THIRD EDITION

HUMAN ANATOMY AND PHYSIOLOGY

THIRD EDITION

HUMAN ANATOMY AND PHYSIOLOGY

ALEXANDER P. SPENCE, Ph.D. • ELLIOTT B. MASON, Ph.D.

State University of New York College at Cortland

ILLUSTRATIONS BY FRAN MILNER

THE BENJAMIN/CUMMINGS PUBLISHING COMPANY, INC.

Menlo Park, California • Reading, Massachusetts
Don Mills, Ontario • Wokingham, U.K. • Amsterdam • Sydney
Singapore • Tokyo • Madrid • Bogota • Santiago • San Juan

Dedication

To my wife Marion and our children Mark, Carol, and Cindy.
Alexander P. Spence

To my wife Marsha and our children Jennifer, Julie, and Jessica.
Elliott B. Mason

Their love, patience, encouragement, and understanding
made this book possible.

Sponsoring Editor: *Andrew Crowley*
Production Supervisor: *Mimi Hills*
Cover and Book Designer: *John Edeen*
Artists: *Fran Milner* with *Barbara Haynes, Cathleen Jackson Miller, Carol Verbeck,* and *Linda McVay*

Library of Congress Cataloging-in-Publication Data

Spence, Alexander P., 1929–
  Human anatomy and physiology.

  (The Benjamin/Cummings series in the life sciences)
  Includes index.
  1. Human physiology. 2. Anatomy, Human.
I. Mason, Elliott B., 1943–   . II. Title.
III. Series.
QP34.5.S72 1985   612   86-18771
ISBN 0-8053-6989-9

  BCDEFGHIJ-DO-8987

The Benjamin/Cummings Publishing Company, Inc.
2727 Sand Hill Road
Menlo Park, California 94025

Magnifications listed with the microscope symbol indicate the final enlargement of a photomicrograph. Magnifications listed in legends indicate the enlargement at which the photomicrograph was shot, with no adjustment for final printed size.

About the Cover

"Hot Leonardo," by Dr. Robin Williams, is a thermograph based on a drawing by Leonardo da Vinci. Thermography is a scanning technique for portraying variations in the heat emitted by different regions of the body. Each color indicates approximately ½°C change in temperature, with dark blue indicating the coldest regions, and red, yellow, and white indicating the hottest. The technique can help detect abnormal conditions and thus acts as an aid in medical diagnosis; for example, cancerous areas show up cooler. Thermography can also show a transplanted organ, help detect vascular disease, and indicate the severity of burns. "Hot Leonardo" © Dr. Robin Williams/Science Photo Library, Science Source/Photo Researchers.

Preface

We are gratified that our textbook **Human Anatomy and Physiology** continues to be so well received that it is necessary to compile a third edition. Our major goal for this edition was the same as for the previous editions: To provide a comprehensive, introductory level text that is useful to students with little background in biology and that integrates anatomy and physiology in a way that reinforces the inseparable relationship between structure and function.

The changes made in this edition make the text even more useful to the nursing, allied-health sciences, physical education, and liberal arts majors enrolled in introductory courses in human anatomy and physiology. For example, we carefully considered whether some areas needed major changes in organization or content. In most cases we felt the basic organization did not need alteration. However, we expanded the coverage of the lymphatic system and placed it in a separate chapter. The defense mechanisms of the body are receiving much current research attention; consequently we significantly updated the discussion of the immune system. We added new sections of special interest to students majoring in physical education, on such topics as the effects of gravity on the cardiovascular system, and cardiovascular adjustments during exercise.

New in This Edition

We made numerous changes in every chapter. The most significant improvements include:

- ### Updated Terminology
  The terminology is revised to conform with the fifth edition of *Nomina Anatomica*, whenever such a change contributes to the clarity of the discussion. When new terms are used the old terms are generally left in parentheses to assist the reader in the transition. In a few instances old terms are retained because they are still in common use, particularly in the introductory courses for which this book is intended.

- ### Increased Use of Color
  The use of color is greatly expanded throughout the text. Color continues to be used in a *functional* manner and we have worked closely with the artist to make certain that color is used as a learning aid—that is, to focus attention, to differentiate related structures, or to guide the viewer through a complex illustration—rather than simply to make the text more flashy.

- ### New Drawings and Photomicrographs
  In addition to minor changes in many figures, there are a number of new drawings and photomicrographs. For instance, all of the histological photomicrographs in Chapter 4 are new and appear in full color.

- ### Frontiers in Health
  To stimulate the readers' interest, a number of essays are included, each of which discusses the treatment of a health-related topic in which major advances have been made in recent years.

- ### Clinical Correlations
  Perhaps of special interest to students majoring in nursing, premedicine, and the allied-health sciences are new Clinical Correlation boxes, each of which discusses an actual case history of a patient suffering from a condition that illustrates some basic physiological principles.

- ### Regional Anatomy Appendix
  Most human anatomy and physiology textbooks at this level are organized by body systems because such an approach is easier to understand and more relevant to most students. However, understanding regional relationships is also an important aspect of learning anatomy. A brief regional anatomy appendix composed of full-color, labeled photographs of a dissected cadaver appears at the end of the book. It will be valuable to anyone wishing to study regional anatomy, especially to those with access to a cadaver in the laboratory.

Special Features

This new edition retains several special features that were well received in the previous editions and that enhance the text's usefulness as a teaching tool and increase student interest. The special features include:

- ## Integration of Embryology

  We have found that structural and functional relationships within the body are best understood when students have some knowledge of the embryonic development of the structures involved. For this reason the discussion of each system begins with a brief consideration of the development of the system in the embryo. These discussions are self-contained and, if the instructor chooses, may be omitted without detracting from the remainder of the chapter.

- ## The Integration of Art and Text

  Perhaps the most distinctive feature of this text, one that previous users consistently praised, is the careful integration of art and text. The artist, Fran Milner, and the designer, John Edeen, along with the Benjamin/Cummings staff, labored over every page of the book, making certain that each figure is in close proximity to the corresponding text. Comparisons with other texts reveal that we have a much higher ratio of figures appearing where referenced, eliminating the annoying and all too common occurrence of figures falling pages away from the relevant text discussion. The integration of artwork and text is supported by a marginal figure referencing system that helps students correlate figures efficiently with the relevant text discussion. To further enhance the effectiveness of the art program, the artist developed insets to help orient students to particular aspects of many figures, and made certain we included only essential labels on the figures. The careful treatment of art and text make this book a superior teaching and learning tool.

- ## Conditions of Clinical Significance

  Throughout the text the emphasis is on normal human anatomy and physiology, but brief discussions of diseases, dysfunctions, and aging are included when they enhance and reinforce an understanding of normal human anatomy and physiology. These discussions appear in separate display boxes following the discussion of each system. This special treatment allows instructors to emphasize or deemphasize these conditions according to the objectives of their particular courses.

In-Text Learning Aids

Students learn in different ways. We have therefore provided various pedagogical aids to assist the diverse group of students who take courses in human anatomy and physiology. Each chapter includes:

 Numbered Learning Objectives

 Chapter Contents Overview
 Key Words and Derivatives
 Marginal Pronunciation Guide
 End-of-Chapter Study Outlines
 Self-Quizzes (Answers are in Appendix 2)

In addition to the pedagogical aids in each chapter, the book includes the following appendices:

- **Word Roots, Prefixes, Suffixes, and Combining Forms**—This guide is a valuable aid to students throughout the course.
- **Glossary**—The glossary includes over 1,500 definitions and provides a phonetic pronunciation of each term.
- **Metric Appendix**—This appendix provides metric/English conversion constants.

Supplements

- ## Instructor's Guide and Transparencies

  Available upon adoption of the text is an Instructor's Guide (Code #37010), which includes 58 chapter-end quizzes (alternate versions of the self-quizzes found at the end of each chapter), a comprehensive final exam, and numerous references and resource bibliographies. Also available are almost 100 two-color acetate Transparencies (Code #36999) of some of the excellent artwork from the text.

- ## Student Study Guide

  A truly marvelous study guide, *The A & P Workbook* by Elaine Marieb (Code #36731) can be ordered through the college bookstore. It is the first and only book to combine the best features of a study guide/workbook with an entertaining and instructive coloring book.

- ## Lab Manuals

  The publisher is pleased to offer three successful lab manuals, all highly complementary to this textbook. One of the three should meet the needs of your lab course, dependent on the length of the course and the animal of dissection.

  Human Anatomy and Physiology Lab Manual, Brief Version, Second Edition (1987) (Code #36733)

  Human Anatomy and Physiology Lab Manual, Cat Version, Second Edition (1985) (Code #36726)

  Human Anatomy and Physiology Lab Manual, Fetal Pig Version, Second Edition (1985) (Code #36727)

  All three manuals were written by Elaine Marieb, Ph.D., R.N., of Holyoke Community College. Please contact your Benjamin/Cummings representative or write directly to the publisher for more information on these excellent lab manuals.

- Computer Assisted Instruction

  Also available to adopters of the text is *Human Anatomy: A Computerized Review and Coloring Atlas* (Code #35810) by Stephen W. Langjahr and Robert D. Brister. This software package consists of a disk for the instructor and workbooks for students. To see a demonstration, call your Benjamin/Cummings representative.

Acknowledgments

In preparing this third edition we relied heavily on the suggestions and reviews received from students and faculty who were using our text, as well as on solicited reviews from faculty who were using other texts in their courses. Their ideas and advice proved invaluable to us, and many of their suggestions were incorporated into the text. The reviewers of this and previous editions are listed below.

Dr. Daniel D. Chiras of the University of Colorado at Denver added a new dimension to this edition by suggesting and researching topics for the

Frontiers in Health boxes. Many of the topics are at the forefront of medical research and provide stimulating reading. We appreciate his willingness to assume this task.

Contributing greatly to this edition, as they did to the others, are the outstanding artwork and the pleasing design. We remain indebted to Fran Milner, the artist, and John Edeen, the designer, for their ideas, their abilities, and their interest in the book.

Of course, it was the dedicated people at Benjamin/Cummings who did not allow us to miss deadlines and who put the book into final form, and we are very much indebted to them—especially Andy Crowley, sponsoring editor, and Mimi Hills, production supervisor.

Alexander P. Spence
Elliott B. Mason
Department of Biological Sciences
State University of New York College at Cortland
Cortland, New York 13045

List of Reviewers

Thomas Adams Michigan State University
Robert M. Anthony Triton College
Maxine A'Hearn Prince Georges Community College
Franklyn Bolander University of South Carolina
Gordon Bradshaw Phoenix College
Alphonse R. Burdi University of Michigan Medical School
William Camelet Lake Michigan College
Cynthia Carey University of Colorado
Jeans Cons College of San Mateo
David Cox Illinois Central College
Dwayne H. Curtis California State University, Chico
Lillian L. Darago Catonsville Community College
Edward Donovan Avila College
Lawrence M. Elson City College of San Francisco
Steven A. Fink West Los Angeles College
Steven Fisher University of California, Santa Barbara
Susan Foster Mt. Hood Community College
John Frehn Illinois State University
Lewis Greenwald Ohio State University
Ann A. Hagan The American University
John P. Harley Eastern Kentucky University
Margaret Hudson Seattle University
August N. Jaussi Brigham Young University
Jack L. Keyes University of Oregon
Ann Marie Kreuger Bunker Hill Community College
Stephen Langjahr Antelope Valley College
Charles Leavell Fullerton College
Sylvia Lianides West Valley College
Kathryn Malone Westchester Community College

Elaine N. Marieb Holyoke Community College
Allan Markezich Black Hawk College
Constance Martin Hunter College
William Matthai Tarrant County Jr. College, Northeast Campus
Joseph W. McDaniel Norwich University
Francis C. Monette Boston University
A. Kenneth Moore Seattle Pacific University
Patricia O'Mahoney-Damon University of Southern Maine
Dennis Peterson DeAnza College
R. Douglas Power Boston College
Gary Resnick Saddleback Community College
Roger H. Sawyer University of South Carolina
David Saxon Morehead State University
Jane Schneider Westchester Community College
Raymond Sicard Boston College
David Smith San Antonio College
Tom Sourisseau Cabrillo College
Carol Spaulding University of Maryland
Robert J. Stark Indiana University—Purdue University Indianapolis
Pauline Tepe Phoenix College
Martha Van Bolt Mott Community College
Eugene Volz Sacramento City College
Elizabeth Walker West Virginia University
Edward Wallen University of Wisconsin—Parkside
James Waters Humboldt State University
Maxine Waughtell Santa Barbara City College
Charles R. Wayne County College of Morris

Brief Contents

Detailed Contents

4 Tissues 87

5 The Integumentary System 111

6 The Skeletal System 129

9 The Muscular System: Gross Anatomy   245

10 The Nervous System: Its Organization and Components   297

11 Neurons, Synapses, and Receptors   315

18 The Circulatory System: The Heart   527

19 The Circulatory System: Blood Vessels   561

20 The Lymphatic System   609

21 Defense Mechanisms of the Body   621

27 Fluid and Electrolyte Balance and Acid-Base Regulation   793

28 The Reproductive System   809

29 Pregnancy, Embryonic Development, and Inheritance 845

Appendix 1
Regional Anatomy 872

Appendix 2
Self-Quiz Answers 888

Appendix 3
Word Roots, Prefixes, Suffixes, and Combining Forms 892

Appendix 4
Units of the Metric System 895

Glossary 896

Index 913

Photo and Art Acknowledgments 937

LEARNING OBJECTIVES

After completing this chapter, you should be able to:

1. Distinguish between anatomy and physiology.

2. Cite several examples of the interrelationship of structure and function in the body.

3. Name four types of tissues produced by the three embryonic cell layers.

4. Name the ten major organ systems in the human body.

5. Describe common directional and regional terms.

6. Name the planes and cavities of the body.

7. Distinguish between the parietal and the visceral membranes of the ventral body cavities.

8. Describe the mesenteries of the abdominopelvic cavity.

9. Describe what is meant by homeostasis.

10. Explain positive and negative feedback mechanisms.

CHAPTER CONTENTS

FIELDS OF ANATOMY

FIELDS OF PHYSIOLOGY

INTERRELATIONSHIP OF STRUCTURE AND FUNCTION

BASIC STRUCTURAL LEVELS

ANATOMICAL AND PHYSIOLOGICAL TERMINOLOGY

BODY POSITIONS

DIRECTIONAL TERMS

REGIONAL TERMS

BODY PLANES

BODY CAVITIES

MEMBRANES OF THE VENTRAL BODY CAVITIES

HOMEOSTASIS

HOMEOSTATIC MECHANISMS

POSITIVE FEEDBACK

KEY TERMS AND DERIVATIVES

anatomy (*ana* = apart; *tom* = cut) the science of the structure of organic beings

homeostasis (*homeo* = same; *stasis* = arresting) the state when the body organs function together to maintain a stable internal environment for the general well-being of the entire body; a state of body equilibrium

inferior (*inferior* = lower) situated below, or directed downward

lateral (*lateral* = side) away from the midline of the body

medial (*medi* = middle) toward the midline of the body

pericardium (*peri* = around; *cardium* = heart) the closed membranous sac enveloping the heart

physiology (*physio* = nature; *ology* = the study of) the science of the functions of organic beings

superior (*super* = above) refers to the head or upper portion; higher

Introduction to Anatomy and Physiology

One of the most exciting and meaningful accomplishments that a person can strive for is to gain an understanding of his or her body. The main goals of this book are to develop in the reader an understanding of how the human body is constructed and how it functions, and to correlate structure with function. Along with this factual knowledge we hope to instill in the reader an appreciation of what a marvelous organism the human body is. With the rapid advances that have been made in scientific knowledge in recent years—particularly in the health sciences—and the prominent coverage given these advances in newspapers and popular magazines, it has become increasingly important for everyone to know more about the human body.

Anatomy is the study of the *structure* of an organism and the relationship among its parts. The term *anatomy* is derived from the Greek words meaning "apart" and "to cut." As this derivation indicates, anatomy is based largely on dissection of the body. In some of the newer fields of anatomy, however, the use of electron microscopes and other instruments provides valuable supplements to dissection. **Physiology** is the study of the *functions* of a living organism. It attempts to explain in physical and chemical terms the factors and processes involved in these functions.

FIELDS OF ANATOMY

The study of anatomy involves examination of the general structures of the body **(gross anatomy)** as well as those structures that can be seen only with the aid of a microscope **(microscopic anatomy).** Gross anatomy can be studied by regions, such as the head, neck, thorax, abdomen, pelvis, or limbs. This approach, referred to as **regional anatomy,** is often used in dissection, in which all structures in a region are studied simultaneously. For our purposes, however, the most helpful approach is to study anatomy by organ systems that perform common functions **(systemic anatomy),** and this book uses that approach. Microscopic anatomy includes the study of cells *(cytology)* and the study of tissues *(histology)*. When anatomy is studied under the extremely high magnifications possible with the electron microscope, it is referred to as *fine structure* or *ultrastructure*. **Developmental anatomy,** another subdivision of anatomy, focuses on the development of the body from the fertilized egg to the adult form. Developmental anatomy includes *embryology*, which is limited to prenatal development.

Radiographic anatomy is particularly valuable in the diagnosis of disorders and injuries. Until recently, the only tools available for this field were the X-ray and fluoroscope machines. Both instruments produce *roentgen rays* (X rays), which pass through the structure under examination and either expose an X-ray film or illuminate a fluorescent screen. When an exposed film is developed, the result is a photographic image called a *roentgenogram*, popularly known as an X ray. Although roentgenograms are diagnostically useful, they are somewhat limited in that the three-dimensional relationships of the body's parts are lost on the film, where the body image is flat. Moreover, small differences in the densities of tissues are not always detectable on a roentgenogram.

In the 1970s, technological advances made **tomography** possible, thus expanding radiographic anatomy. In tomography, a specific level in the body

(a)

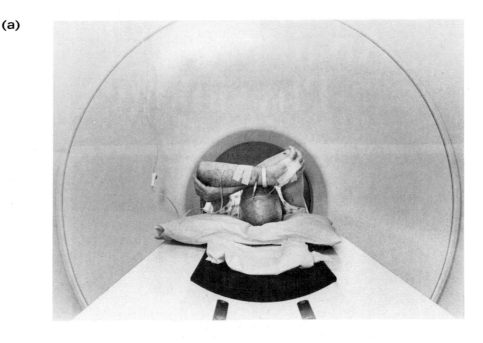

(b)

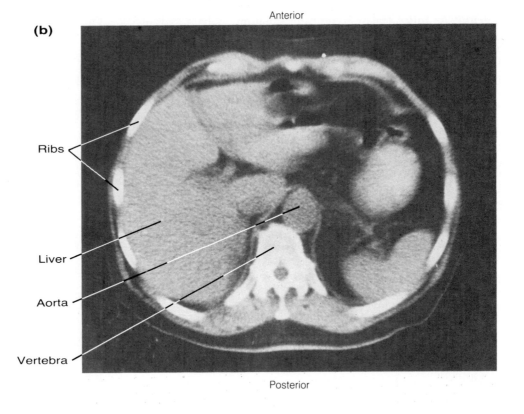

Anterior

Ribs

Liver

Aorta

Vertebra

Posterior

Figure 1.1

(a) Patient inside a CT scanner. The table on which the patient lies slides back and forth as scans are shot in 10–20-mm cross sections. In this instance an intravenous fluid is being used to disperse opaque iodine contrast medium, which will aid in highlighting hard-to-see areas. **(b)** CT scan from the midtrunk region. Density is indicated by the degree of lightness of an area. White areas are the most dense, followed by the grey, and then the black. CT scans are performed primarily as an accurate and rapid means of searching for tumors or other space-occupying lesions. (For additional information on CT scans, see Box 12.1)

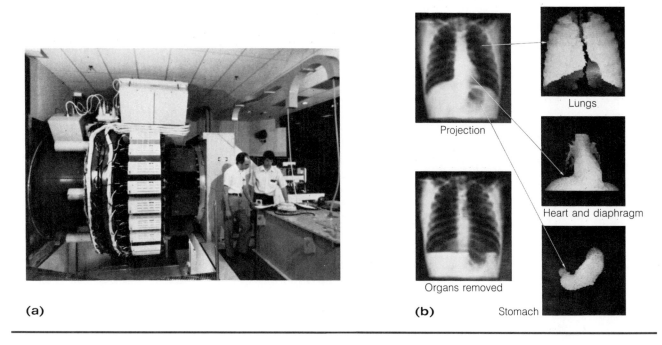

(a)

Projection

Organs removed

(b)

Lungs

Heart and diaphragm

Stomach

is radiographed, whereas structures above and below this level are blurred. This effect is achieved by rotating the X-ray tube and the film around the selected level while repeatedly exposing this region of the body to X rays. The technique, in effect, produces a cross section of the body at the selected level (Figure 1.1). Since the rotation of the X-ray tube and film are coordinated by computer, the technique is referred to as *computed tomography (CT scan)* or *computed axial tomography (CAT scan)*. A CT scan not only shows the three-dimensional relationships of the body parts but differentiates among tissues more clearly than X rays can.

    The recent development of a complex X-ray machine called the *dynamic spatial reconstructor (DSR)* represents a significant advance in radiographic anatomy. The DSR uses multiple X-ray tubes that rapidly revolve around the patient, producing thousands of cross sections within a time span of a few seconds. This is much faster than a CT scanner, which would produce only one section during the same time span. The DSR produces moving three-dimensional images of an organ that can be rotated and tipped in any direction, allowing all sides of the organ to be viewed (Figure 1.2). A CT scan, in contrast, produces only cross-sectional slices of the body. In addition, the image produced by the DSR can be "sliced open" on the screen so that the interior of an organ can be viewed.

    Nuclear magnetic resonance (NMR) is another recent advance in radiographic anatomy. NMR has been used primarily as a research instrument, but is beginning to be used for diagnostic purposes. It has the advantage of using non-ionizing radiation, which is less damaging to cells than are X rays. In NMR the patient is placed in a body-size chamber within a large magnet. The magnetic field causes hydrogen nuclei, as well as certain other nuclei, throughout the body to become aligned. By adjusting the magnetic energy generated by the magnet it is possible to detect the amount of energy absorbed by the various nuclei. This information, when fed into a computer, can be used to plot the distribution of the nuclei and thus provide images of body organs.

    The use of *ultrasonography* provides another means of obtaining images of the body organs. In this procedure, sound waves are sent into the body and the echoes they make when they strike various tissues are used to define the boundaries of the organs. Sound waves appear not to have any significant harmful side effects on cells; therefore ultrasonography can be used safely to produce images of unborn fetuses. It can also produce images of moving objects, such as the blood flowing within a vessel.

F1.1

F1.2

Figure 1.2
(a) Dynamic spatial reconstructor (DSR).
(b) Three-dimensional images of lungs, heart, diaphragm, and stomach produced by a DSR.

Positron-emission tomography (PET) is a very specialized radiographic anatomy process that provides information concerning organ functioning. Whereas CT scans produce images that show what the organs look like, PET, which thus far has only been used to view the brain, produces images that give a measure of the activity of the cells within the organ. During a PET examination, radioactive glucose is injected into the patient's bloodstream. The glucose concentrates in the cells of the most metabolically active regions of the brain. The patient's head is inserted into the PET machine, where detectors absorb the radiation and send signals to a computer. The computer combines the signals into a video image, with bright spots showing where the radioactive glucose has accumulated, thus providing a functional map of the patient's brain.

FIELDS OF PHYSIOLOGY

Physiology is a vast field, with many subdivisions. The subdivisions are often concerned with different levels of organization—from the subcellular to the multicellular. The subdivisions include specialties such as *viral physiology, bacterial physiology, cellular physiology, plant physiology,* and *animal physiology.* In this text we are primarily interested in physiology at the organ level in humans. To study *human physiology,* however, even at the organ level, it is necessary to study certain aspects of cellular physiology.

Scientists whose work is primarily concerned with physiology are referred to as *physiologists.* Human physiologists tend to concentrate their studies on one or another of the body systems. Thus, there are *renal physiologists, reproductive physiologists, neurophysiologists,* and so on. We follow that same approach in this book; the physiology—as well as the anatomy—of each organ system is described as the particular system is studied.

INTERRELATIONSHIP OF STRUCTURE AND FUNCTION

When you study anatomy and physiology at the same time, as in this text, you will benefit by keeping in mind the interrelation of structure and function. By describing how a particular body structure is suited to its function, the study of anatomy helps make the physiological processes of the body more meaningful.

Just as the structures, shapes, and organization of the parts of a machine—such as an automobile—are appropriate to their functions, so the structures, shapes, and organization of the parts of the body are intimately associated with their functions. This interrelation of structure and function is evident at all levels of body organization. At the whole-body level, for example, the structure of the joints in the human hand makes possible an opposable thumb that is essential for efficiently grasping and manipulating objects. At the cellular level, one example of the interrelation of structure and function is the nerve cell, which has long, thin processes extending from the cell body. These processes are well suited to the cell's function of transmitting information in the form of nerve impulses from one body region to another. Even at the molecular level, structure is critical to function. Enzymes, for example, which are protein molecules that speed up the rates at which chemical reactions occur in the body, can act only if they have shapes that specifically "fit" the shapes of the reacting molecules.

BASIC STRUCTURAL LEVELS

There are four structural levels in the body: cells, tissues, organs, and systems. Each body structure has specific functions that contribute to the general well-being of the entire body, not just of the structure itself.

Cells

At its simplest structural level, the body is composed of cells, and even at this level there are structural differences that are closely related to the physiology of each particular cell type. We consider these cellular differences in Chapter 4.

Tissues

Only the simplest animals are able to exist as single cells. In most animals, including humans, groups of similar cells join together to form **tissues.** In the early embryo, where tissue formation first occurs, similar cells group together into three layers: the **ectoderm,** which forms both the outer covering of the body and the nervous tissue; the **endoderm,** which forms the inner lining of the digestive tube and its associated structures; and the **mesoderm,** the layer located between the ectoderm and endoderm tissues, which forms the skeleton and the muscles of the body (Figure 1.3). These three embryonic cell layers—the ectoderm, endoderm, and mesoderm—give rise to four types of tissues, which are briefly discussed here and examined in detail in Chapter 4.

Epithelial Tissues

Epithelial tissues cover the surface of the body and line the various body cavities, ducts, and vessels. The epithelial tissue that forms the outer protective layer of the body, the epidermis, is derived from embryonic ectoderm. The rest of the epithelial tissues originate from either the mesoderm or the endoderm of the embryo.

Muscular Tissues

Muscular tissues are composed of specialized cells that are capable of contracting and thereby decreasing in length. These tissues move the skeleton, propel the blood throughout the body, and aid in digestion by moving food through the digestive tract. As we will explain in Chapter 8, there are three types of muscle tissues: *skeletal, cardiac,* and *smooth.* Each type is derived from embryonic mesoderm.

Nervous Tissues

Nervous tissues, which form the brain, spinal cord, and nerves, consist of cells with long protoplasmic extensions. These nerve cells (neurons) transmit messages throughout the body. They originate from the ectodermal layer of the embryo.

Connective Tissues

Several types of cells are involved in the formation of connective tissues. These tissues, most of which are derived from mesoderm, are used for support (bones and cartilage), for the attachment of other tissues (tendons, ligaments, and fascia), or for other specialized functions (such as blood).

Organs

Tissues allow the body to perform more complex physiologic activities than is possible for individual cells. Physiologic processes that are even more complex are made possible when two or more tissues combine to form an **organ.** The stomach, for instance, is lined with epithelial tissue and its walls are formed by muscular tissue. These tissues are held together as a discrete structure by various connective tissues and are innervated by nervous tissue. Each of these types of tissues contributes in a specific manner to the functioning of the stomach, and without such tissue combinations it would not be possible to process large particles of complex foods. This same principle holds true for all organs: Each organ is a specialized physiologic center for the body.

Systems

The ability of the organs to function for the general well-being of the body is enhanced by the fact that certain organs work together as a system, each

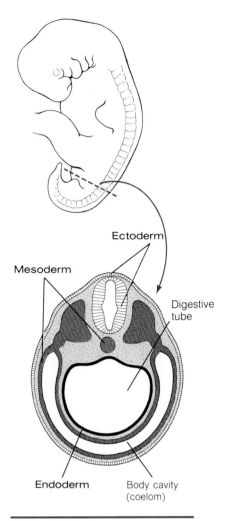

Figure 1.3
Schematic cross section of an embryo showing the location of endoderm, mesoderm, and ectoderm. The dotted line on the embryo indicates the site of the cross section.

organ within the system performing a specific part of a general body function. For example, one shared function is obtaining energy from food for use throughout the body: Food is prepared and partially digested in the mouth; it is transported by the esophagus to the stomach, where it is further prepared and digested; it is absorbed into the blood vessels through the walls of the intestines; and finally, the residue that is not absorbed is eliminated from the body through the rectum.

Organs that function cooperatively to accomplish a common purpose (such as the digestion and absorption of food) are said to be part of a body **system.** There are ten major systems in the human body: *integumentary, skeletal, muscular, nervous, endocrine, circulatory, respiratory, digestive, urinary,* and *reproductive.* The structure and function of each of these body systems are listed in Table 1.1 and discussed in greater detail in later chapters.

in-teg-u-men'tar-ee

Table 1.1

ANATOMICAL AND PHYSIOLOGICAL TERMINOLOGY

Every branch of science has its own special terminology, and anatomy and physiology are no exceptions. Although you will find many terms in this text that are new to you, the terms themselves are not new. A sizable number of them originated centuries ago and have Greek or Latin origins. The terms may appear formidable, but they are quite descriptive if you understand their roots. For example, *ilio* refers to the hip bone (ilium), and *costal* refers to the ribs. Therefore, the *iliocostalis* muscle is clearly a muscle that passes from the ilium to the rib cage.

A knowledge of prefixes and suffixes is also helpful in understanding anatomical and physiological terms. For example, the prefix *endo* means "within" and is used in many scientific terms, including the following:

en-do-kar'-di-um   **endocardium**   (*cardium* refers to the heart)   the inner lining of the heart

en-do-kar-di'-tis   **endocarditis**   (*itis* refers to an inflammation)   an inflammation of the inner lining of the heart

en-doj'-en-us   **endogenous**   (*genous* refers to producing)   produced within the organism

en-do-me'-tree-um   **endometrium**   (*metra* refers to the uterus)   the inner lining of the uterus

The meanings of new prefixes, suffixes, and roots will be explained as they are introduced in the text, and key terms and their roots are listed at the beginning of each chapter. A more thorough list that will be helpful in understanding new words is given at the back of the book (see Appendix 3).

BODY POSITIONS

While studying the detailed description of each body structure, you must also understand the positional relationships among body structures. Thus, you must become familiar with the terms used to describe these relationships.

If the body is lying horizontally with the face downward, the body is in the *prone position.* If the body is on its back, with the face upward, it is in the *supine position.* The relationships of the various body structures to each other differ in these positions. To communicate effectively concerning human anatomy, therefore, we always refer to the body as if it were in a standard position, so that the structural relationships are clear and consistent. This standard position is referred to as the **anatomical position** (Figure 1.4a). In this position, the body is erect, with the feet together. The upper limbs hang at the side, with the palms of the hands facing forward, the fingers extended, and the thumbs pointing away from the body. With the hands in this position, the bones of the hands and fingers are exposed, and their relationships, there-

soo'-pine

F1.4a

Table 1.1   Organ Systems of the Body

System	Major Components	Representative Functions
Integumentary	Skin and associated structures such as hair and nails	Protects internal body structures against injury and foreign substances; prevents fluid loss (dehydration); important in temperature regulation
Skeletal	Bones	Supports and protects soft tissues and organs
Muscular	Skeletal muscles	Moves body and its parts
Nervous	Brain, spinal cord, nerves, special sense organs	Controls and integrates body activities; responsible for "higher functions" such as thought and abstract reasoning
Endocrine	Hormone-secreting glands such as the pituitary, thyroid, parathyroid, adrenals, pancreas, and gonads	Controls and integrates body activities; function closely allied with that of the nervous system
Circulatory	Heart, blood and lymphatic vessels, blood, lymph	Links internal and external environments of the body; transports materials between different cells and tissues
Respiratory	Nose, trachea, lungs	Transfers oxygen from the atmosphere to the blood and carbon dioxide from the blood to the atmosphere
Digestive	Mouth, esophagus, stomach, small intestine, large intestine; accessory structures include salivary glands, pancreas, liver, gallbladder	Supplies body with substances (food materials) from which energy for activity is derived and from which components for synthesis of required substances are obtained
Urinary	Kidneys, ureters, urinary bladder, urethra	Eliminates variety of metabolic end products such as urea; conserves or excretes water and other substances as required
Reproductive	Male: seminal vesicles, testes, prostate gland, bulbourethral glands, penis, associated ducts	Produces male gametes (sperm); provides method for introducing sperm into the female
	Female: ovaries, uterine tubes, uterus, vagina, mammary glands	Produces female gametes (ova); provides proper environment for development of fertilized ovum

fore, are easily described. Moreover, when the palms are facing forward, the bones of the forearm are uncrossed. *Unless stated otherwise, all anatomical descriptions refer to a body in the anatomical position.*

DIRECTIONAL TERMS

The terms used to denote direction come in pairs, each indicating an opposite direction (Figures 1.4b and c). **Anterior** (or **ventral**) refers to the front, whereas its opposite, **posterior** (or **dorsal**) refers to the back. **Superior** (or

F1.4b, F1.4c

kaw'dal

Table 1.2

cranial) means "toward the head"; **inferior** (or **caudal**) means "away from the head." It should be noted that these terms are used differently when they refer to humans than when they refer to four-legged animals. Since humans stand upright, the direction indicated by the terms is different. In this text we are interested only in their meaning as they relate to humans. The directional pairs are listed in Table 1.2.

REGIONAL TERMS

F1.4a

In addition to directional terms, several frequently used terms refer only to special areas of the body (Figure 1.4a):

cervical refers to the neck

thoracic the portion of the body between the neck and the abdomen that is commonly referred to as the chest (thorax)

lumbar the portion of the back between the thorax and the pelvis

sacral the lower portion of the back, just superior to the buttocks

plantar the sole of the foot; the top of the foot is the *dorsal* surface

Table 1.2 **Directional Terms**

Term	Definition	Example
Anterior (ventral)	Situated in front of; the front of the body	The chest is on the anterior surface of the body.
Posterior (dorsal)	Situated in back of; the back of the body	The buttocks are on the posterior surface of the body.
Superior (cranial)	Toward the head; relatively higher in position	The eyebrows are superior to the eyes.
Inferior (caudal)	Away from the head; relatively lower in position	The mouth is inferior to the nose.
Medial	Toward the midline of the body	The breast is medial to the armpit.
Lateral	Away from the midline of the body	The hip is on the lateral surface of the body.
Proximal	Closer to any point of reference, such as the attached end of a limb, the origin of a structure, or the center of the body	The arm is proximal to the forearm.
Distal	Farther from any point of reference, such as the attached end of a limb, the origin of a structure, or the center of the body	The hand is distal to the wrist.
Superficial (external)	Located close to or on the body surface	The skin is superficial to the muscles.
Deep (internal)	Located further beneath the body surface than superficial structures	The muscles are deep to the skin.

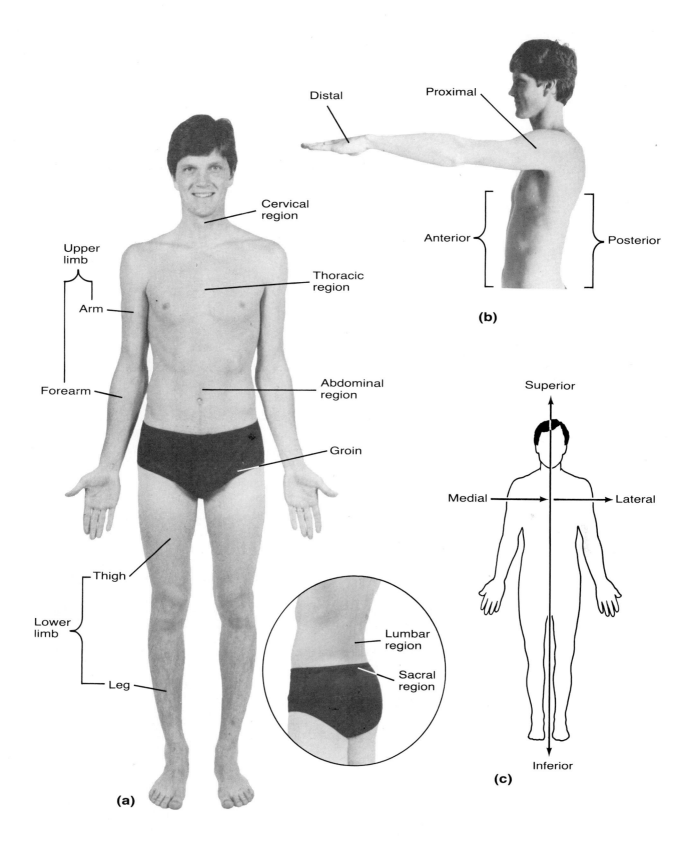

(a)

Distal

Proximal

Anterior

Posterior

(b)

Cervical region

Upper limb

Arm

Thoracic region

Forearm

Abdominal region

Groin

Thigh

Lower limb

Leg

Lumbar region

Sacral region

Superior

Medial

Lateral

Inferior

(c)

Figure 1.4
(a) Anatomical position and regions of the body. **(b, c)** Directional terms.

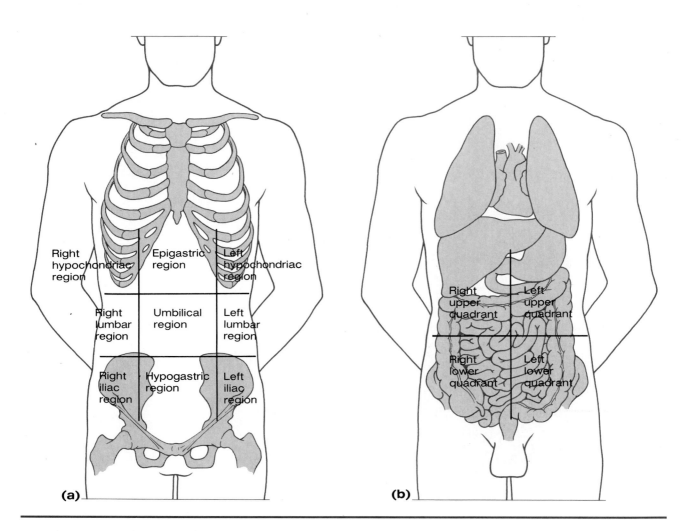

Figure 1.5
(a) Abdominal regions. The top horizontal line passes along the lower edge of the rib cage. The lower horizontal line passes across the upper margins of the hip bones. The vertical lines pass through the midpoints of the clavicles and the inguinal ligaments.
(b) The abdominal wall and abdominopelvic cavity subdivided into four quadrants.

F1.5a

palmar   the anterior surface of the hand; the posterior surface of the hand is the *dorsal* surface

axilla (armpit)   the depression on the inferior surface of the attachment of the upper limb and the body trunk

groin (inguinal region)   the junction of the thigh wall with the abdominal wall

arm   the portion of the upper limb between the shoulder and the elbow

forearm   the portion of the upper limb between the elbow and the wrist

thigh   the portion of the lower limb between the hip and the knee

leg   the portion of the lower limb between the knee and the ankle

To make it easier to describe the location of the abdominal organs, this cavity is divided into nine regions using four imaginary lines: two vertical lines that bisect the clavicles, or collarbones; and two horizontal lines, one along the lower edge of the rib cage and another across the upper edges of the hip bones (iliac crests) (Figure 1.5a). These abdominal regions are:

umbilical   located centrally, surrounding the *umbilicus* (navel)

lumbar   the regions to the right and left of the umbilical region

epigastric   (*epi* means "on or above"; *gastric* refers to the stomach)   the midline region superior to the umbilical region. As the name implies, most of the stomach is located in this region

hypochondriac   (*hypo* means "beneath or under"; *chondral* refers to cartilage)   the regions to the right and left of the epigastric region. The name indicates that the hypochondriac regions are located beneath the cartilage of the rib cage

hypogastric   the midline region directly inferior to the umbilical region

iliac   the regions on either side of the hypogastric region. The name is derived from the iliac (hip) bones that form the lateral boundaries of the regions. These areas are also referred to as the *inguinal regions* because their lower margins end at the inguinal ligament, which follows the fold of the groin

In practice, it is more common to divide the abdominopelvic cavity into four quadrants by means of an imaginary horizontal plane that passes through the umbilicus and a vertical midsagittal plane (Figure 1.5b). These two intersecting planes divide the abdominopelvic cavity into a **right upper (superior) quadrant;** a **right lower (inferior) quadrant;** a **left upper (superior) quadrant;** and a **left lower (inferior) quadrant.**

F1.5b

BODY PLANES

In the study of anatomy, it is useful to visualize the body as cut or sectioned through various planes of reference (Figure 1.6). A **sagittal plane** is a longitudinal section that divides the body or any of its parts into right and left portions. If the section passes through the midline of the body, it is referred to as a **median sagittal (midsagittal) section.** Such a section divides the body into *equal* right and left halves. Sagittal sections other than the median sagittal section are often referred to as **parasagittal sections.** Parasagittal sections divide the body into *unequal* right and left portions. The **frontal (coronal) plane** is also a longitudinal section, but it runs at right angles to the sagittal plane, dividing the body into anterior and posterior portions. A **transverse plane (cross section** or **horizontal section)** divides the body or any of its parts into superior and inferior portions.

saj'-i-tal
F1.6

BODY CAVITIES

The body contains two main cavities: the **dorsal (posterior) cavity** and the **ventral (anterior) cavity** (Figure 1.7). Each of these cavities is lined with membranes and contains a small amount of fluid surrounding the organs that fill the cavities. The dorsal cavity has two subdivisions: the **cranial cavity,** which houses the brain, and the **spinal (vertebral) cavity,** which contains the spinal cord. The spinal cavity communicates with the cranial cavity through the *foramen magnum,* a large opening in the base of the skull. The membranes associated with these dorsal cavities are examined in greater detail in Chapter 12. For now, it is sufficient to call the membranes that cover the brain and the spinal cord the *meninges.* The fluid found in these dorsal cavities is the *cerebrospinal fluid.* It too is considered in more detail with the nervous system.

F1.7

men-in'-jeez

The ventral cavity of the body also has two subdivisions. It is divided by a muscle called the *diaphragm* into an upper **thoracic cavity** and a lower **abdominopelvic (peritoneal) cavity.** Each of these cavities is, in turn, further subdivided. The thoracic cavity is divided into a **pericardial cavity,** which surrounds the heart, and right and left **pleural cavities,** each of which encompasses a lung. The portion of the thoracic cavity between the two pleural cavities is called the **mediastinum.** The pericardial cavity and the heart are

mee-dee-as-tigh'num

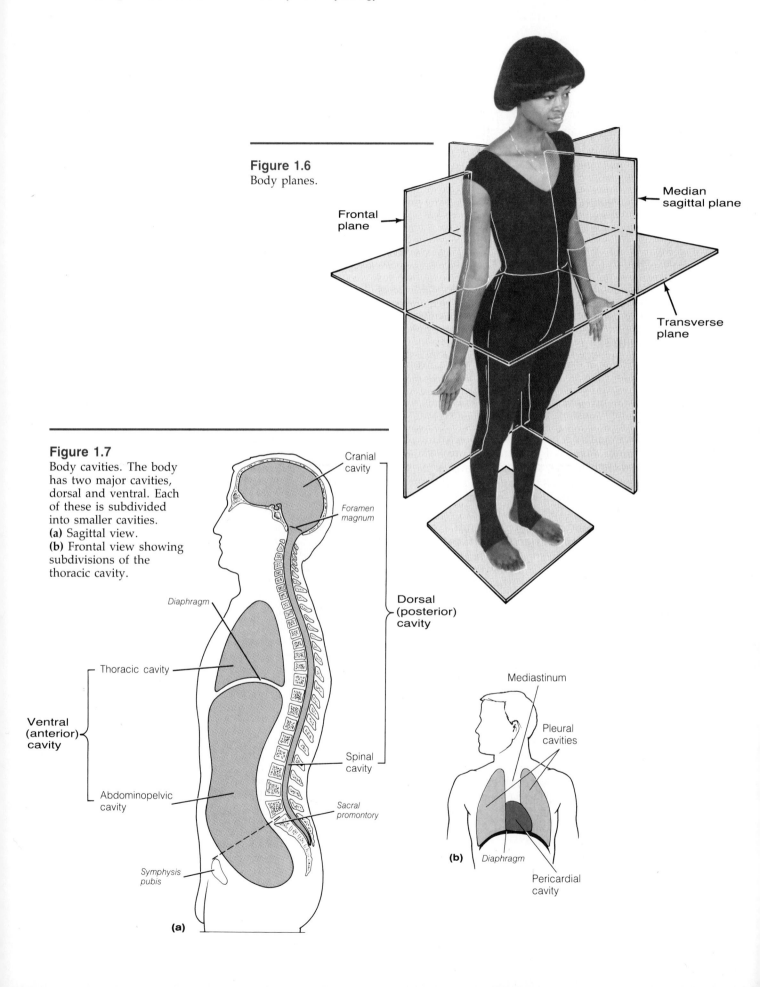

Figure 1.6
Body planes.

Frontal plane

Median sagittal plane

Transverse plane

Figure 1.7
Body cavities. The body has two major cavities, dorsal and ventral. Each of these is subdivided into smaller cavities.
(a) Sagittal view.
(b) Frontal view showing subdivisions of the thoracic cavity.

Cranial cavity

Foramen magnum

Diaphragm

Thoracic cavity

Dorsal (posterior) cavity

Ventral (anterior) cavity

Spinal cavity

Abdominopelvic cavity

Sacral promontory

Symphysis pubis

(a)

Mediastinum

Pleural cavities

(b) *Diaphragm*

Pericardial cavity

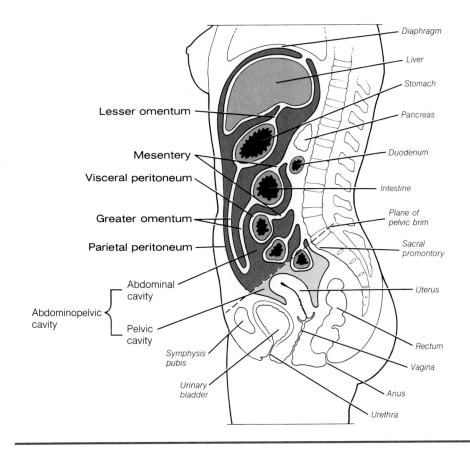

Lesser omentum

Mesentery

Visceral peritoneum

Greater omentum

Parietal peritoneum

Abdominal cavity

Abdominopelvic cavity

Pelvic cavity

Symphysis pubis

Urinary bladder

Diaphragm

Liver

Stomach

Pancreas

Duodenum

Intestine

Plane of pelvic brim

Sacral promontory

Uterus

Rectum

Vagina

Anus

Urethra

Figure 1.8
Sagittal section of the body showing membrane relationships of the abdominopelvic (peritoneal) cavity.

located in the mediastinum. The trachea, esophagus, thymus gland, and several major blood vessels are also located within, or pass through, the mediastinum.

The abdominopelvic cavity, the lower subdivision of the ventral body cavity, is divided for descriptive purposes into a superior **abdominal cavity** and an inferior **pelvic cavity (true pelvis)** by an imaginary oblique plane that passes from the superior margin of the *symphysis pubis* anteriorly to the *sacral promontory* posteriorly (Figure 1.8). The circumference of this plane is called the **pelvic brim.** The pelvic cavity is completely surrounded by the bones of the pelvis. In contrast, the lower portion of the abdominal cavity is bounded posteriorly by the flat hip bones, but its anterior wall is formed by the abdominal wall. This region, which is located just above the pelvic brim, is the **false pelvis.** The abdominal cavity contains the stomach, spleen, liver, gallbladder, pancreas, and the small and large intestines. The pelvic cavity contains the lower part of the digestive system (rectum), the urinary bladder, and, in the female, the internal reproductive organs.

F1.8

MEMBRANES OF THE VENTRAL BODY CAVITIES

To understand the membranes associated with the ventral body cavities, imagine your fist being thrust into an inflated balloon, pushing in one side, as in Figure 1.9a. Notice that the inner wall of the balloon (that is, the wall that has been pushed in) lies close against the fist, whereas the outer wall is separated from the inner wall by the air in the balloon. Now suppose that your fist is an organ. The membranes of the ventral body cavities have the same relationship with the organs in the cavities, except that they are kept apart by fluid rather than air.

F1.9a

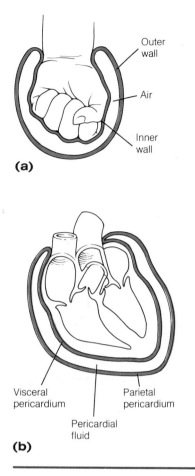

Figure 1.9
Membrane relationships in the ventral body cavities. **(a)** Schematic representation using the analogy of a fist thrust into a balloon. **(b)** Membranes surrounding the heart.

F1.8

F1.10

In the pericardial cavity, the heart (rather than a fist) pushes in one side of a membranous sac (rather than a balloon) (Figure 1.9b). The membrane that lies against the heart (the inner wall of the balloon) is the **visceral pericardium.** The membrane that covers both the heart and the visceral pericardium (the outer wall of the balloon) is the **parietal pericardium** (*parietal* refers to the walls of a body cavity). The parietal pericardium is separated from the visceral pericardium by **pericardial fluid,** which is secreted by the cells of the pericardial membranes. In other words, the outer wall of the pericardial cavity is *lined* with parietal pericardium and the heart is *covered* by visceral pericardium. However, these are both simply different regions of the same membrane. The amount of pericardial fluid that separates these two membranes is very small—just enough to reduce friction and maintain the tissues in a healthy state. If the membranes of the heart become inflamed, the condition is known as *pericarditis.*

The membrane relationships of the pleural cavities are similar to those of the pericardial cavity. The membrane that lies tightly against the surface of the lungs is the **visceral pleura.** The outer walls of the pleural cavities are lined by **parietal pleura.** These two membranes are separated by the **pleural fluid** they secrete. An inflammation of these membranes may result in the secretion of excessive amounts of pleural fluid into the pleural cavity. Prolonged inflammation may cause the visceral and parietal layers of the pleura to adhere to each other. This condition, which makes breathing painful, is called *pleurisy.*

The membrane relationships within the abdominopelvic cavity are similar to those of the thoracic cavity, but here the membrane is called the **peritoneum.** The organs of this cavity are covered with **visceral peritoneum** and the outer walls of the cavity are lined with **parietal peritoneum.** The space between these two membranes is filled with **peritoneal fluid,** which is secreted by the peritoneum. The peritoneum can become inflamed, causing a very serious condition called *peritonitis.*

Most of the organs in the abdominopelvic cavity are suspended from the posterior wall of the cavity by a double-layered membrane of peritoneum (Figure 1.8). These membrane supports are called **mesenteries.** The mesenteries that support particular organs or structures are given specific names such as mesocolon (mesentery of the large intestine), mesoappendix (mesentery of the appendix), mesovarium (mesentery of the ovary), and so forth. The mesenteries not only hold the organs in position, but they also provide a pathway through which blood vessels, lymphatic vessels, and nerves can reach the organs. As the membranes that form the mesenteries continue over the organ that they suspend, they become visceral membranes. Some structures, such as the kidneys, do not hang into the cavity on mesenteries. Instead, they are located outside the cavity, between the body wall and the parietal peritoneum (Figure 1.10). These structures are *retroperitoneal*—that is, they are located *behind* the peritoneum.

HOMEOSTASIS

The body's cells can survive and function efficiently only under relatively constant conditions of temperature, pressure, acidity, and so forth. However, the body as a whole is surrounded by an *external environment* in which there are rather wide fluctuations in temperature, humidity, and other factors. If a person is to survive, therefore, his or her cells must be protected from the variability and extremes of this environment.

The body's cells *are* protected from the variability of the external environment because they exist in an aqueous *internal environment*, which is made up of the fluid portion of the blood and the interstitial fluid that continually bathes the cells (Figure 1.11). The body maintains relatively constant chemical and physical conditions within this environment, and the existence of a relatively constant internal environment is referred to as **homeostasis.**

F1.11

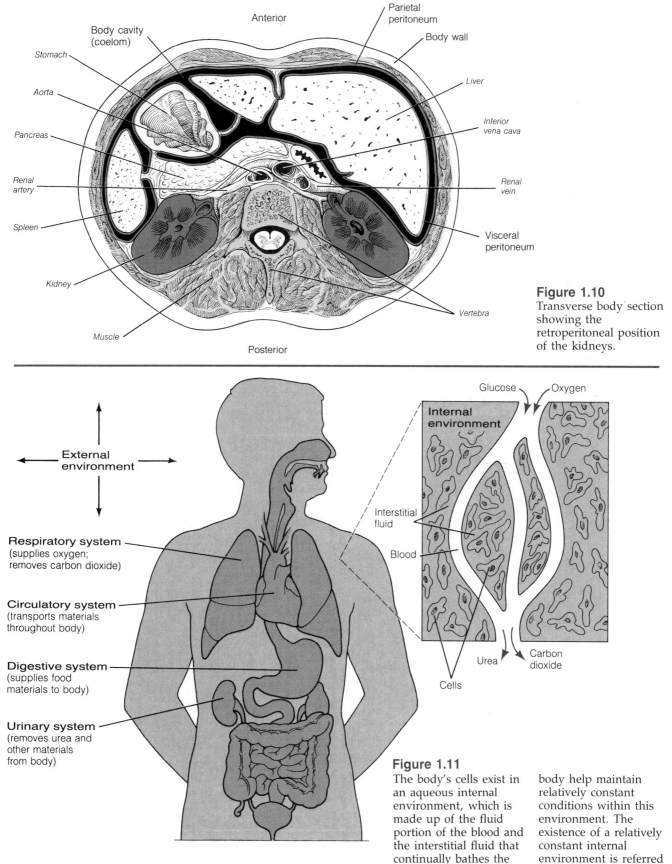

Figure 1.10
Transverse body section showing the retroperitoneal position of the kidneys.

Anterior

Body cavity (coelom)

Stomach

Aorta

Pancreas

Renal artery

Spleen

Kidney

Muscle

Posterior

Parietal peritoneum

Body wall

Liver

Inferior vena cava

Renal vein

Visceral peritoneum

Vertebra

External environment

Respiratory system
(supplies oxygen; removes carbon dioxide)

Circulatory system
(transports materials throughout body)

Digestive system
(supplies food materials to body)

Urinary system
(removes urea and other materials from body)

Glucose   Oxygen

Internal environment

Interstitial fluid

Blood

Urea   Carbon dioxide

Cells

Figure 1.11
The body's cells exist in an aqueous internal environment, which is made up of the fluid portion of the blood and the interstitial fluid that continually bathes the cells. Essentially all the organ systems of the body help maintain relatively constant conditions within this environment. The existence of a relatively constant internal environment is referred to as *homeostasis*.

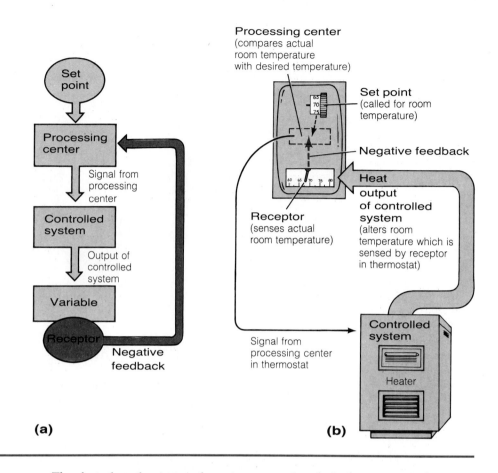

Figure 1.12
(a) Schematic representation of the components of a negative feedback mechanism.
(b) A thermostatically controlled heater is an example of a negative feedback mechanism.

The fact that the internal environment is relatively constant does not imply that it is static or unchanging. Rather, a variety of occurrences continually tend to cause changes in this environment. For example, the activities of the body's cells remove materials such as glucose and oxygen from the internal environment and add materials such as urea and carbon dioxide. As a result, homeostasis can be maintained only if materials are added to the internal environment as rapidly as they are removed or removed as quickly as they are added. The internal environment, therefore, is not static, but rather exists in a dynamic steady state in which the input and output of materials are balanced.

Essentially all the organ systems of the body contribute to the maintenance of homeostasis. For example, as the cells remove glucose and oxygen from the internal environment, the digestive and respiratory systems replace them. In addition, certain materials produced by the cells and added to the internal environment are removed by the urinary system. The circulatory system transports needed materials from areas such as the gastrointestinal tract or the lungs to the cells, and it carries materials produced by the cells to organs such as the kidneys for removal from the body.

As long as the various organ systems function properly, the relative constancy of the internal environment is maintained, and the cells can survive and function efficiently. If the functioning of the organ systems is upset, however, the composition of the internal environment can change and become incompatible with survival.

HOMEOSTATIC MECHANISMS

To maintain homeostasis, the body must be able to sense changes in the internal environment, and it must be able to compensate for the changes.

Thus, the body must be able to control the organ systems concerned with maintaining the composition of the internal environment. The nervous and endocrine systems are the body's principal sensing and controlling systems.

Negative Feedback

The body uses a regulatory principle known as negative feedback to maintain relatively constant, or stable, conditions in the internal environment.

Negative feedback mechanisms have several components (Figure 1.12a). One component, the *controlled system*, is a system whose activity is regulated to maintain the appropriate level of a particular variable—for example, temperature or oxygen. A second component, the *set point*, is a reference that calls for or indicates the level at which the variable is to be maintained. A third component, the *receptor*, monitors the variable and transmits information—referred to as feedback—to a fourth component, the *processing center*. The processing center compares and integrates information from the receptor about the actual level of the variable with information from the set point about the level of the variable called for. If necessary, the processing center increases or decreases the activity of the controlled system in order to bring the actual level of the variable to the level called for by the set point.

A common example of a negative feedback mechanism is the operation of a thermostatically controlled heater, which keeps the temperature in a room comfortable and relatively constant when the temperature outside the room is low (Figure 1.12b). The heater is the controlled system whose activity—heat production—is regulated in order to maintain the appropriate level of the variable, the room temperature. The set point level of the variable is the temperature called for by setting the thermostat at a particular value. The thermostat contains a receptor that provides information about the actual room temperature and a processing center that compares this information with information about the set point temperature. If the actual temperature differs from the set point temperature, the thermostat turns the heater on or off as necessary to bring the actual room temperature to the set point temperature.

Negative feedback mechanisms are called negative because the feedback tends to cause the level of a variable to change in a direction opposite to that of an initial change. In the case of a thermostatically controlled heater, for example, an increase in room temperature above the set point level leads to increased feedback, which acts in an inhibitory fashion to turn the heater off and thereby allow the room temperature to decrease toward the set point level. Conversely, when the room temperature falls below the set point level, the inhibitory feedback decreases, the heater turns on, and the heat produced raises the room temperature toward the set point level. Thus, negative feedback mechanisms minimize the difference between the actual level and the set point level of a variable and, consequently, tend to maintain relatively constant and stable conditions under circumstances that would otherwise cause the conditions to change.

Many negative feedback mechanisms are present in the body. For example, the body temperature control mechanism is believed to include a set point that calls for the maintenance of a particular body temperature. The body contains receptors that monitor the actual body temperature and transmit feedback information to the brain, which contains a processing center. The processing center compares and integrates the information from the receptors about the actual body temperature with information from the set point about the called-for temperature, and if necessary initiates appropriate action to bring the actual body temperature to the called-for temperature. For example, if the actual body temperature is below the set point temperature, shivering may be initiated. The muscular contractions of shivering produce heat that tends to raise the body temperature to the set point level. Conversely, if the body temperature is above the set point temperature, sweating may be stimulated. The evaporation of the water in the sweat cools the body surface and tends to lower the body temperature to the set point level.

F1.12a

F1.12b

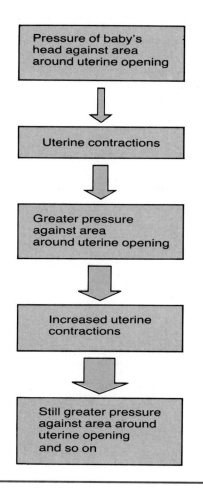

Figure 1.13
During birth, a positive feedback type of response stimulates increased uterine contractions.

POSITIVE FEEDBACK

In contrast to negative feedback mechanisms, positive feedback mechanisms maximize, rather than minimize, the difference between the actual and the set point level of a variable. They are called positive because the feedback tends to cause the level of the variable to change in the same direction as an initial change. With a positive feedback mechanism, an increase in the level of a particular variable above the set point level leads to increased feedback, which acts in a stimulatory fashion to increase the system's activity even more and thereby further increase the level of the variable. This leads to still greater feedback, which in turn leads to a further increase in the activity of the system, and so on and so on. Thus, positive feedback produces a cyclical effect in which a change in the level of a variable leads to further change in the same direction. In a positive feedback situation, however, the level of a variable does not necessarily continue to change indefinitely. For example, the level to which a variable can rise may ultimately be limited by the controlled system's maximum level of activity or by the amount of energy or raw materials available to the system.

Positive feedback does not lead to the maintenance of stable, homeostatic conditions, and consequently it does not occur in the body as frequently as negative feedback. However, positive feedback responses are occasionally evident. For example, during the birth of a baby, the pressure of the baby's head against the area around the opening of the mother's uterus stimulates the contraction of the uterine muscles (Figure 1.13). The contractions, in turn, increase the pressure of the head against the area around the uterine opening, which further stimulates the contraction of the uterine muscles. In this in-

F1.13

stance, a positive feedback type of response is clearly useful in promoting the expulsion of the baby from the uterus.

This chapter has provided a general background in the anatomy of the human body and an introduction to the mechanisms by which the internal environment of the body is kept relatively constant, even under fluctuating external conditions. In the forthcoming chapters we will expand on these topics as we examine the body in greater detail.

STUDY OUTLINE

FIELDS OF ANATOMY Subdivided into gross anatomy and microscopic anatomy. Gross anatomy can be studied by regions or by systems; microscopic anatomy can involve the study of cells or tissues. **pp. 3–6**

FIELDS OF PHYSIOLOGY Physiology is a vast field with many subdivisions, often concerned with different levels of organization. **p. 6**

INTERRELATIONSHIP OF STRUCTURE AND FUNCTION Structures, shapes, and organization of the parts of the body are closely associated with their functions. **p. 6**

BASIC STRUCTURAL LEVELS There are four structural levels in the body: cells, tissues, organs, and systems. **pp. 6–8**

CELLS The simplest structural level of the body.

TISSUES Three embryonic cell layers—the ectoderm, endoderm, and mesoderm—give rise to four tissue types:

EPITHELIAL TISSUES Cover the body surface and line body cavities, ducts, and vessels; they develop from embryonic ectoderm, mesoderm, and endoderm.

MUSCULAR TISSUES Move the skeleton, pump blood, and move food through the digestive tract; they develop from embryonic mesoderm.

NERVOUS TISSUES From the brain, spinal cord, and nerves; they develop from embryonic ectoderm.

CONNECTIVE TISSUES Are used for support and the attachment of other tissues; they develop from embryonic mesoderm.

ORGANS Tissues combine to form organs, such as the stomach.

SYSTEMS Ten major organ systems make up the body: integumentary, skeletal, muscular, nervous, endocrine, circulatory, respiratory, digestive, urinary, reproductive.

ANATOMICAL AND PHYSIOLOGICAL TERMINOLOGY Understanding word roots, prefixes, and suffixes is helpful in understanding anatomical terms; for example, the prefix *endo* means "within," as in *endocardium*, the inner lining of the heart. **p. 8**

BODY POSITIONS Anatomical position is achieved when body stands erect, feet together, upper limbs hanging at sides, palms forward, fingers extended, thumbs pointing away from body. The prone position is lying face down; the supine position, lying face up. **pp. 8–9**

DIRECTIONAL TERMS Denote direction in pairs of opposites, such as medial (toward midline of body) and lateral (away from midline of body). **pp. 9–10**

REGIONAL TERMS Refer to special areas, such as the cervical (neck), thoracic (chest), and plantar (sole of foot). **pp. 10–13**

ABDOMINAL CAVITY Divided into nine regions by two vertical lines and two horizontal lines; it is also divided into four quadrants by a horizontal plane and a vertical plane.

BODY PLANES **p. 13**

SAGITTAL PLANE The longitudinal section that divides the body, or its parts, into right and left parts.

FRONTAL PLANE A longitudinal section that runs at right angles to the sagittal plane, dividing the body into anterior and posterior parts.

TRANSVERSE PLANE Divides the body or its parts into superior and inferior parts.

BODY CAVITIES The body contains two main cavities. **pp. 13–15**

DORSAL (POSTERIOR) CAVITY Has two subdivisions:

CRANIAL CAVITY Houses the brain.

SPINAL (VERTEBRAL) CAVITY Contains the spinal cord.

VENTRAL (ANTERIOR) CAVITY Has two subdivisions:

THORACIC CAVITY Divided into:

Pericardial Cavity

Pleural Cavities (right and left, around lungs)

ABDOMINOPELVIC CAVITY Divided into:

Abdominal Cavity

Pelvic Cavity

MEMBRANES OF THE VENTRAL BODY CAVITIES Both thoracic and abdominopelvic cavities have visceral membranes that cover organs and parietal membranes that line the outer walls of the cavities. Space between two membrane regions is filled with fluid. **pp. 15–16**

HOMEOSTASIS   Refers to the relatively constant chemical and physical conditions that are maintained within the internal environment of the body.   **pp. 16–18**

HOMEOSTATIC MECHANISMS   Means by which the body senses and compensates for changes in the internal environment. The nervous and endocrine systems are the body's principal sensing and controlling systems.   **pp. 18–19**

NEGATIVE FEEDBACK   Used to maintain homeostasis. It involves several components: a controlled system, set point, receptor, and processing center.

POSITIVE FEEDBACK   Maximizes, rather than minimizes, the difference between the actual and the set point level of a variable. It is called positive because the feedback tends to cause the level of a variable to change in the same direction as an initial change. Not used by body to maintain homeostasis, but is used in particular circumstances, such as childbirth.   **pp. 20–21**

SELF-QUIZ

1. When studying anatomy and physiology at the same time, one should keep in mind the interrelation of: (a) nerves and muscles; (b) nerves and the skeleton; (c) structure and function.

2. Each different function of each body structure is usually for the well-being of the structure itself rather than for the general well-being of the entire body. True or False?

3. Nervous tissue is derived from the embryonic: (a) ectoderm; (b) endoderm; (c) mesoderm.

4. The various body cavities are lined with: (a) muscular tissue; (b) epithelial tissue; (c) connective tissue.

5. The component of the word *endocarditis* that signifies "inflammation" is: (a) endo; (b) card; (c) itis.

6. Match the following types of tissues with the appropriate lettered descriptions.

   Epithelial tissues
   Muscular tissues
   Nervous tissues
   Connective tissues

   (a) Derived from embryonic mesoderm
   (b) Derived for the most part from embryonic mesoderm
   (c) Derived from embryonic ectoderm, mesoderm, and endoderm
   (d) Form the brain, spinal cord, and nerves
   (e) Cover the surface of the body and line the body cavities
   (f) Derived from the embryonic ectoderm
   (g) Aid in digestion by moving food through the digestive tract
   (h) Used for support or attachment

7. Match the following body position and directional terms with the appropriate description.

   Prone position
   Anterior (ventral)
   Posterior (dorsal)
   Anatomical position
   Superior (cranial)
   Medial
   Supine position
   Inferior (caudal)
   Lateral
   Proximal
   Cervical
   Plantar
   Distal
   Lumbar
   Palmar

   (a) Toward the head
   (b) Away from the midline of the body
   (c) Lying face up
   (d) Away from the head
   (e) Toward the attached end of a limb
   (f) Sole of the foot
   (g) Front
   (h) Standing erect with feet together, arms at side, palms forward
   (i) Away from the attached end of a limb
   (j) The part of the back between the thorax and pelvis
   (k) Lying face down
   (l) Back
   (m) Neck
   (n) Toward the midline of the body
   (o) Anterior surface of hands

8. Which of the following directional terms are paired correctly? (a) superficial and deep; (b) medial and distal; (c) proximal and lateral.

9. The midline region superior to the umbilical region is called the epigastric region, and it contains most of the stomach. True or False?

10. The midline region directly inferior to the umbilical region is the: (a) hypochondriac; (b) lumbar; (c) hypogastric.

11. All sagittal sections other than the median sagittal section are: (a) intersagittal; (b) parasagittal; (c) intrasagittal.

12. The fluid associated with the body's dorsal cavities is the cerebrospinal fluid. True or False?

13. The portion of the thoracic cavity between the two pleural cavities is called the: (a) mediastinum; (b) meninges; (c) foramen magnum.

14. If the membranes of the heart become inflamed, the condition is known as: (a) pericarditis; (b) peritonitis; (c) pleuritis.

15. Which of the following does the abdominal cavity *not* contain? (a) spleen; (b) liver; (c) lungs.

16. Which of the following does the pelvic cavity *not* contain? (a) urinary bladder; (b) liver; (c) female internal reproductive organs.

17. Most of the organs within the abdominopelvic cavity are suspended from the anterior wall of the cavity by a single-layered membrane of peritoneum. True or False?

18. The mesenteries: (a) support the kidneys; (b) are mostly associated with the dorsal cavities; (c) are double-layered membranes.

19. The existence of a relatively constant internal environment around the cells of the body is referred to as *homeostasis*. True or False?

20. The homeostasis of the body is generally maintained by means of: (a) positive feedback mechanisms; (b) negative feedback mechanisms; (c) a combination of a and b.

LEARNING OBJECTIVES

After completing this chapter, you should be able to:

1. Define the term *chemical element,* and describe the three principal particles that make up an atom.

2. Distinguish between polar and nonpolar covalent bonds, and give an example of each.

3. Describe the composition of carbohydrates, lipids, proteins, and nucleic acids, and give an example of each.

4. Describe how enzymes catalyze metabolic reactions, and cite two examples of the regulation of enzymatic activity.

5. Distinguish between solutions, colloids, and suspensions.

6. Explain three different ways in which the concentrations of solutions are expressed.

7. Define an acid and a base.

8. Explain the use of pH units in the measurement of the hydrogen ion concentration of a solution.

9. Distinguish between diffusion and osmosis, and cite an example of each.

10. Explain the process of dialysis, and describe its application in the artificial kidney.

11. Distinguish between filtration and bulk flow, and cite an example of each.

CHAPTER CONTENTS

CHEMICAL ELEMENTS AND ATOMS

ELECTRON ENERGY LEVELS

ATOMIC NUMBER, MASS NUMBER, AND ATOMIC WEIGHT

ISOTOPES

RADIATION

BIOLOGICAL USES OF RADIOACTIVE ISOTOPES

CHEMICAL BONDS

CARBON CHEMISTRY

ENZYMES AND METABOLIC REACTIONS

SOLUTIONS

SUSPENSIONS

COLLOIDS

ACIDS, BASES, AND pH

DIFFUSION

OSMOSIS

DIALYSIS

BULK FLOW

FILTRATION

KEY TERMS AND DERIVATIVES

aerobic (*aer* = air) requiring oxygen to live or grow

covalent (*co* = with; *valent* = having power) *covalent bonds* are chemical bonds based on electron sharing between atoms

enzyme (*zyme* = causing to ferment) a substance formed by living cells that acts as a catalyst in bodily chemical reactions

isotopes (*iso* = equal) a different form of a given element; isotopes have the same atomic number but different mass numbers

The Chemical and Physical Basis of Life

Although the organs and tissues of the human body differ from one another in both form and function, they are composed of the same basic materials. The body is made up of chemical elements that interact with one another to form the anatomical structures and carry out the physiological processes characteristic of a living organism.

CHEMICAL ELEMENTS AND ATOMS

A **chemical element** is a substance that cannot be broken down into simpler material by chemical means. Each chemical element has its own name and a one- or two-letter symbol. For example, the symbol for the element oxygen is O, and the symbol for the element sodium is Na. At the present time, approximately 109 chemical elements are recognized, but only about 24 of these are normally found in the body (Table 2.1).

A chemical element is made up of extremely small units of matter called **atoms,** which are themselves composed of even smaller particles (Figure 2.1). Positively charged particles called **protons** are located in the central area, or nucleus, of an atom. Uncharged particles called **neutrons,** if present, are also located in the nucleus. Negatively charged particles called **electrons** are in constant motion around the nucleus.

The number of protons and the number of electrons in an atom—and, therefore, the number of positive charges and the number of negative charges—are equal. Consequently, an atom has no overall electrical charge and is electrically neutral.

Protons, neutrons, and electrons, like the atoms they compose, are matter, and, like all matter, they occupy space and possess mass. Each proton or neutron, however, has over 1800 times the mass of an electron. Thus, most of the mass of an atom is concentrated in the nucleus.

ELECTRON ENERGY LEVELS

Energy is the capacity to do work, and electrons possess different amounts of energy. In fact, the present view of atomic structure suggests that the electrons of an atom should be assigned to particular energy levels, which are numbered, starting with the lowest as one. The numbers are called **principal quantum numbers** (n). The lowest energy level ($n = 1$) can contain a maximum of 2 electrons, the second energy level ($n = 2$) can contain 8 electrons,

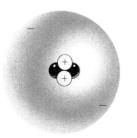

Figure 2.1

An atom of the element helium. There are two protons (plus signs) and two neutrons (black spheres) in the nucleus and two electrons (minus signs) in constant motion around the nucleus. The electrons do not follow fixed pathways as they move around the nucleus, but the color area indicates the region where they are located most of the time.

25

Table 2.1 Chemical Elements of the Human Body

Element	Symbol	Representative Functions
Carbon	C	A primary constituent of organic molecules, such as carbohydrates, lipids, and proteins.
Hydrogen	H	A component of organic molecules and water. As an ion (H^+), it affects the pH of body fluids.
Nitrogen	N	A component of amino acids, proteins, and nucleic acids.
Oxygen	O	A component of many molecules, including water. As a gas, it is important in cellular respiration.
Calcium	Ca	A component of bones and teeth. It is required for proper muscle activity and blood clotting.
Chlorine	Cl	Ionic chlorine (Cl^-) is one of the major anions of the body.
Iodine	I	A constituent of the thyroid hormones thyroxine and triiodothyronine.
Iron	Fe	A constituent of the hemoglobin molecule. It is also a component of a number of respiratory enzymes.
Magnesium	Mg	Found in bone. It is also an important coenzyme in a number of reactions.
Phosphorus	P	A component of bones, teeth, many proteins, and nucleic acids. It is also a constituent of energy compounds such as adenosine triphosphate (ATP) and creatine phosphate (CP).
Potassium	K	As an ion, potassium (K^+) is the major intracellular cation. Potassium is important in the conduction of nerve impulses and in muscle contraction.
Sodium	Na	As an ion, sodium (Na^+) is the major extracellular cation. It is important in water balance and in the conduction of nerve impulses.
Sulfur	S	A component of many proteins, particularly the contractile protein of muscle.
Chromium Cobalt Copper Fluorine Manganese Molybdenum Selenium Silicon Tin Vanadium Zinc	Cr Co Cu F Mn Mo Se Si Sn V Zn	These substances are required by the body in very small amounts. They are referred to as *trace elements*.

the third energy level ($n = 3$) can contain 18 electrons, and the fourth energy level ($n = 4$) can contain 32 electrons (Figure 2.2). Above the fourth energy level are still higher energy levels. (The electron energy levels are sometimes referred to as shells, with the K shell being equivalent to the $n = 1$ energy level, the L shell to the $n = 2$ energy level, the M shell to the $n = 3$ level, and so on.)

Electrons do not follow fixed pathways as they move around the nucleus of an atom, and it is impossible to determine the precise position of a specific electron at any one moment. However, electrons occupying higher energy levels are generally located farther from the nucleus than electrons occupying lower energy levels.

ATOMIC NUMBER, MASS NUMBER, AND ATOMIC WEIGHT

The number of protons in an atom is given by the atom's **atomic number.** The combined number of protons and neutrons is given by the atom's **mass number.** The number of neutrons in an atom is equal to the difference between the atom's atomic number and mass number. The atomic number of an atom is often indicated by a subscript preceding the symbol of the atom's chemical element, and the mass number is indicated by a superscript preceding the symbol. For example, an atom of oxygen with an atomic number of 8 and a mass number of 16 is written as $^{16}_{8}O$.

The total mass of an atom is called its **atomic weight.** The atomic weight of an atom is almost but not exactly equal to the sum of the masses of its constituent protons, neutrons, and electrons. The discrepancy is due to the fact that when protons, neutrons, and electrons combine to form an atom, some of their mass is converted to energy and is given off. The atomic numbers, mass numbers, and atomic weights of atoms of the chemical elements present in the body are given in Table 2.2.

ISOTOPES

All the atoms of a particular chemical element have the same number of protons. However, they can have different numbers of neutrons. These different forms of a given element are called **isotopes.** Isotopes have the same atomic number (number of protons) but different mass numbers (due to different numbers of neutrons). For example, the most common isotope of the element oxygen has eight protons and eight neutrons in its nucleus ($^{16}_{8}O$). However, less common isotopes of oxygen containing eight protons and nine neutrons ($^{17}_{8}O$) and eight protons and ten neutrons ($^{18}_{8}O$) also exist. In nature, oxygen occurs as a mixture of these isotopes (99.76% $^{16}_{8}O$; 0.039% $^{17}_{8}O$; and 0.20% $^{18}_{8}O$).

i'-so-tope

RADIATION

A number of isotopes are radioactive; that is, they emit various kinds of radiation. Radioactive emissions can be either particlelike or they can take the form of electromagnetic rays.

Particulate Radiation

Alpha Radiation

An alpha particle is essentially a helium nucleus (composed of two protons and two neutrons) that is positively charged. Alpha particles are emitted by some radioactive isotopes, such as uranium ($^{238}_{92}U$).

Beta Radiation

A beta particle has essentially the same mass as an electron, but it can be either positively or negatively charged. Positive beta particles, which are called positrons, are emitted by such isotopes as $^{30}_{15}P$. Negative beta particles, which are called negatrons, are emitted by $^{14}_{6}C$ and $^{136}_{53}I$.

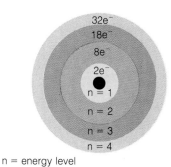

n = energy level
e⁻ = number of electrons
      energy level can hold

Figure 2.2
Diagrammatic representation of the first four electron energy levels ($n = 1$, $n = 2$, $n = 3$, and $n = 4$). The sphere in the center is the nucleus. The maximum number of electrons that can occupy each energy level is given.

Table 2.2

Table 2.2 Atomic Numbers, Mass Numbers, and Atomic Weights of Atoms of Body Elements*

Element	Symbol	Atomic Number	Mass Number	Atomic Weight
Calcium	Ca	20	40	40.08
Carbon	C	6	12	12.011
Chlorine	Cl	17	35	35.453
Chromium	Cr	24	52	51.996
Cobalt	Co	27	59	58.933
Copper	Cu	29	63	63.546
Fluorine	F	9	19	18.998
Hydrogen	H	1	1	1.008
Iodine	I	53	127	126.905
Iron	Fe	26	56	55.847
Magnesium	Mg	12	24	24.305
Manganese	Mn	25	55	54.938
Molybdenum	Mo	42	98	95.94
Nitrogen	N	7	14	14.007
Oxygen	O	8	16	15.999
Phosphorus	P	15	31	30.974
Potassium	K	19	39	39.098
Selenium	Se	34	80	78.96
Silicon	Si	14	28	28.086
Sodium	Na	11	23	22.99
Sulfur	S	16	32	32.064
Tin	Sn	50	120	118.69
Vanadium	V	23	51	50.942
Zinc	Zn	30	64	65.38

*The mass numbers are those of the most common isotope of the element. The atomic weights (expressed in atomic mass units) are weighted averages of the atomic weights of the naturally occurring isotopes of the element.

Electromagnetic Radiation: Gamma Radiation

A gamma ray can be regarded as a bundle of energy that is similar to a high-energy X ray. It has no detectable mass or electrical charge. Gamma ray emissions frequently accompany positive or negative beta emissions.

Transmutation

When a radioactive element emits radiation, it is itself altered. The radioactive element disappears and a different element appears. Thus, atoms of one element spontaneously change into atoms of another element. This change of one element into another is called **transmutation** (radioactive decay). For example, the emission of an alpha particle ($^4_2$He) by uranium ($^{238}_{92}$U) forms thorium ($^{234}_{90}$Th), as follows:

$$^{238}_{92}\text{U} \rightarrow \,^{234}_{90}\text{Th} + \,^4_2\text{He}$$

BIOLOGICAL USES OF RADIOACTIVE ISOTOPES

Radioactive isotopes have proven very useful in studies of living organisms and the chemical reactions that occur within them. The radioactive emissions of different isotopes can be measured by a variety of means. Consequently, radioactive carbon and other radioactive isotopes are used as tracers that can be introduced into the body by ingestion, inhalation, or injection and followed through a variety of physiological processes. For example, the thyroid gland normally removes iodine from the blood and uses it in the formation of thyroid hormones. Thus, radioactive iodine, which is also taken up by the thyroid gland, can be used to study thyroid function.

Since radioactive emissions can damage tissues, high levels of radiation are used to treat certain disease conditions. In some forms of cancer, radiation treatments are used to destroy actively dividing cancer cells.

CHEMICAL BONDS

When atoms are close enough to one another, the outer electrons of one atom may interact with those of others. As a result, attractive forces can develop between atoms that are strong enough to hold the atoms together. These attractive forces are called **chemical bonds.**

In general, many atoms in the human body appear to be particularly stable when their highest electron energy levels are either filled or contain eight electrons, and much chemical bonding results in atoms that have either filled highest electron energy levels or highest levels that contain eight electrons.

Covalent Bonds and Molecules

In many cases, chemical bonds result from atoms sharing electrons. Bonds based on electron sharing are called **covalent bonds,** and two or more atoms held together as a unit by covalent bonds are known as a **molecule.** Many covalent bonds are *single covalent bonds*, in which two atoms share a pair of electrons, with one electron being provided by each atom. However, *double covalent bonds*, in which two atoms share two pairs of electrons, and *triple covalent bonds*, in which two atoms share three pairs of electrons, also occur.

Nonpolar Covalent Bonds

A covalent bond in which there is an equal sharing of electrons between two atoms, and in which one atom does not attract the shared electrons more strongly than the other atom, is called a **nonpolar covalent bond.** For example, the hydrogen atom possesses one proton in its nucleus and one electron in its $n = 1$ energy level. When two hydrogen atoms combine to form a hydrogen molecule, their single electrons are shared equally between the two nuclei so that each electron spends equal time in the vicinity of each nucleus (Figure 2.3). Although neither atom gains complete possession of the other's electron, this sharing allows both to fill their highest ($n = 1$) electron energy levels. F2.3

Polar Covalent Bonds

A covalent bond in which there is an unequal sharing of electrons between two atoms, and in which one atom attracts the shared electrons more strongly

Figure 2.3
(a) Covalent bonding of two hydrogen atoms to form a hydrogen molecule. In this situation, the electrons (minus signs) are shared equally between the two nuclei (plus signs), resulting in a nonpolar covalent bond.
(b, c) Alternative methods of representing the composition of a molecule (in this case a hydrogen molecule) that will be used in later illustrations. In molecular diagrams (b) a shared pair of electrons (a single covalent bond) is represented by a straight line connecting the two bonded atoms. In molecular formulas (c) the number of atoms of each element that are present in a molecule is indicated, but there is little or no information as to how the atoms are connected.

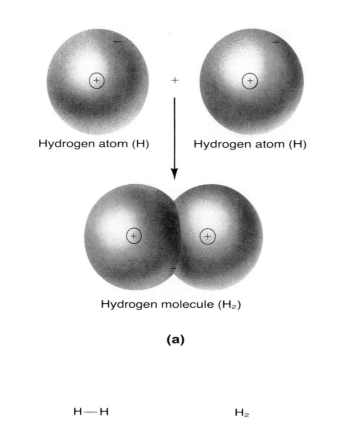

Hydrogen atom (H)   Hydrogen atom (H)

Hydrogen molecule (H_2)

(a)

H — H H_2

(b)   **(c)**

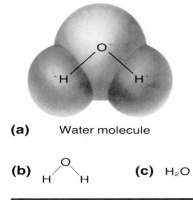

(a)   Water molecule

(b) $H \diagdown O \diagup H$ **(c)** H_2O

Figure 2.4
(a) Two hydrogen atoms (H) bonded covalently to an oxygen atom (O) to form a molecule of water. The shared electrons in this situation are attracted more strongly to the oxygen nucleus than to the hydrogen nuclei, resulting in polar covalent bonds. (b) A molecular diagram of a water molecule. (c) The molecular formula of a water molecule.   F2.5a

than the other atom, is called a **polar covalent bond.** There can be varying degrees of polarity of such bonds, depending on how strongly one atom is able to attract shared electrons from another atom.

The unequal electron sharing that occurs in polar covalent bonds gives rise to polar molecules that have both positive and negative areas. For example, when two hydrogen atoms each bond covalently to an oxygen atom to form water, the shared electrons are more strongly attracted to the oxygen of the water molecule than to the hydrogens (Figure 2.4). As a result, the shared electrons spend more time in the vicinity of the oxygen nucleus than in the vicinities of the hydrogen nuclei. Since electrons are negatively charged, the oxygen portion of the molecule becomes somewhat negative, and the hydrogen portions, with the positively charged protons of their nuclei less balanced by the presence of electrons, become somewhat positive.

Ionic Bonds and Ions

Often, the attraction of one atom for the electrons of another is so strong that electrons are not shared but are actually transferred from one atom to another; that is, they spend essentially all of their time in the vicinity of one nucleus and none in the vicinity of the other. This leaves one atom negatively charged (the one that gains electrons) and one atom positively charged (the one that loses electrons). Such charged atoms (or aggregates of atoms) are called **ions.** Positively charged ions are called **cations,** and negatively charged ions are called **anions.**

Opposite charges attract one another, and oppositely charged ions can be held together by this attraction to form electrically neutral substances known as **salts.** Such attractions are called **ionic attractions** or **ionic bonds.** For example, in the reaction between sodium and chlorine, each sodium atom loses the single electron in its highest ($n = 3$) electron energy level to a chlorine atom (Figure 2.5a). This produces positively charged sodium ions, which have their

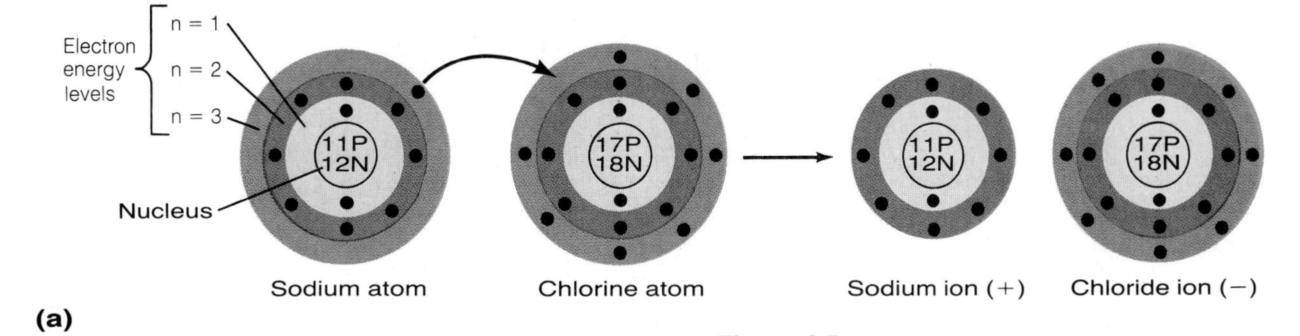

Electron energy levels

$n = 1$

$n = 2$

$n = 3$

Nucleus

Sodium atom   Chlorine atom   Sodium ion (+)   Chloride ion (−)

(a)

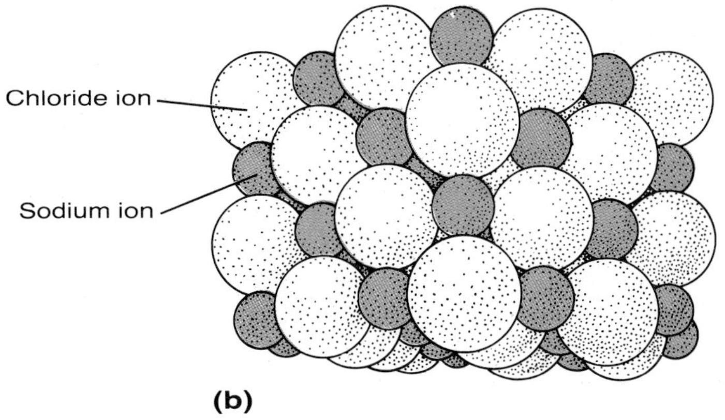

Chloride ion

Sodium ion

(b)

Figure 2.5

(a) In the reaction between sodium and chlorine, each sodium atom loses the single electron (black dot) in its highest ($n = 3$) electron energy level to a chlorine atom. This produces positively charged sodium ions, which have their highest ($n = 2$) electron energy levels filled, and negatively charged chloride ions, which have a stable eight electrons in their highest ($n = 3$) electron energy levels. The number of protons (P) and neutrons (N) in the different nuclei is indicated. **(b)** Positively charged sodium ions and negatively charged chloride ions attract one another, forming crystals of solid sodium chloride (table salt). In a sodium chloride crystal, each sodium ion is surrounded by six chloride ions, and each chloride ion is surrounded by six sodium ions.

highest ($n = 2$) electron energy levels filled, and negatively charged chloride ions, which have a stable eight electrons in their highest ($n = 3$) electron energy levels. The positively charged sodium ions and negatively charged chloride ions attract one another, forming crystals of solid sodium chloride (table salt). In a sodium chloride crystal, sodium ions and chloride ions are packed into a three-dimensional lattice in such a way that each positive sodium ion is surrounded on four sides and top and bottom by negative chloride ions, and each chloride ion is similarly surrounded by six sodium ions (Figure 2.5b). This is a particularly stable arrangement of positive and negative charges, and it occurs in many salts. Strictly speaking, there are no molecules in salts; there are only ordered arrays of ions in which no one positively charged ion belongs to any one negatively charged ion.

Hydrogen Bonds

Oppositely charged regions of polar molecules can attract one another, and an attraction of this sort that involves hydrogen and certain other atoms such as oxygen or nitrogen is called a **hydrogen bond.** For example, as noted previously, in the polar water molecule, the hydrogen portions of the molecule are somewhat positive, and the oxygen portion is somewhat negative. Consequently, the hydrogen portions of a water molecule can attract the oxygen portion of nearby water molecules, forming hydrogen bonds (Figure 2.6). Hydrogen bonds also occur in proteins and other large molecules found in the body.

F2.6

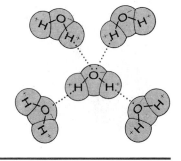

Figure 2.6

Hydrogen bonds (dotted lines) between polar water molecules.

CARBON CHEMISTRY

A tremendous number of molecules, especially those found in living organisms, contain the element carbon. A carbon atom has four electrons in its highest ($n = 2$) electron energy level, which it can share with other atoms.

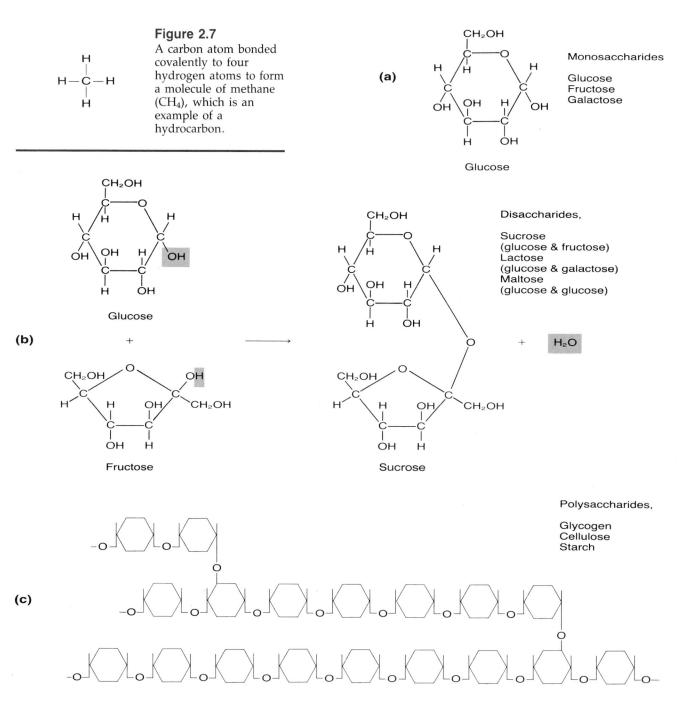

Figure 2.7

A carbon atom bonded covalently to four hydrogen atoms to form a molecule of methane (CH_4), which is an example of a hydrocarbon.

(a) Monosaccharides

Glucose
Fructose
Galactose

Glucose

(b)

Glucose

+

Fructose

Disaccharides,

Sucrose
(glucose & fructose)
Lactose
(glucose & galactose)
Maltose
(glucose & glucose)

Sucrose

+ H_2O

(c) Polysaccharides,

Glycogen
Cellulose
Starch

Portion of a glycogen molecule

Figure 2.8

Carbohydrates. **(a)** The monosaccharide glucose. **(b)** Monosaccharides are joined by bonds in a reaction that generally involves the removal of a molecule of water (a dehydration synthesis reaction, as shown here by the color areas). **(c)** A portion of the poly-saccharide glycogen; the ring structures represent glucose molecules.

F2.7

Carbon atoms, therefore, can each form four covalent bonds, and they commonly bond with hydrogen atoms or with other carbon atoms. Molecules composed of only carbon and hydrogen are called **hydrocarbons** (Figure 2.7). Molecules that contain carbon, hydrogen, and additional elements are called **hydrocarbon derivatives.** Together, hydrocarbons and hydrocarbon derivatives constitute the class of molecules known as **organic molecules.** (Organic molecules are so named because at one time they were thought to be made only by living organisms.) All other molecules are classified as **inorganic molecules.**

Although both organic and inorganic molecules are essential to life, much of the body's chemistry is organic in nature. Several major groups of organic molecules are of particular importance to the body.

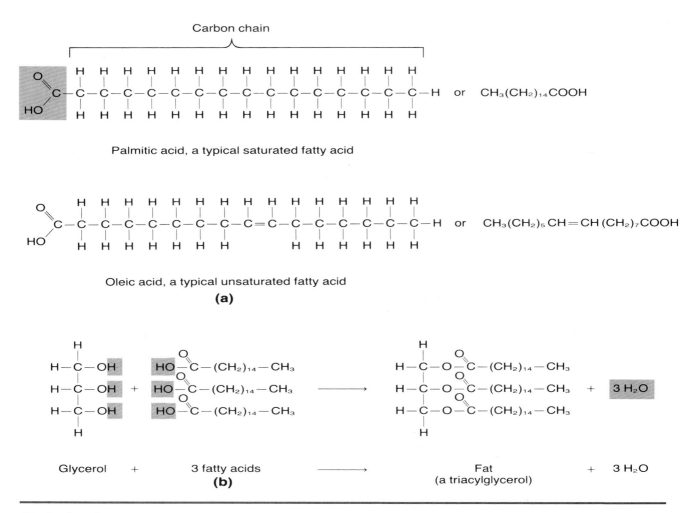

Palmitic acid, a typical saturated fatty acid

Oleic acid, a typical unsaturated fatty acid

(a)

Glycerol   +   3 fatty acids $\longrightarrow$ Fat   +   3 H_2O
(a triacylglycerol)
(b)

Carbohydrates

Carbohydrates, which include sugars and starch, are a major energy source for the body. In general, carbohydrates are composed of carbon, hydrogen, and oxygen, and the hydrogen and oxygen are frequently present in the same 2:1 ratio as they are in water. An important group of carbohydrates is the **monosaccharides,** or single-unit sugars, such as glucose ($C_6H_{12}O_6$) (Figure 2.8). Monosaccharide molecules can be linked together into larger molecules by synthetic reactions that generally involve the removal of a molecule of water (dehydration) at each linkage. The combination of two monosaccharide units produces a molecule called a **disaccharide.** Sucrose, or table sugar, is a disaccharide formed by bonding a glucose molecule to another monosaccharide called fructose. Much larger carbohydrate molecules can be formed by linking together many monosaccharide units. These molecules are called **polysaccharides.** Glycogen, a storage form of body carbohydrates, is a polysaccharide formed from thousands of glucose units bonded together into a single large molecule.

Lipids

Lipids are stored by the body as energy reserves, and they are utilized as structural components. Lipids are almost insoluble in water but very soluble in organic solvents such as benzene, ether, and chloroform. The major lipids are the fatty acids and fats, which, like the carbohydrates, are composed of the elements carbon, hydrogen, and oxygen. However, fatty acids and fats contain only small amounts of oxygen.

Fatty acids consist of chains of carbon atoms with an acid carboxyl group (COOH) at one end of the chain (Figure 2.9a). The most important of the fatty acids are: stearic ($C_{17}H_{35}COOH$); palmitic ($C_{15}H_{31}COOH$); oleic

F2.8

F2.9a

Figure 2.9
Fatty acids and fats.
(a) Saturated fatty acids contain only single covalent bonds between carbon atoms. Unsaturated fatty acids contain at least one double covalent bond between carbon atoms (a double covalent bond is one in which two pairs of electrons are shared between atoms, and it is indicated by two straight lines connecting the bonded atoms).
(b) Glycerol and fatty acids combine to form fat.

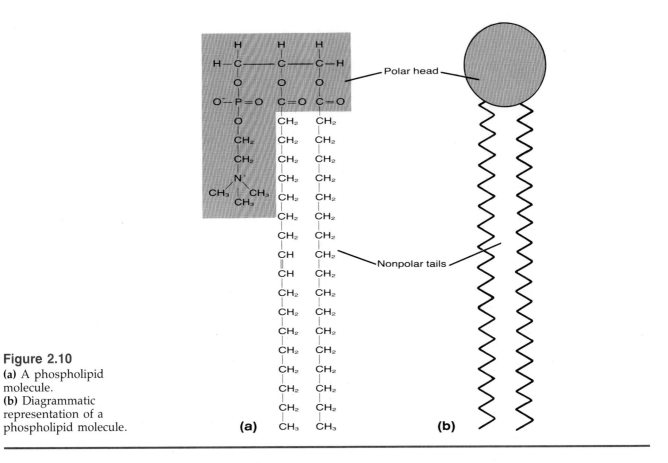

Figure 2.10
(a) A phospholipid molecule.
(b) Diagrammatic representation of a phospholipid molecule.

(a)

(b)

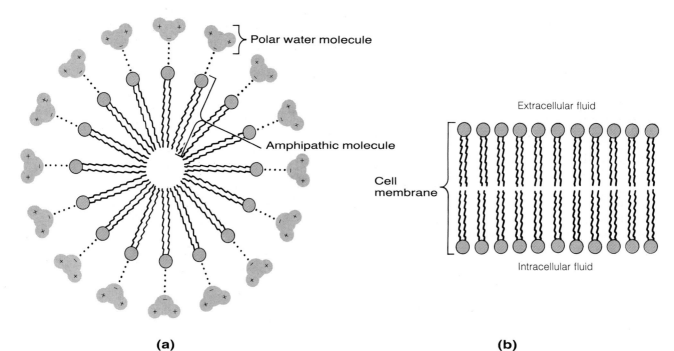

(a)

(b)

Figure 2.11
(a) In an aqueous environment, amphipathic molecules become organized into micelles. **(b)** In cellular membranes, amphipathic phospholipid molecules are organized into a bimolecular layer in which the polar regions of the molecules are located at the membrane surfaces and the nonpolar regions are oriented toward the center of the membrane.

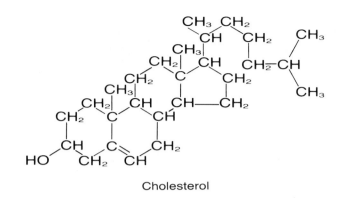

Cholesterol

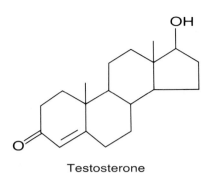

Testosterone

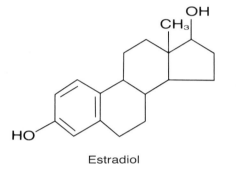

Estradiol

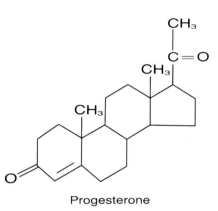

Progesterone

(C$_{17}$H$_{33}$COOH); linolenic (C$_{17}$H$_{29}$COOH); linoleic (C$_{17}$H$_{31}$COOH); and arachidonic (C$_{19}$H$_{31}$COOH).

Fats are formed from the union of fatty acids with the alcohol glycerol (Figure 2.9b). The combination occurs at the carboxyl-group end of the fatty acid by the removal of a molecule of water, which commonly occurs when the organic molecules of the body join with one another. If a single fatty acid molecule attaches to the glycerol molecule, the product is called a **monoacylglycerol (monoglyceride).** If two fatty acids attach to the glycerol molecule, the product is a **diacylglycerol (diglyceride),** and if three fatty acids attach, a **triacylglycerol (triglyceride)** is formed. Triacylglycerols are stored by the body as energy reserves.

Another group of lipids, the **phospholipids (phosphoglycerides),** are important components of cellular membranes. A phospholipid consists of a glycerol molecule to which two fatty acids are attached (Figure 2.10). In addition, a third molecule, containing a phosphate group and, usually, nitrogen, is attached to the glycerol. The phosphate and nitrogen can ionize. As a result, a phospholipid has a polar region at the end of the molecule where the glycerol and ionized phosphate and nitrogen are located, and a nonpolar region where the fatty acid chains of carbon atoms extend from the glycerol.

A molecule such as a phospholipid molecule that has polar or ionized groups at one end and a nonpolar region at the opposite end is called **amphipathic.** In an aqueous (water) environment, amphipathic molecules form spherical clusters known as **micelles** (Figure 2.11a). The polar regions of the molecules are located at the surface of the micelle, where they associate (in part by hydrogen bonds) with water molecules, and the nonpolar regions are oriented toward the center of the micelle. In cellular membranes, amphipathic phospholipid molecules are organized in a similar manner, with the polar regions of the molecules located at the membrane surfaces and the nonpolar regions oriented toward the center of the membrane (Figure 2.11b).

An additional group of lipids are the steroids. **Steroids** basically consist of four interconnected rings of carbon atoms that have few polar groups attached (Figure 2.12). Cholesterol and some hormones—for example, the sex hormones—are steroids that are important in the body.

F2.9b

F2.10

F2.11a

F2.11b

F2.12

Figure 2.12
Steroids. Testosterone is a male sex hormone. Estradiol and progesterone are female sex hormones. As can be seen from the rings of carbon atoms, carbon atoms and the hydrogen atoms bound to them are often not specifically indicated by the letters C or H, respectively, in the representation of the structure of a carbon-containing molecule. This method of representing molecular structure will be encountered in later figures.

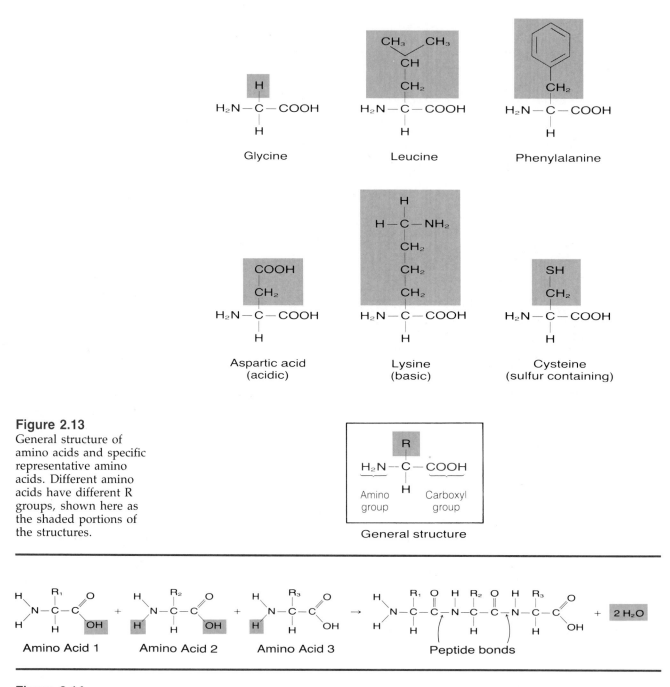

Figure 2.13
General structure of amino acids and specific representative amino acids. Different amino acids have different R groups, shown here as the shaded portions of the structures.

Figure 2.14
Peptide bonds link the amino group of one amino acid to the acid carboxyl group of another.

Proteins

Proteins are components of many body structures, and certain proteins—the enzymes—play critical roles in the chemical reactions that occur within the body. **Proteins** are large, complex molecules that are formed from smaller molecules called **amino acids.** Generally, amino acids have a central or alpha carbon to which is attached a hydrogen atom (H), an acid carboxyl group (COOH), an amino group (NH_2), and a fourth group that differs from one

F2.13

amino acid to another and is often indicated by the letter R (Figure 2.13). Thus, amino acids, and consequently proteins, contain nitrogen (from the amino group) in addition to carbon, hydrogen, and oxygen. They may also contain other elements, such as sulfur, depending on the constitution of the individual R groups. Approximately 20 different amino acids (different because they possess different R groups) are commonly found in the proteins of

Table 2.3   **The Twenty Amino Acids Found in Proteins**

Name	Three-Letter Abbreviation	One-Letter Abbreviation
Alanine	Ala	A
Arginine	Arg	R
Asparagine	Asn	N
Aspartic acid	Asp	D
Cysteine	Cys	C
Glutamic acid	Glu	E
Glutamine	Gln	Q
Glycine	Gly	G
Histidine	His	H
Isoleucine	Ile	I
Leucine	Leu	L
Lysine	Lys	K
Methionine	Met	M
Phenylalanine	Phe	F
Proline	Pro	P
Serine	Ser	S
Threonine	Thr	T
Tryptophan	Trp	W
Tyrosine	Tyr	Y
Valine	Val	V

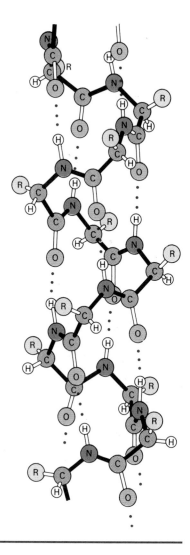

Figure 2.15
Secondary alpha helical structure of proteins. Dotted lines indicate hydrogen bonds.

the human body (Table 2.3). However, no one protein necessarily has all these different amino acids in its structure.

Proteins are formed from amino acids by reactions that bond the amino group of one amino acid to the acid carboxyl group of another, with the simultaneous loss of a molecule of water (Figure 2.14). This bond is called a **peptide bond.** Two amino acids joined together by a peptide bond form a **dipeptide.** Approximately ten or more amino acids linked into a chain by peptide bonds form a **polypeptide.** A **protein** is a chain of approximately 100 or more amino acids linked by peptide bonds.

The sequence of amino acids in a polypeptide chain or protein constitutes what is called the **primary structure** of the molecule. Hydrogen bonds that occur principally between the constituents of the different peptide bonds of the linked amino acids provide the polypeptide or protein with a **secondary structure.** For example, the hydrogen bonds cause some amino acid chains to form a coiled, helical structure called the alpha helix (Figure 2.15). Interactions between atoms of the R groups of different amino acids of an amino acid chain also occur. These interactions cause the amino acid chain (which may be in a helical configuration) to fold into a particular three-dimensional configuration. This folding provides the polypeptide or protein with a **tertiary structure** (Figure 2.16). Protein molecules can consist of a single amino acid chain, or several chains may link together through R-group interactions to form a multichain protein molecule. The interactions between the different amino acid chains of a multichain protein molecule provide still another level of structural organization to protein molecules—the **quaternary structure** (Figure 2.17).

F2.15

F2.16

F2.17

Nucleic Acids

Nucleic acids store and transmit information that is needed to synthesize the particular polypeptides and proteins present in the body's cells. Nucleic acids are complex molecules composed of structures known as purine and pyrimidine bases, five-carbon sugars (pentoses), and phosphate groups (which contain phosphorus and oxygen). A single base–sugar–phosphate unit is called a

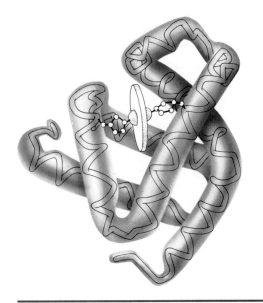

Figure 2.16
Tertiary structure of a molecule of the protein myoglobin. Areas of helical secondary structure are also evident. The tube (color) that appears to enclose the secondary helical structure has been drawn here to make it easier to visualize the molecule's tertiary structure.

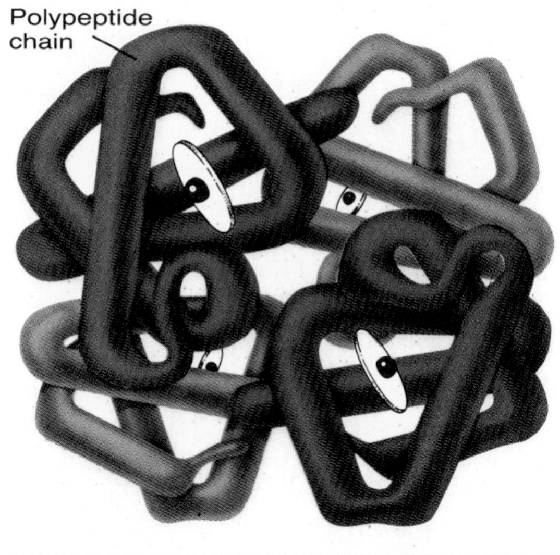

Figure 2.17
Quaternary structure of the protein hemoglobin. A single hemoglobin molecule is composed of four polypeptide chains linked to one another.

The particular configuration of the linked chains is called the quaternary structure of the protein.

Polypeptide chain

F2.18a
F2.18b

dee-ox-i-rye-bo-nu-klee'-ik

F2.19

F2.19

Table 2.4

nucleotide (Figure 2.18a). Individual nucleotides are linked together into a polynucleotide chain by bonds between the phosphate group of one nucleotide and the sugar of the next (Figure 2.18b). If the nucleotides in the polynucleotide chain contain the sugar ribose, the chain is called **ribonucleic acid,** or **RNA.** If the sugar is deoxyribose, the chain constitutes one portion of the two-chain molecule **deoxyribonucleic acid,** or **DNA.** A complete DNA molecule consists of two polynucleotide chains that run in opposite directions to one another. The purine and pyrimidine bases that are opposite one another in each polynucleotide chain link together by hydrogen bonds, and the two linked chains form a double spiral coil known as a *double helix* (Figure 2.19). The complete, two-chain DNA structure is commonly called a DNA molecule even though the two polynucleotide chains are held together by hydrogen bonds rather than by covalent bonds. The purine bases of DNA are adenine and guanine, and the pyrimidine bases are cytosine and thymine. Because of structural and bonding considerations, when the two polynucleotide chains of DNA link with one another, an adenine of one chain always bonds with a thymine of the other chain and vice versa, and cytosine always bonds with guanine and vice versa (Figure 2.19). This is called **complementary base pairing.** As a result of complementary base pairing, if the base sequence of one chain is known, the base sequence of the other can be predicted. The same bases are also found in RNA, with the exception that the base uracil substitutes for thymine. DNA is the genetic material of the cell, and it makes up a major portion of structures called chromosomes. DNA contains coded messages within its base sequence that instruct cells to synthesize particular polypeptides or proteins. RNA is involved in the transmission of the information of DNA to the active synthetic areas of the cells. DNA and RNA molecules are compared in Table 2.4 and are discussed further in Chapter 3.

Adenosine Triphosphate

a-den'-o-sene tri-fos'-fate

A substance called **adenosine triphosphate (ATP)** is the immediate source of energy for organismal activity—for example, muscle contraction. ATP is composed of the nitrogenous base adenine, the five-carbon sugar ribose, and

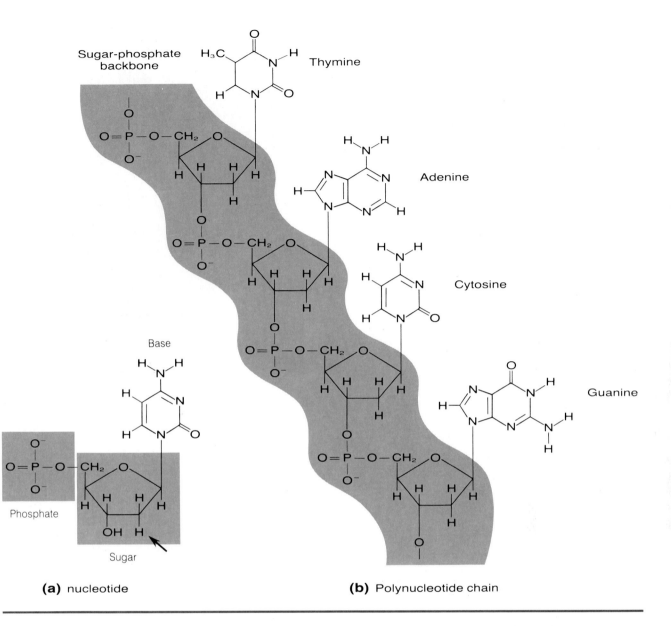

(a) nucleotide

(b) Polynucleotide chain

Figure 2.18
(a) The nucleotide pictured here contains the base cytosine and the sugar deoxyribose. The sugar ribose has the same structure, with the exception that the hydrogen indicated by the arrow is replaced by a hydroxyl (OH) group. **(b)** Four nucleotides linked in a polynucleotide chain. Adenine and guanine are purine bases and cytosine and thymine are pyrimidine bases.

Table 2.4   Comparison of DNA and RNA Molecules

Component or Characteristic	DNA	RNA
Purine bases	Adenine Guanine	Adenine Guanine
Pyrimidine bases	Cytosine Thymine	Cytosine Uracil
Sugar	Deoxyribose	Ribose
Number of strands	Double stranded	Single stranded

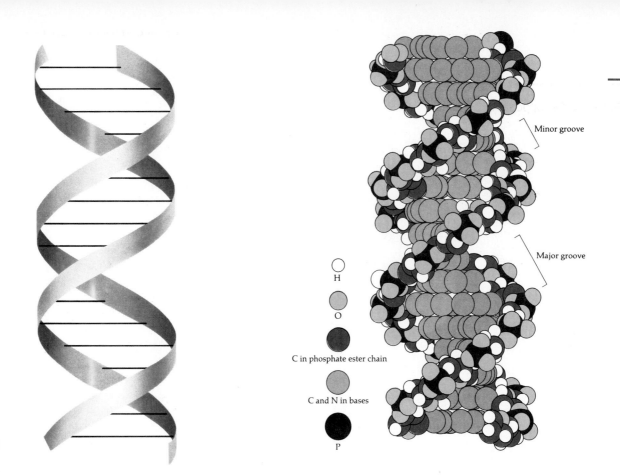

(a)

H

O

C in phosphate ester chain

C and N in bases

P

Minor groove

Major groove

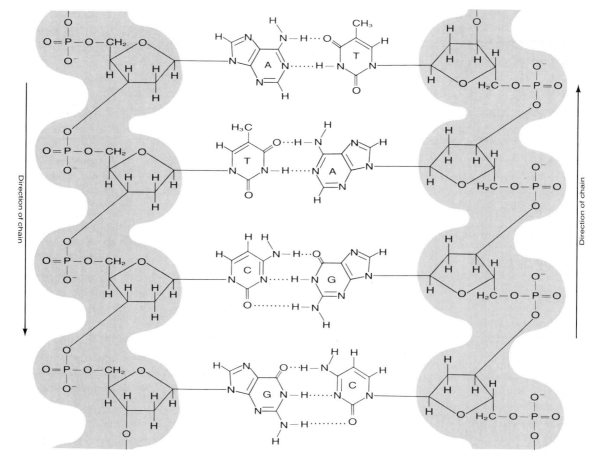

(b)

Direction of chain

Direction of chain

Figure 2.19 (opposite)

(a) The DNA double helix. The ribbons of the model on the left and the strings of dark and colored atoms in the space-filling model on the right represent the sugar–phosphate "backbones" of the two polynucleotide chains. The bases are stacked in the center of the molecule between the two backbones. The bases are 0.34 nm apart (nm = nanometer; 1 nm = 10^{-9} m). (b) A portion of a two-chain DNA molecule. Complementary base pairs (A–T; G–C) are held together by hydrogen bonds (dotted lines). The sugar–phosphate "backbone" of each chain is in the colored region. The phosphate groups are shown in ionized form. Note that the two chains run in opposite directions—that is, they are antiparallel.

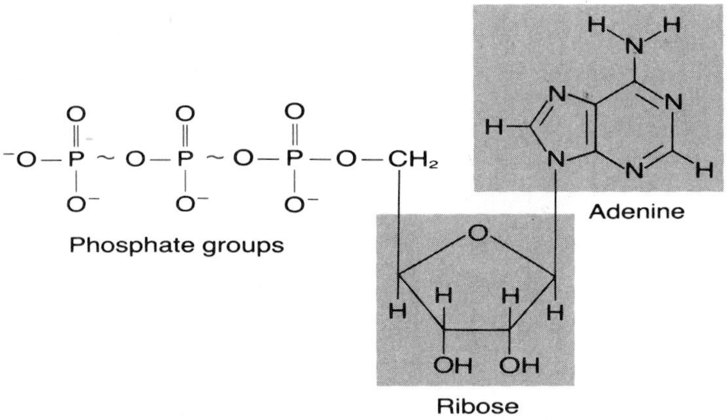

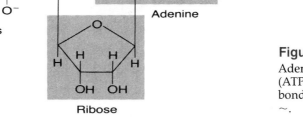

Phosphate groups

Adenine

Ribose

Figure 2.20

Adenosine triphosphate (ATP). High-energy bonds are indicated by ~.

three phosphate groups (Figure 2.20). The phosphate groups are linked by high-energy chemical bonds that, when broken, provide energy to support the activities of the body. For example, when the terminal phosphate group is split away from a molecule of ATP, a molecule of adenosine diphosphate (ADP) is produced and energy is released.

F2.20

$$\text{ATP} \rightleftharpoons \text{ADP} + \text{Phosphate} + \text{Energy}$$

Once ADP has been formed, it can be resynthesized into ATP, provided energy is available. As will be considered later (see Chapter 25), the breakdown of various food materials by chemical reactions that occur in the body releases energy that is utilized in ATP synthesis. In this way, energy contained within the chemical bonds of food materials is made available to the body in a useable form as ATP.

ENZYMES AND METABOLIC REACTIONS

The chemical reactions that constantly occur within the body are lumped together under the classification of **metabolism.** Metabolic reactions, in turn, are subdivided into anabolic, or synthesis, reactions that build up body structure, and catabolic, or decomposition, reactions that break down materials for various purposes such as the supply of energy. Metabolic reactions produce heat, and this heat can be of value because humans must maintain a constant

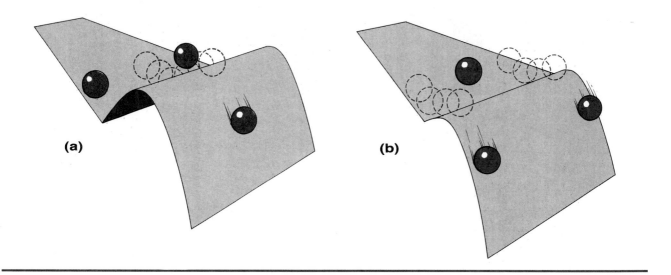

Figure 2.21

Diagrammatic representation of activation energy. **(a)** Just as the balls must have sufficient energy to roll up the small slope before they can roll down the large one, the atoms and molecules of the body must have sufficient energy before they can react with one another. Under normal body conditions, very few atoms and molecules have the energy required to react. **(b)** In the presence of an enzyme, the amount of energy required to react is lowered (the small uphill slope gets smaller); more atoms and molecules will have this lower amount of required energy to react with one another.

body temperature. When humans are exposed to cold, metabolic rates increase, and the heat generated is important in maintaining body temperature.

At normal body temperature, most metabolic reactions do not occur fast enough to benefit the body. Consequently, special catalysts—that is, substances that accelerate chemical reactions without undergoing any net chemical change during the reactions—are utilized to increase the rates of metabolic reactions to levels that can meet the body's needs. These biological catalysts are collectively termed **enzymes.** As a general rule, enzymes are protein in nature. However, in at least some cases, RNA can act as a biological catalyst for certain chemical reactions.

Action of Enzymes

F2.21   Enzymes increase reaction rates by lowering the **activation energy** required for metabolic reactions to occur (Figure 2.21). Under normal body conditions, few of the atoms and molecules that participate in a particular metabolic reaction have the necessary amount of energy to react with one another. In the presence of the proper enzyme, however, more of these atoms and molecules have the necessary energy to react because the enzyme lowers the amount of energy required. As a result, reactions that would otherwise proceed very slowly occur rapidly enough to be useful to the body.

Enzymes act by forming a temporary union with the reacting molecules, which are called **substrates.** This union is called an **enzyme–substrate (ES) complex.** The particular portion of an enzyme molecule with which a substrate combines is called the **active site** of the enzyme. Enzymes are very specific, and each catalyzes only individual reactions or limited classes of reactions. This specificity is due to the fact that a given enzyme has particular characteristics (such as its three-dimensional structure, or shape, and its electrical charge), and only certain substrates have the necessary complementary characteristics that allow them to unite with the enzyme. Therefore, only these substrates can form enzyme–substrate complexes with the enzyme and

F2.22   react (Figure 2.22).

Since the specific characteristics, such as three-dimensional shape and electrical charge, of a protein that acts as an enzyme are essential to its ability

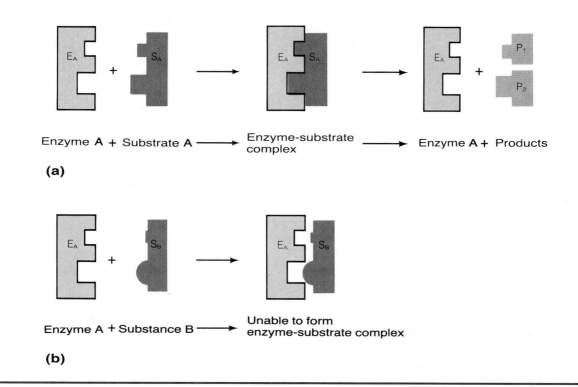

Enzyme **A** + Substrate **A** ⟶ Enzyme-substrate complex ⟶ Enzyme **A** + Products

(a)

Enzyme **A** + Substance **B** ⟶ Unable to form enzyme-substrate complex

(b)

to form an enzyme–substrate complex, factors that disrupt these characteristics can inactivate the enzyme and destroy its catalytic ability. Among the factors that can inactivate enzymatic proteins are variations in such internal environmental conditions of the body as temperature and acidity. The internal environment of the body, therefore, must be relatively constant—that is, homeostasis must be maintained—if the chemical reactions required for survival are to proceed in a stable fashion. Generally, enzymes are not destroyed during the course of the reactions they catalyze, and they appear at the conclusion of the reactions in the same states as when they entered. Many of the body's enzymes are initially produced in inactive forms (precursors) that must be activated before they will be effective catalysts.

Regulation of Enzymatic Activity

Because enzymes increase the rates of metabolic reactions, the amounts and types of different enzymes present in the body at any given moment play an important role in determining how rapidly different reactions can occur. Therefore, the regulation of enzymatic activity provides a method of regulating metabolic reactions.

Control of Enzyme Production and Destruction

One means of regulating enzymatic activity is to control the rates of production and destruction of particular enzymes, and thereby control the total amount of those enzymes present. If a particular enzyme is present in relatively large amounts, the reaction it catalyzes may proceed at a rapid rate, and much product may form. If only relatively small amounts of the enzyme are present, the reaction may proceed very slowly and only little product may form.

Control of Enzyme Activity

A second means of controlling enzymatic activity is to inhibit or enhance the activities of particular enzymes. In one type of inhibition, called **competitive inhibition,** an inhibitor molecule rather than a substrate molecule attaches reversibly to the active site of an enzyme molecule. Although the inhibitor

Figure 2.22
Diagrammatic representation of enzyme action. **(a)** Only certain substrates are able to unite with the active site of a given enzyme (E_A), and only these substrates (S_A) will form enzyme–substrate complexes and react. **(b)** Other materials (S_B) will not be able to form an enzyme–substrate complex with the particular enzyme illustrated.

molecule is generally similar in structure to the substrate molecule, it does not react to form a product as the substrate molecule does. Because both the inhibitor molecule and the substrate molecule can combine reversibly with the active site of the enzyme molecule, there is a competition between them for the active site. If a great number of inhibitor molecules are present, many of them occupy the active sites of the enzyme molecules, and relatively few substrate molecules are able to form enzyme–substrate complexes, and react.

Another type of enzyme inhibition, called **noncompetitive inhibition,** can take several forms. In some cases, a noncompetitive inhibitor molecule combines irreversibly with the active site of an enzyme molecule. As a result, substrate molecules cannot combine with the active site. In other cases, a noncompetitive inhibitor molecule combines with an enzyme molecule at a site other than the active site. As a result of this combination, the structure of the enzyme molecule is altered so that it is less able to form an effective enzyme–substrate complex with its substrate.

Enzymes subject to regulation by small molecules have special binding sites called allosteric effector sites to which regulatory molecules attach by weak bonds. The combination of a regulatory molecule with an allosteric site alters the structure of the enzyme molecule and either activates or inhibits the enzyme. Often, in what is essentially a negative feedback response, the final product of a series of enzymatically catalyzed reactions allosterically inhibits the first enzyme in the series. Thus, as the amount of product increases, the activity of the system that produces the product declines. Allosteric inhibition is a type of noncompetitive inhibition.

Some enzyme molecules are activated or inhibited by the chemical addition or removal of a phosphate (or other) group. The regulation of enzyme activity by phosphorylation or dephosphorylation differs from allosteric regulation in that it involves changes in covalent bonds in enzyme molecules, whereas allosteric regulation involves only patterns of weak bonds. The addition or removal of covalently bound groups requires the intervention of still other enzymes; kinase enzymes add phosphate groups, whereas phosphatase enzymes remove them. Since kinase and phosphatase enzymes are themselves subject to regulation—frequently by feedback mechanisms—enzyme regulation can involve a series of interacting events.

Cofactors

Often, enzymes require the presence of nonprotein structures called **cofactors** to actively catalyze reactions. A cofactor may be either a metal ion or a complex organic molecule called a **coenzyme.** Many vitamins, for example, act as coenzymes.

SOLUTIONS

Water is the medium in which all living processes occur, and life as we know it would be inconceivable in the absence of this molecule. In fact, the chemical reactions that occur continuously within the body involve, for the most part, reactants that are in aqueous (water) solutions. A **solution** is a homogeneous mixture of two or more components that can be gases, solids, or liquids. The components of a true solution cannot be distinguished in the mixture, and they do not settle out at an appreciable rate. If a beam of light is passed through a true solution, the light path will not be visible. The particles dispersed within a true solution are very small—generally in the atomic and molecular size range. For example, sodium chloride dissolved in water forms a true solution. When dealing with solutions, the material present in the greatest amount is generally called the **solvent,** whereas substances present in smaller amounts are generally called **solutes.**

Water as a Solvent

In the body, water is the principal solvent, and solutions most commonly encountered in living organisms result from dissolving gases, liquids, or solids in water. Water is an ideal solvent for living organisms for several reasons:

1. Water is a polar liquid whose chemical properties are such that many different materials can dissolve in it. For example, many salts dissolve easily in water because the positive and negative charges on the polar water molecules can substitute for the positive and negative charges on the ions that compose the crystal lattices of the salts. When a salt crystal dissolves in water, each positively charged ion becomes surrounded by water molecules that have their negative oxygen portions turned toward the ion, and each negatively charged ion becomes surrounded by water molecules that have their positively charged hydrogen portions closest (Figure 2.23). In this condition the ions from the salt crystal are said to be **hydrated.** Thus, when a salt crystal dissolves in water, it does not simply come apart into ions, but it is taken apart by the water molecules. Some physiologically important salts and the ions into which they separate or dissociate when they dissolve in water are listed in Table 2.5.

2. Water has a high specific heat. This means that, compared with other liquids, water requires a good deal of heat to raise its temperature. Thus, the heat produced by metabolism does not affect body temperature as much as if some other solvent were present.

3. Water has a high latent heat of vaporization. This means that, compared with other liquids, water requires a good deal of heat to change it from the liquid to the vapor state. Thus, the evaporation of water from body surfaces carries away large amounts of heat and provides the body with an effective cooling mechanism.

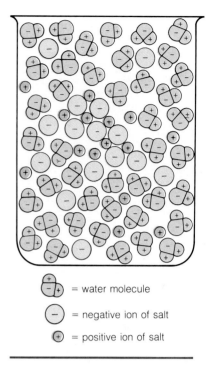

= water molecule

= negative ion of salt

= positive ion of salt

Figure 2.23
Breakup of a salt crystal by water molecules, with hydration of ions. Each salt ion in solution is surrounded by polar water molecules with the opposite charge to that of the ion turned toward it.

Table 2.5   Some Salts of Importance to the Body*

Salt	Ions
Sodium chloride (NaCl) → Sodium (Na^+) and chloride (Cl^-)	
Potassium chloride (KCl) → Potassium (K^+) and chloride (Cl^-)	
Calcium chloride ($CaCl_2$) → Calcium (Ca^{+2}) and chloride ($2\ Cl^-$)	
Magnesium chloride ($MgCl_2$) → Magnesium (Mg^{+2}) and chloride ($2\ Cl^-$)	
Calcium carbonate ($CaCO_3$) → Calcium (Ca^{+2}) and carbonate (CO_3^{-2})	
Calcium phosphate ($Ca_3[PO_4]_2$) → Calcium ($3\ Ca^{+2}$) and phosphate ($2\ PO_4^{-3}$)	
Sodium sulfate (Na_2SO_4) → Sodium ($2\ Na^+$) and sulfate (SO_4^{-2})	

*The arrows indicate the ions into which these salts dissociate when they are dissolved in water.

Methods of Expressing the Concentration of a Solution

It is often necessary to know the concentration of a solution, and concentrations are expressed in a number of ways. One way is simply to indicate the percent of solute in a solution by weight (wt./wt.), by volume (v./v.), or by a combination of the two (wt./v.). For example, a 10% solution by weight can be made by dissolving 10 grams of a solute (such as glucose) if enough solvent (such as water) to make 100 grams of solution. Similarly, a 10% solution by

Table 2.6 Charges of Common Body Ions

Ion	Charge
Bicarbonate (HCO_3^-)	−1
Calcium (Ca^{+2})	+2
Chloride (Cl^-)	−1
Hydrogen (H^+)	+1
Magnesium (Mg^{+2})	+2
Phosphate (PO_4^{-3})	−3
Potassium (K^+)	+1
Sodium (Na^+)	+1
Sulfate (SO_4^{-2})	−2

volume can be made by mixing 10 milliliters of a solute (such as ethyl alcohol) with enough solvent (water, for example) to make 100 milliliters of solution. Alternatively, a 10% solution by weight/volume can be made by dissolving 10 grams of a solute (such as glucose) in enough solvent (such as water) to make 100 milliliters of solution. When the concentration of a solution is expressed in percent, the measurements employed—wt./wt., v./v., or wt./v.—should always be indicated.

A second way of expressing the concentration of a solution is in terms of its **molarity.** In this method the amount of solute is not expressed in terms of weight or volume but in terms of moles. A **mole** of a substance is the amount of the substance in grams that is equal to the molecular weight of the substance. The **molecular weight** of a substance can be determined by adding together the atomic weights of the atoms that make up a molecule of the substance. For example, a molecule of glucose has a chemical formula of $C_6H_{12}O_6$, and is composed of 6 carbon atoms, 12 hydrogen atoms, and 6 oxygen atoms. The atomic weight of carbon is 12.011, the atomic weight of hydrogen is 1.008, and the atomic weight of oxygen is 15.999. A mole of glucose would therefore weigh:

$$
\begin{aligned}
6 \times 12.011 &= 72.066 \quad \text{(for carbon)} \\
12 \times 1.008 &= 12.096 \quad \text{(for hydrogen)} \\
6 \times 15.999 &= \underline{95.994} \quad \text{(for oxygen)} \\
&\ 180.156 \ \text{grams}
\end{aligned}
$$

A mole of any substance contains the same number of molecules as a mole of any other substance. This number is called *Avogadro's number*, and it is 6.022×10^{23}.

Expressing the concentration of a solution in terms of molarity indicates the number of moles of solute in a liter of solution. For example, a one molar (mol/L) solution can be made by adding one mole of a solute to enough solvent to make one liter (1000 milliliters) of solution. If the molecules of the solute remain intact in the solution, the solution will contain 6.022×10^{23} molecules of the solute.

The term *mole* can be applied to atoms and ions as well as to molecules. For example, a mole of potassium is equal to the atomic weight of potassium in grams (39.098 grams). A mole (39.098 grams) of potassium contains 6.022×10^{23} potassium atoms. In the case of salts, which are made up of ions, the chemical formula of a salt can be used to calculate the number of grams in a mole of the salt. Sodium chloride, for example, has the chemical formula NaCl, and a mole of sodium chloride is 58.443 grams of sodium chloride (because the atomic weights of sodium and chlorine are 22.99 and 34.453, respectively). A mole of sodium chloride contains 6.022×10^{23} sodium ions (one mole of sodium ions) and 6.022×10^{23} chloride ions (one mole of chloride ions). Similarly, a mole of the salt calcium chloride ($CaCl_2$) contains 6.022×10^{23} calcium ions (one mole of calcium ions) and $2 \times 6.022 \times 10^{23} = 12.044 \times 10^{23}$ chloride ions (2 moles of chloride ions).

Frequently, the amounts of ionized inorganic constituents in the body fluids are expressed in terms of **equivalents (Eq)** per liter of solution rather than in terms of moles per liter of solution. For present purposes, the number of grams in 1 equivalent of a particular ion can be said to be equal to the sum of the atomic weights of the atoms that compose the ion, divided by the charge of the ion, without regard for the sign (+ or −) of the charge (Table 2.6). For example, 1 equivalent of phosphate ions (PO_4^{-3}) is equal to:

$$
\begin{aligned}
1 \times 30.974 &= 30.974 \quad \text{(for phosphorus)} \\
4 \times 15.999 &= \underline{63.996} \quad \text{(for oxygen)} \\
94.970 \div 3 &= 31.657 \ \text{grams of phosphate ions}
\end{aligned}
$$

Similarly, 1 equivalent of calcium ions (Ca^{+2}) is equal to 40.08/2 = 20.04 grams of calcium ions.

The solutions encountered in the body generally have low concentrations, and the amounts of solute present are often expressed in terms of millimoles (1 millimole = 1/1000th of a mole) or milliequivalents (1 milliequivalent = 1/1000th of an equivalent), rather than in terms of moles or equivalents. The concentrations of solutions considered in this manner are then expressed as millimoles per liter (millimolar; mmol/L) or milliequivalents per liter (mEq/L).

SUSPENSIONS

Other types of mixtures besides true solutions are possible. Among these are **suspensions.** In a suspension, the dispersed particles are so large that they can be kept dispersed only by constant agitation. The components of the suspension remain distinct from one another, thus creating a heterogeneous mixture. If left to stand, the dispersed particles settle. A mixture of sand in water, for example, is an obvious suspension; in the body, blood cells are suspended in the fluid portion of the blood.

COLLOIDS

The transition from the homogeneity of true solutions to the heterogeneity of obvious suspensions is not a sudden one, and there are many gradations in between. **Colloids** are intermediate between true solutions and obvious suspensions. Colloids consist of particles that are dispersed in a medium, much like the composition of obvious suspensions. However, the particles are small enough that they do not readily settle out if left to stand. Colloidal systems can be distinguished from true solutions by passing a beam of light through them. If the system is a colloid, the beam will be scattered by the dispersed particles, and the light path will be visible when viewed from the side. In nature, colloids are very common. Milk is a colloid, and the protoplasm of living cells is considered to have colloidal properties.

ACIDS, BASES, AND pH

Water molecules exist mostly in an undissociated state, with the two hydrogens chemically bonded to the oxygen of the molecule. However, a very small percentage of water molecules (0.0000002%) dissociate into hydrogen ions (H^+) and hydroxide ions (OH^-). Although only about 1 in 500 million water molecules actually dissociates, this small amount of dissociation is one of the most important properties of water, and many metabolic reactions are critically dependent on the hydrogen ion concentration of the solution in which they occur.

Substances that alter the hydrogen ion concentration can be added to water. Substances that increase the hydrogen ion concentration are called **acids,** and substances that decrease the hydrogen ion concentration are called **bases.** Alternatively, acids may be defined as *proton donors* (a hydrogen ion is equivalent to a proton), whereas bases are *proton acceptors*. For example, hydrochloric acid (HCl) is an acid because it can dissociate into hydrogen ions (H^+) and chloride ions (Cl^-) and thereby serve as a hydrogen ion (proton) donor that can increase the hydrogen ion concentration of a solution:

$$HCl \rightleftharpoons H^+ + Cl^-$$

Conversely, ammonia (NH_3) is a base because it can accept hydrogen ions (protons) to form ammonium ions (NH_4^+) and thereby decrease the hydrogen ion concentration of a solution:

$$NH_3 + H^+ \rightleftharpoons NH_4^+$$

The hydrogen ion concentration of the body fluids must be maintained within narrow limits, and it is often necessary to know the hydrogen ion

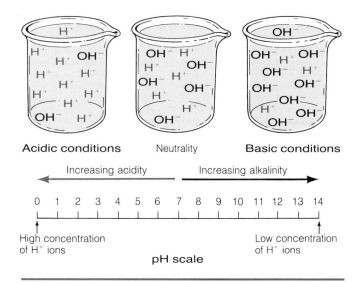

Acidic conditions   Neutrality   Basic conditions

←— Increasing acidity   Increasing alkalinity —→

0 1 2 3 4 5 6 7 8 9 10 11 12 13 14

↑ High concentration of H⁺ ions

Low concentration of H⁺ ions ↑

pH scale

Figure 2.24
The pH scale. pH values below 7 indicate acidic conditions, and values above 7 indicate basic, or alkaline, conditions.

Table 2.7 Typical pH Values of Body Fluids and Common Aqueous Solutions

Substance	Typical pH
1 molar hydrochloric acid	0
0.1 molar hydrochloric acid	1
Gastric juice	1.4–1.8
Lemon juice	2.1–2.3
Vinegar	2.4–3.4
Orange juice	2.8
Soda water	3.8
Tomato juice	4.1–4.2
Black coffee	5.0
Milk	6.3–6.9
Urine	4.8–7.4
Saliva	6.0–7.0
Pure water	7.0
Intestinal juice	6.5–7.5
Venous blood	7.35
Arterial blood	7.4
Bile	7.8–8.6
Pancreatic juice	8.0
Seawater	8.4
Milk of magnesia	10.5
Household ammonia	11.5–11.9
0.1 molar sodium hydroxide	13.0
1 molar sodium hydroxide	14.0

concentration of a solution. In pure water the small dissociation of water molecules results in a hydrogen ion concentration of 0.0000001 (1×10^{-7}) moles per liter. When expressed in this manner, the hydrogen ion concentration is such a small number that it is difficult to work with. Consequently, the hydrogen ion concentration of a solution is commonly expressed as a logarithmic value called the **pH,** which is determined according to the following relationship:

$$pH = \log_{10}\frac{1}{[H^+]}$$

where $[H^+]$ is the molar concentration of hydrogen ions in the solution. The pH of pure water is 7, and a solution with a pH of 7 is considered to be a neutral solution. As the hydrogen ion concentration of a solution increases, the pH value drops. Each drop of one unit on the pH scale indicates a tenfold increase in the hydrogen ion concentration. A solution with a pH of 6, therefore, has ten times the hydrogen ion concentration of a solution with a pH of

7, and a solution with a pH of 5 has 10 × 10 = 100 times the hydrogen ion concentration of a solution with a pH of 7. The lower the pH, therefore, the more acidic the solution (Figure 2.24). Similarly, pH values above 7 indicate progressively more basic, less acidic solutions. Typical pH values for several body fluids as well as for a number of other aqueous solutions are listed in Table 2.7.

DIFFUSION

Atoms, molecules, and ions are constantly in motion. Thus, they possess kinetic energy, or energy of motion. The velocity at which an atom, molecule, or ion moves is a function of temperature—the higher the temperature, the greater the velocity of movement.

Diffusion is the movement of atoms, molecules, or ions from one location to another as a consequence of their thermal motion. For example, if a cube of sugar is placed in a beaker of water, the sugar will dissolve, and the sugar molecules will eventually diffuse throughout the water (Figure 2.25). In the diffusion process, individual sugar molecules move in a random fashion. However, since initially there are more sugar molecules in the area that surrounds the sugar cube (where the sugar molecules are entering solution) than in areas of the water farther from the cube, it is probable that more sugar molecules will move away from the area around the sugar cube than will move toward the area. Thus, although individual sugar molecules move at random, the net movement of sugar molecules by diffusion (that is, the net diffusion) is from regions of high concentrations of sugar molecules to regions of low concentrations of sugar molecules (provided the temperature and pressure throughout the system are constant). **Net diffusion,** then, is the movement of a substance from a region of higher to a region of lower concentration as a consequence of the thermal motion of the atoms, molecules, or ions of the substance when the temperature and pressure throughout the system are constant.

In the preceding example, the movement of sugar molecules by diffusion will eventually result in a uniform distribution of sugar molecules throughout the water, and no differences in concentration will exist. When this occurs, the system is said to be in equilibrium. A state of equilibrium, however, does not imply a state where there is no longer any movement. Rather, it means that as many atoms, molecules, or ions of a substance—in this case, sugar molecules—enter a particular area at any one time as leave it. Thus, although the same atoms, molecules, or ions of a substance may not be in a given area, the same number of atoms, molecules, or ions of the substance will be. As a result, there is no net change in concentration.

Water molecules can also diffuse, and they can exhibit net diffusion from regions of higher to regions of lower concentration. But how can there be different concentrations of water? Consider the following: If one beaker is filled with pure water, there will be 100% water in the beaker, and the entire volume of the beaker will be occupied by water molecules. If an identical second beaker is filled with sugar solution, some of the volume will be occupied by water molecules and some by sugar molecules. Therefore, there will not be 100% water molecules in this second beaker, and the concentration of water in this beaker will be lower than the concentration of water in the first beaker (Figure 2.26).

If, instead of keeping the pure water and the sugar solution in separate beakers, they are placed on separate sides of a single container that has a removable barrier between the sides, the following will occur when the barrier is removed (Figure 2.27): The sugar molecules will show a net diffusion from their region of higher concentration (the sugar solution side) to their region of lower concentration (the pure water side). Likewise, water molecules will exhibit a net diffusion from their region of higher concentration (the pure water side) to their region of lower concentration (the sugar solution side). Both processes will continue until equilibrium is reached—that is, until

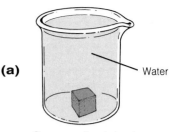
(a) Water

Sugar cube intact

(b) Sugar molecules

Sugar cube partially dissolved

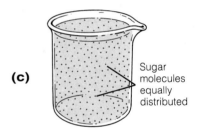
(c) Sugar molecules equally distributed

Sugar cube dissolved showing state of equilibrium

Figure 2.25
Diffusion. **(a)** Initially, all the sugar molecules are within the sugar cube. **(b)** As the cube dissolves, sugar molecules disperse in the water. Note that their concentration is highest near the sugar cube, where they are entering solution. **(c)** When all of the sugar has dissolved and the system is at equilibrium, sugar molecules are dispersed evenly throughout the water as a result of their random movement.

Figure 2.26
When solute molecules (in this case, sugar) are added to water, the concentration of water molecules in a given volume decreases. There can be different concentrations of water (solvent), just as there can be different concentrations of solute in a solution.

● Water molecule
● Sugar molecule

Pure water
(high water
concentration)

Sugar solution
(lower water
concentration)

Semipermeable
membrane

Pure
water

Sugar
solution

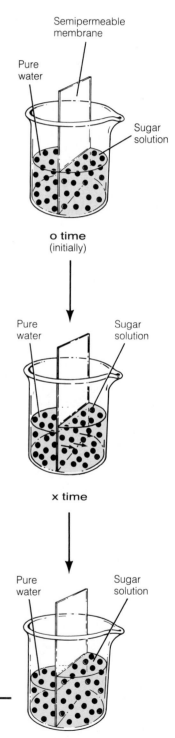

o time
(initially)

Pure
water

Sugar
solution

x time

Figure 2.27
Redistribution of water and sugar molecules when a barrier to molecular movement is removed. Sugar molecules show a net diffusion from their region of high concentration to their region of low concentration. Likewise, water molecules show a net diffusion from their region of high concentration to their region of low concentration. At equilibrium, both sugar and water molecules are randomly and equally distributed throughout the container.

Barrier

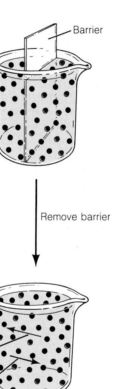

Remove barrier

Sugar

Water

Pure
water

Sugar
solution

Figure 2.28
The process of osmosis. See the text for a detailed discussion.

y time
(at equilibrium)

both sugar molecules and water molecules are equally distributed throughout the system. At this point, no regions of different concentration exist and thus no further net diffusion of either sugar molecules or water molecules will occur.

OSMOSIS

Consider another situation. Suppose pure water placed on one side of a container is separated by a membrane from a sugar solution placed on the other side of the container. Suppose further that the membrane is a **semipermeable membrane** that allows the passage of solvent (water molecules) but not solutes (sugar molecules) (Figure 2.28). In this situation, water will exhibit a net movement from its region of higher concentration through the membrane to its region of lower concentration. Sugar molecules, however, will not be able to move through the membrane. As a result, there will be a net movement of water into the sugar solution, but no movement of sugar into the pure water. The movement of water that takes place across the semipermeable membrane in this instance is an example of osmosis. More generally, **osmosis** is the movement of solvent through any membrane in response to a concentration difference (a concentration gradient) across the membrane.

F2.28

In the preceding example, the occurrence of osmosis will cause the volume of pure water on the one side of the membrane to decrease gradually, and the volume of sugar solution on the other side to increase gradually. However, there will always be a higher concentration of water on the pure water side of the membrane than on the sugar solution side. Nevertheless, the net movement of water into the sugar solution does not go on indefinitely. Eventually, the pressure of the additional volume of fluid on the sugar solution side (the **hydrostatic pressure**) rises to a point at which it is able to balance the force tending to move water into the sugar solution by osmosis, and no further net movement of water into the sugar solution occurs. At this point, the hydrostatic pressure that is exerted against the membrane on the sugar solution side is great enough to force water molecules across the membrane from the sugar solution into the pure water as fast as they move from the pure water into the sugar solution. The result is an equilibrium in which there are not equal concentrations of sugar molecules and water molecules in all parts of the system, but one in which opposing forces prevent the net movement of water molecules and a membrane prevents any movement of sugar molecules.

The pressure required to prevent the net movement of pure water into a solution when the water is separated from the solution by a semipermeable membrane is a measure of the solution's **osmotic pressure.** The osmotic pressure of a solution, which is expressed in units called osmoles or milliosmoles, indicates the tendency of water to move by osmosis into the solution. The osmotic pressure of a solution depends basically on the number of solute particles present and not on their nature. The greater the number of solute particles in a given volume of solution, the greater the osmotic pressure of the solution. If two solutions have the same osmotic pressure, they are said to be **isosmotic** (*iso* = same). If two solutions have different osmotic pressures, the one with the higher osmotic pressure is said to be **hyperosmotic** to the one with the lower osmotic pressure, and the solution with the lower osmotic pressure is said to be **hypoosmotic** to the solution with the higher osmotic pressure (*hyper* = above; *hypo* = below).

The membrane that surrounds living cells is not a simple semipermeable membrane but behaves like a **selectively permeable membrane.** A selectively permeable membrane is a membrane that does not permit the free, unhampered movement of all solutes present, but maintains a differential concentration (a concentration gradient) of at least one solute across itself. Thus, a selectively permeable membrane is not equally permeable to all solute particles present. Osmosis can occur across a selectively permeable membrane in response to different concentrations of water on either side of the membrane.

DIALYSIS

Some membranes are permeable to water molecules and small particles—for example, sodium ions and chloride ions—that may be present in the water, but they are not permeable to large particles such as protein molecules. If such a membrane is placed between a sodium chloride–protein solution on one side and pure water on the other, then water molecules, sodium ions, and chloride ions will be able to pass through the membrane but protein molecules will not. As a result there will be a net diffusion of sodium ions and chloride ions from the solution into the pure water, and water will exhibit a net movement into the sodium chloride–protein solution. If the pure water side is constantly drained away and replenished so that no equilibrium is established, the sodium chloride can be removed from the sodium chloride–protein solution. This process of selectively separating substances in a liquid by taking advantage of their differing diffusibilities through porous membranes is called **dialysis.**

Dialysis achieves its most dramatic application in the artificial kidney. In this device, blood is passed through a membranous tube immersed in a bathing medium of known composition. The membranous tube is permeable to water molecules and other small particles, but it is not permeable to large particles such as protein molecules. Since the atoms, molecules, or ions of most wastes are small particles, these can be removed from the blood while vital protein molecules are retained. Substances required by the body that are also composed of small particles can be retained by having them present in the bathing fluid in the same concentration as they are in the blood. Thus, there will be no concentration difference of these substances on either side of the membrane and, therefore, no net diffusion.

BULK FLOW

Bulk flow is the movement of atoms, molecules, ions, or other particles as a unit in one direction as the result of forces that push them from one point to another. For example, when a sodium chloride solution is pumped through a pipe, the ions and molecules of the solution travel in one direction as a unit.

The bulk flow of a liquid or gas depends on an inequality of pressure that acts on the liquid or gas. If different regions of a liquid or gas are subjected to unequal pressures, the liquid or gas will flow from the region of higher pressure to the region of lower pressure. The flow of blood within the blood vessels and the movement of air into and out of the lungs during respiration are examples of bulk flow that occur in the body.

FILTRATION

Filtration is a process that separates one or more components of a mixture from the other components. In this process, the mixture is forced through a porous filter or membrane by mechanical forces such as hydrostatic pressure. The direction of movement is from a region of higher pressure to a region of lower pressure, and large particles that are present in the mixture may not be able to pass through the filter or membrane. For example, consider a glucose solution to which sand has been added. If this mixture is forced through a membrane containing pores that are too small to pass the sand, water molecules and glucose molecules will move through the membrane, but sand particles will remain behind.

Within the body, a filtration process that allows the passage of only small particles commonly occurs. The blood pressure within blood vessels called capillaries acts to force the fluid portion of the blood, including small dissolved particles such as glucose molecules or the ions of salts, across the capillary walls and out of the circulatory system. However, blood cells and protein molecules that are too large to leave the vessels remain behind in the blood.

STUDY OUTLINE

CHEMICAL ELEMENTS AND ATOMS p. 25

CHEMICAL ELEMENT A substance that cannot be broken down into simpler material by chemical means.

ATOMS Make up chemical elements and are composed of: protons (in nucleus; have positive charge); neutrons (in nucleus; electrically neutral); electrons (move around nucleus; have negative charge).

ELECTRON ENERGY LEVELS Present view of atomic structure suggests that electrons should be assigned to particular energy levels. The statistical distribution of electrons is such that electrons occupying higher energy levels will generally be found farther from the nucleus of an atom than electrons occupying lower energy levels. pp. 25–27

ATOMIC NUMBER, MASS NUMBER, AND ATOMIC WEIGHT p. 27

ATOMIC NUMBER Number of protons in an atom.

MASS NUMBER Combined number of protons and neutrons in an atom.

ATOMIC WEIGHT Total mass of an atom.

ISOTOPES Atoms of an element may differ from one another in numbers of neutrons; these different forms of given elements are called isotopes. Isotopes have same number of protons (atomic number) but different number of neutrons (mass numbers differ). p. 27

RADIATION A number of isotopes emit radiation in form of particles or electromagnetic rays. pp. 27–29

PARTICULATE RADIATION

ALPHA RADIATION Particle is essentially a helium nucleus (two protons, two neutrons); has positive charge.

BETA RADIATION Particle has mass of electron but may have positive or negative charge.

ELECTROMAGNETIC RADIATION: GAMMA RADIATION Gamma ray can be regarded as a bundle of energy similar to a high-energy X ray.

TRANSMUTATION As radioactive element emits radiation, the element disappears and a different element appears.

BIOLOGICAL USES OF RADIOACTIVE ISOTOPES Radioactive isotopes used as tracers can be followed through physiological processes. Radiation is used to treat diseases such as cancer. p. 29

CHEMICAL BONDS Attractive forces between atoms that are strong enough to hold atoms together. pp. 29–31

COVALENT BONDS AND MOLECULES Bonds based on electron sharing are called covalent bonds, and two or more atoms held together by covalent bonds are known as a molecule.

NONPOLAR COVALENT BONDS Involve equal sharing of electrons between atoms.

POLAR COVALENT BONDS Involve unequal sharing of electrons between atoms; condition gives rise to polar molecules that have both positive and negative areas.

IONIC BONDS AND IONS Transfer of electrons from one atom to another creates charged atoms (or aggregates of atoms) called ions; ionic bonds are attractions between oppositely charged ions that can hold ions together to form electrically neutral substances known as salts.

HYDROGEN BONDS Attractions between oppositely charged regions of polar molecules that involve hydrogen and certain other atoms such as oxygen or nitrogen.

CARBON CHEMISTRY Molecules composed only of carbon and hydrogen are called hydrocarbons; molecules containing carbon, hydrogen, and other elements are called hydrocarbon derivatives; hydrocarbons and hydrocarbon derivatives constitute organic molecules; all other molecules are called inorganic. pp. 31–41

CARBOHYDRATES Are composed of carbon, hydrogen, oxygen, with hydrogen and oxygen frequently in a 2:1 ratio.

MONOSACCHARIDES Single-unit sugars, such as glucose.

DISACCHARIDES Two monosaccharide units, for example, sucrose.

POLYSACCHARIDES Many monosaccharide units, for example, glycogen.

LIPIDS Several major types.

FATTY ACIDS Chains of carbon atoms with acid carboxyl group (COOH) at one end.

FATS Union of fatty acids with glycerol.

PHOSPHOLIPIDS Glycerol united with two fatty acid molecules and a third molecule containing a phosphate group and, usually, nitrogen; important components of cellular membranes.

STEROIDS Basically consist of four interconnected rings of carbon atoms that have few polar groups attached; examples are cholesterol and sex hormones.

PROTEINS Large, complex molecules formed from amino acids. About 20 different amino acids form human proteins. Proteins are formed from amino acids by linking the amino acids with peptide bonds.

NUCLEIC ACIDS   Composed of purine and pyrimidine bases, five-carbon sugars, and phosphate groups; single base–sugar–phosphate unit called a nucleotide.

RNA   Formed when the sugar in a polynucleotide chain is ribose.

DNA   Formed when the sugar in two polynucleotide chains is deoxyribose; makes up major portion of chromosomes.

ADENOSINE TRIPHOSPHATE   Composed of adenine, ribose, and three phosphate groups; serves as immediate source of energy for organismal activity.

ENZYMES AND METABOLIC REACTIONS

Sum total of chemical reactions in body is called metabolism. Metabolic reactions are subdivided into anabolic, or synthesis, reactions and catabolic, or decomposition, reactions. Enzymes are biological catalysts that speed up metabolic chemical reactions by lowering activation energies.   **pp. 41–44**

ACTION OF ENZYMES   Enzyme forms enzyme–substrate complex, with substrate at enzyme active site.

REGULATION OF ENZYMATIC ACTIVITY   Provides a method of regulating metabolic reactions.

CONTROL OF ENZYME PRODUCTION AND DESTRUCTION   Controls amounts of particular enzymes present.
1. Competitive inhibition–inhibitor molecule binds reversibly to active site of enzyme.
2. Noncompetitive inhibition–inhibitor molecule binds irreversibly to active site or to site other than active site of enzyme.

COFACTORS   Substances often needed by enzymes to catalyze reactions.

SOLUTIONS   Homogeneous mixtures of two or more

components that can be gases, solids, liquids; material present in a solution in greatest amount is generally called the solvent; materials present in lesser amounts are generally called solutes.   **pp. 44–47**

WATER AS A SOLVENT   Water is ideal solvent for living organisms because:
1. Many different materials dissolve in water.
2. Water has a high specific heat.
3. Water has a high latent heat of vaporization.

METHODS OF EXPRESSING THE CONCENTRATION OF A SOLUTION
1. As a percent of a solute by weight, by volume, or by a combination of the two.
2. As molarity (moles of a solute per liter of solution).
3. As equivalents per liter of solution (for ionized inorganic constituents of body fluids).

SUSPENSIONS   Mixtures in which the dispersed

particles are so large that they tend to settle out.   **p. 47**

COLLOIDS   Between true solutions and obvious sus-

pensions; dispersed particles are small enough that they do not readily settle out.   **p. 47**

ACIDS, BASES, AND pH   In pure water there is

a balance between hydrogen ions and hydroxide ions, since dissociation of a water molecule produces one ion of each type.   **pp. 47–49**

ACIDS   Substances that increase hydrogen ion concentration of an aqueous solution (proton donors).

BASES   Substances that decrease the hydrogen ion concentration of an aqueous solution (proton acceptors).

pH SCALE   A logarithmic mathematical scale used to indicate the hydrogen ion concentration of a solution.

DIFFUSION   The movement of atoms, molecules, or

ions from one location to another as a consequence of their thermal motion. Net diffusion is the movement of a substance from a region of higher concentration to a region of lower concentration as a consequence of the thermal motion of the atoms, molecules, or ions of the substance.   **pp. 49–51**

OSMOSIS   The movement of solvent through any

membrane in response to a concentration difference across the membrane.   **p. 51**

OSMOTIC PRESSURE   The tendency of water molecules to enter a solution in response to a concentration gradient.

ISOSMOTIC   Describes two solutions having the same osmotic concentration.

HYPEROSMOTIC AND HYPOOSMOTIC   Describe two solutions of different osmotic concentrations where the more concentrated one is hyperosmotic to the less concentrated, and the less concentrated is hypoosmotic to the more concentrated.

DIALYSIS   A process that selectively separates sub-

stances in a liquid by taking advantage of their differing diffusibilities through porous membranes.   **p. 52**

APPLICATION IN ARTIFICIAL KIDNEY   Blood is passed through a membranous tube immersed in bathing medium of known composition; wastes that are small particles are removed; vital protein molecules are retained; essential small particles are also retained by having them at same concentration in bathing medium as in blood, hence no net diffusion across membrane.

BULK FLOW   The movement of atoms, molecules,

ions, or other particles as a unit in one direction as the result of forces that push them from one point to another.   **p. 52**

FILTRATION   A process that separates one or more

components of a mixture from the others. In this process, the mixture is forced through a porous filter or membrane by mechanical forces such as hydrostatic pressure.   **pp. 52–53**

SELF-QUIZ

1. At the present time, approximately 109 elements are recognized but only about 24 of these are normally found in the body. True or False?

2. Most of the mass of an atom is concentrated in the atom's: (a) outer electron energy levels; (b) ionic bonds; (c) nucleus.

3. A neutral atom contains the same number of electrons as it does: (a) protons; (b) neutrons; (c) energy levels.

4. The number of neutrons contained in $^{17}_{8}O$ is: (a) 8; (b) 9; (c) 17.

5. Salts are made up of positive and negative ions that form nonpolar covalent bonds with one another. True or False?

6. Molecules composed of only carbon and hydrogen are called: (a) hydrocarbons; (b) carbohydrates; (c) inorganic molecules.

7. Fats are formed from the union of fatty acids with the alcohol glycerol. True or False?

8. Proteins are formed from amino acids by linking the amino group of one amino acid to the acid carboxyl group of another with: (a) an ionic bond; (b) a peptide bond; (c) a disulfide bond.

9. Match the terms with the appropriate lettered descriptions.

Proteins	(a) The sequence of amino acids in a polypeptide chain.
Amino acids	
Peptide bond	
Dipeptide	
Polypeptide	(b) The means by which an amino group of one amino acid is joined to the acid carboxyl group of another amino acid.
Primary structure	
Secondary structure	
Tertiary structure	(c) Composed of 100 or more amino acids linked into a chain.
Quaternary structure	

   (d) The interactions of different amino acid chains of multichain protein molecules.

   (e) Two amino acids bonded chemically.

   (f) Result of hydrogen bonds that occur principally between the constituents of a protein's peptide bonds.

   (g) These compounds are the basic units of proteins.

   (h) The folding of an amino acid chain into a particular three-dimensional configuration.

   (i) Ten amino acids linked into a chain.

10. RNA makes up a major portion of chromosomes, which contain hereditary genetic information for directing the activities of the body's cells. True or False?

11. Enzymes are very specific, and each will catalyze only individual reactions or limited classes of reactions. True or False?

12. Generally, enzymes are destroyed during the course of the reactions they catalyze. True or False?

13. The type of enzyme inhibition in which an inhibitor molecule rather than a substrate molecule combines reversibly with an enzyme molecule at the active site of the enzyme is called: (a) noncompetitive inhibition; (b) competitive inhibition; (c) allosteric inhibition.

14. Match the terms with the appropriate lettered descriptions.

Solvent	(a) In a solution, the substance present in the smaller amount is generally this substance.
Solution	
Solute	
Suspension	(b) The dispersed particles tend to settle out in this mixture.
Colloid	

    (c) In the body, water is this substance.

    (d) A homogeneous mixture of two or more components: gases, solids, or liquids.

    (e) In this mixture, dispersed particles tend not to settle out, and a beam of light passing through the mixture is scattered by the particles.

15. Substances that increase the hydrogen ion concentration of a solution are called: (a) salts; (b) bases; (c) acids.

16. Net diffusion can be viewed as the movement of a substance from a region of lower concentration to a region of higher concentration. True or False?

17. The movement of solvent through any membrane in response to a concentration difference across the membrane is termed: (a) diffusion; (b) osmosis; (c) dialysis.

18. If two solutions have the same osmotic concentration, they are said to be: (a) isosmotic; (b) hyperosmotic; (c) hypoosmotic.

19. The movement of atoms, molecules, ions, or other particles as a unit in one direction as the result of forces that push them from one point to another is called: (a) osmosis; (b) diffusion; (c) bulk flow.

LEARNING OBJECTIVES

After completing this chapter you should be able to:

1. Describe the structure and function of the plasma membrane.

2. Distinguish in detail between mediated and nonmediated transport.

3. Describe the ways in which a cell takes in and expels substances by means of membrane-bounded vesicles.

4. Cite and explain the main functions of DNA.

5. Distinguish among the functions of m-RNA, r-RNA, and t-RNA.

6. Describe the functions of ribosomes, the endoplasmic reticulum, and the Golgi apparatus, indicating the relationship among these structures.

7. Describe the functions of cilia, flagella, basal bodies, and centrioles, indicating the relationship among these structures.

8. Distinguish between mitosis and meiosis.

9. Cite two occurrences during meiosis that lead to genetic diversity, and explain how each occurs.

CHAPTER CONTENTS

CELL COMPONENTS

EXTRACELLULAR MATERIALS

CELL DIVISION

CONDITIONS OF CLINICAL SIGNIFICANCE: THE CELL

KEY TERMS AND DERIVATIVES

chromosome (*chrom* = colored; *some* = body: so named because chromosomes stain deeply) the structures in the nucleus that carry the hereditary factors (genes)

hypertonic (*hyper* = above; *ton* = strength) having an excessive or above-normal tone or tension

hypotonic (*hypo* = below; *ton* = strength) having a below-normal tone or tension

mitosis (*mit* = thread; *osis* = a process: the term is probably derived from the fact that chromosomes first appear as threadlike bodies when mitosis begins) the process of redistributing the genetic material of the cell nucleus into two new nuclei that each contain the same genetic information as the original nucleus; often followed by division of the cytoplasm of a cell

nucleolus (*nucle* = kernel, pit; *olus* = small) organelle located within the nucleus; has a role in the production of ribosomes

nucleus (*nucle* = kernel, pit) large organelle that contains the chromosomes; the control center of the cell

osmosis (*osmos* = pushing) the movement of solvent through a membrane in response to a concentration difference across the membrane

phagocytosis (*phag* = eat; *cyt* = cell; *osis* = a process) the ingestion by cells of large particles suspended in the aqueous medium of the extracellular fluid; the ingestion occurs by vesicle formation

pinocytosis (*pin* = drink; *cyt* = cell; *osis* = a process) the ingestion by cells of materials in solution or very small particles in suspension; the ingestion occurs by vesicle formation

The Cell

The cell is the basic structural and functional unit of the human body, and the cellular nature of the body's tissues, organs, and systems is evident upon microscopic examination. Consequently, a knowledge of cellular function is essential to an understanding of human anatomy and physiology.

Human cells are specialized both anatomically and physiologically. Muscle cells, for example, have a well-developed property of **contractility** (the ability to move or contract), whereas nerve cells are specialized for **conductivity** (the ability to transmit impulses). Other cells exhibit highly developed properties of **metabolism** (the ability to process foods, obtain energy, and synthesize products), **irritability** (the capacity to respond to stimuli), or **reproduction** (the ability to duplicate themselves). It is important to remember, however, that these properties are present to some degree in all cells and as such may be regarded as general cell characteristics.

CELL COMPONENTS

Cells are highly organized units made up of many different components (Figure 3.1). They contain a variety of structures, collectively called **organelles** (little organs), that are responsible for specific cellular functions. Cells also contain chemical substances, such as glycogen granules or lipid droplets, that are collectively called **inclusions.**

A cellular component called the **nucleus** is the control center of the cell. The nucleus is surrounded by a region known as the **cytoplasm,** which contains most of the cell organelles. A membrane called the **plasma membrane** (or **cell membrane**) surrounds the cytoplasm and forms the limiting boundary of the cell. Membranes also surround many organelles. Membranes are selectively permeable barriers that regulate the movement of different materials into and out of cells, as well as into and out of different organelles. They also provide points of attachment for various cell components. For example, the contractile elements of muscle cells are attached to the plasma membrane, and the enzymes that mediate certain chemical reactions are bound to the membranes of particular organelles. Although some differences are evident, the basic organization and fundamental properties of the various cell membranes are believed to be similar to those of the plasma membrane.

Plasma Membrane

The **plasma membrane** is composed of a bimolecular layer of lipid (particularly phospholipid and cholesterol), and various proteins are associated with the lipid (Figure 3.2). Proteins called *integral proteins* are embedded in the lipid, and proteins called *peripheral proteins* are loosely bound to the membrane surface, primarily to integral proteins on the inner surface of the membrane. Carbohydrate molecules are often covalently linked to lipid and protein molecules at the extracellular surface of the membrane. (Proteins that have carbohydrates attached to them are called *glycoproteins*, and lipids that have carbohydrates attached are called *glycolipids*.) In some cases, integral proteins extend completely through the membrane, forming aqueous channels or pores that connect the interior of the cell with the external environ-

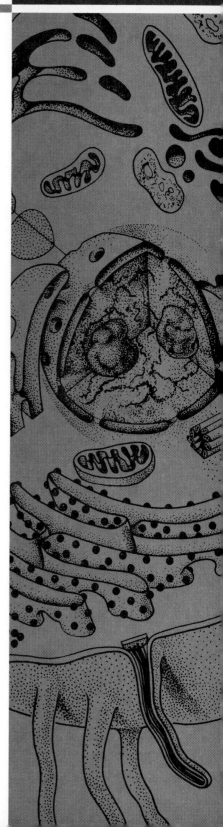

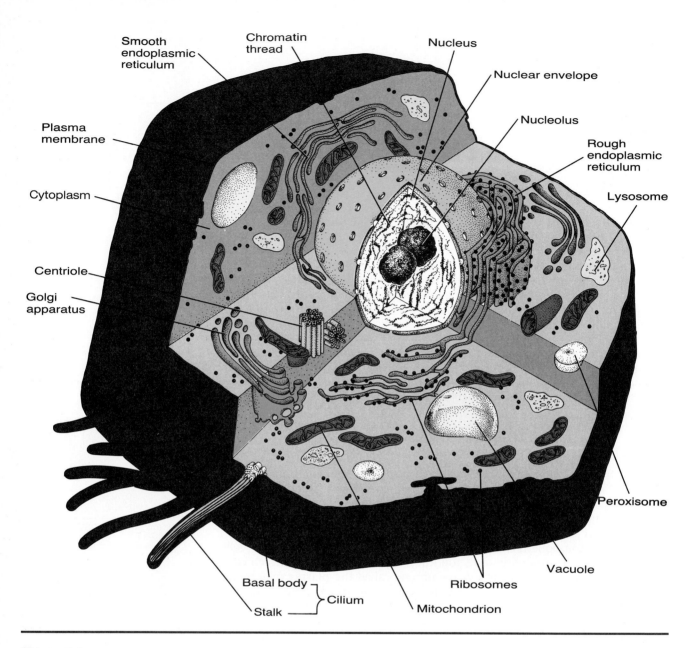

Smooth endoplasmic reticulum

Chromatin thread

Nucleus

Nuclear envelope

Nucleolus

Rough endoplasmic reticulum

Plasma membrane

Lysosome

Cytoplasm

Centriole

Golgi apparatus

Peroxisome

Vacuole

Basal body
Stalk
Cilium

Ribosomes

Mitochondrion

Figure 3.1
A "typical" cell showing subcellular organelles. There is probably no actual cell that can be considered "typical" in all respects.

ment. Moreover, some membrane proteins and glycoproteins serve as receptors to which chemical messengers such as certain hormones can attach and thereby influence cellular function.

Movement of Materials Across the Plasma Membrane

All materials that enter or leave a cell must pass either through the plasma membrane or through channels in the membrane. Thus, the properties of the membrane (as well as the properties of the penetrating materials) are important in determining the ease with which substances enter or leave cells.

DIFFUSION THROUGH THE LIPID PORTION OF THE MEMBRANE Many nonpolar substances that are soluble in lipids move easily through the plasma membrane by simple diffusion. However, water-soluble polar substances that are not very lipid-soluble diffuse through the membrane only with difficulty, if at all. Thus, the lipid portion of the membrane is a selectively permeable barrier to substances entering or leaving cells.

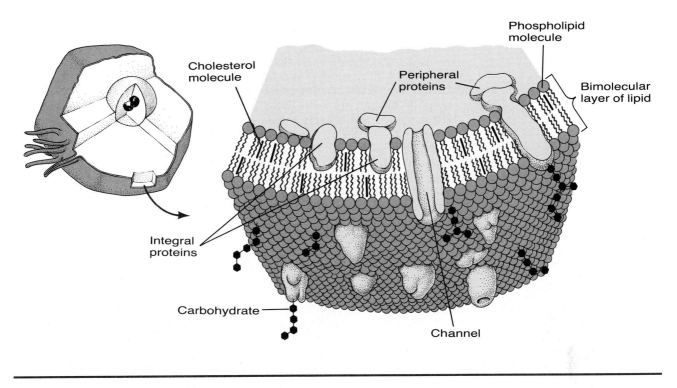

Figure 3.2
The structure of the
plasma membrane.

MOVEMENT THROUGH MEMBRANE CHANNELS The membrane channels provide an alternative route through the membrane for polar molecules and ions. However, this route is limited to substances small enough to fit through the channels, which are estimated to be about 0.8 nanometer (nm) in diameter (1 nm = 1×10^{-6} mm). In addition, the movement of ions through the channels is highly specific, and certain ions pass most easily through certain channels. For example, sodium ions pass easily through channels called *sodium channels*, and potassium ions pass easily through channels called *potassium channels*.

Various signals can open or close membrane channels, and thereby alter membrane permeability. The signals, which are frequently chemical or electrical in nature, are believed to open or close channels by causing changes in the conformations (shapes) of the proteins that form the channels.

MEDIATED TRANSPORT Many polar molecules larger than 0.8 nm in diameter enter cells readily, even though they are quite insoluble in lipids and are too large to fit through membrane channels. Various forms of mediated transport account for the ability of these substances to cross the plasma membrane. **Mediated transport** makes use of molecules called *carrier molecules*, which are part of the membrane itself. (Researchers believe that certain membrane proteins serve as carrier molecules.) Substances to be transported across the membrane by mediated transport attach to specific binding sites on the carrier molecules. This attachment enables the transported substance to cross the membrane and, depending on the direction of transport, either enter or leave the cell.

The exact mechanisms by which carrier molecules enable substances to pass through the membrane are not known. However, one theory proposes that carrier molecules are fixed in place within the structure of the membrane (Figure 3.3). Each carrier possesses a binding site that is oriented so that a substance to be transported that is located on one side of the membrane can attach reversibly to the site. Once the attachment occurs, the carrier molecule undergoes a conformational change that moves the binding site and the attached substance to the other side of the membrane, where the substance is released.

F3.3

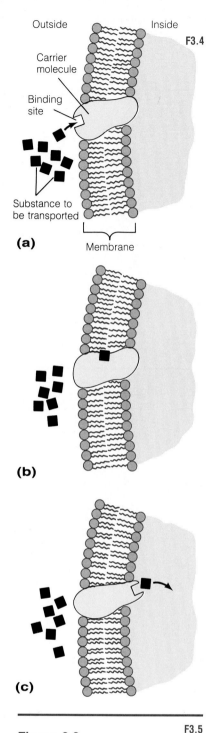

(a)

(b)

(c)

Figure 3.3
A proposed mechanism
of mediated transport.
See text for details.

fag-o-sigh-to'-sis

pi-no-sigh-to'-sis

Mediated transport systems exhibit the characteristics of specificity, saturation, and competition (Figure 3.4). **Specificity** means that only certain substances are carried by mediated transport systems. Each carrier molecule binds only with a select group of substances, and only substances that can attach to the various carrier molecules can cross the membrane by mediated transport. **Saturation** means there is a limit to the amount of a substance that can cross the membrane by mediated transport in a given time. When all of the carrier molecules for a given substance are being utilized, the addition of more of the substance will not increase its rate of transport. **Competition** means that different substances can compete for the services of the same carrier. Suppose, for example, that substance A and substance B are both transported across the membrane by attaching to the same binding site on carrier X. If a substantial amount of substance A but no substance B is present, then all of the carrier X molecules will be utilized in transporting substance A. If substance B is added, it will compete with substance A for the services of the carrier X molecules. In this situation, some substance A and some substance B will be transported, and the transport rate of substance A will decrease compared to the rate when it was the only substance present.

FACILITATED DIFFUSION One type of mediated transport is called **facilitated diffusion.** In this type of transport, carrier molecules move materials across the membrane equally well in either direction, into or out of the cell. However, if the concentration of a substance to be transported is higher outside the cell than inside, more transfer will occur from outside to inside than in the opposite direction, resulting in a net inward movement of the substance. Thus, the net movement in facilitated diffusion is from a region of high concentration to a region of low concentration. When equal concentrations of the substance are reached on either side of the membrane, there will be equal transport in both directions and, therefore, no net movement in either direction. Facilitated diffusion is basically a passive process that requires no expenditure of cellular energy.

ACTIVE TRANSPORT Another type of mediated transport is called **active transport.** In this type of transport, carrier molecules move a substance across the membrane from one side to the other regardless of the substance's concentration on either side of the membrane. Active transport systems can move materials across the membrane from regions of low concentration to regions of high concentration, against normal concentration gradients (or against electrical or pressure gradients). Consequently, active transport systems can accumulate material on one side of the membrane at a concentration that is many times the concentration of the material on the opposite side of the membrane. Energy must be expended to move substances against concentration gradients (or against electrical or pressure gradients), and active transport depends on the metabolic processes of the cell to supply the needed energy. If a cell is unable to supply the required amount of energy, active transport cannot occur.

ENDOCYTOSIS AND EXOCYTOSIS Some cells form vesicles that enclose a small volume of extracellular material (Figure 3.5). In this process, a cell surrounds the material with a section of its plasma membrane. This section of the membrane then separates from the plasma membrane and moves into the interior of the cell. The general term for this process is **endocytosis.** If the vesicle contains large particles suspended in the aqueous medium of the extracellular fluid, the process of vesicle formation is called **phagocytosis** (cell eating). If the contents are materials in solution or very small particles in suspension, the process is called **pinocytosis** (cell drinking). The mechanisms that regulate endocytosis are not fully understood, but in at least some cases, endocytosis is stimulated by the binding of specific particles to sites on the surface of the plasma membrane.

Substances such as cell products and secretions that are contained within membrane-bounded vesicles leave cells by **exocytosis** (Figure 3.6). In this

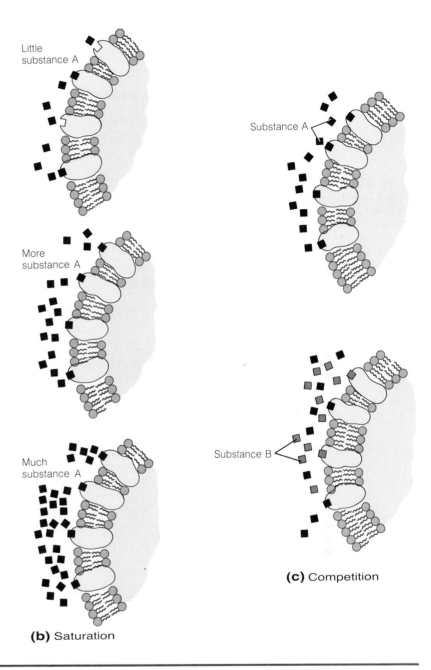

Little
substance A

More
substance A

Much
substance A

Substance A

Substance B

(b) Saturation

(c) Competition

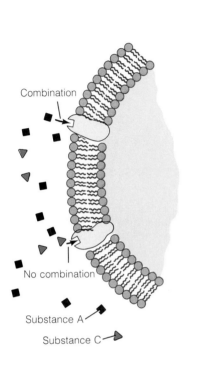

Combination

No combination

Substance A

Substance C

(a) Specificity

Figure 3.4
Mediated transport
systems exhibit
characteristics of
(a) specificity,
(b) saturation, and
(c) competition. See text
for details.

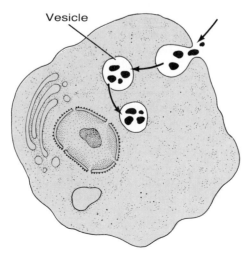

Vesicle

Figure 3.5
The process of
endocytosis. A cell
surrounds a small
volume of extracellular
material with a section of
its plasma membrane.
This section of the
membrane then separates
from the plasma
membrane, giving rise to
a vesicle that moves to
the interior of the cell.

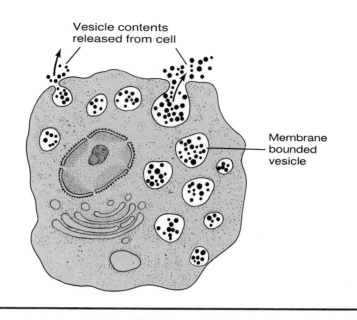

Figure 3.6
The process of exocytosis. A membrane-bounded vesicle fuses with the plasma membrane, and the contents of the vesicle are released from the cell.

process, a membrane-bounded vesicle within a cell fuses with the plasma membrane, and the contents of the vesicle are released from the cell. Exocytosis can be triggered by specific stimuli that cause an increase in the concentration of free calcium ions within cells. The increased concentration of calcium ions leads to the fusion of intracellular vesicles with the plasma membrane, perhaps by activating contractile proteins in the membrane and surrounding cytoplasm, which act upon the vesicles. Both endocytosis and exocytosis are active, energy-requiring processes.

When considering the different ways in which materials enter or leave cells, it is important to remember that a given substance may enter or leave a cell by several different routes. For example, sodium ions and potassium ions can move through membrane channels, and they are also actively transported.

Tonicity

The plasma membrane is permeable to water. When a cell is immersed in an aqueous solution, the movement of water across the plasma membrane can influence the state of tension or tone of the cell, and the cell may swell due to water entry or shrink due to water loss. Consequently, a solution can be described in terms of its **tonicity**—that is, in terms of its ability to influence the state of tension or tone of cells immersed in it, as a result of the movement of water into or out of the cells. An **isotonic** solution is a solution in which cells maintain their normal tone and in which there is no net movement of water into or out of the cells. A **hypotonic** solution is a solution that produces a change in the tone of cells immersed in it as a result of the net movement of water from the solution into the cells. A **hypertonic** solution is a solution that produces a change in the tone of cells immersed in it as a result of the net movement of water out of the cells and into the solution.

The properties of the plasma membrane are important in determining the tonicity of a solution. Consequently, a solution that is isosmotic to particular cells may or may not be isotonic. For example, the plasma membrane of red blood cells is permeable to water and urea. However, the membrane prevents the net entry of sodium chloride into the cells, and it prevents the net exit of osmotically active solute particles from the cells. It is possible to make solutions of urea and solutions of sodium chloride that are isosmotic to one another and also to red blood cells; that is, all three have the same osmotic concentration. If red blood cells are placed in the sodium chloride solution, they neither shrink due to water loss nor swell due to water entry because no osmotic difference exists on either side of the plasma membrane, and the

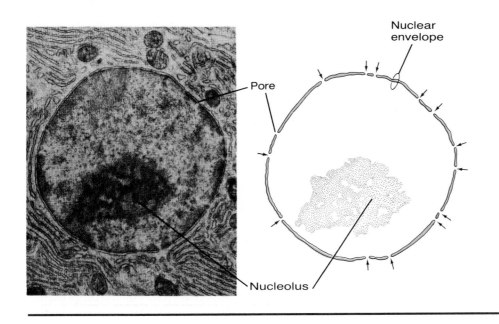

membrane prevents the net exit of osmotically active solute particles from the cells as well as the net entry of sodium and chloride ions from the sodium chloride solution. Thus, the red blood cells maintain their normal tone, and the sodium chloride solution is both isosmotic and isotonic to the cells.

If red blood cells are placed in the urea solution, a different result is observed. Since the plasma membrane of red blood cells is permeable to urea and since the urea is at greater concentration outside the cells than inside them, there will be a net movement of urea into the cells. This increases the number of osmotically active solute particles within the cells (and decreases the concentration of water within the cells). It also decreases the number of osmotically active solute particles outside the cells (and increases the concentration of water outside the cells). As a result, there will be a net movement of water into the cells by osmosis and the cells will swell and perhaps burst. Thus, in spite of the fact that initially the urea solution had the same osmotic concentration as red blood cells, it was not isotonic to the cells since it did not maintain the normal tone of the cells.

The Nucleus

The **nucleus** is a large organelle that is separated from the rest of the cell by a structure called the **nuclear envelope** (Figure 3.7). The nuclear envelope, which is about 40 nm thick, consists of two parallel membranes separated by a *perinuclear space*. Numerous octagonal *nuclear pores* are present in the nuclear envelope. These pores range from 30 nm to 100 nm in diameter, and electron micrographs suggest that the openings of many pores are partially or completely closed by thin diaphragms. Most cells contain a single nucleus. However, some cells contain two or more nuclei, and mature red blood cells contain no nucleus.

F3.7

The nucleus contains the genetic material of the cell in the form of chromatin threads composed of DNA combined with protein and a small amount of RNA. DNA contains coded messages that instruct the cell to synthesize particular polypeptides or proteins. Once the polypeptides or proteins are synthesized, they act as enzymes, as structural proteins, or in other ways to accomplish the work of the cell.

All cells except the reproductive cells contain a person's full genetic complement of DNA. However, all of the possible polypeptides and proteins the DNA can instruct a cell to synthesize are not found at all times in all cells. Various control mechanisms regulate the production of polypeptides and proteins by cells, and a particular polypeptide or protein may be produced by one type of cell but not by another. This is important because a cell's functions depend in large measure on the proteins it produces. For example, if a cell is to carry out a specific sequence of metabolic reactions, it must produce the particular enzymatic proteins required for the reactions. Thus, thyroid gland cells must produce the enzymes needed to synthesize thyroid hormones, and adrenal gland cells must produce the enzymes required to manufacture adrenal hormones. Because DNA provides the instructions for synthesizing particular polypeptides or proteins, it plays a central role in determining cell function.

RNA Synthesis (Transcription)

The DNA in the nucleus is unable to pass through the nuclear envelope into the cytoplasm; yet it is in the cytoplasm that polypeptides and proteins are synthesized. As a result of this situation, DNA must transfer its instructions for synthesizing particular polypeptides or proteins to molecules that carry the instructions from the nucleus to the cytoplasm. It does this by a process called **transcription,** in which DNA serves as a template for the assemblage of molecules of RNA.

As discussed in Chapter 2, both DNA and RNA are made up of base–sugar–phosphate units called *nucleotides,* and a DNA molecule is composed of two polynucleotide chains. The two chains are joined by hydrogen bonds between the bases of the different chains, and the bases always pair with one another in a predictable, complementary fashion. Adenine always links with thymine and vice versa, whereas cytosine always links with guanine and vice versa.

When DNA serves as a template for RNA synthesis, the two linked chains of DNA separate from one another for some distance, exposing part of the base sequence of the DNA (Figure 3.8). An enzyme called RNA *polymerase* links RNA nucleotides into an RNA chain in the order dictated by the sequence of bases in the template DNA. This occurs according to the complementary base-pairing rules—with the exception that in RNA the base uracil substitutes for thymine (and the sugar unit of RNA nucleotides is ribose rather than deoxyribose as in DNA). Thus, RNA polymerase incorporates an adenine-containing nucleotide into RNA wherever the template chain of DNA has thymine, a guanine-containing nucleotide wherever the DNA has cytosine, a uracil-containing nucleotide wherever the DNA has adenine, and a cytosine-containing nucleotide wherever the DNA has guanine. Following this, the newly formed RNA chain separates from the DNA template as an independent RNA molecule.

Usually, only one of the two chains of any given segment of DNA serves as a template for RNA synthesis. However, one DNA chain may serve as the template at some sites, and the other DNA chain may be the template at other sites.

Once RNA molecules have been synthesized using DNA chains as templates, they generally undergo various types of processing. For example, groups of nucleotides can be removed from the center of an RNA molecule, and the remaining portions of the molecule can be joined together to produce a shorter RNA molecule. Ultimately, three major types of RNA molecules—*messenger RNA, transfer RNA,* and *ribosomal RNA*—are produced.

Messenger RNA

The RNA molecules that carry instructions from DNA in the nucleus to the cytoplasmic sites of polypeptide and protein synthesis are called **messenger RNA (m-RNA)** molecules. Each m-RNA molecule contains a coded message that specifies the sequence in which different amino acids are to be joined to form a specific polypeptide or protein. The message is contained within the base sequence of the m-RNA, and this base sequence is ultimately determined by the base sequence of the DNA that serves as a template for RNA synthesis. Each sequence of three bases along an m-RNA molecule constitutes a **codon,** or "word," of the message. For example, the three-base sequence guanine–cytosine–uracil is a codon that specifies the amino acid alanine of a polypeptide or protein.

F3.8

The four principal bases of m-RNA—adenine (A), cytosine (C), guanine (G), and uracil (U)—can form 64 different three-base codons. Since only about 20 amino acids are found in polypeptides and proteins, there are more than enough codons to specify each amino acid. In fact, several different codons can specify the same amino acid (Table 3.1). For example, the codons GCU, GCC, GCA, and GCG all specify the amino acid alanine. In addition, some base sequences serve as "punctuation," indicating termination points for reading the m-RNA message.

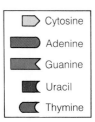

Cytosine
Adenine
Guanine
Uracil
Thymine

Table 3.1 The Genetic Code*

		Second Position			
	U	**C**	**A**	**G**	
U	UUU ⎤ Phe UUC ⎦ UUA ⎤ Leu UUG ⎦	UCU ⎤ UCC ⎥ Ser UCA ⎥ UCG ⎦	UAU ⎤ Tyr UAC ⎦ UAA End UAG End	UGU ⎤ Cys UGC ⎦ UGA End UGG Trp	U C A G
C	CUU ⎤ CUC ⎥ Leu CUA ⎥ CUG ⎦	CCU ⎤ CCC ⎥ Pro CCA ⎥ CCG ⎦	CAU ⎤ His CAC ⎦ CAA ⎤ Gln CAG ⎦	CGU ⎤ CGC ⎥ Arg CGA ⎥ CGG ⎦	U C A G
A	AUU ⎤ AUC ⎥ Ile AUA ⎦ AUG Met	ACU ⎤ ACC ⎥ Thr ACA ⎥ ACG ⎦	AAU ⎤ Asn AAC ⎦ AAA ⎤ Lys AAG ⎦	AGU ⎤ Ser AGC ⎦ AGA ⎤ Arg AGG ⎦	U C A G
G	GUU ⎤ GUC ⎥ Val GUA ⎥ GUG ⎦	GCU ⎤ GCC ⎥ Ala GCA ⎥ GCG ⎦	GAU ⎤ Asp GAC ⎦ GAA ⎤ Glu GAG ⎦	GGU ⎤ GGC ⎥ Gly GGA ⎥ GGG ⎦	U C A G

First Position (left side) / Third Position (right side)

*The three bases in an m-RNA codon are designated, respectively, as the first position, second position, and third position of the codon. See Table 2.3 for amino acid abbreviations. (Three-base sequences designated "End" indicate termination points for reading the m-RNA message.)

DNA chains   RNA chain

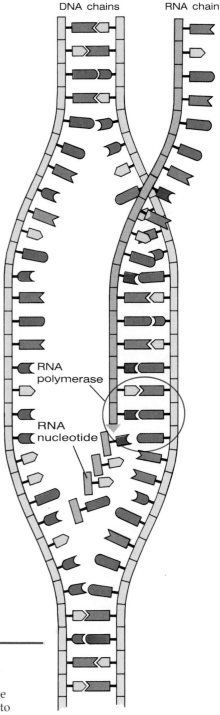

RNA polymerase

RNA nucleotide

Figure 3.8
The enzyme RNA polymerase catalyzes the addition of nucleotides to the growing RNA chain in the sequence dictated by the order of bases in the DNA template.

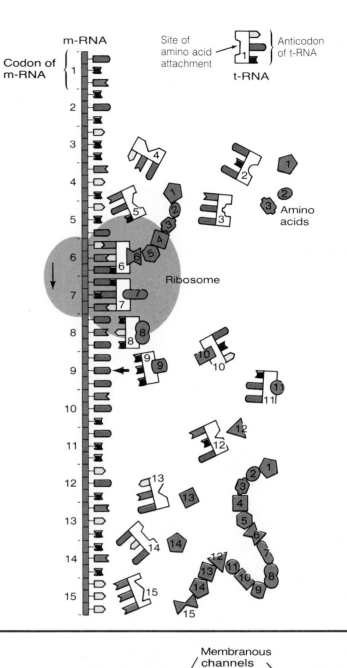

Figure 3.9

A ribosome "reading" the code of m-RNA and assembling a polypeptide or protein chain of amino acids. Molecules of t-RNA bring amino acids to the ribosome for incorporation into the growing chain.

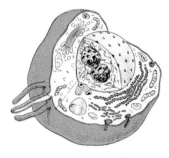

Figure 3.10

An electron micrograph of rough endoplasmic reticulum. The granules that are attached to the reticulum are ribosomes.

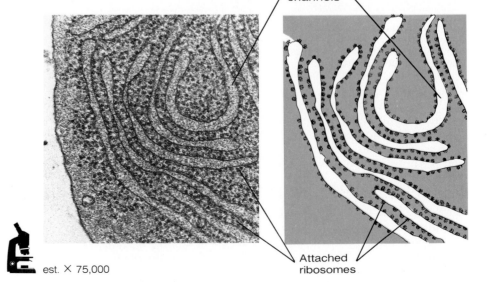

est. × 75,000

Transfer RNA

Transfer RNA (t-RNA) molecules are RNA molecules that combine with amino acids in the cytoplasm and carry the amino acids to the sites where the amino acids are incorporated into a specific polypeptide or protein.

Ribosomal RNA

Ribosomal RNA (r-RNA) molecules are present in the cytoplasm in the form of r-RNA–protein complexes that become organized into structures called *ribosomes*. Precursors of these r-RNA–protein complexes originate in a region of the nucleus called the **nucleolus** (Figure 3.7). The nucleolus is the site of synthesis (on DNA templates) of precursor RNA molecules that are ultimately processed into ribosomal RNA molecules, and in the nucleolar region the precursor RNA combines with protein.

F3.7

Ribosomes

Ribosomes are small cytoplasmic organelles. A single ribosome consists of two subunits: a larger 60S subunit and a smaller 40S subunit (60S and 40S refer to the relative rates of sedimentation of the subunits in a centrifugal field). Each subunit is composed of ribosomal RNA and protein.

Protein Synthesis (Translation)

It is at a ribosome that amino acids are linked together in the sequence specified by messenger RNA to form a polypeptide or protein (Figure 3.9). In this process, which is called **translation,** an m-RNA strand that carries coded instructions for synthesizing a specific polypeptide or protein becomes attached to a 40S ribosomal subunit, which then binds to a 60S subunit to form a fully functional ribosome. The ribosome "reads" the m-RNA message and assembles the particular polypeptide or protein specified by the m-RNA. The m-RNA is always "read" in the same direction from the starting point, and a polypeptide or protein is synthesized from the end with the free amino group to the end with the free acid carboxyl group. Frequently, a number of ribosomes attach to the same strand of m-RNA, forming what is known as a *polyribosome* or *polysome*. Each ribosome assembles a polypeptide or protein according to the instructions of the m-RNA.

F3.9

The amino acids that are incorporated into a growing polypeptide chain are brought to a ribosome from the cytoplasm by transfer RNA (t-RNA) molecules, and there is at least one different t-RNA molecule for each different amino acid. A specific t-RNA molecule contains within its base sequence a region called the **anticodon,** which can attach to a particular m-RNA codon. This allows the amino acid the t-RNA carries to be inserted into a growing polypeptide at the position specified by m-RNA. For example, the m-RNA codon GCU specifies the amino acid alanine. A t-RNA molecule that carries the amino acid alanine to a ribosome has as its anticodon a base sequence that can pair with the GCU codon of m-RNA. When a ribosome encounters a GCU codon, a t-RNA molecule that carries alanine attaches to the ribosome, and an enzyme that is associated with the ribosome joins the alanine to the growing polypeptide. Another t-RNA molecule then attaches according to the next codon of the m-RNA message, and the amino acid it carries is joined to the alanine of the polypeptide. The t-RNA that carries alanine is then released.

When a ribosome has "read" an entire m-RNA message, the newly formed polypeptide or protein is released. The ribosome dissociates into its 60S and 40S subunits, and the m-RNA is released. The ribosomal subunits can re-form ribosomes that "read" other m-RNA molecules and assemble new polypeptides or proteins. The m-RNA can be utilized by other ribosomes.

Endoplasmic Reticulum

A membranous network of tubular or saclike channels called the **endoplasmic reticulum** (*endo* = within; *plasm* = cytoplasm; *reticulum* = network) extends throughout much of the cytoplasm of the cell and is almost always interconnected with the nuclear envelope (Figure 3.10). The walls of the endoplasmic reticulum contain enzymes that play a role in fatty acid and steroid synthesis,

en-do-plaz'-mik re-tik'-u-lum

F3.10

and an extensive endoplasmic reticulum is found in cells that secrete steroid hormones.

Ribosomes are often attached to the endoplasmic reticulum. Endoplasmic reticulum that has ribosomes attached is called **rough** endoplasmic reticulum, and endoplasmic reticulum that does not have attached ribosomes is called **smooth** endoplasmic reticulum. Proteins synthesized by the ribosomes of rough endoplasmic reticulum can pass into the channels of the reticulum, and some are inserted into the endoplasmic reticulum membrane. Within the endoplasmic reticulum, carbohydrates are added to many proteins, forming glycoproteins. Portions of the reticulum can pinch off or break away, giving rise to membrane-bounded vesicles that contain protein.

Golgi Apparatus

gol'-jee

F3.11

The **Golgi apparatus** is a cytoplasmic organelle that consists of a series of flattened, membranous sacs called *cisternae* that are frequently located near the nucleus of the cell (Figure 3.11). Vesicles from the endoplasmic reticulum that contain proteins (largely glycoproteins) can fuse with the membranes of the Golgi apparatus. Within the Golgi apparatus, the proteins are processed and modified. For example, enzymes within the Golgi apparatus can add carbohydrates to or remove them from glycoproteins. Enzymes can also add phosphate groups, sulfate groups, and even fatty acids to protein molecules. Also within the Golgi apparatus, the proteins are sorted for delivery to different destinations. Ultimately, membrane-bounded vesicles pinch off from the Golgi membranes. Protein-containing vesicles called *secretory vesicles* or *secretory granules* migrate from the Golgi apparatus to the plasma membrane. The vesicles fuse with the membrane, and their contents are released from the cell by exocytosis. Other vesicles from the Golgi apparatus fuse with cytoplasmic organelles called *lysosomes* (see below), and still others fuse with the plasma membrane, thereby generating additional membrane.

Lysosomes

F3.12

F3.13

Lysosomes (*lyso* = dissolution; *soma* = body) are membrane-bounded cytoplasmic organelles that appear granular during inactivity but assume the appearance of vesicles when active (Figure 3.12). Lysosomes contain strong digestive enzymes that are capable of breaking down proteins, lipids, certain carbohydrates, DNA, and RNA. Lysosomes form digestive vacuoles by attaching to vesicles that are present within the cell as the result of phagocytosis (Figure 3.13). After attachment, the lysosomes release their enzymes into the vacuoles, and the enzymes digest the phagocytized material, often bacteria, converting the material to products that can enter the cytoplasm and be utilized by the cell. Undigested material remains within the vacuoles, which are then called *residual bodies*. The residual bodies are thought to be removed from the cell by exocytosis. Lysosomes are particularly abundant in certain white blood cells whose principal activity is the phagocytosis of foreign materials in the body.

Parts of a cell itself are sometimes broken down within lysosomal vacuoles that are called *autophagic vacuoles*. Such an event can occur during starvation, allowing the cell to use a part of its own substance as an energy source without doing itself irreparable harm. After a cell has been severely injured or has died, lysosomal membranes may rupture. The enzymes released then digest the material of the cell itself. In addition, a controlled destruction of cells by their lysosomes seems to play an important role in normal embryonic development and in the regression of the mother's mammary glands when her infant is no longer nursing. In both cases there is an excess of cells that must be eliminated. The destruction of apparently normal cells by their lysosomes seems to perform a vital role in these cases of cell death.

Peroxisomes

per-ox'-i-some

Peroxisomes are membrane-bounded cytoplasmic organelles that are very similar to lysosomes in microscopic appearance. Peroxisomes differ from lysosomes in that lysosomes contain digestive enzymes, whereas peroxisomes

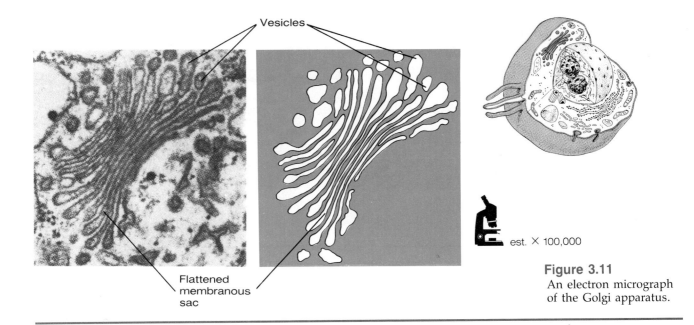

Vesicles

Flattened
membranous
sac

est. × 100,000

Figure 3.11
An electron micrograph
of the Golgi apparatus.

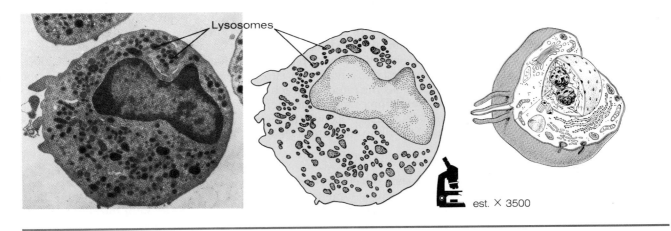

Lysosomes

est. × 3500

Figure 3.12
An electron micrograph
of lysosomes in a white
blood cell. The large
structure is the nucleus.

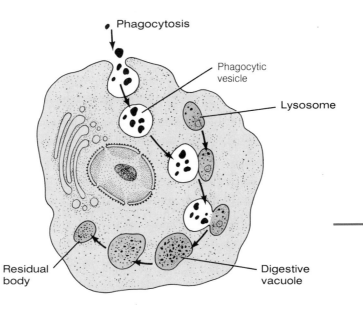

Phagocytosis

Phagocytic
vesicle

Lysosome

Residual
body

Digestive
vacuole

Figure 3.13
Enzymatic digestion of
phagocytized substances
by lysosomes.

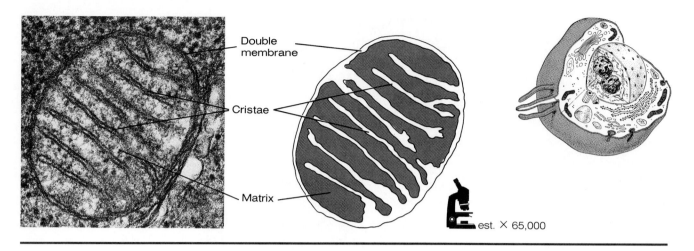

Double membrane

Cristae

Matrix

est. × 65,000

Figure 3.14

An electron micrograph of a mitochondrion.

contain enzymes that produce hydrogen peroxide (H_2O_2). Peroxisomes also contain the enzyme catalase, which splits hydrogen peroxide into water and oxygen.

Mitochondria

F3.14

Mitochondria (*mito* = thread; *chondros* = granule) are cytoplasmic organelles that are bounded by a double membrane (Figure 3.14). The outer membrane is smooth and surrounds the mitochondrion itself. The inner membrane folds at intervals into the central portion of the mitochondrion, forming partitions known as *cristae*. A semisolid substance called the *matrix* lies between the cristae and fills the interior of each mitochondrion. Mitochondria are involved in the generation of metabolic energy for cellular activities, and the metabolic reactions that take place within them are discussed in Chapter 25. Mitochondria contain ribosomes and DNA, and they are capable of self-duplication.

Cytoskeleton

The cytoplasm of cells contains a **cytoskeleton** (cell skeleton) that consists of an intricate, three-dimensional network of filamentous structures. These structures not only maintain the shape and organization of the cell, but they are also involved in various cellular activities, including cellular movement, the transport of substances within the cell, and cell division. Prominent among the components of the cytoskeleton are structures called *microtubules, intermediate filaments,* and *microfilaments*. In addition, the cytoplasm also contains an interconnecting network of thin, filamentous bridges called the **microtrabecular lattice,** which links microtubules (and perhaps other filamentous structures) with one another.

Microtubules

F3.15

Microtubules are small, hollow, cylindrical, unbranched tubules about 25 nm in diameter (Figure 3.15). They are composed primarily of a protein called *tubulin*. Microtubules are involved in cell movement processes such as the translocation of organelles from one place to another, and they play a structural role in the development and maintenance of cell shape.

Intermediate Filaments

Intermediate filaments, which are about 10 nm in diameter, are believed to play a structural role in the cell, perhaps serving as a sort of scaffold for the cytoskeletal framework. Intermediate filaments may be important in maintaining the position of the nucleus within the cell, and they may be involved in the transport of materials into and out of the nucleus. Intermediate filaments form a ring around the nucleus, with branches extending outward

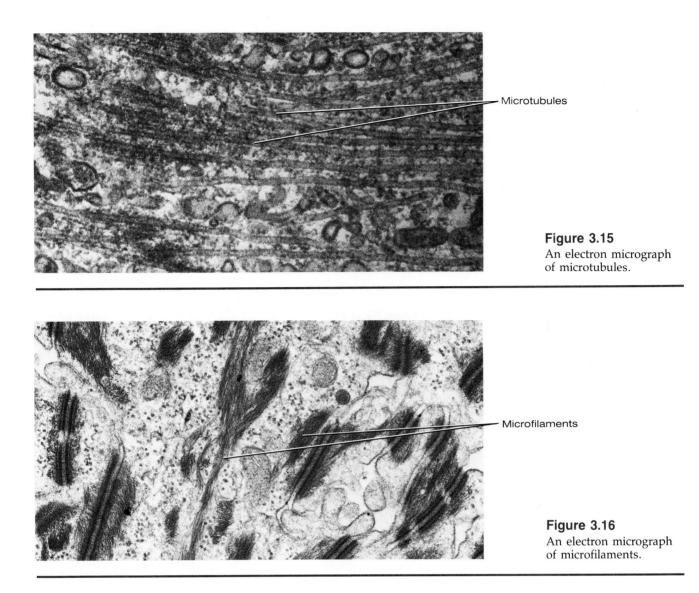

Figure 3.15
An electron micrograph of microtubules.

Figure 3.16
An electron micrograph of microfilaments.

through the cytoplasm, and some filaments extend into the pores of the nuclear envelope.

Microfilaments

Microfilaments are very small fibrils abut 7 nm in diameter (Figure 3.16). They are composed primarily of a protein called *actin*, and they can occur in bundles or other groupings rather than singly. Actin filaments can form connections with the plasma membrane of a cell, and they may play a role in determining cell shape. Microfilaments are associated with contractile activities involved in cell movement, and muscle cells possess an extensive array of microfilaments.

F3.16

Cilia, Flagella, and Basal Bodies

Many cells have one or more thin, cylindrical structures that are motile (capable of movement) projecting from their surfaces. These structures can move substances over the cell surface, or they can move an entire cell through a liquid medium. If the structures are short and numerous, they are called **cilia.** If they are longer and fewer, they are called **flagella.** Both cilia and flagella have the same basic organizational pattern, and both originate from cytoplasmic structures called **basal bodies.** Cilia and flagella consist of a membranous sheath enclosing a series of microtubules. The sheath is continuous with the

fla-jel'-ah

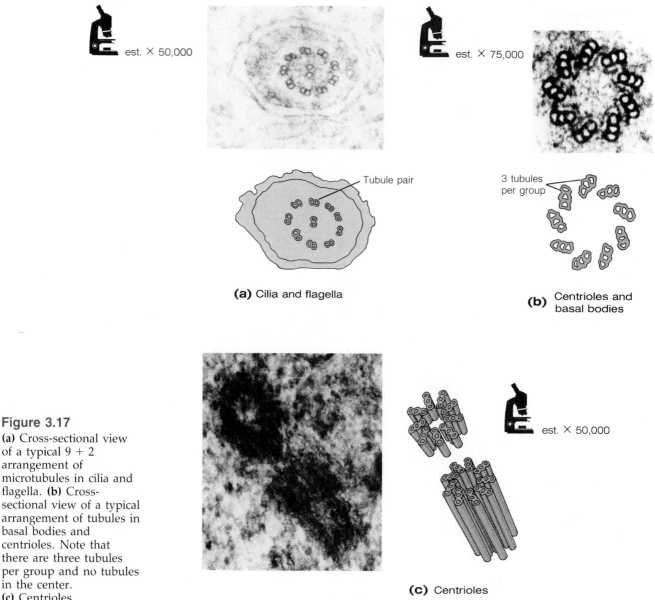

Figure 3.17
(a) Cross-sectional view of a typical 9 + 2 arrangement of microtubules in cilia and flagella. **(b)** Cross-sectional view of a typical arrangement of tubules in basal bodies and centrioles. Note that there are three tubules per group and no tubules in the center. **(c)** Centrioles.

est. × 50,000

est. × 75,000

Tubule pair

(a) Cilia and flagella

3 tubules per group

(b) Centrioles and basal bodies

est. × 50,000

(c) Centrioles

F3.17a

F3.17b

plasma membrane, and the microtubules are arranged in a characteristic circular pattern of nine groups of tubules, with two tubules per group (Figure 3.17a). Two additional microtubules are found in the center of this circular pattern. Basal bodies also have a characteristic circular pattern of groups of microtubules, but basal bodies have three tubules per group, and they do not have the two central tubules (Figure 3.17b).

Centrioles

F3.17b,c

Structures called **centrioles,** which are involved in cell division, are found near the nucleus of the cell in a region called the *centrosome* or *centrosphere.* Centrioles normally occur in pairs with each member of the pair oriented at right angles to the other. Centrioles are cylindrical in shape, and they contain a series of microtubules that are arranged in the same pattern as the microtubules of basal bodies (Figures 3.17b and c). In fact, basal bodies and centrioles are believed to have common origins, and they may even be identical structures.

Inclusions

Cells contain a wide variety of chemical substances that are collectively called **inclusions.** The hemoglobin molecules of red blood cells, which transport oxygen and carbon dioxide, are inclusions. So are particles of the pigment melanin, which is found in some cells of the eyes, the skin, and the hair. Several metabolically important substances are also found in cells as inclusions. For example, the polysaccharide glycogen, a storage form of carbohydrate, is particularly evident in liver and muscle cells, and fats are stored in adipose tissue cells. When a cellular inclusion is a liquid that could mix with the cytoplasm of a cell, the inclusion is often surrounded by a membrane, forming a structure called a *vacuole.*

EXTRACELLULAR MATERIALS

Many body substances are found outside of cells rather than within them. These are collectively called **extracellular materials.** Extracellular materials include the body fluids and the extracellular framework in which many cells are embedded. Many extracellular materials are products of the cells themselves. Among these are chondroitin sulfate, which is a jellylike substance found in bone, cartilage, and heart valves; and hyaluronic acid, which is a viscous, fluidlike substance present in a number of tissues. A variety of fibrous materials, such as the proteins collagen and elastin, also occur extracellularly. Connective tissue is particularly rich in extracellular materials.

CELL DIVISION

Many of the body's cells (for example, liver, intestinal, bone marrow, and epidermal cells) are able to divide and reproduce themselves. The processes by which such cells divide encompass several basic events. One is the replication of the genetic material within the nucleus of the cell. A second is the redistribution of the genetic material into two new nuclei. The process of redistributing the genetic material into two new nuclei that each contain the same genetic information as the original nucleus is called **mitosis.** A third event is the division of the cell's cytoplasm into two new cells (called daughter cells), each with its own nucleus. This process is called **cytokinesis.** Usually, mitosis and cytokinesis occur simultaneously.

Interphase: Replication of DNA

The period between active cell divisions is called **interphase.** During interphase, the DNA molecules that comprise the genetic material of the cell (and the protein and RNA associated with the DNA) appear only as indistinct chromatin threads within the nucleus. During this period, DNA molecules serve as templates for the replication of additional DNA molecules (Figure 3.18). In this process, the two linked chains of a DNA molecule separate from

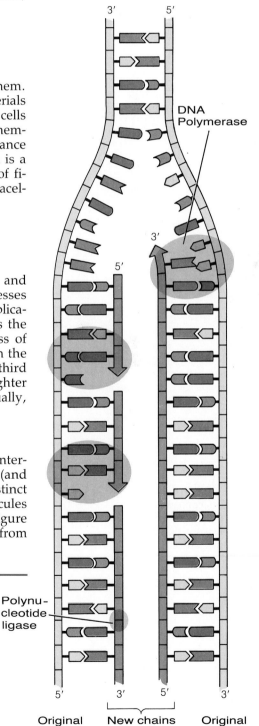

DNA Polymerase

Polynucleotide ligase

Cytosine

Adenine

Guanine

Thymine

5' — Original chain

3' — New chains forming

5' — Original chain

3'

Figure 3.18
Replication of DNA. The original two chains of a DNA molecule separate, each chain then serving as the template for the assembly of a new complementary chain. This results in two DNA molecules, each exactly like the original.

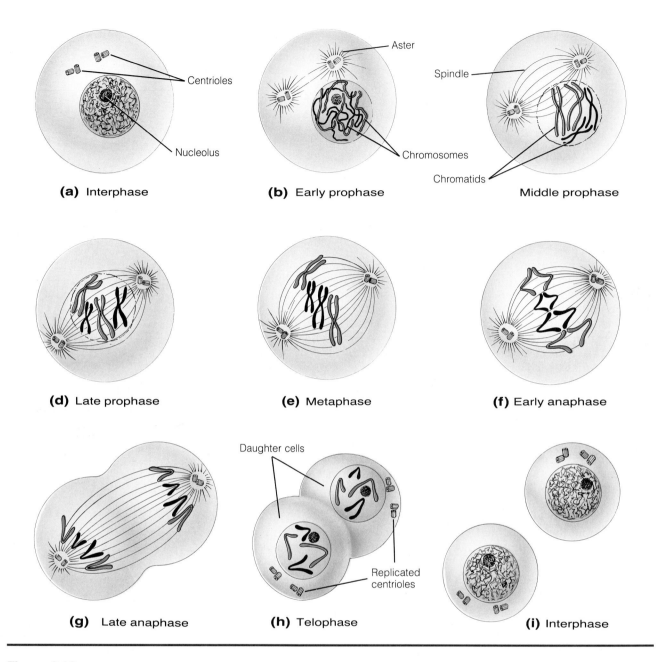

(a) Interphase   **(b)** Early prophase   Middle prophase

(d) Late prophase   **(e)** Metaphase   **(f)** Early anaphase

(g) Late anaphase   **(h)** Telophase   **(i)** Interphase

Figure 3.19

Interphase and the phases of mitosis. See text for detailed discussion.

one another for some distance, and each chain acts as a template that specifies the order in which individual DNA nucleotides are to be incorporated into a new DNA chain. DNA polymerase and polynucleotide ligase enzymes link the nucleotides into the new DNA chain according to the complementary base-pairing pattern. That is, an adenine-containing nucleotide is incorporated into the new DNA chain wherever the template chain has thymine; a guanine-containing nucleotide is incorporated wherever the template DNA has cytosine; a thymine-containing nucleotide, wherever the template has adenine; and a cytosine-containing nucleotide, wherever the template has guanine. The end result is a new DNA chain that is the complement of the original template DNA chain. The newly replicated chain of DNA remains attached to the original template DNA, thereby forming a complete, two-chain DNA molecule exactly like the original. Thus, each new two-chain DNA consists of one chain from the original DNA, which acted as a template, and one newly synthesized chain. By the end of interphase, the cell therefore contains twice the amount of DNA as when it entered interphase.

Because the activities that occur in the cell change as the cell approaches mitosis, interphase is sometimes divided into three phases: G_1 (first gap), S (DNA synthesis), and G_2 (second gap):

1. The **G_1 phase** immediately follows the completion of cell division. During this phase, active synthesis of RNA and protein occurs, the nucleus and the cytoplasm enlarge, and there is increased pinocytotic activity by the cell.

2. The **S phase** follows the G_1 phase. Pinocytotic activity decreases during this phase; however, the most notable event is the synthesis of DNA molecules.

3. The **G_2 phase** follows the completion of DNA synthesis. During this phase the metabolic activities of the cell decrease as changes occur in preparation for mitosis (which is sometimes called the **M phase**).

Mitosis

Several phases in the overall process of mitosis (nuclear division) are recognized (Figure 3.19). However, it must be emphasized that mitosis itself is a *continuous* event and not a series of discrete steps.

Prophase

The initial phase of mitosis is called **prophase** (Figures 3.19b–d). Early in prophase, two pairs of centrioles are present. As prophase proceeds (and in some cells beginning in interphase), one pair of centrioles moves toward one end, or pole, of the cell, and the other pair moves toward the opposite end. Microtubules that project in all directions from the regions of the centrioles become visible, the nucleolus disappears, and the nuclear envelope begins to disintegrate. Also during prophase, the chromatin threads of DNA, protein, and RNA become tightly coiled and visible as structures called **chromosomes** (Figure 3.20). Each chromosome is made up of two separate strands called **chromatids,** and at one point along its length each chromatid has a special region called a **centromere.** The two chromatids are held together at their centromere regions by forces as yet unidentified. One chromatid includes a new, two-chain DNA molecule that was replicated during interphase using one chain of an original DNA molecule as a template. The second chromatid includes the new DNA molecule that was formed using the other chain of the original DNA as a template.

By the end of prophase, the centrioles have nearly reached opposite poles of the cell, and the chromosomes have moved toward a position at the middle or equator of the cell halfway between the two centriole pairs. Some of the microtubules that radiate from the centriole regions end blindly. These are known as *astral fibers,* and they form an aster about each centriole pair. Other microtubules, called *spindle fibers,* form an organized array known as a **spindle** between one centriole pair and the other. The spindle is composed of at least two kinds of spindle fibers. One kind extends from one pole of the cell to the center of the spindle, where it overlaps with similar fibers coming from the opposite pole. A second kind of spindle fiber also extends from a pole of the cell toward the center of the spindle; but it eventually attaches to a chromosome.

Metaphase

By the beginning of **metaphase** the nuclear envelope has completely disappeared (Figure 3.19e). As metaphase proceeds, the chromatids of each chromosome can be seen to be attached by their centromeres to spindle fibers along the central or equatorial plate of the spindle. At the conclusion of this stage, the chromatids of each chromosome uncouple and each of the former chromatids becomes an independent, single-stranded chromosome.

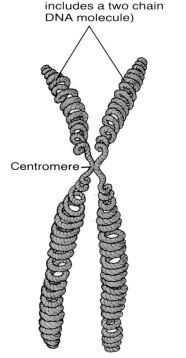

Chromatids (each includes a two chain DNA molecule)

Centromere

Figure 3.20
A chromosome as it appears during late prophase.

F3.20

F3.19e

FRONTIERS IN HEALTH

Doctoring Our Genes

Kevin is 21 years old. Unlike other young men his age, who have graduated from college and are busy starting families and new jobs, Kevin sits at home under his mother's care. With his hands strapped to his wheelchair or bed, Kevin passes his hours watching television. He has never gone to school and he knows practically nothing of the world outside his home. At times Kevin shakes out of control, and his mind drives him toward self-destruction. During one of his fits, he may bite a hand with enough force to amputate a finger; hence the straps.

Sad as it may sound, Kevin has virtually no hope of leaving his bed or wheelchair; no hope of throwing off the straps that bind his hands every moment of his life. Kevin, like 2000 other people in the United States, is afflicted with a rare genetic disorder called Lesch-Nyhan syndrome. This disease, which is detected in about 200 children every year, has turned Kevin's body into a living prison.

A single defective gene has made the difference between a normal life and a life of agony for this young man. The defective gene synthesizes an enzyme, HPRT (hypoxanthine guanine phosphoribosyl transferase), that plays a critical role in the body's metabolism. In its absence, uric acid builds up in the body's tissues, causing the symptoms seen in Kevin, as well as gout and recurrent kidney stones. New drugs can help reduce the buildup of uric acid, and permit victims a somewhat longer life. Most victims now live well into their 20s, whereas previously they generally died in early childhood. However, these drugs do nothing to block the episodes of self-destructiveness.

Curing diseases of this nature has previously seemed impossible, because the only way to reverse such a disease would be to replace the defective genes with normal ones, a task well beyond the scope of medicine—until recently, that is.

New advances in genetic engineering may someday provide a cure for Kevin and thousands of other victims of single-gene defects. Many medical scientists believe that normal genes can be administered to victims of such defects. To do this it is necessary first to isolate the gene in question. Once it is isolated, new copies of the gene are made, usually in a host bacterium. These copies then can be inserted into the genetic material of special viruses that infect human cells but do not cause any undesirable effects. These viruses, carrying the inserted genes, can be injected into the body. If all goes as planned, the viruses will infect body cells and the DNA of the injected gene will be incorporated into the genetic material of the cell. The defective gene remains—it was not functioning anyway—but the new gene becomes active, producing the missing enzymes.

The prospects of gene therapy are exciting, because there are at least 1600 single-gene defects known to medical science. However, a major obstacle to be overcome is perfecting the insertion steps; that is, getting the gene from the virus into the cells of the body, or at least into the cells that require the normal gene.

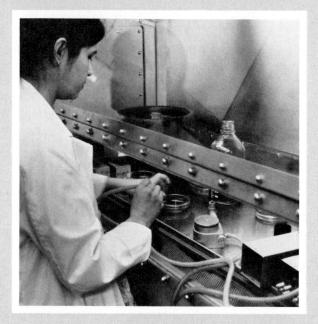

Cell tissue experiment at cancer research lab.

Encouraging results have come from a team of scientists headed by University of California, San Diego geneticist Dr. Theodore Friedman and Dr. Inder Verma of the Salk Institute. After years of tedious work, they have succeeded in transferring the HPRT gene to cultured white blood cells taken from victims of Lesch-Nyhan syndrome. The transplanted genes have raised the enzyme levels to one-quarter of the normal levels. The researchers plan to reinject these white blood cells into Lesch-Nyhan syndrome patients with the hope that the genes will be transferred to other body cells. Even though they are producing only a fraction of the normal levels of HPRT, the transplanted genes may provide enough enzyme to reverse the disease.

It is questionable, however, whether the genes will transfer from the white blood cells to the rest of the body, and it has been suggested that a better option would be to attempt to get the genes directly into every cell of the body, especially those cells of the brain where many of the symptoms originate. Such procedures raise serious questions. For example, how can researchers be certain that the gene will insert properly? What if it interferes with another gene and causes a harmful condition, such as cancer, or turns out to be lethal?

Many medical researchers believe that the possibility of opening the door to a healthy, productive life for the thousands of victims who suffer from genetic diseases far outweighs the risks involved. The future for people suffering from genetic disorders thus appears a bit more hopeful.

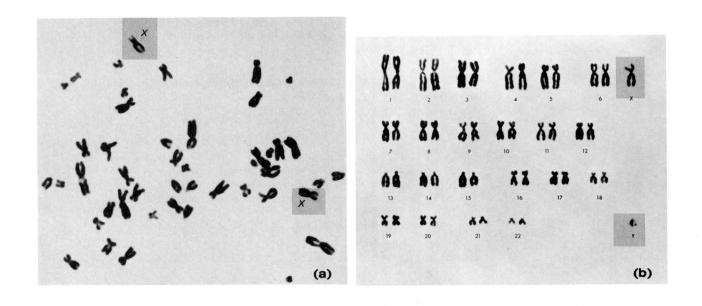

(a)

(b)

Anaphase

During **anaphase,** the single-stranded chromosomes separate and move toward opposite poles of the cells (Figures 3.19f, g). By the end of anaphase, the single-stranded chromosomes (each of which includes a complete, two-chain DNA molecule) have reached the poles of the cell. When mitosis and cytokinesis occur together, the beginning of cytokinesis is generally evident during anaphase as an inward pinching of the cell membrane in the equatorial region.

F3.19f,g

Telophase

In **telophase,** a new nuclear envelope forms, presumably from endoplasmic reticulum, and the nucleolus reappears (Figure 3.19h). When mitosis and cytokinesis are occurring together, cytokinesis is completed when the continued inward pinching of the cell membrane separates the cell into two daughter cells. Also during telophase, the spindle disappears, and the chromosomes become less distinct, gradually assuming their interphase appearance of chromatin threads. In many cells, the centrioles are replicated during this phase, but in other cells they are replicated at other times. By the end of telophase, the two daughter cells have assumed the interphase appearance, and the division cycle is complete.

F3.19h

Meiosis

There are 46 chromosomes in each of the human somatic cells (all cells except the reproductive cells). Two of these are **sex chromosomes** (two X chromosomes in females, and one X and one Y chromosome in males). The remaining 44 chromosomes are called **autosomes.** The 44 autosomes consist of 22 pairs of similar-appearing chromosomes (Figure 3.21). One member of each pair contains genetic information derived from the individual's father, and the other member of each pair contains information derived from the individual's mother. Each pair makes up a set of **homologous chromosomes.** The two sex chromosomes of the female (XX) are also homologous, but the two sex chromosomes of the male (X and Y) are not.

F3.21

ho-mol'-o-gus

Homologous chromosomes each possess genetic information that controls the same functions or characteristics. Often the genetic information for a particular function or characteristic on one chromosome of an homologous pair takes precedence over the corresponding information on the other chromosome of the pair. If the genetic information derived from the individual's

Figure 3.21
Human chromosomes.
(a) Chromosomes of a female, with X sex chromosomes indicated.
(b) Chromosomes of a male, with homologous chromosomes arranged in pairs. X and Y are sex chromosomes. Note X and Y chromosomes in color boxes.

father takes precedence over the genetic information derived from the individual's mother, the paternal function or characteristic will be **dominant** and the individual will display the paternal function or characteristic. In such a case the maternal function or characteristic is said to be **recessive.** If the maternal genetic information takes precedence over the corresponding paternal information, the individual will display the maternal function or characteristic. If neither the paternal nor the maternal genetic information for a particular function or characteristic takes precedence, the individual may display some intermediate function or characteristic. (For a more detailed discussion of human genetics, refer to Chapter 29.)

The 46 chromosomes of human somatic cells actually consist of two 23-chromosome sets (22 autosomes and 1 sex chromosome per set), with one set having been derived from the individual's father and one from the individual's mother. Thus, the **gametes,** or reproductive cells, of the male (the sperm from the testes) and female (the ova from the ovaries) each contain only 23 chromosomes. When a sperm cell fertilizes an ovum, each contributes its 23 chromosomes, thereby establishing the full 46 chromosomes of the new individual.

gam'-eet

Cells with two complete sets of chromosomes (46 chromosomes) are known as **diploid cells.** The formation of gametes, however, must result in the formation of cells that have only one set of chromosomes (23 chromosomes rather than 46 chromosomes). Such cells are called **haploid cells,** and the normal processes of mitosis do not produce such cells. A second type of cell division, known as **meiosis,** is responsible for the production of haploid reproductive cells.

my-o'-sis

F3.22

Two successive division sequences occur in meiosis (Figure 3.22). In the first sequence, prophase occurs essentially as in mitosis, with the exception that all 46 chromosomes do not move separately toward the spindle fibers. Rather, homologous chromosomes pair (synapse) with one another (as do the nonhomologous sex chromosomes of the male). The paired chromosomes move toward the spindle fibers as two chromosome units. During metaphase, the chromatids of the synapsed chromosomes do not uncouple. At anaphase, the paired homologous chromosomes simply move apart, with one member of the pair moving to one pole of the cell and the other member of the pair moving to the opposite pole. Thus, 23 double-stranded chromosomes are moved to each pole during the first division sequence of meiosis, rather than the 46 single-stranded chromosomes that were moved to each pole in mitosis. The remaining events of first anaphase and first telophase occur as in mitosis (with cytokinesis), resulting in two daughter cells that each contain only 23 chromosomes, but the chromosomes are double-stranded rather than single-stranded. Following the first division sequence of meiosis, a short period called **interkinesis** occurs. During this period, which is similar to the interphase period between mitotic divisions, the 23 double-stranded chromosomes of the daughter cells do not duplicate themselves.

Following the interkinesis period, the second division sequence of meiosis occurs. Each of the two 23-chromosome daughter cells undergoes a typical mitotic division with cytokinesis. The result of this second division sequence is four haploid cells, each of which contains 23 single-stranded chromosomes.

Meiosis allows a great deal of genetic diversity in the makeup of reproductive cells (sperm and ova). The chromosomes that contain genetic information derived from the individual's male parent, for example, do not necessarily all line up toward one centriole pair, nor do the chromosomes that contain genetic information derived from the individual's female parent all line up toward the other centriole pair during the synapsis of chromosomes that occurs in the first division sequence of meiosis (Figure 3.23). Rather, a mixture of positions occurs, so that the resulting daughter cells each receive, in random assortment, some chromosomes that contain genetic information derived from the individual's male parent and some that contain genetic information derived from the female parent.

F3.23

Further genetic diversity can result from the process of **crossing over,** which takes place occasionally during the first stage of meiosis, while the

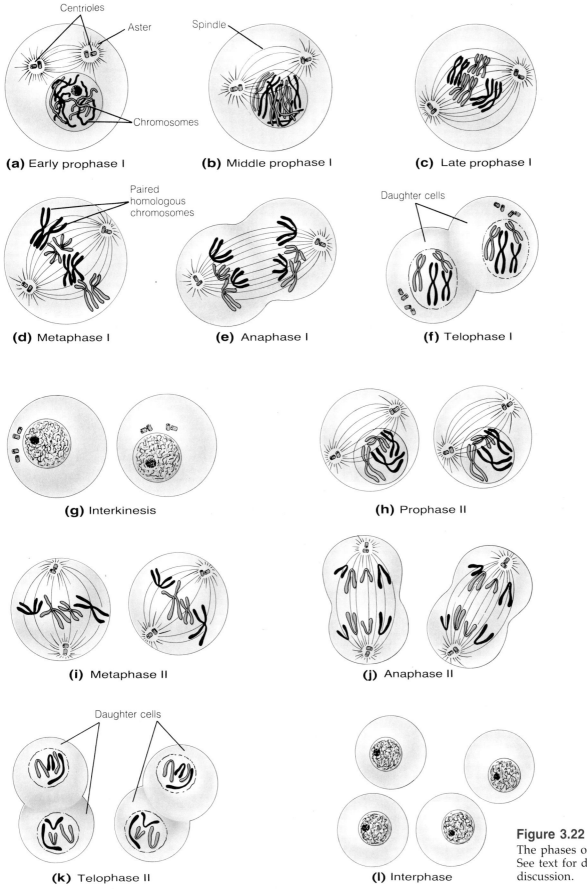

(a) Early prophase I

(b) Middle prophase I

(c) Late prophase I

Centrioles

Aster

Chromosomes

Spindle

(d) Metaphase I

(e) Anaphase I

(f) Telophase I

Paired homologous chromosomes

Daughter cells

(g) Interkinesis

(h) Prophase II

(i) Metaphase II

(j) Anaphase II

(k) Telophase II

(l) Interphase

Daughter cells

Figure 3.22
The phases of meiosis. See text for detailed discussion.

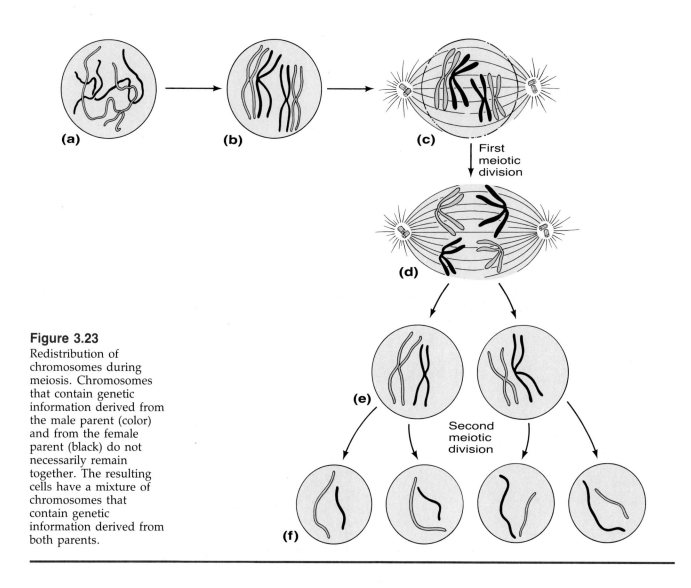

Figure 3.23
Redistribution of chromosomes during meiosis. Chromosomes that contain genetic information derived from the male parent (color) and from the female parent (black) do not necessarily remain together. The resulting cells have a mixture of chromosomes that contain genetic information derived from both parents.

First meiotic division

Second meiotic division

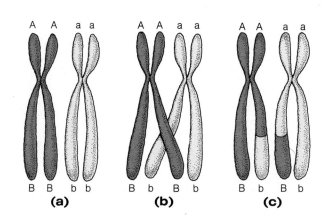

Figure 3.24
The process of crossing over.

chromosomes are synapsed (Figure 3.24). In this process a chromatid of one   **F3.24**
chromosome of a synapsed pair breaks, and a corresponding break occurs in a
chromatid of the other chromosome of the pair. The two free chromatid fragments then exchange places so that when the breaks are repaired, the frag-
ment originating from the chromatid of the one synapsed chromosome is
attached to the chromatid of the other chromosome and vice versa. When the
synapsed chromosomes move apart and daughter cells are ultimately produced, the daughter cells can have a different genetic composition than if
crossing over had not occurred. Thus, the gametes that are ultimately produced from different cells differ from one another genetically depending on
the particular chromosomal redistribution that occurs during meiosis.

CONDITIONS OF CLINICAL SIGNIFICANCE

The Cell

Cancer

In its essence, cancer is simply an uncontrolled division
of cells that gives rise to abnormal growths called *malignant tumors*. Sometimes the growths are localized and, if
discovered early enough, can be removed surgically.
Cancer cells, however, are not very adhesive, and they
may spread (metastasize) throughout the body and
lodge in various organs and tissues. Such an occurrence
renders the cancer inoperable, and other means of treatment must be employed. Radiation therapy is sometimes used in such cases because rapidly dividing cells
are very susceptible to radiation. Various drugs and chemotherapeutic agents that attack rapidly growing cells
are also employed. One problem with radiation therapy
and chemotherapy, however, is that cells such as white
blood cells, which protect the body from foreign organisms, also come from dividing cells that may be injured
by these therapies, leaving the individual less able to
ward off other illnesses and infections.

Although a number of carcinogens—cancer-causing
agents—have been discovered, the exact nature of the
disease remains a mystery. Cancer may be due to
changes that remove cells from whatever controls exist
to regulate their division. Certain viruses have been
shown to cause cancer in some animal species, and there
is some evidence that viruses can also be carcinogenic in
humans.

Genetic Disorders

Many diseases and abnormalities can be traced to genetic defects. Occasionally, incorrect bases are incorporated into DNA chains, and because of complementary
base pairing, RNA molecules produced using these
DNA chains as templates will also contain incorrect
bases. In m-RNA, three-base codons containing incorrect bases may specify amino acids that are different
from those that would normally have been called for.
When ribosomes "read" these erroneous m-RNA molecules, incorrect amino acids are inserted into the proteins being formed. These incorrect proteins may not
function properly, and disease can be one result.

Such is the case in the genetic disease sickle-cell
anemia, which occurs when an incorrect amino acid becomes incorporated into one of the polypeptide chains
of the oxygen-transporting molecule hemoglobin. Red
blood cells that contain the abnormal hemoglobin molecules are fragile and misshapen. They rupture within
and block blood vessels, leading to an impaired delivery
of oxygen to the body's cells.

Also, since many proteins function as enzymes that
catalyze cellular chemical reactions, errors in their construction can make it impossible for certain reactions to
occur. The genetic disease phenylketonuria results from
the lack of a functional enzyme required to convert the
amino acid phenylalanine to tyrosine. The inability of
the cells to carry out this reaction efficiently can ultimately produce mental retardation.

Genetic disorders range from those in which a single amino acid is incorrectly incorporated into a polypeptide chain to those in which entire polypeptides or
proteins are not manufactured at all. Alterations in the
genetic material of the cell that lead to such occurrences
are called **mutations**. In most instances, mutations are
found to be harmful, or, at best, neutral to cell function.
Occasionally, however, they may lead to an improved
condition as far as the organism is concerned. Mutations
that occur in reproductive cells can be preserved and
passed from parent to offspring. Genetic disorders are
discussed further in Chapter 29.

Aging

From the moment of fertilization, the human organism
grows and develops until full maturity is reached. After
this time, the individual gradually ages. The aging processes themselves are currently attracting a great deal of
interest, and several theories have been proposed to
explain what actually causes aging at the cellular level.
These theories range from ideas that external events
such as X rays or cosmic radiation—or the more nebu-
lous "stresses of life"—gradually take their toll on cells,
to ideas more internally oriented, such as that the
changes of aging are built into the genetic apparatus of
the cell.

STUDY OUTLINE

CELL COMPONENTS Nucleus surrounded by cytoplasm, which contains numerous structures called organelles. Limiting boundary of cell is plasma membrane.   **pp. 57–73**

PLASMA MEMBRANE A highly selective barrier that affects the movement of substances into or out of cells. Membrane is lipid bilayer with protein interspersed; channels in membrane connect cell interior with outside. Materials that enter or leave cell must pass through membrane or channels.

MOVEMENT OF MATERIALS ACROSS THE PLASMA MEMBRANE

Diffusion Through the Lipid Portion of the Membrane Many nonpolar substances soluble in lipids move easily through plasma membrane by simple diffusion.

Movement Through Membrane Channels Small polar molecules and ions move through channels.

Mediated Transport Carrier molecules in plasma membrane transport substances across membrane. Mediated transport systems exhibit characteristics of specificity, saturation, and competition.

Facilitated Diffusion Carrier molecules move material across the membrane equally well in either direction. Net movement is from region of high concentration to region of low concentration; does not require cellular energy.

Active Transport Carrier molecules can move materials across the membrane from regions of low concentration to regions of high concentration, against normal concentration gradients; requires cellular energy.

Endocytosis and Exocytosis Some cells take in substances by forming vesicles that enclose small portions of the external environment, a process called endocytosis. Substances leave cell by the fusion of membrane-bounded vesicles with plasma membrane, a process called exocytosis.

TONICITY Refers to the tendency of a solution to influence the state of tension or tone of cells placed in it due to water movement into or out of the cells. The tonicity of a solution is influenced by the properties of a cell's plasma membrane.

THE NUCLEUS
1. Contains the genetic material of the cell in the form of chromatin threads (DNA combined with protein and RNA).
2. DNA instructs cells to synthesize particular polypeptides or proteins.
3. Polypeptides or proteins synthesized according to DNA specifications act as enzymes, as structural proteins, or in other ways to accomplish the work of the cell.

RNA SYNTHESIS (TRANSCRIPTION) DNA serves as the template for RNA synthesis.
1. m-RNA contains coded information, derived ultimately from DNA, that is used to direct protein synthesis.

2. The principal bases of m-RNA—adenine, cytosine, guanine, uracil—can be arranged to form 64 different three-base codons.
3. Codon sequence specifies the amino acid sequence of a polypeptide or protein.

TRANSFER RNA Carries amino acids to sites where the amino acids are incorporated into polypeptides or proteins.

RIBOSOMAL RNA
1. Nucleoli are found in the nucleus and are associated with r-RNA production.
2. r-RNA becomes a part of ribosomes.

RIBOSOMES Cytoplasmic organelles that consist of r-RNA and protein.

PROTEIN SYNTHESIS (TRANSLATION) At ribosomes, amino acids are linked together in the sequence specified by m-RNA to form polypeptides and proteins.

ENDOPLASMIC RETICULUM Membranous network of tubular or saclike channels; walls contain enzymes that play a role in fatty acid and steroid synthesis. Proteins synthesized by attached ribosomes can enter channels, and protein-containing vesicles can break away.

GOLGI APPARATUS Flattened membranous sacs with which vesicles from endoplasmic reticulum can fuse. Proteins are processed, modified, and sorted in Golgi apparatus; secretory vesicles containing processed proteins pinch off from Golgi membranes.

LYSOSOMES Membrane-bounded cytoplasmic organelles containing digestive enzymes that act on proteins, lipids, certain carbohydrates, DNA, and RNA.

PEROXISOMES Contain oxidative enzymes that produce hydrogen peroxide as well as the enzyme catalase, which splits hydrogen peroxide into water and oxygen.

MITOCHONDRIA Involved in the generation of metabolic energy for cell activity.

CYTOSKELETON Consists of an intricate network of filamentous structures.

MICROTUBULES Small, hollow, cylindrical, unbranched tubules involved in cell movement processes and in the development and maintenance of cell shape.

INTERMEDIATE FILAMENTS May be involved in maintaining the position of the nucleus and in transporting materials into and out of the nucleus.

MICROFILAMENTS Can occur in bundles or other groupings; associated with contractile activities involved in cell movement.

CILIA, FLAGELLA, AND BASAL BODIES Cilia and flagella are motile projections from the plasma membrane that can move substances over cell surfaces, or move entire cells about. Both are believed to arise from basal bodies. Cilia, flagella, and basal bodies contain microtubules.

CENTRIOLES Similar in structure to basal bodies; involved in cell division.

INCLUSIONS Many chemical substances in cells, such as hemoglobin.

EXTRACELLULAR MATERIALS Include body fluids and substances in which many cells are embedded. Connective tissues are particularly rich in extracellular materials. p. 73

CELL DIVISION Several basic events are involved, including replication of genetic material in nucleus, redistribution of this material into two new nuclei, and the division of cytoplasm into two new cells, each with its own nucleus. pp. 73–81

INTERPHASE Period between active cell divisions. DNA replication occurs.

MITOSIS A continuous event, although four phases are observable.

PROPHASE Chromosomes become visible; centrioles move toward opposite poles of cell; spindle forms.

METAPHASE Chromatids attached by centromeres to spindle fibers; chromatids uncouple and each becomes a separate, single-stranded chromosome.

ANAPHASE Single-stranded chromosomes separate and move to opposite poles of cell.

TELOPHASE Spindle disappears and chromosomes assume interphase appearance as indistinct chromatin threads.

MEIOSIS Cell division process that produces haploid cells; involves two successive division sequences.

FIRST SEQUENCE

Prophase Essentially as in mitosis, except that homologous chromosomes pair and move toward spindle fibers as two chromosome units.

Metaphase Chromatids of synapsed chromosomes do not uncouple.

Anaphase Paired chromosomes move apart, one member of the pair moves to one pole of cell and the other member to the opposite pole.

Telophase Cytokinesis produces two new cells, each with 23 double-stranded chromosomes.

SECOND SEQUENCE Each of the 23-chromosome daughter cells undergoes mitosis with cytokinesis. Result is four haploid cells, each with 23 single-stranded chromosomes.

CONDITIONS OF CLINICAL SIGNIFICANCE: THE CELL p. 81

CANCER Uncontrolled division of cells. May be treated surgically if discovered early enough; radiation therapy and chemotherapy are other forms of treatment.

GENETIC DISORDERS Many diseases and abnormalities are traceable to genetic defects. Genetic disorders can cause incorrect protein synthesis; erroneous protein might then lead to disease. Alterations in the genetic material of the cell are called mutations; in most cases mutations are harmful.

AGING X rays, cosmic radiation may gradually take their toll on cells, and thus cause aging. Aging changes may be programmed into genetic apparatus of cells.

SELF-QUIZ

1. Many nonpolar substances that are soluble in lipids move relatively easily through the plasma membrane by: (a) exocytosis; (b) active transport; (c) simple diffusion.

2. Match the following terms associated with the plasma membrane with the appropriate descriptions.
   Specificity
   Saturation
   Competition
   Facilitated diffusion
   Active transport
   (a) Carrier molecules move material across the membrane equally well in either direction; net movement is from region of high concentration to region of low concentration.
   (b) Each carrier molecule binds with only a select group of substances.
   (c) Carrier molecules move materials across the plasma membrane against a concentration gradient.
   (d) The amount of a substance that can cross the plasma membrane by mediated transport in a given time is limited.
   (e) A particular binding site on a carrier molecule can bind with more than one substance.

3. The term for the process by which a cell takes in material from the external environment by means of membrane-bounded vesicles is: (a) exocytosis; (b) endocytosis; (c) cytokinesis.

4. A solution that produces a change in the tone of cells immersed in it as a result of the net movement of water from the solution into the cells is a(an): (a) isotonic solution; (b) hypotonic solution; (c) hypertonic solution.

5. The nuclear envelope is a pore-containing double membrane. True or False?

6. t-RNA contains coded messages that instruct a cell to synthesize particular polypeptides or proteins. True or False?

7. A three-base sequence in m-RNA that specifies a particular amino acid of a polypeptide or protein is called a(an): (a) codicil; (b) codon; (c) anticodon.

8. Ribosomes are formed from r-RNA–protein complexes that migrate out of the nucleus and into the cytoplasm. True or False?

9. Enzymes that play a role in fatty acid and steroid synthesis are associated with: (a) endoplasmic reticulum; (b) ribosomes; (c) lysosomes.

10. Enzymes capable of digesting proteins, lipids, DNA, and RNA are associated with: (a) ribosomes; (b) microtubules; (c) lysosomes.

11. Organelles involved in the generation of metabolic energy for cellular activities are: (a) peroxisomes; (b) mitochondria; (c) lysosomes.

12. Thin, short, and numerous projections that can move substances over the surface of a cell are known as: (a) cilia; (b) flagella; (c) basal bodies.

13. Cilia, flagella, basal bodies, and centrioles all contain: (a) mitochondria; (b) microtubules; (c) peroxisomes.

14. Match the following cytoplasmic organelles with their related functions.

Ribosomes	(a)	Organelles that contain the enzyme catalase
Golgi apparatus		
Peroxisomes	(b)	Structures from which cilia and flagella arise
Basal bodies		
Microfilaments	(c)	Structures that are associated with contractile activities involved in cell movement
	(d)	The site of polypeptide and protein synthesis
	(e)	A membranous organelle that gives rise to secretory vesicles

15. Mitosis and cytokinesis always occur simultaneously. True or False?

16. The DNA molecules that comprise the genetic material of the cell appear only as indistinct chromatin threads within the nucleus during: (a) metaphase; (b) anaphase; (c) interphase.

17. During both RNA production and DNA replication, nucleotides attach to an exposed template DNA chain according to a complementary base-pairing pattern. True or False?

18. Match the following items associated with meiosis with the appropriate lettered descriptions.

Autosomes	(a)	XX
XY	(b)	The stage between the first and second meiotic divisions
Homologous chromosomes		
46	(c)	Cells with two complete sets of chromosomes are said to be this type
Diploid		
23	(d)	This process contributes to genetic diversity
Haploid		
Interkinesis	(e)	This cell type is characteristic of gametes
Crossing over		
	(f)	The total number of chromosomes in a human somatic cell
	(g)	Male sex chromosomes
	(h)	Chromosome classification that does not include sex chromosomes
	(i)	The number of chromosomes contained in each of the human male and female gametes

19. A doctor would be likely to use surgery but not radiation or chemotherapeutic agents to treat a patient in whom cancer cells had metastasized. True or False?

20. In most instances, mutations are found to be: (a) neutral; (b) helpful; (c) harmful.

LEARNING OBJECTIVES

After completing this chapter, you should be able to:

1. Name the four primary tissues, and cite one example of each.

2. Describe the specializations by which adjacent cells may be attached to one another.

3. List three means of classifying epithelial tissue, and cite at least one example of each tissue.

4. Describe the shape that characterizes each of these cell types: squamous, cuboidal, and columnar.

5. Distinguish between simple epithelium and stratified epithelium.

6. Distinguish between exocrine glands and endocrine glands.

7. Classify three types of glands by mode of secretion, and describe how each type functions.

8. List the types of connective tissue, and state at least one function of each.

9. Cite several structural differences between bone and cartilage.

10. Name the three main types of muscle tissue, and describe the appearance of the cells of each type.

11. Cite two kinds of tissues that cannot undergo regeneration in a mature person, and two kinds that can.

CHAPTER CONTENTS

EPITHELIAL TISSUES

CONNECTIVE TISSUES

MUSCLE TISSUE

NERVOUS TISSUE

TISSUE REPAIR

KEY TERMS AND DERIVATIVES

adipose tissue (*adip* = fat; *ose* = having or characterized by) a type of connective tissue characterized by the presence of many fat cells

chondrocyte (*chondro* = cartilage; *cyte* = cell) a mature cartilage cell

epithelial (*epi* = upon; *thel* = delicate) one of the primary tissues; covers the surface of the body and lines the body cavities, ducts and vessels

fibroblast (*fibr* = fiber; *blast* = a formative cell) connective tissue cell that produces fibers and portions of matrix

macrophage (*macro* = large; *phage* = eater) a cell type common in connective tissue that removes debris, bacteria, and other materials by phagocytosis

osteocyte (*osteo* = bone; *cyte* = cell) a mature bone cell found in each lacuna

pseudostratified (*pseudo* = false; *strati* = layers) tissue whose cells appear to be arranged in layers, but are not

tissue (*tiss* = weave) a group of similar cells and fibers, forming a distinct structure, that work together to carry out specific functions

Tissues

In the previous chapter you learned that the body is composed of millions of cells and that each cell contains various organelles, which carry on a number of physiological processes. It is important to realize, however, that it is more common for groups of cells to cooperate for the benefit of the organism as a whole, rather than simply for their own individual needs. Groups of cells that are similar in structure, function, and embryonic origin and that are bound together with varying amounts of intercellular material are referred to as **tissues.** There are four primary tissues in the body: *epithelial, connective, muscular,* and *nervous.* Since it is these four tissues that join together to form the organs of the body, an understanding of the structure and function of each tissue type contributes to our understanding of the organ systems.

EPITHELIAL TISSUES

Epithelial tissues are formed of closely joined cells with only a minimum of intercellular material between them. Epithelial cells are always underlain by connective tissue, to which they are attached by a thin layer called the **basement membrane.** The basement membrane consists of two layers; one layer, called the *basal lamina,* is composed of collagen and glycoproteins, and is a product of the epithelial cells; the deeper layer is composed of *reticular fibers* that develop from the connective tissue. Epithelia may develop from either the ectoderm, endoderm, or mesoderm of the embryo.

Epithelia are, by definition, sheets of cells that cover body surfaces and line body cavities. In general, they cover most of the free surfaces of the body, both internal and external. For instance, they form the outer layer of the skin, the lining of the digestive tube, the linings of the ventral body cavities, the lining of the blood vessels, and those glandular ducts and tubules that develop from epithelial linings or coverings. In addition, some epithelial tissues are incorporated into the various glands, where they serve as the functional part of the gland. With such a variety of locations, it is not surprising that epithelial tissues have diverse functions. The epidermis of the skin, for instance, is a *protective* layer that forms a barrier between the organism and its external environment, whereas the linings of the internal body organs are involved in the *absorption* of materials into the body, the *excretion* of waste products, and the *secretion* of special products into the cavities.

Specializations of Epithelial Cell Surfaces

The portion of epithelial cells that forms the surface of the body or lines the cavities and lumina (interior space) of the various tubes in the body is referred to as the **free surface.** The free surfaces of the epithelial cells that line the blood vessels are smooth. The electron microscope shows that other epithelial cells have their free surfaces folded into tiny protoplasmic projections called **microvilli** (Figure 4.1). Because microvilli greatly increase the area of the free surface, they are especially abundant in locations where absorption is the main activity, as in the lining of the digestive tract. Before the electron microscope made it possible to view their structure clearly, these dense groups of microvilli were identified as *striated borders* or *brush borders.* Microvilli line

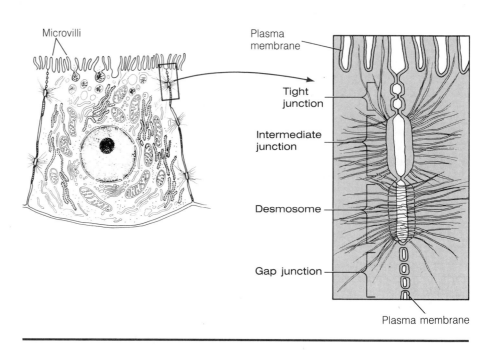

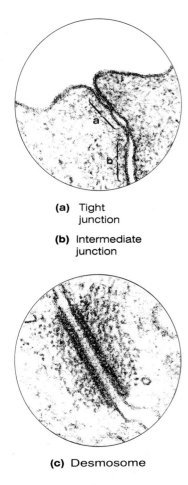

(a) Tight
junction

(b) Intermediate
junction

(c) Desmosome

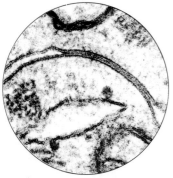

(d) Gap junction

Figure 4.1

Typical epithelial cell as
seen with an electron
microscope. On the right
is a highly magnified
junctional complex
consisting of a tight
junction, an intermediate
junction, and a
desmosome. Parts **(a)**
through **(c)** are
photomicrographs of the
components of a
junctional complex. Part
(d) is a gap junction.

F4.1

some surfaces that are not absorptive, and their function is less well understood there. They may serve to anchor mucus to the cell surface. Microvilli are
not highly motile. In some locations, the free surfaces of epithelial cells are
modified by the presence of **cilia.** Most cilia are motile and move rhythmically, thus serving to propel materials along the epithelial surface.

Specializations for Cell Attachments

We have noted that one characteristic of epithelial tissues is that their cells are
situated close together, with little intercellular material between them. In fact,
adjacent epithelial cells are generally joined together into coherent sheets. It is
these cellular junctions that make epithelial tissues so well suited to cover the
body surfaces and to line the body cavities.

Before the advent of the electron microscope, epithelial cells were thought
to be held together by combinations of cement and bridges between adjacent
cells and by several structures that showed up under the light microscope as
dark spots on the cell boundaries. These spots were called desmosomes or
terminal bars, depending on their location. The electron microscope has clari-
fied the structures of these cellular junctions. In columnar and some cuboidal
epithelium, the intercellular relationships, as revealed by the electron microscope, are referred to as **junctional complexes.**

Junctional Complexes

Every junctional complex generally has three distinct components—a *tight
junction,* an *intermediate junction,* and a *desmosome*—all of which are located on
the lateral cell boundaries (Figure 4.1).

TIGHT JUNCTIONS **Tight junctions (zonula occludens)** are located just
below the free surface of the epithelium. In this region the outer layers of the
cell membranes of adjacent cells fuse in several places, leaving intercellular
separations between the sites of membrane fusion. Tight junctions extend
around the outer margin of the cell like a belt. Tight junctions not only connect adjacent cells but, because they obliterate the intercellular spaces, they
also restrict the movement of substances through the epithelium via the intercellular spaces.

INTERMEDIATE JUNCTIONS **Intermediate Junctions (zonula adherens)**
are located just below tight junctions. In the intermediate junction the cell

membranes of adjacent cells are not modified and are separated by a gap of about 200 Å. There is, however, a mat of filaments located against the inner layer of the cell membrane of each cell. Like the tight junction, the intermediate junction extends like a belt around each cell.

DESMOSOMES The third component of a typical junctional complex is the **desmosome** or **macula adherens.** Each desmosome is an individual point of cell attachment rather than a beltlike zone. The cell membranes remain about 200 Å apart in a desmosome, and the inner layer of each cell membrane is thickened. Cytoplasmic filaments that form part of the cytoskeleton of the cells attach to the thickened cell membranes. Indistinct rodlike proteins called central lamina run across the intercellular space between adjacent desmosomes. It is thought that these rods are related to the cell-to-cell binding at these points. Desmosomes can occur anywhere around the periphery of an epithelial cell. Where the cell membrane contacts connective tissue, as along the basal lamina, half desmosomes **(hemidesmosomes)** are sometimes found. These are particularly prevalent in stratified squamous epithelium, such as is found in the outer layer of the skin.

des'-mo-some
mak'-u-la ad-hear'-uns

Gap Junctions

Apart from the components of the junctional complex is another intercellular specialization called the **gap junction** or **nexus.** In this junction the cell membranes of adjacent cells are separate but very close together—only approximately 20 Å apart. This extremely narrow gap is bridged by small tubular channels that directly link the cytoplasm of adjacent cells. Gap junctions are sites at which small molecules and ions can pass from one cell to another, and they play an important role in the transmission of electrical activity between cells. Unlike junctional complexes, gap junctions are not restricted to epithelia; they are also found in muscle tissue and nerve tissue.

Classification of Epithelia

Epithelial tissues are generally classified on the basis of the *number and arrangement of cell layers* within the tissue and the *shape of the cells at the free surface* of the tissue.

According to Cell Layers

If an epithelium is formed by a single layer of cells, all of which are in contact with the basal lamina, it is called **simple epithelium.** If it has two or more layers of cells, and only the deepest layer is in contact with the basal lamina, it is termed **stratified epithelium.** If the tissue appears to consist of several layers but is actually formed of a single layer with all cells touching the basal lamina, it is **pseudostratified epithelium** (*pseudo* = false). This false impression of stratification occurs because some cells are shorter than others and the taller cells overlap the short ones, preventing them from reaching the free surface of the tissue.

soo-do-strat'-a-fide

According to Cell Shape

The cells that form the free surface of epithelial tissues are of three shapes. **Squamous cells** are flat and thin. **Cuboidal cells** are about as tall as they are wide and therefore appear almost square in vertical section. **Columnar cells** are taller than they are wide and appear rectangular in vertical section. Epithelia can be named according to which of these cell types form their free surfaces.

skway'-mus

General Classification

The general classification of epithelial tissues takes into consideration both the shape of the cells that form the free surface and the number of layers of cells in the tissue.

SIMPLE SQUAMOUS EPITHELIUM **Simple squamous epithelium** is formed of a single layer of squamous cells (Figure 4.2). Since this thin sheet

F4.2

×1250

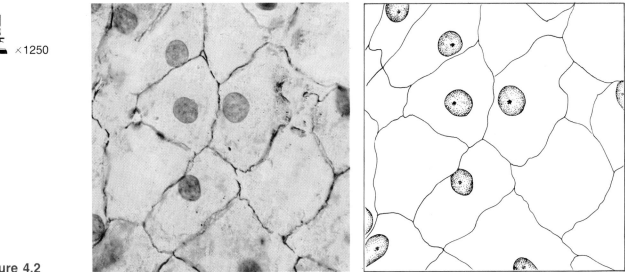

Figure 4.2
Simple squamous
epithelium (surface
view).

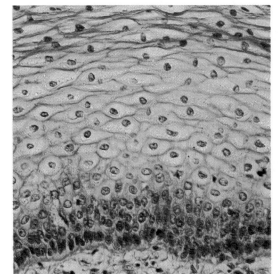

×1250

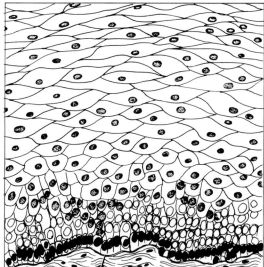

Basal lamina

Figure 4.3
Stratified squamous
epithelium.

does not form a very effective barrier, substances can move easily across it. And the flat cells do not contain enough cytoplasmic inclusions to aid secretion or absorption.

In general, simple squamous epithelium is found in regions where diffusion and filtration occur. Specifically, it lines the heart and the blood vessels and is the only barrier that separates the blood in capillaries from the tissue fluid. The simple squamous lining of the vascular system is called **endothelium.** Similarly, it lines the air sacs (alveoli) of the lungs, where it separates the air from the tissue fluid, and lines the surfaces of body cavities. The simple squamous lining of the body cavities is called **mesothelium.** Simple

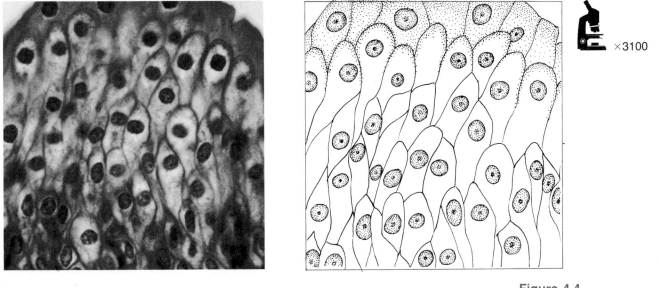

×3100

Figure 4.4
Transitional epithelium.

squamous epithelium also forms the glomerular capsules of the kidneys, the sites where substances are filtered from the blood to form urine.

STRATIFIED SQUAMOUS EPITHELIUM   As the name implies, **stratified squamous epithelium** (Figure 4.3) is composed of many layers, the precise number of which varies in different locations. The deeper cells, close to the basal lamina, tend to be cuboidal, but those against the surface are typical squamous cells. The deeper cells undergo mitosis and thus increase in number. These newly formed cells are pushed toward the surface, where they replace the older surface cells that are being continually sloughed off. Because of the ability of stratified squamous epithelium to replace the cells of the superficial layers, this tissue is capable of compensating for the loss of cells due to such actions as abrasion. Thus, stratified squamous epithelium forms a protective layer on the body surface as the epidermis of the skin, and also in areas that are subjected to friction, such as the linings of the mouth, pharynx, esophagus, anus, and vagina.

F4.3

TRANSITIONAL EPITHELIUM   **Transitional epithelium** is a specialized stratified tissue that lines the urinary bladder and a few other hollow organs (Figure 4.4). The surface cells of transitional epithelium vary between cuboidal and squamous, depending upon whether the bladder is empty or expanded. The deepest cells are columnar and they are overlaid by many layers of cells when the bladder is empty. When the bladder is full and its walls are stretched, the cells flatten and slide over each other, leaving only three or four strata (layers) between the deepest layer and the free surface. In this condition, the surface cells are flattened, like squamous cells. This specialized tissue allows an organ to expand with only minimal resistance from the tissue, thus lessening the chance of the organ rupturing, and reducing the discomfort that occurs as the organ becomes full.

F4.4

SIMPLE CUBOIDAL EPITHELIUM   Cuboidal cells are generally found as a single layer of cells that form **simple cuboidal epithelium** (Figure 4.5). Only rarely are cuboidal cells found in layers (stratified). Some cuboidal cells are capable of forming secretions and consequently are found in glands such as the thyroid, the sweat glands, and the salivary glands. Cuboidal cells also

F4.5

×3100

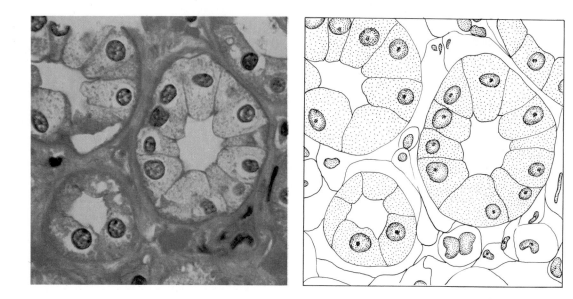

Figure 4.5
Simple cuboidal
epithelium.

×3100

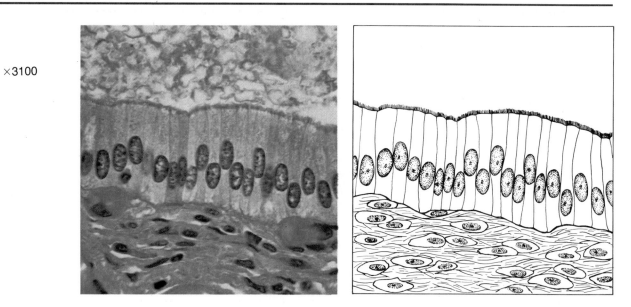

Figure 4.6
Simple columnar
epithelium.

form the ducts of glands and parts of the kidney tubules as well as the outer cell layer covering the ovary.

SIMPLE COLUMNAR EPITHELIUM   Columnar cells, with their large amounts of cytoplasm and cytoplasmic organelles, can perform rather complex chemical reactions and are therefore found in regions where secretion and absorption occur. **Simple columnar epithelium** (Figure 4.6) lines the digestive tube from the stomach to the anal canal. It also forms the ducts of many glands. Cilia are present on the free surfaces of the columnar cells that form the membranes lining the bronchi of the lungs, the nasal cavity, the oviducts, and scattered regions of the uterus. The presence of cilia causes these tissues to be classified as **simple columnar ciliated epithelium.**

F4.6

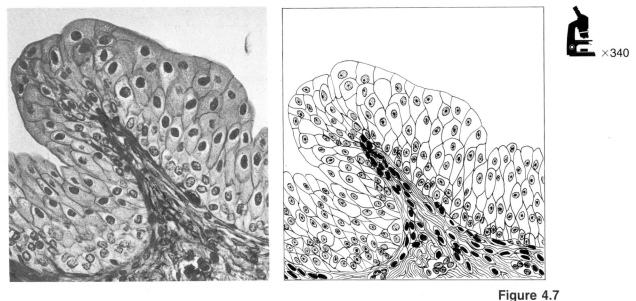

×340

Figure 4.7
Stratified columnar
epithelium.

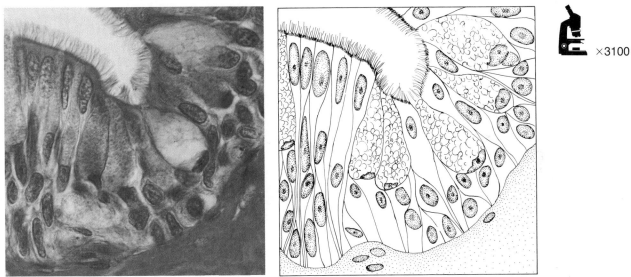

×3100

Figure 4.8
Pseudostratified columnar
ciliated epithelium.

STRATIFIED COLUMNAR EPITHELIUM   There are only a few locations where columnar tissue is truly stratified. In these tissues, the cells next to the basal lamina are small and rounded, and the columnar cells that form the free surface of the tissue do not contact the basal lamina (Figure 4.7). **Stratified columnar epithelium** appears on the epiglottis, in parts of the pharynx and anal canal, and in the male urethra.

F4.7

PSEUDOSTRATIFIED COLUMNAR EPITHELIUM   In **pseudostratified columnar epithelium,** which is much more common than the truly stratified columnar epithelium, all the cells contact the basal lamina but some are shorter than others and do not reach the free surface. The nuclei are also found at different levels within the cells, which adds to the impression of

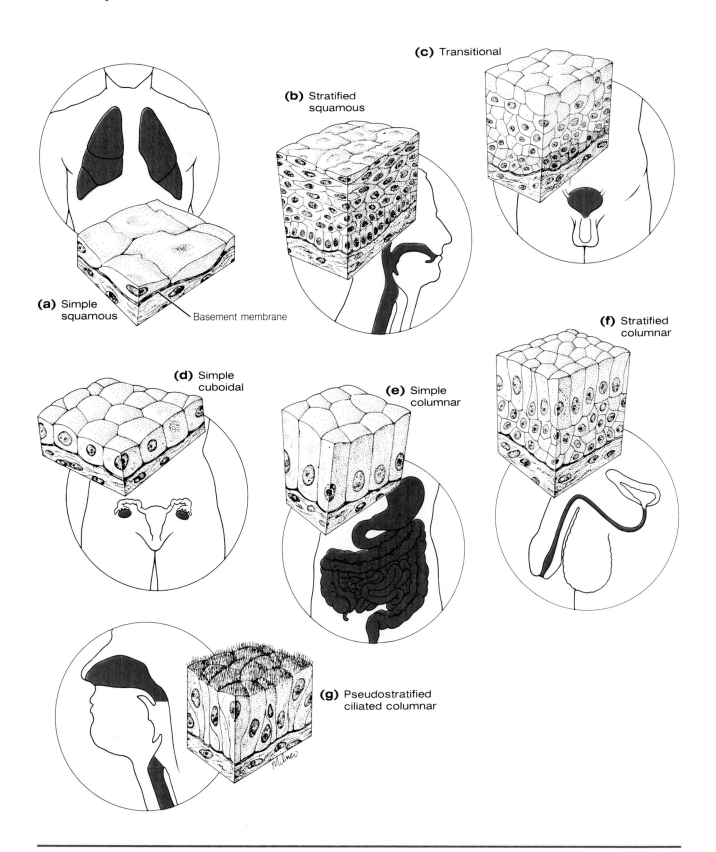

Figure 4.9
Classification of epithelial
tissues according to cell
layers and shape.

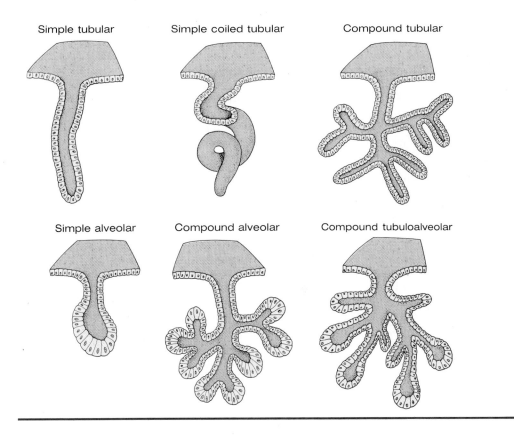

Simple tubular

Simple coiled tubular

Compound tubular

Simple alveolar

Compound alveolar

Compound tubuloalveolar

Figure 4.10
Classification of exocrine glands according to structure.

stratification. This epithelium is found in the large ducts of some glands, such as the parotid gland, and in regions of the male urethra. The free surface of pseudostratified tissue often has cilia, and the tissue is then called **pseudostratified columnar ciliated epithelium** (Figure 4.8). These ciliated tissues line the mucous membranes of the respiratory passageways and the auditory (eustachian) tubes. F4.8

A summary of the classification of epithelial tissues according to cell layers and shape, as well as a typical location for each type of tissue, is given in Figure 4.9. F4.9

Glandular Epithelium

Most glands of the body are composed of epithelial cells that produce a specific secretion (such as sweat, milk, a hormone, or an enzyme) or excrete certain waste products (such as bile pigments). The mucus-secreting **goblet cells** of the respiratory and digestive tracts are examples of *individual cells* that function as glands. Most glands, however, are *multicellular*—that is, they are formed of clusters of cells.

Embryologically, all glands originate from an epithelium. And most glands retain their connection with the epithelium—a connection that serves as a duct through which the secretions of the gland are carried to a particular site. Such glands are called **exocrine glands.** Some glands, however, lose their connection with the epithelium and empty their secretions directly into the blood. These are the **endocrine glands** and their secretions are **hormones.**

The multicellular exocrine glands are classified according to (1) their structure and (2) the manner in which they produce their secretions. Structurally, the ducts of the glands may be *unbranched* or *branched*. Glands whose ducts do not branch are called *simple glands;* glands whose ducts branch repeatedly are called *compound glands*. Simple and compound glands can be further subdivided according to whether their secreting portions are (1) *tubular*, (2) composed of small sacs (*alveoli* or *acini*), or (3) a combination of blindly ending tubules and alveoli (*tubuloalveolar*) (Figure 4.10). Compound tubuloalveolar glands are the most common type of exocrine gland, being present in the pancreas, prostate, salivary glands, and mammary glands. *al-ve'-o-lie* F4.10

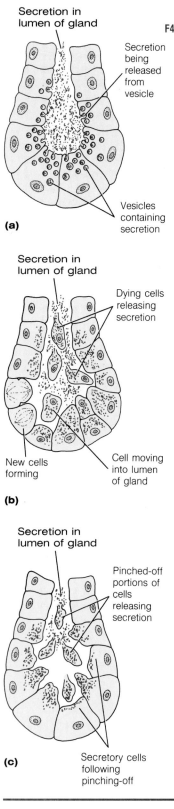

(a)

Secretion in lumen of gland

F4.11

Secretion being released from vesicle

Vesicles containing secretion

Secretion in lumen of gland

Dying cells releasing secretion

New cells forming

Cell moving into lumen of gland

(b)

Secretion in lumen of gland

Pinched-off portions of cells releasing secretion

(c)

Secretory cells following pinching-off

Figure 4.11
Classification of glands according to mode of secretion: **(a)** merocrine, **(b)** holocrine, **(c)** apocrine.

When glands are classified according to their *mode of secretion* there are three different types: merocrine, holocrine, and apocrine (Figure 4.11).

1. **Merocrine glands** produce secretions that do not accumulate significantly in the gland cells. Rather, the secretions pass through the cell membrane within membranous vesicles by the process of exocytosis. In merocrine glands there is no destruction of glandular cells during secretion. The pancreas, the salivary glands, and most sweat glands are merocrine glands. In fact, most exocrine glands are of this type.

2. **Holocrine glands** accumulate their secretions in their cells and discharge them only when the cells rupture and die. New cells then form to replace those that have died. About the only example of holocrine glands are the sebaceous (oil) glands of the skin.

3. **Apocrine glands** produce secretions that accumulate toward the outer ends of the gland cells. The secretions are released and a small amount of cytoplasm is lost when these ends pinch off. But rather than dying, as happens in holocrine glands, the cell is only slightly damaged and it repeats the accumulation of secretory products. Some sweat glands are apocrine glands and the mammary glands are generally considered to be apocrine glands, although they are actually mixed glands as some of their secretion is of a merocrine type.

Epithelial Membranes

There are two important types of body membranes—*mucous membranes* and *serous membranes*—that, while not actually epithelial tissues, are composed of an epithelial layer on their free surface and an underlying connective tissue layer. For this reason they are often referred to as *epithelial membranes;* we will consider them here with the epithelial tissues.

MUCOUS MEMBRANES   **Mucous membranes** are moist epithelial membranes forming the linings of the digestive, respiratory, urinary, and reproductive tracts—all of which open to the exterior of the body. Their free surfaces vary in type, but they are usually either stratified squamous epithelium (mouth and esophagus) or simple columnar epithelium (stomach and intestine).

Mucous membranes are absorptive and secretory, making them particularly well suited to function in digestion and respiration. For example, food substances are absorbed into the body through the mucous membranes of the digestive tract. These membranes also secrete **mucin,** a viscous fluid that moistens the free surface of the membranes and lubricates food as it passes through the digestive tract.

Most mucous membranes have a layer of loose connective tissue—the **lamina propria**—deep to the innermost epithelial layer. Beneath the lamina propria there is often a thin layer of smooth muscle called the **muscularis mucosae.** Because mucous membranes are composed of several types of tissues, they are often considered to be simple organs.

SEROUS MEMBRANES   The ventral body cavities, which are not open to the exterior of the body, are lined with **serous membranes.** Serous membranes form both the parietal and the visceral portions of the pleura, the pericardium, and the peritoneum. The serous membranes consist of a thin layer of loose connective tissue covered by a surface layer of simple squamous epithelium called **mesothelium.** Mesothelium is derived from embryonic mesenchyme (mesoderm). The cells of the mesothelium secrete a clear, watery fluid, called **serous fluid,** that keeps the membranes moist. Serous membranes, like mucous membranes, are composed of more than one type of tissue and may therefore be considered simple organs.

CONNECTIVE TISSUES

Connective tissues vary considerably in form as well as function. Some serve as the framework upon which epithelial cells cluster to form organs; others bind various tissues and organs together, supporting them in their proper locations; some contain the media (tissue fluid) through which nutrients and wastes pass while traveling between blood and body cells; others serve as storage sites for excess food materials in the form of fat; and still others form the rigid skeletal framework of the body.

As you have seen, epithelial tissues are formed of closely packed cells that have very little material (*intercellular matrix*) between adjoining cells. In contrast, connective tissues are characterized by abundant intercellular matrix that surrounds relatively few cells.

Several types of cells are associated with the connective tissues, but fibroblasts and macrophages are the most common. **Fibroblasts** are spindle-shaped cells that form the various fibers characteristic of connective tissues. Fibroblasts that are in a resting or less active phase are called **fibrocytes.** **Macrophages,** which are generally not as abundant as fibroblasts, are active **phagocytes.** They move through the loose connective tissues of the body by means of ameboid movement and engulf foreign matter as well as dead or dying cells.

mak-ro-faj-ez

The activities of macrophages are so important in protecting the body against invasion by microorganisms that they are often referred to collectively as the **macrophage system** (or *reticuloendothelial system*), even though they do not form a discrete system and are distributed widely and rather randomly throughout the body. For instance, macrophages are found in many tissues throughout the body, including the loose connective tissue, lymphatic tissues, mesenteries of the digestive tract, bone marrow, spleen, adrenal gland, and pituitary gland. In some locations, macrophages are given special names—such as *Kupffer cells* that line the blood sinusoids of the liver, the *dust cells* of the lungs, and the *microglia* of the central nervous system. Whereas all these cells have a similar function—phagocytosis—macrophages in specific tissues or organs are often selective as to what they ingest. The macrophages of the spleen and the liver, for example, are particularly active in breaking down aging red blood cells. Macrophages are also involved in the activities of the body's immune system. These selective processes, in addition to their phagocytic role, make the cells of the macrophage system a major defense mechanism of the body.

Intercellular Material

The **intercellular material (matrix)** of connective tissue is formed of *ground substance* and *fibers*. The ground substance is a homogeneous product of the connective tissue cells that it surrounds. It varies in consistency from fluid to a semisolid gel. The fibers, which are also produced by the connective tissue cells (fibroblasts), are found in varying amounts within the ground substance. There are three types of fibers: *collagenous*, *elastic*, and *reticular*.

Collagenous Fibers

Collagenous fibers, the most abundant type, appear as wavy bands under the microscope. Each fiber is made up of bundles of smaller fibrils. Collagenous fibers are very strong and inelastic and are composed primarily of the protein *collagen*. Collagenous fibers that are closely packed have a white color and are sometimes referred to as white fibers.

Elastic Fibers

Elastic fibers are long, threadlike branching fibers that often form interwoven networks. Their main protein, called *elastin*, gives the fibers the capacity of returning to their original lengths after being stretched. They therefore function to give resilience to connective tissue. Large masses of elastic fibers have a slightly yellow color.

×1250

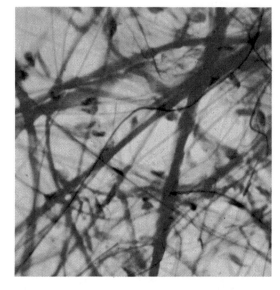

Reticular   Fibroblast   Collagenous
fibers   fibers

Figure 4.12
Typical composition of
loose connective tissue.

×950

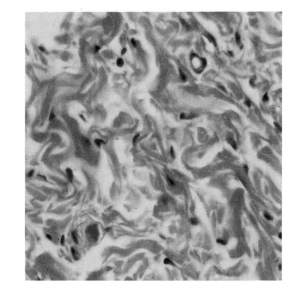

Fibroblasts   Collagenous
                          fibers

Figure 4.13
Dense irregular
connective tissue.

Reticular Fibers

Reticular fibers are short and very thin. They branch freely, forming a tight network called a **reticulum.** These fibers often form a gland's internal framework *(stroma),* to which the epithelial cells that make up the bulk of the gland are attached. Reticular fibers are inelastic and composed primarily of a type of collagen called *reticulin.*

Types of Connective Tissue

Connective tissues are classified according to the nature of the ground substance and the types and organization of the fibers in the ground substance.

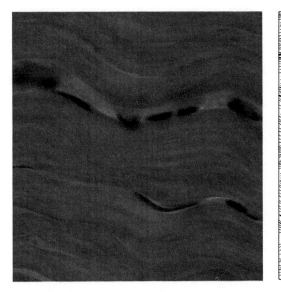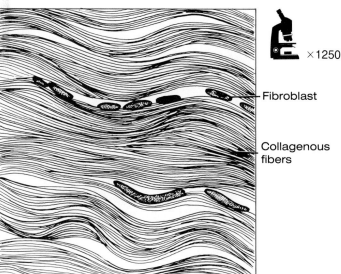

×1250

Fibroblast

Collagenous fibers

Figure 4.14
Dense regular connective tissue.

Loose Connective Tissue

Because the unorganized arrangement of the fibers in **loose connective tissue** leaves many spaces between them, it is also called **areolar connective tissue** (*areolar* = space). Most of the fibers in loose connective tissue are collagenous, but elastic and reticular fibers are also present (Figure 4.12). Loose connective tissue contains several different types of cells, but fibroblasts and macrophages are the most common. It is the most widespread connective tissue of the body, being used to (1) attach the skin to the underlying tissue *(subcutaneous tissue),* (2) fill the spaces between the various organs and thus hold them in place, and (3) surround and support the blood vessels. Because of the large spaces between cells and fibers, loose connective tissue contains a large amount of intercellular fluid (tissue fluid), which enters the spaces from capillaries. Tissue fluid is used to carry nutrients to, and waste products away from, the cells. If excessive fluid accumulates in these spaces, the affected area becomes swollen—a condition called *edema*. The presence of macrophages in loose connective tissue provides the body with a widespread defense against microorganisms.

F4.12

e-dee'-mah

Dense Irregular Connective Tissue

The dense connective tissues are distinguished by an abundance of collagen fibers, which provide the capacity to resist exceptional degrees of tension. **Dense irregular connective tissue** (Figure 4.13) is essentially a dense areolar tissue that contains all the same elements as loose connective tissue but has fewer cells and more numerous collagenous fibers. The fibers are closely interwoven, forming a compact tissue with fewer spaces. Because the tensions that need to be resisted by these tissues come from all directions, the fibers are oriented randomly. An example of dense irregular connective tissue is the dermis layer of the skin.

F4.13

Dense Regular Connective Tissue

Dense regular connective tissue (Figure 4.14) is characterized by a predominance of collagenous fibers that are tightly packed in parallel bundles. In this tissue the tensions to be resisted come from a single direction, parallel to the orientation of the fibers. Because of the prevalence of collagenous fibers, this tissue is sometimes referred to as *white fibrous connective tissue*. The only cells

F4.14

×550

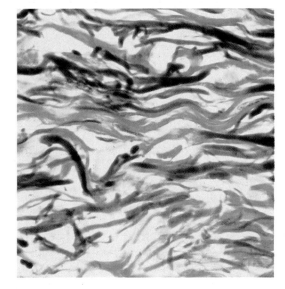

Elastic fibers

Figure 4.15
Elastic connective tissue.

×1250

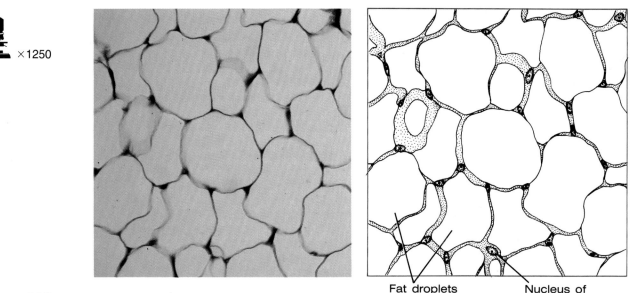

Fat droplets

Nucleus of
fat cell

Figure 4.16
Adipose tissue.

present are fibroblasts, which are located between the fiber bundles. The abundance of fibers gives this tissue great strength. It forms the tendons of muscles, the ligaments of joints (which also contain some elastic fibers), and various fibrous membranes, such as fascia and aponeuroses. **Fascia** surrounds the organs and the muscles; **aponeuroses** are broad sheets that function as thin tendons, attaching muscles to other structures.

fash'-ee-ah
ap-o-noo-ro'-sees

Elastic Connective Tissue

F4.15

In contrast to dense regular connective tissue, **elastic connective tissue** (Figure 4.15) contains more elastic fibers than collagenous fibers. Although this tissue is quite strong, it allows some stretching—which dense regular tissue does not. Elastic tissue is found in the walls of arteries, in the trachea and bronchi, and in the vocal cords, as well as in the walls of some hollow organs.

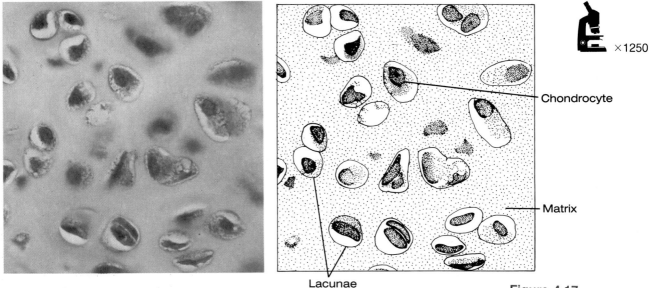

Figure 4.17
Hyaline cartilage.

Adipose Tissue

Adipose tissue (Figure 4.16) is essentially composed of fat cells dispersed in loose connective tissue. Each cell contains a large droplet of fat that squeezes and flattens the nucleus and forces the cytoplasm into a thin ring around the cell's periphery. Adipose tissue serves as a storage site for fats, and it pads and protects certain regions of the body.

ad'-i-pos
F4.16

Cartilage

Cartilage is a specialized fibrous connective tissue that contains numerous collagenous fibers embedded in a firm matrix of **chondrin,** a protein-carbohydrate complex. The connective tissues that we have studied up to this point all have fluid or, at most, semisolid matrices. With a firm matrix, such as chondrin, the connective tissue is able to function as a structural support. At the same time, the presence of fibers in the matrix imparts a certain amount of flexibility to cartilage.

The fibers and matrix of cartilage are formed by cells called **chondroblasts.** Each chondroblast becomes surrounded by fibers and matrix that it produces. As a result, the cartilage-forming cells eventually occupy small spaces called **lacunae.** When cartilage formation is complete, the chondroblast produces only enough matrix to maintain the cartilage. These mature cells are then called **chondrocytes.** The chondrin matrix is nonvascular—that is, it contains no blood vessels. The only blood supply to cartilage is provided by blood vessels found in the inner layer of the **perichondrium.** The perichondrium is a fibrous connective-tissue membrane that covers the external surfaces of all cartilaginous structures (with the exception of the articular cartilages of joints) and is vitally important in the growth of cartilage. Because there is no direct blood supply to chondrin, the nourishment of chondrocytes depends on the diffusion of nutrients through the chondrin of the matrix from capillaries located in the perichondrium or from the synovial fluid of joint cavities. Similarly, waste materials must diffuse from the cells to the vascular perichondrium.

kon'-dro-blast

la-ku'-ni

per-i-kon'-dree-um

Cartilage is especially prevalent in the embryo, but it also forms many adult structures. It is divided into three types according to variations in its fibrous structure: *hyaline, elastic,* and *fibrocartilage.*

HYALINE CARTILAGE   Cartilage that contains many closely packed collagenous fibers dispersed throughout the matrix is called **hyaline cartilage** (Figure 4.17). Since the fibers have the same refractive index and staining properties

high'-a-lin
F4.17

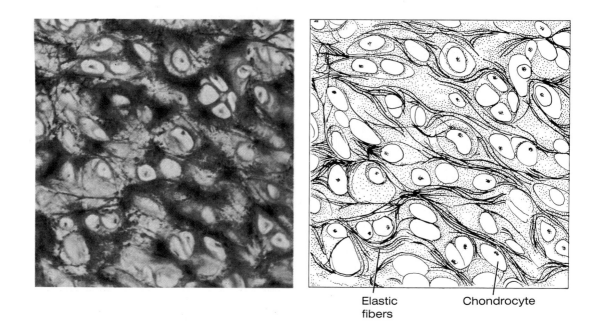

Elastic
fibers   Chondrocyte

Figure 4.18
Elastic cartilage.

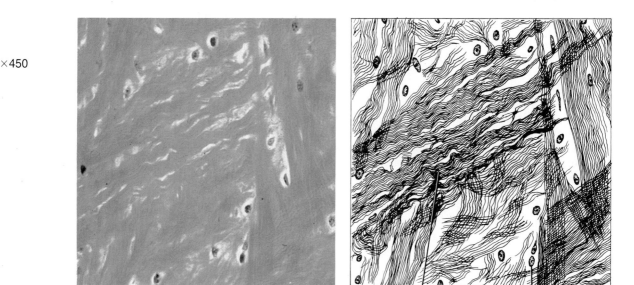

Collagenous   Chondrocyte
fibers

Figure 4.19
Fibrocartilage.

as the matrix, they are not distinguishable by ordinary microscopic examination. Hyaline cartilage is semitransparent and is smooth and firm, but flexible. This type of cartilage, the most abundant in the body, is found primarily in places where strong support is needed but some flexibility is desirable. Hyaline cartilage forms most of the embryonic skeleton, which is gradually replaced by bone. It also makes up the costal cartilages, which attach the ribs to the sternum (breastbone) and allow the thorax to expand during respiration. The cartilage rings of the trachea are also hyaline cartilage, as are the articular cartilages on the ends of bones where two bones meet to form a movable joint. These joint cartilages provide smooth, moist surfaces that permit body movement with a minimum of friction.

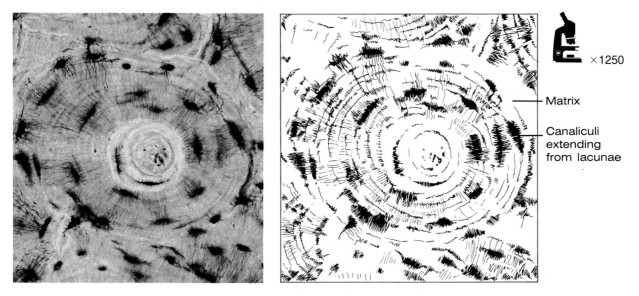

Matrix

Canaliculi
extending
from lacunae

×1250

Figure 4.20
Microscopic structure of
bone showing lacunae
and canaliculi.

ELASTIC CARTILAGE Some body structures must furnish firm but elastic support, and this is the function of **elastic cartilage** (Figure 4.18). Elastic cartilage contains collagenous fibers like hyaline cartilage, but the fibers are not so closely packed. Moreover, elastic cartilage contains a generous network of elastic fibers, which are stainable and therefore show up under the microscope. This cartilage forms the external ear, the epiglottis, and the auditory tubes.

F4.18

FIBROCARTILAGE The construction of **fibrocartilage** (Figure 4.19) differs from that of hyaline cartilage in that its collagenous fibers are arranged in thick, parallel bundles that give the matrix a coarse appearance. Actually, fibrocartilage resembles dense regular connective tissue. Because the fibers are not compacted as much as those in hyaline cartilage, fibrocartilage is slightly compressible—which makes it beneficial in regions that support the body weight or that must withstand heavy pressure. It occurs in the intervertebral discs, which provide cushions between the vertebrae; the articular discs, which are located in the knee joint; and the pad of the pubic symphysis, which creates a partially movable joint between the two sides of the pelvis.

F4.19

Bone

Because it has become mineralized, the matrix of **bone** (Figure 4.20) is even harder than the chondrin of cartilage. Like cartilage, bone contains collagenous fibers, but its rigidity and strength are greatly increased because of the deposition of inorganic salts among the fibers by cells called **osteoblasts.** There are two other structural differences between bone and cartilage: (1) bone is well supplied by blood vessels throughout its matrix and (2) its **lacunae** are interconnected by very small canals called **canaliculi.** Bone forms the major portion of the adult skeleton. Its structure is considered in greater detail in Chapter 6, where the skeletal system is discussed.

F4.20

kan-al-ik'-u-lie

Blood

Because the cells of blood are interspersed in abundant matrix it is often considered to be a type of connective tissue. The matrix of blood, which surrounds the blood cells, is a fluid called **plasma.** Both the cells and the fluid matrix of blood are discussed in Chapter 17.

×850

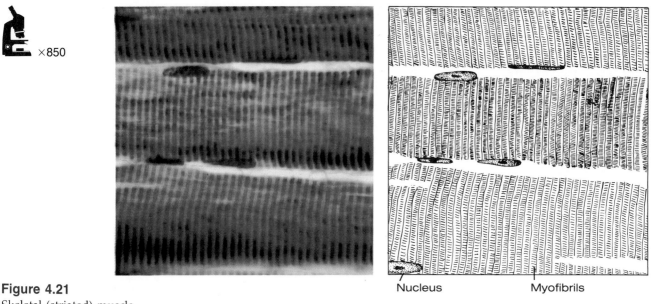

Figure 4.21
Skeletal (striated) muscle
tissue.

Nucleus   Myofibrils

×3100

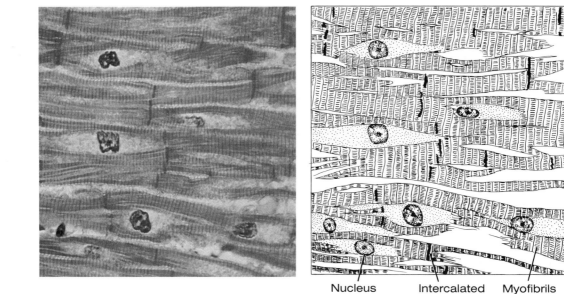

Figure 4.22
Cardiac muscle tissue.

Nucleus   Intercalated   Myofibrils
disc

MUSCLE TISSUE

The long, thin cells of **muscle tissue** are called **fibers.** It is important to realize that muscle cells are living cells and are in no way similar to the fibers of connective tissue. Muscle fibers are highly contractile. There are three structurally different types of muscle tissue: *skeletal, cardiac,* and *smooth.*

Skeletal Muscle

F4.21 **Skeletal muscle** (Figure 4.21) is attached to various bones of the skeleton. The cells of skeletal muscle are long and cylindrical—in fact, some skeletal muscle cells are thought to extend the entire length of the muscle. Running longitudi-

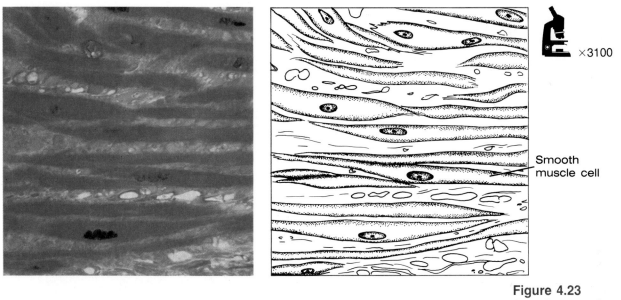

×3100

Smooth
muscle cell

Figure 4.23
Smooth (visceral) muscle
tissue.

nally throughout the skeletal muscle cells are regularly ordered threadlike
arrays of proteins called **myofibrils.** Transverse light and dark bands that
alternate along the myofibrils give skeletal muscle cells a characteristic *striated*
appearance. Each skeletal muscle cell is multinucleate—that is, it has more
than one nucleus. These nuclei are located on the periphery of the cell, just
inside the cell membrane.

my-o-figh'-bril

Cardiac Muscle

Cardiac muscle (Figure 4.22) forms the wall of the heart. The cells of cardiac
muscle, unlike those of skeletal muscle, form branching networks throughout
the tissue. Where adjoining cells meet end to end, their junctions form structures called *intercalated discs* that are visible under the microscope and are
unique to cardiac muscle. Cardiac muscle cells contain myofibrils that are
arranged in a pattern similar to that of skeletal muscle and give cardiac muscle
cells the same *striated* appearance. The nuclei of cardiac muscle cells, in contrast to those of skeletal muscle, are centrally located.

F4.22

Smooth Muscle

Smooth muscle (Figure 4.23) is so named because its cells do not have the
striated appearance of skeletal and cardiac muscle cells. Smooth muscle is also
called *visceral muscle* because it is located in the walls of hollow internal structures, such as ducts, blood vessels, and the digestive tract, as well as in numerous other locations. Each cell of smooth muscle is shaped like a long
spindle, with each end of the cell tapering to a point. A smooth muscle cell
contains a single, centrally located nucleus. Muscular tissue is considered
further in Chapter 8.

F4.23

vis'-ser-al

NERVOUS TISSUE

Nervous tissue is composed of **neurons**—highly specialized cells capable of
receiving and transmitting impulses very rapidly—plus supportive cells, including **neuroglia** and **Schwann cells.** The structure of the neuron is adaptive
to its function. Each neuron consists of a cell body with two or more thin
cytoplasmic extensions. Because of the anatomical arrangement of nervous

FRONTIERS IN HEALTH

Can Aging Be Delayed?

The oldest living things on earth are trees, which can live thousands of years. The life-span of animals is much shorter. The tortoise holds the record for longevity, living to be 200 years old. Humans are the only other animals that can live to be over 100, and few of us live that long. Disease and old age cause most of us to die in our 70s. Could more of us live to be 100?

The most common measure of longevity is life expectancy at birth—that is, the number of years the average person can expect to live. Of course, not everyone will live that long. Some will die younger; others will die older.

Life expectancy has increased significantly in the past 85 years. In 1900 the average white American female could expect to live only 50 years; today she can expect to live 78.3 years. The prospects for males have increased correspondingly. In 1900 the average white American male could expect to live 47 years; today he can expect to live 70.9 years.

These numbers are deceptive, however. They would seem to indicate that we are living longer—that is, that the human life-span has been extended and the aging process delayed. In reality, though, this is not the case. What has happened to produce these numbers is that there has been a reduction in infant deaths, so more people are living past the first year of life. This makes it appear that the aging problem is being solved, but this is unfortunately not the case.

To illustrate this point, consider an example: suppose ten people are born in a given year on a small island. If five of these people die in the first year and the other five live until they are 70 years old, the average life expectancy on the island is 35 years. Now suppose a doctor moves onto the island and is able to reduce the death rate so that only one child out of every ten dies, while the rest of the people continue to live to be 70. The average life expectancy is now 63 years. In this example, the island's residents have not conquered aging, and life-span has not really increased. What has raised the average life expectancy is that more children live past the dangerous first year of life.

This is what is happening in the United States. The gain in life-span is largely due to the significant drop in infant death that has occurred since the early 1900s (from more than 100 per thousand to about 12 per thousand), thanks to improved hygiene and advances in medical practice.

Increased infant survival is not the only factor producing increased longevity. Another way of producing a longer life is to delay death in later life. Improvements in diagnosis and treatment of heart disease have been one of the most important contributors to extended life.

Although such improvements come closer to actually increasing our life-span, they have done nothing to alter the aging process itself. Most medical scientists agree that although it has become possible to maximize the longevity potential of individual people, there has been no significant increase in life-span—that is, no actual slowing of the aging process.

Two young boys suffering from progeria, a disease that causes premature aging.

The search for ways to delay the aging process has been a frustrating one for medical scientists, because no one actually knows what causes aging. There are a number of theories, none of which is completely satisfactory. Some researchers suggest that aging results from a buildup of toxic substances in the body's tissues. These toxins may be the by-products of enzymatic reactions or irradiation, both natural and human produced. Other medical scientists think that cells are preprogrammed to divide only a certain number of times. When they have divided that many times, they die. Still others suggest that aging is largely the result of gradual physiological deterioration of the immune system.

One of the newest discoveries concerning aging is a protein called stomatin. Stomatin was first detected in cultured fibroblasts that had stopped dividing. Since then it has also been found in other cells that do not divide, such as skeletal muscle cells and certain cells of taste buds. It is thought that stomatin may actually cause a cell to stop dividing. If this is the case, then the gene that controls the production of stomatin may affect cellular aging.

The discovery of stomatin could be a significant step in the control of the aging process. Suppose, for instance, that the stomatin gene can be inactivated. Tissues may then be able to continue to regenerate beyond the normal 70-odd years, and humans, like trees, could conceivably live for many, many years.

Despite extensive research, however, there is no indication that we can slow down, much less prevent, aging. The death rate in the aged and the rate of infant death can be reduced by advances in medicine, but the basic process of cellular aging is still beyond our understanding.

tissue, certain of these cytoplasmic extensions transmit impulses toward the cell body of the neuron while others carry impulses away from the cell body—either to another neuron or to a specific structure. Nervous tissue is studied in greater detail with the nervous system in Chapter 10.

TISSUE REPAIR

In the embryo, the cells of all tissues are capable of dividing by mitosis, thus enabling the tissues to grow and to repair damage. As the body continues to develop following birth, however, the ability of the cells of certain tissues to divide is greatly reduced or lost completely. Thus the capability of postembryonic tissues to undergo growth and repair depends on the tissue involved. Cells of nervous tissue and muscle tissue generally become mitotically inactive once the tissues have completed their development. In contrast, cells of the epithelial tissues—including those of the skin, digestive tract, respiratory tract, urogenital tract, and various glands and organs—remain mitotically active and are thus capable of undergoing repair. Fibroblasts also retain the capacity to divide; therefore, like epithelial tissue, connective tissue is able to undergo repair.

STUDY OUTLINE

EPITHELIAL TISSUES occur as coverings for most free surfaces of the body—internal and external.
pp. 87–96

SPECIALIZATIONS OF EPITHELIAL CELL SURFACES

MICROVILLI are tiny protoplasmic projections that increase the area of the free surface; not highly motile.

CILIA are motile processes that move rhythmically to propel materials along the epithelial surface.

SPECIALIZATIONS FOR CELL ATTACHMENT

JUNCTIONAL COMPLEXES attach adjacent cells in columnar and some cuboidal epithelia. Each junctional complex consists of a *tight junction (zonula occludens),* an *intermediate junction (zonula adherens),* and a *desmosome (macula adherens).*

GAP JUNCTIONS hold adjacent cells together in muscle and nerve tissue as well as in epithelia.

CLASSIFICATION OF EPITHELIA epithelia can be classified according to (1) the number and arrangement of cell layers and (2) the shape of cells on the free surface.

ACCORDING TO CELL LAYERS

Simple epithelium has a single layer of cells, all contacting basal lamina.

Stratified epithelium has two or more layers, only deepest layer contacting basal lamina.

Pseudostratified epithelium appears multilayered, but is actually only a single layer with all cells touching basal lamina.

ACCORDING TO CELL SHAPE

Squamous epithelium cells are flat and thin.

Cuboidal epithelium cells are about as tall as they are wide; appear almost square in vertical section.

Columnar epithelium cells are taller than they are wide; appear rectangular in vertical section.

GENERAL CLASSIFICATION is based both on the shape of cells that form free surface and on the number of layers.

Simple squamous epithelium a single layer of squamous cells; found in regions where diffusion and filtration occur; includes endothelium that lines the blood vessels and mesothelium that lines the body cavities.

Stratified squamous epithelium consists of multiple layers; forms protective layer on body surface (epidermis) and other sites of abrasion.

Transitional epithelium specialized stratified tissue lining urinary bladder and certain other hollow organs.

Simple cuboidal epithelium generally a single layer of cuboidal cells; occurs in many glands, such as the salivary glands.

Simple columnar epithelium single layer of columnar cells; capable of performing complex reactions, such as secretion and absorption; lines digestive tube.

Stratified columnar epithelium truly stratified columnar tissue; found on epiglottis and parts of pharynx, anus, and urethra.

Pseudostratified columnar epithelium columnar tissue in which all cells contact basal lamina but not all reach the free surface; found in large ducts of some glands, such as parotid; is often ciliated, as in respiratory tract.

GLANDULAR EPITHELIUM consists of cells that secrete various substances or excrete wastes such as bile pigments.

> *Gland types*
> 1. *Merocrine glands:* produce secretions that do not accumulate in the gland cell; secrete with no glandular destruction; most common type.
> 2. *Holocrine glands:* accumulate secretions in their cells, discharging the secretions only when cells rupture and die.
> 3. *Apocrine glands:* accumulate secretions in outer ends of gland cells; released when end of cell pinches off.

EPITHELIAL MEMBRANES composed of a surface layer of epithelial cells underlain with loose connective tissue.

> *Mucous membranes* absorptive and secretory tissue that line digestive, respiratory, urinary, and reproductive tracts; surface layer of stratified squamous or simple columnar epithelium.

> *Serous membranes* line central body cavities; surface layer of simple squamous epithelium called mesothelium.

CONNECTIVE TISSUES vary considerably in form

and function; all have abundant intercellular matrix; serve as internal framework of organs; bind tissues and organs together for support; some provide media through which nutrients and wastes pass between blood and body cells; serve as food storage sites; form rigid skeletal framework of body.   **pp. 97–103**

INTERCELLULAR MATERIAL

GROUND SUBSTANCE varies from fluid to gel; produced by connective tissue cells.

MACROPHAGES active phagocytes; found throughout body; comprise the macrophage system.

FIBROBLASTS form various fibers characteristic of connective tissues.

> *Collagenous fibers* inelastic fibers composed of bundles of strong fibrils.

> *Elastic fibers* long, threadlike, branching fibers that often form interwoven networks.

> *Reticular fibers* short, thin fibers that branch freely, forming tight, inelastic networks.

TYPES OF CONNECTIVE TISSUE

LOOSE CONNECTIVE TISSUE has unorganized fiber arrangement; is most widespread type. Used to attach skin to underlying tissue; fill spaces between organs, holding them in place; surround and support blood vessels.

DENSE IRREGULAR CONNECTIVE TISSUE dense areolar tissue but contains more randomly oriented collagenous fibers than loose tissue does; forms dermis layer of skin.

DENSE REGULAR CONNECTIVE TISSUE very strong because of numerous parallel collagenous fibers in tightly packed bundles; forms tendons of muscles, ligaments of joints.

ELASTIC CONNECTIVE TISSUE allows some stretching; found for instance, in walls of arteries and trachea.

ADIPOSE TISSUE consists of fat cells dispersed in loose connective tissue; serves as storage site for fats and also pads certain body regions.

CARTILAGE has firm matrix of chondrin; functions as structural support but is somewhat flexible; divided into three types:

> *Hyaline cartilage* most abundant type; contains tightly packed collagenous fibers; provides strong support, as in costal cartilage and rings of trachea.

> *Elastic cartilage* contains elastic fibers; furnishes firm but elastic support, as in outer ear and epiglottis.

> *Fibrocartilage* tough and slightly compressible; found in intervertebral discs of vertebral column and articular discs of knee joint.

BONE has great strength and rigidity provided by inorganic salts; osteoblasts located in lacunae; forms major part of adult skeleton.

BLOOD the matrix that surrounds blood cells is fluid plasma.

MUSCLE TISSUE living cells of muscle tissue are

called fibers.   **pp. 104–105**

SKELETAL MUSCLE attached to skeleton and moves it.

CARDIAC MUSCLE forms walls of heart.

SMOOTH MUSCLE located in walls of hollow internal structures (organs, ducts, and blood vessels).

NERVOUS TISSUE composed of neurons—highly

specialized cells capable of receiving and transmitting impulses; also contains supportive cells called neuroglia and Schwann cells.   **pp. 105–107**

TISSUE REPAIR embryonic cells of all tissues are

capable of mitosis. Some postembryonic tissues are also able to grow and repair damage. With continued development, cells of certain tissues are rendered incapable of mitosis, or mitotic activity is greatly reduced.   **p. 107**

SELF-QUIZ

1. In which one of the following are the cell membranes of adjacent cells closest together? (a) desmosomes; (b) zonula occludens; (c) zonula adherens.

2. Which cell junction seems to allow for the passage of ions and small molecules? (a) gap junction; (b) zonula occludens; (c) zonula adherens.

3. Epithelial tissues are generally classified on the basis of the number and arrangement of cell layers within the tissue, the shape of the cells on the free surface, and/or the location or function of certain epithelia. True or false?

4. Epithelia with two or more layers of cells, with only the deepest layer in contact with the basal lamina, are: (a) stratified epithelia; (b) simple epithelia; (c) pseudostratified epithelia.

5. Match the following terms associated with epithelial cells with the appropriate description.

Squamous cells	(a)	Cells taller than they are wide.
Stratified epithelium	(b)	A single layer of squamous cells.
Simple columnar	(c)	Cells that are flat and thin.
Columnar cells	(d)	Cells about as tall as they are wide.
Cuboidal cells	(e)	A single layer of epithelial cells all in contact with the basal lamina.
Simple squamous epithelium	(f)	Epithelium with two or more layers of cells, only the deepest layer in contact with the basal lamina.

6. Specialized tissue that allows for the expansion of an organ with only minimal resistance from the tissue is composed of: (a) simple cuboidal epithelium; (b) transitional epithelium; (c) stratified squamous epithelium.

7. Which one of these epithelial tissue types has ciliated cells that line the mucous membranes of the respiratory passageways? (a) simple columnar; (b) stratified columnar; (c) pseudostratified columnar.

8. Mucous membranes are particularly well suited to function in digestion and respiration. True or false?

9. Match the following items with the appropriate description.

Mucous membranes	(a)	Tissue lining the wall of the heart.
Lamina propria	(b)	Tissue whose cells secrete serous fluid.
Serous membranes	(c)	Loose connective tissue that is part of the mucous membranes.
Mucin	(d)	Absorptive and secretory tissue lining the respiratory tract.
Mesothelium	(e)	Function as glands and secrete mucus.
Endothelium	(f)	Tissue lining the ventral body cavity.
Goblet cells	(g)	Viscous fluid that lubricates mucous membranes.

10. These glands accumulate their secretions and release them only when the individual cells that store the secretions rupture and die: (a) merocrine; (b) holocrine; (c) apocrine.

11. Match the following terms associated with connective tissues with the appropriate description.

Fibroblasts	(a)	Fiber cells in a resting phase.
Macrophages	(b)	Fibers that form the stroma of glands.
Ground substance	(c)	Structures that occur in the ground substance.
Fibrocytes	(d)	Homogeneous product of the connective tissue cells that it surrounds.
Fibers	(e)	Long, threadlike, branching fibers yellow in color.
Collagenous	(f)	Active phagocytes of loose connective tissue.
Elastic	(g)	Cells that form various fibers of connective tissues.
Reticular	(h)	Most common type of fiber; composed of bundles of strong fibrils.

12. Loose connective tissue is the most common type and forms the dermis layer of the skin. True or false?

13. This connective tissue forms the tendons of muscles and the ligaments of joints: (a) dense irregular; (b) areolar; (c) dense regular.

14. This connective tissue contains numerous collagenous fibers embedded in a firm matrix of chondrin: (a) cartilage; (b) elastic; (c) adipose.

15. The original fibers and matrix of cartilage are formed by: (a) chondrocytes; (b) osteoblasts; (c) chondroblasts.

16. The cartilage that forms the intervertebral discs is termed: (a) hyalin; (b) elastic; (c) fibrocartilage.

17. Although all the cells of the macrophage system have a similar function (phagocytosis), macrophages in specific tissues or organs are often selective as to what they ingest. True or false?

18. Intercalated discs are found in cardiac, skeletal, and smooth muscle tissues alike. True or false?

19. With continued development, the ability of the cells of certain tissues to divide is greatly reduced or is completely lost. True or false?

20. Which one of the following contains cells that remain mitotically active and thus capable of undergoing repair during a human's life span? (a) muscle tissue; (b) nervous tissue; (c) epithelial tissue.

LEARNING OBJECTIVES

After completing this chapter, you should be able to:

1. Name the four body structures that are a part of the integumentary system.

2. Name the two layers that form the skin and describe the composition of each.

3. Describe the process by which a cell becomes cornified.

4. Describe the factors responsible for skin color.

5. Name and describe the two layers that compose the dermis.

6. Distinguish between eccrine sweat glands and apocrine sweat glands, and cite one example of the latter.

7. List five functions of the integument.

8. Cite three ways in which the skin plays a major role in homeostasis.

9. Describe the "rule of nines."

CHAPTER CONTENTS

EPIDERMIS

DERMIS

HYPODERMIS

GLANDS OF THE SKIN

HAIR

NAILS

FUNCTIONS OF THE INTEGUMENTARY SYSTEM

CONDITIONS OF CLINICAL SIGNIFICANCE: THE INTEGUMENTARY SYSTEM

KEY TERMS AND DERIVATIVES

dermis (*dermis* = skin) the layer of skin below the epidermis; composed of connective tissue

epidermis (*epi* = upon; *dermis* = skin) the outer layer of skin; consists of stratified squamous epithelial tissue

keratin (*kerat* = bone) a fibrous insoluble protein produced as epidermal cells die and harden; found in tissues such as hair and nails

melanin (*melan* = dark) a group of dark pigments produced by certain cells (melanocytes) in the epidermis

pore (*por* = channel) opening by which a sweat gland communicates with the surface

sebaceous gland (*seb* = grease) a gland in the skin that secretes sebum, an oily material that lubricates the skin surface

subcutaneous layer (*sub* = under; *cutan* = skin) the layer of loose connective tissue and adipose tissue beneath the skin

The Integumentary System

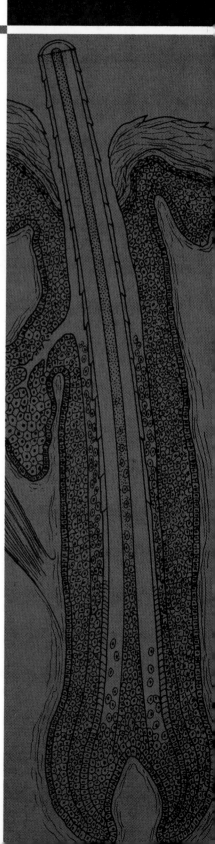

In Chapter 4, we discussed the organization of individual cells into tissues. We noted that when two or more tissues join together, as occurs in the formation of serous membranes and mucous membranes, an organ is formed. This chapter describes another combination of tissues into a simple organ—the **skin.** Although the skin is not often viewed as an organ, it is, in fact, one of the larger organs of the body in terms of surface area and weight. The skin and its accessory structures—hair, nails, and glands—comprise the **integumentary system.**

The skin forms the entire external covering of the body. It is continuous with, but differs structurally from, the mucous membranes lining the external openings of the respiratory, digestive, and urogenital systems (mouth, nose, anus, urethra, and vagina). The skin is composed of two main layers: (1) the *epidermis*, a surface layer of closely packed epithelial cells, and (2) the *dermis*, a deeper layer of dense, irregular connective tissue. The dermis is connected to the underlying fascia of the muscles by a layer of loose connective tissue called the *hypodermis*. In many areas, fat is deposited in the loose connective tissue, thus forming adipose tissue. The hypodermis connects the skin and the underlying fascia of the muscles only loosely, thus allowing the muscles to contract without pulling on the skin. In some areas, where muscles do not lie beneath the skin, there is only a small amount of hypodermis present and the integument is more tightly attached. For example, on the shins the skin is connected directly to the membrane (periosteum) that covers the bone.

EPIDERMIS

The outer portion of the skin is called the **epidermis** (Figure 5.1). The epidermis develops from the single layer of surface ectoderm of the embryo. By the time of birth, it consists of several layers of squamous cells that form a stratified squamous epithelium. The epidermis is generally quite thin, not exceeding 0.12 mm over most of the body. However, it is considerably thicker in areas that are subjected to constant pressure or friction, such as the soles of the feet and the palms of the hands. Continued pressure at a particular location causes the epidermis to thicken, forming calluses and corns.

Epidermal Layers

Where the epidermis is thick, it is possible to identify four layers, or **strata.** The innermost layer is the *stratum germinativum*. This layer is overlaid by the *stratum granulosum, stratum lucidum*, and the *stratum corneum*—in that order. In regions where the epidermis is thin, the stratum lucidum is often absent.

Stratum Germinativum

The **stratum germinativum** (*germinate* = having the capacity to develop) is the deepest layer of the epidermis, lying directly on the dermis (Figures 5.1 and 5.2). As its name indicates, it is within this layer that mitosis occurs, furnishing cells to replace those lost from the more superficial strata of the epidermis. The cells of the stratum germinativum are attached to one another by desmosomes and contain bundles of microfibrils called *tonofibrils* in their cytoplasm.

111

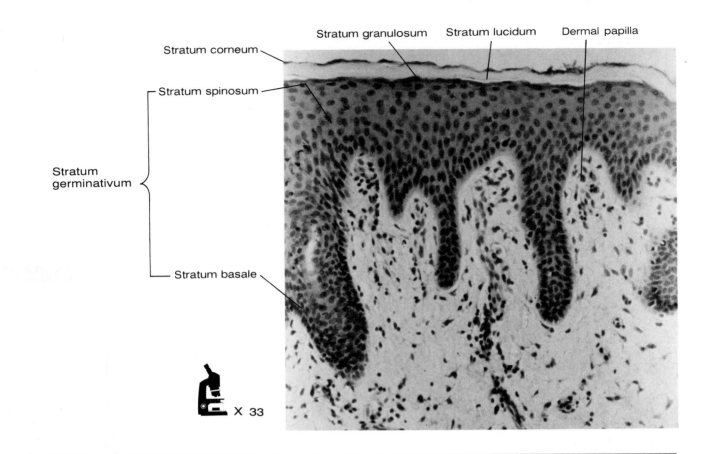

Figure 5.1
Photomicrograph of the
epidermis of thick skin.

The cells in the deepest layer of the stratum germinativum, which contacts the dermis, are columnar cells. It is within this deep layer (the **stratum basale**) of columnar cells that most mitoses take place. Above the basal layer, the cells become somewhat flattened and polyhedral in shape. Under the microscope there appear to be cytoplasmic extensions that connect adjacent cells. Because of these extensions, which are actually artifacts caused by the processes involved in preparing microscope slides, the layers of cells in the stratum germinativum superficial to the stratum basale are sometimes listed as a separate **stratum spinosum** (*spine* = projection).

Stratum Granulosum

The cells of the **stratum granulosum** (*granulosum* = granular) are flattened and are arranged in about three layers just superficial to the stratum germinativum (Figures 5.1 and 5.2). This stratum derives its name from the presence of granules of **keratohyalin** within the cytoplasm of its cells. As the granules increase in size, the nucleus disintegrates, which results in the outermost cells of the stratum granulosum being dead.

F5.1, F5.2

Stratum Lucidum

The **stratum lucidum** (*lucid* = clear) is a clear band superficial to the stratum granulosum (Figures 5.1 and 5.2). It consists of several layers of flattened, closely packed cells, most of which have only indistinct outer boundaries and have lost all of their cytoplasmic inclusions except for keratin fibrils and some droplets of a substance called **eleidin**. Eleidin is transformed into keratin as the cells of the stratum lucidum become part of the outer stratum corneum. The stratum lucidum is most prominent in areas of thick skin and is absent in some locations.

F5.1, F5.2

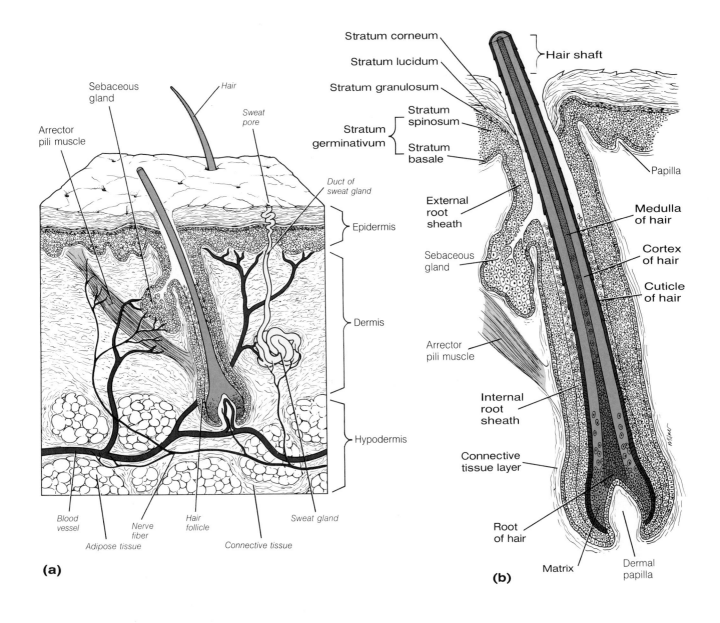

(a)

(b)

F5.1, F5.2

Figure 5.2
(a) Structure of the epidermis and dermis layers of the skin, and the hypodermis.
(b) Vertical section through a hair follicle showing the structure of a hair and the relationship of a sebaceous gland and arrector pili muscle to the follicle.

Stratum Corneum

The **stratum corneum** (*cornu* = horn) is the most superficial layer of the epidermis. It is formed of varying numbers of flat, closely packed cells (Figures 5.1 and 5.2). These cells are dead, their cytoplasm having been replaced by a fibrous protein called **keratin.** Such cells are referred to as being *cornified*, or *horny*. The cornified cells form a covering over the entire body surface that not only protects the body against invasion by substances in the external environment, but also helps to restrict the loss of body water. The cells of the stratum corneum are held together by modified desmosomes that include dense extracellular material. The outermost layers of the stratum corneum are constantly being lost as the result of abrasion—caused by, for example, friction with clothing. However, the lost cells are constantly being replaced by cells from the deeper layers of the epidermis.

Nourishment of the Skin

As is typical of all epithelia, there are no blood vessels within the epidermis, although the underlying dermis is well vascularized. As a result of this situation, the only method by which cells of the epidermis can obtain nourishment is by diffusion from the capillary beds of the dermis. This method is sufficient for those cells closest to the dermis, but as cells divide and some are forced toward the body surface—and thus farther away from the source of nourishment—they die. Their cytoplasm is gradually replaced with keratin, thus forming the structures typical of the outer layers of the epidermis.

Skin Color

me-lan'-o-sites

Skin color is primarily determined by the presence and the distribution of a dark pigment called **melanin.** Melanin is produced by cells called *melanocytes,* which migrate into the epidermis and transfer the pigment to cells of the stratum germinativum and the underlying dermis. There are no great differences in the number of melanocytes found in the skin of the various human races. Differences in skin color are due primarily to the amount of melanin produced by the cells and its distribution. Dark-skinned people have appreciable amounts of melanin in all layers of the epidermis. In light-skinned people, relatively little melanin is distributed among the layers of the epidermis, except in heavily pigmented areas such as the nipples of the breasts. The presence of the yellow pigment **carotene** in the strata of the epidermis, in combination with melanin, produces the yellowish hue that is typical of Oriental people. Skin color is also influenced by a reddish hue that results from the blood vessels of the dermis being visible through the epidermis. Whereas the amount of carotene in the skin is relatively constant for each person, a change in the amount of blood in the capillaries of the dermis—or in the amount of oxygen carried within the blood of these capillaries—can cause intense temporary changes in skin color. For example, *blushing* is caused by an expansion of the capillaries, whereas the bluish skin characteristic of *cyanosis* results from a decrease in the amount of oxygenated hemoglobin in the blood within these capillaries.

DERMIS

F5.2

Lying deep to the stratum germinativum is a layer of dense irregular connective tissue called the **dermis** (Figure 5.2). In contrast to the epidermis, the dermis develops from the mesoderm of the embryo, as do the muscles and the skeleton of the body. The dermis contains some elastic and reticular fibers as well as many collagenous fibers. It is well supplied with blood vessels, lymph vessels, and nerves. It also contains specialized glands and sense organs. The thickness of the dermis varies in different locations, but it averages about 2 mm. The dermis is composed of two indistinctly separated layers, the *papillary layer* and the *reticular layer.*

Papillary Layer

pa'-pil-lar-ee

The outer **papillary layer** fits closely against the basal layer of the stratum germinativum. This layer is so named because it has many **papillae** (projections) that protrude into the epidermal region. On the palms and the soles, these papillae are in the form of curving parallel ridges that cause the overlying epidermis to form the characteristic fingerprint and footprint patterns. Many papillae contain capillary loops, whereas others contain specialized sensory receptors that react to external stimuli such as temperature and pressure changes. These receptors are described in Chapter 10.

Reticular Layer

The deeper **reticular layer** of the dermis consists of dense bundles of collagenous fibers that run in various directions (thus forming a reticulum). The fibers are continuous with the fibers of the hypodermis.

FRONTIERS IN HEALTH

New Hope for Burn Victims

When 6-year-old Glen and his 5-year-old brother Jamie sneaked into an empty house near their home one day they had no idea that their mischief—and misfortune—would make medical history. Joined by a friend, the boys entered the house, where they found cans of paint. They pried open the lids and began splattering paint on the walls and floor and, inadvertently, on themselves. When the fun was over, the boys began to clean themselves with a solvent that was stored with the cans of paint. For some unknown reason, one of the boys struck a match.

The room instantly exploded in fire. The boys raced out of the house in flames. Within a few days of the accident, the friend died. Jamie and Glen, both severely burned over 80% of their bodies, were rushed to Boston. There, plastic surgeon Dr. G. Gregory Gallico III began the long task of stabilizing the boys' condition and replacing the skin that had been burned.

Because of the severity of the burns, Dr. Gallico tried a new technique developed by Dr. Howard Green of the Harvard Medical School. He took postage-stamp-sized pieces of skin from the boys' armpits and groins—about the only skin left intact on their bodies—diced it, and used an enzyme to separate the epidermal cells. The cells were then cultured in special flasks. After several weeks, the skin cells had multiplied thousands of times.

The cultured epidermal cells were next spread over thin gauze pads the size of playing cards and sewn into place on the boys' bodies. In 4 weeks the patches had grown together and had developed the full thickness of normal epidermis. Below the epidermis a thin dermal layer had developed, consisting only of blood vessels, fibroblasts, and collagen fibers. Elastic fibers, which give normal skin its flexibility, were lacking.

Eventually more than 50% of the boys' bodies was covered by new skin. The remaining burned areas either healed on their own or received skin transplants from less severely burned areas that had healed by themselves.

This new epidermal culture and graft technique could help save the lives of 10,000 or more of the 100,000 people who are hospitalized in the United States each year with severe burns. Still, there are some drawbacks to the procedure. While new skin is being grown, burned areas are covered with skin from cadavers, and gauze and antiseptic ointments are applied to help reduce fluid loss and prevent infection. These substances all have potential problems associated with their use.

Medical researchers have therefore sought new ways to speed the regrowth of skin. One of the most promising methods is the use of artificial skin. Researchers at the Massachusetts Institute of Technology have developed an artificial skin that can be applied to newly burned areas to help reduce fluid loss, prevent infection, and promote faster recovery.

The artificial skin consists of two layers that resemble the body's natural epidermal and dermal coverings. The artificial dermis is a spongy layer made of collagen fibers extracted from cowhide and another substance

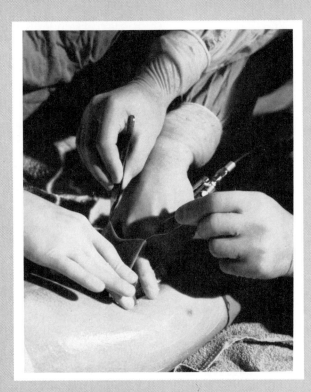

Dermatone used to harvest skin for split skin grafting at the Brooke Burn Center.

extracted from the cartilaginous skeleton of sharks. The artificial epidermal layer is made of plastic.

In severely burned patients, physicians clean away the burned flesh down to the muscle fascia. The artificial skin is then sewn into place. Within a short time fibroblasts and blood vessels invade the artificial dermis, and over a period of several months, new collagen fibers produced by the patient replace the fibers of the artificial skin.

The plastic epidermis is peeled off the dermis within a few weeks, after the dermis has revascularized and shown signs of regrowth. A new epidermis is then reconstructed, using some of the patient's own epidermal cells. Grafted in thin sheets onto the vascularized dermis, the epidermal cells gradually replace the plastic layer.

Artificial skin has been used on patients with moderate burns covering up to 95% of their body surface. In severely burned patients, physicians have successfully used artificial skin to cover 60% of the body.

Artificial skin is relatively easy to produce and can be sterilized and stored at room temperature. Therefore, it can be available for immediate grafting. So far, medical researchers have found no evidence of rejection and virtually no infection associated with the use of artificial skin. Moreover, the new skin grows in without the severe scarring that accompanies more conventional burn treatment.

HYPODERMIS

The **hypodermis** (*hypo* = beneath) is not a part of the skin, but it is important because it attaches the skin to the underlying structures. As was mentioned earlier, the hypodermis is composed of loose connective tissue, often having fat cells deposited among its fibers. This tissue is also referred to as *subcutaneous tissue* or the *superficial fascia.* In some regions, such as over the abdomen and the buttocks, the accumulation of fat within the subcutaneous tissue can become quite extensive. The hypodermis is well supplied by blood vessels and nerve endings.

GLANDS OF THE SKIN

Two types of glands have a widespread distribution in the skin: the sweat glands and the sebaceous glands. In addition, the ceruminous (wax) glands of the external ear canal, the ciliary and meibomian glands of the upper eyelids, and the mammary glands are specialized skin glands. The skin glands begin their embryonic development as solid downgrowths from the ectoderm. The downgrowing cords become hollow tubes as they continue to develop and extend into the dermis, forming the skin glands and their associated ducts.

Sweat Glands

F5.2a

The **sweat glands,** which are also called **sudoriferous glands** (*sudor* = sweat), are distributed over most of the body surface (Figure 5.2a). In only a few places, such as the lips, the nipples, and portions of the skin of the genital organs, are they absent. Typical sweat glands—the *eccrine sweat glands*—are merocrine glands, each in the form of a simple tubule that becomes coiled within the dermis. Stimulation of the sympathetic nerves to these glands causes them to secrete a watery solution of sodium chloride, with traces of urea, sulfates, and phosphates. The amount of sweat secreted depends on such factors as environmental temperature and humidity, amount of muscular activity, and various conditions that cause stress. Sweat glands that are located in the axilla, around the anus, on the scrotum, and on the labia majora of the female external genitalia are unusually large and extend into the subcutaneous tissue. Glands in these locations often empty into a hair follicle rather than directly onto the surface of the skin. These large glands are *apocrine sweat glands*—that is, part of the cytoplasm of the secreting cells is included within the secretion, which is thicker and more complex than true sweat. In the female, these glands periodically become enlarged and hyperactive in conjunction with the menstrual cycle. The **ceruminous glands,** which produce "wax" **(cerumen)** in the ear canal, are also apocrine glands that are considered to be modified sweat glands.

se-roo'-men

Sebaceous Glands

se-bay'-shus
F5.2a

Most **sebaceous glands** develop from, and empty their secretions into, hair follicles (Figure 5.2a). Their secretion **(sebum),** an oily substance rich in lipids, travels along the shaft of the hair to the surface of the skin. Sebum not only serves to oil the skin and the hair, preventing them from drying, but also contains substances that are toxic to certain bacteria. The sebaceous glands, which are known to be stimulated by the presence of sex hormones (mainly testosterone), are particularly active during adolescence. If their secretion accumulates within the duct of the gland, it forms a white pimple. This blocked sebum may become oxidized, darken, and form a "blackhead." Most hairless regions of the body, such as the palms of the hands and the soles of the feet, lack sebaceous glands. However, some areas that lack hair, such as the lips, the glans penis, and the labia minora, do have sebaceous glands. In these regions, the glands empty their secretions directly onto the surface of the epidermis.

Structurally, the typical sebaceous glands are of the simple alveolar type, although there are some that are compound alveolar (for instance, the *meibo-*

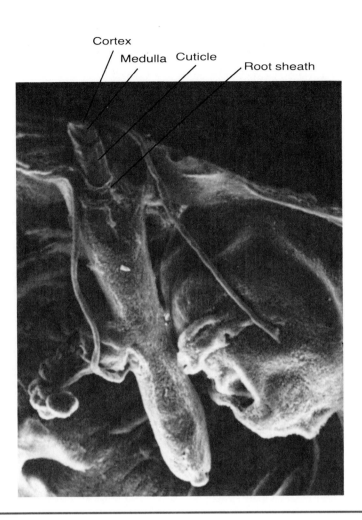

Cortex
Medulla Cuticle
Root sheath

Figure 5.3
Scanning electron micrograph of a hair within a hair follicle (×60).

mian glands of the upper eyelids). Functionally, all sebaceous glands are holocrine glands.

HAIR

Although **hair** is most obvious on the head and in the axillary and pubic regions, it is also present—though much less conspicuous—over most of the body. The only hairless skin is on the lips, the palms of the hands, the soles of the feet, the nipples, and parts of the external genitalia. Hair grows as the result of the mitotic activity of epidermal cells at the bottom of a **hair follicle.** The follicles extend from the epidermis into the dermis (Figure 5.2b). The outermost layer of the follicle, the **external root sheath,** is a downgrowth of the epidermis. From the bottom of the follicle up to the level of the sebaceous glands, the follicles are lined by the **internal root sheath,** which consists of several layers of keratinized cells. Covering the follicle is an outer layer of connective tissue that develops from the dermis. A portion of the dermis protrudes into the bottom of each follicle, forming a **papilla.** Papillae contain blood capillaries that nourish the overlying follicle cells and permit them to undergo repeated mitosis. Emptying into each hair follicle is one or more sebaceous glands, whose secretions help to soften the hair.

Each hair is essentially a column of keratinized cells. The mitotically active cells that cover the papilla are called the **matrix.** The part of the hair that rests upon the matrix is the **root.** The **shaft** of the hair develops from the cells of the matrix; the free end of the shaft extends beyond the surface of the skin. The **medulla,** the central core of the hair shaft, consists of loosely connected horny cells with air spaces between them (Figure 5.3). The **cortex,** which surrounds

F5.2b

F5.3

the medulla, is formed of tightly compressed keratinized cells. Outside the cortex is a **cuticle** of very hard keratinized cells. Straight hairs are cylindrical or oval; curly hairs are somewhat flattened.

Hair follicles exhibit cyclic activity, having active periods alternating with periods of inactivity. During the active periods, cells in the matrix of a follicle undergo mitosis, pushing the older cells upward and causing the hair to elongate. The cells die and become keratinized as they are pushed farther from the nourishment provided by the blood vessels in the papillae. The keratinized cells are incorporated into the hair. During inactive periods, when the matrix cells are not undergoing mitosis, the root of a hair becomes de-tached from the matrix and the hair gradually moves up the follicle. The detached hair may be pulled from the follicle by brushing or combing, or it may remain within the follicle until the next active period, when the new hair produced by matrix cells pushes it out.

Hair follicles in different parts of the body follow different patterns of cyclic activity. For instance, in the scalp, individual follicles can remain active and cause their hairs to elongate continuously for several years before becom-ing inactive for a period of months. In other regions of the body, the follicles may be active for only a few months before entering an inactive phase. Cut-ting or shaving hair does not affect the cyclic activity of the follicles, and therefore has no effect on the growth of hair.

Because of the repeated formation of new hairs during the active periods, normal hair loss does not generally lead to baldness. Baldness is a genetic trait that requires the presence of the male hormones—the androgens—for the hereditary tendency to become effective. For that reason, baldness is more common in males than in females, who may have inherited the trait but lack the androgens necessary to activate it.

Although there are considerable variations in the color of hair, only three different pigments are present—black (melanin), brown, and yellow. Varying combinations of these three pigments produce the different hair colors, al-though melanin is the primary determiner of hair color. Melanin is formed by melanocytes in the follicle and becomes located in the cortex and medulla of each hair. As people age, their hair tends to gradually become gray. This process is due to a decrease in the amount of pigments present, possibly as a result of a decline in the level of a specific enzyme necessary for the produc-tion of melanin. In the complete absence of pigments, the hair appears white.

The hair follicles are generally at an oblique angle to the surface of the skin, as are the hairs themselves. Running diagonally from the connective tissue covering of each follicle to the papillary layer of the dermis is a smooth muscle: the **arrector pili** (Figure 5.2). Contraction of this muscle pulls the follicle and causes the hair to "stand up"—that is, to be perpendicular to the skin surface—and causes the skin to bulge in front of the follicle, producing the "goose pimples" that form in response to cold or to frightening situations. In animals whose bodies are heavily covered with hairs, the erection of hairs traps air between them and the body surface, thus producing an insulating effect and reducing the loss of body heat. This response is probably of little importance in humans, on whom body hair is generally quite sparse.

ar-rek'-tor pih'-lee
F5.2

NAILS

F5.4

ep-o-neech'-ee-um

On the dorsal surfaces of the distal portions of the fingers and toes, the outer two epidermal layers—the strata corneum and lucidum—are heavily corni-fied, forming the **nails** (Figure 5.4). The **nail bed,** upon which the nail rests, is formed by the stratum germinativum. The germinativum is thickened under the proximal end of the nail, forming a whitish area called the **lunula** (*luna* = moon), which is visible through the nail. The region of thickened stratum germinativum is called the **nail matrix.** It is within the nail matrix that mitosis occurs, pushing forward the previously formed cells that have cornified and thus causing the nail to grow. At the proximal end of the nail a narrow fold of epidermis extends onto the free surface, forming the **eponychium** (cuticle). Below the free edge of the nail the stratum corneum is thickened and is called

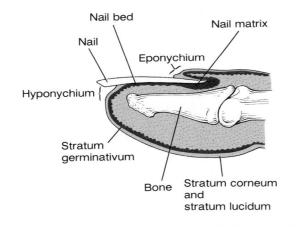

Nail bed

Nail matrix

Nail

Eponychium

Hyponychium

Stratum germinativum

Bone

Stratum corneum and stratum lucidum

Figure 5.4
Structure of a nail.

the **hyponychium.** Nails generally have a pink coloration because the capillary network beneath them is visible through the cornified cells.

high-po-neech'-ee-um

FUNCTIONS OF THE INTEGUMENTARY SYSTEM

Having an understanding of the structure of the skin and of its associated organs should make it easier to understand the various functions the skin performs. These functions can be grouped into several categories: protection, regulation of body temperature, excretion, sensation, and production of vitamin D.

Protection

The skin forms a physical barrier that prevents microorganisms and other foreign substances (including water) from invading the body. It also protects against excessive ultraviolet radiation and greatly reduces the loss of body water to the environment. The surface of the skin is coated with a thin, liquid film that tends to be acidic (pH 4–6.8). Because of its acidity, the film may act as an antiseptic layer and retard the growth of microorganisms on the surface of the skin. When subjected to repeated trauma, the skin—particularly the stratum corneum—becomes thickened, forming calluses in more severe instances.

Body Temperature Regulation

Even under conditions of high environmental temperature or during exercise, the body temperature remains almost normal, in part because considerable heat is lost through the skin. As body temperature begins to rise, the arterioles in the dermis dilate, bringing a greater volume of blood to the body surface and thus allowing more internal heat to be lost to the environment. At the same time, the body surface may become wet because of increased secretory activity by the sweat glands. The evaporation of this sweat further facilitates the loss of body heat. In a similar manner, under cold conditions, body heat can be conserved by the constriction of the dermal arterioles. This constriction reduces the amount of blood that flows to the body surfaces so that less heat will be lost to the external environment. The role of the skin in regulating body temperature is discussed further in Chapter 25.

Excretion

In addition to its cooling effect, the secretion of sweat functions, to a limited extent, as a means of excretion. Small amounts of nitrogenous waste products and sodium chloride leave the body via the sweat. Both the volume and the composition of sweat vary according to the changing needs of the body.

Sensation

Because of the presence within it of nerve endings and specialized receptors, the skin provides the body with much information concerning the external environment. Events such as temperature change, light touch, pressure, or painful trauma all stimulate integumentary receptors. These receptors, in turn, alert the central nervous system to the particular event, thus enabling appropriate action to be taken. This action might be simple and automatic, such as withdrawing the hand from a harmful situation; or it might require a more complicated act, such as deciding that a warmer coat should be worn.

Vitamin D Production

The skin also is involved in the production of vitamin D. In the presence of sunlight or ultraviolet radiation, one of the sterols (7-dehydrocholesterol) found within the skin is altered in such a way that it forms vitamin D_3 (chole-calciferol). After being metabolically transformed, vitamin D_3 assists in the absorption of calcium and phosphate from ingested food. Vitamin D_3, therefore, is important in maintaining the calcium and phosphate levels of the body at optimum levels, thus facilitating the normal growth of bones and their repair following a fracture.

Because of its role in such activities as protection of the deeper-lying structures of the body, regulation of body temperature, and prevention of excessive loss of water to the environment, the skin plays a major role in maintaining the internal homeostasis of the body, and in ensuring the continued normal activity of individual cells.

CONDITIONS OF CLINICAL SIGNIFICANCE

The Integumentary System

The importance of the skin in preventing the invasion of the body by microorganisms is apparent from the abundance of such organisms that are normally present, even on healthy skin, and yet that cause no bodily harm unless the epidermis is damaged, thus allowing them to enter the body. The secretions of the sweat glands and the sebaceous glands provide ample nutrients as well as a favorable environment in which these microorganisms can thrive. The **fungi** that cause *athlete's foot* are often present on the soles of the feet and between the toes without causing harm. Then, due to some change in their environment, the fungi rapidly proliferate and cause the disease condition. There are also some **yeasts** that live harmlessly on healthy skin. By far the most abundant microorganisms on the skin are **bacteria.** These include the rod-shaped forms and the spherical cocci. Most cocci are harmless, but one species, *Staphylococcus aureus*, can cause pimples, boils, and other more serious infections. However, even these powerful disease-producing microorganisms, which are normally present on certain areas of the skin, do not cause skin diseases unless the epidermis is penetrated.

The number of bacteria that are present on the skin varies in different regions of the body as well as from person to person. The largest bacterial populations are found on the face and neck, and in the axillae and the groin. Reported population densities range from 2.41 million bacteria per square centimeter of epidermis in the male axillae, to 314 bacteria per square centimeter on the back.

There are many recognized diseases of the integumentary system, but we will consider only a few of the more common pathologies.

Acne

This is an inflammatory disease caused by a rod-shaped bacterium (*Corynebacterium acnes*). These microorganisms provoke excessive secretion by the sebaceous glands, which, in turn, causes the formation of pimples, blackheads, and dandruff. Acne is most prevalent during puberty because of the hormonal changes that occur during that period. After several years, the skin usually becomes adapted to the higher levels of sex hormones and the condition disappears.

Warts

The common wart is the result of a viral invasion of the skin. This condition is most common in adolescents and young adults. Warts are often found in groups because they are capable of spreading to adjacent areas. Warts that occur on the sole of the foot—*plantar warts*—are particularly painful because they are almost constantly subjected to pressure.

Dermatitis and Eczema

These are general terms that refer to many inflammatory skin conditions. Also included within this category are nonspecific allergic responses of the skin to many different substances.

Psoriasis

This fairly common condition is characterized by small reddish-brown elevations and patches that are covered by layers of silvery scales. When the patches are scraped away, bleeding occurs from minute points that correspond to the tops of the papillae of the dermis. Tiny abscesses form under the stratum corneum, producing an exudate. The cause of psoriasis is unknown.

Impetigo

Impetigo is a highly contagious skin infection that is most common in children. It results from the invasion of the epidermis by various strains of *Staphylococci* and *Streptococci* bacteria. Pus-filled sacs *(pustules)* form beneath the stratum corneum, causing inflammation and swelling. The pustules rupture and form a crust.

Moles

Moles, which are elevations of the skin that are generally pigmented, are very common. Almost everyone has at least one mole, and the average person has about 20 on various locations of the body. Moles are considered to be congenital, although they often do not appear until adulthood. It has been suggested that their eruption may be stimulated by steroid hormones. Most moles are *benign*—that is, they do not develop into tumors. They grow slowly over a period of time, remain stable for a long period, and then gradually diminish in size (atrophy). A few, however, may become *malignant* or capable of spreading to other parts of the body. This change is indicated by an increase in size and pigmentation, a reddening of the skin around the mole, and itching.

Herpes Simplex

This condition is commonly called a *fever blister* or a *cold sore*. It occurs when a particular virus, having been dormant in a spinal nerve, travels along the processes of the nerve cells and becomes mitotically active on the skin and mucous membranes. The active virus causes clusters of watery blisters to form. The blisters generally occur on the lips or the external genitalia. They are often associated with any disease that causes an elevated body temperature.

Herpes Zoster (Shingles)

Like a cold sore, *shingles* results from the invasion of the body by a virus that at first remains dormant in the spinal nerves, generally in the thoracic region. Once the virus becomes active, it affects the sensory nerves of that region, causing an aching pain that follows the paths of the nerves. Groups of small vesicles develop in the skin that overlies the nerve paths. It is currently believed that the same virus that causes the vesicles of shingles also is responsible for the skin vesicles of chickenpox.

Cancers

Numerous types of tumors arise within the skin. Some originate in the various layers of the epidermis, some in the dermis, and others in the sweat glands and the sebaceous glands. Most of these tumors are benign and do not spread to other parts of the body. Warts are an example of benign tumors. Other tumors are malignant and have the capability of spreading *(metastasizing)* to other regions of the body. It is these latter tumors that are generally called *cancers*.

The cause of most skin tumors is not known. However, prolonged overexposure to the ultraviolet rays of sunlight appears to be directly related to the development of many of them. There is a greater incidence of skin tumors in farmers and others whose occupations require that they work outdoors over a period of years. A higher incidence is also noted in the southern United States as compared to the northern regions. Skin cancers are seldom found in dark races, where the skin is heavily pigmented.

Burns

While burns cannot be considered to be pathological conditions of the skin, they disrupt the homeostasis of the body so drastically that we consider them here.

The seriousness of burns results from the destruction of the skin, and clearly demonstrates the skin's importance to the other body systems. When the skin is destroyed, there is a large loss of body water **(tissue fluid)** and blood plasma. Plasma proteins and mineral salts leave the body with these fluids. The loss of plasma proteins upsets the osmotic equilibrium of the body, and the loss of salts produces an electrolyte imbalance (discussed in Chapter 27). The results are dehydration, kidney malfunction, and shock. In addition, with the protection of the skin gone, it is very easy for infectious agents to invade the body.

Burns are classified according to their severity. In *first-degree burns*, only the epidermal layers of the skin are damaged. Symptoms include localized pain, redness, and swelling. Sunburn is usually a first-degree burn. In *second-degree burns*, there is damage to both the epidermis and the dermis. However, the damage is not severe enough to prevent the skin from regenerating quickly. In *third-degree burns*, both the epidermis and the dermis are so severely damaged that they can regenerate only from the edges of the wound. If the burned area is extensive, this regeneration can be a slow process, during which body fluids are constantly being lost from the damaged area and the possibility of infection is high. In addition, such wounds can result in extensive scar tissue formation, which is not only disfiguring but can also restrict the movement of the damaged part. To hasten healing (and thus reduce the loss of body fluids) and to minimize scar formation, large burn areas are often covered with *skin grafts* taken from other regions of the body (Figure 5.5).

In one method of skin grafting—called *split skin grafting*—skin taken from one part of a person's body is used to cover another part. Split skin grafting is useful because it eventually increases the total amount of skin on the body surface. With this method of skin grafting, the outer portion of the skin (the epidermis and perhaps half the dermis) is removed from an undamaged area and placed over a region where serious skin damage has occurred (Figure 5.6a). The cells of the graft are nourished by interstitial fluid from the damaged surface. Gradually, connective tissue cells form new intercellular substance that attaches the graft in place, and eventually the graft becomes vascularized.

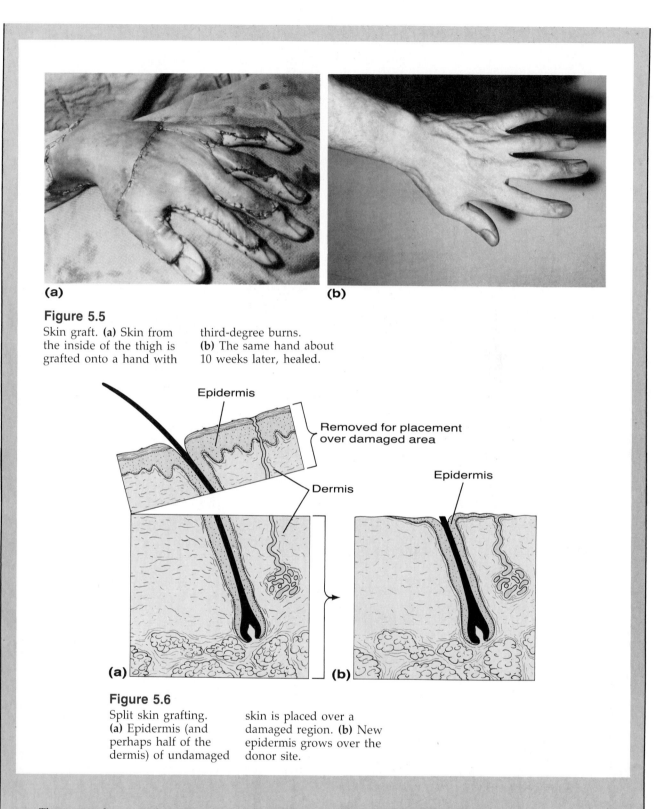

Figure 5.5

Skin graft. **(a)** Skin from the inside of the thigh is grafted onto a hand with third-degree burns. **(b)** The same hand about 10 weeks later, healed.

Figure 5.6

Split skin grafting. **(a)** Epidermis (and perhaps half of the dermis) of undamaged skin is placed over a damaged region. **(b)** New epidermis grows over the donor site.

The exposed, nonepidermal surface in the region from which the graft was taken becomes covered with new epidermis, which grows over the surface from the remaining hair follicles and the ducts of sweat glands (which originate embryonically from the same tissue that gives rise to the epidermis) (Figure 5.6b).

If a patient has suffered extensive burns, removing skin to perform a skin graft such as that described above and shown in Figure 5.5 is ill-advised. Instead, artificial skin made of silicone, collagen, and a polysaccharide is sometimes used in place of real skin.

Because the treatment of burns depends to some degree on the amount of body surface area that has been damaged, it is useful to be able to estimate quickly the extent of a burn. There are methods by which rather precise estimates can be obtained, but a less exact

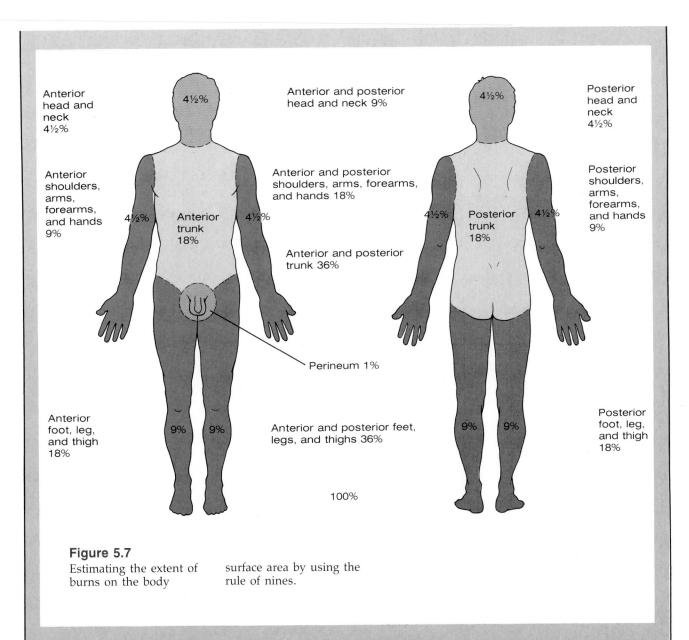

Anterior
head and
neck
4½%

Anterior and posterior
head and neck 9%

Posterior
head and
neck
4½%

Anterior
shoulders,
arms,
forearms,
and hands
9%

Anterior and posterior
shoulders, arms, forearms,
and hands 18%

Posterior
shoulders,
arms,
forearms,
and hands
9%

4½%   Anterior
trunk
18%   4½%

4½%   Posterior
trunk
18%   4½%

Anterior and posterior
trunk 36%

Perineum 1%

Anterior
foot, leg,
and thigh
18%

9%   9%

Anterior and posterior feet,
legs, and thighs 36%

9%   9%

Posterior
foot, leg,
and thigh
18%

100%

Figure 5.7
Estimating the extent of
burns on the body

surface area by using the
rule of nines.

method is commonly used because it is so easy to apply. This method is called the **rule of nines** (Figure 5.7). In this estimation, the body surface area is divided in the following way: each upper limb is considered to have 9% of the body surface area; each lower limb 18%; the anterior and posterior trunk regions also each have 18%; the head and neck together have 9%; and the perineum has the remaining 1%.

Healing of Cut Skin

When the skin is cut and the edges of the wound are drawn together by sutures or other means, a V-shaped slit extends from the surface to the subcutaneous tissue (Figure 5.8a). Soon after the skin is cut, a small amount of a fibrous substance called *fibrin* forms near the bottom of the slit. Fibrin results from the process of blood coagulation. Gradually, epidermis from the surface grows down into the slit, adhering to sound tissue on either side. After about one week, the epidermis extends down into the dermis (Figure 5.8b). The epidermis continues its downward growth, adhering to healthy dermis and passing beneath any fibrin that is present. Epidermal continuity is restored after about two weeks, when the epidermis growing down one side of the slit meets the epidermis growing down the opposite side near the bottom of the slit (Figure 5.8c). While the epidermis is growing down the sides of the slit, fibroblasts and blood vessels from the subcutaneous tissue are engaged in repairing the connective tissue of the skin. At the junction of the dermis and the subcutaneous tissue in the region of the wound, an abundant growth of fibroblasts and small blood vessels called capillaries forms a ridge of new tissue (Figure 5.8c). As this tissue grows, it bulges up at the bottom of the epidermis-lined slit and pushes the slit upward until it becomes level with the surface (Figure 5.8d). Thus, the area previously occupied by the slit becomes occupied by new connective tissue that is derived chiefly from subcutaneous tissue and is covered by a thin epidermis. This epidermis remains thin for some time.

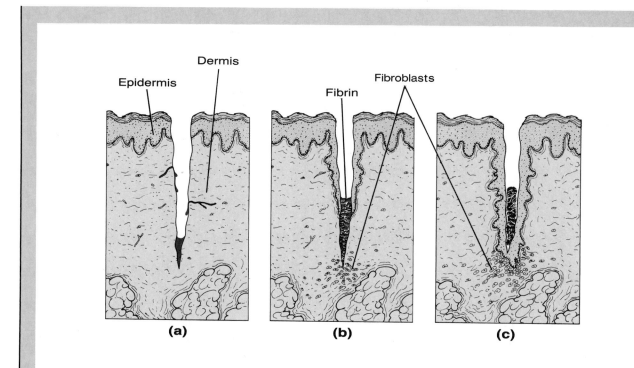

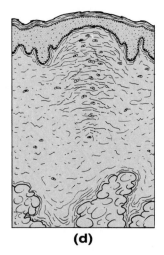

Figure 5.8
Stages of tissue repair (wound healing).
(a) Damage to skin also damages blood vessels. **(b)** Fibrin forms near the bottom of the wound, and the epidermis extends down into the dermis. **(c)** Epidermis extends across the bottom of the wound; fibroblasts and capillaries form a ridge of new tissue. **(d)** New tissue fills in the wound.

Effects of Aging

With aging, the skin tends to become thinned, somewhat wrinkled, dry, and occasionally scaly. Although the actual thickness of the stratum corneum does not seem to change greatly, it may become more permeable, allowing substances to pass through it more readily. Moreover, with aging, the collagen fibers in the dermis become thicker, the elastic fibers less elastic, and there is a gradual decrease in the underlying subcutaneous fat. There is also a decrease in the number and activity of the hair follicles, sweat glands, and sebaceous glands. Consequently, aging is often accompanied by a loss of hair, reduced sweating, and decreased oil (sebum) production. Melanocytes tend to atrophy with age, so there is often a graying of the hair and a mottling of the skin as pigment production decreases in certain areas.

Skin that has been exposed to sunlight over a lifetime will show changes that are more severe than those due to aging alone. Such skin shows more marked wrinkling and furrowing, and may develop nodules of an abnormal type of collagen. Moreover, aging skin that has been exposed to large amounts of sunlight tends to develop more cutaneous cancers than skin that has had less exposure.

STUDY OUTLINE

EPIDERMIS Outermost layer of skin; develops from embryonic ectoderm; lacks blood vessels; skin color is primarily determined by a dark pigment called melanin, but it is also influenced by the yellow pigment carotene and by dermal blood vessels; generally thin, but can thicken as calluses; four epidermal layers. **pp. 111–114**

EPIDERMAL LAYERS

STRATUM GERMINATIVUM Deepest layer; mitosis occurs in basal layer (stratum basale) of cells, thereby supplying epidermis with new cells.

STRATUM GRANULOSUM Composed of cells that contain granules of keratohyalin within cytoplasm; as granules expand, cell nucleus dies, so outermost cells of this layer are dead.

STRATUM LUCIDUM Clear band superficial to the stratum granulosum; cells of this tissue continuously become part of stratum corneum through the presence of eleidin, which is transformed into keratin.

STRATUM CORNEUM Outermost layer; composed of closely packed dead cells filled with fibrous protein, keratin.

NOURISHMENT OF THE SKIN Epidermis obtains nourishment by diffusion from capillary beds of dermis.

SKIN COLOR Determined by presence and distribution of melanin.

DERMIS Lies deep to the stratum germinativum; second main layer of skin; well supplied with blood vessels, lymph vessels, nerves, glands, sense organs; has two indistinctly separated layers. **p. 114**

PAPILLARY LAYER Next to basal layer of stratum germinativum; contains specialized sensory receptors and capillary loops.

RETICULAR LAYER Deep layer consisting of bundles of collagenous fibers, continuous with the deeper hypodermis layer.

HYPODERMIS Not part of skin, but important because it attaches skin to underlying structures; composed of loose connective tissue. **p. 116**

GLANDS OF THE SKIN **pp. 116–117**

SWEAT GLANDS Also called sudoriferous glands; distributed over most of body surface.

ECCRINE SWEAT GLANDS Coiled tubules within dermis; secrete a watery solution of salt, with traces of urea, sulfates, and phosphates.

APOCRINE SWEAT GLANDS Secrete part of their cell contents, so secretion is more complex than true sweat.

CERUMINOUS GLANDS Produce "wax" in ears; modified sweat glands.

SEBACEOUS GLANDS Empty their secretion (sebum) into hair follicles; serve to oil skin and hair. Especially active in adolescence. In regions of skin lacking hair, glands empty secretions onto epidermis surface.

HAIR Covers almost entire body; its growth is due to mitotic activity of epidermal cells at bottom of hair follicles. **pp. 117–118**

HAIR FOLLICLES Extend from epidermis into dermis; composed of two layers: (1) inner layer gives rise to hair; (2) outer layer of connective tissue develops from dermis.

PAPILLAE At bottom of hair follicles; contain blood capillaries for nourishment and mitosis.

ARRECTOR PILI MUSCLE Pulls on follicle; causes hair to "stand up."

SINGLE HAIR Consists of root (part within follicle), shaft (part above skin surface); shaft has central core (medulla) of loose horny cells, cortex of tightly compressed keratinized cells that surround medulla, and outside cuticle of hard keratinized cells. Hair color primarily due to melanin.

NAILS Heavily cornified layers of strata corneum and lucidum. Each nail rests on nail bed of stratum germinativum. Mitosis, which produces nail growth, occurs in thickened matrix under proximal end of nail. **pp. 118–119**

FUNCTIONS OF THE INTEGUMENTARY SYSTEM **pp. 119–120**

PROTECTION Skin forms physical barrier against invasion of body by foreign substances; protects against ultraviolet radiation; reduces water loss.

BODY TEMPERATURE REGULATION

OVERHEATING OF BODY Prevented as capillaries in dermis dilate and bring greater volume of blood to body surface to lose heat to the environment; body surface also becomes wet, providing additional cooling by evaporation.

HEAT CONSERVATION Accomplished during cold by constriction of dermal capillaries.

EXCRETION Some nitrogenous wastes and salt leave the body via sweat.

SENSATION Nerve endings and specialized receptors in skin provide body with much information, such as temperature change, increased pressure.

VITAMIN D PRODUCTION Occurs in skin in presence of sunlight, ultraviolet radiation; helps maintain optimum levels of calcium and phosphate.

CONDITIONS OF CLINICAL SIGNIFICANCE: THE INTEGUMENTARY SYSTEM

Fungi, yeasts, and bacteria live on body skin, yet cause no harm unless the epidermis is damaged, allowing them to enter the body. **pp. 120–124**

ACNE Inflammatory disease caused by bacteria that provoke excessive secretion by sebaceous glands; results in pimples, blackheads, dandruff; most frequent during puberty.

WARTS Caused by viral invasion of skin; most common in adolescents and young adults.

DERMATITIS AND ECZEMA General terms for numerous inflammatory skin conditions.

PSORIASIS Common condition of small reddish-brown elevations and patches covered by layers of silvery scales; accompanied by bleeding and tiny abscesses; cause unknown.

IMPETIGO Highly contagious bacterial infection common in children; pustules that form beneath stratum corneum cause inflammation and swelling.

MOLES Common pigmented elevations of skin; considered to be congenital; may become malignant, but most grow, stabilize, and finally atrophy.

HERPES SIMPLEX Fever blister, or cold sore, caused by viral activity.

HERPES ZOSTER (SHINGLES) Caused by viral activity in spinal nerves; most often affects sensory nerves of thoracic region; cause vesicle formation and pain.

CANCERS Numerous types of tumors in epidermis, dermis, and skin glands; most are nonspreading (benign), but some do spread (malignant); cause of most skin tumors unknown, but some may be caused by prolonged overexposure to ultraviolet radiation.

BURNS The seriousness of burns results from destruction of skin, which can drastically disrupt the homeostasis of the body. Classified according to severity.

HEALING OF CUT SKIN Cut skin is healed by new connective tissue derived chiefly from subcutaneous tissue.

EFFECTS OF AGING Skin tends to become thinned, wrinkled, dry, and sometimes scaly.

SELF-QUIZ

1. The subcutaneous connective tissue is called the: (a) dermis; (b) hypodermis; (c) stratum corneum.

2. In regions where the epidermis is thin, the stratum lucidum is often absent. True or False?

3. Match the following terms associated with the epidermis to the appropriate lettered description:

Basal layer (stratum basale)	(a) The layer that helps restrict loss of body water
Stratum germinativum	(b) The substance that is transformed into keratin
Keratohyalin	
Melanin	(c) A yellow pigment
Stratum corneum	(d) A clear band of several layers of flattened, closely packed cells
Eleidin	
Stratum granulosum	(e) Where mitosis takes place in the stratum germinativum
Carotene	
Stratum lucidum	(f) The deepest layer of the epidermis
Keratin	(g) A fibrous protein
	(h) Granules associated with the disintegration of the nucleus
	(i) The stratum deriving its name from the presence of keratohyalin
	(j) A dark pigment

4. Skin color is determined primarily by the presence and distribution of: (a) carotene; (b) melanin; (c) hemoglobin.

5. The dermis develops from embryonic mesoderm, as do the muscles and the skeleton of the body. True or False?

6. Those glands that empty their secretions into hair follicles are termed: (a) sebaceous; (b) ceruminous; (c) sudoriferous.

7. Hair grows due to the mitotic activity of epidermal cells at the bottom of the arrector pili. True or False?

8. Nail growth occurs at the: (a) nail tips; (b) nail bed; (c) matrix.

9. The skin tends to be slightly basic, with a pH of 6.8. True or False?

10. Body heat is *not* conserved when the dermal capillaries: (a) dilate; (b) constrict.

11. Vitamin D_3 assists in maintaining the calcium and phosphate levels of the body at optimum levels. True or False?

12. The most abundant organisms on the skin are: (a) yeasts; (b) bacteria; (c) fungi.

13. Acne is an inflammatory disease caused by: (a) bacteria; (b) virus; (c) fungi.

14. The common wart is the result of a viral invasion of the skin and is most common in adolescents and young adults. True or False?

15. The cause of which of these disorders is unknown? (a) warts; (b) eczema; (c) psoriasis.

16. Which one of the following conditions is congenital? (a) herpes simplex; (b) warts; (c) moles.

17. Match the terms associated with common pathologies of the integumentary system with the appropriate lettered description:

Acne	(a) Caused when a dormant virus becomes activated on the skin and mucous membranes
Warts	
Psoriasis	
Impetigo	
Moles	
Herpes simplex	(b) The name for any number of inflammatory skin conditions
Herpes zoster	
Cancers	
Eczema	(c) Pigmented elevations of the skin that are congenital

    (d)   The formation of pimples, blackheads, and dandruff due to bacteria

    (e)   Metastasizing tumors

    (f)   Small reddish-brown patches covered by layers of silvery scales

    (g)   Caused by a viral invasion of the skin

    (h)   A highly contagious skin infection common in children

    (i)   Aching pain and formation of vesicles along paths of thoracic sensory nerves

18. A burn victim who has experienced damage to both the epidermis and dermis but whose lost tissue will quickly be regenerated is suffering what degree burn? (a) first; (b) second; (c) third.

19. A burn of the anterior trunk region involves about the same amount of body surface area as does a burn of the anterior surfaces of both lower limbs. True or False?

LEARNING OBJECTIVES

After completing this chapter, you should be able to:

1. List the functions of the skeleton.

2. Cite the main components of bone that provide it with its strength.

3. List and define the major kinds of bones in the human skeleton.

4. Describe the two methods by which bone develops in the embryo.

5. Distinguish between the axial skeleton and the appendicular skeleton, and name the components of each.

6. Name the bones that form the various regions of the skull.

7. Name the components of the skeleton of the thorax.

8. Distinguish between the functions of the pectoral and pelvic girdles, and name the major components of each.

9. Describe the differences between the various types of vertebrae.

10. Describe the differences between the pelvises of males and females.

CHAPTER CONTENTS

FUNCTIONS OF THE SKELETON

CLASSIFICATION OF BONES

STRUCTURE OF BONE

DEVELOPMENT OF BONE

CONDITIONS OF CLINICAL SIGNIFICANCE: THE SKELETAL SYSTEM

INDIVIDUAL BONES OF THE SKELETON

AXIAL SKELETON

APPENDICULAR SKELETON

KEY TERMS AND DERIVATIVES

axial (*ax* = an axis) axial skeleton: upright portion of the skeleton that supports the head, neck, and trunk

coracoid (*corac* = beaklike) coracoid process: the beaklike process of the scapula

crista (*crista* = ridge) crista galli: a bony ridge that projects upward into the cranial cavity

hemopoiesis (*hemo* = blood; *poie* = making) the process by which blood cells are formed

lacuna (*lacuna* = lake) a small depression or space; in bone or cartilage, lacunae are occupied by cells

osteoclast (*osteon* = bone; *klasein* = break) a large multinuclear cell associated with physiologic bone destruction, reabsorption, and remodeling

The Skeletal System

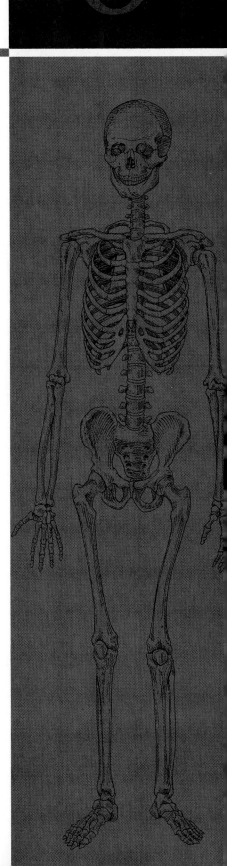

The human skeleton is an endoskeleton—that is, it lies within the soft tissues of the body. It is a living structure capable of growth, adaptation, and repair. An endoskeleton differs greatly from the exoskeleton of arthropods like beetles and crayfish. Because an exoskeleton is a nonliving structure located on the outside of the body, an animal that has an exoskeleton must shed its exterior skeletal structure and form a new, larger one if it is to continue to grow. As you know from your own growth, your skeleton, in contrast, has grown at the same time as the rest of your body structures.

FUNCTIONS OF THE SKELETON

The skeleton performs several important functions.

Support
The skeleton acts as the framework of the body, giving support to the soft tissues and providing points of attachment for most of the body muscles.

Movement
Because many of the body muscles attach to the skeleton, and many of the bones meet (*articulate*) in movable joints, the skeleton plays an important role in determining the kind and extent of movement of which the body is capable.

Protection
Many of the vital internal organs are protected from injury by the skeleton. The brain is encased within the cranial cavity of the skull, the spinal cord is within the canal formed by the vertebrae, the thoracic organs are protected by the rib cage, and the urinary bladder and internal reproductive organs are protected by the bony pelvis.

Mineral Reservoir
Calcium, phosphorus, sodium, potassium, and other minerals are stored within the bones of the skeleton. These minerals can be mobilized and distributed by the blood vascular system to other regions as they are required by the body. During pregnancy, for instance, calcium is removed from the mother's skeleton and used in the development of the baby's bones if the mother's diet does not include enough calcium. Because of the large mineral content of bones, they can remain intact for many years after death.

Hemopoiesis (Blood-Cell Formation)
Following birth, the red marrow within certain bones produces the blood cells that are found within the circulatory system.

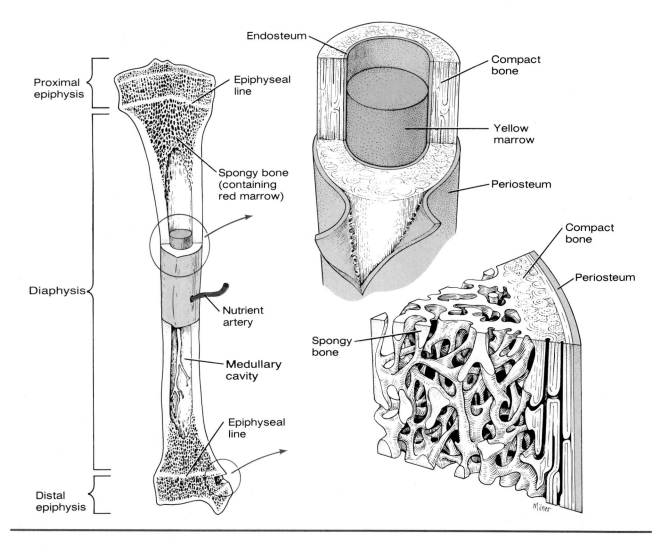

Figure 6.1
Structure of a long bone—longitudinal section and insets of higher magnification.

CLASSIFICATION OF BONES

Bones can be classified according to their shape.

Long Bones

Most of the bones of the upper and lower limbs have a long axis; that is, they are longer than they are wide. These are classified as long bones (humerus, radius, ulna, femur, tibia, fibula, phalanges).

Short Bones

Bones that do not have a long axis, such as those of the wrist (carpals) and ankle (tarsals) are called short bones.

Flat Bones

The rather thin bones that form the roof of the cranial cavity, the ribs, and the sternum (the breastbone) are flat bones.

Irregular Bones

Bones of various shapes that do not fit any of the other categories are classed as irregular bones. Some skull bones, the vertebrae, and the bones of the pectoral and pelvic girdles are examples of irregular bones.

STRUCTURE OF BONE

It is instructive to study the structure of bone at three different levels; at the gross level—where no microscope is used; at the microscopic level; and at the chemical level.

Gross Anatomy

A typical long bone has a shaft, called a **diaphysis,** and two ends, called proximal and distal **epiphyses** (Figure 6.1). The diaphysis is formed of a hollow cylinder of **compact bone** that surrounds a **medullary cavity.** The medullary cavity, which is used as a fat storage site, is also called the **yellow bone marrow cavity.** It is lined by a thin connective tissue layer called the **endosteum.** The outer surfaces of the epiphyses are also formed of compact bone. However, their central regions are filled with interconnecting plates of **spongy (cancellous) bone.** The cavities between the bony plates of spongy bone are lined with endosteum. The spongy bone in the epiphyses of certain bones contains **red bone marrow.** In children and young adults, the diaphysis and epiphysis are separated by an **epiphyseal cartilage** or **plate** that provides the means for the bone to increase in length. In the adult, when skeletal growth has been completed, the epiphyseal cartilage is replaced by bone, firmly uniting the epiphysis with the rest of the bone. This bony junction is called the **epiphyseal line.**

There is no medullary cavity in a flat bone. This kind of bone is formed of spongy bone called **diploe,** which is sandwiched between two surface layers of compact bone (Figure 6.2). The spongy bone contains red marrow.

Bones are covered with a double layer of dense connective tissue called the **periosteum.** There is no periosteum in joints where the bone is covered with an articular cartilage. The outer layer of the periosteum is well supplied with blood vessels and nerves, some of which enter the bone. The inner layer is anchored to the bone by collagenous bundles **(Sharpey's fibers)** that penetrate the bone.

Microscopic Anatomy

When examined under the microscope, compact bone is seen to be composed of many organized systems of interconnecting canals (Figures 6.3 and 6.4). The unit of structure of adult compact bone is the **haversian system,** or **osteon.** Each haversian system has a central **haversian canal** that is surrounded by concentrically arranged **lamellae** (layers) of bone. Because the haversian systems generally run parallel to the long axis of bone, in longitudinal sections the canals appear as long tubes. This orientation of haversian systems contributes to the capacity of bone to resist compressive forces. Located between adjacent lamellae in a haversian system are small cavities called **lacunae.** Each lacuna contains a cell called an **osteocyte.** All of the lacunae within each haversian system are interconnected by tiny canals called **canaliculi.** The osteocytes have tiny protoplasmic processes on their surfaces that enter the canaliculi and contact cell processes of osteocytes located in adjacent lacunae. At the point of contact within the canaliculi, the cell processes of adjacent osteocytes are joined together by gap junctions that make possible the rapid passage of nutrients and wastes from one osteocyte to another (Figure 6.5).

Each haversian canal contains at least one blood capillary, which provides a source of nutrients and a means of waste removal for those osteocytes that are embedded within the lacunae. The nutrients and wastes need to diffuse only a short distance through the tissue fluid within the lacunae and canaliculi from or to the haversian canal. After entering an osteocyte, the nutrients from the blood vessels are distributed to adjacent osteocytes by means of the protoplasmic processes within the canaliculi. The blood vessels reach the canals from larger vessels that are located either on the surface of the bone (that is, in the vascular layer of the periosteum) or within the marrow cavity. Blood vessels, as well as lymph vessels and nerves, enter and leave the marrow

F6.5

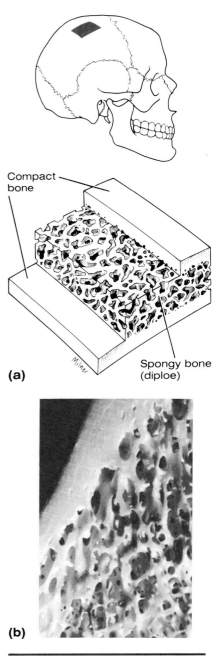

(a)

Compact bone

Spongy bone (diploe)

(b)

Figure 6.2
(a) Cross section showing the structure of a flat bone.
(b) Photomicrograph of spongy bone.

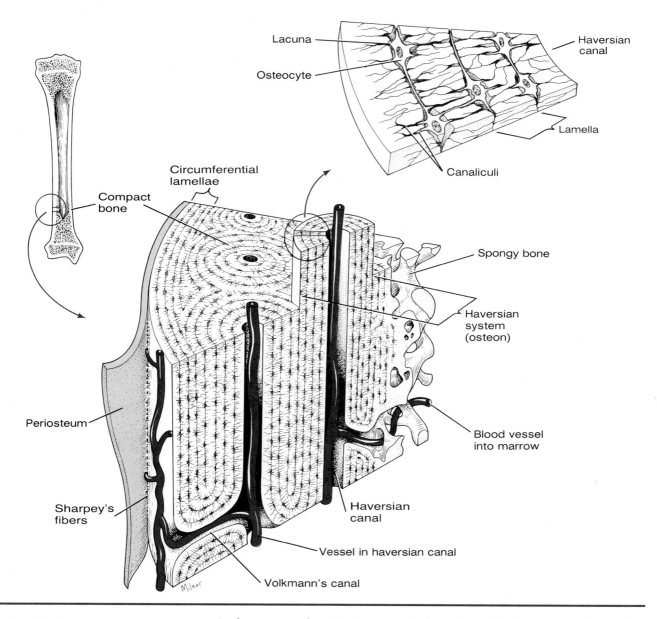

Figure 6.3
Diagram of magnified haversian systems as seen in compact bone tissue. The periosteum has been pulled back to show a blood vessel entering the haversian systems through a Volkmann's canal. The inset is a highly magnified sketch showing osteocytes within lacunae. Note that the lacunae are interconnected by canaliculi.

cavity by means of **nutrient canals** that penetrate the bone from the surface and communicate with the marrow cavity. Blood vessels from either of these sources reach the haversian canals through **Volkmann's canals,** which run at right angles to the haversian canals. At the external surface of a bone, just beneath the periosteum, there may be several **circumferential lamellae,** which follow the circumference of the shaft rather than surrounding a haversian canal.

Spongy bone does not show the organization that is characteristic of compact bone. The osteocytes are embedded within lacunae, and the lacunae intercommunicate via canaliculi, as in compact bone. However, the lamellae are not arranged in concentric layers. Rather, they are arranged in various directions that correspond with the lines of maximum pressure or tension. Blood capillaries reach the vicinity of the osteocytes by passing within the bone marrow spaces between the plates of bone that are formed by the lamellae.

From this consideration of the microscopic anatomy of bone, it is apparent that the skeletal system is a living system, well supplied with blood vessels and nerves. As such, it is capable of performing the dynamic functions of hemopoiesis and of serving as a mineral reservoir, in addition to the static functions of support, movement, and protection.

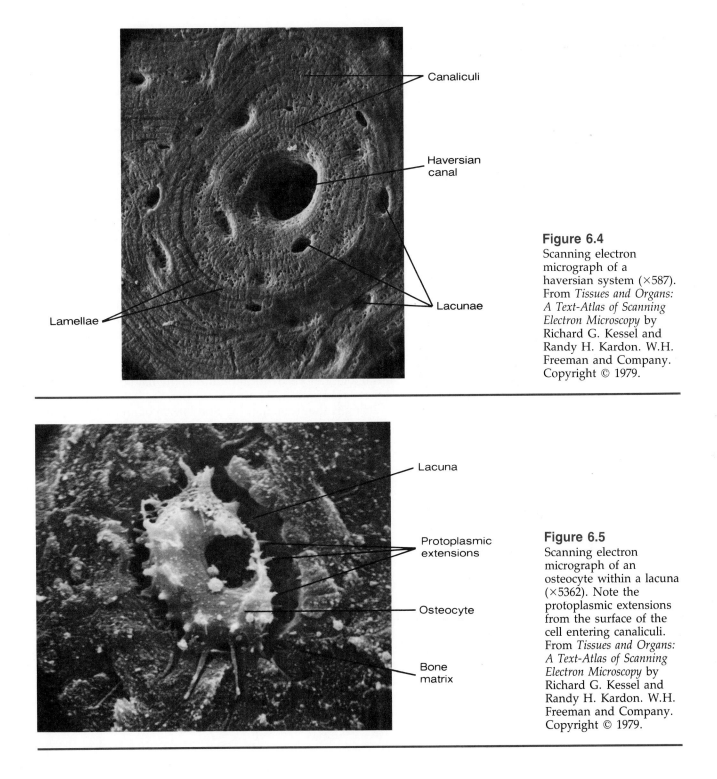

Figure 6.4
Scanning electron micrograph of a haversian system (×587). From *Tissues and Organs: A Text-Atlas of Scanning Electron Microscopy* by Richard G. Kessel and Randy H. Kardon. W.H. Freeman and Company. Copyright © 1979.

Figure 6.5
Scanning electron micrograph of an osteocyte within a lacuna (×5362). Note the protoplasmic extensions from the surface of the cell entering canaliculi. From *Tissues and Organs: A Text-Atlas of Scanning Electron Microscopy* by Richard G. Kessel and Randy H. Kardon. W.H. Freeman and Company. Copyright © 1979.

Composition

The intercellular substance (*matrix*) of bone is composed of two main components—an **organic framework** and **inorganic salts.** The organic framework, in which the inorganic salts become embedded, is formed by collagenous fibers similar to those found within other connective tissues. Surrounding these fibers is a homogenous ground substance. The inorganic salts of bone are composed principally of *calcium* and *phosphate*.

Collagen fibers provide bone with great tensile strength since they are capable of resisting stretching and twisting. The salts allow bone to withstand compression. This combination of fibers and salts makes bone exceptionally strong without being brittle. This same principle is used in reinforced concrete, where steel rods provide tensile strength, and cement, sand, and gravel give compressional strength.

DEVELOPMENT OF BONE

Early Development of Bone

Box 6.1

en-do-kon'-dral

The skeleton develops by the transformation of embryonic connective tissues into bone (see Box 6.1). The connective tissues that give rise to most bones are derived from cells of the mesodermal layer of the embryo. If the embryonic tissues that are transformed into bone are undifferentiated mesoderm (mesenchyme), a relatively simple process called **intramembranous ossification** occurs. If the mesodermal cells transform into cartilage-producing cells before bone formation begins, the process is more complicated and is called **endochondral (intracartilaginous) ossification.** In this process the skeletal structure starts out as cartilage, which is then replaced by bone. The only difference between intramembranous and endochondral bone is the tissue that is replaced. The bone that results from either of these transformations has the same composition.

Intramembranous Ossification

The process of intramembranous ossification forms the flat bones of the roof of the skull and some facial bones. The first indication that intramembranous ossification is occurring is the formation (by fibroblasts) of an organic matrix that has collagenous fibrils extending throughout the tissues. Some of the cells within the matrix increase in size and begin forming bony spicules by calcifying the interstitial substance between the fibrils. These bone-forming

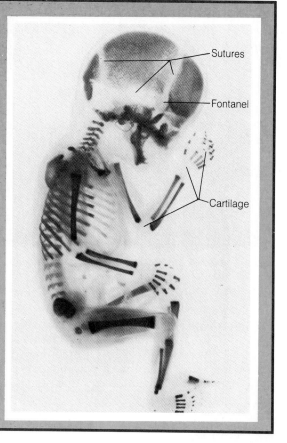

Box 6.1
Fetal Skeleton

Developing bones of a human fetus at about 12 weeks. The darker areas represent ossification (mineral deposition) of the bones. Bones first begin to appear in the fifth week of embryonic development. The clavicle is usually the first bone to ossify. The light areas indicate cartilage that has not yet been replaced by bone. Notice the suture lines separating the individual bones of the skull. Where the sutures meet with one another, a soft spot called a *fontanel* is formed. Some fontanels do not ossify until 18 months after birth. Some bones in the human body are not fully ossified until puberty.

Photograph by Carolina Biological Supply Company.

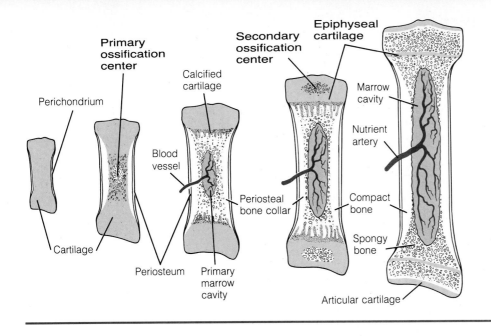

Figure 6.6
Stages of endochondral ossification as seen in a long bone. Cartilage is shown in color. Notice that cartilage remains in the epiphyseal plates and on the articular surfaces. Eventually, the cartilage of the epiphyseal plates will be completely replaced by bone.

cells are **osteoblasts.** Osteoblasts arise from undifferentiated cells called **osteoprogenitor cells,** which develop from embryonic mesenchymal cells. As more layers of bone are deposited, the spicules become heavier and trap the osteoblasts within lacunae. After the osteoblasts are enclosed within lacunae, their activity slows, and they become mature bone cells called **osteocytes.**

Well-developed spicules of bone are known as **trabeculae** ("little beams"). The trabeculae radiate in all directions, uniting with one another to form a network of spongy bone. In areas that will eventually form compact bone, the trabeculae continue to thicken as additional bone is deposited. Gradually, the spaces between the trabeculae are narrowed and bone replaces the spaces, forming compact bone.

tra-bek'-u-lah

During development, most of the external surfaces of bones become covered with a periosteum. This membrane contains osteoprogenitor cells that can give rise to osteoblasts that begin forming bone. Increased activity of these peripheral osteoblasts is responsible for the increase in thickness of the bone.

Even after being calcified, these bones must of necessity be capable of undergoing extensive remodeling in order to meet the changing dimensions of a growing body. This is accomplished through the reabsorption of previously laid down bone by large cells called **osteoclasts,** and the deposition of new bone (by osteoblasts) in patterns that conform to the growth requirements. Osteoclasts, like osteoblasts, arise from osteoprogenitor cells.

Endochondral Ossification

Most bones form by the ossification of hyaline cartilage models that are formed during the early development of the embryo (Figure 6.6). The cartilage models resemble the shape of the future bones. Each model is surrounded by a fibrous connective tissue membrane called the **perichondrium.** During endochondral ossification, the cartilage of the model degenerates, chondrocytes are replaced by osteoblasts, and bone develops.

F6.6

per-i-kon'-dree-um

Endochondral bone development begins with the transformation of the perichondrium into a bone-producing **periosteum.** This change is brought about by the gathering of osteoblasts on the inner surface of the perichondrium. As a result of this process, the cartilage of the diaphysis becomes encased by a collar of compact bone that is laid down by the cells of the periosteum. While bone is forming on the periphery of the diaphysis, the cartilage cells within the diaphysis enlarge, as do the lacunae that surround the cells. As a result of these enlargements, the cartilage matrix that lies deep to the site of the peripheral ossification becomes reduced to thin partitions and spicules. This matrix then begins to calcify, forming a type of tissue called *calcified cartilage,* which makes diffusion of nutrients to the cartilage cells impossible. The cartilage cells therefore die and the calcified matrix begins to degenerate, leaving spaces within the calcified cartilage matrix. Blood vessels enter the spaces within the matrix and form capillary loops throughout the matrix. The blood vessels transport osteoprogenitor cells and osteoblasts from the periosteum to the spaces within the calcified matrix. The osteoblasts form trabeculae of spongy bone on what is left of the calcified matrix. The region within the diaphysis where this occurs is called the **primary ossification center.** In most bones, the primary ossification center appears in the third month of embryonic life. The only difference between the formation of trabeculae in endochondral ossification and their formation in intramembranous ossification is that in intramembranous ossification the trabeculae form around collagenous fibers, whereas in endochondral bones they form around calcified cartilage spicules. As the primary ossification center enlarges within the diaphysis, some of the newly formed spongy bone is broken down by osteoclasts, forming the marrow cavity.

At birth, most bones of the body are at this primary stage of development—that is, they have a diaphysis of compact bone that surrounds remnants of spongy bone and the marrow cavity, while the epiphyses remain as hyaline cartilage. Shortly after birth, the cartilage cells within the epiphyses enlarge, the matrix calcifies, and **secondary ossification centers** appear within both epiphyses. The secondary ossification centers form in the same manner as the primary center, except that there is no marrow cavity formed. These new ossification centers increase in size and eventually occupy the entire epiphyses. The only cartilage that remains is a thin surface layer (which will become the articular cartilage) and a thicker layer that separates each epiphysis from the bony diaphysis. This latter cartilage is called the **epiphyseal cartilage** or **plate.**

Increase in Bone Length and Diameter

As long as the epiphyseal plate remains present, it is possible for the bone to increase in length. It does so in the following manner. Cartilage cells undergo mitosis and therefore tend to increase the size of the epiphyseal plate. At the same time, the side of the plate toward the diaphysis is being replaced by bone. Under normal conditions of growth, these two processes balance one another, so that the diaphysis increases in length while the epiphyseal plate remains a constant thickness. As the bone increases in length, it also undergoes a remodeling process by means of selective bone resorption and formation, which maintains the epiphyses at a relatively constant size. These conditions prevail until the late teen years, when the rate of cartilage growth slows and is gradually overtaken by the continuing ossification on the diaphyseal side of the plate. By age 25 in males (several years earlier in females), the cartilage within the epiphyseal plate is completely replaced by bone, leaving only an **epiphyseal line** to mark its previous location. When this has occurred, the bone is no longer able to increase in length. It does, however, retain its ability to increase in diameter.

Bone increases in diameter as a result of the presence of osteoblasts located in the inner layer of the periosteum. These osteoblasts deposit concentric layers of new bone around the diaphysis, just beneath the periosteum. At the same time as the diameter of the diaphysis is being increased by the addition of these bony layers on the surface, bone is being resorbed from within—beneath the endosteum. Since this resorption increases the volume of the medullary cavity, the thickness of compact bone in the diaphysis may not increase much while the total diameter of the bone does increase.

Theories of Bone Formation

Many of the actual chemical events of bone formation are still incompletely understood. In general, the osteoblasts secrete the organic framework of bone, which includes collagen fibers and glycoproteins such as chondroitin sulfate. The inorganic salts are deposited in the organic framework. It appears that salts such as calcium phosphate $[Ca_3(PO_4)_2]$ are formed initially, and then, through a process of substitution and addition of ions, these salts are converted to the inorganic material of bone, which is composed of highly insoluble crystals of hydroxyapatite $[3Ca_3(PO_4)_2 \cdot Ca(OH)_2]$.

Although it is not known what causes mineralization during bone growth and maintenance, one theory is that bone cells concentrate large quantities of calcium and phosphate and that the cells subsequently release calcium phosphate compounds into the extracellular fluid. The small areas of calcium phosphate salts that result may serve as sites for further salt deposition. An alternative theory suggests that the initial formation of bone mineral takes place in membrane-bounded vesicles called matrix vesicles, which are synthesized and secreted into the extracellular fluid by bone cells.

One interesting property of bone is that when it is deformed, a voltage may be recorded between two opposite surfaces. In fact, when compressed or bent, the convex surface becomes negatively charged. New bone deposition occurs at this negative surface, where it will strengthen the bone most effectively.

Factors That Affect Bone Development

A number of factors can greatly influence the development of bone. Among the more important factors are *stress*, the amounts of certain *hormones* present in the blood, and the *nutrition* of the individual in which bone is developing.

Stress

Bone is a living tissue that is capable of adjusting its strength in proportion to the degree of stress to which it is subjected. Increased amounts of collagen fibers and inorganic salts can be deposited in a bone in response to prolonged heavy loads. Conversely, if a bone is not subjected to stress, salts are withdrawn from bone.

As new stress patterns occur, the orientation of the collagen fibers in a bone may change so that the fibers are aligned in such a manner as to provide maximal tensile strength to withstand the new stress patterns.

Bone is normally subjected to two major kinds of stresses: *gravitational forces*, such as those that result from supporting the weight of the body, and *functional forces*, such as those that result from the pull exerted on the bone by contracting muscles. There is no consensus as to which of these forces is more important in affecting the form and structure of bone. However, it has been repeatedly demonstrated that bone does not develop normally in the absence of either of these forces. For instance, when gravitational forces are removed for extended periods, such as under the weightless conditions of space travel, or when functional forces are greatly reduced, such as occurs when a limb is paralyzed or immobilized by a cast, the bones in the affected areas do not grow, and may actually atrophy (that is, degenerate). In addition, there are some indications that bone growth is promoted by intermittent forces such as would occur during muscular exercise.

Because of these effects of gravitational and functional forces, regular exercise can alter the form of the skeleton. As weight lifters strengthen their muscles so they can lift heavier weights, the bony skeleton is also strengthened. If this bone strengthening did not occur, a greatly strengthened muscle could break the bones to which it is attached. In a similar manner, the skeleton of an obese person becomes heavier because of the increased forces to which it is constantly subjected.

Another effect of the stresses of exercise on bone involves the presence of the epiphyseal cartilages. The chondrocytes within these cartilages remain active for approximately 20 years, and are susceptible to injury and unusual stress during that time. Injury to the epiphyseal cartilages could interrupt the normal rate of bone growth, as well as the pattern of growth. It has not been proven conclusively that contact sports and repeated specialized stresses actually do adversely influence these centers of growth. However, their very presence is reason enough to warrant caution when recommending strenuous exercises for young people.

Hormones

Hormones of the parathyroid and thyroid glands are particularly influential in bone development. Increased levels of the parathyroid hormone **parathor-**

kal-si-to'-nin

mone lead to an increase in both the number and the activity of osteoclasts within bone, thus increasing the resorption of bone. This increased resorption, in turn, releases clacium ions from the bone that pass into the blood and raise the plasma calcium level. The hormone **calcitonin** from the thyroid gland has an effect opposite to that of parathormone. Calcitonin decreases the resorptive activity of osteoclasts and lowers the blood calcium level, particularly in young individuals. Calcitonin may also stimulate the formation of new bone. Any remodeling of bone that occurs involves the interactions of these two hormones. These hormones are discussed in greater detail in Chapter 16.

Nutrition

For normal bone development to occur, it is necessary to follow a balanced diet, that is, one that provides the body with a variety of essential substances. Of particular importance is vitamin D, which is necessary for the proper absorption of calcium into the bloodstream from the gastrointestinal tract. As you already know, calcium is an important constituent of the inorganic portion of bone matrix.

CONDITIONS OF CLINICAL SIGNIFICANCE

The Skeletal System

Fractures

Even though the composition of bones is such that they are well suited to withstand twisting and compressional forces, it is possible to break them. Broken bones are referred to as *fractures*.

Types of Fractures

Fractures are named according to various conditions at the site of the break. The following are the more common types of fractures.

SIMPLE FRACTURE   The broken ends of the bone do not penetrate through the skin.

COMPOUND FRACTURE   The broken ends of the bone protrude through the skin.

COMMINUTED FRACTURE   Rather than being broken in a single plane, the bone is splintered into many fragments at the site of the break.

DEPRESSED FRACTURE   The broken region is pushed inward, as often occurs in fractures of the flat skull bones that form the roof over the brain.

IMPACTED FRACTURE   The broken ends of the bone are driven into each other. Such fractures occur in falls in which the person lands on the ends of the bone.

Healing of Fractures

Fractures undergo a series of progressive changes during the healing process (Figure 6.7).

FORMATION OF A PROCALLUS   When a bone is broken, bleeding occurs from the vessels of the haver-

sian systems and the periosteum. This bleeding produces a swelling and forms a blood clot that is called a **procallus.**

FORMATION OF A FIBROCARTILAGINOUS CALLUS   Fibroblasts invade the procallus and form fibers. Within a few days, the procallus is replaced with a **fibrocartilaginous callus,** which develops between the broken ends of the bone and eventually forms a bridge that reunites the fragments. The part of the callus that extends beyond the usual surface of the bone is the **external callus.** The part of the callus between the broken ends of the bone, including the medullary cavity, is called the **internal callus.**

FORMATION OF A BONY CALLUS   Initially the outer portion of the external callus consists of cartilage that is formed by chondroblasts and chondrocytes that invade the area. Gradually, however, osteoblasts derived from the inner layer of the periosteum and the endosteum form a **bony callus** that knits the ends of the bone firmly together. At first, the bony callus is spongy bone, but it is slowly remodeled to form compact bone. The formation of compact bone is partially under the influence of stress as the repaired bone is again used for body support and movement.

Because of the importance of stress in strengthening the bone, prolonged immobilization of a broken bone may be detrimental to the healing process. For this same reason, pins are sometimes used to hold the ends of a broken bone together, thus allowing the bone to support weight almost immediately. The stresses of use increase the activity of osteoblasts and facilitate healing.

During the months that follow the formation of a bony callus, osteoclasts reabsorb the bone of the external callus and the part of the internal callus that blocks the medullary cavity so that eventually a slight external en-

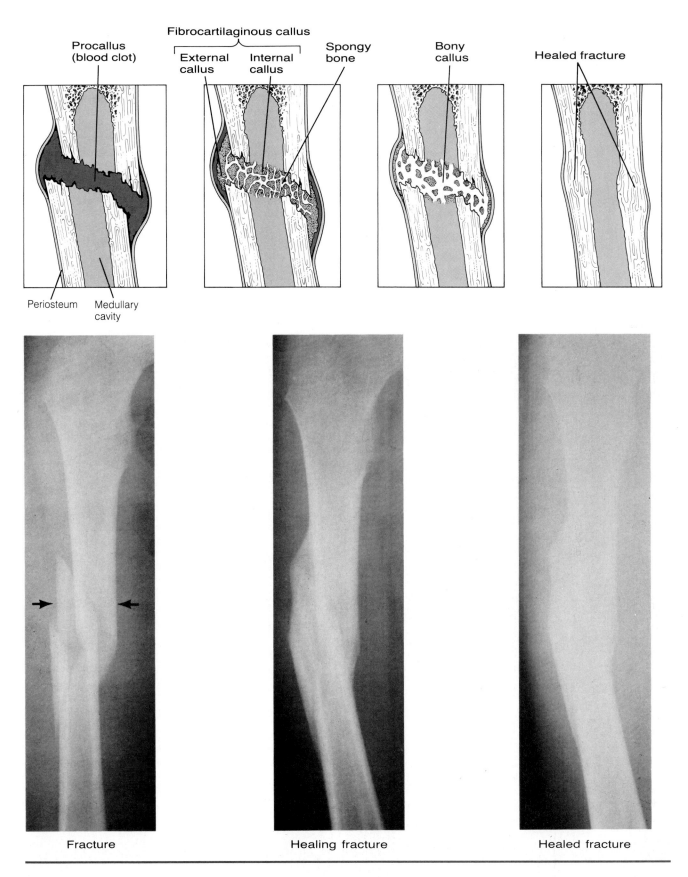

Figure 6.7

Healing of a fracture. The initial repair begins with the formation of a blood clot called a *procallus*. Connective tissue invades the procallus, replacing it with a fibrocartilaginous callus. This fibrous callus is eventually replaced by bone that develops from cells of the periosteum.

largement is all that is left to mark the location of the break.

Metastatic Calcification

The deposition of calcium in tissues that do not usually become calcified is called *metastatic calcification*. It often results if blood calcium levels become too high, as occurs during some decalcifying diseases of bone or when the amount of parathyroid hormone is increased. The kidney is the most common site of metastatic calcification (which results in kidney stones), but it occurs in many other tissues as well.

Spina Bifida

Occasionally, during embryonic development, the posterior portions of the vertebrae of the spinal column fail to form a complete bony arch around the spinal cord. This condition, which is called *spina bifida*, is most common in the lumbar and sacral regions of the spinal column but may occur in a vertebra of any region. If the defect is large enough, the coverings of the spinal cord and the wall of the spinal cord itself may protrude.

Osteoporosis

Osteoporosis, or "porous bone," is a common condition in older people. It also occurs within the bones of paralyzed or immobilized limbs. Osteoporosis is believed to result from a gradual reduction in the rate of bone formation while the rate of bone absorption remains normal. This causes the bone to become porous, fragile, and relatively easily broken.

Osteomyelitis

In *osteomyelitis*, the periosteum, the contents of the marrow cavity, and the bone tissue become infected. Since the causative agent is usually *Staphylococcus aureus*, a bacterium that enters the body through a boil or some other break in the skin, osteomyelitis may follow trauma. The initial damage to the bone has the appearance of an abscess. The abscess spreads throughout the bone, converting the fatty tissue of the marrow cavity into pus and destroying the bony tissue. The infection can pass from the shaft of the bone, where it generally appears first, to the epiphyses, where it perforates the articular cartilage and enters the joint cavity.

Before the discovery of antibiotics, osteomyelitis had a high mortality rate. Because the disease responds well to antibiotic therapy, death from it is now rare.

Tuberculosis of Bone

Tuberculosis of bone is a type of osteomyelitis that is caused by another bacterium, *Mycobacterium tuberculosis*. The bacterium is generally carried by the bloodstream from an infection in the lung or lymph nodes. It is characterized by excessive bone destruction.

Rickets and Osteomalacia

Both *rickets* and *osteomalacia* result from the demineralization and subsequent softening of bone. Rickets occurs in children, while osteomalacia refers to the softening of adult bones. These conditions are due to deficiencies of calcium, phosphorus, or vitamin D, or to a lack of sunlight. Metabolically transformed vitamin D_3 facilitates the absorption of calcium and phosphate from the intestine into the bloodstream, making them available for bone formation. Ultraviolet rays in sunlight provide the body with vitamin D_3 by converting sterols in the skin into vitamin D_3. In children, the softened bones develop bends, such as bowed legs (femur) and pigeon breast (sternum).

Tumors of Bone

Many types of tumors, both benign and malignant, have been identified in bone. Some malignant bone tumors have traveled via the bloodstream from the lungs, breast, prostate, and other structures. Tumors weaken the bone by destroying the tissue. Their presence may be detected for the first time when a patient is x-rayed following a fracture.

Abnormal Growth Patterns

The amount of growth hormone secreted by the anterior pituitary gland can have a dramatic effect on bone development. The presence of excessive pituitary growth hormone delays the ossification of the epiphyseal cartilage. As a consequence, bone development continues for a longer period than normal, producing a *pituitary giant*. Conversely, a deficiency of growth hormone, below the level needed to maintain active epiphyseal cartilages, results in the early replacement of those cartilages by bone. This closure of the epiphyseal plates causes a halt in the development of most bones early in life, producing a midget referred to as a *pituitary dwarf*.

If excessive pituitary growth hormone is secreted after the epiphyseal cartilages have been replaced by bone, an abnormal pattern of bone growth occurs, particularly in the hands, face, and feet. This condition is called *acromegaly*.

Sometimes, under the influence of hereditary factors as well as hormone levels, the epiphyseal cartilages of the bones of the limbs function for only a short time, while the bones of the rest of the body continue developing fairly normally. This condition, called *achondroplasia*, results in a dwarf with short arms and legs but normal trunk and head. Dachshunds are achondroplastic dogs that have been selectively bred.

Effects of Aging

The loss of calcium from the bones, which is the major effect of aging on the skeletal system, is more severe in women than in men. In women, the amount of calcium in the bones steadily decreases after the age of 40, so that by the age of 70 as much as 28% of the calcium in the skeletal system has been lost. Men basically have higher bone calcium levels to start with, and generally do not begin to lose calcium until after the age of 60. The loss of calcium can cause osteoporosis. The cause of the calcium loss is not known, and there are no certain methods of preventing the loss.

Normally the matrix of bone is constantly being broken down and replaced by new matrix. In elderly people, however, along with the calcium loss, the rate of protein formation may be so slow that the organic portion of the matrix is not replaced as rapidly as it is broken down. As a consequence, the matrix of bone gradually comes to contain a greater proportion of inorganic salts. This may cause the bones of elderly people to become brittle and fracture rather easily.

FRONTIERS IN HEALTH

Prescription for Healthy Bones

As Helen, age 62, gets out of bed and steps onto the floor, she fractures a vertebra. However, she is unaware of the fracture, which is, in fact, her seventh in 2 months. All she notices is a nagging pain in her back—a complaint she ascribes to old age. In a matter of months Helen's spinal column will begin to shrink, and she may develop an unsightly hunchback. Like 17 million other Americans, Helen is suffering from osteoporosis, a progressive loss of bone calcium that leads to increased bone brittleness and a marked increase in bone fractures.

One of every two women in the United States will suffer from postmenopausal osteoporosis, but most will not be aware of their problem until they fracture a hip or vertebra. Each year more than 55,000 Americans—mostly women—die from complications of this disease. Hemorrhage, fat embolisms, and shock are three of the most common causes of death.

New advances in prevention and treatment have begun to offer a ray of hope for millions of potential victims. It has been learned that osteoporosis begins to develop much earlier than previously thought. The disease often begins by the time a woman reaches her mid-20s. By age 30, many women have lost one-third of their bone calcium. The gradual demineralization continues until the bones become so brittle that normal activities, such as getting out of bed or dancing, cause tiny fractures throughout the bones of the body.

Why does calcium loss begin so early? The reasons appear to be mostly dietary. Many women become weight conscious in their mid-20s and avoid fatty foods, such as milk and cheese, that happen to be rich in calcium. Many adults also develop an intolerance to milk and other dairy products. Therefore, even though their bodies require over 1000 milligrams of calcium every day, many women consume as little as 450 milligrams.

It is possible to diet and still not suffer from osteoporosis, however, if dietary supplements are taken or if the intake of calcium is increased. Vitamin D, which increases the intestinal absorption of calcium, can also help prevent osteoporosis. Scientific studies also suggest that weight-bearing exercise, such as aerobics, running, walking, and tennis, can help prevent the disease. Instead of taking the elevator, walk up the stairs in order to keep bones from growing soft. Instead of spending the weekend in front of the television, take a brisk walk, or jog in the park, or play a game of tennis.

But what if the disease has already struck? Is there any way to arrest or reverse it? Studies show that osteoporosis can be halted and reversed, even after it has reached an advanced stage. For example, recent work by Dr. Everitt L. Smith and his colleagues at the University of Wisconsin in Madison demonstrated that 45 minutes of moderate exercise three times a week greatly slows the loss of calcium in older women and, after a year, reverses the demineralization. In this study, an inactive control group of older subjects lost 7.5% of their bone calcium over a 3-year period. A second group, which exercised three times a week, lost 3.8% of their bone calcium the first year, but *gained* calcium the second and third years, giving them a 3-year loss of only 1%. Thus,

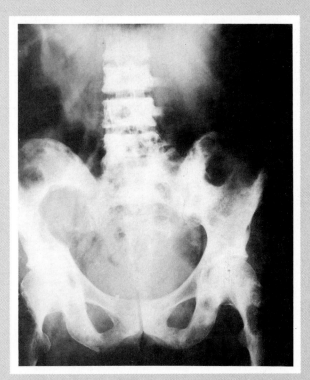

X ray of a person with osteoporosis. The dark mottled areas are regions from which calcium has been withdrawn from the bones.

over the duration of the experiment, the gains nearly offset the first-year losses. These results suggest that continued exercise may yield a long-term increase in bone calcium and, with it, a decrease in bone fractures.

One of the more conventional treatments for osteoporosis in postmenopausal women is the administration of the sex steroid estrogen. Produced by the ovaries until menopause, estrogen promotes bone growth. Fairly low doses of estrogen can stop bone demineralization altogether and may actually promote bone formation. However, women who are given estrogen are at an increased risk of developing endometrial cancer. Physicians therefore often prescribe a mixed dose of estrogen and progesterone. The progesterone lessens the likelihood of cancer.

A better understanding of the prevention and treatment of osteoporosis and the dissemination of this knowledge to the general public will help reduce much of the suffering caused by this disease. For Helen and millions of women in her age group, the goal is to stop the disease before it progresses too far. But for millions of younger women, early detection and sound preventative measures, such as exercise and vitamin D and calcium dietary supplements, can prevent the disease altogether.

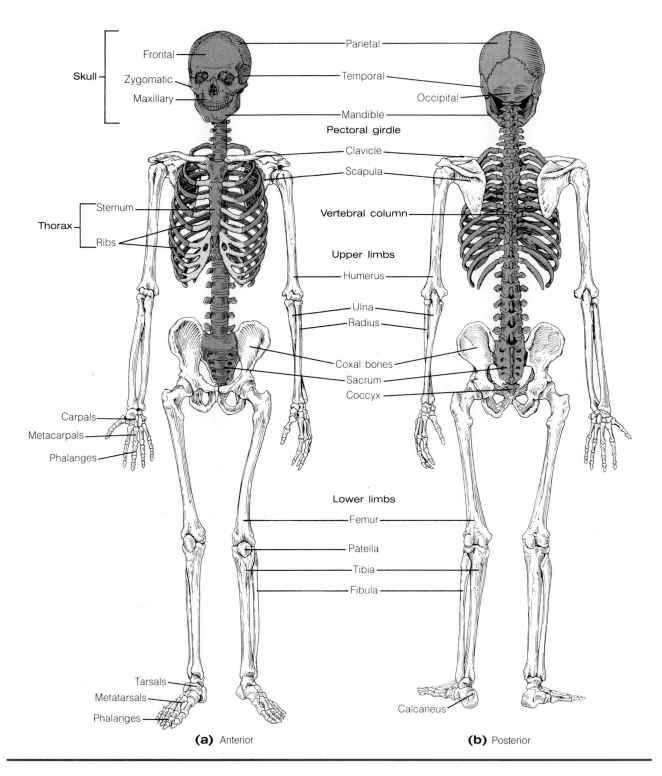

Skull
- Frontal
- Zygomatic
- Maxillary
- Parietal
- Temporal
- Occipital
- Mandible

Pectoral girdle
- Clavicle
- Scapula

Thorax
- Sternum
- Ribs

Vertebral column

Upper limbs
- Humerus
- Ulna
- Radius

Coxal bones
Sacrum
Coccyx

- Carpals
- Metacarpals
- Phalanges

Lower limbs
- Femur
- Patella
- Tibia
- Fibula

- Tarsals
- Metatarsals
- Phalanges

Calcaneus

(a) Anterior   **(b)** Posterior

Figure 6.8
The human skeleton, anterior and posterior views. Axial skeleton in color. Appendicular skeleton in white.

Table 6.1
F6.8

INDIVIDUAL BONES OF THE SKELETON

The human skeleton consists of 206 bones. The bones can be grouped into the axial skeleton and the appendicular skeleton (Table 6.1). The two groups are shown in different colors in Figure 6.8. Your study of the skeletal system will be easier if you first familiarize yourself with some of the more common terms

Table 6.1   Divisions of the Skeleton

Category		Number of Bones
Axial skeleton		80
Skull	29	
Vertebral column	26	
Thorax (ribs and sternum)	25	
Appendicular skeleton		126
Pectoral girdle	4	
Upper limbs	60	
Pelvic girdle	2	
Lower limbs	60	
Total		**206**

Table 6.2   Common Skeletal Structure Terms

Crest   a sharp, prominent bony ridge.

Condyle   a rounded prominence that articulates with another bone.

Epicondyle   a small projection located on or above a condyle.

Facet   a smooth, nearly flat articular surface.

Fissure   a narrow cleftlike passage.

Foramen   a hole.

Fossa   a depression; often used as an articular surface.

Fovea   a pit; generally used for attachment rather than for articulation.

Head   generally, the larger end of a long bone; often set off from the shaft of the bone by a constricted neck.

Line   a slight bony ridge.

Meatus   a canal.

Process   a prominence or projection.

Ramus   a projecting part or elongated process.

Spine   a slender pointed projection.

Sulcus   a groove.

Trochanter   a large, somewhat blunt process.

Tubercle   a nodule or small, rounded process.

Tuberosity   a broad process, larger than a tubercle.

used to describe the structural features of bones. These terms are listed in Table 6.2. Throughout the chapter, each bone is discussed in a general way, and the reader is usually provided with both a figure that illustrates the bone and a table that gives more specific information about it.

Table 6.2

AXIAL SKELETON

The axial skeleton consists of the bones that form the skull, the vertebral column, and the thorax. This portion of the skeleton provides the main axial support for the body and protects the central nervous system and the organs of the thorax.

Skull

The skull is formed of 29 bones, 11 of which are paired. With the exception of the mandible (lower jaw) and three small bones (ossicles) within each middle-

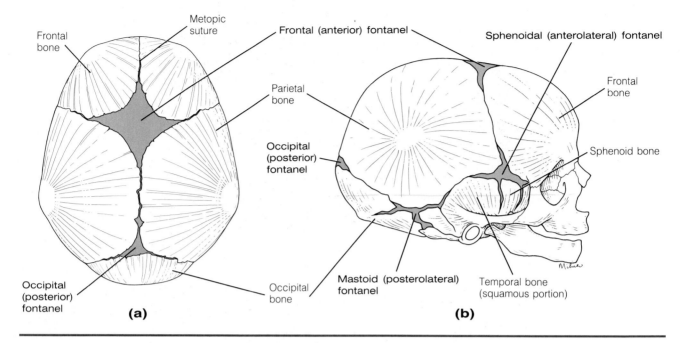

Figure 6.9
Fetal skull, showing the
fontanels. **(a)** Superior
view. **(b)** Lateral view.

ear cavity, all of the bones of the adult skull are joined together in immovable joints called **sutures.** At birth and for some years afterwards, most of these sutures are held together by fibrous connective tissue rather than by bone and therefore are capable of undergoing some movement. This permits the skull to narrow somewhat during birth by allowing the **calvarium** (roofing bones) to override one another when subjected to pressures within the birth canal. The presence of connective tissue sutures also allows for further growth of the skull to accommodate the normal development of the brain. At some points of junction of two or more sutures there are fibrous membrane areas that remain prominent for up to 18 months after birth. These "soft spots" of the skull are called **fontanels** (Figure 6.9).

fon-tan-els'
F6.9

Calvarium

The calvarium is formed of the *frontal, parietal,* and *occipital* bones.

F6.10, F6.11

FRONTAL BONE   The **frontal** is a single bone that forms the anterior superior region of the skull (Figures 6.10 and 6.11). This bone begins its development as two separate bones that meet in a midline **metopic suture.** This suture is generally not distinguishable in the adult. The frontal bone forms the forehead and the roof of the orbital cavities. Inside of the bone, just above its junction with the nasal bones, are the **frontal sinuses.** These, like the other sinuses of the skull, are air spaces that are lined with mucous membrane. Posteriorly, the frontal bone joins with the two parietal bones, forming the **coronal suture.** Table 6.3 lists the features associated with the frontal bone.

Table 6.3

F6.10, F6.11

PARIETAL BONES   Posterior to the frontal bone are the two **parietal bones,** which form most of the calvarium (Figures 6.10 and 6.11). The parietals meet in the midline, forming the **sagittal suture.** The parietal bones form the **coronal suture** across the top of the skull, where they meet with the frontal bone; the **lambdoidal suture** at the back of the skull, where they meet with the occipital bones; and the **squamosal sutures** along the lower sides of the skull, where they meet with the temporal bones. In a young child's skull, the fibrous fontanels occupy the point of junction of these sutures (see Figure 6.9). The **frontal (anterior) fontanel** is located at the junction of the sagittal and coronal sutures. The **occipital (posterior) fontanel** is located at the junction of

F6.9

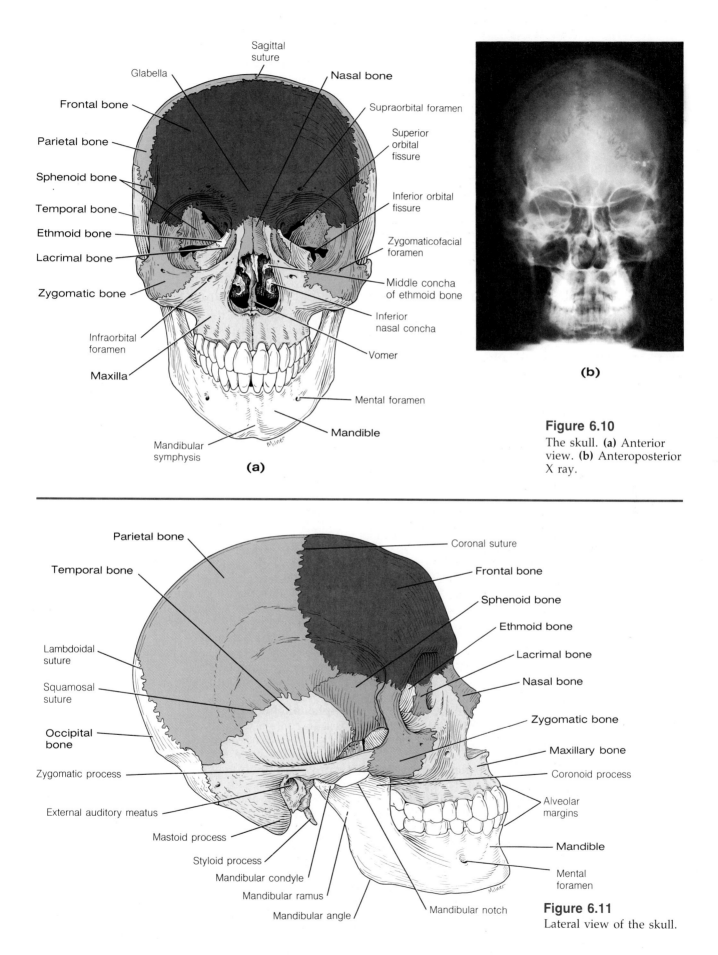

Figure 6.10
The skull. **(a)** Anterior view. **(b)** Anteroposterior X ray.

Figure 6.11
Lateral view of the skull.

the sagittal and lambdoidal sutures. There are right and left **sphenoidal (anterolateral) fontanels** at the junction of the coronal and squamosal sutures, and **mastoid (posterolateral) fontanels** where the squamosal suture meets the lambdoidal suture.

F6.10, F6.11

OCCIPITAL BONE   The single **occipital bone** forms the lower posterior wall of the calvarium, as well as the posterior portion of the floor of the cranial cavity (Figures 6.10 and 6.11). Its most obvious landmark is the large **foramen magnum,** by which the cranial cavity communicates with the vertebral canal. The medulla oblongata of the brain passes through this foramen. Anterior and lateral to the foramen magnum are two **occipital condyles,** one on each side. These condyles articulate with the superior surface of the first cervical vertebra. This is the only connection between the skull and the vertebral column. In addition to articulating with the parietal bones (lambdoidal suture), the occipital bone also meets the temporal and sphenoid bones. Table 6.3 lists the features of the occipital bone.

Table 6.3

Bones That Form the Floor of the Cranial Cavity

F6.12–F6.14

The floor of the cranial cavity, upon which the brain rests, is formed by six bones: the midline *frontal, ethmoid, sphenoid, occipital,* and the paired *temporals* (Figures 6.12, 6.13, and 6.14). We have already discussed the frontal and occipital bones. The frontal bone (together with the ethmoid) forms the **anterior cranial fossa;** the occipital bone forms the **posterior cranial fossa.** We will now consider the ethmoid, the sphenoid, and the temporal bones. The sphenoid and temporal bones form the **middle cranial fossa.**

F6.13–F6.15

ETHMOID BONE   The **ethmoid bone** is situated in the middle of the floor of the anterior cranial fossa, where it forms most of the walls of the upper portion of the nasal cavity (Figures 6.13, 6.14, and 6.15). The ethmoid bone is lightweight and delicate, containing many air sinuses. It has four parts: the **horizontal (cribriform) plate,** with the midline **perpendicular plate,** and two **lateral masses** that project downward from the horizontal plate.

The horizontal plate joins with the frontal bone to form the floor of the anterior cranial fossa. This plate is called the cribriform plate (*cribriform* = sievelike) because it is perforated by many tiny **olfactory foramina.** The olfactory nerves pass through these foramina in traveling between the mucous membranes of the nasal cavity and the olfactory bulbs of the brain. The perpendicular plate forms the major part of the nasal septum, which divides the nose into right and left nasal cavities. On the medial surfaces of the lateral masses there are projections called **superior** and **middle conchae (turbinates),** which form the sidewalls of the nasal cavities; the smooth lateral surfaces of the lateral masses are referred to as the **lamina orbitalis,** because they form part of the medial walls of the orbital cavities. Table 6.3 lists additional information concerning the ethmoid bone.

Table 6.3

F6.12, F6.13, F6.16

SPHENOID BONE   The **sphenoid bone** extends completely across the floor of the middle cranial fossa (Figures 6.12, 6.13, and 6.16). The sphenoid is surrounded on all sides by other bones, articulating posteriorly with the basioccipital portion of the occipital bone, laterally with the temporal and parietal bones, and anteriorly with the frontal and ethmoid bones.

ter'-a-goid

The sphenoid has a complex shape, with a central **body,** from which pairs of **small (lesser) wings, great wings,** and **pterygoid processes** project. The anterior surfaces of the great wings form most of the posterior walls of the orbital cavities. The **optic foramina,** located in the bases of the small wings, provide for the passage of the optic nerves from the eyes to the base of the brain. The superior surface of the body of the sphenoid contains a deep depression called the **sella turcica** (Turk's saddle). The sella turcica houses the pituitary gland. The sella turcica is bounded posteriorly by a ridge of bone called the **dorsum sellae.** Table 6.3 lists additional information concerning the sphenoid bone.

Table 6.3

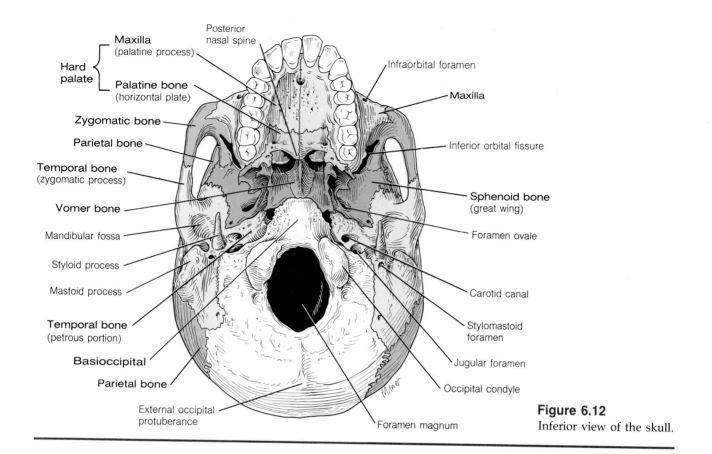

Maxilla (palatine process)
Hard palate
Palatine bone (horizontal plate)
Zygomatic bone
Parietal bone
Temporal bone (zygomatic process)
Vomer bone
Mandibular fossa
Styloid process
Mastoid process
Temporal bone (petrous portion)
Basioccipital
Parietal bone
External occipital protuberance
Posterior nasal spine
Infraorbital foramen
Maxilla
Inferior orbital fissure
Sphenoid bone (great wing)
Foramen ovale
Carotid canal
Stylomastoid foramen
Jugular foramen
Occipital condyle
Foramen magnum

Figure 6.12
Inferior view of the skull.

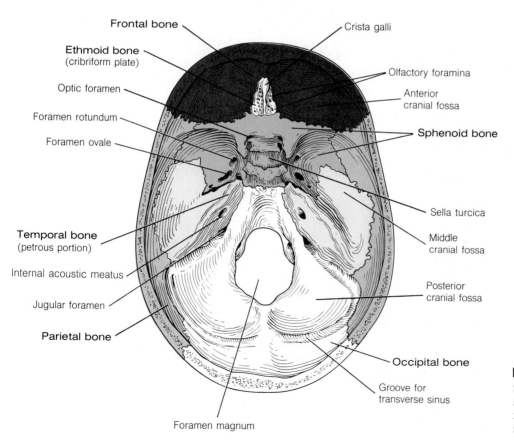

Frontal bone
Ethmoid bone (cribriform plate)
Optic foramen
Foramen rotundum
Foramen ovale
Temporal bone (petrous portion)
Internal acoustic meatus
Jugular foramen
Parietal bone
Foramen magnum
Crista galli
Olfactory foramina
Anterior cranial fossa
Sphenoid bone
Sella turcica
Middle cranial fossa
Posterior cranial fossa
Occipital bone
Groove for transverse sinus

Figure 6.13
Superior view of the skull with the calvarium removed, showing the floor of the cranial cavity.

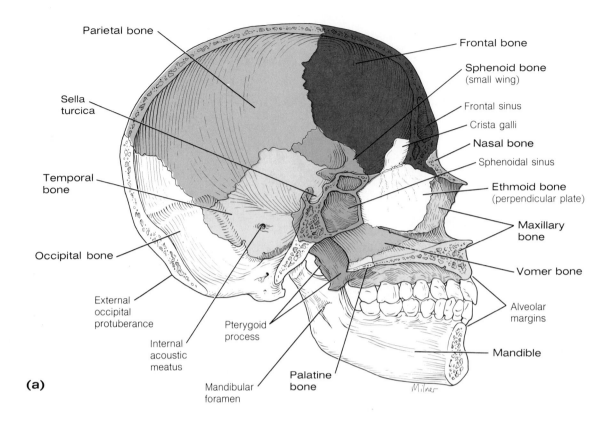

(a)

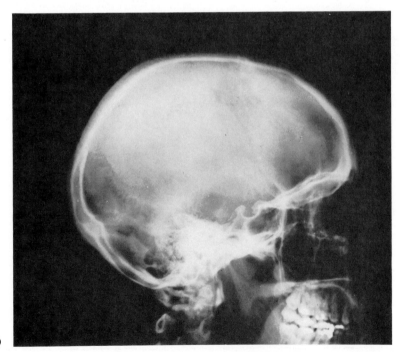

(b)

Figure 6.14

The skull. **(a)** Sagittal
view. **(b)** Lateral X ray.

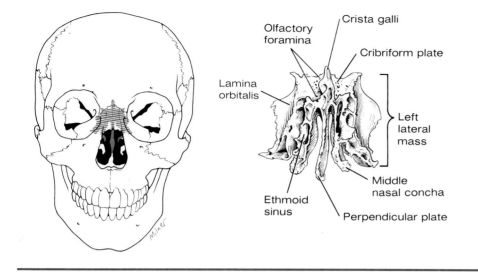

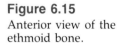

Figure 6.15
Anterior view of the ethmoid bone.

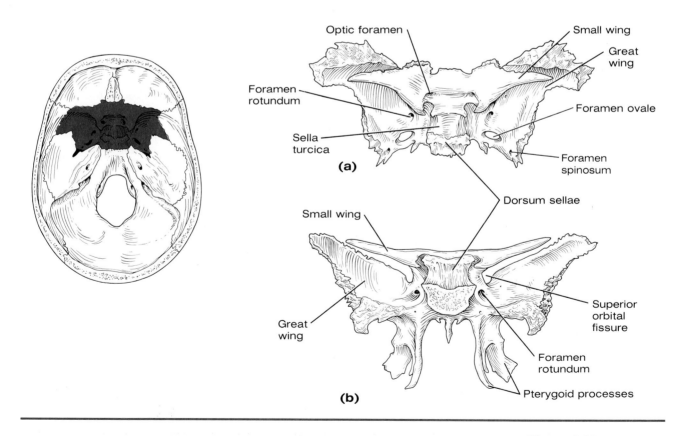

Figure 6.16
Sphenoid bone.
(a) Superior view.
(b) Posterior view.

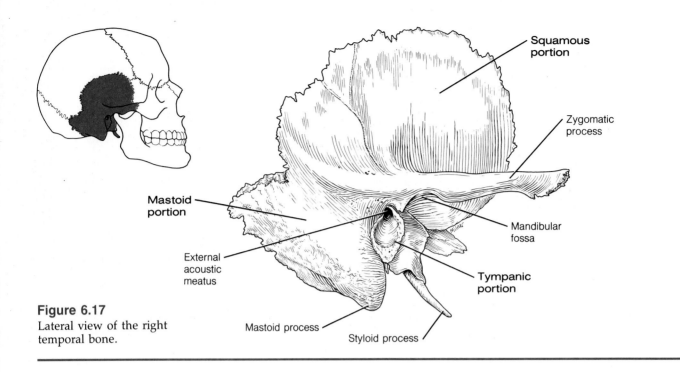

Squamous portion

Zygomatic process

Mastoid portion

Mandibular fossa

External acoustic meatus

Tympanic portion

Mastoid process

Styloid process

Figure 6.17
Lateral view of the right temporal bone.

TEMPORAL BONES The two **temporal bones,** together with the sphenoid bone, form the *middle cranial fossa* (Figures 6.11, 6.12, 6.13, and 6.14). Each bone consists of four regions (Figure 6.17):

F6.11–F6.14
F6.17

1. The thin **squamous portion** projects upward, articulating with the parietal bone in the squamous suture.

2. The **tympanic portion** forms the walls of the external auditory meatus and the region of the bone that closely surrounds the meatus.

F6.12, F6.13
3. The **petrous portion** (Figures 6.12 and 6.13) projects medially between the sphenoid and occipital bones. It contains the middle- and inner-ear cavities.

.4. The **mastoid portion** is located posterior to the external auditory meatus.

Table 6.3
Table 6.3 lists additional information concerning the temporal bones.

Facial Skeleton

F6.10
Ten bones form most of the facial skeleton (Figure 6.10): the unpaired *frontal* and *mandible* bones and the paired *maxillary, zygomatic, nasal,* and *lacrimal* bones. We have previously described the frontal bone.

MAXILLARY BONES The **maxillary bones (maxillae)** form the central part of the facial skeleton (Figure 6.18). With the exception of the mandible, all of the facial bones articulate directly with the maxillae. The two maxillary bones join in the midline to form the upper jaw. In addition, each assists in forming the roof of the mouth, the floor and lateral wall of the nasal cavity, and the floor of the orbit. The large maxillary sinuses are within the body of the bone.
F6.18

Table 6.3
Table 6.3 lists additional information concerning the maxillary bones.

ZYGOMATIC BONES The **zygomatic (malar) bones** articulate with the maxillary and temporal bones to form the prominences of the cheek (Figure 6.10; Figure 6.11). They also articulate with the frontal and sphenoid (great wing) to form part of the floor and the lateral wall of the orbit.
F6.10
F6.11

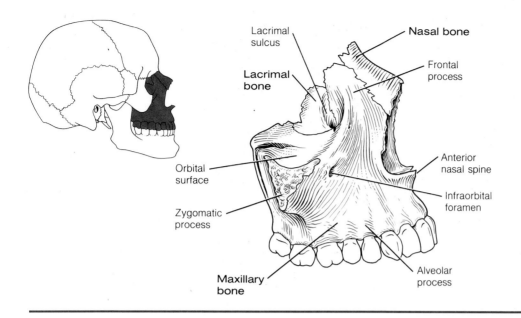

Figure 6.18
Lateral view of the right maxillary, nasal, and lacrimal bones.

NASAL BONES   The **nasal bones** (Figure 6.18) are two small oblong bones that meet at the midline of the face to form the bridge of the nose. In addition, they articulate with the frontal, ethmoid (perpendicular plate), and maxillary bones (frontal process).

F6.18

LACRIMAL BONES   The right and left **lacrimal bones** (Figure 6.18) are small delicate bones that help to form the medial surface of the orbital cavity. They articulate above with the frontal bone, behind with the ethmoid (orbital surfaces of the lateral masses), and in front with the maxillary bones (frontal process). Near their anterior edges, each lacrimal bone has a **lacrimal sulcus** for the lacrimal sac and the nasolacrimal duct. The lacrimal sac and the nasolacrimal duct transport tear fluid from the surface of the eye to the nasal cavity.

F6.18

MANDIBLE   The facial skeleton is completed by the **mandible** (Figure 6.11), which forms the lower jaw. The mandible consists of a horizontal horseshoe-shaped **body** and two perpendicular **rami.** The condyloid processes located on the superior margins of the rami form movable joints with the mandibular fossae of the temporal bones. Table 6.3 lists additional information concerning the mandible.

F6.11

Table 6.3

Bones That Form the Nasal Cavity

We have already described how most of the nasal septum is formed by the perpendicular plate of the ethmoid. Much of the lateral walls of the nasal cavity are formed by the superior and middle conchae of the ethmoid bone (see Table 6.3). Two additional bones also contribute to the formation of the nasal cavity: the *vomer* and the *inferior nasal conchae.*

Table 6.3

The **vomer** (Figure 6.14) is a thin quadrangular bone that forms the posterior inferior portion of the nasal septum. Its superior border articulates with the sphenoid, between the pterygoid processes; its inferior border articulates with the upper surface of the hard palate (maxillae and palatine bones). The upper part of the vomer's anterior border articulates with the perpendicular plate of the ethmoid, and the lower part is in contact with the cartilage portion of the nasal septum.

F6.14

The paired **inferior nasal conchae** (Figure 6.19) form elongated shelves that protrude medially from the lateral walls of the nasal cavity. They are located just below the middle concha of the ethmoid bone. The inferior nasal conchae articulate with the maxillae, lacrimals, ethmoid, and palatine bones.

F6.19

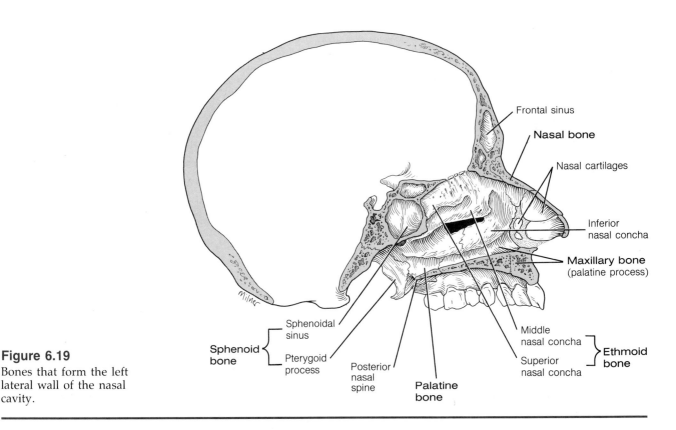

Figure 6.19

Bones that form the left lateral wall of the nasal cavity.

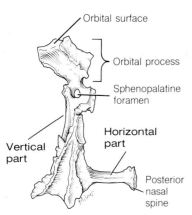

Figure 6.20

Left palatine bone (posterior view).

Bones That Form the Hard Palate

F6.12, F6.19

The hard palate forms the roof of the mouth (Figures 6.12 and 6.19). The anterior portion of the hard palate is formed by the **palatine processes** of the **maxillary bones.** The posterior portion is formed by the **horizontal plates** of the two **palatine bones.**

Each palatine bone is L-shaped, with a horizontal and a vertical portion (Figure 6.20). On the posterior edge of the horizontal plates, at their midline point of junction, is a sharp projection called the **posterior nasal spine.** This spine serves for the attachment of the uvula, a small fleshy mass that hangs from the soft palate. The vertical portion of each palatine bone forms part of the posterior lateral wall of the nasal cavities. A small portion of the vertical plate contributes to the formation of the orbital cavity.

Bones That Form the Orbital Cavity

We have already described the contribution of the various bones to the formation of the orbit (Figure 6.21). These are summarized in Table 6.4.

Paranasal Sinuses

F6.14, F6.19

F6.15

F6.14, F6.19

Located within the *frontal, ethmoid, maxillary,* and *sphenoid* bones are a series of air spaces called the **paranasal sinuses** (Figure 6.22). The paranasal sinuses, which drain into the nasal cavity, are lined with ciliated epithelium on a mucous membrane that is continuous with the mucous membrane of the nasal cavity. The **frontal sinuses** (Figures 6.14 and 6.19) are found above the medial ends of the orbits in the region of the glabella. The **ethmoid sinuses** (Figure 6.15) are a series of small spaces located in the lateral masses of the bone. The **maxillary sinuses,** the largest of the paranasal sinuses, occupy much of the bone from the orbit to the alveolar processes. The **sphenoidal sinuses** (Figures 6.14 and 6.19) are contained within the body of the bone.

Auditory Ossicles

F6.23

Three tiny bones called **auditory ossicles** are located in the *middle-ear (tympanic) cavities* (Figure 6.23). The cavities are inside the petrous portion of each

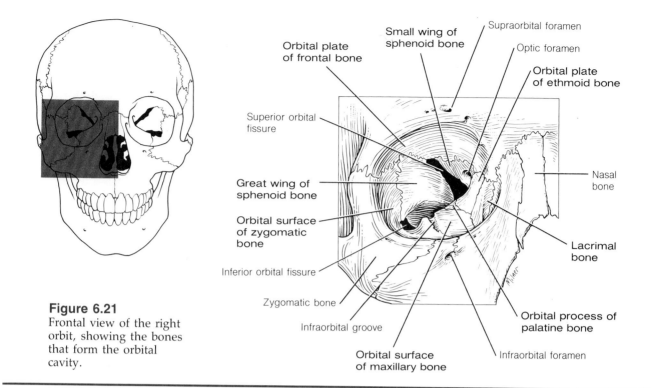

Small wing of
sphenoid bone

Supraorbital foramen

Optic foramen

Orbital plate
of frontal bone

Orbital plate
of ethmoid bone

Superior orbital
fissure

Nasal
bone

Great wing of
sphenoid bone

Orbital surface
of zygomatic
bone

Lacrimal
bone

Inferior orbital fissure

Zygomatic bone

Infraorbital groove

Orbital process of
palatine bone

Orbital surface
of maxillary bone

Infraorbital foramen

Figure 6.21
Frontal view of the right
orbit, showing the bones
that form the orbital
cavity.

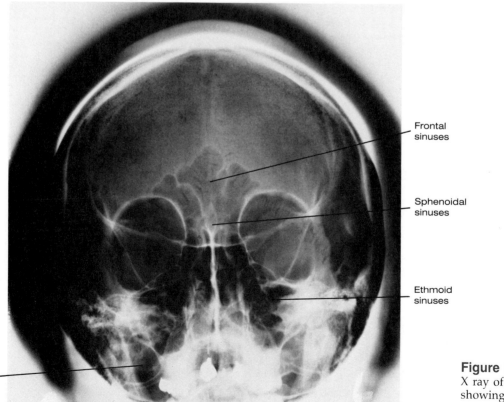

Frontal
sinuses

Sphenoidal
sinuses

Ethmoid
sinuses

Maxillary
sinuses

Figure 6.22
X ray of the skull,
showing several
paranasal sinuses.

temporal bone. The ossicles form a bridge across the tympanic cavity from the eardrum (tympanic membrane) to a membrane (oval window) that separates the middle ear from the inner ear.

The **malleus** (hammer) is attached to the inside surface of the tympanic membrane. The **stapes** (stirrup) fits against the oval window. The **incus** (anvil) forms a connection between the malleus and the stapes. The role played by these bones in transmitting sound in discussed in Chapter 15.

Table 6.3 Summary of Specific Features of Individual Skull Bones

FRONTAL BONES   *F6.10; F6.11; F6.13; F6.14; F6.21*

Metopic suture   The line of junction between the two separate embryonic ossification centers. Generally not present in the adult skull.

Frontal sinuses   *F6.14* Mucous-membrane-lined air cavities located within the bone, close to the orbital cavities.

Supraorbital foramina or notches   *F6.10* Openings for blood vessels and nerves located just above the orbital cavities. They may appear as holes (foramina) or notches.

Glabella   *F6.10* The smooth area located between the two orbital cavities just above the nose.

OCCIPITAL BONE   *F6.11; F6.12; F6.13; F6.14*

Foramen magnum   *F6.12; F6.13* The opening through which the medulla oblongata of the brain stem leaves the skull to become continuous with the spinal cord.

Condyles   *F6.12* Smooth convex external projections on either side of the foramen magnum. They articulate with the first cervical vertebra.

Basioccipital   *F6.12* A narrow portion that extends anteriorly from the foramen magnum. It articulates with the sphenoid bone.

External occipital protuberance   *F6.12; F6.14* A midline prominence on the outer surface, a short distance above the foramen magnum.

Nuchal lines*   Slight ridges on the external surface. *Medial nuchal line*   Runs vertically between the external occipital protuberance and the foramen magnum. *Superior nuchal line*   Extends laterally from the external occipital protuberance. *Inferior nuchal line*   Extends laterally from the medial nuchal line at about its midpoint.

Internal occipital protuberance   A prominence on the inner surface of the bone. This marks the confluence of grooves for the sagittal, transverse, and occipital venous blood sinuses of the brain.

ETHMOID BONE   *F6.13; F6.14; F6.19; F6.21*

Horizontal (cribriform) plate   *F6.13; F6.15* The transverse portion that forms the roof of nasal cavity and floor of the anterior cranial cavity. The plate is perforated by the olfactory foramina to allow for the passage of the olfactory nerves (first cranial nerve).

Crista galli   *F6.13; F6.14; F6.15* A midline projection from the horizontal plate into the cranial cavity. It serves as the anterior point of attachment for the **falx cerebri,** a midline connective tissue septum that anchors the brain within the anterior cranial fossa.

Perpendicular plate   *F6.14; F6.15* A downward projection from the midline of the undersurface of the horizonal plate. It forms the upper portion of the nasal septum. The remainder of the septum is formed by the vomer bone and hyaline cartilage.

Lateral masses   *F6.15* Thin-walled processes that extend downward from the lateral margins of the horizontal plate. They contain the **ethmoid sinuses,** which are mucous membrane–lined air cavities. The smooth lateral surfaces (*lamina orbitalis*) of the lateral masses form the medial walls of the orbital cavities.

Superior and middle conchae (turbinates)   *F6.19* Thin plates of bone that form the medial surfaces of the lateral masses. They also form part of the lateral walls of the nasal cavity. Recesses called **superior, middle,** and **inferior meatuses** are located beneath the shelves of the conchae.

SPHENOID BONE   *F6.10; F6.11; F6.12; F6.13; F6.14; F6.16*

Body   *F6.16* The central portion of the bone. It contains a large mucous-membrane-lined air sinus.

Sella turcica   *F6.13; F6.16* A saddle-shaped depression on the superior surface of the body, bounded posteriorly by the **dorsum sellae.** It serves as the protective cavity for the pituitary gland.

Small wings   *F6.14; F6.16* Sharp lateral projections from the superior portion of the body of the sphenoid. They form part of the posterior walls of the orbital cavities.

Optic foramina   *F6.13; F6.16* Openings through the bases of each small wing for the passage of the optic nerves (second cranial nerves) into the orbital cavities.

Great wings   *F6.12; F6.16* Large lateral projections from the body of the sphenoid. They form most of the posterior wall of the orbital cavity.

Superior orbital fissures *F6.10; F6.16; F6.21* Slitlike openings between the great and small wings. They allow for the passage of the third, fourth, part of the fifth (ophthalmic division), and the sixth cranial nerves from the brain into the orbital cavity.

Foramen rotundum *F6.13; F6.16* The opening through the base of each of the great wings for the passage of the maxillary division of the fifth cranial nerves.

Foramen ovale *F6.12; F6.13; F6.16* The opening through the base of each of the great wings for the passage of the mandibular division of the fifth cranial nerves.

Pterygoid processes *F6.14; F6.19* Two downward projections from the region where the great wings unite with the body. Each process consists of **medial** and **lateral plates.** The processes articulate anteriorly with the palatine bones.

TEMPORAL BONE *F6.10; F6.11; F6.12; F6.13; F6.14; F6.17*

Squamous portion *F6.13; F6.17* The thin vertical projection that forms the anterior and superior portion of the bone. It meets with a parietal bone to form the squamous suture.

Zygomatic process *F6.17* The anterior projection from the squamous portion. It articulates with the zygomatic (malar) bone to form the cheek (zygomatic arch).

Mandibular fossa *F6.17* An oval depression on the inferior surface of the base of the zygomatic process. It articulates with the condyle of the mandible to form the temporomandibular joint.

Tympanic portion *F6.17* Forms and surrounds the external acoustic meatus.

External acoustic meatus *F6.17* The opening that leads into the middle-ear cavity from the exterior of the skull.

Petrous portion *F6.12; F6.23* A medial wedge of bone that forms the floor of the middle cranial fossa between the sphenoid and the occipital bones. It houses the middle- and inner-ear structures.

Internal acoustic meatus *F6.13; F6.14* The opening on the posterior surface of the petrous portion. It transmits the seventh cranial nerve as it travels to the facial structures, and the eighth cranial nerve as it travels to the inner ear.

Styloid process *F6.11; F6.12; F6.17* A sharp spine that projects from the inferior lateral surface of the petrous portion. It serves as a point of attachment for the hyoid bone and for several ligaments and muscles of the pharynx and tongue.

Carotid canal *F6.12* The passageway for the internal carotid artery as it travels through the petrous portion.

Jugular fossa *F6.12* The depression for the internal jugular vein on the inferior surface of the petrous portion.

Stylomastoid foramen *F6.12* The opening between the styloid process and mastoid process through which the seventh cranial nerve leaves the skull. (The nerve enters through the internal acoustic meatus.)

Jugular foramen *F6.12; F6.13* The large opening that allows for passage of several blood vessels, and the ninth, tenth, and eleventh cranial nerves. It is located at the junction of the petrous portion with the occipital bone.

Mastoid process *F6.11; F6.12; F6.17* A prominent downward projection from the mastoid portion, just posterior to the external acoustic meatus.

Mastoid sinuses* Mucous-membrane-lined air spaces within the mastoid process. These sinuses, which communicate with the middle-ear cavity, are the only cranial sinuses that do not drain into the nasal cavity.

MAXILLARY BONE (MAXILLA) *F6.10; F6.12; F6.14; F6.18*

Maxillary sinus *F6.22* A large mucous-membrane-lined cavity within the bone.

Frontal process *F6.18* The vertical process that forms part of the bridge of the nose. It articulates above with the frontal bone, anteriorly with the nasal, and posteriorly with the lacrimal.

Zygomatic process *F6.18* A rough triangular eminence that articulates with the zygomatic (malar) bone.

Alveolar process *F6.11; F6.18* The inferior border that holds the teeth. When the two maxillae are articulated with each other, their alveolar processes together form the alveolar arch.

Palatine process *F6.12; F6.19* The medial horizontal shelf that runs from the inner surface of the alveolar process. It joins with the palatine process of the other maxillary bone to form most of the hard palate.

Anterior nasal spine *F6.18* A pointed process just below the nasal cavity. It joins with the nasal spine of the other maxillary bone to form a point of attachment for the cartilage portion of the nasal septum.

Infraorbital foramen *F6.10; F6.18; F6.21* The opening just below the margin of the orbit. It transmits blood vessels and nerves.

Orbital surface *F6.21* The smooth, flat surface that forms the floor of the orbit.

Table 6.3   Summary of Specific Features of Individual Skull Bones (continued)

MANDIBLE   *F6.10; F6.11; F6.14*

Body   *F6.11* The curved, horizontal portion that forms the chin.

Rami   *F6.11* Two perpendicular projections that join the posterior lateral margins of the body at approximately right angles.

Mandibular symphysis   *F6.10* The vertical midline fusion between the two embryonic ossification centers that form the body.

Alveolar border   *F6.11* The superior edge of the body that contains the sockets for the teeth.

Mental foramina   *F6.10; F6.11* Two openings on the external surface of the body that allow for the passage of blood vessels and nerves.

Angle   *F6.11* A sharp curve on the posterior inferior portion of the ramus.

Mandibular foramina   *F6.14* Openings on the inner surfaces of each ramus for the passage of blood vessels and nerves.

Coronoid processes   *F6.14* Thin upward projection on the anterior surface of each ramus. They provide attachment for the temporalis muscle.

Mandibular condyles   *F6.11* Smooth convex surface on the superior borders of each ramus. They articulate with the mandibular fossae of the temporal bones.

Mandibular notches   *F6.11* Deep depression between the coronoid process and the condyle of each ramus.

*Not illustrated.

Hyoid Bone

F6.24   The **hyoid bone** (Figure 6.24) is a U-shaped bone suspended by ligaments from the styloid processes of the temporal bones. It is located just above the larynx. The hyoid consists of a central **body** and pairs of **greater** and **lesser cornua (horns).** It serves as points of attachment for muscles of the tongue and throat.

Vertebral Column

The embryonic vertebral column consists of 33 vertebrae. The vertebrae are separated into five different types, depending on the regions of the body in which they are located. The upper 7 are **cervical** vertebrae; these are followed by 12 **thoracic,** 5 **lumbar,** 5 **sacral,** and 4 **coccygeal** vertebrae. In the adult, the sacral vertebrae fuse into a single **sacrum,** and the coccygeal vertebrae fuse to form a **coccyx.** Therefore, the adult vertebral column has 26 separate bones (Figure 6.25).

kok-sij'-ee-al

F6.25

Curvatures of the Vertebral Column

When viewed from the lateral aspect, the vertebral column of a newborn infant has a single curve, which is convex posteriorly. As the child begins to raise its head, a *cervical curve* develops that is convex anteriorly. In a similar manner, a secondary *lumbar curve* develops as the child begins to walk. If the anterior lumbar curve is excessive, it is called a *lordosis* (swayback). If the posterior *thoracic curve,* which remains from the primary curve of the newborn, is excessive, it is called a *kyphosis* (hunchback). The vertebral column is normally straight, without any lateral curvatures. If a lateral curve does exist, it is called a *scoliosis.*

Functions of the Vertebral Column

The vertebral column is the main axial support for the body, providing attachments for the skull, the thorax, and the pelvic girdle. Although it is a major support structure, its construction is such that it permits the trunk of the body to have appreciable flexibility. In addition, the vertebral column protects the spinal cord while providing openings between adjacent vertebrae for the passage of spinal nerves.

Table 6.4   Bones that Form the Orbit

Roof of orbit
 Frontal
 Sphenoid (small wing)

Medial wall of orbit
 Maxilla (frontal process)
 Lacrimal
 Ethmoid

Lateral wall of orbit
 Zygomatic
 Sphenoid (great wing)

Floor of orbit
 Maxilla
 Palatine

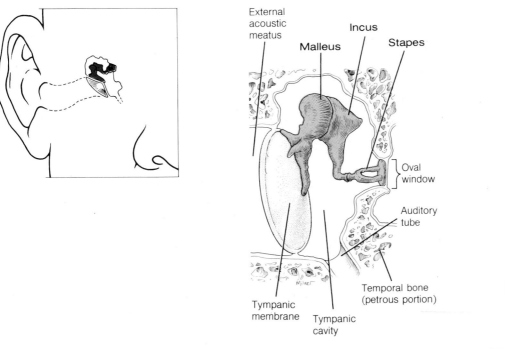

Figure 6.23
The middle-ear cavity
and the auditory ossicles
(color).

Characteristics of a Typical Vertebra

Although there are differences among the vertebrae of the various regions of
the spinal column, enough similarities exist so that it is possible to describe a
typical vertebra (Figure 6.26).

F6.26

    A typical vertebra has a thick anterior **centrum** (body), with a **neural
(vertebral) arch** that arises from the posterior surface of the centrum. The
centra of adjacent vertebrae are separated by fibrocartilaginous **intervertebral
discs.** Each neural arch combines with the posterior surface of the centrum
and encloses a **vertebral foramen.** The vertebral foramina of adjacent vertebrae are aligned to form a **vertebral canal** through which the spinal cord
passes. **Transverse processes** extend laterally from each neural arch. Projecting from the posterior region of the neural arch is the midline **spinous process.** The spinous and transverse processes allow for the attachment of muscles and ligaments. That portion of the neural arch between the centrum and
the transverse process is the **pedicle.** The portion between the transverse
process and the spinous process is the **lamina.** Projecting upward from each
side of the neural arch is a pair of **superior articulating processes;** their articular surfaces face posteriorly. Projecting downward is a pair of **inferior articulating** processes; their articular surfaces face anteriorly. The smooth articular
surface of each process meets with the process of the vertebra above or below
it, thereby increasing the rigidity of the vertebral column. The **intervertebral
foramina,** through which the spinal nerves pass, are located between the
pedicles of adjacent vertebrae (Figure 6.25).

Regional Differences in Vertebrae

The vertebrae of each region have specific characteristics that vary from the
"typical" vertebra and that enable them to be easily identified. These variations are illustrated in Figures 6.27 through 6.31 and are discussed in Table
6.5.

Thorax

The skeleton of the thorax (Figure 6.32) is formed by the *sternum,* the *ribs,* and
the *costal cartilages.* The thoracic vertebrae form its posteriormost portion.

F6.32

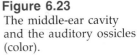

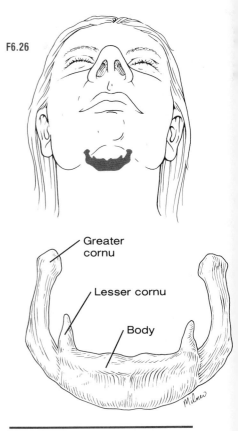

Figure 6.24
Anterior view of the
hyoid bone.

FRONTIERS IN HEALTH

Treatment of Scoliosis

Jennifer was a normal child. She ate well and played actively; but at age 10 something troublesome began to occur. From a point midway up her back, Jennifer's spinal column began to curve unnaturally to the right. Her lower spine twisted to the left.

Fortunately for Jennifer, she lived in one of many school districts that have begun to check its students for this disease, which is called scoliosis. Scoliosis, a lateral curvature of the spine, occurs in about 10% of all children. Every year more than 220,000 individuals are diagnosed as having the disease, which appears most frequently in children between the ages of 10 and 15, when the bones are growing rapidly. Mild forms afflict boys and girls in equal numbers, but the more severe cases are more prevalent in girls.

Scoliosis can be crippling if left untreated, bending the spine into an S-shaped or C-shaped form, and its cosmetic effects are demoralizing. Making matters even worse, as the spine curves it also twists, which throws the rib cage out of normal position. The ribs are pushed together on the concave side and separated on the convex side. In severe cases, the distortion of the rib cage can impair breathing.

Treatment for this disease, whose cause is unknown, varies with the severity of the curvature. In mild cases, a body brace is worn 22 hours a day throughout adolescence to hold the spine straight. This can mean wearing a heavy, unsightly brace for 4 or 5 years. The discomfort and appearance of these braces spurred medical researchers to seek a more comfortable alternative.

Orthopedic surgeon Dr. Walter Bobechko of Toronto, Canada, devised a small electrical muscle stimulator that is used by victims of scoliosis only at night. Electrodes are placed on the skin of the child's back and attached to a power pack. Weak electrical impulses travel to the electrodes and cause the underlying muscles to contract several times a minute throughout the night, without interfering with the child's sleep. In the morning, the child can go to school free of any cumbersome brace. Stimulation of the muscles on the convex side of the curvature straightens the back over time.

Transcutaneous muscle stimulation has been very successful. In one clinical study, Dr. Jens Axelgaard and Dr. John Brown of Rancho Los Amigos Hospital in Downey, California, found that the electrical muscle stimulator stopped the curvature in 95 of every 100 patients if used every night as prescribed. Even more important, the device helped return the spine to its correct position, and 2 years after treatment was stopped the spine remained straight. The muscle stimulator halted the disfiguring bending and twisting of the spine regardless of whether the curvature was slight or large, or whether the rate of bending was fast or slow. Furthermore, no side effects were noted. Muscle on the stimulated side appeared normal. The heart seemed totally unaffected, as did the lungs and skin. The muscle stimulator allowed Jennifer to continue normal activities. Only a handful of friends even knew she had the condition.

In severe cases of scoliosis, corrective surgery may be necessary. Surgeons straighten the back and then

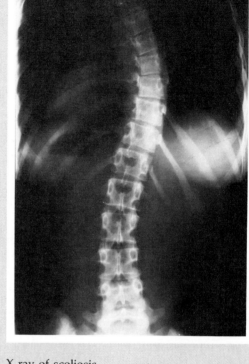

X ray of scoliosis.

place one or two metal rods along the laminae of the vertebrae. The rods are hooked to the laminae and immobilize the straightened spine. To ensure lasting rigidity, surgeons also fuse the vertebrae together by using bone fragments, usually taken from the patient's ilium. After surgery, the patients are encased in a thick cast weighing about 20 pounds and extending from the neck to the thighs. After 6 months, a lighter brace is used to provide support for the next 3 months.

A new technique, devised by a Mexican physician, Dr. Eduardo Luque, promises to reduce the size of the cast. Instead of hooking the metal rods in place, Dr. Luque secures them with wire wrapped tightly around the laminae. This holds the rods more firmly in place. Patients then get all the support they need from a ⅛-inch-thick cast that weighs only 3 pounds and is hardly noticeable beneath their clothing. Dr. Luque's technique cuts the recovery time by a month or more. Recovery is also more comfortable, and the lightweight brace is removable, unlike the heavier body casts that are worn for 6 months and prohibit bathing during that time.

Early detection is the first line of defense in scoliosis. Physicians report that screening programs throughout the country are already proving effective in cutting down on the severity of the disease. It is hoped that screening and electrical muscle stimulation may someday eliminate the need for costly, painful back surgery sometimes necessary to correct scoliosis.

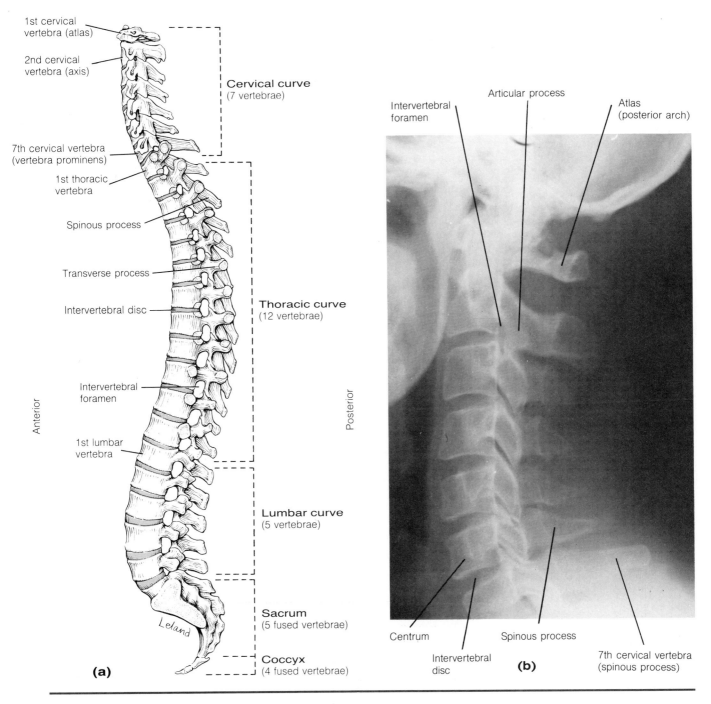

(a)

1st cervical vertebra (atlas)

2nd cervical vertebra (axis)

Cervical curve (7 vertebrae)

7th cervical vertebra (vertebra prominens)

1st thoracic vertebra

Spinous process

Transverse process

Intervertebral disc

Thoracic curve (12 vertebrae)

Intervertebral foramen

1st lumbar vertebra

Lumbar curve (5 vertebrae)

Sacrum (5 fused vertebrae)

Coccyx (4 fused vertebrae)

Anterior

Posterior

Leland

(b)

Intervertebral foramen

Articular process

Atlas (posterior arch)

Centrum

Intervertebral disc

Spinous process

7th cervical vertebra (spinous process)

Figure 6.25
(a) Lateral view of the vertebral column. **(b)** Lateral X ray of the cervical vertebrae.

Sternum

The **sternum** is an elongated flat bone that forms the midline portion of the anterior wall of the thorax (Figure 6.32). It is composed of three parts—the **manubrium,** the **body (gladiolus),** and the **xiphoid process.** The superior portion of the manubrium articulates with the medial end of each clavicle (collar bone). Its lateral margins articulate with the costal cartilages of the first ribs and part of the second ribs. The body of the sternum articulates at its lateral margins with the costal cartilages of the second ribs (which it shares with the manubrium) through the seventh ribs. The small xiphoid process does not articulate with the ribs. It serves as a point of attachment for several ligaments and muscles, including the rectus abdominis muscle. The linea alba, which marks the midline of the abdomen, is also attached to it.

F6.32
zif'-oid

Table 6.5   Identifying Features of Specific Vertebrae

CERVICAL VERTEBRAE   *F6.25; F6.27; F6.28*

Transverse foramina   *F6.27; F6.28* The openings in the transverse processes of each cervical vertebra. They allow for the passage of vertebral arteries and veins to and from the brain.

Bifurcated spinous processes   *F6.27; F6.28* The spinous processes of the cervical vertebrae have a double tip (with the exception of the first and the seventh).

Vertebra prominens   *F6.25* The seventh cervical vertebra, so named because its long, prominent spinous process protrudes beyond those of the other cervical vertebrae, making it useful as a landmark in counting the other spinous processes.

Atlas   *F6.25; F6.28* The first cervical vertebra, which articulates with the occipital condyles of the skull, has no centrum or spinous process. It is ringlike, consisting of anterior and posterior arches.

Axis   *F6.25; F6.28* The second cervical vertebra has a vertical projection called the **odontoid process** or **dens** that arises from the superior surface of its centrum. This process provides a pivot around which the atlas rotates.

THORACIC VERTEBRAE   *F6.25; F6.29*

Spinous processes   *F6.29* Long and slender protuberances that project sharply downward. This is not as noticeable in the lower thoracic vertebrae.

Facets and demifacets   *F6.29* Articular surfaces for the ribs on the transverse processes and the bodies of all thoracic vertebre. (The eleventh and twelfth vertebrae are exceptions because they do not have articular facets on their transverse processes.)

LUMBAR VERTEBRAE   *F6.25; F6.30*

Centra   *F6.30* Larger and heavier than the centra in other regions.

Spinous processes   *F6.30* Short and blunt as compared to the spinous processes in other regions.

Articular processes   *F6.30* The superior articular processes face inward rather than posteriorly; the inferior articular processes face outward rather than anteriorly. This positioning locks the vertebrae together by preventing rotation.

SACRAL VERTEBRAE   *F6.25; F6.31* In the adult, the 5 sacral vertebrae are fused into a single triangular **sacrum.** The transverse lines of fusion are visible on its anterior surface. The spinous processes form the **median sacral crest** on its posterior surface. The fused transverse process form the **alae** (wings), which articulate with the pelvic bones. The **sacral foramina** represent the intervertebral foramina. The superior edge of the ventral border of the first sacral vertebra forms a projection called the **sacral promontory.**

COCCYX   *F6.25; F6.31* The fused coccygeal vertebrae. It articulates with the apex of the sacrum.

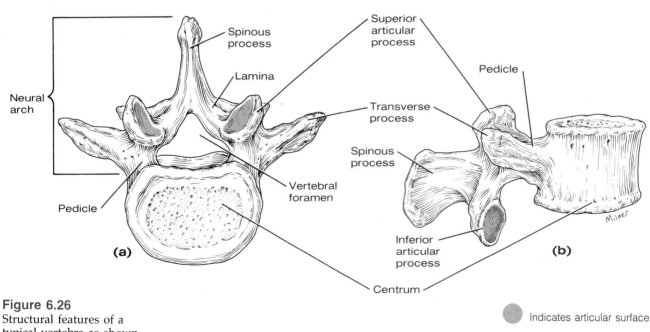

Figure 6.26
Structural features of a typical vertebra as shown in **(a)** superior view and **(b)** lateral view.

Indicates articular surface

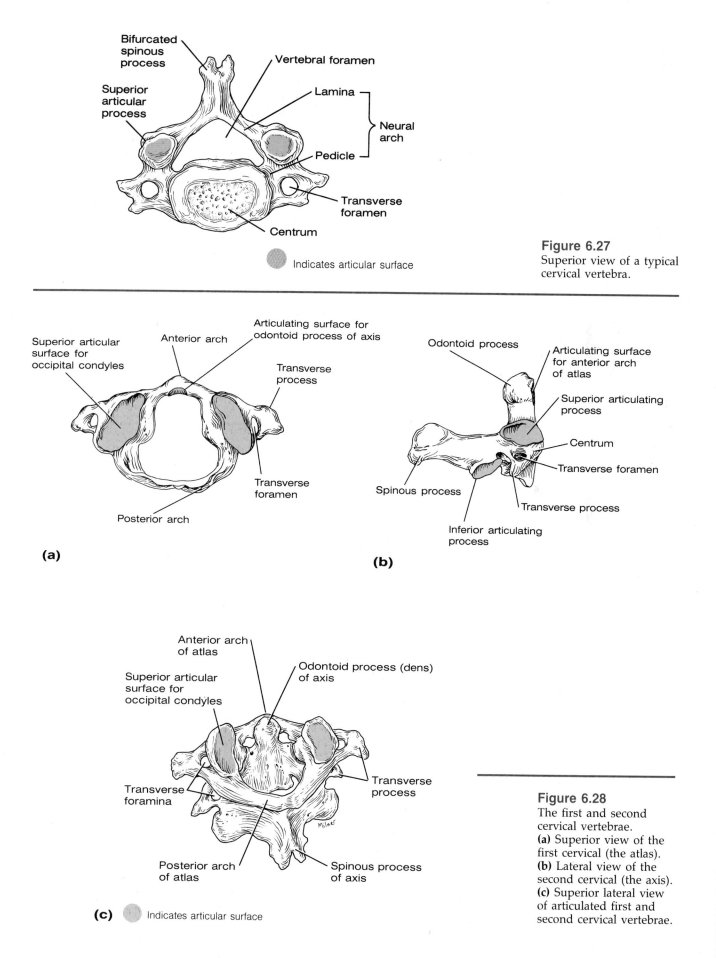

Figure 6.27
Superior view of a typical cervical vertebra.

Indicates articular surface

(a)

(b)

(c) Indicates articular surface

Figure 6.28
The first and second cervical vertebrae.
(a) Superior view of the first cervical (the atlas).
(b) Lateral view of the second cervical (the axis).
(c) Superior lateral view of articulated first and second cervical vertebrae.

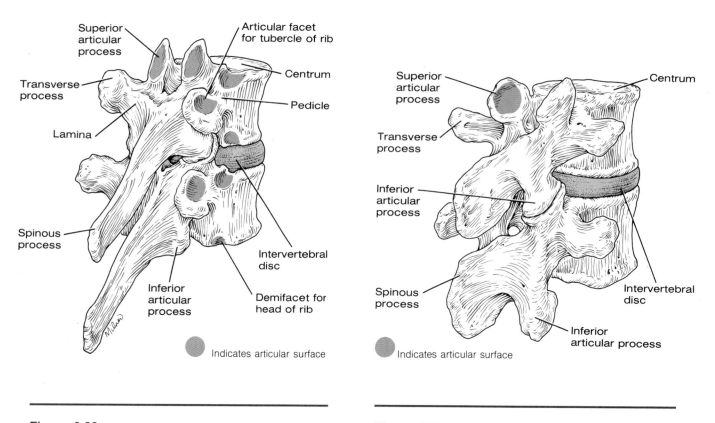

Figure 6.29
Two typical thoracic
vertebrae.

Figure 6.30
Two typical lumbar
vertebrae.

Ribs

F6.32 There are twelve pairs of **ribs** (Figure 6.32). The first seven pairs articulate posteriorly with the thoracic vertebrae and anteriorly with the sternum, through the costal cartilages. These are the **true** or **vertebrosternal ribs.** The remaining five pairs are called **false ribs.**

The first three pairs of false ribs (that is, the eighth, ninth, and tenth ribs) have their costal cartilages attached to the cartilages of the rib above, rather than directly to the sternum. These are called **vertebrochondral ribs.** The costal cartilages of the eleventh and twelfth ribs are short and have no anterior articulation. For this reason these are called **floating** or **vertebral ribs.**

F6.33 The head of a typical rib articulates with the demifacets of two adjacent thoracic vertebra (Figure 6.33). However, the heads of the first, tenth, eleventh, and twelfth ribs each articulate entirely on the facets of one vertebra. A
too'-bur-kul short distance from the head is a **tubercle,** which articulates with the transverse process of a thoracic vertebra. Between the head and the tubercle is a constricted **neck.** Curving anteriorly from the neck is the **shaft** or body of the rib.

Costal Cartilages

The costal cartilages are composed of hyaline cartilage. They strengthen the
F6.32 thorax by serving as the anterior anchors for most of the ribs (Figure 6.32). At the same time, because they are cartilage, they provide flexibility that allows the rib cage to expand during respiration.

Table 6.6 A summary of the skeletal features of the thorax is listed in Table 6.6.
Table 6.7 Table 6.7 summarizes the bones of the axial skeleton.

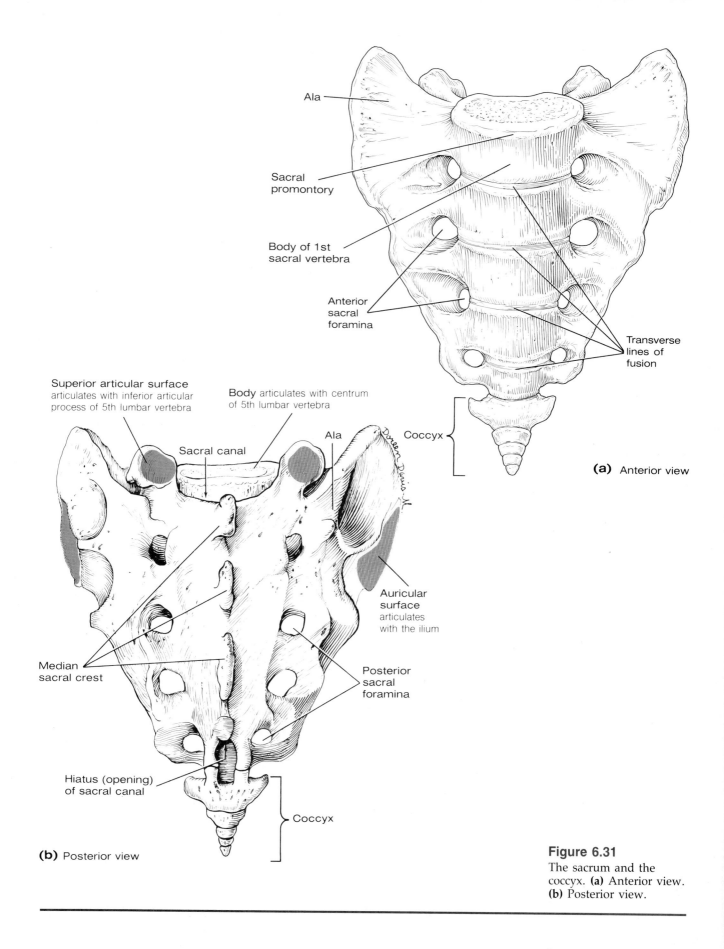

Ala

Sacral promontory

Body of 1st sacral vertebra

Anterior sacral foramina

Transverse lines of fusion

Coccyx

(a) Anterior view

Superior articular surface articulates with inferior articular process of 5th lumbar vertebra

Body articulates with centrum of 5th lumbar vertebra

Sacral canal

Ala

Auricular surface articulates with the ilium

Median sacral crest

Posterior sacral foramina

Hiatus (opening) of sacral canal

Coccyx

(b) Posterior view

Figure 6.31
The sacrum and the coccyx. **(a)** Anterior view. **(b)** Posterior view.

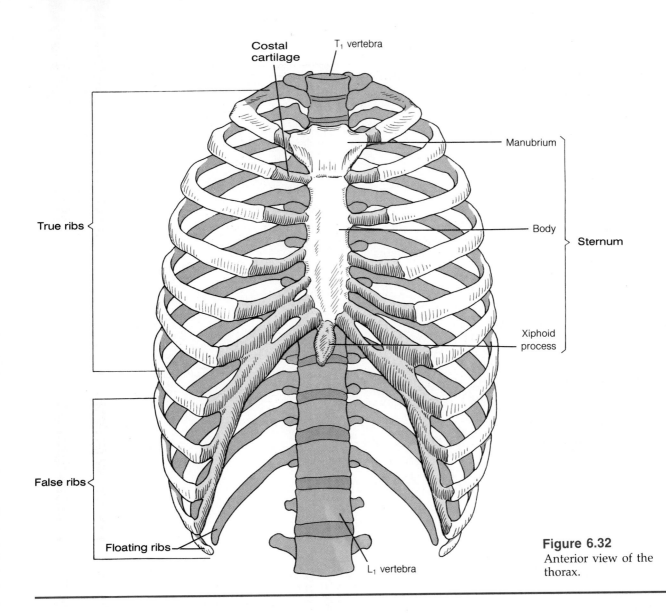

Costal cartilage

T₁ vertebra

Manubrium

True ribs

Body

Sternum

Xiphoid process

False ribs

Floating ribs

L₁ vertebra

Figure 6.32
Anterior view of the thorax.

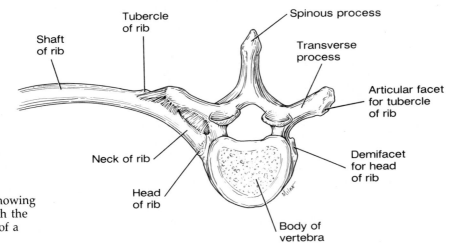

Shaft of rib

Tubercle of rib

Spinous process

Transverse process

Articular facet for tubercle of rib

Neck of rib

Demifacet for head of rib

Head of rib

Body of vertebra

Figure 6.33
Superior view of a thoracic vertebra showing its articulations with the head and tubercle of a rib.

Table 6.6 Summary of Specific Features of the Thoracic Skeleton

STERNUM *F6.32*

Manubrium The broad upper segment that articulates with the medial ends of each clavicle, the costal cartilages of the first pair of ribs, and part of the second pair of ribs. Has a small depression called the *jugular notch* on its superior border.

Body (Gladiolus) The elongated middle segment to which the costal cartilages of the second through the seventh ribs attach. Forms the *sternal angle* at its junction with the manubrium.

Xiphoid process A small inferior cartilage that serves for the attachment of several ligaments and muscles.

RIBS *F6.32*

Head The posterior, medial end that articulates with the bodies of the thoracic vertebrae.

Neck The constricted portion just lateral to the head.

Tubercle A small projection just beyond the neck that articulates with the transverse process of a thoracic vertebra. It is not present in the tenth, eleventh, and twelfth ribs.

TRUE RIBS

Vertebrosternal ribs The first through seventh pairs. They attach directly to the sternum.

FALSE RIBS

Vertebrochondral ribs The eighth, ninth, and tenth pairs. They attach to the costal cartilages of the rib above.

Vertebral (floating) ribs The eleventh and twelfth pairs. They have no anterior attachment.

Table 6.7 Summary of Bones That Form the Axial Skeleton

	Number of Bones		Number of Bones
SKULL	29	AUDITORY OSSICLES	6
CRANIUM *(calvarium and floor of cranial cavity)*	8	*Malleus* 2 *Incus* 2 *Stapes* 2	
Parietal 2		HYOID	1
Temporal 2			
Frontal 1		**VERTEBRAL COLUMN**	26
Occipital 1			
Ethmoid 1		*Cervical* 7	
Sphenoid 1		*Thoracic* 12	
		Lumbar 5	
FACE AND NASAL CAVITY*	14	*Sacrum* 5 fused to form 1	
		Coccyx 4 fused to form 1	
Maxillary 2			
Zygomatic 2		**THORAX**	25
Lacrimal 2			
Nasal 2		*Sternum* 1	
Inferior nasal concha 2		*Ribs* 24	
Palatine 2			
Mandible 1		**TOTAL AXIAL SKELETON BONES**	80
Vomer 1			

*The frontal and ethmoid bones also contribute to the face but are counted under the cranium.

APPENDICULAR SKELETON

The appendicular skelton includes the bones of the upper and lower limbs, and those bones by which these limbs articulate with the axial skeleton—that is, the pectoral girdle and the pelvic girdle.

The pectoral girdle does not provide very firm support, being attached to the axial skeleton only at the sternum. This support is sufficient, however, because the upper limbs are not involved in supporting the body weight. The pectoral girdle does allow for a wide range of movements at the shoulder.

The pelvic girdle, in contrast, does support the body weight. To accomplish this it not only has more extensive attachments to the axial skeleton through its articulation with the sacrum, but also, the two sides of the girdle attach to each other at the pubic symphysis. In addition, the pelvic girdle is aligned with the bones of the lower limbs in such a manner that it transfers much of the weight it supports to the skeleton of the lower limbs.

Upper Limbs

Table 6.8   The bones of the upper limbs are listed in Table 6.8.

Pectoral Girdle

F6.34   The **pectoral girdle** (Figure 6.34) straddles the upper part of the thorax. Its only joint with the axial skeleton is where the medial end of each clavicle articulates with the manubrium of the sternum. The lateral end of each clavicle articulates with the acromion process of a scapula. The scapulae are attached to the posterior thorax by muscles, but they do not contact the ribs directly, being separated from them by other muscles.

F6.34   CLAVICLE   The **clavicle** (Figure 6.34) is an S-shaped bone that serves as a brace for the scapula. Its **sternal (medial) end,** which articulates with the manubrium of the sternum, is enlarged and blunt. Its **acromial (lateral) end,** which articulates with the acromion process of the scapula, is flattened. Through its articulations with the sternum and the scapula, the clavicle holds the shoulder away from the rib cage, thus allowing the arms to swing freely, without first necessitating the lifting of the arm away from the body wall.

Table 6.9   When a clavicle is broken, the entire shoulder collapses. Table 6.9 lists the terms associated with the clavicle.

F6.34   SCAPULA   The **scapula** (Figure 6.34) is a thin, flat, triangular bone that lies over the posterior surfaces of the second to the seventh ribs. The **superior border** forms the base of the triangle, with the **vertebral (medial) border** and the **axillary (lateral) border** joining at the **inferior angle.** The **lateral angle** forms the **glenoid fossa,** which articulates with the humerus. On the superior border, just medial to the glenoid fossa, is the hooked **coracoid process** (*coracoid* = beaklike). The dorsal surface is divided into upper and lower regions by a ridge, the **spine** of the scapula. The spine terminates laterally in a flat **acromion process,** which articulates with the lateral end of the clavicle.

Table 6.9   The features of the scapula are summarized in Table 6.9.

Arm

F6.35   The **humerus** is the only bone in the arm (Figure 6.35). Its proximal epiphysis, which is called the **head,** is smooth and round. The head articulates with the glenoid fossa of the scapula. The **anatomical neck** is a slight constriction below the head. There are two projections just distal to the anatomical neck—the lateral **greater tubercle** and the anterior **lesser tubercle.** The tubercles are separated from each other by an **intertubercular (bicipital) groove.** The distal end of the humerus is flattened, with prominent **medial** and **lateral epicondyles.** Between the epicondyles are two articular surfaces—the lateral rounded **capitulum,** which articulates with the radius, and the medial, deeply grooved **trochlea,** which articulates with the ulna. The features of the hu-

Table 6.9   merus are summarized in Table 6.9.

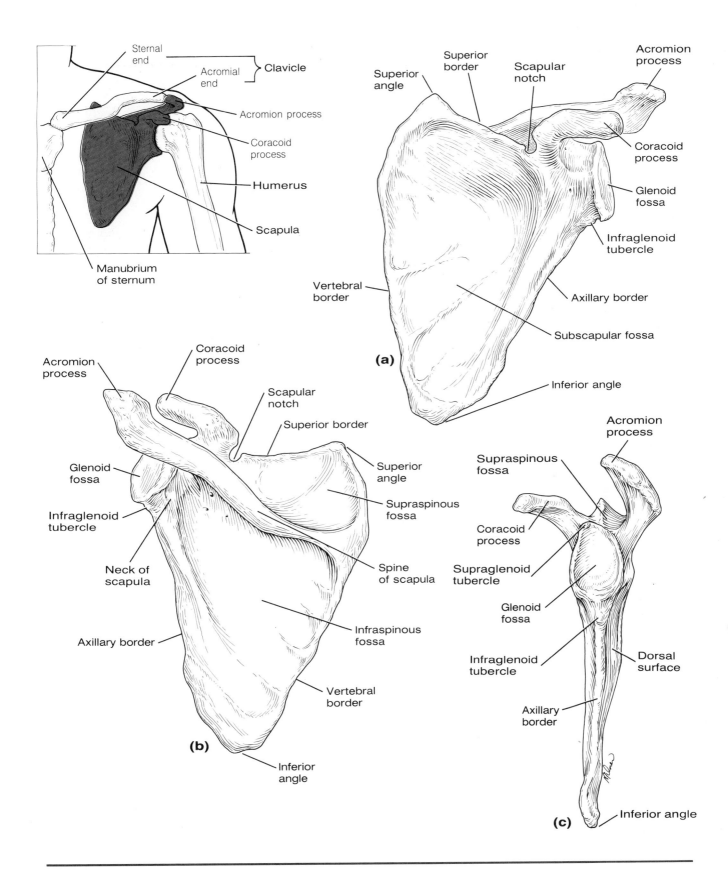

Figure 6.34
The insert is an anterior view of the pectoral girdle with the rib cage removed. **(a)** Anterior view of the left scapula; **(b)** posterior view; **(c)** lateral view.

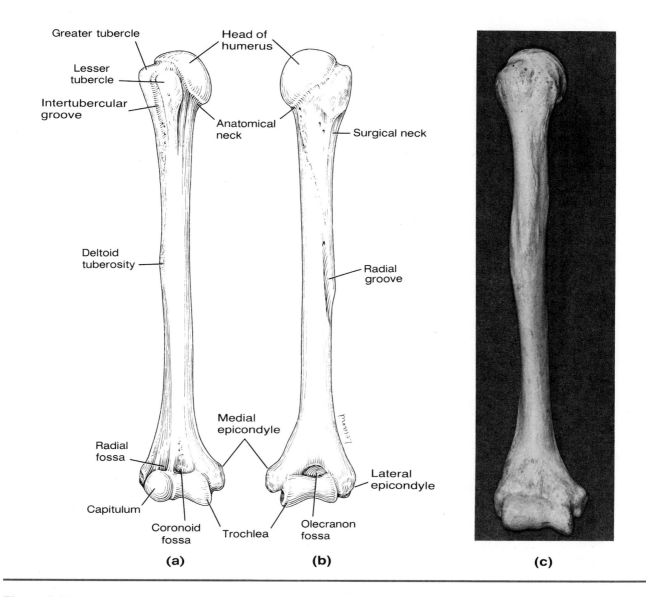

Greater tubercle

Lesser tubercle

Intertubercular groove

Head of humerus

Anatomical neck

Surgical neck

Deltoid tuberosity

Radial groove

Medial epicondyle

Radial fossa

Lateral epicondyle

Capitulum

Coronoid fossa

Trochlea

Olecranon fossa

(a) (b) (c)

Figure 6.35
The right humerus.
(a) Anterior view.
(b) Posterior view.
(c) Anterior photograph.

F6.36
Table 6.9

F6.36

F6.36

too-bur-os'-i-tee

Forearm

There are two parallel bones in the forearm. In the anatomical position, the *ulna* is medial and the *radius* is lateral (Figure 6.36). The features of the radius and ulna are summarized in Table 6.9.

ULNA   The proximal end of the ulna has two prominent processes: the large posterior **olecranon process,** and the smaller anterior **coronoid process** (Figure 6.36). A smooth, concave surface, the **trochlear (semilunar) notch,** lies on the anterior surface of the olecranon process and extends onto the superior surface of the coronoid process. The trochlear notch articulates with the trochlea of the humerus. The **radial notch** is a smooth surface on the lateral side of the coronoid process. It articulates with the edge of the head of the radius. Distally, the ulna has a small rounded **head,** and a posterior medial **styloid process.**

RADIUS   In the anatomical position, the radius lies lateral to the ulna (Figure 6.36). Its small cylindrical proximal epiphysis is the **head.** The head articulates with the capitulum of the humerus and the radial notch of the ulna. On the medial surface, a short distance below the head, is the **radial tuberosity.** The

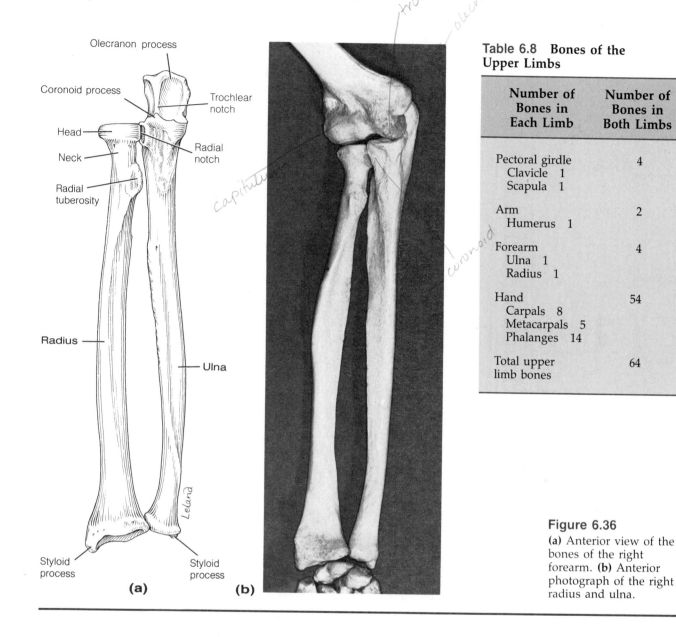

Figure labels (handwritten): trochlea, olecranon, capitulum, coronoid, Leland

Figure labels (a): Olecranon process, Coronoid process, Head, Neck, Radial tuberosity, Trochlear notch, Radial notch, Radius, Ulna, Styloid process, Styloid process, **(a)**, **(b)**

Table 6.8 Bones of the Upper Limbs

	Number of Bones in Each Limb	Number of Bones in Both Limbs
Pectoral girdle		4
Clavicle	1	
Scapula	1	
Arm		2
Humerus	1	
Forearm		4
Ulna	1	
Radius	1	
Hand		54
Carpals	8	
Metacarpals	5	
Phalanges	14	
Total upper limb bones		64

Figure 6.36
(a) Anterior view of the bones of the right forearm. (b) Anterior photograph of the right radius and ulna.

distal end of the radius is broad, and it articulates medially with the ulna and distally with the carpal bones of the wrist. It has a conical **styloid process** projecting from its lateral margin. The space between the radius and ulna is occupied by a strong interosseus membrane.

Hand

The skeleton of the hand consists of *carpal bones, metacarpal bones,* and *phalanges.* The proximal portion of the hand, toward the wrist, is composed of eight **carpal bones** arranged in two transverse rows of four bones each (Figure 6.37). The bones of the proximal row, from lateral to medial, are the **scaphoid, lunate, triquetrum,** and **pisiform.** Those of the distal row, from lateral to medial, are **trapezium, trapezoid, capitate,** and **hamate.** The scaphoid and lunate articulate with the distal end of the radius to form the wrist joint.

    Five **metacarpal bones** form the skeleton of the palm of the hand (Figure 6.37). Rather than being named, they are numbered from lateral (thumb) to medial. Proximally, the metacarpals articulate with the distal row of the carpal bones and with each other. Distally, each articulates with the proximal end of a phalanx.

F6.37

F6.37

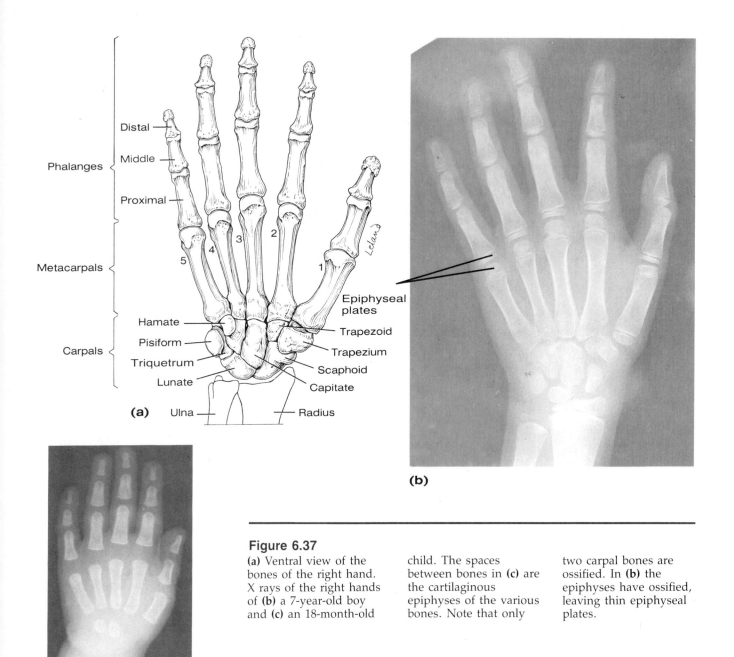

Figure 6.37
(a) Ventral view of the bones of the right hand. X rays of the right hands of (b) a 7-year-old boy and (c) an 18-month-old child. The spaces between bones in (c) are the cartilaginous epiphyses of the various bones. Note that only two carpal bones are ossified. In (b) the epiphyses have ossified, leaving thin epiphyseal plates.

Table 6.9   Summary of Specific Features of the Bones of the Upper Limbs

CLAVICLE   *F6.34*

Sternal end   The blunt medial end that articulates with the manubrium of the sternum.

Acromial end   The flattened lateral end that articulates with the acromion process of the scapula.

SCAPULA   *F6.34*

Superior border   The upper horizontal margin.

Vertebral (medial) border   The vertical margin just lateral to the vertebral column.

Axillary (lateral) border   The thicker oblique lateral margin.

Inferior angle   The most inferior point of the bone. It marks the junction of the vertebral and axillary borders.

Superior angle   The junction of the vertebral and superior borders.

Lateral angle   The junction of the superior and axillary borders. It contains the glenoid fossa.

Glenoid fossa   A shallow depression on the lateral angle. It articulates with the humerus.

Supraglenoid tubercle   A slight elevation just above the glenoid fossa. This is the point of attachment of the long head of the biceps brachii muscle.

Infraglenoid tubercle   The roughened area just below the glenoid fossa. The long head of the triceps brachii muscle originates here.

Coracoid process   The projection that hooks anteriorly from the superior border. It provides for the attachment of ligaments and muscles.

Scapular notch   A deep notch in the superior border at the base of the coracoid process. It allows for the passage of the suprascapular nerve.

Spine   A prominent ridge that runs horizontally across the posterior surface.

Acromion process   The flattened lateral end of the spine. It articulates with the clavicle, thus bracing the scapula.

Supraspinous fossa   The dorsal surface above the spine.

Infraspinous fossa   The dorsal surface below the spine.

Subscapular fossa (costal surface)   The slightly concave ventral surface.

HUMERUS   *F6.35*

Head   The rounded proximal epiphysis. It articulates with the glenoid fossa to form the shoulder joint.

Anatomical neck   A shallow constriction that circles the bone just below the head.

Greater tubercle   The rounded projection from the lateral margin of the bone just distal to the anatomical neck.

Lesser tubercle   The rounded projection from the anterior surface of the bone just distal to the anatomical neck.

Intertubercular (bicipital) groove   A deep groove between the greater and lesser tubercles. The long head of the biceps brachii muscle passes through the groove to reach the supraglenoid tubercle.

Surgical neck   A slightly constricted region just inferior to the tubercles. This is frequently the site of fracture.

Deltoid tuberosity   A triangular roughened area on the anterior lateral surface near the middle of the shaft. The deltoid muscle inserts here.

Radial groove (sulcus)   An oblique groove on the posterior surface just below the deltoid tuberosity. It marks the path of the radial nerve.

Epicondyles (medial and lateral)   The projections from the margins of the distal epiphysis.

Capitulum   The lateral convex portion of the distal condyles. It articulates with the head of the radius.

Radial fossa   A slight depression on the anterior surface above the capitulum. It receives the margin of the head of the radius when the elbow is flexed (bent).

Trochlea   The medial concave portion of the distal condyles. It articulates with the semilunar notch of the ulna.

Coronoid fossa   A small depression on the anterior surface above the trochlea. It receives the coronoid process of the ulna when the elbow is flexed (bent).

Olecranon fossa   A deep depression on the posterior surface above the trochlea. It receives the olecranon process of the ulna when the elbow is extended (straightened).

ULNA   *F6.36*

Olecranon process   The thick posterior projection from the proximal end that forms the point of the elbow. It is received by the olecranon fossa of the humerus when the elbow is extended (straightened).

Coronoid process   The anterior projection from the proximal end. It is received by the coronoid fossa of the humerus when the elbow is flexed (bent).

Semilunar notch   A curved depression formed by the olecranon and coronoid processes. It articulates with the trochlea of the humerus.

Radial notch   A small depression on the lateral side of the coronoid process. It articulates with the margins of the head of the radius, allowing the forearm to rotate, turning the palm down (pronate).

Head   The small distal end that articulates with the fibrocartilaginous disc of the wrist joint.

Styloid process   The posterior medial projection from the distal end. It serves as the point of attachment for the ulnar collateral ligament of the wrist.

RADIUS   *F6.36*

Head   The proximal disc-shaped end. The superior surface articulates with the capitulum of the humerus; the edges articulate with the radial notch of the ulna.

Neck   The constriction just distal to the head.

Radial tuberosity   A flat projection on the medial side, distal to the neck. The biceps brachii muscle inserts here.

Styloid process   The lateral downward projection from the distal end. It is the point of attachment for the radial collateral ligament and the brachioradialis muscle.

Ulnar notch   A depression on the medial margin of the distal end. It articulates with the ulna.

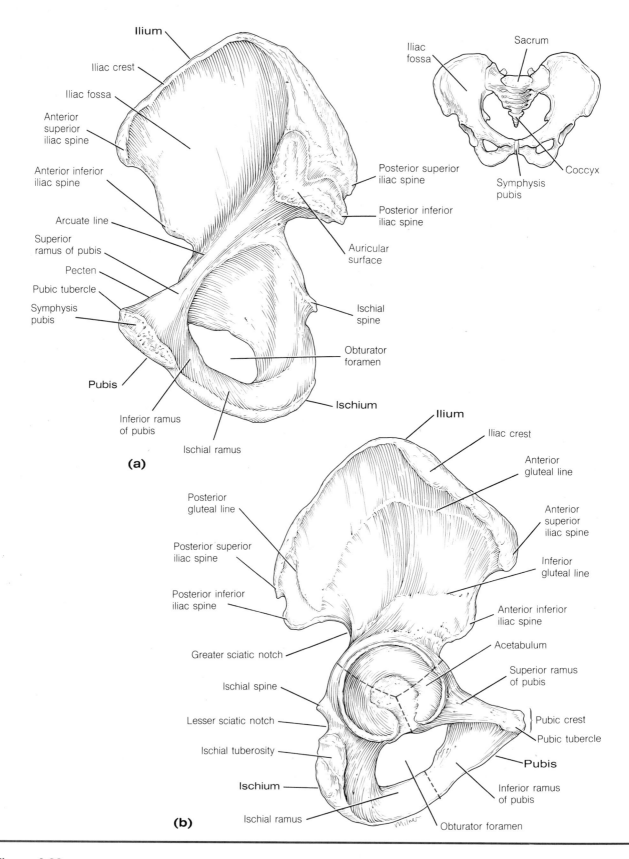

Figure 6.38

Right coxal bone.
(a) Internal surface.
(b) External surface. The
dotted lines indicate the
approximate junctions
between the ilium,
ischium, and pubis.

The skeleton of the fingers is formed by 14 **phalanges.** They are numbered from lateral to medial, in the same manner as the metacarpals. Each finger contains three phalanges, named **proximal, middle,** and **distal.** An exception is the first digit (thumb), which has only proximal and distal phalanges.

fay-lan'-jeez

Lower Limbs

The bones of the lower limbs are listed in Table 6.10.

Table 6.10

Table 6.10   Bones of the Lower Limbs

Number of Bones in Each Limb	Number of Bones in Both Limbs
Pelvic Girdle	2
Coxal bone	
Ilium ⎫	
Ischium ⎬ fused to form 1	
Pubis ⎭	
Thigh	2
Femur 1	
Leg	6
Tibia 1	
Fibula 1	
Patella 1	
Foot	52
Tarsals 7	
Metatarsals 5	
Phalanges 14	
Total lower limb bones	62

Pelvic Girdle

The pelvic girdle is formed by a pair of **coxal bones,** or **ossa coxae** (*ossa* = bones; *coxa* = hip) (Figure 6.38). These bones are also commonly referred to as pelvic bones or innominate bones. The two coxal bones are firmly braced through posterior articulations with the sacrum (the sacroiliac joint) and an anterior articulation with each other (the symphysis pubis). Each coxal bone is a single bone formed by the fusion of three separate embryonic bones: the *ilium, ischium,* and *pubis.* In the adult bone, these individual names are retained for their respective parts. On the lateral surface of the coxal bone, where the ilium, ischium, and pubis bones meet, is a deep cup, the **acetabulum.** The head of the femur articulates with the acetabulum. Below the acetabulum is a large **obturator foramen.** The features of the coxal bones are summarized in Table 6.12.

F6.38

Table 6.12

ILIUM   The **ilium** is a broad, expanded portion of the coxal bone that extends upward from the acetabulum (Figure 6.38). Its superior border is called the **iliac crest.** This crest ends anteriorly at the **anterior superior iliac spine.** A short distance below this spine is the **anterior inferior iliac spine.** The iliac crest ends posteriorly at the **posterior superior iliac spine.** A short distance below this spine is the **posterior inferior iliac spine.** Below the posterior inferior iliac spine is the deep **greater sciatic notch.** The **iliac fossa** is the smooth, slightly concave internal surface. Behind the fossa is a roughened area called the **auricular surface** that articulates with the sacrum. Running diagonally downward and forward from this articular surface, and demarking the lower boundary of the iliac fossa, is the **arcuate line.**

F6.38

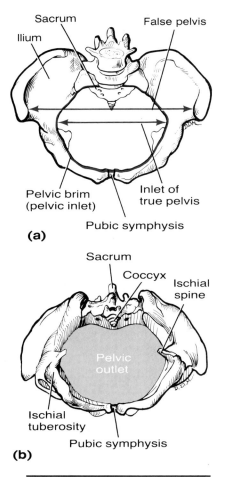

Sacrum

Ilium

False pelvis

Pelvic brim
(pelvic inlet)

Inlet of
true pelvis

Pubic symphysis

(a)

Sacrum

Coccyx

Ischial
spine

Pelvic
outlet

Ischial
tuberosity

Pubic symphysis

(b)

Figure 6.39
Pelvic cavities.
(a) Superior view
showing the expanded
false pelvis separated
from the true pelvis by
the pelvic brim.
(b) Inferior view showing
the pelvic outlet.

Table 6.12

Table 6.11

F6.40

ISCHIUM The **ischium** forms the posterior inferior portion of the coxal bone and part of the acetabulum (Figure 6.38). On the posterior margin of the ischium, below the greater sciatic notch, is the **ischial spine.** The **lesser sciatic notch** is below the spine. The notch is bounded inferiorly by a prominent **ischial tuberosity,** which supports the body weight in the sitting position. The **ischial ramus** is an anterior projection from the tuberosity. It joins with the inferior ramus of the pubis to form the lower border of the obturator foramen.

PUBIS The **pubis** is the anterior part of the coxal bone. It forms the anterior inferior portion of the acetabulum (Figure 6.38). Its **superior ramus** is supported against the superior ramus of the opposite side, forming the **symphysis pubis.** A projection called the **pubic tubercle** is located on the superior ramus, close to the symphysis pubis. A ridge called the **pecten** extends along the superior ramus, from the pubic tubercle to the arcuate line of the ilium. A short distance lateral to the symphysis pubis, the **inferior ramus** of the pubis extends downward and posteriorly to join with the ischial ramus. The junction of the two inferior rami at the symphysis pubis forms the **pubic arch.**

Pelvic Cavities

The cavity of the pelvis is divided into two parts by a horizontal plane that passes from the sacral promontory to the upper margin of the symphysis pubis, following the arcuate lines on the inner surface of the ilium. The circumference of this plane is the **pelvic brim** (Figure 6.39a). The expanded cavity above the pelvic brim is the **greater,** or **false, pelvis.** The **lesser,** or **true, pelvis** is the cavity below the pelvic brim. The false pelvis is bounded posteriorly by the iliac fossa. Anteriorly, it is bounded by the abdominal wall, and is therefore capable of expansion. In contrast, the cavity of the true pelvis is much more restricted, being surrounded on all sides by bone (ilium, ischium, pubis, sacrum, and coccyx). The superior circumference of the true pelvis is called the **pelvic inlet** because it marks the superior entrance to the true pelvis. The margin of the pelvic inlet coincides with the pelvic brim. The **pelvic outlet** (Figure 6.39b) is the lower circumference of the true pelvis. It is bounded posteriorly by the coccyx and two ischial spines and tuberosities, and anteriorly by the lower margin of the symphysis pubis. The features of the pelvic cavities are summarized in Table 6.12.

SEXUAL DIFFERENCES IN THE PELVIS There are several structural differences between the male and the female pelvis, most of which are related to childbearing. Because during birth the fetus must pass from the false pelvis through the true pelvis, the measurements of the pelvic inlet and outlet are of particular importance in the female. Some sexual differences in the pelvis are listed in Table 6.11 and illustrated in Figure 6.40.

Table 6.11   Some Structural Differences Between Female Pelvis and Male Pelvis

Characteristic	Female Pelvis	Male Pelvis
General structure	More delicate	More massive
Anterior iliac spines	More widely separated	Less widely separated
Pelvic inlet	Larger; circular	Heart-shaped
Pelvic outlet	Wider; ischial tuberosities farther apart	Narrower
Pubic arch	Obtuse (greater than 90°)	Acute (less than 90°)
Obturator foramen	Triangular	Oval
Acetabulum	Faces more anteriorly	Faces laterally

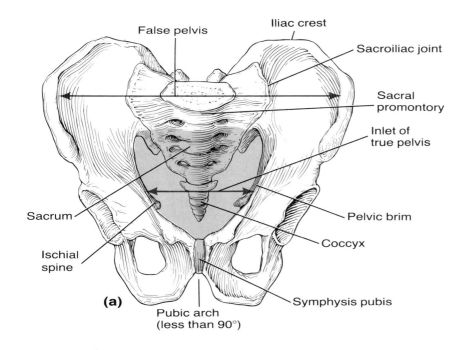

False pelvis

Iliac crest

Sacroiliac joint

Sacral promontory

Inlet of true pelvis

Sacrum

Pelvic brim

Ischial spine

Coccyx

Symphysis pubis

(a)

Pubic arch (less than 90°)

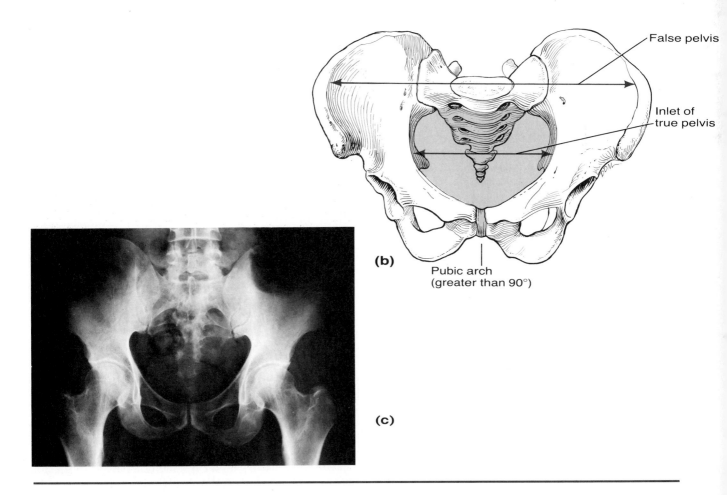

False pelvis

Inlet of true pelvis

(b)

Pubic arch (greater than 90°)

(c)

Figure 6.40

Anterior view of **(a)** the male pelvis and **(b)** the female skelvis. **(c)** Anteroposterior X ray of the female pelvis.

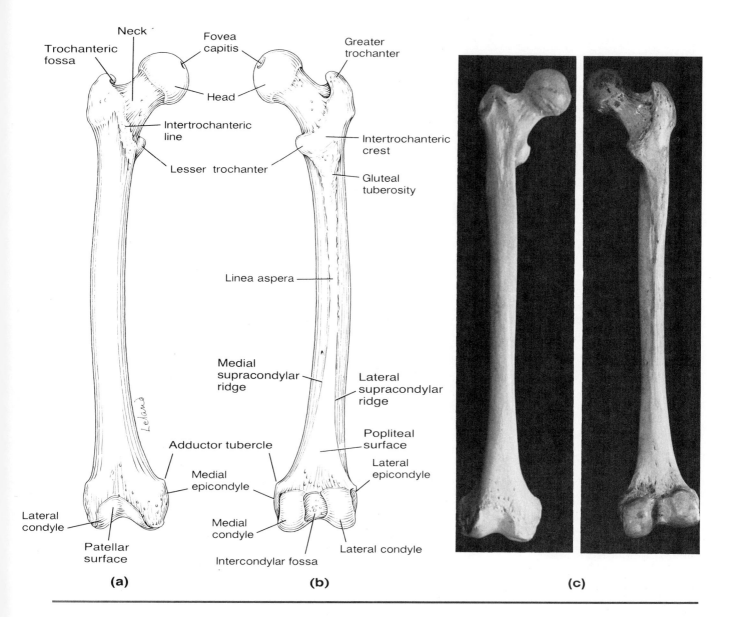

Figure 6.41

Right femur. **(a)** Anterior view. **(b)** Posterior view. **(c)** Anterior (left) and posterior (right) photographs.

F6.41

Thigh

The **femur,** which forms the skeleton of the thigh, is the longest bone of the body (Figure 6.41). The **head** of the femur is the spherical proximal epiphysis, which is directed medially and is received into the acetabulum of the coxal bone. The surface of the head is smooth, except for a central pit, the **fovea capitis.** A strong ligament (*ligamentum teres*) is attached to the fovea and to the acetabulum, helping to maintain the integrity of the hip joint. A constricted **neck** joins the head with the shaft of the bone. The neck forms an angle of about 125° with the shaft. At the point where the neck joins with the shaft, there are two large processes that serve as sites of muscle attachment. The larger lateral process is the **greater trochanter;** the smaller process, which projects medially and posteriorly, is the **lesser trochanter.** On the posterior surface of the shaft there is a distinct longitudinal ridge, the **linea aspera.** At its distal end, the linea aspera divides into **medial** and **lateral supracondylar ridges,** which enclose a flat triangular **popliteal surface** between them.

The distal end of the femur is enlarged into **medial** and **lateral condyles.** The smooth surfaces of the condyles articulate with the tibia. Between the condyles, on the posterior surface, is a deep **intercondylar fossa.** Anteriorly,

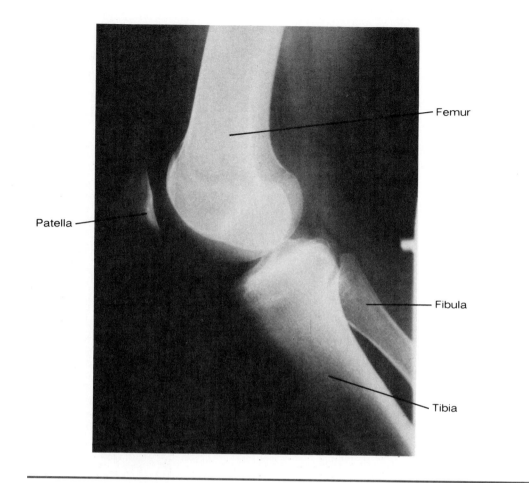

Femur

Patella

Fibula

Tibia

Figure 6.42
Lateral X ray of the knee.

there is a smooth **patellar surface** between the condyles. This surface articulates with the patella when the leg is extended. The features of the femur are summarized in Table 6.12.

Table 6.12

Leg

The skeleton of the leg consists of a strong medial **tibia** and a slender lateral **fibula.** Protecting the knee joint, between the thigh and the leg, is the **patella** (kneecap). The features of the tibia and fibula are summarized in Table 6.12.

Table 6.12

PATELLA   The **patella** (Figure 6.42) is a *sesamoid bone,* which means that it forms within the tendons of muscles and is not firmly anchored to the skeleton. It is located in front of the knee joint. In addition to protecting the knee joint, it improves the leverage of the quadriceps femoris muscle group, in whose tendon it is embedded. The posterior surface of the patella has a smooth articular surface for contact with the patellar surface of the femur.

F6.42

TIBIA   The **tibia** (Figure 6.43), which is the medial bone of the leg, supports the body weight transmitted to it from the femur. Its proximal end is expanded into **medial** and **lateral condyles.** The superior surfaces of the condyles are smooth and flattened. They articulate with the condyles of the femur. Between these articular surfaces is a prominence called the **intercondylar eminence** or **spine.** The **tibial tuberosity** is a prominent elevation on the anterior surface of the bone, just below the condyles. It provides an anchoring point for the tendons of the muscles of the anterior thigh—that is, those that ensheath the patella and form the *ligamentum patellae.* A sharp **anterior crest** extends almost the entire length of the shaft. The distal end of the tibia has a

F6.43

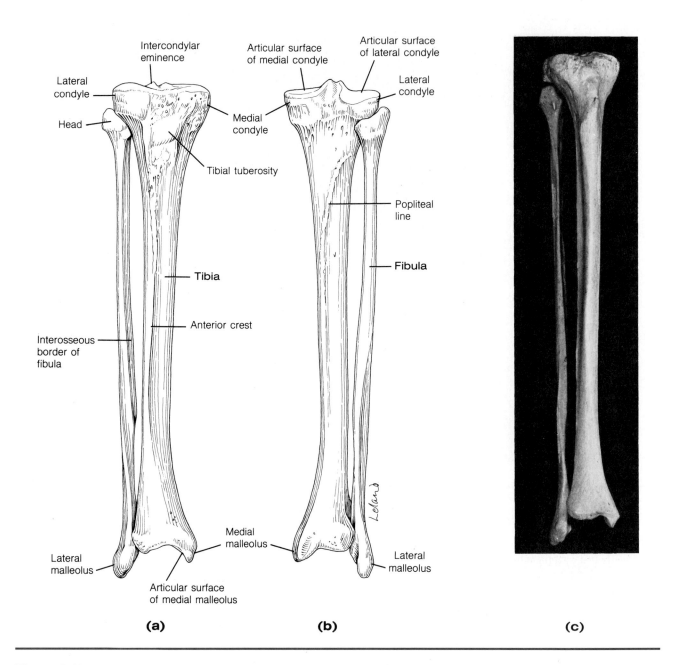

Figure 6.43
Bones of the right leg.
(a) Anterior view.
(b) Posterior view.
(c) Anterior photograph.

downward projection on its inner side called the **medial malleolus.** This is the prominent lump that can be felt on the medial side of the ankle. The distal surface of the tibia is flattened for articulation with the talus, which is a tarsal bone.

F6.43   FIBULA   The **fibula** is a slender bone that is lateral to the tibia (Figure 6.43). Its proximal end is expanded into a **head.** It articulates with the lateral condyle of the tibia. The distal end of the fibula is flattened, forming the **lateral malleolus,** which can be felt on the lateral side of the ankle. The lateral malleolus articulates with the talus and, together with the medial malleolus, forms the ankle joint. Just above the lateral malleolus, the fibula articulates with the distal end of the tibia. The tibia and the fibula are tightly bound together by an interosseous membrane.

Foot

The skeleton of the foot consists of *tarsal bones, metatarsal bones,* and *phalanges.* The proximal portion of the foot, toward the ankle, is composed of seven

FRONTIERS IN HEALTH
Rebuilding Broken Bones

Robert left a local restaurant and drove toward his home in Dumont, New Jersey. He fell asleep while driving, and his car swerved off the road and overturned. When he awoke the next morning surgeons had removed a 4-inch segment of his shattered humerus.

In other times, Robert's arm would have been amputated. Traditional bone grafts would not suffice to rebuild the shattered segment of bone. But today, Robert does yard work and lifts weights with his arm, thanks to the pioneering work of a group of surgeons.

In Robert's case, physicians surgically removed a section of the fibula, the relatively thin bone that lies alongside the tibia. The fibula, which bears very little body weight in humans, has proved useful for replacing segments of other bones. This technique succeeds where conventional bone grafts would have failed because, along with the bone itself, physicians remove the blood vessels that nourish the bone. The graft is then placed in the shattered region of the bone and the blood vessels of the fibula are carefully reconnected with existing arteries and veins by microsurgical techniques, thus ensuring an uninterrupted flow of blood. Ordinary bone grafts often failed because blood flow could not be immediately reestablished.

Supplied with a steady flow of blood, the grafted fibula knits itself into place and gradually thickens, becoming almost indistinguishable in thickness from the bone on either side. Immediately after transplantation, however, the bone must be secured by an external fixator. After 6 to 8 weeks, the fixator is replaced by a cast. Later, a brace is all that is needed to support the bone graft, and eventually even that is no longer necessary.

The success of this technique, which is called the free vascularized fibular graft, has been exciting. Two American teams who have used it on well over 200 patients report outstanding results.

Fibular transplants have been used to repair crushed bones or replace bone segments lost to infection or cancer. The technique was even successfully used to construct a radius in an infant who had been born without one.

Another innovative technique that complements free vascularized fibular grafting is based on the pioneering work of orthopedic surgeon Dr. Marshall Urist, who is developing a method for building new bones from old ones. He obtains bones from human cadavers or even animals and treats them with chemicals to remove the mineral matter. What is left is a rubbery substance, called demineralized bone matter (DBM). Made mostly of collagen, DBM can be surgically implanted in the body to replace lost segments of bone, in much the same way that fibular transplants are done. The implanted DBM is injected with bone cells taken from chips of the patient's own bone. Over time, the rubbery implant is converted into solid bone that is virtually indistinguishable from the normal bone.

Medical scientists are now using DBM to repair congenital birth defects and to fill in cavities formed in bone

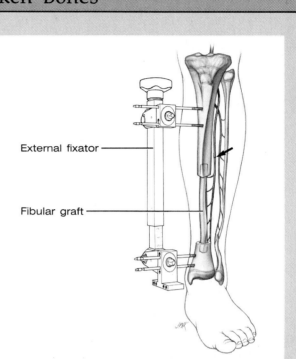

An external fixator stabilizing a fractured tibia, with a fibular graft in place, in a patient with pseudarthrosis (false joints). The arrow points to the anastomosis between the blood vessels of the tibia and the graft.

during surgery to remove cysts, cancer, or serious infections. Oral surgeons are using it to reconstruct jaws damaged in accidents or to rebuild eroded jaw bones in patients with periodontal disease. There is hope that surgeons may be able to use the technique to lengthen arms and legs whose growth was affected by hormonal deficiencies. Perhaps the most widely acclaimed success for DBM was the case of a 6-year-old boy suffering from a rare congenital disease, called cloverleaf deformity, in which bones of the skull fail to grow to accommodate the growing brain. Without treatment the boy would have died. Dr. John Mulliken, a surgeon at Boston's Childrens Hospital, constructed an entirely new skull for the patient out of DBM. Three years later the boy is alive and healthy.

DBM has proved itself in hundreds of cases. Even 5 years after it has been put in place, the new bone grown from DBM appears healthy and strong. The use of DBM has several important advantages. First, it can be produced in large quantities and stored for later use, offering physicians a ready supply of inexpensive, transplantable material. Second, conventional bone grafting and free vascularized fibular grafts require two surgical operations: one to remove the bone and one to put it in its new location. For patients this means more time in surgery, more risk of complications, more pain, higher medical bills, and a longer recovery time. DBM eliminates the need for the extra surgery.

*Illustration by John W. Karapelou. Courtesy of Harold M. Dick, M.D. New York Orthepedic Hospital, CPMC, New York, NY.

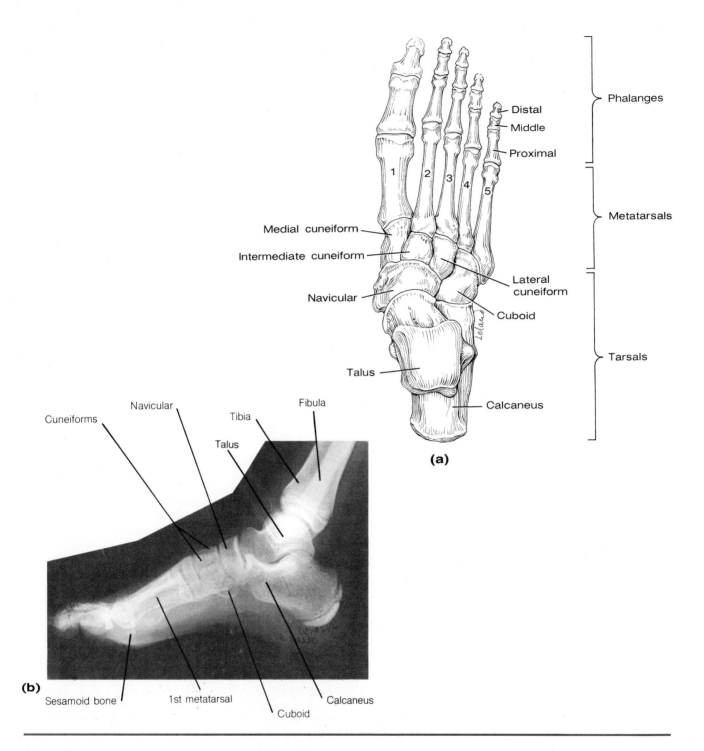

(a)

(b)

Figure 6.44
(a) Superior view of the bones of the right ankle and foot. (b) Lateral X ray of the foot.

F6.44

tarsal bones (Figure 6.44): the **talus, calcaneus, navicular, cuboid, medial cuneiform, intermediate cuneiform,** and **lateral cuneiform.**

The calcaneus forms the prominent heel bone, which provides attachment for several muscles of the calf. Resting on the superior surface of the calcaneus is the talus. The talus is located between the malleoli, where it articulates with the tibia and fibula. In this location, the talus receives the entire weight of the body, which it distributes to the other tarsal bones. The three cuneiforms and the cuboid articulate with the proximal ends of the metatarsal bones.

F6.44

Five **metatarsal bones** form the skeleton of the intermediate region of the foot (Figure 6.44). They are numbered from medial to lateral. Proximally, they articulate with the tarsal bones and with each other. Distally, each articulates with the proximal end of a phalanx.

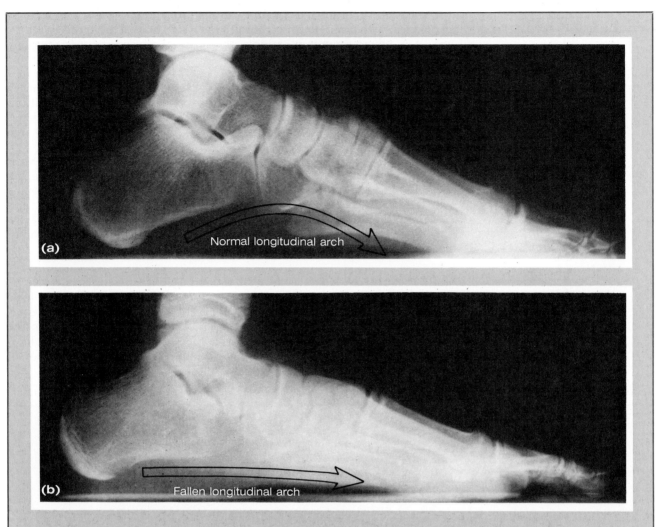

Normal longitudinal arch

Fallen longitudinal arch

Box 6.2
Fallen Arches

These two photos demonstrate the difference between **(a)** a normal longitudinal arch and **(b)** a fallen longitudinal arch. In a foot with a normal longitudinal arch the body weight moves along a precise path during walking, beginning with the heel. Then, with the arch used as a spring, the weight is shifted to the ball of the foot. As weight passes over the arch, the fascia connecting the heel and the metatarsal bones flattens out a bit to accommodate the added pressure and then spreads the weight forward to the ball and metatarsal bones.

A fallen longitudinal arch has lost this flexibility and springlike motion. An arch "falls" because of improper locomotion—that is, the weight is distributed unevenly along this bridgelike path, causing the arch to stretch disproportionately and to become flaccid. Without proper arch support, the foot becomes pronated, with the weight rolling to the inside of the foot instead of on this straight path.

Although anyone can suffer from fallen arches regardless of the level of physical activity, this problem often afflicts runners who have improperly fitted shoes or an incorrect running style. As a runner increases distance and speed, the stress on the feet is increased. The day-to-day jolt of running, especially on hard surfaces, can weaken improperly supported arches.

The skeleton of the toes is similar to that of the fingers. It is formed of 14 **phalanges,** with three (*proximal, middle,* and *distal*) in each digit, except for the great toe, where there is no middle phalanx. The phalanges of the toes are shorter than those of the fingers.

ARCHES OF THE FOOT   The tarsal and metatarsal bones are joined in such a manner that they form three arches. There are **medial** and **lateral longitudinal arches,** which have the calcaneus as their posterior pillar and the metatarsals and other tarsals as their anterior pillar. The **transverse (metatarsal) arch** is formed mostly by the bases of the metatarsal bones.

Box 6.2   The arches are supported primarily by ligaments, with some support gained from muscles and their tendons. They function to distribute the weight of the body fairly evenly between the heel and the metatarsals. People whose longitudinal arches have collapsed suffer from *flat feet* (see Box 6.2) and often find that their feet tire easily. They may also experience low-back discomfort because the arches normally absorb many of the shocks that occur during walking.

Sesamoid Bones

In certain tendons that are subjected to compression or to unusual tensile stress, small bones called **sesamoid bones** form. Sesamoid bones are generally located around a joint. The patella is the largest sesamoid bone in the body. Although the locations of sesamoid bones vary, the most common locations include the areas around the metacarpophalangeal and the metatarsophalangeal joints.

Table 6.12   Summary of Specific Features of the Bones of the Lower Limbs

COXAL BONE (os coxae) *F6.38*

ILIUM *F6.38* The upper, flat expanded portion of the coxal bone.

Iliac crest   The superior margin of the ilium.

Iliac spines
ANTERIOR SUPERIOR   A projection at the anterior end of the iliac crest.

ANTERIOR INFERIOR   A rounded projection just below the anterior spine.

POSTERIOR SUPERIOR   A projection at the posterior end of the iliac crest.

POSTERIOR INFERIOR   A projection just below the posterior superior spine.

Greater sciatic notch   A deep notch just below the posterior inferior spine. It provides passage for the sciatic nerve, as well as other nerves, vessels, and the piriformis muscle.

Gluteal lines   Three arched lines—posterior, anterior, and inferior—on the outer surface of the ilium. The origins of the three gluteus muscles are located between these lines.

Auricular surface   Surface that articulates with the sacrum.

Arcuate line   A slight ridge on the internal surface of the ilium. It extends forward and downward from the top of the sacrum to the pecten of the pubis. The line forms the internal margins of the pelvic inlet.

Iliac fossa   The smooth, concave internal surface of the ilium, above the arcuate line.

ISCHIUM   *F6.38* The posterior portion of the coxal bone.

Ischial spine   A triangular projection from the posterior border behind the acetabulum.

Ischial tuberosity   A roughened enlargement, on the posterior inferior margin.

Lesser sciatic notch   A small indentation that separates the ischial spine and tuberosity. It transmits the tendon of the obturator internus muscle, as well as nerves and blood vessels.

Ischial ramus   The flattened anterior projection that arises from the tuberosity. It joins with the inferior ramus of the pubis to form the lower border of the obturator foramen.

PUBIS   *F6.38* The anterior inferior portion of the coxal bone.

Superior ramus   Extends anteriorly from the acetabulum to form the upper border of the obturator foramen.

Symphysis pubis   The midline joint between the superior rami of both pubic bones.

Pubic tubercle   The projection from the superior ramus just lateral to the symphysis pubis.

Pecten   A ridge that extends upward and laterally along the superior ramus. It runs from the pubic tubercle to the arcuate line.

Pubic crest   A ridge that extends medially from the pubic tubercle to the symphysis pubis.

Inferior ramus   The projection that passes downward and backward from the symphysis pubis. It joins with the ischium to form the lower border of the obturator foramen.

Pubic arch   Formed by the convergence of the inferior rami.

Acetabulum   *F6.38* A cup-shaped depression formed by the junction of ilium, ischium, and pubis. It receives the head of the femur to form the hip joint.

Obturator foramen   *F6.38* The large opening in the inferior region of coxal bone. It allows for the passage of the obturator nerves and blood vessels.

Pelvic brim (inlet)   *F6.39* The boundary of the opening that leads into the true pelvis. It is formed by the sacral promontory, arcuate lines, and the superior margin of the symphysis pubis.

False pelvis   *F6.39* The expanded space above the pelvic inlet. It is actually a portion of the abdominopelvic cavity.

True pelvis   *F6.39* The smaller cavity below the pelvic inlet, bounded by the coxal bones, sacrum, and coccyx.

Pelvic outlet   *F6.39* The lower margin of the true pelvis. It is bounded by the coccyx and the lower parts of the coxal bones.

FEMUR   *F6.41*

Head   The rounded proximal end that fits into the acetabulum.

Fovea capitis   The pit on the head where the ligamentum teres is attached, which in turn is attached to the acetabulum.

Neck   The constriction that connects the head with the shaft.

Greater trochanter   The large lateral process just below the neck.

Lesser trochanter   The smaller medial process just below the neck.

Intertrochanteric crest   A ridge on the posterior surface that connects the two trochanters.

Intertrochanteric line   A slight line on the anterior surface that connects the two trochanters.

Gluteal tuberosity   A small projection just below the greater trochanter.

Trochanteric fossa   A depression on the medial surface of the greater trochanter.

Linea aspera   A longitudinal ridge that extends along the middle third of the posterior surface of the shaft.

Supracondylar ridges   The medial and lateral ridges that are formed by the divergence of the linea aspera at its distal end.

Popliteal surface   A smooth triangular surface on the posterior surface. It is bounded above by the medial and lateral supracondylar ridges.

Condyles   Rounded medial and lateral enlargements at the distal end of the shaft. They have smooth surfaces for articulation with the tibia.

Epicondyles   Roughened prominences on the lateral surfaces of the condyles. They provide points of attachment for the medial and lateral collateral ligaments of the knee joint.

Adductor tubercle   A small projection just above the medial condyle, at the termination of the medial supracondylar ridge.

Intercondylar fossa   The deep notch that separates the condyles posteriorly. It provides points of attachment for the anterior and posterior cruciate ligaments of the knee joint.

Patellar surface   The smooth anterior surface above the condyles. It articulates with the patella when the leg is extended (straightened).

TIBIA   *F6.43*

Condyles   Flattened enlargements at the proximal end of the shaft. The upper surfaces are smooth for articulation with the condyles of the femur.

Intercondylar eminence or **spine**   The vertical projection from the superior surface, between the articular surfaces of the condyles.

Tibial tuberosity   The midline projection from the anterior surface just below the condyles. It serves as the point of attachment for the ligamentum patellae.

Anterior crest   A sharp longitudinal ridge on the anterior surface.

Popliteal line   A ridge on the posterior surface that extends downward and medially from the lateral condyle. It marks the junction between the insertion of the popliteus muscle and the origin of the soleus muscle.

Medial malleolus   A downward projection from the medial side of the distal end of the tibia. It articulates with the talus.

FIBULA   *F6.43*

Head   The proximal, expanded portion of the bone. It articulates with the lateral condyle of the tibia.

Styloid process   A rough vertical prominence on the head of the fibula.

Lateral malleolus   A triangular expansion at the distal end of the tibia and with the talus.

STUDY OUTLINE

FUNCTIONS OF THE SKELETON p. 129

SUPPORT Body framework.

MOVEMENT Muscle attachment to skeleton; movable joints.

PROTECTION Of vital internal organs.

MINERAL RESERVOIR Storage of calcium phosphorus, sodium, potassium.

HEMOPOIESIS Red marrow of certain bones produces the blood cells of adult.

CLASSIFICATION OF BONES p. 130

LONG BONES Have long axis; most bones of upper and lower limbs.

SHORT BONES Lack a long axis; carpals, tarsals.

FLAT BONES Thin bones; ribs.

IRREGULAR BONES Vertebrae.

STRUCTURE OF BONE Studied at gross, microscopic, and chemical levels. pp. 131–134

GROSS ANATOMY

DIAPHYSIS Bone shaft.

Medullary Cavity Fat-storage site.

EPIPHYSIS End of bone; red bone marrow.

EPIPHYSEAL CARTILAGE OR PLATE Separates diaphysis and epiphysis in children; permits increase in bone length.

PERIOSTEUM Double-layered connective tissue that covers bones.

MICROSCOPIC ANATOMY

COMPACT BONE
1. Unit of structure is haversian system, or osteon.
2. Haversian canals surrounded by concentric lamellae.
3. Osteocytes embedded within lacunae.
4. Lacunae interconnected by canaliculi.

SPONGY BONE Osteocytes embedded within lacunae; lamellae not arranged in concentric layers but according to lines of tension.

COMPOSITION

ORGANIC FRAMEWORK Collagenous fibers in homogeneous ground substance; provides tensile strength.

INORGANIC SALTS Calcium and phosphate; allow bone to withstand compression.

DEVELOPMENT OF BONE pp. 134–138

EARLY DEVELOPMENT OF BONE Embryonic connective tissue is transformed into bone by two methods.

INTRAMEMBRANOUS OSSIFICATION
1. Flat bones of cranial roof, certain facial bones.

2. Formation of osteoblasts from undifferentiated mesoderm.
3. Osteoblasts form networks of bony spicules (trabeculae) around collagenous fibers.
4. Trabeculae radiate in all directions, uniting with one another to form a network of spongy bone.
5. Osteoclasts reabsorb previously laid-down bone to permit osteoblasts to deposit new bone for growth.

ENDOCHONDRAL OSSIFICATION
1. Most bones of the body.
2. Formation of bone from hyaline cartilage models.
3. Perichondrium develops into periosteum that contains osteoblasts.
4. Periosteum produces collar of bone that covers surface of diaphysis.
5. Cartilage matrix of diaphysis calcifies.
6. Osteoblasts form spicules of bone around calcified cartilage spicules.
7. Primary ossification center in diaphysis.
8. Secondary ossification center in epiphysis.

INCREASE IN BONE LENGTH AND DIAMETER

CARTILAGE CELLS UNDERGO MITOSIS Tends to increase size of epiphyseal plate.

DIAPHYSIS SIDE OF EPIPHYSEAL PLATE Undergoes calcification, thus increasing the length of the bone.

INCREASE IN DIAMETER Due to bone deposition on outer surface of bone, beneath periosteum.

THEORIES OF BONE FORMATION
1. Osteoblasts secrete organic framework for bone (collagen fibers and glycoproteins); inorganic salts (calcium phosphate) deposited in organic framework.
2. Calcium and phosphate thought to be concentrated in bone cells and released directly into extracellular fluid or else released in membrane-bounded vesicles.

FACTORS THAT AFFECT BONE DEVELOPMENT

STRESS Bone capable of adjusting its strength in proportion to degree of stress to which it is subjected. Increased stress results in formation of more collagen fibers and inorganic salts; salts withdrawn in absence of stress. Bone normally subjected to gravitational and functional forces.

HORMONES Parathormone from parathyroid gland increases resorption of bone by osteoclasts. Calcitonin from thyroid stimulates formation of new bone.

NUTRITION Well-balanced diet required for normal bone development. Vitamin D especially important because of its role in proper absorption of calcium into bloodstream from gastrointestinal tract.

CONDITIONS OF CLINICAL SIGNIFICANCE: THE SKELETAL SYSTEM pp. 138–140

FRACTURES

TYPES OF FRACTURES

Simple Bone ends do not penetrate skin.

Compound Broken bone ends penetrate skin.

Comminuted Bone splintered at site of break.

Depressed Broken region pushed inward.

Impacted Broken ends of bone driven into each other.

HEALING OF FRACTURES

Formation of a Procallus (Blood Clot); Formation of a Fibrocartilaginous Callus; Formation of a Bony Callus

METASTATIC CALCIFICATION Calcium deposits in tissues that are not normally calcified; kidney is a common site.

SPINA BIFIDA Failure of posterior portions of vertebrae to form bony arch around spinal cord; most common in lumbosacral area.

OSTEOPOROSIS Reduced bone formation rate, normal bone absorption rate.

OSTEOMYELITIS Infection of marrow-cavity contents and bone tissue, usually by *Staphylococcus aureus*.

TUBERCULOSIS OF BONE Infection of marrow-cavity contents and bone tissue caused by *Mycobacterium tuberculosis*.

RICKETS AND OSTEOMALACIA Both due to deficiencies of calcium, phosphorus, vitamin D, or lack of sunlight.

RICKETS Demineralization and bone softening in children.

OSTEOMALACIA Softening of adult bones.

TUMORS OF BONE Benign or malignant tumors.

ABNORMAL GROWTH PATTERNS

PITUITARY GIANT Excessive pituitary growth hormone delays ossification of epiphyseal cartilage.

PITUITARY DWARF Growth hormone deficiency resulting in early replacement of epiphyseal cartilages by bone.

ACROMEGALY Excess growth hormone secreted after epiphyseal cartilages replaced.

ACHONDROPLASIA Epiphyseal cartilages function for short time only, resulting in shortened arms and legs.

EFFECTS OF AGING Gradual loss of calcium from bone; more severe in women. Matrix not replaced as rapidly as it is broken down.

INDIVIDUAL BONES OF THE SKELETON   pp. 142–143

AXIAL SKELETON   pp. 143–165

SKULL

CALVARIUM

Frontal Bone Anterior superior region of skull.

Two Parietal Bones Form most of calvarium.

Occipital Bone Posterior floor of cranial cavity.

BONES THAT FORM FLOOR OF CRANIAL CAVITY

Frontal, Occipital, Ethmoid, and Sphenoid Bones Form midline.

Two Temporal Bones Form side of floor.

Sphenoid Bone and Paired Temporals Form middle cranial cavity.

FACIAL SKELETON

Frontal Forehead; *Paired Maxillary* Upper jaw; *Paired Zygomatic* Cheek prominences; *Paired Nasal Bones* Bridge of nose; *Paired Lacrimal* Medial orbital cavity; *Mandible* Lower jaw.

BONES THAT FORM NASAL CAVITY

Ethmoid Perpendicular plate forms part of septum; superior and middle conchae form lateral walls of nasal cavity.

Vomer Posterior inferior nasal septum.

Paired Inferior Nasal Conchae Form elongated shelves from lateral walls of nasal cavity.

BONES THAT FORM THE HARD PALATE Palatine processes of maxillary bones; horizontal processes of palatine bones.

BONES THAT FORM THE ORBITAL CAVITY Frontal, sphenoid, maxilla, lacrimal, ethmoid, zygomatic, palatine.

PARANASAL SINUSES Mucous-membrane-lined air spaces in the frontal, ethmoid, maxillary, and sphenoid bones.

AUDITORY OSSICLES

Malleus (Hammer), Stapes (Stirrup), Incus (Anvil)

HYOID BONE U-shaped; attachment point of tongue and throat muscles.

VERTEBRAL COLUMN

CURVATURES OF VERTEBRAL COLUMN

Infant Single curve, convex posteriorly.

Cervical Curve Convex anteriorly; develops with head raising.

Thoracic Curve Convex posteriorly; remains from primary curve of newborn.

Lumbar Curve Convex anteriorly; develops as child walks.

Abnormalities

Lordosis (Swayback) Excessive anterior lumbar curve.
Kyphosis (Hunchback) Excessive posterior thoracic curve.
Scoliosis Lateral curvature of spine.

FUNCTIONS OF THE VERTEBRAL COLUMN

Support and Flexibility

Spinal Cord Protection Intervertebral foramina provide passage for spinal nerves.

CHARACTERISTICS OF A TYPICAL VERTEBRA
1. Thick anterior centrum with neural arch that arises posteriorly.

2. Spinous process extends from neural arch.
3. Vertebral foramen—passageway for spinal cord.
4. Transverse processes.
5. Superior and inferior articulating processes.

REGIONAL DIFFERENCES IN VERTEBRAE summarized in Table 6.5.

THORAX

STERNUM Elongated, flat bone; midline anterior thorax wall; rib articulation.

RIBS 12 pairs.

True Ribs: Vertebrosternal Ribs Pairs 1 through 7; articulate anteriorly with sternum through costal cartilage; articulate posteriorly with thoracic vertebrae.

False Ribs: Vertebrochondral Ribs Pairs 8 through 10; costal cartilages attach to cartilage of rib above. *Vertebral (Floating) Ribs* Pairs 11 and 12; short costal cartilages with no anterior articulation.

COSTAL CARTILAGES Hyaline cartilage; strengthen thorax and provide flexibility.

APPENDICULAR SKELETON   pp. 166–182

UPPER LIMBS

PECTORAL GIRDLE Straddles upper thorax.

Clavicle S-shaped bone bracing scapula; holds shoulder away from rib cage.

Scapula Thin, flat, triangular bone over posterior surfaces of rib pairs 2 through 7.

ARM Humerus is the only bone.

FOREARM 2 bones.

Ulna Medial to radius in anatomical position.

Radius Lateral to ulna in anatomical position.

HAND 8 carpal bones in 2 transverse rows of 4 each toward wrist.

Proximal Row Lateral to medial—scaphoid, lunate, triquetrum, pisiform.

Distal Row Lateral to medial—trapezium, trapezoid, capitate, hamate.

PALM OF HAND 5 *metacarpal bones.*

FINGERS 14 *phalanges.*

LOWER LIMBS

PELVIC GIRDLE Formed by a pair of coxal bones.

Ilium, Ischium, Pubis Individual bones during embryonic development; fuse to form each adult coxal bone.

PELVIC CAVITIES

Greater or False Pelvis Above pelvic brim; expansible.

Lesser or True Pelvis Cavity below pelvic brim; restricted.

Sexual Differences in Pelvis Most are related to childbearing.

THIGH Femur forms thigh skeleton.

LEG: Tibia Strong, medial. *Fibula* Slender, lateral. *Patella* Protects knee joint between thigh and leg.

FOOT

7 tarsal bones toward ankle: talus, calcaneus, navicular, cuboid, medial cuneiform, intermediate cuneiform, and lateral cuneiform.

5 metatarsal bones.

TOES 14 *phalanges*

ARCHES OF FOOT 3 arches formed: 2 longitudinal, 1 transverse; supported by ligaments as well as muscles and their tendons; distribute weight evenly between heel and metatarsals.

SESAMOID BONES Form in tendons subjected to compression; generally around a joint; e.g., patella.

SELF-QUIZ

1. Bones may act as a storehouse for: (a) calcium; (b) phosphorus; (c) both calcium and phosphorus.

2. Flat bones lack: (a) periosteum; (b) a medullary cavity; (c) diploe.

3. The combination of collagen fibers and inorganic salts makes bone exceptionally strong without being brittle. True or False?

4. The connective tissues that give rise to bone are derived from cells of the ectodermal layer of the embryo. True or False?

5. The reabsorption of previously laid-down bone is accomplished by: (a) osteoblasts; (b) osteoprogenitor cells; (c) osteoclasts.

6. The marrow cavity of a long bone enlarges as the bone grows due to the action of: (a) fibroblasts; (b) osteocytes; (c) osteoclasts.

7. Calcitonin: (a) decreases the resorptive activity of osteoclasts and lowers the blood calcium level; (b) increases the resorptive activity of osteoclasts and raises the blood calcium level; (c) releases calcium ions from the bone, which pass to the blood and raise the plasma calcium level.

8. A gradual reduction in the rate of bone formation while the rate of bone absorption remains normal results in: (a) spina bifida; (b) osteoporosis; (c) osteomyelitis.

9. The effect of vitamin D is to: (a) soften the matrix of bone; (b) remove calcium from the matrix of bone; (c) increase the absorption of calcium from food by the wall of the intestine.

10. The single occipital bone forms the lower posterior wall of the calvarium as well as the anterior portion of the floor of the cranial cavity. True or False?

11. An abnormal lateral curve that occurs in the vertebral column is called a: (a) scoliosis; (b) lordosis; (c) kyphosis.

12. Match the common skeletal structure terms with the appropriate description.

Process
Trochanter
Tubercle
Condyle
Sulcus
Crest
Facet
Fossa
Fovea
Foramen
Meatus

(a) A smooth, nearly flat articular surface
(b) A pit; generally used for attachment rather than for articulation
(c) A hole
(d) A rounded prominence that articulates with another bone
(e) A canal
(f) A large, somewhat blunt process
(g) A depression; often used as an articular surface
(h) A sharp, prominent bony ridge
(i) A prominence or projection
(j) A nodule or small rounded process
(k) A groove

13. Match the structures with their appropriate lettered description.

Fontanels
Frontal bone
Sinuses
Parietal bones
Occipital bone
Ethmoid bone
Temporal bones
Maxillary bones
Zygomatic bones
Lacrimal bones
Mandible
Vomer bone
An auditory ossicle
Hyoid bone
Cervical vertebrae
Thoracic vertebrae
Sacral vertebrae

(a) Form most of the calvarium
(b) One of the bones of the floor of the cranial cavity
(c) Form the central part of the facial skeleton
(d) Air spaces in bone lined with mucous membranes
(e) The lower jaw
(f) "Soft spots" of the skull
(g) Forms the posterior inferior portion of the nasal septum
(h) Its most obvious landmark is the large foramen magnum
(i) Form the prominences of the cheek
(j) Single bone that forms the anterior superior region of the skull
(k) Help form the medial surface of the orbital cavity
(l) Form part of the middle cranial cavity
(m) Form the spinal column in the neck
(n) Serves for attachment of muscles of the tongue and throat
(o) Are fused and form a wedge between the coxal bones
(p) Malleus
(q) There are 12 of these

14. The xiphoid process is a component of the: (a) sternum; (b) ribs; (c) thoracic vertebrae.

15. Those ribs that have their costal cartilages attached to the cartilages of the rib above are called: (a) true ribs; (b) vertebrosternal ribs; (c) false ribs.

16. The first seven pairs of ribs articulate anteriorly with the sternum through the: (a) costal cartilages; (b) tubercles; (c) xiphoid process.

17. Which of the following are part of the appendicular skeleton? (a) pectoral girdle; (b) pelvic girdle; (c) mandible; (d) both a and b.

18. The scaphoid, lunate, triquetrum, and pisiform are bones of the: (a) foot, (b) hand; (c) tibia-fibula complex.

19. Which of the following forms the junction of the superior and axillary borders of the scapula and contains the glenoid fossa? (a) inferior angle; (b) superior angle; (c) lateral angle.

20. The coxal bone includes which of the following? (a) ischium; (b) fibula; (c) patella.

21. The lesser or true pelvis is the cavity: (a) above the pelvic brim; (b) anterior to the pelvic brim; (c) below the pelvic brim.

22. Compared with the male pelvis, the female pelvis: (a) is more massive; (b) is narrower at the pelvic outlet; (c) has an obtuse pubic angle.

23. Match the following terms with the appropriate lettered description.

Ilium
Ischium
Pubis
Acetabulum
False pelvis
Pelvic outlet
Femur
Greater trochanter
Condyles
Adductor tubercle
Tibia
Anterior crest
Fibula

(a) Forms anterior portion of each coxal bone
(b) Expanded space above the pelvic inlet, actually part of abdominopelvic cavity
(c) Rounded medial and lateral enlargements at distal end of femur shaft
(d) The medial bone of the leg
(e) Bone that forms the thigh
(f) Sharp longitudinal ridge on anterior surface of the tibia
(g) Small portion just above medial condyle of femur
(h) Upper flat expanded portion of each coxal bone
(i) Large lateral process just below the neck of the femur
(j) An opening bounded by coccyx and lower parts of coxal bones
(k) Posterior inferior portion of each coxal bone
(l) A slender bone lateral to the tibia
(m) Cup-shaped depression formed by junction of ilium, ischium, and pubis

LEARNING OBJECTIVES

After completing this chapter, you should be able to:

1. State two especially useful criteria for classifying body joints.

2. Name and distinguish between the two types of fibrous joints, and give an example of each.

3. Name and distinguish between the two types of cartilaginous joints, and give an example of each.

4. Describe the distinguishing features of synovial joints.

5. Name the four types of synovial joints, and give examples of each.

6. Name the ligaments of the shoulder, hip, and knee joints.

7. Describe several common joint disorders.

CHAPTER CONTENTS

FIBROUS JOINTS

CARTILAGINOUS JOINTS

SYNOVIAL JOINTS

CONDITIONS OF CLINICAL SIGNIFICANCE: ARTICULATIONS

KEY TERMS AND DERIVATIVES

bursa (*bursa* = purse or sac) a small sac or cavity filled with fluid and located at friction points

condyle (*condyl* = knob) a rounded projection at the end of a bone that articulates with another bone

extrinsic (*exterus* = exterior; *secus* = alongside) originating from outside an organ or part

glenoid (*glen* = joint socket) glenoid cavity: a depression in the scapula that articulates with the head of the humerus

symphysis (*symphusis* = a growing together) a relatively immovable joint in which the involved bones are usually connected by fibrocartilage

synchondrosis (*syn* = together; *chondrum* = cartilage) a relatively immovable joint in which the involved bones surfaces are connected by hyaline cartilage

syndesmosis (*syndesmos* = ligament) a relatively immovable joint in which the involved bone surfaces are joined by connective tissue fibers or connective tissue sheets

synostosis (*syn* = together; *ostosis* = bone) a union of originally separate bones by osseous material

synovial fluid (*syn* = like; *oon* or *ovum* = egg) a fluid that is secreted by the synovial membrane and that lubricates joint surfaces and nourishes articular cartilages

Articulations

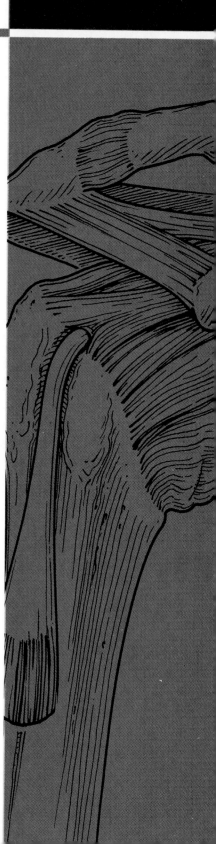

In the previous chapter, we were primarily interested in the support function of the skeleton. Here we will study how the individual bones of the skeleton join, or **articulate.** While some joints are rather rigid and permit little, if any, movement, most joints allow the bones to move in relation to each other. Various criteria can be used to classify the large numbers of joints within the body. Two of the most useful are (1) according to the *material that connects the joints* and (2) according to the *movement* allowed by the joints.

FIBROUS JOINTS

The **fibrous joints** include all those articulations in which the bones are held tightly together by fibrous connective tissue. Because of its role in strengthening the joint, this connective tissue is referred to as the *sutural ligament.* Very little material separates the ends of the bones, and no appreciable movement is allowed. For this reason the fibrous joints are also classified as **synarthroses** (*syn* = together; *arthron* = joint—that is, a nonmovable joint). However, slight movement is, in fact, permitted in some synarthroses. There are two types of fibrous joints, *sutures* and *syndesmoses,* depending in part on the length of the connective tissue fibers that hold the bones together.

Sutures

In **sutures,** the edges of the bones have interdigitations, or grooves, that fit closely and firmly together (Figure 7.1). Consequently, the connecting fibers spanning the small gap between the bones are very short. This type of joint is found only between the flat bones of the skull. In early adulthood, the fibers of the suture begin to be replaced by bone. Eventually, if the fibers are completely replaced, the bones on either side of the sutures become firmly fused together. This condition is called *synostosis* (that is, held together by bone).

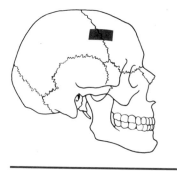

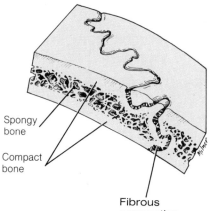

Spongy bone

Compact bone

Fibrous connective tissue

Figure 7.1
Suture. Note the short connective tissue fibers joining the two bones.

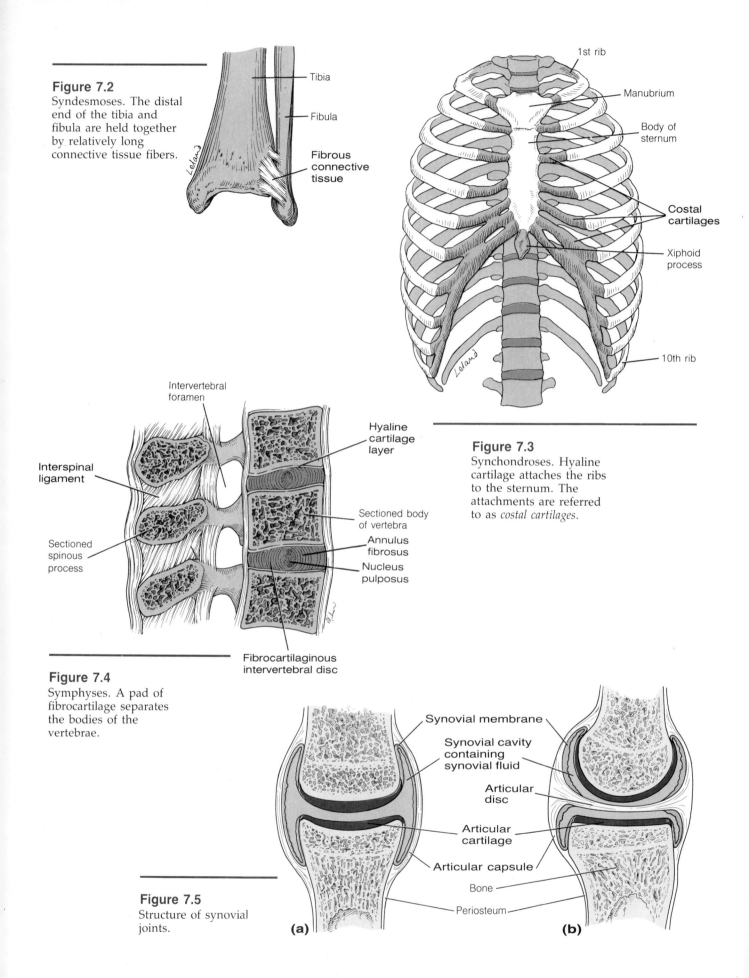

Figure 7.2
Syndesmoses. The distal end of the tibia and fibula are held together by relatively long connective tissue fibers.

Tibia

Fibula

Fibrous connective tissue

1st rib

Manubrium

Body of sternum

Costal cartilages

Xiphoid process

10th rib

Figure 7.3
Synchondroses. Hyaline cartilage attaches the ribs to the sternum. The attachments are referred to as *costal cartilages*.

Intervertebral foramen

Hyaline cartilage layer

Interspinal ligament

Sectioned body of vertebra

Annulus fibrosus

Nucleus pulposus

Sectioned spinous process

Fibrocartilaginous intervertebral disc

Figure 7.4
Symphyses. A pad of fibrocartilage separates the bodies of the vertebrae.

Synovial membrane

Synovial cavity containing synovial fluid

Articular disc

Articular cartilage

Articular capsule

Bone

Periosteum

Figure 7.5
Structure of synovial joints.

(a)   (b)

Syndesmoses

Like a suture, a **syndesmosis joint** (*syndesmosis* = together by bands) is held together by fibrous connective tissue (Figure 7.2). However, the ends of the bones are farther apart in a syndesmosis than in a suture. Consequently, the connective tissue fibers joining the bones are longer. The bones joined by syndesmoses are not held as firmly as those joined by sutures. And although syndesmoses can permit a kind of movement best described as "give," they still do not allow any true movement; therefore, these joints are considered to be synarthroses. The joint between the distal ends of the tibia and fibula and the mid–radius/ulna and mid–tibia/fibula joints—where the bones are held together by interosseous membranes—are examples of syndesmoses.

F7.2

CARTILAGINOUS JOINTS

In **cartilaginous joints,** the bones are united by cartilage. Because slight movement is possible in these articulations they are also classified as **amphiarthroses** (*amphi* = on both sides). There are two types of cartilaginous joints—*synchondroses* and *symphyses.*

am-fee-ar-throw'-seez

Synchondroses

The bones in a **synchondrosis joint** (*synchondrosis* = together by cartilage) are held together by hyaline cartilage (Figure 7.3). Many synchondroses are temporary joints, with the cartilage eventually being replaced by bone. This occurs between the epiphyses and the diaphysis of long bones (where the epiphyseal cartilages are replaced) and between certain of the skull bones. The joints formed between the first ten ribs and their costal cartilages are permanent synchondroses.

F7.3

Symphyses

The articular surfaces of bones that are joined by a **symphysis** are covered with a thin layer of hyaline catilage (Figure 7.4). Separating the bones within the joint is a fibrocartilaginous pad, which is the distinguishing feature of symphyses. These pads, or discs, are compressible, allowing the symphyses to serve as shock absorbers. The junction of the two pubic bones and the junctions between the bodies of adjacent vertebrae are examples of symphyses. The pads between the vertebrae are called *intervertebral discs.* Each disc is composed of a firm outer portion called the *annulus fibrosus* and a softer central portion, the *nucleus pulposus.* During development, the two halves of the mandible are joined by a midline symphysis; however, this joint becomes completely ossified by adulthood.

F7.4

an'-yoo-lus figh-bro'-sus

SYNOVIAL JOINTS

Most joints of the body are **synovial joints,** which are characterized by being freely movable. The movement of synovial joints is limited only by ligaments, muscles, tendons, or adjoining bones. Because of this freedom, synovial joints are also referred to as **diarthroses.** The term *diarthrosis* means "through joint," indicating that there are only relatively slight limitations to the movement of such joints. Another characteristic of synovial joints is the presence of a fluid-filled joint cavity.

sa-no'-vee-al

dye-ar-throw'-sis

Synovial joints have four distinguishing features: an *articular cartilage;* an *articular capsule;* a *synovial membrane;* and *synovial fluid* (Figure 7.5a). The **articular cartilage** is a thin layer of hyaline cartilage that covers the smooth articular surfaces of the bones. The **articular capsule** is a double-layered membrane that surrounds and encloses the joint. The outer layer of the capsule is composed of dense fibrous connective tissue whose fibers are firmly joined to the

F7.5a

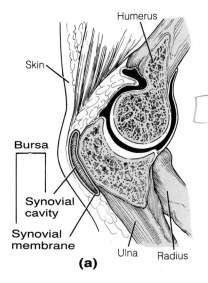

(a)

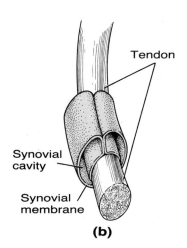

(b)

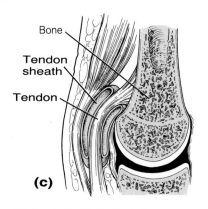

(c)

Figure 7.6
(a) Subcutaneous bursa of the elbow. **(b)** Structure of a tendon sheath. **(c)** Longitudinal section showing the position of a tendon sheath.

periosteum of the bones. Parallel bundles of fibers in the outer layer form ligaments that strengthen the joint. We will name some of these ligaments later in the chapter. The inner layer of the articular capsule is referred to as the **synovial membrane.** The synovial membrane consists of loose connective tissue whose inner surface is well supplied with capillaries. The membrane, which is often thrown into folds that project into the joint cavity, lines the entire joint cavity. However, it does not cover the surfaces of the articular cartilages or the articular disc. The synovial membrane produces a thick fluid called **synovial fluid.** Synovial fluid provides nourishment to the articular cartilages while it lubricates the joint surfaces. In fact, synovial fluid serves as a weight-bearing element in the joint as it keeps the articular cartilages of the bones that form the joint separate, preventing the cartilages from contacting one another. Normally, only enough synovial fluid is secreted to form a thin film over the surfaces within the joint. However, fluid production may be stimulated in a joint that is injured or becomes inflamed, and enough fluid may accumulate to cause swelling and discomfort. The joint capsule is well supplied with nerve fibers, which not only makes the perception of pain possible but also provides constant information concerning movement and position of the joint.

In addition to these four features, some synovial joints have **articular discs** (Figure 7.5b), or in the knee, **menisci,** of fibrocartilage, which extend inward from the articular capsule. Articular discs divide the synovial cavity into two separate cavities. In this form of joint, the synovial membrane lining the cavities extends only a short distance onto the surfaces of the disc. The jaw, the sternoclavicular joint, and the distal radioulnar joint contain articular discs.

In addition to the strengthening provided by the ligaments from the fibrous layer of the articular capsule, various muscles and their tendons, which cross the joints, serve to stabilize the joints while still permitting them to move.

Bursae and Tendon Sheaths

Synovial membranes form two other structures that, while not actually part of the synovial joints, are often associated with them. These are *bursae* and *tendon sheaths*. Both of these structures contain synovial fluid and function to reduce the friction that would occur during movement between a structure—such as skin, muscle, tendons, or ligaments—and the bone.

Bursae are small sacs lined with synovial membranes. Because they are filled with synovial fluid, they act as cushions between the structures they separate. There are many bursae distributed throughout the body. Some are subcutaneous, lying between the bone and the skin, such as the bursa separating the olecranon process of the elbow from the skin (Figure 7.6a). Most bursae, however, are located between the tendons and the bone. In some cases, bursae may be continuous with a joint cavity through small membrane tunnels.

Tendon sheaths (Figure 7.6b and c) are found where tendons cross joints and would, without the sheaths, be subjected to constant friction against the bones, as in the wrist and fingers. The sheaths are cylindrical synovial sacs, similar to bursae. They wrap around the tendons, forming fluid-filled, double-walled cushions for the tendons to slide through.

Movements of Synovial Joints

Synovial joints, unlike fibrous and cartilaginous joints, are not classified according to the material that connects the bones. Rather, they are named on the basis of the movements they permit. The shapes of the bony structures that surround a joint, and often the articular surface itself, generally limit the movements that are possible by any one joint. Many joints have *axes of rotation* that allow bones to move in various planes. A particular plane of movement is generally perpendicular to the axis. For instance, in the movement of the elbow, the axis is a horizontal line that passes through the joint from side to side. The bones rotate around this fulcrum (pivot point) in a vertical plane.

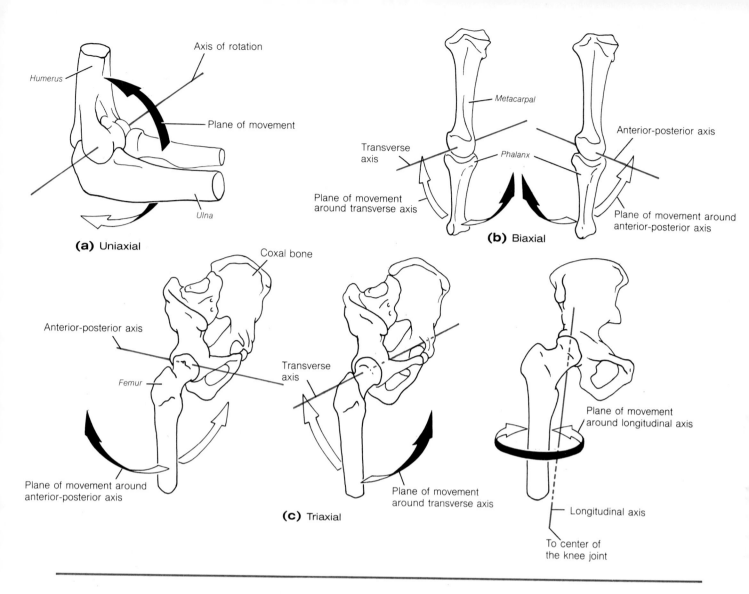

(a) Uniaxial

(b) Biaxial

(c) Triaxial

F7.7a
Box 7.1
F7.7b
F7.7c

Figure 7.7
Movements typical of synovial joints.

Joints that have only one axis of motion and can therefore move in only one plane, such as the elbow, are called **uniaxial joints** (Figure 7.7a). Note the movement of the knee, a uniaxial joint, in Box 7.1. Some joints have two axes, thus allowing movement in two planes that are at right angles to each other. Such joints are called **biaxial joints** (Figure 7.7b). Still other joints have more than two axes and permit movement in three planes. These are called **triaxial joints** (Figure 7.7c). The movement in many of the smaller joints is not restricted by the shapes of the articular surfaces, and slight movement is possible in any direction. Because their movements do not follow particular axes, these joints are referred to as **nonaxial.**

The general movements allowed in synovial joints can be placed into four groups: *gliding, angular, circumduction,* and *rotation.* In addition, several synovial joints allow movements unique to those particular joints. These unique movements are considered in the section on "Special Movements."

Gliding

The simplest and most common type of motion that can occur in a synovial joint is **gliding.** In this motion, the surfaces of adjoining bones move back and forth in relation to one another. In many cases the articulating surfaces are flat or slightly concave, but gliding can occur between any two adjoining surfaces regardless of their forms. The joints between the heads of the ribs and the bodies of the vertebrae and between the tubercles of the ribs and the transverse processes of the vertebrae allow gliding movement, as do numerous other joints.

Angular Movements

Angular movements increase or decrease the angle between two adjoining bones by moving in a single plane. Four kinds of angular movements may occur in various synovial joints: *flexion, extension, abduction,* and *adduction* (Figure 7.8).

flek'-shun
F7.8

FLEXION   When a bone is moved in an anterior-posterior plane in such a manner as to *decrease* the angle between it and the adjoining bone, **flexion** occurs. Examples include bending the elbow, bringing the thigh up toward the abdomen, and bringing the calf of the leg up toward the back of the thigh. Pulling the heel upward, thereby lowering the toe region of the foot, is referred to as **plantar flexion.**

EXTENSION   Extension is the opposite of flexion. It causes the angle between adjoining bones to *increase*. Extension occurs when a flexed joint is moved back to the anatomical position, such as straightening the arm, thigh, or knee. **Hyperextension** occurs when the part is moved beyond the straight position, such as arching the back or bringing the limbs posteriorly beyond the plane of the body. Although raising the toe region toward the shin is often considered to be extension of the foot, this movement is called **dorsiflexion.**

ABDUCTION   When a part such as a limb is moved away from the midline of the body, **abduction** occurs. In the case of the fingers, abduction involves moving them away from the midline of the hand. Abduction of the toes is accomplished by moving them away from the longitudinal axis of the second toe.

ADDUCTION   **Adduction,** the opposite of abduction, involves the movement of a part toward the midline of the body, back toward the anatomical position. In the case of the fingers, the movement is toward the midline of the hand. Adduction of the toes is accomplished by moving them toward the longitudinal axis of the second toe.

Circumduction

F7.8

The joint motion known as **circumduction** delineates a cone (Figure 7.8). The base of the bone is outlined by the movement of the distal end of the bone, with the apex of the cone lying in the articular cavity. The movement is actually a sequential combination of flexion, abduction, extension, and adduction. Circumduction is common at the hip and the shoulder joints and is possible in other joints also.

Rotation

F7.8

The motion of a bone around a central axis without any displacement of that axis is **rotation** (Figure 7.8) If the anterior surface of a bone such as the humerus or femur moves inward, it is called *inward (medial) rotation*. When the anterior surface turns outward it is *outward (lateral) rotation*.

SUPINATION   The term used to describe the outward rotation of the forearm, causing the palm to face upward or forward and the radius and the ulna to be parallel, is **supination.** The forearms are supinated in the anatomical position.

soo-pa-nay'-shun

PRONATION   The term used to describe the inward rotation of the forearm, causing the radius to cross diagonally over the ulna and the palm to face downward or backward, is **pronation.**

Special Movements

Some synovial joints allow special movements that cannot be described by any of the previously mentioned movements. These movements are *elevation, depression, inversion, eversion, protraction,* and *retraction.*

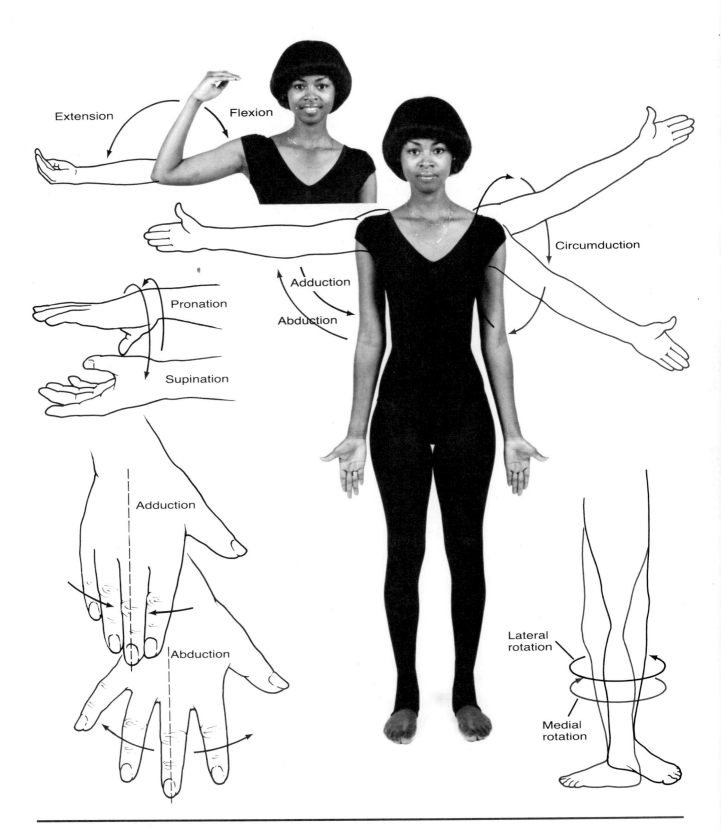

Extension

Flexion

Circumduction

Pronation

Adduction

Supination

Abduction

Adduction

Lateral
rotation

Abduction

Medial
rotation

Figure 7.8
Angular and circular
movements of synovial
joints.

ELEVATION   The motion that raises a part is **elevation.** This term is most commonly used to refer to the raising of the scapula, as when shrugging the shoulders, or raising the mandible, as when closing the mouth.

DEPRESSION   The motion that lowers a part is **depression.** This term is often used to refer to the lowering of the scapula or the mandible.

INVERSION   The twisting of the foot so that the sole faces inward with its inner margin raised is **inversion.**

EVERSION   The twisting of the foot so that the sole faces outward with its outer margin raised is **eversion.**

PROTRACTION   The motion that moves a part, such as the mandible, forward is **protraction.**

RETRACTION   The motion that returns a protracted part to its usual position is **retraction.**

Types of Synovial Joints

On the basis of the movements allowed and the shapes of the articular surfaces involved, it is possible to separate the synovial joints into six types. These types can be grouped according to whether they are *nonaxial, uniaxial, biaxial,* or *triaxial.*

Nonaxial Joints

GLIDING (ARTHRODIAL) JOINTS   The **gliding joints** are formed primarily by the apposition of flat, or only slightly curved, articular surfaces. Movement is allowed in any direction, being limited only by ligaments or bony processes that surround the articulation. Gliding joints are found between the articular process of vertebrae, and between most carpal and tarsal bones.

Uniaxial Joints

HINGE (GINGLYMUS) JOINTS   In **hinge joints,** the articular surfaces are shaped such that the only movements possible are flexion and extension. The elbow joint, the knee, and the joints between the phalanges of the fingers and toes (interphalangeal joints) are examples of hinge joints.

PIVOT (TROCHOID) JOINTS   The only movement allowed in a **pivot joint** is rotation around the longitudinal axis of the bone. Examples are the rotation of the first cervical vertebra (atlas) around the odontoid process of the second cervical vertebra (axis), and the proximal articulations between the radius and the ulna. In the atlas/axis joint, the odontoid process is held against the inside surface of the anterior arch of the atlas by a transverse ligament that passes behind the process while connecting the two sides of the anterior arch. In the radial/ulnar joint, the head of the radius is held firmly against the radial notch of the ulna by a strong annular ligament that encircles its head. The radius rotates within the annular ligament, allowing the forearm to pronate and supinate.

Biaxial Joints

CONDYLOID (ELLIPSOID) JOINTS   **Condyloid joints** have one articular surface slightly concave and the other slightly convex; thus, movement is allowed in two planes that are at right angles to each other. Flexion, extension, abduction, and adduction can occur in condyloid joints. Circumduction is possible also, but axial rotation is not. The articulations between the radius and the carpals, the occipital condyles of the skull on the first cervical verte-

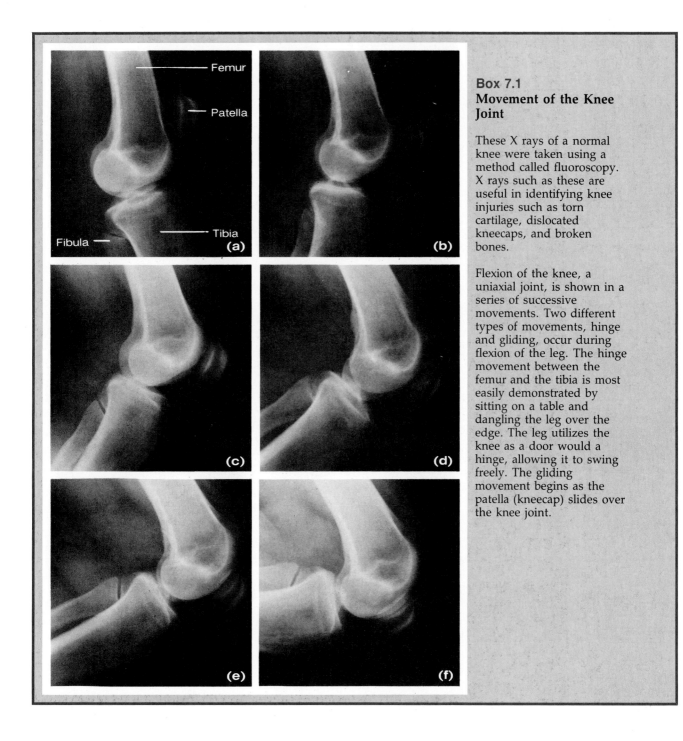

Box 7.1
Movement of the Knee Joint

These X rays of a normal knee were taken using a method called fluoroscopy. X rays such as these are useful in identifying knee injuries such as torn cartilage, dislocated kneecaps, and broken bones.

Flexion of the knee, a uniaxial joint, is shown in a series of successive movements. Two different types of movements, hinge and gliding, occur during flexion of the leg. The hinge movement between the femur and the tibia is most easily demonstrated by sitting on a table and dangling the leg over the edge. The leg utilizes the knee as a door would a hinge, allowing it to swing freely. The gliding movement begins as the patella (kneecap) slides over the knee joint.

bra, the metacarpophalangeal, and the metatarsophalangeal joints are condyloid joints.

SADDLE JOINTS   **Saddle joints** allow the same movements as the condyloid joints—flexion, extension, abduction, adduction, and circumduction. The articular surface of each bone is concave in one direction and convex in the other; therefore, the bones fit together just as two saddles would if the riding surface of one saddle were rotated 90° in relation to the other and the two surfaces were placed on top of one another. The only true saddle joint in the body is the carpometacarpal joint of the thumb.

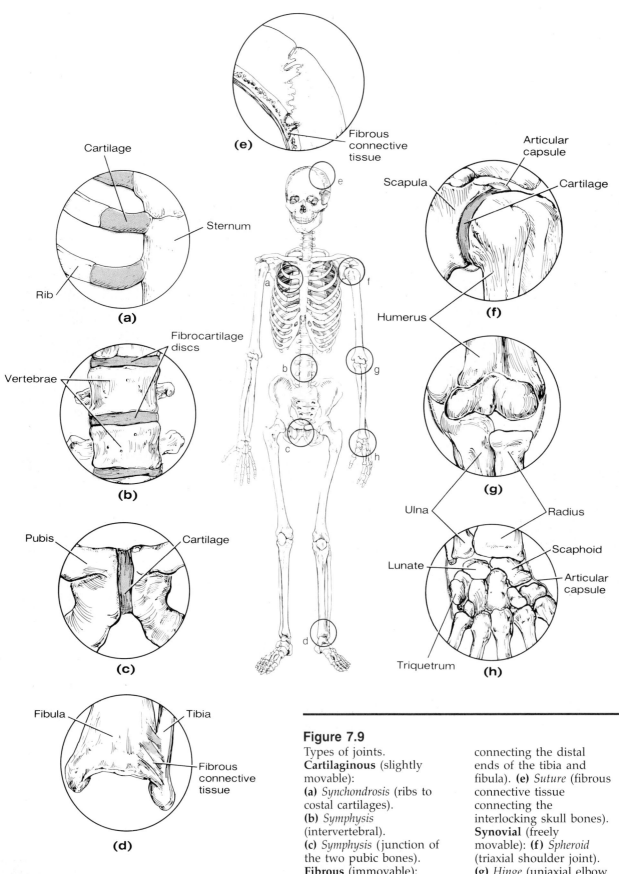

Figure 7.9
Types of joints.
Cartilaginous (slightly movable):
(a) *Synchondrosis* (ribs to costal cartilages).
(b) *Symphysis* (intervertebral).
(c) *Symphysis* (junction of the two pubic bones).
Fibrous (immovable):
(d) *Syndesmosis* (fibrous connective tissue connecting the distal ends of the tibia and fibula). **(e)** *Suture* (fibrous connective tissue connecting the interlocking skull bones).
Synovial (freely movable): **(f)** *Spheroid* (triaxial shoulder joint).
(g) *Hinge* (uniaxial elbow joint). **(h)** *Gliding* (nonaxial intercarpal joints).

Triaxial Joints

SPHEROID (BALL AND SOCKET) JOINTS **Spheroid joints** are formed by a spherical head of one bone fitting into a cup-shaped cavity on the other. Such joints allow movement around an indefinite number of axes. In addition to flexion, extension, abduction, adduction, and circumduction, spheroid joints allow medial and lateral rotation to occur. There are only two spheroid joints in the body—the shoulder and the hip.

Figure 7.9 illustrates most of the types of joints; Table 7.1 summarizes the main joints of the body.

F7.9,
Table 7.1

Table 7.1   Summary of the Main Joints of the Body

Joint	Type	Movement
Between the cranial bones	Fibrous (suture)	No appreciable movement
Between the distal tibia and the fibula	Fibrous (syndesmosis)	Slight movement ("give")
Between the mid-radius and the ulna (interosseous membrane)	Fibrous (syndesmosis)	Slight movement
Between the ribs and the sternum (sternocostal)	Cartilaginous (synchondrosis)	Slight movement
Between the two pubic bones	Cartilaginous (symphysis)	Slight movement
Between the bodies of the vertebrae	Cartilaginous (symphysis)	Slight movement
Between the sacrum and the ilium	Partly cartilaginous (synchondrosis) and partly synovial (gliding)	Generally no movement, but slight gliding movement is possible. In older people, fibers may hold the two bones firmly together.
Between the articular processes of the vertebrae	Synovial (gliding)	Nonaxial; gliding
Between the head of the rib and the body of the vertebra	Synovial (gliding)	Nonaxial; gliding
Between the occipital and the atlas	Synovial (condyloid)	Biaxial; flexion, extension, abduction, adduction, circumduction
Between the atlas and the odontoid process of the axis	Synovial (pivot)	Uniaxial; pivoting around the odontoid process
Between the sternum and the clavicle	Synovial (gliding)	Nonaxial; gliding
Between the acromion of the scapula and the clavicle	Synovial (gliding)	Nonaxial; gliding and rotation of the scapula upon the clavicle
Between the humerus and the scapula	Synovial (spheroid)	Triaxial: flexion, extension, abduction, adduction, circumduction, medial and lateral rotation
Between the ulna and the humerus	Synovial (hinge)	Uniaxial; flexion and extension
Between the head of the radius and the ulna	Synovial (pivot)	Uniaxial; pivoting longitudinal axis, as in pronation and supination
Between the radius and the carpals (scaphoid, lunate)	Synovial (condyloid)	Biaxial; flexion, extension, abduction, adduction, circumduction
Between the carpals	Synovial (gliding)	Nonaxial; gliding

Table 7.1   Summary of the Main Joints of the Body (continued)

Joint	Type	Movement
Between the first metacarpal and carpal	Synovial (saddle)	Biaxial; flexion, extension, abduction, adduction, circumduction
Between the second through fifth metacarpals and the carpals	Synovial (gliding)	Nonaxial; gliding
Between the second through fifth metacarpals and the phalanges	Synovial (condyloid)	Biaxial; flexion, extension, abduction, adduction, circumduction
Between the phalanges (hand and foot)	Synovial (hinge)	Uniaxial; flexion, extension
Between the femur and the coxal bone	Synovial (spheroid)	Triaxial; flexion, extension, abduction, adduction, circumduction, medial and lateral rotation
Between the tibia and the femur	Synovial (hinge)	Uniaxial; flexion, extension (some rotation)
Between the proximal end of the fibula and the tibia	Synovial (gliding)	Nonaxial; gliding
Between the distal ends of the fibula and the tibia and the talus	Synovial (hinge)	Uniaxial; flexion, extension
Between the tarsals	Synovial (gliding)	Nonaxial; gliding
Between the tarsals and the metatarsals	Synovial (gliding)	Nonaxial; gliding
Between the metatarsals and the phalanges	Synovial (condyloid)	Biaxial; flexion, extension, abduction, adduction, circumduction

Ligaments of Selected Synovial Joints

Ligaments play an important role in maintaining the proper positioning of bones that articulate in synovial joints while at the same time allowing relatively free movement of the joints. Therefore, we will consider the ligaments of four of the more important joints: the shoulder, elbow, hip, and knee.

Ligaments of the Shoulder Joint

In the shoulder joint, the head of the humerus is received by the shallow glenoid fossa of the scapula. The joint is loosely constructed, which permits extremely free movement but also allows it to become frequently dislocated. It is protected above by the coracoid process and the acromion process of the scapula.

F7.10
Like all synovial joints, the shoulder joint is enclosed by an **articular capsule** (Figure 7.10). This capsule attaches to the rim of the glenoid fossa and extends outward to the anatomical neck of the humerus. The capsule is strengthened anteriorly by two ligaments: the **coracohumeral ligament,** which extends from the coracoid process to the greater tubercle; and the **glenohumeral ligaments,** which are several thickenings of the lower portion of the capsule itself. The **glenoid labrum** (*labrum* = lip), a rim of fibrocartilage that surrounds the glenoid fossa, adds somewhat to the stability of the joint by deepening the fossa.

In addition to these ligaments, the shoulder joint depends on the surrounding muscles for strength. The biceps brachii muscle has a unique arrangement: The tendon of its long head, which arises from the superior border of the glenoid fossa, passes inside of the joint capsule of the shoulder and

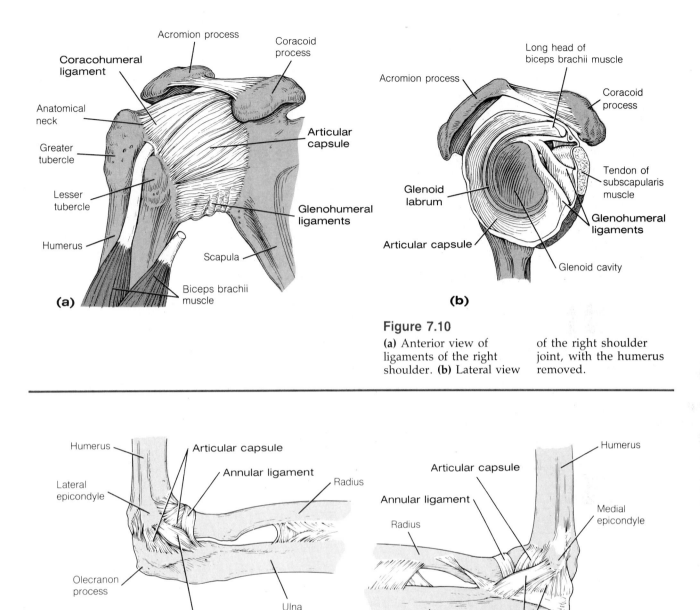

Figure 7.10
(a) Anterior view of ligaments of the right shoulder. **(b)** Lateral view of the right shoulder joint, with the humerus removed.

Figure 7.11
Ligaments of the right elbow joint. **(a)** Lateral view. **(b)** Medial view.

through the intertubercular groove of the humerus. Thus, in effect, it helps to hold the head of the humerus against the scapula.

Ligaments of the Elbow Joint

The elbow joint is a hinge joint where the trochlea of the humerus is received into the trochlear notch of the ulna. Closely associated with the elbow joint is a gliding joint between the capitulum of the humerus and the superior surface of the head of the radius. These two articulations share a common joint cavity and are enclosed by a single fibrous **articular capsule** (Figure 7.11).

Medial and lateral thickenings of the articular capsule serve to stabilize the elbow joint. The medial thickening, the **ulnar collateral ligament,** passes from the medial epicondyle of the humerus to the medial surface of the ulna between the olecranon process and the coronoid process. The lateral thickening, the **radial collateral ligament,** passes from the lateral epicondyle of the

F7.11

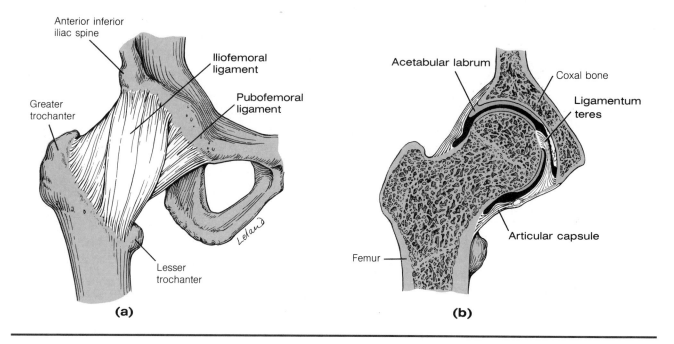

Anterior inferior
iliac spine

Iliofemoral
ligament

Pubofemoral
ligament

Greater
trochanter

Lesser
trochanter

(a)

Acetabular labrum

Coxal bone

Ligamentum
teres

Articular capsule

Femur

(b)

Figure 7.12
(a) Anterior view of ligaments of the right hip joint. **(b)** Frontal section through the right hip joint.

F7.12

is-kee-o-fem'-a-ral

humerus to the annular ligament and to the lateral surface of the ulna. The *annular ligament* is a strong band of fibers in the distal margin of the articular capsule of the elbow joint. The annular ligament does not strengthen the elbow joint to any great degree. Instead, it encircles the head and the upper part of the neck of the radius as it passes between the anterior and posterior margins of the trochlear notch of the ulna. In this manner, the radius is held tightly against the ulna and yet is allowed to rotate freely, as occurs during pronation and supination.

Ligaments of the Hip Joint

The head of the femur fits into the deep acetabulum of the coxal bone (Figure 7.12), making the hip joint a more stable joint than the shoulder. The **articular capsule,** which extends from the margin of the acetabulum to the anatomical neck of the femur, completely encloses the joint. The capsule is strengthened anteriorly by the **iliofemoral** and the **pubofemoral ligaments.** On its posterior surface, the capsule is strengthened by the **ischiofemoral ligament.**

The acetabulum is surrounded by a fibrocartilaginous rim called the *acetabular labrum.* The acetabular labrum is incomplete at its inferior margin, which gives it a horseshoe shape. Like the glenoid labrum of the shoulder joint, this labrum deepens the joint cavity. A unique ligament called the **ligamentum teres** extends through the joint cavity from the fovea on the head of the femur to the gap at the lower portion of the acetabular labrum. Because the ligamentum teres is slack during most movements of the hip, it is believed not to contribute significantly to the strength of the joint.

Ligaments of the Knee Joint

F7.13

The knee joint (Figure 7.13) is a complicated joint that is vulnerable to injury. It is classified as a hinge joint because its movements are restricted by the surrounding ligaments to, for the most part, flexion and extension. It has the structure, however, of a condyloid joint, with the condyles of the femur articulating with the slightly concave condyles of the tibia. The articular surface on the medial condyle of the femur is somewhat longer from front to back and is less curved than the articular surface on the lateral condyle. As a consequence of these structural differences, the final phase of complete extension of the knee joint primarily involves movement of the medial condyle of the femur on the tibia. This causes the tibia to undergo some lateral rotation (or the femur to undergo some medial rotation). Similarly, the extended joint must

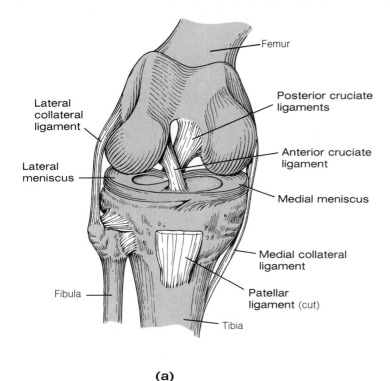

Lateral collateral ligament

Lateral meniscus

Fibula

Femur

Posterior cruciate ligaments

Anterior cruciate ligament

Medial meniscus

Medial collateral ligament

Patellar ligament (cut)

Tibia

(a)

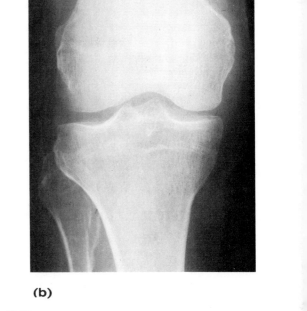

(b)

Figure 7.13

(a) Ligaments of the right knee joint. The femur is flexed slightly to allow the ligaments to be seen. The articular capsule and the patella have been removed.
(b) Anteroposterior X ray of the right knee.

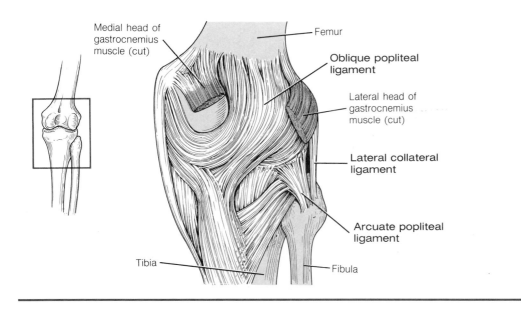

Medial head of gastrocnemius muscle (cut)

Femur

Oblique popliteal ligament

Lateral head of gastrocnemius muscle (cut)

Lateral collateral ligament

Arcuate popliteal ligament

Tibia

Fibula

Figure 7.14

Posterior ligaments of the right knee joint.

be unlocked by a slight medial rotation of the tibia before flexion of the knee can occur. When the knee joint is in a partially flexed position, it is possible for it to undergo even more rotation.

The articulation is completely enclosed by an **articular capsule.** The capsule is strengthened posteriorly by the **oblique popliteal ligament** and the **arcuate popliteal ligament** (Figure 7.14). The oblique popliteal ligament is a broad, flat band attached proximally to the posterior surface of the femur just above the articular surface of the lateral condyle. From here it extends downward and medially to attach to the posterior surface of the head of the tibia.

pop-la-tee'-al
F7.14

203

Figure 7.15
Functions of the cruciate ligaments. **(a)** Knee extended: taut anterior cruciate ligament prevents overextension of the joint. **(b)** Knee flexed: taut posterior cruciate ligament prevents the tibia from slipping posteriorly. (The medial condyle of the femur has been removed to expose the cruciate ligaments.)

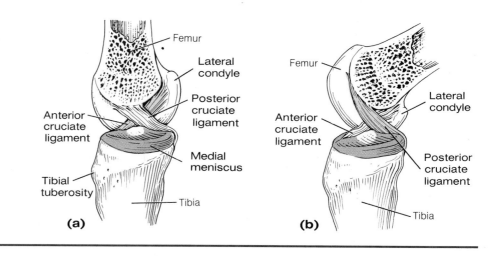

The arcuate popliteal ligament passes from the posterior surface of the lateral condyle of the femur to the styloid process of the head of the fibula. The knee joint is stabilized medially and laterally by very strong **medial** and **lateral collateral ligaments,** which extend from the condyles of the femur to the tibia or the fibula. The collateral ligaments limit the amount of rotation that is possible by the knee joint. The joint is strengthened anteriorly by the **patellar ligament,** which extends from the patella to the tibial tuberosity. This ligament is a continuation of the central tendon of the quadriceps femoris muscles of the anterior thigh.

The flat superior surface of the condyles of the tibia, which are the largest weight-bearing surfaces in the body, are deepened by crescent-shaped cartilages called **medial** and **lateral menisci.** The menisci are attached only at their outer margins, and they frequently become damaged or torn loose in athletic injuries.

Additional stability is added to the knee joint by the presence within the joint cavity of **anterior** and **posterior cruciate ligaments** (*cruciate* = cross-shaped). These ligaments extend diagonally from the superior surface of the tibia to the distal end of the femur, between the condyles. They are called cruciate because their paths cross each other. Because of their unique structural arrangements, the cruciate ligaments perform very specialized functions. When the knee is extended, the anterior cruciate ligament is taut, thus preventing overextension of the joint (Figure 7.15a). When the knee is flexed, the posterior cruciate ligament becomes taut, preventing the tibia from slipping posteriorly (Figure 7.15b).

F7.15a

F7.15b

CONDITIONS OF CLINICAL SIGNIFICANCE
Articulations

Sprains

Sprains result when the twisting or overstretching of a joint causes a ligament to tear or to separate from its bony attachment. If the trauma is severe, excessive tissue fluid may accumulate and cause swelling.

Dislocations

When the articular surfaces of the bones are forcibly displaced, *dislocations* occur. When the dislocations are severe, the bones, as well as the surrounding tendons and ligaments, may be damaged. The most commonly dislocated joints are those of the shoulder, thumb, and fingers.

Bursitis

When one or more of the bursae surrounding a joint become inflamed, the condition is known as *bursitis.* This disorder can result from injury, heavy exercise, or infection. The bursae fill with excessive synovial fluid, causing discomfort and limiting motion in the affected joint.

Tendinitis

Tendinitis is the inflammation of tendon sheaths around a joint. The condition is generally characterized by local tenderness at the point of inflammation and severe pain upon movement of the affected joint. Tendinitis can result from trauma to, or excessive use of, a joint. The wrist, elbow (where it is referred to as "tennis elbow"), and shoulder joints are most often affected.

Slipped Disc

Among the more common discomforts that people endure are those associated with the back. There are numerous causes of back pain; one cause that involves joints is a *slipped disc*. In this condition, which is most common in the lower back, the relatively soft nucleus pulposa within an intervertebral disc is squeezed to one side of the disc. This can result from trauma or from improper distribution of weight along the vertebral column resulting from poor posture or deformity of the vertebrae. The displacement of the nucleus pulposa causes the relatively firm anulus fibrosus of the disc to protrude, or in some cases, to rupture. If the disc protrudes into the vertebral canal it can compress spinal nerves or, in the thoracic or cervical regions, the spinal cord itself. The pressure can cause severe pain along the paths of the nerves, as well as numbness of the regions supplied by the nerves. If the pressure on the spinal nerves continues, actual nerve damage can result. The nerve damage, in turn, can cause weakness and degeneration of muscles supplied by the damaged nerves.

Torn Menisci

The fibrocartilage menisci of the knee joint serve to somewhat adapt the surfaces of the tibia to the shape of the femoral condyles. The menisci are not firmly attached, and during rotation they move slightly on the tibia. Because of this movement, sudden changes of direction while bearing the body weight can cause the menisci to tear loose. Severe pain and swelling of the joint can result.

An innovative surgical technique called *arthroscopic surgery* has greatly increased the success of operations performed to repair damaged menisci as well as other joint injuries. In this technique, a needlelike viewing instrument called an *arthroscope* is inserted into the joint through a tiny incision (Figure 7.16). The arthroscope contains a fiber-optic light source that makes it possible to examine the interior of the joint visually and to observe the positions of cutting instruments inserted through other tiny incisions. This procedure can be performed under local anesthesia, and the patient can often return home from the hospital the same day.

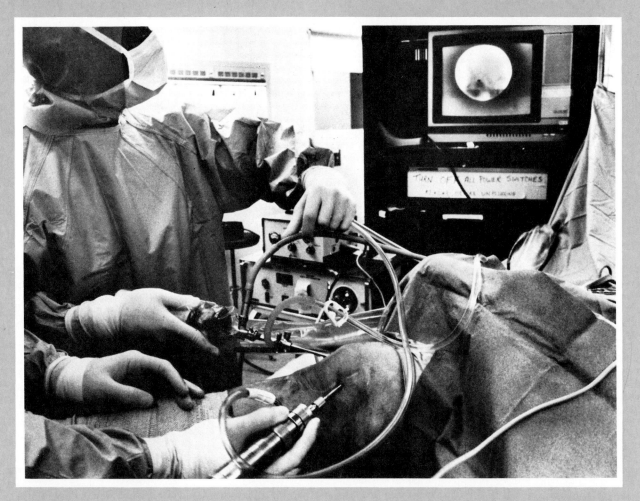

Figure 7.16
Arthroscopic surgery.

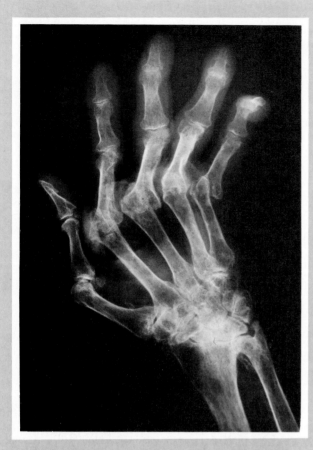

Figure 7.17

X ray of rheumatoid arthritis in a hand. The enlarged joints are a product of an inflamed synovial membrane. In most of the serious cases, the articular cartilage completely erodes, allowing the bones to fuse together. This is visible in the second finger from the right.

Arthritis

Many different types of inflammation of the joints fall under the general term *arthritis*. In these conditions there may be pathological changes in the joint membranes, cartilage, and bone that cause swelling and pain. The causes of arthritis are unknown, but trauma to a joint, bacterial infection (staphylococci, streptococci, and gonococci), and metabolic disorders have been implicated. At least one form of arthritis (rheumatoid) is thought to be caused by an immune response to inflammation of the synovial membrane of the affected joint cavity. There is evidence suggesting that arthritis may be genetically inherited, since some families show a predisposition toward the condition.

Osteoarthritis

Osteoarthritis, the most common form of arthritis, is a chronic inflammation that causes the articular cartilage in the affected joint to gradually degenerate. The inflammation is accompanied by pain, swelling, and stiffness. As the articular cartilage degenerates, bony spurs develop from the exposed ends of the bones that form the joint. These spurs tend to restrict the movement of the joint and often cause discomfort.

Rheumatoid Arthritis

Rheumatoid arthritis is a severely damaging form of the disease that tends to affect principally the small joints of the body, such as those in the hands, feet, knees, ankles, elbows, and wrists (Figure 7.17). It affects women more frequently than men. The condition begins with inflammation of the synovial membrane of the joint, causing swelling and pain. The inflamed synovial membrane produces an abnormal tissue known as a *pannus*, which grows over the surface of the articular cartilage. As the disease progresses, the articular cartilage beneath the pannus, and in some cases the bone itself, is gradually destroyed. Eventually the pannus fills the joint space and becomes invaded by fibrous tissue, thereby restricting joint movement. In severe cases, calcification of the pannus may ankylose (fuse) the joint, making it immovable.

Gouty Arthritis

Gout is a condition characterized by sudden severe pain and swelling of the joints. It affects primarily the toes, insteps, ankles, heels, knees, and wrists. Gout is more common in males. It is due to an inherited genetic defect that causes either an increased production of uric acid or a reduction in the ability of the kidneys to excrete uric acid. In either case the result is an increase in the level of uric acid in the blood *(hyperuricemia)*. Uric acid is an end product of purine metabolism. The excessive level of uric acid in the blood causes the body fluids to become supersaturated and eventually results in the formation of sodium urate crystals in the soft tissues of the body as well as in the joints.

When the joints are affected, the condition is called *gouty arthritis*. As crystals accumulate in the joints and the soft tissues surrounding the joints, they cause an inflammation that eventually may erode the articular cartilage and the underlying bone, causing intense pain and immobility of the joint.

Effects of Aging

With aging, there is a progressive loss of the cartilaginous surface of the joints, which may be accompanied by the appearance of bony, spurlike overgrowths in the joint. This often produces a condition known as *degenerative osteoarthropathy*.

The effects of aging on the joints—particularly of degenerative osteoarthropathy—vary greatly from individual to individual and are influenced both by genetic factors and by environmental factors such as the heavy use of certain joints. There is a progressive increase in degenerative osteoarthropathy starting as early as the age of 20. By the age of 80, virtually everyone has some degenerative osteoarthropathy of the knee and the elbow joints, with somewhat lower frequencies for the hip and the shoulder joints. Women often develop bony swellings (called Heberden's nodes) in the terminal phalanges. In men, the spine is the most common site of degenerative osteoarthropathy. Here the intervertebral discs degenerate, which results in new bone formation between the bodies of the vertebrae. This degeneration can produce a narrowing of the intervertebral foramina, which causes chronic and painful pressure on the nerve roots.

FRONTIERS IN HEALTH

New Hope for the Arthritic

Arthritis is this nation's number one crippler, afflicting more than 31 million Americans, or one in seven. Arthritis strikes people of all ages, although most cases occur in people older than 30. The disease claims a million new victims each year and costs the nation more than $14 billion a year in lost wages, medical bills, and disability claims.

Pauline is one of those victims. For her, each day is a long episode of pain and discomfort. Bathing, eating, writing, and even dressing are difficult and painful. She cannot open a screw-top jar or pick up coins. The disfigured, swollen joints of her hand have become practically useless. For Pauline a relaxing walk in the park is out of the question too, as her knee joints have been painfully affected by this incurable disease.

Arthritis literally means inflammation of the joints. According to the Arthritis Foundation, what we call arthritis actually encompasses at least 100 different rheumatic diseases—the painful diseases that affect the joints and are characterized by inflammation and stiffness.

Osteoarthritis and **rheumatoid arthritis** are the most common arthritic diseases. Osteoarthritis, which causes a degeneration of the articular cartilages and the development of bony spurs from the ends of the bones, is believed to result from excessive stress on the joints. It especially affects joints that have been abused by improper exercise or constant demanding work. Obesity contributes to the excessive stress.

Rheumatoid arthritis, whose cause remains obscure, results from an erosion of the articular cartilage in joints and its eventual replacement with fibrous tissue. Rheumatoid arthritis affects mainly the joints of the hands, hips, knees, and feet. Enzymes from phagocytic cells in the synovial fluid and synovial membranes appear to be responsible for the erosion. Inflammation occurs along with the erosion and weakens the joint capsule and supporting ligaments. The affected joint slowly degenerates and becomes stiff, swollen, and sore.

For Pauline and millions of Americans like her, new hope has emerged in recent years with the perfection of artificial joints. One of the most successful of these is the metacarpophalangeal joint. In a recent study carried out by scientists in the United States and the Soviet Union, 100 patients were fitted with artificial metacarpophalangeal joints made of a special type of plastic. The results of this work are very encouraging. All the recipients felt that the newly constructed joints noticeably improved the appearance of their hands. Even more encouraging, the new joints allowed renewed use of the crippled digits: nearly all the patients reported that their ability to grasp objects improved after the operation. Common chores that previously were difficult to perform or required assistance became easier, and many of the tasks made impossible by arthritis were once again accomplishable. The surgery also relieved the constant pain that accompanies arthritis.

Arthroplasty—the surgical replacement of a joint—has improved tremendously in recent years, thanks

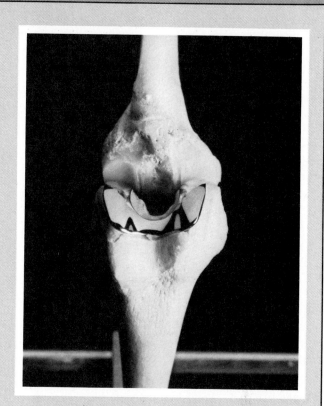

Photo of artificial knee joint.

largely to new glues and new materials that make artificial joints more durable. The computer has also contributed.

Until recently most artificial joints were manufactured by hand. In the first step of this complex process, physicians would X ray a patient's joint. The X rays were used to make blueprints for the replacement joint, and the blueprints were then sent to a laboratory, which constructed the replacement joints by hand. This process has now undergone considerable change, thanks to medical researchers at New York's Hospital for Special Surgery, who are working on a computer-assisted design, computer-assisted manufacturing technique called CAD-CAM.

With CAD-CAM, physicians first make numerous X rays of the joint, so that a complete three-dimensional picture can be composed. This information is fed into a computer, which takes the patient's age and other factors into consideration to construct a workable joint design. The information is then fed into a sophisticated set of lathes and milling machines that build the prescribed joint from a block of titanium.

For the more than 150,000 Americans who receive artificial joints each year, this cooperation of science and technology offers joints that move and a relief from pain.

STUDY OUTLINE

FIBROUS JOINTS (SYNARTHROSES) Articulations in which bones are held together tightly by fibrous connective tissue; nonmovable joints; two types classed by length of fibers that hold bone together.
pp. 189–191

SUTURES Grooves on edges of bones fit closely together; short connecting fibers; found only between flat bones of skull.

SYNOSTOSIS In adulthood, suture fibers replaced by bone; sutures fused together.

SYNDESMOSES Joint fibers are longer, called ligaments; allow some "give" movement; example is joint between distal ends of tibia and fibula.

CARTILAGINOUS JOINTS (AMPHI-ARTHROSES) Bones united by cartilage; slight movement   p. 191

SYNCHONDROSES Joints in which bones are held together by hyaline cartilage.

TEMPORARY SYNCHONDROSES Cartilage replaced by bone; for example, epiphyses of long bones.

PERMANENT SYNCHONDROSES Between first ten ribs and costal cartilages.

SYMPHYSES Articular surfaces of bones covered with thin layer of hyaline cartilage; fibrocartilage pad separates bones within the joint; shock-absorbing action; junction of pubic bones and junctions between adjacent vertebrae are examples.

SYNOVIAL JOINTS (DIARTHROSES) Freely movable, with movement limited only by ligaments, muscles, tendons, adjoining bones.   pp. 191–204

FLUID-FILLED JOINT CAVITY

ARTICULAR CARTILAGE Thin layer of hyaline cartilage that covers smooth articular bone surfaces.

ARTICULAR CAPSULE Dense fibrous connective tissue enclosing joint; joined to periosteum.

SYNOVIAL MEMBRANE Inner surface of articular capsule; vascular.

SYNOVIAL FLUID Clear, viscous secretion of synovial membrane; lubricates and nourishes.

ARTICULAR DISCS Present in some synovial joints; fibrocartilage that extends inward from capsule, dividing synovial cavity in two; for example, jaw, distal radioulnar joints.

BURSAE AND TENDON SHEATHS Formed by synovial membranes; reduce friction.

BURSAE Sacs lined with synovial membranes, filled with synovial fluid; subcutaneous or between tendon and bone.

TENDON SHEATHS Cylindrical synovial sacs around tendons; found where tendons cross joints.

MOVEMENTS OF SYNOVIAL JOINTS

GLIDING Bones move back and forth upon one another.

ANGULAR MOVEMENTS Increase or decrease angle between two adjoining bones by moving in a single plane.

Flexion Bone moved in anterior-posterior plane to decrease angle between it and adjoining bone.

Extension Increases angle between adjoining bones.

Abduction Body part moved away from midline.

Adduction Body part moved toward midline.

CIRCUMDUCTION Delineates a cone; base outlined by movement of distal end of bone and apex in articular cavity; hip and shoulder joints are examples.

ROTATION Motion of bone around a central axis.

Supination Outward rotation of forearm; palm anterior.

Pronation Inward rotation of forearm; palm posterior.

SPECIAL MOVEMENTS

Elevation Raising a body part.

Depression Lowering a body part.

Inversion Twisting foot so that sole faces inward.

Eversion Twisting foot so that sole faces outward.

Protraction Moving a part forward.

Retraction Returning protracted body part to usual position.

TYPES OF SYNOVIAL JOINTS Classed by movements allowed and shapes of articular surfaces.

NONAXIAL JOINTS

Gliding (Arthrodial) Joints

ARTICULAR SURFACE Formed by apposition of flat or slightly curved surfaces.

MOVEMENT Any direction.

EXAMPLE Found between vertebrae, carpal, and tarsal bones.

UNIAXIAL JOINTS

Hinge (Ginglymus) Joints

MOVEMENT Flexion and extension.

EXAMPLE Proximal articulations of radius and ulna.

Pivot (Trochoid) Joints

MOVEMENT Rotation around longitudinal axis of bone.

EXAMPLE Proximal articulations of radius and ulna.

BIAXIAL JOINTS

Condyloid (Ellipsoid) Joints

ARTICULAR SURFACE One slightly concave; other slightly convex.

MOVEMENT In two perpendicular planes (flexion, extension, abduction, adduction, circumduction).

EXAMPLE Radiocarpal articulations.

Saddle Joints

ARTICULAR SURFACE For each bone, is concave in one direction and convex in the other.

MOVEMENT Same as condyloid joints.

EXAMPLE Carpometacarpal joint of thumb.

TRIAXIAL JOINTS

Spheroid (Ball and Socket) Joints

ARTICULAR SURFACE Spherical head of one bone fits in cup-shaped socket on second bone.

MOVEMENT Allows medial and lateral rotation plus all condyloid movements.

EXAMPLE shoulder and hip joints.

LIGAMENTS OF SELECTED SYNOVIAL JOINTS

LIGAMENTS OF THE SHOULDER JOINT

Articular Capsule Encloses joint.

Coracohumeral Ligament Strengthens joint anteriorly; extends from coracoid process to greater tubercle.

Glenohumeral Ligaments Several thickenings of lower portion of capsule; glenoid labrum deepens fossa.

Biceps Brachii Muscle Tendon enhances joint strength.

LIGAMENTS OF THE ELBOW JOINT

Articular Capsule Encloses joint.

Ulnar Collateral Ligament From medial epicondyle to ulna.

Radial Collateral Ligament From lateral epicondyle to annular ligament and ulna.

LIGAMENTS OF THE HIP JOINT

Articular Capsule Encloses joint.

Iliofemoral and Pubofemoral Ligaments Provide anterior joint strength.

Ischiocapsular Ligament Strengthens posteriorly.

Acetabular Labrum Deepens joint cavity.

Ligamentum Teres Extends through joint cavity; no significant contribution to hip-joint strength.

LIGAMENTS OF THE KNEE JOINT

Articular Capsule Encloses joint.

Oblique and Arcuate Popliteal Ligaments Strengthen capsule posteriorly.

Medial Collateral Ligament Stabilizes joint medially.

Lateral Collateral Ligament Stabilizes joint laterally.

Patellar Ligament Strengthens joint anteriorly.

Medial and Lateral Menisci Cartilages that deepen condyles of tibia.

Anterior and Posterior Cruciate Ligaments Prevent hyperextension and posterior tibial slip.

CONDITIONS OF CLINICAL SIGNIFICANCE: ARTICULATIONS pp. 204–206

STRAINS Joint being overstretched or twisted results in ligament tearing or separation.

DISLOCATIONS Articular surfaces of bones forcibly displaced.

BURSITIS Inflamed bursa resulting from injury, exercise, or infection.

TENDINITIS Inflammation of tendon sheath.

SLIPPED DISC Nucleus pulposa squeezed to cause protrusion from intervertebral disc; causes pressure on spinal nerves.

TORN MENISCI Menisci torn loose; generally occurs during sudden changes of direction.

ARTHRITIS Joint inflammation caused by trauma or bacterial infection, metabolic disorders, or other unknown causes; may be inherited.

OSTEOARTHRITIS Most common form; gradual degeneration of articular cartilage and development of bony spurs.

RHEUMATOID ARTHRITIS Severely damaging form; pannus develops on surfaces of articular cartilage; articular cartilage and bone beneath it often destroyed; joint may fuse.

GOUTY ARTHRITIS Sudden severe pain and swelling; due to excessive production of uric acid or inability to excrete uric acid; sodium urate crystals form in joints and soft tissues; articular cartilage may be eroded.

EFFECTS OF AGING Progressive loss of articular cartilage and growth of bony spurs. Often results in degenerative osteoarthropathy.

SELF-QUIZ

1. The humerus/ulna joint is an example of a fibrous joint. True or False?

2. The joint found between the flat bones of the skull is classed as: (a) syndesmosis; (b) suture; (c) amphiarthrosis.

3. Most joints of the body are: (a) synchondroses; (b) symphyses; (c) synovial.

4. Which of the following exemplifies a symphysis? (a) junction of the two pubic bones; (b) junctions between the costal cartilages; (c) the epiphyses of a long bone to the diaphysis.

5. The term *diarthrosis* refers to a synovial joint that is kept apart by a fluid-filled cavity. True or False?

6. Match the terms with the appropriate lettered description.

Synarthroses	(a)	A joint in which the bones on either side of a suture become firmly fused
Sutures		
Syndesmoses		
Synostosis	(b)	The fibers in this synarthrosis joint are relatively long
Amphiarthroses		
Synchondroses	(c)	These joints usually serve as shock absorbers
Symphyses		
Synovial	(d)	All fibrous joints
	(e)	Most joints of the body are of this type
	(f)	Articulations that permit slight movement
	(g)	The edges of the bones that form these joints have grooves that fit very firmly and closely together
	(h)	Temporary joints in which the original cartilage eventually is replaced by bone

7. The inability to produce the fluid that keeps most joints moist would likely be due to a disorder in the: (a) bursae; (b) synovial membrane; (c) articular cartilage.

8. Bursae and tendon sheaths are part of the synovial joints. True or False?

9. Synovial joints, like fibrous and cartilaginous joints, are classified according to the material that connects the bones. True or False?

10. The elbow is an example of a: (a) uniaxial joint; (b) biaxial joint; (c) triaxial joint.

11. Match the angular movement terms with the appropriate lettered descriptions. There may be more than one answer per term.

    | Flexion | (a) | When you move a toe toward the midline of your foot |
    | Extension | | |
    | Abduction | (b) | Arching the back |
    | Adduction | (c) | When you move a limb away from the midline of your body |
    | Plantar flexion | | |
    | Dorsiflexion | | |
    | | (d) | A bone moved in an anterior-posterior plane, decreasing the angle between it and adjoining bone |
    | | (e) | Lowering the toe region of the foot |
    | | (f) | The opposite of flexion |
    | | (g) | Raising the toe region of the foot toward the shin |

12. *Supination* is the term used to describe the inward rotation of the forearm, which causes the radius to cross diagonally over the ulna. True or False?

13. This movement is characteristic of the hip and shoulder joints: (a) pronation; (b) supination; (c) circumduction.

14. Twisting the foot so the sole faces outward, with its outer margin raised, is called: (a) inversion; (b) eversion; (c) protraction.

15. The joints found between the articular processes of vertebrae and between most carpal and tarsal bones are termed: (a) hinged; (b) gliding; (c) condyloid.

16. Match the joint types listed with the appropriate lettered descriptions. There may be more than one answer per joint type.

    | Nonaxial | (a) | Condyloid (ellipsoid) |
    | Gliding | (b) | Carpometacarpal joint of the thumb |
    | Uniaxial | | |
    | Hinge | (c) | The shoulder and hip joints |
    | Pivot | | |
    | Biaxial | (d) | Gliding (arthrodial) |
    | Condyloid | (e) | The only movement here is rotation around the longitudinal axis of the bone. |
    | Saddle | | |
    | Triaxial | | |
    | Sphenoid | (f) | Hinge (ginglymus) |
    | | (g) | Found between most carpal bones |
    | | (h) | Articulation between the radius and the carpals |
    | | (i) | The elbow joint |

17. The first metacarpal/carpal joint is which of the following? (a) saddle; (b) condyloid; (c) suture.

18. Both the shoulder joint and the hip joint contain: (a) an articular capsule; (b) an iliofemoral ligament; (c) a ligamentum teres.

19. The ligament that prevents the tibia from slipping posteriorly is the: (a) medial meniscus; (b) posterior cruciate; (c) ligamentum patellae.

20. Match the joint types listed with the appropriate lettered descriptions. There may be more than one answer per joint type.

Distal tibia/ fibula	(a) Fibrous (syndesmosis)
Pubic/pubic	(b) Cartilaginous (symphysis)
Ulna/humerus	(c) Synovial (condyloid)
Sternum/clavicle	(d) Synovial (gliding)
Radius/carpals	(e) Nonaxial
Tarsal/tarsal	(f) Uniaxial; flexion; extension
Occipital bone/ atlas	(g) This joint is capable of slight movement ("give")

21. Arthritis refers to many different types of inflammation of the joints. True or False?

22. A painful condition due to an inherited genetic defect that causes an increased level of uric acid in the blood is called: (a) bursitis; (b) arthritis; (c) gouty arthritis.

LEARNING OBJECTIVES

After completing this chapter, you should be able to:

1. Describe three types of muscle.

2. Describe the subcellular structure of a skeletal muscle fiber.

3. Discuss the events involved in excitation–contraction coupling.

4. Explain the mechanism of skeletal muscle contraction.

5. Cite three ways that adenosine triphosphate (ATP) is supplied to a muscle during activity.

6. Describe the influence of length on the development of tension by a skeletal muscle.

7. Explain what is meant when muscles are called prime movers, antagonists, synergists, or fixators.

8. Describe three different types of skeletal muscle fibers.

9. Describe three functional differences between smooth muscle and skeletal muscle.

CHAPTER CONTENTS

KEY TERMS AND DERIVATIVES

antagonist (*ant* = against) a muscle that opposes the action of the agonist under consideration

endomysium (*end* = internal; *mysium* = muscle) the thin connective tissue between the fibers of a muscle bundle

epimysium (*epi* = around; *mysium* = muscle) the sheath of connective tissue surrounding a muscle

fasciculus (*fase* = a bundle) a bundle of muscle fibers

myofibril (*myo* = muscle) a fibril found in the cytoplasm of muscle

myofilament (*myo* = muscle; *fil* = thread) a component of the sarcomeres of skeletal muscle cells

sarcomere (*sarx* = flesh; *meros* = part) one of the functional segments into which a myofibril of striated muscle is divided

synergist (*syn* = together; *ergon* = work) a muscle cooperating with another muscle to produce a movement neither alone can produce

The Muscular System:
General Structure and Physiology

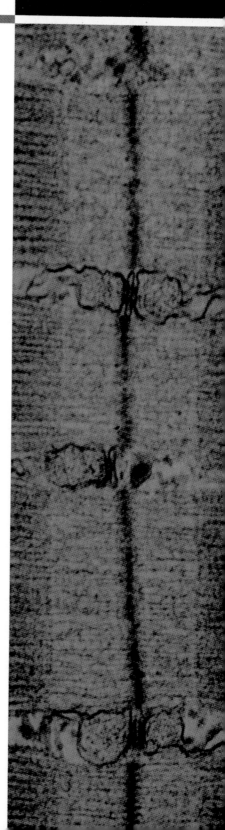

Muscle is composed of contractile cells that actively develop tension and shorten. As a result, muscle is important in the movement of body parts, the alteration of the diameters of tubes within the body, the propulsion of materials within the body, and the excretion of substances from the body. In addition, muscle contractions produce significant amounts of heat that can be used to maintain normal body temperature. Because of its many functions, muscle tissue contributes importantly to the maintenance of homeostasis.

MUSCLE TYPES

The body contains three types of muscle—skeletal muscle, smooth muscle, and cardiac muscle.

Skeletal Muscle

As the name implies, most **skeletal muscle** attaches to the bones of the skeleton. The contractions of skeletal muscle exert force on the bones, thereby moving them. Consequently, skeletal muscle is responsible for activities such as walking and manipulating objects in the external environment.

Skeletal muscle is *voluntary* muscle—that is, its contractions are normally under the conscious control of the individual. Under many conditions, however, skeletal muscle contractions require no conscious thought. For example, a person does not usually have to think about contracting the skeletal muscles involved in maintaining posture. Skeletal muscle contractions are regulated by signals transmitted to the muscle by a portion of the nervous system known as the somatic nervous system.

When viewed microscopically, skeletal muscle cells exhibit alternating transverse light and dark bands that give them a striped, or *striated*, appearance.

Smooth Muscle

Smooth muscle is so named because its cells lack the striations evident in skeletal muscle cells. Smooth muscle is found in the walls of hollow organs and tubes such as the stomach, intestines, and blood vessels, and its contractions govern the movement of materials through these structures.

Smooth muscle is *involuntary* muscle—that is, its contractions are not normally under the conscious control of the individual. However, under appropriate circumstances (see Chapter 14), a person can gain some voluntary control over smooth muscle. Smooth muscle contractions are regulated by factors intrinsic to the muscle itself, by hormones, and by signals transmitted to the muscle by a portion of the nervous system known as the autonomic nervous system.

Cardiac Muscle

Cardiac muscle is a specialized type of muscle that forms the wall of the heart. Cardiac muscle is involuntary muscle, and its contractions are regulated by intrinsic factors, hormones, and the autonomic nervous system. Cardiac muscle cells are striated. The anatomy and the physiology of cardiac muscle are considered in Chapter 18.

213

EMBRYONIC DEVELOPMENT OF MUSCLE

Skeletal Muscle

With the exception of the muscles of the limbs and some of the muscles of the head, skeletal muscles develop embryonically from **somites,** which are masses of mesodermal cells. The somites are located dorsally, along both sides of the axial skeleton. Only cells from one portion of a somite, called the *my'-a-tome* **myotome,** differentiate into muscle cells. The mesodermal cells from the myotomes spread downward between the skin and the body cavity, until the right and left sides meet ventrally in the midline. The skeletal muscles of the trunk develop from this sheet of mesoderm. Typical somites do not form in the head of the embryo; consequently, most of the muscles of the head develop from the general mesoderm of that region. The muscles of the limbs begin development from condensations of mesoderm within the embryonic limb buds. Some of the mesodermal cells that form the limb muscles probably migrate to the area from the myotomes.

Individual muscle cells are called **muscle fibers,** and the immature cells that give rise to skeletal muscle fibers are called *myoblasts.* To form a skeletal muscle fiber, individual myoblasts become aligned into a tubelike structure (myotube). The plasma membranes of adjacent myoblasts then fuse, producing a long tube that contains all of the nuclei from the fused myoblasts. Therefore, each mature skeletal muscle fiber contains several nuclei. With further maturation, contractile proteins and an extensive internal membrane system develop within the myotube.

Mature skeletal muscle fibers are generally incapable of mitosis. However, scattered among the muscle fibers are *satellite cells,* which are considered to be inactive myoblasts—and thus are potentially capable of dividing. Satellite cells are prevalent in the muscles of young children, but as a muscle matures, the number of satellite cells present decreases; in a mature muscle, they represent less than 1% of the tissue. Nevertheless, their presence would seem to represent the potential for formation of new muscle fibers, even within a fully developed muscle.

Smooth Muscle

As the digestive tube and the body organs form within the embryo, mesodermal cells migrate to them and form a thin layer around them. These mesodermal cells develop into the smooth muscles of the body.

Cardiac Muscle

The muscle of the heart forms in a manner similar to that of the smooth muscles. Mesodermal cells migrate to and surround the early heart while it is still in the form of a tubule. Cardiac muscle begins contracting very early in the development of the embryo, even before the peripheral blood vessels have completely formed.

GROSS ANATOMY OF SKELETAL MUSCLE

When examining skeletal muscles without the aid of a microscope, as during gross dissection, it is possible to note various distinguishing features of individual muscles. Among the most obvious features are the connective tissue coverings, the attachments, and the shape of each muscle.

Connective-Tissue Coverings

Each skeletal muscle is composed of many individual muscle fibers (muscle cells) that are held together by thin sheets of fibrous connective tissue called

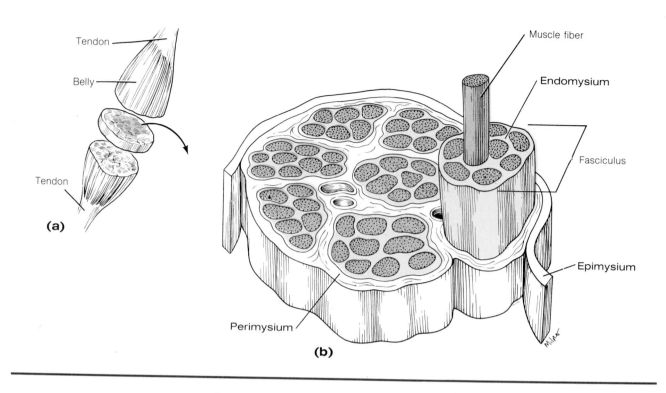

Tendon

Belly

Tendon

(a)

Muscle fiber

Endomysium

Fasciculus

Epimysium

Perimysium

(b)

F8.1

Figure 8.1
Connective tissue of a muscle. **(a)** Entire muscle, with the belly sectioned. **(b)** Enlargement of a cross section of the belly.

fascia. The fascia that invests an entire muscle is called the **epimysium** (Figure 8.1). Fascia also penetrates the muscle, separating the muscle fibers into bundles called **fasciculi.** This fascia is called the **perimysium.** Very thin extensions of the fascia, called the **endomysium,** envelop the cell membrane of each individual muscle fiber. Blood vessels and nerves pass into the muscle with the fascial sheaths to reach the individual muscle cells. Beds of capillaries are located between the muscle cells, and each cell is supplied by a branch of a nerve cell.

Skeletal Muscle Attachments

Skeletal muscles are anchored to the skeleton by extensions of the endomysium, perimysium, and epimysium. These connective tissues continue beyond the end of a muscle and either they attach directly to the periosteum of a bone, as often seen at the proximal attachment of a muscle, or they may blend into a strong fibrous connection called a **tendon,** which then becomes continuous with the periosteum of the bone. Some tendons are quite short, whereas others are more than a foot in length. Tendons that take the form of broad thin sheets are called **aponeuroses.**

en-doo-mis'-ee-um
per-a-mis'-ee-um
e-pa-mis'-ee-um

ap-o-noo-ro'-seez

The attachments of both ends of a skeletal muscle are given specific names. The **origin** is the less movable end and is generally proximal. The **insertion** is the more movable end and is generally distal. The widest portion of a muscle, between the origin and the insertion, is known as the **belly** of the muscle. A muscle is described as *arising* from the origin and *inserting* at the insertion. The origin may be rather broad, arising from several different places on a bone, or even from several different bones. The insertion, in contrast, tends to be much more restricted. With the exception of sphincter muscles that surround body openings and some of the facial muscles that insert on the skin, joints are located between the origins and the insertions of skeletal muscles. When a skeletal muscle contracts, it shortens, using the joint as a pivot point to pull the insertion closer to the origin.

The origins and insertions that are described in Chapter 9 are the most common. However, you should keep in mind that for some muscles it is functionally possible to reverse the origin and the insertion—that is, the more

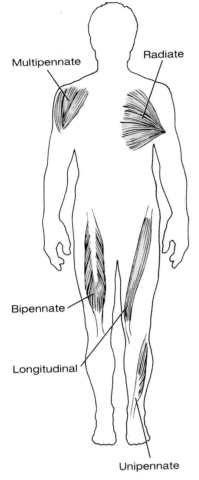

Figure 8.2
Variation in muscle shapes.

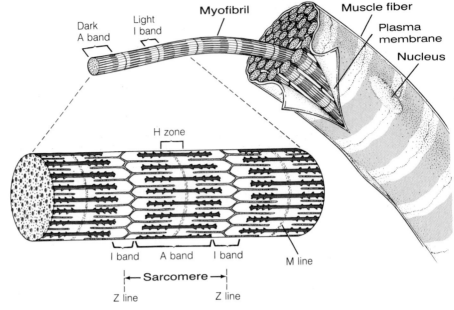

Figure 8.3
Microscopic anatomy of an individual skeletal muscle fiber (cell). Note the striated (striped) appearance of the muscle fiber and the myofibrils.

fixed end, which is normally called the origin, can be used in some actions as the more movable end. For instance, certain of the superficial muscles of the chest are described as having their origins on the thorax and their insertions on the humerus. Clearly, the humerus is generally more movable than the thorax. However, if an individual is doing pull-ups, the thorax is moved towards the humerus and is therefore serving as an insertion.

Skeletal Muscle Shapes

The arrangement of the bundles of muscle fibers (fasciculi) varies within the different skeletal muscles (Figure 8.2). In some muscles, the fasciculi run parallel to the long axis of the muscle, forming straplike muscles. The contraction of muscles that possess such **longitudinally** arranged fasciculi produces considerable movement; however, such muscles do not have much power. Less movement, though greater power, is produced by muscles that have a tendon running their entire length, with the fasciculi inserting diagonally into this tendon. In some muscles of this type, all of the fasciculi insert onto one side of the tendon. This arrangement is called **unipennate. Bipennate muscles** have fasciculi inserting obliquely on both sides of the tendon. The fasciculi of certain muscles have a complex arrangement that involves the convergence of several tendons. These are **multipennate muscles.** In a few muscles, the fasciculi converge from a broad origin into a single narrow tendon, an arrangement known as **radiate** (fan shaped).

F8.2

yoo-na-pen'-it

MICROSCOPIC ANATOMY OF SKELETAL MUSCLE

When a skeletal muscle is examined with the aid of a microscope, it is apparent that the muscle fibers have a regular subcellular structure. Skeletal muscle fibers are multinucleate cells approximately 10 to 100 microns in diameter and

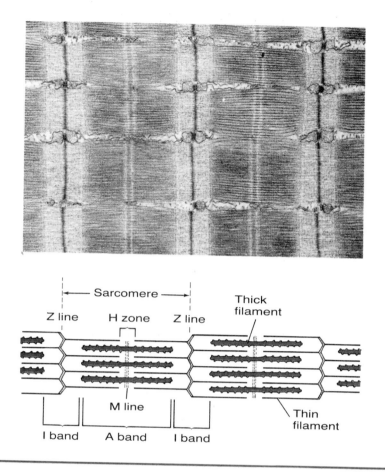

Figure 8.4
Longitudinal view of the structure of sarcomeres. The I band consists only of thin filaments and is divided by the Z line. The A band consists of thick filaments that overlap at either end with the thin filaments. The region where only thick filaments occur is the H zone. A single sarcomere extends from one Z line to the next. Thus, half of the I band is associated with one sarcomere and half with the neighboring sarcomere.

frequently many centimeters long. Each fiber contains several hundred to several thousand regularly ordered, threadlike **myofibrils** that extend lengthwise throughout the cell (Figure 8.3). When highly magnified, a myofibril exhibits alternating transverse light and dark bands, which are responsible for the striated appearance of the cell. The light bands are named **isotropic bands,** or **I bands,** and the dark bands are named **anisotropic bands,** or **A bands.** Crossing the center of each I band is a dense, fibrous **Z line.** The Z lines divide the myofibrils into a series of repeating segments known as **sarcomeres.** In the center of a sarcomere and, therefore, in the center of an A band, is a somewhat less dense region, the **H zone.** A thin, dark **M line** crosses the center of the H zone. F8.3

A sarcomere contains two distinct types of longitudinally oriented **myofilaments:** thick filaments and thin filaments (Figure 8.4). **Thick filaments** occupy the A band; and the H zone contains only thick filaments. The M line is formed by linkages between the thick filaments that hold them in a parallel arrangement. **Thin filaments** occupy the I band and part of the A band. The thin filaments attach to the Z lines. In the region of the A band where thick and thin filaments overlap, there is a hexagonal arrangement of thin filaments around each thick filament (Figure 8.5). F8.4

F8.5

Composition of the Myofilaments

The thick filaments consist mainly of the protein **myosin.** A myosin molecule is made up of two identical subunits, each shaped something like a golf club. The two subunits are tightly wound around each other so that a complete myosin molecule has two rather bulbous heads protruding from one end of a straight shaft. A thick filament contains approximately 200 myosin molecules arranged in such a way that the shafts of the molecules are bundled together

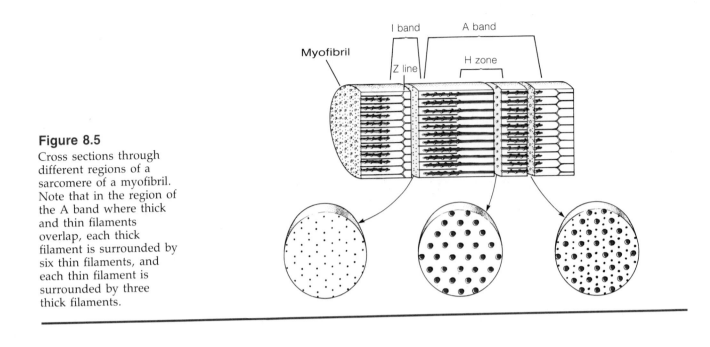

Figure 8.5

Cross sections through different regions of a sarcomere of a myofibril. Note that in the region of the A band where thick and thin filaments overlap, each thick filament is surrounded by six thin filaments, and each thin filament is surrounded by three thick filaments.

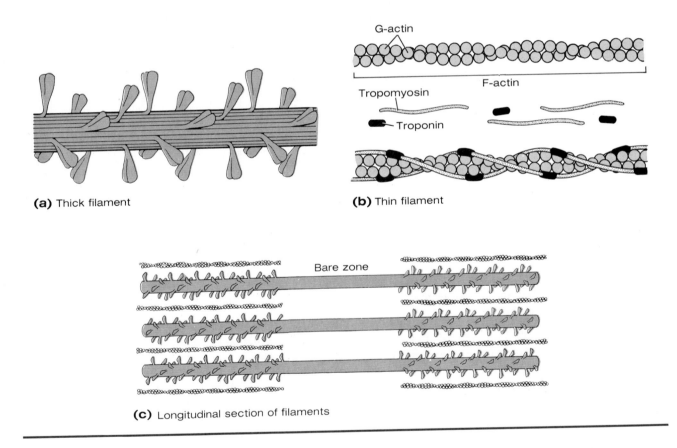

(a) Thick filament

(b) Thin filament

(c) Longitudinal section of filaments

Figure 8.6

Composition of myofilaments. **(a)** The thick filament consists mainly of golf-club–shaped molecules of myosin bundled together so the heads of the molecules project from the shaft of the filament and spiral around the shaft. **(b)** The thin filament. Individual G-actin subunits link into a double chain of F-actin. In a single thin filament, two chains of G-actin subunits twist around one another.

Along the surface of each chain lie threadlike tropomyosin molecules that each cover seven G-actin subunits. Each tropomyosin molecule is attached to a molecule of troponin. Troponin is also attached to actin. **(c)** Longitudinal view of thick and thin filaments as arranged in a sarcomere. Note that the myosin molecules project in opposite directions on either side the bare zone.

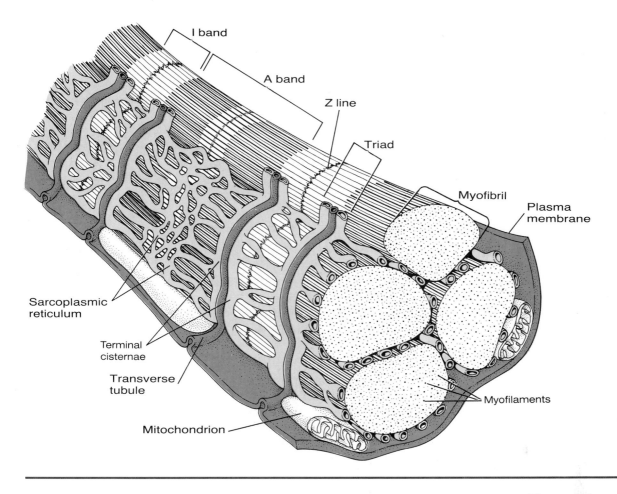

I band

A band

Z line

Triad

Myofibril

Plasma membrane

Sarcoplasmic reticulum

Terminal cisternae

Transverse tubule

Mitochondrion

Myofilaments

Figure 8.7
Schematic representation of transverse tubules and the sarcoplasmic reticulum of skeletal muscle.

with the heads of the molecules (called **cross bridges**) facing outward (Figure 8.6a). The myosin molecules face in opposite directions on either side of the center of a thick filament, with the shafts of the molecules directed toward the center. Because of this arrangement, the central area of the filament contains the shaft portions of myosin molecules but no myosin heads (Figure 8.6c).

The thin filaments consist mainly of the proteins **actin, tropomyosin,** and **troponin** (Figure 8.6b). The actin portion of a thin filament consists of spherical subunits of globular (G) actin that are organized into a double chain of fibrous (F) actin. The F-actin structure resembles two strings of pearls twisted around one another in a spiral, with each pearl being the equivalent of a G-actin subunit. Although the G-actin subunits are spherical, they have a definite polarity, and they link to one another from front to back. Associated with each chain of G-actin subunits are threadlike molecules of tropomyosin. The tropomyosin molecules lie end to end along the surfaces of the actin chains, and each tropomyosin molecule covers approximately seven G-actin subunits. Attached to each tropomyosin molecule and also to actin is a smaller molecule of the protein troponin. The arrangement of thick and thin filaments in a sarcomere is illustrated in Figure 8.6c.

Transverse Tubules and Sarcoplasmic Reticulum

Tubules known as **transverse tubules (t tubules)** pass deep into a skeletal muscle fiber from the plasma membrane (Figure 8.7). In addition, a membranous network, the **sarcoplasmic reticulum,** extends throughout the fiber and surrounds each myofibril. The sarcoplasmic reticulum is in some respects similar to the endoplasmic reticulum of other cells. Elements of the sarcoplasmic reticulum and t-tubule system lie close to one another at the junction of

F8.6a

F8.6c

F8.6b

F8.6c

F8.7

Figure 8.8
Isometric and isotonic muscle contractions. **(a)** In an isometric contraction, the muscle develops tension but does not shorten. **(b)** In an isotonic contraction, the muscle shortens while under a constant load.

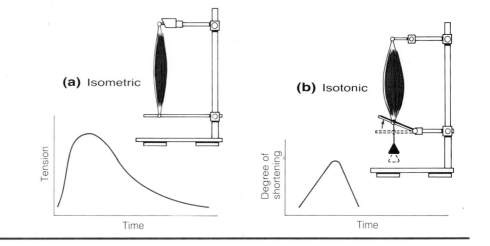

the A and I bands of the sarcomeres. At these locations, structures consisting of three tubules (triads) are formed.

CONTRACTION OF SKELETAL MUSCLE

Experimentally, two basic types of muscle contraction—isometric contraction and isotonic contraction—are frequently employed to study muscle function. An **isometric** (*iso* = equal; *metric* = measure) contraction is a contraction during which the length of a muscle remains constant (Figure 8.8a). In this type of contraction, the muscle actively develops tension and exerts force on an object but does not shorten. In the body, isometric types of muscle contractions occur when a person supports an object in a fixed position or attempts to lift an object that is too heavy to move. An **isotonic** (*iso* = equal; *tonic* = tension) contraction is a contraction during which a muscle shortens while under a constant load (Figure 8.8b). In this type of contraction, the force against which the muscle shortens remains constant even though the length of the muscle changes considerably. In the body, where most skeletal muscles attach to bones and more than one muscle is usually involved in a movement, pure isotonic contractions seldom occur because the load on a muscle frequently changes as the muscle shortens. Nevertheless, contractions during which a muscle shortens, such as the contractions that move the legs during walking or the arms when lifting an object, are often referred to as isotonic contractions. Regardless of whether the load on a muscle changes or not, before a muscle can shorten, it must first develop sufficient tension to equal and overcome the resistance of the load against which it contracts—only then can shortening occur (Figure 8.9).

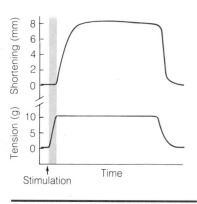

Figure 8.9
Record of a contraction during which a muscle shortens and lifts a 10-gram load. Immediately following its stimulation, there is a period (color) during which the muscle develops sufficient tension to equal and overcome the resistance of the load but does not shorten. Following this the muscle shortens and lifts the load.

F8.9

Contraction of a Skeletal Muscle Fiber

At the cellular level, the same basic events occur during both isometric and isotonic skeletal muscle contractions. A skeletal muscle contracts when stimuli from the nervous system excite the individual muscle fibers. This initiates a series of events that lead to interactions between the thick and thin filaments of the sarcomeres of the fibers. These interactions are responsible for the development of tension and the shortening of the fibers.

The Neuromuscular Junction

Nerve cells known as **motor neurons** supply the neural stimulation that skeletal muscle fibers require in order to contract. These neurons form specialized

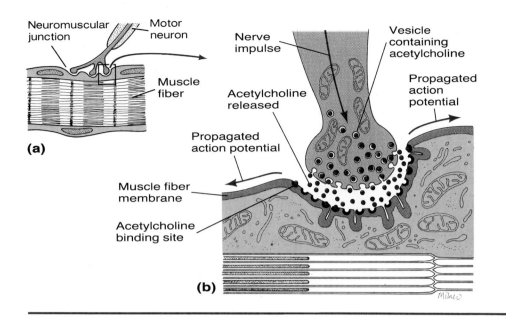

Figure 8.10
A neuromuscular junction.
(a) Diagrammatic representation of the general structure of a neuromuscular junction. **(b)** A more highly magnified section of the junctional area illustrating the events that occur at a neuromuscular junction. When a nerve impulse arrives at a terminal ending of a motor neuron, acetylcholine is released. The acetylcholine binds with receptors on the muscle fiber plasma membrane at the junction. This leads to a change in the membrane's permeability to sodium and potassium ions and produces a propagated action potential that travels along the membrane.

junctions, called **neuromuscular junctions,** with skeletal muscle fibers (Figure 8.10a). At a neuromuscular junction, an ending of a motor neuron closely approaches a specialized point along the plasma membrane of a skeletal muscle fiber, but it does not directly contact the membrane. Instead, a small gap separates the motor neuron ending from the muscle fiber membrane. Most skeletal muscle fibers have only one neuromuscular junction.

Excitation of a Skeletal Muscle Fiber

Motor neurons transmit brief, intermittent electrical signals called **nerve impulses** toward skeletal muscle fibers. The events that occur when a nerve impulse arrives at a neuromuscular junction are discussed in detail in Chapter 11. Here it is sufficient to state that a nerve impulse does not directly stimulate a skeletal muscle fiber because it cannot cross the gap that separates a motor neuron ending from the muscle fiber plasma membrane. Instead, a nerve impulse stimulates a skeletal muscle fiber indirectly, and it does so in the following way. When a nerve impulse reaches a neuromuscular junction, a chemical substance, *acetylcholine,* is released from the motor neuron ending (Figure 8.10b). The acetylcholine diffuses to the muscle fiber plasma membrane, where it attaches to acetylcholine receptors. This causes a change in the membrane's permeability to sodium and potassium ions and produces a stimulatory electrical impulse, or **propagated action potential,** that travels along the membrane of the muscle fiber. The propagated action potential triggers a series of intracellular events that culminate in interactions between the thick and thin filaments of the sarcomeres and the contraction of the fiber.

Excitation–Contraction Coupling

The series of events by which a propagated action potential in the plasma membrane of a skeletal muscle fiber causes interactions between the thick and thin filaments of the sarcomeres and the contraction of the fiber are grouped together under the heading of **excitation–contraction coupling.** From the plasma membrane, the propagated action potential passes along the t tubules into the central areas of the fiber. As it moves along the t tubules, it triggers the release of calcium ions from sites in the sarcoplasmic reticulum called *terminal cisternae* (Figure 8.7). The calcium ions bind to troponin molecules of the thin filaments, leading—as is explained in the following section—to interactions between the thick and thin filaments of the sarcomeres. These interactions are directly responsible for muscle contraction.

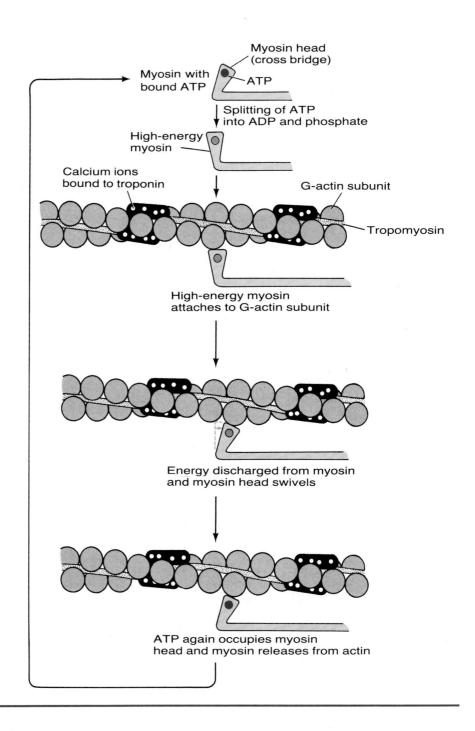

Figure 8.11

Mechanism of muscle contraction. A myosin head of a thick filament has ATP bound to it. The ATP is split, producing a high-energy form of myosin. The high-energy myosin binds to an actin subunit of the thin filament. The energy of the high-energy myosin is discharged and the myosin head swivels, pulling on the actin-containing thin filament. The myosin remains bound to the actin until ATP again occupies the myosin's head. When this occurs, the ATP is split and the cycle is repeated. For this contractile cycle to occur, troponin must bind calcium. This results in the removal of tropomyosin from a position blocking the myosin–actin interaction.

Mechanism of Contraction

F8.11   Muscle contraction requires energy, which is supplied by adenosine triphosphate (ATP). In a muscle fiber, an ATP molecule occupies a binding site on a club-shaped head of a myosin molecule of a thick filament (Figure 8.11). The myosin molecule possesses enzymatic activity, and it splits the ATP into adenosine diphosphate (ADP) and phosphate. This reaction releases energy, which is transferred to the myosin, producing a high-energy form of myosin.

In addition to an ATP binding site, a myosin head contains a binding site that can attach to a complementary site on an actin subunit of a thin filament, and a high-energy form of myosin has a strong tendency to bind to actin. In a relaxed, unstimulated muscle fiber, this binding is prevented by tropomyosin, which lies along the surface of the actin and physically blocks interactions

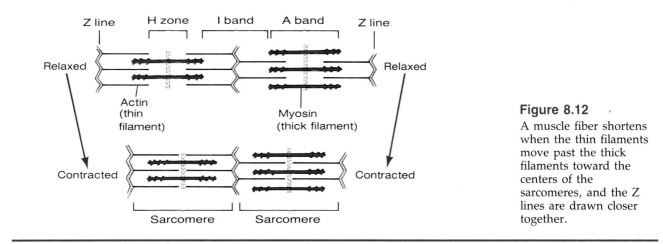

Figure 8.12
A muscle fiber shortens when the thin filaments move past the thick filaments toward the centers of the sarcomeres, and the Z lines are drawn closer together.

between high-energy myosins and actin subunits. However, when a muscle fiber is stimulated and calcium ions are released from the sarcoplasmic reticulum, the calcium ions bind to troponin molecules, which are linked to both actin and tropomyosin. The binding of calcium ions to troponin weakens the linkage between troponin and actin. This allows tropomyosin to move away from its blocking position and deeper into the groove formed by the twisting of the two actin chains around one another. With the tropomyosin out of the way, high-energy myosins can link with G-actin subunits.

When a high-energy myosin combines with an actin subunit, the energy stored within the myosin is discharged, with the resultant production of a force that causes the myosin head (cross bridge) to move. In essence, the myosin head swivels toward the center of the sarcomere, pulling on the actin-containing thin filament. The myosin remains attached to the actin until an ATP molecule again occupies the myosin head. When this occurs, the myosin releases from the actin. It again splits ATP into ADP and phosphate, producing a high-energy form of myosin that attaches to an actin subunit, and the cycle is repeated. At any instant during the contraction of a skeletal muscle fiber, approximately 50% of the myosin heads are attached to actin subunits, and the rest are at intermediate stages of the activity cycle.

The force of the myosin heads of the thick filaments pulling on the actin-containing thin filaments is transmitted to the plasma membrane of the muscle fiber and ultimately to the load. If enough force is developed by the fiber—and by other fibers involved in the muscle contraction—to overcome the resistance of the load, the repeated cycling of the myosin heads pulls the thin filaments past the thick filaments toward the centers of the sarcomeres (Figure 8.12). This draws the Z lines closer together, and the muscle fiber shortens.

Regulation of the Contractile Process

Once a nerve impulse stimulates a skeletal muscle fiber and calcium ions are released from the sarcoplasmic reticulum, why doesn't the formation of attachments between thick and thin filaments and, therefore, the contractile process continue indefinitely? In fact, the calcium ions released in response to a single nerve impulse are free for only a short time. Following their release, an active transport mechanism quickly pumps the calcium ions back into the sarcoplasmic reticulum. With the return of the calcium ions to the sarcoplasmic reticulum, troponin strengthens its connection with actin, pulling the tropomyosin back into its blocking position. When this occurs, no further interactions between high-energy myosins and actin subunits are possible. Consequently, the contractile process ceases, and the muscle fiber relaxes. When another nerve impulse stimulates the muscle fiber and calcium ions are again released from the sarcoplasmic reticulum, the contractile process occurs

Table 8.1   Sequence of Events Involved in the Excitation and Contraction of a Skeletal Muscle Fiber

1. A nerve impulse arrives at a neuromuscular junction. Acetylcholine is released from the motor neuron and binds to receptors on the muscle fiber plasma membrane.

2. A propagated action potential travels over the plasma membrane of the muscle fiber and along the transverse tubules into the interior of the cell.

3. The propagated action potential triggers the release of calcium ions from lateral sacs of the sarcoplasmic reticulum.

4. Calcium ions bind with troponin.

5. Tropomyosin moves away from its blocking position, permitting actin and myosin to interact.

6. High-energy myosins bind with actin subunits of the thin filaments.

7. Energy stored in the high-energy myosins is discharged, and the myosin heads swivel, pulling on the thin filaments.

8. ATP binds with myosin heads, which release from actin subunits.

9. ATP is split into ADP and phosphate, again producing high-energy myosins, and steps 6 and 7 are repeated as long as calcium ions are bound to troponin.

10. When calcium ions are returned to the lateral sacs of the sarcoplasmic reticulum, tropomyosin moves back into its blocking position, preventing further interaction between high-energy myosins and actin subunits.

11. Contraction ceases, and the muscle fiber relaxes.

once more. If many nerve impulses arrive at a skeletal muscle fiber in rapid succession so that calcium ions continue to be available to bind with troponin, the fiber does not relax between successive impulses, and the contractile process continues until the impulses cease and the calcium ions are returned to the sarcoplasmic reticulum.

Table 8.1   The sequence of events involved in the excitation and contraction of a skeletal muscle fiber are outlined in Table 8.1.

Sources of ATP for Muscle Contraction

ATP is the immediate source of energy for muscle contraction. However, the amount of ATP in skeletal muscle fibers is sufficient to support muscle contraction during strenuous exercise for only a few seconds. If muscular activity is to continue for longer periods, additional ATP must be produced.

kree'-uh-tin   Skeletal muscle fibers contain a substance called **creatine phosphate** that provides them with a means of forming ATP rapidly. Creatine phosphate contains phosphate and energy that can be transferred to ADP to produce ATP:

$$\text{Creatine phosphate} + \text{ADP} \rightleftharpoons \text{Creatine} + \text{ATP}$$

During periods of exercise, when ATP is being utilized to provide energy for muscle contraction, this reaction can form additional ATP. Although skeletal muscle fibers contain more creatine phosphate than they do ATP, the utilization of creatine phosphate provides only enough ATP to support muscle contraction during strenuous exercise for a few additional seconds. Nevertheless, creatine phosphate is an important source of ATP during the period immediately following the initiation of muscle contraction, when other sources of F8.13   ATP are not yet operating at high levels (Figure 8.13).

Ultimately, the metabolic breakdown of substances such as glucose, glycogen, and fatty acids by skeletal muscle fibers provides the ATP required to support continued muscular activity. Under conditions of mild-to-moderate exercise, ATP is produced by **aerobic** (that is, oxygen-utilizing) metabolic processes that break down these substances into carbon dioxide and water.

(The oxygen required by these aerobic processes is supplied by the increased breathing that occurs during exercise.)

During periods of intense exercise, oxygen cannot be supplied to many muscle fibers fast enough, and oxidative metabolism cannot produce all the ATP required for muscle contraction. During such periods, additional ATP is produced by an **anaerobic** (that is, non-oxygen-utilizing) metabolic process called *glycolysis*. Glycolysis breaks down glucose and stored glycogen into lactic acid, which can diffuse out of the muscle fibers and into the blood.

Oxygen Debt

The use of creatine phosphate and glycolysis to provide ATP for muscle contraction allows muscles to maintain a high level of activity for a longer period of time than would be possible if only oxidative processes supplied the ATP. However, when muscular activity ceases, the creatine phosphate level within the muscle fibers must be returned to normal (as must the ATP level, if it has dropped), and any lactic acid produced by glycolysis must be metabolized. These activities are accomplished by aerobic metabolic processes. These oxygen-utilizing processes provide ATP for the resynthesis of creatine phosphate (which occurs by a reversal of the creatine phosphate breakdown reaction described above) and, if necessary, for the replenishment of ATP. They also convert a portion of any lactic acid back into glucose or glycogen (this conversion occurs primarily in the liver), and they utilize some of the lactic acid as an energy source for ATP production. To supply the oxygen required by these processes, increased breathing continues for some time after strenuous muscular activity ceases. In essence, an **oxygen debt** is built up during periods of muscular activity, when nonoxidative sources of ATP are used to support muscle contraction. This debt is paid back by the increased breathing that continues after the end of the activity. The increased breathing provides the oxygen required by the aerobic processes that replenish creatine phosphate and ATP supplies and metabolize lactic acid.

Following exercise, it is not only necessary to restore creatine phosphate and ATP levels to normal, but it is also necessary to replace any muscle glycogen that was used for ATP production. The rate at which muscle glycogen stores are replenished is strongly influenced by diet. After exhaustive exercise that severely depletes muscle glycogen stores, a person on a high carbohydrate diet can replenish them in about two days. However, if the person is on a high-fat, high-protein diet, glycogen stores will not be fully replenished even five days after the exercise.

Muscle Fatigue

Intense skeletal muscle activity cannot continue indefinitely, and muscles eventually become fatigued. **Muscle fatigue** can be defined as the inability of a muscle to maintain a particular strength of contraction or tension (that is, a particular power output) over time. The immediate events that cause muscle fatigue are not well understood and are thought to differ with different types of exercise (for example, lifting a heavy weight as opposed to running a marathon). However, it is widely believed that a major factor underlying the occurrence of muscle fatigue is the inability of a muscle to generate energy at a rate sufficient to meet its energy requirements. This inability may be due to a depletion of metabolic reserves of substances such as muscle glycogen or it may be due to a buildup of metabolic by-products of contraction such as lactic acid. A build-up of lactic acid can cause the pH of a muscle to become more acidic and can thereby inhibit the activation of certain enzymes involved in cross-bridge cycling and reduce the activity of certain enzymes involved in energy production. (Note that "psychological fatigue" can cause a person to stop exercising even before any of the factors just discussed create an actual muscle fatigue).

The Motor Unit

There are more muscle fibers in a skeletal muscle than there are neurons to supply the muscle. Consequently, each neuron must branch to supply several

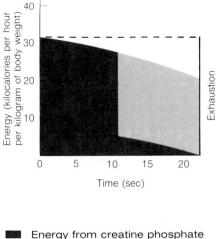

■ Energy from creatine phosphate
☐ Energy from aerobic metabolism
▨ Energy from anaerobic metabolism

Figure 8.13
Energy sources for muscle contraction during strenuous exercise (running at 18 km/hr on a treadmill inclined upward at an angle of 15°). Energy from "phosphagen" (ATP and creatine phosphate) that is present within muscles at the beginning of exercise is the initial and major energy source during superexertion. Energy from aerobic metabolism increases exponentially from the onset of exercise, but the mechanism is sluggish and accounts for only a small amount of the energy used during the first few seconds. The remaining energy is derived from anaerobic glycolysis.

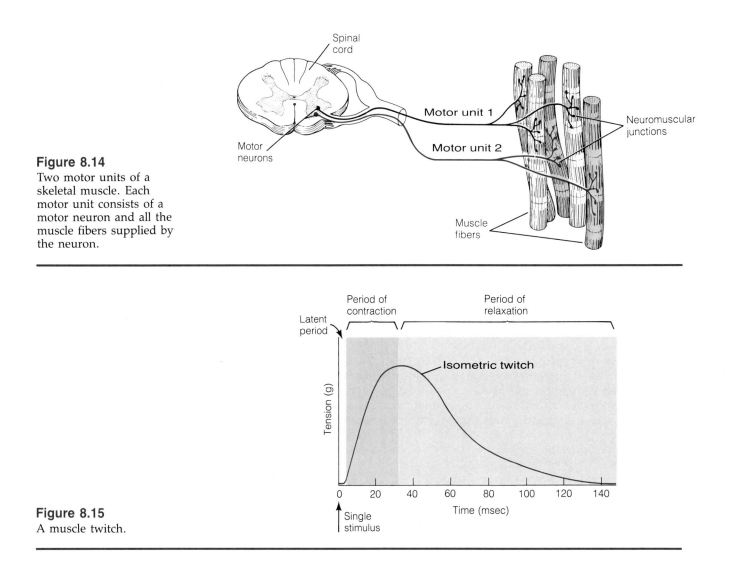

Figure 8.14
Two motor units of a skeletal muscle. Each motor unit consists of a motor neuron and all the muscle fibers supplied by the neuron.

Figure 8.15
A muscle twitch.

muscle fibers, and a nerve impulse transmitted along a particular neuron will reach all the fibers supplied by the neuron.

A single neuron and all the muscle fibers it supplies make up a **motor unit** (Figure 8.14). In the body, the motor unit and not the individual muscle fiber can be thought of as the functional unit of muscle activity because a nerve impulse in a single neuron stimulates all the muscle fibers supplied by the neuron. Muscles used for fine movements over which great control is exercised, such as the muscles of the hand, generally contain motor units in which each neuron supplies a relatively small number of muscle fibers—perhaps only a dozen or so. Muscles whose contractions are less precise, such as the muscles of the back or calf, have motor units in which a single neuron may supply several hundred muscle fibers.

Responses of Skeletal Muscle

The contractions of the muscle fibers of the motor units of a muscle combine to produce contractions of the whole muscle that vary in both duration and strength.

Muscle Twitch

If the fibers of a skeletal muscle are stimulated with a single brief stimulus, the muscle will contract once rapidly and then relax. This response, which is called a **muscle twitch,** shows three distinct phases (Figure 8.15). Immediately following the arrival of the stimulus at the muscle, there is a short *latent*

F8.14

F8.15

period during which no response is seen. During this period, the processes associated with excitation–contraction coupling occur. Following the latent period, a *period of contraction* occurs. During this period, the muscle actively develops tension, and if enough tension develops to overcome the resistance of the load, the muscle shortens. The final phase of the muscle twitch is the *period of relaxation*. During this period, the tension that was actively developed by the muscle diminishes, and if the muscle had shortened, it returns to its original, unstimulated length.

Graded Muscular Contractions

When a skeletal muscle fiber is stimulated and a propagated action potential travels over the plasma membrane and along the t tubules, enough calcium ions are released from the sarcoplasmic reticulum to completely activate the fiber. Consequently, when a nerve impulse triggers a propagated action potential in the membrane of a skeletal muscle fiber, the fiber contracts to the maximum extent possible for the existing conditions. (Note that this does not mean the contraction of a skeletal muscle fiber is exactly the same at all times. The physiological condition of a fiber at the time it is stimulated affects its contraction, and different conditions can exist at different times. For example, a fiber may or may not have contracted previously, or it may contain greater or lesser amounts of the by-products of contraction, such as lactic acid.)

Even though individual skeletal muscle fibers contract to the maximum extent possible for the existing conditions, a muscle as a whole responds in a graded fashion to meet the demands of the task at hand. For example, the biceps brachii muscle of the arm does not contract to the same extent when a person lifts a feather as when the person lifts a bowling ball. A smooth, graded response by a muscle depends on such factors as the number of motor units activated at any particular time, the frequency of nerve impulses, and the asynchronous activation of different motor units.

MULTIPLE MOTOR UNIT SUMMATION   The term **multiple motor unit summation** (spatial summation) refers to the ability of the individual motor units of a muscle to combine their simultaneous activities to influence the degree of contraction of the entire muscle. If only a few motor units are active at any one time, the muscle contraction will be relatively weak. If many motor units are active, the muscle contraction will be relatively strong.

WAVE SUMMATION AND TETANUS   Following the stimulation of a skeletal muscle fiber and the triggering of a propagated action potential that leads to the contraction of the fiber, there is a brief period of time during which a second stimulus will not produce another propagated action potential no matter how strong the stimulus is. This period is called the *absolute refractory period*. Following the absolute refractory period there is another brief interval during which a stimulus stronger than normal is required to produce another propagated action potential. This period is called the *relative refractory period*. The absolute and relative refractory periods usually last only one or two milliseconds, and they end well before the actual contraction of the muscle fiber (which can take 20 to 100 milliseconds) is completed. Consequently, a skeletal muscle fiber can be stimulated to contract a second time before it has completely relaxed from an initial contraction, a third time before it has completely relaxed from the second contraction, and so on.

In the body, the neurons that supply the muscle fibers of the motor units of a skeletal muscle normally do not transmit just a single stimulatory impulse that would produce only a twitch-type response. Instead, the neurons transmit volleys of impulses to the fibers, with one impulse closely following another. Consequently, the muscle fibers are stimulated a second time before they have completely relaxed from their initial contractions, a third time before they have relaxed from their second contractions, and so on. This results in a summation of individual contractions, or twitches, called **wave summation** (temporal summation) that can create a state of more-or-less sustained

Figure 8.16
Record of response of an isolated skeletal muscle stimulated with increasing frequencies of stimuli of sufficient intensity to produce a maximal response from the muscle. **(a)** Indicates single muscle twitches; **(b)** illustrates partial or incomplete tetanus (summation of twitches); **(c)** shows complete tetanus.

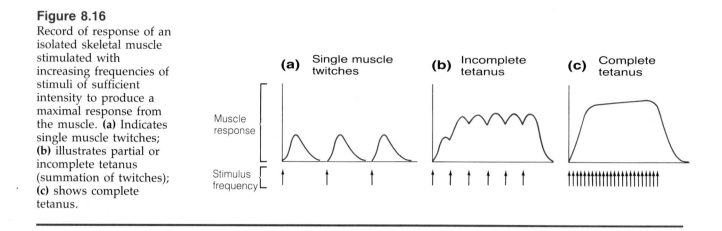

Figure 8.17
The maintenance of a nearly constant tension in an entire muscle is a result of the asynchronous activity of individual motor units of the muscle.

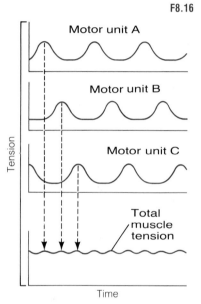

F8.16   contraction called **tetanus** (Figure 8.16). If the successive stimuli arrive far enough apart in time that the muscle is allowed to relax partially between stimuli, a condition of *incomplete tetanus* is seen. If the stimuli arrive so rapidly that no relaxation of the muscle occurs between stimuli, a condition of *complete tetanus* exists.

ASYNCHRONOUS MOTOR UNIT SUMMATION   A single muscle, such as the biceps brachii of the arm, consists of many motor units, and a sustained, coordinated contraction of the muscle is due in part to the asynchronous activation of different motor units (Figure 8.17). Some motor units are activated initially; then they relax while other motor units are activated; then these relax and still other motor units become active. The result is a smooth, sustained contraction of the muscle as a whole.

Factors Influencing the Development of Muscle Tension

The amount of force, or tension, developed by a skeletal muscle is influenced by the composition of the muscle and by the length of the muscle at the time it contracts.

Contractile and Series Elastic Elements

A muscle contains both contractile elements and series elastic elements. The **contractile elements** are those structures actively involved in contraction, such as the thick and thin filaments of the muscle fibers. The **series elastic elements** are structures that resist stretching (but can be stretched) that are located between the contractile elements and the load. They include connective tissue and some parts of the muscle fibers themselves. When a muscle contracts, the force generated by the contractile elements stretches the series elastic elements, which, in turn, exert force on the load.

INFLUENCE OF SERIES ELASTIC ELEMENTS ON MUSCLE TENSION   Because of the presence of series elastic elements, the tension developed by the contractile elements during a muscle contraction—that is, the internal tension or active state of the muscle fibers—is not always equivalent to the external tension exerted on the load. When a muscle is stimulated, the formation of attachments between the thick and thin filaments and the development of internal tension by the contractile elements of the muscle occur rather rapidly. However, following the formation of attachments between the thick and thin filaments and the development of internal tension, some time and effort are required to stretch or take up any slack in the series elastic elements of the muscle. During a single muscle twitch, the internal tension reaches its peak and decreases before the series elastic elements are stretched to a tension

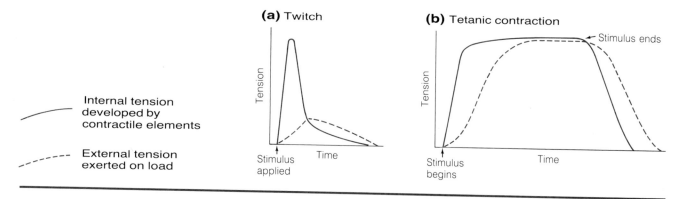

Figure 8.18

(a) During a single muscle twitch, the internal tension developed by the contractile elements (the active state of the muscle fibers) reaches its peak and decreases before the series elastic elements are stretched to a tension equal to the maximum internal tension. As a result, less than the full internal tension is transmitted to the load.
(b) During a tetanic contraction, the repeated stimulation of the muscle maintains the internal tension long enough for the series elastic elements to be stretched to a tension similar to the internal tension. As a result, more of the internal tension is transmitted to the load during a tetanic contraction than during a single muscle twitch, and the response of a muscle during a tetanic contraction is greater than that during a single muscle twitch.

equal to the maximum internal tension. As a result, less than the full internal tension is transmitted to the load (Figure 8.18a). During a tetanic contraction, however, the repeated stimulation of the muscle maintains the internal tension long enough for the series elastic elements to be stretched to a tension similar to the internal tension (Figure 8.18b). As a result, more of the internal tension is transmitted to the load during a tetanic contraction than during a single muscle twitch, and the response of a muscle during a tetanic contraction is greater than that during a single muscle twitch (see also Figure 8.16).

F8.18a

F8.18b

F8.16

ACTIVE AND PASSIVE TENSION   Because a muscle contains both contractile and elastic elements, it can develop two types of tension: active tension and passive tension. **Active tension** is the tension due to the activity of the contractile elements when a muscle is stimulated and contracts. **Passive tension,** in contrast, is a consequence of a muscle's elasticity, and a muscle does not have to contract to exert passive tension. When a skeletal muscle is stretched so that it lengthens, it behaves much like a spring or rubber band. Within limits, the more it is stretched, the greater the passive tension it exerts.

Influence of Length on the Development of Muscle Tension

The amount of active tension developed by a skeletal muscle fiber when it contracts varies with the length of the fiber at the time of contraction (Figure 8.19). This relation between active tension and fiber length is related to the fact that the fiber contracts as the result of the formation of attachments between the thick and thin filaments of its sarcomeres. When the thin filaments completely overlap the portions of the thick filaments that possess cross bridges, more attachments can form than when the muscle fiber is stretched, so there is less overlap. If a muscle fiber is stretched to the point where no overlap occurs, no attachments between thick and thin filaments can form, and the fiber cannot contract. On the other hand, when a muscle fiber is compressed, the thin filaments overlap and interfere with one another, and the thick filaments are forced against the Z lines of the sarcomeres. As a result, the active tension the muscle fiber can develop when the fiber contracts is less than maximal.

A whole skeletal muscle exhibits a similar relationship between its length and the amount of active tension it develops when it contracts, and each

F8.19

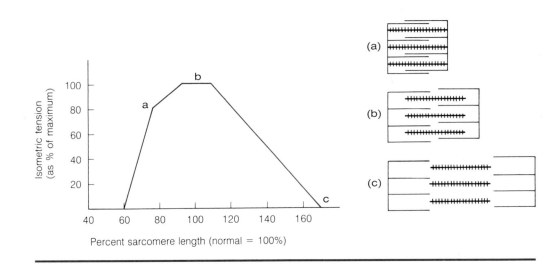

Figure 8.19

Active tension developed by a skeletal muscle fiber at different initial fiber lengths, expressed in terms of percent sarcomere length.

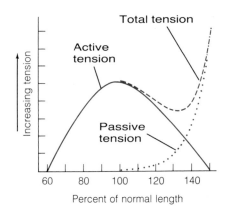

Figure 8.20

Graph indicating the influence of muscle length on the tension developed by a skeletal muscle.

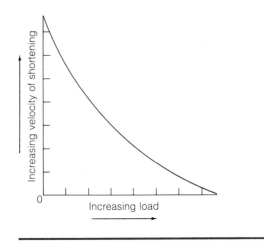

Figure 8.21

Relation of load to velocity of shortening of a skeletal muscle. As the load increases, the velocity of shortening decreases.

F8.20 skeletal muscle has an optimal length from which it can develop the maximum active tension (Figure 8.20). In addition, as a muscle is lengthened— that is, stretched, it exerts increasing amounts of passive tension, and the total tension exerted by a muscle when it contracts is equal to the sum of its active and passive tensions. In the body, the lengths of most relaxed, unstimulated muscles are such that they allow the development of maximal active tension when the muscles are stimulated.

Relation of Load to Velocity of Shortening

F8.21 The greater the load on a muscle, the slower the velocity of shortening (Figure 8.21). With only a small load, the velocity of shortening is relatively fast. As the load increases, the velocity of shortening decreases until a point is reached at which the load is too great for the muscle to move. At this point, the velocity of shortening is zero, and the contraction of the muscle is isometric.

MUSCLE ACTIONS

It is the contraction of skeletal muscles that causes the various movements at the different joints, as described in Chapter 7. Some muscles pass in front of a joint and thus flex the bone to which they are attached; others pass behind a joint and extend the bone to which they are attached. Some muscles move a part away from the midline of the body—or abduct it; others move the part back toward the midline—or adduct it. Some muscles rotate the bones that form certain joints, and so forth.

In order to bring about these movements, muscles usually work in groups rather than individually. Those muscles whose contractions are primarily responsible for a particular movement are called **prime movers.** In any movement there are always some muscles, generally situated on the opposite side of the joint, whose actions oppose the particular movement. These muscles, whose contraction offers resistance to the movement, are called **antagonists.** When prime movers contract and produce a movement, the antagonists are stretched. It is important to realize that a particular muscle is not always either a prime mover or an antagonist; rather, its role changes, depending on the movement that is being produced. For instance, when the forearm is flexed, the anterior muscles of the arm are the prime movers and the posterior arm muscles are the antagonists. When the forearm is brought back to the anatomical position—or extended, the posterior muscles of the arm become the prime movers and the anterior arm muscles are the antagonists.

In addition to prime movers and antagonists, most joint movements involve muscles that act as **synergists.** Synergists are muscles that indirectly aid *sin'-er-jists* in a particular movement by steadying a joint, thus preventing unwanted movements and thereby allowing the prime movers to function more efficiently, or by otherwise aiding in a particular movement. Muscles that function as prime movers often cause actions other than the movement desired. For example, flexion may be accompanied by rotation in the joint. In this case, the contraction of synergistic muscles may assist the prime-mover muscles by opposing an undesired rotation. In a similar manner, if synergistic muscles did not act to immobilize the wrist and thus keep it from moving, the wrist would flex every time a person made a fist. This action would occur because the muscles that flex the fingers also pass anteriorly across the wrist. Therefore, some synergistic muscles may assist a movement by acting as antagonists.

When a synergist acts to immobilize a joint or an individual bone, it is referred to as a **fixator.** The muscles described in the previous paragraph as *fiks'-ay-tor* immobilizing the wrist were functioning as fixators. Many of the muscles that attach the scapula to the axial skeleton have important actions as fixators. The scapula is freely movable, and in order for it to serve as a firm origin for those muscles that move the arm, it must be held steady when the muscles contract. The contractions of fixators hold the scapula firmly against the thorax so that contractions of the arm muscles can move only their insertions, which are on the bones of the arm and forearm.

Although it is the prime movers that cause the actual movements, the contractions of antagonistic and synergistic muscles are necessary to produce the smooth coordinated movements that are typical of a normal person. The strength of an antagonist's contraction affects the strength and the speed of the prime mover's contraction. For instance, if the extensor muscles of the forearm remain partially contracted—and therefore act as antagonists—while the flexor muscles of the forearm are causing the elbow to bend—and therefore are serving as prime movers, the flexor muscles have to contract harder to overcome the opposition, and the joint movement will be slower than it might otherwise be. The actions of antagonists and synergists make very fine and precise movements possible.

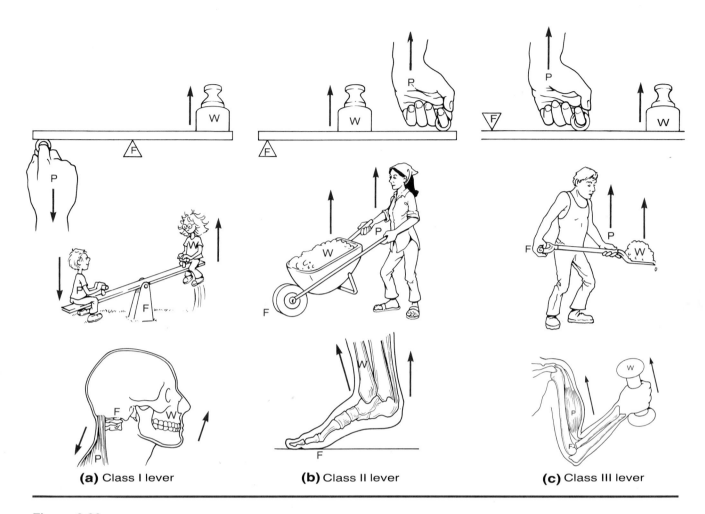

(a) Class I lever **(b)** Class II lever **(c)** Class III lever

Figure 8.22

(a) Class I lever. The fulcrum (F) is located between the weight (W) (or resistance) and the force (P) (or pull). Arrows indicate the direction of movement. **(b) Class II lever.** The weight (W) (or resistance) is located between the fulcrum (F) and the point of force (P) (or pull).

Arrows indicate the direction of movement. **(c) Class III lever.** The force (P) (or pull) is applied between the fulcrum (F) and the weight (W) or resistance. Arrows indicate the direction of movement.

RELATIONSHIP BETWEEN LEVERS AND MUSCLE ACTIONS

The movements brought about by the actions of most skeletal muscles involve the use of levers. A **lever** is a rigid structure capable of moving around a pivot point, called a **fulcrum,** when a force is applied. In the body the bones of the skeleton function as levers, the joints serve as fulcrums, and skeletal muscles provide the force to move the bones. Depending on the location of the fulcrum, a lever can make it possible to move heavier loads than could otherwise be moved, or to alter both the rate of movement and the distance over which a load can be moved.

Classes of Levers

There are three *classes of levers:* Class I, Class II, and Class III.

Class I Levers

In a **Class I lever,** the fulcrum is located between the point at which the force is applied and the weight that is to be moved (Figure 8.22a). A seesaw is a common example of a Class I lever. In the body, this type of lever is used

F8.22a

when the head is tipped back to raise the face: The occipital condyles on the atlas serve as the fulcrum, the facial portion of the skull is the weight, and the force (pull) is applied to the back of the skull by the posterior muscles of the neck.

Class II Levers

In a **Class II lever,** the weight to be moved is between the fulcrum and the point of force (Figure 8.22b). A wheelbarrow involves this type of lever. The best example in the body is raising the body on the toes. In this case, the base of the toes serves as the fulcrum, the toes support the weight, and the contraction of the posterior muscles of the calf causes a force (pull) to be exerted on the calcaneous bone.

F8.22b

Class III Levers

In a **Class III lever,** the weight is at one end, the fulcrum is at the other, and the force is applied between them (Figure 8.22c). Lifting a shovel utilizes this type of leverage. There are many examples of Class III levers in the body, since it is the most common lever system used. One example is flexion of the forearm: The weight is at the wrist, the fulcrum is the elbow joint, and the force (pull) is exerted by the contraction of flexor muscles on the anterior of the arm that insert on the radius or the ulna, between the fulcrum and the weight.

F8.22c

Effects of Levers on Movements

The portion of a lever located between the fulcrum and the point where a force is applied is called the **power arm;** the portion between the fulcrum and the weight is the **weight arm.** When the weight arm is long in relation to the power arm, a weight can be moved rapidly over a considerable distance, but a strong force is required. Conversely, when the weight arm is short in relation to the power arm, the same weight can be moved with less force, but both the speed of movement and the range of movement are reduced. Thus, depending on the arrangement of the particular muscles and bones, the levers of the body may enable muscles to move loads faster over greater distance than would otherwise be possible or they may enable muscles to move heavier loads than they otherwise could.

TYPES OF SKELETAL MUSCLE FIBERS

Not all skeletal muscle fibers are identical. For example, the myosin molecules of some fibers split ATP more rapidly than the myosin molecules of other fibers. Because of this, cross-bridge cycling occurs more rapidly in some fibers than others, and some fibers contract faster than others. In addition, the metabolic processes by which muscle fibers produce the ATP necessary for contraction differ from one fiber to another, and the ability of a fiber to produce ATP influences its resistance to fatigue.

Based on their contraction speed and fatigue resistance, the following three types of skeletal muscle fibers can be distinguished.

Slow Twitch, Fatigue-Resistant Fibers

Slow twitch, fatigue-resistant fibers contain myosin molecules that split ATP at a slow rate. Consequently, cross-bridge cycling occurs slowly, and the fibers contract at a slow speed. Slow twitch, fatigue-resistant fibers have a highly developed capacity for oxidative metabolism. They are surrounded by many blood vessels, and they contain large amounts of an oxygen-binding protein known as **myoglobin.** Myoglobin increases the rate of oxygen diffusion into the fibers, and it provides them with small stores of oxygen. Slow twitch, fatigue-resistant fibers can meet their needs for ATP almost entirely through aerobic processes, and they are extremely resistant to fatigue.

my-o-glo'-bin

Fast Twitch, Fatigue-Resistant Fibers

Fast twitch, fatigue-resistant fibers are also well supplied by blood vessels, contain large amounts of myoglobin, and have a highly developed capacity for oxidative metabolism. However, they contain myosin molecules that split ATP at a rapid rate. Consequently, cross-bridge cycling occurs rapidly, and they contract at a fast speed. The highly developed oxidative processes of these fibers can supply most of their needs for ATP, and they are quite resistant to fatigue (although less so than slow twitch, fatigue-resistant fibers).

Fast Twitch, Fatigable Fibers

Fast twitch, fatigable fibers contain myosin molecules that split ATP rapidly, and they contract at a fast speed. However, these fibers are not well supplied by blood vessels, and they do not contain large amounts of myoglobin. Instead, they contain large stores of glycogen, and they are especially geared to the utilization of anaerobic metabolic processes. These processes are unable to supply the fibers continually with the amount of ATP they require, and they fatigue easily.

UTILIZATION OF DIFFERENT FIBER TYPES

In some skeletal muscles, one type of skeletal muscle fiber predominates. However, most skeletal muscles contain all three fiber types, in proportions that vary from muscle to muscle. Consequently, muscles show a range of contraction speeds and fatigue resistances.

The presence of varying proportions of the three fiber types in different skeletal muscles is consistent with the following facts.

1. Not all skeletal muscles perform the same functions, and different functions often require different types of muscle contraction. For example, the postural muscles of the back and legs, which support the body against the force of gravity, must be able to undergo sustained contractions without fatigue, and the muscles that move the arms must be able to develop large amounts of tension quickly to allow the rapid lifting of heavy objects.

2. The same skeletal muscle often performs different functions at different times. For example, certain muscles of the legs may be involved in supporting the body against the force of gravity at one time, and they may participate in moving the legs during walking or running at another.

Even though most skeletal muscles contain all three fiber types, all the fibers of any one motor unit are of the same type. Moreover, the different fiber types tend to be utilized in characteristic fashions when a muscle contracts. For example, during activities of short duration, if only a weak contraction is required from a muscle, just the slow twitch, fatigue-resistant fibers are activated. If a stronger contraction is required, the fast twitch, fatigue-resistant fibers are also activated, and if a still stronger contraction is necessary, the fast twitch, fatigable fibers are added.

EFFECTS OF EXERCISE ON SKELETAL MUSCLE

Regular exercise can produce increases in muscle size, strength, and endurance. However, different types of exercise cause different kinds of changes in the fibers of a skeletal muscle. Endurance types of exercise such as distance running are associated with a transformation of fast twitch, fatigable fibers into fast twitch, fatigue-resistant fibers, with relatively small increases in the diameters of the fibers or their strength. High-intensity, short-duration types of exercise such as weight lifting, however, cause fast twitch, fatigable fibers to hypertrophy. The fibers increase in diameter, and there is an increased synthesis of actin and myosin filaments that results in a large increase in the strength of the fibers.

The increases in strength that occur with exercise do not always seem to be accounted for simply by increased muscle size. It has been suggested that in addition to increased muscle size, more nerve pathways that activate more motor units may be utilized. Normally, all the motor units of a muscle are not activated simultaneously; if they were, the muscle's tendons could be torn from their attachments. As exercise develops a muscle, however, perhaps more motor units can be activated simultaneously, thereby increasing strength.

Endurance exercises produce cardiovascular and respiratory changes that cause muscles to be better supplied with materials such as oxygen and carbohydrates. Thus, the improved muscular performance that results from exercise is not solely the result of muscular changes but involves other systems as well.

CONDITIONS OF CLINICAL SIGNIFICANCE
Skeletal Muscle

Muscle Atrophy

The shrinkage and death of muscle cells—*muscle atrophy*—cause a reduction in the size of the affected muscles. Atrophy of muscles can be caused by prolonged disuse or by a number of disorders, most of which reduce the blood supply or interfere with the nerve supply to the muscle. Muscle atrophy can be widespread or localized. Localized atrophy may not reduce the size of the muscle, however, since the unaffected cells undergo compensatory enlargement (hypertrophy).

Cramps

Cramps are painful, involuntary muscle contractions that are slow to relax. They can occur during exercise or at rest, and their precise cause is not known. Cramps may be caused by conditions within the muscle itself—for example, a low oxygen supply—or by stimulation from the nervous system. There is some evidence that cramps that occur during heavy exercise are caused by low blood levels of sodium and chloride ions, as the result of the loss of those ions by sweating. However, it is not clear whether the depletion of sodium chloride acts on the muscles or on the nervous system.

Muscular Dystrophy

The term *muscular dystrophy* refers to a group of diseases characterized by progressive muscular weakness. The weakness is the result of the degeneration of muscle fibers, an increase in connective tissue within the muscles, and in some forms, the replacement of muscle fibers by fatty tissue. The muscular dystrophies are genetically transmitted. Some forms of muscular dystrophy are fatal, but other forms have a more favorable outlook.

Myasthenia Gravis

Myasthenia gravis is a chronic condition characterized by extreme muscle weakness. It is believed to be caused by an abnormal response of the body's immune system that disrupts acetylcholine receptors on the muscle cell membranes at the neuromuscular junctions. This decreases the responsiveness of the muscle fibers to acetylcholine released from motor neuron endings. About 10% of the people who have this disease die from it. However, if an afflicted individual survives the first three years, there is a good chance that the condition will stabilize, with some degree of recovery.

Effects of Aging

Starting in the middle 20's, a progressive and continuous loss of skeletal muscle mass begins, and much of the loss is replaced by fat. Since fat weighs less than muscle, the normal body weight at age 50 is less than that at age 20. As the muscle mass decreases, so does the maximal strength, which declines by about 50% between ages 20 and 80.

CLINICAL CORRELATION

Skeletal Muscle Contracture During Sustained Activity

CASE REPORT . . . .

THE PATIENT: A 25-year-old male.

PRINCIPAL COMPLAINT: Extreme muscle stiffness during exercise.

HISTORY: The patient's birth and development were normal. His earliest recollection of a disorder of movement was at 5 years of age, when he fell during a foot race because his muscles stiffened. Thereafter, he was aware that muscle stiffness occurred during vigorous exercise or sudden, rapid movements. Strenuous exercise caused painless stiffness within seconds, and the stiffness was of such a magnitude that the exercise could not be continued. However, the stiffness disappeared after only a few seconds of rest. Despite his problem, the patient was able to carry out daily activities, provided he paced his movements. If he was careful, he could even play tennis and swim. He managed to qualify for the college wrestling team by using "brute strength," but when speed was required, he lost competitive matches. Cold temperatures led to the rapid occurrence of stiffness, and the patient often noted clumsiness in his hands under these conditions. He had been accepted into military service, but subsequently was discharged because he could not perform "double time" marches. The patient denied weakness and had no history of myoglobinuria (myoglobin in the urine). He had no siblings, and no one in his known ancestry had a similar condition.

CLINICAL EXAMINATION: At examination, the patient appeared healthy and well developed. Abnormal findings were limited to the skeletal muscles. Under gross examination, the muscles were normal in contour, tone, and strength, but a progressive lengthening of the relaxation time was observed during vigorous exercise. After 10 to 15 seconds of repetitive contraction with maximal effort, the arm muscles became paralyzed in a contracted state, which was recognizable grossly. After 5 to 15 seconds of rest, the muscles relaxed again, and exercise could be resumed. The momentary "contracture"* was painless unless the patient continued his efforts to contract the shortened muscles. This phenomenon was identified in muscles of the limbs, the face, and the jaws. There were no symptoms of uncoordination or disturbance of gait, and passive movements did not induce the persistent shortening in any muscles. Electrical studies using external electrodes showed that the velocities of nerve impulse transmission were normal and that the muscle responded normally to either single or repetitive excitation. During the contracture induced by exercise, no electrical activity could be detected in the muscle.

A hollow needle was used to obtain samples of muscle tissue for study. Examination of samples by both light and electron microscopy disclosed no abnormal morphology, and biochemical analysis revealed a normal concentration of ATP. A sample of muscle was homogenized, and a cell fraction containing sarcoplasmic reticulum was obtained from the sample by ultracentrifugation. The uptake of calcium by the fraction was measured using radioactive calcium (^{45}Ca). The calcium uptake was found to be significantly less than the calcium uptake of control tissues.

COMMENT: The studies that were done suggest a defect in the uptake of calcium ions by the sarcoplasmic reticulum. Apparently calcium ions accumulate in the cytoplasm of the muscle cells during repeated contractions until contracture is produced. The condition is made worse by cold, which further decreases the rate of uptake of calcium ions by the sarcoplasmic reticulum. At present, the cause of the abnormal function of the sarcoplasmic reticulum is not known.

OUTCOME: The patient's condition has not progressed, and despite the lack of any useful therapy, he lives a reasonably normal life. He accepts his limitations and avoids muscular activity that is rapid or prolonged enough to induce contracture.

Contracture: a retarded relaxation of a muscle due to maintained force in the contractile apparatus of the muscle cells that is independent of any electrical activity in the plasma membranes of the cells.

SMOOTH MUSCLE

Smooth muscle fibers are uninucleate, spindle-shaped cells that are considerably smaller than skeletal muscle fibers. Smooth muscle fibers have a well-developed capacity for anaerobic metabolism, but their overall metabolic machinery is not as highly developed as that of skeletal muscle fibers.

Smooth muscle fibers possess thick filaments that contain myosin, and smaller proteins called *myosin light chains* are associated with myosin. The fibers also possess thin filaments that contain actin and tropomyosin. However, the thick and thin filaments of smooth muscle fibers are not organized into regularly ordered sarcomeres, and smooth muscle fibers are not striated. Nevertheless, smooth muscle uses cross-bridge movements between myosin and actin to generate force, and it is generally believed that smooth muscle contraction occurs by a sliding-filament mechanism similar to that in skeletal muscle.

Smooth muscle contraction is triggered by calcium ions obtained from two sources—the sarcoplasmic reticulum and the extracellular fluid, and most smooth muscle fibers rely on both sources to some extent. When a smooth muscle fiber is stimulated, calcium ions are released from the sarcoplasmic reticulum. Moreover, the plasma membrane becomes more permeable to calcium ions, which enter the fiber from the extracellular fluid. The calcium ions are believed to trigger smooth muscle contraction by binding to a cytoplasmic protein called *calmodulin,* which then activates an enzyme called *myosin light chain kinase.* The activated myosin light chain kinase phosphorylates (adds phosphate groups to) myosin light chains associated with myosin molecules of the thick filaments. This phosphorylation leads to the interaction of the thick myosin-containing filaments and the thin actin-containing filaments, and the muscle fiber contracts.

Smooth muscle relaxation occurs when the calcium ions are pumped back into the sarcoplasmic reticulum or across the plasma membrane and out of the fiber. The removal of calcium ions leads to the inactivation of the myosin light chain kinase, with the result that myosin light chains cease being phosphorylated. Moreover, another enzyme within the muscle fiber, called *myosin light chain phosphatase,* removes phosphate groups from myosin light chains. This dephosphorylation prevents the interaction of the thick myosin-containing filaments and the thin actin-containing filaments, and the muscle fiber relaxes.

Overall, the contraction and relaxation of smooth muscle fibers depend on the relative activities of the enzymes myosin light chain kinase and myosin light chain phosphatase. When a smooth muscle fiber is stimulated and calcium ions are present in relatively high concentrations, the phosphorylating activity of myosin light chain kinase is greater than the dephosphorylating activity of myosin light chain phosphatase. Consequently, myosin light chains tend to have phosphate groups attached, and the muscle fiber contracts. Following stimulation, when calcium ions are present in lower concentration and the myosin light chain kinase is inactive, the dephosphorylating activity of myosin light chain phosphatase is dominant. As a result, myosin light chains lose their phosphate groups, and the muscle fiber relaxes.

Stress-Relaxation Response

Smooth muscle can be stretched much more than can skeletal muscle before exhibiting any marked changes in tension. When a smooth muscle is suddenly stretched, it initially displays an increased tension. However, this tension begins to decrease almost immediately, and within a few minutes it has returned to its original level. This phenomenon, which is called **stress-relaxation,** is important because many smooth muscles are located in the walls of hollow structures such as the stomach, intestines, or urinary bladder. As these structures become filled, they enlarge, stretching their walls and the smooth muscles associated with them. Within limits, however, the stress-relaxation response of smooth muscle allows these hollow structures to enlarge with little appreciable change in the pressures exerted on their contents.

Ability to Contract When Stretched

Smooth muscle is able to contract and actively generate tension under greater stretching than skeletal muscle. This ability is at least partly explained by the fact that the filaments of smooth muscle fibers are not organized into any regular pattern of sarcomeres but rather exist in a somewhat irregular array. Because of this, the problem of stretching smooth muscle fibers to the point where filaments no longer overlap one another is less serious than in skeletal muscle cells. At many degrees of stretch, some filaments still overlap one another and the fibers can still contract.

The ability to contract when stretched is important to smooth muscles that are associated with hollow structures. During the filling and distension of these structures, the smooth muscles associated with them undergo considerable stretching. Despite this stretching, the muscles maintain the ability to contract and actively generate tension.

Degree of Shortening During Contraction

When smooth muscles contract, they can shorten far more as a percentage of their length than can skeletal muscles. The useful distance of contraction for a skeletal muscle is about 25% to 35% of the length of the muscle, whereas a smooth muscle can contract from twice its normal length to one-half its normal length. This allows structures such as the stomach, intestines, or urinary bladder to vary the diameters of their lumens (cavities) from considerably large values to practically zero.

Smooth Muscle Tone

When compared with skeletal muscle, smooth muscle contracts rather slowly. Smooth muscle, however, is quite resistant to fatigue, and many smooth muscles undergo sustained, long-term (tonic) contractions that consume relatively small amounts of energy. The tonic contraction of smooth muscle (*smooth muscle tone*) is important in the circulatory system, where the smooth muscles of blood vessels called arterioles must contract for long periods. Also, the tonic contraction of smooth muscles of the intestine exerts a steady pressure on the intestinal contents.

Types of Smooth Muscle

Two basic arrangements of smooth muscle occur in the body: single unit and multiunit. Although some intergrading between the two types is evident, most smooth muscles adhere basically to one pattern or the other.

Single-Unit Smooth Muscle

The cells of single-unit smooth muscle are connected by gap junctions at which electrical impulses can spread from cell to cell. As a result, many cells respond as a unit to stimulation. Single-unit smooth muscle is self-excitable, and it contracts without external stimulation. Occasionally, one cell of the unit becomes spontaneously stimulated, and this stimulus passes to other cells, causing the entire unit to contract. This type of muscle is sensitive to stretch, and a quick stretch of the muscle can produce a contraction. Single-unit smooth muscle is found in small arteries and veins, the intestine, and the uterus.

Multiunit Smooth Muscle

The individual cells of multiunit smooth muscle are usually sufficiently separated from one another so that a stimulus to one cell is not transferred to neighboring cells. Thus, each cell responds independently to stimulation. Multiunit smooth muscle is generally not self-excitable, and it requires external stimulation to initiate a contraction. This type of smooth muscle is not responsive to stretch. Multiunit smooth muscle is found in large arteries and the large airways to the lungs.

Influence of External Factors on Smooth Muscle Contraction

Smooth muscle contraction is influenced by external factors that include neural activity and chemical substances such as hormones. As just mentioned, external factors are generally required to initiate the contraction of multiunit smooth muscle. Moreover, external factors modulate—that is, enhance or inhibit—the frequency and intensity of the spontaneous contractions of single-unit smooth muscle. The responses of different smooth muscles to particular factors vary greatly, and a factor that stimulates one smooth muscle may inhibit another. As particular smooth muscles are encountered in later portions of the text, their responses to specific substances will be considered.

STUDY OUTLINE

MUSCLE TYPES   p. 213

SKELETAL MUSCLE   Most attached to skeleton; striated cells; voluntary control; regulated by somatic nervous system.

SMOOTH MUSCLE   In walls of hollow organs and tubes; cells lack striations; involuntary control; regulated by factors intrinsic to muscle itself, hormones, and autonomic nervous system.

CARDIAC MUSCLE   Forms wall of heart; striated cells; involuntary control; regulated by factors intrinsic to muscle itself, hormones, and autonomic nervous system.

EMBRYONIC DEVELOPMENT OF MUSCLE   p. 214

SKELETAL MUSCLE
1. Except for muscles of head and limbs, skeletal muscles develop from embryonic masses of mesoderm called somites—located dorsally along axial skeleton.
2. Myotome—that portion of a somite that differentiates into muscle cells.
3. Head musculature develops from general mesoderm of that region.
4. Limb musculature develops from mesodermal condensations within embryonic limb buds.
5. Individual muscle cells are called muscle fibers.
6. Myoblasts give rise to skeletal muscle fibers.

SMOOTH MUSCLE   Mesodermal cells migrate to embryonic digestive tube and body organs and surround them in a thin layer.

CARDIAC MUSCLE   Mesodermal cells migrate to and surround early tubular heart.

GROSS ANATOMY OF SKELETAL MUSCLES   pp. 214–216

CONNECTIVE-TISSUE COVERINGS   Skeletal muscle is composed of muscle fibers held together by thin sheets of fibrous connective tissue called fascia.

EPIMYSIUM   Fascia that encases entire muscle.

PERIMYSIUM   Penetrates muscle; separates fibers into bundles called fasciculi.

ENDOMYSIUM   Thin extensions of fascia; envelop plasma membrane of each muscle fiber.

SKELETAL MUSCLE ATTACHMENTS   Extensions of endomysium, perimysium, and epimysium may directly attach to bone or may blend into strong fibrous connection called a tendon.

TENDONS   Extensions of connective tissue beyond end of muscle; length varies; continuous with periosteum.

APONEUROSES   Broad, thin tendon sheets.

ORIGIN   Less movable end of muscle; usually proximal.

INSERTION   More movable end of muscle; usually distal.

SKELETAL MUSCLE SHAPES   Fasciculi may run parallel to long axis of muscle, producing considerable movement but little strength; or fasciculi may insert diagonally into a tendon running length of muscle, producing less movement but greater power.

UNIPENNATE   All fasciculi insert on one side of tendon.

BIPENNATE   Fasciculi insert on both sides of tendon.

MULTIPENNATE   Convergence of several tendons.

RADIATE   Fasciculi converge from broad origin to single narrow tendon.

MICROSCOPIC ANATOMY OF SKELETAL MUSCLE   Skeletal muscle fibers (cells) contain: pp. 216–220

MYOFIBRILS   Longitudinal threadlike arrays of proteins; cross-striated because of alternating light and dark bands.

ANISOTROPIC (A, DARK) BANDS   Have less dense H zone in center; H zone is crossed by M line.

ISOTROPIC (I, LIGHT) BANDS   Have dense Z line crossing center; Z lines divide myofibrils into sarcomeres.

SARCOMERES   Repeating units of myofibrils; contain filamentous structures called myofilaments.

THICK FILAMENTS   Only in A band; H zone has only thick filaments; each thick filament surrounded by six thin filaments.

THIN FILAMENTS   I band; part of A band; attach to Z lines.

COMPOSITION OF THE MYOFILAMENTS
1. **Thick filaments** Composed of protein myosin.
2. **Thin filaments** Composed of proteins actin, tropomyosin, and troponin.

TRANSVERSE TUBULES AND SARCOPLASMIC RETICULUM

TRANSVERSE TUBULES   (t tubules) Run deep into muscle cell from plasma membrane.

SARCOPLASMIC RETICULUM   Membranous network that runs throughout cell and surrounds myofibrils.

CONTRACTION OF SKELETAL MUSCLE   pp. 220–231
1. **Isometric** Muscle actively develops tension but does not shorten.
2. **Isotonic** Muscle actively develops tension and shortens.

CONTRACTION OF A SKELETAL MUSCLE FIBER   At cellular level, same basic events occur during both isometric and isotonic skeletal-muscle contractions.

THE NEUROMUSCULAR JUNCTION   Ending of a motor neuron approaches specialized point along plasma membrane of skeletal-muscle fiber, forming neuromuscular junction.

EXCITATION OF A SKELETAL MUSCLE FIBER
Nerve impulse reaches neuromuscular junction; acetylcholine is released, changes permeability of muscle-fiber plasma membrane at junction; propagated action potential produced.

EXCITATION–CONTRACTION COUPLING Propagated action potential moves along t tubules to fiber interior; calcium ions released from reticular sites called terminal cisternae; calcium ions bind to troponin molecules of thin filaments of sarcomeres; binding leads to interactions between thick and thin filaments and to muscle contraction.

MECHANISM OF CONTRACTION
1. High-energy myosin formed when ATP is split.
2. When calcium ions bind to troponin molecules, tropomyosin molecules move aside.
3. High-energy myosin links with G-actin subunit.
4. Energy discharged from high-energy myosin; myosin head swivels and pulls on actin-containing filament.
5. When ATP again occupies myosin head, cycle repeats.

REGULATION OF THE CONTRACTILE PROCESS
Active transport mechanism returns calcium ions to sarcoplasmic reticulum; when calcium ions return to sarcoplasmic reticulum, tropomyosin returns to blocking position.

SOURCES OF ATP FOR MUSCLE CONTRACTION
Creatine phosphate can rapidly produce ATP. Aerobic metabolism and glycolysis also produce ATP.

OXYGEN DEBT Built up during periods of muscular activity when nonoxidative sources of ATP are used to support muscle contraction; oxygen debt repaid by continued increased breathing after activity.

MUSCLE FATIGUE Can be defined as inability of muscle to maintain particular strength of contraction or tension (that is, particular power output) over time.

THE MOTOR UNIT Single neuron and all muscle cells it supplies.

RESPONSES OF SKELETAL MUSCLE Contractions of muscle fibers of motor units combine to produce contractions of entire muscle.

MUSCLE TWITCH
1. **Latent period** Following arrival of stimulus at muscle.
2. **Period of contraction** Active development of tension and, possibly, shortening.
3. **Period of relaxation.**

GRADED MUSCULAR CONTRACTIONS When action potential occurs in membrane of skeletal-muscle fiber, fiber contracts to maximum extent possible for existing conditions; however, entire muscle responds in graded fashion to meet demands of particular tasks.

FACTORS INFLUENCING THE DEVELOPMENT OF MUSCLE TENSION
1. **Contractile elements:** Those structures actively involved in contraction, such as thick and thin filaments of muscle fibers.

2. **Series elastic elements:** Structures that resist stretch (but can be stretched) located between contractile elements and load.

INFLUENCE OF LENGTH ON THE DEVELOPMENT OF MUSCLE TENSION Greatest active tension develops at optimal muscle length, decreases when a muscle is stretched or compressed. Within limits, passive tension increases with stretch of the muscle.

RELATION OF LOAD TO VELOCITY OF SHORTENING The greater the load on a muscle, the slower the velocity of shortening.

MUSCLE ACTIONS p. 231

PRIME MOVERS Those muscles whose contractions are primarily responsible for a particular movement.

ANTAGONISTS Those muscles whose contraction offers resistance to the movement; generally on opposite side of joint.

SYNERGISTS Those muscles that indirectly aid a particular movement by preventing unwanted movements—for example, steadying a joint.

FIXATOR A synergist acting to immobilize.

RELATIONSHIP BETWEEN LEVERS AND MUSCLE ACTIONS pp. 232–233

LEVER Rigid structure capable of moving around pivot point (fulcrum) when force is applied.

CLASSES OF LEVERS

CLASS I LEVERS Fulcrum is between point of force and weight to be moved; for example—a seesaw, head tipping back to raise face.

CLASS II LEVERS Weight to be moved is between fulcrum and point of force; for example—a wheelbarrow, raising body on toes.

CLASS III LEVERS Weight at one end, fulcrum at other, point of force between them; for example—shovel lifting, flexion of forearm.

EFFECTS OF LEVERS ON MOVEMENTS

POWER ARM Portion of lever between fulcrum and point of force.

WEIGHT ARM Portion of lever between fulcrum and weight.

TYPES OF SKELETAL MUSCLE FIBERS
Three types, based on contraction speed and fatigue resistance. pp. 233–234

SLOW TWITCH, FATIGUE-RESISTANT FIBERS Myosin splits ATP at slow rate; slow speed of contraction; many blood vessels, much myoglobin; well-developed oxidative metabolic processes; very fatigue-resistant.

FAST TWITCH, FATIGUE-RESISTANT FIBERS Myosin splits ATP at rapid rate; fast speed of contraction; many blood vessels; much myoglobin; well-developed oxidative metabolic processes; quite fatigue-resistant.

FAST TWITCH, FATIGABLE FIBERS Myosin splits ATP at rapid rate; fast speed of contraction; few blood vessels; little myoglobin; geared to anaerobic metabolic processes; easily fatigued.

UTILIZATION OF DIFFERENT FIBER TYPES
Most skeletal muscles contain all three fiber types; each type tends to contribute in characteristic fashion to a muscle contraction. All fibers of any one motor unit are of same type. **p. 234**

EFFECTS OF EXERCISE ON SKELETAL MUSCLE **pp. 234–235**
1. Muscle size and strength can increase with exercise.
2. Endurance-type exercise causes transformation of fast twitch, fatigable fibers into fast twitch, fatigue-resistant fibers with small increase in fiber mass.
3. High-intensity, short-duration exercise causes hypertrophy of fast twitch, fatigable fibers with large increase in fiber mass and strength.

CONDITIONS OF CLINICAL SIGNIFICANCE: SKELETAL MUSCLE **p. 235**

MUSCLE ATROPHY Reduction in size of muscles due to shrinkage and death of muscle cells; may be widespread or localized.

CRAMPS Involuntary, painful muscle contractions, slow to relax; may be caused by low oxygen supply in muscles or by nervous system stimulation; those that occur during heavy exercise may be due to low levels of sodium and chloride ions in blood.

MUSCULAR DYSTROPHY Genetically transmitted progressive muscular weakness resulting from muscle-cell degeneration; increase in connective tissue or replacement of muscle cells by fatty tissue.

MYASTHENIA GRAVIS Rare chronic condition of extreme muscle weakness; related to inability of muscle-fiber plasma membrane at neuromuscular junction to respond to acetylcholine.

EFFECTS OF AGING Progressive and continuous loss of muscle mass.

SMOOTH MUSCLE **pp. 236–238**
1. Differs from skeletal muscle.
2. Smaller cells; no regular cross striations.
3. No regularly ordered myofibrils, but does possess thin filaments that contain actin and tropomyosin, and thick filaments that contain myosin; myosin light chains are associated with myosin.
4. Calcium source extracellular as well as from sarcoplasmic reticulum; calcium enters, is then pumped out of cell.

STRESS-RELAXATION RESPONSE Does not increase tension when greatly stretched.

ABILITY TO CONTRACT WHEN STRETCHED Can contract and actively develop tension when stretched.

DEGREE OF SHORTENING DURING CONTRACTION Can shorten by relatively large amounts.

SMOOTH MUSCLE TONE Generally slower response than skeletal muscle; can undergo long-term, sustained contractions.

TYPES OF SMOOTH MUSCLE

SINGLE-UNIT SMOOTH MUSCLE
1. Cells connected by gap junctions; impulses spread from cell to cell.
2. Self-excitable; spontaneous contractions.
3. Found in small arteries, veins, intestines, uterus.

MULTIUNIT SMOOTH MUSCLE
1. Cells separated; stimulus to one cell not transferred to others; cells generally require external stimulation.
2. Found in large arteries; large airways to lungs.

INFLUENCE OF EXTERNAL FACTORS ON SMOOTH MUSCLE CONTRACTION Neural activity and chemical factors such as hormones can influence contraction.

SELF-QUIZ

1. Skeletal muscles are voluntary muscles. True or False?

2. Skeletal muscle fibers: (a) exhibit cross striations; (b) are innervated by the autonomic nervous system; (c) are uninucleate.

3. Match the following terms with the appropriate lettered descriptions:

Epimysium	(a) The less movable end of a skeletal muscle
Perimysium	
Endomysium	(b) When all the fasciculi insert onto one side of the tendon
Aponeuroses	
Origin	
Insertion	(c) The fascia that envelops an entire muscle
Unipennate	
Bipennate	(d) Tendons that take the form of long, thin sheets
Multipennate	
	(e) When the fasciculi of certain muscles have a complex arrangement

involving the convergence of several tendons

(f) Thin extensions of the fascia that envelop the plasma membrane of each muscle fiber

(g) The more movable end of a skeletal muscle

(h) Fascia that penetrates a muscle, separating the fibers into bundles

(i) Muscles that have fasciculi inserting obliquely on both sides of the tendon

4. The thick filament of a sarcomere of a skeletal muscle cell contains: (a) actin; (b) troponin; (c) myosin.

5. The threadlike molecule that lies along the surface of actin is: (a) myosin; (b) tropomyosin; (c) troponin.

6. Structures that run from the plasma membrane into the interior of a skeletal-muscle fiber form the: (a) t tubule network; (b) sarcoplasmic reticulum; (c) myofibers.

7. Acetylcholine causes a change in the permeability of the plasma membrane of a skeletal-muscle fiber at the neuromuscular junction. True or False?

8. The membrane system of skeletal muscle fibers that contains calcium ions necessary for contraction is the: (a) sarcomere; (b) sarcoplasmic reticulum; (c) myofibril.

9. During a skeletal muscle contraction, ATP occupies: (a) actin; (b) myosin; (c) troponin.

10. Actin acts as an enzyme to split ATP into ADP and phosphate. True or False?

11. In an unstimulated skeletal-muscle fiber, the interaction of actin and myosin is believed to be directly blocked by: (a) calcium ions; (b) t tubules; (c) tropomyosin.

12. During a skeletal muscle contraction, calcium ions bind with: (a) tropomyosin; (b) myosin; (c) troponin.

13. During a contraction in which a skeletal-muscle fiber shortens: (a) troponin shortens; (b) the Z lines are drawn closer together; (c) calcium binds with myosin.

14. During sustained, intense, rapid exercise, a skeletal muscle produces: (a) acetylcholine; (b) glycogen; (c) lactic acid.

15. Which substance provides skeletal muscle fibers with a means of rapidly forming ATP immediately following the initiation of muscular activity? (a) fatty acids; (b) creatine phosphate; (c) kinase.

16. Within a skeletal muscle: (a) there are more neurons than muscle fibers; (b) there are no neuromuscular junctions between nerve endings and muscle fibers; (c) each neuron branches to supply a number of muscle fibers.

17. An individual neuron, together with all of the skeletal-muscle fibers it supplies, makes up a: (a) sarcomere; (b) motor unit; (c) myofibril.

18. When a nerve impulse triggers a propagated action potential in the membrane of a skeletal-muscle fiber, the fiber contracts to the maximum extent possible for the existing conditions. True or False?

19. Fast twitch, fatigable skeletal-muscle fibers are: (a) well supplied by blood vessels; (b) myoglobin rich; (c) geared for the utilization of anaerobic metabolic processes.

20. With training of a high-intensity, short-duration type: (a) skeletal muscle strength decreases; (b) fast twitch, fatigable fibers hypertrophy; (c) slow twitch, fatigue-resistant fibers are converted to fast twitch, fatigable fibers.

21. Smooth muscle fibers are: (a) multinucleate; (b) found in the walls of internal or visceral organs; (c) principally under voluntary control.

22. Which type of muscle is likely to become spontaneously stimulated? (a) single-unit smooth muscle; (b) skeletal muscle; (c) multiunit smooth muscle.

23. The muscular dystrophies are generally considered to be genetically transmitted. True or False?

LEARNING OBJECTIVES

After completing this chapter, you should be able to:

1. Cite several criteria used to name muscles, and give an example of each.

2. Distinguish between intrinsic and extrinsic muscles, citing several examples of each.

3. Name the major muscles of the head and neck, describing the origin, insertion, action, and innervation of each.

4. Contrast the general functions of the neck muscles that are found in the posterior and anterior triangles.

5. Name the major muscles of the trunk, describing the origin, insertion, action, and innervation of each.

6. Name the major muscles that move the vertebral column.

7. Name the major muscles of the upper limbs, describing the origin, insertion, action, and innervation of each.

8. Name the major muscles of the lower limbs, describing the origin, insertion, action, and innervation of each.

CHAPTER CONTENTS

MUSCLES OF THE HEAD AND NECK

MUSCLES OF THE TRUNK

MUSCLES OF THE UPPER LIMBS

MUSCLES OF THE LOWER LIMBS

KEY TERMS AND DERIVATIVES

adductor (*ad* = toward) an adductor moves a part closer to the midline; the adductors magnus, longus, brevis, and pectineus all adduct the femur

deltoid (*delta* = triangular) large triangular muscle that forms a cap over the shoulder

gastrocnemius (*gaster* = belly; *kneme* = leg) one of three superficial, posterior muscles that form the bulge of the calf

latissimus dorsi (*latissimus* = widest; *dorsum* = back) a large muscle arising from the vertebrae of the lower back to insert on the anterior surface of the humerus; it extends the arm

masseter (*maseter* = chewer) a muscle arising from the zygomatic arch, which assists the temporalis in raising the mandible.

orbicularis (*orb* = circular) a type of sphincter muscle, which is a ring-shaped muscle that surrounds certain body openings

sternocleidomastoid (*sternum* = breastbone; *cleido* = clavicle) the muscle that runs diagonally across the lateral margins of the neck from the mastoid process of the temporal bone to the sternum and the clavicle

The Muscular System:
Gross Anatomy

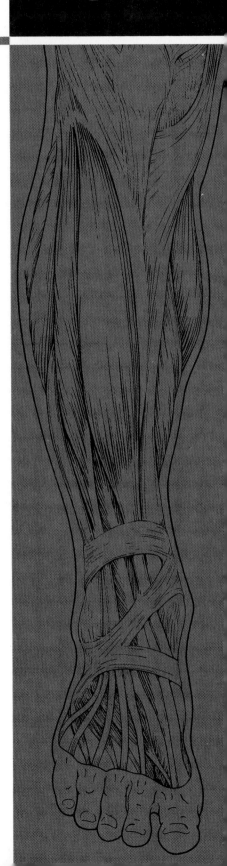

There are over 600 muscles within the human body (Figures 9.1 and 9.2). Only the more commonly studied muscles are included in this chapter. Several criteria are used to name muscles; each describes a particular characteristic of the muscle being named, such as its shape, action, or location. You will find it quite useful in your study of the muscles to familiarize yourself with these criteria:

SHAPE   The names of some muscles include references to their shape. For example, the trapezius muscles are shaped like trapezoids and the rhomboideus muscles resemble rhomboids.

ACTION   Various muscle names include references to the actions of the muscle by using the terms flexor, extensor, adductor, or pronator. For example, the flexor carpi radialis muscles flex the hands, and the extensor digitorum longus muscles extend the toes.

LOCATION   It is possible to locate certain muscles by their names. For example, the intercostal muscles (*inter* = between; *costal* = rib) are located between the ribs, and the tibialis anterior muscles lie alongside the anterior margin of each tibia.

ATTACHMENTS   The attachments of a muscle to the skeleton are included in some names. For example, the sternocleidomastoid muscles have origins on the sternum and clavicles and insert on the mastoid processes of the temporal bones; the coracobrachialis muscles have their origins on the coracoid processes of the scapulae and insert on each brachium—which refers to the arm (humerus).

NUMBER OF DIVISIONS   Some muscles are separated into two, three, or four divisions, and this is indicated in their names. For example, the biceps brachii muscles have two divisions, the triceps brachii have three, and the quadriceps femoris muscles have four.

SIZE RELATIONSHIPS   Terms referring to size are often included in muscle names—for example, the gluteus maximus and gluteus minimus muscles of the buttocks are large and small, respectively, and the peroneus longus and peroneus brevis muscles of the leg are long and short, respectively.

In many cases muscle names include more than one of these criteria. For example, the name of the flexor digitorum longus muscle indicates the muscle's action (flexion), its insertion (digits), and its size relationship (long, in comparison to the flexor digitorum brevis muscle).

To organize their study, we will consider the muscles in various groups (but note that some muscles belong to more than one group). Each group of muscles is discussed in a general way, and the reader is provided with figures that illustrate the muscles as well as tables that give a more detailed description of each muscle, including information on each muscle's origin, insertion, principal actions, and nerve innervation.

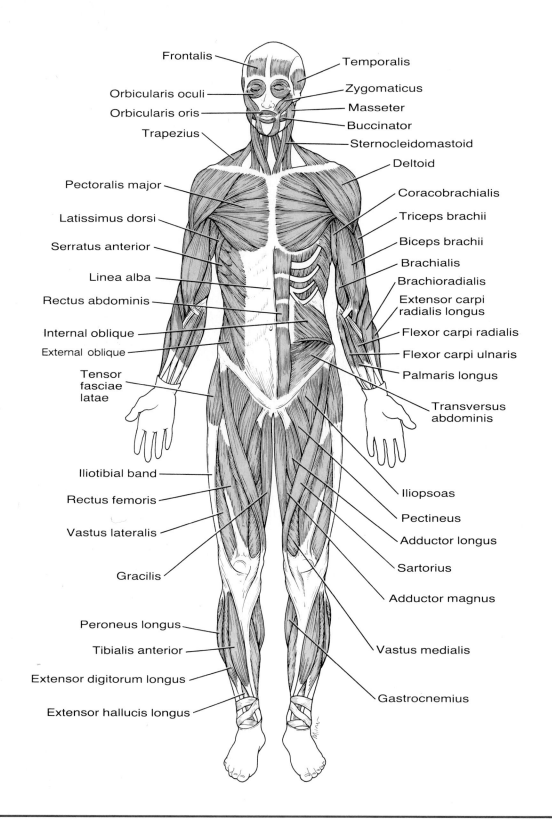

Frontalis

Temporalis

Orbicularis oculi

Zygomaticus

Orbicularis oris

Masseter

Buccinator

Trapezius

Sternocleidomastoid

Deltoid

Pectoralis major

Coracobrachialis

Latissimus dorsi

Triceps brachii

Serratus anterior

Biceps brachii

Linea alba

Brachialis

Rectus abdominis

Brachioradialis

Internal oblique

Extensor carpi radialis longus

External oblique

Flexor carpi radialis

Tensor fasciae latae

Flexor carpi ulnaris

Palmaris longus

Transversus abdominis

Iliotibial band

Iliopsoas

Rectus femoris

Pectineus

Vastus lateralis

Adductor longus

Gracilis

Sartorius

Adductor magnus

Peroneus longus

Tibialis anterior

Vastus medialis

Extensor digitorum longus

Extensor hallucis longus

Gastrocnemius

Figure 9.1

Anterior view of the muscles of the body. The left external oblique muscle has been removed.

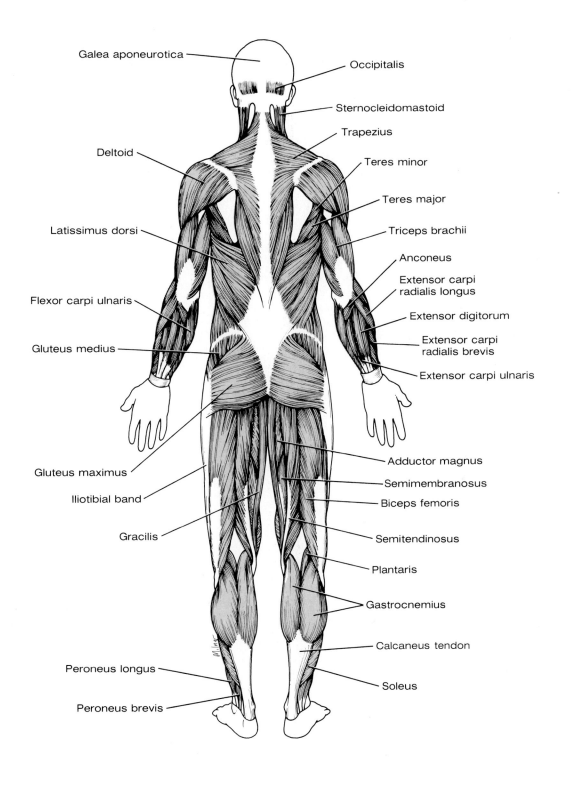

Galea aponeurotica

Occipitalis

Sternocleidomastoid

Trapezius

Deltoid

Teres minor

Teres major

Triceps brachii

Latissimus dorsi

Anconeus

Extensor carpi
radialis longus

Flexor carpi ulnaris

Extensor digitorum

Extensor carpi
radialis brevis

Gluteus medius

Extensor carpi ulnaris

Adductor magnus

Semimembranosus

Biceps femoris

Gluteus maximus

Semitendinosus

Iliotibial band

Plantaris

Gracilis

Gastrocnemius

Calcaneus tendon

Peroneus longus

Soleus

Peroneus brevis

Figure 9.2

Posterior view of the
muscles of the body.

Table 9.1   Muscles of the Face [F9.3]

Muscle	Origin	Insertion	Action	Innervation
Buccinator	Alveolar process of the mandible and the maxillary bone	Orbicularis oris and skin at the angle of the mouth	Compresses cheek; pulls corner of the mouth laterally	Facial (cranial nerve VII)
Corrugator	Frontal bone, lateral to the glabella	Skin of the eyebrows	Draws the eyebrows together, as in frowning	Facial
Depressor anguli oris	Body of the mandible, below the mental foramen	Skin and muscles at the angle of the mouth	Pulls the angle of the mouth downward	Facial
Depressor labii inferioris	Body of the mandible between the symphysis and the mental foramen	Skin and muscles of the lower lip	Pulls the lower lip downward	Facial
Epicranius *Frontalis*	Galea aponeurotica	Skin and muscles of the forehead	Raises the eyebrows; wrinkles the skin of the forehead	Facial
Occipitalis	Occipital bone (superior nuchal line)	Galea aponeurotica	Draws the scalp posteriorly	Facial
Levator labii superioris	Lower margin of orbit (maxillary and zygomatic bones)	Skin and muscles of the upper lip, and wing of the nose	Raises the upper lip; dilates the nares (nostrils)	Facial
Mentalis	Mandible, near the symphysis	Skin of the chin	Raises and protrudes the lower lip	Facial
Orbicularis oculi	Frontal and maxillary bones; medial palpebral ligament	Circles the orbit and extends within the eyelids	Closes the eyelids; tightens the skin of the forehead	Facial
Orbicularis oris	Muscles surrounding the mouth	Skin surrounding the mouth	Closes and protrudes the lips	Facial
Platysma	Fascia over the pectoralis major and the deltoid muscles	Lower border of the mandible, and the skin of the chin and cheek	Depresses the mandible; draws the angle of the mouth downward; tightens and wrinkles the skin of the neck	Facial
Procerus	Lower portion of the nasal bone; upper part of the lateral nasal cartilage	Skin between the eyebrows	Wrinkles the skin between the eyebrows	Facial
Risorius	Fascia of the masseter muscle	Skin at the angle of the mouth	Pulls the angle of the mouth backward	Facial
Zygomaticus major and zygomaticus minor	Zygomatic bone	Skin and muscles above the angle of the mouth	Raise the angle of the mouth	Facial

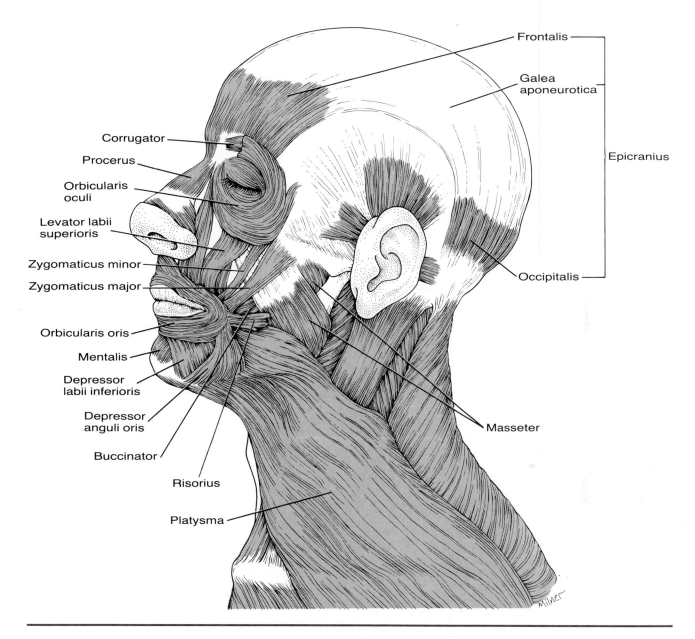

Figure 9.3
Muscles of the face and
neck. (The temporalis
muscle has been
removed.)

MUSCLES OF THE HEAD AND NECK

Muscles of the Face

Because of their actions, the muscles of the face (Figure 9.3; Table 9.1) are also
referred to as the muscles of facial expression. Whereas some facial muscles
arise from the bones of the skull, others arise from the superficial fascia of the
face. Most of them insert into the skin of the region, and therefore serve to
move the skin rather than a joint. Among the unusual types of muscles in this
group are the *sphincters*, which are ring-shaped muscles that surround body
openings. They can enlarge or close the opening by relaxing or contracting.
The **orbicularis oculi** is a sphincter used in closing the eye, winking, and
squinting. Contraction of the sphincter muscle named **orbicularis oris** closes
the mouth and purses the lips. Several of the other facial muscles insert onto
the fascia that covers the orbicularis oris.

F9.3, Table 9.1

sfink'-ters

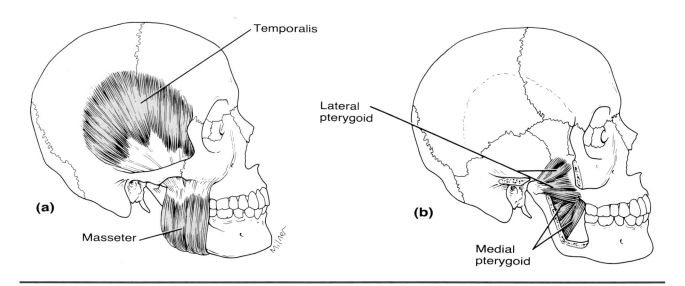

Figure 9.4

Muscles of mastication.
(a) The temporalis and masseter muscles are the strongest masticatory

muscles. **(b)** The temporalis and masseter muscles have been removed, and the

zygomatic arch and mandible have been sectioned to reveal the pterygoid muscles.

Table 9.2 Muscles of Mastication [F9.4]

Muscle	Origin	Insertion	Action	Innervation
Temporalis	Temporal fossa	Coronoid process and ramus of the mandible	Raises the mandible, closing the jaws; retracts the mandible	Trigeminal (cranial nerve V)
Masseter	Zygomatic arch	Angle and ramus of the mandible	Raises the mandible, closing the jaws	Trigeminal
Medial pterygoid	Medial surface of the lateral pterygoid plate of the sphenoid bone, and the tuberosity of the maxillary bone	Inner surface of the mandible, at the angle	Closes the jaws; together with the lateral pterygoid, it aids in sideways movement of the jaws	Trigeminal
Lateral pterygoid	Lateral surface of the lateral pterygoid plate and the great wing of the sphenoid bone	Mandible just below the condyle	Opens and protrudes the mandible; moves the mandible from side to side	Trigeminal

Another unusual facial muscle is the **epicranius.** This muscle has two parts—the anterior **frontalis** and the posterior **occipitalis.** These two muscular portions are connected by a broad, flat tendon, the **galea aponeurotica,** which lies tight against the top of the skull. Contraction of one or the other of the muscular portions pulls the scalp forward or backward.

While the **platysma** is not actually a facial muscle, we will consider it here because its main actions are on the mandible and the skin around the mouth. It is a superficial sheetlike muscle that covers the ventral surface of the upper thorax and the neck and extends over the chin to the region of the mouth. Contraction of the platysma lowers the mandible, the lower lip, and the corners of the mouth, as well as tightening the skin of the neck.

Muscles of Mastication

Four pairs of muscles are involved in biting and chewing. The large fan-shaped **temporalis,** which passes deep to the zygomatic arch of the cheek, and the quadrilateral-shaped **masseter,** which arises from the zygomatic arch,

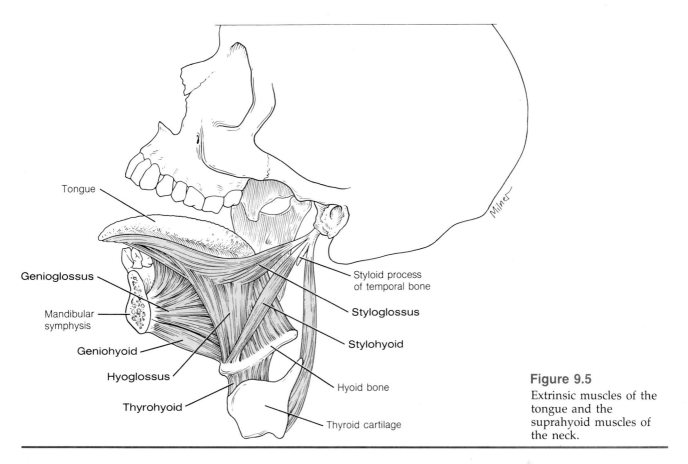

Figure 9.5
Extrinsic muscles of the tongue and the suprahyoid muscles of the neck.

Table 9.3   Extrinsic Muscles of the Tongue [F9.5]

Muscle	Origin	Insertion	Action	Innervation
Genioglossus	Internal surface of the mandible, near the symphysis	Undersurface of the tongue; body of the hyoid	Protracts, retracts, and depresses the tongue	Hypoglossal (cranial nerve XII)
Hyoglossus	Body and greater cornu of the hyoid bone	Side of the tongue	Depresses the tongue; draws its sides down	Hypoglossal
Styloglossus	Styloid process of the temporal bone	Side of the tongue	Retracts and elevates the tongue	Hypoglossal

both serve to raise the mandible (Figure 9.4; Table 9.2). These muscles can be felt when the teeth are forcibly clenched. The other two pairs of muscles involved in mastication are the medial and lateral **pterygoid** muscles, which move the mandible sideways in grinding movements, as well as assisting in opening and closing the mouth.   F9.4, Table 9.2

Muscles of the Tongue

The tongue is a muscular organ covered with mucous membrane. Some of the muscles lie entirely within the tongue. These are called **intrinsic muscles.** The fibers of the intrinsic muscles are arranged in longitudinal, vertical, and horizontal planes; consequently, when they contract, they squeeze, fold, and curl the tongue. These actions are particularly useful in speaking and manipulating food within the mouth.

The **extrinsic muscles** (Figure 9.5; Table 9.3) anchor the tongue to the skeleton (hyoid, mandible, and temporal bones) and control the protrusion, retraction, and sideward movement of the tongue.   F9.5, Table 9.3

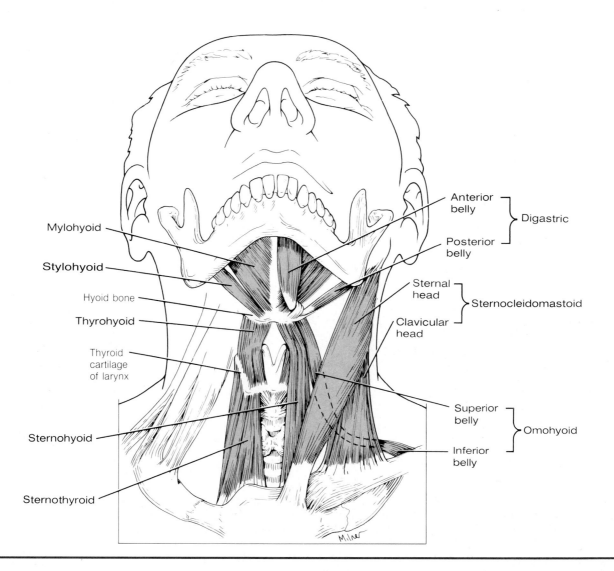

Figure 9.6
The suprahyoid and infrahyoid muscles of the neck. The sterno-cleidomastoid and digastric muscles have been removed on the right.

F9.6,
Table 9.4

Table 9.5

F9.5, Table 9.4

Muscles of the Neck

The muscles of the neck (Figure 9.6; Table 9.4) are often described as being located within one of two triangles. Those within the anterior triangle are separated from those within the posterior triangle by the **sternocleidomastoid** muscle. This muscle runs diagonally across the lateral margins of the neck, from the mastoid process of the temporal bone to the sternum and the clavicle. It is beyond the scope of this text to describe all of the muscles within these two triangles. Rather we will consider at this time only selected muscles of the anterior triangle—in particular, the muscles of the throat. The muscles of the posterior triangle are included with those muscles that move the vertebral column and the head (Table 9.5).

Muscles of the Throat

The muscles of the throat are the deep muscles of the anterior triangle (Figure 9.5; Table 9.4). They help to form the floor of the oral cavity and are attached to the hyoid bone. Because the tongue is also attached to the hyoid bone, these muscles are involved with movements of the tongue. In addition, some of the throat muscles are attached to the larynx and therefore aid in swallowing. These muscles are often divided into two groups:

1. The *suprahyoid muscles:* **digastric, stylohyoid, mylohyoid,** and **geniohyoid.** As a group, these muscles raise the hyoid bone during swallowing and lower the jaw when the hyoid bone is fixed.

Table 9.4 Muscles of the Anterior Triangle of the Neck [F9.6]

Muscle	Origin	Insertion	Action	Innervation
Sternocleidomastoid	By two heads: the manubrium of the sternum, and the medial portion of the clavicle	Mastoid process of the temporal bone	Both muscles acting together flex the cervical vertebral column; acting singly, each rotates head to the opposite side	Accessory (cranial nerve XI) and upper cervical spinal nerves
SUPRAHYOID MUSCLES				
Digastric	*Anterior belly:* inner surface of the mandibular symphysis *Posterior belly:* mastoid process of the temporal bone	Hyoid bone, via the intermediate tendon	Raises the hyoid and assists in lowering the jaw	Trigeminal (anterior belly); facial (posterior belly)
Stylohyoid	Styloid process of the temporal bone	Hyoid bone	Raises the hyoid and pulls it backward	Facial
Mylohyoid	Inner surface of the mandible, from the symphysis to the angle	Hyoid bone	Raises the hyoid and the floor of the mouth	Trigeminal
Geniohyoid	Inner surface of the mandibular symphysis	Hyoid bone	Pulls the hyoid anteriorly	C_1 (through hypoglossal)
INFRAHYOID MUSCLES				
Sternohyoid	Manubrium and the medial end of the clavicle	Hyoid bone	Pulls the hyoid inferiorly	C_1–C_3 (through ansa cervicalis—see p. 399)
Sternothyroid	Manubrium	Thyroid cartilage of the larynx	Pulls the larynx inferiorly	C_1–C_3 (through ansa cervicalis)
Thyrohyoid	Thyroid cartilage of the larynx	Hyoid bone	Pulls the hyoid inferiorly and raises the larynx	C_1 (through hypoglossal)
Omohyoid	Superior border of the scapula	Hyoid bone	Pulls the hyoid inferiorly	C_1–C_3 (through ansa cervicalis)

2. The *infrahyoid muscles:* **sternohyoid, sternothyroid, thyrohyoid,** and **omohyoid.** These muscles pull down on the larynx and hyoid, returning them to their normal positions after swallowing.

MUSCLES OF THE TRUNK

The muscles of the trunk include those that are associated with the vertebral column, the back, the thorax, the floor of the pelvic cavity, and the wall of the abdomen. Trunk muscles have various actions, depending on their locations. Some move the vertebral column, others move the head; some are involved in respiratory movements, others function to move the upper limbs; and so forth. We will study them in groups according to their actions.

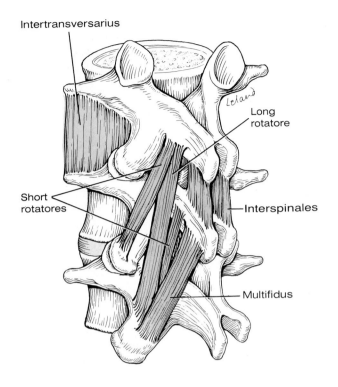

Figure 9.7
Muscles of the vertebral
column.

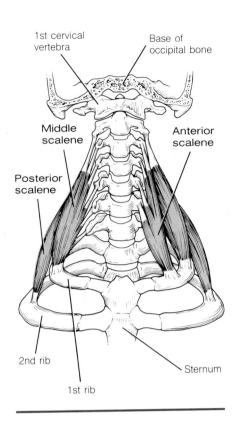

Figure 9.8
The scalene muscles
viewed from the front.
The right anterior scalene
muscle has been
removed.

Muscles of the Vertebral Column

Most of the muscles that move the vertebral column are located on the posterior surface of the spine (Table 9.5). A few, such as the *splenius,* have fibers that insert onto the skull and therefore move the head as well as the vertebral column. The deepest of these muscles are located medially and travel only a few segments superiorly before inserting onto the transverse processes or spinous processes of the vertebrae (Figure 9.7). These include the **multifidus, rotatores, interspinales,** and **intertransversarii,** and the **semispinalis thoracis, cervicis,** and **capitis.** The **scalenes** (anterior, middle, and posterior) pass from the transverse processes of the cervical vertebrae to the upper two ribs (Figure 9.8).

Lateral to these muscles, located in the depression between the spinous processes and the transverse processes and the ribs, is a longitudinal muscle mass that extends from the sacrum to the skull. This is the **sacrospinalis (erector spinae)** muscle (Figure 9.9). The sacrospinalis has three subdivisions in the form of columns. The **iliocostalis,** which is the most lateral column, inserts on the ribs; the medial column is the **spinalis,** the fibers of which insert on the vertebrae; the **longissimus** subdivision is located between the other two columns. Each of these columns is further separated into **lumborum, thoracis, cervicis,** and/or **capitis** parts, which are named according to their points of insertion. In Table 9.5 the origins, insertions, and actions of these parts have been combined for each subdivision of the sacrospinalis.

All the muscles associated with the vertebral column act to extend the spine, and when acting on one side only, they bend the vertebral column to that side and may assist in its rotation. These muscles associated with the vertebral column are not the only ones that move the vertebral column. Some

Margin references:
Table 9.5
F9.7
F9.8
F9.9
Table 9.5

Table 9.5   Muscles That Move the Vertebral Column

Muscle	Origin	Insertion	Action	Innervation
Semispinalis [F9.9] thoracis cervicis capitis	Transverse processes of the thoracic and the seventh cervical vertebrae	Spinous processes of the second cervical through the fourth thoracic vertebrae, and the occipital bone	Extend the vertebral column and the head (capitis); rotate them to the opposite side	Branches of the spinal nerves
Multifidus [F9.7]	Posterior surface of the sacrum and the ilium, and the transverse processes of the lumbar, thoracic, and lower cervical vertebrae	Spinous processes of the lumbar, thoracic, and cervical vertebrae	Extends the vertebral column; rotates it towards the opposite side	Branches of the spinal nerves
Rotatores (long and short) [F9.7]	Transverse processes of all the vertebrae	Base of the spinous process of the vertebra above the vertebra of origin (short) or the second vertebra above (long)	Extend the vertebral column; rotate it towards the opposite side	Branches of the spinal nerves
Interspinales [F9.7]	Superior surface of all the spinous processes	Inferior surface of the spinous process of the vertebra above the vertebra of origin	Extend the vertebral column	Branches of the spinal nerves
Scalenes [F9.8]	Transverse process of cervical vertebrae	Upper two ribs	Flex and rotate the neck; assist in inspiration	Branches of the lower cervical nerves
Intertransversarii [F9.7]	Transverse processes of all the vertebrae	Transverse processes of the vertebra above the vertebra of origin	Bend the vertebral column laterally	Branches of the spinal nerves
Splenius [F9.9] capitis cervicis	Spinous processes of the upper thoracic and the seventh cervical vertebrae, and from the ligamentum nuchae	Occipital bone, mastoid process of the temporal bone, and the transverse processes of the upper three cervical vertebrae	Acting together, they extend the head and the neck; acting singly, they abduct and rotate the head towards the same side	Branches of the spinal nerves

SACROSPINALIS (ERECTOR SPINAE) [F9.9]

Muscle	Origin	Insertion	Action	Innervation
Iliocostalis lumborum thoracis cervicis	Crest of the sacrum; spinous processes of the lumbar and lower thoracic vertebrae; iliac crests; angles of the ribs	Angles of the ribs; transverse processes of the cervical vertebrae	Extend the vertebral column and bend it laterally	Branches of the spinal nerves
Longissimus thoracis cervicis capitis	Transverse processes of the lumbar, thoracic, and lower cervical vertebrae	Transverse processes of the vertebra above the vertebra of origin, and the mastoid process of the temporal bone (capitis)	Extend the vertebral column and head; rotate the head towards the same side	Branches of the spinal nerves
Spinalis thoracis cervicis	Spinous process of the upper lumbar, lower thoracic, and seventh cervical vertebrae	Spinous processes of the upper thoracic and the cervical vertebrae	Extend the vertebral column	Branches of the spinal nerves

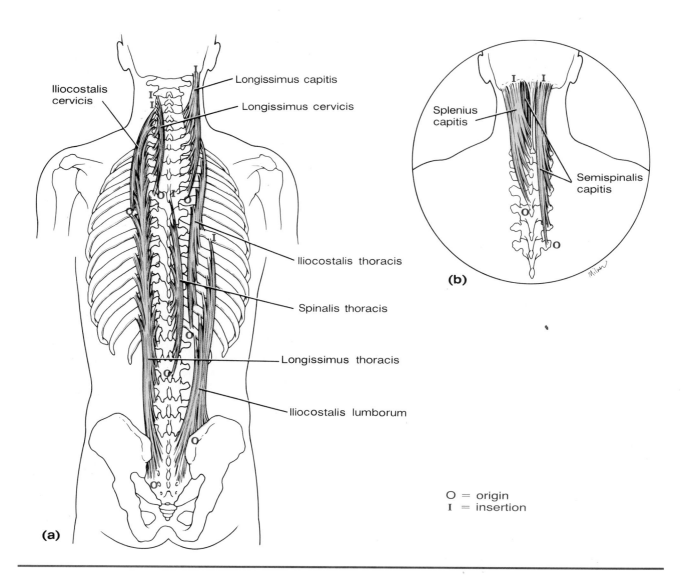

Figure 9.9
(a) The sacrospinalis muscles showing the longissimus, iliocostalis, and spinalis portions. (b) The splenius capitis and semispinalis capitis muscles.

Table 9.7
Table 9.17

of the muscles of the abdominal wall, such as the **rectus abdominis** and the **quadratus lumborum** (Table 9.7) also act on the vertebral column. In addition, the psoas major (Table 9.17), which generally acts on the hip joint, can cause the vertebral column to flex if the thighs are fixed.

Deep Muscles of the Thorax

Table 9.6

Table 9.6

Most of the deep muscles of the thorax insert on the ribs and assist in breathing by drawing the ribs together or by elevating or depressing the rib cage (Table 9.6). The ribs slope downward as they pass forward; therefore, any muscle that elevates them increases the volume of the thoracic cavity, causing inspiration. Conversely, muscles that depress the ribs back to their usual positions decrease the volume of the thoracic cavity, forcing air from the lungs in expiration. The muscles described in Table 9.6 are those involved in normal, quiet respirations. In forced breathing, when overexpansion of the thoracic cage is beneficial, additional muscles may be involved, such as the **scalenes** (Table 9.5), the **sternocleidomastoid** (Table 9.4), and the **quadratus lumborum** (Table 9.7). Because these muscles have other actions that are more commonly performed, they are listed within other groups, as their table references indicate.

Table 9.5, Table 9.4
Table 9.7

F9.10

The spaces between adjacent ribs are reinforced primarily by **external** and **internal intercostal** muscles (Figure 9.10). In addition, small **innermost intercostal** muscles lie deep to the internal intercostals. The fibers of the external

Table 9.6   Deep Muscles of the Thorax (Respiratory Muscles)

Muscle	Origin	Insertion	Action	Innervation
Diaphragm [F9.11]	The xiphoid process; inner surfaces of lower six ribs; and the lumbar vertebrae	Central tendon of the diaphragm	Pulls central tendon downward, increasing the size of the thoracic cavity and therefore causing inspiration	Phrenic
External intercostals [F9.10]	Inferior border of the ribs and the costal cartilages	Superior border of the rib below the rib of origin	Draw ribs together, aiding respiration	Intercostal
Internal intercostals [F9.10]	Inner surface of the ribs and the costal cartilages	Superior border of the rib below the rib of origin	Draw ribs together, aiding respiration	Intercostal
Subcostales*	Inner surface of the ribs, near their angles	Inner surface of the second or third rib below the rib of origin	Draw ribs together, aiding expiration	Intercostal
Transversus thoracis*	Inner surface of the sternum and the xiphoid process	Inner surface of the costal cartilages	Draws anterior portion of the rib cage downward, aiding expiration	Intercostal

*Not illustrated.

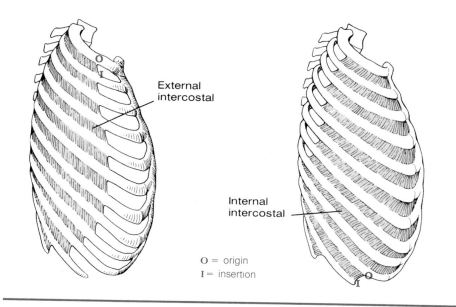

External intercostal

Internal intercostal

O = origin
I = insertion

Figure 9.10
The external and internal intercostal muscles.

intercostal muscles run at right angles to the fibers of the internal and innermost muscles, as in a bias-ply automobile tire, thus providing a strong muscular wall between the ribs, without requiring heavy musculature.

The **diaphragm** is the muscle chiefly responsible for quiet breathing. Dome-shaped, it separates the thoracic cavity from the abdominopelvic cavity. The upper surface of the diaphragm is in contact with the heart and lungs; the lower surface contacts the liver, the stomach, the spleen, and the pad of fat that surrounds the suprarenal glands and kidneys.

The muscle fibers of the diaphragm are divisible into sternal, costal, and lumbar portions (Figure 9.11). The small *sternal* portion arises from the inner surface of the xiphoid process; the *costal* fibers arise from the inner surfaces of the seventh, eighth, and ninth ribs and the distal ends of the last three ribs;

F9.11

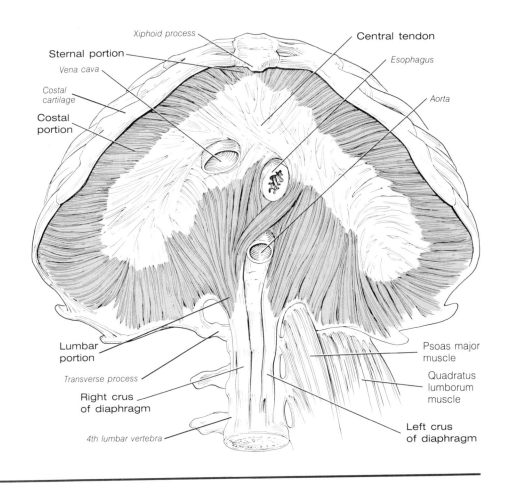

Xiphoid process

Sternal portion

Vena cava

Costal
cartilage

Costal
portion

Central tendon

Esophagus

Aorta

Lumbar
portion

Transverse process

Right crus
of diaphragm

4th lumbar vertebra

Psoas major
muscle

Quadratus
lumborum
muscle

Left crus
of diaphragm

Figure 9.11
The abdominal surface of
the diaphragm.

the *lumbar* portion arises from the front of the lumbar vertebrae by two tendinous bands called crura. The muscle fibers from these three portions insert on a common *central tendon*, which they surround. When the diaphragm contracts, its dome is pulled downward, flattening the muscle and increasing the volume of the thoracic cavity.

The diaphragm is pierced by a number of openings that permit the passage of structures between the thorax and abdomen. The largest openings are for the aorta, the inferior vena cava, and the esophagus.

The mechanics of respiration are considered in greater detail in Chapter 22.

Muscles of the Abdominal Wall

Because there are no skeletal supports within the walls of the abdominal cavity, the abdominal cavity derives its strength entirely from muscles. There are three layers of muscles in the wall (Figure 9.12; Table 9.7). The fibers of each of the muscles run in different directions, providing additional strength. The outermost layer is the **external abdominal oblique** muscle, whose fibers pass medially and downward as a continuation of the external intercostal muscles. The **internal abdominal oblique** lies just beneath the external oblique. Its fibers run upward and medially, becoming continuous over the ribs with the internal intercostal muscles. Deep to both of the oblique muscles is a thin muscle whose fibers run horizontally, encircling the abdominal cavity. This is the **transversus abdominis** muscle. The tendons of the three abdominal-wall muscles pass medially in the form of broad aponeuroses that insert on a midline **linea alba** ("white line"). The linea alba is a fibrous band that extends from the xiphoid process of the sternum to the symphysis pubis.

F9.12, Table 9.7

Table 9.7   Muscles of the Abdominal Wall [F9.12]

Muscle	Origin	Insertion	Action	Innervation
External abdominal oblique	External surface of the lower eight ribs	Linea alba and the anterior half of the iliac crest	Compresses the abdominopelvic cavity; assists in flexing and rotating the vertebral column	Intercostal, iliohypogastric, and ilioinguinal
Internal abdominal oblique	Inguinal ligament, the iliac crest, and the lumbodorsal fascia	Linea alba, the pubic crest, and the lower four ribs	Compresses the abdominopelvic cavity; assisting in flexing and rotating the vertebral column	Intercostals, iliohypogastric, and ilioinguinal
Transversus abdominis	Inguinal ligament, the iliac crest, the lumbodorsal fascia, and the costal cartilages of the last six ribs	Linea alba, and the pubic crest	Compresses the abdominopelvic cavity	Intercostals, iliohypogastric, and ilioinguinal
Rectus abdominis	Pubic crest	Xiphoid process and the costal cartilages of the fifth through the seventh ribs	Compresses the abdominopelvic cavity; flexes the vertebral column	Intercostals
Quadratus lumborum	Iliac crest, and the iliolumbar ligament	Lower border of the twelfth rib; the transverse processes of the upper lumbar vertebrae	Pulls the thoracic cage toward the pelvis; bends the vertebral column laterally toward the side that is being contracted	Twelfth thoracic and first lumbar

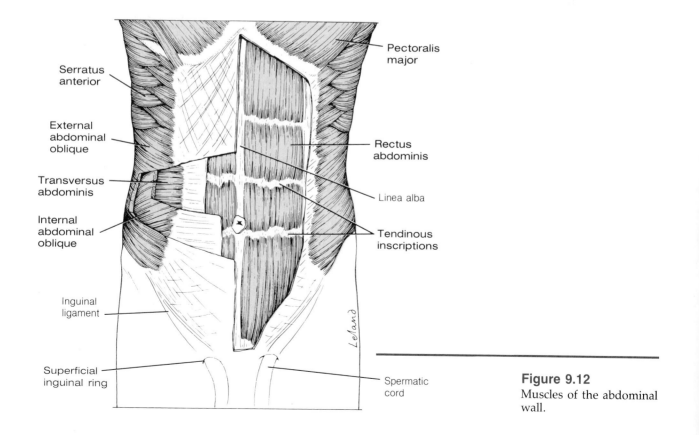

Figure 9.12

Muscles of the abdominal wall.

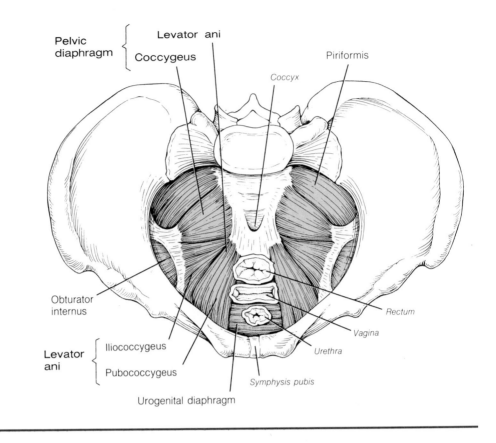

Figure 9.13
Pelvic diaphragm of the female, viewed from the inside of the pelvic cavity.

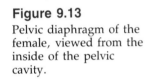

ing'-gwi-nal

At the lower margin of the external abdominal oblique muscle, the fascia of the muscle forms a tendinous border called the **inguinal ligament.** This ligament runs between the pubic tubercle and the anterior superior iliac spine. It marks the separation between the body wall and the thigh. At one point there is an opening between the muscle fascia and the inguinal ligament. This opening, which is called the *superficial inguinal ring*, is the external opening of the **inguinal canal.** The canal passes laterally, above and parallel with the inguinal ligament. About midway between the anterior superior iliac spine and the pubic tubercle, the canal opens into the abdominal cavity through an aperture in the fascia of the transversus abdominis muscle. This opening is called the *deep inguinal ring*. In the male, the spermatic cord passes through the canal. In the female, the canal provides for passage of the round ligament of the uterus. If the inguinal canal is weak, increased abdominal pressure can force some of the abdominal contents into the canal. This condition is called a *hernia*, or *rupture*. Inguinal hernias are more common in males because during embryonic development their inguinal canals are expanded and weakened as a result of the passage of the testes through them into the scrotal sacs.

The **rectus abdominis** is a narrow, flat muscle on the ventral aspect of the abdominal wall. Its fibers run vertically from the pubis to the rib cage, alongside the linea alba. Each rectus abdominis is completely ensheathed by the fascia of the oblique and transversus abdominal muscles. The fasciae separate in various combinations to pass superficially and deep to the rectus muscle. Each rectus is crossed by three transverse fibrous bands called the **tendinous inscriptions.** In a person who has developed his or her rectus muscle through exercise, the portions of the muscle between these inscriptions enlarge, causing the inscriptions to appear through the skin as horizontal depressions.

The three oblique muscles and the rectus abdominis compress the abdominal cavity, assisting in such actions as forced expiration, defecation, and urination.

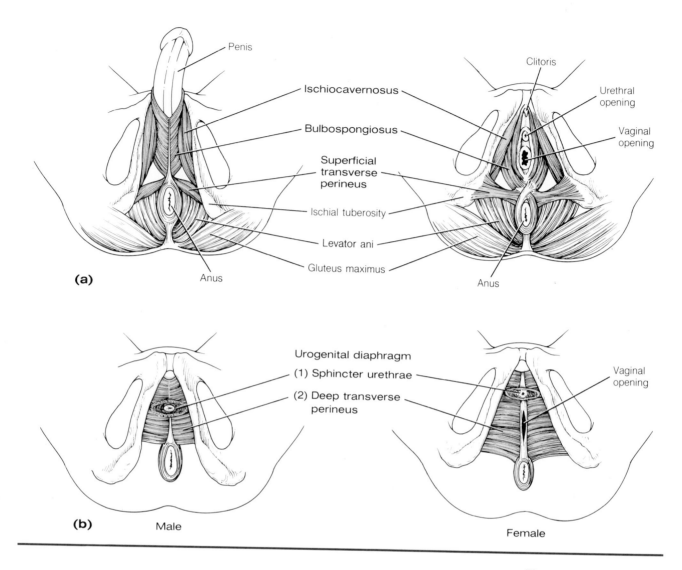

Figure 9.14
Muscles of the perineum.
(a) Superficial muscles.
(b) Deep muscles forming
the urogenital
diaphragm.

F9.28

Table 9.17

Most of the posterior portion of the abdominal wall is formed by the **quadratus lumborum** muscle (Figure 9.28). This is a broad quadrilateral muscle that runs from the posterior region of the iliac crest to the twelfth rib. The psoas major muscle also forms part of the posterior wall, but it acts primarily on the femur and so is described with that group of muscles (Table 9.17). A small psoas minor muscle that acts on the lumbar vertebral column is sometimes present ventral to the psoas major.

Muscles That Form the Floor of the Abdominopelvic Cavity

The viscera of the abdominopelvic cavity are supported by a muscular floor called the **pelvic diaphragm** (Figure 9.13; Table 9.8). Two muscles, the **levator ani** and the **coccygeus,** form the pelvic diaphragm. The levator ani is composed of several parts, of which the **pubococcygeus** muscle and the **iliococcygeus** muscle are the most prominent. There are openings through these muscles for the rectum, the urethra, and in the female, the vagina.

F9.13, Table 9.8

Muscles of the Perineum

The **perineum** is the lower end of the trunk between the thighs (Figure 9.14; Table 9.9). It is bounded anteriorly by the pubic arch, posteriorly by the coccyx, and laterally by the ischiopubic rami and the sacrotuberous ligaments,

F9.14

Table 9.9

Table 9.8   Muscles That Form the Floor of the Abdominopelvic Cavity [F9.13]

Muscle	Origin	Insertion	Action	Innervation
PELVIC DIAPHRAGM				
Levator ani	Inner surface of the superior ramus of the pubic bone, the lateral pelvic wall, and the spine of the ischium	Inner surface of the coccyx	Supports the pelvic viscera	Third through fifth sacral
Coccygeus	Spine of the ischium and the sacrospinous ligament	Sides of the coccyx and the sacrum	Supports the pelvic viscera	Fourth and fifth sacral

Table 9.9   Muscles of the Perineum [F9.14]

Muscle	Origin	Insertion	Action	Innervation
Ischiocavernosus	Tuberosity and rami of ischium	Crus of penis or clitoris	Retards return of blood through veins, thereby maintaining erection of penis or clitoris	Pudendal (second through fifth sacral)
Bulbospongiosus (Bulbocavernosus)	*Male:* from ventral median raphe on base of penis	Encircles base of penis and joins with fibers from opposite side on dorsum of penis	Empties urethral canal; assists in erection of penis	Pudendal
	Female: central tendinous point of perineum near anus	Pass on either side of vagina to insert on base of clitoris	Assists in erection of clitoris	Pudendal
Superficial transverse perineus	Tuberosity of ischium	Central tendinous point of perineum	Fixes (tightens) central tendinous point	Pudendal
Urogenital diaphragm *deep transverse perineus*	Inferior rami of ischium	Median raphe where it joins with corresponding muscle from opposite side	Both muscles of urogenital diaphragm act as constrictors of urethra	Pudendal
sphincter urethrae	Encircle membranous portion of urethra			

which run between the lateral margins of the sacrum and coccyx and the ischial tuberosities.

The perineal muscles are located inferior to the pelvic diaphragm. They consist of superficial muscles associated with the external genital organs and deeper muscles that form the **urogenital diaphragm.** The urogenital diaphragm is located just below the anterior portion of the pelvic diaphragm, to which it adds support. The urogenital diaphragm is composed primarily of the **deep transverse perinei** and **sphincter urethrae** muscles and various fascial sheets. The superficial perineal muscles include the **ischiocavernosus, bulbospongiosus (bulbocavernosus),** and **superficial transverse perinei** muscles.

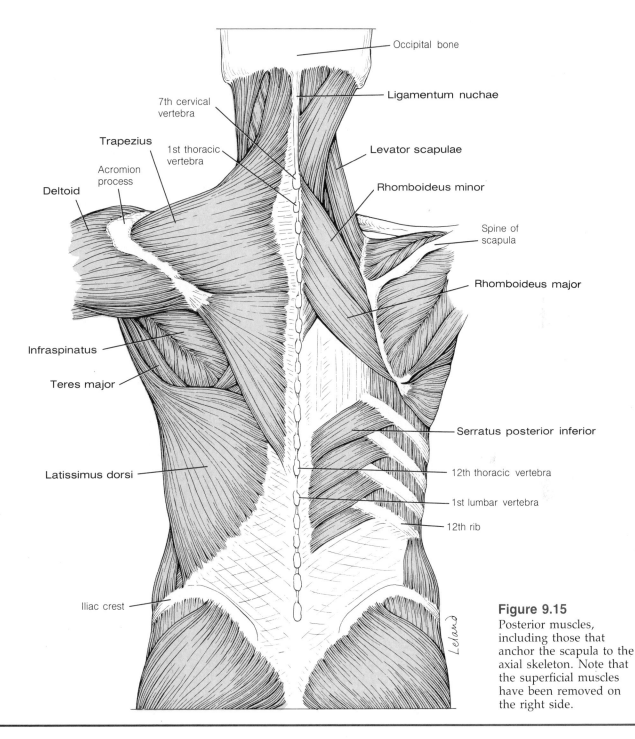

Occipital bone

Ligamentum nuchae

7th cervical vertebra

Levator scapulae

Trapezius

1st thoracic vertebra

Rhomboideus minor

Acromion process

Deltoid

Spine of scapula

Rhomboideus major

Infraspinatus

Teres major

Serratus posterior inferior

12th thoracic vertebra

Latissimus dorsi

1st lumbar vertebra

12th rib

Iliac crest

Figure 9.15
Posterior muscles, including those that anchor the scapula to the axial skeleton. Note that the superficial muscles have been removed on the right side.

MUSCLES OF THE UPPER LIMBS

Muscles That Act on the Scapula

Included in the muscles of the upper limbs are those muscles of the pectoral girdle that anchor the scapula (and to a lesser extent, those muscles that anchor the clavicle) to the axial skeleton (Table 9.10). Although it is possible for the scapula to move over the ribs, these muscles act primarily as fixators of the scapula. When it is immobilized by these muscles, the scapula serves as a stable point of origin for most of the muscles that move the arm.

Four posterior muscles anchor the scapula (Figure 9.15): the **trapezius,** the **rhomboideus major** and **minor,** and the **levator scapulae.** The trapezius

Table 9.10

F9.15

Table 9.10   Muscles That Act on the Scapula [*F9.15, F9.16*]

Muscle	Origin	Insertion	Action	Innervation
Trapezius	Occipital bone, the ligamentum nuchae, and the spinous processes of the seventh cervical and all of the thoracic vertebrae	Lateral third of the clavicle, the acromion process, and the spine of the scapula	Elevates (upper portion) or depresses (lower portion), rotates, adducts, and stabilizes the scapula	Accessory (cranial nerve XI)
Rhomboideus major	Spinous processes of the second through the fifth thoracic vertebrae	Vertebral border of the scapula, below the spine of the scapula	Adduct, stabilize, and rotate the scapula, lowering its lateral angle	Dorsal scapular (fifth cervical)
Rhomboideus minor	Spinous processes of the seventh cervical and first thoracic vertebrae	Vertebral border of scapula, at the base of the spine of the scapula		
Levator scapulae	Transverse processes of the upper four cervical vertebrae	Vertebral border of the scapula, above the spine of the scapula	Elevates scapula and bends the neck laterally when the scapula is fixed	Dorsal scapular
Pectoralis minor	Anterior surface of the third through the fifth ribs	Coracoid process of the scapula	Depresses the scapula and pulls it anteriorly	Medial pectoral (eighth cervical and first thoracic)
Serratus anterior	Outer surface of the first nine ribs	Entire length of the ventral surface of the vertebral border of the scapula	Stabilizes, abducts, and rotates the scapula upward	Long thoracic (fifth through seventh cervical)
Subclavius	Outer surface of the first rib	Inferior surface of the lateral portion of the clavicle	Stabilizes and depresses the pectoral girdle	Fifth and sixth cervical

and rhomboids are named according to their shapes, the levator according to its action.

Because of the shape of the *trapezius,* its fibers pull in various directions, and its actions depend on which portion of the muscle contracts. If the upper portion contracts, the scapula is elevated, as in shrugging the shoulders. If the lower portion contracts, the scapula is depressed. If the entire muscle contracts, the scapula is pulled toward the vertebral column—that is, it is adducted. If the scapula is fixed, the trapezius muscle assists in moving the head posteriorly.

The *rhomboids* insert on the vertebral border of the scapula and act to pull the scapula medially as well as downwardly rotating it, thus lowering the lateral angle. The *levator scapulae,* which runs from the upper cervical vertebrae to the superior angle of the scapula, elevates the scapula, acting as a synergist to the upper portion of the trapezius muscle.

F9.16   Two anterior muscles anchor the scapula to the thorax (Figure 9.16). The **pectoralis minor** pulls the scapula anteriorly and downward, lowering the lateral angle. In this manner, it acts antagonistically to the trapezius, rhomboid, and levator scapulae muscles. The **serratus anterior** derives its name from its notched origin on the anterior surfaces of the ribs. From this origin it passes posteriorly, between the dorsal surface of the ribs and the subscapular fossa of the scapula, to insert on the vertebral border of the scapula. The

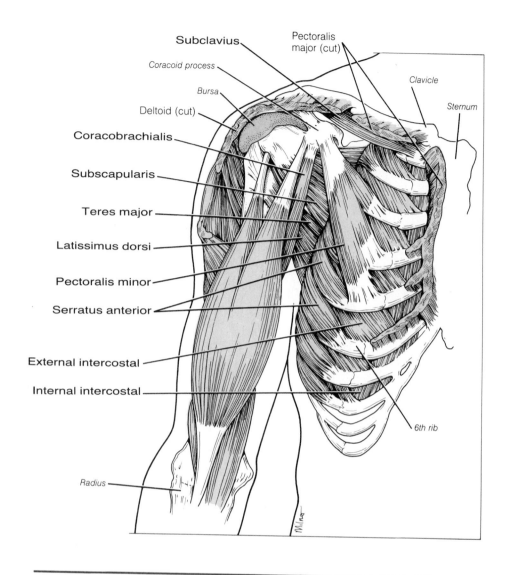

Subclavius

Pectoralis major (cut)

Coracoid process

Clavicle

Bursa

Sternum

Deltoid (cut)

Coracobrachialis

Subscapularis

Teres major

Latissimus dorsi

Pectoralis minor

Serratus anterior

External intercostal

Internal intercostal

6th rib

Radius

Figure 9.16
Deep anterior muscles, including those that anchor the scapula and the clavicle to the thorax.

serratus anterior pulls the scapula laterally; therefore, it acts antagonistically to the rhomboid muscles.

In addition to these muscles, the **subclavius** muscle anchors the pectoral girdle to the thoracic cage through its insertion onto the clavicle. For this reason it is included in Table 9.10, even though it does not act directly on the scapula.

Table 9.10

Muscles That Act on the Arm (Humerus)

Nine muscles cross the shoulder joint and insert on the humerus (Table 9.11). Seven of these muscles arise from the scapula, indicating how important it is for the muscles discussed in Table 9.10 to fix the scapula. The remaining two muscles arise from the axial skeleton and have no attachments to the scapula. These are the *pectoralis major* and the *latissimus dorsi* muscles. The actions of all these muscles are summarized in Table 9.12.

Table 9.11

Table 9.10

Table 9.12

F9.17

The **pectoralis major** (Figure 9.17) is a large fan-shaped chest muscle that completely covers the smaller pectoralis minor. The pectoralis major passes from the thoracic cage to the humerus, forming the anterior border of the axilla. It acts to adduct, to flex, and—because it passes anteriorly to the shoulder joint—to rotate the humerus medially.

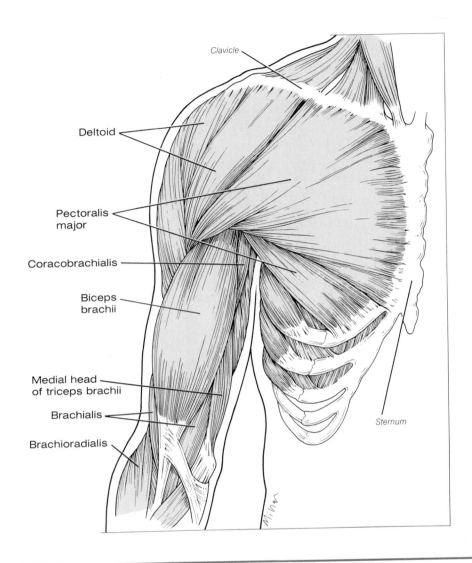

Figure 9.17
Superficial muscles of the chest, shoulder, and anterior arm.

F9.15   The **latissimus dorsi** (Figure 9.15) arises from the vertebrae of the lower back and from the pelvis. It twists upon itself as it passes between the humerus and the scapula to insert on the anterior surface of the humerus. In doing so, it forms the posterior border of the axilla. The latissimus dorsi extends the arm, pulling it downward or backward, thereby acting antagonistically to the pectoralis major. At the same time, it acts synergistically to the pectoralis major in adducting and—because of the manner in which it wraps around the humerus to insert on the anterior surface—in rotating the arm medially.

F9.15, F9.17   The **deltoid** (Figures 9.15 and 9.17) is a large muscle that forms a cap over the shoulder. It arises anteriorly from the clavicle and posteriorly from the scapula. Some of its fibers, therefore, pass in front of the shoulder joint; some pass behind; and some pass directly over the lateral surface of the joint. Because of the different positions of their fibers, the anterior and posterior parts of the muscle have actions that are antagonistic to each other. If the entire muscle contracts, it abducts the arm (antagonistic to the pectoralis major and the latissimus dorsi). The anterior fibers act to flex and medially rotate the arm (synergistic to the pectoralis major). The posterior fibers extend and laterally rotate the arm (antagonistic to the action of the anterior fibers).

F9.16   The origins of the **supraspinatus, infraspinatus,** and **subscapularis** muscles cover most of the ventral and dorsal surfaces of the scapula (Figures 9.16

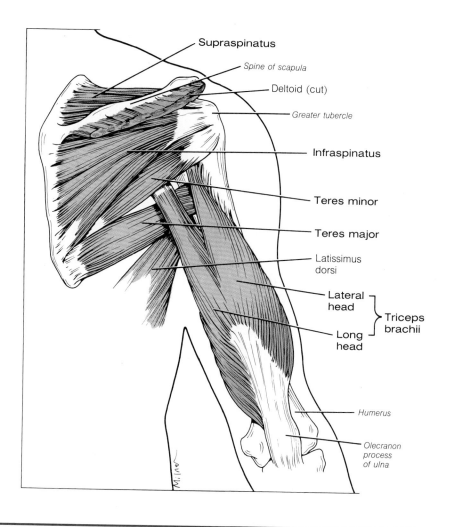

Supraspinatus
Spine of scapula
Deltoid (cut)
Greater tubercle
Infraspinatus
Teres minor
Teres major
Latissimus dorsi
Lateral head ⎤
⎥ Triceps brachii
Long head ⎦
Humerus
Olecranon process of ulna

Figure 9.18
Deep posterior muscles that attach the arm to the scapula. The deltoid muscle has been removed.

and 9.18). They derive their names from the scapular fossae from which they arise. The *supraspinatus* crosses the upper part of the shoulder joint, allowing it to serve as an abductor of the arm. The *infraspinatus* passes posteriorly to the shoulder joint, acting as a lateral rotator of the arm. The *subscapularis*, whose origin is on the ventral surface of the scapula, separates the scapula from the serratus anterior muscle. It inserts on the anterior surface of the humerus; consequently, it acts as a medial rotator of the arm. In addition to their prime actions, all these deep muscles that cross the shoulder joint assist in strengthening and stabilizing the joint.

The **teres major** and **teres minor** muscles both originate from the axillary border of the scapula (Figure 9.18). The teres major is the longer, its origin being from the inferior angle. The two muscles bracket the humerus, the teres major inserting on the anterior surface, the teres minor inserting on the posterior surface. Because of these insertions, the major is a medial rotator of the arm, and the minor is a lateral rotator.

The **coracobrachialis** (Figures 9.16 and 9.20) is the remaining muscle of scapular origin. As the name indicates, it arises from the coracoid process of the scapula and inserts onto the humerus. It is an anterior muscle that assists in flexion and adduction of the arm.

Two additional muscles, the *biceps brachii* and the *triceps brachii*, have origins on the scapula and pass into the arm. However, because their major actions are on the elbow joint, they are discussed with that group of muscles (Table 9.13).

F9.18

F9.18

F9.16, F9.20

bray'-kee-eye

Table 9.13

Table 9.11   Muscles That Act on the Arm (Humerus)

Muscle	Origin	Insertion	Action	Innervation
ORIGIN ON AXIAL SKELETON				
Pectoralis major [F9.17]	Medial half of the clavicle, the sternum, the costal cartilages of the upper six ribs, and the aponeurosis of the external oblique muscle	Greater tubercle of the humerus	Flexes, adducts, and medially rotates the arm	Medial and lateral pectoral
Latissimus dorsi [F9.15]	Spinous processes of the lower six thoracic and the lumbar vertebrae, the sacrum, the posterior iliac crest—all via the lumbodorsal fascia	Medial margin of the intertubercular groove of the humerus	Extends, adducts, and medially rotates the arm	Thoracodorsal
ORIGIN ON SCAPULA				
Deltoid [F9.15]	Lateral third of the clavicle, the acromion process, and the spine of the scapula	Deltoid tuberosity of the humerus	Abducts arm; anterior fibers flex and medially rotate the arm; posterior fibers extend and laterally rotate the arm	Axillary
Supraspinatus [F9.18]	Supraspinatus fossa of the scapula	Greater tubercle of the humerus	Abducts the arm; slight lateral rotation	Suprascapular
Infraspinatus [F9.18]	Infraspinatus fossa of the scapula	Greater tubercle of the humerus (posterior to the supraspinatus)	Rotates the arm laterally; slight adduction	Suprascapular
Subscapularis [F9.16]	Subscapular fossa of the scapula	Lesser tubercle of the humerus	Rotates the arm medially	Subscapular
Teres major [F9.18]	Dorsal surface of the inferior angle of the scapula	Lesser tubercle of the humerus	Adducts, extends, and medially rotates the arm	Subscapular
Teres minor [F9.18]	Axillary border of the scapula	Greater tubercle of the humerus (posterior to the infraspinatus)	Rotates the arm laterally; weakly adducts and extends the arm	Axillary
Coracobrachialis [F9.20]	Coracoid process of the scapula	Middle of the humerus, medial surface	Flexes and adducts the arm	Musculocutaneous

Table 9.12   Summary of Muscle Actions on the Arm

Flexion	Extension	Adduction	Abduction	Medial rotation	Lateral rotation
Deltoid	Deltoid	Pectoralis major	Deltoid	Pectoralis major	Deltoid
Pectoralis major	Latissimus dorsi	Latissimus dorsi	Supraspinatus	Latissimus dorsi	Supraspinatus
Coracobrachialis	Teres major	Teres major Coracobrachialis Infraspinatus		Deltoid Subscapularis Teres major	Infraspinatus Teres minor

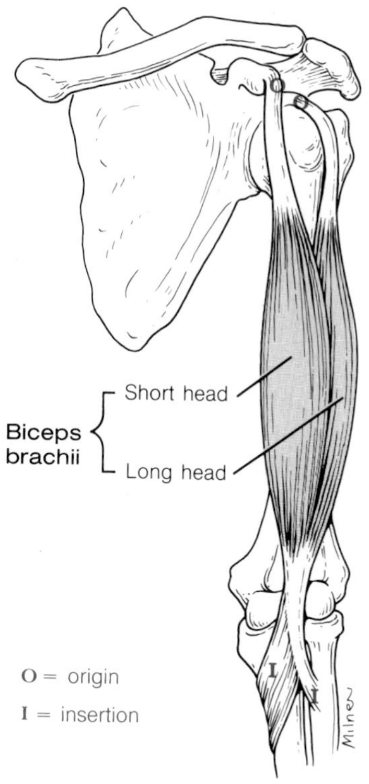

Biceps brachii { Short head, Long head

O = origin

I = insertion

Figure 9.19
The biceps brachii
muscle.

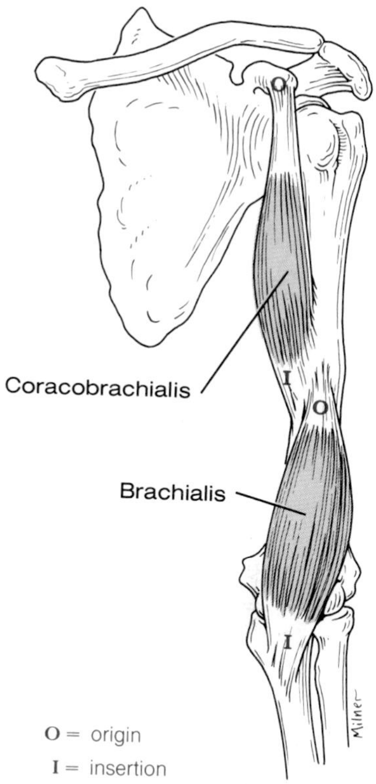

Coracobrachialis

Brachialis

O = origin
I = insertion

Muscles That Act on the Forearm (Radius and Ulna)

The more powerful of the muscles that move the elbow and/or the proximal radial/ulnar joint (producing supination and pronation) are located in the arm (Table 9.13). They are assisted, however, by several muscles whose bellies lie in the forearm. Most of the latter muscles have their prime actions on the hand and are described with that group (Table 9.14).

The two anterior muscles of the arm, the *biceps brachii* and the *brachialis*, flex the forearm. A third anterior arm muscle, the *coracobrachialis*, has its prime action on the shoulder joint; it was described earlier with that group of muscles (Table 9.11).

As the name indicates, the **biceps brachii** (Figures 9.17 and 9.19) has two heads, both of which have their origins on the scapula. The tendon of the *long head* passes over the top of the shoulder joint and travels through the intertubercular groove on the humerus. The two heads blend into the thick belly of the muscle. The main insertion of the biceps brachii is on the tuberosity of the radius. When the forearm is supinated, the biceps flexes it. However, when the forearm is pronated with the tuberosity of the radius rotated toward the ulna, the biceps acts to supinate the forearm. In addition, it strengthens the shoulder joint and assists in flexion of the joint.

The **brachialis** (Figures 9.17 and 9.20), which is deep to the biceps, is also a strong flexor of the forearm. Because it arises on the humerus and inserts on the ulna, it acts only on the elbow joint.

There are two muscles that act to extend the forearm. The first, the **triceps brachii** (Figure 9.18), is the only muscle in the posterior compartment of the arm. One of its heads—the *long head*—arises from the scapula; the other two—the *lateral* and *medial heads*—arise from the humerus. All three heads insert by a common tendon onto the olecranon process of the ulna. Because the long head, which passes between the teres major and minor muscles, crosses the shoulder joint, it acts as a weak synergist to the latissimus dorsi muscle to extend and adduct the arm. Its main action, however, is extension of the forearm.

Figure 9.20
The brachialis and coracobrachialis muscles. The more superficial biceps brachii muscle has been removed.

Table 9.13

Table 9.14

Table 9.11
F9.17, F9.19

F9.17, F9.20

F9.18

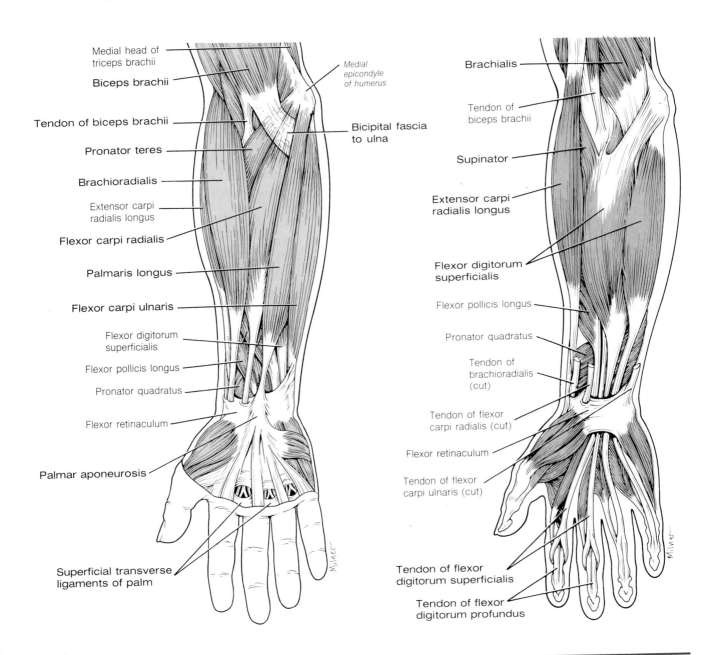

Figure 9.21
Superficial anterior
muscles of the right
forearm and hand.

Figure 9.22
Second layer of anterior
muscles of the right
forearm. The superficial
muscles have been
removed.

F9.24, F9.25   The second extensor of the forearm is the **anconeus** (Figures 9.24 and
9.25), a small muscle that appears to be a lateral continuation of the triceps
brachii. It runs from the lateral epicondyle of the humerus to the lateral side of
the olecranon process. Although it is considered to be a muscle of the forearm, the anconeus does not act on the wrist; its only action is to assist the
triceps in extending the forearm.

There is one additional muscle that is located in the forearm but acts to
F9.21   flex the forearm. This is the **brachioradialis** muscle (Figure 9.21). It runs from
the distal end of the humerus to the distal end of the radius, forming the
lateral surface of the forearm.

Table 9.13 **Muscles That Act on the Forearm (Radius and Ulna)**

Muscle	Origin	Insertion	Action	Innervation
Biceps brachii [F9.19]	*Long head:* Supraglenoid tubercle of the scapula *Short head:* Coracoid process of the scapula	Tuberosity of the radius	Flexes the forearm and the arm; supinates the forearm	Musculocutaneous
Brachialis [F9.20]	Anterior surface of the distal half of the humerus	Coronoid process of the ulna	Flexes the forearm	Musculocutaneous
Triceps brachii [F9.18]	*Long head:* Infraglenoid tubercle of the scapula *Lateral head:* Posterior surface of the humerus above the radial groove *Medial head:* Posterior surface of the humerus below the radial groove	Olecranon process of the ulna	Extends the forearm; long head also extends the arm	Radial
MUSCLES OF FOREARM Anconeus [F9.24]	Lateral epicondyle of the humerus	Lateral surface of the olecranon process of the ulna	Extends the forearm	Radial
Brachioradialis [F9.20]	Lateral supracondylar ridge of the humerus	Styloid process of the radius	Flexes the forearm	Radial

Muscles That Act on the Hand and Fingers

Most of the muscles that form the bulge of the proximal end of the forearm have at least a part of their origins on the distal end of the humerus. As a result, they cross the elbow joint as well as the wrist. However, their actions on the elbow joint are very slight. Their prime actions are on the hand and fingers.

Although this is a complex group of muscles with formidable-sounding names, most of the names indicate the muscle's action, origin, or insertion. These muscles can be divided into two groups on the basis of location and function. The muscles of the *anterior group* serve as flexors or pronators. Most of these anterior muscles have their origins on the medial epicondyle of the humerus. A few of them insert on the radius, but most insert on the carpals, metacarpals, or phalanges. The *posterior group* of forearm muscles serve as extensors and supinators. Most of these muscles have their origins on the lateral epicondyle of the humerus and insert on the metacarpals or phalanges. Both the anterior and posterior groups can be divided further into superficial and deep muscles. The tendons of these forearm muscles are held down at the wrist by heavy thickenings of the fascia called **flexor** and **extensor retinacula** (*retinaculum* = halter). If these transverse bands of fascia were not present, the tendons would protrude when the hand is flexed or extended.

Although the muscles that act on the hand and fingers are not described here individually, they are illustrated in Figures 9.21 through 9.25, and their precise locations and actions are listed in Table 9.14. In addition, Table 9.15 summarizes their actions and makes it possible to easily identify synergists and antagonists. Notice that while the three muscles that extend the wrist are antagonistic to the flexors of the forearm, the **extensor carpi radialis longus** acts synergistically with the **flexor carpi radialis** to abduct the hand. In a similar manner, the **extensor carpi ulnaris** acts synergistically with the **flexor carpi ulnaris** to adduct the hand.

F9.21–F9.25
Table 9.14, Table 9.15

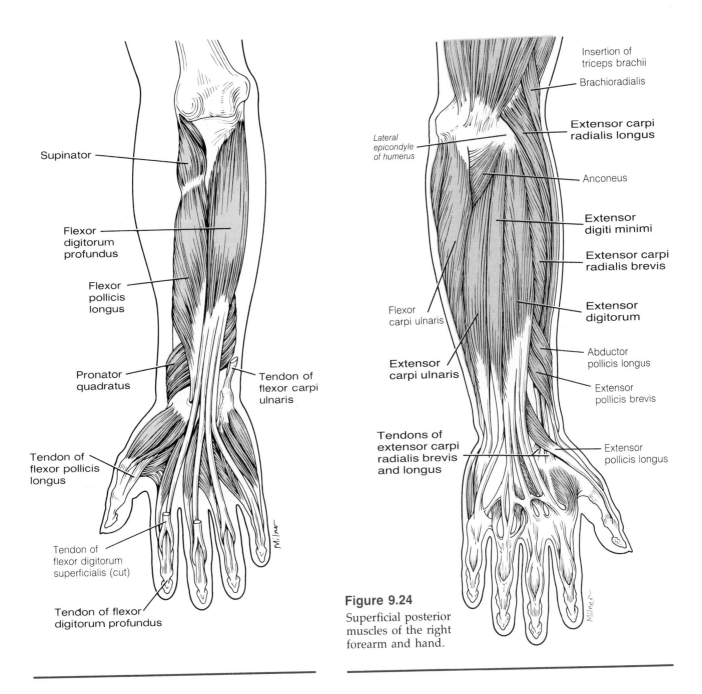

Supinator

Flexor digitorum profundus

Flexor pollicis longus

Pronator quadratus

Tendon of flexor pollicis longus

Tendon of flexor digitorum superficialis (cut)

Tendon of flexor digitorum profundus

Tendon of flexor carpi ulnaris

Insertion of triceps brachii

Brachioradialis

Extensor carpi radialis longus

Lateral epicondyle of humerus

Anconeus

Extensor digiti minimi

Extensor carpi radialis brevis

Flexor carpi ulnaris

Extensor digitorum

Extensor carpi ulnaris

Abductor pollicis longus

Extensor pollicis brevis

Tendons of extensor carpi radialis brevis and longus

Extensor pollicis longus

Figure 9.24
Superficial posterior muscles of the right forearm and hand.

Figure 9.23
Deep anterior muscles of the right forearm. The more superficial muscles that are illustrated in Figures 9.21 and 9.22 have been removed.

Intrinsic Muscles of the Hand

We have seen that several of the muscles of the forearm have long tendons that reach the phalanges and serve to move the fingers. Apart from these forearm muscles, there are several groups of small muscles whose origin and insertion are both in the hand. As we learned earlier, such muscles are called *intrinsic* muscles. The intrinsic muscles make possible the fine and precise movements that are typical of the fingers.

F9.26

The intrinsic muscles of the hand (Figure 9.26) are divided into three groups. Those that act on the thumb form the **thenar eminence** at the base of the thumb. Those that act on the little finger form the **hypothenar eminence** on the medial side of the hand. The intermediate, or **midpalmar,** muscles act on all the phalanges except the thumb. These intrinsic muscles of the hand

Table 9.16 are described in Table 9.16. Notice that there are no intrinsic muscles on the dorsum of the hand, since the dorsal interossei are located between the metacarpal bones.

Table 9.14   **Muscles That Act on the Hand and Fingers**

Muscle	Origin	Insertion	Action	Innervation
ANTERIOR GROUP				
SUPERFICIAL MUSCLES (LISTED FROM LATERAL TO MEDIAL)				
Pronator teres [F9.21]	Medial epicondyle of the humerus and the coronoid process of the ulna	Middle of the lateral surface of the shaft of the radius	Pronates and weakly flexes the forearm	Median
Flexor carpi radialis [F9.21]	Medial epicondyle of the humerus	Ventral surface of the second and third metacarpals	Flexes and abducts the hand; aids in flexion and pronation of the forearm	Median
Palmaris longus [F9.21]	Medial epicondyle of the humerus	Palmar aponeurosis	Flexes the hand	Median
Flexor carpi ulnaris [F9.21; F9.24]	Medial epicondyle of the humerus, olecranon process, and the proximal two-thirds of the posterior surface of the ulna	Pisiform, hamate, and fifth metacarpal	Flexes and adducts the hand	Ulnar
Flexor digitorum superficialis (beneath the other superficial muscles) [F9.21; F9.22]	Medial epicondyle of the humerus, coronoid process of the ulna, and the anterior surface of the radius	Ventral surface of the middle phalanges of the second through the fifth fingers	Flexes the phalanges and the hand	Median
DEEP MUSCLES				
Flexor digitorum profundus (*profundus* = deep) [F9.23]	Medial epicondyle and the coronoid process of the humerus, the interosseus membrane, and the ventral surface of the ulna	Ventral surface base of the distal phalanges of the second through the fifth fingers	Flexes the phalanges and the hand	Median and ulnar
Flexor pollicis longus (*pollex* = thumb) [F9.21; F9.22; F9.23]	Ventral surface of the radius and the interosseus membrane	Ventral surface base of the distal phalanx of the thumb	Flexes the thumb; aids in flexing the hand	Median
Pronator quadratus [F9.21; F9.22; F9.23]	Distal ventral surface of the ulna	Distal ventral surface of the radius	Pronates the forearm	Median
POSTERIOR GROUP				
SUPERFICIAL* MUSCLES (LISTED FROM LATERAL TO MEDIAL)				
Extensor carpi radialis longus [F9.21; F9.22; F9.24]	Lateral supra-condylar ridge of the humerus	Dorsal surface of the base of the second metacarpal	Extends and abducts the hand	Radial
Extensor carpi radialis brevis (*brevis* = short) [F9.24]	Lateral epicondyle of the humerus	Dorsal surface of the base of the third metacarpal	Extends the hand	Radial
Extensor digitorum [F9.24]	Lateral epicondyle of the humerus	Dorsal surface of the phalanges of the second through the fifth fingers	Extends the fingers and the hand	Radial

Table 9.14   Muscles That Act on the Hand and Fingers (continued)

Muscle	Origin	Insertion	Action	Innervation
Extensor digiti minimi (= little finger) [F9.24]	Tendon of the extensor digitorum	Tendon of the extensor digitorum on the dorsum of the little finger	Extends the little finger	Radial
Extensor carpi ulnaris [F9.24]	Lateral epicondyle of the humerus	Base of the fifth metacarpal	Extends and adducts the hand	Radial
POSTERIOR GROUP DEEP MUSCLES (LISTED LATERAL TO MEDIAL) Supinator [F9.22; F9.23; F9.25]	Lateral epicondyle of the humerus	Proximal end of the lateral surface of the shaft of the radius	Supinates the forearm	Radial
Abductor pollicis longus [F9.22; F9.24; F9.25]	Posterior surface of the middle of the radius and ulna, and the interosseus membrane	Base of the first metacarpal	Extends the thumb and abducts the hand	Radial
Extensor pollicis brevis [F9.24; F9.25]	Posterior surface of the middle of the radius, and the interosseus membrane	Base of the first phalanx of the thumb	Extends the thumb and abducts the hand	Radial
Extensor pollicis longus [F9.24; F9.25]	Posterior surface of the middle of the ulna, and the interosseus membrane	Base of the last phalanx of the thumb	Extends the thumb and abducts the hand	Radial
Extensor indicis [F9.25]	Posterior surface of the distal end of the ulna, and the interosseus membrane	Tendon of the extensor digitorum to the index finger	Extends the index finger	Radial

*The brachioradialis is located with the posterior superficial muscles but was described in Table 9.13 because it is a flexor of the forearm.

Table 9.15   Summary of Muscle Actions on the Hand

Flexion	Extension	Adduction	Abduction
Flexor carpi radialis	Extensor carpi radialis longus	Flexor carpi ulnaris	Flexor carpi radialis
Palmaris longus	Extensor carpi radialis brevis	Extensor carpi ulnaris	Extensor carpi radialis longus
Flexor carpi ulnaris			Abductor pollicis longus
Flexor digitorum superficialis	Extensor carpi ulnatis		Extensor pollicis brevis
Flexor digitorum profundus	Extensor digitorum		Extensor pollicis longus
Flexor pollicis longus			

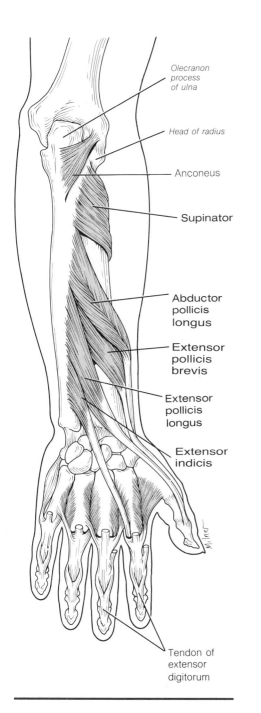

Olecranon process of ulna

Head of radius

Anconeus

Supinator

Abductor pollicis longus

Extensor pollicis brevis

Extensor pollicis longus

Extensor indicis

Tendon of extensor digitorum

Figure 9.25
Deep posterior muscles of the right forearm and hand. The superficial muscles have been removed.

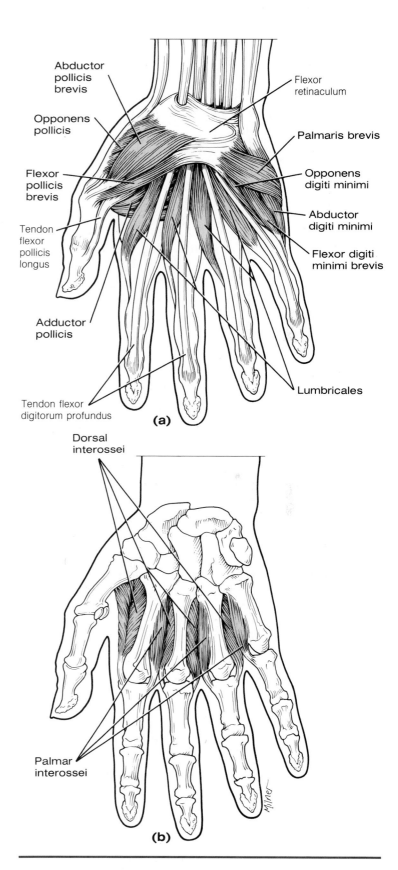

Abductor pollicis brevis

Opponens pollicis

Flexor pollicis brevis

Tendon flexor pollicis longus

Adductor pollicis

Tendon flexor digitorum profundus

Flexor retinaculum

Palmaris brevis

Opponens digiti minimi

Abductor digiti minimi

Flexor digiti minimi brevis

Lumbricales

(a)

Dorsal interossei

Palmar interossei

(b)

Figure 9.26
Palmar view of the intrinsic muscles of the hand. **(a)** Superficial muscles. **(b)** Interossei. The superfical muscles have been removed.

Table 9.16 Intrinsic Muscles of the Hand

Muscle	Origin	Insertion	Action	Innervation
THENAR MUSCLES				
Abductor pollicis brevis [F9.26a]	Flexor retinaculum, scaphoid, and trapezium	Proximal phalanx of the thumb	Abducts the thumb	Median
Opponens pollicis (*opponens* = one that opposes) [F9.26a]	Flexor retinaculum and trapezium	Lateral border of the metacarpal of the thumb	Pulls the thumb in front of the palm to meet the little finger	Median
Flexor pollicis brevis [F9.25a]	Flexor retinaculum, trapezium, and first metacarpal	Base of the proximal phalanx of the thumb	Flexes and adducts the thumb	Median and ulnar
Adductor pollicis [F9.26a]	Capitate, and second and third metacarpals	Proximal phalanx of the thumb	Adducts the thumb	Ulnar
HYPOTHENAR MUSCLES				
Palmaris brevis [F9.26a]	Flexor retinaculum	Skin on the ulnar border of the hand	Pulls the skin toward the middle of the palm	Ulnar
Abductor digiti minimi [F9.26a]	Pisiform, and the tendon of the flexor carpi ulnaris	Base of the proximal phalanx of the little finger	Abducts the little finger	Ulnar
Flexor digiti minimi brevis [F9.26a]	Flexor retinaculum and hamate	Base of the proximal phalanx of the little finger	Flexes the little finger	Ulnar
Opponens digiti minimi [F9.26a]	Flexor retinaculum and hamate	Metacarpal of the little finger	Brings the little finger out to meet the thumb	Ulnar
MIDPALMAR MUSCLES				
Lumbricales [F9.26a]	Tendons of the flexor digitorum profundus	Tendons of the extensor digitorum	Flex the proximal phalanx and extend the middle and distal phalanges of the second through fifth fingers	Median and ulnar
Dorsal interossei (4) [F9.26b]	Adjacent sides of all of the metacarpals	Proximal phalanx of second, third, and fourth fingers	Abduct the fingers from the middle finger; flex the proximal phalanx	Ulnar
Palmar interossei (3) [F9.26b]	Medial side of the second metacarpal, and lateral side of the fourth and fifth metacarpals	Proximal phalanx of the same finger	Adduct the fingers toward the middle finger; flex the proximal phalanx	Ulnar

MUSCLES OF THE LOWER LIMBS

When compared to the muscles of the upper limbs, those of the lower limbs tend to be bulkier and more powerful. The versatile movements that are characteristic of the upper limbs are somewhat sacrificed in the lower limbs in favor of strength, stability, and locomotion. Many of the muscles of the lower

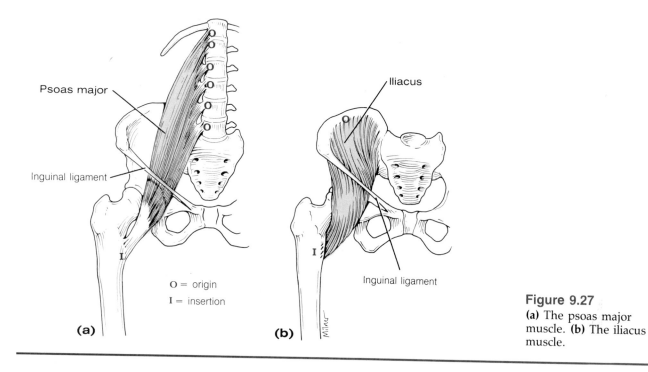

O = origin
I = insertion

(a)   **(b)**

Figure 9.27
(a) The psoas major muscle. **(b)** The iliacus muscle.

limbs are used in maintaining the upright posture and therefore must constantly resist the pull of gravity. Unlike the pectoral girdle, the pelvis does not rely entirely on muscles to stabilize or fix it. The only movement possible in the bony pelvis is a slight gliding between the sacrum and the ilium. Many of the muscles of the lower limbs cross two joints—either the hip and the knee, or the knee and the ankle—and act equally strongly on both joints. These double actions are listed in Tables 9.17–9.19.

Tables 9.17–9.19

Muscles That Act on the Thigh (Femur)

Most of the muscles that act on the femur arise from the pelvis. One muscle, the **psoas major** (Figures 9.27a and 9.28), arises from the lumbar vertebrae. Because the psoas major muscle has a common insertion with, and acts synergistically with, the **iliacus** muscle (Figures 9.27b and 9.28), the two muscles are often referred to as the **iliopsoas** muscle. The tendon of the iliopsoas passes beneath the inguinal ligament to reach the femur, which it flexes. When the lower limbs are fixed, the psoas muscles flex the vertebral column, as when bending over.

so'-az
F9.27a, F9.28
F9.27b, F9.28

Three large gluteal muscles (Figures 9.29 and 9.30) give shape to the buttocks and serve as powerful mobilizers of the hip joint. The largest and most superficial is the **gluteus maximus**. The gluteus maximus covers the posterior third of the smaller **gluteus medius;** deep to the gluteus medius is the still smaller **gluteus minimus**. The broad tendon of the gluteus maximus passes behind the hip joint, causing it to extend and laterally rotate the femur. In contrast, the tendon of the gluteus medius passes above, and that of the minimus passes in front of, the hip joint, causing them to abduct and medially rotate the femur. In the rotation of the thigh, therefore, the gluteus maximus is an antagonist to the two smaller gluteal muscles. Acting synergistically with the gluteus maximus in rotating the thigh laterally are deep muscles that extend from the sacrum and the coxal bone to the posterior surface of the proximal end of the femur. These lateral rotators of the thigh are included in Table 9.17.

F9.29, F9.30

The **tensor fasciae latae** (Figure 9.28) is a lateral hip muscle that inserts on a strong band of connective tissue called the **iliotibial tract** of the **fascia lata** ("broad fascia"). The fascia lata invests all of the muscles of the thigh, but it is especially thick laterally—thus forming the iliotibial tract. The tensor serves

Table 9.17
F9.28

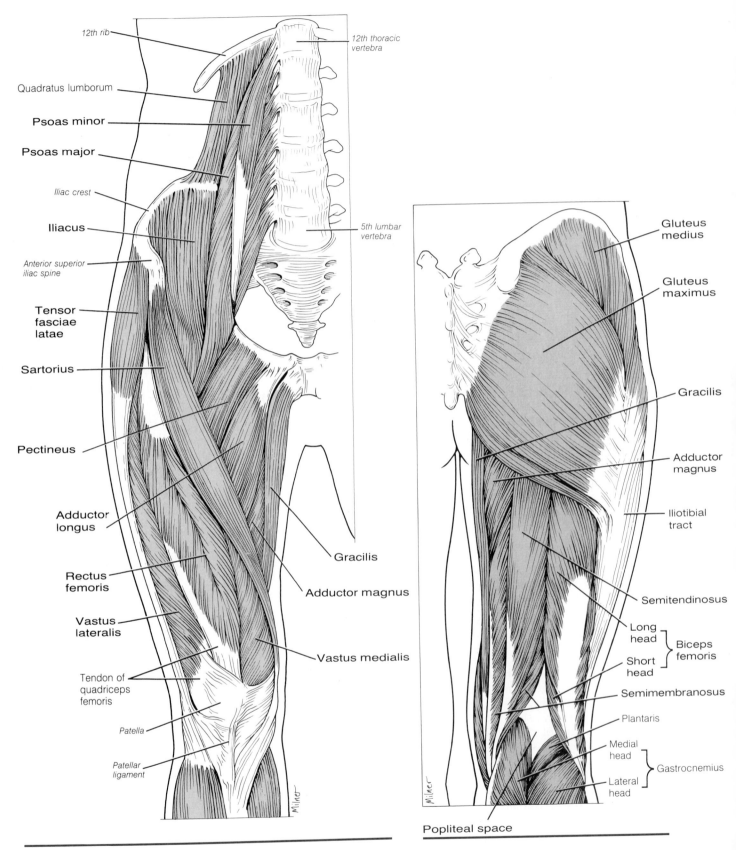

Figure 9.28
Anterior view of the muscles that attach the femur to the pelvis and the lumbar vertebrae.

Figure 9.29
Superficial muscles of the posterior hip and thigh.

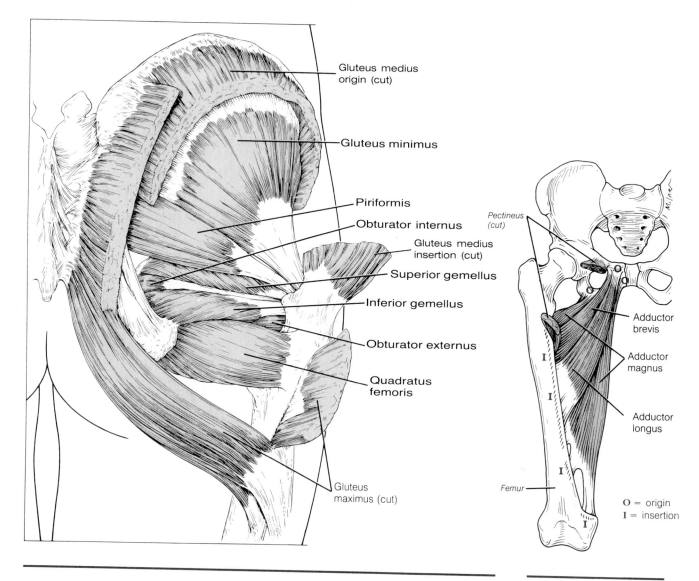

Figure 9.30

Deep muscles of the posterior hip. The gluteus maximus and gluteus medius have been cut to expose the deep muscles.

Figure 9.31

Anterior view of the right thigh showing the major muscles of the medial (adductor) compartment of the thigh. The gracilis muscle, which is also in the medial compartment, is illustrated in Figure 9.33.

primarily to stabilize the knee by tightening the fascia lata, but it also assists in flexing the thigh.

Three other large muscles that act on the femur are the medial muscles of the thigh. All of these muscles originate from the pubis and serve to pull the femur toward the midline. They are therefore named *adductors* (Figure 9.31). Because they are inserted onto the posterior surface of the femur, the adductor muscles also rotate the femur laterally, acting synergistically with the gluteus maximus and the six small, deep muscles just mentioned. The *pectineus* (Figure 9.28), which is included with the adductors in the medial group of thigh muscles, acts synergistically with the adductor muscles.

Muscles That Act on the Leg (Tibia and Fibula)

The muscles that act on the leg are located in the thigh, with their tendons crossing the knee joint. The thigh muscles are grouped by connective tissue sheets into *anterior, posterior,* and *medial compartments* (Figure 9.32; Table 9.18).

F9.31

F9.28

F9.32, Table 9.18

Table 9.17 Muscles That Act on the Femur (Thigh)

Muscle	Origin	Insertion	Action	Innervation
Iliopsoas *PSOAS MAJOR* [F9.28]	Transverse processes and bodies of the last thoracic and all of the lumbar vertebrae	Lesser trochanter of the femur	Flex the thigh; flex the trunk on the femur	Femoral and first lumbar
ILIACUS [F9.28]	Iliac crest and fossa			
Gluteus maximus [F9.29]	Posterior gluteal line of the ilium, and the posterior surface of the sacrum and the coccyx	Gluteal tuberosity of the femur; iliotibial band	Extends and laterally rotates the thigh	Inferior gluteal
Gluteus medius [F9.30]	Outer surface of the ilium, between the posterior and the anterior gluteal lines	Lateral surface of the greater trochanter of the femur	Abducts and medially rotates the thigh	Superior gluteal
Gluteus minimus [F9.30]	Outer surface of the ilium, between the anterior and the inferior gluteal lines	Anterior surface of the greater trochanter of the femur	Abducts and medially rotates the thigh	Superior gluteal
Tensor fasciae latae [F9.28]	Anterior portion of the iliac crest, and the anterior superior iliac spine	Iliotibial tract of the fascia lata	Tenses the fascia lata; assists in flexion, abduction, and medial rotation of the thigh	Superior gluteal
Piriformis [F9.30]	Anterior surface of the sacrum	Superior border of the greater trochanter of the femur	Rotates the thigh laterally; assists in extending and abducting the thigh	Second sacral
Obturator internus [F9.30]	Inner surface of the obturator membrane and the bony margins of the obturator foramen	Greater trochanter of the femur	Rotates the thigh laterally	Fifth lumbar, and first and second sacral
Obturator externus [F9.30; F9.33]	Outer surface of the obturator membrane and the bony margins of the obturator foramen	Trochanteric fossa of the femur	Rotates the thigh laterally	Obturator
Superior gemellus [F9.30]	Ischial spine	Greater trochanter of the femur	Rotates the thigh laterally	Fifth lumbar, and first and second sacral
Inferior gemellus [F9.30]	Ischial tuberosity	Greater trochanter of the femur	Rotates the thigh laterally	Fourth and fifth lumbar and first sacral
Quadratus femoris [F9.30]	Ischial tuberosity	Shaft of the femur just below the greater trochanter	Rotates the thigh laterally	Fourth and fifth lumbar

Table 9.17   Muscles That Act on the Femur (Thigh) (continued)

Muscle	Origin	Insertion	Action	Innervation
*Adductor magnus [F9.31]	Inferior ramus of the pubis and the ischium, and the ischial tuberosity	Most of the length of the linea aspera, and the adductor tubercle of the femur	Adducts and laterally rotates the thigh; assists in extending the thigh	Obturator and sciatic
Adductor longus [F9.31]	Crest and the symphysis of the pubis	Middle third of the linea aspera of the femur	Adducts and laterally rotates the thigh; assists in flexion of the thigh	Obturator
Adductor brevis [F9.31]	Inferior ramus of the pubis	Upper part of the linea aspera of the femur	Adducts and laterally rotates the thigh	Obturator
Pectineus [F9.28]	Superior ramus of the pubis	Posterior surface of the femur just below the lesser trochanter	Adducts, flexes, and laterally rotates the thigh	Obturator and femoral

*The three adductor muscles and the pectineus are in the medial compartment (F9.32) of the thigh musculature but are included in the table because they act upon the femur.

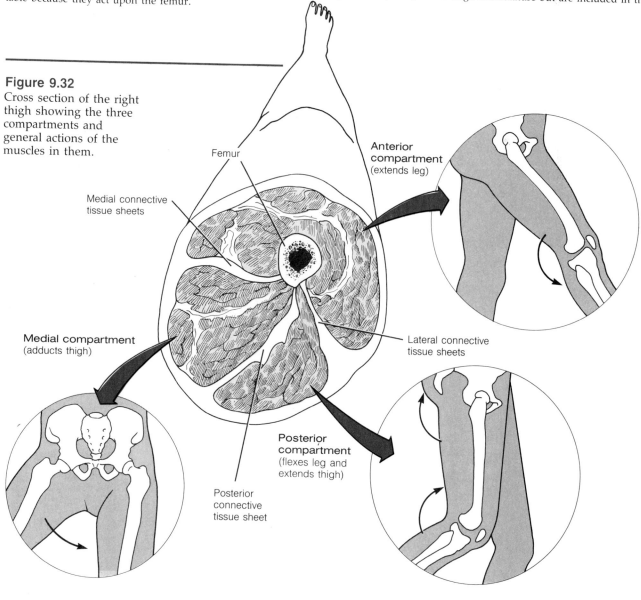

Figure 9.32
Cross section of the right thigh showing the three compartments and general actions of the muscles in them.

Femur

Medial connective tissue sheets

Anterior compartment (extends leg)

Medial compartment (adducts thigh)

Lateral connective tissue sheets

Posterior compartment (flexes leg and extends thigh)

Posterior connective tissue sheet

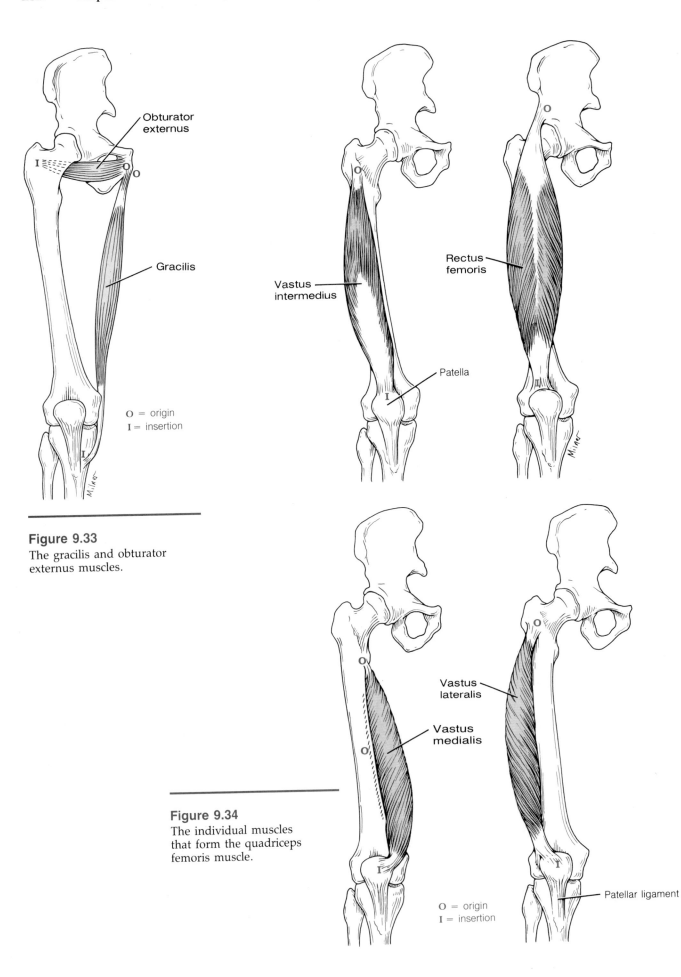

Figure 9.33
The gracilis and obturator externus muscles.

Figure 9.34
The individual muscles that form the quadriceps femoris muscle.

The *medial compartment* (Figure 9.31) is referred to as the *adductor compartment* because the muscles within this compartment function to adduct the femur. The **adductors magnus, longus,** and **brevis,** and the **pectineus** muscles are included in Table 9.17 because they act entirely on the femur. The one remaining muscle of the medial compartment of the thigh is the **gracilis** (Figure 9.33). In addition to adducting the thigh, the gracilis assists in flexing the leg; therefore it is listed in Table 9.18.

The *anterior compartment* of the thigh is called the *extensor compartment* because the muscles that it contains function primarily to extend the leg. There are five muscles within the anterior compartment. Four of them, although individually named—**rectus femoris, vastus intermedius, vastus lateralis,** and **vastus medialis**—are often grouped as the **quadriceps femoris** muscle (Figure 9.34). The four heads of the quadriceps muscle have a common insertion on the patella, or kneecap, which is a large sesamoid bone lying within the tendon of the quadriceps as it crosses in front of the knee joint. A strong band of connective tissue called the **patellar ligament** (ligamentum patellae) extends from the patella to the tibial tuberosity. Therefore, functionally the four heads of the quadriceps insert on the tibial tuberosity. The **sartorius** (Figure 9.35) is the remaining muscle of the anterior compartment. It is a long, narrow muscle that forms a band diagonally across the thigh from the ilium to the medial side of the tibia. It crosses both the hip and the knee, producing flexion at both joints. It is unusual for an anterior thigh muscle to flex the leg, but the sartorius does this because it passes posteriorly to the medial condyle of the femur.

The muscles within the *posterior,* or *flexor, compartment* of the thigh function to flex the leg. They also extend the thigh. The three muscles within this compartment (Figure 9.29)—**biceps femoris, semimembranosus,** and **semitendinosus**—are known as the **hamstring muscles.** They have a common origin from the ischial tuberosity and insert on the tibia and fibula. Consequently, they act on the hip as well as on the knee joint. The tendons of insertion of the semimembranosus and the semitendinosus pass medially behind the knee; that of the biceps femoris passes laterally. Between these tendons, on the posterior surface of the knee, is a triangular *popliteal space.*

In Table 9.18, notice that the muscles in a compartment of the thigh are all innervated by branches from the same nerve: those in the anterior compartment by the femoral nerve; in the medial compartment by the obturator nerve; and in the posterior compartment by the sciatic nerve.

Table 9.19 summarizes the actions of various muscles on the thigh.

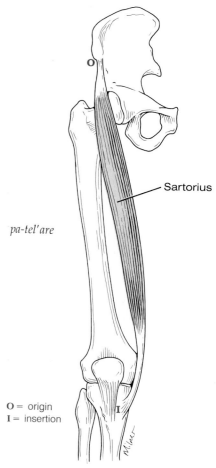

pa-tel'are

O = origin
I = insertion

Figure 9.35
The sartorius muscle.

Table 9.19

Table 9.18   Muscles That Act on the Leg (Muscles of the Thigh)

Muscle	Origin	Insertion	Action	Innervation
MEDIAL COMPARTMENT				
Adductor magnus Adductor longus Adductor brevis Pectineus	These muscles act only on the femur. They are illustrated in Figures 9.28 and 9.31 described in Table 9.17.		Adduct and laterally rotate the thigh	Obturator
Gracilis [F9.33]	Symphysis pubis and the pubic arch	Medial surface of the tibia just below the condyle	Adduct the thigh; flex the leg	Obturator
ANTERIOR COMPARTMENT				
Sartorius [F9.35]	Anterior superior iliac spine	Proximal medial surface of the tibia, below the tuberosity	Flex the thigh and the leg; laterally rotate the thigh	Femoral

Table 9.18   Muscles That Act on the Leg (Muscles of the Thigh) (continued)

Muscle	Origin	Insertion	Action	Innervation
Quadriceps femoris				
Rectus femoris [F9.34]	Anterior inferior iliac spine and just above the acetabulum of the coxal bone			
Vastus lateralis [F9.34]	Greater trochanter and lateral lip of the linea aspera of the femur	Tibial tuberosity, via the patella and the patellar ligament	Extend the leg; the rectus femoris also flexes the thigh	Femoral
Vastus medialis [F9.34]	Medial lip of the linea aspera of the femur			
Vastus intermedius [F9.34]	Anterior surface of the shaft of the femur			
POSTERIOR COMPARTMENT				
Hamstrings				
Biceps femoris [F9.29]	*Long head:* ischial tuberosity *Short head:* lateral lip of the linea aspera	Lateral surface of the head of the fibula, and the lateral condyle of the tibia	Flexes the leg; long head extends the thigh	Sciatic
Semitendinosus [F9.29]	Ischial tuberosity	Medial surface of the proximal end of the tibia	Flexes the leg; extends the thigh	Tibial
Semimembranosus [F9.29]	Ischial tuberosity	Medial surface of the proximal end of the tibia	Flexes the leg; extends the thigh	Tibial

Table 9.19   Summary of Muscle Actions on the Thigh (Including Muscles of the Thigh)

Flexion	Extension	Adduction	Abduction	Medial rotation	Lateral rotation
Illiopsoas	Gluteus maximus	Adductor magnus	Gluteus medius	Gluteus medius	Gluteus maximus
Sartorius	Biceps femoris	Adductor longus	Gluteus minimus	Gluteus minimus	Piriformis
Rectus femoris	Semitendinosus	Adductor brevis	Tensor fasciae latae	Tensor fasciae latae	Obturator internus
Pectineus	Semimembranosus	Pectineus	Piriformis		Obturator externus
Adductor longus	Piriformis	Gracilis			Superior and inferior gemelli
Tensor fasciae latae	Adductor magnus				Quadratus femoris Adductor magnus Adductor longus Adductor brevis Pectineus

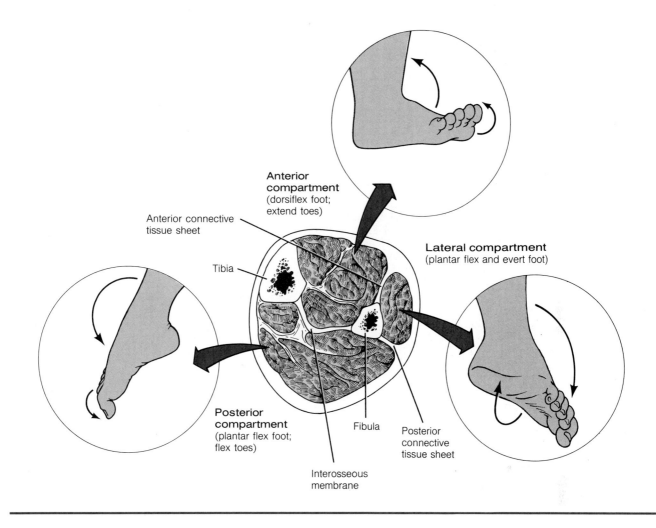

Figure 9.36
Cross section of the right leg showing the three compartments and general actions of the muscles in them.

Muscles That Act on the Foot and Toes

The muscles of the leg, like the muscles of the thigh, are grouped into three compartments (Figure 9.36; Table 9.20). The interosseous membrane between the tibia and fibula, and connective tissue sheets that extend anteriorly and posteriorly from the fibula, separate the muscles into *anterior, posterior,* and *lateral compartments.*

The muscles of the *anterior compartment* (Figures 9.37 and 9.38) act to extend the toes and/or dorsiflex the foot. The tendons of these anterior muscles are held firmly to the ankle by **superior** and **inferior extensor retinaculae,** in a manner similar to the tendons at the wrist.

The *lateral* or *peroneal compartment* (Figures 9.39 and 9.40) contains two of the three peroneal muscles. The **peroneus tertius** is fused to the extensor digitorum longus muscle and is included in the anterior compartment. The tendons of the **peroneus longus** and **brevis** pass behind the lateral malleolus of the fibula to insert on the plantar and lateral surfaces of the metatarsals. The peroneus longus tendon crosses the sole of the foot obliquely, from the lateral edge to the medial tarsal and metatarsal bones. Using the lateral malleolus as a pulley, the peroneal muscles of the lateral compartment act to plantar flex and evert the foot.

The muscles of the *posterior compartment* are grouped into three superficial posterior muscles (Figure 9.41)—the **gastrocnemius, soleus,** and **plantaris**—and four deep posterior muscles (Figures 9.42 and 9.43)—the **popliteus, flexor hallucis longus, flexor digitorum longus,** and the **tibialis posterior.** The tendons of the latter three muscles pass behind the medial malleolus of the tibia and use it as a pulley. The gastrocnemius and the underlying soleus

F9.37, F9.38

F9.39, F9.40

gas-trok-nee'-mee-us
F9.41, F9.42, F9.43

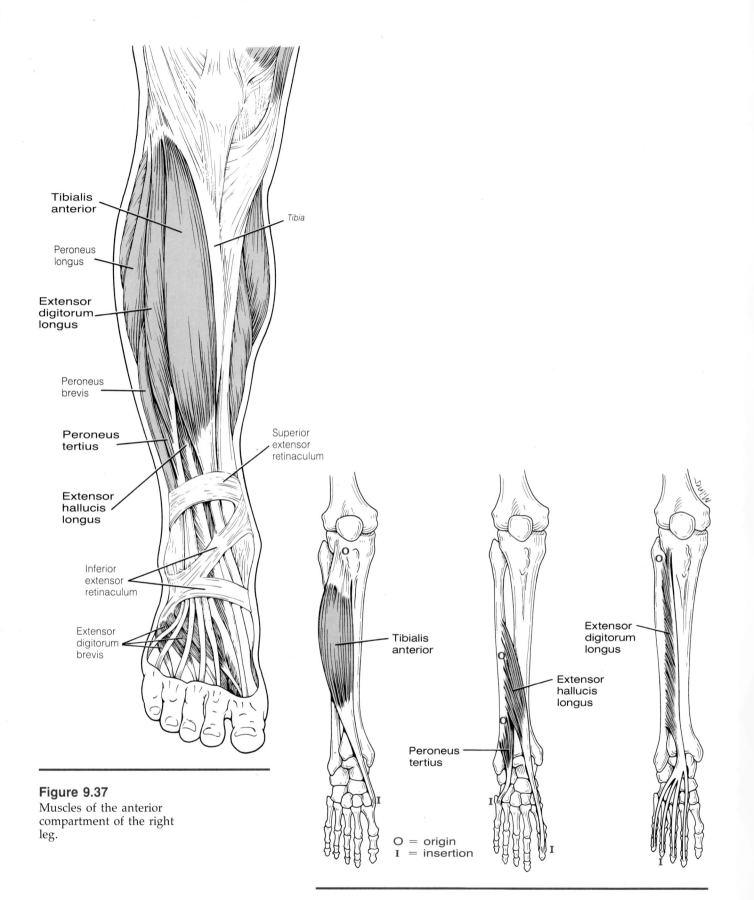

Figure 9.37
Muscles of the anterior compartment of the right leg.

Figure 9.38
The individual muscles of the anterior compartment of the right leg.

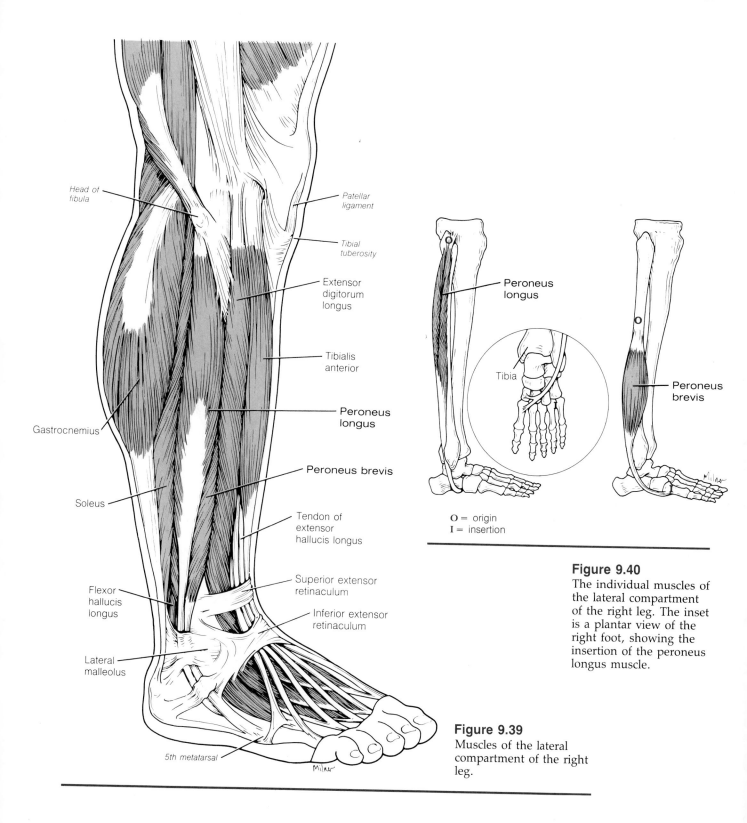

Figure 9.40
The individual muscles of the lateral compartment of the right leg. The inset is a plantar view of the right foot, showing the insertion of the peroneus longus muscle.

O = origin
I = insertion

Figure 9.39
Muscles of the lateral compartment of the right leg.

form the characteristic bulge of the calf. They share a common tendon, the **calcaneus (Achilles) tendon,** which inserts onto the calcaneus. With the exception of the popliteus, which flexes and medially rotates the leg, the muscles of the posterior compartment flex the toes and/or plantar flex the foot.

As is the case in the thigh, the muscles within each compartment of the leg are innervated by branches from the same nerve (Table 9.20). The muscles in the anterior compartment are innervated by the deep peroneal nerve; those

Table 9.20

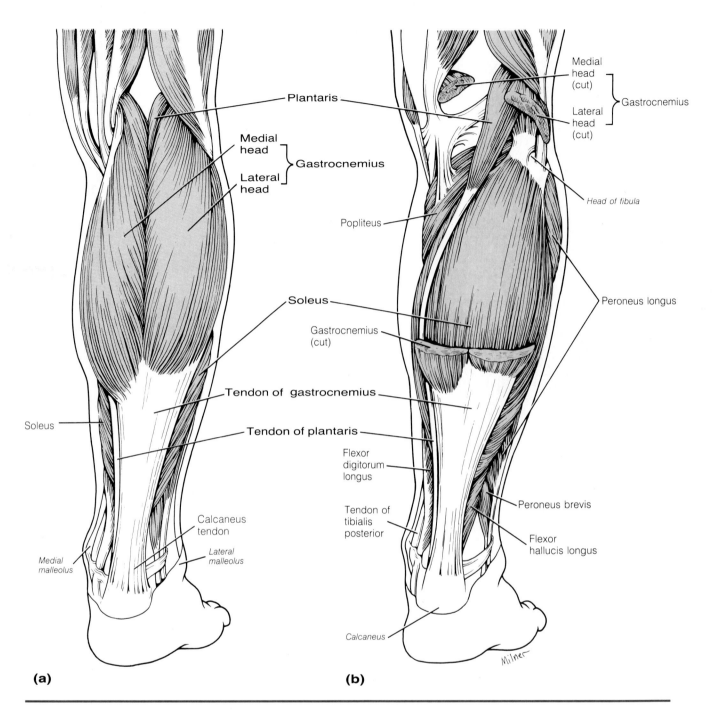

(a)   **(b)**

Figure 9.41

Superficial muscles of the
posterior compartment of   **Table 9.21**
the right leg.
(a) Superficial layer.
(b) The origin of the
gastrocnemius has been
removed to reveal the
underlying muscles.

in the lateral compartment by the superficial peroneal nerve; and those in the
posterior compartment by the tibial nerve.

Table 9.21 summarizes the muscle actions on the foot.

Intrinsic Muscles of the Foot

The skeletal structure of the foot is very similar to that of the hand; however,
it performs quite different functions. Whereas the hand has great versatility,
being capable of grasping and other fine movements, the foot is adapted for
support and locomotion. Consequently, the muscles of the foot tend to be
heavier than those of the hand, thus making it possible for them to support
the arches of the foot. The intrinsic (and extrinsic) muscles are aided in this
support by a tough fibrous **plantar aponeurosis** that extends from the calcaneus to the phalanges. Another difference in the musculature of the foot as

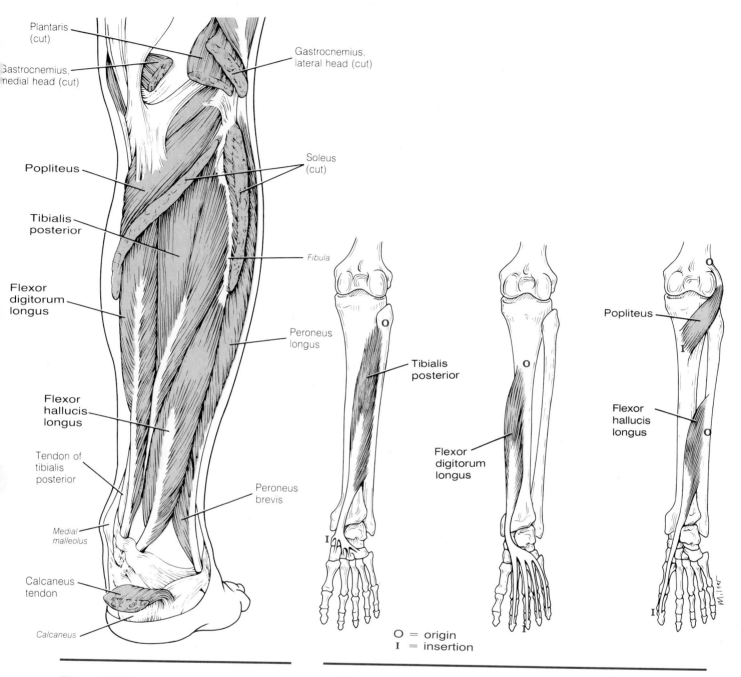

Plantaris (cut)

Gastrocnemius, medial head (cut)

Gastrocnemius, lateral head (cut)

Popliteus

Soleus (cut)

Tibialis posterior

Fibula

Flexor digitorum longus

Peroneus longus

Flexor hallucis longus

Tendon of tibialis posterior

Peroneus brevis

Medial malleolus

Calcaneus tendon

Calcaneus

Tibialis posterior

Flexor digitorum longus

Popliteus

Flexor hallucis longus

O = origin
I = insertion

Figure 9.42
Deep muscles of the posterior compartment of the right leg. The gastrocnemius and the soleus have been removed.

Figure 9.43
The individual deep muscles of the posterior compartment of the right leg.

compared to that of the hand is the presence of one intrinsic muscle, the **extensor digitorum brevis,** on the dorsum of the foot.

The remaining intrinsic muscles are on the plantar surface. They are illustrated in Figure 9.44 and described in Table 9.22. These muscles are separated into four layers (as illustrated), and they can be divided functionally into those that act on the big toe, those that move only the small toe, and those that move all the toes except the big toe.

F9.44, Table 9.22

Table 9.20 Muscles That Act on the Foot and Toes (Muscles of the Leg)

Muscle	Origin	Insertion	Action	Innervation
ANTERIOR COMPARTMENT				
Tibialis anterior [F9.37]	Lateral condyle and proximal two-thirds of the shaft of the tibia, and the interosseous membrane	Medial surface of first cuneiform and first metatarsal	Dorsiflexes and inverts foot	Deep peroneal
Extensor hallucis longus (hallus = great toe) [F9.38]	Anterior surface of the middle of the fibula, and the interosseous membrane	Dorsal surface of the distal phalanx of the great toe	Dorsiflexes and inverts foot; extends the great toe	Deep peroneal
Extensor digitorum longus [F9.38]	Lateral condyle of the tibia, proximal three-fourths of the anterior surface of the fibula, and the interosseous membrane	Dorsal surface of the phalanges of the second through fifth toes	Dorsiflexes and everts the foot; extends the toes	Deep peroneal
Peroneus tertius (tertius = third) [F9.38]	Distal third of the anterior surface of the fibula, and the interosseous membrane	Dorsal surface of the fifth metatarsal	Dorsiflexes and everts the foot	Deep peroneal
LATERAL COMPARTMENT				
Peroneus longus [F9.37]	Proximal two-thirds of the lateral surface of the fibula	Ventral surface of the first metatarsal and the medial cuneiform	Plantar flexes and everts the foot	Superficial peroneal
Peroneus brevis [F9.37]	Distal two-thirds of the fibula	Lateral side of the fifth metatarsal	Plantar flexes and everts the foot	Superficial peroneal
POSTERIOR COMPARTMENT				
SUPERFICIAL MUSCLES				
Gastrocnemius [F9.39; F9.41a]	Medial and lateral condyles of the femur	Calcaneus, via the calcaneus tendon	Flexes the leg; plantar flexes the foot	Tibial
Soleus [F9.39; F9.41]	Posterior surface of the proximal third of the fibula, and the middle third of the tibia	Calcaneus, via the calcaneus tendon	Plantar flexes the foot	Tibial
Plantaris [F9.41b]	Posterior surface of the femur above the lateral condyle	Calcaneus	Flexes the leg; plantar flexes the foot	Tibial
DEEP MUSCLES				
Popliteus [F9.41b]	Lateral condyle of the femur	Proximal portion of the tibia	Flexes the leg and rotates it medially	Tibial
Flexor hallucis longus [F9.41b]	Lower two-thirds of the fibula	Distal phalanx of the great toe	Flexes the great toe; plantar flexes and inverts the foot	Tibial
Flexor digitorum longus [F9.41b]	Posterior surface of the tibia	Distal phalanx of the second through fifth toes	Flexes the toes; plantar flexes and inverts the foot	Tibial

Table 9.20   Muscles That Act on the Foot and Toes (Muscles of the Leg) (continued)

Muscle	Origin	Insertion	Action	Innervation
Tibialis posterior [F9.42]	Posterior surface of the interosseous membrane, the tibia, and the fibula	Navicular, cuneiforms, cuboid; second through fourth metatarsals	Plantar flexes and inverts the foot	Tibial

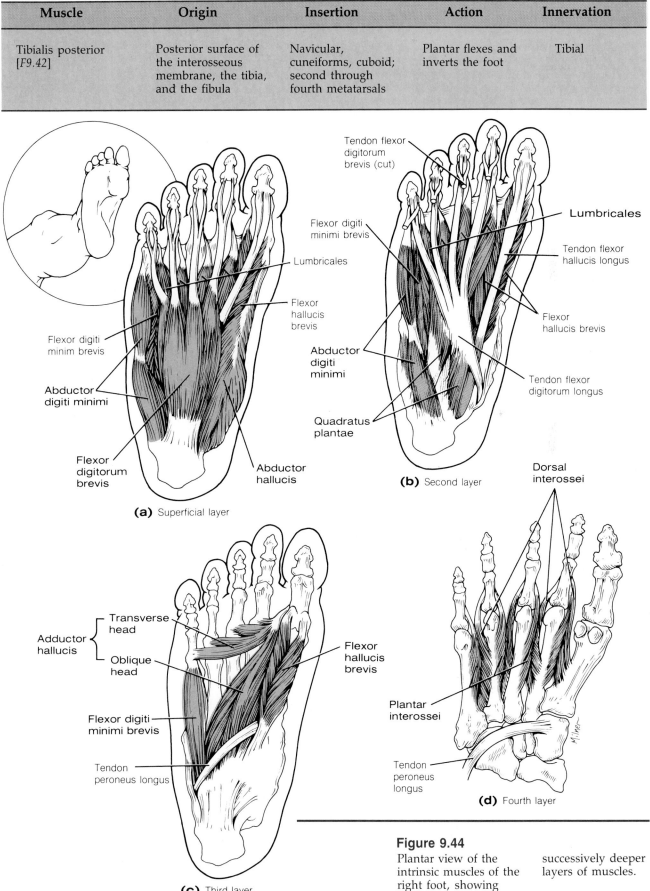

(a) Superficial layer

(b) Second layer

(c) Third layer

(d) Fourth layer

Figure 9.44
Plantar view of the intrinsic muscles of the right foot, showing successively deeper layers of muscles.

Table 9.21   Summary of Muscle Actions on the Foot

Dorsiflex	Plantar flex	Adduction Plus Inversion	Abduction Plus Eversion
Tibialis anterior	Peroneus longus	Tibialis anterior	Peroneus tertius
Extensor hallucis longus	Peroneus brevis	Tibialis posterior	Peroneus longus
Extensor digitorum longus	Gastrocnemius	Extensor hallucis longus	Peroneus brevis
Peroneus tertius	Soleus	Flexor hallucis longus	Extensor digitorum longus
	Plantaris	Flexor digitorum longus	
	Tibialis posterior		
	Flexor hallucis longus		
	Flexor digitorum longus		

Table 9.22   Intrinsic Muscles of the Foot

Muscle	Origin	Insertion	Action	Innervation
DORSAL MUSCLE Extensor digitorum brevis [F9.37]	Lateral surface of the calcaneus	Proximal phalanx of the great toe and the tendons of the extensor digitorum longus	Extends the first through fourth toes	Deep peroneal
PLANTAR MUSCLES **SUPERFICIAL LAYER** Abductor hallucis [F9.44a]	Calcaneus	Proximal phalanx of the great toe (with the tendon of the flexor hallucis brevis)	Abducts the great toe	Medial plantar
Flexor digitorum brevis [F9.44a]	Calcaneus and plantar aponeurosis	Middle phalanx of the second through the fifth toes	Flexes the second through fifth toes	Medial plantar
Abductor digiti minimi [F9.44a]	Calcaneus and plantar aponeurosis	Proximal phalanx of the small toe	Abducts the small toe	Lateral plantar
SECOND LAYER Quadratus plantae [F9.44b]	Calcaneus	Into tendons of the flexor digitorum longus	Aids in flexing the second through fifth toes by straightening the pull of the flexor digitorum longus	Lateral plantar
Lumbricales (4) [F9.44b]	From tendons of the flexor digitorum longus	Into tendons of the extensor digitorum longus	Flex the proximal phalanx; extend the middle and distal phalanges of the second through fifth toes	Medial and lateral plantar
THIRD LAYER Flexor hallucis brevis [F9.44c]	Cuboid and lateral cuneiform	Proximal phalanx of the great toe	Flexes the great toe	Medial plantar
Adductor hallucis [F9.44c]	*Oblique head:* second, third, and fourth metatarsals *Transverse head:* ligaments of the metatarsophalangeal joints	Proximal phalanx of the great toe	Adducts the great toe	Lateral plantar

Table 9.22 Intrinsic Muscles of the Foot (continued)

Muscle	Origin	Insertion	Action	Innervation
Flexor digiti minimi brevis [F9.44c]	Fifth metatarsal	Proximal phalanx of the small toe	Flexes the small toe	Lateral plantar
FOURTH LAYER Plantar interossei (3) [F9.44d]	Third, fourth, and fifth metatarsals	Proximal phalanx of the same toe	Adduct the toes toward the second toe	Lateral plantar
Dorsal interossei (4) [F9.44d]	Bases of the adjacent metatarsals	Proximal phalanges; both sides of the second toe; lateral side of the third and fourth toes	Abduct the toes from the second toe; move the second toe medially and laterally	Lateral plantar

STUDY OUTLINE

CRITERIA USED TO NAME MUSCLES
p. 245

MUSCLE SHAPE For example: trapezius, rhomboid.

MUSCLE ACTION For example: flexor, extensor, adductor, pronator.

MUSCLE LOCATION For example: intercostal, tibialis anterior.

MUSCLE ATTACHMENTS For example: sternocleidomastoid, coracobrachialis.

MUSCLE DIVISIONS For example: biceps brachii, triceps brachii, quadriceps femoris.

MUSCLE SIZE RELATIONSHIPS For example: gluteus maximus, gluteus minimus, peroneus longus, peroneus brevis.

MUSCLES OF THE HEAD AND NECK
pp. 249–253

MUSCLES OF THE FACE (Table 9.1) Also called muscles of facial expression. Some arise from skull bones, others from superficial fascia of the face. Most insert into skin of the region and thus serve to move the skin rather than joints. Facial muscles control mouth and lips; compress cheeks; control eyebrow, eyelid, and scalp movement; dilate nares; tighten and wrinkle neck and forehead skin; and depress mandible.

MUSCLE OF MASTICATION (Table 9.2) Four pairs of muscles involved in biting and chewing. Two pairs raise mandible. Two pairs move mandible sideways and assist in opening and closing mouth.

MUSCLES OF THE TONGUE Intrinsic muscles of tongue squeeze, fold, and curl tongue; extrinsic muscles of tongue (Table 9.3) anchor it to skeleton and control tongue's protrusion, retraction, and sideways movement.

MUSCLES OF THE NECK Include muscles grouped within one of two triangles—anterior (Table 9.4) or posterior—that are separated by sternocleidomastoid muscle.

MUSCLES OF THE THROAT Consist of deep muscles of anterior triangle; help form floor of oral cavity and are attached to hyoid bone; aid in movements of tongue and in swallowing.

MUSCLES OF THE TRUNK Include muscles of vertebral column, the back, thorax, abdominal wall, and muscles that form floor of abdominopelvic cavity.
pp. 253–262

MUSCLES OF THE VERTEBRAL COLUMN (Table 9.5) Most are located on posterior surface of spine. Some have fibers that insert onto skull and thus also move the head. Others (some that comprise abdominal wall and those that act on hip) also act on vertebral column.

DEEP MUSCLES OF THE THORAX (Table 9.6) Most insert on ribs and assist in breathing by elevating or depressing rib cage; spaces between ribs are reinforced by double layer of intercostal muscles; diaphragm is muscle chiefly responsible for quiet breathing.

MUSCLES OF THE ABDOMINAL WALL (Table 9.7) Wall of abdominal cavity lacks skeletal support and thus derives strength entirely from muscles, which are composed of three layers.
1. **External abdominal oblique:** Outermost layer, with fibers that pass medially and downward.
2. **Internal abdominal oblique:** Layer immediately beneath external oblique; has fibers that run upward and medially.
3. **Transversus abdominis muscle:** Deep to both oblique muscles; a thin muscle whose fibers run horizontally and encircle abdominal cavity.

   RECTUS ABDOMINIS MUSCLE: Ensheathed by fasciae of oblique and transversus muscles; passes vertically from pubic symphysis to rib cage.

   QUADRATUS LUMBORUM MUSCLE: Forms most of posterior abdominal wall.

MUSCLES THAT FORM THE FLOOR OF THE ABDOMINOPELVIC CAVITY (Table 9.8) Viscera of abdominopelvic cavity are supported by muscular floor called pelvic diaphragm, which is composed of two principal muscles: levator ani and coccygeus.

MUSCLES OF THE PERINEUM (Table 9.9) Five muscles in two layers; located just below pelvic diaphragm; urogenital diaphragm formed by two of the muscles.

MUSCLES OF THE UPPER LIMBS Include muscles that act on scapula; clavicle; arm and forearm; hand and fingers. **pp. 263–276**

MUSCLES THAT ACT ON THE SCAPULA (Table 9.10) Six muscles—four posterior and two anterior—act directly on scapula and anchor it to thorax.

MUSCLES THAT ACT ON THE ARM (HUMERUS) (Tables 9.11 and 9.12) Nine muscles cross shoulder joint and insert on humerus; seven of these arise from scapula.

MUSCLES THAT ACT ON THE FOREARM (RADIUS AND ULNA) (Table 9.13) Elbow joint and/or proximal radial/ulnar joints are moved by powerful muscles in arm, but are assisted by muscles in forearm.

MUSCLES THAT ACT ON THE HAND AND FINGERS (Tables 9.14 and 9.15) Moved by muscles that form bulge of proximal end of forearm; anterior group of muscles serves as flexors and/or pronators; the posterior group serves as extensors and/or supinators.

INTRINSIC MUSCLES OF THE HAND (Table 9.16) Several muscles of forearm move fingers; intrinsic finger muscles produce fine, precise finger movement.

MUSCLES OF THE LOWER LIMBS Include muscles that act on thigh (femur), leg (tibia and fibula), foot and toes, and intrinsic muscles of foot. Lower-limb muscles tend to be bulkier and more powerful than those of upper limbs because the former are concerned with support and locomotion. Many muscles of lower limbs cross and act on two joints, either hip and knee or knee and ankle. **pp. 276–293**

MUSCLES THAT ACT ON THE THIGH (FEMUR) (Tables 9.17 and 9.19) Most arise from pelvis.

MUSCLES THAT ACT ON THE LEG (TIBIA AND FIBULA) (Table 9.18) Located in thigh; their tendons cross knee; grouped into anterior (extensor), posterior (flexor), and medial (adductor) compartments.

MUSCLES THAT ACT ON THE FOOT AND TOES (Tables 9.20 and 9.21) Most located in leg; grouped into anterior (dorsiflex ankle and/or extend toes), posterior (plantar flex foot and/or flex toes), and lateral (plantar flex and evert foot) compartments.

INTRINSIC MUSCLES OF THE FOOT (Table 9.22) Four layers on plantar surface and one intrinsic muscle on dorsum of foot; tend to be heavier than hand muscles because foot is adapted for support and locomotion.

SELF-QUIZ

1. Match each muscle with the correct criterion or criteria on which the name is based.

   Intercostalis
   Rhomboideus
   Pronator
   Coracobrachialis
   Gluteus
     maximus
   Trapezius
   Rectus
     abdominis
   Pectoralis minor
   Flexor carpi
     radialis
   Biceps brachii
   Quadriceps
     femoris

   (a) Shape
   (b) Size relationship
   (c) Attachment
   (d) Location
   (e) Number of divisions
   (f) Action

2. Most of these muscles insert into the skin of their particular region: (a) intrinsic muscles of the hand; (b) facial muscles; (c) intrinsic muscles of the foot.

3. Match each facial muscle with the correct statement about its action.

   Depressor labii
     inferioris
   Epicranius
     occipitalis
   Corrugator
   Orbicularis oculi
   Orbicularis oris
   Procerus
   Risorius

   (a) Pulls angle of mouth backward
   (b) Draws eyebrows together in frowning
   (c) Draws scalp posteriorly
   (d) Pulls lower lip down
   (e) Wrinkles skin between eyebrows
   (f) Closes eyelids; tightens forehead skin
   (g) Closes and protrudes lips

4. Which muscle retracts and elevates the tongue? (a) styloglossus; (b) hypoglossus; (c) genioglossus.

5. The muscles of the throat are the deep muscles of the anterior triangle. True or False?

6. Raising the hyoid, pulling it backward and anteriorly, and raising the floor of the mouth are actions performed by the: (a) infrahyoid muscles; (b) suprahyoid muscles; (c) sternothyroid muscle.

7. The diaphragm is the muscle chiefly responsible for quiet breathing. True or False?

8. Match each vertebral column muscle with its correct statement.

   Multifidus
   Interspinales
   Splenius capitis
     and cervicis
   Spinalis thoracis
     and cervicis

   (a) *Origin:* spinous processes of upper lumbar, lower thoracic, and 7th cervical vertebrae
   (b) *Insertion:* occipital bone, mastoid process of temporal bone, and transverse processes of upper three cervical vertebrae
   (c) *Action:* extends vertebral column; rotates it toward opposite side

(d) *Origin:* superior surface of all spinous processes.

9. Select the correct statement about the temporalis muscle. (a) It originates in the zygomatic arch. (b) It opens and protrudes the mandible. (c) It is innervated by the trigeminal nerve.

10. The internal intercostal muscles, the external intercostal muscles, and the transversus muscles are all involved in respiration. True or False?

11. Inguinal hernias are more common in females than in males. True or False?

12. Which of these muscles of the abdominal wall pulls the thoracic cage toward the pelvis and bends the vertebral column toward the side being contracted? (a) quadratus lumborum; (b) internal abdominal oblique; (c) rectus abdominis.

13. A muscle that forms the floor of the abdominopelvic cavity is the: (a) transversus abdominis; (b) levator ani; (c) internal abdominal oblique.

14. Match the following muscles that act on the scapula with the correct statement about each muscle's action.

Subclavius	(a)	Depresses the scapula and pulls it anteriorly
Rhomboideus major	(b)	Stabilizes and depresses the pectoral girdle
Pectoralis minor	(c)	Adducts, stabilizes, and rotates the scapula, lowering its lateral angle

15. Match the muscle action with the appropriate muscles. (*Note:* A given muscle may have more than one action associated with it.)

Deltoid	(a)	Flexion
Latissimus dorsi	(b)	Extension
Teres minor	(c)	Adduction
Teres major	(d)	Abduction
Coracobrachialis	(e)	Medial rotation
Infraspinatus	(f)	Lateral rotation

16. Which of the following muscles *flex* the forearm? (a) biceps brachii; (b) brachialis; (c) triceps brachii; (d) anconeus; (e) brachioradialis.

17. Match the muscle action with the appropriate hand or finger muscle.

Flexor carpi radialis	(a)	Flexes the hand and phalanges
Extensor indicis	(b)	Extends and abducts the hand
Extensor carpi ulnaris	(c)	Extends the little finger
Extensor pollicis longus		
Extensor carpi radialis longus	(d)	Extends the index finger
Extensor digiti minimi	(e)	Extends and adducts the hand
Flexor carpi ulnaris	(f)	Extends the thumb and abducts the hand
Flexor digitorum superficialis	(g)	Flexes and adducts the hand
	(h)	Flexes and abducts the hand

18. Which two of the following muscles are involved when you move your little finger in to meet your thumb, and when you move your thumb in to meet your little finger? (a) flexor digiti minimi brevis; (b) opponens digiti minimi; (c) flexor pollicis brevis; (d) palmaris brevis; (e) opponens pollicis.

19. Match the muscle action with the appropriate muscles.

Gluteus maximus	(a)	Abducts and medially rotates the thigh
Adductor magnus	(b)	Adducts, flexes, and laterally rotates the thigh
Pectineus		
Adductor longus	(c)	Adducts and laterally rotates the thigh; assists in flexion of the thigh
Gluteus medius		
	(d)	Extends and laterally rotates the thigh
	(e)	Adducts and laterally rotates the thigh; assists in extension of the thigh

20. Those muscles of the thigh that flex the leg and extend the thigh are located in the: (a) medial compartment; (b) anterior compartment; (c) posterior compartment.

21. What is the action of the muscles that are located in the anterior compartment of the leg? (a) flex the leg; (b) dorsiflex foot and/or extend toes; (c) plantar flex foot and flex toes.

22. Match the muscle action with the appropriate leg muscle that acts on the foot and/or toes.

Tibialis anterior	(a)	Flexes the great toe; plantar flexes and inverts the foot
Peroneous longus	(b)	Plantar flexes and everts the foot
Extensor digitorum longus	(c)	Dorsiflexes and inverts the foot
Flexor hallucis longus	(d)	Flexes the leg and rotates it medially
Popliteus	(e)	Dorsiflexes and everts the foot; extends the toes

LEARNING OBJECTIVES

After completing this chapter, you should be able to:

1. Distinguish between receptors and effectors.

2. Name the two divisions of the nervous system according to structural location, and list the principal components of each.

3. Name the divisions of the nervous system according to function.

4. Distinguish between the somatic nervous system and the autonomic nervous system, and cite the principal function of each.

5. Differentiate between myelinated and unmyelinated nerve fibers.

6. Identify the connective-tissue sheaths that cover a nerve.

7. Distinguish between a dendrite and an axon.

8. Name the types of neuroglia and state the principal function of each.

9. Name the types of neurons according to structural and functional classification.

10. Cite the various ways receptors may be classified.

11. Describe the specialized peripheral neuron endings.

CHAPTER CONTENTS

ORGANIZATION OF THE NERVOUS SYSTEM

EMBRYONIC DEVELOPMENT OF THE NERVOUS SYSTEM

COMPONENTS OF THE NERVOUS SYSTEM

KEY TERMS AND DERIVATIVES

astrocyte (*astr* = starlike) star-shaped neuroglial cells that provide most of the structural support for the CNS

axon (*axon* = axle) the conductive process of a neuron; transmits the nerve impulses

dendrite (*dendr* = tree) a branching neuron process where electrical signals originate

ependymal cells (*ependym* = tunic) neuroglial cells that line spaces within the brain and spinal cord

ganglion (*gangl* = mass) a group of nerve cell bodies, usually located in the peripheral nervous system

neurilemma (*neuro* = nerve; *lemma* = rind or peel) composed of the outermost parts of Schwann cells that were pushed peripherally as the cells wrapped around an axon

neuroglia (*neuro* = nerve; *glia* = glue) specialized supporting cells of the nervous system that are also involved in the transfer of nutrients from blood vessels to neurons and in waste-product removal

oligodendrocyte (*oligo* = few) a small neuroglial cell with few cellular processes

The Nervous System:
Its Organization and Components

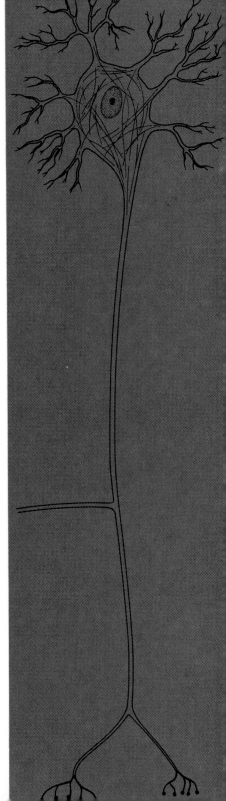

The survival of any multicellular organism depends on its having some means of regulating and coordinating the activities of its cells. If there were no coordination, each cell would function just as genetically programmed, without regard to the functions other cells were performing. At best, such haphazard organization would be inefficient; at worst, it could be fatal.

The human organism, which is composed of billions of cells, has two systems that serve primarily as means of *internal communication* among the cells—the nervous system and the endocrine system. Because both systems cooperate to provide the body's internal communications, they can be viewed as closely allied networks operating together to ensure proper body functioning. By monitoring and regulating how the various body functions interrelate, the nervous system and the endocrine system help maintain the homeostasis of the body. In Chapters 10 through 15 we consider the nervous system. The endocrine system is discussed in Chapter 16.

The nerve cells of the nervous system carry messages in the form of electrical signals. A particularly important kind of electrical signal carried by nerve cells is called a **nerve impulse.** Nerve impulses often originate within a nerve cell as a result of the activity of sensory structures called **receptors.** Receptors are generally activated by changes in either the internal or external environments of the body. The changes that activate the receptors are called **stimuli.** When a stimulus activates a receptor, nerve impulses are initiated within **sensory nerve cells.** These impulses are carried by the sensory nerve cells to the spinal cord and brain (Figure 10.1), where other nerve cells may be activated. And these nerve cells may then conduct nerve impulses to various other locations within the spinal cord and brain. Ultimately, nerve impulses carried by **motor nerve cells** leave the brain and the spinal cord and bring about responses to environmental changes by selectively activating various **effectors.** Effectors capable of responding to nerve impulses include muscle cells and the secretory cells of glands and organs. In addition to responding to environmental stimuli, the human nervous system has the ability to integrate and store the information that it receives, thus providing for such capacities as memory, abstract reasoning, and conceptualizing.

ORGANIZATION OF THE NERVOUS SYSTEM

Although there is actually only one nervous system, it can be separated conceptually into various divisions based on either structural locations or functional characteristics. Keep in mind, however, that these divisions—which are themselves called *nervous systems*—are all integral parts of a single nervous system. Structurally, the nervous system may be divided into two parts: the *central nervous system* and the *peripheral nervous system* (Figure 10.2).

Central Nervous System

The **central nervous system (CNS)** consists of the brain and the spinal cord. It is completely encased within bony structures—the brain within the cranial cavity of the skull and the spinal cord within the vertebral canal of the spinal

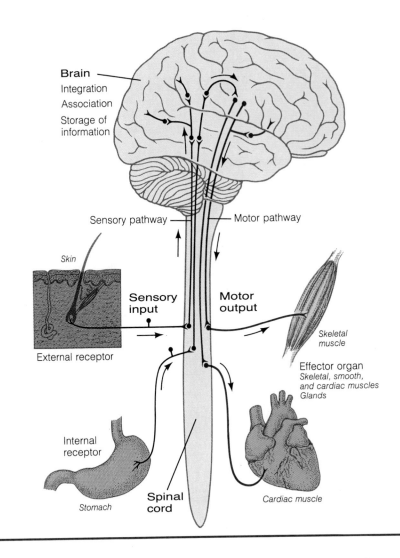

Figure 10.1
Schematic representation of the basic organization of the nervous system. The arrows follow a typical pathway of nerve impulses from receptor to effector.

column. The central nervous system is the *integrative and control center* of the nervous system. It receives sensory input from the peripheral nervous system and formulates responses to this input.

Peripheral Nervous System

per-if'-er-al   The **peripheral nervous system (PNS)** is composed of all nervous structures located outside of the central nervous system. Specifically:

1. **nerves** that connect the outlying parts of the body and their receptors with the central nervous system

gan'-glee-uh   2. **ganglia** (groups of nerve cell bodies) associated with the nerves

The peripheral nervous system includes 12 pairs of **cranial nerves,** which arise from the brain and the brain stem and leave the cranial cavity through foramina in the skull, and 31 pairs of **spinal nerves,** which arise from the spinal cord and leave the vertebral canal through the intervertebral foramina. The paired spinal nerves include 8 cervical, 12 thoracic, 5 lumbar, 5 sacral, and 1 coccygeal nerve (Figure 10.2). The peripheral nervous system can be divided functionally into afferent (sensory) and efferent (motor) divisions.

F10.2

Afferent Division

af'-er-ent   The **afferent division** of the peripheral nervous system includes *somatic sensory* nerve cells, which carry impulses to the central nervous system from

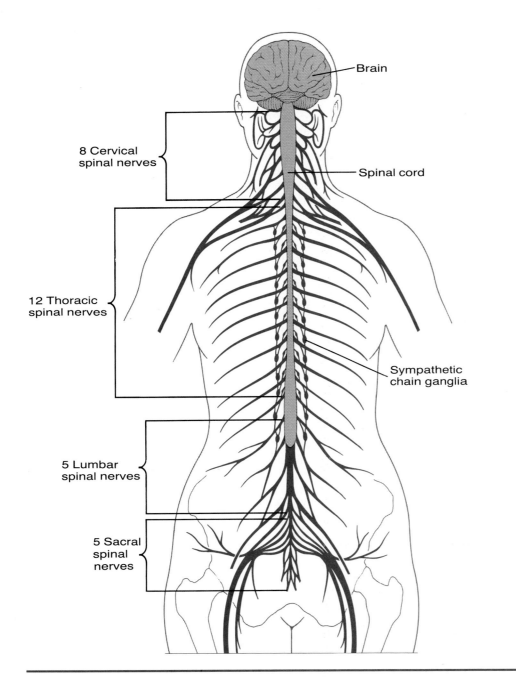

8 Cervical spinal nerves

12 Thoracic spinal nerves

5 Lumbar spinal nerves

5 Sacral spinal nerves

Brain

Spinal cord

Sympathetic chain ganglia

Figure 10.2
The central nervous system and the proximal portions of the peripheral nervous system.

receptors located in the skin, in the fascia, and around the joints, and *visceral sensory* nerve cells, which carry impulses from the viscera of the body to the central nervous system.

Efferent Division

The **efferent division** of the peripheral nervous system is divided into the *somatic nervous system* and the *autonomic nervous system*.

ef'-er-ent

1. The **somatic nervous system** is also called the **voluntary nervous system,** because its motor functions may be consciously controlled. It includes *somatic motor* nerve cells, which carry impulses from the central nervous system to the skeletal muscles. The impulses carried by somatic motor nerve cells produce contractions of the skeletal muscles. Muscle contractions brought about by the somatic nervous system may be under

so-mat'-ik

the conscious control of the individual, or in the case of reflex responses, they may not be consciously controlled.

aw-to-nom'-ik

2. The **autonomic nervous system**—or **involuntary nervous system**—in contrast to the somatic nervous system, is composed of *visceral motor* nerve cells, which transmit impulses to smooth muscles, cardiac muscle, and glands. Visceral motor impulses generally cannot be consciously controlled.

sim-pa-thet'-ik
par-a-sim-pa-thet'-ik

The autonomic nervous system may be subdivided functionally into **sympathetic** and **parasympathetic divisions,** each of which is studied in more detail in Chapter 14. As noted earlier, the autonomic and somatic nervous "systems" are actually overlapping divisions of the body's single nervous system. Moreover, although the motor neurons of both the somatic and the autonomic divisions are generally considered to be parts of the peripheral nervous system, they are under the control of centers located in the central nervous system. The organization of the nervous system is summarized in

F10.3   Figure 10.3.

EMBRYONIC DEVELOPMENT OF THE NERVOUS SYSTEM

The nervous system develops from ectodermal cells that, by the second week of gestation, form a flat thickening on the dorsal surface of the embryo. This thickening, called the **neural plate,** gives rise to all of the nerve cells of the nervous system. In addition, neural plate cells give rise to most of the supportive cells *(neuroglial cells)* of the nervous system.

As development continues, the midline of the neural plate sinks inward. At the same time, the proliferation of cells along the margins of the plate produces elevations. The result is a **neural groove** flanked by **neural folds**

F10.4   that extend the entire length of the embryo (Figure 10.4).

The neural groove deepens as the neural folds increase in height. Eventually, the folds meet and fuse in the midline, converting the groove into the

F10.5   **neural tube** (Figure 10.5). The neural tube separates from the surface ectoderm and subsequently develops into the brain and the spinal cord. In addition, cells within the neural tube send out processes to peripheral structures. These cells and their processes form the motor nerve cells of both the somatic and the autonomic nervous systems.

While the neural folds are fusing, ectodermal cells from the tops of the folds move laterally to form columns of cells on each side of the neural tube. The cells within these columns are called **neural crest cells.** These neural crest cells develop connections with the central nervous system and with peripheral structures, thus forming sensory nerve cells within cranial nerves and spinal nerves. Some neural crest cells migrate to other locations and develop into motor nerve cells of the sympathetic nervous system, into Schwann cells that wrap around nerve cells, or into other structures not directly associated with the nervous system.

The anterior end of the neural tube undergoes the most rapid growth and gives rise to the brain. By the fourth week of gestation, the brain is in the form

pross-en-sef'-uh-lon   of three fluid-filled enlargements, or vesicles: the **prosencephalon (fore-**
mes-en-sef'-uh-lon   **brain),** the **mesencephalon (midbrain),** and the **rhombencephalon (hind-brain).** During subsequent development, this region undergoes several bendings or flexures and the vesicles further subdivide. As a result, by the fifth week, the brain consists of five vesicles tightly curved upon themselves

F10.6   at the anterior end of the embryo (Figure 10.6). The prosencephalon subdi-
tel-en-sef'-uh-lon   vides into two vesicles—an anterior **telencephalon** and, just behind it, the
dye-en-sef'-uh-lon   **diencephalon.** The mesencephalon remains unchanged, but the rhomben-
met-en-sef'-uh-lon   cephalon divides into the **metencephalon** and the most posterior brain vesi-
mile-len-sef'-uh-lon   cle, the **myelencephalon.** The cavities in these brain vesicles will become the
ser-ee-bro-spy'-nul   *ventricles* of the adult brain, and the fluid in them is the **cerebrospinal fluid.** Brain development is discussed further in Chapter 12.

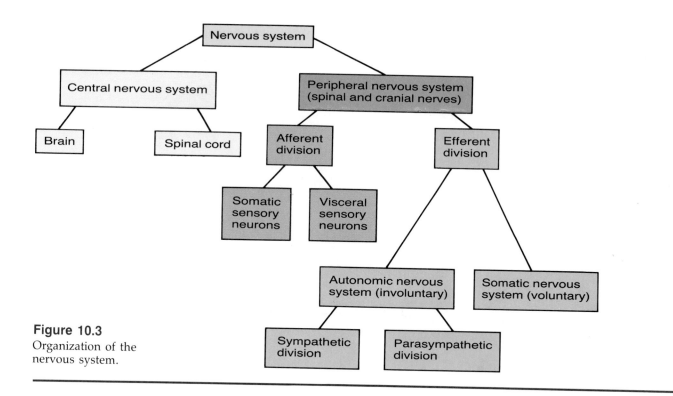

Figure 10.3
Organization of the nervous system.

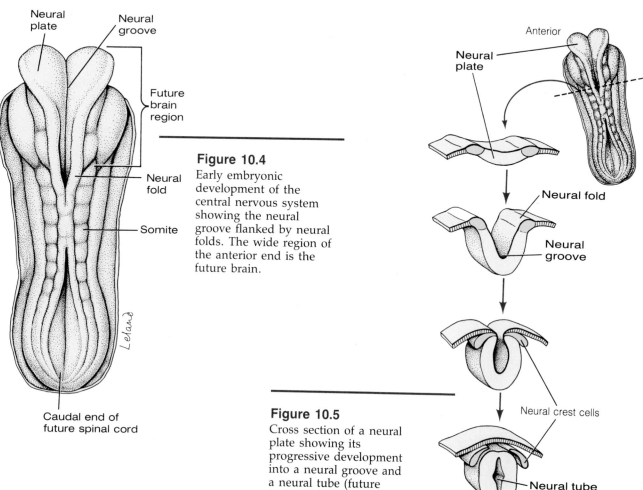

Figure 10.4
Early embryonic development of the central nervous system showing the neural groove flanked by neural folds. The wide region of the anterior end is the future brain.

Figure 10.5
Cross section of a neural plate showing its progressive development into a neural groove and a neural tube (future brain and spinal cord).

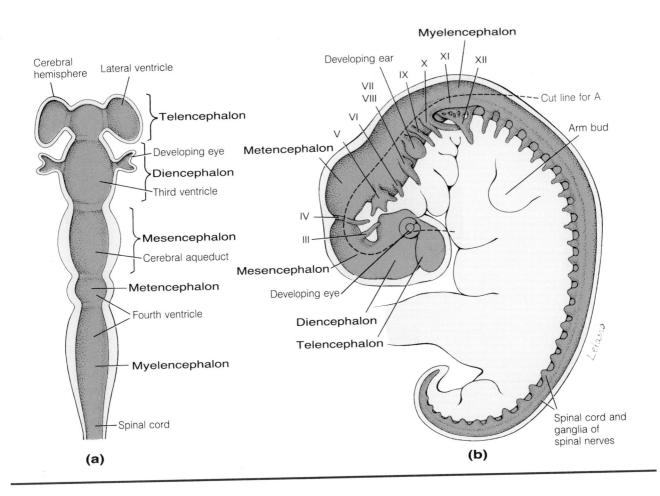

(a)

(b)

Figure 10.6

Development of the brain (at five weeks) into telencephalon, diencephalon, mesencephalon, metencephalon, and myelencephalon. **(a)** Frontal section. **(b)** Lateral view. Roman numerals indicate cranial nerves.

noo-rog'-lee-uh

The region of the neural tube posterior to the myelencephalon gives rise to the spinal cord. The development of the spinal cord is also discussed in Chapter 12.

COMPONENTS OF THE NERVOUS SYSTEM

Nervous tissue is composed of three types of cells that differ structurally and functionally: (1) **neurons (nerve cells),** which transmit nerve impulses; (2) **Schwann cells (neurolemmocytes),** which form a segmented covering around the processes of many of the neurons of the peripheral nervous system; and (3) **neuroglia (glial cells),** which are specialized to serve as support tissue between the neurons in the central nervous system. The nerves of the peripheral nervous system, therefore, are formed by neurons and Schwann cells, and the tracts of the brain and spinal cord are formed by neurons and neuroglial cells. In the peripheral nervous system, some neurons are closely associated with receptors.

Neurons

The **neuron (nerve cell)** is the basic structural and functional component of the nervous system. It has the ability to respond to stimulation by initiating and conducting electrical signals. Certain other cells, such as muscle fibers, are also capable of conducting electrical signals, but the unique shape of the nerve cell makes it especially well suited for transmitting information from one point to another. Neurons have processes that can be quite long. For instance, a single neuron—that is, a single nerve cell—may extend from the spinal cord to the tip of a toe!

Mature neurons are unable to undergo mitosis. For this reason, once embryonic development is complete, nerve cells cannot be replaced when they die or are destroyed. Under the proper conditions, however, peripheral neuronal processes that have been damaged or severed can be repaired or regenerated, thus reestablishing the nerve supply to the affected structures.

Structure of a Neuron

Each neuron is composed of a **cell body (soma,** or **perikaryon)** and one or more processes containing cytoplasm that extend from the cell body (Figure 10.7). Within the cytoplasm of the cell body is a large nucleus containing a prominent nucleolus. As in most cells, the cytoplasm of neurons also contains mitochondria and Golgi apparatus. In addition, the cytoplasm contains a dark-staining **chromatophilic substance** *(Nissl bodies)* and slender fibrils called **neurofibrils.** Under the electron microscope, the chromatophilic substance appears to be parallel layers of rough endoplasmic reticulum, and the neurofibrils are seen to be composed of microfilaments. Microtubules are also present. It is thought that the fibrils provide support to the cell, whereas the microtubules may be involved in the transport of materials within the cell.

Locations of the Cell Bodies of Neurons

The cell bodies of most neurons are located in the central nervous system, although there are large numbers located in the peripheral system. Nerve cell bodies in the CNS tend to be located in clusters called **nuclei.** (Note that this use of the term *nucleus* has no relation to the nucleus of a cell.) A nucleus or group of nuclei whose neurons have related functions is referred to as a **center.** Nerve cell bodies located outside the CNS, in the peripheral nervous system, are generally found in groups called **ganglia.**

Processes of Neurons

The processes associated with a neuron are very thin extensions of the cell. There are two types of processes: *dendrites* and *axons.* According to classical usage of the terms, dendrites are the neuronal process or processes that conduct electrical signals *toward* a cell body; an axon (of which there is only one per cell) is the neuronal process that conducts electrical signals *away* from a cell body. It is becoming more common, however, to differentiate dendrites and axons in the following way:

DENDRITES The **dendritic zone** is the *receptive* portion of a neuron, where electrical signals originate. The dendritic zone can include the cell body as well as branching processes, or **dendrites,** that are either direct extensions of the cell body or more remote branchings separated from the cell body by a length of axon. The number, length, and extent of branching of the dendrites vary in different types of neurons. Dendrites possess many of the structures found in the cell body, including filaments and microtubules that are oriented parallel to the long axis of the dendrite.

AXONS The **axon (nerve fiber)** is the *conductive* process of the neuron—that is, the part that transmits the electrical signals. In motor neurons, the cell body is located between the axon and the dendrites, but in sensory neurons it is located to one side of the axon. The axon frequently arises from a cone-shaped process on the cell body called the **axon hillock.** The lengths of axons vary considerably. They may be quite short, traveling only a short distance in the central nervous system, or they may be long, traveling a considerable distance within the CNS. Some axons extend a meter or more out into the peripheral nervous system. Each neuron has only one axon, but each axon generally has several branches called **collaterals.** Axons contain mitochondria, microtubules, and neurofibrils, but they lack chromatophilic substance. An axon and its collaterals end by separating into many fine branches called **telondendria.** The distal end of each telodendron expands

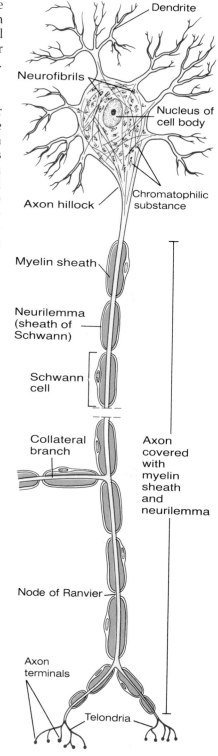

Figure 10.7
Structure of a typical motor neuron.

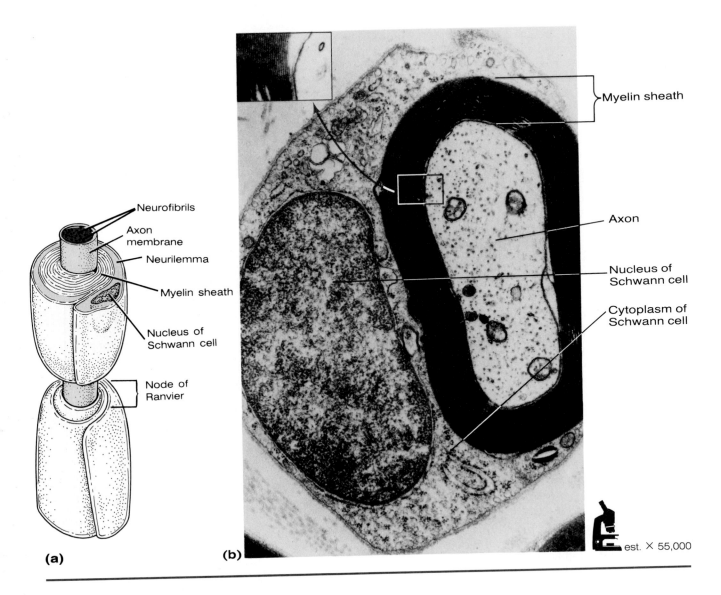

(a)

(b)

est. × 55,000

Figure 10.8
The ultrastructure of
myelinated neurons.
(a) Schematic.
(b) Photomicrograph of a
cross section of a nerve
fiber surrounded by a
myelin sheath. The inset
is a higher magnification
of a portion of the
myelin sheath.

my'-uh-li-nay-ted

my'-uh-lin

F10.8

F10.7, F10.8a

into small bulblike structures called **axon terminals,** or **synaptic knobs.**

Formation of Myelinated Neurons

Most axons are covered with a fatty substance called **myelin.** Such axons are called **myelinated fibers;** axons not covered with myelin are called **unmyelinated fibers.** Because of the lipid content of myelin, myelinated fibers appear white; when traveling in groups they form white pathways in the nervous system.

When myelinated axons of the peripheral nervous system are viewed under the light microscope, there appears to be a thin membrane between the myelin and the connective tissue sheath (endoneurium) surrounding the axon. This membrane is called the **neurilemma,** or **sheath of Schwann** (Figure 10.8). The neurilemma and the myelin of these myelinated axons are interrupted at regular intervals along the length of the axon. Each of these points of interruption is called a **node of Ranvier** (Figures 10.7 and 10.8a).

Both myelinated and unmyelinated axons of the peripheral nervous system are surrounded by **Schwann cells,** which are arranged sequentially along the length of the axon. During embryonic development, Schwann cells, which are derived from embryonic neural crest cells, migrate along the axon and envelop it. We consider the myelinated axons first.

Under the electron microscope, it can be seen that the myelin sheath of myelinated axons is formed when each Schwann cell wraps itself around the axon several times, pushing the cytoplasm and nucleus of the Schwann cell peripherally (Figure 10.8). Consequently, the myelin sheath is composed of alternating layers of the proteins and lipids that comprise the plasma membrane of the Schwann cell. The outermost part of the Schwann cell, which contains the cytoplasm and the nucleus that were pushed peripherally as the Schwann cell wrapped around the axon, is the neurilemma (sheath of Schwann). The nodes of Ranvier are the junctions between two successive Schwann cells. It is at the nodes of Ranvier that any collateral branches of the axon occur. The nodes are also instrumental in increasing the rate of transmission of the nerve impulse, as is discussed in Chapter 11.

Not all axons are myelinated, however. As shown in Figure 10.9, it is not unusual for the axons of several neurons to become embedded within the cytoplasm of a single Schwann cell. The Schwann cell may simply envelop some of these axons without forming a myelin sheath around them. Such axons are unmyelinated. The unmyelinated axons of the peripheral nervous system are covered by a neurilemma formed by the Schwann cells. However, the unmyelinated axons lack the additional spiraled wrappings of the Schwann cells that form the myelin covering.

Many neurons in the central nervous system are myelinated, but there are no Schwann cells in the CNS. Instead, the central nervous system contains a type of neuroglial cell called an **oligodendrocyte** that sends out processes that spiral around axons, thus forming a myelin covering. There is no neurilemma in the central nervous system, however. Myelinated axons form the white matter of the central nervous system.

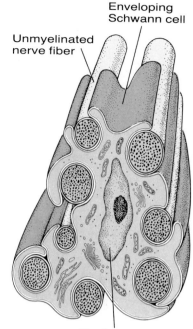

Enveloping Schwann cell

Unmyelinated nerve fiber

Nucleus of Schwann cell

Figure 10.9
A single Schwann cell encompassing eight unmyelinated nerve fibers.

Types of Neurons

Neurons can be classified according to their form, or structure, and according to their function—that is, the role they perform in the nervous system.

CLASSIFICATION ACCORDING TO STRUCTURE   Structurally, neurons can be classified into three types, based on the number of processes that extend from the cell body:

1. **Bipolar neurons** (Figure 10.10a) have two processes, one extending from each end of the cell body. There are very few examples of bipolar neurons in the body.   F10.10a

2. **Unipolar neurons** (Figure 10.10b) are formed during embryonic development when the two processes of certain bipolar neurons fuse together so that there is only a single process arising from the cell body. Beyond this point of fusion, however, the two processes remain separate.   F10.10b

3. **Multipolar neurons** (Figures 10.10c and 10.11), the most common type of neuron, have one long process that arises from the cell body and functions as an axon; numerous other processes that arise from the cell body function as dendrites.   F10.10c, F10.11

CLASSIFICATION ACCORDING TO FUNCTION   Functionally, there are also three types of neurons:

1. **Motor (efferent) neurons** transmit impulses away from the CNS to an effector, or from a higher center within the CNS to a lower center.

2. **Sensory (afferent) neurons** carry impulses from receptors to the CNS, or from a lower center within the CNS to a higher center.

3. **Internuncial neurons** (*interneurons*), when present, transmit impulses from one neuron to another.

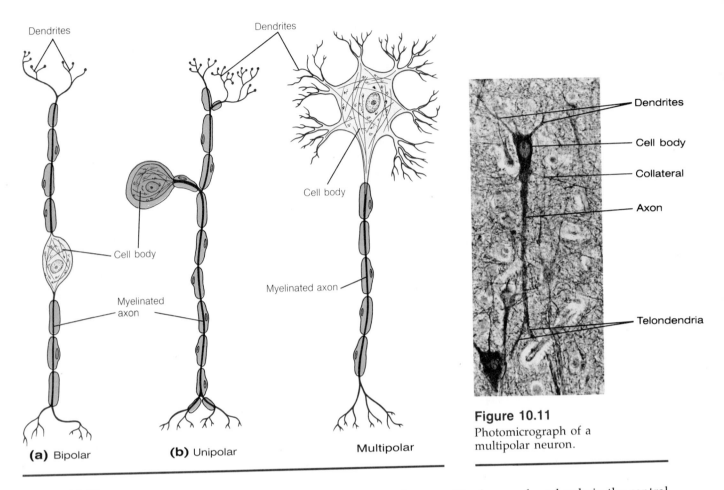

Figure 10.10
Types of neurons.

Figure 10.11
Photomicrograph of a multipolar neuron.

Internuncial neurons, which are multipolar, are found only in the central nervous system. Motor neurons are also multipolar. Most sensory neurons are unipolar, but those that carry impulses from the retina of the eye, inner ear, taste buds, and the olfactory epithelium are bipolar.

Nerves

F10.12

A **nerve** is composed of the processes of many neurons held together by connective-tissue sheaths (Figure 10.12). Each nerve fiber is individually wrapped in a thin connective-tissue sheath called the **endoneurium.** The processes of the neurons are separated into groups within the nerve by another connective-tissue wrapping, the **perineurium.** Each bundle of nerve fibers surrounded by perineurium is called a **fasciculus.** Several fasciculi surrounded by a connective-tissue sheath called the **epineurium** constitute a single nerve. Blood vessels and lymphatic vessels travel within the connective-tissue sheaths to supply the neurons.

Nerves are found only in the peripheral nervous system, and they vary in size and composition. Nerves that contain only the processes of sensory neurons and carry nerve impulses to the CNS are **sensory nerves.** A few of the

Table 10.1   Comparison of Neurons, Nerve Fibers, and Nerves

Neuron	A nerve cell
Nerve Fiber	Any long process of a neuron. The term usually refers to axons, but also includes the peripheral processes of sensory neurons.
Nerve	A collection of nerve fibers in the peripheral nervous system.

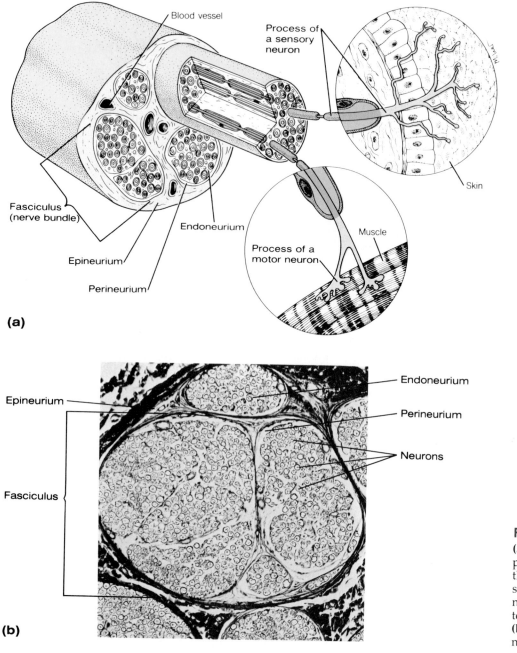

(a)

(b)

Figure 10.12
(a) Cross section of a peripheral nerve showing the connective-tissue sheaths that surround the neurons and bind them together to form a nerve. **(b)** Photomicrograph of a nerve.

cranial nerves are sensory nerves. Most nerves, however, are **mixed nerves**— that is, they contain the processes of both sensory and motor neurons. As a result of their composition, nerve impulses within mixed nerves travel both to and from the central nervous system.

To help you understand the terminology used for the main components of the nervous system, Table 10.1 provides a reference for distinguishing between a neuron, a nerve fiber, and a nerve.

Table 10.1

Specialized Peripheral Neuron Endings

The endings of neuronal processes in the peripheral regions of the body are generally specialized structures. They range from simple free nerve endings to complex encapsulated structures (Figure 10.13).

F10.13

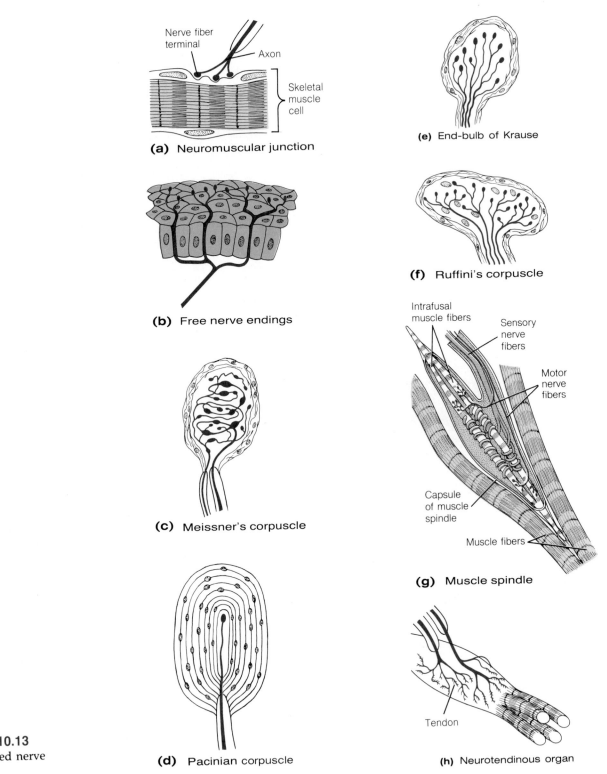

Figure 10.13
Specialized nerve endings.

(a) Neuromuscular junction

(b) Free nerve endings

(c) Meissner's corpuscle

(d) Pacinian corpuscle

(e) End-bulb of Krause

(f) Ruffini's corpuscle

(g) Muscle spindle

(h) Neurotendinous organ

Motor Neuron Endings

noo-ro-mus'-kyoo-lar

F10.13a

Motor neurons (**somatic motor neurons,** or **efferent neurons**) form a **neuromuscular** or **myoneural junction** with the skeletal muscles they supply (Figure 10.13a). The axon of the somatic motor neuron divides into many terminal branches that end in neuromuscular junctions at skeletal muscle cells. The myelin sheath of the axon does not extend to the terminal end of the process;

thus, the bare axon, which is covered on its outer surface with neurilemma, is able to come very close to the membrane of the muscle cell. The process by which a nerve impulse crosses a neuromuscular junction is described in Chapter 11.

Sensory Neuron Endings

The terminal portions of the peripheral processes of sensory neurons are dendrites. The terminal ends of the dendrites of many sensory neurons are sensitive to changes in their environment. For this reason, the dendritic endings of many sensory neurons function as **receptors.** In the skin, for example, there are modified dendrites that serve as receptors associated with the various cutaneous senses.

The cutaneous senses include those of pain, touch, pressure, heat, and cold. For many years, it has been believed that the body's ability to distinguish one cutaneous sense from another is due to the presence of structurally distinct and functionally specialized receptors—a different type for the reception of each type of sensation. There is now some doubt about the accuracy of this one-to-one structural-functional specificity. It may be that each anatomically distinct type of receptor in the skin is, in fact, sensitive to a variety of stimuli rather than to only one basic type of stimulus. According to this view, one type of receptor may have a somewhat different sensitivity to a variety of stimuli than another type. Although the following discussion indicates that each type of receptor may be responsible for sensing a specific kind of stimulus, bear in mind that this notion may require modification as additional information becomes available. Regardless of their precise function, however, the presence of structurally distinct types of receptors in the skin has been clearly established.

FREE NERVE ENDINGS   **Free nerve endings** are the least modified receptors. They consist of bare dendrites (Figure 10.13b). The free nerve endings   F10.13b
branch between epithelial cells, connective-tissue cells, muscle cells, cells of mucous membranes, and so forth. They are thought to serve primarily as *pain* receptors of the body, although free nerve endings that surround hair follicles are believed to be important *touch* receptors, and some free nerve endings serve as heat and cold receptors.

ENCAPSULATED SENSORY ENDINGS   The rest of the general sensory receptors are surrounded by connective-tissue capsules that contribute to the activity of the receptors in ways that are not completely understood. These encapsulated receptors are Meissner's corpuscles, Pacinian corpuscles, end bulbs of Krause, Ruffini's corpuscles, muscle spindles, and neurotendinous organs.

Meissner's corpuscles are small elliptical connective-tissue capsules that surround a spiraled ending of a dendrite (Figure 10.13c). These receptors are   F10.13c
thought to be particularly sensitive to *light touch* and are abundant in the dermal papillae of the skin (especially the fingertips), in the mucous membranes of the tongue, and in other sensitive regions of the body.

Located deeper than the Meissner's corpuscles, **Pacinian corpuscles** (Figure 10.13d) are found in the deeper layers of the skin, in the mesenteries, and   F10.13d
in loose connective tissue. They are composed of a dendrite and several concentric layers of connective tissue surrounding the dendrite. Because they are surrounded by a comparatively heavy capsule, Pacinian corpuscles are thought to be sensitive to heavier *pressure* but not to light touch.

End-bulbs of Krause (Figure 10.13e), which are quite common through-   F10.13e
out the body, are thought to serve as *cold* receptors. Oval capsules called
Ruffini's corpuscles (Figure 10.13f) are found primarily in subcutaneous tis-   F10.13f
sue and are believed to function as *heat* receptors.

Muscle spindles are complex capsules found in skeletal muscles (Figure   F10.13g
10.13g). Within the capsules are thin skeletal muscle fibers called *intrafusal fibers.* The intrafusal fibers are supplied by sensory neurons, some of which spiral around the fibers. When the muscle is stretched, the intrafusal fibers

are also stretched. This increases the number of nerve impulses carried back to the spinal cord by the neurons that surround the intrafusal muscle fibers. In response to the sensory nerve impulses produced by the stretching of the skeletal muscle, there is an increase in the frequency of motor nerve impulses to the same muscle. In this way, the stretch is reflexly resisted.

Neurotendinous organs *(Golgi tendon organs)* function in close association with the muscle spindles (Figure 10.13h). They are composed of dendrites that divide into many small branches in a tendon, near its junction with a muscle. The contraction of the muscle causes varying amounts of tension to be exerted on the tendon. The development of excessive tension by the muscle activates the neurotendinous organs. Sensory impulses from the organs are then carried back to the spinal cord, where motor neurons to the same muscle are inhibited, thus relaxing the muscle.

Types of Receptors

Receptors are structures that are generally activated by changes (stimuli) in either the internal or external environment of the body. As a result of the activity of the receptors, nerve impulses are initiated within sensory nerve cells. Receptors may be the endings of peripheral sensory neurons such as just described, or they may be specialized cells associated with such peripheral endings. Receptors can be classified in two ways: according to the *location* of stimulus or the *type* of stimulus.

Classification According to Location of Stimulus

One means of classification is according to the location of the stimuli to which the receptors respond. Thus, **exteroceptors** respond to stimuli from the body surface, including touch, pressure, pain, temperature, light, and sound. **Interoceptors (visceroceptors)** are sensitive to pressure, pain, and chemical changes in the internal environment of the body. Much of the information transmitted over interoceptors does not reach the level of consciousness. Therefore, a person normally is not aware of small changes in blood pressure, the amount of gases carried within the blood, or contractions of the smooth muscles of the organs. A type of interoceptor called a **proprioceptor** provides information concerning the position of body parts, without the necessity of visually observing the parts. Thus, for example, it is possible to know the position of the fingers even with the eyes closed. The muscle spindles and the neurotendinous organs are proprioceptors. There are also sensory receptors that monitor the degree of stretch in joint capsules. Since these stretch receptors provide information concerning the position of joints, they, too, serve as proprioceptors.

Classification According to Type of Stimulus

Receptors can also be classified according to the type of stimulus to which they are sensitive. **Mechanoreceptors** are sensitive to physical deformation, such as pressure or stretch. These receptors include Pacinian corpuscles, muscle spindles, neurotendinous organs, and perhaps Meissner's corpuscles. Receptors that respond to changes in temperature are known as **thermoreceptors.** Ruffini's corpuscles and the end-bulbs of Krause, as well as some free nerve endings, are thought to be thermoreceptors. **Chemoreceptors** are stimulated by various chemicals in food, in the air, or in the blood. Thus, the senses of taste and smell rely upon chemoreceptors. The receptors that monitor the pH and the levels of gases in the blood are also chemoreceptors. Specialized receptors that are sensitive to light energy are called **photoreceptors.** The only photoreceptors in the body are located in the retina of the eyes. These receptors, along with the chemoreceptors associated with the senses of taste and smell, are considered to be organs of special senses; they are discussed in Chapter 15.

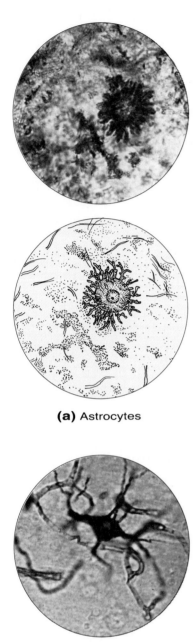

(a) Astrocytes

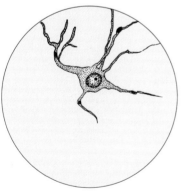

(b) Oligodendrocytes

Figure 10.14
Types of neuroglial cells.

Neuroglia

There are billions of neurons in the central nervous system, but there is an even greater number of supportive cells distributed among the neurons. These cells are called **neuroglia (glial cells).** Some neuroglia provide structural support for the neurons; others are involved in the transfer of nutrients from the blood vessels to the neurons and assist in the removal of waste products from the neurons. Still others serve as active phagocytes within the central nervous system. One type of neuroglial cell forms the myelin covering of axons. In contrast to neurons, neuroglia are capable of undergoing mitotic division. Included in the neuroglia are *astrocytes, oligodendrocytes, microglia,* and *ependymal cells.* With the exception of the microglia, all of the neuroglial cells are of ectodermal origin. Microglia originate from the mesoderm that forms connective tissues.

There are also two types of cells present in the peripheral nervous system that are often considered to be neuroglia. These are *satellite cells,* which are found in the capsule that surrounds the body of sensory neurons, and *Schwann cells,* which surround axons of peripheral nerves.

Astrocytes

Astrocytes are rather large cells with star-shaped bodies from which numerous processes radiate outward (Figure 10.14a). They are the most numerous of the neuroglia, and they provide most of the structural support for the CNS. Moreover, many of the processes of astrocytes contact blood vessels within the CNS. Because of this contact, the astrocytes are thought to be involved in the formation of the *blood–brain barrier,* which surrounds the capillaries of the CNS. In this role, astrocytes are thought to regulate the transport of substances between capillaries and neurons.

The concept of a blood–brain barrier is based on the observation that only water, oxygen, and carbon dioxide can readily enter or leave the capillaries of the CNS. All other substances that move across the wall of the capillaries of the central nervous system do so at a slower rate. This impediment to exchange in the brain is in contrast to other regions of the body, where many substances move rather freely and quickly across capillary walls. Apparently, the role of the blood–brain barrier is to prevent sudden and extreme fluctuations in the composition of the tissue fluid of the CNS, thus protecting the neurons, which are unreplaceable.

Oligodendrocytes

Oligodendrocytes (oligodendroglia) are smaller than astrocytes and have fewer processes (Figure 10.14b). Some of these processes wrap around axons within the central nervous system, forming a myelin sheath, much like the Schwann cells do in the peripheral nervous system.

Microglia

Microglia are the smallest type of neuroglial cells (Figure 10.14c). They are phagocytic and therefore function in the removal of dead tissue or foreign materials from the CNS. For this reason, microglia are considered to be a part of the macrophage system (see Chapter 4).

Ependymal Cells

Ependymal cells are neuroglial cells that line the ventricles (cavities) of the brain and the central canal of the spinal cord (Figure 10.14d). Some of these cuboidal epithelial cells are ciliated. The ependymal cells in the roofs of the ventricles of the brain are further modified, having microvilli extending from their free surfaces. These modified ependymal cells cover vascular networks called *choroid plexuses.* The choroid plexuses produce cerebrospinal fluid, which fills the ventricles.

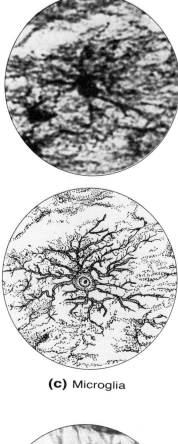

(c) Microglia

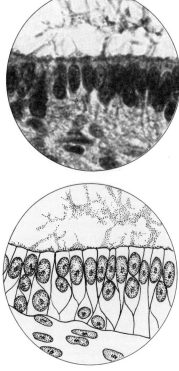

(d) Ependymal cells

Figure 10.14 *continued*
Types of neuroglial cells.

STUDY OUTLINE

ORGANIZATION OF THE NERVOUS SYSTEM pp. 297–300

CENTRAL NERVOUS SYSTEM (CNS) Brain and spinal cord.

PERIPHERAL NERVOUS SYSTEM (PNS) Spinal nerves, cranial nerves, ganglia, and receptors.

AFFERENT DIVISION Includes somatic and visceral sensory nerve cells.

EFFERENT DIVISION Includes somatic and visceral motor nerve cells—*somatic nervous system (voluntary), autonomic nervous system (involuntary).*

EMBRYONIC DEVELOPMENT OF THE NERVOUS SYSTEM pp. 300–302

ECTODERMAL CELLS THICKEN Form neural plate.

NEURAL GROOVE BECOMES NEURAL TUBE Develops into brain and spinal cord.

NEURAL CREST CELLS Form sensory nerve cells in cranial and spinal nerves, sympathetic motor nerves, and Schwann cells.

ANTERIOR END OF NEURAL TUBE Forms brain.

COMPONENTS OF THE NERVOUS SYSTEM pp. 302–311

NEURONS

STRUCTURE OF A NEURON
1. *Cell Body (Soma, Perikaryon)* And one or more processes containing cytoplasm.
2. *Large Nucleus* Prominent nucleolus.
3. *Chromatophilic Substance* Parallel layers of rough endoplasmic reticulum.
4. *Neurofibrils* Fibrils for cell support and microtubules for transport within cell.

LOCATIONS OF THE CELL BODIES OF NEURONS
1. *Nuclei* Cell body clusters in CNS.
2. *Center* Group of nuclei with same function.
3. *Ganglia* Groups of nerve cell bodies in PNS.

PROCESSES OF NEURONS
1. *Dendrites (Dendritic Zone)* Receptive portion of neuron where electrical signal originates.
2. *Axon* Conducting process of neuron along which transmission of electric signal occurs; one axon per neuron; axon branches called collaterals; an axon plus its covering is called a nerve fiber; lack chromatophilic substance.

FORMATION OF MYELINATED NEURONS
1. *Unmyelinated Fibers* Axons not covered with myelin; are covered by Schwann cells.
2. *Myelinated Fibers* Axons covered with myelin, a fatty substance that causes them to appear white; spiral wrappings of Schwann cells form myelin.

TYPES OF NEURONS
Classification According to Structure
1. *Bipolar neurons* Two processes; found mostly in the embryo.
2. *Unipolar neurons* Single process that arises from cell body.
3. *Multipolar neurons* One axon, several dendrites.

Classification According to Function
1. *Motor neurons (efferent)* Transmit from CNS to effector or to lower center in CNS; multipolar.
2. *Sensory neurons (afferent)* Transmit from receptor to CNS or to higher center in CNS; most unipolar, some bipolar.
3. *Internuncial neurons (interneurons)* Transmit from neuron to neuron; multipolar; only in CNS.

NERVES
1. Individual neuron (nerve cell) wrapped in connective-tissue endoneurium.
2. Neuron processes in bundles wrapped in perineurium; bundle termed a fasciculus.
3. Nerve is a group of fasciculi surrounded by connective-tissue epineurium.
4. Sensory nerves contain only sensory neurons; transmit *to* CNS.
5. Mixed nerves contain sensory and motor neuron processes; transmit *to* and *from* CNS.

SPECIALIZED PERIPHERAL NEURON ENDINGS

MOTOR NEURON ENDINGS
1. *Neuromuscular or Myoneural Junctions* Where terminal branches of axon meet but do not touch skeletal muscles.
2. *Myelin* Not present at end of axon.

SENSORY NEURON ENDINGS Dendrites function as receptors, of which there are histologically distinct types. Each type of receptor may or may not be responsible for sensing just one specific kind of stimulus.
1. *Free Nerve Endings* Bare dendrites; primarily pain receptors; also touch and temperature.
2. *Encapsulated Sensory Endings* Surrounded by connective-tissue capsules.
    Meissner's Corpuscles; Pacinian Corpuscles; End-Bulbs of Krause; Ruffini's Corpuscles; Muscle Spindles; Neurotendinous Organs

TYPES OF RECEPTORS

CLASSIFICATION ACCORDING TO LOCATION OF STIMULUS
1. *Exteroceptors* Respond to body-surface stimuli, such as touch, light.
2. *Interoceptors (Visceroceptors)* Respond to pressure, pain, and chemical changes in internal body environment.
3. *Proprioceptors* Respond to position of body parts, for example, muscle spindles and neurotendinous organs.

CLASSIFICATION ACCORDING TO TYPE OF STIMULUS
1. *Mechanoreceptors* Sensitive to physical deformations such as pressure or stretch; Pacinian corpuscles, muscle spindles, neurotendinous organs.
2. *Thermoreceptors* Sensitive to temperature changes; Ruffini's corpuscles, end-bulbs of Krause, and some free nerve endings.
3. *Chemoreceptors* Sensitive to chemicals; taste, smell, and pH.
4. *Photoreceptors* Sensitive to light; retina of the eyes.

NEUROGLIA (GLIAL CELLS) Structural support for neurons; nutrient transfer; phagocytosis; insulation of electrical activity; undergo mitotic divisions; all are ectodermal in origin, except for microglia, which are mesodermal. Satellite cells and Schwann cells considered to be peripheral neuroglia.

ASTROCYTES Large star-shaped bodies with many processes. CNS structural support; blood–brain barrier.

OLIGODENDROCYTES Form myelin sheath in CNS.

MICROGLIA Considered part of macrophage system; remove dead tissue and foreign materials from CNS.

EPENDYMAL CELLS Line ventricles of brain and central canal of spinal cord; some in roof of ventricles have microvilli.

SELF-QUIZ

1. There are two systems that serve primarily as means of internal communications—the nervous system and the endocrine system. True or False?

2. The central nervous system includes: (a) ganglia; (b) an autonomic division; (c) the spinal cord.

3. The peripheral nervous system includes: (a) the somatic nervous system; (b) the brain; (c) sensory input centers.

4. Impulses that produce contraction of skeletal muscles are carried by: (a) somatic motor nerves; (b) visceral motor nerves; (c) visceral sensory nerves.

5. Both the somatic and the autonomic divisions are under the control of centers located in the peripheral nervous system. True or False?

6. The brain develops from embryonic: (a) mesoderm; (b) endoderm; (c) ectoderm.

7. Nerve impulses in mixed nerves travel: (a) both to and from the central nervous system; (b) only from the CNS; (c) only to the CNS.

8. Mature neurons are generally able to undergo mitosis. True or False?

9. Clusters of nerve cell bodies in the central nervous system are generally called: (a) ganglia; (b) soma; (c) nuclei.

10. Match the following nervous system components with the appropriate lettered item.

Nuclei	(a) Groups of nerve cell bodies in the peripheral nervous system
Soma	
Center	
Ganglia	(b) The portion of the neuron in which an electrical impulse originates
Dendrite	
Axon	
Nerve fiber	(c) An axon together with certain sheaths
Sheath of Schwann	
Myelin	(d) Clusters of cell bodies of neurons in the CNS

    (e) A thin membrane between the myelin and the endoneurium of myelinated axons of the peripheral nervous system

    (f) The name of the cell body of each neuron

    (g) A fatty substance that covers most axons; formed by spiraling of Schwann cells

    (h) A group of nuclei whose neurons all have a specific function

    (i) A neuronal process that conducts a nerve impulse

11. Axon processes in the peripheral nervous system may have which of the following associated with them? (a) neurilemma; (b) nodes of Ranvier; (c) both.

12. Schwann cells are not found in the central nervous system. True or False?

13. The most common type of neurons are those called: (a) unipolar; (b) bipolar; (c) multipolar.

14. These efferent neurons transmit impulses away from the central nervous system to an effector: (a) motor; (b) sensory; (c) internuncial.

15. The terminal portions of the peripheral processes of sensory neurons are dendrites. True or False?

16. Match the following terms related to specialized nerve endings with the appropriate lettered item.

Motor neuron endings	(a) Modified dendrites that serve as receptors associated with cutaneous senses
Sensory neuron endings	
Pacinian corpuscles	(b) These receptors are stimulated by heavy pressure
End-bulbs of Krause	
Ruffini's corpuscles	(c) These are the least modified receptors, consisting of bare dendrites
Muscle spindles	
Neurotendinous organs	(d) Heat receptors
Free nerve endings	(e) Form a neuromuscular junction with the muscles
Meissner's corpuscles	(f) Cold receptors

    (g) These receptors are particularly sensitive to light touch

    (h) Complex capsules in skeletal muscles

    (i) Function in close association with muscle spindles

17. These receptors provide information concerning the position of body parts without an individual having to observe the parts: (a) exteroceptors; (b) proprioceptors; (c) mechanoreceptors.

18. Ruffini's corpuscles and the end-bulbs of Krause are thought to be: (a) thermoreceptors; (b) chemoreceptors; (c) photoreceptors.

LEARNING OBJECTIVES

After completing this chapter, you should be able to:

1. Describe the movements of potassium, sodium, and chloride ions across an unstimulated neuronal membrane.

2. Describe the role of active transport mechanisms in maintaining the resting membrane potential of a neuron.

3. Describe the movements of sodium and potassium ions during an action potential.

4. Explain the all-or-none nature of an action potential and a nerve impulse.

5. Cite two factors that influence the velocity at which an axon conducts a nerve impulse.

6. Discuss the events involved in the transfer of information from one neuron to another at a synapse.

7. Distinguish between an excitatory postsynaptic potential and an inhibitory postsynaptic potential.

8. Distinguish between temporal summation and spatial summation.

9. Describe the characteristics of a generator potential.

10. Describe the process of adaptation.

CHAPTER CONTENTS

BASIC ELECTRICAL CONCEPTS

MEMBRANE POTENTIALS

SYNAPSES

NEURAL INTEGRATION

RECEPTORS

EFFECTORS

KEY TERMS AND DERIVATIVES

saltatory (*saltator* = a dancer) *saltatory conduction* is a nerve impulse conduction in which the impulse seems to jump from node to node along the nerve fiber

synapse (*syn* = together; *apse* = to join) the region of communication between two neurons

Neurons, Synapses, and Receptors

11

Chapter 10 described the general organization of the nervous system and examined the structure of the basic functional component of the nervous system—the nerve cell, or neuron. This chapter focuses on the physiology of the neuron. It describes how a neuron transmits information from one point to another in the form of electrical signals, and it examines the ways in which one neuron communicates with another at specialized junctions known as *synapses*. It also considers the function of *receptors*, which are structures that convert information about conditions in the body's internal or external environments into neural signals.

BASIC ELECTRICAL CONCEPTS

As just mentioned, a neuron transmits information from one point to another in the form of electrical signals. Therefore, in order to understand the function of a neuron, it is necessary to understand the basic principles of electricity.

There are two types of electrical charge: positive and negative. The total amount of positive charge in the universe is believed to equal the total amount of negative charge; consequently the universe is electrically neutral. However, limited areas within the universe—and within the human body as well—can have more positive than negative charge, or vice versa. These areas are said to be either positively or negatively charged, depending on which charge predominates.

Like charges repel one another; that is, positive charge repels positive charge, and negative charge repels negative charge. However, an electrical force occurs between opposite—that is, positive and negative—charges that attracts them to one another and tends to draw them together. This force increases as the amount of charge increases and as the distance between the charges decreases.

Because of the attractive force between positive and negative charges, energy must be expended and work must be done to separate them. Conversely, if positive and negative charges are allowed to come together, energy is liberated, and this energy can be used to perform work. Thus, when positive and negative charges are separated from one another, they have the *potential* to perform work. The measure of this potential is known as **voltage,** and the units of measurement are volts or millivolts (one millivolt = 0.001 volt). Voltage is always measured between two points in a system, and it is often referred to as the potential difference, or potential, between the points.

The actual movement, or flow, of electric charge from one point to another is known as **current.** The amount of charge that moves between two points depends on the voltage and on the hindrance to the movement of charge offered by the material between the points. This hindrance is called **resistance.**

The relation between voltage (*E*), current (*I*), and resistance (*R*) is expressed by Ohm's law:

$$I = \frac{E}{R}$$

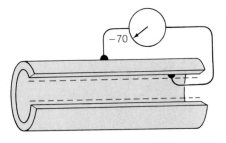

Figure 11.1
Like all body cells, an unstimulated neuron exists in an electrically polarized state. Unstimulated neurons exhibit voltage differences across their cell membranes of about −70 to −85 millivolts (the inside of the cell is negative relative to the outside).

Thus, when the resistance is constant, the current flow between two points increases when the voltage between the points increases; and, when the voltage is constant, the current flow decreases when the resistance increases.

Ions are electrically charged particles, and if positively charged ions are separated from negatively charged ions, a voltage develops. Moreover, the movement of ions from one point to another produces a current. Many ions are present within the body—particularly in the intracellular and extracellular fluids—and the distribution and movements of ions give rise to electrical phenomena that are important in the functioning of neurons and other cells.

MEMBRANE POTENTIALS

All body cells are electrically polarized, with the inside of the cells negative relative to the outside. This polarity can be measured as a difference in electrical potential, or voltage, between the inside and the outside of a cell, and this voltage is known as the **membrane potential.** Changes in the membrane potential convey important information to cells, and a change called an *action potential* is the basis of the transmission of nerve impulses by neurons.

Resting Membrane Potential

F11.1

The membrane potential of an unstimulated neuron, which is called the **resting membrane potential,** is about −70 to −85 millivolts (Figure 11.1). (The minus sign indicates that the inside of the neuron is negative relative to the outside.) The resting membrane potential is a result of differences in the ionic compositions of the intracellular and extracellular fluids (see, for example,

Table 11.1

Table 11.1). These differences are due to the characteristics and function of the plasma membrane of the neuron, which include the following:

1. The plasma membrane contains energy-utilizing, active transport mechanisms that "pump" certain ions into cells and other ions out of cells. For example, the neuronal membrane contains a *sodium–potassium pump* that transports sodium ions outward and potassium ions inward. As a result, the intracellular fluid of an unstimulated neuron contains a higher concentration of potassium ions and a lower concentration of sodium ions than the extracellular fluid.

2. The plasma membrane is not equally permeable to all ions. For example, the plasma membrane of a neuron contains a number of permanently open channels that are quite permeable to potassium ions but relatively impermeable to sodium ions. Consequently, the unstimulated neuronal membrane is 50 to 100 times more permeable to potassium ions than to sodium ions.

As a consequence of the differences in the ionic compositions of the intracellular and extracellular fluids that result from the characteristics and function of the plasma membrane, a very slight excess of positively charged ions accumulates immediately outside the membrane, and an equal number of negatively charged ions accumulates immediately inside the membrane. This separation of positive and negative charges across the membrane gives rise to the resting membrane potential. The development of this potential involves only a minute quantity of the total ions present, and it does not significantly affect the overall electrical neutrality of the intracellular or extracellular fluids.

Development of the Resting Membrane Potential

Of particular importance in establishing the resting membrane potential are the facts that: (1) the intracellular fluid of an unstimulated neuron contains a higher concentration of potassium ions and a lower concentration of sodium ions than the extracellular fluid, and (2) the unstimulated neuron is 50 to 100 times more permeable to potassium ions than to sodium ions. As a result, potassium ions can diffuse through the neuronal membrane from the intracellular fluid, where they are in high concentration, to the extracellular fluid, where they are in lower concentration, relatively easily. However, sodium ions have difficulty diffusing from the extracellular fluid, where they are in high concentration, to the intracellular fluid, where they are in lower concentration. Since both potassium ions and sodium ions are positively charged, the overall result is that, initially, a somewhat greater number of positively charged ions move out of the neuron than into it, leaving behind a slight excess of negatively charged ions (particularly, negatively charged proteins). This occurrence contributes to the separation of positive and negative charges across the neuronal membrane that gives rise to the resting membrane potential. However, like charges repel and unlike charges attract one another. Thus, as the resting membrane potential develops and the inside of the neuron becomes negative relative to the outside, an electrical force begins to oppose the outward movement of positively charged ions such as potassium and tends to attract positively charged ions into the neuron. Ultimately, a steady state is reached at which the forces tending to move potassium ions—and also sodium and other ions—out of the neuron are balanced by forces tending to move the ions into the neuron. In this state, which is the normal state of unstimulated neurons in the body, the neuron is polarized at its resting membrane potential.

Movement of Ions Across the Unstimulated Neuronal Membrane

The movements of potassium, sodium, and chloride ions across the plasma membrane of an unstimulated neuron that is in a steady state and polarized at its resting membrane potential are influenced by both concentration and electrical forces as well as by active transport mechanisms.

POTASSIUM IONS   The plasma membrane of an unstimulated neuron is relatively permeable to potassium ions, and a concentration force favors a net diffusion of potassium ions from the intracellular fluid, where they are in high concentration, to the extracellular fluid, where they are in lower concentration (Table 11.1). However, in an unstimulated neuron that is polarized at its resting membrane potential, the inside of the neuron is negative relative to the outside. Consequently, an electrical force opposes a net outward movement of positively charged ions such as potassium ions and tends to attract positively charged ions into the neuron. This electrical force is not sufficient by itself to prevent a net outward diffusion of potassium ions, but the concentration force tending to move potassium ions outward is balanced by a combination of this electrical force and the active transport of potassium ions from the extracellular fluid to the intracellular fluid by the sodium–potassium pump (Figure 11.2). Thus, equal numbers of potassium ions move across the membrane in each direction, and there is no net movement in either direction.

Table 11.1

F11.2

Table 11.1   Representative Concentrations of Selected Ions in the Extracellular Fluid and the Intracellular Fluid of an Unstimulated Neuron.*

Ion	Extracellular Fluid	Intracellular Fluid
Potassium (K$^+$)	5	150
Sodium (Na$^+$)	150	15
Chloride (Cl$^-$)	125	10

*Values are in milliequivalents per liter. Many other ions are also present in the extracellular and intracellular fluids, but potassium, sodium, and chloride ions are particularly important to neural function.

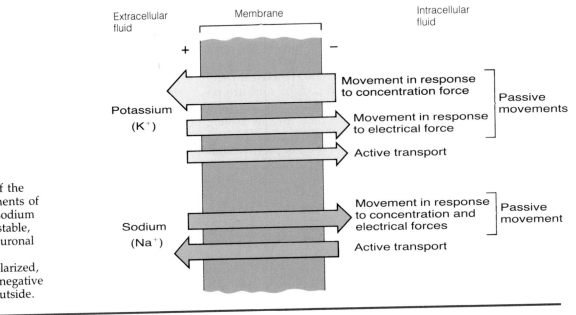

Figure 11.2
Diagrammatic representation of the balanced movements of potassium and sodium ions across the stable, unstimulated neuronal membrane. The membrane is polarized, with the inside negative relative to the outside.

Table 11.1

F11.2

SODIUM IONS In an unstimulated neuron that is polarized at its resting membrane potential, a concentration force favors a net diffusion of sodium ions from the extracellular fluid, where they are in high concentration, to the intracellular fluid, where they are in lower concentration (Table 11.1). In addition, the electrical force resulting from the polarity of the neuron also favors a net inward movement of positively charged sodium ions (Figure 11.2). Nevertheless, there is no net movement of sodium ions across the neuronal membrane because the membrane is relatively impermeable to sodium ions, and the sodium–potassium pump within the membrane actively transports any sodium ions that enter the neuron back into the extracellular fluid.

Table 11.1

CHLORIDE IONS The plasma membrane of an unstimulated neuron that is polarized at its resting membrane potential is permeable to chloride ions, and a concentration force favors a net diffusion of chloride ions from the extracellular fluid, where they are in high concentration, to the intracellular fluid, where they are in lower concentration (Table 11.1). This force is opposed by the electrical force due to the polarity of the neuron, which favors a net outward movement of negatively charged ions. Until recently it was thought that the concentration and electrical forces balanced one another, and chloride ions were passively distributed—that is, in electrochemical equilibrium—across the membrane. Now it is known that in many neurons the electrical force is not sufficient by itself to prevent a net entry of chloride ions. Consequently, an active transport mechanism called a chloride pump, which moves chloride ions outward, is believed to be important in preventing a net entry of chloride ions into at least some unstimulated neurons.

Role of Active Transport Mechanisms in Maintaining the Resting Membrane Potential

As previously indicated, the resting membrane potential is a result of differences in the ionic compositions of the extracellular and intracellular fluids, and these differences are maintained by active transport mechanisms. For example, if the sodium–potassium pump stops operating, a net movement of potassium ions out of the neuron occurs in response to a concentration force that is not balanced by an opposing electrical force, and a net movement of sodium ions into the neuron occurs in response to both concentration and electrical forces. These movements alter the ionic compositions of the extracellular and intracellular fluids, and this, in turn, alters the resting membrane potential.

Local Potential

The plasma membrane of an axon contains selectively permeable *voltage-sensitive channels,* which open and close in response to alterations in membrane polarity. When an axon is stimulated so that it undergoes some degree of depolarization—that is, its resting membrane potential changes toward zero— a number of voltage-sensitive channels that are selectively permeable to sodium ions open. Consequently, in the stimulated area the permeability of the membrane to sodium ions increases, and there is a net movement of sodium ions across the membrane from the extracellular fluid to the intracellular fluid. (Recall that both a concentration force and an electrical force favor this movement.) Since the axon is polarized, with the inside negative relative to the outside, and since sodium ions are positively charged, the net movement of sodium ions into the axon in the stimulated area tends to enhance the depolarization due to the stimulus. However, if the depolarization resulting from the stimulus is slight, the increased permeability of the membrane to sodium ions is only slight, and the net movement of sodium ions into the axon is quickly balanced by a net outward movement of potassium ions. (Note that the electrical force opposing the net outward movement of potassium ions in the unstimulated axon is diminished by the depolarization of the axon.) In this situation, only a transient local depolarization called a *local potential* occurs in the area of stimulation, and the axon returns to its unstimulated, fully polarized state when the stimulus is removed. The local potential is a graded potential—that is, its magnitude increases with increasing strengths of stimuli (Figure 11.3).

Action Potential

If a stimulus of sufficient intensity is applied to an axon, the local depolarization in the area of stimulation reaches a critical level called **threshold.** (Generally, threshold is reached when an axon depolarizes by 15 to 20 millivolts from its resting level.) Once threshold is reached, the axon continues to depolarize without further stimulation. The membrane potential quickly decreases to zero and then reverses so that the inside of the axon becomes positive relative to the outside (Figure 11.4). This rapid depolarization and polarity reversal is called an **action potential,** and it is unique to nerve and muscle cells. The next two sections describe how an action potential develops and subsides in one area of the plasma membrane of an axon. The section following these explains how an action potential travels along the membrane as a nerve impulse.

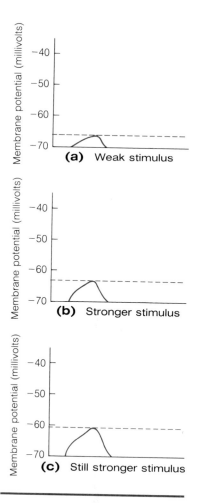

(a) Weak stimulus

(b) Stronger stimulus

(c) Still stronger stimulus

Figure 11.3
Record of the response of the axon of a nerve cell to locally applied stimulation. Note that the response is graded: With increased strengths of stimuli, greater local potentials occur. These potentials are transient, and the axon returns to its unstimulated state when the stimulus is removed.

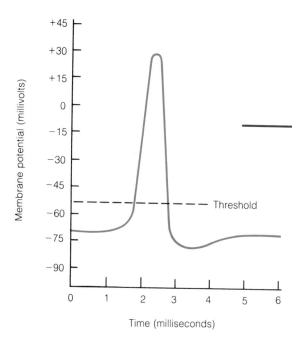

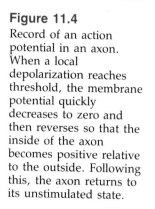

Figure 11.4
Record of an action potential in an axon. When a local depolarization reaches threshold, the membrane potential quickly decreases to zero and then reverses so that the inside of the axon becomes positive relative to the outside. Following this, the axon returns to its unstimulated state.

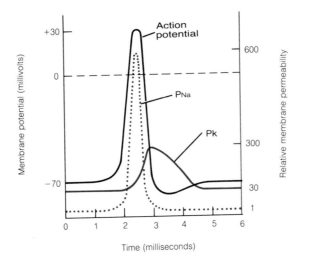

Figure 11.5
Change in the permeability of the plasma membrane of an axon to sodium ions (P_{Na}) that is associated with an action potential. Also shown is the change in permeability of the membrane to potassium ions (P_K) that occurs in unmyelinated axons.

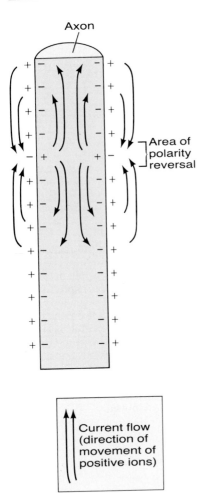

Figure 11.6
Current flow between an area of polarity reversal and adjacent areas of the axonal membrane of an unmyelinated axon during an action potential.

Depolarization and Polarity Reversal

As a local depolarization of an axon approaches threshold, more and more voltage-sensitive sodium channels open; and the plasma membrane in the depolarized area becomes increasingly permeable to sodium ions. Consequently, there is an increasing net movement of sodium ions into the axon. At threshold, the opening of voltage-sensitive sodium channels has increased the sodium permeability of the membrane to the point where the net movement of sodium ions into the axon is too great to be balanced by an equal net movement of potassium ions outward. As a result, the net inward movement of positively charged sodium ions increases the depolarization of the axon, which in turn causes more voltage-sensitive sodium channels to open. The opening of these channels further increases the permeability of the membrane to sodium ions and leads to a net movement of even more sodium ions into the axon. The net inward movement of more sodium ions further depolarizes the axon, causing still more voltage-sensitive sodium channels to open, and so on. Thus, when an axon is depolarized to threshold, a regenerative, positive-feedback cycle occurs that no longer requires the participation of the stimulus. The changing permeability of the membrane to sodium ions and the increased net movement of sodium ions into the axon during an action potential continue the depolarization of the axon, independent of the stimulus. As a result, the original polarity of the axon in the region of depolarization decreases to zero and then reverses so that the inside of the axon becomes positive relative to the outside.

Return to the Unstimulated Condition

The voltage-sensitive sodium channels remain open for only a brief interval, and the actual time of increased permeability to sodium ions at any one point on the axonal membrane during an action potential is quite short—on the order of a millisecond (Figure 11.5). Moreover, following the depolarization and polarity reversal due to the net movement of sodium ions into the axon, there is a significant net movement of positively charged potassium ions out of the axon. (Note that the axonal membrane is relatively permeable to potassium ions and that both concentration and electrical forces favor a net outward movement of potassium ions when the polarity reverses during an action potential.) The net outward movement of potassium ions causes the inside of the axon to become less positive and then more negative compared to the outside. The rapid decrease in membrane permeability to sodium ions and the net outward movement of potassium ions reestablish the original unstimulated polarity of the axon, with the inside being negative relative to the outside. In unmyelinated axons, but apparently to a much smaller degree or not at all in myelinated axons, the net outward movement of potassium ions is aided by the opening of voltage-sensitive channels that are selectively permeable to potassium ions. These channels open a short time after the voltage-sensitive sodium channels open, and they remain open somewhat longer than the voltage-sensitive sodium channels. While the voltage-sensitive potassium channels are open, the permeability of the axonal membrane to potassium ions is increased, and this increased permeability aids in the net movement of potassium ions out of the axon.

During the occurrence of an action potential and the return of the axon to its original unstimulated polarity, only relatively small numbers of sodium ions actually enter the axon, and only relatively small numbers of potassium ions actually leave. Considered alone, these ionic movements have essentially no effect on the sodium ion and potassium ion concentrations of the fluids within and around the axon (the intracellular and extracellular fluids). However, if thousands of action potentials were to occur, the sodium ion and potassium ion concentrations of these fluids could become altered to such a degree that no additional action potentials could be generated. Even though numerous action potentials can occur in an axon, the sodium ion and potassium ion concentrations of the intracellular and extracellular fluids are maintained at normal levels by the activity of the sodium–potassium pump, which moves sodium ions out of the axon and into the extracellular fluid and potassium ions into the axon and out of the extracellular fluids. Thus, this active transport system compensates for the movements of these ions that occur during action potentials and the return of the axon to its original, unstimulated polarity. Note, however, that the sodium–potassium pump is not directly involved in the occurrence of an action potential or in the return of the axon to its original, unstimulated polarity.

The Nerve Impulse

In an unmyelinated axon, when one area of the axonal membrane reverses its polarity during an action potential, the voltage difference between that area and the immediately adjacent area of the membrane leads to a local current flow between the areas (Figure 11.6). The current is carried by ions and, by convention, the direction of current flow is taken to be the direction of movement of positively charged ions. Along the inside of the membrane, positively charged ions are repelled from the area of polarity reversal and negatively charged ions are attracted toward the area. Along the outside of the membrane, positively charged ions are attracted toward the area of polarity reversal and negatively charged ions are repelled from the area. Consequently, current flows away from the area of polarity reversal along the inside of the membrane and toward the area of polarity reversal along the outside of the membrane. The current flow depolarizes the area of the membrane immediately adjacent to the area of polarity reversal, and this area becomes more permeable to sodium ions. When the depolarization of the area reaches threshold, an action potential occurs. A local current flow between this area and the next adjacent area of the membrane depolarizes the next adjacent area, and an action potential occurs there. This activity continues along the length of the membrane, producing a **propagated action potential,** or **nerve impulse,** that moves along the axon.

In a myelinated axon, a type of nerve-impulse conduction occurs that is called **saltatory conduction** (Figure 11.7). Myelin is a good insulator, with a high resistance to current flow, and in a myelinated axon the local current that gives rise to a propagated action potential flows from one node of Ranvier, where the myelin is interrupted, to the next. Thus, action potentials do not occur all along the axon, but only at the nodes of Ranvier. As a result, a nerve impulse in a myelinated axon "jumps" quickly along the axon from node to node. Because of this saltatory (or "jumping") conduction, a nerve impulse travels faster in a myelinated axon than it would if the axon was unmyelinated.

The saltatory conduction of nerve impulses in myelinated axons is energetically more efficient than the conduction of impulses in unmyelinated axons. In a myelinated axon, where action potentials occur only at the nodes

F11.7

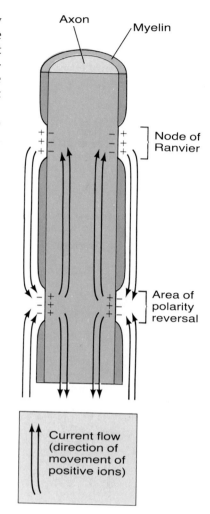

Figure 11.7
In a myelinated axon, the local current that gives rise to a propagated action potential flows from one node of Ranvier to the next.

of Ranvier, fewer sodium ions enter the axon and fewer potassium ions leave during the conduction of a nerve impulse than in an unmyelinated axon, where action potentials occur all along the axon. Consequently, less metabolic energy is required to restore the resting distribution of ions in a myelinated axon than in an unmyelinated axon.

Refractory Periods

The movements of sodium and potassium ions across any given region of an axonal membrane during an action potential require only a few milliseconds. During a short period that basically corresponds to the period of sodium permeability changes, no additional stimulus applied to a region of the membrane that is undergoing these changes can evoke another action potential, regardless of how strong that stimulus is. This period is called the **absolute refractory period,** and its length limits the number of action potentials—and therefore the number of nerve impulses—that can be produced in a given period of time. Intact nerve cells can generally produce action potentials at frequencies between 0 and 500 per second, although some cells are capable of much higher frequencies for brief periods of time.

During a brief period following the absolute refractory period, a stimulus stronger than that normally required to reach threshold may be able to initiate another action potential. This period is known as the **relative refractory period.**

All-Or-None Response

As is evident from the events involved in the generation of a propagated action potential, the conduction of a nerve impulse depends on the properties of the nerve cell membrane (for example, the membrane's ability to change its permeability to sodium ions). Once an axon is depolarized to threshold and a nerve impulse is triggered, the impulse travels along the axon with a magnitude that is characteristic of the particular neuron being stimulated, and independent of the strength of the initial stimulus. Thus, a nerve impulse is not a graded response, but instead has the same magnitude (under similar physiological conditions) whether it is triggered by a weak stimulus that just causes the axon to reach threshold or by a much stronger stimulus. This type of response is called an *all-or-none response.*

Direction of Nerve Impulse Conduction

If nerve impulses are triggered experimentally in the middle of an axon, they travel away from their site of origin in both directions toward the two ends of the axon. In the body, however, the initial action potentials and polarity reversals that give rise to nerve impulses normally occur at one end of an axon. Consequently, nerve impulses travel in one direction toward the other end of the axon. Note, however, that this unidirectional transmission of nerve impulses is due to their site of origin and not to any inability of an axon to conduct impulses in the opposite direction.

Conduction Velocities

Both the diameter of an axon and the presence or absence of myelin influence the velocity at which the axon conducts a nerve impulse. In general, the larger the diameter of an axon, the faster it will conduct a nerve impulse. Also, as previously indicated, a nerve impulse travels faster in a myelinated axon than it would if the axon was unmyelinated.

SYNAPSES

For information to be transmitted throughout the nervous system, not just one neuron but chains of them must be traversed. This activity requires a means of passing information from neuron to neuron as well as the ability to transmit information along the axon of a single neuron in the form of nerve impulses.

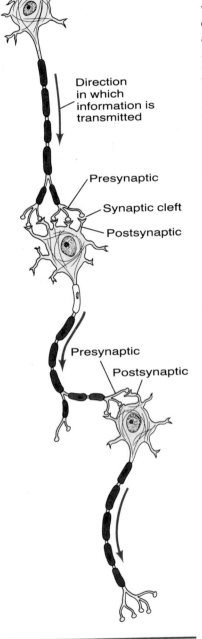

Figure 11.8
A single neuron can be a presynaptic neuron at one synapse and a postsynaptic neuron at another.

Direction in which information is transmitted

Presynaptic

Synaptic cleft

Postsynaptic

Presynaptic

Postsynaptic

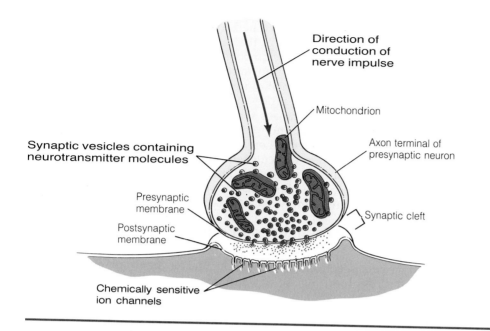

Direction of
conduction of
nerve impulse

Mitochondrion

Synaptic vesicles containing
neurotransmitter molecules

Axon terminal of
presynaptic neuron

Presynaptic
membrane

Postsynaptic
membrane

Synaptic cleft

Chemically sensitive
ion channels

Figure 11.9
A chemical synapse.
When a nerve impulse
arrives at an axon
terminal, chemical
neurotransmitter
molecules are released.
The molecules diffuse
across the synaptic cleft
and attach to receptors
on the membrane of the
postsynaptic neuron.
This attachment initiates
a series of events that
influence the activity of
the postsynaptic neuron.

Information is transferred from one neuron to another at neuronal junctions called **synapses.** A neuron that transmits information toward a synapse is called a **presynaptic neuron,** and a neuron that transmits information away from a synapse is called a **postsynaptic neuron.** A single neuron can be a presynaptic neuron at one synapse and a postsynaptic neuron at another (Figure 11.8).

sin'-ap-siz

F11.8

Near its end, an axon branches into numerous **axon terminals,** and most synapses occur between an axon terminal of a presynaptic neuron and a dendrite or cell body of a postsynaptic neuron. However, in some cases, synapses occur between two axons, two dendrites, or a dendrite and a cell body.

There are two types of synapses: electrical and chemical.

Electrical Synapses

At an **electrical synapse,** a presynaptic and postsynaptic neuron are connected by a gap junction (see Chapter 4). This connection allows local electrical currents resulting from action potentials in the presynaptic neuron to pass directly to the postsynaptic neuron and influence its activity.

Chemical Synapses

At a **chemical synapse,** an axon terminal of a presynaptic neuron closely approaches a dendrite or cell body of a postsynaptic neuron (Figure 11.9). The two neurons, however, remain separated by a small **synaptic cleft,** which prevents a nerve impulse in the presynaptic neuron from crossing directly to the postsynaptic neuron. Instead, when a nerve impulse arrives at an axon terminal of a presynaptic neuron, chemical **neurotransmitter** molecules are released from the terminal. The neurotransmitter molecules diffuse across the synaptic cleft and attach to receptors on the membrane of the postsynaptic neuron. This attachment initiates a series of events that influence the activity of the postsynaptic neuron.

F11.9

In the human nervous system, chemical synapses occur much more often than electrical synapses. The following sections deal with the function of chemical synapses.

Release of Neurotransmitters

A number of different substances serve as neurotransmitters, and these substances are manufactured by the neurons that release them. After their manufacture, neurotransmitter molecules are stored in small membrane-bounded sacs called **synaptic vesicles.** When a nerve impulse arrives at an axon terminal of a neuron, the permeability of the terminal to calcium ions increases, and there is a net movement of calcium ions into the cell. The entry of calcium ions triggers the release of neurotransmitter molecules from the axon terminal. The generally accepted method of neurotransmitter release is essentially a process of exocytosis. In this process, the calcium ions that enter the cell initiate a series of events that causes some synaptic vesicles to fuse with the membrane of the axon terminal and release their neurotransmitter molecules. In at least some cases, however, there is evidence that the entry of calcium ions causes neurotransmitter molecules present in the cytoplasm of axon terminals to be released directly across the terminal membrane. In such cases, neurotransmitter molecules within synaptic vesicles may be released into the cytoplasm to replenish the supply of cytoplasmic neurotransmitter molecules.

Actions of Neurotransmitters on Postsynaptic Neurons

Once they are released from a presynaptic neuron, neurotransmitter molecules diffuse across the synaptic cleft and attach to receptors on the membrane of a postsynaptic neuron. Depending on the effect this attachment has on the postsynaptic neuron, a chemical synapse can be either excitatory or inhibitory.

EXCITATORY SYNAPSES   At an excitatory synapse, the binding of neurotransmitter molecules to receptors on the membrane of the postsynaptic neuron increases the likelihood that the neuron will transmit a nerve impulse. In general, the combination of neurotransmitter molecules with receptors at an excitatory synapse produces changes in the permeability of the membrane of the postsynaptic neuron that lead to a depolarization of the neuron called an **excitatory postsynaptic potential (EPSP).** At many excitatory synapses, the neurotransmitter–receptor combination triggers the opening of specific *chemically sensitive channels* in the membrane of the postsynaptic neuron that increase the permeability of the membrane to potassium ions and sodium ions. This increased permeability results in a net outward movement of some potassium ions and a net inward movement of a greater number of sodium ions, which depolarizes the cell. At other excitatory synapses, permeability changes to ions such as calcium ions may contribute to the depolarization of the postsynaptic neuron and the production of an excitatory postsynaptic potential.

An excitatory postsynaptic potential is a graded potential that varies in magnitude with the degree of stimulation of the postsynaptic neuron by neurotransmitter molecules. If the stimulation is sufficiently intense, the axon of the postsynaptic neuron depolarizes to threshold, and a nerve impulse is triggered.

INHIBITORY SYNAPSES   At an inhibitory synapse, the binding of neurotransmitter molecules to receptors on the membrane of a postsynaptic neuron decreases the likelihood that the neuron will transmit a nerve impulse. In general, the combination of neurotransmitter molecules with receptors at an inhibitory synapse triggers the opening of chemically sensitive channels in the membrane of the postsynaptic neuron that increase the permeability of the membrane to potassium ions and/or chloride ions. This increased permeability usually results in a net outward movement of potassium ions and/or a net inward movement of chloride ions, which causes an increase in the resting polarity of the neuron called an **inhibitory postsynaptic potential (IPSP).** In some cases, however, the permeability change leads to a stabilization of the membrane potential without a hyperpolarization. In any event, a stronger-than-normal excitatory stimulus is required to trigger a nerve impulse in the postsynaptic neuron.

Synaptic Delay

A neuron can conduct a nerve impulse quite rapidly (speeds up to 250 miles per hour have been recorded), but the speed of transmission at chemical synapses is much slower. The time required to cross a chemical synapse is called the **synaptic delay.** This delay—which is approximately 0.5 to 1 millisecond—is primarily due to the time required for the release of neurotransmitter molecules from the presynaptic ending upon the arrival of a nerve impulse.

Removal of Neurotransmitters

In many cases, the attachment of neurotransmitter molecules to receptors on the membrane of a postsynaptic neuron causes permeability and postsynaptic potential changes that last only a few milliseconds. However, in some cases, the changes last hundreds of milliseconds and perhaps longer. Depending on the particular neurotransmitter substance and synapse involved, the neurotransmitter molecules themselves are ultimately removed from the region of the receptors by enzymatic inactivation, by diffusion away from the region, or by being taken up by the neuron that released them, by other neurons, or by neuroglial cells.

Kinds of Neurotransmitters

More than 30 different substances are known or suspected to serve as neurotransmitters. The following are some of the more common and widely accepted neurotransmitter substances.

ACETYLCHOLINE   Acetylcholine is a neurotransmitter released by neurons   *a-see-til-ko'-leen*
of the parasympathetic division of the autonomic nervous system at their junctions with effectors. Acetylcholine is also released by certain neurons in the brain and by motor neurons at their junctions with skeletal muscles (that is, at neuromuscular junctions).

Acetylcholine is synthesized in the neurons from which it is released. Within the neurons, molecules of acetylcoenzyme A react with choline molecules to form acetylcholine. This reaction is catalyzed by the enzyme choline acetyltransferase.

$$\text{Acetylcoenzyme A + choline} \xrightleftharpoons[]{\substack{\text{Choline}\\\text{acetyltransferase}}} \text{Acetylcholine + coenzyme A}$$

The acetylcholine released from an axon terminal of a neuron upon the arrival of a nerve impulse appears to be cytoplasmic acetylcholine. Acetylcholine present in synaptic vesicles is believed to be used to replenish the cytoplasmic supply. Following its release, acetylcholine is rapidly broken down into acetate and choline by the enzyme acetylcholinesterase, which is located on the outer surfaces of cell membranes as well as in the synaptic cleft. The choline can be taken up by the axon terminal and used in the synthesis of new acetylcholine.

NOREPINEPHRINE   The neurotransmitter **norepinephrine** is released by   *nor-ep-i-nef'-rin*
most neurons of the sympathetic division of the autonomic nervous system at their junctions with effectors. Norepinephrine is also released by a number of neurons in the brain, and it is thought to be a neurotransmitter in certain nerve pathways involved with the maintenance of arousal, in the brain system of reward, in dreaming sleep, and in the regulation of mood.

Norepinephrine is synthesized from the amino acid tyrosine (Figure   F11.10
11.10). Once it is released, norepinephrine is removed primarily by being taken up by the axon terminals that released it. Norepinephrine can also be inactivated by the enzyme catechol-O-methyltransferase (COMT), which is associated with the membranes of postsynaptic cells. However, enzymatic inactivation is not a major way that norepinephrine is removed from synaptic junctions. Norepinephrine taken up by axon terminals can eventually be released again. Also, some of it may be metabolized in a reaction sequence involving the enzyme monoamine oxidase (MAO), which is associated with mitochondria present in the axon terminals.

CLINICAL CORRELATION

Myasthenia Gravis

CASE REPORT . . . .

THE PATIENT   A 34-year-old woman

PRINCIPAL COMPLAINT   Muscle weakness

HISTORY   About two months before the current admission, the patient began to notice episodes of double vision, drooping of the eyelids, difficulty chewing and swallowing, and generalized weakness. The severity of the symptoms increased with activity, decreased with rest, and varied with time. About one month before the current admission, she experienced a particularly severe episode of muscular weakness, during which she was unable to walk. The patient was seen by a physician, who suspected that she had myasthenia gravis. She was given 0.5 mg of neostigmine methylsulfate intramuscularly, which inhibits acetylcholinesterase, thereby prolonging the action of the acetylcholine released at synapses and effector sites. Her condition temporarily improved. Another severe attack resulted in her present admission.

CLINICAL EXAMINATION   The patient was breathing with difficulty, and both eyelids were drooping. She had some loss of ability to move her eyes and turn her head, inability to close her mouth completely, and weakness of her arms and legs. Repeated attempts to open her eyes widely or to clench her teeth produced greater fatigue. Because of the fatigue of the respiratory muscles, an endotracheal tube was inserted and mechanical respiration was provided.

TREATMENT   Neostigmine methylsulfate was administered intravenously in increments of 0.125 mg/5 min until a satisfactory response was obtained (indicated by strength of hand-grip). The endotracheal tube was removed, and neostigmine bromide was given orally in increments of 7.5 mg until a satisfactory response was obtained. Pyridostigmine bromide, which also inhibits acetylcholinesterase and prolongs the action of the acetylcholine released at synapses and effector sites, in sustained-release tablets (6–8 hours duration) was used at night.

COMMENT   The muscular weakness of myasthenia gravis is caused by a decreased responsiveness of skeletal muscle cells to the acetylcholine released from motor neuron endings at neuromuscular junctions. The condition appears to be due to an abnormal response of the body's immune system that disrupts acetylcholine receptors on skeletal muscle cell membranes at the junctions. (Antireceptor antibodies have been identified in patients who have myasthenia gravis.) Exacerbations and remissions of the myasthenic condition occur frequently, sometimes in response to such variables as upper respiratory tract infection, loss of sleep, intake of alcohol, or menstruation.

Drugs that inhibit acetylcholinesterase benefit the patient by decreasing the rate of destruction of acetylcholine. Presumably, the inhibition of acetylcholine destruction results in levels of acetylcholine that are high enough to stimulate the skeletal muscle cells. The fluctuations in the myasthenic condition require that patients learn to modify the dosage schedule as their needs change.

OUTCOME   The patient was stabilized on neostigmine bromide by day and sustained-release pyridostigmine bromide at night. Her condition has not increased in severity, but the prognosis (probable outcome) with myasthenia gravis is uncertain.

---

ep-i-nef'-rin

EPINEPHRINE   The chemical substance **epinephrine** is very closely related to norepinephrine. It is present in certain neurons of the brain stem, and it is believed to be a neurotransmitter in some nerve pathways concerned with behavior, mood, and perhaps emotions.

F11.10

Like norepinephrine, epinephrine is synthesized from tyrosine (Figure 11.10). Following its release, epinephrine is removed in the same manner as norepinephrine.

DOPAMINE   **Dopamine** is a chemical substance found in the brain that is widely accepted as a neurotransmitter. It has been implicated in the regulation of emotional responses and in the control of complex movements.

F11.10

Dopamine is similar to norepinephrine in chemical structure, and it is synthesized from tyrosine by the same series of reactions that produce norepinephrine (Figure 11.10). Dopamine is also removed from synaptic junctions in the same manner as norepinephrine. Because of their chemical structures, norepinephrine, epinephrine, and dopamine are referred to as *catecholamines*.

kat-i-kol'-uh-meenz

SEROTONIN   **Serotonin** (5-hydroxytryptamine) is another chemical substance found in the brain that is widely considered to be a neurotransmitter. It

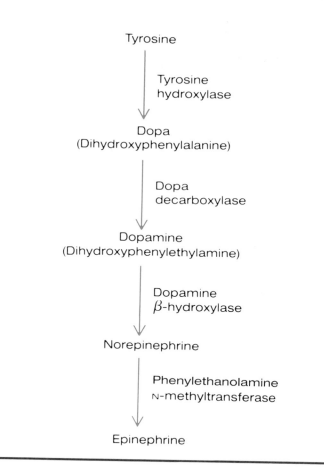

Tyrosine

Tyrosine
hydroxylase

Dopa
(Dihydroxyphenylalanine)

Dopa
decarboxylase

Dopamine
(Dihydroxyphenylethylamine)

Dopamine
β-hydroxylase

Norepinephrine

Phenylethanolamine
N-methyltransferase

Epinephrine

Figure 11.10
Reactions that synthesize dopamine, norepinephrine, and epinephrine from tyrosine.

is believed to play a role in temperature regulation, in sensory perception, and in the onset of sleep.

Serotonin is synthesized from the amino acid tryptophan. Following its release, serotonin can be removed by being taken up by axon terminals.

AMINO ACIDS   Several amino acids are also believed to function as neurotransmitters. Gamma-aminobutyric acid (GABA) is a common inhibitory transmitter in the brain. Glutamic acid and aspartic acid powerfully excite many neurons, and glycine has an inhibitory effect in the spinal cord. Once they are released, the centrally active amino acid neurotransmitters are removed by a combination of uptake and enzymatic inactivation.

Neuromodulators

Neuromodulators are chemical substances that alter neuronal activity by altering neurons directly or by influencing the effectiveness of neurotransmitters. For example, a neuromodulator may alter the synthesis or release of a neurotransmitter from a presynaptic neuron or it may increase or decrease the sensitivity of a postsynaptic neuron to a neurotransmitter. A number of hormones are believed to function as neuromodulators, and many substances currently believed to be neuromodulators also exert nonneural effects. Moreover, it is possible that a particular substance may act as a neurotransmitter in one region of the nervous system and as a neuromodulator in another.

Neuropeptides

A group of substances known collectively as **neuropeptides** are present in the nervous system. These substances are believed to function as neuromodulators or neurotransmitters. Neuropeptides are chains of amino acids, and essentially all neuropeptides discovered to date also have nonneural effects. In fact, many of them were originally identified in nonneural tissues. Table 11.2 lists some neuropeptides and describes some of their neural effects.

Table 11.2

Table 11.2 **Representative Neuropeptides**

Name	Neural Effects
Substance P	Stimulates spinal cord neurons that respond to painful stimuli; may be a neurotransmitter involved in the transmission of pain signals; may play a role in emotional behavior
Enkephalins	Appear to inhibit the release of sensory pain neurotransmitters such as substance P in spinal cord
Leu-enkephalin	May be involved in regulating emotional behavior
Met-enkephalin	May be involved in integrating sensory perception; may be principal mediator of analgesic effects of enkephalins
Neurotensin	May play a role in pain perception; has potent analgesic effect
Cholecystokinin	May be involved in pain-integrating functions; is powerful excitant of cerebral cell firing
Vasoactive intestinal polypeptide	May activate and synchronize neuronal activity of certain cerebral cortical cells; is a potent and rapid excitant of neurons in hippocampus of brain
Bradykinin	May be involved in blood pressure regulation and in pain perception

NEURAL INTEGRATION

The nervous system integrates information from many different sources in order to produce useful, coordinated responses by the body to a wide variety of conditions. At the cellular level, this integrative ability depends on such occurrences as divergence, convergence, summation, and facilitation.

Divergence and Convergence

F11.11 The axon of a presynaptic neuron may branch many times and, thus, may synapse with many postsynaptic neurons (Figure 11.11). This phenomenon, which is called **divergence,** permits a nerve impulse in a single presynaptic neuron to affect many postsynaptic neurons. Conversely, axon terminals of many different presynaptic neurons may all synapse with a single postsynaptic cell. This phenomenon is called **convergence.** Divergence and convergence allow for a wide variety of neuronal interactions and information transfers within the nervous system.

Summation

The postsynaptic potentials produced by the release of neurotransmitter molecules at chemical synapses can add together, or **summate,** to influence the activity of a postsynaptic neuron. In fact, summation is normally required to trigger a nerve impulse in a postsynaptic neuron because the excitatory postsynaptic potential (EPSP) produced by the neurotransmitter released in response to the arrival of a single nerve impulse at a single excitatory synapse is rarely, if ever, large enough to depolarize the axon of the postsynaptic neuron to threshold. There are two types of summation: temporal and spatial.

Temporal Summation

Temporal summation is the summation that occurs when many nerve impulses arrive at a single synapse within a short period of time. For example, the arrival of one nerve impulse at an excitatory synapse leads to the release of some neurotransmitter molecules, which produce a relatively small EPSP in the postsynaptic neuron. However, if a second nerve impulse arrives and a second EPSP is produced before the initial EPSP dies away, the two EPSPs

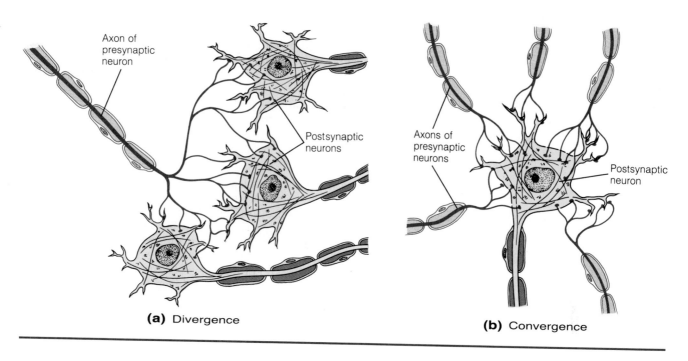

(a) Divergence

(b) Convergence

Figure 11.11
Divergence and
convergence.
(a) Neuronal processes of
a single cell diverge to
a number of other cells.
(b) Many neuronal
processes converge on a
single cell.

can add together, or summate. This summation produces a greater depolarization of the postsynaptic neuron than would result from either EPSP alone. If enough nerve impulses arrive at the synapse close enough together in time, the summation of the EPSPs that are produced can depolarize the axon of the postsynaptic neuron to threshold and trigger a nerve impulse in it.

Spatial Summation

Spatial summation is the summation that occurs when nerve impulses arrive very close together in time at a number of synapses between different presynaptic axon terminals and the same postsynaptic neuron. For example, as previously indicated, the EPSP produced by the neurotransmitter released in response to the arrival of a single nerve impulse at a single excitatory synapse is not normally very great. But, if nerve impulses arrive at many excitatory synapses at about the same time, the EPSPs produced at each synapse can summate and depolarize the axon of the postsynaptic neuron to threshold.

Facilitation

Quite often, as a result of spatial summation, the axon of a postsynaptic neuron is depolarized toward threshold by EPSPs produced at some excitatory synapses. Consequently, less depolarization is required for the axon to reach threshold when additional EPSPs occur. This phenomenon is called **facilitation** because the initial EPSPs that depolarize the axon toward threshold facilitate the attainment of threshold and the generation of a nerve impulse in the axon when additional EPSPs occur.

Determination of Postsynaptic Neuron Activity

As is the case for EPSPs, inhibitory postsynaptic potentials (IPSPs) produced by neurotransmitter molecules released at inhibitory synapses can also undergo temporal and spatial summation. Moreover, a postsynaptic neuron often receives thousands of presynaptic inputs, some of which may form excitatory synapses with the neuron and some of which may form inhibitory synapses.

Whether or not the axon of a postsynaptic neuron is depolarized to threshold depends on the relationship between the excitatory and inhibitory

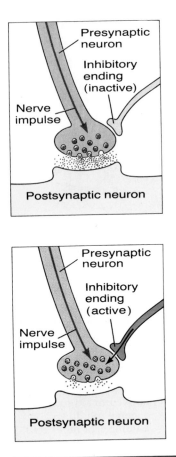

Figure 11.12
Presynaptic inhibition. An inhibitory neuronal ending makes synaptic contact with the ending of the presynaptic neuron at an excitatory synapse. When the inhibitory neuronal ending is activated, less neurotransmitter is released from the ending of the presynaptic neuron when a nerve impulse arrives than would otherwise be the case. As a consequence, the stimulation of the postsynaptic neuron is reduced.

events influencing it at any moment. If excitatory events are sufficiently dominant, the axon may reach threshold and transmit a nerve impulse. If inhibitory events predominate, the axon will be unlikely to transmit a nerve impulse. Thus, the activity of a postsynaptic neuron is determined by the integrated activities of many presynaptic inputs from various sources. Integrative processes of this sort, involving many neurons in multineuronal pathways, allow the nervous system to generate appropriate responses to many different circumstances.

Presynaptic Inhibition

In some cases, the amount of chemical neurotransmitter released when a nerve impulse arrives at an excitatory synapse is reduced by the process of **presynaptic inhibition.** In this process, an inhibitory neuronal ending makes synaptic contact with the ending of a presynaptic neuron at an excitatory synapse (Figure 11.12). When the inhibitory neuronal ending is activated, less neurotransmitter is released from the ending of the presynaptic neuron when a nerve impulse arrives than would otherwise be the case. As a consequence, the stimulation of the postsynaptic neuron is reduced. The occurrence of presynaptic inhibition provides the nervous system with a means of regulating the influence that nerve impulses in a particular presynaptic neuron will have on a postsynaptic neuron.

RECEPTORS

Receptors are structures that convert information about conditions in the body's internal or external environments into neural signals. Some receptors, such as those in the nose for smell, are specialized endings of neurons. Others, such as those of the ears for sound, are separate cells that are synaptically connected to neurons. Regardless of their nature, however, receptors act as *transducers*—that is, they convert various forms of energy (light, heat, pressure, sound, and so on) into the ionic-electrical phenomena that ultimately result in the generation of neural signals.

The environmental factors that activate receptors are called *stimuli*. However, not all receptors respond to all stimuli. The receptors of the eyes, for example, are normally activated by light but not by sound, and the receptors of the ears are usually activated by sound but not by light. The particular stimuli that activate a given receptor are referred to as *adequate stimuli* for that receptor.

Generator Potentials

When an adequate stimulus is applied to a receptor that is a specialized neuronal ending, the receptor responds with a general increase in permeability to all small ions that produces a depolarization called a **generator potential** (Figure 11.13). A generator potential is a graded potential. The stronger the stimulus applied to a receptor ending, the greater will be the magnitude of the generator potential. In addition, the rate of application or removal of a stimulus can affect the magnitude of a generator potential. Thus, a stimulus that rapidly builds or diminishes in intensity usually produces a greater generator potential than a stimulus that only slowly builds or diminishes in intensity. A generator potential generally lasts longer than the 1 or 2 milliseconds of an action potential, and it does not exhibit a refractory period. Consequently, if a second stimulus is applied before a generator potential resulting from an initial stimulus disappears, a second generator potential can add to the first, producing an even greater depolarization. Thus, a *summation* of generator potentials can occur. A generator potential of sufficient magnitude depolarizes the axon of the neuron to threshold and triggers a nerve impulse in the axon.

Receptor Potentials

When an adequate stimulus is applied to a receptor that is a separate cell synaptically connected to a neuron, the receptor responds with a depolarization that is called a **receptor potential.** A receptor potential is a graded potential that exhibits characteristics similar to those of a generator potential. The receptor potential causes the release of a chemical neurotransmitter from the receptor cell, and the neurotransmitter alters the permeability and polarity of the associated neuron. If the axon of the associated neuron is depolarized to threshold, a nerve impulse is triggered.

Discrimination of Differing Stimulus Intensities

The fact that a nerve impulse is conducted in an all-or-none fashion regardless of the intensity of the stimulus that triggers it raises some question about neural function. If the magnitude of a nerve impulse does not vary with the intensity of the stimulus, how does the nervous system transmit information about stimulus intensity? For example, how can a light touch on the hand be distinguished from a much stronger touch? One way is to have more neurons activated by strong stimuli than by weak stimuli. Another, less obvious, way is to have different frequencies of nerve impulses generated in a single neuron in response to varying intensities of stimuli. The frequency of nerve impulses in a neuron associated with a receptor is related to the magnitude of the generator or receptor potential produced by a stimulus. A strong stimulus produces a large generator or receptor potential, which, in turn, triggers a high frequency of nerve impulses in the neuron; and a weak stimulus produces a smaller generator or receptor potential, which, in turn, triggers a lower frequency of nerve impulses in the neuron. In this way, nerve impulse frequency provides the nervous system with a means of distinguishing stimulus intensity.

Adaptation

When a stimulus is continuously applied, the frequency of nerve impulses in a neuron may diminish with time, even though the intensity of the applied stimulus remains the same. This phenomenon is called **adaptation.** In some cases, adaptation is due to the fact that the responsiveness of the receptor membrane diminishes with time, causing the magnitude of the generator or receptor potential evoked by the stimulus to diminish also. Consequently, a lower frequency of nerve impulses is triggered in the neuron. Adaptation may explain the observation that many sensations—smell, for instance— diminish in intensity when stimuli are applied for long periods.

Some sensations—pain, for one—do not diminish in intensity when stimuli are applied for long periods. A pain sensation that does not diminish in intensity with time can be beneficial to a person because it may warn of a potentially harmful situation. Unfortunately, the person cannot always control the situation—or the pain.

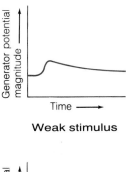

Weak stimulus

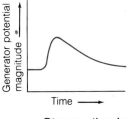

Strong stimulus

Figure 11.13
Adequate stimuli applied to neuronal endings that act as receptors produce generator potentials. Generator potentials are graded potentials—that is, the stronger the stimulus, the greater the generator potential produced by the stimulus.

EFFECTORS

Ultimately, neural signals are transmitted to structures called **effectors** to bring about responses to the various stimuli received by the nervous system. Effectors capable of responding to neural signals include muscle cells and the secretory cells of glands and organs. A neuron–effector junction is similar to a chemical synapse between two neurons, and information is transmitted from neurons to effectors by chemical neurotransmitters. When a nerve impulse arrives at the ending of a neuron at a neuron–effector junction, chemical neurotransmitter molecules are released. The neurotransmitter molecules diffuse to the effector cell and alter its activity. For example, at a *neuromuscular*

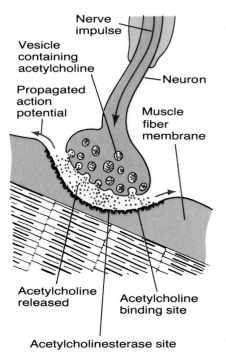

Figure 11.14

A neuromuscular
junction. When a nerve
impulse arrives at the
ending of the neuron,
acetylcholine is released.
The acetylcholine binds
with receptors on the
muscle cell membrane at
the junction. This
binding leads to a change
in the membrane's
permeability to sodium
and potassium ions and
produces a propagated
action potential that
travels along the muscle
cell membrane.

junction between a neuron and skeletal muscle cell, an ending of a neuron closely approaches the membrane of the muscle cell (Figure 11.14). When a nerve impulse arrives at the junction, the neurotransmitter *acetylcholine* is released from the neuronal ending. The acetylcholine diffuses to the plasma membrane of the muscle cell, where it attaches to acetylcholine receptors. This attachment leads to an increase in the permeability of the membrane to both sodium and potassium ions, and the muscle cell depolarizes in the region of the junction. This depolarization, which is called an *end-plate potential*, is sufficient to produce a propagated action potential that travels along the muscle cell membrane. The propagated action potential triggers the intracellular events that lead to the contraction of the muscle cell.

The effective combination of acetylcholine with the receptors of the muscle cell membrane at the neuromuscular junction lasts only a few milliseconds. Following its release from a neuronal ending at a neuromuscular junction, acetylcholine is rapidly inactivated by the enzyme *acetylcholinesterase*, which is located on the muscle cell membrane very close to the acetylcholine receptor sites. The inactivation of acetylcholine terminates its action on the muscle cell.

Effects of Aging on Neuromuscular Junctions

As a person ages, his or her physical strength generally declines. Recent evidence from animal experiments indicates that the reduced physical strength associated with aging may be due in part to changes that occur at neuromuscular junctions.

Accompanying the aging process is a reduction in the amount of the neurotransmitter acetylcholine present in nerve endings at neuromuscular junctions. This reduction is thought to be due to an increased leakage of acetylcholine from the nerve endings. The amount of acetylcholine released from a nerve ending in response to a nerve impulse does not change with age. However, because of the leakage, less acetylcholine is present in the nerve endings of older animals in comparison to younger animals. As a result, older animals have a tendency to run out of acetylcholine faster. Moreover, there are two known types of acetylcholine receptors on skeletal muscle cells, and many of the receptors on the muscle cells of older animals are of the less effective of the two types. The reduction of acetylcholine and a proportional increase of less effective receptors are thought to contribute to the decline of physical strength typical of aging.

STUDY OUTLINE

BASIC ELECTRICAL CONCEPTS pp. 315–316
1. Two types of electrical charge: positive and negative.
2. Like charges repel one another; unlike charges attract one another.
3. *Voltage:* A measure of the potential of separated positive and negative charges to perform work.
4. *Current:* Actual movement of electric charge from one point to another.
5. *Resistance:* Hindrance to the movement of charge.
6. *Ions:* Electrically charged particles; their distribution and movements in body give rise to electrical phe-

nomena important in functioning of neurons and other cells.

MEMBRANE POTENTIALS All body cells are
electrically polarized; voltage between inside and outside of cell is the *membrane potential*. pp. 316–322

RESTING MEMBRANE POTENTIAL Result of differences in ionic composition of intracellular and extracellular fluids due to characteristics and function of plasma membrane of neuron.
1. Plasma membrane contains active transport mechanisms that pump ions into and out of cells.
2. Plasma membrane is not equally permeable to all ions.

3. Differences in ionic compositions of intracellular and extracellular fluids result in accumulation of positively charged ions outside membrane and negatively charged ions inside membrane that gives rise to resting membrane potential.

DEVELOPMENT OF THE RESTING MEMBRANE POTENTIAL Of particular importance are the facts that

1. Intracellular fluid of unstimulated neuron contains higher concentration of potassium ions and lower concentration of sodium ions than extracellular fluid.
2. Unstimulated neuron is 50 to 100 times more permeable to potassium ions than to sodium ions.

MOVEMENT OF IONS ACROSS THE UNSTIMULATED NEURONAL MEMBRANE Movements of potassium, sodium, and chloride ions across plasma membrane of an unstimulated neuron that is in steady state and polarized at resting membrane potential are influenced by both concentration and electrical forces as well as by active transport mechanisms.

Potassium Ions Concentration force favoring net outward movement of potassium ions balanced by electrical force favoring net inward movement of positively charged ions and by inward pumping of potassium ions.

Sodium Ions Electrical and concentration forces favoring net inward movement of sodium ions balanced by outward pumping of sodium ions; membrane not very permeable to sodium ions.

Chloride Ions Concentration force favoring net inward movement of chloride ions balanced by electrical force favoring net outward movement of negatively charged ions and in at least some neurons by outward pumping of chloride ions.

ROLE OF ACTIVE TRANSPORT MECHANISMS IN MAINTAINING THE RESTING MEMBRANE POTENTIAL Active transport mechanisms maintain differences in ionic compositions of the extracellular and intracellular fluids that give rise to resting membrane potential.

LOCAL POTENTIAL Axon of nerve cell stimulated so it undergoes some degree of depolarization. Membrane becomes more permeable to positively charged sodium ions, and sodium ions enter cell. If depolarization resulting from stimulus is slight, increased sodium permeability is only slight and net inward movement of sodium ions is balanced by net outward movement of potassium ions. Cell returns to unstimulated, fully polarized state when stimulus removed. Local potential is graded potential—magnitude increases with increasing strengths of stimuli.

ACTION POTENTIAL When axon reaches threshold, membrane potential quickly decreases to zero, then reverses so inside of axon is positive relative to outside.

DEPOLARIZATION AND POLARITY REVERSAL Due to entry of sodium ions into axon through voltage-sensitive channels.

RETURN TO THE UNSTIMULATED CONDITION Due to decreased entry of sodium ions into axon

and movement of potassium ions out of axon. Sodium–potassium pump restores normal distribution of ions.

THE NERVE IMPULSE In unmyelinated axon, axonal membrane adjacent to area of initial action potential is depolarized and generates action potential. Action potentials continue to be generated along axon, producing propagated action potential, or nerve impulse. In myelinated axon, saltatory conduction occurs.

REFRACTORY PERIODS
1. **Absolute refractory period.** Time after action potential when no additional stimulus can evoke another action potential.
2. **Relative refractory period.** Period when stimulus greater than that normally required can initiate action potential.

ALL-OR-NONE RESPONSE Under similar physiological conditions, nerve impulse has same magnitude regardless of stimulus strength.

DIRECTION OF NERVE IMPULSE CONDUCTION In the body, nerve impulse normally originates at end of axon and is conducted in one direction toward other end of axon.

CONDUCTION VELOCITIES
1. In general, the larger the diameter of an axon, the faster it conducts a nerve impulse.
2. Nerve impulse travels faster in myelinated axon than it would if the axon was unmyelinated.

SYNAPSES Junctions between neurons at which information is transferred from one neuron to another. **pp. 322–327**

ELECTRICAL SYNAPSES Presynaptic and postsynaptic neurons connected by gap junction. Electrical currents in presynaptic neuron can pass directly to postsynaptic neuron.

CHEMICAL SYNAPSES Axon terminal of presynaptic neuron is separated from postsynaptic neuron by synaptic cleft. Axon terminal releases neurotransmitter molecules that diffuse across synaptic cleft and attach to receptors on postsynaptic neuron.

RELEASE OF NEUROTRANSMITTERS May be released from synaptic vesicles or directly from cytoplasm.

ACTIONS OF NEUROTRANSMITTERS ON POSTSYNAPTIC NEURONS Can be excitatory or inhibitory.

Excitatory Synapses The combining of neurotransmitter molecules with receptors on plasma membrane of postsynaptic neuron produces changes in permeability of membrane that lead to depolarization of neuron called an *excitatory postsynaptic potential.*

Inhibitory Synapses The combining of neurotransmitter molecules with receptors on plasma membrane of postsynaptic neuron produces changes in permeability of membrane that usually cause an increase in resting polarity of neuron called an *inhibitory postsynaptic potential.*

SYNAPTIC DELAY Time required to cross chemical synapse.

REMOVAL OF NEUROTRANSMITTERS Depending on particular neurotransmitter substance and synapse involved, neurotransmitter molecules are ultimately removed by enzymatic inactivation, by diffusion away, or by being taken up by neuron that released them, by other neurons, or by neuroglial cells.

KINDS OF NEUROTRANSMITTERS Over 30 different substances known or suspected of serving as neurotransmitters, including: *acetylcholine, norepinephrine, epinephrine, dopamine, serotonin,* and *amino acids.*

NEUROMODULATORS Chemical substances that alter neuronal activity by altering neurons directly or by influencing effectiveness of neurotransmitter.

NEUROPEPTIDES Chains of amino acids believed to function as neuromodulators or neurotransmitters.

NEURAL INTEGRATION Nervous system integrates information from many sources to produce useful, coordinated responses by the body to wide variety of conditions. pp. 328–330

DIVERGENCE AND CONVERGENCE
1. **Divergence:** Presynaptic neuron processes branch many times to synapse with many postsynaptic neurons.
2. **Convergence:** Processes of many presynaptic neurons synapse with single postsynaptic cell.

SUMMATION Postsynaptic potentials produced by release of neurotransmitter molecules at chemical synapses can add together, or summate, to influence activity of postsynaptic neuron.

TEMPORAL SUMMATION Occurs when many nerve impulses arrive at single synapse within short period of time.

SPATIAL SUMMATION Occurs when nerve impulses arrive close together in time at a number of synapses between different presynaptic axon terminals and same postsynaptic neuron.

FACILITATION Initial depolarization of axon of postsynaptic neuron by excitatory postsynaptic potentials produced at some excitatory synapses; facilitates attainment of threshold and generation of nerve impulse when additional excitatory postsynaptic potentials occur.

DETERMINATION OF POSTSYNAPTIC NEURON ACTIVITY Determined by integrated activities of many presynaptic inputs from various sources.

PRESYNAPTIC INHIBITION Can alter transmission of information from one neuron to another at synapses.

RECEPTORS Act as transducers that convert various forms of environmental energy into ionic-electrical phenomena that result in generation of neural signals. pp. 330–331

GENERATOR POTENTIALS Depolarization in response to adequate stimulation of neuronal ending that acts as receptor; a graded potential with longer duration than action potential and no refractory period.

RECEPTOR POTENTIALS Depolarization in response to adequate stimulus of separate receptor cell synaptically connected to neuron. Similar to generator potential, but receptor cell releases chemical transmitter.

DISCRIMINATION OF DIFFERING STIMULUS INTENSITIES Strong stimuli may activate more neurons than weak stimuli; different frequencies of nerve impulses in single neuron in response to varying intensities of stimuli.

ADAPTATION A phenomenon that occurs when stimulus is applied continuously; but frequency of nerve impulses diminishes with time even though intensity of applied stimulus is constant.

EFFECTORS Muscle cells and secretory cells of glands and organs. pp. 331–332

SELF-QUIZ

1. In an unstimulated neuron: (a) the inside is negative relative to the outside; (b) the inside is positive relative to the outside; (c) the sodium–potassium pump does not operate.

2. The resting, unstimulated neuron has: (a) a greater concentration of sodium ions inside than outside; (b) a greater concentration of potassium ions inside than outside; (c) an equal concentration of both sodium and potassium ions inside and outside.

3. In an unstimulated neuron, a concentration force favors the net movement of: (a) potassium ions into the neuron; (b) sodium ions out of the neuron; (c) chloride ions into the neuron.

4. In an unstimulated neuron, an electrical force favors the movement of: (a) potassium ions out of the neuron; (b) sodium ions into the neuron; (c) chloride ions into the neuron.

5. The membrane of an unstimulated neuron is: (a) essentially impermeable to chloride ions; (b) quite permeable to potassium ions; (c) extremely permeable to sodium ions.

6. During a local potential in an axon there is: (a) a net movement of potassium ions into the axon; (b) an increased permeability of the axonal membrane to sodium ions; (c) a net movement of sodium ions out of the axon.

7. When the axon of a neuron reaches threshold and an action potential is generated: (a) potassium ions rapidly enter the axon; (b) the membrane permeability to both sodium and potassium ions decreases substantially; (c) sodium ions rapidly enter the axon.

8. The actual time of increased permeability to the entrance of sodium ions at any one point on the axo-

nal membrane at the onset of an action potential is quite long and generally lasts more than 5 seconds. True or False?

9. Which of the following is an all-or-none response? (a) action potential; (b) excitatory postsynaptic potential; (c) generator potential.

10. Saltatory, or "jumping," conduction of nerve impulses occurs in: (a) large-diameter myelinated axons; (b) large-diameter unmyelinated axons; (c) small-diameter unmyelinated axons.

11. The axons of different neurons all transmit nerve impulses at the same velocity. True or False?

12. At a chemical synapse the plasma membranes of the presynaptic and postsynaptic neurons are: (a) fused with one another; (b) separated by a synaptic cleft; (c) connected by a gap junction.

13. At an excitatory synapse, the combination of neurotransmitter molecules with receptors: (a) depolarizes the postsynaptic neuron; (b) increases the resting polarity of the postsynaptic neuron; (c) stabilizes the polarity of the postsynaptic neuron at the normal resting membrane potential.

14. Acetylcholine is a chemical transmitter substance released by neurons at their junctions with skeletal muscle cells. True or False?

15. Substance P is believed to be involved in the transmission of neural signals associated with painful stimuli. True or False?

16. The postsynaptic potentials produced by the release of neurotransmitter molecules at chemical synapses can add together, or summate, to influence the activity of a postsynaptic neuron. True or False?

17. Generator potentials: (a) can summate; (b) have a long absolute refractory period; (c) generally are of shorter duration than action potentials.

18. Differing stimulus intensities may lead to different: (a) magnitudes of nerve impulses in a given neuron; (b) velocities of conduction of nerve impulses in a given neuron; (c) frequencies of nerve impulses in a given neuron.

19. The sensation of pain adapts very rapidly. True or False?

20. Muscle cells are effectors that are unable to respond to neural signals. True or False?

LEARNING OBJECTIVES

After completing this chapter, you should be able to:

1. Name the types of tracts in the brain's white matter, and state the function of each.

2. Describe the ventricles of the brain and the function and flow of cerebrospinal fluid.

3. Distinguish between the white and gray matter of the spinal cord in terms of structure and function.

4. Name and describe the ascending and descending tracts of the spinal cord.

5. Name and describe the meninges.

6. Describe the formation of a typical spinal nerve.

7. Distinguish between a stretch reflex and a tendon reflex.

8. Describe possible mechanisms involved in the sleep–wakefulness cycle.

9. Correlate the symptoms of several dysfunctions of the central nervous system with the region affected.

CHAPTER CONTENTS

THE BRAIN

THE SPINAL CORD

NEURON POOLS

INTEGRATIVE FUNCTIONS OF THE NERVOUS SYSTEM

CONDITIONS OF CLINICAL SIGNIFICANCE: THE CENTRAL NERVOUS SYSTEM

KEY TERMS AND DERIVATIVES

cerebral cortex (*cortex* = bark) the outer gray layer of the cortex

choroid plexus (*plex* = interweaving) a mass of specialized capillaries associated with ventricles in the brain

funiculus (*funi* = small cord or fiber) cordlike structures; anterior, posterior, and lateral divisions of white matter in the spinal cord

hypothalamus (*hypo* = under; *thalamus* = inner chamber) the region of the diencephalon forming the floor of the third ventricle of the brain

medulla oblongata (*medulla* = innermost part; *oblongata* = oblong) most inferior portion of the brain; controls heartbeat, respiration, blood pressure, etc.

meninges (*mening* = membrane) the membranes that cover the brain and spinal cord

pons (*pons* = bridge) any bridgelike structure or part; the structure connecting the cerebellum with the brain stem and providing linkage between upper and lower levels of the central nervous system

pyramidal tracts (*pyramidal* = relating to a pyramid; *tract* = pathway) also known as corticospinal tracts; these motor tracts descend from the cerebral cortex and cross in the pyramids of the medulla, extending into the spinal cord; they regulate the activity of skeletal muscles

thalamus (*thalamus* = inner chamber) a mass of gray matter that forms part of the lateral wall of the third ventricle of the brain; relays information between the spinal cord and the cerebrum

ventricle (*ventr* = hollowed part) the cavities in the interior of the brain

Table 12.1 Subdivisions of the Neural Tube and the Major Adult Structures Derived from Each

Primary Division	Subdivision	Adult Brain Structures	Neural Canal Region
Prosencephalon (forebrain)	Telencephalon	Cerebral hemispheres (cerebrum)	Lateral ventricles and upper portion of the third ventricle
		Cerebral cortex	
		Basal ganglia	
		Olfactory bulbs and tracts	
	Diencephalon	Epithalamus	Most of the third ventricle
		Thalamus	
		Hypothalamus	
Mesencephalon (midbrain)	Mesencephalon	Corpora quadrigemina	Cerebral aqueduct
		Cerebral peduncles	
Rhombencephalon (hindbrain)	Metencephalon	Cerebellum	Fourth ventricle
		Pons	
	Myelencephalon	Medulla oblongata	Part of the fourth ventricle
Spinal cord	Spinal cord	Spinal cord	Central canal

Box 12.1
CT Scan of the Human Brain

This is a CT scan of the central portion of the human brain. The thick white area indicates the skull, and the outer thin white area shows the scalp. Gray represents the brain. Symmetric darkened areas represent the ventricular system. The whitened areas within the brain indicate that there is some calcification of the blood vessels, which may result in blood vessel constriction. This is a scan of an older person who may have atherosclerosis, a vascular disease. If these whitish areas of calcified vessels appear in the brain, it is likely they are present throughout the entire body.

CT scans, which are primarily used to determine the presence of tumors, lesions, and vascular disease, reduce the need for exploratory surgery in certain situations. For instance, in the case of a head injury where there may be some concern regarding the possible presence of internal bleeding, a CT scan is an efficient, rapid, and low-risk way of ascertaining whether or not bleeding is present.

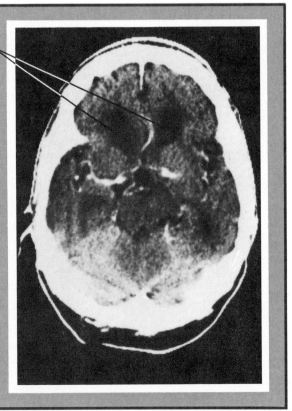

Ventricles

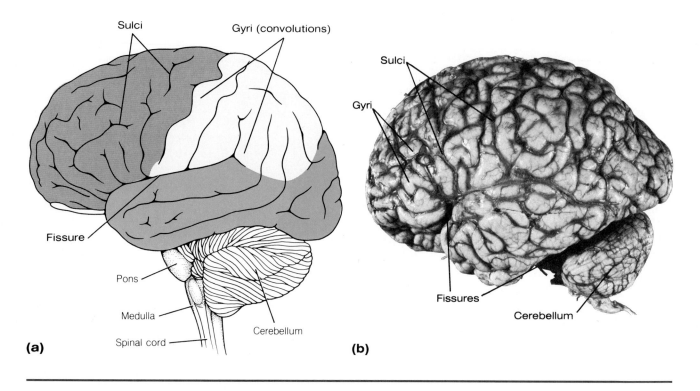

Figure 12.2
Left cerebral hemisphere.
(a) Lateral view.
(b) Photograph. The
surface of each cerebral
hemisphere has
numerous convolutions
separated by either sulci
or fissures.

recovery, the person appears to function normally. With special testing procedures, however, it is possible to show that following the cutting of the commissure, a task learned with one hand cannot be performed by the other hand unless the task is relearned with this hand. A person whose corpus callosum is intact can generally perform a task with either hand, although perhaps not with equal dexterity. Therefore, it appears that the corpus callosum makes possible the transfer of information between cerebral hemispheres; and if the commissure is severed, information learned by one cerebral hemisphere is unavailable to the other.

Gyri, Fissures, Sulci, and Lobes of the Cerebrum

The surface of the cerebrum has many rounded ridges called **convolutions,** or **gyri** (singular: *gyrus*) (Figure 12.2). Separating the gyri are furrows. The deeper furrows are called **fissures;** the shallower ones are **sulci** (singular: *sulcus*). The folding of the cortex that produces the gyri and sulci makes the surface area of the cerebral cortex much greater than it would be if the brain's surface were smooth. As it is, a significant percentage of the cerebral cortex is located in the fissures and sulci and is not visible from the surface.

jigh'-rus
F12.2
sul'-kus

The patterns of the gyri and fissures or sulci vary somewhat from one brain to another. Nevertheless, the locations of certain fissures and sulci are constant enough to serve as surface landmarks by which each hemisphere can be divided into *frontal, parietal, temporal,* and *occipital* lobes (Figure 12.3). Each lobe is located in the same general region as the correspondingly named skull bones.

F12.3

The **longitudinal fissure** (Figure 12.1) is a deep furrow that extends down to the corpus callosum in the central region of the cerebrum. It runs anteriorly and posteriorly, dividing the cerebrum into right and left hemispheres. Each hemisphere is further divided into a **frontal lobe** and a **parietal lobe** by the **central sulcus** *(fissure of Rolando)* (Figure 12.3), which runs at right angles to the longitudinal fissure. Two gyri run parallel to the central sulcus: The one anterior to the sulcus is the **precentral gyrus,** and the one posterior to the sulcus is the **postcentral gyrus.** The functional significance of these gyri is explained in the next section.

F12.1

F12.3

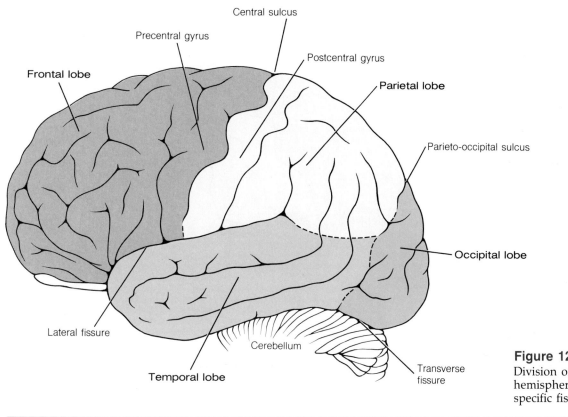

Central sulcus

Precentral gyrus

Postcentral gyrus

Frontal lobe

Parietal lobe

Parieto-occipital sulcus

Occipital lobe

Lateral fissure

Cerebellum

Transverse fissure

Temporal lobe

Figure 12.3
Division of a cerebral hemisphere into lobes by specific fissures and sulci.

The parietal lobe is separated posteriorly from the **occipital lobe** by an indistinct **parieto-occipital sulcus.** The **temporal lobes** extend forward along the lateral sides of the cerebral hemispheres. Each temporal lobe is separated from the lower portions of the frontal and parietal lobes by a deep **lateral fissure** *(fissure of Sylvius)* (Figure 12.3). A portion of the cerebral cortex called the **insula,** which is considered to be a fifth lobe of the cerebrum, is located deep within the lateral fissure (Figure 12.1). The insula is covered by portions of the frontal, parietal, and temporal lobes. The cerebrum is completely separated posteriorly from the cerebellum by a deep **transverse fissure** (Figure 12.3).

F12.3

F12.1

F12.3

Functional Areas of the Cerebral Cortex

On the basis of the effects of electrical stimulation of specific areas of the cerebral cortex in humans, from observing the clinical manifestations of brain disease or damage in humans, and from the results of detailed experiments on other mammals, it has been determined that certain areas of the cortex are related to specific functions. Some of these areas have been precisely mapped and numbered in a system called the *Brodmann classification,* but for our purposes it is sufficient to consider only the general locations of the major functional areas (Figure 12.4). Keep in mind, however, that Figure 12.4 represents an oversimplification of a very complex organ: No area of the brain functions alone. Because of extensive interconnections between various cortical areas by commissural and association fibers, any function attributed to a specific cortical area actually probably involves several cortical areas.

F12.4

PRIMARY MOTOR AREA   The **primary motor area** is located in the precentral gyrus of the frontal lobe just anterior to the central sulcus. Since the neurons in this gyrus control the conscious and precise voluntary contractions of skeletal muscles, this area is also referred to as the **primary somatic motor area.** The neurons of the primary motor area are distributed in an

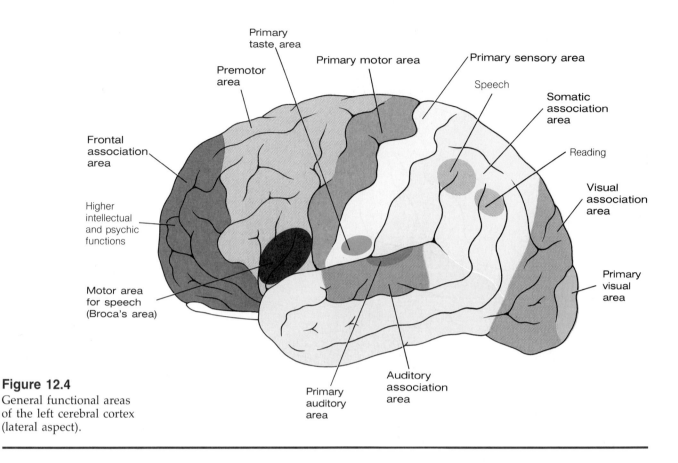

Figure 12.4
General functional areas of the left cerebral cortex (lateral aspect).

F12.5

organized manner. Neurons involved in controlling toe movements are located medially, deep in the longitudinal fissure; neurons involved in controlling the movements of other body parts are located in a regular but disproportionate sequence laterally along the gyrus (Figure 12.5). Originating in the precentral gyrus are descending motor nerve tracts called **pyramidal tracts** (so named because the cell bodies of the motor neurons whose processes form these tracts are pyramid-shaped).

PREMOTOR AREA   Located just anterior to the primary motor area is a region referred to as the **premotor area.** The neurons in the premotor area cause groups of muscles to contract in a specific sequence, thereby producing stereotyped movements. These repetitive movements are involved in learned activities such as playing a musical instrument and typing. The motor neurons that have their cell bodies located in the premotor area travel within **extrapyramidal tracts.**

At the lower margin of the premotor area is a motor area associated with the ability to speak. This region, called **Broca's area,** seems to be located only in the left cerebral hemisphere in most individuals, whereas the other functional areas that are identifiable are found in both hemispheres.

PRIMARY SENSORY AREA   Located just posterior to the central sulcus in the postcentral gyrus of the parietal lobe is the **primary sensory area (primary somatic sensory area).** Within this area are the terminations of pathways that transmit general sensory information concerning temperature, touch, pressure, pain, and proprioception from the body to the cerebral cortex. The pathways cross from one side of the nervous system to the other as they ascend to the cortex. Thus, the primary sensory area of the right cerebral hemisphere receives information from the left side of the body, and the primary sensory area of the left cerebral hemisphere receives information from the right side of the body. The neurons of the primary sensory area are organized sequentially

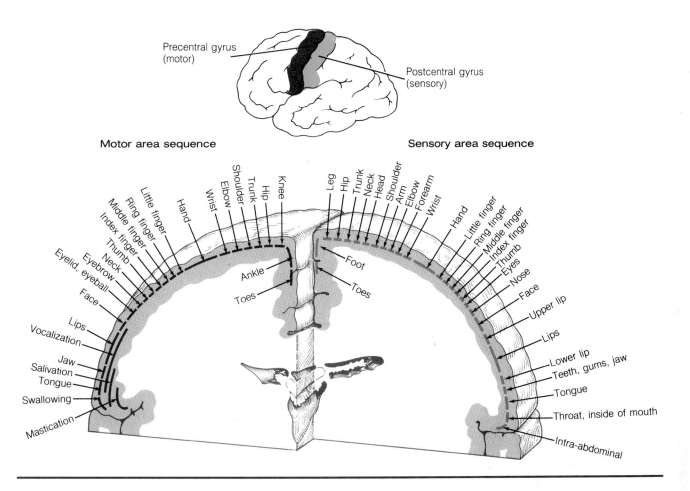

within the primary sensory area in a manner similar to the motor neurons of the primary motor area (Figure 12.5).

SPECIAL SENSES AREA   The **primary visual area** is in the posterior portion of the occipital lobe. Located along the upper margin of the temporal lobe is the **primary auditory area,** which receives nerve impulses associated with hearing. The area concerned with the sense of smell, the **primary olfactory area,** is located on the medial surface of the temporal lobe. The **primary taste area** is located in the parietal lobe, near the bottom of the postcentral gyrus.

When dealing with sensory areas, it is important to realize that only nerve impulses, not sensations, are transmitted from receptors to the brain. It is in the brain that nerve impulses are consciously interpreted as particular sensations (pain, touch, sound, taste, and so on). The brain then projects the sensations—usually with considerable accuracy—to the locations of the stimuli that are activating the receptors.

ASSOCIATION AREAS   Several **association areas** surround the sensory and motor areas of the cerebral cortex. These areas are involved in processing, integrating, and interpreting sensory information and in formulating patterns of motor responses.

The **frontal association area,** located anterior to the premotor area, is considered to be the site of origin of the higher intellectual activities characteristic of humans. These activities include foresight, the ability to make judgments, and the capacity to select behavior for a variety of circumstances.

The **somatic association area** is located on the parietal lobe, posterior to the primary sensory area. This integration and interpretation center makes it possible to determine an object's shape and texture without viewing it and provides information about the positional relationships of body parts.

Figure 12.5

Frontal section of the cerebrum. *Left half:* through the precentral gyrus, showing the locations of neurons within the cerebral cortex that control voluntary motor movement of specific structures. *Right half:* through the postcentral gyrus, showing the locations of regions of the cerebral cortex that receive sensory nerve impulses from specific body structures.

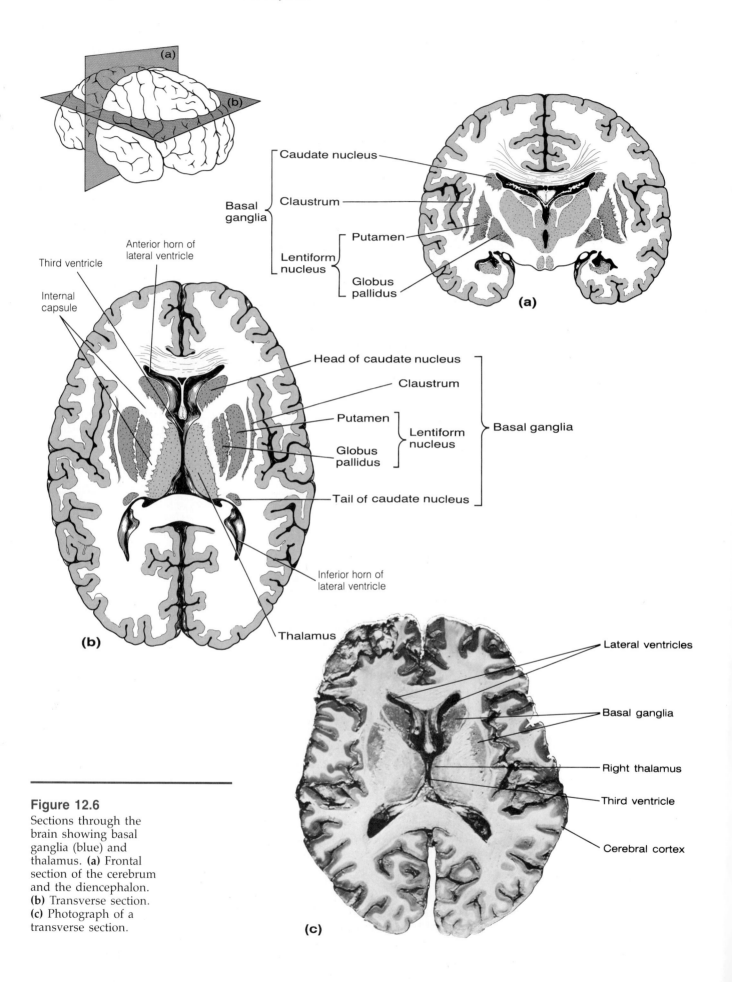

Figure 12.6
Sections through the brain showing basal ganglia (blue) and thalamus. **(a)** Frontal section of the cerebrum and the diencephalon. **(b)** Transverse section. **(c)** Photograph of a transverse section.

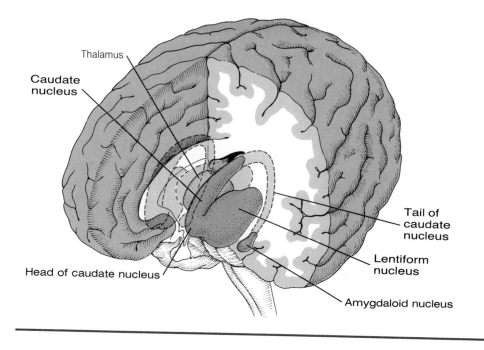

Thalamus

Caudate
nucleus

Tail of
caudate
nucleus

Lentiform
nucleus

Head of caudate nucleus

Amygdaloid nucleus

Figure 12.7
Three-dimensional
relationships among the
structures that comprise
the basal ganglia.

Located posterior to the somatic association area is a **visual association area,** and in the temporal lobe is an **auditory association area.** These areas contribute to the interpretation of visual and auditory experiences.

Basal Ganglia

Located deep within each cerebral hemisphere are several masses of gray matter known collectively as the **basal ganglia** *(basal nuclei)* (Figure 12.6). The ganglia, which are surrounded by white matter, are composed of groups of nerve cell bodies. Included in the basal ganglia are: the long arching **caudate nucleus;** the **amygdaloid nucleus,** which is located at the tip of the tail of the caudate nucleus; the **lentiform nucleus,** which is subdivided into the **puta-men** and the **globus pallidus;** and the **claustrum,** a thin layer of gray matter just deep to the cortex of the insula (Figure 12.7). The band of white matter located between the basal ganglia and the thalamus is called the **internal capsule.** The internal capsule is composed of projection fibers of the major motor and sensory tracts as they pass to and from the cerebral cortex. Because of their appearance, the caudate nucleus, the internal capsule, and the len-tiform nucleus are sometimes referred to as the **corpus striatum** ("striped body").

The basal ganglia, like the neurons of the primary motor area, are in-volved in controlling skeletal muscle activity. Because they are located outside the precentral gyrus, the basal ganglia are part of the *extrapyramidal system.* In other words, the somatic motor activities of the body are controlled both by the pyramidal neurons of the cerebral cortex and by motor neurons located elsewhere in the brain (extrapyramidal system), including the basal ganglia and the premotor area. In contrast to the pyramidal tracts, however, the neu-rons of the basal ganglia act in part to *inhibit* muscle contraction. This inhibi-tion, together with the stimulatory effects of the pyramidal system, provides a means by which muscular movements can be precisely controlled. Disorders of the basal ganglia result in involuntary contractions of skeletal muscles, such as the muscular rigidity and persistent tremors of the limbs associated with Parkinson's disease.

Olfactory Bulbs

On the ventral surface of each cerebral hemisphere is a small **olfactory bulb** and its associated **olfactory tract** (Figure 12.8). These structures, which are

F12.6

F12.7

F12.8

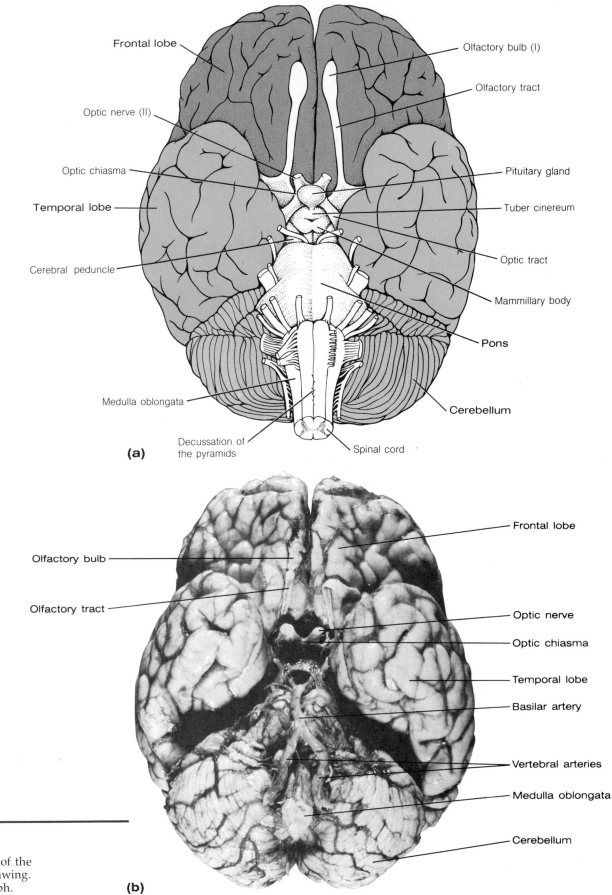

Figure 12.8
Ventral view of the brain. **(a)** Drawing. **(b)** Photograph.

associated with the sense of smell, are located in the portion of the brain called the *rhinencephalon.* The neurons of the **olfactory nerve (cranial nerve I)** pass from the nasal mucosa, through the cribriform plate of the ethmoid bone, and into the olfactory bulb, where they synapse with neurons of the olfactory tract. The neurons of the tracts pass to the olfactory area of the cortex on the medial surface of the temporal lobe. In addition to its olfactory function, the rhinencephalon is thought to be involved, in some obscure manner, with certain emotional and behavioral responses. Those portions of the rhinencephalon that are not involved with olfaction are considered to be part of the *limbic system,* which we will consider shortly.

rye-nen-sef'-a-lon

Hemispheric Specialization

Each cerebral hemisphere is somewhat specialized for carrying out certain kinds of mental processes. For example, language ability tends to be localized in the left cerebral hemisphere. In general, the left hemisphere is concerned with verbal and sequential processes and behaviors such as writing business letters and solving simple equations. This hemisphere excels at performing rational, linear, verbally oriented tasks, and it seems to process information in a fragmentary or analytical way. In contrast, the right cerebral hemisphere is primarily concerned with the recognition of complex visual patterns or with mentally picturing objects in three-dimensional space. The right hemisphere is also more involved in the expression and recognition of emotion and in certain musical or artistic abilities, such as identifying a theme in an unfamiliar piece of music. The information processing in the right hemisphere tends to be holistic and unitary rather than fragmentary and analytical.

The ability to recognize faces is well localized within the brain without being a function of primarily one cerebral hemisphere or the other. Damage to the medial undersides of both occipital lobes and the ventromedial surfaces of the temporal lobes produces a failure to recognize a person by sight. People with brain damage in these areas can usually correctly name objects but cannot identify faces. Such individuals may even fail to recognize their parents, spouses, or children by sight. Nevertheless, when a familiar but unrecognized person speaks, the person with brain damage can immediately recognize the other person's voice and can then name him or her.

Diencephalon

The second subdivision of the forebrain is the **diencephalon.** Since the cerebral hemispheres extend downward and almost completely surround the diencephalon, it is not visible from the exterior of the brain, except for a portion that can be seen when the brain is viewed ventrally. The **third ventricle** forms a midplane cavity within the diencephalon (Figure 12.9). The most important parts of the diencephalon are the *thalamus, hypothalamus,* and *epithalamus.*

F12.9

Thalamus

The **thalamus** consists of two oval masses of nerve-cell bodies (gray matter) that form the lateral walls of the third ventricle (Figure 12.6). A small bridge of commissural fibers called the **intermediate mass** (*massa intermedia*) passes across the third ventricle and connects the two thalamic masses. Each thalamic mass is deeply embedded in a cerebral hemisphere and is bounded laterally by the internal capsule. The thalamus contains over 20 functionally separate nuclei (cell-body masses). Functionally, the thalamus acts as a major sensory relay and integrating center of the brain. Except for the tracts associated with olfaction, all sensory fiber tracts that transmit nerve impulses from receptors to the cerebral cortex synapse within one of the thalamic nuclei. From the thalamus, nerve impulses are relayed to specific regions of the cerebral cortex as well as to subcortical areas such as the basal ganglia and hypothalamus. Apart from its sensory role, the thalamus is also involved with some of the motor tracts that leave the cerebral cortex.

thal'-a-muss
F12.6

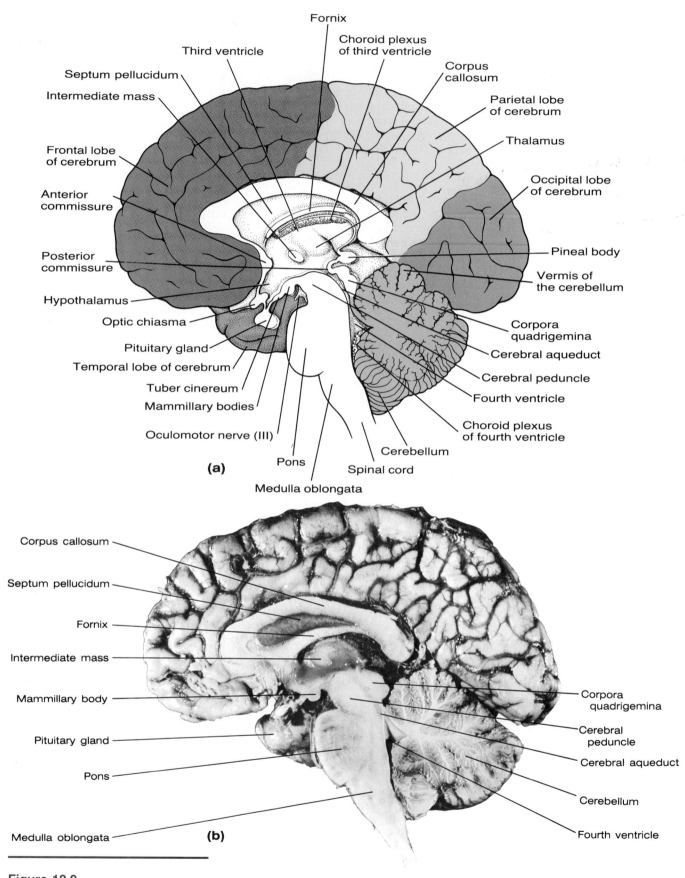

Figure 12.9
Midsagittal section of the
brain and brain stem.
(a) Drawing. (b) Photograph.

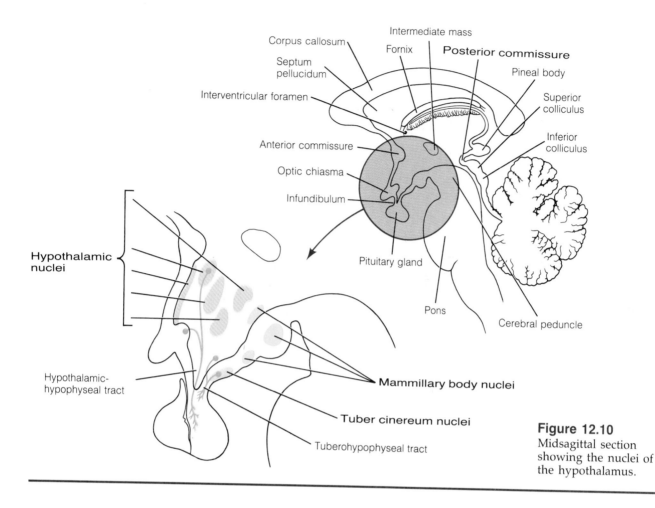

Figure 12.10
Midsagittal section showing the nuclei of the hypothalamus.

Hypothalamus

As the name indicates, the **hypothalamus** lies below the thalamus, where it forms part of the walls and floor of the third ventricle. Like the thalamus, the hypothalamus is composed of several nuclei, each of which is involved with specific functions (Figure 12.10). Several hypothalamic structures are visible externally, including the *mammillary bodies, tuber cinereum, infundibulum,* and the *optic chiasma* (or *chiasm*).

F12.10

The two **mammillary bodies** are small, round nuclear masses that form external bulges from the undersurface of the brain posterior to the infundibulum. The mammillary bodies function as relay stations for olfactory neurons and are involved in olfactory reflexes.

Located just anterior to the mammillary bodies is the **tuber cinereum,** which contains neurons that transport regulatory hormones (or factors) from the hypothalamus to the infundibulum. These neurons of the tuber cinereum form the *tuberohypophyseal tract.* In the infundibulum, the hormones from the tuberohypophyseal tract enter blood vessels that transport them to the adenohypophysis of the pituitary gland (Chapter 16).

Extending downward from the tuber cinereum is the stalklike **infundibulum.** Nerve fibers from some of the hypothalamic nuclei pass through the infundibulum on their way to the pars nervosa of the posterior lobe of the pituitary gland. These neurons, which form the *hypothalamic-hypophyseal tract,* transport hormones (oxytocin and ADH) synthesized in hypothalamic nuclei to the posterior pituitary, from which they are released.

in-fun-dib'-yoo-lum

Anterior to the infundibulum is the **optic chiasma,** which is formed by the decussation (crossing) of some of the neurons in the optic nerves.

The hypothalamus controls many vital processes, most of them associated with the autonomic nervous system. Some of the hypothalamic nuclei

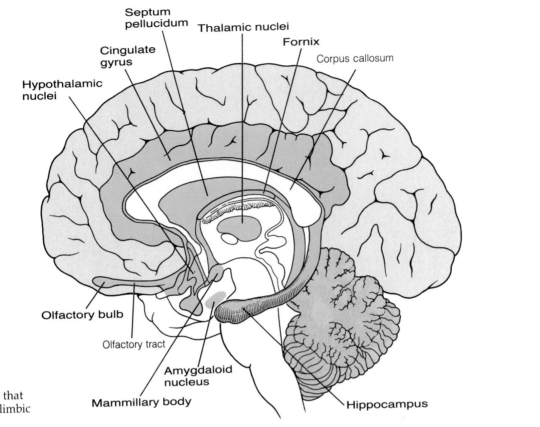

Figure 12.11

The structures that constitute the limbic system.

have been shown experimentally to regulate sympathetic activity; others control parasympathetic functions. The hypothalamus is involved in regulating body temperature, water balance, appetite, gastrointestinal activity, sexual activity, and even emotions such as fear and rage. The hypothalamus also regulates the release of the hormones of the pituitary gland; and thus it greatly affects the endocrine system (see Chapter 16).

Epithalamus

The **epithalamus,** the most dorsal portion of the diencephalon, forms a thin roof over the third ventricle. The roof has a vascular choroid plexus located on its internal surface. A small mass called the **pineal body (epiphysis)** extends outward from the posterior end of the epithalamus. The possible neuroendocrine function of the pineal body is discussed in Chapter 16. The **posterior commissure** is located just ventral to the pineal body.

The Limbic System

Although emotions are influenced by the hypothalamus, emotional responses involve a complex interaction of structures in several different regions of the brain, including the cerebrum and the diencephalon. A group of structures collectively referred to as the **limbic system** are particularly important in emotional responses (Figure 12.11). The limbic system includes the **olfactory bulbs;** the **septum pellucidum;** a band of fibers called the **fornix,** which passes from beneath the corpus callosum to the mammillary bodies of the hypothalamus; a cerebral gyrus called the **cingulate gyrus,** located just above the corpus callosum; parts of the **basal ganglia,** including the **amygdaloid nucleus;** the **hippocampus,** which is a part of the cerebrum located in the floor of the lateral ventricle close to the amygdaloid nucleus; the **mammillary bodies;** and various **thalamic** and **hypothalamic nuclei.**

In animal experiments it is possible to electrically stimulate various centers within the limbic system. When certain centers are stimulated, the animal

F12.11

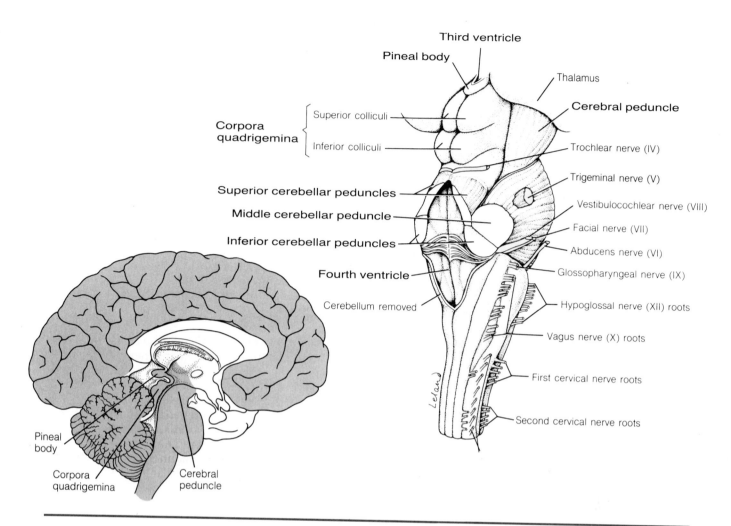

Third ventricle

Pineal body

Thalamus

Cerebral peduncle

Corpora quadrigemina
{
Superior colliculi

Inferior colliculi
}

Trochlear nerve (IV)

Trigeminal nerve (V)

Superior cerebellar peduncles

Middle cerebellar peduncle

Inferior cerebellar peduncles

Vestibulocochlear nerve (VIII)

Facial nerve (VII)

Abducens nerve (VI)

Glossopharyngeal nerve (IX)

Fourth ventricle

Cerebellum removed

Hypoglossal nerve (XII) roots

Vagus nerve (X) roots

First cervical nerve roots

Second cervical nerve roots

Pineal body

Corpora quadrigemina

Cerebral peduncle

Figure 12.12
Dorsolateral view of the brain stem. The cerebellum has been removed.

responds to the stimulation as if it were pleasant ("pleasure centers"). When other centers are stimulated, the animal responds to the stimulation as if it were unpleasant ("punishment centers"). A wide variety of emotional behavior patterns can be produced by stimulating or removing specific regions of the limbic system. It is assumed therefore that the structures of the limbic system play an important role in regulating emotional behavior.

Mesencephalon

The **mesencephalon (midbrain)** is a short, constricted region between the forebrain and the hindbrain. Within the mesencephalon is a small tunnel called the **cerebral aqueduct** (*aqueduct of Sylvius*) (Figure 12.14), which connects the third ventricle (located in the diencephalon) with the fourth ventricles (located in the metencephalon).

F12.14

Cerebral Peduncles

On the ventral surface of the mesencephalon are two cylindrical bulges called the **cerebral peduncles** (Figure 12.12). The peduncles are composed of motor nerve fibers that travel from the primary motor area of the cerebral cortex to the pons and spinal cord and sensory nerve fibers that travel from the spinal cord to the thalamus. The **oculomotor nerves (cranial nerve III)** emerge between the peduncles. Deeper within the mesencephalon, between the peduncles and cerebral aqueduct, is a small island of gray matter called the **red nucleus.** The cell bodies of neurons that compose the *rubrospinal tract* (discussed later in the chapter) are located in the red nucleus. The red nucleus serves as a relay station that coordinates impulses between the cerebellum and the cerebral hemispheres, thereby contributing to the coordination of movements and to the sense of balance.

F12.12

Corpora Quadrigemina

F12.12

The dorsal surface of the mesencephalon, which forms the roof of the cerebral aqueduct, consists of four rounded prominences called the **corpora quadrigemina** ("four twin bodies") (Figure 12.12). The upper pair of prominences are called the **superior colliculi.** Some neurons of the optic tracts from the retina of the eyes travel to the superior colliculi, where they participate in activities concerned with certain reflex responses to visual stimuli. The lower pair of prominences, the **inferior colliculi,** serve as relay stations and reflex centers for auditory stimuli. The **trochlear nerves (cranial nerve IV)** emerge from the roof of the mesencephalon just below the inferior colliculi.

Metencephalon

The major structures of the **metencephalon,** which is the most superior portion of the hindbrain, are the *cerebellum* and the *pons.* The cerebral aqueduct of the mesencephalon expands into the **fourth ventricle** in the metencephalon. The inferior portion of the fourth ventricle extends into the myelencephalon. As is true of all the ventricles of the brain, there is a vascular choroid plexus in the thin membrane that forms the roof of the fourth ventricle.

Cerebellum

ser-a-bel'-lum
F12.8, F12.9

Projecting from the dorsal surface of the metencephalon, the **cerebellum** (Figures 12.8 and 12.9) is separated from the cerebral hemispheres by a strong membrane called the *tentorium cerebelli.* The tentorium lies within the transverse fissure of the brain and supports the occipital lobes of the cerebrum, thus minimizing the pressure that the lobes exert on the cerebellum.

The cerebellum is composed of two lateral **cerebellar hemispheres** connected in the midline by a structure called the **vermis.** The surface of the cerebellum consists of a thin cortex of gray matter. The cortex dips deeply below the apparent surface of the cerebellum in a manner similar to the fissures and sulci of the cerebrum, although the cerebellar fissures are more parallel, giving the appearance of a series of flattened plates. The ridges between the fissures are called **folia.**

F12.12

The cerebellum is connected to the mesencephalon by a pair of nerve-fiber tracts called the **superior cerebellar peduncles;** to the pons by a pair of **middle cerebellar peduncles;** and to the medulla oblongata by a pair of **inferior cerebellar peduncles** (Figure 12.12). The superior cerebellar peduncles are composed principally of efferent nerve fibers from the cerebellum; the middle and inferior cerebellar peduncles are composed mostly of afferent nerve fibers that transmit impulses from the pons and the medulla oblongata to the cerebellum. These extensive interconnections with other regions of the central nervous system provide the cerebellum with widespread input and output capabilities.

The cerebellum coordinates the activities of the skeletal muscles through sensory information carried to it from receptors for proprioception, equilibrium, and balance. Moreover, the cerebellum receives some sensory information concerning touch, vision, and sound. Further coordination occurs by way of motor impulses sent from the cerebellum to higher brain centers. In particular, nerve impulses from the cerebellum may dictate specific movement sequences to the primary motor area of the cerebral cortex, which then initiates the nerve impulses necessary to carry out the movements. A person whose cerebellum has been damaged experiences muscular weakness, a loss of muscle tone, and uncoordinated movements. All the functions with which the cerebellum is concerned remain below the level of consciousness. Thus, the cerebellum is able to mediate certain responses without having them reach the conscious level.

Pons

F12.8, F12.9

The **pons** ("bridge"), which is located on the ventral surface of the metencephalon, consists of bands of nerve-fiber tracts and several nuclei (Figures 12.8 and 12.9). The tracts in the pons are both transverse and longitudinal.

The transverse tracts consist of neurons that enter the cerebellar hemispheres through the middle cerebellar peduncles. The longitudinal tracts are composed of neurons that travel between the brain stem (see the next section) and the cerebrum. The pons, therefore, functions primarily to connect the cerebellum with the cerebrum and the brain stem, thus providing connections between upper and lower levels of the central nervous system. The cerebral cortex, in particular, achieves most of its connections to the cerebellum by way of nerve-fiber tracts that pass through the pons. In addition, the stimulation of nuclei within the pons affects the rate of respiration. The nuclei of the **trigeminal (V)**, **abducens (VI)**, **facial (VII)**, and **vestibulocochlear (VIII)** cranial nerves are located in the pons.

Myelencephalon

The **myelencephalon**, the most inferior division of the brain, is also known as the **medulla oblongata**. The medulla, the pons, and the mesencephalon together form the **brain stem**. At its lower end, the medulla is continuous with the spinal cord. The cavity in the medulla forms the lower portion of the fourth ventricle and continues into the spinal cord as the central canal of the cord.

On the ventral surface of the medulla are two large columns of nerve-fiber tracts called the **pyramids.** The pyramids contain the same motor tracts found in the cerebral peduncles. Therefore, the tracts in the pyramids carry the voluntary motor output from the primary motor area of the cerebral cortex. The tracts of the pyramids originate from cell bodies in the precentral gyrus of the cerebral cortex and continue into the spinal cord as the *corticospinal tracts* (discussed later in this chapter). Some of the nerve tracts in the pyramids cross from one pyramid to another. This crossing, called the *decussation of the pyramids,* is visible on the ventral surface of the medulla in the groove that separates the pyramids (Figure 12.8). As a consequence of the decussation of these nerve tracts, motor areas located on one side of the cerebral cortex can control muscular movements on the opposite side of the body.

The medulla oblongata also contains nuclei that give rise to the last four cranial nerves: the **glossopharyngeal (IX)**, the **vagus (X)**, the cranial portion of the **accessory (XI)**, and the **hypoglossal (XII)**.

Located in the medulla are *medullary centers,* which are groups of neurons that are involved in the control of a variety of vital functions, such as heart rate, respiration, dilation and constriction of blood vessels, coughing, swallowing, and vomiting.

Reticular Formation

Inside the medulla is a region of gray matter containing a network of interlacing nerve fibers called the **reticular formation.** The reticular formation extends throughout the brain stem and up into the diencephalon. In addition to receiving nerve impulses from the cerebellum, from the basal ganglia, and from various other nuclei in the brain, the reticular formation also receives input from all the sensory tracts as they ascend through the medulla. Selected impulses passing through the reticular formation are relayed to the cerebral cortex, thus activating it (Figure 12.13). Because it exerts this control over the cerebral cortex, the reticular formation is considered to be an activating or arousal system that is essential in maintaining wakefulness and alertness. For this reason it is also referred to as the **reticular activating system (RAS).** Injury or diseases that affect the reticular activating system often produce coma.

Ventricles of the Brain

The **ventricles** of the brain (Figure 12.14) develop from expansions of the lumen of the embryonic neural tube and form a continuous fluid-filled system in the brain. The roof of each ventricle is thin and contains no neurons. Each roof does, however, have a network of capillaries called a **choroid plexus** associated with it. These plexuses, together with the ependymal cells that

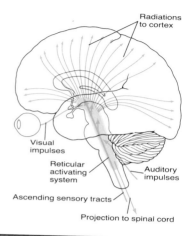

Figure 12.13
Schematic representation of the reticular formation. The arrows indicate input to and output from the reticular activating system.

F12.8

F12.13

F12.14

ko'-roid

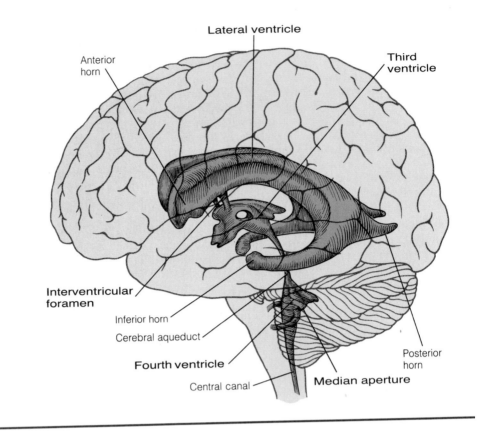

Figure 12.14
Ventricles of the brain viewed as if they could be seen from the surface of the brain.

cover them, are the production sites of **cerebrospinal fluid.** The fluid fills the ventricles of the brain, the central canal of the spinal cord, and the subarachnoid space, which surrounds the brain and spinal cord (Figure 12.15).

F12.15

If air is injected into the ventricles, they become distinguishable on an X ray. This procedure is used to detect brain damage or the presence of tumors that distort the normal outlines of the ventricles.

Lateral Ventricles

Within each cerebral hemisphere is a **lateral ventricle** that has its major portion located in the parietal lobe. Extensions from this portion protrude into the frontal lobe *(anterior horn)*, the occipital lobe *(posterior horn)*, and the temporal lobe *(inferior horn)*. The lateral ventricles are separated from each other medially by a thin vertical partition called the **septum pellucidum** (Figure 12.9). Each lateral ventricle communicates with the third ventricle by a small opening called the **interventricular foramen** *(foramen of Monroe)*.

F12.9

Third Ventricle

The **third ventricle** is a narrow midline chamber in the diencephalon. The right and left masses of the thalamus form most of its lateral walls. A commissure called the *intermediate mass* passes through the ventricle. The third ventricle opens into the fourth ventricle by means of the **cerebral aqueduct** of the mesencephalon.

Fourth Ventricle

The **fourth ventricle** is a pyramidal cavity located in the hindbrain just ventral to the cerebellum. There are two openings in the lateral walls of the fourth ventricle that are called the **lateral apertures** *(foramina of Luschka)* (Figure 12.15). In the roof is a single opening, the **median aperture** *(foramen of Magendie)*. The ventricles communicate through these three openings with a *subarachnoid space* that surrounds the brain and spinal cord. Inferiorly, the fourth ventricle is continuous with the narrow *central canal* that extends the length of the spinal cord.

F12.15

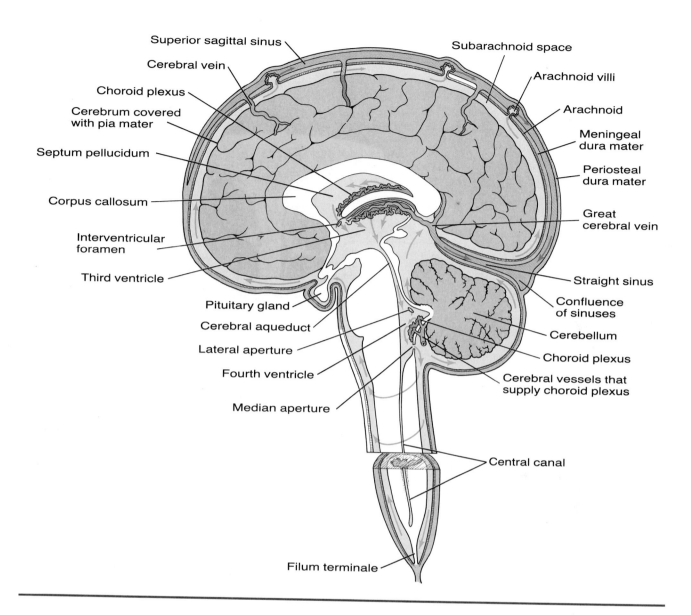

Superior sagittal sinus
Cerebral vein
Choroid plexus
Cerebrum covered with pia mater
Septum pellucidum
Corpus callosum
Interventricular foramen
Third ventricle
Pituitary gland
Cerebral aqueduct
Lateral aperture
Fourth ventricle
Median aperture
Filum terminale

Subarachnoid space
Arachnoid villi
Arachnoid
Meningeal dura mater
Periosteal dura mater
Great cerebral vein
Straight sinus
Confluence of sinuses
Cerebellum
Choroid plexus
Cerebral vessels that supply choroid plexus
Central canal

Figure 12.15
Location of the cerebrospinal fluid (blue) that surrounds the brain and spinal cord. The arrows indicate the direction of the fluid's flow. Blood is shown in rust.

F12.15, F12.16

The Meninges

The entire central nervous system is covered by three layers of connective tissue called the **meninges** (singular: *meninx*). The meninges are composed of the *dura mater*, the *arachnoid*, and the *pia mater*.

Dura Mater

The **dura mater** is the outermost meninx. It is a strong membrane composed of fibrous connective tissue. Around the brain, the dura mater is a double-layered structure (Figures 12.15 and 12.16). The outer layer of the dura mater adheres closely to the bones of the skull, serving as the periosteum of the cranial bones. The inner layer of the dura mater is continuous with the dura mater of the spinal cord. The two layers of the dura that surround the brain are fused together over most of the brain. In certain regions, however, the layers are separated, forming venous sinuses *(dural sinuses)* that carry blood to the internal jugular veins of the neck. The inner layer of the dura dips into the longitudinal fissure that separates the cerebral hemispheres, forming a strong septum called the **falx cerebri** that is anchored anteriorly to the crista galli of the ethmoid bone. The inner layer of the dura also forms the **falx cerebelli,** which is a septum located between the cerebellar hemispheres. Another extension of the dura mater passes transversely in the fissure that separates the

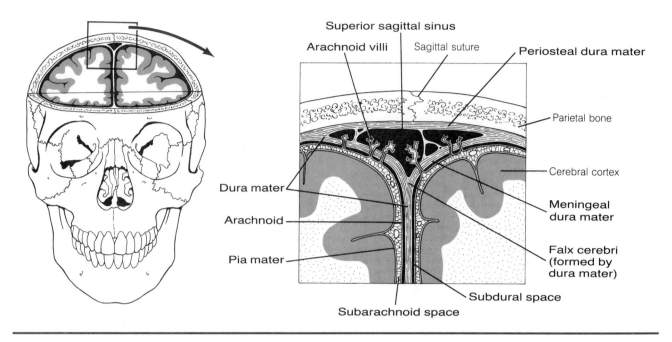

Figure 12.16
Frontal section showing the relationships of the dural venous sinuses, meninges, and the subarachnoid space.

cerebrum and the cerebellum, where it forms a septum called the **tentorium cerebelli.** All of these dural extensions anchor the brain to the inside of the cranial cavity.

Arachnoid

a-rak'-noid
F12.16

The middle of the three meninges is the **arachnoid** meninx (Figure 12.16), which is located deep to the dura mater. The arachnoid, a delicate membrane, is rather closely adherent to the inner surface of the dura mater, with only a very narrow **subdural space** separating the two membranes. Between the arachnoid and the deepest meninx, the pia mater, is a wider separation called the **subarachnoid space.** The subarachnoid and the subdural spaces contain cerebrospinal fluid. The subarachnoid space is bridged by weblike strands of the arachnoid.

Pia Mater

The innermost meninx is the **pia mater.** A delicate vascular membrane of loose connective tissue, the pia mater adheres closely to the brain and spinal cord, dipping deeply into the fissures and the sulci. In the roofs of the ventricles, the pia mater plus associated ependymal cells become modified and contribute to the formation of the choroid plexuses.

Cerebrospinal Fluid

Cerebrospinal fluid is a watery fluid that serves as a cushion for the entire central nervous system, protecting the soft tissue from jolts and blows. Besides filling the ventricles of the brain, the fluid also surrounds the brain and spinal cord, so the central nervous system actually floats in the fluid and is effectively lightened by it. Cerebrospinal fluid is actively secreted into the ventricles by the capillaries of the choroid plexuses. It is also secreted by ependymal cells that surround the blood vessels of the cerebrum and line the central canal of the spinal cord. Because it is a selective secretion, the composition of cerebrospinal fluid differs from that of the blood plasma. For example, proteins, potassium, and calcium are present in lower concentrations in the cerebrospinal fluid than in the plasma, and sodium and chloride are present in higher concentrations.

There is normally a slight pressure in the ventricles, and the cerebrospinal fluid circulates slowly from the lateral ventricles into the third and then the fourth ventricle. From the fourth ventricle, some of the cerebrospinal fluid

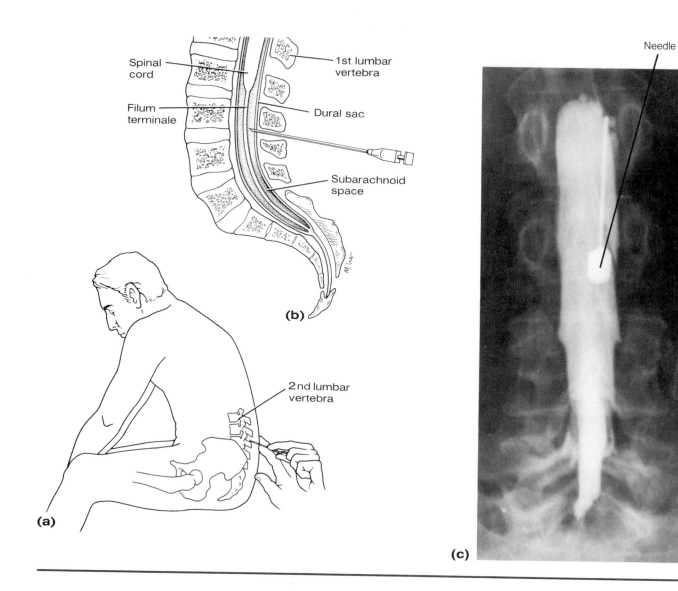

Spinal cord

Filum terminale

1st lumbar vertebra

Dural sac

Subarachnoid space

(b)

2nd lumbar vertebra

(a)

Needle

(c)

flows into the central canal of the spinal cord, but most of it passes through the apertures (one medial; two lateral) in the roof of the fourth ventricle and enters the subarachnoid space of the meninges. In the subarachnoid space, the cerebrospinal fluid circulates slowly down the posterior surface of the spinal cord, around the cord, and ascends in the anterior portion of the space to reach the brain (Figure 12.15).

If cerebrospinal fluid were allowed to accumulate, it would exert enough pressure to compress and damage the brain. If the circulation of cerebrospinal fluid is blocked during infancy, before the skull bones have united firmly, the head enlarges as the pressure within the brain increases. This condition is called *hydrocephalus.* Normally, however, cerebrospinal fluid is reabsorbed into the blood at the same rate at which it is formed. This reabsorption is accomplished through thin projections of the arachnoid meninx called **arachnoid villi,** which project into the dural venous sinuses of the brain (Figure 12.16). Cerebrospinal fluid therefore is formed from the blood, and after circulating through and around the central nervous system, it returns to the blood.

Substances such as nutrients and the end products of metabolism can move between the neural tissue and the cerebrospinal fluid, and an examination of cerebrospinal fluid provides a means of determining whether infectious organisms are present in the central nervous system. Samples of cerebrospinal fluid are withdrawn for diagnostic purposes by inserting a needle between the third and fourth lumbar vertebrae into the subarachnoid space— a procedure called a **lumbar puncture** (Figure 12.17). By inserting the needle

F12.15

F12.16

F12.17

Figure 12.17
(a) Lumbar puncture technique. **(b)** Position of the needle within the subarachnoid space below the termination of the spinal cord. **(c)** Lumbar myelogram. The white material is contrast medium that was injected into the subarachnoid space through the needle.

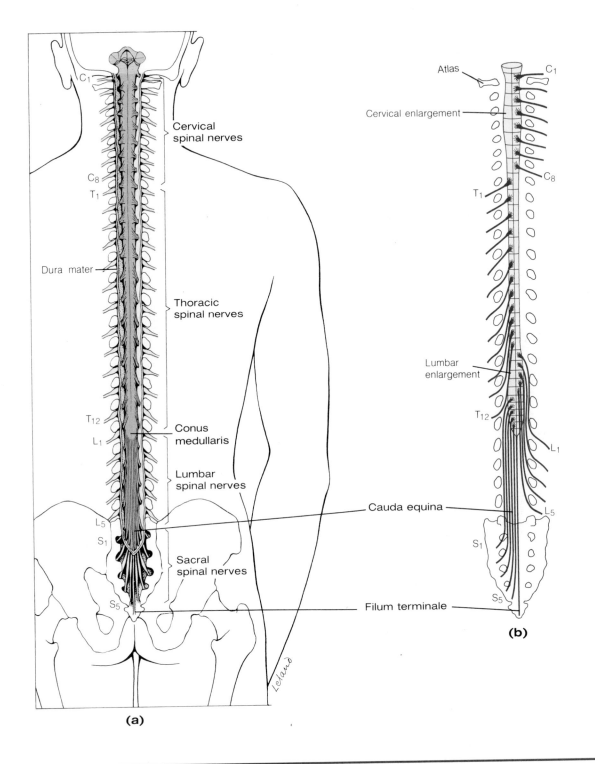

(a)

(b)

Figure 12.18

(a) The spinal cord and the proximal portions of the spinal nerves in their normal positions, viewed with the neural arches of the vertebrae removed and the dura mater that surrounds the cord cut open. **(b)** Schematic representation of the spinal cord and spinal nerves to illustrate the cauda equina and the cervical and lumbar enlargements. The letters indicate specific spinal nerves: C = cervical; T = thoracic; L = lumbar; S = sacral.

that far down the spinal column, there is little danger of damaging the spinal cord, which ends at the level of the first or second lumbar vertebra. Spinal anesthesias *(spinal blocks)* are sometimes administered in a similar manner. To identify damaged intervertebral discs or the presence of structures such as tumors, which may cause pressure on the spinal cord, contrast medium that appears opaque on an X ray is sometimes injected into the subarachnoid space. The X ray taken after the injection of the contrast medium is referred to as a *lumbar myelogram* (Figure 12.17c). Any obstruction of the contrast medium may be due to constriction of the subarachnoid space by a ruptured disc or by a tumor.

F12.17c

THE SPINAL CORD

Below the level of the medulla, the central nervous system continues as the **spinal cord.** The spinal cord performs two main functions: (1) It *conducts* nerve impulses to and from the brain, and (2) it *processes* sensory information in a limited manner, making it possible for the cord to initiate stereotyped reflex actions (called *spinal reflexes*) without input from higher centers in the brain.

General Structure of the Spinal Cord

The spinal cord passes through the vertebral canal of the vertebrae. It extends from the foramen magnum of the skull to the level of the first or second lumbar vertebra (Figure 12.18a). Until the third month of fetal development, the spinal cord is as long as the vertebral column. As the embryo continues to develop, however, the vertebral column grows at a faster rate than the spinal cord. As a result, the spinal cord does not extend the entire length of the vertebral column in the adult. A thin fibrous filament of the spinal meninges called the **filum terminale** extends from the tip of the spinal cord **(conus medullaris)** to the coccyx.

F12.18a

Thirty-one pairs of spinal nerves arise from the spinal cord and pass through the intervertebral foramina between adjacent vertebrae of the vertebral column. The spinal nerves that leave the vertebral column between the cervical vertebrae are called *cervical* nerves; those that leave the vertebral column between the thoracic vertebrae are called *thoracic* nerves. Similarly, *lumbar, sacral,* and *coccygeal* nerves leave the vertebral column between the lumbar, sacral, and coccygeal vertebrae, respectively. Each portion of the cord that gives rise to a pair of spinal nerves is called a *spinal segment*. Because the vertebral column grows at a faster rate than the spinal cord, the spinal nerves are pulled downward as the column lengthens. As a result, the roots of the lower spinal nerves pass some distance inferiorly before reaching the appropriate intervertebral foramina (Figure 12.18b). At the end of the spinal cord, the mass of descending lumbar and sacral nerve roots has the appearance of a horse's tail and is therefore called the **cauda equina.**

F12.18b

The spinal cord displays two prominent enlargements (Figure 12.18b). The *cervical enlargement* is located in the portion of the cord that gives rise to the spinal nerves that supply the upper limbs. These nerves form the brachial plexus (see Chapter 13). The *lumbar enlargement* is in the region of the cord that gives rise to the nerves that innervate the lower limbs. These nerves form the lumbosacral plexus.

F12.18b

In Chapter 10 we discussed how the spinal cord and brain develop embryonically from a neural groove into a neural tube. With further development, the lateral walls of the neural tube (and brain stem) show greater development than the roof or floor of the tube (Figure 12.19). These lateral thickenings become separated into dorsal and ventral portions by a groove **(sulcus limitans)** along each wall of the central canal. The two dorsal thickenings are called **alar plates.** The neurons in the alar plates are *internuncial neurons* that receive *sensory* and *coordinative* information from afferent neurons. The two ventral thickenings of the developing neural tube are called **basal plates.** The neurons in the basal plates develop into *motor* neurons. As the

F12.19

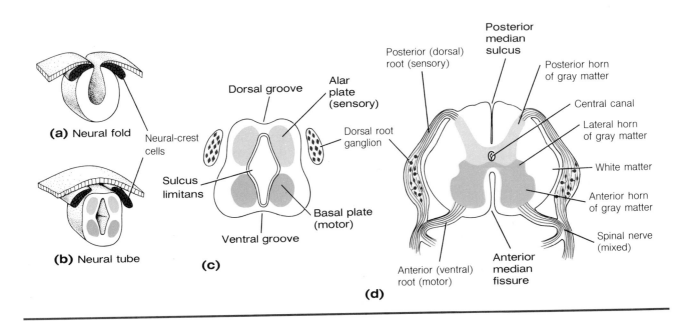

Figure 12.19
Development of alar and basal plates within the spinal cord. (a) Neural fold stage. (b) Neural tube stage. (c) Separation of alar and basal plates by the sulcus limitans. (d) Fully developed spinal cord with a pair of spinal nerves.

lateral regions of the neural tube develop, grooves form along the roof and floor of the tube. A fairly deep longitudinal fissure called the **anterior median fissure** forms on the ventral surface of the spinal cord (Figure 12.20). A shallow longitudinal groove called the **posterior median sulcus** is located on the posterior surface of the spinal cord.

Meninges of the Spinal Cord

The spinal cord is covered with the same three meninges that cover the brain: the dura mater, the arachnoid, and the pia mater (Figure 12.20).

The spinal **dura mater** is continuous at the foramen magnum with the inner layer of the dura mater of the brain. Unlike the situation in the skull, the spinal dura does not fuse to the bone of the surrounding vertebrae. Therefore, there is a small epidural space between the spinal dura and the vertebral column. The spinal dura extends beyond the lower end of the cord, enclosing the cauda equina (Figure 12.18a). In the sacral region, it forms a covering around the filum terminale. The dura extends laterally to blend with the connective tissue that covers each spinal nerve.

As is the case in the coverings of the brain, the spinal **arachnoid** forms a close lining of the dura mater. The subarachnoid space, which contains cerebrospinal fluid, is largest in the region of the cauda equina.

The innermost spinal meninx is the **pia mater.** It adheres closely to the surface of the cord and the roots of the spinal nerves. The pia mater contains a rich network of blood vessels.

The spinal cord is held in a somewhat fixed position within the meninges by fibrous bridges that cross the subarachnoid space, joining the pia mater with the arachnoid and dura mater. The heaviest of the bridges are the **denticulate ligaments** (Figure 12.21), located along the lateral margins of the spinal cord.

Composition of the Spinal Cord

The spinal cord, like the brain, consists of areas of white matter and gray matter (Figure 12.22). As in the brain, the white matter is composed primarily of the myelinated processes of neurons, and the gray areas are composed primarily of nerve-cell bodies and unmyelinated internuncial nerve fibers. Neuroglia are present in both the white and gray matter. As we have seen, the gray matter in the cerebrum and cerebellum of the brain forms the surface layer (cortex). By contrast, the gray matter of the spinal cord is centrally located and is surrounded by the white matter (Figures 12.20, 12.21, and 12.22).

F12.18a

F12.21

F12.22

F12.20–F12.22

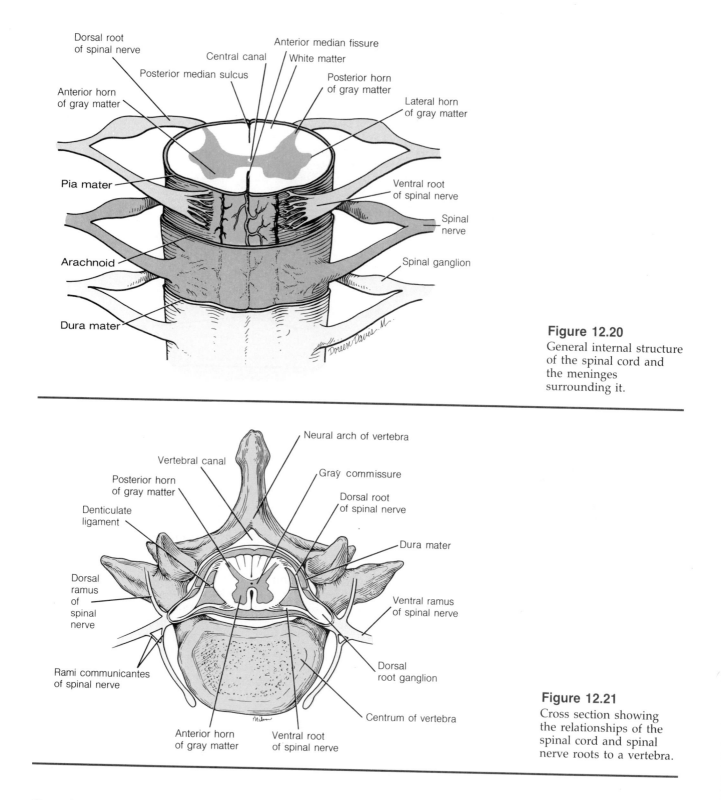

Figure 12.20
General internal structure of the spinal cord and the meninges surrounding it.

Figure 12.21
Cross section showing the relationships of the spinal cord and spinal nerve roots to a vertebra.

Gray Matter of the Spinal Cord

The gray matter of the spinal cord is roughly in the form of the letter H. The transverse bar of gray matter that connects the two lateral gray areas is the **gray commissure** (Figure 12.21). Within the gray commissure is the narrow fluid-filled **central canal,** which is continuous with the fourth ventricle. The vertical bars of the gray H, on either side of the gray commissure, are separated into a pair of **posterior (dorsal) horns,** or **columns,** and a pair of **anterior (ventral) horns,** or **columns.** In the thoracic and upper lumbar regions, the

F12.21

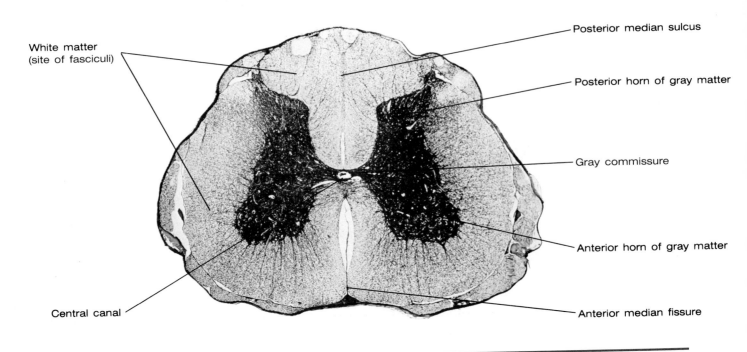

White matter (site of fasciculi)

Posterior median sulcus

Posterior horn of gray matter

Gray commissure

Anterior horn of gray matter

Anterior median fissure

Central canal

Figure 12.22
Photograph of cross section of a spinal cord.

spinal cord also has a pair of **lateral horns,** or **columns,** of gray matter located between the other two gray horns.

The posterior horns of gray matter develop from the alar plates (sensory) of the embryonic neural tube and are composed of axons of sensory neurons from the spinal nerves and internuncial neurons that transmit sensory information within the central nervous system. Some of the axons of the sensory neurons enter the white matter of the posterior area of the spinal cord. These axons then travel to higher levels of the cord or to the brain. Other sensory axons enter the gray substance and synapse either directly with neurons in the anterior horns (thus forming a spinal reflex arc) or with internuncial neurons. The internuncial neurons, in turn, may synapse with anterior horn motor neurons at the same level, pass to higher or lower levels within the spinal cord, or travel all the way up to various regions in the brain.

The anterior and lateral horns of gray matter develop from the basal plates (motor) of the embryonic neural tube. The anterior horns contain the cell bodies of somatic motor (voluntary) neurons whose axons leave the cord and enter a spinal nerve. The cell bodies of visceral motor (involuntary) neurons are found in the lateral horns (see Chapter 14).

Dorsal and Ventral Roots of Spinal Nerves

Groups of nerve fibers called **dorsal roots** enter the spinal cord where the tips of the posterior horns of gray matter come close to the surface of the cord. Similarly, groups of nerve fibers called **ventral roots** leave the spinal cord where the tips of the anterior horns of gray matter come close to the surface of the cord. The dorsal root and ventral root on each side of each spinal segment unite to form a **spinal nerve** (Figures 12.20 and 12.21). F12.20, F12.21

The dorsal roots contain only axons of sensory neurons (somatic and visceral) that pass from the spinal nerve into the posterior horn of gray matter in the spinal cord. The cell bodies of these sensory neurons are located outside the spinal cord within enlargements of the dorsal roots. These enlargements, which lie in the intervertebral foramina, are called **dorsal root ganglia,** or **spinal ganglia** (Figure 12.21). F12.21

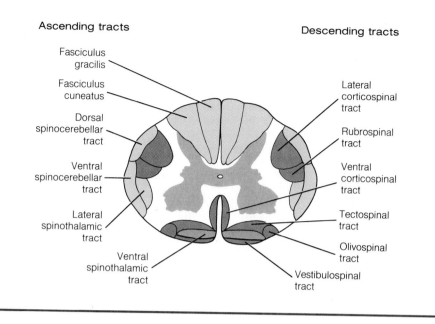

Ascending tracts

Descending tracts

Fasciculus gracilis

Fasciculus cuneatus

Dorsal spinocerebellar tract

Ventral spinocerebellar tract

Lateral spinothalamic tract

Ventral spinothalamic tract

Lateral corticospinal tract

Rubrospinal tract

Ventral corticospinal tract

Tectospinal tract

Olivospinal tract

Vestibulospinal tract

Figure 12.23
Main fasciculi of the spinal cord. The ascending (sensory) tracts are shown in blue and are labeled only on the left side. The descending (motor) tracts are shown in rust and are labeled only on the right side.

The ventral roots are formed by the axons of neurons located in the anterior and lateral horns of gray matter. The ventral roots therefore contribute processes of both somatic motor and visceral motor (autonomic) neurons to the spinal nerves.

White Matter of the Spinal Cord

The white matter of the spinal cord completely surrounds the gray matter. It is composed primarily of myelinated axons. These axons travel in three directions: (1) up the spinal cord to higher levels in the cord or brain; (2) down the spinal cord from the brain or higher levels of the cord; or (3) across the cord, transmitting impulses from one side to the other.

In each half of the spinal cord the white matter is divided by the gray matter into three areas: the **posterior funiculus** (cord), the **lateral funiculus,** and the **anterior funiculus.** Within the funiculi are smaller bundles of nerve fibers called **tracts,** or **fasciculi** (bundles) (Figure 12.23). The tracts are composed of the processes of neurons that carry similar types of impulses to a specific destination. Some tracts are *ascending (sensory) tracts,* carrying impulses to the brain. (The impulses reach the spinal cord by way of afferent neurons of a spinal nerve.) Other tracts are *descending (motor) tracts,* carrying impulses from the brain down to the motor neurons in the anterior or lateral gray horns of the spinal cord. The tracts are not visibly discernible, but their locations have been determined by experimental methods. Most of the spinal tracts have descriptive names that indicate where they begin and where they terminate (Figure 12.23).

fah-sik'-yoo-lye
F12.23

F12.23

ASCENDING (SENSORY) SPINAL TRACTS   The ascending tracts of the spinal cord carry afferent (sensory) impulses from peripheral sensory receptors to various centers in the brain. These tracts generally contain three successive neurons called first-order, second-order, and third-order neurons. A **first-order neuron** is a sensory neuron that has the peripheral portion of its nerve fiber in the spinal nerve and its cell body in a dorsal root ganglion. A **second-order neuron** is a neuron whose cell body is located in the spinal cord or the medulla. A second-order neuron connects a first-order neuron to a **third-order neuron,** whose cell body is located in the thalamus. The nerve fiber of a third-order neuron extends to the cerebral cortex. All of the ascending spinal tracts cross to the other side of the central nervous system—either at the level of entry into the spinal cord or a few segments above the entry level or within the medulla. As a result, sensory information received by receptors on the right side of the body is interpreted in the primary sensory area of the left cerebral cortex and sensory information from the left side of the body is interpreted in the primary sensory area of the right cerebral cortex. The major ascending (sensory) spinal tracts are the:

1. fasciculus gracilis
2. fasciculus cuneatus
3. spinothalamic tracts
4. spinocerebellar tracts

F12.24   The **fasciculus gracilis** and **fasciculus cuneatus** are two tracts in the posterior funiculus that carry similar types of sensory information concerning *joint sense, muscle-position sense,* and *fine-touch localization* from different parts of the body (Figure 12.24). Information carried by these tracts makes it possible to know the position of a body part without having to see it. This information also enables a person to locate where an object is touching the body and helps identify the object by shape, texture, weight, and so on. The fasciculi gracilis and cuneatus receive impulses generated by proprioceptors in the muscles and joints and mechanoreceptors in the skin. The fasciculus cuneatus, which is located lateral to the fasciculus gracilis, transmits these impulses from the upper limbs, the trunk, and the neck. The fasciculus gracilis transmits impulses that arise from receptors in the lower limbs and lower trunk.

Nerve fibers of the fasciculus gracilis synapse in a center in the medulla called the **nucleus gracilis.** Nerve fibers of the fasciculus cuneatus synapse in a medullary center called the **nucleus cuneatus.** The processes of the second-order neurons that leave the nucleus gracilis or nucleus cuneatus cross to the other side of the medulla and give rise to a tract called the *medial lemniscus.* The nerve fibers in the lemniscus synapse in the thalamus with third-order neurons that ascend to the cortex of the post-central gyrus.

F12.25   In the spinothalamic tracts, most of the second-order neurons cross within the cord, after synapsing with a sensory neuron from a spinal nerve, and ascend in the white matter of the opposite side as the lateral and ventral spinothalamic tracts. The nerve fibers of the spinothalamic tracts synapse in the thalamus. From the thalamus, third-order neurons ascend to the postcentral gyrus of the cerebral cortex. The **lateral spinothalamic tracts** (Figure 12.25) convey impulses concerned with *pain* and *temperature.* The **ventral spinothalamic tracts** carry impulses from receptors sensitive to *touch* and *pressure.*

F12.26   There are four spinocerebellar tracts: a pair of **dorsal** and a pair of **ventral spinocerebellar tracts.** The spinocerebellar tracts carry information concerning unconscious *proprioception* from neuromuscular receptors. As the name of the tracts indicates, their nerve fibers end in the cerebellum. The nerve fibers of these tracts do not synapse with higher-order neurons that pass to the cerebral cortex, or even to the thalamus (Figure 12.26). Nerve fibers of the spinocerebellar tracts do, however, synapse with neurons in the cerebellum that can cause contraction of skeletal muscles (by way of the red nucleus of the midbrain). Some of the nerve fibers in the spinocerebellar tracts cross to the opposite side of the cord, whereas others remain on the side of the cord they enter and project directly to the cerebellum. The nerve fibers of the tracts enter the cerebellum by way of the inferior cerebellar peduncles.

DESCENDING (MOTOR) SPINAL TRACTS   The descending spinal tracts carry impulses from the brain to lower motor neurons that regulate the activity of skeletal muscles. All of these tracts cross from one side of the central nervous system to the other, and they all contain two or three consecutive neurons. There are two types of descending (motor) spinal tracts:

1. pyramidal tracts
2. extrapyramidal tracts

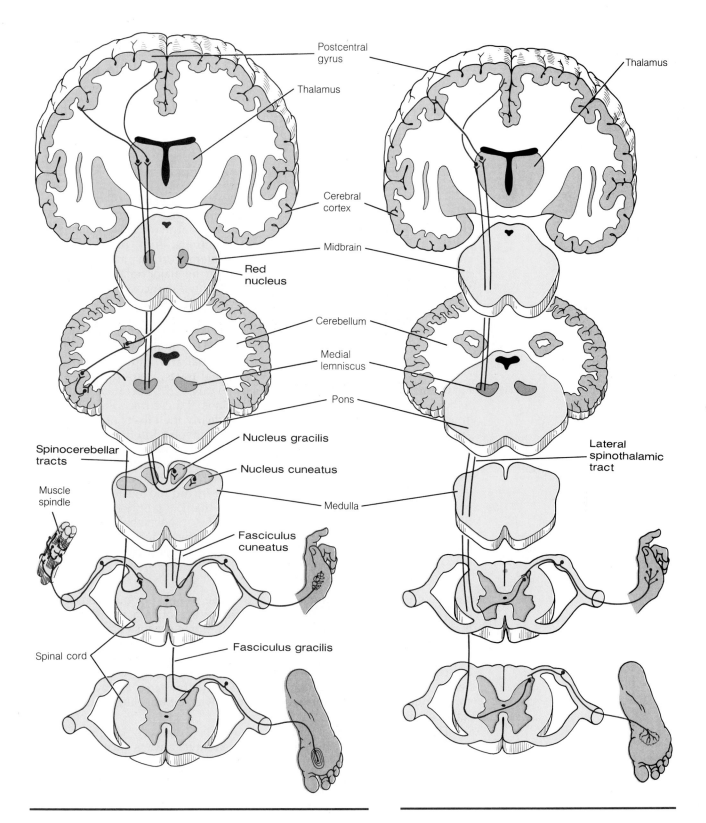

Figure 12.24

Sensory pathways for touch, pressure, and proprioception (conscious and unconscious) within the fasciculi gracilis, the fasciculi cuneatus, and the spinocerebellar tracts.

Figure 12.25

Sensory pathways for pain and temperature within the lateral spinothalamic tracts.

Pyramidal tracts are motor tracts that originate primarily from large cells (pyramidal cells) in the cortex of the precentral gyrus and travel through the cerebral peduncles of the midbrain and the pyramids of the medulla. The pyramidal tracts are also called **corticospinal tracts,** a name that indicates their origin (cerebral cortex) and termination (spinal cord). From the cerebral cortex these tracts descend through the internal capsule, midbrain, pons and pyramids of the medulla. The nerve fibers of these tracts (which are called **upper motor neurons**) synapse primarily with motor neurons in the anterior horns of the gray matter of the spinal cord, although some of them synapse

F12.27 with motor neurons associated with certain cranial nerves (Figure 12.27). Most of the upper motor neurons in the corticospinal tracts cross to the opposite side of the cord in the medulla and form the **lateral corticospinal tracts.** The remaining upper motor neurons descend uncrossed as the **ventral corticospinal tracts.** Some of the nerve fibers in the ventral tracts cross at the levels at which they synapse with **lower motor neurons.** Both corticospinal tracts carry *motor stimuli* to *skeletal muscles.* The lateral corticospinal tracts, which extend the entire length of the spinal cord, are the major tracts involved in the voluntary control of skeletal muscles. The ventral corticospinal tracts generally do not extend below the thoracic level of the cord.

The remaining motor tracts, which originate from various regions of the cerebral cortex and subcortical areas, are referred to as **extrapyramidal tracts.** These tracts descend from various nuclei in the brain stem and influence muscular actions, coordination, balance, visual and auditory stimuli, and other functions. They include the **rubrospinal tracts, vestibulospinal tracts,**

F12.23 **tectospinal tracts,** and **olivospinal tracts** (Figure 12.23). Because there seems to be considerable overlap between the actions of these tracts and the corticospinal tracts, it is not possible to separate clearly the effects of each. In general, the pyramidal tracts control the muscles involved in fine movements of the body, whereas the extrapyramidal tracts tend to modify muscular contractions related to *posture* and *balance.* Some extrapyramidal tracts have been found to be inhibitory of movements rather than excitatory.

The rubrospinal tracts, which originate in the red nucleus of the mid-

F12.26 brain, are illustrated in Figure 12.26. Notice that the proprioceptive information that is carried to the cerebellum over the spinocerebellar tracts is ultimately conveyed to the rubrospinal tracts. In this manner, the rubrospinal tracts help coordinate the reflexes concerned with postural adjustment.

The Spinal Reflex Arc

Not all of the afferent impulses carried to the spinal cord over sensory neurons enter one of the ascending tracts of the spinal cord and thus reach a higher center of the central nervous system. Some sensory neurons synapse directly or through internuncial neurons with motor neurons in the anterior horn of the gray matter of the spinal cord at the same level at which they enter the cord. Other neurons travel only a few segments up or down the cord before synapsing with a motor neuron. The neural pathways by which sensory impulses from receptors reach effectors without traveling to the brain are called **spinal reflex arcs.**

The presence of spinal reflex arcs makes possible automatic, stereotyped reactions to stimuli in which a particular stimulus always elicits a particular response. Such reactions are called **reflexes.** Since spinal reflexes occur at the level of the spinal cord without involving the pyramidal tracts from the cortex of the brain, they are involuntary responses, even though they often involve skeletal muscles. At the same time that a reflex is occurring, however, information about the stimulus that initiated the reflex may also be transmitted to the brain, where a conscious sensation is elicited. When you touch something hot with your hand, for example, you become aware of it through impulses carried to the brain over the lateral spinothalamic tracts. By the time you feel the burning sensation, however, you have already withdrawn your hand from the hot object as a result of reflex action. If the spinal cord is severed in the cervical region, thus making it impossible for impulses to reach the brain

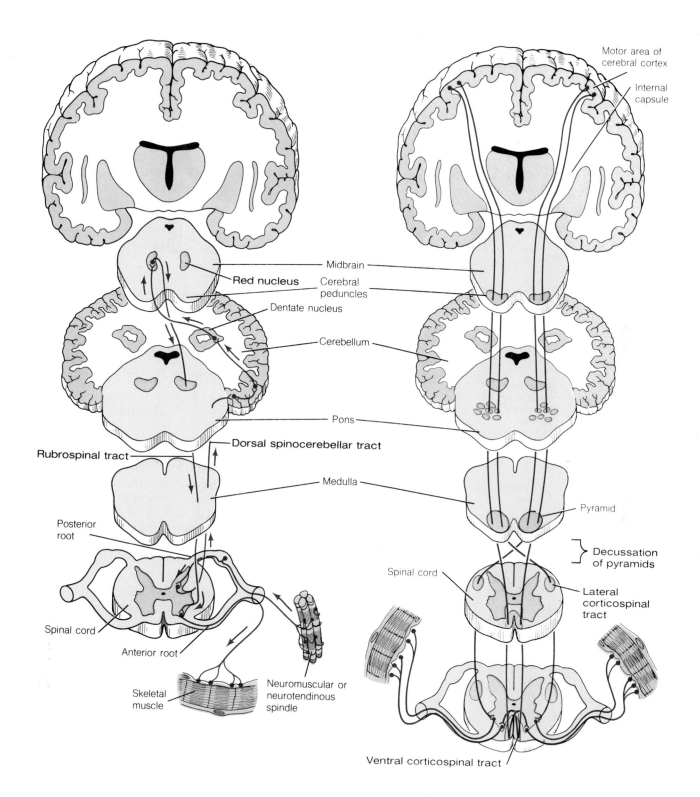

Motor area of
cerebral cortex

Internal
capsule

Midbrain

Red nucleus

Cerebral
peduncles

Dentate nucleus

Cerebellum

Pons

Dorsal spinocerebellar tract

Rubrospinal tract

Medulla

Pyramid

Posterior
root

Decussation
of pyramids

Spinal cord

Lateral
corticospinal
tract

Spinal cord

Anterior root

Skeletal
muscle

Neuromuscular or
neurotendinous
spindle

Ventral corticospinal tract

Figure 12.26
Sensory and motor
pathways for
unconscious muscle
movement within the
dorsal spinocerebellar
and rubrospinal tracts.

Figure 12.27
Pathways of the
pyramidal tracts (lateral
and ventral corticospinal
tracts) carrying motor
impulses to skeletal
muscles.

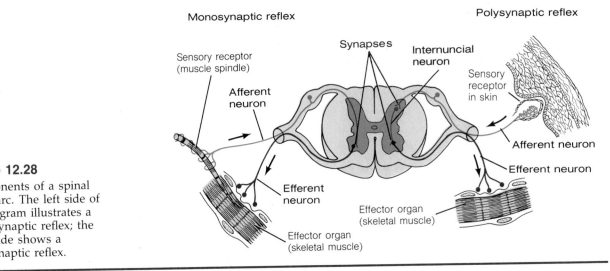

Monosyntaptic reflex

Polysynaptic reflex

Figure 12.28
Components of a spinal reflex arc. The left side of the diagram illustrates a monosynaptic reflex; the right side shows a polysynaptic reflex.

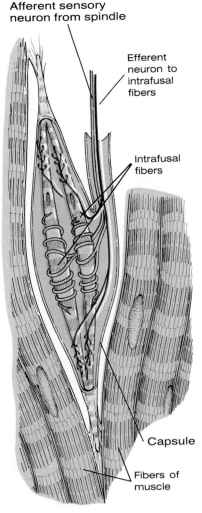

Figure 12.29
A muscle spindle.

from the spinal cord, the hand would still be withdrawn as a result of reflex activity that occurs in the spinal cord itself. In this instance, however, there would be no sensation of pain.

A spinal reflex arc has at least five components (Figure 12.28):

1. A **receptor,** which can be a peripheral ending of a sensory neuron or a specialized cell associated with a peripheral ending of a sensory neuron.

2. A **sensory (afferent) neuron,** which carries impulses through a spinal nerve from the receptor to the spinal cord.

3. A **synapse,** within the spinal cord between a sensory neuron and a motor neuron. If a sensory neuron synapses directly with a motor neuron, the arc formed is known as a **monosynaptic reflex arc.** If there are one or more internuncial neurons between a sensory and a motor neuron, requiring more than one synapse, the arc is a **polysynaptic reflex arc.** Most CNS reflexes are polysynaptic.

4. A **motor (efferent) neuron,** which transmits the nerve impulse from the anterior horn of the gray matter to an effector.

5. An **effector,** which responds to the efferent nerve impulses. Muscle cells (skeletal, smooth, or cardiac) and secretory cells serve as effectors.

Two important spinal reflexes—the stretch reflex and the tendon reflex—influence the contractions of skeletal muscles.

Stretch Reflex

The muscle **stretch reflex** is initiated by skeletal muscle receptors called **muscle spindles** (Figure 12.29). The muscle spindles provide information about the lengths of skeletal muscles, and they respond to both the rate and the magnitude of a length change. Within a muscle spindle are three to ten specialized muscle cells, called *intrafusal fibers,* that differ from the cells of the muscle as a whole *(extrafusal fibers).* The intrafusal fibers are contractile only at their ends, and the ends of the fibers are supplied by efferent neurons called *gamma efferent neurons.* The noncontractile, central portions of the intrafusal fibers are supplied by afferent neurons that spiral around the intrafusal fibers. The frequency of nerve impulses in the afferent neurons increases in response to both the sudden and maintained stretch of the central areas of the intrafusal fibers and decreases when the central areas of the fibers are compressed. A muscle spindle is attached in parallel with the extrafusal fibers of a skeletal

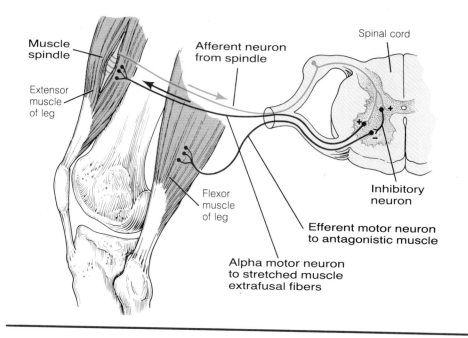

Muscle spindle

Extensor muscle of leg

Flexor muscle of leg

Afferent neuron from spindle

Spinal cord

Inhibitory neuron

Efferent motor neuron to antagonistic muscle

Alpha motor neuron to stretched muscle extrafusal fibers

Figure 12.30
Pathways involved in the stretch reflex. Excitation is indicated by a +; inhibition is indicated by a −.

muscle so that stretching the muscle stretches the muscle spindle (and its intrafusal fibers), and contracting the muscle compresses the muscle spindle.

In the stretch reflex, the stretching of a muscle results in an increased frequency of nerve impulses in the afferent neurons associated with the muscle spindles (Figure 12.30). Within the spinal cord, the afferent neurons synapse with efferent neurons called *alpha motor neurons* that supply the extrafusal fibers of the muscle. As a result, the increased frequency of nerve impulses in the afferent neurons increases the stimulation of the alpha motor neurons, causing the muscle to contract and resist the stretch. The afferent neurons also synapse with inhibitory neurons that, in turn, synapse with efferent neurons controlling muscles whose activities oppose the contraction of the stretched muscle (that is, antagonistic muscles). Thus, when the stretch reflex stimulates the stretched muscle to contract, antagonistic muscles that oppose the contraction are inhibited. This occurrence is called *reciprocal inhibition,* and the neuronal mechanism that causes it is called *reciprocal innervation.*

The *patellar reflex* (knee jerk) is a common example of a stretch reflex. Striking the patellar tendon at the knee pulls on the tendon and thus stretches the quadriceps muscles of the anterior thigh. If all the components of the reflex arc are intact, the muscle spindles within the quadriceps muscles initiate a stretch reflex that causes the quadriceps muscles to contract and swing the leg forward.

In addition to their synapses with neurons involved in the stretch reflex, the afferent neurons from muscle spindles also synapse with neurons that relay impulses to the brain. As a result, nerve impulses from muscle spindles provide information about the state of stretch or contraction (that is, the length) of skeletal muscles that is important in maintaining body posture and in coordinating muscular activity.

The activity and sensitivity of a muscle spindle is influenced by nerve impulses carried to the intrafusal fibers of the spindle by the gamma efferent neurons. The contraction of the ends of the intrafusal fibers due to gamma efferent stimulation stretches the central, noncontractile portions of the fibers. This stretching increases the activity of the afferent neurons and enhances the response of the spindle to further stretch. Thus, the excitability of the muscle spindle and the amount of muscle stretch necessary to initiate a stretch reflex can be altered by altering the degree of contraction of the intrafusal fibers.

Nerve impulses to skeletal muscles from higher centers often simultaneously stimulate both the extrafusal fibers of a muscle (by way of the alpha

F12.30

FRONTIERS IN HEALTH

Electric Pain Relief

People who experience pain can tell you that it exists in dozens of forms, from intense to subtle, from nagging to stabbing, from chronic to acute. Pain is a leading affliction in the United States, with nearly one in every three Americans suffering from chronic pain. It has been estimated that pain costs $70 billion a year in medical costs and lost working days. The personal cost of chronic pain, of course, is immeasurable.

Ironically, pain is one of the least understood of all medical conditions. The study of pain falls outside the realm of internal medicine, surgery, or any of the other traditional subdivisions of medicine. But pain is now being studied more than it has been, and several recent techniques offer sufferers of chronic pain new hope.

Pain has traditionally been treated with painkillers, with acupuncture, biofeedback, and hypnosis to complement these medications. For severe pain, surgeons have resorted to cutting nerves or destroying areas of the brain that perceive pain. However, a large percentage of patients who have undergone such treatment report the recurrence of pain, usually after about a year, and often with greater intensity.

One technique that offers great promise is deep brain stimulation. Surgeons find that electrodes implanted in certain regions of the brainstem and midbrain can block out pain impulses that are transmitted to the brain. After the electrodes are inserted through a tiny hole in the skull of pain victims, they are pushed into specific regions of the brain and connected to a small battery worn on the belt, or sometimes implanted under the skin. When pain begins, the patient turns on the current, giving himself or herself a small electric shock that blocks out the pain. Deep brain stimulation successfully blocks out a wide spectrum of painful stimuli without noticeably affecting other brain functions.

Scientists at the Johns Hopkins University Laboratory of Applied Physics, working with Pacesetter Systems in Sylmar, California, have developed an implantable battery pack that can be used in deep brain stimulation. No bigger than a deck of cards, the stimulator is implanted beneath the skin on the lower part of the rib cage. Once they become accustomed to its presence, patients hardly know the device is there. Fine wires run from the unit to the brain or to nerve bundles. According to the developers, the stimulator can last for 10 years with only monthly recharging, which can be done transcutaneously by simply holding an alternating magnetic field over the device.

Deep brain stimulation requires surgery, and with that come potential complications. But with a success rate of 75% most patients are willing to accept the risks in order to rid themselves of the excruciating and often disabling pain that racks their bodies day and night.

For many people, relief may come in a less costly and simpler form of treatment, called transcutaneous electrical nerve stimulation (TENS). In this method of pain control, patients wear a small battery pack, which is connected by wires to small electrodes. Rather than being implanted, however, the electrodes are located on

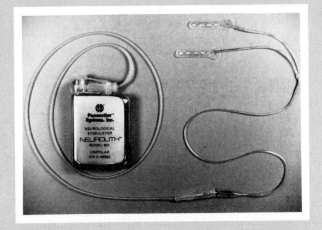

This cerebellar stimulator was designed by Pacesetter Systems for patients with cerebral palsy. Dr. Joseph H. Schulman of the Neurodyne Corporation is monitoring clinical testing, under the authorization of the U.S. F.D.A.

the body's surface, over nerve fibers that transmit pain. When a patient feels pain, he or she simply turns on the electricity. The small current generated by the battery stimulates the skin and blocks the pain fibers.

Medical researchers are not entirely certain how TENS works, but many think that it blocks pain by overloading the neuronal circuitry. Two types of nerve fibers are thought to carry sensory information to the spinal cord. Small-diameter fibers carry pain of many varieties, while larger-diameter fibers carry other forms of sensory information, such as pressure and light touch, from receptors in the skin. Both types of sensory fibers converge in the spinal cord. Nerve impulses traveling from the pain receptors are thought to be blocked by simultaneous stimulation of the larger fibers, either with acupuncture or TENS.

TENS has been used successfully to reduce the pain experienced after major surgery. In one study, throughout the postoperative period, patients using TENS needed only one-third as much narcotic pain killer as patients not using the method. Moreover, patients using TENS were able to leave the hospital several days sooner than those who were treated conventionally.

Alleviation of postoperative pain is only one use for this relatively inexpensive and safe treatment. Dentists have found that TENS can virtually eliminate the need for local anesthesia in routine dental work. It could also find an important niche in obstetrics by reducing the pain of childbirth. Athletic injuries, arthritic pain, low-back pain, neck pain, and painful joints are additional applications for this promising form of electric pain relief.

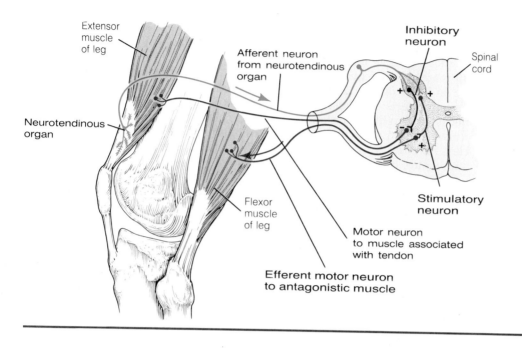

Extensor
muscle
of leg

Afferent neuron
from neurotendinous
organ

Inhibitory
neuron

Spinal
cord

Neurotendinous
organ

Flexor
muscle
of leg

Stimulatory
neuron

Motor neuron
to muscle associated
with tendon

Efferent motor neuron
to antagonistic muscle

Figure 12.31
Pathways involved in the
tendon reflex. Excitation
is indicated by a +;
inhibition is indicated by
a −.

motor neurons) and the intrafusal fibers of the spindles (by way of the gamma efferent neurons). This simultaneous stimulation maintains the responsiveness of the spindles when the muscle shortens, and it may allow the basic stretch reflex mechanism to assist in attaining the proper degree of muscular contraction. For example, suppose picking up an object requires a certain degree of contraction by a muscle. If both the extrafusal fibers of the muscle and the intrafusal fibers of the muscle spindles are stimulated, and if the contraction of the muscle is equal to the contraction of the intrafusal fibers of the spindles, the degree of stimulation of the muscle spindles will not change as a result of the contraction of the muscle. However, if for some reason (for example, fatigue) the stimulation of the extrafusal fibers results in a muscle contraction that is less than the contraction of the intrafusal fibers of the spindles, the central portions of the intrafusal fibers will be stretched. As a result, the activity of the afferent neurons from the spindles increases. This increased activity further stimulates the alpha motor neurons to the extrafusal fibers and thereby helps achieve the needed degree of muscular contraction (perhaps by activating additional motor units). In addition, nerve impulses from the muscle spindles are relayed to higher brain centers, and these centers may also alter their output to the alpha motor neurons in order to attain the proper degree of muscular contraction.

Tendon Reflex

The **tendon reflex** helps protect tendons and their associated muscles from the damage that could result from excessive tension. The receptors for this reflex are **neurotendinous organs,** which are encapsulated structures located within tendons near the junction of a tendon with a muscle. Unlike muscle spindles, which are sensitive to muscle length, neurotendinous organs are sensitive to tension. Within a neurotendinous organ are afferent neuron endings wrapped around small bundles of collagen fibers. When the tension applied to a tendon increases, so does the frequency of nerve impulses in the afferent neurons; when the tension decreases, so does the afferent impulse frequency. Because of their locations within tendons, the neurotendinous organs are in series with the muscles themselves, and they respond to changes in tension that result from either passive stretch or muscular contraction.

In the tendon reflex, an increase in the tension applied to a tendon—most commonly as a result of muscular contraction—increases the frequency of nerve impulses in the afferent neurons associated with the tendon organs (Figure 12.31). Within the spinal cord, the afferent neurons synapse with inhibitory neurons that, in turn, synapse with alpha motor neurons that supply the muscle associated with the tendon. As a result, the increased frequency of nerve impulses in the afferent neurons from the tendon organs ultimately diminishes the stimulation of the alpha motor neurons to the muscle and may even completely inhibit muscular contraction. Thus, the tendon reflex protects the muscle and tendon from excessive tension. The afferent neurons from the neurotendinous organs also synapse with stimulatory neurons that, in turn, synapse with motor neurons controlling antagonistic muscles. As a result, the tendon reflex is generally accompanied by a reciprocal stimulation of the antagonistic muscles.

In addition to synapsing with neurons involved in the tendon reflex, the afferent neurons from the neurotendinous organs also synapse with neurons that relay impulses to the brain. Thus, just as nerve impulses from muscle spindles provide the brain with information about the length of a skeletal muscle, nerve impulses from neurotendinous organs provide information about the tension developed by the muscle.

NEURON POOLS

Within the central nervous system, neurons are organized into functional groups called **neuron pools,** and information transmitted from a receptor is often received by a particular neuron pool. Within the pool, this information is processed and integrated with information received from other sources. It is then transmitted from the pool to various destinations. Within a neuron pool, the processing of information involves integrative activities such as divergence, convergence, summation, and facilitation (see Chapter 11).

Neuron pools occur at all levels of the central nervous system, including the cerebral cortex, and there are numerous interconnections among the various pools. Although the neural pathways involved in particular activities are commonly depicted as simple neuron-to-neuron chains, the performance of even the simplest actions generally requires the participation of many neurons whose activities are coordinated in complex circuits in neuron pools. For example, the alpha motor neurons to a skeletal muscle are influenced by nerve impulses arriving from muscle spindles, neurotendinous organs, and higher brain centers. As a result, the degree of activity of the alpha motor neurons at any one moment is determined by a combination of influences.

INTEGRATIVE FUNCTIONS
OF THE NERVOUS SYSTEM

The nervous system brings together information from many different sources, and it coordinates and regulates numerous interrelated activities that take place within the body. Consequently, the nervous system is an integrative system that is vital in maintaining homeostasis.

The brain is an extremely complex organ. It has been studied intensively for many years, but researchers have so far achieved only enticing bits of understanding about its function. Bear in mind, then, that much of the following discussion of brain activity is largely theoretical. As more information becomes available, some of these ideas may have to be abandoned in favor of new ones.

The Electroencephalogram

F12.32

ee-lek-tro-en-sef'-a-lo-gram

The degree and pattern of electrical activity that occur within the brain can be detected by electrodes placed on the head (Figure 12.32). The record of the electrical waves produced during brain activity is called an **electroencephalogram (EEG).** The electroencephalogram is essentially a record of waves of different amplitudes (that is, heights) and frequencies that arise from the cerebral cortex, perhaps at times as a result of the influence of some subcorti-

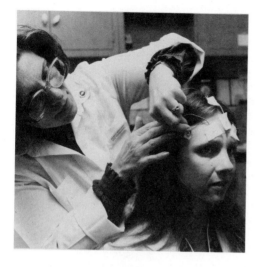

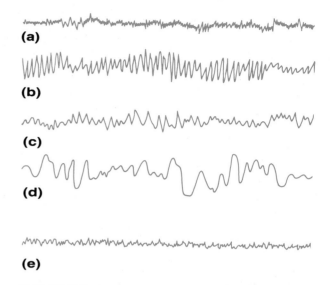

Figure 12.32
Using an electroencephalograph, a machine that graphically records the electrical activity of the brain, researchers are able to record brain waves.

Figure 12.33
Typical electroencephalograms showing various states of consciousness. **(a)** Awake and alert. **(b)** Relaxed with eyes closed. **(c)** Drowsy. **(d)** Asleep, slow-wave sleep. **(e)** Asleep, paradoxical sleep.

cal center such as the thalamus. These "brain waves," as they are called, are always present, even during unconsciousness, indicating that the brain is continually active as long as it is alive.

There is some degree of correlation between brain waves and body activity (Figure 12.33). For example, when a person is awake, brain wave recordings show low-amplitude, high-frequency waves. When a person falls asleep, the frequency of the brain waves slows, but the amplitude increases. A person who is anesthetized also exhibits slow waves.

The Aroused Brain

When the brain is aroused or awake, it is in a state of readiness and is able to react consciously to stimuli. The attainment of this state seems to depend in large part on nerve impulses sent throughout the brain by the reticular activating system, or RAS (see p. 353). Stimulation of the RAS in a sleeping animal produces an EEG like that of an aroused brain and causes the animal to awaken. Conversely, destruction of the RAS produces a permanent coma, with an EEG characteristic of the sleeping state.

The activity of the RAS is influenced by nerve impulses from cutaneous, visual, auditory, muscular, and visceral receptors. These impulses, which are called *arousal signals*, stimulate the RAS to send impulses throughout the brain to arouse it. Consequently, general sensory input plays an important role in maintaining the arousal of the brain.

Sleep

Sleep can be defined as a state of altered consciousness or partial unconsciousness from which a person can be aroused by appropriate stimuli. In contrast, a *coma* is a state of unconsciousness from which a person cannot be aroused.

As a person becomes drowsy and falls asleep, there is a gradual shift in the EEG toward higher amplitude, slower (lower-frequency) waves. The state of sleep characterized by high-amplitude, low-frequency EEG waves is called **slow-wave sleep.** During deep slow-wave sleep, the respiratory cycle is regular and deep, the heart rate is rhythmic and slow, and the blood pressure is below waking levels.

Figure 12.34

These photographs were taken using time-lapse photography during a period of paradoxical, or rapid-eye-movement, sleep. Note how the eyes have moved from side to side behind the eyelids. Although the reason for this movement is not known for certain, some researchers hypothesize that the sleeper is surveying dream imagery, perhaps giving the imagery logic and direction by supplying movement and depth to the characters and objects in the dream.

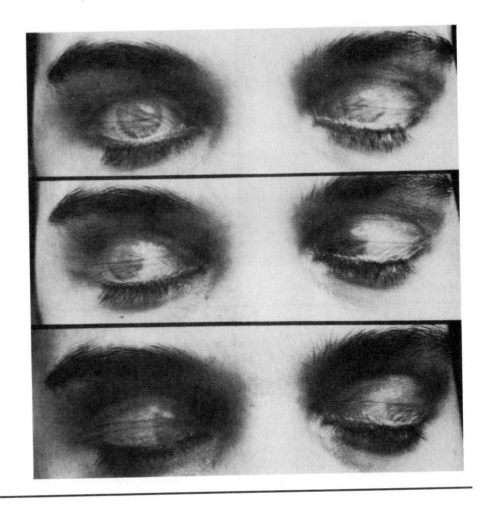

Approximately once every 90 minutes during a normal night of sleep, the EEG of slow-wave sleep is interrupted by episodes (lasting 5–20 minutes) during which the EEG resembles that of an aroused brain. This state of sleep is called **paradoxical sleep.** Because the eyes move rapidly behind the closed eyelids during paradoxical sleep, it is also referred to as **rapid-eye-movement (REM)** sleep.

Paradoxical sleep occurs in conjunction with slow-wave sleep, and the first paradoxical sleep episode normally takes place 80 to 100 minutes after a person falls asleep. A specific region of the pons called the locus caeruleus appears to induce paradoxical sleep, and damage to this region prevents its occurrence. Neurons of the area contain an abundance of norepinephrine, suggesting that release of norepinephrine leads to paradoxical sleep.

During paradoxical sleep, muscle tone throughout the body is greatly depressed, although periodic twitching of the facial muscles and limbs occurs. In addition, the respiration and heart rate are irregular, and the blood pressure may rise or fall. If a person is awakened every time a period of paradoxical sleep begins so that a paradoxical-sleep deficit develops, the person spends a greater proportion of the total sleep time in paradoxical sleep when allowed to sleep undisturbed. Most dreaming occurs during paradoxical sleep (Figure 12.34).

Sleep–Wakefulness Cycle

The normal sleep–wakefulness cycle is one of the most obvious human rhythms, and both mental and physical benefits are derived from alternating periods of sleep and wakefulness. Although the processes that underlie this cycle are not fully understood, awakening appears to occur when signals

F12.34

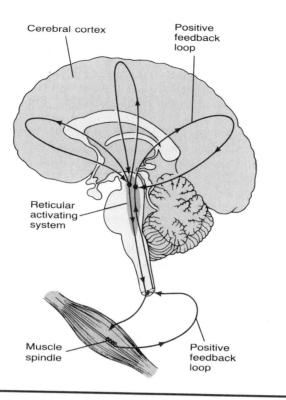

Cerebral cortex

Positive feedback loop

Reticular activating system

Muscle spindle

Positive feedback loop

Figure 12.35
Positive-feedback loops from the reticular activating system to the brain and skeletal muscles.

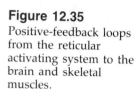

from the RAS reach a sufficient intensity to arouse the brain. Conversely, sleep seems to result when the activity of the RAS declines to a level no longer adequate to maintain arousal.

The RAS not only transmits nerve impulses throughout the brain, but—in what is basically a positive feedback response—brain areas such as the cerebral cortex send stimulatory signals to the RAS (Figure 12.35). In addition, a second positive feedback response occurs in which the RAS not only receives nerve impulses from receptors in the peripheral musculature, but it also transmits signals to the muscles. Thus signals from the RAS that enhance the activity of the brain or the muscles lead to the transmission of stimulatory impulses from these structures to the RAS. These positive feedback responses are believed to be part of a neural wakefulness circuit that operates as follows.

When the relatively dormant RAS of a sleeping individual is stimulated by arousal signals from various receptors, the RAS sends more intense signals to the brain and muscles. When this occurs, the positive feedback responses just described further stimulate the RAS. The RAS, in turn, further stimulates the brain, and so on and so on. When the signals from the RAS to the brain become sufficiently intense, the brain is aroused and the individual awakens. The wakefulness circuit contains numerous individual positive-feedback loops, and the degree of wakefulness at any moment depends on the number of active loops and on their level of activity. If many loops are activated simultaneously in a sleeping person, the positive feedback aspect of the wakefulness circuit produces a rapidly increasing response that could explain the often-sudden transition between the sleeping and the waking states. The ultimate level of activity of the positive feedback loops is limited by the capacities of the neurons involved to transmit nerve impulses. Therefore, once particular loops are activated, their level of activity does not continue to increase indefinitely but finally stabilizes.

With the passage of time, the excitability of the wakefulness circuit is believed to diminish, and consequently the intensity of the signals from the RAS to the brain declines. As a result, the individual becomes drowsy and falls asleep. During sleep, the excitability of the wakefulness circuit increases, and arousal signals again set in motion the positive feedback events that lead

F12.35

to awakening. Because of the cyclical variation in the excitability of the wakefulness circuit, arousal signals that are not strong enough to produce wakefulness early in the sleep period may be sufficient to awaken the individual later in the period. Moreover, the system exerts selectivity as to the types of signals that produce arousal. For example, a mother may quickly awaken if her child cries, but she may not be awakened by the sounds of traffic, trains, or airplanes.

Researchers have proposed a number of theories to explain the cyclical variation in the excitability of the wakefulness circuit. One suggestion is that, with continued activity during the waking period, neurons of the positive-feedback loops become depleted of neurotransmitter substances. Consequently, the feedback diminishes, the activity of the RAS declines, and the person becomes drowsy and falls asleep. During the sleep period, the neurotransmitter substances are replenished and the feedback loops recover their excitability.

Another suggestion is that the accumulation or removal of various chemical substances influences the excitability of the wakefulness circuit. For example, inhibitory chemical substances could gradually build up during the waking period and, through their influence on the RAS, lead to drowsiness and sleep. During the sleep period, the substances are removed, leading to an increased excitability of the RAS and the arousal of the brain. There is some evidence that supports this possibility. If dialyzed blood or whole cerebrospinal fluid from animals that have been kept awake for several days is injected into the brain ventricle systems of other animals, it can put them to sleep. Moreover, neurons within a region of the brain stem called the median raphe secrete serotonin into the RAS, and damage to these neurons induces a state of relative sleeplessness. Perhaps, then, sleep is caused by the secretion of serotonin from these neurons.

Researchers have proposed that a feedback relationship exists between the serotonin-secreting brain-stem neurons that may cause sleep and the neurons of the pons that appear to induce paradoxical sleep. According to this proposal, the brain-stem neurons facilitate the neurons of the pons, whose activity leads to paradoxical sleep. These neurons, in turn, stimulate the brain-stem neurons to take up serotonin that was previously secreted. This uptake decreases the extracellular serotonin concentration and lessens the inhibition of the RAS, permitting a return to the waking state. As a result, periods of paradoxical sleep would be important because they would remove an inhibitory chemical substance from the system.

Memory

Memory refers to the ability to store experiences, thoughts, and sensations for later recall. There are many unanswered questions about the processes that make memory possible, but it appears that there are at least two basic types of memory: short-term and long-term.

Different physical-chemical mechanisms are believed to be responsible for the two types of memory. In short-term memory, a bioelectrical process is thought to be involved, whereas in long-term memory, structural as well as biochemical changes in neurons or synapses are thought to occur.

Short-Term Memory

Short-term memory allows the recall of information for only short periods following its initial presentation. Theorists believe that short-term memory decays in a matter of seconds unless the information within the short-term memory is rehearsed. Rehearsal, which is essentially the process of keeping one's attention on the information to be remembered, holds information within short-term memory for an indefinite period of time. Rehearsal is also believed to be important in the transfer of information from short-term to long-term memory.

Some researchers suggest that short-term memory depends on the activation of *reverberating (oscillating) circuits* of neurons within the brain by signals

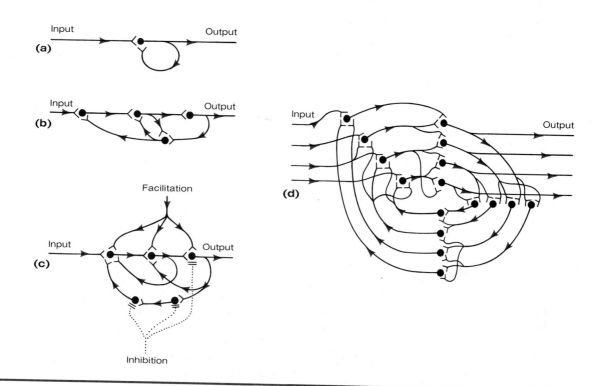

Figure 12.36
Reverberating circuits.
(a) and **(b)** Simple
reverberating circuits
that include collateral
branches from neu-
rons that restimulate
themselves. **(c)** A more
complex reverberating
circuit that includes both
facilitatory and inhibitory
fibers. **(d)** A complex
reverberating circuit that
includes many parallel
fibers with collateral
branches.

from various receptors (Figure 12.36). The components of the reverberating circuits repeatedly reactivate each other for a period of time, thereby retaining information within the circuits. Without rehearsal, the activity of the reverberating circuits is believed to diminish, the short-term memory fades, and the person can no longer remember the information.

It has also been proposed that short-term memory depends on a temporary facilitation of nerve impulse transmission at the synapses of particular neuronal circuits. This facilitation could be due to an increase in the release of neurotransmitters at synapses following the repeated activation of the neurons of the circuit, or to facilitatory changes in the resting membrane potentials of the neurons. In any event, without rehearsal, the facilitation—and the memory—are thought to decline.

Long-Term Memory

The ability of the nervous system to recall information long after it was first presented is **long-term memory.** The consolidation of a bit of information within the long-term memory appears to take some time. During this time, various events can occur that interfere with the entrance of the information into the long-term memory. To some investigators this possibility for interference suggests that information must remain within the short-term memory while the activities necessary to incorporate it into the long-term memory take place.

Theorists do not believe that long-term memory is due to the same type of neural activity responsible for short-term memory. Rather, they believe that long-term memory is the result of changes in the structure or biochemistry of neurons or synapses. Some investigators suggest that long-term memory involves alterations in the functional contacts between neurons in the brain (for example, changes in the number or area of synaptic junctions between neurons). Recent evidence suggests that mature neurons are constantly branching and establishing connections with new cells; at the same time they are discontinuing connections with other cells. Such changes are thought to be part of the physiological basis of learning and memory.

Other investigators have suggested that the excitability of synapses is altered by a change in the ability of presynaptic neurons to secrete transmitter substances or by an alteration in the sensitivity of postsynaptic neurons. Moreover, some investigators have proposed that changes in protein synthesis—perhaps because of changes in gene expression—are important for the development of long-term memory. These changes are believed to influence the structures or activities of neurons.

Alterations in the structure or biochemistry of neurons or synapses, such as those just discussed, are thought to provide for memory by producing long-term or permanent facilitation of nerve impulse transmission at the synapses of particular neuronal circuits, thus allowing the circuits to be easily re-excited by incoming signals at later dates. Such hypothetical facilitated neuronal circuits are called *memory traces (memory engrams).*

Locomotion

The alternating flexion of one limb and extension of the other that occurs during locomotion (walking or running) depends on rhythmic patterns of activity occurring within the spinal cord. These patterns, it is thought, make use of reverberating circuits of interneurons to stimulate alternately the flexor and extensor muscles of each leg, while reciprocal inhibition mechanisms keep one leg extended when the other is flexed. These activity patterns are believed to be coordinated by the integrative actions of higher centers, including the cerebral cortex, the basal ganglia, the cerebellum, and the reticular formation. The higher centers control these activities according to the desires of the individual.

The Cerebellum and the Coordination of Movement

The cerebellum is an especially important controller of muscular activities, particularly very rapid activities such as typing or running. During movement, the cerebellum receives information about the signals being sent to the muscles from areas of the motor system such as the cerebral cortex, the basal ganglia, and the reticular formation. The cerebellum also receives information from muscle spindles, neurotendinous organs, and joint receptors (and also from the eyes and the vestibular apparatus of the ears, which are discussed in Chapter 15) about the position, rate of movement, inertia, momentum, and so forth of body parts. The cerebellum compares the information about the movement called for by the motor system with information about the actual status of the body parts. If necessary, it sends signals back to the motor system to bring the actual movement into line with the intended one.

The cerebellum appears to play an important role in providing the central nervous system with the ability to predict the future position of a body part in the next few hundredths of a second during a movement. Some time before a moving body part reaches its intended position, signals are sent from the cerebellum to slow the moving part and stop it at its intended point. This activity involves the excitation of antagonistic muscles near the end of a movement coupled with the inhibition of the muscles that started the movement.

The cerebellum is also important in maintaining body equilibrium. When the direction of movement changes, signals transmitted to the cerebellum from receptors such as the vestibular apparatus of the ears help a person anticipate an impending loss of equilibrium and take corrective action before the equilibrium loss actually occurs.

Skilled Movements

Almost all skilled movements (such as the hand movements involved in typing or playing a violin) are learned movements whose mastery requires repetition. Such movements are usually slow and awkward at first, but their speed and efficiency generally improve with practice. It has been proposed that as a skilled movement is learned, the pattern of neural activity necessary to perform the movement becomes stored within the brain, perhaps in the form of a facilitated neuronal circuit (that is, as a memory trace). When the learned, preprogrammed movement is to be repeated, this stored pattern is called forth, and the muscle contractions necessary to perform the movement are repeated in the orderly sequence dictated by the pattern. The precise parts of the brain in which such patterns may be stored are not known. However, the motor cortex, the sensory cortex, and deeper centers such as the basal ganglia and even the cerebellum may be involved. Of particular interest is a

---

CLINICAL CORRELATION
Alzheimer's Disease

CASE REPORT ....

THE PATIENT A 61-year-old woman

PRINCIPAL COMPLAINT Loss of memory and disorientation

HISTORY The patient was reported by her sister to have shown a progressive decline in her mental capabilities over the past 5 years. Her family had attributed the mental changes to grief after the death of the patient's husband, but the increasing severity of the symptoms prompted them to seek medical evaluation. The family described the patient as very forgetful, and reported that she became confused and disoriented in places other than her home and the small food market that she and her husband have owned for years. The patient was no longer capable of handling the bookkeeping for the market or of balancing her checkbook, although she had managed the market's finances for more than 30 years.

CLINICAL EXAMINATION The patient was aware of her problems with memory and appeared concerned about the changes in her mental abilities. During an interview, she could not correctly state the current month, and although she correctly named the current president of the United States, she could not remember who preceded him. She had no difficulty with language, either receptive or expressive, no weakness of arms or legs, and no sensory deficits. Her gait was normal, and no abnormal movements were observed. Neuropsychological testing revealed a profound deficit in short-term memory, impaired visuospatial skills (manifested by the inability to duplicate three-dimensional constructions with blocks), and mild psychomotor slowing (decreased reaction time, decreased rate of finger-tapping). A standard adult-intelligence test revealed that the patient's IQ was approximately 20 points lower than would be predicted by her performance in high school and her vocational history. Pertinent laboratory findings included a CT scan consistent with mild, diffuse cerebral atrophy, an electroencephalogram (EEG) diffusely slow, with no focal abnormality, normal thyroid function tests, and normal electrocardiogram (ECG) and blood pressure.

COMMENT Clinically, *dementia* refers to a generalized diminution of mental abilities that consists mainly of memory loss, confusion and/or disorientation, personality changes, and generally slowed "mental processing." The term *dementia* usually is reserved for an irreversible condition; similar cognitive changes caused by reversible processes (such as drug effects, toxins, or hypoxia) are referred to as *delirium*.

Dementia can be caused by brain tumors, vascular disease, nutritional deficits, or hydrocephalus and can accompany many specific neurological degenerative diseases. The most common cause, however, accounting for more than 40% of geriatric patients who have dementia, is Alzheimer's disease.

Patients in the early stages of Alzheimer's disease generally have impairments of recent memory. As the disease progresses, higher cognitive functions deteriorate, and the abilities to read, write, calculate, and use language appropriately are lost. Alzheimer's patients may also become irritable, display emotional lability, and have hallucinations. The diagnosis of Alzheimer's disease, except at autopsy, is made by excluding other possible causes of the patient's symptoms, especially in the early stages. Other causes, particularly those that are treatable, must be considered first.

A number of changes occur in the brains of Alzheimer's patients. These changes include the occurrence of *neuritic plaques*, which consist of abnormal axons (primarily axon terminals) associated with a core of extracellular amyloid (a complex protein that has a hyaline, structureless nature). Also present are *neurofibrillary tangles*, which are bundles of paired helical filaments such as cross-linked polypeptides that accumulate within the cell bodies of neurons. Neuritic plaques and neurofibrillary tangles are prominent in the cerebral cortex and hippocampus, and generally the more severe a patient's cognitive defects, the greater will be the density of neuritic plaques in the patient's cortex (as determined at autopsy). Alzheimer's patients have a particular deficiency of acetylcholine-releasing neural endings in the cerebral cortex and hippocampus, and it appears that this deficiency is due at least in part to the degeneration of certain neurons whose cell bodies are located in the basal forebrain. (The axons of these neurons normally extend into the cerebral cortex and the hippocampus.) In addition, there is evidence that a degeneration of other neural systems can also occur in Alzheimer's patients. Although the cause of Alzheimer's disease is unknown, there is evidence that the disease has a genetic component.

OUTCOME Although there is no known effective treatment for Alzheimer's disease, the patient was given the tranquilizer diazepam (valium), 2 mg twice a day, to make her and her family more comfortable. Because of the progressive nature of the disorder, this patient probably will have to be placed in a special-care facility when the disease becomes more advanced.

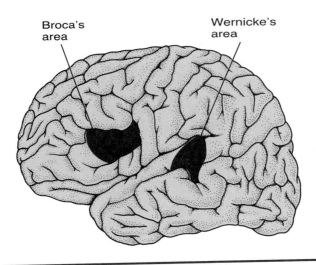

Broca's area   Wernicke's area

Figure 12.37
Brain areas associated with language.

cortical area in front of the primary motor cortex called the premotor area. Electrical stimulation of this area sometimes produces skilled patterns of movement such as hand movements, and this area is thought to be particularly involved in the occurrence of learned movements that can be rapidly performed.

As a skilled movement is being learned, it is usually performed slowly and deliberately. During this time, feedback from receptors in muscles and joints, from cutaneous receptors, and from the eyes is important in indicating the effectiveness and degree of success of the movement and in establishing a pattern of learned motor functions in the brain. Even after a precision movement has been mastered, feedback is important in monitoring and, if necessary, correcting the movement—if it is performed slowly. However, once they are learned, many skilled movements are performed so rapidly that there is no time for feedback control or correction to occur. Nevertheless, feedback is important in determining if the movement has been correctly performed and in correcting the skilled movement pattern if necessary so that the movement is executed properly the next time it is performed. The feedback signals are believed to be sent not only to the cerebellum but also to the sensory areas of the cerebral cortex, which, in turn, relays signals to the motor cortex.

The primary motor area of the cerebral cortex was long believed to be the area principally responsible for initiating voluntary motor activity. However, more recent studies have indicated that this is not the case. When monkeys were taught to perform simple hand movements in response to cues, it was found that neurons in the motor cortex, the cerebellum, and the basal ganglia all discharged impulses prior to the occurrence of the movement. Moreover, other studies indicated that cerebellar neurons become active before neurons of the motor cortex. These findings have led to the proposal that subcortical structures such as the cerebellum and basal ganglia are involved in initiating activity in motor cortex neurons. According to this view, signals related to the performance of learned, preprogrammed movements reach the motor cortex by way of the cerebellum and the basal ganglia. Moreover, it has been suggested that the cerebellum is particularly involved in the performance of rapid movements, the basal ganglia are important in the performance of slower movements, and the motor cortex is involved in the more precise integration of both rapid and slow movements.

Language

Language is a complex form of communication in which spoken or written words represent objects or concepts. One person communicates with or transfers information to another person primarily by language.

The areas of the cerebral cortex concerned with language are usually located in only one cerebral hemisphere—the left in 95% of people (Figure 12.37). A portion of the left frontal lobe known as *Broca's area* is concerned with the motor processes of word formation. If Broca's area is damaged, the person's speech is slow and labored, and he or she has difficulty articulating words. The person can decide what he or she wants to say but cannot make the vocal system perform smoothly.

A portion of the left temporal lobe known as *Wernicke's area* is responsible for choosing appropriate words to express a person's thoughts and for attaching meanings to words. If Wernicke's area is damaged, the person can still articulate words, but the words are often inappropriate or even nonsense words. Moreover, a person with extensive damage to Wernicke's area loses all comprehension of spoken language.

Broca's area and Wernicke's area cooperate to produce spoken language. Wernicke's area determines the appropriate choice and sequence of words to express a person's thoughts, and transmits signals to Broca's area. In Broca's area, the incoming signals are processed and a detailed motor program for the muscular movements of vocalization is generated. This program is transmitted to the primary motor cortex, which, in turn, sends nerve impulses to the speech muscles.

CONDITIONS OF CLINICAL SIGNIFICANCE

The Central Nervous System

Because of the complex structure of the brain and spinal cord, innumerable abnormalities can occur in the central nervous system as the result of either trauma or disease. Central nervous system dysfunctions are particularly serious because the neurons in the CNS are not capable of effective regeneration once they have been damaged. Consequently, complete recovery from CNS dysfunction is seldom achieved.

In the following sections we discuss some of the more common and more illustrative abnormalities of the CNS. However, we will begin by discussing the sensation of pain. Although this sensation itself is not generally considered to be an abnormality, it is associated with many abnormalities and dysfunctions, and it is certainly of great clinical significance.

Pain

Stimuli strong enough to cause tissue damage commonly give rise to a sensation of pain. Stimuli such as excessive mechanical stress, extremes of heat or cold, and various chemical substances stimulate pain receptors. This stimulation gives rise to neural signals that are transmitted to brain areas, including the reticular formation, the thalamus, and the somatic sensory areas of the cerebral cortex. In addition to eliciting a sensation of pain, the transmission of pain signals to the brain leads to emotional reactions (crying, anxiety, fear, for example) and behavioral responses (such as withdrawal or defensive responses).

Referred Pain

A person can often identify the location of a stimulus that produces pain because the brain usually projects the sensation of pain to the site of the stimulus. In some instances, however, signals from pain receptors in internal organs are incorrectly interpreted by the brain as coming from areas quite distant from the actual sites of stimulation—particularly as coming from sites on the body surface (Figure 12.38). This phenomenon is called **referred pain.** The locations of some referred pains are so consistent that they are used by physicians in diagnosing visceral dysfunction. For example, a heart attack frequently causes referred pain in the skin over the heart, on the left shoulder, and down the medial surface of the left arm.

One attempt to explain the cause of referred pain suggests that sensory neurons that transmit pain signals from a particular area of the body surface and sensory neurons that transmit pain signals from an internal organ connect with the same ascending neurons within the spinal cord. Thus, these ascending neurons carry pain signals to the brain from both the particular body surface area and the internal organ. Because cutaneous pain is much more common than visceral pain, pain signals carried over the ascending neurons are interpreted by the brain as having originated in the skin rather than in the viscera, and the pain sensation is projected to the skin site.

Phantom Pain

Phantom pain is the phenomenon whereby a person who has undergone an amputation continues to feel pain that he or she perceives as coming from the amputated body part. Phantom pain, like referred pain, is a case of inaccurate projection of the pain sensation by the brain. The neurons that supplied the affected structure are, of course, severed as a result of the amputation. However, the remaining portions of the neurons may continue to send nerve impulses to the same area of the brain as previously. The brain continues for some time to interpret impulses from the severed neurons as originating from the same body region as they did before the amputation. As a result, the sensations evoked in the brain are projected to that region. In this manner, pain (and other sensations) may still be "felt," for example, in the toes, even after the foot has been amputated.

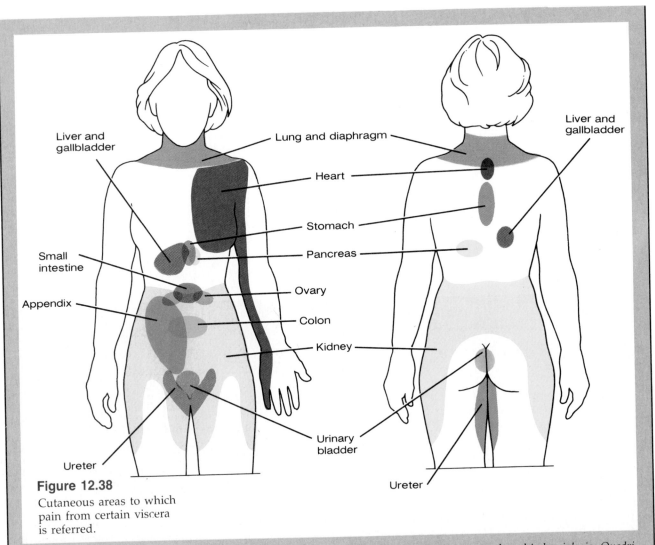

Figure 12.38
Cutaneous areas to which
pain from certain viscera
is referred.

Anesthesia

Various anesthetics are commonly used to reduce a person's sensitivity to painful stimuli. A number of local anesthetics (for example, procaine, tetracaine) exert their effects in circumscribed areas by decreasing the permeability of nerve-cell membranes to sodium ions. This decreased sodium permeability reduces the excitability of the neurons so that they do not transmit nerve impulses. General anesthetics reduce sensitivity to pain by rendering a person unconscious. This result depends at least partially on the ability of the general anesthetic to inhibit or depress conduction in the reticular activating system.

Spinal Cord Dysfunctions

Damage to tissues or organs that results in impairment or loss of function are referred to as *lesions*. Lesions of the spinal cord, either from injury or from disease, generally involve the motor and sensory fiber tracts of which the cord is composed.

Paralysis

Lesions of the motor spinal tracts cause *paralysis* of the structures supplied by the neurons involved. Paralysis of both of the lower limbs is called *paraplegia*. The condition in which both the upper limb and the lower limb on one side of the body are paralyzed is *hemiplegia*. *Quadriplegia* is the paralysis of all four limbs. Paralysis may be *flaccid*, when the affected muscles lose their normal slight contraction (tonus) and reflexes are absent; or it may be *spastic*, with increased tonus and reflexes. Spastic paralysis generally results from lesions of descending motor tracts from the brain. These lesions are often in the extrapyramidal system, which eliminates the normal inhibitory effects of the higher centers on the spinal reflexes. Flaccid paralysis generally results when the pyramidal tracts or the lower motor neurons are affected.

Lesions of Sensory Tracts of the Spinal Cord

Generally, lesions of the spinal cord tracts are not so precisely located that they involve only one tract. Damage to sensory tracts can be located by identifying the combination of sensory losses that are experienced. Damage to the tracts in the posterior funiculi (fasciculi gracilis and cuneatus) results in a loss of proprioceptive ability and touch discrimination. A loss of thermal and pain sensations from a particular region of the body indicates damage to the lateral spinothalamic tracts above the affected region. Damage to the ventral spinothalamic tracts causes a slight decrease in the sense of touch.

Specific Dysfunctions of the Spinal Cord

TABES DORSALIS   *Tabes dorsalis* is a condition caused by the progressive degeneration of the posterior funiculi of the spinal cord and the dorsal roots of the spinal nerves. It is a result of invasion of the CNS by the spirochete bacteria of syphilis. Because of the loss of the proprioceptive pathways in the posterior columns and dorsal roots, tabes dorsalis causes a loss of muscular coordination. Consequently, people affected with tabes dorsalis walk unsteadily *(ataxia)* and must watch the ground in order to maintain their balance. There may be severe leg and abdominal pain in the early stages of the disease due to the irritation of the sensory neurons in the dorsal roots. As these dorsal roots degenerate, the pain diminishes along with the sense of touch.

POLIOMYELITIS   The disease *poliomyelitis* is caused by a virus that destroys primarily the motor nerve-cell bodies in the anterior horns of the spinal cord, especially those in the cervical and lumbar enlargements. This degeneration produces fever, severe headache, and a flaccid paralysis of the muscles supplied by these neurons. After several weeks of paralysis, the muscles begin to atrophy, and eventually the muscle tissue may be almost completely replaced by connective tissue and adipose tissue. Death can result from respiratory failure or heart failure if the virus invades the nerve cells in the regulatory centers of the medulla.

SYRINGOMYELIA   The condition *syringomyelia* occurs when small fluid-filled sacs (cysts) form in the gray matter of the spinal cord and brain stem, and the neuroglia that line the central canal proliferate. Numerous commissural pathways are interrupted in syringomyelia, producing various sensory dysfunctions and muscular weakness accompanied by muscle atrophy. For example, the lateral spinothalamic tracts are often destroyed as they cross within the gray commissure of the cord. As a result, impulses concerned with the sensations of pain and temperature may not be transmitted to the brain from a particular region of the body.

MULTIPLE SCLEROSIS   *Multiple sclerosis* is a chronic condition that results in widespread destruction of the myelin sheaths of the nerve fibers in the spinal cord and brain. The destroyed myelin sheaths are replaced by hardened plaques that interfere with the normal transmission of nerve impulses. The cause of the destruction is not known; however, a virus is suspected by some investigators. Multiple sclerosis causes a great variety of symptoms, involving both motor and sensory tracts, depending on the areas of the CNS where plaque formation occurs. Common symptoms include abnormal sensations, spastic paralysis, and exaggerated reflexes. Although multiple sclerosis is a chronic condition, it is not unusual for the symptoms of the disease to disappear for years at a time. However, with each recurrence, there is permanent damage to the neurons of the CNS, thus causing progressive incapacitation.

Dysfunctions of the Brain Stem

All the sensory and motor tracts of the central nervous system travel through the brain stem (medulla, pons, and mesencephalon). The brain stem also contains various centers that control the autonomic nervous system, and most cranial nerves originate there. Consequently, lesions in this region of the CNS can produce a complex mixture of motor and sensory symptoms. Moreover, a condition such as tumor, hemorrhage, or trauma can affect the reticular activating system in the brain stem, producing a *coma*.

Dysfunctions of the Cerebellum

Lesions of the cerebellum or its pathways can cause dysfunctions in the smoothly coordinated actions that normally occur among groups of muscles. These dysfunctions may be evident as disturbances of gait and posture, as the inability to perform smoothly a task that involves the movement of several joints, or as an inability to estimate the range of movement and thus to stop a movement at a particular point.

Dysfunctions of the Basal Ganglia

The basal ganglia, which are part of the extrapyramidal system, are small islands of gray matter deep within the cerebral hemispheres. The neurons in the basal ganglia regulate skeletal muscle activity, in part by inhibiting contractions. This inhibition is in contrast to the excitatory effects of the neurons in the pyramidal tracts. For this reason, lesions of the basal ganglia generally result in spastic movements that appear to be voluntary but are actually beyond the individual's control. The person's limbs may be flung wildly about, writhe continuously, or move in a rapid, jerky manner.

Parkinsonism

Parkinsonism is the most common dysfunction of the basal ganglia. Since its onset is gradual, the disease primarily affects people over age 50. A number of factors are thought to produce Parkinson's syndrome, all of which cause the degeneration of cells in the basal ganglia. The condition is characterized by useless contractions of skeletal muscles, causing tremor and rigidity of the muscles, as well as by a decrease in muscular movements, such as swinging the arms while walking and changing facial expressions related to emotions.

There is impressive evidence that parkinsonism may be related to the abnormal metabolism of the neurotransmitter *dopamine* by the cells of the basal ganglia. In Parkinson's disease, the level of dopamine in the basal ganglia is found to be substantially reduced. It is not known how the reduced dopamine level brings about the symptoms of Parkinson's disease. Since dopamine itself is unable to effectively cross the blood–brain barrier (see page 311), the administration of supplementary dopamine is ineffective in controlling parkinsonism. However, *L-dopa*, a precursor (forerunner) of dopamine, will pass from the blood to the brain cells and has proved effective in treating the condition.

Huntington's Chorea

Another dysfunction that results from degeneration of

cells in the basal ganglia is *Huntington's chorea.* The symptoms of this progressive hereditary disorder usually do not appear until about age 40. Huntington's chorea is characterized by occasional involuntary flailing movements of the arms and legs and writhing movements of the hands, head, trunk, and feet. As the condition progresses, mental deterioration occurs, often resulting in personality changes. In contrast to parkinsonism, Huntington's chorea is treated with drugs that block the action of dopamine.

Inflammatory Diseases of the Central Nervous System

Encephalitis and Myelitis

The invasion of nervous tissue by viruses (and in some instances by bacteria, fungi, and other agents) can produce inflammation of the CNS. Inflammation of the brain is called *encephalitis.* Inflammation of the spinal cord is known as *myelitis.* Both conditions display a variety of possible motor and sensory symptoms, including paralysis, coma, and death.

Meningitis

Meningitis is an inflammation of the meninges that cover the brain and spinal cord. It is caused by a number of microorganisms, the most common being the meningococci, streptococci, pneumococci, and tubercle bacilli. The organisms are thought to enter the body through the nose and throat. Meningitis produces a high fever with a severe headache and stiffness of the neck. In the more serious cases, meningitis can cause coma and death.

Tumors of the Central Nervous System

Most tumors of the CNS develop from glial cells, including the oligodendrocytes that form the neurilemma in the CNS, but it is not uncommon for cells of the meninges to form tumors also. Only rarely do the neurons themselves give rise to tumors. The presence of tumors in the brain and spinal cord can produce a variety of dysfunctions, depending on their location. The symptoms of a tumor can include headache, convulsions, a change in behavior patterns, pain, and paralysis. These dysfunctions result from the destruction of nervous tissue by the tumor, increased pressure in the skull, or edema of the nervous tissue. Tumors may be treated by surgical removal or destroyed by chemical and radiation therapy.

STUDY OUTLINE

THE BRAIN pp. 337–359

TELENCEPHALON Cerebrum in adult; contains two lateral ventricles.

Cerebral Cortex Outer surface of gray matter; unmyelinated neurons.

Basal Ganglia Gray matter structures deep inside cerebral hemispheres.

White Matter Myelinated nerve-fiber tracts.

MYELINATED NERVE-FIBER TRACTS
1. *Projection tracts:* carry motor or sensory impulses from one level of brain or spinal cord to another.
2. *Association tracts:* connect areas in cortex of same cerebral hemisphere.
3. *Commissural tracts:* connect left and right hemispheres.

GYRI, FISSURES, SULCI, AND LOBES OF THE CEREBRUM

Gyri Convolutions on surface of cerebrum.

Fissures Deeper furrows between gyri.

Sulci Shallower furrows between gyri.

Lobes of cerebrum Division of fissures and sulci into frontal, parietal, temporal, and occipital lobes; insula located deep in lateral fissure.

FUNCTIONAL AREAS OF THE CEREBRAL CORTEX

Primary Motor (Primary Somatic Motor) Area In precentral gyrus of frontal lobe; involved in controlling voluntary contractions of skeletal muscle.

Premotor Area Anterior to primary motor area; causes groups of muscles to contract, producing stereotyped movements.

Primary Sensory (Primary Somatic Sensory) Area In postcentral gyrus of parietal lobe; receives sensory information concerning temperature, touch, pain, proprioception.

Special Senses Areas
1. *Primary visual area:* posterior occipital lobe.
2. *Primary auditory area:* upper margin of temporal lobe.
3. *Primary olfactory area:* medial surface of temporal lobe.
4. *Primary taste area:* parietal lobe, near bottom of postcentral gyrus.

Association Areas Interconnect motor and sensory areas.
1. *Frontal association area:* site of intellectual activities.
2. *Somatic association area:* integration and interpretation center.
3. *Visual and auditory association areas:* contribute to interpretation of visual and auditory experiences.

BASAL GANGLIA Involved in control of skeletal muscle activity; part of extrapyramidal system; act in part to inhibit muscle contraction.

OLFACTORY BULBS On ventral surface of cerebral hemisphere; smell.

HEMISPHERIC SPECIALIZATION Each cerebral hemisphere is somewhat specialized for carrying out certain kinds of mental process.

DIENCEPHALON

THALAMUS Two oval masses of gray matter forming walls of third ventricle; sensory relay and integrating center of brain; some motor involvement.

HYPOTHALAMUS Located below thalamus; regulates:
1. *Sympathetic and parasympathetic activity* (for example, body temperature, water balance, appetite, gastrointestinal activity, sexual activity).
2. *Emotions* (fear and rage, for example).
3. *Pituitary hormone release.*

Mammillary bodies Olfactory reflexes.

Tuber cinereum Contains neurons that transport regulatory hormones (or factors) from hypothalamus to pituitary gland.

Infundibulum Hormone transport to posterior lobe of pituitary gland.

Optic chiasma Point of crossing of some optic nerve fibers.

EPITHALAMUS Forms roof of third ventricle; includes pineal body, which has possible neuroendocrine function.

THE LIMBIC SYSTEM Includes structures that affect emotional responses.

MESENCEPHALON Between forebrain and hindbrain.

CEREBRAL AQUEDUCT Connects third and fourth ventricles.

CEREBRAL PEDUNCLES Nerve fiber tracts from primary motor area of cerebral cortex to pons and spinal cord, and sensory nerve fibers from spinal cord to thalamus.

RED NUCLEUS Coordinates cerebellum and cerebral hemispheres.

CORPORA QUADRIGEMINA Visual and auditory functions.

METENCEPHALON Superior portion of hindbrain; contains fourth ventricle.

CEREBELLUM Concerned with functions below conscious level; directs precise, smooth movements and maintains equilibrium.

PONS Consists of nerve fiber tracts that connect cerebellum with brain stem; connects cerebellum with various levels of central nervous system; also affects respiration rate.

MYELENCEPHALON (MEDULLA OBLONGATA) Includes motor nerve tracts called *pyramids* on ventral surface and *medullary centers* that control vital functions such as heart rate, respiration, blood vessel dilation and constriction, coughing, swallowing, and vomiting.

RETICULAR FORMATION Receives input from several areas of brain; sensory input from spinal tracts; activates cerebral cortex; maintains wakefulness; also called *reticular activating system (RAS).*

VENTRICLES OF THE BRAIN Four interconnected spaces filled with cerebrospinal fluid.

CHOROID PLEXUSES In roof of each ventricle; form cerebrospinal fluid.

THE MENINGES Three layers of connective tissue that cover entire CNS:

DURA MATER

ARACHNOID

PIA MATER

CEREBROSPINAL FLUID Cushions CNS; formed from blood via choroid plexuses; circulates in and around CNS; returns to blood via arachnoid villi.

THE SPINAL CORD pp. 359–372

GENERAL STRUCTURE OF THE SPINAL CORD Gives rise to 31 pairs of spinal nerves.

CERVICAL ENLARGEMENT Forms brachial plexus; supplies upper limbs.

LUMBAR ENLARGEMENT Forms lumbosacral plexus; supplies lower limbs.

MENINGES OF THE SPINAL CORD Same as brain: dura mater, arachnoid, pia mater.

COMPOSITION OF THE SPINAL CORD White myelinated areas surround gray, unmyelinated central area.

GRAY MATTER OF SPINAL CORD H-shaped.

Posterior Horns Axons of sensory neurons from spinal nerves; develop from alar plate.

Anterior Horns Cell bodies of somatic motor neurons whose axons leave cord for a spinal nerve; develop from basal plate.

Lateral Horns Only in thoracic and upper lumbar regions; cell bodies of visceral motor neurons; develop from basal plate.

DORSAL AND VENTRAL ROOTS OF SPINAL NERVES Dorsal and ventral roots unite on each side of each spinal segment, forming a spinal nerve.

Dorsal Roots Enter spinal cord at tips of posterior gray horns; sensory nerves; contain dorsal root ganglia.

Ventral Roots Leave spinal cord at tips of anterior gray horns; somatic and visceral motor nerves.

WHITE MATTER OF SPINAL CORD Surrounds gray matter.

Funiculi Posterior, lateral, and anterior regions of white matter in each half of spinal cord.

Tracts Small bundles of neuron processes within funiculi:
1. *Ascending (sensory) spinal tracts:* carry afferent sensory impulses from peripheral sensory receptors to brain centers; all cross over in CNS.
   a. *Fasciculus gracilis* and *fasciculus cuneatus:* awareness of joint sense, muscle position sense, and fine-touch localization.
   b. *Spinothalamic tracts:* pain, temperature, touch, pressure.
   c. *Spinocerebellar tracts:* unconscious proprioception.
2. *Descending (motor) spinal tracts:* carry impulses from brain to lower motor neurons that regulate skeletal muscle.
   a. *Pyramidal tracts* (corticospinal tracts): originate from pyramidal cells in cortex of precentral gyrus and travel through medulla pyramids; voluntary control of skeletal muscles, especially fine movements.
   b. *Extrapyramidal tracts:* originate from various nuclei in brain stem; modify muscular contractions for posture and balance.

THE SPINAL REFLEX ARC Neural pathway by which sensory impulses from receptors reach effectors without traveling to brain.

RECEPTOR → SENSORY NEURON → SYNAPSE → MOTOR NEURON → EFFECTOR.

STRETCH REFLEX Initiated by muscle spindles that respond to stretch; stretch of spindle increases activity of afferent neurons from spindle, which increases stimulation of alpha motor neurons to muscle (for example, patellar reflex). Muscle spindles provide information about lengths of skeletal muscles that is important in coordination of muscular activity. Activity of muscle spindles may assist in attaining proper degree of muscular contraction.

TENDON REFLEX Neurotendinous organs respond to increased tension with increased afferent neuron activity that ultimately inhibits alpha motor neurons to muscle associated with tendon. Neurotendinous organs provide information about tension developed by muscle.

NEURON POOLS Functional groups of neurons within central nervous system; divergence, convergence, summation, and facilitation occur within pools; even simple actions require participation of many neurons, whose activities are coordinated in complex circuits in neuron pools. **p. 372**

INTEGRATIVE FUNCTIONS OF THE NERVOUS SYSTEM pp. 372–381

THE ELECTROENCEPHALOGRAM Record of electrical waves produced during brain activity; spontaneous waves present, even during unconsciousness; some correlation between brain waves and body activity.

THE AROUSED BRAIN Able to consciously react to stimuli; reticular activating system important in arousing brain; general sensory input plays important role in maintaining brain arousal.

SLEEP State of partial or complete unconsciousness from which a person can be aroused by appropriate stimuli; *slow-wave sleep:* high-amplitude, low-frequency EEG; *paradoxical sleep:* rapid-eye-movement (REM) sleep; dreams.

SLEEP–WAKEFULNESS CYCLE Awakening appears to occur when impulses to cerebral cortex from RAS reach sufficient intensity to arouse brain.
1. Positive feedback responses involving brain areas such as cerebral cortex as well as peripheral musculature contribute to the activity of the RAS leading to wakefulness.
2. Over time, excitability of wakefulness circuit diminishes, intensity of signals from RAS declines, and individual becomes drowsy and falls asleep.

MEMORY Ability to store experiences, thoughts, and sensations for later recall.

SHORT-TERM MEMORY Decays rapidly unless rehearsed; may depend on activation of reverberating circuits of neurons within the brain by signals from receptors.

LONG-TERM MEMORY May involve actual alterations in structure or biochemistry of neurons or synapses; may involve changes in number or area of synaptic junctions between neurons; may involve facilitated neuronal circuits called memory traces (memory engrams).

LOCOMOTION
1. Believed to depend on rhythmic patterns of activity occurring within spinal cord that involve reverberating neuronal circuits and reciprocal inhibition.
2. Rhythmic activity patterns are coordinated to produce effective locomotion by integrative actions of higher centers (such as cerebral cortex, basal ganglia, cerebellum, reticular formation).
3. Higher centers control activities according to desires of the individual.

THE CEREBELLUM AND THE COORDINATION OF MOVEMENT
1. During movement, receives signals from motor areas about intended movement and from periphery about actual movement.
2. Compares information and, if necessary, sends signals to motor areas to bring actual movement into line with intended movement.
3. Appears to play important role in enabling CNS to predict future position of a body part during movement.
4. Important in maintaining equilibrium.

SKILLED MOVEMENTS
1. Almost all are learned movements that require repetition to master.
2. As a movement is learned, pattern of neural activity necessary to perform it may become laid down in brain; when the movement is to be repeated, the pattern is called forth and the necessary muscular contractions are repeated.
3. Premotor area of the cortex may be important in pattern storage.
4. Sensory feedback important in learning skilled movements and establishing patterns of learned motor function in brain; also involved in monitor-

ing and perhaps correcting pattern of movement after movement is mastered.

5. Cerebellum and basal ganglia appear to be involved in initiating activity in neurons of motor cortex during learned, preprogrammed movements; cerebellum particularly involved in rapid movements, basal ganglia in slow movements.

LANGUAGE Complex form of communication in which spoken or written words represent objects or concepts; *Broca's area* concerned with motor processes of word formation; *Wernicke's area* responsible for choosing appropriate words to express thoughts and for attaching meanings to words.

CONDITIONS OF CLINICAL SIGNIFICANCE: THE CENTRAL NERVOUS SYSTEM pp. 381–384

PAIN Signals to brain from pain receptors elicit sensation of pain and also lead to such activities as emotional reactions and behavioral responses.

REFERRED PAIN Inaccurate projection of pain sensation to site other than stimulus site.

PHANTOM PAIN Person feels pain from amputated body part.

ANESTHESIA Anesthetics reduce sensitivity to painful stimuli. Local anesthetics can decrease permeability of nerve-cell membranes to sodium ions so neurons do not transmit impulses. General anesthetics render person unconscious, at least partially by inhibiting or depressing conduction in the reticular activating system.

SPINAL CORD DYSFUNCTIONS

PARALYSIS Caused by lesions of motor spinal tracts.

LESIONS OF SENSORY TRACTS OF THE SPINAL CORD Various sensory losses, depending on lesion site.

SPECIFIC DYSFUNCTIONS OF THE SPINAL CORD

Tabes Dorsalis Progressive degeneration of posterior funiculi and dorsal roots of spinal nerves caused by spirochete bacteria of syphilis.

Poliomyelitis Caused by virus that destroys nerve-cell bodies in anterior horns of spinal cord.

Syringomyelia Cyst formation in gray matter of cord and brain stem, with proliferation of neuroglia of central canal.

Multiple Sclerosis Chronic, widespread destruction of myelin sheaths of neurons in spinal cord and brain.

DYSFUNCTIONS OF THE BRAIN STEM Lesions, tumors, hemorrhage, or trauma can produce variety of motor and sensory symptoms; damage to reticular activating system can produce coma.

DYSFUNCTIONS OF THE CEREBELLUM Lesions cause dysfunction in coordinated actions between muscle groups.

DYSFUNCTIONS OF THE BASAL GANGLIA Lesions result in spastic movements.

PARKINSONISM May be related to abnormal metabolism of dopamine by basal ganglia.

HUNTINGTON'S CHOREA Treated with drugs that block dopamine.

INFLAMMATORY DISEASES OF THE CENTRAL NERVOUS SYSTEM

ENCEPHALITIS Brain inflammation.

MYELITIS Spinal cord inflammation.

MENINGITIS Inflammation of meninges that cover brain and spinal cord.

TUMORS OF THE CENTRAL NERVOUS SYSTEM Symptoms include headaches, convulsions, behavior change, pain, paralysis; tumors destroy nervous tissue, increase pressure within skull, or cause edema of nervous tissue; treatments include surgical removal or chemical and radiation therapy.

SELF-QUIZ

1. The fibers that connect the left and right cerebral hemispheres form the: (a) projection tracts; (b) association tracts; (c) commissural tracts.

2. The cerebrum is divided into left and right hemispheres by the: (a) longitudinal fissure; (b) occipital lobe; (c) parietal lobe.

3. Each hemisphere of the cerebrum is divided by the central sulcus into: (a) frontal and parietal lobes; (b) parietal and occipital lobes; (c) occipital and frontal lobes.

4. The taste area is located in the parietal lobe, deep in the longitudinal fissure. True or False?

5. The control of emotions is influenced largely by the: (a) thalamus; (b) hypothalamus; (c) epithalamus.

6. "Pleasure centers" and "punishment centers" of the brain are located in the: (a) mammillary bodies; (b) limbic system; (c) intermediate mass.

7. The cerebellum is part of the: (a) mesencephalon; (b) telencephalon; (c) metencephalon.

8. The part of the brain that directs precise, smooth movements and the maintenance of equilibrium is the: (a) cerebrum; (b) cerebellum; (c) medulla oblongata.

9. The myelencephalon is the most inferior division of the brain. True or False?

10. The control of heart rate, coughing, and swallowing is a function of the: (a) reticular activating system; (b) pons; (c) medullary centers.

11. The vagus cranial nerve is number: (a) IX; (b) X; (c) XI.

12. The roof of each ventricle is thin and contains a plexus of capillaries. True or False?

13. Cushioning the CNS against jolts and blows is a function of the: (a) meninges; (b) cerebrospinal fluid; (c) ventricles of the brain.

14. The portion of the spinal cord that gives rise to the spinal nerves supplying the upper limbs is the: (a) cervical enlargement; (b) lumbar enlargement; (c) lumbosacral plexus.

15. The white matter of the brain and spinal cord is composed of: (a) meninges; (b) myelinated processes of neurons; (c) nerve-cell bodies.

16. Match the terms associated with the brain with the appropriate description.

Cerebrum
Projection tracts
Basal ganglia
Cerebral hemisphere
Insula
Primary motor area
Visual area
Olfactory area
Thalamus
Cerebral peduncles
Metencephalon
Pons
Hypothalamus
Ventricles
Telencephalon
Reticular formation

(a) This structure is divided into frontal, parietal, temporal, and occipital lobes
(b) Located in the posterior portion of the occipital lobe
(c) Acts as the sensory relay and integrating center of the brain
(d) Connects the cerebellum with the brain stem
(e) Carry motor and/or sensory nerve impulses from one level of the CNS to another
(f) Contains gyri, fissures, and sulci
(g) Controls many vital processes, including appetite and sexual activity
(h) A fluid-filled system of cavities in the brain
(i) The structure that envelops the diencephalon and obscures much of the rhombencephalon
(j) Network of interlacing nerve fibers that exerts control over the cerebral cortex
(k) A portion of the cerebral cortex located deep in the lateral fissure
(l) Located on the medial surface of the temporal lobe
(m) Gray-matter structures located deep in the cerebral hemispheres
(n) Two cylindrical nerve tracts on the ventral surface of the mesencephalon
(o) Contains the cerebellum
(p) Located in the precentral gyrus

17. The ascending tracts of the spinal cord carry afferent (sensory) impulses from peripheral receptors to various centers in the brain. True or False?

18. Sensory information received by receptors on the right side of the body is transmitted to the: (a) left cerebral cortex; (b) right cerebral cortex; (c) both.

19. Match the items associated with the spinal cord with the appropriate description.

Cauda equina
Meninges
White matter
Gray matter
Ventral roots
Fasciculus gracilis
Spinothalamic tracts
Spinocerebellar tracts
Pyramidal tracts
Extrapyramidal tracts

(a) Contains myelinated axons that travel up or down the spinal cord to higher or lower levels in the CNS
(b) Convey impulses concerned with pain and temperature
(c) These structures contribute processes of both somatic motor neurons and visceral motor neurons to the spinal nerves
(d) Dura mater, arachnoid, pia mater
(e) Modify muscular contractions related to posture and balance
(f) Major structures involved in the voluntary control of skeletal muscles
(g) A mass of descending nerve roots below the end of the spinal cord
(h) Carry information concerning unconscious proprioception from neuromuscular receptors
(i) Enables you to locate where an object is touching your body
(j) Substance whose anterior horns contain the cell bodies of somatic motor neurons

20. Increasing the tension applied to a neurotendinous organ within a tendon: (a) decreases the activity of the afferent neurons from the tendon organ; (b) decreases the stimulation of alpha motor neurons to the muscle associated with the tendon; (c) increases the contraction of the intrafusal fibers of the tendon organ.

21. Electroencephalogram records during slow-wave sleep show waves of: (a) low amplitude; (b) low frequency; (c) no amplitude or frequency.

22. Electroencephalogram records of an alert or awake brain show waves of: (a) low amplitude and high

frequency; (b) high amplitude and low frequency; (c) high amplitude and high frequency.

23. Destruction of the reticular activating system produces: (a) temporary loss of memory; (b) permanent coma; (c) an inability to sleep soundly.

24. General sensory input may play an important role in maintaining the arousal of the brain. True or False?

25. REM sleep is the same as: (a) slow-wave sleep; (b) drowsy sleep; (c) paradoxical sleep.

26. Parkinsonism is a dysfunction of the: (a) basal ganglia; (b) brain stem; (c) cerebellum

LEARNING OBJECTIVES

After completing this chapter, you should be able to:

1. State the function of the peripheral nervous system, and list its two divisions.

2. Name the 12 pairs of cranial nerves, and specify the functions of each.

3. Describe the distribution patterns of the spinal nerves.

4. Distinguish between dorsal rami and dorsal roots and between ventral rami and ventral roots.

5. List the three major nerve plexuses, and state the body regions they supply.

6. Name and describe the specific distribution of each nerve formed by the major nerve plexuses.

7. Distinguish between the somatic and autonomic nervous systems by describing the function of each.

8. Cite the most common disorders of the peripheral nervous system.

CHAPTER CONTENTS

CRANIAL NERVES

SPINAL NERVES

CONDITIONS OF CLINICAL SIGNIFICANCE: THE PERIPHERAL NERVOUS SYSTEM

KEY TERMS AND DERIVATIVES

chiasma (*chiasm* = across) an x-shaped crossing; the optic chiasma is formed by the crossing of the optic nerves

dermatome (*derma* = skin; *tom* = cut) the specific area of skin innervated by a specific spinal nerve

peripheral nervous system (*peri* = all around) a system of nerves that connect the outlying parts of the body and their receptors with the central nervous system

plexus (*plexus* = braid) a general term for a network, such as a network of nerves

somatic nervous system (*somatic* = referring to the body wall, in contrast to the internal organs) a division of the peripheral nervous system that includes the sensory receptors on the body surface and within the muscles, and the nerves that link them with the CNS

The Peripheral Nervous System

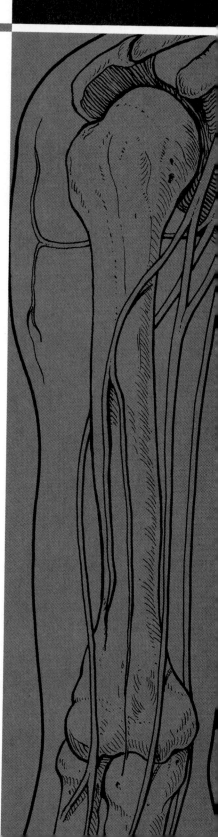

The **peripheral nervous system (PNS)** includes all neurons except those that are restricted to the brain and spinal cord—that is, all except the neurons in the central nervous system. It consists of pathways of nerve fibers between the central nervous system and all outlying structures of the body. Included in the peripheral nervous system are 12 pairs of **cranial nerves** and 31 pairs of **spinal nerves.** In terms of function, the peripheral nervous system is divided into the:

1. **Afferent (sensory) division,** whose nerve fibers relay impulses from all areas of the body, including the viscera, to the central nervous system.

2. **Efferent (motor) division,** which is subdivided into the:
   a. **Somatic nervous system,** whose fibers carry motor impulses between the central nervous system and the skeletal muscles.
   b. **Autonomic nervous system,** which connects motor fibers from the central nervous system with smooth muscles, cardiac muscle, and glands.

CRANIAL NERVES

Twelve pairs of cranial nerves arise from the brain (Figure 13.1). Most of the cranial nerves are mixed nerves, being composed of both motor and sensory neurons, although a few cranial nerves carry only sensory impulses. Previously it was thought that some cranial nerves transmitted only motor impulses. It is now known, however, that the cranial nerves formerly considered to be entirely motor actually contain some sensory neurons from proprioceptors in the muscles they innervate. Thus, they convey information to the central nervous system concerning the lengths of specific muscles and the tension generated when the muscles contract or are stretched.

In the cranial nerves, as in the spinal nerves, the cell bodies of the sensory neurons are in ganglia located *outside* the central nervous system. The cell bodies of the motor neurons are located in nuclei *within* the central nervous system. Some motor neurons in the cranial nerves supply skeletal muscles. These neurons, which are under conscious control, are called **somatic motor neurons.** Other motor neurons in the cranial nerves are not under conscious control. These supply smooth muscles, cardiac muscle, or glands and are called **visceral motor neurons.** The visceral motor neurons in the cranial nerves are part of the parasympathetic division of the autonomic nervous system, which is discussed in Chapter 14. With the exception of the vagus nerve, the cranial nerves supply only structures in the head and neck. The cranial nerves are numbered (using Roman numerals) from superior to inferior, in the order in which they leave the brain. Table 13.1 (pp. 400–401) summarizes the cranial nerves and their functions.

I: Olfactory Nerves

The first pair of cranial nerves, the **olfactory nerves,** arise from receptor cells in the epithelial lining of the nose—that is, the nasal mucosa (Figure 13.2;

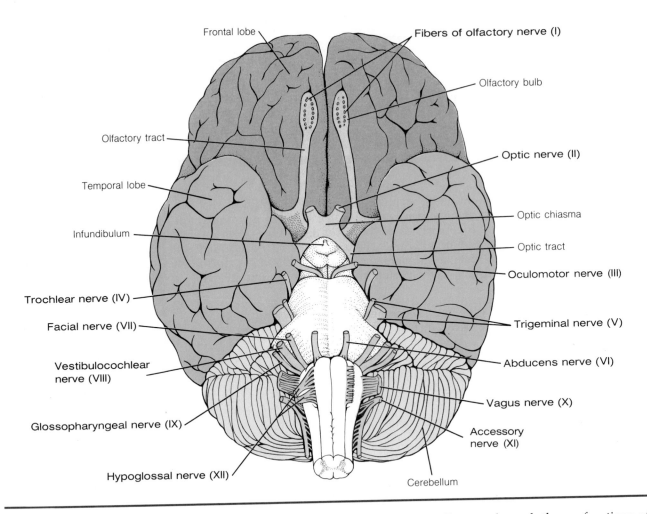

Frontal lobe
Fibers of olfactory nerve (I)
Olfactory bulb
Olfactory tract
Optic nerve (II)
Temporal lobe
Optic chiasma
Infundibulum
Optic tract
Oculomotor nerve (III)
Trochlear nerve (IV)
Trigeminal nerve (V)
Facial nerve (VII)
Vestibulocochlear nerve (VIII)
Abducens nerve (VI)
Vagus nerve (X)
Glossopharyngeal nerve (IX)
Accessory nerve (XI)
Hypoglossal nerve (XII)
Cerebellum

Figure 13.1
The ventral surface of the brain showing the cranial nerves.

op'-tik kigh-az'-mah

F13.3,
Table 13.1

Table 13.1). Processes of these receptor cells pass through the perforations of the *cribriform plate* of the ethmoid bone and enter the olfactory bulbs of the telencephalon portion of the brain. In the olfactory bulbs, the nerve fibers synapse with neurons that pass posteriorly in the olfactory tracts. The fibers of the olfactory tracts enter the brain, and many of them travel to the cerebral cortex of the medial sides of the temporal lobes. The olfactory nerves are entirely sensory, carrying impulses associated with the sense of smell.

II: Optic Nerves

The **optic nerves** carry impulses associated with vision. Like the olfactory nerves, they are entirely sensory. The optic nerves are actually brain tracts rather than true nerves, since they are formed from outgrowths of the embryonic diencephalon.

The optic nerves originate in the retina of the eyes, on which images are focused. After leaving the posterior surface of the eyeball, each optic nerve exits from the orbital cavity and enters the cranial cavity through an *optic foramen* in the sphenoid bone. Shortly after entering the cranium, the two optic nerves meet in the **optic chiasma,** just anterior to the pituitary gland (Figure 13.3; Table 13.1). In the optic chiasma, nerve fibers from the medial half of each retina cross to the opposite side, and those from the lateral half of each retina remain on the same side. The fibers then continue to the brain as the **optic tracts.** Because of the crossing of nerve fibers at the optic chiasma, each optic tract consists of fibers from the retinas of both eyes.

Some nerve fibers in the optic tracts terminate in the superior colliculi of the midbrain, where they function in subconscious visual reflexes. However, most of the fibers in the optic tracts travel to the lateral geniculate bodies of the thalamus, where they synapse with neurons that form pathways called **optic radiations.** The neurons of the optic radiations, which are third-order

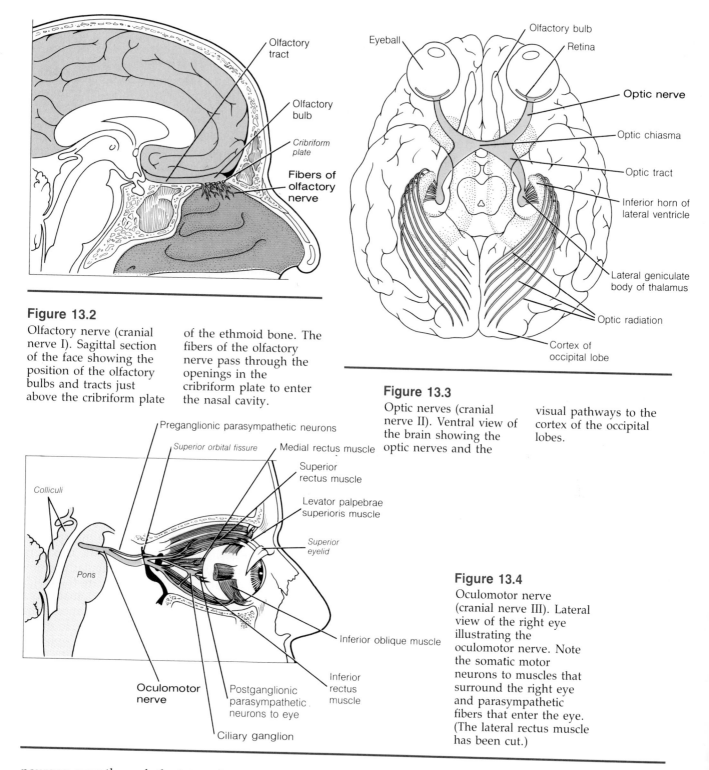

Figure 13.2

Olfactory nerve (cranial nerve I). Sagittal section of the face showing the position of the olfactory bulbs and tracts just above the cribriform plate of the ethmoid bone. The fibers of the olfactory nerve pass through the openings in the cribriform plate to enter the nasal cavity.

Figure 13.3

Optic nerves (cranial nerve II). Ventral view of the brain showing the optic nerves and the visual pathways to the cortex of the occipital lobes.

Figure 13.4

Oculomotor nerve (cranial nerve III). Lateral view of the right eye illustrating the oculomotor nerve. Note the somatic motor neurons to muscles that surround the right eye and parasympathetic fibers that enter the eye. (The lateral rectus muscle has been cut.)

neurons, pass through the internal capsule and terminate in the visual cortex of the occipital lobes.

III: Oculomotor Nerves

The **oculomotor nerves** emerge from the midbrain, just superior to the pons, and enter the orbits through the *superior orbital fissures*, which are located between the small wings and great wings of the sphenoid bone. The oculomotor nerves consist of somatic motor neurons traveling to, and proprioceptive sensory neurons traveling from, four of the six extrinsic muscles that move the eyeball (Figure 13.4; Table 13.1). Specifically, the oculomotor nerves

F13.4,
Table 13.1

innervate the superior rectus, medial rectus, inferior rectus, and inferior oblique muscles of the eyeball. In addition, the oculomotor nerves supply the levator palpebrae superioris muscles, which function to elevate the upper eyelids.

The oculomotor nerves also contain neurons of the parasympathetic division of the autonomic nervous system (Chapter 14). The efferent pathways of the autonomic nervous system have two neurons between the central nervous system and the innervated structures. The first neuron, called a **preganglionic (presynaptic) neuron,** always travels from the central nervous system to a ganglion located outside the central nervous system. Within the ganglion, the preganglionic neuron synapses with second neurons, called **postganglionic (postsynaptic) neurons,** which leave the ganglion and travel to the innervated structures. Preganglionic parasympathetic neurons in the oculomotor nerves synapse in ganglia called **ciliary ganglia,** which are located behind the eyeballs. From a ciliary ganglion, postganglionic neurons enter the eye and innervate the intrinsic smooth muscles that regulate the size of the pupil and shape of the lens. Therefore, apart from regulating the voluntary movement of the eyeballs, the oculomotor nerves are also involved in the reflex adjustments of the eyes to varying intensities of light and in focusing the eyes for near and far vision.

pree-gang-lee-on'-ik

post-gang-lee-on'-ik

IV: Trochlear Nerves

The **trochlear nerves** are small nerves that arise below the inferior colliculi on the dorsal surface of the midbrain. These nerves, which are the only cranial nerves to exit from the dorsal surface of the brain, curve around the lateral sides of the brain and enter the orbits through the *superior orbital fissures* along with the oculomotor nerves (Figure 13.5; Table 13.1). The trochlear nerves carry somatic motor neurons to, and proprioceptive neurons from, one of the extrinsic muscles of the eyes, the superior oblique muscles. Thus, the trochlear nerves aid in the voluntary movements of the eyeballs.

F13.5, Table 13.1

V: Trigeminal Nerves

The large **trigeminal nerves** emerge from the lateral sides of the pons. As their name indicates, each trigeminal nerve has three divisions: the **ophthalmic, maxillary,** and **mandibular nerves** (Figure 13.6; Table 13.1). The three divisions exit the skull through openings in the sphenoid bone: the ophthalmic division leaves the skull through the *superior orbital fissure;* the maxillary division passes through the *foramen rotundum;* and the mandibular division exits the skull through the *foramen ovale.* The trigeminal nerves are the major sensory nerves of the face. They contain sensory neurons that originate from the skin of the face and anterior scalp and from the mucous membranes of the nasal cavity and mouth. The cell bodies of these sensory neurons are found in large **trigeminal (semilunar) ganglia** located at the points where the ophthalmic, maxillary, and mandibular nerves join before entering the brain.

of-thal'-mik

F13.6, Table 13.1

The trigeminal nerves also contain motor neurons that travel within the mandibular nerve to the muscles of mastication (medial and lateral pterygoids, masseter, and temporalis), to the mylohyoid muscle, and to the anterior belly of the digastric muscle.

VI: Abducens Nerves

The sixth cranial nerves originate in the metencephalon and exit the brain stem just below the pons (Figure 13.7; Table 13.1). The **abducens nerves** enter the orbits through the *superior orbital fissures* along with the oculomotor (III) and trochlear (IV) cranial nerves. The abducens nerves carry somatic motor neurons to, and proprioceptive sensory neurons from, the remaining extrinsic muscle of each eye, the lateral rectus muscles.

F13.7, Table 13.1

Eye movements involve the coordinated contraction of the extrinsic muscles of both eyes. This coordination, in turn, requires a synchronization of the nerve impulses carried by the oculomotor (III), trochlear (IV), and abducens (VI) cranial nerves.

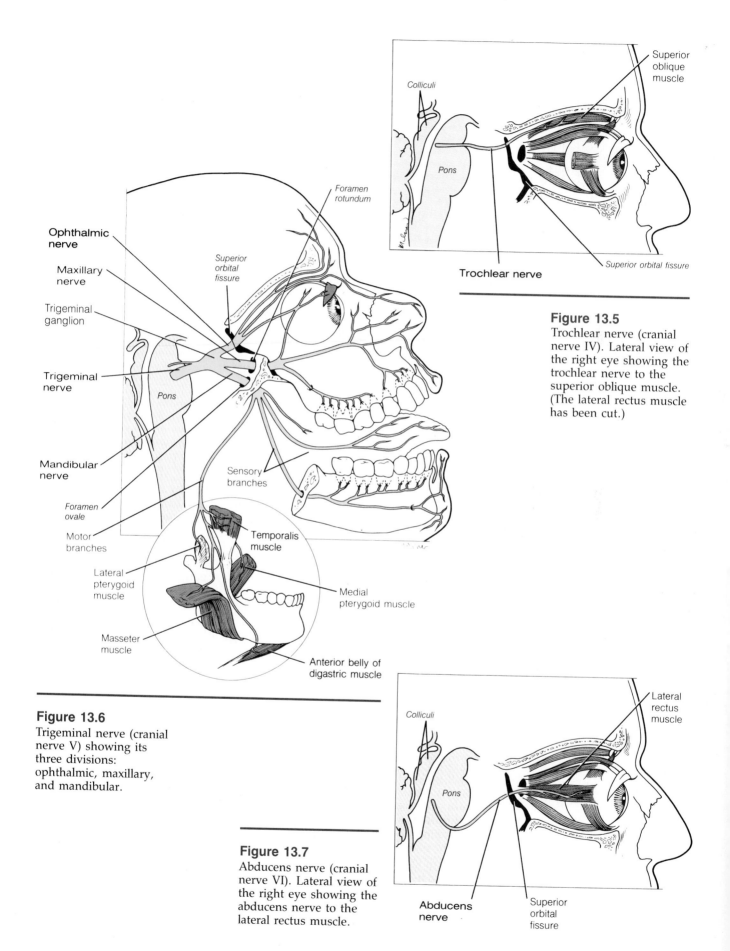

Ophthalmic nerve

Maxillary nerve

Trigeminal ganglion

Trigeminal nerve

Pons

Mandibular nerve

Foramen ovale

Motor branches

Lateral pterygoid muscle

Masseter muscle

Foramen rotundum

Superior orbital fissure

Sensory branches

Temporalis muscle

Medial pterygoid muscle

Anterior belly of digastric muscle

Colliculi

Pons

Superior oblique muscle

Superior orbital fissure

Trochlear nerve

Figure 13.5
Trochlear nerve (cranial nerve IV). Lateral view of the right eye showing the trochlear nerve to the superior oblique muscle. (The lateral rectus muscle has been cut.)

Figure 13.6
Trigeminal nerve (cranial nerve V) showing its three divisions: ophthalmic, maxillary, and mandibular.

Colliculi

Pons

Lateral rectus muscle

Abducens nerve

Superior orbital fissure

Figure 13.7
Abducens nerve (cranial nerve VI). Lateral view of the right eye showing the abducens nerve to the lateral rectus muscle.

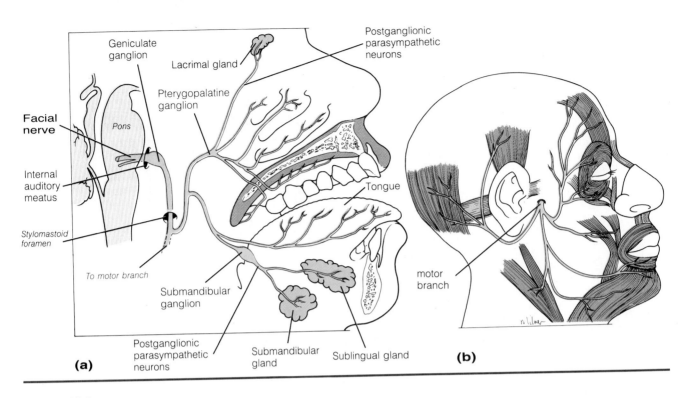

Figure 13.8
Facial nerve (cranial nerve VII). **(a)** Sensory and parasympathetic neurons. **(b)** Voluntary motor branches.

F13.8,
Table 13.1

VII: Facial Nerves

The **facial nerves** leave the metencephalon at the lower border of the pons, just lateral to the abducens nerves, and enter the petrous portions of the temporal bones through the *internal auditory meatus*. After traveling through the temporal bones, the facial nerves leave the skull by way of the *stylomastoid foramina*. They then pass forward across the cheek through the parotid salivary glands and divide into numerous branches that supply motor neurons to the muscles of the face and scalp (Figure 13.8; Table 13.1).

The facial nerves also contain sensory neurons that originate from the taste buds on the anterior two-thirds of the tongue. The cell bodies of these sensory neurons are located in the **geniculate ganglia,** which lie within the petrous portions of the temporal bones.

Parasympathetic neurons to the lacrimal (tear) glands, to mucous glands in the nasal cavity, and to the submandibular and sublingual salivary glands are also carried in the facial nerves. Preganglionic parasympathetic neurons to the lacrimal glands synapse in the **pterygopalatine ganglia** with postganglionic parasympathetic neurons that then pass to the lacrimal glands. The preganglionic parasympathetic neurons to the submandibular and sublingual glands synapse in the **submandibular ganglia** with postganglionic parasympathetic neurons that then travel to the glands.

VIII: Vestibulocochlear Nerves

koak'-lee-er
F13.9,
Table 13.1

The eighth cranial nerves have two separate divisions: the **cochlear** and the **vestibular nerves** (Figure 13.9; Table 13.1). Both of these divisions are entirely sensory and originate from inner-ear receptors located in the petrous portions of the temporal bones. The two divisions join to form a common trunk, the **vestibulocochlear nerve,** which leaves the temporal bones through the *internal auditory meatus* and enters the brain stem just below the pons.

The cochlear divisions transmit impulses related to hearing from the spiral organ located in the cochlea of the ear (Chapter 15). The cell bodies of the cochlear nerves lie in **spiral ganglia** located within the cochlea.

The vestibular divisions are concerned with equilibrium. Their receptors are located in the ampullae of the semicircular canals and in the saccule and utricle of the vestibule of the ear (Chapter 15). The cell bodies of the vestibular nerves are located in the **vestibular ganglia.**

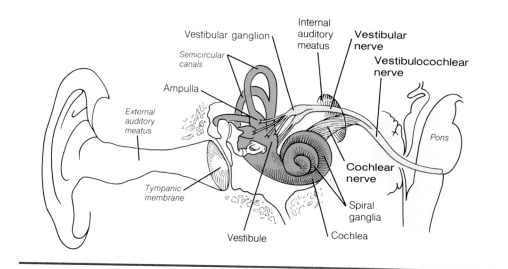

Figure 13.9
Vestibulocochlear nerve (cranial nerve VIII), showing the vestibular nerve that supplies the vestibule and ampullae and the cochlear nerve that supplies the cochlea. All these structures are within the inner ear.

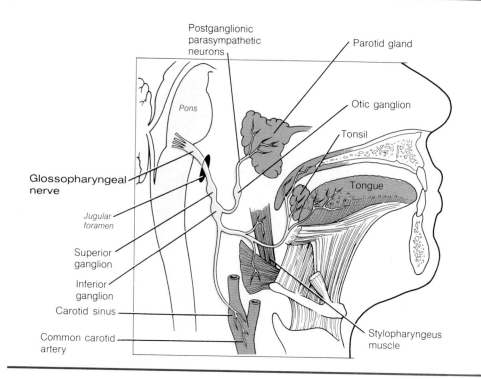

Figure 13.10
Glossopharyngeal nerve (cranial nerve IX) to the tongue, throat, and parotid gland.

IX: Glossopharyngeal Nerves

The **glossopharyngeal nerves** are mixed nerves, carrying both motor and sensory impulses. As the name indicates, these nerves supply the tongue and the pharynx. They emerge from the medulla and leave the skull through the *jugular foramina* of the temporal bone (Figure 13.10; Table 13.1).

Sensory neurons in the glossopharyngeal nerves carry impulses from the taste buds of the posterior third of the tongue; from the mucous membranes of the pharynx, tonsils, and middle-ear cavity; from receptors sensitive to changes in blood levels of oxygen and carbon dioxide in the carotid body; and from receptors that monitor the blood pressure in the carotid sinus (Chapter 19). The cell bodies of these sensory neurons are located in the **superior** and **inferior ganglia** of the nerves.

The glossopharyngeal nerves also contain motor neurons that supply the stylopharyngeus muscles of the pharynx, which are involved in swallowing. In addition, some motor neurons of the glossopharyngeal nerves intermix with the vagus (X) and accessory (XI) cranial nerves to innervate several other pharyngeal muscles.

F13.10,
Table 13.1

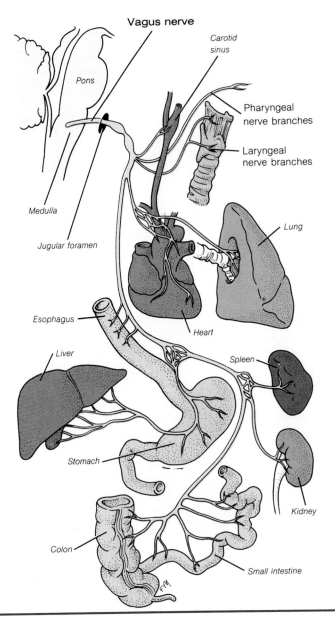

Figure 13.11
Vagus nerve (cranial nerve X), showing its distribution to the neck, thoracic cavity, and abdominal cavity.

Preganglionic parasympathetic neurons that travel in the glossopharyngeal nerves synapse in the **otic ganglia,** from which postganglionic parasympathetic neurons travel to the parotid salivary glands.

X: Vagus Nerves

The **vagus nerves** are the only cranial nerves that are not restricted to the head and neck regions. They leave the sides of the medulla by several rootlets, pass through the *jugular foramina* of the temporal bones, and descend along the pharynx close to the common carotid arteries and the internal jugular veins (Figure 13.11; Table 13.1). After leaving the neck, the vagus nerves enter the thorax and abdomen.

F13.11, Table 13.1

The vagus nerves carry motor impulses to the voluntary muscles of the pharynx and larynx, and sensory impulses from taste receptors located toward the base of the tongue and from the skin of the external ear. Moreover, they have a broad parasympathetic distribution, carrying preganglionic neurons to the involuntary muscles of the thoracic and abdominal viscera as far caudally as the transverse colon of the large intestine. Located close to, or in, the walls of the innervated structures are small ganglia from which short postganglionic parasympathetic neurons supply the structures. The vagus nerves also carry sensory fibers from the viscera. The sensory input from the viscera generally does not reach the conscious level; rather, it automatically

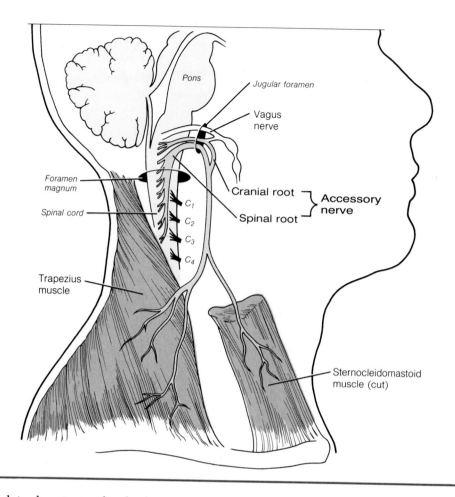

Figure 13.12
Accessory nerve (cranial nerve XI), with cranial and spinal portions separated.

regulates heart rate, depth of respiration, blood pressure, digestive processes, and so forth. Under certain conditions these visceral sensations can reach conscious levels, however, as evidenced by sensations of distension and nausea.

XI: Accessory Nerves

The **accessory nerves** are composed of motor neurons to voluntary muscles and some sensory neurons from proprioceptors. Each accessory nerve is actually formed by two nerves. One arises from the medulla and is thus a true cranial nerve, whereas the other arises from the cervical region of the spinal cord and is actually a spinal nerve (Figure 13.12; Table 13.1). The spinal portions pass upward along the sides of the spinal cord and enter the skull through the *foramen magnum* of the occipital bone. In the cranial cavity, the spinal and cranial portions join and leave the skull through the *jugular foramina* along with the glossopharyngeal (IX) and vagus (X) cranial nerves.

F13.12, Table 13.1

The fibers of the cranial portions intermix with the vagus nerves to supply the muscles of the larynx and pharynx. The fibers of the spinal portions supply the trapezius and sternocleidomastoid muscles.

XII: Hypoglossal Nerves

The **hypoglossal nerves** leave the anterior surface of the medulla as a series of rootlets and pass out of the skull through the *hypoglossal canals* of the occipital bone. As the name indicates, they are located beneath the tongue, where they travel anteriorly in close relationship with the first cervical spinal nerve (Figure 13.13; Table 13.1). The hypoglossal nerves consist of motor neurons that supply the intrinsic and extrinsic muscles of the tongue, and some sensory neurons from proprioceptors. The first three cervical nerves send motor fibers to some of the muscles of the neck through branches called the *ansa cervicalis*, which are so closely associated with the hypoglossal nerves that they appear to arise from them.

F13.13, Table 13.1

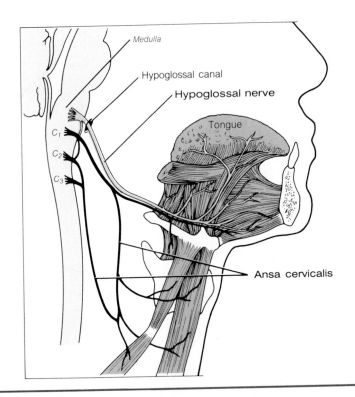

Figure 13.13
Hypoglossal nerve (cranial nerve XII), which supplies the muscles of the tongue. Branches of the first three cervical spinal nerves interconnect with the hypoglossal nerve to supply certain muscles of the throat.

Table 13.1   Summary of the Cranial Nerves

Nerves	Site of Exit from Brain	Site of Exit from Skull	Functions
I: Olfactory [F13.2] *sensory*	Telencephalon (cerebral hemisphere)	Cribriform plate of ethmoid	*Sensory* olfaction (sense of smell)
II: Optic [F13.3] *sensory*	Diencephalon	Optic foramen	*Sensory* vision
III: Oculomotor [F13.4] *motor and proprioception*	Mesencephalon (midbrain)	Superior orbital fissure	*Motor* levator palpebrae superioris and the external eye muscles, except superior oblique and lateral rectus *Proprioception* from the innervated muscles *Parasympathetic* ciliary muscle of the lens and the sphincter of pupil
IV: Trochlear [F13.5] *motor and proprioception*	Mesencephalon (midbrain)	Superior orbital fissure	*Motor* superior oblique muscle of the eye *Proprioception* from the superior oblique muscle
V: Trigeminal [F13.6] *mixed*	Metencephalon (pons)		
Ophthalmic division		Superior orbital fissure	*Sensory* cornea; skin of nose, forehead, and scalp
Maxillary division		Foramen rotundum	*Sensory* nasal cavity, palate, and upper teeth, skin of cheek, and upper lip
Mandibular division		Foramen ovale	*Sensory* tongue, lower teeth, skin of chin, lower jaw, and temporal regions

Table 13.1   Summary of the Cranial Nerves   (continued)

Nerves	Site of Exit from Brain	Site of Exit from Skull	Functions
			Motor muscles of mastication *Proprioception* from muscles of mastication
VI: Abducens [F13.7] *motor and proprioception*	Metencephalon (pons)	Superior orbital fissure	*Motor* lateral rectus muscle of the eye *Proprioception* from lateral rectus muscle
VII: Facial [F13.8] *mixed*	Metencephalon (pons)	Stylomastoid foramen	*Motor* muscles of facial expression *Proprioception* from muscles of facial expression *Sensory* taste from the anterior two-thirds of tongue *Parasympathetic* sublingual and submandibular salivary glands; lacrimal glands; mucous glands of nasal cavity
VIII: Vestibulocochlear [F13.9] *sensory*	Metencephalon (pons)		
Vestibular division		Internal auditory meatus	*Sensory* equilibrium
Cochlear division		Internal auditory meatus	*Sensory* hearing
IX: Glossopharyngeal [F13.10] *mixed*	Myelencephalon (medulla)	Jugular foramen	*Motor* stylopharyngeus muscle; other pharyngeal muscles via cranial nerves X and XI *Proprioception* from innervated muscles *Parasympathetic* parotid salivary glands *Sensory* taste and general sensation from the posterior one-third of the tongue; pharynx, middle-ear cavity, carotid sinus
X: Vagus [F13.11] *mixed*	Myelencephalon (medulla)	Jugular foramen	*Motor* muscles of the pharynx and the larynx *Proprioception* from innervated muscles *Sensory* skin of the external ear; taste from the rear of the tongue; visceral sensory from the thoracic and the abdominal organs *Parasympathetic* organs of the thoracic and abdominal cavities
XI: Accessory [F13.12] *motor and proprioception*	Myelencephalon (medulla)	Jugular foramen	*Motor* trapezius and sternocleidomastoid muscles *Proprioception* from innervated muscles
XII: Hypoglossal [F13.13] *motor and proprioception*	Myelencephalon (medulla)	Hypoglossal canal	*Motor* intrinsic and extrinsic muscles of the tongue *Proprioception* from innervated muscles

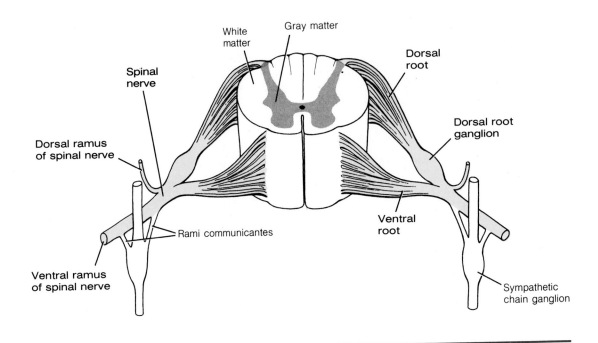

Figure 13.14
A segment of the spinal cord showing the formation of a pair of spinal nerves from dorsal and ventral roots.

SPINAL NERVES

There are 31 pairs of **spinal nerves,** including 8 cervical, 12 thoracic, 5 lumbar, 5 sacral, and 1 coccygeal. With the exception of the first pair of cervical nerves, the spinal nerves leave the vertebral canal by passing through the *intervertebral foramina.* The first pair of cervical nerves exit between the occipital bone and the atlas. The second through the seventh pairs of cervical nerves emerge *above* the vertebrae for which they are named. The eighth pair of cervical nerves emerge between the seventh cervical and first thoracic vertebrae. All remaining pairs of spinal nerves pass *below* the vertebrae for which they are named.

Formation of Spinal Nerves

F13.14

The spinal nerves are formed from the union of **ventral** and **dorsal roots** that leave or enter the spinal cord (Figure 13.14). The ventral roots contain axons of motor neurons that leave the anterior and lateral gray horns of the spinal cord. The cell bodies of these motor neurons are located in the spinal cord. The dorsal roots contain axons of sensory neurons that enter the posterior horns of the gray matter. The cell bodies of these sensory neurons lie outside the spinal cord in **dorsal root ganglia (spinal ganglia)** on each dorsal root. After the roots join, all spinal nerves are mixed nerves containing the processes of motor (somatic and visceral) and sensory neurons.

Branches of Spinal Nerves

F13.15

Soon after passing through the intervertebral foramina, each spinal nerve divides into two branches: a **dorsal ramus** and a **ventral ramus** (Figure 13.15). The dorsal rami pass posteriorly to supply the skin and muscles of the back. The ventral rami are longer, and their distribution varies in different body regions. Like the spinal nerves, both rami are mixed, containing motor and sensory fibers. In the thoracic region, the ventral rami travel in the intercostal spaces between the ribs to supply the skin and muscles of the lateral and anterior body walls. In the cervical, lumbar, and sacral regions, the ventral rami of successive spinal nerves unite to form *plexuses* (networks) that give rise to nerves that supply the skin, muscles, and joints of the upper and lower limbs.

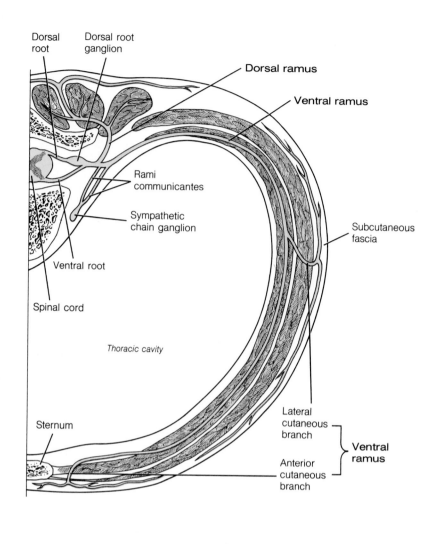

Figure 13.15
The pathways of the ventral and dorsal rami of a thoracic spinal nerve.

Distribution of Spinal Nerves

The rami of the spinal nerves are distributed throughout the body in a fairly systematic pattern, as evidenced by the distribution of sensory fibers to the skin in uniform regions called **dermatomes** (Figure 13.16). The pattern is particularly regular in the trunk region, where the rami of each spinal nerve supply a horizontal band of skin. The regular pattern of nerve distribution is also present, although somewhat modified, in the skin of the limbs. In the upper limbs, the ventral rami of cervical nerves supply the posterior and anterolateral surfaces of the entire limb, as well as the ventral surfaces of the hands. The ventral ramus of the first thoracic nerve (T_1) supplies the antero-medial surfaces of the limbs (in the anatomical position). In the lower limbs, the ventral rami of the lumbar nerves supply the anterior surfaces, and the ventral rami of the sacral nerves supply the posterior surfaces.

F13.16

Plexuses and Peripheral Nerves

The peripheral nerves that are formed by the intermixing of the ventral rami of the spinal nerves in plexuses have specific names. These nerves primarily supply the skin and the underlying muscles of the limbs. Since each of these named peripheral nerves is formed of fibers from more than one spinal nerve, their distribution patterns (Figure 13.16) do not duplicate the cutaneous distribution patterns of the spinal nerves. The main nerve plexuses, all of which are paired, are the *cervical, brachial,* and *lumbosacral* plexuses.

F13.16

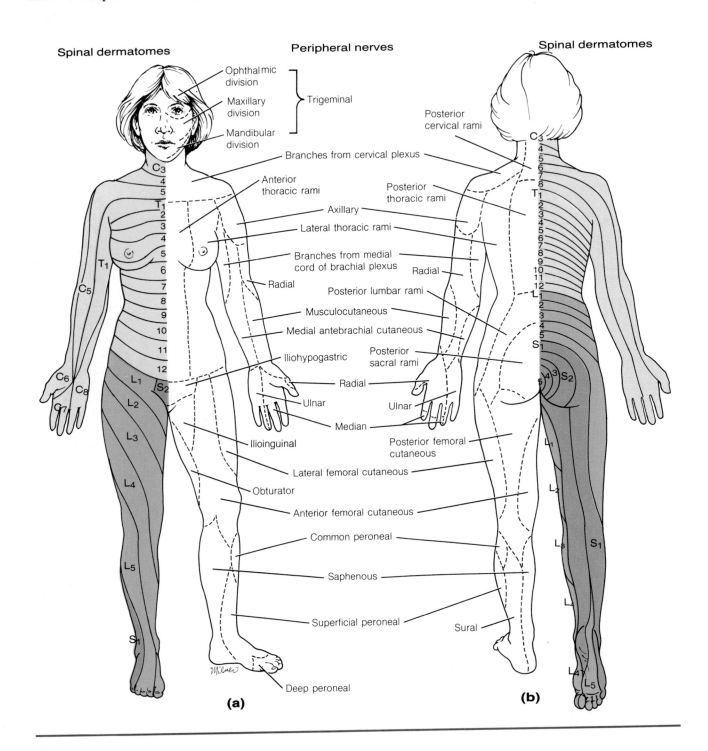

Spinal dermatomes   **Peripheral nerves**   **Spinal dermatomes**

Ophthalmic division
Maxillary division   } Trigeminal
Mandibular division

Posterior cervical rami

Branches from cervical plexus

Anterior thoracic rami

Posterior thoracic rami

Axillary

Lateral thoracic rami

Branches from medial cord of brachial plexus

Radial

Radial

Posterior lumbar rami

Musculocutaneous

Medial antebrachial cutaneous

Iliohypogastric

Posterior sacral rami

Radial

Ulnar

Ulnar

Median

Ilioinguinal

Posterior femoral cutaneous

Lateral femoral cutaneous

Obturator

Anterior femoral cutaneous

Common peroneal

Saphenous

Superficial peroneal

Sural

Deep peroneal

(a)   **(b)**

Figure 13.16

Dermatome and peripheral distribution of spinal nerve innervations. **(a)** Anterior surface of the body. **(b)** Posterior surface of the body.

Table 13.2

Cervical Plexus

Each **cervical plexus** is formed by the ventral rami of the first four cervical nerves (Figure 13.17). Branches from the plexus supply the muscles and skin of the neck, shoulder, and chest. Branches of the cervical plexus also interconnect with cranial nerves X (vagus), XI (accessory), and XII (hypoglossal). One branch from each of the cervical plexuses, the **phrenic nerve,** passes through the thorax to supply the diaphragm. Although this innervation may seem unusual for a cervical nerve, during embryonic development the diaphragm arises from cervical myotomes. As the diaphragm assumes its adult position in the lower portion of the thoracic cavity, it retains its embryonic nerve supply. The major branches from the cervical plexus are summarized in Table 13.2.

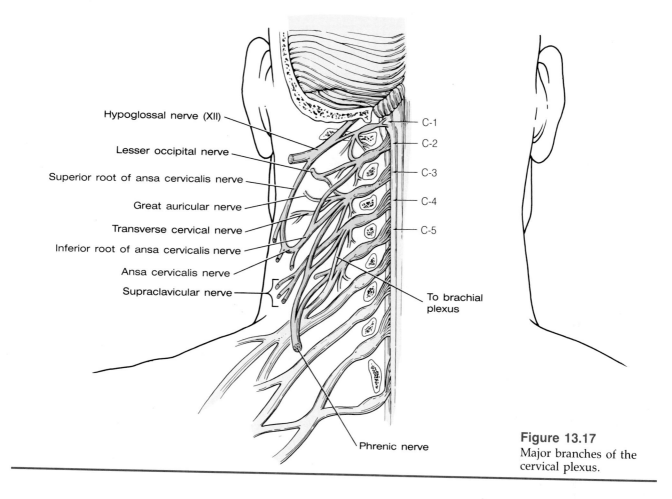

Hypoglossal nerve (XII)

Lesser occipital nerve

Superior root of ansa cervicalis nerve

Great auricular nerve

Transverse cervical nerve

Inferior root of ansa cervicalis nerve

Ansa cervicalis nerve

Supraclavicular nerve

C-1
C-2
C-3
C-4
C-5

To brachial plexus

Phrenic nerve

Figure 13.17
Major branches of the cervical plexus.

Table 13.2   The Cervical Plexus [F13.16; F13.17]

Nerve	Spinal Nerves Involved (Ventral Rami)	Distribution
SUPERFICIAL BRANCHES		
Lesser occipital	C_2	*Skin* of scalp behind and above ear
Greater auricular	C_2 through C_3	*Skin* over parotid gland, over mastoid process, and on back of ear
Transverse cutaneous	C_2 through C_3	*Skin* over side and front of neck
Supraclavicular	C_3 through C_4	*Skin* over upper region of shoulder and chest
DEEP (MUSCULAR) BRANCHES		
Ansa cervicalis (leaves the hypoglossal nerve)		
Superior root	C_1 through C_2	Deep *muscles* of neck, including geniohyoid and thyrohyoid
Inferior root	C_2 through C_3	Infrahyoid *muscles*, including sternohyoid and sternothyroid
Phrenic	C_3 through C_5	*Diaphragm*
Muscular Branches	C_2 through C_4	*Muscles* sternocleidomastoid, trapezius, levator scapulae, and scalenus medius

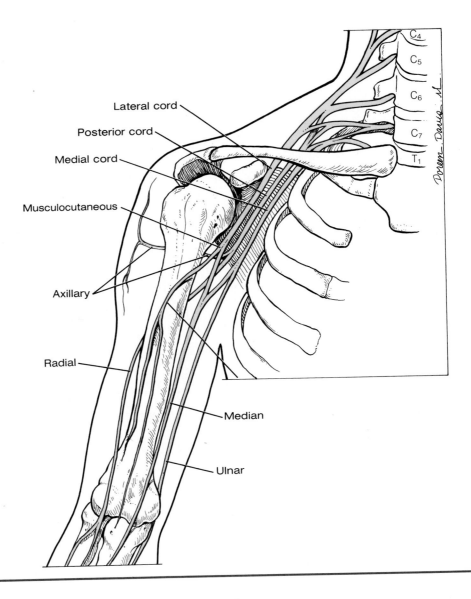

Figure 13.18
The brachial plexus and the nerves of the upper limb.

Brachial Plexus

Each **brachial plexus** is formed by the ventral rami of the last four cervical nerves (C_5 to C_8) and the first thoracic nerve (T_1). The brachial plexus extends downward and laterally, passing behind the clavicle, to enter the axilla (Figure 13.18). The *roots* of the plexus, which are the ventral rami of the spinal nerves, unite to form a *superior trunk* (C_5–C_6), a *middle trunk* (C_7), and an *inferior trunk* (C_8–T_1). Branching off of the roots are a *dorsal scapular nerve* to the levator scapulae and rhomboid muscles, and a *long thoracic nerve* to the serratus anterior muscle.

Each trunk divides into *anterior* and *posterior divisions*. Branching off of the trunks are a *suprascapular nerve* to the supraspinatus and infraspinatus muscles and a *nerve to the subclavius* muscle. The trunks, in turn, separate into *posterior*, *lateral*, and *medial* cords. Five major nerves, as well as several smaller ones, arise from these cords and provide the entire nerve supply to the skin and muscles of the upper limbs (Figure 13.19). The major branches of the brachial plexus are summarized in Table 13.3.

POSTERIOR CORD   The **axillary nerve** passes laterally from the posterior cord to supply the skin and muscles of the shoulder. The **radial nerve,** the main branch of the posterior cord, passes behind the humerus and curves around it (in the radial groove) to supply the skin on the posterior surface of

F13.18

F13.19
Table 13.3

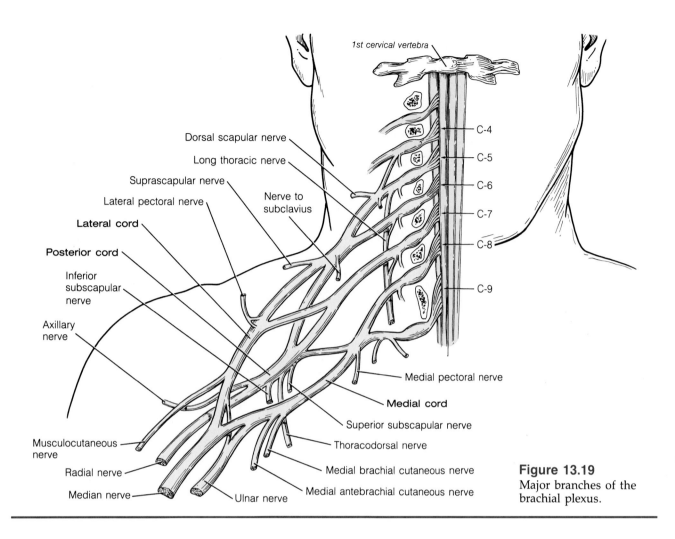

Figure 13.19
Major branches of the brachial plexus.

the arm, forearm, and hand as well as the extensor muscles of the arm and forearm. Smaller branches of the posterior cord include the *superior* and *inferior subscapular nerves* to the subscapularis and teres major muscles and the *thoracodorsal nerve* to the latissimus dorsi muscle.

LATERAL CORD   The **musculocutaneous nerve** from the lateral cord supplies the skin on the lateral surface of the forearm and several anterior muscles of the arm. The *lateral pectoral nerve* branches off of the lateral cord and supplies the pectoralis major muscle.

MEDIAL CORD   The **ulnar nerve** leaves the medial cord and passes behind the medial epicondyle of the humerus to supply the skin on the medial surface of the hand and some flexor muscles of the forearm as well as several intrinsic muscles of the hand. Other branches from the medial cord include the *medial pectoral nerve* to the pectoralis major and minor muscles, the *medial brachial cutaneous nerve* to the skin of the medial side of the arm, and the *medial antebrachial cutaneous nerve* to the skin on the medial side of the forearm.

MEDIAN NERVE   The **median nerve** is formed by branches from the medial and lateral cords of the brachial plexus. It supplies the skin and muscles of the lateral portion of the palmar surface of the hand and the flexor muscles of the forearm.

Lumbosacral Plexus

The nerves of each **lumbosacral plexus** supply the skin and muscles of the buttocks, pelvis, lower abdomen, and lower limbs (Figure 13.20). The plexus   F13.20

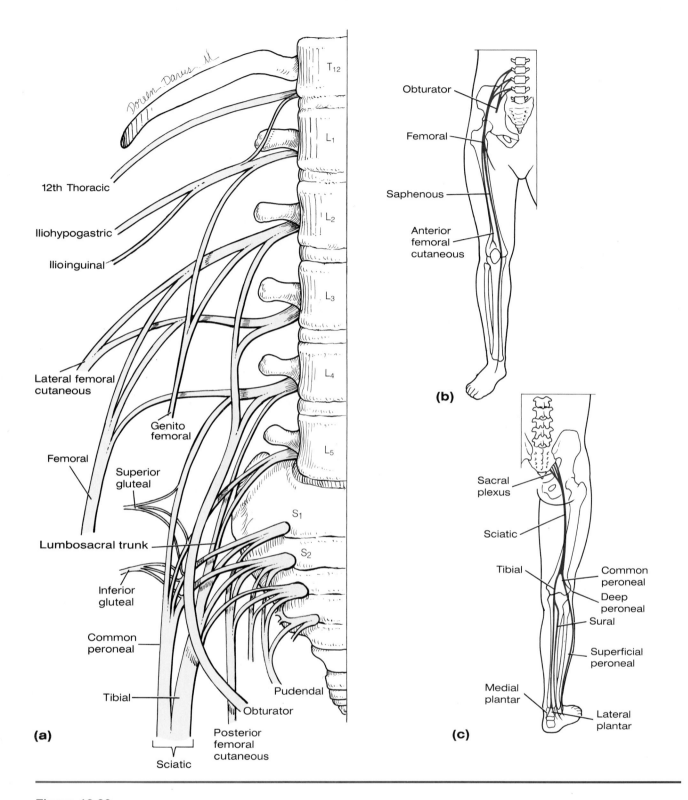

Figure 13.20
(a) Major branches of the lumbosacral plexus.
(b) Peripheral distribution of the femoral nerve and its branches to the anterior surface of the limb.
(c) Peripheral distribution of the sciatic nerve and its branches to posterior and lateral surfaces of limb.

is divisible into two sections, the *lumbar plexus* and the *sacral plexus*, which are connected by a *lumbosacral trunk* of nerves.

LUMBAR PLEXUS   Each **lumbar plexus** is formed from the ventral rami of the first four lumbar nerves and some nerve fibers from the twelfth thoracic nerve. The nerves that arise from the lumbar plexus supply the lower abdomen and the anterior and medial portions of the lower limb (Figure 13.16). The major nerves that arise from the lumbar plexus are the femoral and the obturator nerves. Smaller branches of this plexus are described in Table 13.4.

Table 13.3 Major Branches of the Brachial Plexus
[F13.16; F13.18; F13.19]

Nerve	Spinal Nerves Involved (Ventral Rami)	Distribution
Posterior cord Axillary	C_5 and C_6	*Skin* of the shoulders *Muscles* teres minor and deltoid
Radial	C_5 through C_8; T_1	*Skin* of posterior lateral surface of arm, forearm, and hand *Muscles* triceps brachii, supinator, anconeus, brachioradialis, extensor carpi radialis brevis, extensor carpi radialis longus, extensor carpi ulnaris, and several muscles that move the fingers (see Table 9.14)
Lateral cord Musculocutaneous	C_5 through C_7	*Skin* of the lateral surface of the forearm *Muscles* brachialis, biceps brachii, coracobrachialis
Medial cord Ulnar	C_8 and T_1	*Skin* of the medial third of the hand *Muscles* flexor carpi ulnaris, flexor digitorum profundus ($\frac{1}{2}$), and several intrinsic muscles of the hand (see Table 9.16)
Median nerve	C_5 through C_8; T_1	*Skin* of the lateral two-thirds of the hand *Muscles* pronator teres, pronator quadratus, palmaris longus, flexor carpi radialis, flexor pollicis longus, flexor digitorum superficialis, flexor digitorum profundus ($\frac{1}{2}$), and several intrinsic muscles of the hand (see Tables 9.14 and 9.16)

Table 13.4   The Lumbar Plexus [F13.16, F13.20]

Nerve	Spinal Nerves Involved (Ventral Rami)	Distribution
Iliohypogastric	T_{12} and L_1	*Skin and muscles* of lower back, hip, and lower abdomen
Ilioinguinal	L_1	*Skin* of upper medial thigh and external genitalia
		Muscles of lower abdominal wall
Genitofemoral	L_1 and L_2	*Skin* of anterior thigh, almost to knee, and external genitalia
Lateral femoral cutaneous	L_2 and L_3	*Skin* of lateral thigh
Femoral	L_2 through L_4	*Skin* of anterior, medial surface of thigh, leg, and foot through anterior femoral cutaneous and saphenous branches
		Muscles sartorius, iliopsoas, quadriceps femoris, pectineus
Obturator	L_2 through L_4	*Skin* of medial thigh
		Muscles adductor longus, adductor magnus, adductor brevis, gracilis, pectineus, obturator externus

F13.20b   The **femoral nerve** passes underneath the inguinal ligament to supply the anterior muscles of the thigh (Figure 13.20b). Two superficial branches of the femoral nerve, the **anterior femoral cutaneous nerve** and the **saphenous nerve,** innervate the skin of the anterior medial surface of the thigh, leg, and foot. The **obturator nerve** leaves the pelvis through the obturator foramen and supplies the skin on the medial surface of the thigh and the muscles of the medial (adductor) compartment of the thigh.

F13.16
Table 13.5   SACRAL PLEXUS   The **sacral plexus** is formed from the ventral rami of the last two lumbar spinal nerves (L_4 and L_5) and the first four sacral spinal nerves (S_1–S_4). The fibers of the lumbar neurons reach the plexus by way of the **lumbosacral trunk.** The nerves from the sacral plexus supply the lower back, the pelvis, and the posterior surface of the thigh and leg, as well as the dorsal and ventral surfaces of the foot (Figure 13.16). The nerves that form the sacral plexus are summarized in Table 13.5.

F13.20c   The **sciatic nerve,** which is the largest nerve in the body, is the main branch of the sacral plexus. It leaves the pelvis through the greater sciatic notch and travels down the posterior surface of the thigh, innervating the muscles and skin of the region (Figure 13.20c). The sciatic nerve is actually two nerves wrapped in a common sheath. In the lower thigh, these two nerves separate into the common peroneal nerve and the tibial nerve. The **common peroneal nerve** passes obliquely along the lateral side of the popliteal fossa to supply muscles in the anterior and lateral compartments of the leg and the skin on the anterior surface of the leg and the dorsum of the foot. The common peroneal nerve has branches called the **superficial** and **deep peroneal nerves.** The superficial peroneal supplies the muscles of the lateral

Table 13.5   The Sacral Plexus [F13.16, F13.20]

Nerve	Spinal Nerves Involved (Ventral Rami)	Distribution
Superior gluteal	L_4, L_5; S_1	*Muscles* gluteus minimus, gluteus medius, and tensor fasciae latae
Inferior gluteal	L_4, L_5; S_1	*Muscles* gluteus maximus
Posterior femoral cutaneous	S_1 through S_3	*Skin* on posterior surface of the thigh
Pudendal	S_2 through S_4	*Skin and muscles* of perineum External genitalia
Sciatic		
Tibial *Sural*	L_4, L_5; S_1 through S_3	*Skin* of posterior surface of leg and sole of foot
Medial and lateral plantar		*Muscles* biceps femoris, semimembranosus, semitendinosus, flexor digitorum longus, flexor hallucis longus, tibialis posterior, popliteus, and intrinsic muscles of foot (see Table 9.22)
Common peroneal *Superficial and deep peroneal*	L_4, S_5; S_1, S_2	*Skin* of anterior surface of leg and dorsum of foot *Muscles* peroneus brevis, peroneus longus, peroneus tertius, tibialis anterior, extensor hallucis longus, extensor digitorum longus, extensor digitorum brevis

compartment of the leg; the deep peroneal supplies the muscles of the anterior compartment of the leg. The **tibial nerve** supplies the muscles and skin on the posterior surface of the leg and the sole of the foot. The tibial nerve gives rise to the **sural nerve** (which supplies the skin on the back of the leg) and ends on the sole of the foot as the **medial** and **lateral plantar nerves.**

CONDITIONS OF CLINICAL SIGNIFICANCE

The Peripheral Nervous System

Injury and Regeneration of Peripheral Nerves

The most common disorders involving neurons of the peripheral nervous system are those associated with damage or inflammation. When a peripheral nerve is severely damaged or severed, the portion of the nerve that is distal to the injury undergoes degenerative changes (Figure 13.21). Within a few days following the injury, the nerve fibers and their myelin sheaths are broken down by macrophages from the endoneurium. Cells of the sheath of Schwann seem to assist in the degeneration by exerting phagocytic action themselves.

Following the injury, the Schwann cells proliferate and form cords in the endoneurial tubes. In a matter of days, sprouts form on the stumps of the damaged nerve fibers. Some of these sprouts grow into the endoneurial tubes, and if there are no obstructions (such as scar tis-

sue) in the tubes, the fibers may again grow out to the periphery and eventually innervate the structures that were separated from their nerve supply. As the new fibers from the proximal stump of the neuron grow along the endoneurial tubes, they are surrounded by Schwann cells. Because nerve fibers regenerate at a rate of from 1 to 4 mm per day, it is possible to estimate the length of time required for the nerve supply to return to the denervated structure.

The first indication that a peripheral nerve has reached the vicinity of a denervated structure is an improved blood supply to the area. This stage is followed by the return of sensory function to the structure. In the case of a paralyzed skeletal muscle, motor function is the last function to return.

Numerous surgical techniques have been used in an attempt to rejoin the severed portions of a damaged nerve. Unfortunately, the success rate of these opera-

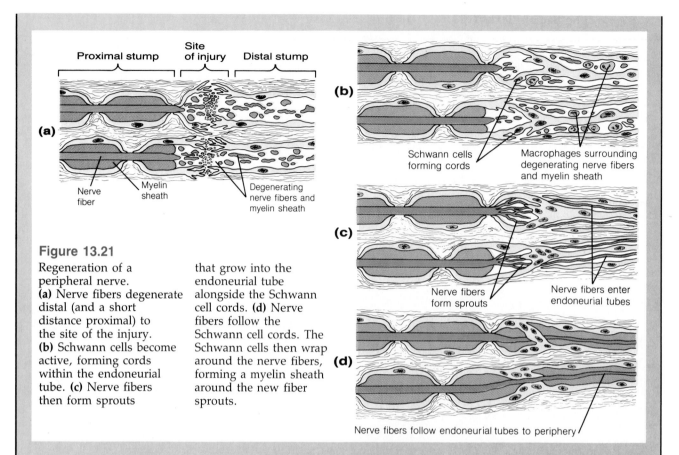

Figure 13.21

Regeneration of a peripheral nerve. **(a)** Nerve fibers degenerate distal (and a short distance proximal) to the site of the injury. **(b)** Schwann cells become active, forming cords within the endoneurial tube. **(c)** Nerve fibers then form sprouts that grow into the endoneurial tube alongside the Schwann cell cords. **(d)** Nerve fibers follow the Schwann cell cords. The Schwann cells then wrap around the nerve fibers, forming a myelin sheath around the new fiber sprouts.

tions has been only about 15%. Two new procedures have been very successful in reconnecting severed nerves in animals, and they hold great promise for use in humans.

One of these procedures uses a silicone tube filled with a collagen-carbohydrate material to bridge the gap between the ends of a severed nerve. This material provides support for new growths from the proximal portions of the severed nerve fibers. By the time the carbohydrate material degrades six weeks later, some nerve fibers have bridged the gap between the severed ends. Critical to the usefulness of any neurosurgical technique, however, is the number of nerve fibers that reinnervate the same structure they innervated prior to being severed. Whether enough valid nerve fiber reconnections occur when this procedure is used has not yet been satisfactorily determined.

The second procedure attempts to increase the number of valid nerve fiber reconnections between proximal and distal ends of a severed nerve by eliminating the gap that occurs between the nerve endings at the site of injury. After freezing the severed ends of the nerve, soaking them in a solution with the same ionic concentration as the cytoplasm of the nerve cells, and cutting the severed ends of the nerve with a vibrating blade so that they are very smooth, the nerve is held firmly together with a specially designed rubber support. The reported success rates of this procedure have been very high, and the hope is that it will prove just as successful in humans.

Neuritis

The term *neuritis* means "inflammation of a nerve," although many of the conditions referred to as neuritis are more degenerative than inflammatory. Neuritis is characterized by a range of sensations—from mild tingling to sharp, stabbing pains. It can result from a number of conditions, including mechanical damage to the involved nerves, prolonged pressure on the nerves, vascular disorders involving the nerves, and direct invasion of the nerves by pathological organisms.

Bell's palsy is a term used to describe peripheral inflammation of the facial nerve (cranial nerve VII). In Bell's palsy, all the muscles of facial expression are paralyzed on the affected side, causing them to sag. This paralysis causes difficulties in speaking and eating, and it results in the lower portion of the face being pulled to the unaffected side when the person smiles.

Neuralgia

Neuralgia refers to attacks of severe pain along the path of a peripheral nerve. There are many varieties of neuralgia, most of which have unknown causes. In *trigeminal neuralgia (tic douloureux)*, there are short attacks of excruciating pain along the course of the maxillary and mandibular divisions of the trigeminal nerve (cranial nerve V). The attacks of pain are short at first, and they may be followed by periods of several weeks without pain. With time, however, the attacks generally become more frequent and of longer duration. The cause of trigeminal neuralgia is unknown.

Herpes Zoster (Shingles)

The disease *herpes zoster*, also known as *shingles*, is a viral infection of the dorsal root ganglia of the spinal nerves. The infection causes pain and produces fluid-filled vesicles of the skin along the path of the peripheral sensory neurons that are infected.

STUDY OUTLINE

PERIPHERAL NERVOUS SYSTEM Includes 12 pairs of cranial nerves and 31 pairs of spinal nerves; divided into afferent (sensory) and efferent (motor) divisions; efferent division divided functionally into somatic and autonomic nervous systems. **p. 391**

CRANIAL NERVES 12 pairs. **pp. 391–401**

I: OLFACTORY NERVES Arise from receptor cells in nasal mucosa; entirely sensory; control sense of smell.

II: OPTIC NERVES Entirely sensory; carry impulses concerned with vision; arise from retina of eye.

OPTIC CHIASMA Where the two optic nerves meet.

OPTIC TRACT Beyond optic chiasma; each optic tract consists of fibers from retinas of both eyes.

OPTIC RADIATIONS Neurons from thalamus to visual cortex of occipital lobe.

III: OCULOMOTOR NERVES
1. Emerge from midbrain, superior to pons.
2. Innervate four extrinsic muscles of eyeball, which assist in voluntary movements of eyeball, and also the levator palpebrae superioris muscle, which elevates upper eyelids.
3. Also contain neurons of parasympathetic nervous system that cause reflex adjustments of eyes to varying light intensities and focus eyes for near and far vision.

IV: TROCHLEAR NERVES
1. Arise below inferior colliculi on dorsal surface of midbrain.
2. Motor neurons to, and proprioceptive neurons from, superior oblique eye muscle.
3. Assists in voluntary movements of eyeball by controlling contraction of superior oblique muscle.

V: TRIGEMINAL NERVES Ophthalmic, maxillary, and mandibular nerves.
1. Emerge from lateral sides of pons.
2. Major sensory nerves of face.
3. Motor neurons to muscles of mastication.

VI: ABDUCENS NERVES
1. Originate from metencephalon below pons.
2. Motor neurons to, and proprioceptive neurons from, lateral rectus eye muscle.
3. Coordinate with oculomotor and trochlear nerves to cause contraction of extrinsic muscles of both eyes.

VII: FACIAL NERVES
1. Arise from metencephalon at lower border of pons.
2. Motor neurons to face and scalp muscles.
3. Carry sensory neurons from taste buds on anterior two-thirds of tongue.
4. Carry parasympathetic neurons to lacrimal glands and submandibular and sublingual salivary glands.

VIII: VESTIBULOCOCHLEAR NERVES Cochlear and vestibular nerves.
1. Originate from inner-ear receptors; entirely sensory.
2. Cochlear division transmits impulses related to hearing.

3. Vestibular division concerned with equilibrium.

IX: GLOSSOPHARYNGEAL NERVES
1. Emerge from medulla.
2. Mixed nerves—motor and sensory.
3. Sensory—impulses from taste buds of posterior third of tongue, from membranes of pharynx and tonsils, from receptors in carotid sinus and carotid bodies.
4. Motor—innervate pharyngeal muscles.
5. Parasympathetic neurons to parotid salivary glands.

X: VAGUS NERVES Only cranial nerves not restricted to head and neck regions.
1. Arise from sides of medulla.
2. Motor neurons to, and sensory neurons from, pharynx and larynx.
3. Sensory input from viscera regulates heart rate, respiration depth, blood pressure, digestion, etc.
4. Parasympathetic neurons to thoracic and abdominal viscera and blood vessels.

XI: ACCESSORY NERVES
1. Composed of a true cranial nerve (which arises from medulla) and a cervical spinal nerve.
2. Cranial portions supply pharynx and larynx muscles.
3. Spinal portions supply trapezius and sternocleidomastoid muscles.

XII: HYPOGLOSSAL NERVES
1. Arise from anterior surface of medulla.
2. Supply intrinsic and extrinsic tongue muscles.
3. Sensory neurons from proprioceptors.

SPINAL NERVES 31 pairs. **pp. 402–411**

FORMATION, BRANCHES, AND DISTRIBUTION
1. Union of ventral and dorsal roots.
2. All are mixed (motor and sensory) nerves.
3. Each spinal nerve divides into dorsal and ventral rami.
4. Sensory nerves in rami are distributed in bands called dermatomes.

PLEXUSES AND PERIPHERAL NERVES

CERVICAL PLEXUS Formed by ventral rami of first four cervical nerves; supply muscles and skin of neck and shoulder; phrenic nerve supplies diaphragm.

BRACHIAL PLEXUS Formed by ventral rami of last four cervical nerves and first thoracic nerve. Dorsal scapular and long thoracic nerves branch from roots; suprascapular nerve and nerve to subclavius muscle branch from the trunks of plexus.

Posterior Cord
1. *Axillary nerve:* skin and muscles of shoulder.
2. *Radial nerve:* skin on posterior surface of arm, forearm, and hand; extensor muscle of arm and forearm.
3. *Superior* and *inferior subscapular nerves* and *thoracodorsal nerves* branch from posterior cord.

Lateral Cord
1. *Musculocutaneous nerve:* skin of lateral surface of forearm; anterior muscles of arm.
2. *Lateral pectoral nerve.*

Medial Cord
1. *Ulnar nerve:* skin of medial surface of hand; forearm; intrinsic hand muscles.
2. *Medial pectoral nerve; medial brachial cutaneous nerve; medial antebrachial cutaneous nerve.*

Median Nerve From branches of *medial* and *lateral* cords; flexor muscles of forearm; skin and muscles of lateral palmar portion of hand.

LUMBOSACRAL PLEXUS Skin and muscles of buttocks, pelvis, lower abdomen, lower limbs.

Lumbar Plexus ventral rami L_1–L_4.
1. *Femoral nerve:* anterior thigh muscles, skin of anterior medial thigh, leg, and foot; has anterior femoral cutaneous and saphenous branches.
2. *Obturator nerve:* skin of medial thigh; adductor muscles of thigh.

Sacral Plexus Ventral rami L_4, L_5, and S_1–S_4.
1. *Sciatic nerve:* muscles and skin of posterior surface of thigh.
2. *Common peroneal nerve:* muscles of anterior and lateral compartments of leg; skin on anterior surface of leg and dorsum of foot; has superficial and deep peroneal branches.
3. *Tibial nerve:* muscles and skin on posterior surface of leg and sole of foot; has sural as well as medial and lateral plantar branches.

CONDITIONS OF CLINICAL SIGNIFICANCE: THE PERIPHERAL NERVOUS SYSTEM pp. 411–412

INJURY AND REGENERATION OF PERIPHERAL NERVES
1. Macrophages break down nerve fibers and myelin sheath distal to site of injury.
2. Schwann cells proliferate, forming cords.
3. Buds form on stumps of nerves.
4. New fibers develop from buds and grow toward periphery.
5. Schwann cells surround new fibers.

NEURITIS
1. Degenerative or inflammatory condition.
2. Tingling to sharp pain.
3. Results from mechanical damage, prolonged pressure, vascular disorders, pathological organism invasion.
4. Bell's palsy is neuritis of facial nerve.

NEURALGIA Attacks of severe pain along peripheral nerve path.

TRIGEMINAL NEURALGIA (TIC DOULOUREUX) Short attacks of severe pain along maxillary and mandibular divisions of trigeminal cranial nerve.

HERPES ZOSTER (SHINGLES) Viral infection of dorsal root ganglia of spinal nerves.

SELF-QUIZ

1. The peripheral nervous system includes the: (a) brain; (b) spinal cord; (c) spinal nerves.

2. Which motor neurons are not under conscious control, are within the cranial nerves, and supply smooth or cardiac muscle? (a) visceral motor neurons; (b) somatic motor neurons; (c) sympathetic neurons.

3. The olfactory nerves are: (a) motor and leave the mesencephalon; (b) sensory and enter the pons; (c) sensory and enter the telencephalon.

4. The optic nerves are: (a) sensory and leave the skull through the optic foramen; (b) motor and leave the telencephalon; (c) motor and leave the skull through the superior orbital fissure.

5. The oculomotor nerves: (a) include proprioceptor fibers; (b) are located in the diencephalon; (c) leave the skull through the foramen rotundum.

6. The trochlear nerves: (a) arise in the diencephalon; (b) aid in voluntary movements of the eyeball; (c) are large nerves.

7. The trigeminal nerves are the: (a) major sensory nerves of the face; (b) principal nerves that control eyeball movement; (c) nerves that control the tongue.

8. The abducens nerves: (a) originate from the telencephalon; (b) are sensory nerves; (c) leave the skull through the superior orbital fissure.

9. The facial nerves: (a) leave the skull through the foramen magnum; (b) also supply the viscera of the thorax; (c) are mixed nerves.

10. The vestibulocochlear nerves: (a) are concerned with equilibrium; (b) are motor nerves; (c) arise in the telencephalon.

11. The glossopharyngeal nerves: (a) leave the skull through the stylomastoid foramen; (b) carry motor and sensory impulses; (c) regulate contractions of the abdominal viscera.

12. The vagus nerves: (a) supply most organs of the thorax and abdomen; (b) are restricted to the head and neck region; (c) carry sensory impulses to the pharynx.

13. The accessory nerves are composed of: (a) somatic motor neurons; (b) sensory neurons from proprioceptors; (c) parasympathetic neurons; (d) both a and b.

14. The hypoglossal nerves: (a) supply the intrinsic and extrinsic muscles of the pharynx; (b) arise in the telencephalon; (c) consist of motor neurons that supply the tongue.

15. The spinal nerves are formed from the union of ventral and dorsal roots that leave or enter the spinal cord. True or False?

16. After the ventral and dorsal roots join, all the spinal nerves are mixed nerves containing the pro-

cesses of motor neurons (somatic and visceral) and sensory neurons. True or False?

17. The cell bodies of the motor neurons are located outside the spinal cord, whereas those of the sensory neurons lie within the spinal cord. True or False?

18. From which plexus does the phrenic nerve pass through the thorax to supply the diaphragm? (a) lumbosacral; (b) brachial; (c) cervical.

19. The skin on the posterior surface of the arm, forearm, and hand, and the extensor muscles of the arm and forearm are supplied by the: (a) radial nerve; (b) phrenic nerve; (c) musculocutaneous nerve.

20. The palmaris longus and flexor carpi radialis muscles are supplied by the: (a) ulnar nerve; (b) radial nerve; (c) median nerve.

21. The sciatic nerve, which is the largest nerve in the body, is the main branch of the sacral plexus. True or False?

22. It is possible for a peripheral nerve that has been severed to regenerate and innervate the same structures it supplied before it was damaged. True or False?

LEARNING OBJECTIVES

After completing this chapter, you should be able to:

1. State the function of the autonomic nervous system.

2. Distinguish between the sympathetic and parasympathetic divisions of the autonomic nervous system in terms of structure and function.

3. List the paths that the preganglionic sympathetic axons can follow upon entering the chain ganglia.

4. Describe biofeedback as it relates to the autonomic nervous system.

CHAPTER CONTENTS

ANATOMY OF THE AUTONOMIC NERVOUS SYSTEM

NEUROTRANSMITTERS OF THE AUTONOMIC NERVOUS SYSTEM

RECEPTORS FOR AUTONOMIC NEUROTRANSMITTERS

FUNCTIONS OF THE AUTONOMIC NERVOUS SYSTEM

CONDITIONS OF CLINICAL SIGNIFICANCE: THE AUTONOMIC NERVOUS SYSTEM

KEY TERMS AND DERIVATIVES

autonomic nervous system (*auto* = self; *nom* = governing)   the division of the nervous system that functions involuntarily and innervates cardiac muscle, smooth muscle, and glands

cranial nerves (*cranial* = pertaining to the upper portion of the head)   the 12 pairs of nerves that emerge from the brain and transmit information directly between certain sensory receptors and the brain and between the brain and certain effectors

parasympathetic nervous system (*para* = beside)   a division of the autonomic nervous system; also referred to as the *craniosacral division*

postganglionic neuron (*post* = after; *gangl* = mass)   a neuron of the autonomic nervous system having its cell body in a ganglion and its axon extending to an organ or tissue

preganglionic neuron (*pre* = before; *gangl* = mass)   a neuron of the autonomic nervous system having its cell body in the brain or spinal cord and its axon terminating in a ganglion

rami (*ramus* = branch)   a branch of a nerve, vein, or bone, especially a primary division

sympathetic nervous system (*syn* = together with; *pathos* = feeling)   a division of the autonomic nervous system; also referred to as the *thoracolumbar division*

The Autonomic Nervous System

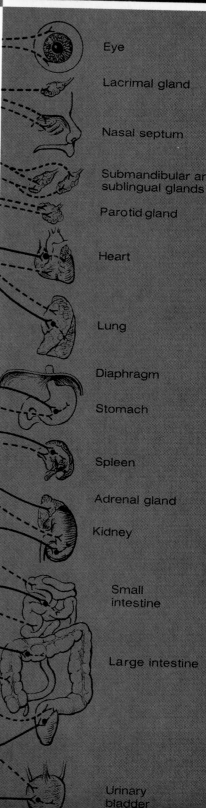

Eye

Lacrimal gland

Nasal septum

Submandibular and sublingual glands

Parotid gland

Heart

Lung

Diaphragm

Stomach

Spleen

Adrenal gland

Kidney

Small intestine

Large intestine

Urinary bladder

The **autonomic nervous system (ANS)** is a part of the efferent division of the peripheral nervous system. It is composed entirely of visceral motor (efferent) neurons that innervate and thus regulate the activity of cardiac muscle, smooth muscle, and the glands of the body. This system is normally an involuntary system, functioning below the conscious level.

We will consider the autonomic nervous system separately from the peripheral nervous system, but keep in mind that the ANS is structurally and functionally an integral part of the body's single nervous system. In fact, many nerve fibers of the ANS travel in the spinal nerves and certain cranial nerves. Even though the autonomic nervous system has only motor functions, visceral sensory fibers of the afferent division of the peripheral nervous system travel along the same pathways as the visceral motor fibers of the autonomic nervous system. The cell bodies of these visceral sensory fibers are located in the dorsal root ganglia of the spinal nerves or in one of the outlying ganglia of certain cranial nerves.

ANATOMY OF THE AUTONOMIC NERVOUS SYSTEM

The efferent pathways of the autonomic nervous system that run from the central nervous system to the effectors are composed of two neurons. One of these neurons, called a **preganglionic (presynaptic) neuron,** has its cell body in the central nervous system. The axon of the preganglionic neuron travels to an **autonomic ganglion** located outside the central nervous system; in the autonomic ganglion it synapses with other neurons called **postganglionic (postsynaptic) neurons.** Shortly after leaving an autonomic ganglion, the axons of postganglionic neurons generally form into branching networks known as **autonomic plexuses** and then travel to the various effectors. This two-neuron chain is in contrast to the somatic nervous system, where a single motor neuron travels from the central nervous system to the structure innervated.

The autonomic nervous system can be separated both structurally and functionally into two divisions: the *sympathetic division* and the *parasympathetic division* (Figure 14.1).

Sympathetic Division

The cell bodies of the preganglionic neurons of the **sympathetic division** of the autonomic nervous system are located in the lateral horns of the gray matter of the spinal cord, from the first thoracic segment (T_1) through the second lumbar segment (L_2). For this reason, the sympathetic division is also called the **thoracolumbar division.** The axons of these visceral motor neurons leave the spinal cord through the ventral roots, along with somatic motor axons, and enter the dorsal and ventral rami of the spinal nerves of the various segments. After going only a short distance in the ventral rami of the spinal nerves, all the preganglionic sympathetic nerve fibers leave the ventral rami and enter one of a series of interconnected **chain (paravertebral or sympathetic) ganglia** (Figure 14.2). The chain ganglia form longitudinal pathways, called **sympathetic trunks,** on both sides of the bodies of the vertebrae

417

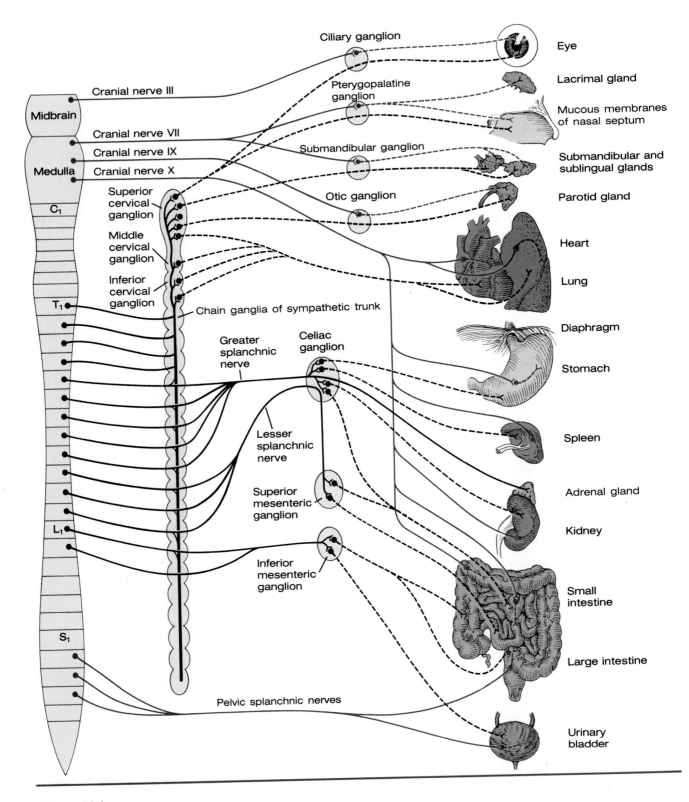

Figure 14.1

The autonomic nervous system. The parasympathetic division is shown in rust; the sympathetic division is shown in black. The solid lines indicate preganglionic nerve fibers; the dashed lines indicate postganglionic nerve fibers. Postganglionic nerve fibers that originate from the thoracic and lumbosacral chain ganglia are not shown.

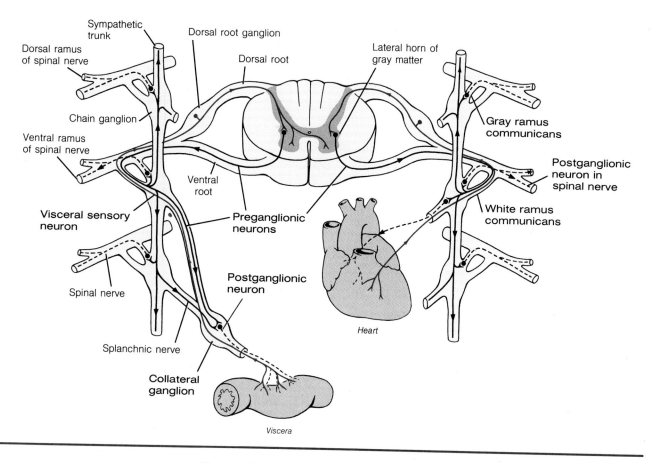

Figure 14.2

The sympathetic division of the autonomic nervous system. The central connections of the sympathetic neurons are shown as seen in one segment of the thoracic spinal cord. Solid black lines indicate sympathetic preganglionic neurons; dotted black lines indicate sympathetic postganglionic neurons; colored lines indicate sensory neurons. The arrows show the direction of the nerve impulse.

along the entire length of the vertebral column, including the cervical and sacral regions. Except in the cervical region, where several chain ganglia fuse together to form three or four larger ganglia, the ventral ramus of each spinal nerve generally has a chain ganglion associated with it. Since most of the preganglionic sympathetic nerve fibers are myelinated, the short pathways they form in passing from the ventral ramus of the spinal nerve to the sympathetic trunks appear white and are called **white rami communicantes**. There are 14 pairs of white rami communicantes connecting the first thoracic through the second lumbar spinal nerves to the chain ganglia.

Upon entering the chain ganglia, the preganglionic sympathetic axons follow one of three courses:

1. *Preganglionic sympathetic axons may snyapse with the cell bodies of postganglionic neurons in the chain ganglion located at the same level at which the preganglionic fibers entered the chain.* The axons of the postganglionic neurons return directly to the spinal nerve and travel to the periphery in the dorsal and ventral rami of the spinal nerve. These axons innervate effectors in the skin, including smooth muscles in the walls of the blood vessels of the skin, sweat glands, and the arrector pili muscles of the hairs. Because most postganglionic axons are unmyelinated, the pathways they form as they pass from the chain ganglia to the spinal nerves appear gray, and they are called **gray rami communicantes.**

2. *Preganglionic sympathetic axons may travel up or down within the sympathetic trunk before synapsing with postganglionic neurons at a higher or lower level.* The axons of some of these postganglionic neurons enter the cervical and sacral spinal nerves through gray rami communicantes and ultimately innervate the skin (blood vessels, sweat glands, arrector pili muscles) of the regions supplied by those nerves. Notice that, although preganglionic axons enter the chain ganglia through white rami from only the first thoracic through the second lumbar segments, every chain ganglion is connected to a spinal nerve by a gray ramus. Thus, because preganglionic axons may travel to higher or lower levels in the sympathetic trunk, every spinal nerve receives axons of postganglionic sympathetic neurons. However, some postganglionic axons in the thoracic and cervical ganglia pass directly from the ganglia to innervate the thoracic viscera and structures in the head, rather than entering the spinal nerves.

3. *Preganglionic sympathetic axons may pass through the chain ganglia without synapsing.* Some preganglionic axons from the thoracic region form pathways, called **greater** and **lesser splanchnic nerves,** that pass through the diaphragm and lead to **collateral (prevertebral) ganglia** located on the front of the abdominal aorta (which is the large blood vessel that carries blood from the heart and distributes it to the body). The major collateral ganglia, which are named for the vascular branches of the aorta nearest them, are the **celiac, superior mesenteric,** and **inferior mesenteric.** Within the collateral ganglia, the preganglionic axons synapse with postganglionic neurons. The axons of the postganglionic neurons leave the collateral ganglia, interconnect to form an autonomic plexus, and supply the viscera of the abdominopelvic cavity.

Preganglionic sympathetic neurons that innervate the adrenal medulla travel in the splanchnic nerves and do not synapse before reaching the gland. Thus, there are no postganglionic sympathetic neurons innervating the adrenal medulla—the only exception to the usual two-neuron chain in autonomic efferent pathways. The reason for this exception is explained later in the chapter when we discuss the differences between the divisions of the autonomic nervous system.

Accompanying the sympathetic motor fibers that supply the viscera are sensory fibers of the afferent division of the peripheral nervous system returning from the viscera. These visceral sensory fibers travel, without synapsing, from the innervated structure, through the chain ganglia and the white rami communicantes, to their cell bodies in the dorsal root ganglia. Therefore, the cell bodies of both visceral and somatic sensory neurons are located in the same ganglia.

Parasympathetic Division

The cell bodies of the preganglionic neurons of the **parasympathetic division** of the autonomic nervous system are located either within nuclei in the brain stem or within the lateral portions of the gray matter of the spinal cord in the second, third, and fourth sacral segments (Figure 14.1). Because of these origins, the parasympathetic division of the autonomic nervous system is also referred to as the **craniosacral division.** The distribution of the parasympathetic division differs from that of the sympathetic division in that parasympathetic fibers do not travel within the rami of the spinal nerves. Consequently, the sweat glands, arrector pili muscles, and cutaneous blood vessels do not receive parasympathetic innervation. In fact, with few exceptions, the parasympathetic division does not innervate blood vessels anywhere in the body.

Parasympathetic axons whose cell bodies are located in nuclei of the brain stem travel to the viscera of the head, thorax, and abdomen within the cranial

F14.1

nerves—specifically, the oculomotor, facial, glossopharyngeal, and vagus nerves. (The specific distribution of the parasympathetic axons in these cranial nerves is discussed in Chapter 13 and summarized in Table 13.1.) Preganglionic parasympathetic axons in the four cranial nerves synapse with postganglionic neurons in ganglia (ciliary, pterygopalatine, otic, submandibular, and terminal ganglia) that are located close to the structures innervated by the postganglionic neurons.

Table 13.1

Preganglionic parasympathetic axons whose cell bodies are located in the sacral region of the spinal cord leave the cord in the ventral roots of the sacral spinal nerves. The parasympathetic axons then leave the ventral roots and join together to form the **pelvic splanchnic nerves,** which interconnect in the hypogastric plexus and supply the viscera of the pelvic cavity. The preganglionic axons of the sacral parasympathetic neurons synapse with postganglionic neurons in terminal ganglia located close to the organs they innervate.

Anatomical Differences Between the Divisions

The sympathetic and parasympathetic divisions of the autonomic nervous system differ not only in the locations of the cell bodies of their preganglionic neurons but also in the lengths of their fibers. In the sympathetic division, most preganglionic axons are relatively short, synapsing in the chain ganglia close to the vertebral column. The postganglionic axons are long, extending from the chain ganglia to the structures they innervate. In contrast, the parasympathetic preganglionic axons are relatively long, passing uninterrupted from their origins in the central nervous system to terminal ganglia located on or close to the walls of the organs they supply. The postganglionic axons of the parasympathetic division are short, extending from the terminal ganglia to the organs innervated.

NEUROTRANSMITTERS OF THE AUTONOMIC NERVOUS SYSTEM

Parasympathetic preganglionic and postganglionic fibers, as well as sympathetic preganglionic fibers, release *acetylcholine*—the same neurotransmitter substance released by somatic efferent neurons (Figure 14.3). Therefore, these fibers are called **cholinergic fibers.** In contrast, most sympathetic postganglionic fibers secrete *norepinephrine (noradrenaline).* Consequently, these fibers are called **adrenergic fibers** (or sometimes, noradrenergic fibers). Some postganglionic sympathetic fibers, however, secrete acetylcholine. For example, the sympathetic postganglionic fibers that innervate the sweat glands are cholinergic.

F14.3
ko-lin-ur'-jick

ad-ren-ur'-jick

Norepinephrine (as well as the closely related substance epinephrine) is also secreted by the **adrenal medulla,** an endocrine gland. The adrenal medulla possesses physiological and biochemical properties similar to those of the sympathetic nervous system, and it is innervated by cholinergic preganglionic sympathetic neurons that do not synapse before reaching the gland. In fact, the adrenal medulla is essentially a modified sympathetic ganglion. Norepinephrine has the same effects, whether released into the bloodstream by the adrenal medulla or secreted directly onto an organ by a sympathetic nerve fiber. However, when norepinephrine is released into the bloodstream, it is carried to all parts of the body, and thus its effects may be more widespread.

RECEPTORS FOR AUTONOMIC NEUROTRANSMITTERS

Once they are released, acetylcholine, norepinephrine, and epinephrine influence the activity of postganglionic neurons and effector cells by combining with receptors that are present on the neurons and effector cells. There are a

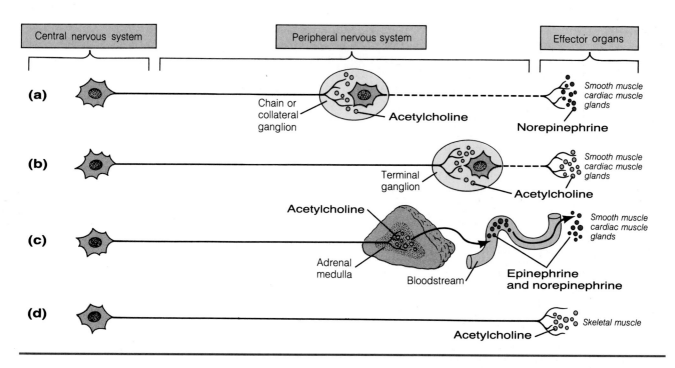

Figure 14.3
Neurotransmitters of the efferent division of the peripheral nervous system. **(a)** Preganglionic neurons (solid line) of the sympathetic division of the autonomic nervous system release acetylcholine at their synapses with postganglionic neurons (dashed line). Although exceptions occur, the postganglionic neurons release mainly norepinephrine at their junctions with effectors. **(b)** Preganglionic neurons (solid line) of the parasympathetic division of the autonomic nervous system release acetylcholine at their synapses with postganglionic neurons (dashed line), and the postganglionic neurons also release acetylcholine at their junctions with effectors. **(c)** The adrenal medulla is supplied by preganglionic sympathetic neurons that release acetylcholine. The adrenal medulla releases epinephrine and norepinephrine into the bloodstream, which carries the secretions to effectors. **(d)** Somatic efferent neurons release acetylcholine at their junctions with effectors.

number of different types of receptors, and a given postganglionic neuron or effector cell can possess more than one receptor type.

Acetylcholine receptors are described as being either *muscarinic* or *nicotinic*. This is because the substance muscarine mimics the effects of acetylcholine at some receptors, and the substance nicotine mimics the effects of acetylcholine at other receptors. Muscarinic receptors are present on many of the effectors supplied by parasympathetic postganglionic fibers. Nicotinic receptors are present on postganglionic neurons within the ganglia of the autonomic nervous system (and also on skeletal muscle cells at neuromuscular junctions). The effective combination of acetylcholine with muscarinic receptors can be blocked by the substance *atropine*, which competes with acetylcholine for the receptor sites. The effects of acetylcholine at nicotinic receptors on postganglionic neurons within autonomic ganglia can be blocked by the substance *tetraethylammonium*, which prevents changes in cellular permeability. The effective combination of acetylcholine with nicotinic receptors at neuromuscular junctions can be blocked by the substance *curare*, which competes with acetylcholine for skeletal muscle receptor sites. Thus, there appear to be two types of nicotinic receptors.

There are four principal receptors with which norepinephrine or epinephrine can combine. They are designated $alpha_1$, $alpha_2$, $beta_1$, and $beta_2$. $Alpha_1$ receptors are the most common alpha receptors on effector cells, but the cells of some organs such as the uterus and brain appear to contain both $alpha_1$ and $alpha_2$ receptors. In most instances, the combination of norepinephrine or epinephrine with $alpha_1$ receptors results in a stimulatory response, and the combination of norepinephrine or epinephrine with $alpha_2$

Table 14.1   Effects of the Autonomic Nervous System

Structure	Effects of Sympathetic Stimulation	Effects of Parasympathetic Stimulation
Heart	Increase rate	Decrease rate
Lungs		
Bronchioles	Dilation	Constriction
Bronchial glands	Possible inhibition of secretion	Stimulation of secretion
Salivary glands	Secretion of viscous fluid	Secretion of watery fluid
Stomach		
Motility	Decreased	Increased
Secretion	Possible inhibition	Stimulation
Intestine		
Motility	Decreased peristalsis	Increased peristalsis
Secretion	Possible inhibition	Stimulation
Pancreas (Exocrine Portion)		Stimulation of secretion
Liver	Increased release of glucose	
Eye		
Iris	Dilation of pupil (contraction of radial muscles)	Constriction of pupil (contraction of sphincter muscles)
Ciliary muscle	Slight relaxation	Contraction (accommodates for near vision)
Sweat glands	Stimulation of secretion (cholinergic)	
Adrenal medulla	Stimulation of secretion (cholinergic preganglionic neurons)	
Urinary bladder	Relaxation	Contraction
Blood vessels of:		
Skin	Constriction	
Salivary glands	Constriction	
Abdominal viscera	Constriction	
External genitalia	Constriction	Dilation

receptors results in an inhibitory response. Beta$_2$ receptors are the most common beta receptors on effector cells, but the cells of some organs, such as the heart and the kidneys, possess beta$_1$ receptors.

Norepinephrine combines effectively with alpha$_1$, alpha$_2$, and beta$_1$ receptors, but it combines either weakly or not at all with certain beta$_2$ receptors. Epinephrine combines effectively with all four receptor types. These differences in the effectiveness of the combination of norepinephrine and epinephrine with the different receptors are at least in part responsible for the fact that norepinephrine and epinephrine sometimes have different actions. For example, both norepinephrine and epinephrine stimulate the heart, which has mainly beta$_1$ receptors, and constrict certain blood vessels, which have alpha$_1$ receptors. Yet, only epinephrine dilates certain other blood vessels, which have beta$_2$ receptors.

FUNCTIONS OF THE
AUTONOMIC NERVOUS SYSTEM

Many body organs (for example, the heart, the intestines, and the lungs) are supplied by both the sympathetic and parasympathetic divisions of the autonomic nervous system. In such cases, the two divisions frequently (but not always) cause opposite responses. For example, if one division increases the activity of an organ, the other may decrease it. Although most organs are predominantly controlled by one division or the other, the dual innervation of an organ by both divisions of the autonomic nervous system contributes to the precise control of the organ's activity. The effects of sympathetic and parasympathetic stimulation on a number of body organs are summarized in Table 14.1.

Table 14.1

There is no generalization that indicates whether sympathetic or parasympathetic stimulation will excite or inhibit a particular organ. However, when viewed in broad terms, parasympathetic stimulation tends to produce responses that are primarily concerned with maintaining bodily functions under relatively quiet conditions. For example, parasympathetic stimulation decreases the heart rate and promotes digestive activities. Sympathetic stimulation, in contrast, tends to produce responses that prepare a person for strenuous physical activity, such as may be required in an emergency or in situations that lead to aggressive or defensive behavior. In fact, emotional states such as rage and fear are generally accompanied by a widespread activation of the sympathetic division of the autonomic nervous system. This broad sympathetic activity produces a group of responses—such as increased heart rate and dilation of the bronchii of the lungs—that increase the capability of the body to perform vigorous muscular activity. These responses are particularly beneficial to a person who must defend against or flee from a physical threat or challenge; consequently, they are frequently called "fight or flight" responses.

CONDITIONS OF CLINICAL SIGNIFICANCE

The Autonomic Nervous System

Biofeedback

The fact that the autonomic nervous system normally functions below the conscious level implies that a person has no control over the activities governed by this system. However, this is not entirely true.

Normally, a person receives only limited information at the conscious level about what is occurring within the body. For example, blood pressure may fluctuate or brain wave patterns change without the person becoming aware of the changes. Consequently, no conscious effort is made to react to or to control such changes. However, the technique of **biofeedback** provides a person with conscious information about body events that usually go unnoticed. Biofeedback utilizes electronic instruments to monitor some of the normally subconscious activities that occur within the body and raise them to the conscious level. The instruments provide information about such events as temperature changes, changes in heart rate, and variations in nerve-impulse patterns. With such conscious knowledge of previously subconscious body activities, it has become possible in

some cases for people to learn to control certain of these activities. For example, using biofeedback techniques, people have learned to lower their heart rate, lower their blood pressure, increase the circulation of blood through their limbs, relieve migraine headaches by reducing the blood pressure within the vessels of the head, and control epileptic seizures. Biofeedback, then, shows considerable promise as a self-administered therapeutic technique with broad applications.

Raynaud's Disease

Raynaud's disease is characterized by episodes of pallor or cyanosis of the extremities—especially the fingers and toes, and less frequently, the tip of the nose and the ears. It is thought to be the result of exaggerated vasomotor responses, both central and local, by the sympathetic division of the autonomic nervous system. These responses cause episodes of vasoconstriction of the blood vessels of the affected regions.

The episodes are generally first noticed in cold weather and may be infrequent. The course of Raynaud's disease is variable; it often remains as nothing more than a nuisance for years and in some cases subsides spontaneously. Occasionally, however, the condition becomes progressive and produces ulcerations and areas of gangrene of the fingertips.

Achalasia

Achalasia, or *cardiospasm*, is characterized by difficulty in swallowing accompanied by a feeling that food is sticking in the esophagus. It is the result of uncoordinated and ineffectual peristalsis of the esophagus and persistent contraction of the esophagus where it enters the cardiac region of the stomach. These conditions produce a functional obstruction of the esophagus. Achalasia is probably caused by a number of factors. Emotions and a hypersensitivity to the hormone gastrin are implicated, but there may also be structural or functional disorders of the portion of the parasympathetic nervous system that innervates the esophagus.

Hirschsprung's Disease

Hirschsprung's disease, or *megacolon*, is somewhat similar to achalasia, except that the functional obstruction occurs in the distal portion of the colon and the rectum. In response to this obstruction, the colon above the level of the obstruction dilates greatly (megacolon). This disorder is thought to be caused by a reduction in the parasympathetic innervation to the affected structures. The reduction of parasympathetic innervation allows the sympathetic neurons to inhibit peristalsis and maintain a chronic contraction in the affected region.

STUDY OUTLINE

AUTONOMIC NERVOUS SYSTEM p. 417

VISCERAL MOTOR (EFFERENT) NEURONS Innervate and regulate cardiac muscle, smooth muscle, and glands; control involuntary body processes.

ENTIRELY MOTOR FUNCTIONS Part of efferent division of peripheral nervous system.

ANATOMY OF THE AUTONOMIC NERVOUS SYSTEM pp. 417–421

EFFERENT PATHWAYS Composed of two neurons.

PREGANGLIONIC (PRESYNAPTIC) NEURON Cell body in CNS; axon synapses with postganglionic neurons.

POSTGANGLIONIC (POSTSYNAPTIC) NEURON Located outside CNS; axons travel to effectors.

SYMPATHETIC DIVISION (THORACOLUMBAR DIVISION) Preganglionic neuron cell bodies in lateral horns of spinal cord gray matter from T_1 through L_2. Fibers leave ventral rami and enter a series of chain ganglia that form longitudinal pathways along both sides of vertebral column. *White rami communicantes* (14 pairs) are short pathways formed by myelinated preganglionic nerve fibers as they pass from ventral ramus to chain ganglia. Preganglionic sympathetic neurons follow one of three courses upon entering chain ganglia:
1. May synapse with postganglionic neurons in chain ganglia at same level. Postganglionic axons return to the spinal nerve to innervate effectors located in skin. *Gray rami communicantes* are pathways formed by unmyelinated postganglionic axons as they pass from chain ganglia to spinal nerves.
2. May travel up or down within the sympathetic trunks before synapsing with postganglionic neurons that supply effectors in skin, head, or thorax.
3. May pass through chain ganglia without synapsing and synapse with postganglionic neurons in collateral ganglia; postganglionic neurons from collateral ganglia supply viscera of abdominopelvic cavity.

PARASYMPATHETIC DIVISION (CRANIOSACRAL DIVISION)
1. Preganglionic neuron cell bodies are located within nuclei in the brain or in the lateral portions of the gray matter of the spinal cord from S_2 through S_4.
2. Fibers do not travel through rami of spinal nerves.
3. Preganglionic parasympathetic axons in four cranial nerves (III, VII, IX, X) synapse with postganglionic neurons in ganglia close to structures innervated.
4. Sacral preganglionic parasympathetic axons leave ventral roots of spinal nerves and form a pelvic nerve that supplies viscera of the pelvic cavity.

ANATOMICAL DIFFERENCES BETWEEN THE DIVISIONS

LOCATION OF PREGANGLIONIC NEURON CELL BODIES

Sympathetic Lateral horns of spinal cord gray matter from T_1 through L_2.

Parasympathetic Brain stem and lateral horns of spinal cord gray matter from S_1 through S_4.

FIBER LENGTH Short sympathetic preganglionic axons; long sympathetic postganglionic axons. Long parasympathetic preganglionic axons; short parasympathetic postganglionic axons.

NEUROTRANSMITTERS OF THE AUTONOMIC NERVOUS SYSTEM Most postganglionic sympathetic fibers secrete norepinephrine; postganglionic parasympathetic fibers secrete acetylcholine. p. 421

RECEPTORS FOR AUTONOMIC NEUROTRANSMITTERS Acetylcholine receptors are described as being either muscarinic or nicotinic; there are four principle receptors with which norepinephrine or epinephrine can combine. pp. 421–423

FUNCTIONS OF THE AUTONOMIC NERVOUS SYSTEM
Many organs are innervated by both sympathetic and parasympathetic fibers, which generally cause opposite responses; parasympathetic stimulation tends to produce responses primarily concerned with maintaining bodily functions under relatively quiet conditions; sympathetic stimulation tends to produce responses that prepare a person for strenuous physical activity. **p. 424**

CONDITIONS OF CLINICAL SIGNIFICANCE: THE AUTONOMIC NERVOUS SYSTEM **pp. 424–425**

BIOFEEDBACK Technique by which a person made aware of normally subconscious activities can exert some voluntary control over them.

RAYNAUD'S DISEASE Characterized by pallor or cyanosis of the extremities; thought to be caused by vasoconstriction of blood vessels in the area by the sympathetic nervous system.

ACHALASIA Also called *cardiospasm*; characterized by difficulty in swallowing, with food remaining in esophagus; peristalsis of esophagus is uncoordinated and ineffectual, and esophagus is constricted where it enters stomach; structural or functional disorders of parasympathetic nervous system probably involved.

HIRSCHSPRUNG'S DISEASE Also called *megacolon*; result of functional obstruction in distal portion of colon and rectum; probably caused by reduction in parasympathetic innervation to these structures; proximal portion of colon dilates greatly in response to obstruction.

SELF-QUIZ

1. The autonomic nervous system regulates the activity of: (a) cardiac muscle; (b) skeletal muscles; (c) smooth muscle; (d) both a and c.

2. The autonomic nervous system is normally a voluntary system that functions at the conscious level. True or False?

3. The autonomic nervous system has both motor and sensory functions. True or False?

4. The efferent pathways of the autonomic nervous system, which run from the central nervous system to the effectors, are composed of how many neurons? (a) 1; (b) 2; (c) 4.

5. The ventral ramus of each spinal nerve generally has associated with it: (a) a collateral ganglion; (b) a chain ganglion; (c) a dorsal root ganglion.

6. Preganglionic sympathetic axons may synapse with the cell bodies of postganglionic neurons in the chain ganglion located at the same level at which the preganglionic axons entered the ganglion. True or False?

7. Most postganglionic axons are myelinated. True or False?

8. Preganglionic sympathetic axons may pass through the chain ganglia without synapsing. True or False?

9. Preganglionic sympathetic axons may travel up or down in the sympathetic trunks before synapsing with: (a) other preganglionic neurons; (b) postganglionic neurons; (c) sensory neurons.

10. The distribution of the sympathetic division differs from that of the parasympathetic division in that its fibers do not travel through the rami of the spinal nerves and, therefore, the cutaneous structures do not receive sympathetic innervation. True or False?

11. The parasympathetic division of the autonomic nervous system is also referred to as: (a) the thoracolumbar division; (b) the craniosacral division; (c) neither a nor b.

12. The sympathetic and parasympathetic divisions of the autonomic nervous system differ in: (a) the location of the cell bodies of their preganglionic neurons; (b) the length of their preganglionic fibers; (c) both a and b.

13. Most sympathetic postganglionic fibers secrete: (a) norepinephrine; (b) acetylcholine; (c) neither a nor b.

14. The adrenal medulla causes reactions similar to those caused by the sympathetic nervous system. True or False?

15. The "fight-or-flight" responses are produced by widespread activation of the: (a) parasympathetic division; (b) somatic nervous system; (c) sympathetic division.

16. Parasympathetic stimulation tends to produce responses that are primarily concerned with maintaining bodily functions under relatively quiet conditions. True or False?

17. Under controlled conditions it may be possible for a person to learn to control such responses of the autonomic nervous system as heart rate and blood pressure. True or False?

LEARNING OBJECTIVES

After completing this chapter, you should be able to:

1. Name and describe the functions of the basic layers of the eye and their main component parts.

2. Describe the focusing of images on the retina.

3. Describe the structure of the retina.

4. Describe the responses of the rods to light.

5. Explain how colors are perceived.

6. Distinguish among presbyopia, astigmatism, and emmetropia.

7. Name the basic structural compartments of the ear and describe their main components.

8. Explain how a sound wave is converted into a perceived sound.

9. Describe mechanisms by which the source of a sound is localized.

10. Describe the function of the utricle.

11. Describe the function of the semicircular ducts.

12. On a neurophysiological level, explain how odors are detected.

13. On a neurophysiological level, explain how substances are tasted.

CHAPTER CONTENTS

THE EYE—VISION

CONDITIONS OF CLINICAL SIGNIFICANCE: THE EYE

THE EAR—HEARING AND HEAD POSITION AND MOVEMENT

CONDITIONS OF CLINICAL SIGNIFICANCE: THE EAR

SMELL (OLFACTION)

TASTE (GUSTATION)

KEY TERMS AND DERIVATIVES

auditory (*aud* = to hear) pertaining to the sense of hearing

cochlea (*cochlea* = snail) the spiral-shaped hearing organ of the inner ear

cornea (*corn* = horn) the transparent anterior portion of the eyeball

endolymph (*endo* = within; *lympha* = lymph) the fluid contained in the membranous labyrinth of the ear

labyrinth (*labyrinth* = maze) the complex system of interconnecting chambers and tubes of the inner ear

ossicles (*ossiclum* = a small bone) the three bones of the middle ear: malleus, stapes, and incus

olfactory (*olfact* = to smell) pertaining to the sense of smell

sclera (*scler* = hard) the firm, fibrous outer layer of the eyeball; functions for protection and maintenance of eyeball shape

tympanic membrane (*tympano* = drum) the eardrum

vitreous humor (*vitre* = glass) the transparent, jellylike substance within the posterior cavity of the eye

The Special Senses

The body possesses specialized receptors for *vision* (sight), *hearing* (audition), *head position and movement* (the labyrinthine or vestibular sensations), *smell* (olfaction), and *taste* (gustation). These receptors are located in specific structures. The receptors for vision are located in the eyes; those for hearing and head position and movement are located in the ears; the receptors for smell are located in the nose; and the receptors for taste are found in the mouth and throat.

THE EYE—VISION

The **eye** is the sense organ for vision. It is housed in a bony skull cavity called the *orbit*, which protects much of the eye from physical injury.

Embryonic Development of the Eye

By the fourth week of embryonic development, lateral outgrowths from either side of the diencephalon have formed structures called **optic vesicles.** As the optic vesicles enlarge, the distal portion of each invaginates, forming a double-layered **optic cup** (Figure 15.1). The optic cup is destined to become the retina of the eye. The internal layer of the cup thickens to form the nervous-tissue layers of the retina, which contain receptor cells and neural elements. The external layer develops into the pigmented layer of the retina. The proximal portion of each optic vesicle narrows into an **optic stalk,** which will become incorporated into the optic nerve.

As the optic cups approach the underside of the surface ectoderm of the embryo, the ectoderm thickens into a **lens placode** directly over each cup. As the cavity of each optic cup deepens, the lens placodes invaginate to form **lens vesicles,** each of which is surrounded by one of the optic cups. With further development, the lens vesicles separate from the surface ectoderm and form a rounded body within the opening of each optic cup. The lens of each adult eye develops from these lens vesicles. Ectodermal cells superficial to the lens vesicles develop into the **cornea** of the eye.

On the undersurface of each optic cup and optic stalk is a groove called the **optic fissure.** The optic fissure serves as a path that directs the fibers of the optic nerve from the inner layer of the retina to the brain and that provides a means by which blood vessels are able to supply the interior of the eye. By the time of birth, the optic fissures have closed, completely encompassing the optic nerves and blood vessels.

Loose mesenchymal (mesoderm) cells accumulate around the outside of each optic cup. These cells differentiate into two connective tissue layers of the eye: the fibrous tunic and the vascular tunic.

Structure of the Eye

The eye is essentially a spherical structure composed of three basic coats, or layers: the *fibrous tunic*, the *vascular tunic*, and the internal tunic, or *retina* (Figure 15.2).

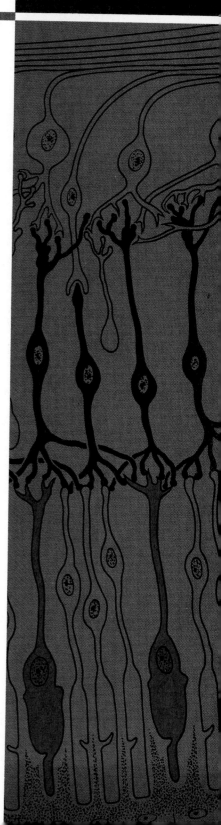

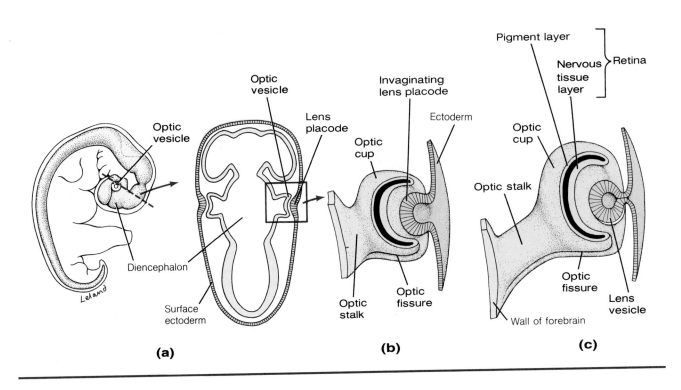

Figure 15.1

Embryonic formation of the eye. **(a)** Section through diencephalon and optic vesicle. (The dotted line indicates the plane of the section.) The optic vesicle grows out from the diencephalon and contacts the surface ectoderm. The surface ectoderm then thickens to form a lens placode.

(b) The lens placode invaginates, forming a hollow lens vesicle. The optic vesicle surrounds the lens vesicle, forming an optic cup. **(c)** Saggital section of the optic cup and lens vesicle.

Fibrous Tunic

skleh´-rah

The outermost layer of the eye is called the **fibrous tunic.** The posterior five-sixths of the fibrous tunic, the **sclera,** is white and opaque. The sclera, composed of dense connective tissue, aids in protecting the inner structures of the eye and helps maintain the shape of the eye. The anterior one-sixth of the fibrous tunic is clear and is called the **cornea.** The cornea is composed primarily of dense connective tissue, with an outer layer of stratified squamous epithelium. It has a greater curvature than the sclera, which causes it to protrude somewhat from the sclera. As light enters the eye, it passes through the cornea.

Vascular Tunic

Internal to the fibrous tunic is a layer called the **vascular tunic,** which is composed of the *choroid*, the *ciliary body*, and the *iris*.

The **choroid,** which lines most of the internal region of the sclera, is darkly pigmented and contains many blood vessels. Around the edge of the cornea the choroid forms the **ciliary body,** which contains smooth muscles called **ciliary muscles.**

The anterior portion of the vascular tunic is the **iris.** The iris, which is largely a continuation of the choroid, is a thin, muscular diaphragm whose pigmentation is responsible for eye color. In the center of the iris is a rounded opening, the **pupil,** through which light enters the interior regions of the eye.

Retina

Box 15.1

The innermost layer of the eye is the internal tunic, or **retina** (see Box 15.1). The retina consists of an outer **pigmented layer** and inner **nervous tissue layers.** The outer pigmented layer is composed of a single layer of heavily pigmented cuboidal epithelial cells that lie in contact with, and are tightly

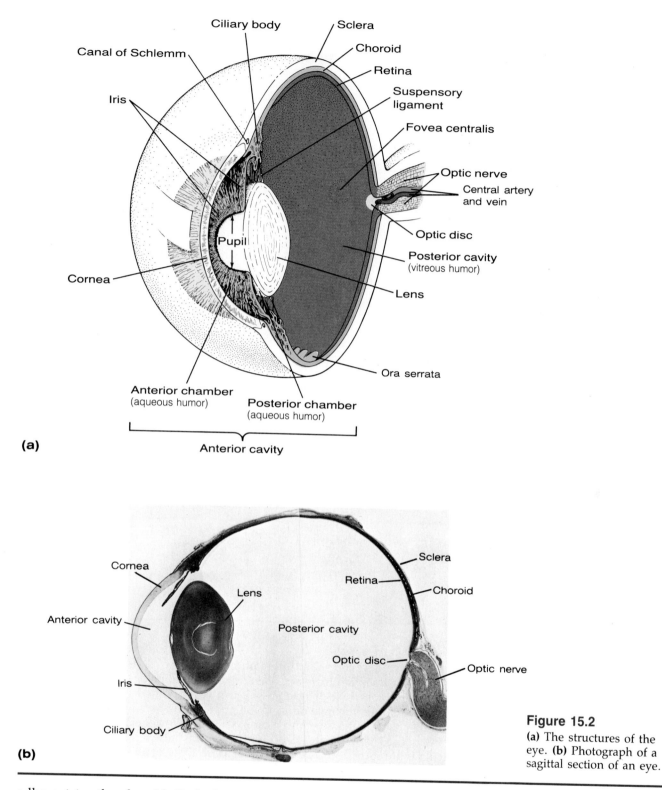

(a)

(b)

Figure 15.2
(a) The structures of the eye. (b) Photograph of a sagittal section of an eye.

adherent to, the choroid. Both the pigmented layer of the retina and the choroid contain the brown-black pigment melanin. The dark pigmentation of these structures reduces the reflection of the light that enters the eye.

The nervous tissue layers of the retina contain the actual receptors for light—photoreceptors called **rods** and **cones**—as well as numerous neural interconnections. After several synapses in the retina, nerve fibers converge and exit through the rear of the eye, slightly medial to the posterior pole of the eye, as the **optic nerve.** No photoreceptors are present at the point of exit,

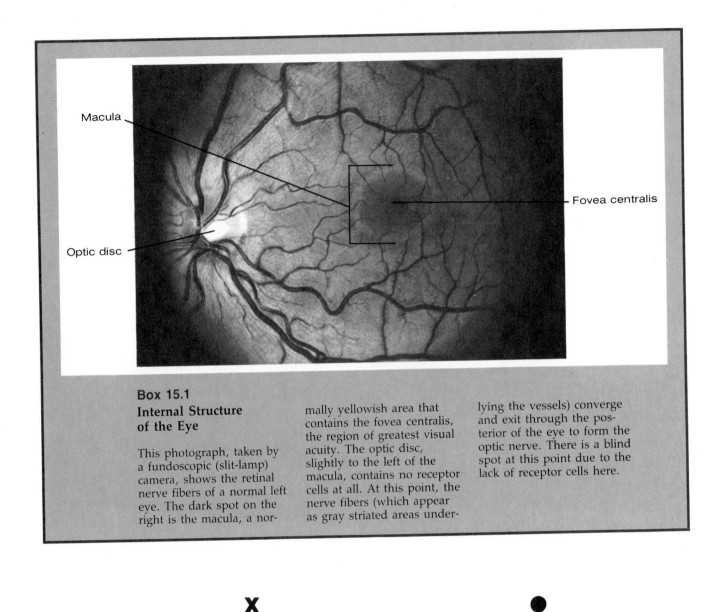

Macula

Optic disc

Fovea centralis

Box 15.1

**Internal Structure
of the Eye**

This photograph, taken by
a fundoscopic (slit-lamp)
camera, shows the retinal
nerve fibers of a normal left
eye. The dark spot on the
right is the macula, a normally yellowish area that
contains the fovea centralis,
the region of greatest visual
acuity. The optic disc,
slightly to the left of the
macula, contains no receptor
cells at all. At this point, the
nerve fibers (which appear
as gray striated areas underlying the vessels) converge
and exit through the pos-
terior of the eye to form the
optic nerve. There is a blind
spot at this point due to the
lack of receptor cells here.

X   ●

Figure 15.3
Demonstration of the
blind spot. Close your
left eye and focus your
right eye on the X. Move
the page toward or away
from you until the black
spot disappears. When
this occurs, the image of
the spot has fallen on the
blind spot (the optic disc)
of the retina.

and this area is called the **blind spot,** or **optic disc** (Figure 15.3). The *central
retinal* artery and vein that supply and drain the retina enter and leave the eye
through the optic disc. These vessels can be examined with an ophthalmo-
scope, providing valuable information about a person's health. Lateral to the
optic disc is a slightly yellow region known as the **macula.** The macula is
located almost exactly at the posterior pole of the eye. In the center of the
macula is a depression called the **fovea centralis** ("central depression"). The
fovea, which is the portion of the retina where visual acuity is greatest, contains no rods but does contain very densely packed cones. When you look
directly at an object, the image of the object generally is focused on the fovea.

The nervous tissue layers appear to end anteriorly, near the ciliary
body, in a scalloped margin called the *ora serrata*. However, they actually
continue forward onto the inner surface of the ciliary body and the iris. The
nervous tissue layers are attached to the pigmented layer of the retina only
around the optic nerve and at the ora serrata. Because the connection is loose,
it is possible for the nervous tissue layers of the retina to become detached
from the pigmented layer. Such a separation can be repaired surgically, or the
layers can be fused together using a laser.

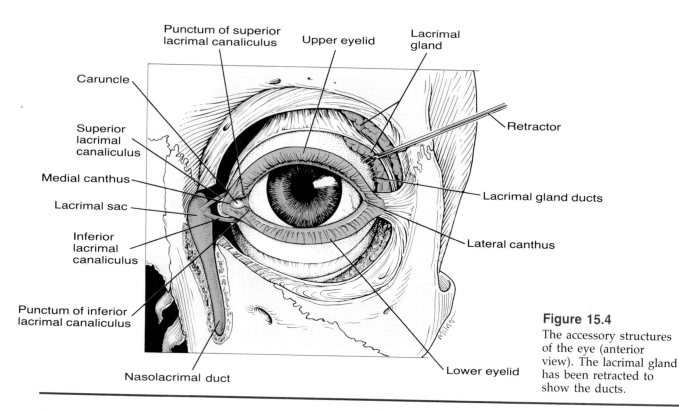

Punctum of superior lacrimal canaliculus

Upper eyelid

Lacrimal gland

Caruncle

Superior lacrimal canaliculus

Medial canthus

Lacrimal sac

Inferior lacrimal canaliculus

Punctum of inferior lacrimal canaliculus

Nasolacrimal duct

Retractor

Lacrimal gland ducts

Lateral canthus

Lower eyelid

Figure 15.4
The accessory structures of the eye (anterior view). The lacrimal gland has been retracted to show the ducts.

Lens

Behind the pupil is a clear **lens,** which is held in position by a fibrous **suspensory ligament** that attaches to the ciliary body. The lens is a biconvex structure that is relatively elastic. By changing shape, the lens aids in properly focusing on the retina the light that enters the eye. The lens has no blood supply. It receives its nourishment from substances called the aqueous and vitreous humors.

Cavities and Humors

The lens separates the interior of the eye into two cavities. In front of the lens is an **anterior cavity;** behind the lens is a **posterior cavity.** The anterior cavity can be subdivided into two chambers. The **anterior chamber** is located in front of the lens and iris and behind the cornea. The **posterior chamber** is located between the iris and the suspensory ligament. The anterior cavity contains a clear fluid, the **aqueous humor,** and the posterior cavity contains a transparent, semifluid, jellylike substance called the **vitreous humor.** The aqueous humor is produced at a rate of about 5 to 6 ml per day by folds called ciliary processes, which project from the ciliary body. The aqueous humor drains into a canal (venous sinus) called the **canal of Schlemm** and eventually reaches the bloodstream. The canal of Schlemm is located near the junction of the cornea and the iris. The rates of production and removal of aqueous humor are such that a relatively constant pressure of about 15 mm of mercury (the *intraocular pressure*) is maintained in the eye.

Accessory Structures of the Eye

Several structures associated with the eye contribute in various ways to its functioning. These include the *eyelids,* the *conjunctiva,* the *lacrimal apparatus,* and the *extrinsic eye muscles.*

Eyelids

Skin-covered structures called upper and lower **eyelids (palpebrae)** can be drawn over the anterior surface of the eye to protect the exposed portion (Figure 15.4). When closed, the eyelids prevent about 99% of the incident light from entering the eye. The movements of the eyelids are controlled by

pal´-pe-bral

F15.4

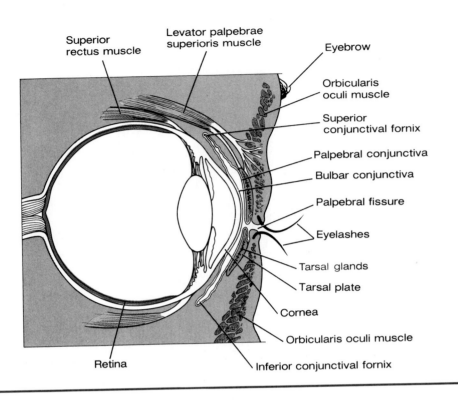

Figure 15.5

Accessory structures of the eye (sagittal section).

skeletal muscles. Fibers of the **orbicularis oculi** muscle extend into both eyelids and serve to close the lids. The **levator palpebrae superioris** muscle is inserted into the upper lid and functions to raise it (Figure 15.5). Much of the activity of these muscles occurs as the result of reflexes that produce an eyeblink response. This response may occur spontaneously about 25 times per minute when a person is awake, and it is particularly evident if a foreign object comes into contact with the surface of the eye or with the hairs **(eyelashes)** projecting from the borders of the eyelids.

Within each eyelid is dense connective tissue that forms a structure called a **tarsal plate.** The tarsal plates are important in maintaining the shape of the eyelids. Sebaceous glands called **tarsal (meibomian) glands** are located close to the inner surfaces of the eyelids and are embedded in the tarsal plates. The ducts of the tarsal glands open onto the margins of the eyelids, and their secretions onto these margins help prevent tears from overflowing between the eyelids. An infection of the tarsal glands can produce a *cyst* on the eyelid. The eyelids also possess modified sweat glands called **ciliary glands,** as well as sebaceous glands located at the bases of the hair follicles of the eyelashes. Occasionally, these sebaceous glands may become infected, thereby producing a *sty.* Between the upper and lower eyelids of each eye is a gap through which the eye is exposed—the **palpebral fissure.** The angle formed by the lateral junction of the upper and lower eyelids is called the **lateral canthus;** the angle formed by the medial junction of the upper and lower eyelids is called the **medial canthus** (Figure 15.4). A small, reddish mound of tissue at the medial canthus of each eye, the **caruncle,** contains a few sebaceous glands.

Conjunctiva

Both the insides of the eyelids and the anterior surface of the eye itself are covered with a thin protective layer of epithelium that forms a mucous membrane called the **conjunctiva.** The portion of the conjunctiva associated with the inner surface of the eyelids is called the **palpebral conjunctiva,** and the portion associated with the surface of the eye is called the **bulbar conjunctiva** (Figure 15.5). The palpebral conjunctiva of the upper eyelid is reflected at its upper margin onto the anterior surface of the eye as the bulbar conjunctiva.

F15.5

F15.4

kon-junk-tigh´-vah

F15.5

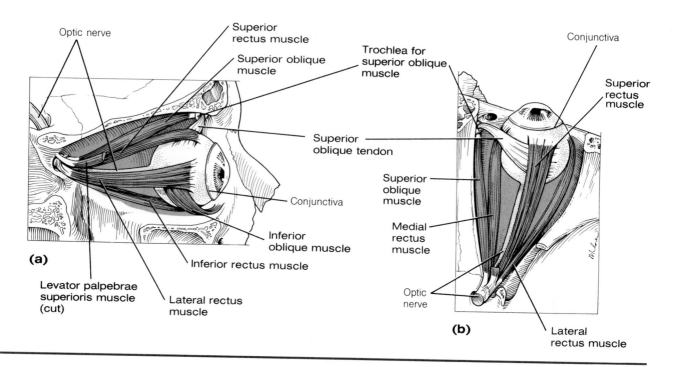

Figure 15.6
The extrinsic muscles of the eye viewed **(a)** from the side and **(b)** from above.

The angle formed by the palpebral and bulbar portions of the conjunctiva at the line of reflection is called the **superior conjunctival fornix.** Similarly, the palpebral conjunctiva of the lower eyelid is reflected at its lower margin onto the anterior surface of the eye, and an **inferior conjunctival fornix** is located at the line of reflection. An inflammation of the conjunctiva is called *conjunctivitis.*

Lacrimal Apparatus

In the superior lateral region of the orbit is a gland called the **lacrimal gland** (Figure 15.4). Six to 12 ducts arise from the lacrimal gland and carry its secretions onto the surface of the conjunctiva, primarily near the lateral portion of the superior conjunctival fornix. The lacrimal gland produces a watery secretion called **tears** that continually bathes the surface of the eye. Tears—which contain salts, some mucin, and a bacteriocidal enzyme called lysozyme—flow over the bulbar conjunctiva from their area of secretion to the medial canthus. Most tears simply evaporate. But excess fluid sometimes overflows from the eye, through the palpebral fissure, or it can drain via ducts into the nasal cavity, which is why it is generally necessary to blow one's nose when crying.

lak´-ra-mal
F15.4

Toward the medial end of each eyelid is a small projection called a **lacrimal papilla.** The lacrimal papilla contains the opening, or **punctum,** of a **lacrimal canaliculus.** The lacrimal canaliculi of the upper and lower eyelids converge medially from their openings and connect to a **lacrimal sac.** The inferior portion of the lacrimal sac opens into a **nasolacrimal duct,** which, in turn, opens into the nasal cavity beneath the inferior nasal conchae (turbinates). Normally, the secretions of a lacrimal gland amount to less than 1 ml per day; but these secretions increase in amount during certain emotional states (such as crying) and when foreign objects or other irritants contact the eye.

Extrinsic Eye Muscles

Six straplike skeletal muscles that originate from the orbit and insert on the connective tissue of the eye are responsible for eye movement (Figure 15.6). Originating from the rear of the orbit and inserting on the lateral surface of the eye is the **lateral rectus** muscle, whose contraction principally turns the eye laterally (outward). The **medial rectus** muscle, which originates from the

F15.6

rear of the orbit and inserts on the medial surface of the eye, primarily turns the eye medially (inward). Originating from the rear of the orbit and inserting on the superior surface of the eye is the **superior rectus** muscle, which principally turns the eye upward and somewhat medially. Originating from the rear of the orbit and inserting on the inferior surface of the eye is the **inferior rectus** muscle, which principally turns the eye downward and somewhat medially. The **inferior oblique** muscle originates from the medial surface of the front of the orbit and inserts on the inferior surface of the eye; it primarily turns the eye upward and laterally. The **superior oblique** muscle originates from the posterior portion of the orbit, with the rectus muscles, and runs along the medial surface of the orbit. Anteriorly, its tendon passes through a pulleylike loop, the **trochlea,** and turns to insert on the superior surface of the eye. The superior oblique muscle principally turns the eye downward and laterally.

The extrinsic eye muscles are among the most rapidly acting and precisely controlled skeletal muscles in the body. The motor units of the extrinsic eye muscles each contain only a few muscle fibers, and relatively high frequencies of stimuli are required to produce tetanus (350 stimuli per second versus 100 stimuli per second for some other skeletal muscles).

Light

Light is the small portion of the electromagnetic spectrum to which the receptors of the eye are sensitive (Figure 15.7). Light possesses both particlelike and wavelike properties. From a particlelike viewpoint, light is propagated in small, discrete packets of energy called photons. The travel of light, however, is often described in terms of wavelike properties such as wavelengths and frequencies (Figure 15.8).

The eye is stimulated by those wavelengths of electromagnetic radiation that range from about 400 to 700 nanometers (1 nanometer = one-billionth of a meter) (Figure 15.9). Within this range, the eye is able to distinguish an immense number of different wavelengths that are interpreted by the brain as different colors. The shorter wavelengths are interpreted as blues and violet, and the longer wavelengths are sensed as oranges and reds. A mixture of all the wavelengths of light is interpreted as white, and the absence of light is interpreted as black.

Optics

Light rays traveling in a straight line through a uniform medium of a given optical density (such as air) travel at a uniform speed. If the rays enter a second medium, with a different optical density (such as glass), their speed is altered. Moreover, if they strike the surface of the second medium at an angle other than perpendicular to the medium's surface, they are bent (refracted) (Figure 15.10). The greater the deviation from the perpendicular at which the rays strike the second medium, the greater the bending. If the optical density of the second medium is higher than that of the first (for example, when light rays traveling in air strike glass), the rays are bent toward the perpendicular. If the optical density of the second medium is lower than that of the first (for example, when light rays traveling through glass strike air), the rays are bent away from the perpendicular.

Suppose that light rays traveling through air from a point of light strike a spherical, biconvex glass lens (Figure 15.11). Rays from the point of light that strike the lens at angles other than the perpendicular are bent as they enter and leave the lens. Moreover, light rays striking the periphery of the lens are bent more than rays striking the central areas. This difference in bending occurs because rays striking the peripheral areas enter and leave the lens at angles more divergent from the perpendicular than rays striking the central areas. As a result of this differential refraction, the rays converge and come to a focus at a single point behind the lens called the *focal point*. The distance from the lens to the focal point increases as the curvature of the lens diminishes and also as the distance from the light source to the lens diminishes.

F15.7

F15.8

F15.9

F15.10

F15.11

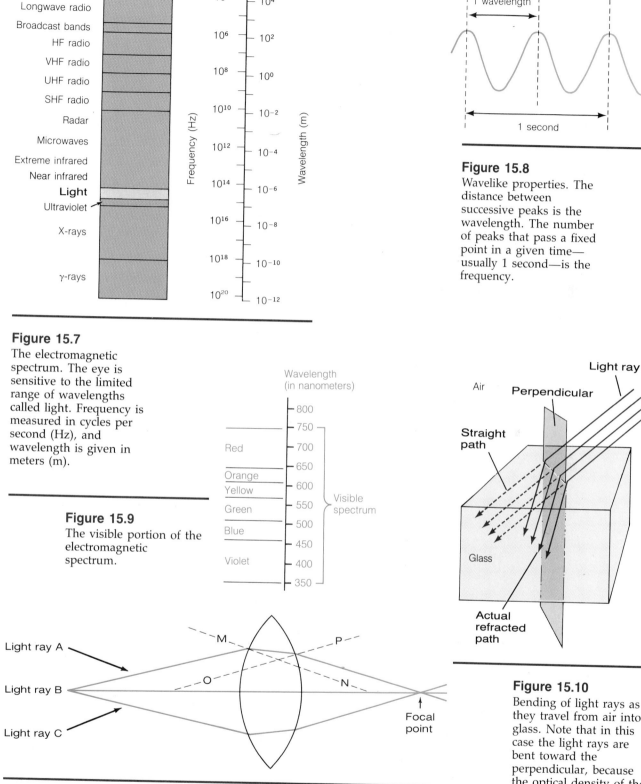

Longwave radio
Broadcast bands
HF radio
VHF radio
UHF radio
SHF radio
Radar
Microwaves
Extreme infrared
Near infrared
Light
Ultraviolet
X-rays
γ-rays

Frequency (Hz)

10^4
10^6
10^8
10^{10}
10^{12}
10^{14}
10^{16}
10^{18}
10^{20}

Wavelength (m)

10^4
10^2
10^0
10^{-2}
10^{-4}
10^{-6}
10^{-8}
10^{-10}
10^{-12}

Figure 15.7
The electromagnetic spectrum. The eye is sensitive to the limited range of wavelengths called light. Frequency is measured in cycles per second (Hz), and wavelength is given in meters (m).

Figure 15.8
Wavelike properties. The distance between successive peaks is the wavelength. The number of peaks that pass a fixed point in a given time— usually 1 second—is the frequency.

1 wavelength

1 second

Wavelength (in nanometers)

Red
Orange
Yellow
Green
Blue
Violet

— 800
— 750
— 700
— 650
— 600
— 550
— 500
— 450
— 400
— 350

Visible spectrum

Figure 15.9
The visible portion of the electromagnetic spectrum.

Air
Light ray
Perpendicular
Straight path
Glass
Actual refracted path

Figure 15.10
Bending of light rays as they travel from air into glass. Note that in this case the light rays are bent toward the perpendicular, because the optical density of the glass is greater than that of the air.

Light ray A
Light ray B
Light ray C
M
O
P
N
Focal point

Figure 15.11
Bending of light rays from a point of light as they strike a biconvex lens. Line M–N is the perpendicular for ray A as it enters the lens. Line O–P is the perpendicular for ray A as it leaves the lens. Note that ray B, which strikes the very center of the lens perpendicular to its surface, is not bent.

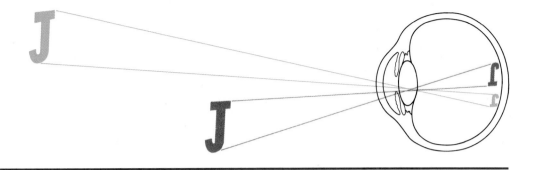

Figure 15.12
Retinal images are reversed (left to right) and upside down. The farther away the object, the smaller is its image on the retina.

F15.12

If the light source is an object instead of a point, an image of the object forms behind the lens. The image is inverted (upside down), and its right and left are reversed. Usually, the image is reduced in size, and the farther away the object, the smaller its image (Figure 15.12).

Focusing of Images on the Retina

When light rays strike the cornea, they are bent (refracted) in much the same way as light rays striking the lens in the previous example. In fact, when light rays enter the eye, the greatest bending occurs when the rays pass into the cornea. Additional bending occurs as the light rays pass from the cornea to the aqueous humor and also when the rays enter and leave the lens. The lens carries out the focusing adjustments required to ensure that images are formed sharply and clearly on the retina.

Emmetropia

The intraocular pressure exerted by the fluid within the eye forces the walls of the eye apart. Because the lens is attached to the walls by way of the suspensory ligament that connects to the ciliary body, the lens is under tension. Since the lens is elastic, the tension flattens it somewhat, giving it a less pronounced curvature than it might otherwise assume. In this condition, the normal, or **emmetropic,** eye focuses sharply onto the retina parallel light rays from distant points (light rays from points farther away than 20 feet are considered parallel). Divergent light rays entering the eye from points closer than 20 feet are focused behind the retina.

Accommodation

The focusing system of the eye must adjust in order to focus onto the retina light rays from objects closer than 20 feet. The adjustment process is referred to as **accommodation,** and it is accomplished by the ciliary muscles of the ciliary body, which contract in response to parasympathetic stimulation. The contraction of the ciliary muscles pulls the ciliary body slightly forward and inward, narrowing the ring of the ciliary body. This action lessens the tension on the suspensory ligament and permits the elastic lens to assume a more pronounced curvature. The greater curvature results in a greater bending of light rays that strike the lens at angles other than the perpendicular, enabling the lens to focus onto the retina divergent light rays from points closer than 20 feet. Under such conditions, light rays from distant points are focused in front of the retina.

Information that reaches the visual association areas of the occipital cortex from the retina appears to be important in accommodation. When the image on the retina is out of focus, the visual association areas transmit signals to the ciliary muscles to alter the curvature of the lens and thus bring the image into focus. These focusing adjustments usually require less than a second.

During accommodation, the pupil constricts, eliminating the most divergent light rays that would otherwise pass through the most peripheral portions of the lens. The elimination of these rays is beneficial because even a good lens may not be perfect, and light rays that pass through the peripheral regions of the lens do not always focus at exactly the same point as those that pass through the central areas. Thus, the constriction of the pupil aids in the formation of a sharp image on the retina (and also reduces the amount of light entering the eye).

Another event associated with accommodation is the convergence of the eyes. That is, when viewing close objects, the eyes turn inward so the image falls on the fovea of each retina, which is the portion of the retina where visual acuity is greatest.

Near and Far Points of Vision

The closer an object is to the eye, the more divergent are the light rays from the object and the greater is the accommodation required to focus the light rays onto the retina. However, the ability of the eye to accommodate for viewing near objects is limited. The distance from the eye to the nearest point whose image can be focused on the retina is called the **near point of vision.** The near point of vision varies with age: It is close to the eye in youth and recedes farther and farther from the eye as a person gets older.

The **far point of vision** is the distance from the eye to a point whose image can be focused on the retina without accommodation. The far point of vision for the normal emmetropic eye is infinity (any point beyond 20 feet).

Control of Eye Movements

The movements of the eyes, which are caused by the extrinsic eye muscles, are synchronized with one another so that both eyes move in a coordinated manner when viewing an object. When a person wants to look at something, he or she moves the eyes voluntarily to find the object and fix upon it. These *voluntary fixation movements* are controlled by a small area of the cerebral cortex located bilaterally in the premotor regions of the frontal lobes. Once the object is found, *involuntary fixation movements* keep the image of the object fixed on the foveal portions of the retinas. The involuntary fixation movements are controlled by the occipital region of the cortex, particularly the visual association areas. The visual fixation mechanisms depend on feedback from the retinas to keep the image of an object on the foveas. For example, each time the image drifts to the edge of the foveas, sudden flicking movements of the eyes occur. These movements, which are automatic reflex responses controlled by the involuntary fixation mechanisms, prevent the image from leaving the foveas.

Binocular Vision and Depth Perception

Humans have **binocular vision**—that is, each eye views a portion of the external world that overlaps considerably with that viewed by the other eye. However, the visual field of the left eye is not identical with that of the right eye (Figure 15.13). Moreover, the two eyes do not form exactly identical images of an object because they occupy slightly different locations. This *retinal disparity* contributes to a person's ability to judge relative distances when objects are nearby (that is, closer than about 70 meters). Depth perception with only one eye depends partly on the use of learned cues such as the fact that the sizes of objects in the environment appear to diminish with distance.

F15.13

Diplopia (Double Vision)

Light rays coming from an object strike the retinas of both eyes. However, a person normally perceives only one object, not two. This is because the retinas possess *corresponding points* that, when stimulated, result in the perception of a single object. Normally, when an individual looks at an object, the eyes move in such a manner that light rays from the object fall on corresponding points of the foveas of the retinas. However, if the extrinsic eye muscles are weak or paralyzed, or if the muscles of the two eyes act in an unequal or uncoordinated fashion, light rays from an object may not fall on corresponding points of the retinas. As a result, the individual may develop **diplopia** (double vision) and perceive two objects instead of one. Transient diplopia

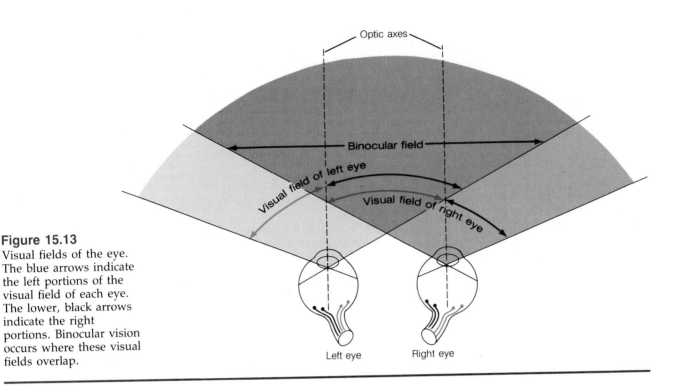

Figure 15.13
Visual fields of the eye.
The blue arrows indicate
the left portions of the
visual field of each eye.
The lower, black arrows
indicate the right
portions. Binocular vision
occurs where these visual
fields overlap.

can develop in acute alcoholic intoxication due to the partial paralysis of the
extrinsic eye muscles.

Strabismus

stra-biz´-mus

In the condition known as **strabismus** (squint, cross-eyedness), the movements of the two eyes are not properly coordinated with one another when a
person looks at an object. In some cases of strabismus, the eyes alternate in
fixing on an object. In others, the same eye is used all the time. In such cases,
the image formed by the unused eye can eventually become suppressed, and
if the condition is not corrected, the repressed eye may become functionally
blind.

Photoreceptors of the Retina

F15.14

The outermost portion of the nervous tissue layers of the retina (the portion
closest to the pigmented layer of the retina and the choroid) contains the
light-sensitive photoreceptors: the **rods** and **cones** (Figure 15.14). The rods
and cones contain photopigments whose configurations are altered when
light strikes them and they absorb the light. These alterations lead to changes
in the polarity of the photoreceptors, ultimately resulting in the transmission
of neural signals from the retina to the brain that are interpreted as visual
events.

There are four different photopigments, each consisting of a protein,
called an *opsin*, to which a chromophore molecule called *retinal* is attached.
The opsins differ from pigment to pigment, and they confer specific light-
sensitive properties on each pigment. Retinal is produced from vitamin A_1,
and retinal and vitamin A_1 can be interconverted. The pigmented layer of the
retina contains stores of vitamin A_1.

When light of the proper wavelength strikes a photopigment, the chemical configuration of the retinal is altered, and the retinal breaks away from the
opsin. The altered photopigment activates a protein that can bind with the
substance guanosine triphosphate (GTP). The GTP-binding protein, in turn,
activates the enzyme phosphodiesterase, which breaks down the substance
cyclic guanosine monophosphate (cyclic GMP). In the photoreceptors, cyclic
GMP acts on ion channels (particularly sodium channels) in the plasma membrane to keep them open. When cyclic GMP is broken down, the channels

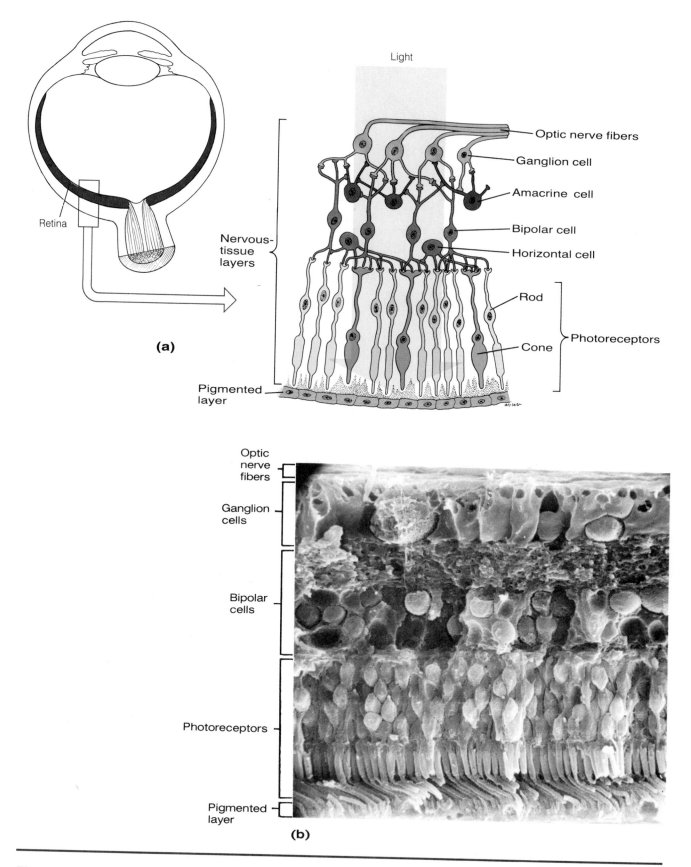

Figure 15.14
The structure of the retina. **(a)** Schematic view. **(b)** Scanning electron micrograph (×1500). (From *Tissues and Organs: A Text-Atlas of Scanning Electron Microscopy* by Richard G. Kessel and Randy H. Kardon. W.H. Freeman and Company. Copyright © 1979.)

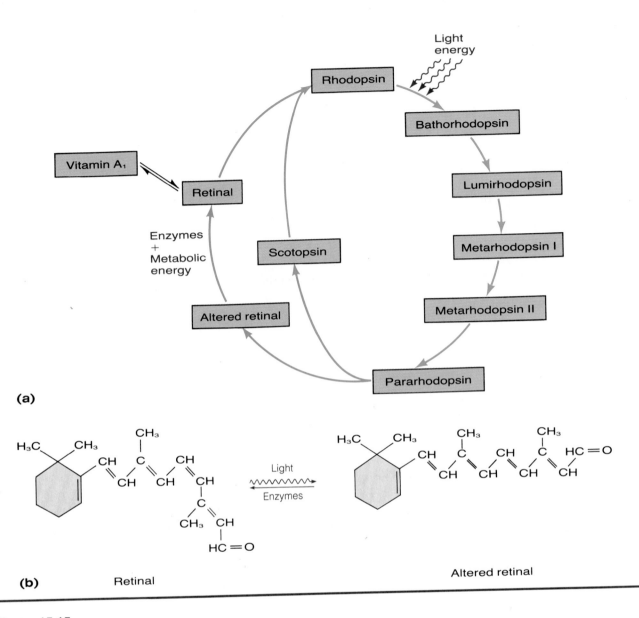

(a)

(b)   Retinal   Altered retinal

Figure 15.15

(a) The rhodopsin–retinal chemical cycle. Light energy alters the chemical configuration of the retinal of rhodopsin, leading to the formation of bathorhodopsin. Note that retinal can be formed from vitamin A_1. **(b)** Retinal, indicating the change in chemical configuration brought about by light. Enzymes and metabolic energy are required to convert altered retinal back to retinal.

F15.15

close, resulting in a hyperpolarization of the cell. Consequently, the photoreceptors respond to an increase in the intensity of light of the proper wavelength by hyperpolarizing, and they respond to a decrease in the intensity of the light by depolarizing. After the light-induced breakdown of the photopigment, the altered retinal is rearranged and rejoined to the opsin to restore the photopigment.

Rods

The photopigment contained within the rods is called *rhodopsin*. Rhodopsin is composed of retinal and the opsin scotopsin. Vitamin A_1 absorbed by the rods is converted to retinal, which combines with scotopsin to form rhodopsin. When the eyes are not exposed to light, the concentration of rhodopsin can build up to a very high level.

When light strikes the rods, the chemical configuration of the retinal is altered, and the altered retinal begins to pull away from the scotopsin, forming a substance called bathorhodopsin (prelumirhodopsin) (Figure 15.15). Bathorhodopsin is very unstable; it quickly decays into lumirhodopsin, which in turn rapidly decomposes into metarhodopsin I. Metarhodopsin I spontaneously forms metarhodopsin II, which in turn becomes pararhodopsin. Pararhodopsin then splits into altered retinal and scotopsin. Following these

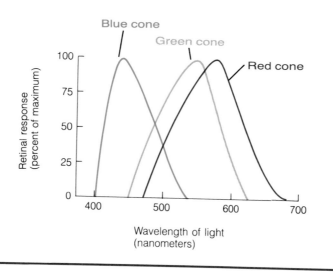

Figure 15.16

Sensitivities of the three different types of cones to light of different wavelengths.

events, the altered retinal is rearranged and rejoined to scotopsin to form rhodopsin. The rearrangement of the altered retinal is enzymatically mediated and requires metabolic energy.

The rods are very light-sensitive, and they can respond to low levels of illumination, such as those present at night or in dimly lit areas. The responses of the rods indicate degrees of brightness. However, the rod system is characterized by a relative lack of color discrimination, and rod responses are interpreted principally as shades of gray.

Cones

The chemical events that occur when light strikes the cones are similar to those that take place when light strikes the rods. However, there are three different types of cones. Each type contains a different photopigment and each is selectively sensitive to particular wavelengths of light (Figure 15.16). Red cones respond more intensely than the others to those wavelengths of light that the brain interprets as red. Green cones respond more intensely than the others to those wavelengths of light that the brain interprets as green. Blue cones respond most intensely to those wavelengths of light that the brain interprets as blue or violet.

The cones are primarily responsible for color vision. The ability to perceive many different colors rather than only three colors (red, green, and blue) is due to the fact that different wavelengths of light striking the retina evoke different ratios of response from the three cone types. These varied responses are interpreted by the brain as a tremendous variety of different colors. For example, when light with a wavelength of 580 nanometers strikes the retina, the red cones respond more intensely than the green cones, and the blue cones do not respond at all. The brain interprets this pattern of responses as the color yellow.

The cones operate only at relatively high levels of illumination, and they are the principal photoreceptors during daylight or in brightly lit areas. The cones are concentrated in the center of the retina and most highly in the fovea, whereas the rods are more numerous in the peripheral retina.

Neural Elements of the Retina

The portions of the nervous tissue layers of the retina closest to the interior of the eye are composed of neural elements (Figure 15.14). The middle portion of the nervous tissue layers contains **bipolar cells,** and the innermost portion contains **ganglion cells.** The rods and cones synapse with bipolar cells, which in turn synapse with ganglion cells.

In general, the neural pathways that transmit the responses of the rods display greater convergence than the pathways that transmit the responses of

F15.16

F15.14

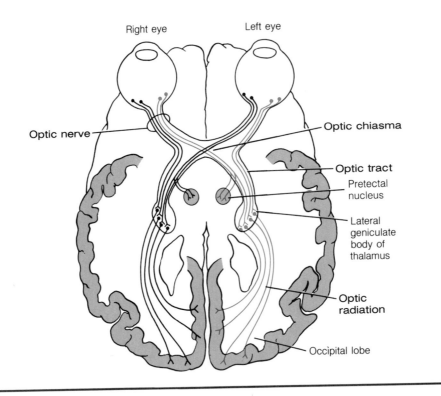

Figure 15.17
Neural pathways for vision (ventral view of brain).

the cones. In the peripheral regions of the retina, for example, many rods synapse with a single bipolar cell, and many bipolar cells synapse with a single ganglion cell. Within the fovea, in contrast, there are approximately equal numbers of cones, bipolar cells, and ganglion cells, and relatively little convergence occurs. Consequently, the cones in the fovea provide more precise information about the area of the retina stimulated than do the rods in the peripheral retina. As a result, vision is more precise in the fovea than in the peripheral retina. On the other hand, the convergence that occurs along the visual pathways of the rods provides numerous opportunities for spatial summation, and this, together with the differences in light sensitivity of the rods and cones themselves, contributes to the fact that rod vision is more effective in dim light than cone vision.

F15.14   In addition to bipolar cells and ganglion cells, the retina contains neural cells called **horizontal cells** and **amacrine cells** (Figure 15.14). These cells are important in lateral interactions that occur within the retina. For example, rods and cones respond to an increase in the intensity of light of the proper wavelength by hyperpolarizing and diminishing their release of neurotransmitter molecules. Since the neurotransmitter molecules released by rods and cones cause bipolar cells to hyperpolarize, this response is in essence stimulatory to the bipolar cells, and the bipolar cells, in turn, stimulate ganglion cells with which they synapse. However, when the rods or cones of a given area of the retina respond to an increase in light intensity, their responses are transmitted to horizontal cells as well as to bipolar cells. The horizontal cells act in an inhibitory fashion on bipolar cells adjacent to the bipolar cells directly affected by the rods and cones. This *lateral inhibition* enhances the signal contrast between strongly stimulated regions of the retina and adjacent, more weakly stimulated regions.

Visual Pathways

The axons of the retinal ganglion cells come together and leave the eye as the **optic nerve,** and each optic nerve contains about one million nerve fibers. The two optic nerves (one from each eye) meet at a structure called the **optic** F15.17   **chiasma,** located just anterior to the pituitary gland (Figure 15.17). Within the optic chiasma, ganglion cell axons from the medial half of each retina cross to the opposite side, and those from the lateral half of each retina remain on the

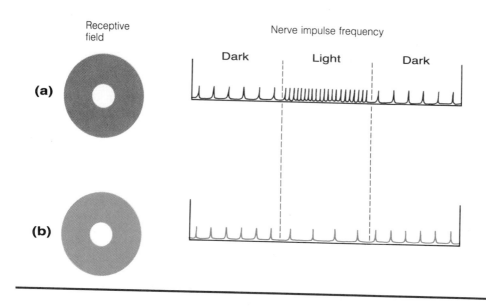

Receptive field

Nerve impulse frequency

Figure 15.18
Ganglion cell responses.
(a) When the central area of the receptive field of an on-center type of ganglion cell is illuminated by a spot of light, the frequency of nerve impulses transmitted by the cell increases. **(b)** When the central area of the receptive field of an off-center type of ganglion cell is illuminated, the frequency of nerve impulses transmitted by the cell decreases.

same side. From the optic chiasma, the axons continue as the **optic tracts.** Thus, the left optic tract consists of ganglion cell axons from the lateral half of the retina of the left eye and the medial half of the retina of the right eye (the right portion of the visual field of each eye), and the right optic tract consists of axons from the lateral half of the retina of the right eye and the medial half of the retina of the left eye (the left portion of the visual field of each eye).

Most of the ganglion cell axons within the optic tracts travel to the lateral geniculate bodies of the thalamus. There they synapse with neurons that form pathways called **optic radiations,** which terminate in the visual cortex of the occipital lobes. Some of the axons within the optic tracts travel to midbrain nuclei called pretectal nuclei, where they are involved in a reflex called the pupillary light reflex (see page 446).

Processing of Visual Signals

The response of a visual pathway cell is commonly described in terms of its receptive field. The *receptive field* of a cell is the area of the retina that, when illuminated, influences the activity of the cell. In essence, it is an area of photoreceptors that provide input to the cell. This input is often provided indirectly by way of other cells, and it can be inhibitory as well as stimulatory. For example, a photoreceptor may provide input to a bipolar cell by way of an inhibitory horizontal cell.

Because of such occurrences as convergence and lateral interactions, the processing of visual signals begins in the retina, and the ganglion cells transmit to the brain a coded message rather than a simple mosaic representation of the image on the retina. The receptive field of a ganglion cell is circular and is composed of essentially two parts: a small central area and a ringlike peripheral area (Figure 15.18). In the dark, a ganglion cell continually transmits nerve impulses to the brain. However, when the receptive field of a ganglion cell is illuminated, the frequency of nerve-impulse transmission changes. In one type of ganglion cell, called an *on-center cell*, positioning a spot of light on the central area of a cell's receptive field increases the frequency of nerve-impulse transmission. Conversely, illuminating the peripheral area decreases the frequency of nerve-impulse transmission. In a second type of ganglion cell, called an *off-center cell*, positioning a spot of light on the central area of a cell's receptive field decreases the frequency of nerve-impulse transmission, and illuminating the peripheral area increases the frequency of nerve-impulse transmission. In either cell type, nerve-impulse frequency varies with the degree of illumination of the central area of a cell's receptive field compared to the degree of illumination of the peripheral area. That is, when both areas are illuminated simultaneously, the activity of the ganglion cell is a reflection of both stimulatory and inhibitory influences. Consequently, the signals transmitted to the brain by a ganglion cell indicate such things as the light level in

F15.18

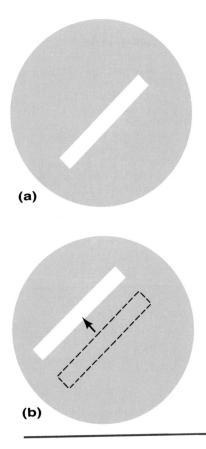

(a)

(b)

Figure 15.19

Responses of the cells of the visual cortex. **(a)** A simple cell responds when a line is oriented at a precise angle in a particular portion of the cell's receptive field. **(b)** A complex cell responds when the line is oriented at a precise angle regardless of its position in the cell's receptive field. Thus, the cell continues to respond as a properly oriented line moves across its receptive field. (Some complex cells respond better when the line moves in one direction than in the other.)

one small area of the visual scene compared to the average illumination of the immediately surrounding region.

The neurons of the lateral geniculate bodies of the thalamus that transmit nerve impulses to the visual cortex have receptive fields that are similar to those of ganglion cells—that is, their receptive fields are circular, with either an excitatory central area and an inhibitory peripheral region or vice versa.

In the visual cortex there is a hierarchy of cells with receptive fields that vary widely in organization. Some cortical cells have circularly symmetrical receptive fields, and their responses resemble those of ganglion cells and lateral geniculate neurons. However, the receptive fields of other cortical cells are organized so the cells respond best not to spots of light but to specifically oriented line segments (such as narrow slits of light) positioned on the retina. For example, cells called *simple cells* respond only when the line is oriented at a precise angle in a particular portion of a cell's receptive field (Figure 15.19a). The responses of these cells provide information about lines and borders. Other cells, called *complex cells*, respond when the line is oriented at a precise angle regardless of its position in a cell's receptive field (Figure 15.19b). The responses of these cells help indicate the movement of a visual stimulus. Thus, the processing of visual signals that begins in the retina continues in other parts of the visual pathway. As a result, the components of the pathway transmit a coded message about certain aspects of the image on the retina rather than a mosaic pattern of the image. The message contains information about such characteristics of the image as contrast, movement, and color, and it is this information that the brain utilizes in developing a visual representation of the image.

Light and Dark Adaptation

When a person moves from a dimly lit area to a brightly lit area, the person's eyes often feel uncomfortable, and vision is poor for several minutes. With time, however, **light adaptation** occurs and vision improves. Similarly, when a person moves from a well-lit area to a dimly lit area, the person initially cannot see well. However, as time passes, **dark adaptation** occurs and vision gets better. Several mechanisms contribute to light and dark adaptation. These mechanisms include changes in the concentration of photopigments, the pupillary light reflex, and neural adaptation.

Photopigment Concentration

Relatively small changes in the concentration of rhodopsin within the rods can greatly influence their sensitivity to light. For example, with a decrease in rhodopsin content of only 0.6% from the maximum, rod sensitivity declines about 3000 times. A similar situation is believed to exist with respect to the cones and their photopigments.

When a person moves from a dimly lit area to a brightly lit area, significant amounts of photopigments (particularly rhodopsin) are broken down by the light. This occurrence, which requires about five to ten minutes for completion, decreases the sensitivity of the eyes and raises the visual threshold so that a greater intensity of light is required for effective vision.

When a person moves from a well-lit area to a dimly lit or dark area, the photopigments of the rods and cones that were broken down in bright light are regenerated. The regeneration of the photopigments lowers the visual threshold and increases the sensitivity of the eyes to light. Although this process can continue for hours, a considerable increase in sensitivity occurs within 20 to 40 minutes. The increase in visual sensitivity that occurs during the first ten minutes or so is relatively slight and has been attributed primarily to the cones. This increase is followed by a much greater increase in sensitivity that is due to the more slowly adapting rods, which are primarily responsible for vision in dim light.

Pupillary Light Reflex

The **pupillary light reflex** constricts (reduces the diameter of) the pupils in bright light and helps regulate the amount of light entering the eyes. In this reflex, light striking the retina initiates neural signals that are sent to the

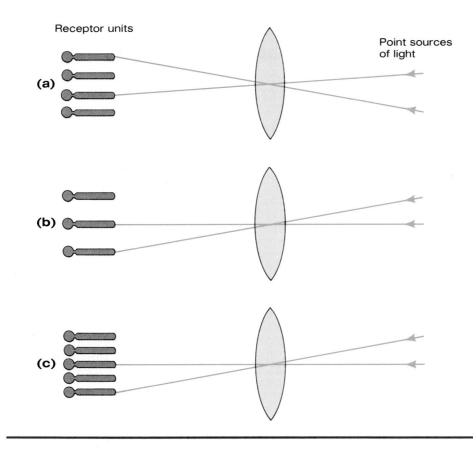

Figure 15.20
Resolving power of the eye. In order for two point sources of light to be discriminated as separate, light rays from one point must stimulate a different receptor unit of the retina than light rays from the second point, with at least one unstimulated or only weakly stimulated unit between the two. The more closely the receptor units are packed, the more likely this will occur. In **(a)**, two widely spaced points may be detected as separate. In **(b)**, the two points will not be discriminated as separate. In **(c)**, two closely spaced points may be detected as separate.

pretectal nuclei of the midbrain. These signals ultimately lead to an increased stimulation of the circular smooth muscles of the iris by neurons of the parasympathetic division of the autonomic nervous system. The contraction of the smooth muscles constricts the pupils. In dim light, the stimulation of the circular smooth muscles decreases, and the pupils dilate (increase their diameter).

If bright light shines into only one eye, the pupils of both eyes will constrict. However, the constriction of the pupil of the eye into which the light shines will be greater than the constriction of the pupil of the other eye. The reflex leading to pupillary constriction in the eye into which the light shines is called the *direct light reflex*, and the reflex leading to pupillary constriction in the other eye is called the *consensual light reflex*.

Neural Adaptation

Neurons of the visual pathways in the retinas undergo rapid neural adaptation. When a person moves from a dimly lit area to a brightly lit area, the signals transmitted by the bipolar cells, horizontal cells, amacrine cells, and ganglion cells are initially very intense. However, within a fraction of a second, the intensity of the signals declines severalfold. The cone system contributes to light adaptation by way of neural interconnections that inhibit rod function in bright-light conditions appropriate for cone function.

Visual Acuity

Visual acuity is the ability of the eye to distinguish detail. It is related to resolving power, which is the eye's ability to distinguish two closely spaced points as separate. For two closely spaced points to be distinguished as separate, it is believed, light rays entering the eye from one of the two points must stimulate a different receptor unit in the retina than light rays from the other point, and there must be at least one unstimulated or only weakly stimulated unit between the two receptor units (Figure 15.20). Whether or not this pattern of stimulation occurs depends on both the focusing system of the eye and

F15.20

CONDITIONS OF CLINICAL SIGNIFICANCE

The Eye

Many disorders can affect the eyes. Focusing problems are among the most common.

Myopia

Myopia is nearsightedness. In this condition, light rays from distant objects are focused in front of the retina, and only light rays from close objects can be focused accurately onto the retina. Myopia is caused by a focusing system that has too great a refractive power with respect to the position occupied by the retina. Most commonly this is due to an elongated eyeball, which can result from a weakness of the coats of the eye (Figure 15.21). In myopia, the near point of vision is closer than normal and the far point of vision is closer than 20 feet. In all but the severest cases, myopia can be corrected by eyeglasses with concave lenses. Such lenses cause light rays to diverge slightly as they enter the eye, helping the refractive system of the eye focus them onto the retina.

Hypermetropia

Hypermetropia (hyperopia) is farsightedness. In this condition, light rays from distant objects are focused behind the retina when the eye is at rest, and accommodation is necessary in order to focus the rays onto the retina. Since the eyes can accommodate only so much, a hyper-

metropic individual has trouble viewing close objects. Moreover, accommodation is required to view any object, whether close or far. Thus, in hypermetropia there is no far point of vision, and the near point lies farther away than normal. Hypermetropia is most commonly due to a shortened eyeball (Figure 15.22). The condition can usually be corrected with eyeglasses having convex lenses. Such lenses cause light rays to converge as they enter the eye, and they thus assist the eye's refractive system to focus the rays onto the retina.

Astigmatism

Previous considerations of the refractive system of the eye have assumed that all elements of the system possess uniformly curved surfaces (much like a marble). Often, however, the surface of one or more elements of the system is not uniformly curved in all planes. For example, the surface of the cornea may have a different curvature in the horizontal plane than in the vertical plane (much like a chicken's egg). As a result, light rays entering the eye in different planes are focused at different points, creating an out-of-focus image. The condition that results from unequal curvatures of portions of the eye's refractive system is called *astigmatism*. If an astigmatic individual examines a pattern such as that in Figure 15.23, in which straight lines radiate from a cen-

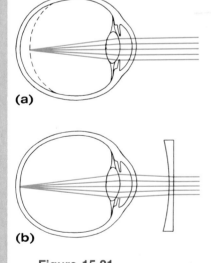
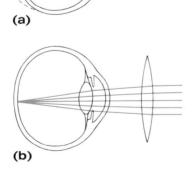
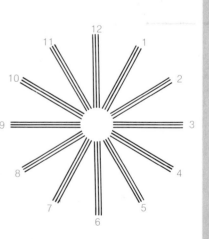

Figure 15.21
Myopia. **(a)** In myopia, light rays from distant objects focus in front of the retina.
(b) Most cases of myopia can be corrected by a concave lens, which causes light rays to diverge as they enter the eye.

Figure 15.22
Hypermetropia. **(a)** In hypermetropia, light rays from distant objects focus behind the retina.
(b) Most cases of hypermetropia can be corrected by a convex lens, which causes light rays to converge as they enter the eye.

Figure 15.23
Astigmatism may be detected by viewing a series of radiating lines. To astigmatic individuals, some lines appear more distinct than others.

tral point, some lines are sharply focused on the retina and are seen clearly, whereas others are focused in front of or behind the retina and are seen indistinctly. All but very severe cases of astigmatism can be corrected by a lens (eyeglass) that has a greater curvature in one plane than another. The lens is oriented in front of the eye so that the differential bending of light rays passing through the lens compensates for the differential bending of light rays by the eye.

Cataract

In some cases, the lens of the eye or a portion of it becomes cloudy or opaque so that vision is impaired. This cloudiness or opacity is known as a *cataract* (Figure 15.24). Often, lenses with cataracts are removed surgi-

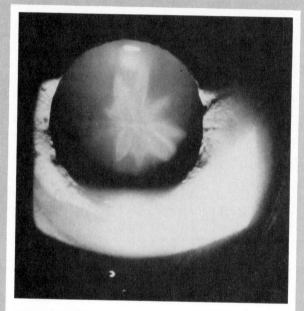

Figure 15.24
Rose petal anterior
cataract.

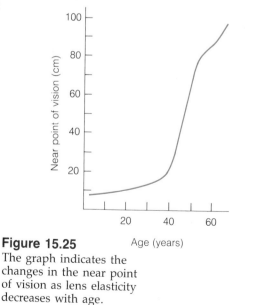
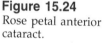

Figure 15.25
The graph indicates the
changes in the near point
of vision as lens elasticity
decreases with age.

cally, and effective vision is restored by the use of lens implants, contact lenses, or special eyeglasses.

Glaucoma

Glaucoma is a condition in which there is an abnormal elevation of the intraocular pressure, often resulting from deficient drainage of the aqueous humor. When severe, the high intraocular pressure of glaucoma squeezes shut blood vessels that supply the eye, leading to the degeneration of the retina and to blindness.

Color Blindness

Color blindness is a deficiency in color perception that ranges from an inability to distinguish certain shades of color to a complete lack of color perception. The condition is caused by a functional deficiency or absence of one or more of the different cone types involved in color vision. Difficulty in distinguishing reds and greens (red–green color blindness) is the most common form of color blindness. There is a strong hereditary component in color blindness, and about 8% of males but less than 1.5% of females are red–green color blind.

Effects of Aging

With aging, the lens of the eye becomes yellowed due to the effects of ultraviolet rays from such sources as sunlight. In addition, the lens is one of the few body structures that exhibits increased cellular growth with age, which may contribute to the development of cataracts.

As an individual ages, the pupil of the eye is no longer able to dilate fully, and the amount of light that reaches the photoreceptors of the retina by age 70 may be only 50% of the amount that reaches the retina during youth. The inability of the pupil to dilate fully can also contribute to poor drainage of the aqueous humor, resulting in increased intraocular pressure and a greater likelihood of glaucoma.

The continuous exposure of the rods and cones to light causes damage to their membranes and generates cellular debris. The debris can be removed by cells of the pigmented layer of the retina, and as long as these cells function properly, no problems occur. With aging, however, these cells can become congested, and debris from the rods and cones may accumulate, thus contributing to a loss of visual acuity. This loss is particularly striking if it occurs in the fovea.

With aging, the lens gradually loses its elasticity, and thereby loses some of its ability to change shape during accommodation for viewing near objects. As a result, the near point of vision recedes farther and farther from the eye. Although the loss of lens elasticity begins early in life, the greatest loss occurs after about the age of 40 (Figure 15.25). As the ability to accommodate for viewing near objects diminishes, it becomes necessary to hold reading materials farther and farther from the eye in order to focus the printed letters onto the retina. Ultimately, books and newspapers may have to be held so far from the eye that the images of the letters on the retina are too small to be recognized. This condition, called *presbyopia*, can be corrected by the use of eyeglasses with convex lenses. The convex lens increases the convergence of the light rays so that the refractive system of the eye can focus them onto the retina when the printed matter is held reasonably close to the eye.

the area of the retina struck by the light rays. Light rays entering the eye from two points have the greatest likelihood of stimulating different receptor units in the fovea, where the cones are packed closely together and each cone has a relatively direct path to the brain. Moreover, in the fovea the inner layers of the retina are generally displaced to one side, allowing light to pass relatively unimpeded to the cones. Therefore, visual acuity is greatest in the region of the fovea.

Visual acuity is often assessed with the aid of eye charts, which frequently consist of a series of letters of varying sizes. A person with normal visual acuity should be able to read the letters on the 20-foot line of the eye chart at a distance of 20 feet. If a person can read only the 40-foot line at 20 feet, the person is said to have 20/40 vision (meaning that the person's eyes cannot distinguish beyond 20 feet what normal eyes can distinguish at 40 feet).

THE EAR—HEARING AND HEAD POSITION AND MOVEMENT

The ear contains receptors for hearing as well as receptors that detect head position and movement. The ear is divided into *external, middle,* and *inner* regions (Figure 15.26). The external ear is essentially a funnel-shaped structure used for collecting sound waves. The middle ear contains three small bones called ossicles, which transmit sound waves from the external ear to the inner ear. The inner ear is composed of a system of fluid-filled semicircular canals and chambers that contain receptors for the detection of sound waves, as well as receptors concerned with head position and movement.

F15.26

Embryonic Development of the Ear

In the embryo, the development of the ear begins with the formation of a thickened ectodermal plate called an **otic placode,** which is located on the side of the head in the vicinity of the hindbrain (Figure 15.27a). Each otic placode invaginates to form an **otic pit** (Figure 15.27b). By the fourth week of development, the otic pit separates from the surface ectoderm and develops into a closed sac called an **otic vesicle** (Figure 15.27c).

plak´-ode
F15.27a
F15.27b

F15.27c

While the otic vesicles are forming, lateral pouches develop from the side of the pharynx. As these **pharyngeal pouches** expand, the surface ectoderm indents toward them, forming the **branchial grooves.**

With further development, the otic vesicle forms the *membranous labyrinth* of the inner ear. While this is occurring, fibers from the vestibulocochlear nerve (cranial nerve VIII), grow toward the otic vesicle and innervate it. The distal end of the first pharyngeal pouch forms the cavity of the middle ear. The proximal portion of the first pharyngeal pouch forms a tube called an *auditory (eustachian) tube,* which connects the middle-ear cavity with the pharynx (Figure 15.27d). The three ossicles form from condensations of mesenchymal cells in the middle-ear cavity. On the surface of the embryo, the branchial groove associated with the first pharyngeal pouch deepens and forms a canal called the *external auditory meatus.* The *tympanic membrane (eardrum)* develops from membrane that separates the first pharyngeal pouch from the floor of the branchial groove. The *external ear (auricle)* forms from the coalescence of a series of elevations that develop around the external auditory meatus.

F15.27d

mee-ay´-tus

Structure of the Ear

The structures of the various regions of the ear are uniquely suited to collect sound waves, to convert the vibrations of sound waves traveling through air into other vibrations—first by small bones in the middle ear and then by fluid in the inner ear—and finally to convert the fluid vibrations into nerve impulses.

External Ear

The **auricle,** or **pinna,** is the most prominent portion of the external ear (Figure 15.26). It consists of an irregularly shaped framework of elastic cartilage

F15.26

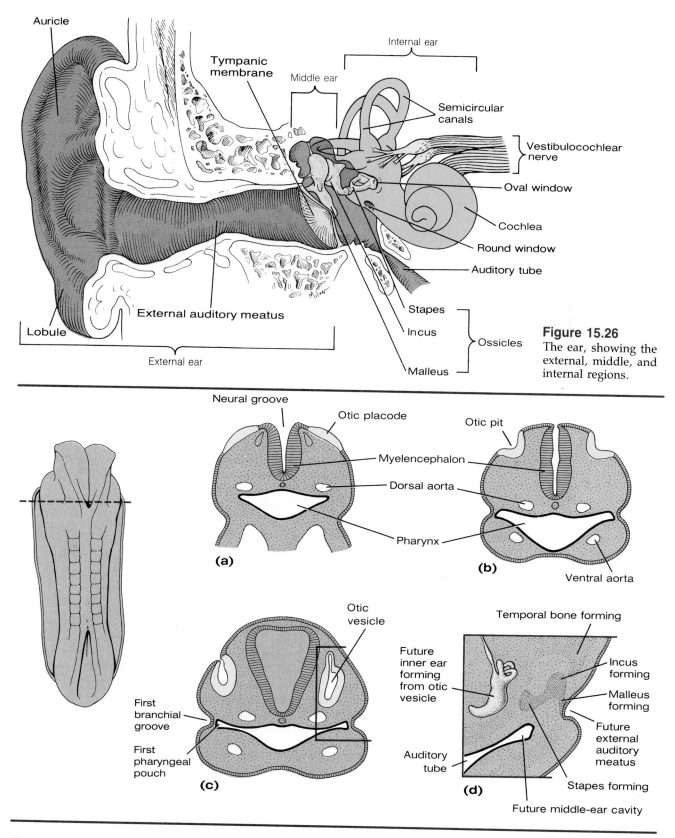

Figure 15.26
The ear, showing the external, middle, and internal regions.

Figure 15.27
Embryonic development of the ear. **(a)** Formation of the otic placodes and pharyngeal pouches. **(b)** Invagination of the otic placodes to form the otic pits, and beginning of the formation of the branchial grooves. **(c)** Otic vesicles form from the otic pits, and the branchial grooves deepen.

(d) Inner-ear structures form from the otic vesicles. Each branchial groove develops into an external auditory meatus, and each pharyngeal pouch becomes an auditory tube and a middle-ear cavity.

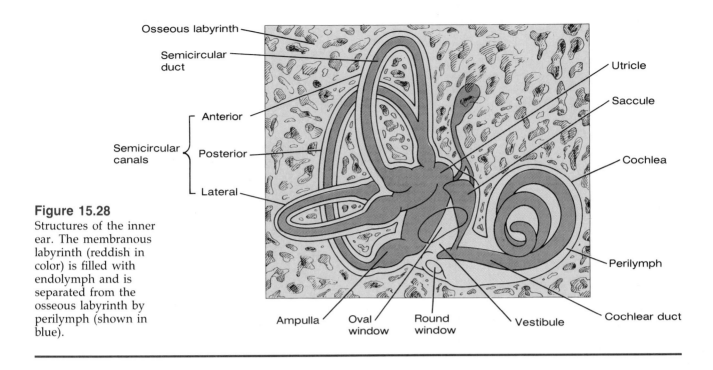

Osseous labyrinth
Semicircular duct
Anterior
Semicircular canals
Posterior
Lateral
Utricle
Saccule
Cochlea
Perilymph
Cochlear duct
Ampulla
Oval window
Round window
Vestibule

Figure 15.28
Structures of the inner ear. The membranous labyrinth (reddish in color) is filled with endolymph and is separated from the osseous labyrinth by perilymph (shown in blue).

covered with skin. The only part of the auricle that is not supported by carti-lage is the **lobule,** a flap of skin-covered connective tissue that extends from the lower margin of the auricle. The auricle directs sound waves into the external auditory meatus.

The **external auditory meatus** (canal) is a curved passageway approxi-mately 2.5 cm long that extends from the auricle to the eardrum. The meatus is lined with skin, and near its entrance are fine hairs and sebaceous glands. It also contains modified sweat glands called **ceruminous glands** that secrete *cerumen (earwax).* The hairs and the cerumen help to prevent small foreign objects from reaching the eardrum.

The external auditory meatus serves as a resonator for the range of sound waves typical of human speech (2500 to 5000 cycles per second). Because of its resonating properties, the meatus increases the sound pressure on the ear-drum for sound waves in this frequency range.

Middle Ear

The middle ear is a small air-filled chamber in the temporal bone. It is sepa-rated from the external auditory meatus by the **tympanic membrane (ear-drum)** and separated from the inner ear by a bony wall in which there are two small membrane-covered openings—the **oval window (fenestra vestibuli)** and the **round window (fenestra cochleae)** (Figure 15.26). An opening in the posterior wall of the middle ear leads to the mastoid sinuses in the mastoid portion of the temporal bone. Another opening connects the middle-ear chamber with the **auditory tube,** which leads to the nasopharynx.

F15.26

The auditory tube provides a means by which the air pressure in the middle-ear chamber remains equalized with atmospheric pressure. When atmospheric pressure is reduced, as at higher altitudes, the tympanic mem-brane would bulge outward if the pressure in the middle-ear chamber were not correspondingly reduced. Not only would this be painful, but it would also impair hearing by interfering with the vibrations of the tympanic mem-brane. The auditory tube, which is closed most of the time in adults, can be opened by swallowing or yawning. This action allows the air pressure in the middle-ear chamber to equalize with the atmospheric pressure.

The mucous membranes that line the middle-ear chamber, the mastoid sinuses, and the auditory tubes are continuous with the mucous membrane of the throat. For this reason, infections of the throat can readily spread to the middle-ear chamber—particularly in young children, whose auditory tubes are straighter than in the adult and tend to remain open most of the time. Because of the opening between the middle-ear cavity and the mastoid sinuses, infections of the middle-ear cavity can spread to the mucous membrane that lines the mastoid sinuses, producing a condition called *mastoiditis*. The mastoid sinuses are separated from the brain only by thin, bony partitions. Therefore, it is also possible for infection to spread from the mastoid sinuses to the meninges of the brain.

The three ear **ossicles,** or middle-ear bones, form a flexible bridge across the middle-ear chamber (Figure 15.26). The handle-shaped portion of the ossicle called the **malleus** (*hammer*) attaches to the inner surface of the tympanic membrane. The foot-plate portion of the ossicle called the **stapes** (*stirrup*) fits against the oval window on the medial wall of the middle-ear chamber. The third ossicle, the **incus** (*anvil*), lies between the malleus and the stapes and articulates with them. Thus, the three ossicles form a bridge between the tympanic membrane and the oval window. The articulations between the ear ossicles are freely movable synovial joints. The ossicles form a lever system that picks up vibrations of the tympanic membrane and transmits them to the oval window, which in turn leads into the inner ear. Two small muscles attach to the ear ossicles: the **stapedius muscle** attaches to the stapes, and the **tensor tympani muscle** attaches to the malleus. These muscles support the ossicles so that movements of the head will not result in substantial movements of the ossicles.

F15.26

Inner Ear

The inner ear is located medial to the middle ear in the petrous portion of the temporal bone (Figure 15.28). It consists of a series of canals called the **osseous labyrinth,** which are hollowed out of the bone. Within the osseous labyrinth, and following its course, is a **membranous labyrinth.** The membranous labyrinth is filled with a fluid called **endolymph,** which has a high concentration of potassium ions and a low concentration of sodium ions. The membranous labyrinth is suspended in a fluid called **perilymph,** which has a low concentration of potassium ions and a high concentration of sodium ions. Perilymph therefore separates the walls of the membranous labyrinth from the osseous labyrinth. The osseous labyrinth is divided into three areas: the *vestibule*, the *semicircular canals*, and the *cochlea*.

F15.28

VESTIBULE   The **vestibule** is a chamber just medial to the middle-ear chamber. Since the oval window forms a membranous partition between the middle-ear chamber and the vestibule of the inner ear, vibrations of the oval window induced by the stapes are transmitted to the perilymph of the vestibule. Within the vestibule are two enlargements of the membranous labyrinth: the **utricle** and the **saccule.** The utricle and the saccule contain receptor cells that detect the position and movement of the head, and information from these receptor cells contributes to the sense of balance. The utricle is connected to the portion of the membranous labyrinth that is located in the osseous semicircular canals; the saccule is connected with the portion of the membranous labyrinth located in the cochlea. The membranous labyrinth, therefore, is a continuous series of ducts that are filled with endolymph.

SEMICIRCULAR CANALS   Within the inner ear are three bony **semicircular canals,** which contain three membranous **semicircular ducts.** The semicircular canals are arranged at right angles to each other, forming anterior, lateral, and posterior canals. The anterior and posterior canals are vertical; the lateral canal is horizontal. Each membranous semicircular duct possesses an enlargement called an **ampulla,** which contains receptor cells that detect certain movements of the head and thereby provide information concerning equilibrium.

COCHLEA   The portion of the inner ear associated with hearing is the **cochlea.** It resembles a snail shell, spiraling $2\frac{1}{2}$ turns around a central bony core

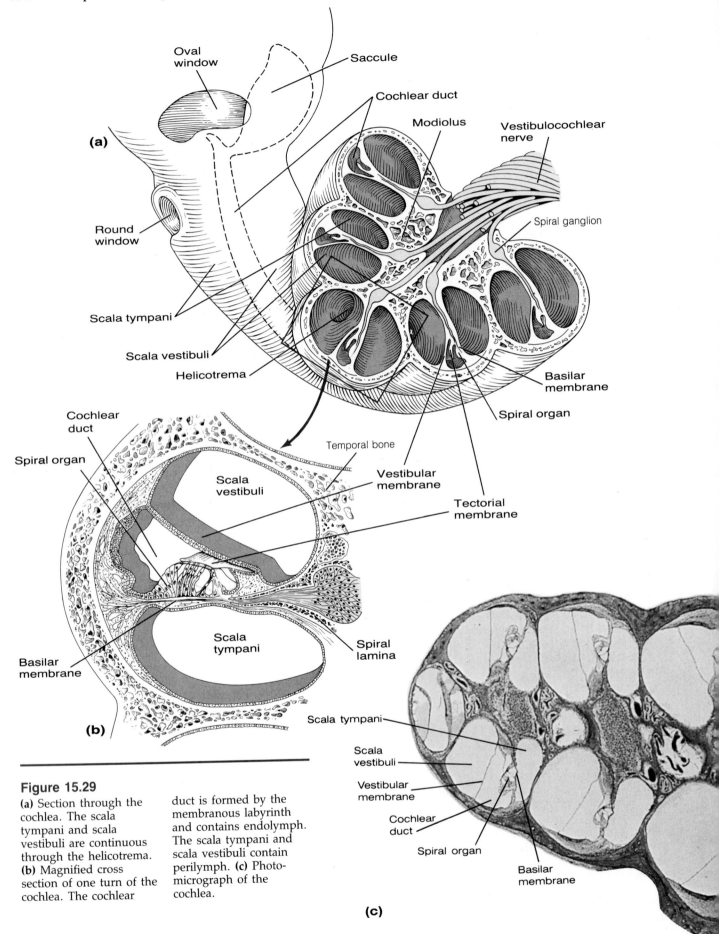

Figure 15.29

(a) Section through the cochlea. The scala tympani and scala vestibuli are continuous through the helicotrema. (b) Magnified cross section of one turn of the cochlea. The cochlear duct is formed by the membranous labyrinth and contains endolymph. The scala tympani and scala vestibuli contain perilymph. (c) Photomicrograph of the cochlea.

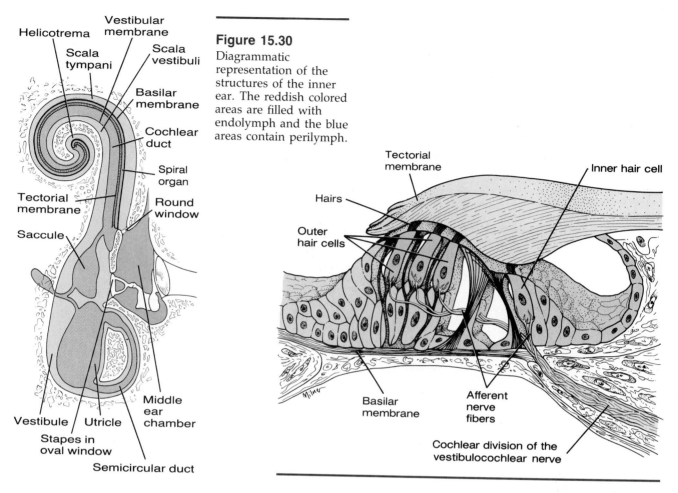

Figure 15.30
Diagrammatic representation of the structures of the inner ear. The reddish colored areas are filled with endolymph and the blue areas contain perilymph.

Figure 15.31
Section of the spiral organ.

called the **modiolus** (Figure 15.29a). A bony shelf called the **spiral lamina** extends into the cochlea from the modiolus (Figure 15.29b). Two membranes, the **vestibular membrane** and the **basilar membrane,** extend across the cochlea from the spiral lamina, dividing the cochlea into three longitudinal tunnels.

The central tunnel between the vestibular and basilar membranes is called the **cochlear duct,** or **scala media.** The cochlear duct is the membranous labyrinth of the cochlea and is filled with endolymph. One tunnel, the **scala vestibuli,** is separated from the cochlear duct by the vestibular membrane. The other tunnel, the **scala tympani,** is separated from the cochlear duct by the basilar membrane. The scala vestibuli and scala tympani both contain perilymph and are continuous with one another at the apex of the cochlea through an opening called the **helicotrema.** The scala tympani terminates at the round window, whereas the scala vestibuli ends at the oval window. The relationships of these structures are more easily understood if the cochlea is visualized as being uncoiled, as shown in Figure 15.30.

The **spiral organ,** which contains the receptors for hearing, is located on the basilar membrane. It consists of a series of sensory hair cells and supporting cells (Figure 15.31). The hair cells are arranged in a single row of about 3500 *inner hair cells* and three or four rows of about 20,000 *outer hair cells* (Figure 15.32). The hair cells are innervated by sensory fibers from the cochlear division of the vestibulocochlear nerve (cranial nerve VIII). These sensory fibers travel through the modiolus and spiral lamina to reach the basilar membrane. Overhanging the spiral organ is a flexible flap of fibrous and gelatinous tissue called the **tectorial membrane** that is anchored to the spiral lamina. Hairs of the sensory cells of the spiral organ are in contact with the tectorial membrane.

F15.30

F15.31

F15.32

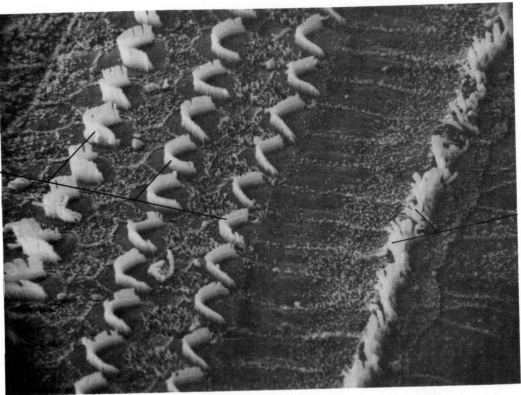

Outer hair cells

Inner hair cells

Figure 15.32

Scanning electron micrograph of the hair cells of the spiral organ (×2420). (From *Tissues and Organs: A Text-Atlas of Scanning Electron Microscopy* by Richard G. Kessel and Randy H. Kardon. W.H. Freeman and Company. Copyright © 1979.)

Mechanisms of Hearing

Hearing is the process of perceiving sound.

Sound Waves

Sound energy (a type of mechanical energy) travels through air in the form of pressure waves that are produced when air molecules are alternately compressed and rarefied (Figure 15.33). For example, when the arm of a vibrating tuning fork moves in one direction, air molecules ahead of the arm are pushed together, or compressed, and the pressure in the area increases. In turn, air molecules in the area of compression bump into other air molecules ahead of them, pushing these air molecules together and creating a new region of compression. Through many repetitions of this process, the initial compression of air molecules by the arm of the tuning fork can be transmitted a considerable distance, even though individual air molecules move only a short distance. Similarly, when the arm of the tuning fork moves back in the opposite direction, the compression of the air molecules and the pressure are decreased even below normal. These changes, too, are transmitted from air molecule to air molecule. Thus, a sound wave consists of regions of compression, in which the air molecules are close together and the pressure is relatively high, alternating with areas of rarefaction, where the molecules are farther apart and the pressure is lower.

The loudness of a sound is related to the amplitude of the sound wave—that is, to the pressure difference between a zone of compression and a zone of rarefaction. The pitch of a sound is related to the frequency of the sound wave (usually expressed in cycles per second, or Hertz, Hz). In general, the greater the amplitude of a particular sound wave, the louder the sound; and the higher the frequency of a sound wave, the higher the pitch of the sound.

The human ear is capable of detecting sound waves with frequencies between about 20 and 20,000 cycles per second. However, the ear is most sensitive to sound waves with frequencies between 1000 and 4000 cycles per second. Sound waves with frequencies outside this range must have greater amplitudes than sound waves with frequencies within the range if they are to be detected.

Transmission of Sound Waves to the Inner Ear

If sounds are to be perceived—that is, if a person is to hear a sound—the vibratory movements of sound waves traveling in air must be transferred to the fluids within the inner ear. However, fluids have much greater inertia than air, and greater pressures are required to cause movements of fluids than are required to cause movements of air. The pressures necessary to move the fluids of the inner ear result in part from the arrangement of the ossicles of the middle ear and in part from the difference in area between the tympanic membrane and the oval window.

A sound wave enters the external auditory meatus and strikes the tympanic membrane. The alternating higher and lower pressure regions of the sound wave cause the tympanic membrane to move forward and backward (vibrate) at the same frequency as the sound wave. The vibratory movements of the tympanic membrane are transmitted across the middle-ear cavity by the malleus, incus, and stapes to the oval window of the inner ear. Because of the arrangement of the lever system of the ossicles, the force exerted on the oval window by the stapes is about 1.3 times greater than the force exerted on the malleus at the tympanic membrane (although the distance of movement of the stapes at the oval window is only about 75% as great as the distance of movement of the malleus at the tympanic membrane). Moreover, the area of the tympanic membrane that the sound wave can strike (about 55 mm^2) is approximately 17 times greater than the area of the oval window (about 3.2 mm^2). Thus, the pressure applied to the perilymph within the cochlea by the movement of the stapes against the oval window is about 22 times (1.3 × 17) greater than the pressure that would be exerted if the sound wave were to strike the oval window directly. It is this higher pressure that acts on the perilymph to cause its movement.

A reflex called the *tympanic* or *sound attenuation reflex* occurs in response to loud sounds. After a latent period of about 40 to 60 milliseconds following the occurrence of a loud sound, there is a reflexive contraction of the stapedius muscle (but probably not the tensor tympani muscle). This contraction dampens the vibrations of the ossicles and decreases the transmission of sound waves (particularly sound waves with frequencies below 1000 cycles per second) through the middle ear and to the cochlea. This reflex provides some protection to the auditory receptors of the cochlea from damage by excessively loud sounds, and it may also mask low-frequency sounds in loud environments. Because of the latent period, however, the reflex provides little protection from sudden, explosive types of loud sounds.

Function of the Cochlea

If the forward movement of the stapes at the oval window is very slow, the pressure exerted on the perilymph pushes the perilymph along the scala vestibuli, through the helicotrema, and into the scala tympani, causing the round window to bulge outward into the middle ear (Figure 15.34a). When the stapes moves backward and exerts less pressure, the perilymph moves back from the scala tympani into the scala vestibuli. Thus, very-low-frequency pressure waves have little effect on the basilar membrane. Because of the inertia of the fluids of the inner ear, however, higher-frequency pressure waves of the sort associated with sound perception do not follow this pathway. Instead, these pressure waves are transmitted from the perilymph of the scala vestibuli near the oval window, through the flexible vestibular membrane, to the endolymph of the cochlear duct. From the endolymph, the waves are transmitted through the basilar membrane to the perilymph of the scala tympani, and finally to the round window. The transmission of the pressure waves through the basilar membrane at the base of the cochlea near the oval and round windows causes this area of the membrane to vibrate (move downward and upward). The vibration of the basilar membrane near

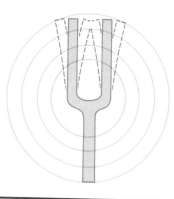

Figure 15.33
Production of a sound wave. As a sound source—here a tuning fork—vibrates, it creates alternating areas of compression (higher pressure), in which the air molecules are close together, and rarefaction (lower pressure), where the molecules are farther apart. This activity produces a pressure wave that radiates outward from the sound source. This pressure wave is called a sound wave.

F15.34a

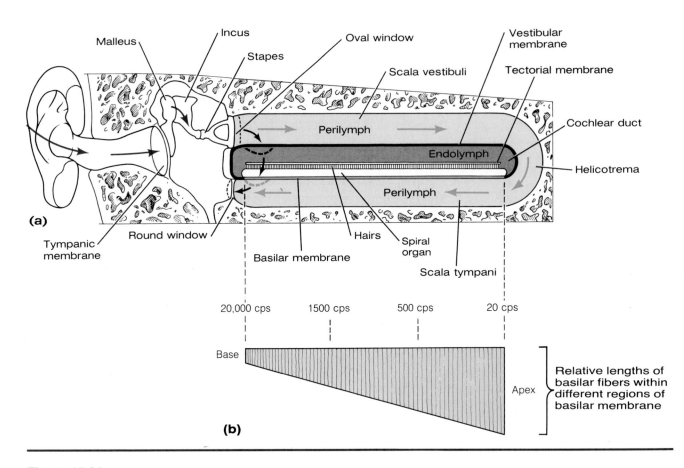

Figure 15.34

(a) Schematic representation of the transmission of pressure waves in the cochlea. Blue/green arrows indicate the transmission pathway when the movement of the stapes is slow. Black arrows indicate the transmission pathway of higher-frequency pressure waves of the sort associated with sound perception. (b) Graphic representation of the basilar membrane showing that the end of the membrane toward the middle ear contains shorter basilar fibers than the end toward the helicotrema. Maximum vibrations occur in the region of the basilar membrane that has the same resonant frequency as the sound wave that causes the pressure wave. Low-frequency sound waves cause maximum vibration toward the apex of the basilar membrane. High-frequency waves have their maximal effect toward the base of the basilar membrane.

F15.34b

the base of the cochlea initiates a wave that travels along the basilar membrane toward the helicotrema.

The basilar membrane contains about 25,000 basilar fibers that project from the spiral lamina toward the outer wall of the cochlea. Although the basilar membrane extends completely across the cochlea from the spiral lamina, its fibers do not. Rather, the distal ends of the fibers are embedded in the basilar membrane. The basilar fibers increase in length and decrease in thickness and rigidity from the base of the cochlea toward the helicotrema (Figure 15.34b). Because of this, the portion of the basilar membrane near the base of the cochlea tends to vibrate at high frequencies, and the portion near the helicotrema tends to vibrate at lower frequencies. Moreover, when a portion of the basilar membrane vibrates, the fluid between the vibrating portion of the membrane and the oval and round windows must also move. Since less fluid mass must move (that is, less inertia must be overcome) when a region of the membrane near the base of the cochlea vibrates than when a portion of the membrane near the helicotrema vibrates, the fluid loading of the basilar membrane also favors high-frequency vibration at the base of the cochlea near the oval and round windows and low-frequency vibration at the apex of the cochlea near the helicotrema. Thus, the sympathetic vibration, or resonance, of the basilar membrane in response to high-frequency sound waves is greatest near the base of the cochlea, and the sympathetic vibration, or resonance,

FRONTIERS IN HEALTH

The Bionic Ear

More than 2 million Americans suffer from profound deafness. Brought on by the destruction of hair cells as a result of loud noises or infections such as meningitis, or resulting from birth defects, profound deafness has been considered virtually untreatable. For the unfortunate victims, police sirens, automobile horns, and all other sounds go unperceived. Living such a life is much like watching television with the sound off, 24 hours a day for an entire lifetime.

Profound deafness can seriously affect the development of children born deaf or deafened before they begin to speak. Banished to a silent world, they often fail to mature emotionally. Learning to communicate is difficult, and reading comprehension is generally impaired. In fact, some never advance beyond the elementary reading level.

Hearing aids are usually of no value to those who are born deaf or to those who have completely lost their hearing from some other cause. For this large group of people, however, there is now some hope. It comes in the form of a cochlear implant—a device that simulates the function of the ear.

Pioneered by Dr. William House of the House Ear Institute in Los Angeles, the cochlear implant comprises several parts. Mounted at ear level is a small microphone, which picks up sound and transmits it to a stimulator unit worn on a belt or on the chest. The stimulator, which is about the size of a deck of cards, processes the signals it receives from the microphone and converts them into electric signals, which are sent to an induction coil attached to the outside of the body over the mastoid bone. Smaller than a quarter, the induction coil beams the signal through the skin to an implanted receiver. The receiver, in turn, picks up the faint signal and generates an electric current. This tiny current travels to the inner ear along an extremely fine implanted electrode, and from there to a ground electrode implanted a short distance away. The current produces an activity in nerve cells of the vestibulocochlear nerve, and this activity is interpreted as sound sensations. In some models the electrodes are imbedded in the vestibulocochlear nerve fibers themselves.

This complex method makes it possible for the deaf to hear, but the hearing is very different from normal hearing. Patients with cochlear implants can detect speech and environmental sounds and can distinguish some words from others, but the fine discriminations that allow people to communicate freely through complex language systems are impossible.

However, even this rudimentary form of hearing is tremendously exciting to people who have been closed off from the sounds of the world. For them, the cochlear implant provides valuable outside stimuli and a connection with the world. It also makes living a little safer. The sounds of horns and sirens, which may be annoying to most of us, serve an important purpose. The implant also helps adults monitor and regulate their own voices. Speech reading (reading lips and matching with sound)

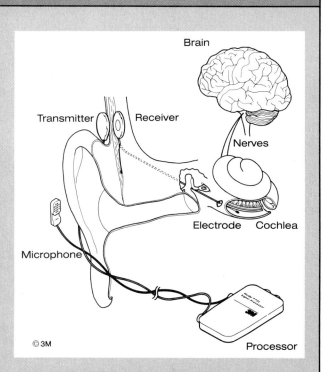

Diagram of the 3M cochlear implant.

becomes easier, for the cochlear implant allows the recipient to follow normal conversation at a comfortable level and to distinguish some words by their sounds. Thus, the minimal gains achieved by cochlear implants can significantly enhance the lives of profoundly deaf adults. For children, minimal hearing afforded by the cochlear implant can mean the difference between learning to speak and read or remaining mute and illiterate— the difference between communication and complete isolation.

The 3M Company of St. Paul, Minnesota manufactures a single-electrode cochlear implant that is being used by hundreds of adults and children. Some patients who were involved in the earliest experiments have been wearing their bionic ears for more than a decade with no significant problems. In 1984, the Food and Drug Administration approved the single-electrode implant for general use.

However, the single-electrode cochlear implant may soon become obsolete as medical researchers continue to search for ways to introduce more electrodes into the inner ear, thereby bringing a wider range of sounds to the organ of hearing. Dr. Robert L. White and his colleagues at Stanford University, among others, are working on multiple-channel cochlear implants. Recipients of these devices may be able to perceive many distinct words, not just the sound of a telephone ringing or a horn blowing. With this development, normal hearing may be a bit closer for those suffering from profound deafness.

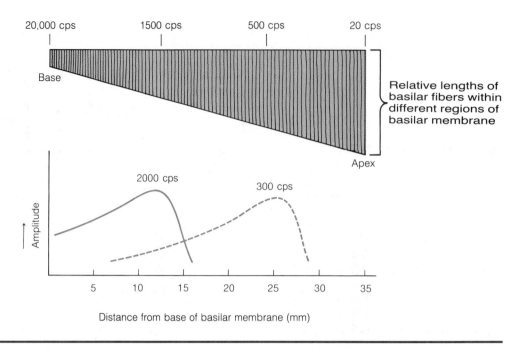

Figure 15.35
Graph indicating relative amplitudes of vibration of portions of the basilar membrane in response to sound waves of different frequencies.

of the basilar membrane in response to low-frequency sound waves is greatest near the apex of the cochlea.

As a particular-frequency wave travels along the basilar membrane from its point of initiation near the oval and round windows, the vibration of the basilar membrane increases in amplitude and is greatest at that portion of the membrane that has the same natural resonant frequency as the frequency of the wave (Figure 15.35). At this point the energy of the wave is dissipated, and beyond this region the wave quickly dies away. Thus, high-frequency sound waves cause maximal vibration of the basilar membrane near the base of the cochlea near the oval and round windows, and low-frequency sound waves cause maximal vibration of the basilar membrane near the apex of the cochlea near the helicotrema.

The spiral organ, which contains receptor cells called hair cells, is located on the basilar membrane (Figure 15.31). When the basilar membrane vibrates, the hairs of the hair cells are displaced, altering the polarity of the cells. When the basilar membrane moves upward toward the scala vestibuli, the hair cells of the affected region depolarize, and afferent nerve impulses are generated in neurons of the cochlear division of the vestibulocochlear nerve that are associated with the hair cells. When the basilar membrane moves downward, the affected hair cells hyperpolarize, and the generation of afferent impulses within the neurons of the cochlear division of the vestibulocochlear nerve decreases.

Auditory Pathways

The neural pathways between the spiral organ and the auditory portion of the cerebral cortex include synapses in the medulla, in the inferior colliculi of the midbrain, and in the medial geniculate bodies of the thalamus (Figure 15.36). Auditory signals from each ear are transmitted on both sides of the brain stem and cortex, and dysfunctions of the auditory pathways on one side do not greatly affect hearing in either ear.

Determination of Pitch

The ability to determine the pitch of most sounds is related to the fact that a sound wave of a particular frequency causes a specific region of the basilar membrane to vibrate more intensely than other regions, and the brain is able to detect which area of the basilar membrane is most stimulated. In fact,

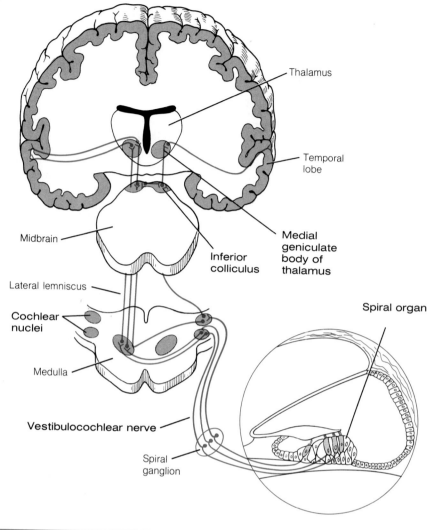

Figure 15.36
Auditory pathways from the spiral organ through the central nervous system to the cortex of the temporal lobe of the brain.

certain neurons within the auditory cortex respond only to a narrow range of sound frequencies. When the maximal vibration of the basilar membrane is near the base of the cochlea, the sounds perceived are interpreted as high-pitched; when the maximal vibration is in the middle of the cochlea, the sounds are interpreted as being of intermediate pitch; and when the maximal vibration is near the apex of the cochlea, the sounds are interpreted as low-pitched.

Determination of Loudness

High-intensity (high-amplitude) sound waves cause a greater amplitude of vibration of the basilar membrane than low-intensity (low-amplitude) waves. As the amplitude of vibration of the basilar membrane increases, hair cells of the spiral organ are more strongly stimulated. In fact, certain hair cells apparently do not become strongly stimulated until the amplitude of vibration of the basilar membrane becomes quite high. The stronger stimulation of the hair cells activates more neurons and increases the frequency of transmission of afferent nerve impulses to the brain. The brain interprets this increased neural activity as an increase in the loudness of the sound.

Sound Localization

Several mechanisms are involved in localizing the source of a sound. The auditory system is able to detect differences in the time of arrival of a sound

---

CLINICAL CORRELATION
Hearing Loss Due to Aging

CASE REPORT . . . .

THE PATIENT   A 76-year-old male

PRINCIPAL COMPLAINT   Progressive hearing loss and difficulty understanding speech

HISTORY   The patient had noticed gradually increasing difficulty with his hearing and a great change in his ability to understand conversations in a background of noise. He had no history of dizziness, middle-ear infections, head trauma, hypertension, or diabetes, and he was taking no medicines routinely.

CLINICAL EXAMINATION   Examination of the head and neck revealed no abnormality. No bruits (sounds caused by turbulent blood flow, due to vascular constriction or obstruction) were heard in the neck. Hearing tests showed a mild-to-moderate bilateral loss of the ability to hear high-frequency sounds. His understanding of monosyllabic, phonetically balanced speech was consistent with the degree of hearing loss. His understanding of nonsense sentences embedded in a background story was poorer than his understanding of single words.

COMMENT   The loss of hearing with aging, which is called *presbycusis*, is characterized by loss in the high-frequency range. The ability to understand simple spoken material is comparable to that of normal young people. When the pattern of speech is interrupted, the rate of speaking is changed, or background noise or unrelated speech interferes, older people do not comprehend as well as young people do. This difference of function may appear in persons as young as 40 to 50 years of age.

The pathophysiology of presbycusis is fourfold: peripheral, central, mechanical, and metabolic. Within the cochlea, hair cells, supporting cells, and the basilar membrane degenerate. Nerve fibers are lost, initially in the basal cochlea and later throughout the cochlea. Centrally, fewer neurons are found at each successive level within the auditory pathways. The elasticity of the basilar and tectorial membranes changes, which causes mechanical differences between a young system and an old one. Metabolically, the walls of certain capillaries in the cochlea thicken, and the nutrient supply to the cochlea is decreased. Moreover, the electrolyte balance of the cochlear fluids may also be altered by increasing age.

OUTCOME   A moderate-amplification, over-the-ear-style hearing aid was fitted to the patient, and his ability to hear was restored to normal.

---

wave at each ear, and this ability provides one sound-localization mechanism. For example, a sound wave coming from a point directly in front of a person reaches both ears at the same time, but a sound wave coming from a point to the person's right reaches the right ear slightly before it reaches the left ear. This aspect of sound localization is most effective at low frequencies (that is, frequencies less than 3000 cycles per second).

The auditory system is also able to detect differences in the intensity of a sound wave at one ear compared to the other, and this ability provides a second sound-localization mechanism. The head acts as a sound-wave barrier, particularly for higher-frequency sound waves. Thus, if a sound wave comes from a point directly in front of a person, both ears will receive the same intensity of stimulation. However, if the sound wave comes from a point to the person's right, the right ear will be more intensely stimulated than the left ear.

Head Position and Movement

In addition to its role in hearing, the inner ear provides information about the position and movement of the head. This information is utilized in the coordination of movements that maintain body equilibrium and balance. Moreover, signals from the inner ear are involved in controlling the extrinsic eye muscles so that the eyes can remain fixed on the same point despite changes in the position of the head. The portion of the inner ear involved in these activities is called the **vestibular apparatus.** Receptor cells of the vestibular apparatus are located within the utricle, the saccule, and the ampulae of the semicircular ducts.

Function of the Utricle

Receptor cells of the utricle provide information about the position of the head with respect to gravity and about the linear acceleration or deceleration of the head (such as would occur when a person starts to run forward in a straight line).

The receptor structure within the utricle is called a **macula.** The macula is composed of groups of hair cells whose protruding hairs are embedded in a gelatinous substance that contains tiny particles of calcium carbonate called **otoliths.** The otoliths make the gelatinous substance heavier than the surrounding endolymph, and under the influence of gravity, the gelatinous substance exerts force on the hair cells. Even when the head is upright, neurons associated with the hair cells transmit a continuous series of impulses to the brain by way of the vestibular division of the vestibulocochlear nerve (Figure 15.37). When the position of the head changes, the direction of the force of the gelatinous substance on the hair cells also changes, causing the hairs to be displaced from their normal position. The displacement of the hairs alters the stimulation of the neurons associated with the hair cells, and a different pattern of nerve impulses is transmitted to the brain. For example, when the head is tilted to the left, the transmission of nerve impulses from the left utricle increases and the transmission of impulses from the right utricle decreases. When the head is upright, it is possible to detect a change in head position of as little as one-half degree.

The hair cells of the macula also detect linear acceleration or deceleration. When the head undergoes linear acceleration, the inertia of the gelatinous substance causes it to lag behind the movement of the head. The lag of the gelatinous substance displaces the hairs of the hair cells, and an altered pattern of nerve impulses is transmitted to the brain. For example, the lag of the gelatinous substance that occurs when a person starts to run forward displaces the hairs of the hair cells to the rear. This is similar to the displacement that would occur if the head was tilted backward so the face was oriented upward. If the linear movement of the head continues at a constant velocity the inertia of the gelatinous substance is overcome, and it eventually moves at the same rate as the head is moving. During linear deceleration, the momentum of the gelatinous substance causes it to continue to move as the head slows and stops. The continued movement of the gelatinous substance also displaces the hairs of the hair cells and alters the pattern of nerve impulses transmitted to the brain. Thus, the hair cells of the macula of the utricle detect linear acceleration or deceleration, but they provide little information during linear motion at a constant velocity.

Function of the Saccule

The receptor cells of the saccule are organized into a macula like that of the utricle, and the saccule is generally considered to function in a similar manner as a component of the vestibular apparatus.

Function of the Semicircular Ducts

The semicircular ducts detect rotational (angular) acceleration or deceleration of the head (such as would occur when the head is suddenly turned to one side). The semicircular ducts are filled with endolymph, and the ampulla of each duct contains a group of hair cells called a **crista** (Figure 15.38). The hairs are embedded in a gelatinous mass called the **cupula.** In general, the hairs of the hair cells within at least one semicircular duct are displaced by rotational acceleration or deceleration in any given plane.

When the head undergoes rotational acceleration, the inertia of the endolymph causes it to lag behind the movement of the semicircular duct. In effect, this response is equivalent to a flow of endolymph in a direction opposite to the movement of the duct—that is, opposite to the direction of the head's rotation. The lag of the endolymph alters the force exerted on the cupula within the ampulla of the duct and displaces the cupula and the hairs of the hair cells from their normal position. The displacement of the hairs

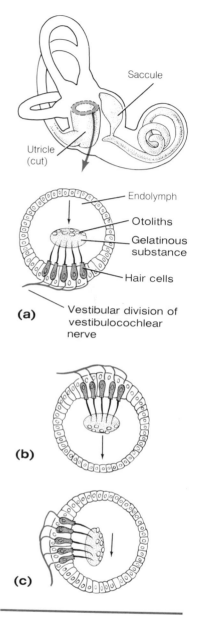

Figure 15.37
Stimulation of the macula within the utricle. The arrows indicate the direction of the force of gravity. **(a)** Shows a macula when the head is in an upright position. **(b)** Illustrates a macula when the head is inverted. **(c)** Shows a macula when the head is in a horizontal position.

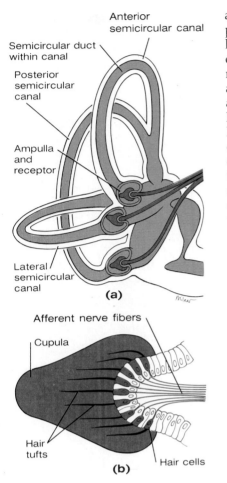

Anterior
semicircular canal

Semicircular duct
within canal

Posterior
semicircular
canal

Ampulla
and
receptor

Lateral
semicircular
canal

(a)

Afferent nerve fibers

Cupula

Hair
tufts

(b)   Hair cells

Figure 15.38
(a) The semicircular canals and ducts. The ampullae of the ducts have been cut open to show the locations of the receptors (cristae) within them. **(b)** Enlargement of a crista.

alters the stimulation of neurons associated with the hair cells, and a different pattern of nerve impulses is transmitted to the brain. If the rotation of the head continues at a constant velocity, the inertia of the endolymph is overcome, and it eventually moves at the same rate as the semicircular duct is moving. Under these conditions, the cupula returns to its normal position and afferent-nerve-impulse transmission returns to the resting pattern within about 20 seconds. During rotational deceleration, the momentum of the endolymph causes it to continue to move as the semicircular duct slows and stops. In effect, this response is equivalent to a flow of endolymph within the semicircular duct in the direction of rotation. The continued movement of the endolymph also alters the force exerted on the cupula within the ampulla of the duct, and displaces the cupula and the hairs of the hair cells so that a different pattern of nerve impulses is transmitted to the brain. Thus, the hair cells of the cristae detect rotational acceleration or deceleration, but they provide little information during periods of rotation at a constant velocity.

Nystagmus

Nystagmus is a characteristic movement of the eyes that is associated with their ability to remain fixed on the same point despite changes in the position of the head. A type of nystagmus called *vestibular nystagmus* occurs even when the eyes are closed. Vestibular nystagmus is produced when the cupulae and the hairs of the hair cells within the ampullae of the semicircular ducts are displaced. For example, if the head undergoes horizontal rotational acceleration (for example, the person suddenly spins around and around), the eyes move relatively slowly in a direction opposite to the direction in which the head is turning (as would occur if the eyes were fixed on some object in the environment). After the eyes have turned far to the side, they jump rapidly back to their forward position, and the movements are repeated. When the head undergoes horizontal deceleration (for example, the person suddenly stops spinning), nystagmus also occurs, but the directions of the eye movements are opposite to those that took place during acceleration (that is, the slow drift of the eyes occurs in the direction in which the rotation took place). These opposite directions of eye movements occur because, during deceleration, the cupulae and the hairs of the hair cells of the cristae are displaced in a direction opposite to that which took place during acceleration; therefore, a different pattern of nerve impulses is transmitted to the brain.

Another type of nystagmus, called *optokinetic nystagmus*, is produced when visual images move across the retinas of the eyes. For example, when the head moves horizontally in relation to the visual scene (such as when a person looks out of a window of a moving train), the eyes fix on an object and follow it, moving slowly until they have turned far to the side. The eyes then jump rapidly back to their forward position and fix upon a new object, and the movements are repeated. Optokinetic nystagmus does not occur when the eyes are closed.

Maintenance of Equilibrium

Nerve impulses transmitted to the brain from receptors that provide information about the position and movement of the body or its parts strongly influence the activity of muscles involved in maintaining equilibrium, or balance. Among these receptors are those of the eyes and the vestibular apparatus of the ears, as well as receptors in the skin, muscles, and joints.

It is difficult to assign varying degrees of importance to the different types of information employed in maintaining equilibrium. Visual information is important, but so are other types of information. For example, receptors in the neck provide information about the position of the head relative to the body. This information is important because the vestibular apparatus of the ears provides information only about the position or movement of the head itself. Thus information from the neck receptors helps the system determine if signals from the vestibular apparatus are due to the movement of the head

CONDITIONS OF CLINICAL SIGNIFICANCE

The Ear

Middle-Ear Infections

Infections of the mucous membrane of the throat can travel through the auditory tube and cause an inflammation of the mucous membrane that lines the middle-ear cavity, including the inner surface of the tympanic membrane. The inflammation may cause fluid to collect within the middle ear, temporarily interfering with the ability to hear.

Deafness

Diseases of the ear or injuries to the ear can cause partial or total deafness. Such hearing losses can be classified as conductive deafness or as nerve deafness.

In *conductive deafness*, there is interference with the transmission of sound waves through the external or middle ear. The interference can be caused by a physical blockage of the external auditory meatus by a foreign object or by cerumen (earwax), inflammation of the eardrum, adhesions between the ossicles, or thickening of the oval window. There is no damage to the receptor cells of the spiral organ or to the nerve pathways in conductive deafness. Therefore, hearing aids that transmit sound waves to the inner ear through the bone of the skull rather than through the middle ear can aid a person suffering from this condition.

In *nerve deafness*, the loss of hearing results from disorders that affect the sound receptors in the inner ear, the vestibulocochlear nerve, or nerve pathways or centers within the central nervous system. Such disorders can be caused by a number of factors, including infections, tumors, and trauma. Hearing aids may not be helpful in cases of nerve deafness in which the hearing loss is due to destroyed receptor cells or nerve pathways.

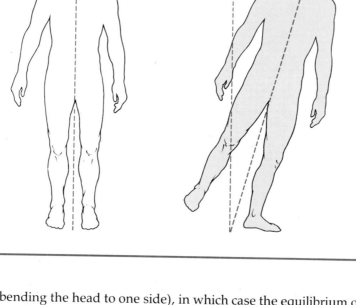

Figure 15.39
Proprioceptive joint receptors in the neck (as well as receptors in other body areas) play an important role in maintaining equilibrium. They provide information to determine whether signals from head-region receptors such as the vestibular apparatus are due to the movement of the head alone or to the movement of the body as a whole.

alone (such as bending the head to one side), in which case the equilibrium of the body may not be endangered, or to the movement of the whole body (such as tilting the entire body to one side), in which case the equilibrium of the body may be endangered and adjustments may be necessary to prevent a fall (Figure 15.39). Signals from the various receptors are integrated within the central nervous system to provide information about the position of the body and its parts in space, and this information helps determine the degree of activity in individual muscles that is required to maintain balance.

F15.39

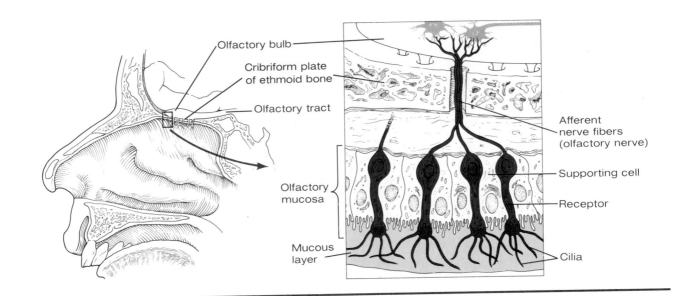

Figure 15.40
Location and structure of
the olfactory receptors.

F15.40

SMELL (OLFACTION)

The receptors for smell (olfaction) are specialized neurons located in the epithelium of the nasal mucosa in the upper portion of the nasal cavity on either side of the nasal septum (Figure 15.40). These neurons have two processes. One of the processes, together with similar processes from other olfactory receptor cells, travels to the *olfactory bulbs* in the brain as a component of the olfactory nerves. Within the olfactory bulbs, the processes of the receptor cells synapse with neurons that leave the bulbs as the *olfactory tracts*. From the olfactory tracts, nerve fibers travel to areas that include the amygdaloid nuclei and the primary olfactory cortex on the medial sides of the temporal lobes.

The second process of an olfactory receptor cell bears many fine projections called olfactory hairs or cilia that extend into the nasal cavity and are embedded in the mucous lining of the cavity.

For an odorous substance to be detected, it must reach the olfactory receptors. Normally, air moving through the nose does not pass through the upper portion of the nasal cavity where the olfactory receptors are located, so odorous substances must diffuse to the region. This process is aided by sniffing. Once in the region of the receptors, an odorous substance must dissolve in the mucous layer that covers the receptors and interact in some way with the receptors. This interaction is believed to depolarize the receptor, resulting in a generator potential that can initiate nerve impulses that are conveyed back to the brain, where they are interpreted as a particular smell.

Tens of thousands of different odors can be discriminated, but the basis for this discrimination is largely unknown. One theory is that a receptor cell membrane possesses 20 to 30 different types of receptor sites where the molecules of various odorous substances can interact and stimulate the depolarization of the cell. Although any one site is thought to be able to interact with a wide variety of odorous molecules, it responds best to molecules of particular sizes, polarities, and shapes. Thus, a particular molecule may interact well at some receptor sites and not as well at others. The better the interaction of the odorous molecule with a specific receptor site, the greater will be the depolarization resulting from the interaction. The depolarizations that occur at the different receptor sites of a single receptor cell are believed to be additive, resulting in a generator potential that determines the firing rate of the receptor cell. Different receptor cells are believed to have different combinations and concentrations of specific types of receptor sites. Hence, a given odorous substance may cause a simultaneous but differential stimulation of different receptor cells. The system is thus provided with the ability to discriminate many different odors.

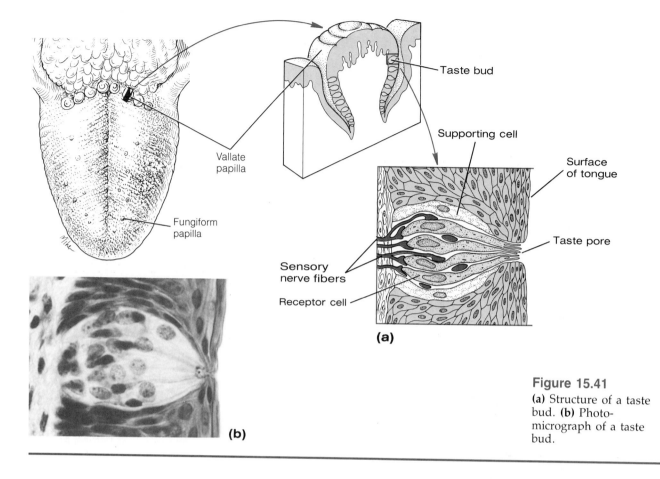

Vallate papilla

Fungiform papilla

Taste bud

Supporting cell

Surface of tongue

Taste pore

Sensory nerve fibers

Receptor cell

(a)

(b)

Figure 15.41
(a) Structure of a taste bud. **(b)** Photomicrograph of a taste bud.

TASTE (GUSTATION)

The receptors for taste (gustation) are called **taste buds.** Most taste buds are located on the surface and the papillae of the tongue, but some are found on the roof of the mouth and in the pharynx and larynx. Two types of cells have been identified within a taste bud. These cells have been called *receptor cells* and *supporting cells*. Although these names imply a specific functional role for each cell type, it has not been possible to confirm functional differences. In fact, there is evidence that the two cell types are actually different stages in the development of a single cell type. The cells of a taste bud are replaced frequently, and the life span of a single taste cell is only about one week. Microvilli from the upper portions of the receptor cells extend into a small opening, called a *taste pore*, at the surface of the taste bud (Figure 15.41). This position allows them to be bathed by the fluids of the mouth.

Sensory nerve fibers contact the receptor cells of the taste buds. A single nerve fiber may contact several different receptor cells, and a single receptor cell may have several different nerve fibers contacting it. These fibers run to the brain stem in the facial nerve (from the anterior two-thirds of the tongue), the glossopharyngeal nerve (from the posterior third of the tongue), and the vagus nerve (a few fibers from the pharyngeal region). The axons in these cranial nerves synapse in the brain stem with interneurons that cross to the opposite side and travel through the medial lemnisci to the thalamus. In the thalamus the interneurons synapse with neurons that are conveyed to the tongue area of the somatosensory cortex in the postcentral gyrus of the parietal lobe.

If a substance is to evoke a taste sensation, it must dissolve in the fluids that bathe the tongue and interact with the receptor cells of the taste buds. Four primary taste sensations have traditionally been identified—sweet, salty, bitter, and sour, with each taste being detected best in specific regions

F15.41

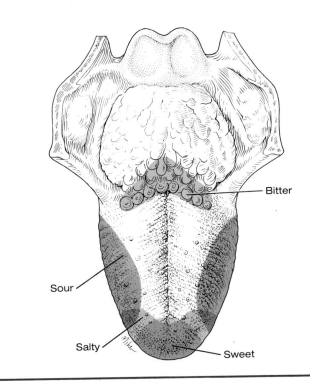

Figure 15.42
Areas of the tongue that
are most sensitive to
particular taste
sensations.

F15.42   of the tongue (Figure 15.42). It appears, however, that there is no corresponding specificity of taste receptor cell types. A given taste receptor cell can interact with and respond to a variety of different substances that belong to more than one of the specific taste categories.

Taste receptor cells have been postulated to possess several different types of receptor sites that can form loose combinations with different types of molecules. The formation of these combinations depolarizes the receptor cell, leading to the generation of nerve impulses in the sensory nerve fibers that contact the receptor cell. Thus, the sensation of a specific taste is a complex phenomenon that involves the relative activities and firing patterns of a number of different sensory neurons from a variety of different taste receptor cells.

Much of what is commonly considered to be "taste" actually involves the stimulation of olfactory receptors. This fact is particularly evident when a person has a cold that produces congestion of the nasal mucosa that blocks olfactory stimulation. Under such conditions, the person often reports an inability to "taste" food, even though his or her actual taste system continues to function normally.

STUDY OUTLINE

THE EYE—VISION pp. 429–450

EMBRYONIC DEVELOPMENT OF THE EYE
1. Outgrowths from lateral diencephalon form optic vesicles; each invaginates, forming optic cup, which becomes retina.
2. Optic stalk becomes incorporated into optic nerve.
3. Ectoderm over optic cup thickens into lens placode; placodes develop into lens vesicles that later separate from ectoderm, eventually forming lens of adult eye.
4. Cornea forms from cells superficial to lens vesicles.

5. Loose mesenchymal cells form fibrous tunic and vascular tunic of eye.

STRUCTURE OF THE EYE
FIBROUS TUNIC Outermost layer.
1. *Sclera* Posterior; white; dense connective tissue; protects eye and maintains its shape.
2. *Cornea* Anterior; clear; allows for light passage.

VASCULAR TUNIC Beneath fibrous tunic.
1. *Choroid* Posterior; dark; contains many blood vessels.
2. *Ciliary Body* Around edge of cornea; contains smooth ciliary muscles; ciliary processes produce aqueous humor.

3. *Iris* Anterior; thin muscular diaphragm; pigmentation responsible for eye color.
4. *Pupil* Opening in center of iris through which light passes.

RETINA Innermost layer of eye (internal tunic) consists of outer pigmented layer and inner nervous tissue layers.
1. *Pigmented Layer* Composed of epithelial cells in contact with choroid.
2. *Nervous Tissue Layers* Contain photoreceptors—rods and cones.
3. *Optic Nerve* Area of exit from eye is blind spot (optic disc).
4. *Macula* Located at posterior pole of eye; at its center is fovea centralis with high concentration of cone photoreceptors.

LENS Aids in focusing light onto retina; relatively elastic, biconvex structure.

CAVITIES AND HUMORS Lens separates eye into two cavities: anterior cavity (contains aqueous humor) and posterior cavity (contains vitreous humor).

ACCESSORY STRUCTURES OF THE EYE

EYELIDS Anterior, skin-covered structures; provide protection and light screening; controlled by skeletal muscles (orbicularis oculi and levator palpebrae superioris).

CONJUNCTIVA Mucous membrane lining that covers inside of eyelids and anterior surface of eye.

LACRIMAL APPARATUS Includes lacrimal gland and associated ducts; gland produces tears.

EXTRINSIC EYE MUSCLES Lateral, medial, superior, and inferior rectus muscles; inferior and superior oblique muscles. Rapid acting; among the most precisely controlled skeletal muscles.

LIGHT Portion of electromagnetic spectrum to which eye receptors are sensitive.

OPTICS Light rays are bent (refracted) when striking a biconvex lens; light rays converge at focal point behind lens.

FOCUSING OF IMAGES ON THE RETINA

EMMETROPIA Lens under tension appears relatively flat; eye focuses parallel light rays from distant points (farther than 20 feet) sharply onto retina.

ACCOMMODATION Focusing adjustments for viewing objects closer than 20 feet; ciliary muscles contract and permit greater lens curvature; pupil constriction eliminates divergent rays; convergence of eyes.

NEAR AND FAR POINTS OF VISION Near point of vision varies with age—close to the eye when young, recedes with age; far point of vision: any object beyond 20 feet for normal eye can be focused onto retina without accommodation.

CONTROL OF EYE MOVEMENTS Movements of eyes are synchronized so both eyes move in a coordinated manner when viewing an object.

VOLUNTARY FIXATION MOVEMENTS Controlled by small area of premotor regions of frontal lobes of cerebral cortex; move eyes voluntarily to find an object and fix on it.

INVOLUNTARY FIXATION MOVEMENTS Controlled by occipital region of cortex; keep image of object focused on foveal portions of retinas.

BINOCULAR VISION AND DEPTH PERCEPTION Retinal disparity enhances person's ability to judge relative distances when objects are nearby.

DIPLOPIA (DOUBLE VISION) May occur when light rays from an environmental object do not fall on corresponding points on retinas.

STRABISMUS Improper coordination of eye movements. Eyes may alternate in fixing on environmental object, or one eye may be used all the time and other eye may become functionally blind.

PHOTORECEPTORS OF THE RETINA Rods and cones contain photopigments. A photopigment consists of a protein called an opsin to which a chromophore molecule called retinal is attached. Opsins differ from pigment to pigment. Light alters chemical configuration of retinal, and retinal breaks away from the opsin, leading to changes in polarity of receptor cells containing photopigment.

RODS Responses indicate degrees of brightness, but rod system is characterized by relative lack of color discrimination; contain rhodopsin; numerous in peripheral retina.

CONES Primarily responsible for color vision; three different cone types: red, green, blue; cones are concentrated in center of retina and fovea.

NEURAL ELEMENTS OF THE RETINA
1. Rods and cones synapse with bipolar cells, which synapse with ganglion cells.
2. Horizontal cells and amacrine cells are involved in lateral interactions that occur within retina.
3. Neural pathways within retina are partially responsible for differences in rod vision and cone vision.

VISUAL PATHWAYS Ganglion cell axons from medial half of each retina cross over at optic chiasma; optic tracts to thalamus; optic radiations to visual cortex.

PROCESSING OF VISUAL SIGNALS Begins in retina and continues throughout visual pathway. Components of pathway transmit coded message about certain aspects of image on retina.

LIGHT AND DARK ADAPTATION Due to several mechanisms.

PHOTOPIGMENT CONCENTRATION Photopigments are broken down in light, regenerated in darkness.

PUPILLARY LIGHT REFLEX Pupils constrict in bright light and dilate in dim light.

NEURAL ADAPTATION When person moves from dimly lit area to brightly lit area, signals transmitted by bipolar cells, horizontal cells, amacrine cells, and ganglion cells are initially very intense. However, within a fraction of a second, the intensity of signals declines severalfold.

VISUAL ACUITY Ability of eye to distinguish detail—greatest in fovea; light rays stimulate different receptor units, with an unstimulated or weakly stimulated unit between them.

CONDITIONS OF CLINICAL SIGNIFICANCE: THE EYE pp. 448–449

MYOPIA Nearsightedness.
1. Light rays from distant objects focused in front of retina.
2. Near point of vision closer than normal, far point of vision closer than 20 feet.
3. Corrected by concave lenses.

HYPERMETROPIA Farsightedness.
1. No far point of vision; near point farther away than normal.
2. Refraction increased with convex lenses.

ASTIGMATISM Unequal curvatures of portions of refraction system of eye. Light rays entering eye in one plane focus at a point different from that of light rays entering eye in another plane.

CATARACT Cloudiness or opaqueness of lens of eye or portion thereof.

GLAUCOMA Abnormal elevation of intraocular pressure.

COLOR BLINDNESS Functional deficiency or absence of certain cones.

EFFECTS OF AGING
1. Lens becomes yellow due to exposure to ultraviolet rays.
2. Potential for development of cataracts.
3. Pupil is less able to dilate, resulting in reduction of light reaching photoreceptors.
4. Poor drainage of aqueous humor increases likelihood of glaucoma.
5. Debris from rods and cones may cause loss of visual acuity.
6. Lens loses elasticity, leading to presbyopia.

THE EAR—HEARING AND HEAD POSITION AND MOVEMENT pp. 450–465

EMBRYONIC DEVELOPMENT OF THE EAR
1. Plate of thickened ectoderm called otic placode invaginates, forming otic pit; otic pit separates from surface ectoderm to become otic vesicle; eventually forms membranous labyrinth of inner ear.
2. Part of first pharyngeal pouch forms eustachian tube.
3. Mesenchymal cells in middle-ear cavity form three ossicles.
4. Branchial groove external to first pharyngeal pouch forms external auditory meatus.
5. Tympanic membrane develops from membrane separating first pharyngeal pouch and branchial groove.

STRUCTURE OF THE EAR

EXTERNAL EAR
1. *Auricle (Pinna)* Directs sound waves.
2. *External Auditory Meatus (Canal)* Contains hairs and cerumen; serves as resonator.

MIDDLE EAR Air chamber in temporal bone; extends from tympanic membrane to oval window.
1. *Auditory Tube* Regulates pressure.
2. *Ossicles (Malleus, Incus, Stapes)* Transmit and amplify vibrations of tympanic membrane to oval window.

INNER EAR Series of fluid-filled (perilymph) canals called osseous labyrinth that contain membranous labyrinth, which is filled with endolymph.
1. *Vestibule* Saccule and utricle; associated with sense of balance.
2. *Semicircular Canals* Equilibrium.
3. *Cochlea* Portion of inner ear for hearing; spiral organ.

MECHANISMS OF HEARING

SOUND WAVES Sound energy travels through air in the form of pressure waves, consisting of regions of compression (air molecules are close together and pressure is relatively high) alternating with areas of rarefaction (molecules are farther apart and pressure is lower). The ear can detect sound waves with frequencies between about 20 and 20,000 cycles per second but is most sensitive to frequencies between 1000 and 4000 cycles per second.

TRANSMISSION OF SOUND WAVES TO THE INNER EAR Sound wave enters external auditory meatus and causes tympanic membrane to vibrate. Vibrations of tympanic membrane are transmitted across middle-ear cavity by malleus, incus, and stapes to oval window of inner ear.

FUNCTION OF THE COCHLEA Pressure waves generated by movement of stapes against oval window cause basilar membrane to vibrate near base of cochlea, initiating a wave that travels along basilar membrane toward helicotrema. Vibration of basilar membrane alters polarity of hair cells of spiral organ (located on membrane), leading to generation of nerve impulses that are transmitted to the brain and interpreted as sounds. High-frequency sound waves cause maximal vibration of basilar membrane near base of cochlea; low-frequency sound waves cause maximal vibration near apex of cochlea.

AUDITORY PATHWAYS Auditory signals from each ear are transmitted on both sides of brain stem and cortex, and dysfunctions of auditory pathways on one side do not greatly affect hearing in either ear.

DETERMINATION OF PITCH Sound wave of particular frequency causes particular region of basilar membrane to vibrate maximally. The brain can detect which area of basilar membrane is most stimulated and interpret this as particular pitch.

DETERMINATION OF LOUDNESS High-intensity sound waves cause greater vibration of basilar membrane than do low-intensity sound waves; the brain detects this difference and interprets greater vibration as louder sound.

SOUND LOCALIZATION
1. Auditory system is able to detect differences in sound-wave arrival time at each ear.
2. Auditory system is able to detect differences in sound-wave intensity at each ear.

HEAD POSITION AND MOVEMENT Vestibular apparatus of inner ear provides information about head position and movement. Information is utilized in coordination of movements that maintain body equilibrium and balance, and in controlling extrinsic eye muscles so eyes can remain fixed on the same point despite change in head position.

FUNCTION OF THE UTRICLE Provides information about position of head with respect to gravity and about linear acceleration or deceleration of head.

Macula Receptor structure composed of hair cells, with hairs embedded in gelatinous substance containing calcium carbonate. Hairs are displaced and stimulation of hair cells is altered by changing position of head or by linear acceleration of deceleration.

FUNCTION OF THE SACCULE Considered to function like the utricle as component of vestibular apparatus.

FUNCTION OF THE SEMICIRCULAR DUCTS Detect rotational acceleration or deceleration. Crista-receptor structure composed of hair cells, with hairs embedded in gelatinous mass called the cupula. Hairs are displaced and stimulation of hair cells is altered by rotational acceleration or deceleration.

NYSTAGMUS Characteristic eye movement associated with ability of eyes to remain fixed on same point despite changes in head position.

Vestibular Nystagmus Produced when cupulae and hairs of hair cells within ampullae of semicircular ducts are displaced.

Optokinetic Nystagmus Produced when visual images move across retinas.

MAINTENANCE OF EQUILIBRIUM Nerve impulses from various receptors provide information about position and movement of the body or its parts. Neural signals are integrated within CNS and help determine degree of activity in individual muscles that is required to maintain balance.

CONDITIONS OF CLINICAL SIGNIFICANCE: THE EAR p. 465

MIDDLE-EAR INFECTIONS Inflammation of mucous membrane of middle ear.

DEAFNESS

CONDUCTIVE DEAFNESS Interference with sound-wave transmission by physical blockage, inflammation of eardrum, ossicle adhesions, or oval window thickening.

NERVE DEAFNESS Loss of hearing that involves damage to sensory cells or nerve pathways.

SMELL (OLFACTION) p. 466
1. Receptors are neurons in epithelium of nasal mucosa in upper nasal cavity.
2. Odorous substances must dissolve in mucous layer that covers receptors, and must interact with receptors.
3. Different receptors have different combinations and concentrations of specific types of receptor sites.

TASTE (GUSTATION) pp. 467–468
1. Receptors are taste buds on tongue, roof of mouth, pharynx, larynx.
2. Much of what is commonly considered to be "taste" actually involves stimulation of olfactory receptors.

SELF-QUIZ

1. The retina develops from the embryonic: (a) telencephalon; (b) diencephalon; (c) mesencephalon.

2. The iris belongs to the: (a) fibrous tunic; (b) internal tunic; (c) vascular tunic.

3. The anterior cavity of the eye is separated from the posterior cavity by the: (a) lens; (b) iris; (c) cornea.

4. The aqueous humor is produced by the: (a) cornea; (b) ciliary processes that project from the ciliary body; (c) vitreous body.

5. The extrinsic eye muscles are the least rapid-acting but among the most precisely controlled skeletal muscles in the body. True or False?

6. The greater the deviation from the perpendicular at which light rays enter a block of glass: (a) the greater the wavelength of the light rays; (b) the less the bending of the light rays; (c) the greater the bending of the light rays.

7. During accommodation for viewing near objects: (a) the pupils constrict; (b) the ciliary muscles relax; (c) the eyes diverge.

8. Those eye receptors most responsible for vision in dim light are: (a) cones; (b) rods; (c) ossicles.

9. Visual acuity is greatest in the: (a) fovea centralis; (b) peripheral retina; (c) ciliary body.

10. The condition in which there is an abnormal elevation of the intraocular pressure is called: (a) strabismus; (b) astigmatism; (c) glaucoma.

11. The loss of lens elasticity that occurs with aging is called: (a) presbyopia; (b) myopia; (c) hyperopia.

12. The middle ear contains the: (a) cochlea; (b) semicircular canals; (c) incus.

13. A membranous partition between the middle-ear chamber and the scala vestibuli of the inner ear is formed by the: (a) oval window; (b) round window; (c) fenestra cochleae.

14. The cochlea is the portion of the inner ear associated with hearing. True or False?

15. The spiral organ is located on the: (a) tympanic membrane; (b) basilar membrane: (c) vestibular membrane.

16. The loudness of a sound is determined primarily by which aspect of a sound wave? (a) amplitude; (b) frequency; (c) pitch.

17. The endolymph within at least one semicircular duct is generally affected by rotational acceleration in any given plane. True or False?

18. Receptor cells of the semicircular ducts of the inner ear are best able to detect: (a) linear motion of the head; (b) rotational motion of the head at a constant velocity; (c) rotational acceleration.

19. A characteristic movement of the eyes that occurs while rotational acceleration of the body is taking place is termed: (a) dizziness; (b) nystagmus; (c) vertigo.

LEARNING OBJECTIVES

After completing this chapter, you should be able to:

1. Describe two ways that hormones exert their effects at the cellular level.

2. List the hormones of the major endocrine glands and describe their effects.

3. Explain how the release of pituitary hormones is influenced by the nervous system.

4. Explain the application of feedback control to endocrine regulation.

5. Describe the synthesis, storage, and release of the thyroid hormones.

6. Describe how the adrenal medulla is controlled.

7. Describe four factors that influence the release of aldosterone.

8. Cite the symptoms of adrenal cortical hypofunction and hyperfunction, and relate them to the effects of the adrenal cortical hormones.

9. Distinguish between the roles of glucagon and insulin in controlling blood-glucose levels.

10. Cite the symptoms of diabetes mellitus, and relate them to the effects of the hormone insulin.

CHAPTER CONTENTS

BASIC ENDOCRINE FUNCTIONS

PITUITARY GLAND

CONDITIONS OF CLINICAL SIGNIFICANCE: PITUITARY GLAND

FEEDBACK CONTROL OF HORMONE RELEASE

THYROID GLAND

CONDITIONS OF CLINICAL SIGNIFICANCE: THYROID DISORDERS

PARATHYROID GLANDS

CONDITIONS OF CLINICAL SIGNIFICANCE: PARATHYROID DISORDERS

ADRENAL GLANDS

CONDITIONS OF CLINICAL SIGNIFICANCE: ADRENAL DISORDERS

PANCREAS

CONDITIONS OF CLINICAL SIGNIFICANCE: PANCREATIC DISORDERS

GONADS

OTHER ENDOCRINE TISSUES AND HORMONES

PROSTAGLANDINS

KEY TERMS AND DERIVATIVES

adrenal cortex (*ad* = near; *renal* = kidney) the outer portion of the adrenal gland; produces steroid hormones

glucagon (*gluc* = glucose; *gon* = produce) hormone produced in the islets of the pancreas; stimulates the breakdown of liver glycogen

hormones (*hormon* = excite) highly potent, specialized organic substances that function as biological regulators

prolactin (*pro* = before; *lact* = milk) hormone secreted by the pituitary gland; involved in initiation and maintenance of milk production in females

The Endocrine System

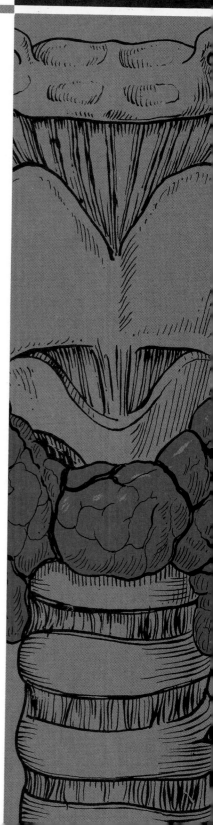

The endocrine system is a regulatory system that coordinates and integrates many body processes by releasing chemical messengers called *hormones*. The endocrine system interacts with all other body systems and affects virtually every type of cell. Consequently, the endocrine system contributes importantly to the maintenance of homeostasis. Some of the specific activities influenced by the endocrine system are listed in Table 16.1.

This chapter provides an overview of the activities of the endocrine system and the major endocrine glands. The activities of individual endocrine structures are considered further in those chapters dealing with the specific body processes they influence.

BASIC ENDOCRINE FUNCTIONS

In general, **endocrine glands** are ductless structures that contain epithelial cells specialized for the synthesis and secretion of specific hormones. The hormones are released from the cells into the surrounding extracellular spaces. From the extracellular spaces, the hormones enter the bloodstream, which transports them throughout the body. (In contrast, exocrine glands typically secrete their products by way of ducts onto body surfaces—either external surfaces like the skin, or internal surfaces like the lining of the digestive tract.) The major endocrine glands of the body are shown in Figure 16.1. Besides the clearly recognized endocrine glands such as the thyroid gland and the adrenal glands, other structures—for example, the gastrointestinal tract— also exhibit endocrine activity.

Hormones

Hormones are highly potent, specialized organic substances that function as biological regulators. Many hormones are derivatives of amino acids. Included among these hormones are the *protein hormones* (for example, prolactin from the pituitary gland), the *peptide hormones* (for example, antidiuretic hormone from the pituitary gland), and the *catecholamine hormones* (for example, epinephrine and norepinephrine from the medulla of the adrenal gland). Other hormones are synthesized from cholesterol. Among these hormones are the *steroid hormones* (for example, testosterone from the testes, estrogens and progesterone from the ovaries, and cortisol and aldosterone from the cortex of the adrenal gland). Table 16.3, which is located at the end of the chapter, provides a summary of the major endocrine glands and the hormones they produce.

Transport of Hormones

Within the bloodstream, many hormones—particularly the steroid hormones—are bound to specific carrier proteins, and the free and bound forms of a hormone exist in equilibrium with one another. The reversible association between a hormone and a carrier protein provides both a storage system and a buffer system for the hormone. Usually only a small fraction of the hormone is present in the free form, and the protein-bound portion is a readily available reserve. The binding of a hormone to a carrier protein can protect the body against excessive concentrations of the hormone, and it can delay the

473

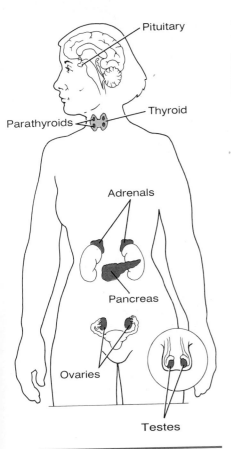

Figure 16.1
The major endocrine glands.

a-den'-o-seen mon-o-fos'-fate

F16.2

Table 16.1   Representative Activities Regulated or Influenced by the Endocrine System

Intracellular and extracellular fluid volumes
Acid–base balance
Carbohydrate, protein, lipid, and nucleic acid metabolism
Food and fluid intake
Digestion, absorption, and nutrient distribution
Reproduction and lactation
Immune processes
Blood-cell formation
Blood pressure and blood flow
Resistance to stress
Adaptation to environmental change

degradation of the hormone. The binding can also protect certain small hormone molecules from being excreted by the kidneys.

Mechanisms of Hormone Action

The bloodstream transports hormones throughout the body. However, a particular hormone does not necessarily affect all the body's cells. Only certain cells, called **target cells,** respond to any given hormone. The target cells of a particular hormone possess receptors to which molecules of the hormone can attach. When the hormone molecules attach to the receptors, a series of events is triggered that influences such cellular activities as the rates of certain reactions or the permeability of plasma membranes, leading ultimately to the observed effects of the hormone on body processes. Receptors for a number of hormones—particularly the protein, peptide, and catecholamine hormones—are located on the plasma membranes of their target cells. Receptors for some hormones—particularly the steroid hormones and possibly several nonsteroid hormones as well—are located within their target cells.

Different hormones affect the activities of their target cells by many different mechanisms. However, two important general mechanisms of hormone action are (1) by utilizing intracellular mediators and (2) by activating genes within cells.

Utilization of Intracellular Mediators

Many hormones (and also other substances) that bind to receptors on plasma membranes influence cellular function by way of intracellular mediators. One intracellular mediator is a substance known as 3',5' *cyclic adenosine monophosphate (cyclic AMP).* Cyclic AMP is formed from adenosine triphosphate (ATP) by the enzyme adenylate cyclase, which is bound to the inner surface of the plasma membrane. The receptors for hormones that utilize cyclic AMP as an intracellular mediator can be either stimulatory receptors or inhibitory receptors (Figure 16.2). The binding of hormone molecules to stimulatory receptors leads to the activation of membrane proteins known as stimulatory G proteins, which bind the substance guanosine triphosphate (GTP). The stimulatory G proteins with bound GTP, in turn, stimulate the activity of adenylate cyclase. The binding of hormone molecules to inhibitory receptors leads to the activation of inhibitory G proteins, which also bind GTP. The inhibitory G proteins with bound GTP inhibit the activity of adenylate cyclase.

Alterations of adenylate cyclase activity lead to increases or decreases in the level of cyclic AMP within cells, which in turn can affect various cellular functions such as enzyme activities, secretory activities, and membrane permeability. The specific effects of changes in cyclic AMP levels on a particular cell depend on the characteristics of the cell. For example, when the pancreatic hormone glucagon binds to stimulatory receptors on liver cells, elevated cyclic AMP concentrations promote the breakdown of glycogen into glucose, and when thyroid-stimulating-hormone from the pituitary gland binds to

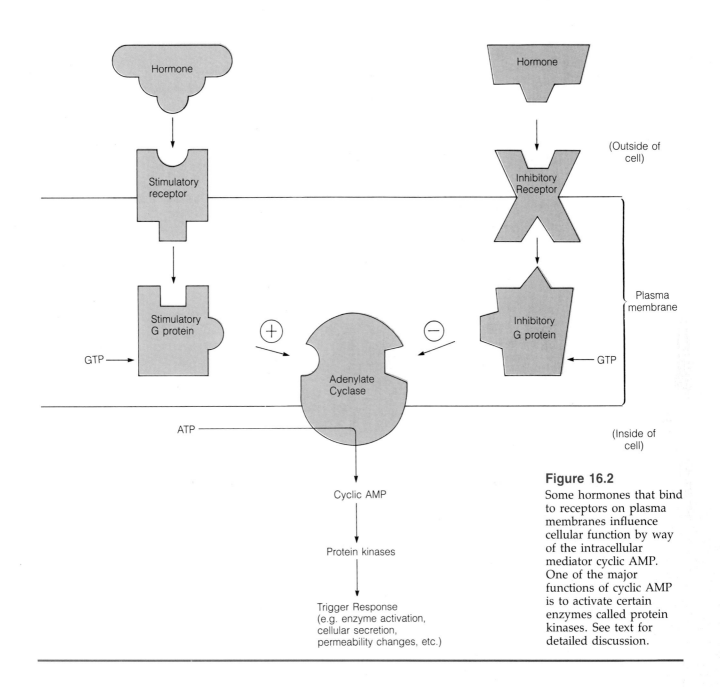

Figure 16.2
Some hormones that bind to receptors on plasma membranes influence cellular function by way of the intracellular mediator cyclic AMP. One of the major functions of cyclic AMP is to activate certain enzymes called protein kinases. See text for detailed discussion.

stimulatory receptors on cells of the thyroid gland, increased cyclic AMP leads to the release of the thyroid hormones.

One of the major functions of cyclic AMP is to activate certain enzymes called *protein kinases*. The protein kinases attach phosphate groups to specific proteins, which are often enzymes themselves. This phosphorylation can activate or inhibit enzymatic proteins (or alter the function of other regulatory molecules), leading to the stimulation or inhibition of various cellular reactions or processes. There are different protein kinases, and there is generally more than one protein kinase in a given type of cell. Each protein kinase phosphorylates particular proteins and therefore influences particular cellular reactions or processes. Consequently, the effects of cyclic AMP in a given cell depend at least in part on the specific protein kinases present and also on the particular proteins present that are phosphorylated by the protein kinases.

In addition to cyclic AMP, there are other intracellular mediators of hormonal effects. In some cases, hormone molecules (or other substances) bind to receptors on plasma membranes, and the binding activates certain G pro-

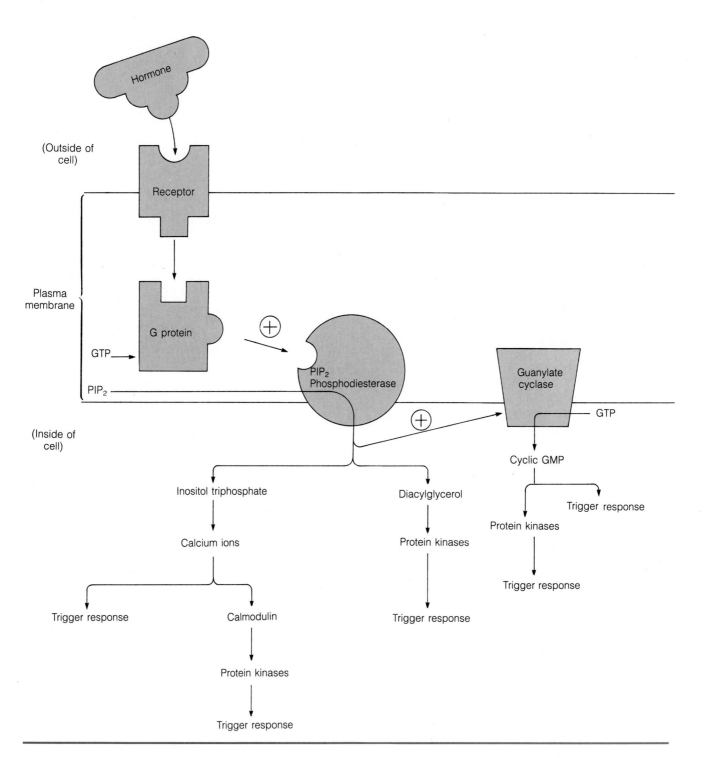

Figure 16.3
Some hormones that bind to receptors on plasma membranes influence cellular function by way of intracellular mediators such as inositol triphosphate, diacylglycerol, cyclic GTP, and/or calcium ions. See text for detailed discussion.

teins, which bind GTP. The G proteins with bound GTP, in turn, activate the enzyme PIP_2 phosphodiesterase, which is bound to the inner surface of the plasma membrane (Figure 16.3). The PIP_2 phosphodiesterase catalyzes the splitting of a membrane phospholipid called phosphatidylinositol 4,5 biphosphate (PIP_2) into diacylglycerol and inositol triphosphate. Often, the activation of this pathway is accompanied by the conversion of GTP into cyclic guanosine monophosphate (cyclic GMP) by the enzyme guanylate cyclase, indicating that the pathway may also induce cyclic GMP formation. Diacylglycerol can activate certain protein kinase enzymes, with results similar to those described for cyclic AMP. Cyclic GMP can also activate protein kinase enzymes and in some cases can act directly on regulatory molecules in cells to influence their functions.

Hormones often influence the concentration of calcium ions within cells, and inositol triphosphate (as well as cyclic AMP and perhaps cyclic GMP) can modulate intracellular calcium-ion levels. For example, inositol triphosphate can stimulate the release of calcium ions into the cytoplasm from the endoplasmic reticulum of certain cells. The ability of hormones—perhaps acting by way of inositol triphosphate and/or cyclic AMP—to influence the levels of calcium ions within cells is important because calcium ions are themselves an important intracellular mediator. In some cases, calcium ions bind directly to proteins, thereby altering their functions. In other cases, calcium ions bind to a protein called *calmodulin,* and the calmodulin with bound calcium activates protein kinase enzymes.

Activation of Genes

Some hormones, particularly the steroid hormones, exert at least some of their effects by activating genes within a cell (Figure 16.4). In this activity, hormone molecules enter a cell and combine with receptors, forming hormone–receptor complexes. The hormone–receptor complexes interact either directly or indirectly with the genetic material to activate certain genes. This gene activation leads to the synthesis of m-RNA and ultimately to the production of proteins (for example, enzymes) that influence cellular reactions or processes.

Hormonal Interrelationships

There are numerous interrelationships between different endocrine glands and hormones. In fact, almost every activity influenced by the endocrine system is affected by groups of hormones acting either together or in sequence. For example, hormones secreted by the pituitary gland, the thyroid gland, the adrenal glands, and the pancreas all influence carbohydrate metabolism.

In some cases, one hormone affects the activity of a second hormone. One way that this occurs is for the first hormone to increase or decrease the number or affinity of the target-cell receptors for the second hormone, and thereby increase or decrease the ability of the second hormone to influence its target cells.

A number of hormones influence the rates of production of other hormones. As discussed in subsequent sections, the pituitary gland produces several hormones that act in this fashion.

Relationship Between the Endocrine System and the Nervous System

The regulatory role of the endocrine system is similar to that of the nervous system. In fact, the endocrine and nervous systems are very closely related. Hormones influence neural functions, and neurotransmitters contribute to the control of hormone secretion. Moreover, in certain cases a single chemical substance can act as a hormone at some sites and as a neurotransmitter or neuromodulator at others. The close relationship between the endocrine system and the nervous system is evident in the control and function of the pituitary gland.

PITUITARY GLAND

The hormones of the **pituitary gland,** or **hypophysis,** regulate several other endocrine glands and affect a number of diverse body activities.

Embryonic Development and Structure

The pituitary gland is located beneath the brain and is surrounded by the *sella turcica* of the sphenoid bone. The opening of the sella turcica is covered over

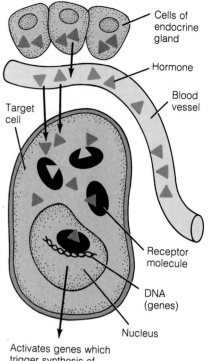

Figure 16.4
Mechanism by which some hormones, particularly the steroid hormones, activate genes within a cell.

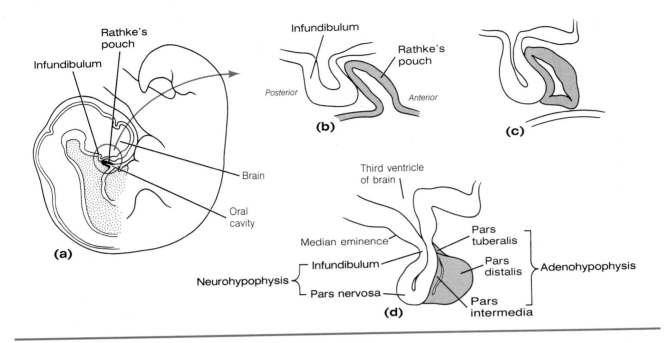

Figure 16.5
Stages in the embryonic development of the pituitary gland. See text for detailed discussion.

by a tough membrane called the *diaphragm sellae*. The pituitary develops embryonically from two different ectodermal regions: the floor of the brain and the roof of the mouth (Figure 16.5).

Neurohypophysis

noo-ro-high-pof'-i-sis

The portion of the pituitary known as the **neurohypophysis** is an outgrowth of nervous tissue from the floor of the brain in the region of the hypothalamus. The fully developed neurohypophysis consists of a structure called the **pars nervosa (lobus nervosus),** which is connected to the brain by a stalklike **infundibulum.** Where the infundibulum connects the pars nervosa to the brain there is a small elevation called the *median eminence* (Figures 16.5 and 16.6).

F16.5, F16.6

Adenohypophysis

The second portion of the pituitary arises from ectodermal tissue from the roof of the mouth (Figure 16.5b). An outpouching from the mouth called **Rathke's pouch** grows toward the developing neurohypophysis. (There is evidence that, during development, Rathke's pouch is invaded by neural cells, which are believed to contribute to endocrine functions.) As it meets the neurohypophysis, Rathke's pouch loses its connection with the mouth (Figure 16.5c). The portion of the pituitary derived from Rathke's pouch becomes the **adenohypophysis** of the fully developed pituitary and includes the **pars distalis, pars tuberalis,** and **pars intermedia** (Figure 16.5d). The pars intermedia, however, is virtually nonexistent in the pituitary of the adult human, although it is present in most other vertebrates and in human fetuses.

F16.5b

F16.5c
ad-e-no-high-pof'-i-sis
F16.5d

Table 16.2

Table 16.2 summarizes the terminology used to identify the divisions and subdivisions of the pituitary gland. Note that the pituitary gland may also be divided into an *anterior lobe,* which includes the pars distalis and pars tuberalis, and a *posterior lobe,* which includes the pars nervosa and, when present, the pars intermedia.

Relationship to the Brain

There is a close relationship between the pituitary and the nervous system. The neurohypophysis is connected to the brain by way of the infundibulum. The adenohypophysis is also closely associated with the brain. Within the median eminence of the lower hypothalamic region of the brain is a capillary

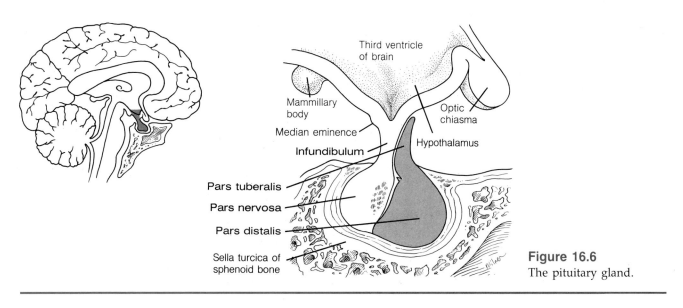

Third ventricle
of brain

Mammillary
body

Optic
chiasma

Median eminence

Hypothalamus

Infundibulum

Pars tuberalis

Pars nervosa

Pars distalis

Sella turcica of
sphenoid bone

Figure 16.6
The pituitary gland.

Table 16.2   Terminology of the Pituitary Gland

Neurohypophysis	Infundibulum	
	Pars nervosa	Posterior lobe
	Pars intermedia	
Adenohypophysis	Pars distalis	Anterior lobe
	Pars tuberalis	

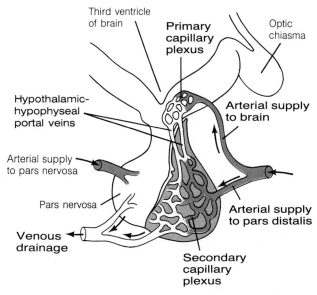

Third ventricle
of brain

Primary
capillary
plexus

Optic
chiasma

Hypothalamic-
hypophyseal
portal veins

Arterial supply
to brain

Arterial supply
to pars nervosa

Pars nervosa

Arterial supply
to pars distalis

Venous
drainage

Secondary
capillary
plexus

Figure 16.7
Diagrammatic
representation of the
circulatory link between
the brain and the adeno-
hypophysis of the
pituitary gland.

bed known as the *primary capillary plexus* (Figure 16.7). This plexus receives blood from the internal carotid arteries and the cerebral arterial circle (see Chapter 19). Blood in the primary capillary plexus flows downward along the stalk of the pituitary by way of *hypothalamic-hypophyseal portal veins* directly to a *secondary capillary plexus* in the adenohypophysis. Thus, the brain and the adenohypophysis are linked by the circulatory system.

Neurohypophyseal Hormones and Their Effects

The pars nervosa releases two peptide hormones: antidiuretic hormone (ADH) and oxytocin (Table 16.3).

Table 16.3

Antidiuretic Hormone

Antidiuretic hormone (ADH), also called **vasopressin,** promotes the reabsorption of water from the urine-forming structures of the kidneys and thus helps retain fluid within the body. At moderate to high concentrations, ADH constricts blood vessels called arterioles. ADH is a peptide hormone composed of nine amino acids.

an-tee-dye-u-re'-tik

Figure 16.8
Neurosecretory cells in the hypothalamic region of the brain manufacture antidiuretic hormone (ADH) and oxytocin. These hormones are transported along the axons of the neurosecretory cells to the pars nervosa of the pituitary. From the pars nervosa, the hormones are released into the circulation.

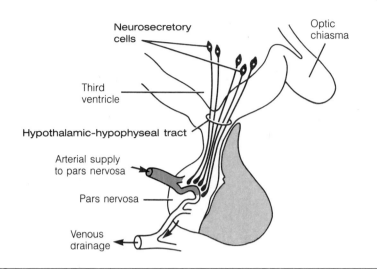

Oxytocin

Oxytocin stimulates the smooth muscles of the uterus. It also promotes the contraction of myoepithelial cells surrounding the saclike alveoli of the mammary glands, resulting in the "let down" of milk during lactation. The functions of oxytocin in males have not as yet been fully defined. Like ADH, oxytocin is a peptide hormone composed of nine amino acids.

Release of Neurohypophyseal Hormones

F16.8

Although both ADH and oxytocin are released from the pars nervosa, neither hormone is manufactured there. Rather, both are manufactured in certain hypothalamic nuclei by specialized neural cells called **neurosecretory cells.** These cells possess long processes (axons) that extend from their cell bodies in the hypothalamic nuclei, along the infundibulum, to the pars nervosa. The cell processes form the **hypothalamic-hypophyseal tracts** in the infundibulum (Figure 16.8). ADH and oxytocin synthesized in the brain by these neurosecretory cells move down the axons to the pars nervosa, from which they are released to enter the bloodstream. Both hormones are transported from the hypothalamus to the pars nervosa attached to carrier proteins called *neurophysins.*

In addition to synthesizing and transporting ADH and oxytocin, the specialized neurosecretory cells also conduct nerve impulses. This activity regulates the release of the neurohypophyseal hormones.

Because ADH and oxytocin are synthesized in the hypothalamus by specialized neurosecretory cells, the synthesis and release of these hormones can be influenced by neural inputs to the brain. Information from appropriate receptors is neurally relayed to the brain to alter the production and release of ADH or oxytocin in accordance with the regulatory requirements of the body.

Adenohypophyseal Hormones and Their Effects

Table 16.3

The adenohypophysis of the pituitary produces and releases several hormones (Table 16.3).

Gonadotropins

go-nad-o-tro'-pinz

Two of the adenohypophyseal hormones are called **gonadotropins** because they particularly affect the gonads (ovaries and testes). In females, **follicle-stimulating hormone (FSH)** stimulates the development of structures called follicles within the ovaries and induces the secretion of the estrogenic female sex hormones. In this activity, FSH works in conjunction with the other adenohypophyseal gonadotropin, **luteinizing hormone (LH).** Surging levels of

LH together with FSH lead to ovulation and the formation of a structure called the corpus luteum from an ovarian follicle. The corpus luteum produces estrogenic hormones, as well as the female sex hormone progesterone.

Both FSH and LH are also present in males. FSH promotes the maturation of cells within the testes called Sertoli cells, which are involved in the development and maturation of sperm. LH, which in males is sometimes called **interstitial-cell-stimulating hormone (ICSH),** stimulates the interstitial cells of the testes to produce the male sex hormones (that is, androgens such as testosterone). Since the male sex hormones are themselves involved in sperm production, LH can be regarded as important in this process as well. Both FSH and LH are glycoproteins (proteins combined with carbohydrate molecules).

in-ter-stish'-al

Thyrotropin

Thyrotropin, also called **thyroid-stimulating hormone (TSH),** stimulates the synthesis and release of the thyroid hormones from the thyroid gland. Like FSH and LH, thyrotropin is a glycoprotein.

Adrenocorticotropin

Adrenocorticotropin, or **adrenocorticotropic hormone (ACTH),** stimulates the release of hormones from the cortical regions of the adrenal glands, particularly cortisol and other glucocorticoid hormones that are active in carbohydrate metabolism. ACTH is a polypeptide hormone composed of 39 amino acids.

a-dree'-no-kort-i-ko-tro'-pik

Growth Hormone

Growth hormone (GH), also called **somatotropin,** stimulates growth in general, and the growth of the skeletal system in particular. GH enhances the entrance of amino acids into cells and favors their incorporation into protein. These actions decrease the concentrations of amino acids in the blood. Growth hormone increases the release of fatty acids from adipose tissue into the blood, and it inhibits fat synthesis. The fatty acids can be taken up from the blood and used as energy sources by most cells. Growth hormone also promotes glucose formation from liver glycogen. Although GH may initially increase the rate of glucose uptake by adipose tissue, heart, and skeletal muscle cells, after some hours glucose uptake and utilization are reduced, an effect that contributes to the maintenance of adequate blood-glucose levels during fasting. In many instances, growth hormone acts synergistically with other hormones to enhance their effects. GH is a protein hormone consisting of 191 amino acids.

A number of growth hormone effects are brought about through the mediation of a group of growth-promoting peptides found in the blood plasma that are collectively called **somatomedins.** For example, the ability of growth hormone to stimulate cartilage and bone growth occurs, at least in part, by way of the somatomedins, which directly stimulate cartilage and bone. Growth hormone stimulates the production of somatomedins by the liver and perhaps by the muscles and kidneys as well.

Although growth hormone produces its most dramatic effects during the period of body growth and development, it is secreted throughout life. The release of growth hormone increases in response to declining concentrations of blood glucose and also in response to elevated blood levels of certain amino acids, particularly arginine. An increased secretion of growth hormone is associated with fasting, hypoglycemia (low blood-glucose levels), exercise, and certain kinds of stress.

Prolactin

Prolactin is involved in the initiation and maintenance of milk production in females. Like oxytocin, the functions of prolactin in males have not as yet been fully defined. The structure of prolactin, which is a protein hormone, resembles that of growth hormone.

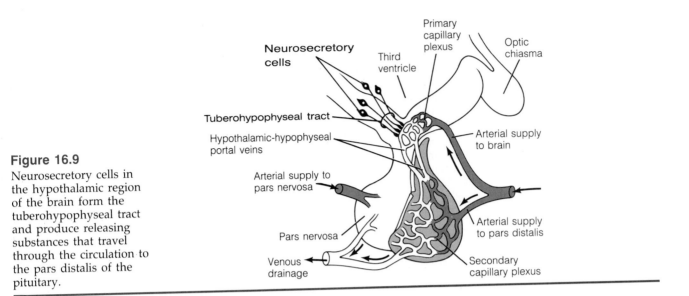

Figure 16.9
Neurosecretory cells in the hypothalamic region of the brain form the tuberohypophyseal tract and produce releasing substances that travel through the circulation to the pars distalis of the pituitary.

Endorphins

In addition to the hormones just discussed, the adenohypophysis contains several forms of **endorphins,** which are peptides that have opiatelike analgesic effects. In fact, the largest of these peptides—beta endorphin—is a much more potent analgesic than morphine. Although endorphins appear to be produced principally in the adenohypophysis, they occur in the medulla oblongata and hypothalamus of the brain and at other sites.

Pro-opiomelanocortin and the Production of Adenohypophyseal Hormones

A number of the peptide and protein hormones of the adenohypophysis, as well as other peptide and protein substances, are believed to be produced initially as parts of large precursor molecules that are subsequently split by enzymes into fragments of precise length and biological activity. For example, cells in the adenohypophysis and also in the hypothalamus of the brain synthesize a protein molecule called *pro-opiomelanocortin* that contains the amino acid sequences of an enkephalin, an endorphin, and ACTH, as well as the sequence of a substance called gamma melanocyte-stimulating hormone (gamma MSH), which may be used as a neurotransmitter by certain hypothalamic cells. Thus, pro-opiomelanocortin is apparently a precursor molecule from which ACTH and other substances can be formed.

It should be noted that not all cells containing a precursor like pro-opiomelanocortin necessarily secrete the same hormones or other substances. In some cells a precursor may be split to produce one substance, and in others it may be split to form another substance. It is also possible that the same cell may produce more than one substance from a precursor molecule, or that it may form one substance at one time and another substance at some other time. In any event, the release of each hormone or other substance is precisely controlled, and there is no massive, undifferentiated secretion of multiple materials. The mechanisms that control the release of the adenohypophyseal hormones are discussed in the following sections.

Release of Adenohypophyseal Hormones

The brain exercises a good deal of control over the synthesis and release of the adenohypophyseal hormones.

Hormone-Releasing and Hormone-Inhibiting Substances

Within the brain, specialized neurosecretory cells manufacture a group of molecules known collectively as **releasing** and **inhibiting substances.** If the chemical identity of a particular releasing or inhibiting substance is known, the substance is referred to as a hormone. If its chemical composition is unknown, it is referred to as a releasing or inhibiting factor. The neurosecretory cells in the brain that manufacture releasing or inhibiting substances form the *tuberohypophyseal tract* as they pass through the tuber cinereum of the hypothalamus to the median eminence. Within the median eminence the neurosecretory cells release their products in the region of the primary capillary plexus (Figure 16.9). The releasing or inhibiting substances travel through the circulation by way of the hypothalamic-hypophyseal portal veins to the adenohypophysis, where they stimulate or inhibit the release of adenohypophyseal hormones.

F16.9

The following releasing or inhibiting substances have been identified: **gonadotropin-releasing hormone (GnRH),** which stimulates the release of both FSH and LH; **thyrotropin-releasing hormone (TRH); corticotropin-releasing hormone (CRH); prolactin-releasing factor (PRF);** and **growth-hormone-releasing hormone (GH-RH).** In addition, there is a **prolactin-inhibiting factor (PIF),** and there is an inhibiting hormone for growth hormone called **somatostatin.**

Nervous System Effects
on Adenohypophyseal Hormone Release

Regardless of whether a particular factor or hormone promotes or inhibits the release of certain adenohypophyseal hormones, the adenohypophyseal hormones are under the influence of the nervous system. For example, information neurally relayed to the brain can alter the secretion of various releasing or inhibiting substances, and an altered secretion of these substances, in turn, influences the release of adenohypophyseal hormones. In addition, since many adenohypophyseal hormones stimulate the release of other hormones (for example, those of the thyroid, adrenals, and gonads), a pathway exists by which neural activity can ultimately influence a wide variety of endocrine functions. Even brain activities of the sort that are responsible for emotions can affect hormonally controlled events. For instance, emotional trauma can upset the menstrual cycle or nursing.

CONDITIONS OF CLINICAL SIGNIFICANCE
Pituitary Gland

Disorders of the glands of the endocrine system generally result in either an underproduction of hormones—a condition referred to as *hyposecretion*—or an overproduction—or *hypersecretion*—of hormones. Pituitary malfunctions can be caused by disorders of the gland itself or by difficulties involving the releasing or inhibiting substances from the brain. Pituitary disorders can involve the functioning of major segments of the gland, or they may be limited to disorders in the release of a single pituitary hormone.

Individual hormone disorders produce a number of observable conditions. Deficiencies of the gonadotropins upset normal gonadal function, and serious deficiencies can lead to gonadal inactivity in males and to the impairment or cessation of menstruation in females. TSH deficiency results in underactivity of the thyroid gland (hypothyroidism). Deficiencies of prolactin may result in a failure of lactation after delivery, whereas prolactin overproduction can lead to lactation in a woman who has not recently given birth to a child. ACTH deficiency leads to adrenal cortical insufficiency, whereas excessive ACTH produces symptoms of adrenal cortical hyperfunction. A deficiency of growth hormone in the young can result in impaired growth and may produce pituitary dwarfs. An excess of growth hormone before the growth in length of the long bones is complete (that is, before the epiphyses have closed) results in great height and gigantism. After this time, an excess secretion of growth hormone leads to acromegaly. In acromegaly, height does not increase as in gigantism, but bones thicken and soft tissues increase in size. Structures such as the jaw, hands, feet, and tongue enlarge, and there is a coarsening of the facial features. A deficiency of antidiuretic hormone results in diabetes insipidus, which is characterized by copious urination.

CLINICAL CORRELATION
Pituitary Gigantism

CASE REPORT . . . .

THE PATIENT   A 19-year-old male

PRINCIPAL COMPLAINT   Extreme growth (gigantism)

HISTORY   The patient has normal parents and siblings. He was normal himself until 10 years of age, when he began to experience abnormally rapid growth. The rate of growth continued at the 99th percentile for two years and then remained between the 50th and 75th percentiles for two more years. At age 14 years, the concentration of growth hormone in the blood plasma was 113 nanograms (ng) per milliliter (normal: 10–20 ng/ml), and the level could not be increased by injection of insulin (which lowers the blood-glucose level) or decreased by injection of glucose. The fasting blood-glucose level was 110 milligrams (mg) per 100 ml (normal: 70–110 mg/100 ml). The blood serum inorganic phosphorus was 8 mg/100 ml (normal: 3–4.5 mg/100 ml). In the initial period of rapid growth the patient was strong and vigorous, but during the last two years he developed muscular weakness and atrophy. Specific weakness developed in his hands, making it difficult for him to hold a pencil and write or to use eating utensils.

CLINICAL EXAMINATION   At admission, all of the limb reflexes were markedly dulled. Cranial tomograms (sectional radiography) showed no significant abnormality of the sella turcica, and no signs of a tumor (such as impaired vision or increased intracranial pressure) were found. Nevertheless, on the basis of the increased levels of growth hormone in the blood plasma and its lack of susceptibility to stimulation or suppression, a pituitary tumor (microadenoma) was diagnosed.

TREATMENT   The pituitary gland was surgically removed, and tissue consistent with a tumor was also removed. Hormone-substitution therapy for pituitary insufficiency (due to removal of the gland) was initiated. Although the level of growth hormone in the blood plasma decreased considerably to 20–25 ng/ml, it was still slightly higher than normal. This level suggests that the tumor may not have been removed completely.

COMMENT   A pituitary tumor that produces one type of hormone often decreases production of other pituitary hormones by compressing pituitary tissue and destroying cells that produce other hormones. For example, the muscular weakness and atrophy that pituitary giants subsequently develop reflect adrenal cortical deficiency to some extent. Lack of secretion of pituitary gonadotropins delays puberty; hence the epiphyses do not close and bone growth continues. The menstrual cycles of adult women who develop such tumors often cease as a consequence of decreased FSH and LH. Neurological symptoms develop as abnormal growth of nerves causes compression in the channels through bones and joints. The weakness of the hands in this patient was caused by pressure on the medial nerve in the space formed by the wrist bones and the transverse carpal ligament ("carpal tunnel syndrome").

OUTCOME   Although substitution therapy with thyroid, adrenal, and growth hormones should bring about normal puberty and the subsequent limitation to linear growth, the continued high levels of growth hormone may produce some degree of acromegaly, and further surgery or radiation of the remaining tumor probably will be necessary.

FEEDBACK CONTROL OF HORMONE RELEASE

F16.10   Depending on the particular gland or hormone, endocrine activity may be stimulated or inhibited by a variety of factors, such as neural activity or the levels of certain chemical substances in the body. In many cases, feedback control—particularly negative feedback (see Chapter 1)—contributes to the regulation of endocrine function. For example, a stimulus may promote an increased secretion of a releasing substance from the hypothalamus (Figure 16.10). The increased secretion of the releasing substance causes an augmented release of a pituitary hormone, which in turn stimulates a target endocrine gland to release more of its hormone. The increased amount of hormone released by the target endocrine gland, in addition to exerting its normal physiological effects, may provide negative feedback to either the hypothalamus to inhibit the secretion of the releasing substance, or to the pituitary gland to inhibit the release of the pituitary hormone. Either occurrence leads to a diminished release of hormone from the target endocrine gland. This particular type of feedback pathway (from target gland to pituitary or brain) is called a *long feedback loop*.

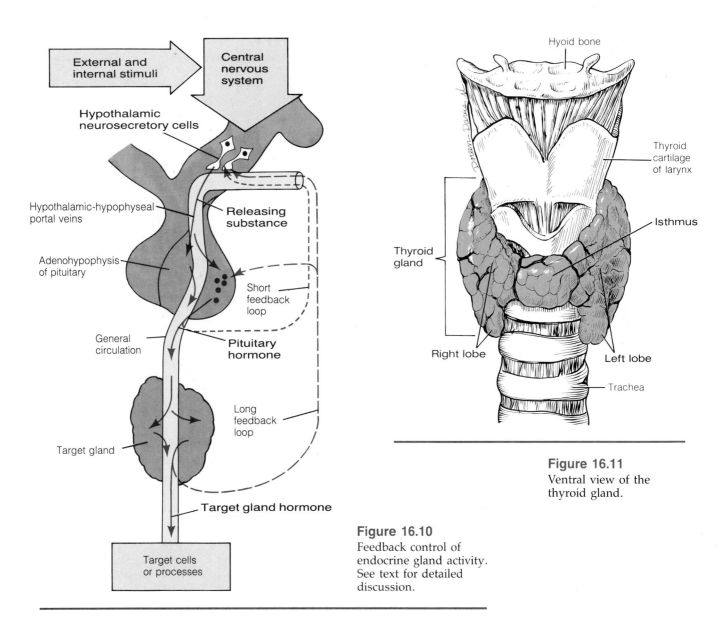

Figure 16.10
Feedback control of endocrine gland activity. See text for detailed discussion.

Figure 16.11
Ventral view of the thyroid gland.

Additionally, the pituitary hormones themselves may feed back to the brain to influence the secretion of releasing or inhibiting substances. This feedback pathway is called a *short feedback loop*. Both the long and short feedback loops provide mechanisms by which endocrine function can influence the activity of the endocrine system itself.

THYROID GLAND

The fully developed thyroid gland is among the largest of the body's endocrine organs. It is located anterior to the upper part of the trachea, near its junction with the larynx.

Embryonic Development and Structure

The **thyroid gland** originates as an epithelial thickening in the floor of the pharynx. The thickening grows outward from the pharynx, eventually loses its connection with the gastrointestinal tract, and comes to occupy a position around the trachea just below the larynx.

The thyroid gland is divided into right and left lobes that are joined across the trachea by a thin band called the **isthmus** (Figure 16.11). The gland is well vascularized, receiving blood from major arteries in the neck region.

F16.11

Follicles
(containing colloid)

Follicular cells

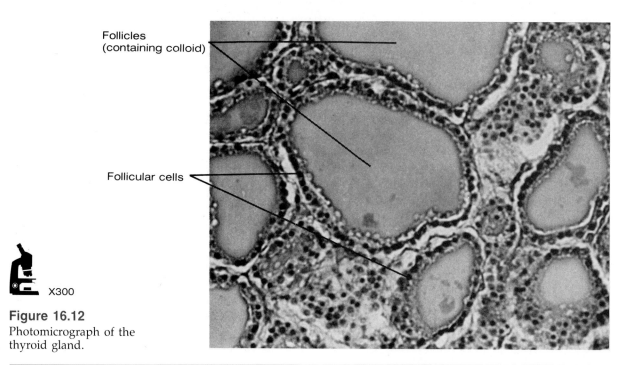

X300

Figure 16.12
Photomicrograph of the
thyroid gland.

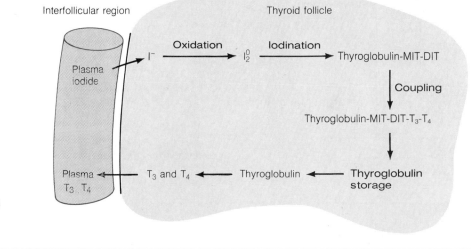

Figure 16.13
Diagrammatic
representation of the
steps involved in the
synthesis of the thyroid
hormones. T_3 is tri-
iodothyronine, and T_4 is
tetraiodothyronine
(thyroxine).

F16.12   The basic internal structure of the thyroid consists of hollow balls of cells
called **follicles** that are bound together with connective tissue (Figure 16.12).
The central region of each follicle contains a protein substance, the *colloid*. The
colloid contains a stored form of thyroid hormones. The thyroid gland is
different from all other endocrine glands in having extracellular storage sites—
the colloid regions of the follicles—for its hormones. The other endocrine
glands store their hormones in the cells of the gland.

The Thyroid Hormones: Triiodothyronine and Thyroxine

The thyroid hormones contain iodine. Most of the body's iodine is obtained
from the diet and appears in the blood as ionic iodide (I^-), which is actively
F16.13   taken up by the thyroid follicles (Figure 16.13). Within the follicles, the iodide
is converted to an active form of iodine, which is attached to the amino acid
F16.14   tyrosine to form monoiodotyrosine (MIT) (Figure 16.14). A second iodine
can also be attached to the tyrosine to form diiodotyrosine (DIT). Two

diiodotyrosines are then coupled to form **tetraiodothyronine (T_4, or thyroxine)**. Alternatively, a monoiodotyrosine and a diiodotyrosine can combine to form **triiodothyronine (T_3).** These synthetic reactions do not occur with the tyrosines in the free state. Rather, the tyrosines are part of a glycoprotein molecule known as thyroglobulin, which is manufactured by the follicle cells and secreted into the colloid area. The iodination of the tyrosines appears to occur at the cell–colloid interface as the thyroglobulin is secreted, and the iodinated tyrosines are then coupled to one another. As a result, thyroglobulin molecules containing MIT, DIT, T_3, and T_4 are produced, and these are stored in the colloid areas of the follicles.

When the thyroid is actively secreting, thyroglobulin molecules are taken into the follicle cells by endocytosis. Within the cells, lysosomes fuse with the endocytic vesicles, and lysosomal enzymes break down the thyroglobulin molecules, freeing MIT, DIT, T_3, and T_4. The MIT and DIT are deiodinated, and the iodine can be recycled for use in other iodinations. The T_3 and T_4, which are collectively referred to as the thyroid hormones, pass out of the follicle cells and enter the bloodstream.

Within the bloodstream almost all of the thyroid hormones are bound to plasma proteins such as *thyroid-binding globulin (TBG)*. The thyroid normally produces about 10% T_3 and 90% T_4. In the tissues, however, much of the T_4 is converted to T_3, and there is increasing evidence that T_3 is the major active form of the thyroid hormones at the cellular level.

Among the most evident effects of the thyroid hormones are those associated with metabolism. The administration of thyroid hormones increases the body's oxygen consumption and heat production (the calorigenic effect). Although most body tissues are responsive to this influence, some tissues— including the spleen, brain, uterus, and testes—are not. When thyroid hormones are administered, the calorigenic effect normally does not become evident for some time, but it may last several days.

In physiological amounts, the thyroid hormones favor protein synthesis, but in excessive amounts they cause protein breakdown. Excessive amounts of the thyroid hormones can lead to muscle wasting and weakness, with the weakness being especially evident in the eye muscles and cardiac muscle.

The thyroid hormones affect almost all aspects of carbohydrate metabolism, and many of their influences depend on or are modified by other hormones, particularly the catecholamines and insulin. The thyroid hormones increase gluconeogenesis (carbohydrate production from noncarbohydrate precursors) and glycolysis. They enhance the rate of intestinal absorption of glucose and the uptake of glucose by muscle and adipose tissue. In general, the thyroid hormones increase the utilization of carbohydrates and accelerate the liberation of energy. In excessive amounts, the thyroid hormones may cause hyperglycemia (high blood-glucose levels).

The thyroid hormones stimulate many phases of lipid metabolism, including synthesis, mobilization, and degradation. Generally, degradation effects predominate, and excess thyroid hormones cause a decrease in lipid stores and in the plasma concentrations of triacylglycerols, phospholipids, and cholesterol.

Release of the Thyroid Hormones

Thyrotropin (TSH) from the pars distalis of the pituitary controls the synthesis and release of the thyroid hormones, and TSH, in turn, is regulated by thyrotropin-releasing hormone (TRH) from the brain. This system is influenced by a negative feedback of the thyroid hormones themselves. Most of the negative feedback effect is manifested at the level of the pituitary gland by changing the sensitivity of the TSH-secreting cells to TRH.

Calcitonin

In addition to the thyroid hormones, the thyroid releases a hormone called **calcitonin,** which is produced by parafollicular cells ("C" cells) located between or adjacent to the thyroid follicles. Calcitonin lowers blood-calcium and

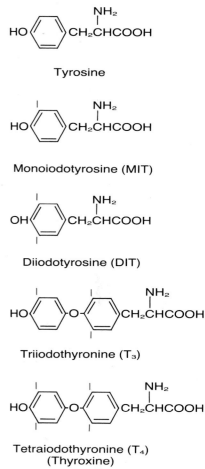

Figure 16.14
Structures of the thyroid hormones and substances important in their synthesis.

kal-si-to'-nin

CONDITIONS OF CLINICAL SIGNIFICANCE
Thyroid Disorders

In children, the thyroid hormones are essential for normal growth and maturation, and growth is retarded by hypothyroidism (lowered thyroid function). In the young, hypothyroidism can lead to an abnormal development of bones, connective tissue, and reproductive structures. In addition, the development of the nervous system may be faulty, leading to mental retardation. The condition caused by severe hypothyroidism beginning in infancy is called *cretinism*. If the hypothyroid state is recognized early and thyroid hormone replacement therapy is begun, the hypothyroid condition can be markedly improved. If treatment is delayed, the mental retardation becomes permanent.

Severe hypothyroidism in the adult is called *myxedema*. This condition, which is considerably more common in women than in men, is characterized by a puffiness of the face and eyelids and a swelling of the tongue and larynx. The skin becomes dry and rough, and the hair becomes scant. The individual has both a low basal-metabolic rate and a low body temperature. The sufferer also has poor muscle tone, lacks strength, and fatigues easily. His or her mental activity is generally sluggish and retarded. Myxedema can be alleviated by the administration of thyroid hormones.

An excess of thyroid hormones acting on the tissues is called *thyrotoxicosis*. One form of thyrotoxicosis is *hyperthyroidism* (thyroid overfunction). Hyperthyroidism is characterized by an elevated basal metabolism, elevated body temperature, and rapid heartbeat. Despite an increased appetite, there may be a large weight loss. The sufferer perspires freely and may be nervous, emotionally unstable, and unable to sleep. Hyperthyroidism can be treated surgically, with drugs that impair thyroid function, or with radioactive iodine that is taken up by the gland and destroys some of the thyroid cells.

A *goiter* is simply an enlargement of the thyroid gland. It may or may not be associated with hypothyroidism or hyperthyroidism. In one form of thyrotoxicosis known as Graves' disease, the thyroid usually exhibits a diffuse enlargement (goiter). Simple goiter is glandular enlargement without hypothyroidism or thyrotoxicosis.

-phosphate levels. It acts on bone cells to suppress bone resorption and enhance bone formation, and it is believed to be involved in the process of bone remodeling and in preventing excessive bone resorption. The plasma-calcium level controls calcitonin secretion: when the concentration of calcium ions in the plasma rises, calcitonin secretion increases. Certain hormones released by the gastrointestinal tract during the digestion and absorption of food (for example, gastrin) also promote calcitonin secretion, and it has been proposed that a release of calcitonin associated with digestion may help the body conserve calcium obtained from the diet by preventing a rise in plasma-calcium levels that could lead to an increased urinary excretion of calcium. Calcitonin is a polypeptide consisting of 32 amino acids.

PARATHYROID GLANDS

F16.15   Embedded on the posterior surface of the lobes of the thyroid gland are four small **parathyroid glands** (Figure 16.15). The blood supply of these glands is the same as that of the thyroid gland, but the parathyroid and thyroid glands differ in their embryonic development, their structure, and their function.

Embryonic Development and Structure

The parathyroid glands develop from the dorsal halves of the third and fourth pairs of embryonic structures called pharyngeal pouches. With continued development they lose their attachments to the pouches and migrate to the neck, where they assume their adult positions on the posterior surfaces of the lateral lobes of the thyroid gland.

There are usually two masses of parathyroid tissue (*superior* and *inferior*) on each of the two thyroid lobes. This tissue is composed of densely packed masses or cords of cells.

Parathyroid Hormones and Their Effects

Parathyroid hormone (PTH, or **parathormone)** is a polypeptide, and it is currently believed that two or possibly three forms of the hormone may appear in the blood. Parathyroid hormone is a principal controller of calcium

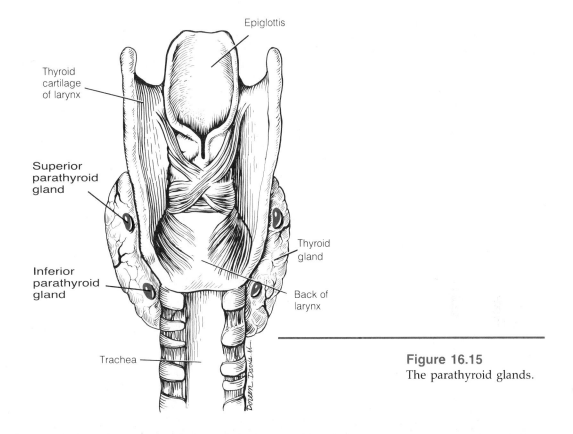

Epiglottis

Thyroid cartilage of larynx

Superior parathyroid gland

Inferior parathyroid gland

Trachea

Thyroid gland

Back of larynx

Figure 16.15
The parathyroid glands.

CONDITIONS OF CLINICAL SIGNIFICANCE
Parathyroid Disorders

Hyperparathyroidism, which results in an excess of parathyroid hormone, causes extensive bone decalcification and may lead to deformities and fractures. The plasma-calcium level rises, and the calcification of soft tissues—especially the kidneys—may occur. Hyperparathyroidism is generally treated by the surgical removal of glandular tissue.

Hypoparathyroidism, which results in a deficiency of parathyroid hormone, leads to a lowered plasma-calcium level, which greatly increases neural and mus-

cular excitability. A common symptom of a low plasma-calcium level is numbness and tingling of the extremities due at least in part to an increased activity of sensory nerve fibers that carry impulses associated with light touch. In addition, skeletal-muscle twitching and cardiac dysrhythmia may occur. If the plasma-calcium level is low enough, skeletal-muscle spasms (tetany) may occur. Hypoparathyroidism is generally treated by administering large doses of calcium salts and/or vitamin D.

and phosphate metabolism, and it is involved in the remodeling of bone. Parathyroid hormone increases the plasma-calcium concentration and decreases the plasma-phosphate concentration. It increases phosphate excretion in the urine and acts on the kidneys to decrease calcium excretion. However, a high plasma-calcium concentration due to the action of parathyroid hormone can often lead to an increased loss of calcium in the urine. Parathyroid hormone acts on bone cells to increase bone resorption. In addition, it enhances a step in the metabolic transformation and activation of vitamin D_3. In the skin, the precursor substance provitamin D (7-dehydrocholesterol) is converted to vitamin D_3 (cholecalciferol) by exposure to ultraviolet radiation (sunlight). In the liver, vitamin D_3 subsequently is converted to 25-hydroxycholecalciferol. In the kidneys, 25-hydroxycholecalciferol is converted to 1,25-dihydroxycholecalciferol, and parathyroid hormone promotes this conversion. 1,25-Dihydroxycholecalciferol is a potent substance that enhances the absorption of calcium from the gastrointestinal tract. Consequently, parathyroid hormone indirectly promotes this absorptive activity. The plasma-calcium level is the major controller of parathyroid hormone secretion: When the plasma-calcium level falls, parathyroid hormone secretion increases.

ADRENAL GLANDS

F16.16

F16.17

Table 16.3

The two **adrenal glands,** or **suprarenal glands,** are pyramid-shaped organs located behind the peritoneum close to the superior border of each kidney (Figure 16.16). Each gland is surrounded by a connective-tissue capsule and embedded in fat. The adrenal glands are well supplied with blood vessels.

Each adrenal gland consists of two separate portions: an inner *medulla* and an outer *cortex* (Figure 16.17). The medulla and the cortex have different embryonic origins and different structures, and the actions of their hormones differ considerably (Table 16.3). Consequently, each adrenal gland is, in effect, actually two distinct endocrine organs. Let us examine these two portions in greater detail.

Adrenal Medulla

The central portion of each adrenal gland, the **adrenal medulla,** is composed of cells arranged in groups or short cords surrounding blood capillaries and venules.

Embryonic Development and Structure

The adrenal medulla arises embryonically from neural-crest cells. Since the neural-crest cells also give rise to postganglionic sympathetic neurons, the adrenal medulla can be regarded as a modified portion of the sympathetic division of the autonomic nervous system. In fact, the cells of the adrenal medulla function in a manner similar to postganglionic sympathetic cells. Because they stain with chromium salts, the cells of the adrenal medulla are called *chromaffin cells.*

CLINICAL CORRELATION
Abnormal Secretion of Parathyroid Hormone

CASE REPORT ....

THE PATIENT   A 34-year-old woman

PRINCIPAL COMPLAINT   Persistent fatigue, fever, weight loss, and frequent headaches

HISTORY   The symptoms had begun to develop about one year before admission, and the patient had been seen first one month before the present admission.

CLINICAL EXAMINATION   Because of the nonspecific nature of the symptoms, general laboratory studies were performed. The only significant abnormal findings were the plasma concentrations of calcium, 13.7 mg/100 ml (normal: 8.5–10.5 mg/100 ml); phosphorus, 1.2 mg/100 ml (normal 3.4–5 mg/100 ml); and the enzyme alkaline phosphatase, 95 units (U) per liter (normal: 13–19 U/liter). These values are consistent with hyperparathyroidism. The concentration of parathyroid hormone in the plasma was determined to be 301 mU/ml (normal preovulatory level: 5–22 mU/ml).

TREATMENT   Two weeks after admission, the patient's neck was explored surgically. The four parathyroid glands appeared normal, and an examination of tissue samples from each revealed no abnormality. Since the evidence suggested a parathyroid adenoma (a benign tumor) in a supernumerary (extra) parathyroid gland, a search for the extra gland was begun. Injection of radiographic contrast material into the left internal mammary artery revealed an area of abnormally high blood supply, which is characteristic of parathyroid adenomas, in the anterior mediastinum. Subsequent surgical exploration of the area disclosed a large parathyroid nodule within the left lobe of the thymus gland.

OUTCOME   The abnormal tissue was removed, the patient's plasma-calcium concentration became normal, and the symptoms of hyperparathyroidism disappeared.

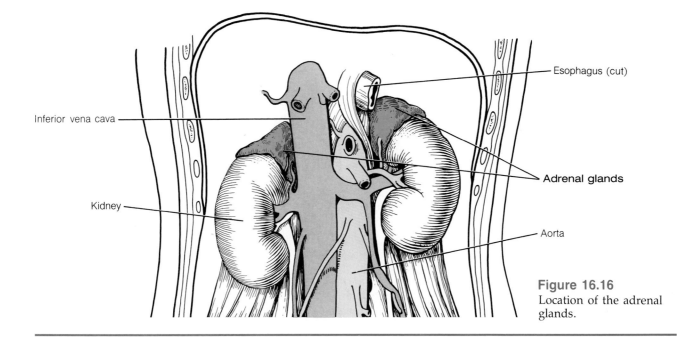

Figure 16.16
Location of the adrenal glands.

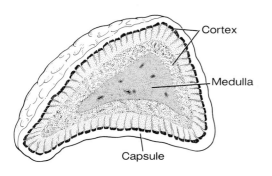

Figure 16.17
An adrenal gland. The gland is a dual structure that consists of a central medulla and a surrounding cortex enclosed in a fibrous capsule.

Figure 16.18
Structures of the adrenal medullary catecholamine hormones epinephrine and norepinephrine.

Adrenal Medullary Hormones and Their Effects

The adrenal medulla produces two catecholamine hormones: epinephrine (adrenaline) and norepinephrine (noradrenaline) (Figure 16.18). (Recall that norepinephrine is also released from the end terminals of the postganglionic neurons of the sympathetic nervous system; thus, it may be present from sources other than the adrenal medulla.) Although epinephrine and norepinephrine are structurally similar molecules that exert a number of common effects, they are not identical in function.

F16.18

EPINEPHRINE **Epinephrine** counteracts the hypoglycemic effects of insulin by elevating blood-glucose levels. It stimulates the production of lactic acid from glycogen in muscle, and the lactic acid can be used by the liver to manufacture new carbohydrate (glucose or glycogen).

Epinephrine increases the rate, force, and amplitude of the heartbeat. It constricts blood vessels in a number of body areas, including the skin, mucous membranes, and kidneys, but it can induce vessels to dilate in some areas, such as skeletal muscles. Epinephrine also causes the dilation of respiratory passageways called bronchioles in the lungs by relaxing bronchiolar smooth muscle.

NOREPINEPHRINE **Norepinephrine** increases the heart rate and the force of contraction of cardiac muscle. It also constricts blood vessels in almost all areas of the body. In addition, norepinephrine promotes the breakdown of fat in adipose tissue.

Release of the Adrenal Medullary Hormones

In general, the adrenal medulla releases a mixture of about 80% epinephrine and 20% norepinephrine, but these percentages vary considerably under different physiological conditions. The release of the catecholamine hormones of the adrenal medulla is controlled by preganglionic neurons to the medulla from the sympathetic division of the autonomic nervous system. A variety of conditions leads to the release of adrenal medullary hormones, including emotional excitement, injury, exercise, and low blood-glucose levels. Because catecholamines are liberated by both the adrenal medulla and the sympathetic nervous system, it is often convenient to consider the two as a single sympatheticoadrenal system. Together the divisions of this system maintain blood pressure and help regulate carbohydrate metabolism.

Adrenal Cortex

The outer portion of each adrenal gland, the *adrenal cortex,* makes up approximately 80% of the total weight of each fully developed gland.

Embryonic Development and Structure

F16.19

The adrenal cortex is derived embryonically from the mesoderm of the region that gives rise to gonadal tissue. The outer portion of the adrenal cortex is surrounded by a connective-tissue capsule. Trabeculae (strands of connective tissue) extend from the capsule into the gland itself. The endocrine cells of the adrenal cortex are organized into three layers (Figure 16.19): the *zona glomerulosa,* the *zona fasciculata,* and the *zona reticularis.*

The **zona glomerulosa** is a relatively thin region in the cortex. It is located directly beneath the capsule and is composed of clusters of cells. Beneath the zona glomerulosa is a thick region called the **zona fasciculata.** The cells of the zona fasciculata are arranged in parallel columns that run at right angles to the surface of the gland. Occupying the deepest region of the adrenal cortex and lying adjacent to the adrenal medulla is the **zona reticularis,** in which the cells are arranged in a network of interconnecting cords.

Adrenal Cortical Hormones and Their Effects

min-er-al-o-kor'-ti-koidz

MINERALOCORTICOIDS The adrenal cortex produces hormones that regulate sodium and potassium metabolism. These are the **mineralocorticoids**—such as **aldosterone,** which is produced by the zona glomerulosa. Aldosterone promotes the reabsorption of sodium and the excretion of potassium by the urine-forming structures of the kidneys.

gloo-ko-kor'-ti-koidz
kor'-ti-sol

GLUCOCORTICOIDS The adrenal cortex also produces hormones that affect carbohydrate metabolism. These are the **glucocorticoids**—such as **cortisol,** which is produced by the zona fasciculata. The glucocorticoids supplement and conserve the energy derived from circulating glucose. In response to the glucocorticoids, glucose utilization in many peripheral tissues is inhibited, fatty acids are mobilized from adipose tissue, and muscle shifts from glucose to fatty acids for much of its metabolic energy. The glucocorticoids promote gluconeogenesis (carbohydrate production from noncarbohydrate precursors) and glycogen deposition in the liver, and the elevation of blood glucose. They also accelerate the breakdown of proteins and inhibit amino acid uptake and protein synthesis by many tissues other than the liver. In the liver, however, amino acid uptake is enhanced, as is the utilization of amino

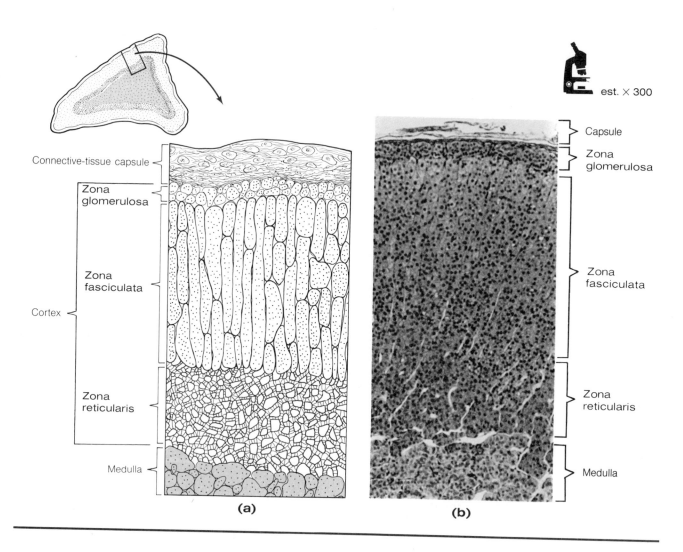

Connective-tissue capsule

Zona
glomerulosa

Zona
fasciculata

Cortex

Zona
reticularis

Medulla

(a)

est. × 300

Capsule

Zona
glomerulosa

Zona
fasciculata

Zona
reticularis

Medulla

(b)

Figure 16.19
Layers of the adrenal
cortex. **(a)** Schematic
representation.
(b) Photomicrograph.

acids for protein and glucose synthesis. At pharmacological concentrations (that is, concentrations higher than those normally present in the body), cortisol has anti-inflammatory effects.

The adrenal cortex also produces some androgenic substances that resemble male sex hormones, as well as small quantities of estrogenic materials that resemble female sex hormones. The adrenal cortical hormones (and the male and female sex hormones) are steroids that are synthesized from cholesterol.

Release of the Adrenal Cortical Hormones

The release of the mineralocorticoids (such as aldosterone) from the adrenal cortex is influenced by several factors. One of these is an enzyme called **renin,** which is released by specialized cells of the kidneys. Renin acts on the precursor substance **angiotensinogen,** which is manufactured by the liver and is present in the blood. Renin converts angiotensinogen to angiotensin I. In turn, angiotensin I is converted into other angiotensins (II, III), but these molecules will be considered simply as **angiotensin.** Among the effects of angiotensin is the stimulation of aldosterone release from the adrenal cortex. Renin secretion is itself controlled by a number of factors, including the activity of sympathetic nerves to the kidneys and the renal arterial blood pressure (decreased pressure increases renin secretion).

Aldosterone release is also stimulated by an elevation of the potassium concentration of the fluids bathing the adrenal glands, as well as by a decrease in the sodium content of the body.

The release of the glucocorticoids (such as cortisol) is controlled primarily by ACTH from the pituitary, and the release of ACTH, in turn, is influenced

CONDITIONS OF CLINICAL SIGNIFICANCE

Adrenal Disorders

Adrenal cortical hypofunction in humans is called *Addison's disease.* Inadequate amounts of aldosterone impair the body's ability to conserve sodium and excrete potassium. This impairment may lead to a decreased extracellular fluid volume, weight loss, decreased plasma volume, low blood pressure, decreased cardiac size and output, general weakness, and shock. A deficiency of glucocorticoids in Addison's disease may result in loss of appetite (anorexia), fasting hypoglycemia, apathy, weakness, and a diminished ability to withstand various types of physiological stress. Hormone administration can alleviate the symptoms of Addison's disease.

Adrenal cortical hyperfunction can result in a number of disorders. Hypercortisolism may produce *Cushing's syndrome,* which is characterized by increased blood-glucose levels, increased protein breakdown, osteoporosis (softening of bones), weakness, and hypertension. Hyperaldosteronism is characterized by potassium depletion and expansion of the extracellular fluid compartment, which may result in hypertension or edema (excessive accumulation of fluid in the tissue spaces). Certain adrenal tumors can secrete excessive quantities of androgenic substances, producing masculinizing effects that are particularly evident in females.

by corticotropin-releasing hormone (CRH) from the hypothalamus of the brain. The glucocorticoids exert an inhibitory influence over ACTH and CRH release by way of negative feedback. Various stressful situations (such as trauma or emotional stress) cause increased release of ACTH and, consequently, increased glucocorticoid secretion. This effect is probably neurally mediated by way of CRH. Once released, cortisol is bound to a plasma glycoprotein called *transcortin* for transport through the circulation.

PANCREAS

The **pancreas,** which is located behind the stomach, between the spleen and the duodenum of the small intestine, is both an endocrine gland that produces hormones and an exocrine gland that produces digestive enzymes. The endocrine portion of the pancreas produces hormones that have a role in regulating the metabolic activities of the body—particularly those associated with carbohydrate metabolism.

Embryonic Development and Structure

The pancreas arises embryonically through a fusion of two outgrowths from the duodenum of the small intestine. The exact origin of the cells that make up the endocrine portion of the pancreas remains uncertain. At least some of the endocrine cells are believed to bud off from the lining of the pancreatic ductules—either from endodermal cells that may have arisen at the site or from neuroectodermal cells that migrated into the intestinal mucosa at an earlier developmental stage.

F16.20   The endocrine portion of the fully developed pancreas consists of aggregations of cells that are clustered into groups called **pancreatic islets (islets of Langerhans)** (Figure 16.20). The islet cells are arranged in irregular cords separated by a rich vascular system of capillary vessels or sinusoids. Nerves from both the sympathetic and parasympathetic divisions of the autonomic nervous system supply the pancreas.

Pancreatic Hormones and Their Effects

The pancreatic islets contain a number of functionally different cell types. *Alpha$_2$ cells* produce the hormone **glucagon,** *beta cells* produce the hormone **insulin,** and *alpha$_1$ cells* produce somatostatin—the same substance that is produced in the brain and inhibits the release of growth hormone. There is some evidence that somatostatin, which is also secreted by the mucosa of the upper gastrointestinal tract and has been proposed as a possible neurotransmitter substance, inhibits the release of glucagon, and high concentrations of

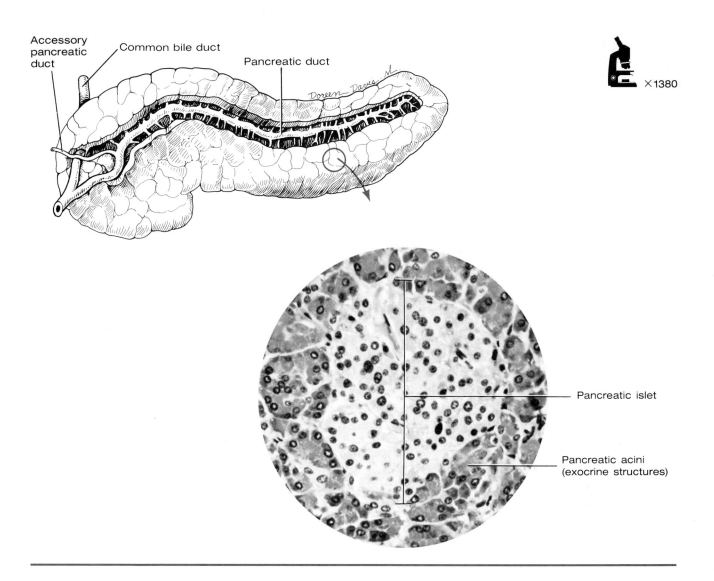

Accessory
pancreatic
duct

Common bile duct

Pancreatic duct

×1380

Pancreatic islet

Pancreatic acini
(exocrine structures)

Figure 16.20
The pancreas. The
photomicrograph shows
details of the pancreatic
cellular structure.

somatostatin appear to inhibit the release of insulin. Thus, somatostatin exemplifies the fact that a given chemical substance may be synthesized at different sites and may perform different functions.

Insulin

Insulin is composed of two linked polypeptide chains. One chain consists of 21 amino acids and the other consists of 30 amino acids.

Insulin facilitates the uptake of glucose and the utilization of glucose as an energy source by many cells, especially skeletal muscle cells. However, it does not increase the rate of glucose uptake by the brain (except, perhaps, by the hypothalamus). In skeletal muscle cells, insulin promotes the formation of glycogen from glucose that is taken up but not used as an energy source, and it prevents excessive glycogen breakdown. In liver cells, insulin promotes the conversion of glucose into both glycogen and fat (glucose is an important precursor of fatty acids and fat). In addition, insulin opposes the breakdown of glycogen and the release of glucose from liver cells into the blood. In adipose tissue cells, insulin enhances glucose uptake and fat synthesis, and it inhibits the breakdown and mobilization of stored fat. The net result of the foregoing actions of insulin is to lower blood-glucose levels. Insulin also favors the synthesis of proteins, in part by facilitating the movement of amino acids into cells.

CONDITIONS OF CLINICAL SIGNIFICANCE
Pancreatic Disorders

The most common pancreatic endocrine disorder is the condition known as *diabetes mellitus*. This condition is caused by a relative insulin deficiency that can be due to a deficient pancreatic secretion of insulin or to a decreased sensitivity of target cells to insulin. There is also some evidence that a relative or absolute excess of glucagon, as well as a deficiency of insulin, occurs in certain forms of diabetes. Diabetes occurs most often in children or adolescents and in adults over 40 years of age, and two types of diabetes are recognized. Juvenile-onset (Type I) diabetes occurs mostly in young people and is due to an insulin deficiency resulting from beta-cell pathology. Some researchers believe that juvenile-onset diabetes may be genetically predetermined, with something being required to trigger it—perhaps a viral infection or a chemical imbalance of some sort. Maturity-onset (Type II) diabetes mainly strikes overweight people who are middle-aged or older. In fact, more than 80% of adult diabetics are or have been overweight. In most forms of maturity-onset diabetes, cells become progressively less sensitive to insulin. This insulin insensitivity, which is especially evident in skeletal muscle and liver cells, appears to have multiple causes, but in some cases various target cells for insulin apparently have a deficiency of insulin receptors. Extra insulin is produced to compensate for the insulin insensitivity, and maturity-onset diabetics may have insulin levels that are above normal. However, over time, the extra activity of the pancreas can cause it to become exhausted.

Diabetes mellitus is characterized by increased levels of blood glucose coupled with a reduced entry of glucose into cells and an impairment of cellular ability to use glucose. Free fatty acids are mobilized and released from adipose tissue in diabetes mellitus, and they provide energy sources for the cells. Such energy sources are necessary because of the deficient cellular uptake and utilization of glucose. The oxidation of fatty acids in the liver produces substances known as *ketone bodies* (for example, acetoacetic acid and beta-hydroxybutyric acid). Ketone bodies may be produced faster than other body tissues can metabolize them, and the level of these sub-stances in the blood and other body fluids can increase considerably. In severe cases of diabetes, the body's buffering and excretory mechanisms are not able to cope with these acidic substances, and the pH of the body fluids falls. The low pH, in turn, can cause altered respiration, nervous system depression, coma, and death.

Protein synthesis decreases and protein breakdown increases in diabetes mellitus. These effects contribute to weight loss, and they can impair the body's ability to combat infections and to repair injured tissues.

As the blood levels of glucose and ketone bodies rise in diabetes mellitus, these substances appear in the urine. Extra water is osmotically required to excrete them, and a large volume of urine is produced (polyuria). As a consequence of the substantial loss of water in the urine, strong sensations of thirst occur, and large volumes of fluid are consumed (polydipsia). If compensation for the water loss is inadequate, dehydration can occur, as can circulatory difficulties such as hypotension.

Many diabetics develop long-term complications 20 or more years after the onset of the disease. The complications are probably the result of long-term high blood-glucose levels, which can damage blood vessels and nerves. Proliferative lesions can occur in blood vessels—particularly in the kidneys, where they can cause renal insufficiency, and in the retina, where they can impair vision and cause blindness. Atherosclerosis is also a frequent complication, and coronary artery disease is the most common cause of death among diabetics.

In the case of a diabetic whose pancreas remains functional but does not produce sufficient insulin, drugs taken orally have been used to stimulate insulin production. However, the effectiveness and safety of these drugs remains controversial, and they are not used as commonly as they once were. If the individual's pancreas cannot secrete insulin, the insulin deficiency must be made up by injection. In many obese maturity-onset diabetics, weight loss can eliminate many of the manifestations of the disease.

Glucagon

The activities of glucagon, which is a polypeptide molecule composed of 29 amino acids, are generally opposite to those of insulin. Glucagon decreases glucose oxidation and promotes hyperglycemia. Its main action seems to be to stimulate the breakdown of liver glycogen. Glucagon also stimulates the formation of carbohydrate in the liver from noncarbohydrate precursors. In addition, it stimulates the breakdown of fat in liver and adipose tissue. This breakdown increases the substances available for carbohydrate production and also results in the formation of substances called ketone bodies. Glucagon may also have a mild stimulating effect on protein breakdown.

Release of Pancreatic Hormones

The release of the pancreatic hormones is under chemical, hormonal, and neural control. The blood-glucose level appears to be the major factor governing insulin release: the higher the blood-glucose level, the greater the insulin release. The blood-glucose level also influences the release of glucagon, but in

FRONTIERS IN HEALTH

Insulin Pumps—A Step Toward an Artificial Pancreas

Lauren is one of those unfortunate people whose lives are ruled by an incurable disease: diabetes mellitus, or sugar diabetes. Three times a day she must inject insulin into her body to maintain blood-glucose levels.

For the most part, Lauren's daily injections work, but occasionally things go awry. Too much insulin or too much exercise can cause her blood-glucose levels to fall, producing hypoglycemia and causing her to feel dizzy and weak. At other times, blood glucose rises to excessive levels, producing hyperglycemia and with it depression, fatigue, irritability, and weakness. Teetering between hyperglycemia and hypoglycemia, Lauren, like many diabetics, lives a life dependent on her daily shots of insulin.

To prevent a variety of long-term complications of diabetes and to better treat the immediate symptoms of the disease as well, physicians have long tried to find ways to mimic the secretion of insulin by the pancreatic islet cells. But simulating the body's intricate and precise glucose-regulating system is not an easy task. The body constantly maintains glucose levels, minute by minute. Even a conscientious diabetic can measure blood glucose only three or four times a day. Adjustments can be made to take into account large meals or extra exercise, but such changes are crude in comparison to the body's elaborate regulating system.

New advances offer some hope for better regulation of insulin levels in diabetics. One of the most promising is the insulin infusion pump. Worn outside the body, this pump delivers tiny amounts of insulin to the body day and night, providing the baseline insulin levels needed to maintain proper blood-glucose concentrations. The pump also delivers a surge of insulin at meal times to take care of the sudden rise in blood glucose that accompanies the digestion of a meal. The diabetic simply presses a button on the pump 30 minutes before a meal to deliver the necessary, preprogrammed surge of insulin. If the meal will be bigger than usual, an additional small amount of insulin can be delivered to protect against hyperglycemia.

Insulin infusion pumps are being used by approximately 8000 Americans. According to Dr. Joseph Lowenstein, an endocrinologist at the Louisiana State University School of Medicine, most of the pumps introduce insulin into the bloodstream by slowly depressing a plunger on a disposable plastic syringe. Insulin is delivered through a small plastic tube connected on one end to the pump and on the other to a needle inserted under the skin of the thigh or the abdomen. The needle and tubing are periodically replaced, usually every three or four days.

New designs for infusion pumps are now under development. Some of them capitalize on computer technology to regulate insulin flow continuously. According to their developers, the new models are simple to operate, yet offer greater flexibility by accommodating exercise and meals of varying size. They even have memories that make it possible to store information on the exact doses given over a period of time. Physicians can use this information to reprogram the pump. Moreover, built-in alarms are designed to detect any discrepancies between the pump's dosage and the preprogrammed levels that were supposed to be administered.

However, the constant presence of tubes, needles, and computerized pumps is cumbersome. Many diabetics give the current pumps low marks for comfort and aesthetics. But the devices win high praise for the level of control over blood glucose they make possible.

For many diabetics, the simplicity of daily injections may outweigh the benefits of the pump. However, for diabetic women who become pregnant, the infusion pump may mean the difference between a normal child and a defective one. Even mild hyperglycemia can seriously affect the fetus. Although an insulin pump may not simplify a diabetic's life, and it may well complicate it, it can provide greater freedom from constant fluctuations in blood-glucose levels.

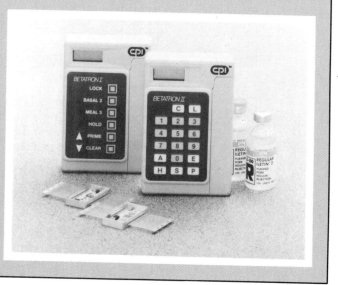

Two models of Betatron insulin infusion pumps manufactured by Cardiac Pacemakers, Inc.

opposite fashion: the lower the blood-glucose level, the greater the glucagon release. Amino acids stimulate the simultaneous secretion of both insulin and glucagon.

kol-e-sis-toe-kine'-in

Several hormones from the gastrointestinal tract (such as secretin, gastrin, and cholecystokinin) have been reported to promote insulin release, and glucagon itself has also been found to promote insulin secretion.

Parasympathetic neural activity stimulates insulin release, but sympathetic neural activity apparently has a dual effect. Adrenergic alpha-receptor stimulation inhibits insulin secretion and adrenergic beta-receptor activity increases insulin secretion. Similarly, adrenergic alpha-receptor stimulation decreases glucagon secretion and adrenergic beta-receptor stimulation increases it. Neural mechanisms are thought to be continuously active in modulating the basal secretion of insulin and glucagon, but they do not appear to be absolutely necessary for normal insulin and glucagon secretion.

GONADS

The male gonads *(testes)* and female gonads *(ovaries)* produce hormones as well as gametes (reproductive cells, that is, sperm or ova). The testes produce the male sex hormones—the **androgens**—and the ovaries produce the female sex hormones—the **estrogens** and **progesterone.** The endocrine as well as the reproductive roles of the gonads are discussed in Chapter 28.

OTHER ENDOCRINE TISSUES AND HORMONES

The digestive tract is the source of a number of hormonal substances (such as gastrin, secretin, cholecystokinin). These are discussed in Chapter 24. The placenta is also a source of hormones, as is discussed in Chapter 29.

The **pineal body** of the brain is widely believed to be an endocrine structure. However, the precise function of the pineal body in humans is currently unclear, and the exact nature of the pineal hormone remains in dispute. Some investigators believe that the principal pineal hormone is a substance called melatonin. Others believe it to be a polypeptide.

The **thymus gland,** which is discussed in relation to immune responses (Chapter 21), is also regarded as a source of hormonal material. A group of polypeptides collectively called thymosin, which appear to be involved in the development of immunologically competent white blood cells, have been isolated from the thymus.

PROSTAGLANDINS

The **prostaglandins** are a specialized group of fatty acids that serve as chemical messengers. They occur in very small quantities in the body, and probably only a few tenths of a milligram of prostaglandins are synthesized per day. Nevertheless, prostaglandins have been isolated from a wide variety of tissues, including intestine, liver, kidney, pancreas, heart, lung, brain, and male and female reproductive structures.

Some investigators believe that prostaglandins travel through the bloodstream and thus act as hormones. However, most researchers believe that the prostaglandins are primarily local messengers that exert their actions in the tissues that synthesize them. In any event, they have a wide range of effects. Prostaglandins PGE_2 and $PGF_{2\alpha}$, for example, stimulate uterine contractility and may be important in the process of giving birth and in the fertilization of the ovum. Prostaglandin $PGF_{2\alpha}$ also inhibits the secretion of progesterone by the corpus luteum of the ovary. Other prostaglandins inhibit gastric secretion, relax smooth muscle in the airways to the lungs, increase urine flow, and increase or lower blood pressure.

Table 16.3   Major Endocrine Glands and Their Hormones

Gland	Hormones	Representative Effects	Selected Disorders
PITUITARY *F16.6*			
Neurohypophysis (hormones are actually manufactured in hypothalamus of the brain)	**Antidiuretic hormone (ADH)**	Promotes reabsorption of water from urine-forming structures of kidneys	Undersecretion leads to diabetes insipidus
	Oxytocin	Stimulates contraction of uterine smooth muscle and myoepithelial cells around alveoli of mammary glands. As such, it is involved in birth processes and milk "letdown" during nursing	
Adenohypophysis	**Follicle-stimulating hormone (FSH) and luteinizing hormone (LH)**	Stimulate gonads to produce gametes and sex hormones	Undersecretion causes gonadal inactivity in males and impairment or cessation of menstruation in females
	Thyrotropin (TSH)	Stimulates thyroid gland to secrete thyroid hormones	Undersecretion leads to symptoms of hypothyroidism
	Adrenocorticotropin (ACTH)	Stimulates adrenal cortex to secrete glucocorticoids (such as cortisol)	Undersecretion leads to symptoms of adrenal cortical insufficiency. Oversecretion leads to symptoms of adrenal cortical hyperfunction
	Growth hormone (GH)	Stimulates growth in general and growth of skeletal system in particular. It also affects metabolic functions	Undersecretion produces pituitary dwarfs. Oversecretion causes gigantism or, in adults, acromegaly
	Prolactin	Involved in milk production in females	Undersecretion may cause failure to lactate after giving birth. Oversecretion may lead to lactation without recently having given birth
THYROID *F16.11*	**Thyroxine (T₄ or tetraiodothyronine) and triiodothyronine (T₃)**	Increase oxygen consumption and heat production (calorigenic effect). Important for normal growth and development. These hormones affect many metabolic processes	Undersecretion leads to symptoms of hypothyroidism, possibly causing cretinism in children or myxedema in adults. Oversecretion leads to hyperthyroidism, possibly causing Graves' disease
	Calcitonin	Lowers blood-calcium and -phosphate levels	
PARATHYROIDS *F16.15*	Parathyroid hormone	Affects calcium and phosphate metabolism to raise plasma-calcium levels and decrease plasma-phosphate levels	Undersecretion leads to nervous excitability and tetanus. Oversecretion leads to bone decalcification, and calcification of soft tissues such as the kidneys may occur

Table 16.3   Major Endocrine Glands and Their Hormones (continued)

Gland	Hormones	Representative Effects	Selected Disorders
ADRENALS *F16.16*			
Medulla	Epinephrine	Affects carbohydrate metabolism; generally tends to raise blood-glucose levels. Constricts vessels in skin, mucous membranes, and kidneys. Dilates respiratory passageways called bronchioles in the lungs	
	Norepinephrine	Increases heart rate and force of contraction of cardiac muscle; constricts blood vessels in almost all areas of the body	
Cortex	Mineralocorticoids (such as aldosterone)	Promote reabsorption of sodium and excretion of potassium from urine-forming structures of kidneys	Undersecretion may lead to decreased fluid volume and circulatory difficulties, and contribute to Addison's disease. Oversecretion may cause increased fluid volume, edema, and hypertension
	Glucocorticoids (such as cortisol)	Affect many aspects of carbohydrate metabolism; generally tend to increase blood-glucose levels	Undersecretion contributes to Addison's disease. Oversecretion leads to Cushing's syndrome
PANCREAS *F16.20*	Insulin	Affects many aspects of carbohydrate metabolism; generally tends to lower blood-glucose levels	Relative deficiency of insulin leads to hyperglycemia and diabetes mellitus
	Glucagon	Affects metabolism in fashion generally opposite of insulin; generally tends to raise blood-glucose levels	
GONADS			
Ovaries	Estrogens Progesterone	The gonadal hormones are involved in the processes of reproduction. Their functions are discussed in Chapters 28 and 29	
Testes	Androgens (such as testosterone)		

STUDY OUTLINE

BASIC ENDOCRINE FUNCTIONS In general, endocrine glands are ductless structures that contain epithelial cells specialized for synthesis and secretion of specific hormones. pp. 473–477

HORMONES Highly potent, specialized organic substances that function as biological regulators.

TRANSPORT OF HORMONES Many hormones, particularly steroid hormones, are bound to specific carrier proteins within bloodstream. Reversible association between a hormone and a carrier protein provides both storage system and buffer system for the hormone.

MECHANISMS OF HORMONE ACTION A particular hormone does not necessarily affect all cells—only its target cells. Target cells of a hormone possess receptors to which molecules of the hormone can attach.

UTILIZATION OF INTRACELLULAR MEDIATORS Binding of some hormone molecules to receptors on plasma membranes alters activity of the enzyme adenylate cyclase, leading to changes in level of cyclic AMP within cells that can affect various cellular functions. Other substances (for example, calcium ions, diacylglycerol, and cyclic GMP) may also mediate hormones' effects at cellular level.

ACTIVATION OF GENES Hormone molecules enter cell and combine with receptors. Hormone–receptor complex interacts directly or indirectly with genetic material to activate certain genes; gene activation leads to m-RNA synthesis, and ultimately to production of proteins (for example, enzymes) that influence cellular reactions or processes.

HORMONAL INTERRELATIONSHIPS Almost every activity influenced by endocrine system is affected by groups of hormones acting together or in sequence. Some hormones influence other hormones' activities or rates of production.

RELATIONSHIP BETWEEN THE ENDOCRINE SYSTEM AND THE NERVOUS SYSTEM Hormones influence neural functions; neurotransmitters contribute to the control of hormone secretion.

PITUITARY GLAND Produces hormones that regulate several other endocrine glands and affect diverse body activities. pp. 477–483

EMBRYONIC DEVELOPMENT AND STRUCTURE

NEUROHYPOPHYSIS From ectodermal tissue of brain floor.

ADENOHYPOPHYSIS From ectodermal tissue of roof of mouth.

RELATIONSHIP TO THE BRAIN Neurohypophysis connected to brain by way of infundibulum; brain and adenohypophysis linked by circulatory system.

NEUROHYPOPHYSEAL HORMONES AND THEIR EFFECTS Pars nervosa releases two peptide hormones:

ANTIDIURETIC HORMONE Promotes reabsorption of water from urine-forming structures of kidneys; constricts arterioles.

OXYTOCIN Stimulates smooth muscles of uterus; promotes contraction of myoepithelial cells that surround alveoli of mammary glands.

RELEASE OF NEUROHYPOPHYSEAL HORMONES Hormones synthesized in brain by neurosecretory cells; transported to pars nervosa; neural inputs to brain influence release.

ADENOHYPOPHYSEAL HORMONES AND THEIR EFFECTS Adenophypophysis releases several hormones:

GONADOTROPINS Follicle-stimulating hormone (FSH) and luteinizing hormone (LH); FSH and/or LH are involved in:
1. Ovarian follicle development and estrogen production.
2. Ovulation and formation of corpus luteum.
3. Spermatogenesis and androgen production.

THYROTROPIN Stimulates synthesis and release of thyroid hormones.

ADRENOCORTICOTROPIN Stimulates release of hormones from cortical region of adrenal gland, particularly cortisol and other glucocorticoids.

GROWTH HORMONE Affects growth of skeletal system; enhances entrance of amino acids into cells, and incorporation into proteins; increases release of fatty acids into blood.

PROLACTIN Involved in initiation and maintenance of milk production in females.

ENDORPHINS Peptides in adenohypophysis that have an opiatelike analgesic effect.

PRO-OPIOMELANOCORTIN AND THE PRODUCTION OF ADENOHYPOPHYSEAL HORMONES Pro-opiomelanocortin is polypeptide molecule containing amino acid sequences of an enkephalin, an endorphin, and ACTH; apparently is precursor molecule from which ACTH and other substances can be split.

RELEASE OF ADENOHYPOPHYSEAL HORMONES Brain exercises good deal of control.

HORMONE-RELEASING AND HORMONE-INHIBITING SUBSTANCES Manufactured in brain by neurosecretory cells; travel through circulation by way of hypothalamic-hypophyseal portal veins to adenohypophysis, where they influence release of adenohypophyseal hormones.

NERVOUS SYSTEM EFFECTS ON ADENOHYPOPHYSEAL HORMONE RELEASE Neural activity influences many endocrine functions.

CONDITIONS OF CLINICAL SIGNIFICANCE: PITUITARY GLAND Can be caused by disorders of pituitary gland itself or by difficulties involving releasing or inhibiting substances from brain. p. 483

FEEDBACK CONTROL OF HORMONE RELEASE Negative feedback inhibits hormone release; long and short feedback loops. pp. 484–485

THYROID GLAND Located anterior to upper part of trachea. pp. 485–488

EMBRYONIC DEVELOPMENT AND STRUCTURE Originates as epithelial thickening in pharynx floor; right and left lobes joined across trachea by isthmus; consists of follicles capable of storing hormone.

THE THYROID HORMONES: TRIIODOTHYRONINE AND THYROXINE Iodine attached to tyrosine molecules; involved in wide range of metabolic activities; increase oxygen consumption and heat production; increase utilization of carbohydrate and accelerate liberation of energy; stimulate many aspects of lipid metabolism.

RELEASE OF THYROID HORMONES TSH from pituitary controls synthesis and release of thyroid hormones; thyrotropin-releasing hormone and negative feedback involved in control.

CALCITONIN From parafollicular cells of thyroid; lowers blood-calcium and -phosphate levels.

CONDITIONS OF CLINICAL SIGNIFICANCE: THYROID DISORDERS p. 488

1. Cretinism—Severe hypothyroidism beginning in infancy.
2. Myxedema—Severe hypothyroidism in the adult.
3. Thyrotoxicosis—Excess of thyroid hormones acting on tissues; hyperthyroidism, for example.
4. Goiter—Enlarged thyroid gland.

PARATHYROID GLANDS Embedded on posterior surface of thyroid gland. pp. 488–490

EMBRYONIC DEVELOPMENT AND STRUCTURE
From dorsal halves of third and fourth pairs of pharyngeal pouches; usually two masses of parathyroid tissue on posterior surface of each lateral lobe of thyroid.

PARATHYROID HORMONES AND THEIR EFFECTS
Two or three forms; control calcium and phosphate metabolism.

CONDITIONS OF CLINICAL SIGNIFICANCE: PARATHYROID DISORDERS p. 489

1. Hyperparathyroidism—Bone decalcification; calcification of soft tissue.
2. Hypoparathyroidism—Lowered plasma-calcium level, increased neuromuscular excitability.

ADRENAL GLANDS Located along superior border of each kidney; composed of inner medulla and outer cortex. pp. 490–494

ADRENAL MEDULLA

EMBRYONIC DEVELOPMENT AND STRUCTURE Arises from neural-crest cells; functions in manner similar to postganglionic sympathetic cells.

ADRENAL MEDULLARY HORMONES AND THEIR EFFECTS Catecholamines released by adrenal medulla and sympathetic nervous system.

Epinephrine (Adrenaline) Tends to elevate blood-glucose levels; affects cardiovascular system.

Norepinephrine (Noradrenaline) Affects cardiovascular system; promotes breakdown of fat in adipose tissue.

RELEASE OF THE ADRENAL MEDULLARY HORMONES Controlled by preganglionic neurons from sympathetic division of autonomic nervous system; may be caused by emotional excitement, injury, exercise, low blood-glucose levels.

ADRENAL CORTEX

EMBRYONIC DEVELOPMENT AND STRUCTURE Derived from mesoderm of region that gives rise to gonadal tissue. Arranged in three layers: zona glomerulosa, zona fasciculata, and zona reticularis.

ADRENAL CORTICAL HORMONES AND THEIR EFFECTS

Mineralocorticoids Such as aldosterone; regulate sodium and potassium metabolism.

Glucocorticoids Such as cortisol; inhibit glucose utilization in many peripheral tissues; promote gluconeogenesis and glycogen deposition in liver and elevation of blood glucose; influence protein metabolism.

RELEASE OF THE ADRENAL CORTICAL HORMONES Renin-angiotensin system and concentration of potassium in fluids bathing adrenal glands influence release of mineralocorticoids, as does sodium content of body; ACTH controls release of glucocorticoids; CRH influences ACTH release.

CONDITIONS OF CLINICAL SIGNIFICANCE: ADRENAL DISORDERS p. 494

1. Addison's disease—Hypofunction of adrenal cortex; inadequate aldosterone; impaired ability to conserve sodium and excrete potassium; inadequate glucocorticoids, fasting hypoglycemia, weakness.
2. Cushing's syndrome—Increased blood-glucose levels; increased protein breakdown.
3. Hyperaldosteronism—Potassium depletion; edema.
4. Adrenal tumors—May produce virilizing or feminizing effects.

PANCREAS Both endocrine and exocrine gland. pp. 494–498

EMBRYONIC DEVELOPMENT AND STRUCTURE
Arises through fusion of two outgrowths from duodenum of small intestine; endocrine portion composed of cell clusters called pancreatic islets.

PANCREATIC HORMONES AND THEIR EFFECTS
Alpha$_2$ cells produce glucagon; beta cells produce insulin; alpha$_1$ cells produce somatostatin.

INSULIN Facilitates uptake of glucose and utilization of glucose as an energy source by many cells; promotes glycogen formation in skeletal muscle cells and liver cells; influences lipid and protein metabolism.

GLUCAGON Activities are generally opposite to those of insulin; stimulates breakdown of liver glycogen.

RELEASE OF PANCREATIC HORMONES Chemical, hormonal, and neural control.
1. Increased plasma-glucose levels increase insulin release and decrease glucagon release.
2. Amino acids stimulate secretion of insulin and glucagon.
3. Several gastrointestinal hormones promote insulin release.
4. Parasympathetic neural activity stimulates insulin release, but sympathetic neural activity apparently has dual effect.

CONDITIONS OF CLINICAL SIGNIFICANCE: PANCREATIC DISORDERS Diabetes mellitus caused by relative insulin deficiency. p. 496

GONADS Testes and ovaries produce hormones as well as gametes. p. 498

OTHER ENDOCRINE TISSUES AND HORMONES Digestive tract and placenta are sources of hormonal substances; *pineal body* believed to be a source of endocrine material; *thymus gland* produces thymosin. p. 498

PROSTAGLANDINS Chemical messengers that stimulate uterine contractility; may function in parturition and ovum fertilization, as well as many other actions. p. 498

SELF-QUIZ

1. Endocrine glands generally are ductless glands composed of epithelial cells that release their secretions directly into the target organ. True or False?

2. The effects of some hormones on cellular function appear to be at least partly mediated intracellularly by cyclic AMP. True or False?

3. ADH and oxytocin are manufactured in the: (a) pars nervosa; (b) hypothalamus; (c) adenohypophysis.

4. Gonadotropins are: (a) produced by the ovaries and testes; (b) neurohypophyseal hormones; (c) adenohypophyseal hormones.

5. Surging levels of LH together with FSH are important in: (a) stimulation of the thyroid gland; (b) ovulation; (c) stimulation of the cortical region of the adrenal glands.

6. A hormone secreted by adenohypophysis of the pituitary gland is: (a) thyrotropin; (b) antidiuretic hormone; (c) oxytocin.

7. Match the hormones with their appropriate effects.

ADH	(a) Stimulates the cortical region of the adrenal gland
Oxytocin	
LH	
TSH	(b) Stimulates the reabsorption of water from the urine-forming structures of the kidneys
ACTH	
GH	
	(c) Stimulates growth
	(d) Stimulates the smooth muscle of the uterus
	(e) Stimulates the synthesis of the thyroid hormones
	(f) Acts on the interstitial cells of the testes to stimulate the production of the male sex hormones

8. Specialized neurosecretory cells within the brain manufacture a group of substances known collectively as: (a) gonadotropins; (b) luteinizing hormone; (c) releasing and inhibiting substances.

9. Brain activities of the sort that are responsible for emotions are generally unable to influence hormonally controlled events. True or False?

10. Hormones that contain iodine are produced by which of these glands? (a) pituitary; (b) adrenal medulla; (c) thyroid.

11. The follicular storage form of the thyroid hormones is: (a) thyroglobulin; (b) thyroid-binding globulin; (c) free thyroxine.

12. The thyroid hormones generally travel freely in the circulation. True or False?

13. Match the pituitary gland disorders with their possible symptoms.

Lack of ACTH	(a) Underactivity of thyroid
Gonadotropin deficiency	
TSH deficiency	(b) Adrenal cortical hyperfunction
ACTH overproduction	(c) Menstrual impairment
Growth hormone deficiency	(d) Pituitary dwarfs
	(e) Acromegaly
	(f) Adrenal cortical insufficiency
Growth hormone overproduction	(g) Diabetes insipidus
Antidiuretic hormone deficiency	

14. In the young, hypothyroidism may lead to: (a) elevated body temperature; (b) elevated basal metabolism; (c) abnormal bone development.

15. Enlargement of the thyroid gland is known as: (a) Addison's disease; (b) folliculitis; (c) goiter.

16. When the plasma-calcium level falls, parathyroid hormone secretion: (a) decreases; (b) remains constant; (c) increases.

17. Hypoparathyroidism: (a) increases neuromuscular excitability; (b) raises plasma-calcium levels; (c) leads to calcification of the kidneys.

18. The release of epinephrine is controlled by nerves to the adrenal medulla from the sympathetic division of the autonomic nervous system. True or False?

19. Epinephrine is able to counteract the hypoglycemic effects of insulin by: (a) elevating blood-glucose levels; (b) lowering renal-potassium levels; (c) raising renal-sodium levels.

20. Steroid hormones: (a) are released only by the cortex of the adrenal glands; (b) may attach to carrier proteins within the bloodstream; (c) exert their effects solely by attaching to receptors at cell membranes.

21. Insulin release is stimulated by: (a) low blood-sugar levels; (b) the activity of sympathetic nerves to the liver; (c) gastrointestinal hormones.

22. A major effect of insulin is to facilitate the uptake of glucose by many cells. True or False?

23. Which hormone would most likely be released in increased amounts in response to elevated blood-glucose levels? (a) epinephrine; (b) glucagon; (c) insulin.

LEARNING OBJECTIVES

After completing this chapter, you should be able to:

1. Describe four functions of the blood.

2. Describe three functions of the plasma proteins.

3. Explain the function of each of the formed elements of the blood.

4. Describe the formation and fate of erythrocytes.

5. Describe the structure and function of hemoglobin.

6. Discuss iron metabolism.

7. Explain the control of erythrocyte production.

8. Compare and contrast anemia and polycythemia.

9. Discuss the formation of a platelet plug as a hemostatic mechanism.

10. Explain the process of blood clotting, and distinguish between the extrinsic and intrinsic clotting mechanisms.

11. Explain why clots do not usually form within normal blood vessels, and describe the difficulties that can result from intravascular clotting.

12. Compare and contrast hemophilia and thrombocytopenia.

CHAPTER CONTENTS

PLASMA

FORMED ELEMENTS

CONDITIONS OF CLINICAL SIGNIFICANCE: ERYTHROCYTES

CONDITIONS OF CLINICAL SIGNIFICANCE: LEUKOCYTES

HEMOSTASIS

CONDITIONS OF CLINICAL SIGNIFICANCE: EXCESSIVE BLEEDING

KEY TERMS AND DERIVATIVES

anemia (*an* = without; *hem* = blood) a condition characterized either by a decreased number of erythrocytes in the blood or a decreased concentration of hemoglobin

erythrocyte (*erythr* = red) red blood cell

erythropoiesis (*erythr* = red; *poiet* = making) the formation process of red blood cells

hematocrit (*hem* = blood; *crit* = separate) the percentage of erythrocytes to total blood volume

hemostasis (*hem* = blood; *stasis* = a halting) arrest of bleeding or stopping or slowing of flow of blood through a vessel or a part

The Circulatory System:
The Blood

The cells and tissues of the body are linked by the blood vessels and the blood that flows within them. The blood transports materials throughout the body. It carries oxygen and food materials to and carbon dioxide away from the cells. It also transports cellular products such as hormones between cells. In addition, the blood and its components help maintain homeostasis by buffering the tissues against extremes of pH, by aiding in the regulation of body temperature, and by protecting the tissues from toxic foreign materials or organisms.

The volume of blood in both lean males and lean females varies almost directly with body weight and averages approximately 79 milliliters per kilogram of body weight ($\pm10\%$). However, fat tissue has little vascular volume, and the volume of blood per unit of body weight declines as the proportion of adipose tissue in the body increases. Since the average female has a greater fat/lean-tissue ratio than the average male, the average female tends to have a lower blood volume per kilogram of body weight than the average male (about 67 ml/kg body weight for average females and about 75 ml/kg body weight for average males). Because of this difference, as well as differences in body size, the general range of total blood volume is 4 to 5 liters in females and 5 to 6 liters in males.

The blood consists of both a liquid component, called *plasma*, and a non-liquid portion whose structures are collectively called *formed elements*.

PLASMA

The liquid portion of the blood, the **plasma,** is approximately 91% water. The portion of the plasma that is not water consists of various dissolved or colloidal materials (Figure 17.1; Table 17.1). Hormones and other cellular products are present in the plasma, as are metabolic end products such as urea. The plasma contains plasma proteins, including various albumins, fibrinogen (which is involved in blood clotting), and globulins (some of which act as antibodies in immune responses, others of which serve as transport molecules). The plasma proteins act as buffers that help stabilize the pH of the internal environment, and they contribute importantly to the osmotic pressure and viscosity of the plasma. Various ions, including sodium (Na^+), chloride (Cl^-), and bicarbonate (HCO_3^-), are present within the plasma. These ions also contribute to plasma osmotic pressure, which averages about 300 milliosmoles per liter. The plasma contains food materials such as carbohydrates (for example, glucose), amino acids, and lipids, as well as gases such as oxygen, nitrogen, and carbon dioxide.

FORMED ELEMENTS

The portion of the blood that is not plasma consists of **formed elements** (Figure 17.1; Table 17.2) The formed elements include erythrocytes, or red blood cells; various types of leukocytes, or white blood cells; and platelets.

Hemocytoblast

Basophilic erythroblasts

Polychromatophilic erythroblast

Normoblast

Reticulocyte

Erythrocytes

505

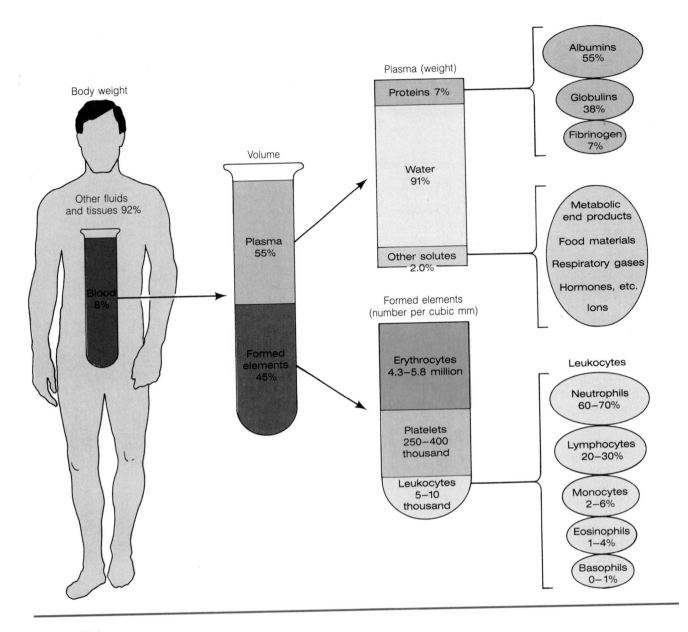

Figure 17.1
Components of blood.
The values indicated are
approximate, and in
many cases there are
individual variations and/
or variations by sex.

hee-mo-sigh'-to-blasts

Hemopoiesis

The formed elements of the blood are produced by a process called **hemopoiesis.** Prior to birth, formed elements are produced in a number of tissues, including the liver, spleen, bone marrow, thymus gland, and lymph nodes. The exact source of formed elements following birth has been a subject of disagreement over the years. However, it is now generally believed that all the formed elements of the blood are derived from primitive stem cells called **hemocytoblasts** that are present in **red bone marrow** (*myeloid tissue*).

The cellular composition of bone marrow varies in different regions of the skeleton. It also varies with the age of the individual. In the fetus, almost all the bones contain red bone marrow (the red color indicates that the marrow is capable of producing the formed elements of the blood). Following birth, in most areas of bone marrow the number of active hemocytoblasts decreases, and they are replaced with fat cells. The abundance of fat cells causes the color of the marrow to change from red to yellow. Therefore, bone marrow in which this replacement has occurred is referred to as *yellow bone marrow.* Yellow bone marrow functions primarily as a site for fat storage and generally is not actively producing formed elements of the blood. In the adult, most bones

Table 17.1   Principal Constituents of Plasma*

Constituent	Approximate Amount
Water	91%
Plasma proteins	
Albumins	3.2 to 5.0 g/100 ml
Fibrinogen	0.20 to 0.45 g/100 ml
Globulins	
Alpha	0.40 to 0.98 g/100 ml
Beta	0.56 to 1.06 g/100 ml
Gamma	0.44 to 1.04 g/100 ml
Glucose	61 to 130 mg/100 ml
Cholesterol	128 to 347 mg/100 ml
Bilirubin (total, indirect reacting)	0 to 1.1 mg/100 ml
Urea	13.8 to 39.8 mg/100 ml
Sodium	310 to 356 mg/100 ml
Potassium	12 to 21 mg/100 ml
Calcium (total)	8.2 to 11.6 mg/100 ml
Iron	0.04 to 0.21 mg/100 ml
Chloride	355 to 381 mg/100 ml

*Plasma also contains a wide variety of other substances too numerous to list here, including vitamins, hormones, enzymes, and metabolic end products.

contain yellow bone marrow. Red bone marrow is present only in certain bones, such as the ribs, the sternum, the vertebrae, and the pelvis. The production of the formed elements of the blood, therefore, normally occurs only in these red-bone-marrow-containing regions.

Red bone marrow consists primarily of precursors of the formed elements of the blood and some fat cells packed within a fibrous meshwork of reticular connective tissue. It is well supplied with blood from the nutrient artery of the bone. Capillaries from this artery form a system of thin-walled spaces called *sinusoids*, which allow for the slow flow of blood throughout the marrow. The lining of the sinusoids permits formed elements that are produced in the marrow to pass through the walls of the sinusoids and enter the blood.

The hemocytoblasts in red bone marrow are immature cells capable of developing into five types of cells that, in turn, give rise to the formed elements of the blood (Figure 17.2). These five types of cells are:

F17.2

1. **Proerythroblasts,** which form *erythrocytes*.

2. **Myeloblasts,** which form three types of leukocytes collectively called *granulocytes*.

3. **Lymphoblasts,** which form a type of leukocyte called *lymphocytes*.

4. **Monoblasts,** which form a type of leukocyte called *monocytes*.

5. **Megakaryoblasts,** which form cells called *megakaryocytes* that give rise to cell fragments called *platelets*.

Erythrocytes

Erythrocytes, or **red blood cells,** are small, circular, biconcave discs approximately 7.5 microns in diameter that have no nuclei (Figure 17.3; Table 17.2). Their principal function is to transport oxygen and carbon dioxide.

e-rith'-ro-sites
F17.3,
Table 17.2

Erythrocytes are the most numerous of the formed elements contained in the blood, and they contribute importantly to total blood viscosity (which is normally 3.5 to 5.5 times that of water). Although their numbers are quite variable, a cubic millimeter of peripheral venous blood usually contains about 5.1 to 5.8 million erythrocytes in males and about 4.3 to 5.2 million erythrocytes in females. The proportion of erythrocytes in a sample of blood is called

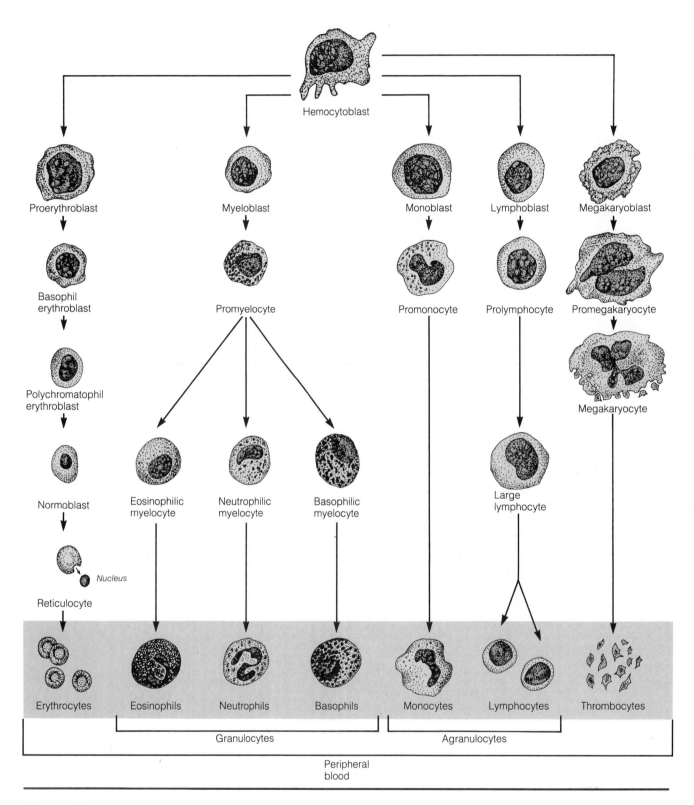

Figure 17.2
Origin and development of the formed elements of the blood.

the **hematocrit.** The hematocrit is determined by first centrifuging a blood sample in a hematocrit tube until the erythrocytes are packed at the bottom of the tube and then measuring the ratio or percentage of the packed erythrocyte volume to the total sample volume. Although there is considerable variability, normal hematocrits of samples of peripheral venous blood average about 46.2% for males and 40.6% for females. The hematocrit of arterial blood is generally slightly lower than that of venous blood, and the hematocrit of

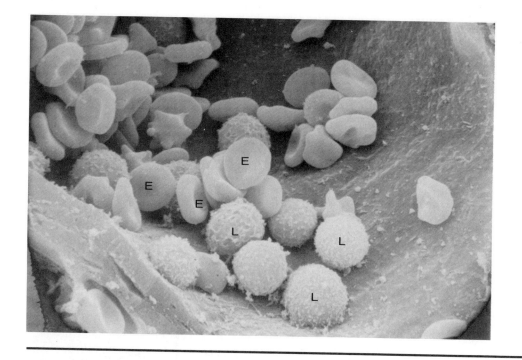

Figure 17.3

Formed elements of the blood within a blood vessel. E: erythrocytes, or red blood cells. L: leukocytes, or white blood cells (×2480). (From *Tissues and Organs: A Text-Atlas of Scanning Electron Microscopy* by Richard G. Kessel and Randy H. Kardon. W.H. Freeman and Company. Copyright © 1979)

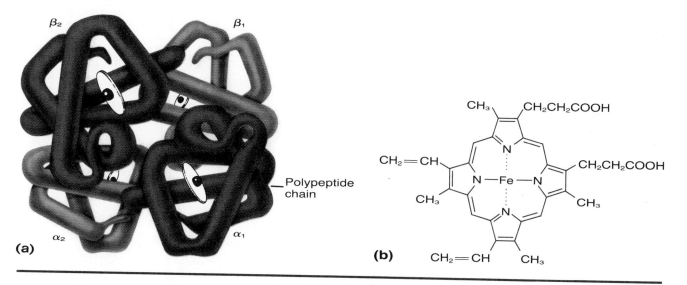

Figure 17.4

Hemoglobin. **(a)** A hemoglobin molecule, showing its four polypeptide chains (2 α chains and 2 β chains). The structure in the center of each chain represents an iron-containing heme group. **(b)** Structure of a single iron-containing heme group.

blood in the very small vessels of the body is considerably lower than that of blood in the large arteries and veins.

Hemoglobin

A substance called **hemoglobin** is essential to the ability of erythrocytes to transport oxygen and carbon dioxide, and a single erythrocyte contains up to 300 million hemoglobin molecules. Males generally have more hemoglobin (14 to 18 grams per 100 ml of peripheral venous blood) than females (12 to 16 grams per 100 ml of peripheral venous blood).

A molecule of hemoglobin is composed of the protein *globin*, combined with four nonprotein groups called *hemes* (Figure 17.4). The globin protein consists of four polypeptide chains, each of which has a heme group bound to it. Each heme group contains an iron atom (Fe^{+2}) that can combine reversibly with one molecule of oxygen. Therefore, a hemoglobin molecule can potentially associate with four oxygen molecules. When hemoglobin is combined with oxygen, it is called *oxyhemoglobin*; when it is not carrying oxygen, it is

F17.4

FRONTIERS IN HEALTH
Artificial Blood

In 1966, Dr. Leland C. Clark of the University of Cincinnati's College of Medicine stunned the medical world by immersing a mouse in a bubbling solution of fluorocarbon. The mouse's lungs filled with the oxygen-rich liquid, but nothing traumatic happened. The mouse's heart kept beating, and the animal showed no signs of anoxia (lack of oxygen). After a while, the mouse was removed from the solution, apparently none the worse for the experience.

What Dr. Clark and his colleague, Dr. Frank Gollan, had done was develop the first prototype of an artificial blood. But was there a need for artificial blood?

Consider the fate of an accident victim in rural America. Miles from a hospital and the closest blood bank, a victim of a severe accident often has little chance of surviving long enough to reach the emergency room. Drs. Clark and Gollan began to work on an artificial blood—or more correctly, a blood substitute—that could be readily available, even in rural areas, and would have the potential of saving thousands of lives each year.

The result of their work was a fluorocarbon emulsion called fluosol. Fluosol has now been approved for use in Canada, Holland, and Italy. The milky white solution contains two fluorocarbons, a number of salts, and water. It carries twice as much oxygen as blood does and contains fine particles one-seventieth the size of erythrocytes. Because the particles are so small, fluosol can pass through narrowed arteries that might not allow the passage of erythrocytes, and victims of stroke and heart attack may benefit from this form of oxygen delivery.

The Japanese have performed much of the pioneering work on human subjects. The first volunteers to receive the blood substitute were Japanese, and the first clinical trial also took place in Japan. Reported in 1980, this test was performed by Dr. Kenji Honda and his colleagues at the Fukushima Medical College. A 65-year-old patient suffering from a massive bleeding ulcer was rushed to surgery to have his stomach removed, but the blood bank could not supply the required blood on time. The patient's blood pressure fell dangerously low. To prevent the patient's death, Dr. Honda administered some fluosol; within minutes the blood pressure rose. The patient's life had been saved. Two hours later, when whole blood became available, surgeons completed the removal of the damaged stomach.

Since that historic event, fluosol has been used in dozens of patients requiring emergency surgery when no whole blood was available or in anemic patients who refused conventional blood transfusions before more routine surgery. There have been very few side effects, and it is possible that the Food and Drug Administration will approve fluosol for widespread use in the United States.

Fluosol's applications may extend beyond emergency surgery or surgery on anemic patients. Consider the benefits of using it to sustain the tissues of a brain-dead person, whose organs can then be made available for transplant in various distant hospitals. Consider the benefits of using artificial blood during a war. Finally, consider the actual case of a patient who had fallen into a deep coma caused by infectious hepatitis, which is a disease that affects the liver. The patient's diseased liver poured toxins into the blood, and the toxins poisoned the liver cells, thus creating a vicious cycle of liver destruction. Dr. Gerald Klebanhoff of Lackland Air Force Base Medical Center used artificial blood to break this cycle. He drained the patient's blood entirely and replaced it with artificial blood. This removed the toxins and allowed the beleaguered liver to begin to recover, while life was sustained by a blood substitute. After a short time, the artificial blood was drained and replaced with whole blood. The comatose patient awoke in the recovery room a few hours after the procedure, attesting to the success of the total blood replacement.

Dr. Anthony Hunt of the University of California, San Francisco, and Dr. Ronald Burnette of the University of Wisconsin, Madison, have recently perfected a new method of producing artificial blood. They have devised a way to make artificial miniature red blood corpuscles, which they call neohemocytes. Neohemocytes, which are microscopic spheres of hemoglobin surrounded by lipids, are capable of carrying oxygen and are proving to be a substitute for erythrocytes.

One of the exciting findings of this research was that neohemocytes are more stable than erythrocytes. They can be stored in refrigerators for two months, which is five weeks longer than the body's erythrocytes can be stored. Because they are smaller than erythrocytes, neohemocytes can pass through narrowed blood vessels that might not allow the passage of erythrocytes. Victims of heart attack and stroke may therefore have faster recoveries and less tissue damage from transfusions with this promising erythrocyte substitute.

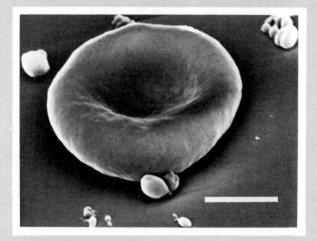

A scanning electron micrograph of a single human erythrocyte surrounded by several neohemocytes. Each neohemocyte encapsulates purified hemoglobin. (The bar is 2.0 microns long.) Courtesy of Dr. C. A. Hunt, University of California, San Francisco.

called *reduced hemoglobin*. Under normal circumstances, a liter of blood traveling to the tissues contains approximately 198 ml of oxygen. Of this amount, about 195 ml are associated with hemoglobin molecules within erythrocytes. The remaining 3 ml are in physical solution dissolved in the plasma. Thus, erythrocytes and hemoglobin account for almost all of the blood's ability to transport oxygen.

Hemoglobin can also combine with carbon dioxide and thus aid in the transport of this substance. However, in contrast to oxygen, which is carried in association with the iron of the heme groups of hemoglobin, carbon dioxide is carried in reversible association with the protein portion of the hemoglobin molecule. The transport of oxygen and carbon dioxide by hemoglobin is discussed further in Chapter 22.

Iron

The production of hemoglobin requires iron, and the body normally contains about 4 grams of iron. Approximately 65% of the body's iron is found in hemoglobin, and about 15 to 30% is stored in the liver, spleen, bone marrow, and elsewhere, mainly in the form of intracellular iron–protein complexes called *ferritin* and *hemosiderin*. In the blood, iron forms a loose combination with a beta globulin called *transferrin*, and it is in this form that iron is transported in the plasma. Iron can be released from transferrin for use or storage.

Small amounts of iron are lost each day through the feces, urine, sweat, and cells sloughed from the skin. In women, additional iron is lost as a result of menstrual bleeding. The average daily loss of iron is about 0.9 mg in males and about 1.7 mg in females.

Iron losses must be replaced if iron homeostasis is to be maintained. Ingested iron is absorbed into the body by active processes, primarily in the upper portion of the small intestine, and total body iron is largely regulated by alterations in the intestinal absorption rate. When the body stores of iron are high, the intestinal absorption rate of iron is low; but when iron stores are depleted, the rate of iron absorption increases. In general, the rate of iron absorption is slow, and even at the maximum absorption rate only a few milligrams of iron can be absorbed per day.

Production of Erythrocytes

During erythrocyte production, some of the primitive stem cells (hemocytoblasts) in the red bone marrow give rise to cells called *proerythroblasts* (Figure 17.2). The proerythroblasts are cells that have just begun to activate the biochemical machinery needed for the synthesis of hemoglobin. As hemoglobin synthesis occurs, the proerythroblasts differentiate into cells called *basophil erythroblasts*. As hemoglobin synthesis continues, the basophil erythroblasts give rise to cells called *polychromatophil erythroblasts*, which in turn differentiate into cells called *normoblasts* (*acidophil erythroblasts*). When the normoblast cytoplasm has attained a hemoglobin concentration of about 34%, the nucleus is pinched off, enclosed in a portion of the cell membrane and a thin layer of cytoplasm. The nonnucleated cells that result are called *reticulocytes* (*polychromatophil erythrocytes*). Reticulocytes are essentially young erythrocytes that contain small numbers of residual ribosomes. It is these cells that are usually released from the bone marrow into the blood. Reticulocytes generally become mature erythrocytes within one or two days after their release from the bone marrow, and the normal proportion of reticulocytes in the blood is less than 0.5 to 1.0%.

The process of erythrocyte formation is called **erythropoiesis.** During erythropoiesis, the various cells continue to divide through the normoblast stage of development so that greater and greater numbers of cells are formed. Occasionally, when erythrocytes are being manufactured very rapidly, nucleated cells appear in the blood, but this is not a usual occurrence.

Among the substances required for the production of erythrocytes is vitamin B_{12}. Vitamin B_{12} is required for DNA formation and, as a result, is necessary for nuclear maturation and division. Folic acid, another vitamin, is also required for DNA formation and thus for erythrocyte production.

F17.2

e-rith-ro-poi-ee'-sis

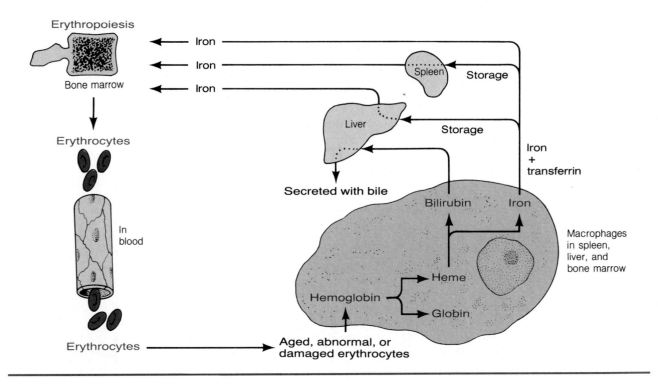

Figure 17.5
Aged, abnormal, or damaged erythrocytes are disposed of by macrophages. The iron that is released may be either stored or reused in the synthesis of new hemoglobin.

Control of Erythrocyte Production

A substance called **erythropoietin** stimulates erythrocyte production. Erythropoietin is a glycoprotein produced in the kidneys and possibly elsewhere. The kidneys are believed to release an enzyme called renal erythropoietic factor (erythrogenin) into the blood. Renal erythropoietic factor is thought then to act on a plasma protein precursor to form erythropoietin.

The production of erythropoietin, and thus erythrocyte production, is regulated by a negative feedback mechanism that is sensitive to the amount of oxygen delivered to the tissues. An inadequate supply of oxygen—particularly to the kidneys—leads to a greater production of erythropoietin and thereby to increased erythrocyte production. Conversely, an increased oxygen supply to the tissues leads to a decrease in erythropoietin levels and, as a result, to a decrease in erythropoiesis. The male sex hormone testosterone enhances erythropoietin production, and the estrogenic female sex hormones tend to depress it. This could account at least in part for the greater numbers of erythrocytes and higher levels of hemoglobin in males than in females.

Fate of Erythrocytes

F17.5   The average life of an erythrocyte is approximately 120 days in males and 109 days in females. Aged, abnormal, or damaged erythrocytes are disposed of by phagocytic cells called *macrophages* (Figure 17.5). These cells are found in the spleen, liver, bone marrow, and other tissues. The spleen is particularly important in the destruction of pathologic or defective erythrocytes, and the number of abnormal erythrocytes circulating in the blood increases considerably if the spleen is removed.

F17.5   During erythrocyte destruction, the macrophages degrade hemoglobin (Figure 17.5). The globin protein is broken down into its component amino acids, and the amino acids can be used in the synthesis of new proteins. The iron is liberated from the nonprotein heme, and the remainder of the heme is *bil-i-ver'-din*   converted to *biliverdin*. Biliverdin is in turn converted to *bilirubin*, which enters *bil-i-roo'-bin*   the blood and binds to albumin. Bilirubin is eventually taken up by the liver, conjugated to glucuronic acid, and secreted in the bile. The liberated iron also enters the blood, where it combines with transferrin. The iron can be used in the synthesis of new hemoglobin during erythropoiesis in the bone marrow or it can be stored.

CONDITIONS OF CLINICAL SIGNIFICANCE
Erythrocytes

Anemia

Anemia is a condition characterized either by a decreased number of erythrocytes in the blood or by a decreased concentration of hemoglobin. Whatever the cause, anemia decreases the blood's ability to transport oxygen to the tissues. Because tissues cannot function at optimum levels without adequate supplies of oxygen, anemia is frequently associated with listlessness, fatigue, and a lack of energy. Anemia can result from a number of conditions, several of which are discussed in the following sections.

HEMORRHAGIC ANEMIA *Hemorrhagic anemia* is the result of the loss of substantial quantities of blood. When blood is lost as the result of such occurrences as large wounds, stomach ulcers, or excessive menstrual bleeding, the lost volume is rather quickly replaced by fluid from the tissue spaces. The lost erythrocytes, however, are replaced by the slower processes of erythropoiesis. Until erythrocyte replacement is complete (which may take several weeks), there are fewer circulating erythrocytes than normal. Repeated or chronic hemorrhage can produce severe anemia, even if there is an increase in erythropoiesis.

IRON DEFICIENCY ANEMIA *Iron deficiency anemia* is due to an excessive loss, deficient intake, or poor absorption of iron. In iron deficiency anemia, the erythrocytes produced are smaller than usual, and they contain less hemoglobin than normal.

PERNICIOUS ANEMIA *Pernicious anemia* is the result of an inability to absorb adequate amounts of vitamin B_{12} from the digestive tract. Vitamin B_{12}, which is required for the normal maturation of erythrocytes, is found in such foods as liver, kidney, milk, eggs, cheese, and meat. The absorption of vitamin B_{12} requires the presence of a glycoprotein substance called *intrinsic factor*, which is produced by cells in the epithelial lining of the stomach. Individuals who suffer from pernicious anemia lack sufficient quantities of functional intrinsic factor and cannot adequately absorb vitamin B_{12}. As a result, they produce fewer erythrocytes than normal, and those that are produced are larger and more fragile than normal. Pernicious anemia can be treated by injections of vitamin B_{12}, thus bypassing the absorption problem.

SICKLE CELL ANEMIA *Sickle cell anemia* is the result of the production of an abnormal hemoglobin (hemoglobin S; HbS) due to a genetic defect. In this condition, the globin portion of the molecule is abnormal. When the abnormal hemoglobin molecules are exposed to low concentrations of oxygen, they form fibrous precipitates within the erythrocytes, distorting them into the sickle shape characteristic of the disease. The misshapen erythrocytes are very fragile and often rupture as they pass through the capillaries, and especially through the spleen. Sickled cells can also become trapped in small blood vessels, impeding blood flow. In its severest form sickle cell anemia can be fatal.

APLASTIC ANEMIA *Aplastic anemia* is the result of an inadequate production of erythrocytes due to the inhibition or destruction of the red bone marrow. Aplastic anemia can be caused by radiation, various toxins, and certain medications.

Polycythemia

Polycythemia is a condition in which there is a net increase in the total circulating erythrocyte mass of the body. There are several types of polycythemia.

PRIMARY POLYCYTHEMIA *Primary polycythemia* (also called *polycythemia vera* or *erythremia*) occurs when excess erythrocytes are produced as a result of tumorous abnormalities of the tissues that produce blood cells. Often, excess white blood cells and platelets are also produced. In primary polycythemia there may be 8 to 9 million and occasionally 11 million erythrocytes per cubic millimeter of blood, and the hematocrit may be as high as 70 to 80%. In addition, the total blood volume sometimes increases to as much as twice normal. The entire vascular system can become markedly engorged with blood, and circulation times for blood throughout the body can increase up to twice the normal value. The increased numbers of erythrocytes can increase the viscosity of the blood to as much as five times normal. Capillaries can become plugged by the very viscous blood, and the flow of blood through the vessels tends to be extremely sluggish.

SECONDARY POLYCYTHEMIA *Secondary polycythemia* is caused by either appropriate or inappropriate increases in the production of erythropoietin that result in an increased production of erythrocytes. In secondary polycythemia there may be 6 to 8 million and occasionally 9 million erythrocytes per cubic millimeter of blood. A type of secondary polycythemia in which the production of erythropoietin increases appropriately is called *physiologic polycythemia*. Physiologic polycythemia occurs in individuals living at high altitudes (4275 to 5200 meters), where oxygen availability is less than at sea level. Such people may have 6 to 8 million erythrocytes per cubic millimeter of blood.

Blood Groups

An *antigen* is a substance that has the capacity to elicit a specific immune response, such as the production of a specialized protein called an *antibody*, and an antibody can react with the specific antigen that stimulated its production. The surfaces of erythrocytes contain antigens that can react with particular antibodies. These reactions form the basis of classifying blood into various groups—most commonly, groups known as type A, type B, type AB, or type O. Because these classifications are based on antigen–antibody reactions, they are considered in detail with the specific immune responses in Chapter 21.

Leukocytes

Leukocytes, or **white blood cells,** are formed elements that defend the body against invasion by foreign organisms or chemicals and remove debris that results from dead or injured cells. Leukocytes act primarily in the loose connective tissues; those within the blood are mainly being transported by the circulation. Leukocytes are present within the blood in much smaller numbers than erythrocytes: a cubic millimeter of blood usually contains between 5000 and 10,000 leukocytes. Many leukocytes are found in lymphoid tissues, such as the thymus, lymph nodes, spleen, and lymphoid areas in the linings of the gastrointestinal tract. There are two major classes of leukocytes: granulocytes and agranulocytes (Figure 17.2; Table 17.2).

F17.2,
Table 17.2

Granulocytes

Granulocytes have clearly evident granules in their cytoplasm. They are formed in the red bone marrow from stem cells called *myeloblasts*. Three types of granulocytes are distinguished, based on their reactions to certain stains. They are neutrophils, eosinophils, and basophils.

NEUTROPHILS **Neutrophils** possess small cytoplasmic granules that appear light pink to blue-black when stained with Wright's stain. (The granules of neutrophils are lysosomes.) Because they typically have a nucleus that varies in shape and consists of two or more lobes connected by narrow strands, neutrophils are also referred to as *polymorphonuclear leukocytes*. However, this term is also sometimes used to refer to all three types of granulocytes, since the shape of the nucleus varies somewhat in all of them.

Neutrophils are the most abundant type of leukocyte, comprising approximately 60–70% of the total leukocyte count. Neutrophils are phagocytic cells that are capable of ameboid movement. They are able to leave the blood vessels and enter the tissues, where they protect the body by ingesting bacteria and other foreign substances.

ee-o-sin'-o-fils   EOSINOPHILS **Eosinophils** possess coarse cytoplasmic granules that appear reddish-orange when stained with Wright's stain. (As is the case for neutrophils, the granules of eosinophils are also lysosomes.) Eosinophils are capable of both phagocytosis and ameboid movement, and they are believed to ingest and destroy antigen–antibody complexes (see Chapter 21). The number of eosinophils in the blood and in the tissues increases during certain parasitic infections and in conditions involving allergic hypersensitivity (for example, asthma and hay fever).

bay'-so-fils   BASOPHILS **Basophils** possess relatively large cytoplasmic granules that appear reddish-purple to blue-black when stained with Wright's stain. The granules of basophils contain a number of chemical substances, including the substances *histamine* and *heparin*. Basophils are active in phagocytosis, but their exact function is still being investigated. Basophils are believed to be functionally similar to cells called *mast cells*, which occur in connective tissue and in the extracellular spaces near blood vessels, particularly in the lungs. Mast cells have granules much like those in basophils, and they also contain histamine and heparin. Histamine causes vascular dilation and increased blood vessel permeability in inflammation and contributes to allergic responses. Heparin can prevent blood coagulation and can also enhance the removal of fat particles from the blood after a fatty meal.

Agranulocytes

Certain leukocytes, called **agranulocytes** or *nongranular leukocytes*, do not have prominent granules in their cytoplasm. There are two types of agranulocytes: monocytes and lymphocytes.

Table 17.2 Formed Elements of the Blood

Formed Element	Approximate Diameter (microns)	Approximate Abundance (per mm³)	Function
ERYTHROCYTES	7.5	Males: 5.1 to 5.8 million Females: 4.3 to 5.2 million	Transport oxygen and aid in the transport of carbon dioxide
PLATELETS	2.5	250,000 to 400,000	Involved in the processes of hemostasis and blood coagulation
LEUKOCYTES **Granulocytes** Neutrophils	12 to 14	3000 to 7000 (60 to 70 percent of total leukocytes)	Phagocytic cells that are capable of ameboid movement
Eosinophils	9	50 to 400 (1 to 4 percent of total leukocytes)	Phagocytic cells that are believed to destroy antigen–antibody complexes
Basophils	12	0 to 50 (0 to 1 percent of total leukocytes)	Believed to release chemicals such as histamine and heparin
Agranulocytes Monocytes	20 to 25	100 to 600 (2 to 6 percent of total leukocytes)	Develop into large phagocytic cells called macrophages within the tissue spaces
Lymphocytes	9 (small) 12 to 14 (large)	1000 to 3000 (20 to 30 percent of total leukocytes)	Involved in specific immune responses, including antibody production

MONOCYTES **Monocytes** are formed in the red bone marrow from *monoblasts*. Monocytes are capable of ameboid movement, and they can leave the blood vessels and enter the tissues, where they develop into large, phagocytic cells called *macrophages* that can ingest bacteria and other foreign substances.

LYMPHOCYTES Only a few of the total number of **lymphocytes** are found in the blood; most are lodged in the lymphoid tissues. Lymphocytes, which are formed from *lymphoblasts*, are important in the body's specific immune responses, including antibody production. On the basis of their size, lymphocytes are classified as small or large. Their development and specific functions are discussed in Chapter 21.

Differential Count

It is important for diagnostic purposes to be able to estimate the relative abundance of each type of leukocyte in the blood. This can be done by a procedure called a **differential count.** A differential count involves staining a blood smear and then identifying and counting the leukocytes under a microscope (Figure 17.6). A differential count can also be obtained by placing a blood sample in a specialized machine that automatically identifies and tabulates the various types of leukocytes. A differential count provides a percentage figure for each type of leukocyte. The percentages of the different types of leukocytes present in the blood change in certain diseases, and a differential count can aid in their diagnosis.

F17.6

Platelets

Platelets are formed elements that are small cytoplasmic fragments about 2.5 microns in diameter having no nuclei and containing many granules. Platelets are formed in the red bone marrow as pinched-off portions of large cells called

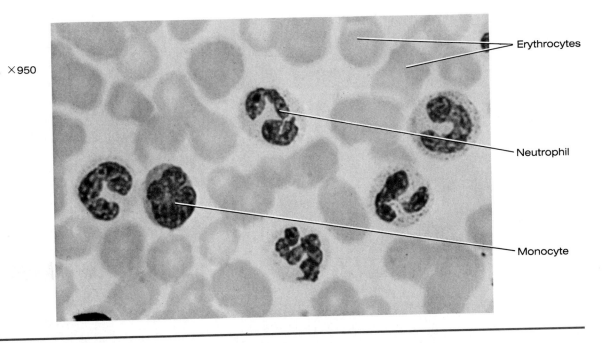

×950

Figure 17.6
Stained leukocytes as they appear during a differential blood count.

megakaryocytes. There are about 250,000 to 400,000 platelets per cubic millimeter of blood. Platelets are involved in blood clotting, and they have also been implicated in other hemostatic processes.

HEMOSTASIS

Hemostasis is the arrest of bleeding or the stopping or slowing of the flow of blood through a vessel or a part. When a blood vessel is severed or damaged, hemostasis is achieved by mechanisms that include vascular constriction, the formation of a platelet plug, and blood clotting.

Vascular Constriction

When a blood vessel is cut or ruptured, one of the first noticeable responses is the constriction of the vessel. This vascular constriction decreases or stops blood flow and allows time for other hemostatic activities to occur. The mechanisms responsible for vascular constriction include a contraction of vascular smooth muscle in direct response to injury and neurally mediated vascular constriction reflexes that occur in response to pain. In general, the greater the damage to a vessel, the greater the degree of constriction. Thus, a vessel ruptured by crushing usually bleeds less than a cleanly cut vessel.

Formation of a Platelet Plug

Within an undamaged vessel, platelets generally circulate freely and do not stick to the normal endothelial cells that line the vessel. However, in an injured vessel where the deeper lying subendothelial tissues are exposed, platelets adhere to the underlying connective tissues—specifically, collagen—and release several chemicals, among which is adenosine diphosphate (ADP). The ADP causes the surfaces of nearby platelets to become "sticky," and as a result, still more platelets adhere, forming a platelet plug or clump. The formation of a platelet plug may slow or completely stop bleeding from a damaged blood vessel that has a relatively small hole in it; but if the hole is large, blood clotting may be necessary to stop the bleeding. The ability of platelets to plug small vascular holes is important in closing minute tears that occur many times daily in the capillaries and other small vessels, and a person with an insufficient number of platelets often develops numerous small hemorrhagic areas under the skin and throughout the internal tissues. In addition to ADP,

Leukemia

Leukemia is a cancerous condition in which an uncontrolled proliferation of leukocytes leads to a diffuse and almost total replacement of the red bone marrow with leukemic cells. If the leukemic cells are from the granulocyte or, in some cases, monocyte cell lines, the leukemia is referred to as *myelogenous leukemia*. If the leukemic cells are from the lymphocyte cell line, the leukemia is called *lymphocytic leukemia*. Leukemic cells often replace the cells that form erythrocytes, and anemia results. In addition, there is frequently a decrease in blood platelets, which are also formed in the red bone marrow. Since platelets are involved in blood clotting and other hemostatic processes, leukemia may be accompanied by bleeding and hemorrhage. In fact, one of the causes of death in leukemia is internal hemorrhage, especially cerebral hemorrhage. Many of the leukocytes that are formed in the red bone marrow in leukemia are immature or abnormal. These abnormal leukocytes are unable to defend the body adequately against invasion by foreign organisms, and leukemia victims can also die from infection. Leukemic cells can leave the bone marrow and infiltrate other tissues or organs in the body, where they can damage or interfere with the functions of the tissues or organs.

Infectious Mononucleosis

Infectious mononucleosis, which occurs mostly in children and young adults, is caused by a virus called the *Epstein-Barr virus*. The disease is characterized by an increase in the relative and absolute numbers of lymphocytes in the blood, and many of the lymphocytes are atypical. The symptoms of infectious mononucleosis include fatigue, sore throat, and slight fever. Affected individuals are usually watched carefully for complications during the course of the disease, which may last several weeks. Recovery is usually complete, without any ill effects.

other chemicals released by platelets during the formation of a platelet plug facilitate vasoconstriction (for example, serotonin) and contribute to blood clotting.

Blood Clotting

Blood clotting, or **coagulation,** is a complex process that involves a number of different factors, many of which are present in the plasma (Table 17.3). The formation of a blood clot requires the conversion of a soluble protein called *fibrinogen* (normally present in the plasm) into an insoluble, threadlike polymer called *fibrin* (Figure 17.7). The insoluble fibrin threads form a network that entraps blood cells, platelets, and plasma to form the clot itself (Figure 17.8). The fibrin threads can adhere to damaged blood vessels and thus anchor the clot in place.

Table 17.3

F17.7

F17.8

The conversion of fibrinogen to fibrin during clot formation normally requires the enzymatic action of a factor called *thrombin*, which is produced from the inactive plasma protein precursor *prothrombin*. Minute amounts of thrombin may be continually produced from prothrombin, but this thrombin is generally inactivated or destroyed relatively rapidly so that its concentration in the blood does not rise high enough to promote clotting. During cell, tissue, or platelet disruption, however, the formation of thrombin increases considerably, and its concentration may rise high enough to lead to clot formation.

The exact sequence of events that leads to the formation of thrombin during clotting is currently an area of much study and debate. However, researchers believe that the production of thrombin from prothrombin requires a *prothrombin activator (prothrombin-converting factor)*. The production of prothrombin activator occurs either as a result of the action of an extrinsic mechanism that is activated when blood vessels are ruptured and tissues are damaged, or as the result of the action of an intrinsic mechanism that is present within the blood itself and that is activated when the blood is traumatized.

Extrinsic Mechanism

The extrinsic mechanism is activated when a substance called *tissue thromboplastin* (actually a complex mixture of lipoproteins that contains one or more phospholipid substances) is released from damaged tissues. Tissue thromboplastin, together with factor VII from the plasma and calcium ions, activates

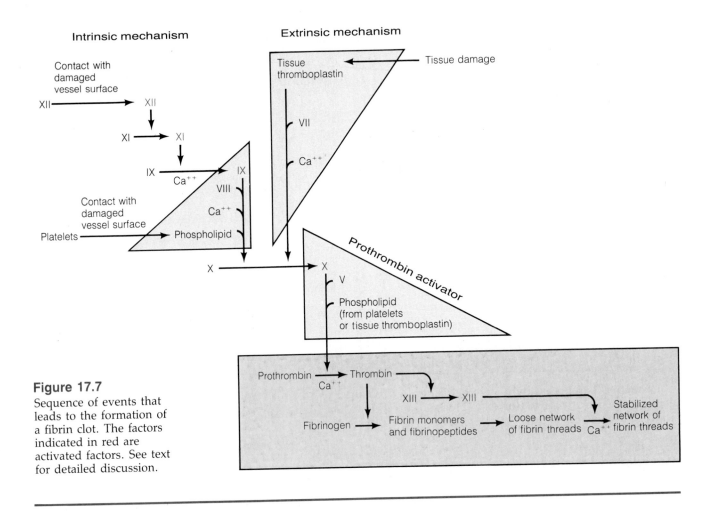

Figure 17.7
Sequence of events that leads to the formation of a fibrin clot. The factors indicated in red are activated factors. See text for detailed discussion.

Table 17.3   factor X (Table 17.3). Activated factor X, together with the effective form of factor V and a phospholipid substance, forms prothrombin activator. Prothrombin activator acts enzymatically to catalyze the formation of thrombin from prothrombin. Thrombin, in turn, converts fibrinogen into fibrin monomers and fibrinopeptides. The fibrin monomers associate (polymerize) with one another, forming a loose network of fibrin threads. Thrombin also activates factor XIII, which augments the bonding between fibrin monomers, thereby strengthening and stabilizing the fibrin network.

Intrinsic Mechanism

The intrinsic mechanism is triggered when inactive factor XII in the plasma is activated by contact with a damaged vessel surface, perhaps by contact with underlying collagen fibers. Activated factor XII activates factor XI, and activated factor XI in turn activates factor IX. Activated factor IX, together with the effective form of factor VIII, calcium ions, and a phospholipid substance, then activates factor X. From this point on, the sequence of events occurs in the same fashion as in the extrinsic mechanism. The intrinsic clotting mechanism, however, is generally a slower-acting mechanism than the extrinsic mechanism. In the intrinsic clotting sequence, the phospholipid substance required as a cofactor for some of the steps becomes exposed on the surfaces of platelets (as platelet factor III) during platelet adhesion and clumping (agglutination), which is initiated when platelets contact collagen fibers.

Many of the reactions of the clotting process occur in what is called a "cascade" fashion. One factor becomes activated, and this factor in turn activates another that activates still another and so on. In this way, inactive factors in the blood are changed into active clotting factors.

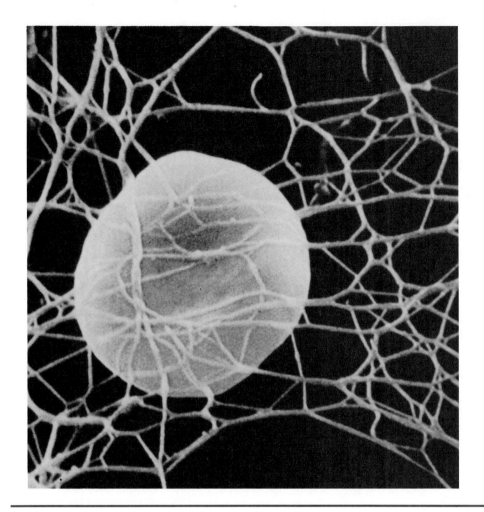

Figure 17.8
Portion of a blood clot showing fibrin network and an entrapped erythrocyte.

Positive Feedback Nature of Clot Formation

Once a blood clot starts to develop, thrombin acts in a positive feedback fashion to promote the development of the clot. For example, thrombin acts enzymatically on prothrombin to generate still more thrombin. In addition, thrombin accelerates the actions of some of the other clotting factors, and it enhances platelet adhesion and agglutination.

Limiting Clot Growth

Despite the positive feedback nature of the clotting process, clot formation normally occurs only locally at a site of damage. In general, once the various clotting factors are activated, they are rapidly inactivated or carried away by the blood so that their overall concentrations do not rise high enough to induce widespread clotting. For example, thrombin is removed from the blood in the region of a clot by being adsorbed onto the fibrin threads and by being inactivated by combination with an alpha globulin fraction of the plasma proteins called *antithrombin III* (*antithrombin–heparin cofactor*).

Clot Retraction

After the formation of a clot, a phenomenon known as *clot retraction* (syneresis) occurs. The fibrin meshwork shrinks and becomes denser and stronger. This activity helps pull the edges of a damaged vessel closer together. Clot retraction requires large numbers of platelets that, during clot formation, become trapped within the fibrin meshwork and attached to the fibrin strands. The platelets contain an actomyosinlike contractile protein, and they contract and pull the fibrin threads closer together.

CLINICAL CORRELATION
Monomyeloblastic Leukemia

CASE REPORT . . . .

THE PATIENT   A 63-year-old woman

PRINCIPAL COMPLAINTS   Weakness, fatigue, and dyspnea (difficult breathing)

HISTORY   The patient appeared to be chronically ill. During the last year, she had gradually lost 25 pounds in weight. The weakness and fatigue had become noticeable several months before she was seen and was progressing. She reported difficulty climbing even a single flight of stairs.

CLINICAL EXAMINATION   The jugular veins of the neck were visible (distended) almost to the angles of the mandible when the patient was sitting at 45° to the horizontal. The heart was moderately enlarged, the heart rate was 198/min (normal, <100/min), and the rate of respiration was 20/min (normal, <17/min). The liver was enlarged. The lower edge of the liver was felt 4–5 cm (normal, 1–2 cm) below the right costal margin (edge of the rib cage). The upper edge of the liver, determined by percussion, was at the normal level of the fifth rib. Pitting edema of the ankles was detected; that is, tissue in the ankle area was so swollen by fluid contained in the extracellular spaces (edema) that it could be indented by pressure with the fingers. The oral temperature was 37.9°C. The arterial blood pressure was 115/60 mm Hg (normal, 90–140/60–90 mm Hg). The hematocrit was 32% (normal, 37–48%). The white cell count was 66,000/mm^3 (normal, 4300 to 10,800/mm^3), with 1% neutrophils, 27% monocytes, 10% lymphocytes, 43% myeloblasts (immature granulocytes), and 26% promonocytes (immature monocytes) (normal: neutrophils, 60–70%; monocytes, 2.4–11.8%; lymphocytes, 20–53%; and immature forms, 0–25%). The blood levels of several enzymes were determined: serum glutamic oxaloacetic transaminase (SGOT) was 20 U/ml (normal, 10–40 U/ml); lactic dehydrogenase (LDH) was 295 U/ml (normal, 60–120 U/ml); and creatine phosphokinase (CPK) was 5 U/ml (normal, 5–35 U/ml). High blood levels of these enzymes are consistent with damage to heart muscle. An electrocardiogram (ECG) was obtained, and it displayed abnormalities that were consistent with acute myocardial infarction (an area of damaged heart muscle like that resulting from a heart attack).

Digoxin, which increases cardiac contractility, was administered. Five days after admission, bone marrow was aspirated from the left posterior iliac crest. Pathological examination revealed large numbers of immature mononuclear cells, consistent with the diagnosis of acute monocyte leukemia. Six days after admission, the hematocrit was 27%, the white cell count was 167,000/mm^3, with 27% monocytes and 36% immature forms (monoblasts). The patient became increasingly weak and dyspneic and died nine days after admission.

COMMENT   Leukemia is generally characterized by an abnormally high concentration of leukocytes without any demonstrable stimulus. The high proportion of immature forms that occurs in many types of leukemia is believed to represent some cellular defect in maturation. The concentrations of normal blood cells usually are decreased. The abnormal leukocytes infiltrate the organs and systems of the body, although the leukemic cells are not invasive in the way that the cells of malignant carcinomas (malignant tumors of epithelial origin) are invasive. Enlarged spleen, liver, and lymph nodes are common. The central nervous system, the heart, the kidneys, and the gastrointestinal system can be affected adversely.

In this patient, infiltration of the heart caused heart failure and ECG abnormalities consistent with myocardial infarction. However, changes of serum enzymes characteristic of infarction (SGOT, LDH, and CPK) were not seen. Postmortem examination revealed leukemic infiltration of the heart, lungs, liver, and colon; the latter contained numerous leukemic ulcerations. Patients who have leukemia may die within a few days of diagnosis; or, with treatment, they may live for years. The anemia and organ degeneration of this patient precluded the usual chemotherapy and contributed to her early demise.

During clot retraction, a fluid known as *serum* is extruded from the fibrin meshwork. Serum is essentially plasma that lacks fibrinogen and some of the other clotting factors that are removed during the clotting process.

Clot Dissolution

An enzyme called **plasmin (fibrinolysin),** which can decompose fibrin and dissolve clots, is present in the blood in an inactive form (plasminogen). Activated factor XII, thrombin, enzymes released from damaged tissues, and substances released from vascular endothelial cells can function as plasmin activators that convert plasminogen to plasmin. Normally, when clotting occurs, the plasmin activators initiate a mechanism for the eventual dissolution of the clot. For example, when a clot occurs around a blood vessel, the adjacent vascular endothelium releases substances that convert plasminogen to plas-

Table 17.3   Factors Involved in Blood Coagulation

Factor Number	Name	Type of Factor and Origin
I	Fibrinogen	A plasma protein produced in the liver
II	Prothrombin	A plasma protein produced in the liver
III	Tissue thromboplastin	A complex mixture of lipoproteins containing one or more phospholipid substances; released from damaged tissues
IV	Calcium ions	An ion in the plasma that is acquired from the diet and from bones
V	Proaccelerin (labile factor, accelerator globulin)	A plasma protein produced in the liver
VI	Not utilized	
VII	Serum prothrombin conversion accelerator (stable factor, proconvertin)	A plasma protein produced in the liver
VIII	Antihemophilic factor (antihemophilic globulin)	A plasma protein produced in the liver
IX	Plasma thromboplastin component (Christmas factor)	A plasma protein produced in the liver
X	Stuart factor (Stuart–Prower factor)	A plasma protein produced in the liver
XI	Plasma thromboplastin	A plasma protein produced in the liver
XII	Hageman factor	A plasma protein
XIII	Fibrin stabilizing factor	A protein present in plasma and in platelets

min and thereby promote clot dissolution. In general, within a few days after blood has leaked into the tissues and clotted, the plasmin activators cause the formation of enough plasmin to dissolve the clot. Plasmin may also remove clots that form spontaneously in small peripheral blood vessels (for example, capillaries), although clots that block large vessels are less likely to be removed.

Prevention of Clotting Within Normal Blood Vessels

Under most circumstances, blood does not clot as it flows through the blood vessels. This is due in part to the fact that the endothelial cells that line normal, undamaged blood vessels form a smooth surface that tends to prevent the contact activation of the intrinsic clotting mechanism. In addition, a monomolecular layer of negatively charged protein is adsorbed onto the endothelial cell lining, and this layer is believed to repel various clotting factors and platelets.

Anticlotting factors within the blood also contribute to clot prevention within normal blood vessels. For example, heparin, which is secreted by mast cells and basophils, inhibits the clotting process by accelerating the activity of antithrombin III, which inactivates thrombin. In addition, plasmin can destroy a number of clotting factors, as well as fibrin. Because of the presence of anticlotting factors, clotting normally does not occur until the rate of formation of thrombin rises above a critical level.

Intravascular Clotting

In spite of the various anticlotting mechanisms, clots occasionally may occur within the vascular system itself. If a clot blocks blood flow in a vessel that supplies tissue critical to survival, the result can be very serious.

If the clot is fixed or adheres to a vessel wall, it is called a *thrombus*. A thrombus that forms in the coronary circulation may block the blood supply to heart tissue, leading to heart damage and ultimately to heart failure and death. The formation of a thrombus in a coronary artery is called a *coronary thrombosis*. Just what causes such clots to form is currently the subject of intense investigation, and several theories have been presented. One line of evidence suggests that continuous, unreversed damage to a blood vessel wall that removes the inner endothelium and exposes deeper-lying tissues establishes conditions favorable to clot formation. Other investigations have revealed alterations in blood clotting characteristics in persons prone to thromboses that suggest a hyperreactivity of the clotting mechanism itself. Still others indicate that changes occur in a blood vessel wall as the result of the activity of viruses, and that the altered vessel wall may promote clot formation at the site. None of these occurrences is mutually exclusive, and all may play a role in clot formation.

If an intravascular clot is not fixed but floats within the blood, it is called an *embolus*. Emboli are potentially dangerous because they can become lodged in smaller vessels, thus blocking them and leading to tissue damage due to blood flow restriction. Emboli that lodge in the vessels of a lung can damage the lung itself and elicit cardiovascular reflexes leading to hypotension (low blood pressure) and death.

CONDITIONS OF CLINICAL SIGNIFICANCE

Excessive Bleeding

One of the factors required in the blood-clotting sequence is factor VIII (antihemophilic factor). An inherited genetic deficiency of an effective form of this factor results in a clotting defect called *hemophilia A*. A genetic deficiency of an effective form of another clotting factor, factor IX (plasma thromboplastin component), results in the less common defect *hemophilia B*. The blood of individuals who suffer from hemophilia does not clot properly, and even minor damage to blood vessels can result in substantial bleeding.

Nutritional deficiencies or liver disease can also lead to clotting problems and excessive bleeding. Vitamin K is required for the normal synthesis of prothrombin in the liver, and a deficiency of this vitamin interferes with prothrombin production. As a result, decreased levels of prothrombin may be present in the blood, leading to clotting difficulties. The liver is also involved in the production of a number of other clotting factors, and diseases that impair liver function can severely affect the clotting system.

Reduced numbers of platelets, a condition known as *thrombocytopenia*, causes bleeding states in which blood loss occurs through capillaries and other small vessels. Low platelet numbers can result from increased platelet destruction or from depressed platelet production due to pernicious anemia, certain drug therapies, or radiation.

STUDY OUTLINE

PLASMA Approximately 91% water; transports hormones and metabolic end products; contains proteins (such as albumin, fibrinogen, globulins), ions (such as sodium, bicarbonate, chloride), food materials, and gases. **p. 505**

FORMED ELEMENTS **pp. 505–515**

HEMOPOIESIS Formed elements of the blood develop from stem cells called hemocytoblasts.

ERYTHROCYTES (RED BLOOD CELLS) Small, circular, biconcave discs with no nuclei that function in transport of oxygen and carbon dioxide; *hematocrit* is proportion of erythrocytes in a sample of blood.

HEMOGLOBIN Substance in erythrocytes that can bind with oxygen and carbon dioxide. Composed of protein called *globin* and four nonprotein groups called *hemes* that each contain an iron atom; oxygen binds reversibly with iron of heme groups; carbon dioxide binds reversibly with globin portion of hemoglobin, which is composed of four polypeptide chains.

IRON Component of hemoglobin. Iron is transported in plasma in loose combination with a beta globulin called *transferrin*. Iron can be stored in the liver, spleen, and elsewhere in form of intracellular iron–protein complexes called *ferritin* and *hemosiderin*. Iron is absorbed into body in upper part of small intestine. Total body iron regulated largely by alterations in intestinal absorption rate.

PRODUCTION OF ERYTHROCYTES Erythrocytes develop from hemocytoblasts, which give rise to proerythroblasts, which become basophil erythroblasts, which differentiate into polychromatophil erythroblasts, which become normoblasts. When hemoglobin content of normoblast cytoplasm reaches about 34%, nucleus is pinched off, and resulting reticulocytes mature into erythrocytes. Erythrocyte formation process (erythropoiesis) requires vitamin B_{12} and folic acid.

CONTROL OF ERYTHROCYTE PRODUCTION Erythropoietin, which stimulates erythrocyte production, is produced in kidneys by action of renal erythropoietic factor on plasma protein precursor. Erythropoietin production, and thus erythrocyte production, is regulated by negative-feedback mechanism that is sensitive to amount of oxygen delivered to tissues; inadequate oxygen supply leads to increased erythropoiesis.

FATE OF ERYTHROCYTES Aged, abnormal, or damaged erythrocytes are disposed of by phagocytic macrophages (especially in spleen, liver, and bone marrow); hemoglobin is degraded. Amino acids of globin can be used in synthesis of new protein. Iron is liberated from heme; remainder of heme is converted to biliverdin, which is in turn converted to bilirubin. Bilirubin is eventually taken up by liver, conjugated to glucuronic acid, and secreted with bile.

BLOOD GROUPS Based on reactions between antigens on erythrocytes and particular antibodies.

CONDITIONS OF CLINICAL SIGNIFICANCE: ERYTHROCYTES p. 513

ANEMIA Condition characterized by either decreased number of erythrocytes in blood or decreased percentage of hemoglobin.

HEMORRHAGIC ANEMIA Result of substantial blood loss and relatively slow replacement of lost erythrocytes.

IRON DEFICIENCY ANEMIA Due to excessive loss, deficient intake, or poor absorption of iron; red blood cells are smaller than normal and contain lower-than-normal level of hemoglobin.

PERNICIOUS ANEMIA Result of inability to absorb adequate amounts of vitamin B_{12} due to insufficient quantities of functional intrinsic factor. Erythrocytes are larger and more fragile than normal.

SICKLE CELL ANEMIA Result of production of abnormal hemoglobin due to genetic defect. Abnormal hemoglobin causes distortion of erythrocytes into sickle shape at low concentrations of oxygen; misshapen erythrocytes can rupture or block blood vessels.

APLASTIC ANEMIA Inadequate production of erythrocytes due to inhibition or destruction of red bone marrow.

POLYCYTHEMIA Condition in which there is a net increase in body's total circulating erythrocyte mass.

PRIMARY POLYCYTHEMIA Production of excess erythrocytes due to tumorous abnormalities of tissues that produce blood cells.

SECONDARY POLYCYTHEMIA Increased production of erythrocytes due to increased levels of erythropoietin formation; *physiologic polycythemia* is seen in people living at high altitudes.

LEUKOCYTES Formed elements of blood that act primarily in loose connective tissues and are mainly being transported by circulation.

GRANULOCYTES Leukocytes with clearly evident granules in their cytoplasm; formed in red bone marrow.

Neutrophils Phagocytic cells; capable of ameboid movement; can leave blood vessels and enter tissues, where they ingest bacteria and other foreign substances.

Eosinophils Ameboid, phagocytic cells; believed to ingest and destroy antigen–antibody complexes.

Basophils May be functionally related to mast cells of connective tissue that contain histamine and heparin.

AGRANULOCYTES Two types.

Monocytes Ameboid cells formed in red bone marrow; can leave blood vessels and enter tissues, where they develop into large, phagocytic cells called macrophages.

Lymphocytes Important in body's specific immune responses, including antibody production; many found in lymphoid tissues.

DIFFERENTIAL COUNT Provides percentage figure for each type of leukocyte in blood.

CONDITIONS OF CLINICAL SIGNIFICANCE: LEUKOCYTES p. 516

LEUKEMIA Cancerous condition involving excessive, uncontrolled proliferation of leukocytes that leads to diffuse and almost total replacement of the red bone marrow with leukemic cells.

INFECTIOUS MONONUCLEOSIS Caused by Epstein-Barr virus; characterized by increase in relative and absolute numbers of lymphocytes in blood.

PLATELETS Small cytoplasmic fragments of megakaryocytes; involved in blood clotting and other hemostatic processes.

HEMOSTASIS Arrest of bleeding or stopping or slowing of blood flow through a vessel or a part.
pp. 515–522

VASCULAR CONSTRICTION Occurs when blood vessel is cut or ruptured; decreases or stops blood flow.

FORMATION OF A PLATELET PLUG In damaged vessel, platelets adhere to exposed collagen and release ADP, which leads to further platelet adhesion and the formation of platelet plug; plug may slow or stop bleeding if vessel damage is not too great; platelets also release chemicals that facilitate vasoconstriction and contribute to blood clotting.

BLOOD CLOTTING Involves conversion of soluble plasma protein called *fibrinogen* into insoluble, thread-like polymer called *fibrin* by enzymatic action of thrombin; fibrin threads entrap blood cells, platelets, and plasma to form clot; thrombin can be formed from inactive plasma protein precursor by two mechanisms:

EXTRINSIC MECHANISM Tissue thromboplastin released from damaged tissues acts with other clotting factors to form prothrombin activator; prothrombin activator catalyzes formation of thrombin from prothrombin.

INTRINSIC MECHANISM Activation of factor XII and platelet adhesion and clumping may occur when inactive factor XII and platelets contact damaged vessel surface; phospholipid of platelets and active factor XII act with other clotting factors to form prothrombin activator.

POSITIVE FEEDBACK NATURE OF CLOT FORMATION Once clot formation begins, thrombin promotes clot development by acting on prothrombin to generate more thrombin, by accelerating the actions of other clotting factors, and by enhancing platelet adhesion and agglutination.

LIMITING CLOT GROWTH Once the various clotting factors are activated, they are rapidly inactivated or carried away by the blood so their overall concentrations do not rise enough to induce widespread clotting.

CLOT RETRACTION After clot formation, fibrin meshwork shrinks and becomes denser and stronger; requires large numbers of platelets that apparently contain actomyosinlike contractile protein.

CLOT DISSOLUTION Enzyme *plasmin* can decompose fibrin and dissolve clots.

PREVENTION OF CLOTTING WITHIN NORMAL BLOOD VESSELS Due in part to smooth, endothelial lining of vessels and to layer of negatively charged protein adsorbed onto surface of endothelial cells; anticlotting factors in blood (such as heparin, antithrombin III, plasmin) may also inhibit clotting.

INTRAVASCULAR CLOTTING Clots may occur within vascular system and block blood flow to tissues critical to survival; clot that is fixed or adheres to vessel wall is called *thrombus* and clot that floats within blood is called *embolus*. Clots may form in blood vessels as result of continuous, unreversed damage to vessel walls, hyperreactivity of the clotting mechanisms, or viral activity.

CONDITIONS OF CLINICAL SIGNIFICANCE: EXCESSIVE BLEEDING Inherited genetic deficiencies of effective forms of factor VIII (antihemophilic factor) or factor IX (plasma thromboplastin component) may result in blood that does not clot properly and substantial bleeding; nutritional deficiencies and liver diseases can also cause clotting problems and excessive bleeding; thrombocytopenia (reduced numbers of platelets) may result in bleeding from capillaries. **p. 522**

SELF-QUIZ

1. As the amount of adipose tissue increases, the volume of blood per unit of body weight: (a) increases; (b) declines; (c) remains constant.

2. The hematocrit of a sample of peripheral venous blood is usually higher in females than in males. True or False?

3. Most of the oxygen that is transported by the blood is in physical solution dissolved in the plasma. True or False?

4. Total body iron is regulated to a large extent by alterations in the intestinal absorption rate of iron. True or False?

5. Pernicious anemia is a condition that results from: (a) an inability to absorb adequate amounts of vitamin B_{12} from the digestive tract; (b) a deficiency of iron in the blood; (c) the destruction of the red bone marrow.

6. Leukocytes act primarily in the loose connective tissues, and those within the blood are mainly being transported by the circulation. True or False?

7. Match the following terms with their appropriate descriptions:

   Bilirubin
   Heme
   Erythropoietin
   Globin
   Ferritin

   (a) A substance that stimulates erythrocyte production
   (b) A component of hemoglobin that contains iron
   (c) A component of hemoglobin that can bind reversibly with carbon dioxide
   (d) A storage form of iron
   (e) A breakdown product of hemoglobin that is taken up by the liver, conjugated to glucuronic acid, and excreted with the bile

8. Which of the following leukocyte types leave the blood vessels and enter the tissues, where they pro-

tect the body by ingesting bacteria and other foreign substances? (a) neutrophil; (b) megakaryocyte; (c) lymphocyte.

9. Mast cells and basophils: (a) contain heparin; (b) produce antibodies; (c) secrete renal erythropoietic factor.

10. In the tissues, monocytes are converted to: (a) mast cells; (b) macrophages; (c) megakaryocytes.

11. The leukocyte type that is involved in specific immune responses, including antibody production, is the: (a) lymphocyte; (b) neutrophil; (c) megakaryocyte.

12. The Epstein-Barr virus is believed to cause: (a) sickle cell anemia; (b) infectious mononucleosis; (c) leukemia.

13. Platelets are formed in the red bone marrow as pinched-off portions of cells called: (a) reticulocytes; (b) mast cells; (c) megakaryocytes.

14. ADP released by platelets facilitates: (a) vasoconstriction; (b) the conversion of prothrombin to thrombin; (c) the formation of a platelet plug.

15. During clot formation, fibrinogen is converted to fibrin by: (a) thrombin; (b) thromboplastin; (c) antithrombin III.

16. Blood clotting cannot occur unless tissue thromboplastin is released from damaged tissues. True or False?

17. Once it is formed, thrombin may contribute to clot development by acting enzymatically to convert serotonin to thromboplastin. True or False?

18. Clot retraction requires large amounts of: (a) antithrombin III; (b) plasmin; (c) platelets.

19. Fibrin can be decomposed and clots dissolved by: (a) thromboplastin; (b) plasmin; (c) fibrinogen.

20. A deficiency of vitamin K interferes with the normal production of prothrombin. True or False?

LEARNING OBJECTIVES

After completing this chapter, you should be able to:

1. Describe the coverings of the heart.

2. Describe the structure of the heart, including its "skeleton."

3. Describe the flow of blood through the heart and the pumping action of the heart during a contractile cycle.

4. Describe the excitation process that occurs in the heart, and explain why the sinuatrial node acts as the pacemaker of the heart.

5. Describe the conduction of a stimulatory impulse through the heart.

6. Describe and explain the pressure changes, volume changes, and valve actions that occur within the left chambers of the heart during a cardiac cycle.

7. Define cardiac output, and discuss the interrelationship of heart rate and stroke volume with reference to cardiac output.

8. Describe the control of heart rate.

9. Describe the control of stroke volume.

10. Discuss the effect of exercise on cardiac function.

11. Describe the process of congestive heart failure.

12. Explain the causes of the normal heart sounds.

13. Describe a normal electrocardiogram, and explain its principal features.

CHAPTER CONTENTS

EMBRYONIC DEVELOPMENT OF THE HEART

POSITION OF THE HEART

COVERINGS OF THE HEART

ANATOMY OF THE HEART

CIRCULATION THROUGH THE HEART

PUMPING ACTION OF THE HEART

CARDIAC MUSCLE

EXCITATION AND CONDUCTION IN THE HEART

THE CARDIAC CYCLE

CARDIAC OUTPUT

FACTORS THAT INFLUENCE CARDIAC FUNCTION

CONDITIONS OF CLINICAL SIGNIFICANCE: THE HEART

HEART SOUNDS

ELECTROCARDIOGRAPHY

KEY TERMS AND DERIVATIVES

bradycardia (*brady* = slow; *cardia* = heart) slowness of the heart rate

diastolic pressure (*diastol* = dilation) blood pressure during the period between heart contractions, when the heart relaxes, dilates, and fills with blood

electrocardiogram (*cardia* = heart; *gram* = something written or recorded) a graphic record of the electric current associated with heartbeats

myocardium (*myo* = muscle) the cardiac muscle layer of the wall of the heart

The Circulatory System:
The Heart

The circulatory system can be separated into two divisions: the **cardiovascular system** and the **lymphatic system.** The cardiovascular system includes the *heart*, which serves as a pump for the blood, and the *blood vessels*, which transport the blood throughout the body. The lymphatic system consists of *organs* (tonsils, thymus, spleen, lymph nodes, and lymphatic nodules) that play a role in the specific immune responses (see Chapter 21), and *vessels* that collect tissue fluid from between the cells of the body and transport it to the cardiovascular system. In this chapter we will study the heart. The blood vessels and the lymphatic system are discussed in Chapters 19 and 20, respectively.

The cardiovascular system is a continuous closed system. Confined to the heart and the numerous vessels, the blood repeatedly travels through the heart, into arteries, then to capillaries, into veins, and back to the heart. Normally, blood does not leave this system, although some of the fluid part of the blood does pass through the walls of the capillaries to join the tissue fluid between the cells. However, even this fluid is returned to the cardiovascular system, either directly or by way of the lymphatic system. The heart is the pump that keeps the blood flowing through the system of vessels.

EMBRYONIC DEVELOPMENT OF THE HEART

Early in embryonic development the future heart is a simple pulsating tube that receives blood from the veins at its posterior end and pumps it into the arterial system through its anterior end (Figure 18.1a). From this simple beginning, the heart must not only respond to the changing needs of the developing embryo but must also be capable of functioning under the vastly different conditions immediately following birth. To accomplish this, the tubular heart must develop into a four-chambered organ complete with valves and a midline partition. And the heart must undergo these alterations without interrupting its delivery of blood to the developing embryo.

By the fifth week of development it becomes apparent that the heart is beginning to undergo changes as it grows rapidly (Figure 18.1b) and evolves into an S-shaped structure (Figure 18.1c). With continued development, the anterior vessel, which carries blood away from the heart, divides into two vessels: the future *pulmonary trunk*, which supplies blood to the lungs, and the *aorta*, which carries blood to the vessels that supply the rest of the body. At the same time, a midline septum (wall) is developing within the heart. When completed, this wall will separate the blood flow through the heart into two channels; the blood on one side of the wall passes through the pulmonary trunk and to the lungs, and that on the other side is directed into the aorta and thus to the rest of the body (Figure 18.1d).

By the seventh week the embryonic heart has further divided into the four chambers it will retain in the adult—two atria and two ventricles. Through a series of changes, including the development of new segments in some veins and the degeneration of portions of other veins, the vessels that enter the posterior (venous) region of the heart develop into *superior* and *inferior vena cavae*, which return all the venous blood from the body to the heart. No other major changes occur in the heart until birth.

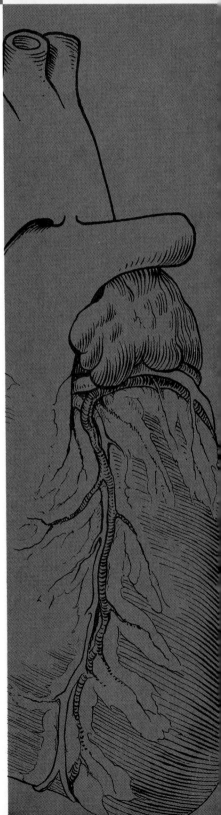

527

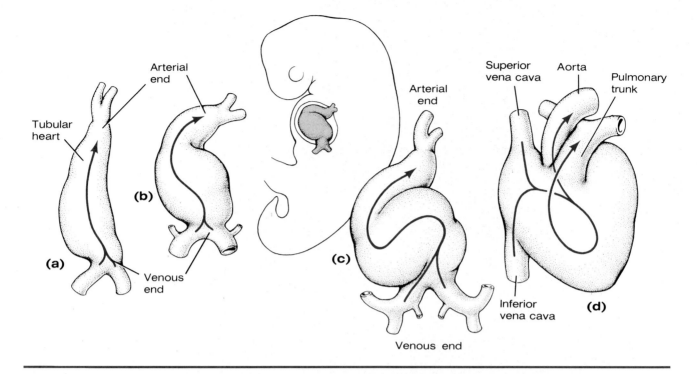

Figure 18.1
Successive stages in the embryonic development of the heart. The arrows indicate the direction of blood flow.

F18.2

POSITION OF THE HEART

The adult heart is a cone-shaped organ about the size of a fist. It is located between the lungs in the space called the **mediastinum** (Figure 18.2). The heart lies obliquely in the mediastinum and is described as having a base and an apex, diaphragmatic and sternocostal surfaces, and four borders.

The *base* of the heart faces posteriorly and is located superiorly and to the right, behind the sternum, at about the level of the second and third ribs. It consists mainly of the left atrium, part of the right atrium, and the proximal portions of the large veins that enter the posterior wall of the heart. From the base the heart projects downward, anteriorly, and to the left, ending in a blunt *apex*. The apex reaches the fifth intercostal space, about 8 cm to the left of the midsternal line. The *diaphragmatic surface* of the heart is that part between the base and the apex that rests upon the diaphragm. It involves the left and right ventricles. The anterior surface of the heart, which is formed mainly by the right ventricle and the right atrium, is referred to as the *sternocostal surface*.

The *superior border* of the heart is formed by both atria and is the region where the great vessels enter and leave the heart. It lies at about the level of the second intercostal space. The *inferior border* extends from behind the lower portion of the sternum to the left fifth intercostal space, where it ends at the apex; it is formed mostly by the right ventricle, plus a small portion of the left ventricle at the apex. The *right border* of the heart is formed by the right atrium and is located about 2.5 cm to the right of the sternum. The *left border* is formed mainly by the left ventricle, with the left atrium forming the upper portion. The left border extends to the apex of the heart, from the level of the junction of the left second rib with its costal cartilage.

COVERINGS OF THE HEART

per-i-kar´-dee-um

The heart is enclosed in a double-walled membranous sac called the **pericardium.** The inner layer of the pericardium, called the **epicardium** or the **visceral pericardium,** is a serous membrane with a surface layer of mesothelium overlying a thin layer of loose connective tissue that adheres to the outer

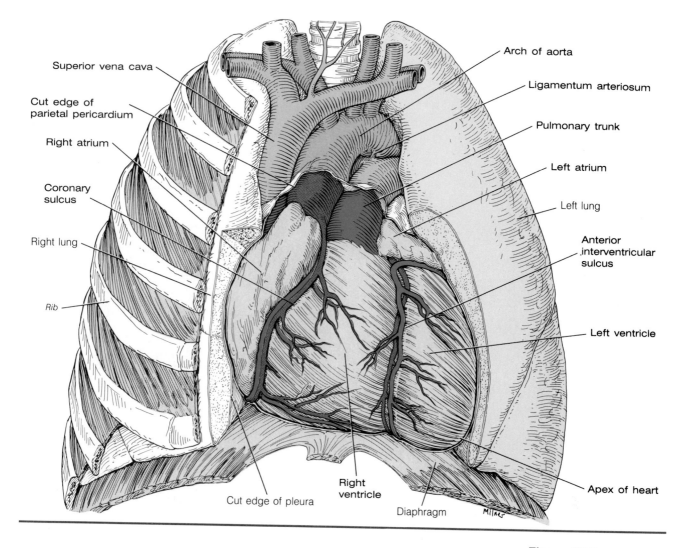

Superior vena cava

Cut edge of
parietal pericardium

Right atrium

Coronary
sulcus

Right lung

Rib

Arch of aorta

Ligamentum arteriosum

Pulmonary trunk

Left atrium

Left lung

Anterior
interventricular
sulcus

Left ventricle

Apex of heart

Cut edge of pleura

Right
ventricle

Diaphragm

Figure 18.2
Frontal view of the
thorax showing the
position of the heart in
the mediastinum.

surface of the heart (Figure 18.3). At the point where the large vessels enter
and leave the heart, the serous layer of the visceral pericardium folds back
and is continuous with the outer layer of the pericardium, the **parietal pericardium.** The parietal pericardium is composed of two layers: an outer *fibrous*
layer, which strengthens it and anchors it within the mediastinum; and an
inner *serous* layer, which lines the inside of the fibrous layer and is continuous
with the serous layer of the visceral pericardium. Between the serous membranes of the visceral and parietal layers is a small space called the **pericardial
cavity.** This cavity contains **pericardial fluid,** which is secreted by the cells of
the serous membrane of the pericardium. The fluid lubricates the membranes, permitting them to slide over one another with a minimum of friction
as the heart beats.

Inflammation of the pericardium, which is referred to as *pericarditis*, can
result from a variety of causes. The amount and character of the pericardial
fluid varies in the different forms of pericarditis. In some cases, the pericardial
fluid is scanty; in others, it is abundant; and some infections produce fibrin,
others produce pus, in the pericardial cavity. In pericarditis, the serous layers
of the pericardium become roughened, which causes pain as they move over
one another and also interferes with the normal filling of the heart chambers.

ANATOMY OF THE HEART

In order to function as a pump the heart must have both receiving chambers
and delivery chambers, valves to direct the flow of blood through the heart, a

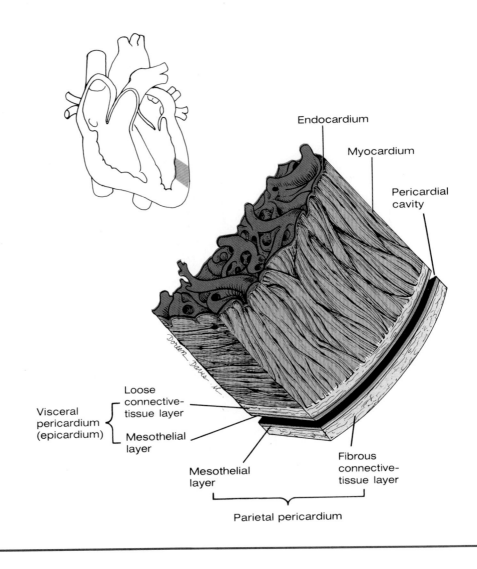

Figure 18.3

Section through the wall of the heart showing the parietal and visceral layers of the pericardium (heart sac), the myocardium (heart muscle), and the endocardium (inner lining of the heart chamber).

wall that is strongly compressible and thus provides the force to propel blood, and vessels to deliver blood to and from the heart.

Chambers of the Heart

ven´-tri-kals
F18.4–F18.6

The heart consists of four chambers: **right** and **left atria** and **right** and **left ventricles** (Figures 18.4–18.6). The atria are small and are located toward the superior region of the heart. The ventricles are larger and comprise the bulk of the heart. Located inferiorly, the ventricles form the apex of the heart. The right ventricle forms most of the heart's anterior surface; the left ventricle forms most of the inferior surface and left margin of the heart. The atria are separated from one another by an **interatrial septum**. The ventricles are separated by an **interventricular septum.**

Vessels Associated with the Heart

F18.6

Several large vessels enter or leave the base and the superior border of the heart (Figure 18.6):

1. **Superior** and **inferior venae cavae,** which return venous blood from the vessels of the body to the right atrium.

2. The **pulmonary trunk,** which divides into **right** and **left pulmonary arteries** and carries blood from the right ventricle to the lungs.

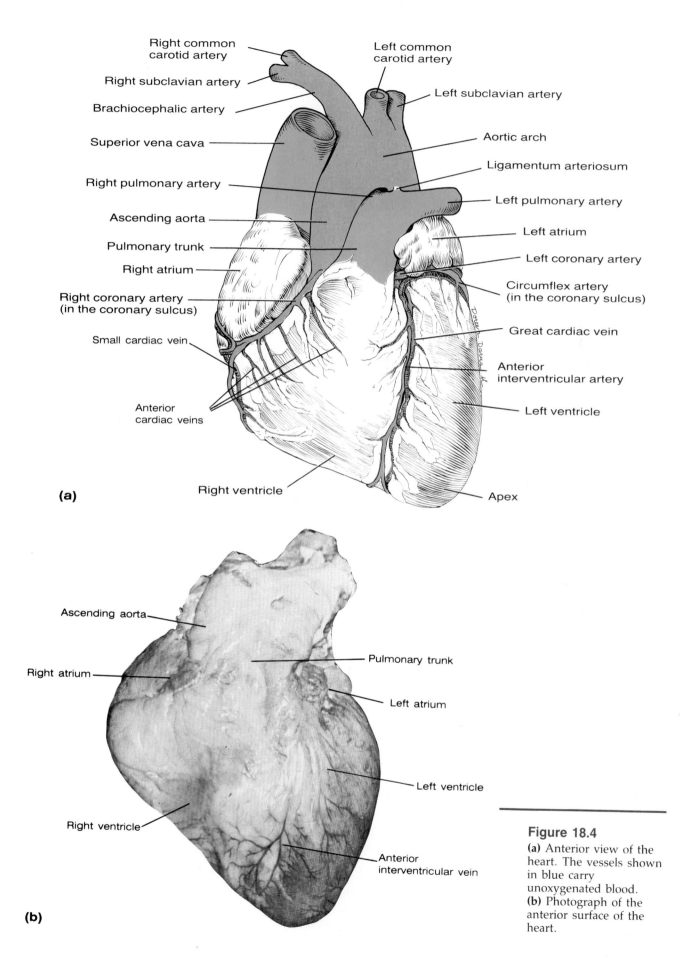

(a)

(b)

Figure 18.4
(a) Anterior view of the heart. The vessels shown in blue carry unoxygenated blood.
(b) Photograph of the anterior surface of the heart.

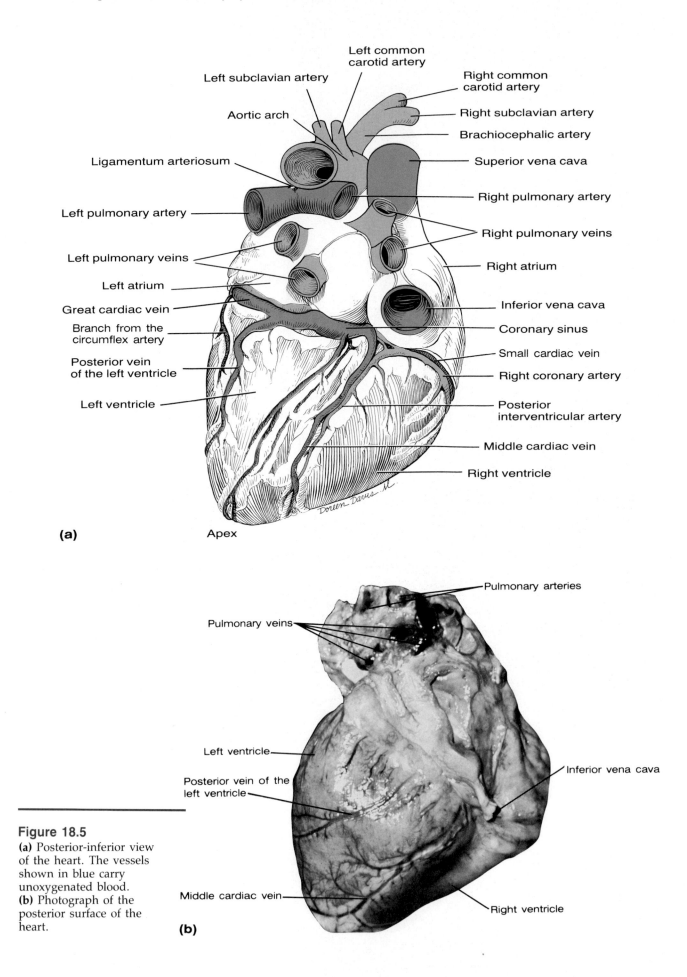

(a)

Left common
carotid artery

Left subclavian artery

Aortic arch

Right common
carotid artery

Right subclavian artery

Brachiocephalic artery

Ligamentum arteriosum

Superior vena cava

Left pulmonary artery

Right pulmonary artery

Left pulmonary veins

Right pulmonary veins

Left atrium

Right atrium

Great cardiac vein

Inferior vena cava

Branch from the
circumflex artery

Coronary sinus

Posterior vein
of the left ventricle

Small cardiac vein

Right coronary artery

Left ventricle

Posterior
interventricular artery

Middle cardiac vein

Right ventricle

Apex

Pulmonary arteries

Pulmonary veins

Left ventricle

Inferior vena cava

Posterior vein of the
left ventricle

Middle cardiac vein

Right ventricle

(b)

Figure 18.5
(a) Posterior-inferior view
of the heart. The vessels
shown in blue carry
unoxygenated blood.
(b) Photograph of the
posterior surface of the
heart.

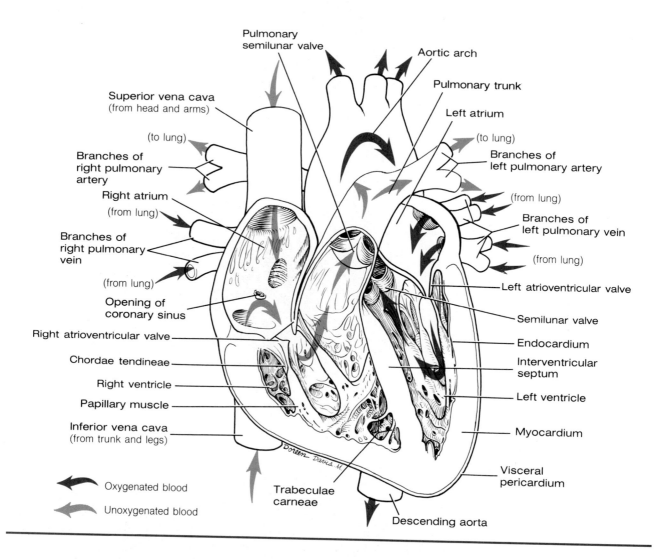

Pulmonary
semilunar valve

Aortic arch

Pulmonary trunk

Superior vena cava
(from head and arms)

Left atrium

(to lung)

(to lung)

Branches of
right pulmonary
artery

Branches of
left pulmonary artery

Right atrium

(from lung)

(from lung)

Branches of
left pulmonary vein

Branches of
right pulmonary
vein

(from lung)

(from lung)

Left atrioventricular valve

Opening of
coronary sinus

Semilunar valve

Right atrioventricular valve

Endocardium

Chordae tendineae

Interventricular
septum

Right ventricle

Left ventricle

Papillary muscle

Myocardium

Inferior vena cava
(from trunk and legs)

Visceral
pericardium

Oxygenated blood

Trabeculae
carneae

Descending aorta

Unoxygenated blood

3. The four **pulmonary veins,** which carry blood from the lungs
   to the left atrium.

4. The **aorta,** which carries blood from the left ventricle into the
   vessels that supply the body.

Wall of the Heart

The heart is composed primarily of cardiac muscle anchored to a fibrous skeleton.

Epicardium, Myocardium, and Endocardium

The wall of the heart is formed of three layers: the epicardium, the myocardium, and the endocardium (Figure 18.3). The **epicardium (visceral pericardium)** is a thin serous membrane that adheres to the outer surface of the heart. The thickest layer of the wall of the heart, the **myocardium,** is composed of cardiac muscle. The myocardium is lined on the inside by the **endocardium,** which is composed of connective tissue and a surface layer of squamous cells. Foldings of the endocardium form the valves that separate the atria from the ventricles—the *atrioventricular valves*—and the ventricles from the aorta and the pulmonary trunk—the *semilunar valves*. The endocardial lining of the heart is continuous with the endothelium that lines all the arteries, veins, and capillaries of the body.

The myocardium varies considerably in thickness from one heart chamber to another. Its thickness is related to the resistance encountered in pumping

Figure 18.6
Frontal section of the heart. The arrows indicate the path of the blood flow through the chambers, the valves, and the major vessels. Note that the branches of the right pulmonary vein pass behind the heart and enter the left atrium.

F18.3

my-o-kar´-dee-um
en-do-kar´-dee-um

a-tree-o-ven-trik´-yoo-lar
sem-i-loo´-ner

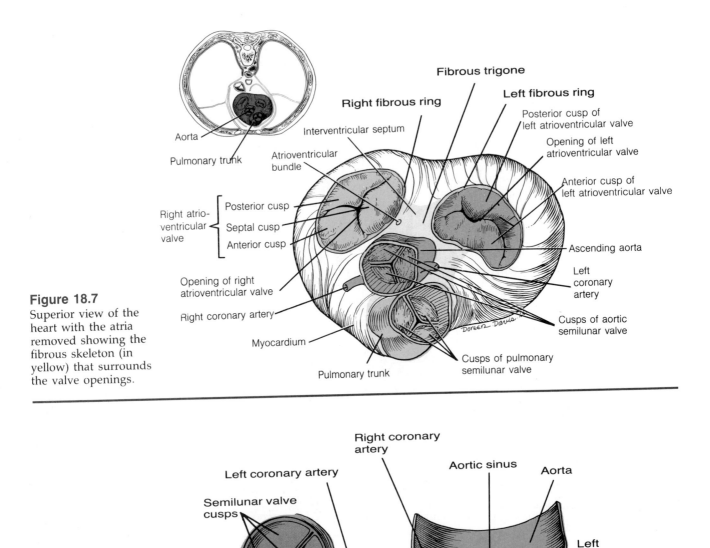

Figure 18.7
Superior view of the
heart with the atria
removed showing the
fibrous skeleton (in
yellow) that surrounds
the valve openings.

Figure 18.8
The origin of the
coronary arteries. **(a)** A
closed aortic valve
viewed from above.
(b) An aortic orifice cut
and opened to show the
semilunar valves.

F18.6 the blood from the different chambers. Since the muscles of the atria meet little resistance in pushing the blood into the ventricles, the walls of the atria are the thinnest part of the myocardium (Figure 18.6). In contrast, the ventricles must move the blood through the blood vessels of either the lungs or the rest of the body and back into a receiving chamber. Consequently, the myocardium of the ventricles is thicker than that of the atria. Moreover, the left ventricle, which propels blood through all parts of the body (other than the lungs) and back into the right atrium, has thicker walls than the right ventricle, which moves the blood only through the blood vessels of the lungs and back into the left atrium.

The inner surface of the myocardium of the ventricles is irregular, having folds and bridges—called **trabeculae carneae**—and cone-shaped **papillary muscles** that project into the lumen (Figure 18.6). Strong fibrous strands called **chordae tendineae** run from the papillary muscles to the cusps (flaps) of the atrioventricular valves.

F18.6

Skeleton of the Heart

Horizontal **fibrous rings** surround the atrioventricular openings and the openings of the aorta and pulmonary trunk. The rings are fused together by additional fibrous tissue called **fibrous trigones** (Figure 18.7). Collectively, these fibrous supports are referred to as the **skeleton of the heart.** This fibrous skeleton not only serves as the attachment for the heart muscle and valves but also helps form the septa that separate the atria from the ventricles.

F18.7

Vessels of the Myocardium

The myocardium receives an abundant blood supply through the **right** and **left coronary arteries** (Figures 18.4 and 18.5; Box 18.1). These arteries arise from the aorta just beyond the point where it leaves the superior border of the heart (Figure 18.7). The coronary arteries obtain their blood from sinuses behind the cusps of the aortic semilunar valve (Figure 18.8).

The **right coronary artery** arises from the right anterior surface of the aorta and passes to the right margin of the heart in a groove called the **coronary sulcus** (Figure 18.4). The coronary sulcus separates the atria from the ventricles. The right coronary artery extends around the margin of the heart to the

F18.4, F18.5,
Box 18.1
F18.7
F18.8

F18.4

Box 18.1
Cast of Blood Vessels of the Heart

This photograph of a cardiac cast shows the network of vessels that carry blood from the chambers of the heart to the heart muscle and back again to the heart chambers. The thick, dark, branchlike structure on the left is the right coronary artery. On the right is the left coronary artery, which branches into two different arteries: the circumflex artery, which appears slightly out of focus at the back of the heart, and the anterior interventricular artery. The large white structure on the left is the coronary sinus, which drains blood into the right atrium from several of the vessels supplying the heart.

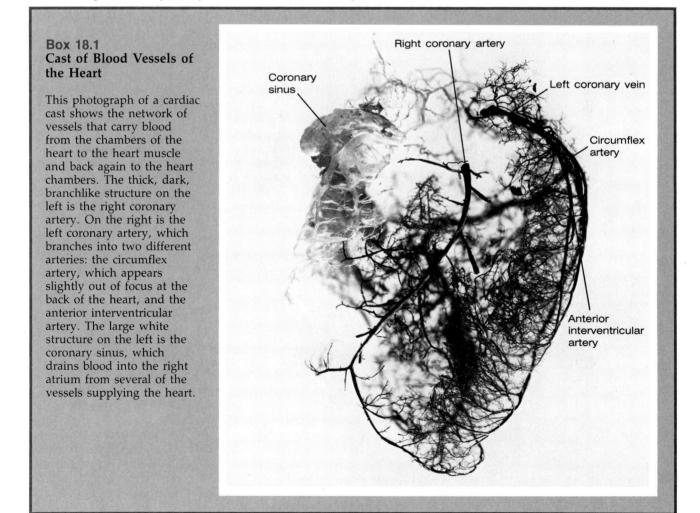

F18.5

posterior surface, sending branches to the atrium and the ventricle. On the posterior surface of the heart, the main branch of the artery turns downward toward the apex of the heart. This branch, the **posterior interventricular artery,** supplies smaller branches to both ventricles (Figure 18.5).

The **left coronary artery** arises from the left anterior surface of the aorta, behind the pulmonary trunk; after passing a short distance toward the left margin of the heart, it divides into anterior interventricular and circumflex branches. The **anterior interventricular artery** travels down the anterior surface of the interventricular septum toward the apex of the heart. It supplies branches to both ventricles. The **circumflex artery** travels in the coronary sulcus, between the left atrium and the left ventricle, to the left margin of the heart. After passing around the left margin, the circumflex artery supplies branches to the posterior surfaces of the left atrium and left ventricle.

Any narrowing or blockage of the coronary arteries interferes with the oxygen supply to the myocardium, which causes regions of the cardiac muscle to die. This condition, which can be disabling or fatal, is commonly called a heart attack. If the blockage of the coronary arteries is only temporary, resulting in an inadequate oxygen supply to the myocardium for only a few seconds, a sharp pain in the chest results, often radiating down the left arm.

an-jee´-nah pek´-tor-is

This condition is called *angina pectoris.*

F18.4, F18.5

After passing through an extensive capillary network, the blood from the coronary arteries enters the **cardiac veins,** which travel alongside the arteries (Figures 18.4 and 18.5). The cardiac veins join together to form an enlarged vessel, the **coronary sinus,** which is located in the coronary sulcus on the posterior surface of the heart, between the atria and ventricles.

The anterior surface of the heart is drained primarily by the **great cardiac vein,** which passes upward alongside the anterior interventricular artery. The great cardiac vein begins at the apex of the heart and ascends to the base of the ventricles, where it becomes continuous with the coronary sinus. The largest veins on the posterior-inferior surface of the heart are the **posterior vein of the left ventricle,** which accompanies the circumflex artery, and the **middle cardiac vein,** which travels alongside the posterior interventricular artery. These veins, like the great cardiac vein, empty into the coronary sinus. The coronary sinus, in turn, empties into the right atrium. Also draining into the coronary sinus is a **small cardiac vein,** which travels along the right margin of the heart and enters the coronary sulcus on the posterior aspect of the heart. In addition to these veins there are several small **anterior cardiac veins,** which drain the ventral surface of the right ventricle directly into the right atrium.

Valves of the Heart

There are four sets of valves that keep the blood flowing in the proper direction through the chambers of the heart—two sets of *atrioventricular valves* and two sets of *semilunar valves.*

Atrioventricular Valves

F18.6, F18.7

Located between the atria and the ventricles, the two **atrioventricular (AV) valves** are flaps of endocardium with an inner framework of fibrous connective tissue (Figures 18.6 and 18.7). The flaps are anchored to the papillary muscles of the ventricles by the chordae tendineae. The papillary muscles, which are extensions of the myocardium, exert tension on the valve cusps, thus preventing them from being forced into the atria while the ventricles are contracting. The right atrioventricular valve, which separates the right atrium from the right ventricle, has three flaps, or *cusps,* and is therefore known as the *tricuspid valve.* The left atrioventricular valve is called *bicuspid* or *mitral valve* because it has only two cusps. Both of the atrioventricular valves are forced shut as the pressure in the ventricles increases, thus preventing the flow of blood back into the atria while the ventricles are contracting.

Semilunar Valves

F18.6, F18.7

Blood is prevented by **semilunar valves** from returning to the ventricles after the ventricles have completed their contractions (Figures 18.6 and 18.7). The

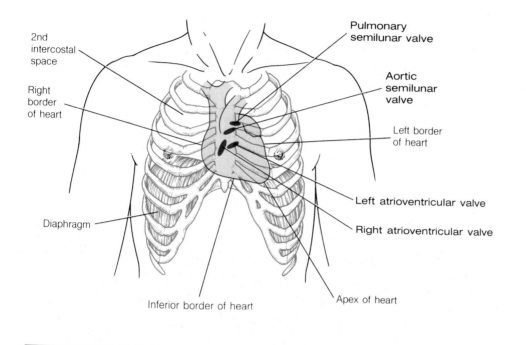

Figure 18.9
Anterior view of the thorax showing the position of the heart and the heart valves in relationship to the ribs, sternum, and diaphragm.

pulmonary semilunar valve is located in the proximal end of the pulmonary trunk; the **aortic semilunar valve** is in the proximal end of the aorta. Both sets of semilunar valves have three cusps. Each cusp resembles a shallow cup that has been cut in half vertically, with the cut edges attached to the walls of the vessel (Figure 18.8). When the ventricles contract, the force of the blood pushes the cusps against the vessel wall. When the ventricles relax, the blood starts to flow back into them. As this occurs, blood fills the cusps and causes their free margins to meet in the middle of the vessel, thus preventing any further backflow.

  F18.8

  The openings of the superior and inferior venae cavae, which empty into the right atrium, and those of the pulmonary veins, which empty into the left atrium, are not guarded by functional valves.

Surface Locations of the Valves

It is helpful to be able to locate the valves of the heart in relationship to the surface of the thorax (Figure 18.9). The semilunar valves are located toward the base of the heart. The aortic semilunar valve is behind the left side of the sternum, opposite the third intercoastal space. The pulmonary semilunar valve lies slightly above and to the left of the aortic semilunar valve, behind the costal cartilage of the third rib. The atrioventricular valves are situated more centrally in the heart than are the semilunar valves. The right atrioventricular valve lies almost directly behind the sternum, extending between the levels where the fourth and fifth costal cartilages join with the sternum. The left atrioventricular valve is located at the level of the fourth costal cartilage, behind the left side of the sternum.

  F18.9

CIRCULATION THROUGH THE HEART

Because the chambers on the right side of the heart are separated from those on the left by septa (the interatrial and interventricular septa), the heart functions as a double pump. Each pump has a receiving chamber (the atrium) and a propulsion chamber (the ventricle).

  The right pump receives blood that has passed through the vessels of the body and sends it to the lungs, through the *pulmonary circuit* (Figure 18.10).

  F18.10

CLINICAL CORRELATION
Aortic Stenosis and Replacement of the Aortic Valve

CASE REPORT . . . .

THE PATIENT   A 53-year-old man

PRINCIPAL COMPLAINT   Fatigue and chest pain during exertion.

HISTORY   At the age of 31 years, a heart murmur was detected during a routine physical examination for an insurance policy. Medical surveillance of the problem was 'not maintained. About a year before the present admission, the patient noted chest pain during exertion. The condition has progressed, with fatigue and chest pain during even slight exertion, and on one occasion the patient temporarily lost consciousness.

CLINICAL EXAMINATION   The patient appeared chronically ill. His arterial pressure was 95/80 mm Hg (normal: 90–140/60–90 mm Hg). His heart rate was 85 beats/min (normal: 60–100 beats/min), and his respiratory rate was 21 breaths/min (normal: 12–19 breaths/min). The ECG suggested left ventricular hypertrophy, but no evidence of myocardial infarction was found. A chest radiograph confirmed that the heart was enlarged and indicated that the lungs were clear. Calcification was indicated in the region of the aortic semilunar valve. A harsh ejection murmur (a murmur occurring

when the ventricles contract and eject blood) was heard, with maximal intensity at the apex of the heart.

CLINICAL COURSE   Digoxin (which increases cardiac contractility) was administered, and cardiac catheterization was performed. Coronary angiography (injection of contrast medium and observation under a fluoroscope) revealed no significant decrease in coronary blood flow. The aortic pressure was 90/50 mm Hg. The orifice of the aortic semilunar valve could not be penetrated. On the following day, the chest was opened, and the aortic semilunar valve was examined. It was heavily calcified, and only a pinhole-size orifice existed. The defective valve was excised, and a prosthetic aortic valve was installed.

OUTCOME   The patient recovered without event and was discharged three weeks after admission, on a regimen of digoxin and warfarin (an anticoagulant). At a follow-up examination six weeks after the operation, the patient felt well, the arterial pressure was 115/75 mm Hg, and no cardiac murmur was heard. The digoxin and warfarin were discontinued, and the patient was continued on aspirin (inhibits the aggregation of platelets) 150 mg twice a day.

Venous blood from the body enters the right atrium through several vessels:

1. The *superior vena cava*, which brings blood from the head, thorax, and upper limbs.

2. The *inferior vena cava*, which returns blood from the trunk, lower limbs, and abdominal viscera.

3. The *coronary sinus* and *anterior cardiac veins* that drain the myocardium.

From the right atrium, the blood enters the right ventricle, which pumps it through the pulmonary trunk and pulmonary arteries to the capillary network of the lungs. In the lungs, the blood gives up carbon dioxide and receives oxygen (see Chapter 22).

The left pump receives newly oxygenated blood from the lungs and sends it to the body, through the *systemic circuit*. Blood from the lungs is returned to the left atrium by pulmonary veins. From the left atrium, the blood enters the left ventricle, which pumps it through the aorta to the body.

PUMPING ACTION OF THE HEART

The right and left pumps of the heart work in unison: When the heart beats, both atria contract simultaneously, and then both ventricles do the same.

The period from the end of one heartbeat to the end of the next is called the *cardiac cycle*. During this cycle, blood from the superior and inferior venae cavae (as well as from the coronary sinus and anterior cardiac veins) moves through the right atrium, past the open right atrioventricular valve, and into the right ventricle (Figure 18.11). At the same time, blood from the pulmonary

F18.11

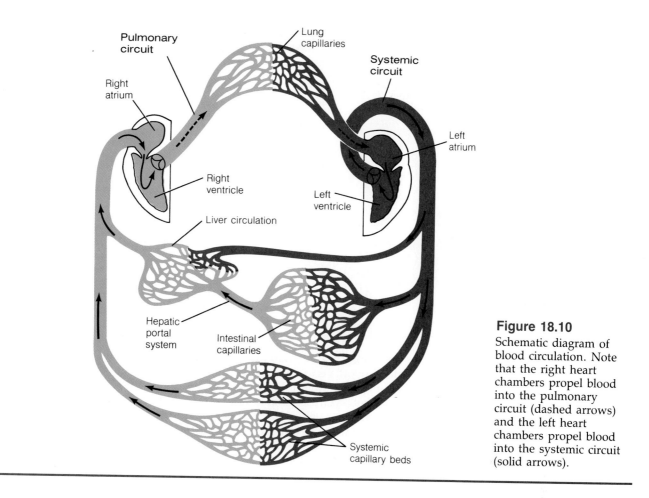

Figure 18.10
Schematic diagram of blood circulation. Note that the right heart chambers propel blood into the pulmonary circuit (dashed arrows) and the left heart chambers propel blood into the systemic circuit (solid arrows).

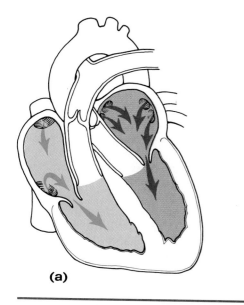 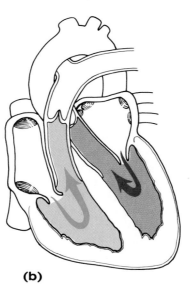 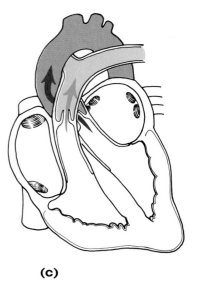

(a)   (b)   (c)

Figure 18.11
Blood flow through the heart during a single cardiac cycle. **(a)** Blood fills both atria and enters both ventricles. **(b)** The atria contract, which squeezes more blood into the ventricles. **(c)** The ventricles contract, which forces blood into the aorta and the pulmonary trunk.

×3100

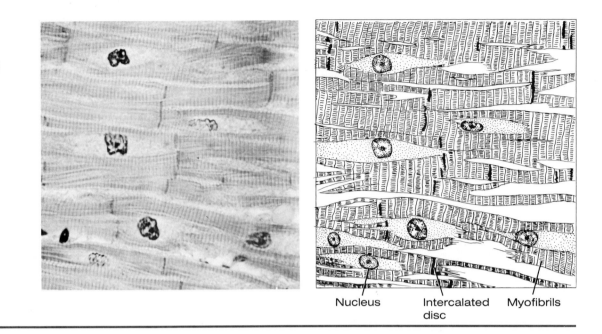

Nucleus   Intercalated   Myofibrils
                         disc

Figure 18.12

Microscopic appearance
of cardiac muscle.

veins moves through the left atrium, past the open left atrioventricular valve, and into the left ventricle. The simultaneous contraction of both atria then squeezes more blood into the ventricles. Subsequently, the simultaneous contraction of both ventricles closes the atrioventricular valves and forces blood past the semilunar valves and into the pulmonary trunk (from the right ventricle) and the aorta (from the left ventricle).

Note from this discussion that the contraction of the atria is not essential for the movement of blood into the ventricles. In fact, even if the atria fail to function, the ventricles can still pump considerable quantities of blood.

CARDIAC MUSCLE

The wall of the heart is composed almost entirely of cardiac muscle, and it is this muscle that is responsible for the pumping action of the heart.

Cellular Organization

F18.12

The individual cells of cardiac muscle interconnect with one another to form branching networks (Figure 18.12). Where adjoining cells meet end to end, their junctions form structures called *intercalated discs*. Within the intercalated discs are two types of cell-to-cell membrane junctions: desmosomes, which attach one cell to another; and gap junctions, which allow electrical impulses to spread from cell to cell. Cardiac muscle cells are cross-striated, and they possess thick myosin-containing filaments and thin actin-containing filaments that are arranged into regularly ordered sarcomeres and myofibrils. The basic contractile events that occur in cardiac muscle cells are believed to be similar to those that occur in skeletal muscle cells (Chapter 8). However, cardiac muscle cells have larger t tubules and a less-developed sarcoplasmic reticulum than skeletal muscle cells, and some of the calcium ions involved in contraction are obtained from the extracellular fluid.

Automatic Contraction

Like some types of smooth muscle, certain cells of cardiac muscle undergo spontaneous, rhythmical self-excitation. Moreover, a stimulatory impulse can spread from cell to cell through gap junctions. Thus, cardiac muscle is self-excitable, and it contracts automatically without external stimulation. However, the spontaneous activity of cardiac muscle is continually influenced by neurons of the autonomic nervous system that supply the heart and by certain chemicals and hormones that circulate within the blood.

Degree of Contraction

In general, when cardiac muscle contracts, it contracts as much as it can for the existing conditions. The individual cells of skeletal muscles react in basically the same way, but a stimulatory impulse that excites one skeletal muscle cell does not spread to adjacent cells. Thus, a skeletal muscle, which is composed of thousands of cells, shows graded contractions, depending on the number of cells stimulated. In contrast, a stimulus that excites a cardiac muscle cell can be transmitted to other cardiac muscle cells through gap junctions. Thus, an entire mass of interconnected cardiac muscle cells responds as a unit, contracting as much as it can for the existing conditions whenever a single cell is stimulated. This does not imply that all contractions of the cardiac muscle of the heart are identical. Indeed, there is considerable variation, depending on the conditions at the time of contraction.

Refractory Period

Following the excitation of a cardiac muscle cell, there is a relatively long period of time—called the *refractory period*—during which the cell cannot normally be re-excited. This period is particularly evident in ventricular cardiac muscle cells, for which the refractory period extends well into the relaxation phase of the contractile cycle (Figure 18.13). The relatively long refractory period normally prevents the heart from undergoing a prolonged tetanic contraction or a spasm, which would halt blood flow and therefore cause death.

Metabolism of Cardiac Muscle

The myocardium can obtain only insignificant amounts of energy—that is, ATP—from anaerobic metabolism. Therefore, cardiac muscle depends on aerobic metabolism for the continuous supply of energy required to support its contractile activity. For this reason, a substantial and continuous delivery of oxygen to the myocardium is essential, and the myocardium receives an abundant blood supply through the coronary arteries.

In a resting individual the myocardium obtains most of its energy from the oxidation of fatty acids and small amounts from the oxidation of lactic acid and glucose. During increased activity—for example, during heavy physical exercise—the myocardium actively removes lactic acid from coronary blood and oxidizes it directly for the production of energy. In this regard, it should be noted that the anaerobic metabolic processes utilized by skeletal muscles during vigorous activity produce lactic acid that is released into the blood.

EXCITATION AND CONDUCTION IN THE HEART

The heart contains specialized cardiac muscle cells that initiate the impulses that cause it to beat (Figure 18.14). It also contains specialized cardiac muscle cells that conduct the impulses throughout the myocardium. This conducting system coordinates the heartbeat, producing an efficient pumping action.

F18.14

Excitation

Like other cells, cardiac muscle cells maintain an unequal distribution of ions on either side of their cell membranes, and they are electrically polarized. If the cell membrane of a cardiac muscle cell becomes depolarized to threshold, an action potential results and the cell is stimulated to contract.

In the wall of the right atrium, near the entrance of the superior vena cava, is a small mass of specialized cardiac muscle cells called the **sinuatrial node (sinoatrial node** or **SA node).** In a resting adult, the cells of this node spontaneously depolarize to threshold and generate an action potential approximately 70 to 80 times each minute (that is, about every 0.8 sec). The reasons for this depolarization are not entirely clear. Some investigators suggest that the membranes of the cells of the sinuatrial node are relatively permeable to sodium and calcium ions and that the entrance of these ions into

sign-you-ay´-tree-al

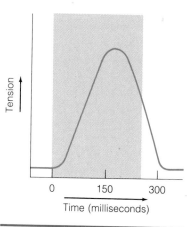

Figure 18.13
Contractile response of a ventricular cardiac muscle cell following its excitation at time 0. The refractory period is indicated by the shaded box.

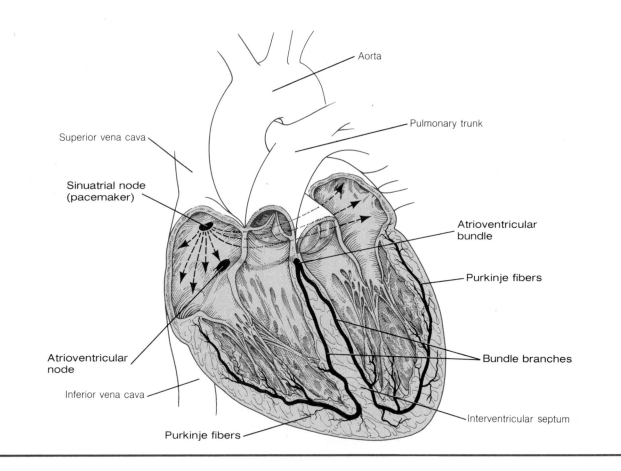

Aorta

Pulmonary trunk

Superior vena cava

Sinuatrial node (pacemaker)

Atrioventricular bundle

Purkinje fibers

Atrioventricular node

Bundle branches

Inferior vena cava

Interventricular septum

Purkinje fibers

Figure 18.14
The conducting system of the heart. The dashed arrows indicate the action potential from the sinuatrial node as it travels through the myocardium of both atria.

the cells gradually decreases the membrane potential (normally about −55 to −70 millivolts, inside negative) to threshold. Others stress the importance of a decreasing membrane permeability to potassium and a consequent diminished outflow of potassium ions to the depolarization process. In any event, the action potentials developed by the cells of the sinuatrial node depend in large measure on movements of sodium, calcium, and potassium ions, and a coupled sodium–potassium pump is operative in these cells.

Cells in other areas of the heart can also undergo spontaneous depolarization to threshold and generate action potentials. However, they do so at slower rates than the cells of the sinuatrial node. As a result, impulses from the sinuatrial node spread to these other cells and stimulate them so frequently that they do not generate action potentials at their own inherent rates. Thus, the rate of discharge of the sinuatrial node sets the rhythm for the entire heart, and it is for this reason that the sinuatrial node is called the *pacemaker* of the heart.

Conduction

An impulse generated in the sinuatrial node spreads from cell to cell (and also along special atrial pathways) throughout the myocardium of the atria, stimulating the atria to contract. However, the fibrous skeleton of the heart that surrounds the openings between the atria and the ventricles as well as the openings of the aorta and the pulmonary trunk (Figure 18.7) cannot depolarize. Therefore, a stimulatory impulse transmitted from the sinuatrial node throughout the atria cannot pass directly to the myocardium of the ventricles. Instead, the impulse reaches the ventricles by way of a specialized conducting system.

A group of specialized cardiac muscle cells called the **atrioventricular node (AV node)** is located within the interatrial septum just above the junction of the atria and the ventricles (Figure 18.14). From the atrioventricular node, a bundle of specialized muscle tissue called the **atrioventricular bundle** (*bundle of His*) passes to the ventricles. The atrioventricular bundle enters the

F18.7

F18.14

interventricular septum, where it divides into **right** and **left bundle branches** (or *crura*) that travel down the septum. Small groups of terminal conducting fibers called **Purkinje fibers** (or *crural rami*) exit from the bundle branches and end on the true cardiac muscle cells of the ventricles. At the apex of the heart, the Purkinje fibers pass to the outer wall of the ventricles and travel back toward the base of the heart.

per-kin´-jee

As a stimulatory impulse from the sinuatrial node spreads throughout the atrial myocardium, it reaches the atrioventricular node. After a delay of approximately 0.1 sec, the impulse is transmitted from the atrioventricular node, through the atrioventricular bundle and the Purkinje fibers, to the cells of the ventricles. One of the reasons for the delay in the transmission of the impulse at the atrioventricular node is that the cells in the region are quite small and thus transmit impulses slowly. This delay is important to a coordinated heartbeat because it ensures that the atria will contract and squeeze blood into the ventricles before the ventricles contract.

In contrast to the delay in the conduction of the stimulatory impulse at the atrioventricular node, the rate of conduction of the impulse from the node to the ventricles along the specialized conducting fibers is quite rapid. Thus, the entire ventricular myocardium is stimulated almost simultaneously, producing a strong, effective pumping action.

THE CARDIAC CYCLE

The main function of the heart is to receive blood from the veins at low pressure and pump it into the arteries at a pressure high enough to propel it through the vessels and back again to the heart.

In order to function as a pump, the heart repeats two alternating phases: (1) The chambers contract, forcing the blood within them out of the chambers. This phase is called **systole.** (2) The chambers relax, allowing them to refill with blood. This phase is known as **diastole.** The cardiac cycle, therefore, consists of a period of contraction and emptying (systole) followed by a period of relaxation and filling (diastole). Although both the atria and the ventricles undergo systole and diastole, the terms usually refer to ventricular contraction or relaxation.

sis´-ta-lee
dye-ass´-ta-lee

The pressure changes, volume changes, and valve actions that occur in the heart of a resting adult during the cardiac cycle are compared in Figure 18.15. Although this figure is concerned with the left chambers of the heart, the right chambers follow a similar pattern. However, the peak pressure within the right ventricle is considerably lower. The systolic pressure in the right ventricle—which only has to pump the blood through the pulmonary circuit—reaches about 30 millimeters of mercury (mm Hg), whereas, as the figure shows, the systolic pressure in the left ventricle reaches 120 mm Hg.

F18.15

In a resting adult, each cardiac cycle lasts approximately 0.8 sec. Atrial systole requires about 0.1 sec, and ventricular systole about 0.3 sec. Thus, during each cardiac cycle the atria are in diastole for 0.7 sec and the ventricles for 0.5 sec. It is important to emphasize that during ventricular diastole, blood flows into the atria and through the open atrioventricular valves into the ventricles prior to the contraction of the atria. In fact, the ventricles are 70% filled before the atria contract. When the atria do contract, they force a small amount of additional blood (approximately 30% of the ventricular capacity) into the ventricles. However, since the ventricles are almost full prior to atrial contraction, defects that impair atrial pumping may not seriously impair cardiac function, at least in resting individuals. (During exercise, when the heart rate is elevated and the diastolic filling period is shortened, atrial contraction contributes more importantly to ventricular filling than it does in a resting individual.)

Pressure Curve of the Left Atrium

During most of ventricular diastole, the pressure within the left atrium is slightly higher than the pressure within the left ventricle. Moreover, the pressure within the left atrium gradually increases as blood flows into it (and

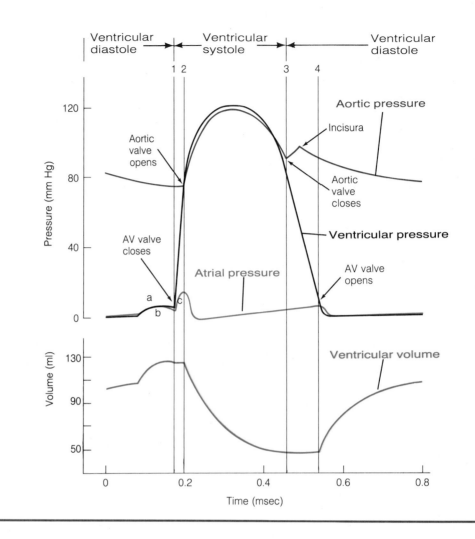

Figure 18.15

Comparisons of pressure
changes, volume
changes, and valve
actions within the left
chambers of the resting
heart during a single
cardiac cycle. Note that
the duration of a
complete cycle is 0.8 sec.

F18.15 continues into the left ventricle) from the pulmonary veins (Figure 18.15).
When the left atrium contracts, there is a sudden further increase in the atrial
pressure (curve a). Shortly thereafter, ventricular systole begins, and the
pressure within the left ventricle rises above that in the left atrium, causing
the left atrioventricular valve to close (line 1). The sudden increase in the
intraventricular pressure is so great that it causes the left atrioventricular
valve to bulge slightly into the left atrium. This activity contributes to a further increase in the atrial pressure (curve c) that lasts only until the aortic
semilunar valve opens (line 2). The pressure within the left atrium, which is
now in diastole, then quickly drops to almost 0 mm Hg. However, as blood
from the pulmonary veins again fills the atrium, the atrial pressure shows a
steady increase until the left atrioventricular valve reopens during the following ventricular diastole (line 4). The reopening of the left atrioventricular
valve allows blood from the left atrium to enter the left ventricle. The flow of
blood from the atrium into the ventricle causes the atrial pressure to drop
again to almost 0 mm Hg. The atrium is now ready to begin another cycle.

Pressure Curve of the Left Ventricle

F18.15 During most of ventricular diastole, the pressure within the left ventricle is
slightly lower than the pressure within the left atrium (Figure 18.15). The
contraction of the left atrium causes a slight increase in the pressure within
the left ventricle (curve b) because atrial systole increases the volume of blood
within the ventricle, as shown in the ventricular volume curve. When ventricular systole begins, the pressure within the left ventricle rises rapidly (lines 1
to 2). When the pressure in the left ventricle exceeds the pressure in the left

atrium, the left atrioventricular valve closes. When the pressure within the left ventricle becomes greater than the pressure within the aorta, the aortic semilunar valve is forced open (line 2), allowing blood to flow into the aorta. The pressure within the ventricle continues to rise to about 120 mm Hg and then begins to fall. At the end of ventricular systole, the intraventricular pressure drops below the pressure in the aorta. As a result, blood from the aorta begins to flow back toward the left ventricle, causing the aortic semilunar valve to close (line 3). The intraventricular pressure continues to fall precipitously until it becomes less than the atrial pressure (line 4), thus allowing blood in the left atrium to push open the left atrioventricular valve and flow into the left ventricle. Notice that the volume of blood within the left ventricle is at its lowest at this point (line 4) and that it immediately begins to increase as the blood that has collected within the left atrium flows into the left ventricle. This increase in blood volume occurs prior to atrial systole. Notice also that there is only a slight increase in ventricular volume following atrial systole.

Pressure Curve of the Aorta

The aorta is an elastic artery. As blood is pumped into the aorta during ventricular systole, the vessel distends, or expands. Then, as the blood flow from the left ventricle into the aorta decreases and stops during late systole and diastole, the aorta recoils and squeezes down on the blood within it. This helps maintain a certain amount of pressure within the vessel (Figure 18.15).   F18.15
If the aorta and the other elastic arteries had rigid walls that were not distensible, the pressure within them would rise very abruptly during ventricular systole. If they were not capable of elastic recoil, the pressure within them would fall just as abruptly during ventricular diastole.

When the left ventricle is in diastole and no blood is being ejected into the aorta, the aortic pressure curve shows a steady but gradual decrease as the blood pumped into the aorta during the previous systole flows into other vessels of the body (Figure 18.15). The elastic recoil of the aortic wall main-   F18.15
tains some pressure on the blood (the aortic blood pressure never drops much below 80 mm Hg), but it is not sufficient to keep the pressure constant. When the left ventricle contracts and the aortic semilunar valve opens (line 2), the aortic pressure increases rapidly as a large volume of blood is pumped into the aorta. The pressure peaks at about 120 mm Hg and then decreases as the flow of blood from the left ventricle into the aorta diminishes. However, the ventricular pressure falls more rapidly than the aortic pressure, and as it drops below the aortic pressure, the aortic semilunar valve closes (line 3). The closure of the aortic semilunar valve produces a brief rise in the aortic pressure, which forms the incisura on the diagram.

CARDIC OUTPUT

Each cardiac cycle moves blood out of the ventricles into the pulmonary circuit and the systemic circuit. The term **cardiac output** is used to indicate the volume of blood pumped per minute by either ventricle; for this reason, cardiac output is sometimes referred to as **cardiac minute output.**

Generally, it is the left ventricle that is referred to when cardiac output is discussed. However, over any period of time, the normal heart ejects the same amount of blood from each ventricle, although some variation may occur from beat to beat. If this was not the case and the left ventricle pumped a greater volume than the right, blood would accumulate in the systemic circuit. Similarly, if the right ventricle pumped more blood than the left, blood would accumulate in the pulmonary circuit. As we discuss later, unequal pumping by the two ventricles can be a problem in certain diseases of the heart.

Cardiac output is equal to the heart rate multiplied by the stroke volume. The **stroke volume** is the amount of blood pumped by one side of the heart per beat. The average stroke volume of a human heart is about 75 ml, and the

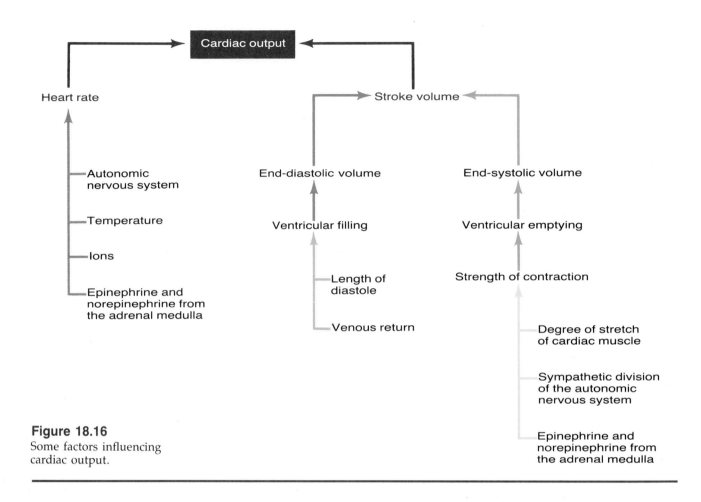

Figure 18.16
Some factors influencing cardiac output.

F18.16

heart rate of a resting individual is generally between 70 and 80 beats per minute. Therefore, the cardiac output under resting conditions is between 5 and 6 liters per minute (for example, 75 ml/beat × 70 beats/minute = 5250 ml/minute = 5.25 liters/minute). Under stressful conditions, the cardiac output can be greatly increased—up to 30 liters per minute in a trained athlete. The difference between the volume of blood moved per minute during rest and the volume that the heart is capable of moving per minute is called the **cardiac reserve.**

Changes in either the heart rate or the stroke volume can alter the cardiac output, and in general an increase in either factor increases the cardiac output. However, at very high heart rates—usually beginning between 170 and 250 beats per minute—cardiac output actually decreases. This decrease is due to the fact that stroke volume declines at very high heart rates, offsetting the effects of the increased rate. A major reason for the decline in stroke volume is that the length of diastole decreases as the heart rate increases, leaving a shorter time for the ventricles to fill with blood. Thus, less blood is present within the ventricles at the end of diastole to be pumped out during the following systole.

The cardiac output changes to meet the varying needs of the body, and since it is equal to the heart rate multiplied by the stroke volume, the mechanisms that control the heart rate and the stroke volume also control the cardiac output (Figure 18.16).

Control of Heart Rate

The most important factor in the control of heart rate is the effect of the autonomic nervous system. The sinuatrial node and the atrioventricular node are particularly well supplied by sympathetic and parasympathetic neurons, and the atria also receive both sympathetic and parasympathetic innervation.

The ventricles are supplied mainly by sympathetic neurons, with far fewer parasympathetic neurons present.

Stimulation of the sympathetic neurons increases the heart rate, and maximal sympathetic stimulation can almost triple the heart rate. The sympathetic neurons release norepinephrine, which increases the rate of discharge of the sinuatrial node, decreases the conduction time through the atrioventricular node, and increases the excitability of all portions of the heart. The exact mechanisms by which norepinephrine exerts its effects are not clearly understood, but it is believed to either decrease membrane permeability to potassium ions or increase membrane permeability to sodium ions and calcium ions, thereby enhancing the depolarization of excitable cells.

Parasympathetic fibers innervate the heart by way of the vagus nerves. The right vagus is distributed predominately to the sinuatrial node, and the left vagus innervates principally the atrioventricular node. Stimulation of the parasympathetic neurons to the heart decreases the heart rate. The parasympathetic neurons release acetylcholine, which decreases the rate of discharge of the sinuatrial node and increases the conduction time through the atrioventricular node. Acetylcholine increases membrane permeability to potassium ions, thereby causing hyperpolarization and making excitable tissue less excitable. Intense parasympathetic stimulation can stop impulse generation in the sinuatrial node and block the transmission of impulses through the atrioventricular node to the ventricles. However, soon after the ventricles stop contracting, some other portion of the conducting system—often in the atrioventricular bundle—begins to discharge spontaneously and stimulate ventricular contraction. This phenomenon is known as *ventricular escape*.

The heart rate at any given moment is largely determined by the balance between the stimulatory effects of norepinephrine released from sympathetic neurons and the inhibitory effects of acetylcholine released from parasympathetic neurons. If the sympathetic nerves to the heart are blocked, the heart rate decreases (due to the unopposed parasympathetic activity), and if the parasympathetic nerves are blocked, the heart rate increases (due to the unopposed sympathetic activity). In a resting individual, parasympathetic activity is dominant, and if all autonomic innervation is blocked, the heart rate increases from the normal 70 to 80 beats per minute to about 100 beats per minute, which is the inherent rate of discharge of the sinuatrial node. During exercise, the activity of the sympathetic neurons increases, the activity of the parasympathetic neurons decreases, and the heart beats faster.

Control of Stroke Volume

The amount of blood that moves out of a ventricle with each beat—that is, the stroke volume—is equal to the difference between: (1) the volume of blood within the ventricle at the end of diastole just as systole begins (the *end-diastolic volume*), and (2) the volume of blood remaining within the ventricle when the semilunar valves close at the end of systole (the *end-systolic volume*). A number of factors influence the end-diastolic and end-systolic volumes and, therefore, the stroke volume.

End-Diastolic Volume

The end-diastolic volume depends on the amount of blood that enters a ventricle during diastole, and this, in turn, is determined by two main factors: the length of diastole and the venous return.

LENGTH OF DIASTOLE   The longer the period of diastole, the more time is available for the ventricles to fill with blood. However, as the heart rate increases, the length of diastole becomes shorter, and less time is available for ventricular filling. In this regard, it should be noted that norepinephrine from the sympathetic nervous system not only increases the heart rate but also accelerates the rates at which cardiac muscle fibers shorten and relax. Thus, as norepinephrine increases the heart rate, the decreased time required for cardiac muscle fibers to shorten and relax provides somewhat more time for ventricular filling during diastole than would otherwise be the case. Moreover, the rapid relaxation of the ventricles after contraction causes the intraventricular pressure to fall rapidly, which enhances the pressure gradient for the flow of blood from the atria into the ventricles. Nevertheless, in a very rapidly beating heart, the length of diastole may be so shortened that ventricular filling is significantly reduced.

VENOUS RETURN   The volume of blood that returns from the veins to one side of the heart in a given time has a major influence on the end-diastolic volume. (The term *venous return* usually refers to the volume of blood flowing from the systemic veins into the right side of the heart in a given time, but under normal circumstances the same volume of blood flows from the pulmonary veins into the left side of the heart.) In general, if the length of diastole is constant, an increased venous return leads to increased ventricular filling and a relatively large end-diastolic volume that stretches the ventricle wall. If the length of diastole decreases (for example, when the heart rate increases), a ventricle may still fill completely if, as is often the case, the venous return increases. As we discuss in Chapter 19, conditions in the peripheral blood vessels have a major influence on the magnitude of the venous return.

End-Systolic Volume

The end-systolic volume depends on the amount of blood ejected from a ventricle during its contraction. A ventricle does not eject all of the blood it contains when it contracts, and the degree to which a ventricle is emptied during systole is determined primarily by the strength of the ventricular contraction.

THE FRANK-STARLING LAW OF THE HEART   The Frank-Starling law of the heart states that, within limits, as cardiac muscle is stretched, its force of contraction increases. Thus, when ventricular filling produces a comparatively large end-diastolic volume that stretches the ventricle wall, the ventricle contracts more forcefully and ejects a larger volume of blood than when ventricular filling produces a smaller end-diastolic volume that does not stretch the ventricle wall as greatly. This phenomenon provides an inherent self-regulating mechanism by which the heart is able to adjust stroke volume to changing end-diastolic volumes. However, if a ventricle is overly stretched (that is, if the end-diastolic volume is too great), the effectiveness of the ventricular contraction diminishes.

The influence of stretch on the force of contraction of cardiac muscle helps ensure that over any period of time the normal heart ejects the same amount of blood from each ventricle, even though some variation may occur beat to beat. For example, suppose the stroke volume of the right ventricle increases while that of the left ventricle remains unchanged. The increased right ventricular stroke volume causes an increased volume of blood to enter the pulmonary circuit, which in turn, leads to an increase in the volume of blood returned to the left ventricle. The increase in the volume of blood returned to the left ventricle produces a greater left ventricular end-diastolic volume, which stretches the ventricle wall. Consequently, the left ventricle contracts more forcefully, and its stroke volume increases, thereby restoring the balance in the amount of blood pumped by each ventricle.

SYMPATHETIC NERVOUS SYSTEM   Norepinephrine from the sympathetic nervous system is an important factor that influences the strength of ventricular contraction. Norepinephrine, which increases heart rate and accelerates the rates at which cardiac muscle fibers shorten and relax, also increases the force of contraction of cardiac muscle. At any particular end-diastolic volume (that is, degree of stretch), a more forceful contraction ejects a larger volume of blood from the ventricles, which in turn decreases the end-systolic volume and increases the stroke volume. Under normal conditions, sympathetic activity maintains the strength of ventricular contraction at a level about 20% greater than would be the case without stimulation, and maximal sympathetic stimulation can increase contractile strength to about 100% greater than normal. Norepinephrine is believed to increase contractile strength at least in part by increasing membrane permeability to calcium ions, which directly enhances the contractile activity of cardiac muscle cells.

In contrast to the effects of sympathetic activity, parasympathetic activity has comparatively little effect on the strength of ventricular contraction, and maximal parasympathetic stimulation decreases ventricular contractile strength by only about 30%.

FACTORS THAT INFLUENCE CARDIAC FUNCTION

Cardiac function can be influenced by a number of factors.

Cardiac Center

Both the sympathetic and parasympathetic neurons that supply the heart are under the control of a center within the brain stem called the *cardiac center*. This center receives neural input from higher centers in the brain and from various receptors associated with the cardiovascular system. In general, when the sympathetic nerves to the heart are stimulated, the parasympathetic nerves are inhibited. Conversely, sympathetic inhibition and parasympathetic stimulation are usually elicited simultaneously. The role of various inputs to the cardiac center in the control of cardiovascular activity is considered in Chapter 19.

Exercise

Chronic endurance-types of exercise generally cause cardiac muscle to hypertrophy and the ventricular chambers to enlarge, leading to an increase in stroke volume. For example, it has been found that the hearts of marathon runners have stroke volumes that are approximately 40 to 50% larger than the stroke volumes of the hearts of untrained people. Compared to a heart with a smaller stroke volume, a heart with a larger stroke volume can pump the same amount of blood per minute with fewer beats, and it can pump more blood when the rates of the two hearts are the same. Thus, both at rest and during exercise, an individual in good physical condition can generally maintain the cardiac output required for a particular activity level with a lower heart rate than an individual in poor physical condition. In addition, the maximum cardiac output and cardiac reserve are greater in a person in good physical condition.

During exercise, cardiac output increases greatly (for example, from a resting value of about 5.25 liters/min to a maximum value of about 30 liters/min in a trained athlete). This increase in cardiac output is due to an increase in both stroke volume and heart rate. In an exercising marathon runner with a cardiac output of 30 liters/min, the stroke volume is about 50% greater than the resting value (162 ml compared to 105 ml), and the heart rate is about 270% greater (185 beats per minute compared to 50 beats per minute). As the cardiac output increases from the resting level to 30 liters/min, the stroke volume reaches its maximum (162 ml) at a cardiac output of approximately 15 liters/min. At this cardiac output, the heart rate is about 92 beats per minute. Increases in cardiac output above this level are achieved by further increases in heart rate.

Temperature

When the heart is warmed, the discharge rate of the sinuatrial node increases, and a rise in body temperature of 1°C increases the heart rate about 12 to 20 beats per minute. This may account for the rapid heart rate that accompanies fever. However, increased activity of the sympathetic nerves that supply the heart may also be involved. When the heart is cooled, the heart rate declines and the heart ultimately stops. During some surgical procedures, the body temperature is artificially lowered, and the reduced heart rate that results makes it possible to perform operations that cannot be performed on a rapidly beating heart.

Ions

The levels of various inorganic ions in the blood and interstitial fluid can influence cardiac function. If the level of potassium ions is increased substantially, the heart rate drops and the heart becomes extremely dilated, flaccid, and weak. The elevated level of potassium ions is believed to decrease membrane potentials, which in turn decreases the intensities of action potentials, thereby weakening heart contractions. When the level of calcium ions is increased excessively, the heart contracts spastically, probably because of the increasing level of direct involvement of calcium ions in the contractile process. A substantial increase in the level of sodium ions slows the heart and depresses cardiac function, presumably by interfering with the normal role of calcium ions in the contractile process. The levels of calcium and sodium ions rarely rise sufficiently to alter cardiac function greatly.

Catecholamines

Epinephrine and norepinephrine from the adrenal medulla increase both the heart rate and the force of contraction, and they can provide a relatively small, slower-acting, but effective adjunct to the sympathetic innervation of the heart.

CONDITIONS OF CLINICAL SIGNIFICANCE

The Heart

Valvular Malfunctions

Valvular malfunctions can interfere with the normal movement of blood through the heart. They can decrease the amount of blood pumped out of a ventricle with each contraction, thereby making it necessary for the heart to work harder to maintain a given cardiac output.

Valvular Regurgitation

Valvular regurgitation occurs when the cusps of a valve do not form a tight seal when the valve is closed. As a result, blood leaks back, or regurgitates, into the chamber from which it came. If an atrioventricular valve does not close completely, blood flows back into the atrium when the ventricle contracts, and less than the normal amount of blood may be moved into the aorta or pulmonary trunk. Similarly, if a semilunar valve does not close completely, blood that moves into the aorta or pulmonary trunk during ventricular contraction flows back into the ventricle when it relaxes. Growths or scar tissue that form on a valve as a result of diseases such as rheumatic fever can prevent the valve from closing securely and cause valvular regurgitation.

Valvular Stenosis

Valvular stenosis is a condition in which the opening of a valve becomes so narrowed that it interferes with the flow of blood through it. If the opening of an atrioventricular valve is too narrow, the ventricle may not fill completely with blood, and a lower-than-normal amount of blood may be pumped out when the ventricle contracts. If the opening of a semilunar valve is too nar-

row, the ventricle may not eject a normal amount of blood into the aorta or pulmonary trunk when it contracts. Growths or scar tissue on a valve can cause valvular stenosis as well as valvular regurgitation. In many cases valvular regurgitation and valvular stenosis occur in the same valve.

Congestive Heart Failure

Congestive heart failure is a condition in which the heart fails to pump enough blood to meet the body's needs. In congestive heart failure there is an abnormal increase in blood volume and interstitial fluid, and the heart is generally dilated with blood (as are the veins and capillaries). Congestive heart failure may be due to impaired contractile ability of cardiac muscle (for example, as a result of a heart attack) or it may be due to an increased workload placed on the heart (for example, as a result of valvular malfunction). In any event, the heart is unable to perform normally the work demanded of it.

In congestive heart failure, compensatory mechanisms may initially allow the cardiac output to be maintained—at least in resting individuals. In response to impaired cardiovascular function, the activity of sympathetic nerves to the heart increases and the kidneys retain fluid in the body. The increased sympathetic activity increases the contractile strength of cardiac muscle, thereby helping to maintain cardiac output. The retention of fluid by the kidneys increases the blood volume (and the interstitial fluid volume). The increased blood volume leads to an increased venous return and end-diastolic volume that stretches the ventricular muscle. As described by the Frank-Starling law of the heart, the stretched muscle contracts more forcefully, increasing

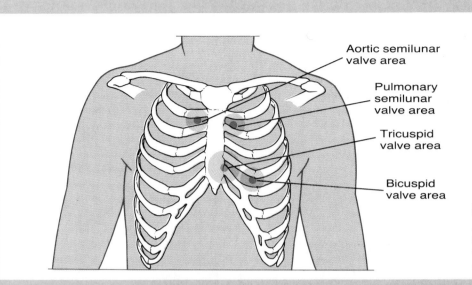

Aortic semilunar valve area

Pulmonary semilunar valve area

Tricuspid valve area

Bicuspid valve area

Figure 18.17
Areas of the chest where sounds associated with the different heart valves can be best detected.

the stroke volume and thereby helping to maintain cardiac output. In addition, the myocardium may hypertrophy (particularly in cases of valvular malfunction), and although the contractile activity per unit weight of hypertrophied muscle may be below normal, the increased muscle mass permits an overall increase in work capacity.

If the heart continues to fail, the compensatory mechanisms may become detrimental. The continued retention of fluid can cause the end-diastolic volume to increase to the point where the cardiac muscle is stretched excessively and its contractile strength declines. In addition, hypertrophied cardiac muscle may not receive a sufficient blood supply to meet its needs. Thus, the stroke volume decreases and the cardiac output declines.

In congestive heart failure, only one side of the heart may fail initially. For example, if the left heart is failing, the right ventricle continues to pump blood normally into the pulmonary circuit, but blood returning from the lungs is not pumped efficiently into the systemic circuit by the left ventricle. As a consequence, blood accumulates in the pulmonary circuit, and the pressure within the lung capillaries can rise to the point where fluid is forced out of the vessels. The accumulation of fluid in the lung tissues can result in potentially fatal pulmonary edema. Moreover, since the cardiovascular system is a closed circuit, the failure of the left side of the heart eventually produces an excessive strain on the right side of the heart that can result in total heart failure.

In addition to correcting the cause—for example, surgically repairing a defective valve—congestive heart failure is treated with drugs that increase the contractile strength of the failing cardiac muscle (for example, digitalis) and eliminate excess fluid (for example, diuretics).

HEART SOUNDS

Normal Heart Sounds

There are two principal sounds that normally occur as blood moves through the heart during a cardiac cycle.

These sounds are best described as "lub-dup." The first heart sound (the "lub") is associated with the closure of the atrioventricular valves at the beginning of ventricular contraction. It is largely due to vibrations of the taut atrioventricular valves immediately after closure and to the vibration of the walls of the heart and major vessels around the heart. The second sound (the "dup") is associated with the closure of the semilunar valves as the ventricles begin to relax following their contraction. This sound is due largely to vibrations of the taut, closed semilunar valves and to the vibration of the walls of the pulmonary artery, the aorta, and to some extent the ventricles. The areas of the chest where a stethoscope can be placed to detect more effectively the sounds associated with the different valves are indicated in Figure 18.17.

Abnormal Heart Sounds

By listening to the heart, a trained person can obtain considerable information about its condition. Abnormal sounds known as *heart murmurs* can be indicative of particular problems. These sounds, which are described as blowing or vibrating sounds, are caused by a turbulent flow of blood as it passes through the heart.

Both valvular regurgitation and valvular stenosis can cause heart murmurs. In valvular regurgitation, the backward movement of the regurgitated blood interferes with the normal pattern of blood flow through the heart, causing a detectable turbulence. In valvular stenosis, there is a rapid, turbulent flow of blood through the narrowed valvular opening. In addition, the walls around a narrowed valve are often roughened, further contributing to the turbulence.

Certain heart murmurs, called *functional murmurs*, are not pathological but are considered to be normal. For example, the rapid movement of blood through the heart during heavy exercise may result in turbulence that produces a functional murmur. Functional murmurs are particularly common in young people.

ELECTROCARDIOGRAPHY

The pattern of electrical activity associated with the contraction of cardiac muscle during a heartbeat can be recorded by **electrocardiography.** This procedure, which produces a recording called an **electrocardiogram (ECG)**, is useful in detecting conditions that interfere with the normal conduction of impulses through the heart.

As an impulse generated by the sinuatrial node travels through the heart, it produces electrical currents that spread through the body fluids surrounding the heart and then continue onward to the body surface. By placing electrodes on the body surface, it is possible to detect and record the electrical potentials generated by the heart.

The electrodes used to obtain electrocardiograms are placed in various positions on the body surface. Commonly, electrocardiograms are obtained using three "standard" limb leads, or locations for electrode placements (Figure 18.18). Although all three leads give similar patterns of recordings, there are differences in the amplitudes of the waves. These differences are often important in diagnosing various heart conditions.

Normal Electrocardiogram

A normal electrocardiogram of a single heartbeat consists of a regularly spaced series of waves designated **P, Q, R, S,** and **T** (Figure 18.19). (The letters simply indicate the order of appearance of the waves.) The P wave is caused by electrical currents that are produced as the atria depolarize prior to contraction. The QRS complex—actually three separate waves: a Q wave, an R wave, and an S wave—is caused by currents that are generated as the ventricles repolarize—that is, as the ventricles recover from being depolarized. The repolarization of the atria occurs during ventricular depolarization, and the atrial recovery wave is generally obscured by the QRS complex.

The time interval between the beginning of the P wave and the beginning of the QRS complex indicates the length of time between the beginning of the contraction of the atria and the beginning of the contraction of the ventricles. This period of time is referred to as either the P–Q interval or the P–R interval (because the Q wave is often absent). In a similar manner, the time between the beginning of the Q wave and the end of the T wave (the Q–T interval) provides a general indication of the duration of ventricular contraction.

Abnormal Electrocardiograms

The appearances of the different waves and the durations of the various intervals between the waves of the electrocardiogram are useful in diagnosing abnormalities that alter the conduction of impulses through the heart. The following are a few of the more common cardiac abnormalities detectable with electrocardiograms.

Abnormal Heart Rates

In a resting adult, the heart normally beats about 70 to 80 times a minute. When the rate drops below about 60 beats per minute, it is referred to as *bradycardia*. Bradycardia is generally not considered to be pathological. Much more serious and often associated with cardiovascular pathology is a resting heart rate of over 100 beats per minute, which is called *tachycardia*. When the heart rate is very fast, the ventricles do not have time to fill

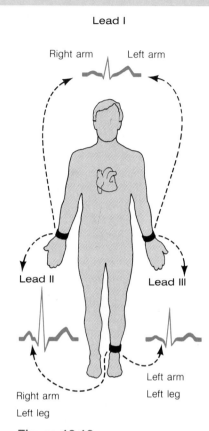

Figure 18.18
Locations of electrodes in the standard limb leads for an electrocardiogram. The leads are designated I, II, and III.

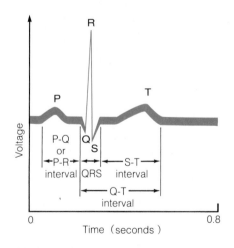

Figure 18.19
A normal electrocardiogram.

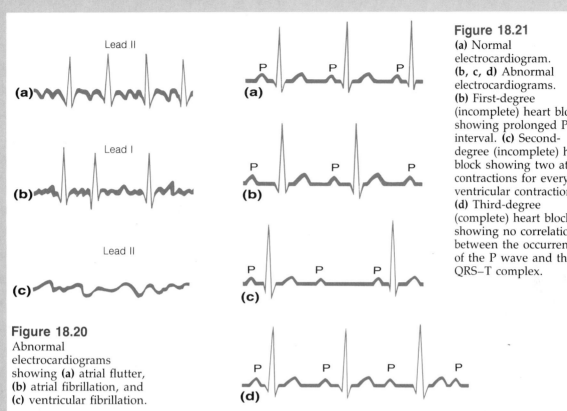

Figure 18.21
(a) Normal electrocardiogram. **(b, c, d)** Abnormal electrocardiograms. **(b)** First-degree (incomplete) heart block showing prolonged P–R interval. **(c)** Second-degree (incomplete) heart block showing two atrial contractions for every ventricular contraction. **(d)** Third-degree (complete) heart block showing no correlation between the occurrence of the P wave and the QRS–T complex.

Figure 18.20
Abnormal electrocardiograms showing **(a)** atrial flutter, **(b)** atrial fibrillation, and **(c)** ventricular fibrillation.

properly, and the movement of blood through the heart is impaired.

Somewhat coordinated atrial contractions that occur at a very rapid rate (often between 200 and 340 per minute) are called *atrial flutter* (Figure 18.20a). During atrial flutter, the atria pump almost no blood into the ventricles. Atrial flutter may occur in rheumatic and coronary heart disease. Extremely rapid, uncoordinated contractions of the atrial myocardium are called *atrial fibrillation* (Figure 18.20b). In atrial fibrillation, numerous impulses spread through the atria in all directions, and no P wave is evident on the electrocardiogram. The uncoordinated contractions of the atrial myocardium during atrial fibrillation are ineffective in pumping blood into the ventricles. Like atrial fibrillation, *ventricular fibrillation* is characterized by rapid, uncoordinated contractions of the ventricular myocardium. During ventricular fibrillation, a total irregularity of the QRS complex is evident on the electrocardiogram (Figure 18.20c). The uncoordinated contractions of the ventricular myocardium are ineffective in pumping blood from the ventricles, and ventricular fibrillation is generally fatal unless emergency measures (such as electrical defibrillation) are undertaken immediately.

Heart Block

Occasionally, the conduction of a stimulatory impulse through the heart is blocked at some point.

ATRIOVENTRICULAR BLOCK   Damage to or depression of the atrioventricular node or the atrioventricular bundle can impair the conduction of stimulatory

impulses from the atria to the ventricles. In first-degree (incomplete) heart block, there is a longer-than-normal delay in the conduction of the impulse from the atria to the ventricles. In first-degree heart block the electrocardiogram of a resting person's heart shows a P–R interval of greater than 0.20 sec (possibly up to between 0.35 and 0.45 sec) compared to a normal P–R interval of approximately 0.16 sec (Figure 18.21b). In second-degree (incomplete) heart block, some of the stimulatory impulses from the sinuatrial node fail to be transmitted to the ventricles. As a result, only every second (or third, or fourth, and so on) atrial contraction is followed by a ventricular contraction (Figure 18.21c). In this condition, the heart retains a definite, though altered, rhythm. In third-degree (complete) heart block, the impairment of conduction is so severe that no stimulatory impulses are conducted from the atria to the ventricles. In this condition, the ventricles contract at a rate that is slower than and completely independent of the rate of the atria. Often, impulses that stimulate the ventricles originate spontaneously in the atrioventricular bundle (whose inherent rhythm is slower than that of the sinuatrial node). Thus, the atrioventricular bundle functions as an ectopic (that is, "out of place") pacemaker for the ventricles in place of the normal pacemaker (the sinuatrial node). In the electrocardiogram of a complete heart block, there is no correlation between the occurrences of the P wave and the QRS–T complex (Figure 18.21d).

BUNDLE BRANCH BLOCK   If there is an impairment of conduction of the stimulatory impulse along one of the major branches of the atrioventricular bundle, the

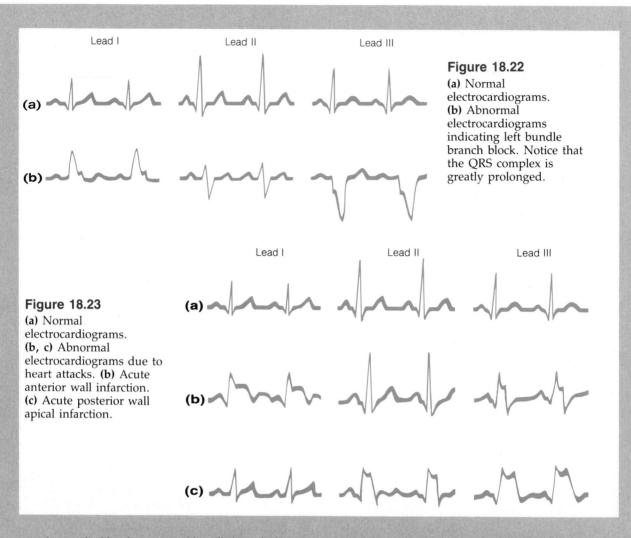

Lead I   Lead II   Lead III

(a)

(b)

Figure 18.22
(a) Normal electrocardiograms. **(b)** Abnormal electrocardiograms indicating left bundle branch block. Notice that the QRS complex is greatly prolonged.

Figure 18.23
(a) Normal electrocardiograms. **(b, c)** Abnormal electrocardiograms due to heart attacks. **(b)** Acute anterior wall infarction. **(c)** Acute posterior wall apical infarction.

Lead I   Lead II   Lead III

(a)

(b)

(c)

ventricle supplied by the impaired bundle branch depolarizes more slowly than the other ventricle. Such a block shows up on the electrocardiogram as a prolonged QRS complex (Figure 18.22). By varying the positions of the leads of the electrocardiograph, it is possible to determine whether the left or right bundle branch is blocked.

Heart Attacks

An insufficient flow of blood through the coronary arteries—due to an obstruction of a vessel, for example—can lead to myocardial damage and, if severe enough, to the deaths of cardiac muscle cells. An area of dead cardiac muscle cells resulting from inadequate blood flow is called a *myocardial infarct*, and a person who suffers a sudden obstruction of a coronary vessel (that is, a heart attack) generally incurs such damage. Because damaged or dead cardiac muscle cells do not conduct impulses normally, the presence of such areas produces altered ECG tracings (Figure 18.23).

FRONTIERS IN HEALTH

New Weapons in the Fight Against Heart Attacks

Jerry collapsed while playing tennis. When he arrived at the hospital his heart was beating, but only weakly, and only because of the heroic efforts of the emergency medical personnel who had transported him to the hospital.

Still dazed and only barely aware of what was going on, the 52-year-old patient was ushered into the cardiology ward. After a series of tests, his physicians made an incision in Jerry's groin and inserted a small plastic catheter into the femoral artery. With the aid of special equipment, they snaked the tube up through the aorta and into the coronary artery until it was stopped by a blood clot. The physician then injected streptokinase, an experimental drug, directly into the clot. This enzyme began to dissolve the clot and within 30 minutes had opened up the blood vessel, restoring blood flow to the oxygen-starved heart muscle.

Serious heart attacks strike thousands of Americans every year, and many of the victims never recover. The most common cause of a heart attack is atherosclerosis, a thickening of the arterial wall that narrows the lumen of the vessel and creates conditions favorable for the formation of blood clots. A heart attack can occur when a clot forms and obstructs blood flow in a coronary vessel, thus depriving the cardiac muscle cells supplied by the vessel of sufficient blood flow to meet their needs. What is responsible for atherosclerosis? The complete answer is not known, but too many cigarettes, too much alcohol, too much animal (saturated) fat, too much cholesterol, and too little physical activity have all been implicated.

Medical scientists are now seeking new ways to reduce the severity of heart attacks. Streptokinase, the enzyme given to Jerry, is one promising answer. It has been shown that, when given within four hours of the onset of an attack, streptokinase opens up clogged arteries and restores blood flow. This treatment reduces damage to heart muscle and hastens a patient's recovery. In fact, in several clinical studies streptokinase reduced the number of deaths by 75%. Recent work shows that streptokinase is effective even when given intravenously. This means that even hospitals without expensive cardiac catheterization facilities can administer the drug and improve the prognosis for hundreds of victims of heart attacks.

Streptokinase is a bacterial enzyme and unfortunately evokes an allergic reaction in many patients. As a result, it cannot be readministered to a patient for at least six months without possible serious consequences. Some patients may react violently to the foreign protein and die of anaphylactic shock. Streptokinase may also be hazardous if surgery is required, since small breaks in arteries may cause uncontrollable hemorrhaging in patients who have received the enzyme.

Because of the allergic response to streptokinase, scientists have been working with new chemicals that they hope will not cause such a response. Urokinase, an enzyme extracted from human tissue, is currently under study. Urokinase dissolves blood clots and, because it is

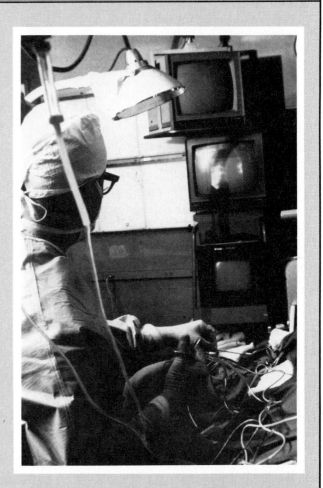

A cardiopulmonary catheter in use at the Beth Israel Hospital.

chemically the same in all humans, will not trigger an immune reaction. In time this substance may prove to be a suitable replacement for streptokinase. Mass production by genetic engineering could help lower its cost, which now is prohibitively high.

Medical researchers are also experimenting with another naturally occurring clot dissolver, called plasminogen activator, or TPA. TPA has already been produced successfully using genetic engineering, and medical researchers have established a cell line that produces large quantities of TPA when maintained in culture. TPA has proved remarkably effective when given to dogs. It works fast, breaking up the clot and re-establishing blood flow in as little as seven minutes. This quick re-establishment of blood flow reduces damage to cardiac muscle by 30–70%. TPA has been tested in humans and appears free of the side effects of streptokinase.

STUDY OUTLINE

EMBRYONIC DEVELOPMENT OF THE HEART p. 527
1. Begins as pulsating tubule.
2. Tubule becomes S-shaped.
3. Anterior vessel becomes pulmonary trunk and aorta.
4. Posterior vessel becomes superior and inferior venae cavae.
5. Four chambers develop: two atria and two ventricles.

POSITION OF THE HEART　Cone-shaped organ in mediastinum; size of closed fist. Base behind sternum at level of second and third ribs; apex to left of midsternal line at level of fifth intercostal space. p. 528

COVERINGS OF THE HEART pp. 528–529
1. Pericardium is a double-walled membranous sac (visceral and parietal layers).
2. Parietal pericardium has fibrous and serous layers.
3. Pericardial cavity contains pericardial fluid.

ANATOMY OF THE HEART pp. 529–537

CHAMBERS OF THE HEART

ATRIA (RIGHT AND LEFT)　Smaller, located toward superior region of heart; separated by interatrial septum.

VENTRICLES (RIGHT AND LEFT)　Larger, located at apex of heart; separated by interventricular septum.

VESSELS ASSOCIATED WITH THE HEART

SUPERIOR AND INFERIOR VENAE CAVAE　Return venous blood from body to right atrium.

PULMONARY TRUNK　From right ventricle to lungs.

PULMONARY VEINS　From lungs to left atrium.

AORTA　From left ventricle to body.

WALL OF THE HEART

EPICARDIUM　Serous membrane that adheres to outer surface of heart.

MYOCARDIUM　Cardiac muscle.

ENDOCARDIUM　Connective tissue, squamous cells; folds to form valves.

SKELETON OF THE HEART　Fibrous rings separating atria from ventricles.

VESSELS OF THE MYOCARDIUM　Supplied by coronary arteries; drained into coronary sinus by cardiac veins.

VALVES OF THE HEART

ATRIOVENTRICULAR VALVES
1. Right and left AV valves; tricuspid valve on right; bicuspid (mitral) valve on left.
2. Prevent blood backflow into atria during ventricular contraction.

SEMILUNAR VALVES
1. Pulmonary and aortic semilunar valves.

2. Prevent return of blood to ventricle from aorta and pulmonary trunk after contraction.

SURFACE LOCATION OF THE VALVES
1. Aortic semilunar valve—opposite left third intercostal space.
2. Pulmonary semilunar valve—behind left third costal cartilage.
3. Right AV valve—behind sternum, at level of fourth and fifth costal cartilages.
4. Left AV valve—at level of left fourth costal cartilage.

CIRCULATION THROUGH THE HEART pp. 537–538
1. Heart functions as a double pump: right pump receives blood from systemic circuit and pumps it into pulmonary circuit; left pump receives blood from pulmonary circuit and pumps it into systemic circuit.
2. Superior and inferior venae cavae, coronary sinus, and anterior cardiac veins return blood from body to right atrium. Blood then moves into right ventricle, which pumps it through pulmonary trunk and pulmonary arteries to capillary network of lungs. Blood from the lungs returns to left atrium by way of pulmonary veins, then moves into left ventricle, which pumps it through aorta to body.

PUMPING ACTION OF THE HEART pp. 538–540
1. Both atria contract simultaneously, followed by simultaneous contraction of both ventricles.
2. Contraction of atria not essential for movement of blood into ventricles; even if atria fail to function, ventricles still pump considerable quantities of blood.

CARDIAC MUSCLE pp. 540–541

CELLULAR ORGANIZATION　Cardiac muscle cells form branching networks and cells are connected end to end by intercalated discs. Contractile events believed similar to those in skeletal muscle cells.

AUTOMATIC CONTRACTION　Cardiac muscle contracts automatically without external stimulation; spontaneous activity is continually influenced by neurons of autonomic nervous system and by certain chemicals and hormones.

DEGREE OF CONTRACTION　When cardiac muscle contracts, it contracts as much as it can for existing conditions.

REFRACTORY PERIOD　Cardiac muscle cells have relatively long refractory periods that normally prevent heart from undergoing tetanic contraction or spasm.

METABOLISM OF CARDIAC MUSCLE　Myocardium can obtain only insignificant amounts of energy from anaerobic metabolism, and cardiac muscle depends primarily on aerobic metabolism for continuous supply of energy required to support its contractile activity.

EXCITATION AND CONDUCTION IN THE HEART pp. 541–543

EXCITATION   Sinuatrial node spontaneously generates stimulatory impulses approximately 70 to 80 times per minute in resting individual. Other areas of myocardium may also exhibit spontaneous activity, but sinuatrial node dominates them and is pacemaker of heart.

CONDUCTION   Stimulatory impulse from sinuatrial node spreads throughout myocardium of atria, stimulating atrial contraction and reaching atrioventricular node. After delay of about 0.1 sec, impulse is transmitted from atrioventricular node through atrioventricular bundle and Purkinje fibers to cells of ventricles.

THE CARDIAC CYCLE   Heart repeats two alternating phases: systole (contraction) and diastole (relaxation).   pp. 543–545

PRESSURE CURVE OF THE LEFT ATRIUM
1. Pressure increases gradually as blood enters from pulmonary veins.
2. Pressure increases suddenly when atrium contracts.
3. Pressure increases when ventricular systole begins and atrioventricular valve bulges into atrium.
4. Pressure then drops quickly.

PRESSURE CURVE OF THE LEFT VENTRICLE
1. Slight pressure increase when atrium contracts.
2. Rapid rise in pressure during ventricular systole.
3. Pressure falls at end of ventricular systole to value below that in aorta.

PRESSURE CURVE OF THE AORTA
1. Pressure decreases gradually during ventricular diastole.
2. Pressure increases rapidly when ventricle contracts and ejects blood into aorta.
3. Pressure then decreases, but rises briefly when aortic semilunar valve closes.

CARDIAC OUTPUT   Equal to heart rate multiplied by stroke volume. Stroke volume—amount of blood pumped by one side of heart per beat—decreases at very high heart rates. A major reason for this decrease is diminished time for ventricular filling due to shortened time of diastole.   pp. 545–549

CONTROL OF HEART RATE   Most important factor is autonomic nervous system: sympathetic stimulation increases heart rate and parasympathetic stimulation decreases heart rate.

CONTROL OF STROKE VOLUME   Stroke volume is equal to difference between end-diastolic volume and end-systolic volume.

END-DIASTOLIC VOLUME   Determined mainly by:

Length of Diastole

Venous Return

END-SYSTOLIC VOLUME   Determined mainly by strength of ventricular contraction, which is influenced by:

The Frank-Starling Law of the Heart   Stretching the ventricle wall increases strength of ventricular contraction.

Sympathetic Nervous System   Sympathetic stimulation increases strength of ventricular contraction.

FACTORS THAT INFLUENCE CARDIAC FUNCTION   pp. 549–550

CARDIAC CENTER   Both sympathetic and parasympathetic neurons to heart are under control of a cardiac center in brain stem.

EXERCISE   Chronic endurance-types of exercise generally cause cardiac muscle to hypertrophy and ventricular chambers to enlarge, leading to increase in stroke volume.

TEMPERATURE   Warming the heart increases heart rate; cooling the heart decreases heart rate.

IONS   When level of potassium ions is increased, heart rate drops and heart becomes extremely dilated, flaccid, and weak. When level of calcium ions is increased excessively, heart exhibits spastic contraction. Increase in level of sodium ions slows heart and depresses cardiac function.

CATECHOLAMINES   Epinephrine and norepinephrine from adrenal medullae can increase heart rate and force of contraction.

CONDITIONS OF CLINICAL SIGNIFICANCE: THE HEART   pp. 550–554

VALVULAR MALFUNCTIONS   Can make it necessary for heart to work harder to maintain a given cardiac output.

VALVULAR REGURGITATION   Occurs when cusps of a valve do not form tight seal when valve is closed. As a result, blood leaks into chamber from which it came.

VALVULAR STENOSIS   Valve opening becomes so narrowed that it interferes with flow of blood through it.

CONGESTIVE HEART FAILURE   Condition in which heart fails to pump enough blood to meet body's needs. In congestive heart failure there is an abnormal increase in blood volume and interstitial fluid, and heart is generally dilated with blood. Kidneys retain fluid, activity of sympathetic nerves to heart increases, and myocardium may hypertrophy.

HEART SOUNDS   p. 551

NORMAL HEART SOUNDS   First heart sound ("lub") is associated with closure of atrioventricular valves at beginning of systole. Second heart sound ("dup") is associated with closure of semilunar valves as ventricles begin to relax following their contraction.

ABNORMAL HEART SOUNDS   Often due to turbulent flow and are called *murmurs*. Both valvular regurgitation and valvular stenosis can cause heart murmurs.

ELECTROCARDIOGRAPHY   Electrodes placed on body surface detect electrical potentials generated by heart.   pp. 551–554

NORMAL ELECTROCARDIOGRAM   Consists of regularly spaced series of waves. P wave represents atrial depolarization. QRS complex represents ventricular depolarization. T wave represents ventricular repolarization. P–R interval represents length of time between beginning of contraction of atria and beginning of contraction of ventricles. Q–T interval provides general indication of duration of ventricular contraction.

ABNORMAL ELECTROCARDIOGRAMS

ABNORMAL HEART RATES *Bradycardia*: below 60 beats per minute; *tachycardia*: above 100 beats per minute. Somewhat coordinated atrial contractions occurring at very rapid rate are called *atrial flutter*. Extremely rapid, uncoordinated atrial contractions are called *atrial fibrillation* (no evident P wave on electrocardiogram). Extremely rapid, uncoordinated ventricular contractions are called *ventricular fibrillation* (total irregularity of QRS complex on electrocardiogram).

HEART BLOCK

Atrioventricular Block May delay or prevent impulse transmission from atria to ventricles. *First-degree block*: delayed impulse transmission (P–R interval exceeds 0.20 sec on electrocardiogram). *Second-degree block*: transmission of some impulses from sinuatrial node to ventricles is prevented (two, three, or more P waves for each QRS complex on electrocardiogram). *Third-degree block*: transmission of impulses from atria to ventricles is prevented (no correlation between P wave and QRS–T complex on electrocardiogram).

Bundle Branch Block Impairment of conduction in one of major branches of atrioventricular bundle (prolonged QRS complex on electrocardiogram).

HEART ATTACKS An area of dead cardiac muscle cells resulting from inadequate blood flow is called a *myocardial infarct*; areas of damaged or dead cardiac muscle cells result in altered electrocardiograms.

SELF-QUIZ

1. The heart is enclosed in a double-walled membranous sac called the: (a) mediastinum; (b) pericardium; (c) epicardium.

2. Match the various terms associated with the heart with the appropriate description.

   Pericardial cavity
   Atria
   Ventricles
   Interatrial septum
   Inferior vena cava
   Aorta
   Epicardium
   Myocardium
   Endocardium
   Trabeculae carneae

   (a) Separates the two atria
   (b) Carries blood from the left ventricle into the systemic circuit
   (c) The muscular layer of the wall of the heart
   (d) Small chambers located toward the superior region of the heart
   (e) Folds and bridges of the inner surface of the myocardium
   (f) Conducts blood to the right atrium
   (g) Space located between the visceral and parietal layers
   (h) Foldings of this structure form the valves that separate the atria from the ventricles
   (i) Large chambers that compose the bulk of the heart
   (j) A thin serous membrane that adheres to the outer surface of the heart

3. Blood is prevented from returning to the ventricles after they have completed their contractions by the: (a) atrioventricular valves: (b) tricuspid valve; (c) semilunar valve.

4. Blood within the pulmonary veins returns to the: (a) right atrium; (b) right ventricle; (c) left atrium; (d) left ventricle.

5. The left atrium contracts at the same time as the: (a) right atrium; (b) right ventricle; (c) left ventricle.

6. The cardiac muscle of the heart: (a) commonly undergoes prolonged tetanic contractions; (b) does not contract unless stimulated by the nervous system; (c) obtains only an insignificant amount of energy from anaerobic metabolism.

7. The sinuatrial node is the only area of the myocardium that can undergo spontaneous depolarization to threshold and generate action potentials. True or False?

8. The stimulatory impulse from the sinatrial node is normally delayed for a short time at the: (a) atrioventricular node; (b) atrioventricular bundle; (c) Purkinje fibers.

9. During left ventricular systole, which event occurs first? (a) the atrioventricular valve closes; (b) the semilunar valve opens; (c) the pressure within the ventricles peaks.

10. In a resting adult, during ventricular diastole, the ventricles are about 70% filled before the atria contract. True or False?

11. During most of ventricular diastole, the pressure within the left ventricle is slightly lower than the pressure within the left atrium. True or False?

12. Cardiac output is equal to the heart rate multiplied by the: (a) aortic pressure; (b) stroke volume; (c) ventricular end-systolic volume.

13. Stimulation of the parasympathetic neurons to the heart: (a) decreases the heart rate; (b) decreases the membrane permeability to potassium ions; (c) decreases the conduction time through the atrioventricular node.

14. Within limits, as cardiac muscle is stretched, its force of contraction increases. True or False?

15. Norepinephrine: (a) decreases the contractile strength of the ventricles; (b) decreases the heart rate; (c) accelerates the rates at which cardiac muscle fibers shorten and relax.

16. Chronic endurance-types of exercise most likely will lead to (a) an increased heart rate at rest; (b) an increased stroke volume; (c) a decreased cardiac reserve.

17. In general, when the level of potassium ions increases excessively, the heart rate increases and the heart exhibits spastic contraction. True or False?

18. Growths or scar tissue that form on a valve as a result of rheumatic fever can cause valvular regurgitation. True or False?

19. In congestive heart failure: (a) the blood volume decreases; (b) the kidneys excrete large volumes of fluid; (c) the activity of the sympathetic nerves to the heart increases.

20. The first heart sound is due largely to the: (a) vibrations of the taut atrioventricular valves immediately after closure; (b) vibrations of the taut, closed semilunar valves; (c) turbulent flow of blood past the open semilunar valves.

21. The P wave of a normal electrocardiogram indicates: (a) atrial depolarization; (b) ventricular depolarization; (c) atrial repolarization; (d) ventricular repolarization.

22. The Q–T interval of a normal electrocardiogram provides a general indication of: (a) the duration of atrial contraction; (b) the duration of ventricular contraction; (c) the length of time between the beginning of atrial contraction and the beginning of ventricular contraction; (d) the length of time that the stimulatory impulse is delayed at the atrioventricular node.

23. A P–R interval of 0.30 sec on an electrocardiogram would most likely indicate: (a) ventricular fibrillation; (b) an atrioventricular block; (c) atrial flutter.

LEARNING OBJECTIVES

After completing this chapter, you should be able to:

1. Describe the structure of arteries, veins, and capillaries.

2. Describe how blood viscosity, vessel length, and vessel radius affect the resistance to blood flow.

3. Discuss the relationship among blood flow, pressure, and resistance.

4. Describe how baroreceptors in the aortic arch and carotid sinus influence arterial pressure.

5. Describe how local autoregulatory mechanisms affect blood flow.

6. Discuss the factors involved in the movement of fluid between the blood and the interstitial fluid.

7. Describe the factors that influence venous return.

8. Distinguish between the pulmonary and the systemic circuits.

9. Identify the principal pulmonary arteries and veins.

10. Identify the principal systemic arteries and veins.

CHAPTER CONTENTS

TYPES OF VESSELS

GENERAL STRUCTURE OF BLOOD VESSEL WALLS

STRUCTURE OF ARTERIES

STRUCTURE OF ARTERIOLES

STRUCTURE OF CAPILLARIES

STRUCTURE OF VENULES

STRUCTURE OF VEINS

PRINCIPLES OF CIRCULATION

ARTERIAL PRESSURE

BLOOD FLOW THROUGH TISSUES

CAPILLARY EXCHANGE

VENOUS RETURN

EFFECTS OF GRAVITY ON THE CARDIOVASCULAR SYSTEM

CIRCULATION IN SPECIAL REGIONS

CARDIOVASCULAR ADJUSTMENTS DURING EXERCISE

CONDITIONS OF CLINICAL SIGNIFICANCE: BLOOD VESSELS

ANATOMY OF THE VASCULAR SYSTEM

KEY TERMS AND DERIVATIVES

angiotensin (*angio* = vessel) a substance found in the blood that constricts blood vessels

hypotension (*hypo* = below; *tension* = pressure) low blood pressure

phlebitis (*phleb* = vein) an inflammation of a vein

The Circulatory System:
Blood Vessels

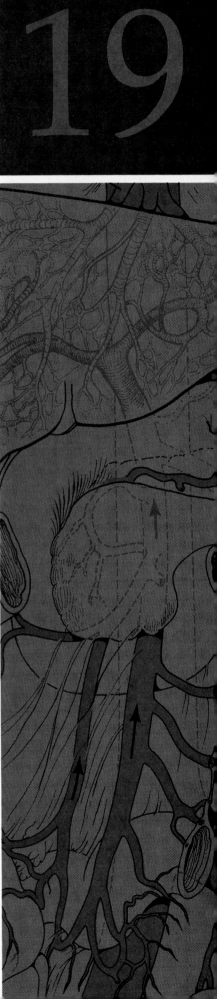

Upon leaving the heart, the blood enters the vascular system, which is composed of numerous *blood vessels.* The vessels transport the blood to all parts of the body; permit the exchange of nutrients, metabolic end products, hormones, and other substances between the blood and the interstitial fluid; and ultimately return the blood to the heart. Both the size of the vessels and the thickness of their walls vary, as does the blood pressure within them.

TYPES OF VESSELS

Large vessels called **arteries** carry the blood *away* from the heart. The major arteries divide into smaller arteries, then into still smaller **arterioles,** and finally into tiny **capillaries.** The capillaries converge into very small vessels called **venules,** which in turn join to form longer vessels called **veins.** The major veins *return* blood to the atria of the heart.

GENERAL STRUCTURE OF BLOOD VESSEL WALLS

Blood vessel walls vary in thickness. This variation is due to the presence or absence of one or more of three layers of tissues and to the differences in their thicknesses (Figure 19.1). The **tunica intima** (*tunic* = coat) is the innermost layer. It is formed of a layer of *simple squamous epithelium* called the **endothelium,** a layer of *connective tissue,* and a *basement membrane.* The endothelium of the tunica intima is the only layer present in vessels of all sizes. Moreover, it is continuous with the endocardium of the heart. Separating the tunica intima from the middle layer, and considered a part of the intima, there is often a thin layer of elastic fibers called the **internal elastic lamina.** The middle layer, the **tunica media,** is generally quite thick and is composed of *smooth muscle fibers* (mostly circularly arranged) mixed with *elastic fibers.* The outer border of the tunica media, where it contacts the outermost layer of the blood vessel wall, is often in the form of a distinct layer of elastic fibers called the **external elastic lamina.** The outermost layer is the **tunica externa** (or **adventitia**). This relatively thin layer of *connective tissue* contains elastic and collagenous fibers that run parallel to the long axis of the vessel. The walls of the larger vessels are too thick to be nourished by diffusion from the blood in the vessel. Instead, they are supplied by their own small nutrient vessels—the **vasa vasorum** ("vessels of the vessels")—which are located in the tunica externa and arise either from the blood vessel itself or from other vessels located close by.

STRUCTURE OF ARTERIES

The composition of the walls of the arteries differs, depending on the size of the vessels.

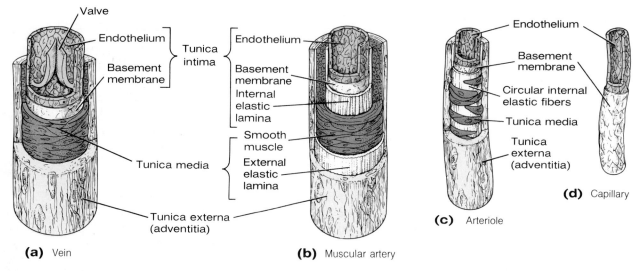

(a) Vein

(b) Muscular artery

(c) Arteriole

(d) Capillary

Figure 19.1
Comparison of the structure of blood vessels.

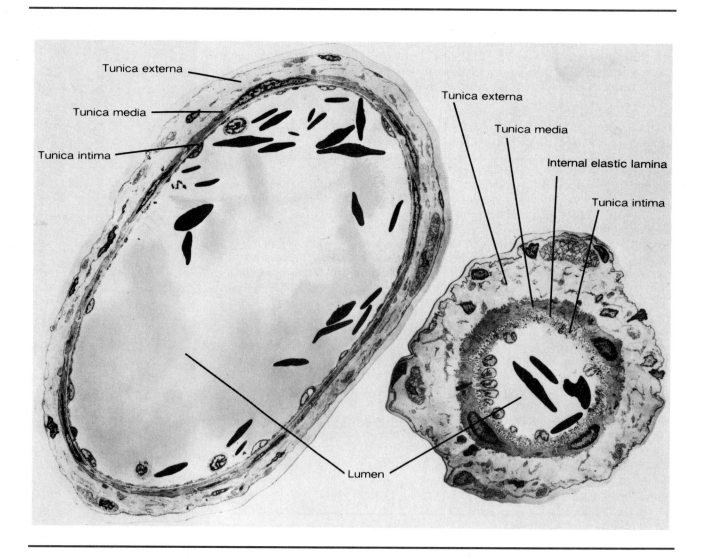

Figure 19.2
Photomicrograph of cross sections of dilated arteriole (left)
and constricted arteriole (right).

Elastic Arteries

The large arteries, such as the aorta and its major branches and the pulmonary trunk, are called **elastic arteries.** The walls of these arteries are composed of the three tunics just described. The tunica media of the large arteries is very thick, and in addition to smooth muscle fibers it also contains many elastic fibers.

During ventricular systole, elastic arteries are stretched as blood is ejected from the heart. During diastole, the recoil of the elastic arteries helps maintain pressure within the vessels. Although these arteries are commonly and conveniently considered to behave like passive elastic tubes, there is some evidence that this is not entirely the case. In animal experiments it has been found that smooth muscle in the walls of large arteries may be stimulated by the cardiac pacemaker during the contraction of the heart, thus producing some resistance to stretch and decreasing the distensibility of the arteries. This response may prevent overdilation of the arteries when blood is pumped into them.

Muscular Arteries

The tunica media of the walls of most smaller arteries consists almost entirely of smooth muscle cells, with relatively few elastic fibers. Such arteries are called **muscular arteries** or—because they carry blood throughout the body— **distributing arteries.** Muscular arteries have well-defined internal and external elastic laminae.

STRUCTURE OF ARTERIOLES

When an arterial vessel has a diameter of less than 0.5 mm, it is referred to as an **arteriole.** Arterioles have a small lumen and a relatively thick tunica media that is composed almost entirely of smooth muscle, with very little elastic tissue. In the smallest arterioles—that is, those closest to the capillaries—the external elastic membrane is lost and the tunica media is gradually reduced until it is composed of only a few scattered smooth muscle cells.

The arterioles play a major role in regulating the flow of blood into the capillaries. When the smooth muscle of the tunica media contracts, the internal cavities, or **lumens,** of the vessels are narrowed—that is, the vessels undergo *vasoconstriction,* which restricts the flow of blood into the capillaries (Figure 19.2). When the muscles relax, the lumens of the arterioles enlarge— that is, the vessels undergo *vasodilation,* which allows the blood to enter the capillaries freely.

STRUCTURE OF CAPILLARIES

In most tissues, a capillary network contains two types of vessels: *preferential channels,* which directly connect arterioles and venules, and *true capillaries,* which branch from and join with the preferential channels (Figure 19.3). A ring of smooth muscle called a *precapillary sphincter* usually surrounds each true capillary at the point where it arises from a preferential channel. The contraction and relaxation of the sphincters help regulate the flow of blood through the capillaries.

Capillaries have extremely thin walls. As a consequence, they are sites at which the exchange of materials between the blood and the interstitial fluid takes place. Capillary structure varies from one part of the body to another, but in general, a capillary consists of a single layer of endothelial cells surrounded by a thin basal lamina of the tunica intima. There is no tunica media or tunica externa present, and a single endothelial cell may form the entire circumference of a capillary (Figure 19.4). Endothelial cells are held to one another by tight junctions. Water-filled clefts occur between adjacent endothelial cells. In some capillaries, the endothelial cells contain small oval windows called *fenestrations,* which are usually covered by a very thin diaphragm.

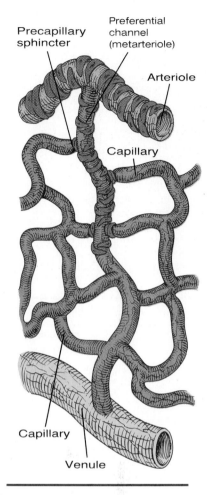

Figure 19.3
A capillary network. Preferential channels directly connect arterioles and venules; true capillaries branch from and join with the preferential channels.

F19.3

F19.4

fen-es-tray'-shuns

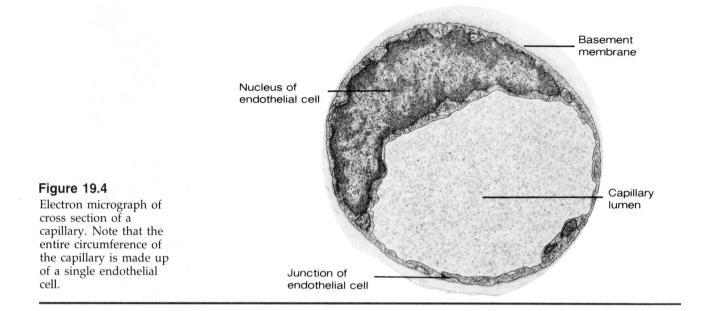

Figure 19.4

Electron micrograph of cross section of a capillary. Note that the entire circumference of the capillary is made up of a single endothelial cell.

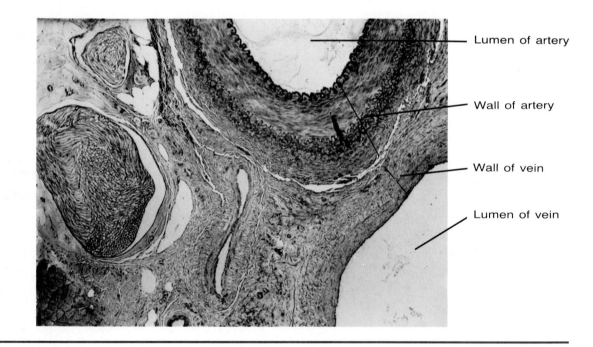

Figure 19.5

Photomicrograph of a portion of an artery and a vein. Notice how much thicker the wall of the artery is as compared to the wall of the vein.

Although a single capillary is only about 0.5–1 mm long and 0.01 mm in diameter, capillaries are so numerous that their total surface area within the body has been estimated to be more than 600 square meters. This provides a large surface across which the exchange of materials can occur. Substances may enter or leave capillaries by four possible routes: (1) through the tight junctions that anchor the endothelial cells together; (2) directly through the cell membrane; (3) within small membrane-bounded vesicles; and (4) in the case of capillaries with fenestrations, through pores between the endothelial cells.

The arterioles in certain structures, rather than connecting with capillaries, empty into relatively large vascular channels that have very thin walls. These channels are **sinusoids.** Sinusoids are so thin-walled that they generally conform in shape to the space in which they are located rather than being cylindrical. In some sinusoids the endothelium that lines them is discontinuous, with gaps present between the cells; in others the endothelium has small pores that are closed by thin diaphragms, as in fenestrated capillaries. Sinusoids are characteristic of the liver, spleen, bone marrow, and some endocrine glands.

STRUCTURE OF VENULES

In the venules closest to the capillaries, the walls have an inner lining composed of the endothelium of the tunica intima, surrounded by a very thin tunica externa. The larger venules that are further from the capillaries are encircled by a few smooth muscle fibers that form a thin tunica media.

STRUCTURE OF VEINS

The veins, which receive blood from the venules, have the same three coats as the arteries. In general, however, the tunica media of the veins is quite thin and has few muscle fibers. The tunica externa forms the greatest part of the wall, often being several times thicker than the media, although there is very little elastic tissue present. Veins have no internal or external elastic laminae. They tend to have a larger lumen and thinner walls than the arteries they accompany (Figure 19.5). The walls of the veins, like the walls of the arteries, receive nourishment through tiny vasa vasorum. Some veins contain valves that allow the one-way flow of blood toward the heart (Figure 19.6). These valves are folds of the tunica intima and have a form similar to the semilunar valves of the heart. When, in response to the pull of gravity, blood in the veins has a tendency to flow backward, away from the heart, the flaps of the valves fill with blood, thereby blocking the vessel. Valves are most common in the veins of the lower limbs, where the movement of blood depends largely on the contraction of the surrounding skeletal muscles.

PRINCIPLES OF CIRCULATION

Efficient circulation requires an adequate volume of circulating blood, blood vessels that are in good condition, and a heart that functions smoothly to pump the blood through the vessels.

Velocity of Blood Flow

In general, the velocity of blood flow in any segment of the cardiovascular system is inversely related to the total cross-sectional area of the vessels of the segment. In other words, the larger the total cross-sectional area, the lower the velocity of flow.

Although the lumen of an arteriole is smaller than the lumen of an artery, there are so many arterioles that their total cross-sectional area exceeds that of the arteries (Figure 19.7). Thus, the velocity of blood flow in the arterioles is lower than that in the arteries. Similarly, the smallest individual vessels, the capillaries, are so numerous that they have the greatest total cross-sectional area of the entire cardiovascular system. As a consequence, the velocity of blood flow in the capillaries is lower than that in any other vessels. The slow blood flow in the capillaries allows adequate time for the exchange of nutrients, metabolic end products, and other substances between the blood and the interstitial fluid.

Pulse

When blood is ejected from the heart during ventricular systole, the pressure within the arteries rises and the arteries expand. During diastole, the arterial pressure falls and the arteries recoil. The expansion and subsequent recoil of the arteries can be felt at various locations on the body surface. This is the **pulse.**

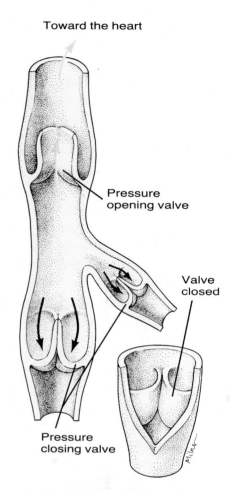

Toward the heart

Pressure opening valve

Valve closed

Pressure closing valve

Figure 19.6
Valves of a vein. As the arrows indicate, the valves are forced open by pressure from below and shut by pressure from above. This arrangement of valves allows blood to move in only one direction—toward the heart.

F19.7

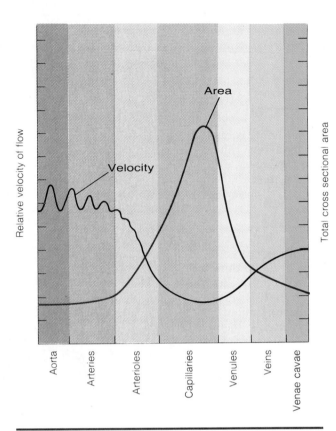

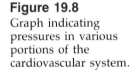

Figure 19.7
Graph indicating the relationship between the velocity of blood flow and the total cross-sectional area of various segments of the cardiovascular system.

Figure 19.8
Graph indicating pressures in various portions of the cardiovascular system.

The pressure fluctuations within the arterioles are less extreme than those within the arteries. Within the capillaries and beyond, the pressures do not rise and fall greatly with the beating of the heart, but remain at relatively constant levels.

Blood Flow

The **blood flow** is the actual volume of blood that passes through a vessel in a given time. Blood flow is a function of the pressure forcing the blood through a vessel, and the resistance of the vessel to the flow of blood through it.

Pressure

The pumping action of the heart imparts energy to the blood. This energy is evident as the **pressure** that drives the blood through the vessels.

Resistance

F19.8

As the blood flows through the vessels, it encounters varying degrees of **resistance,** which is essentially a measure of friction. As a consequence, the energy imparted to the blood by the pumping action of the heart is ultimately dissipated as heat. This dissipation of energy is evident as a progressive drop in pressure as the blood moves through the vessels (Figure 19.8). Thus, the pressure is highest in the arteries, and it gradually drops as the blood flows through the arterioles, capillaries, venules, and veins. In general, the greater the resistance the blood encounters, the harder the heart must pump, and the more pressure it must generate in order to keep the blood circulating.

Factors That Affect Resistance

Several factors affect the magnitude of the resistance the blood encounters as it flows through the vessels.

BLOOD VISCOSITY   The more viscous the blood, the greater the resistance to its flow through any given vessel. The more protein and cells within the blood, the greater its viscosity.

VESSEL LENGTH   The longer a vessel, the greater the resistance encountered as blood flows through it.

VESSEL RADIUS   The smaller the radius of a vessel, the greater its resistance to the flow of blood through it. The resistance of a vessel is inversely related to the fourth power of the vessel's radius:

$$\text{resistance } \alpha \ \frac{1}{\text{radius}^4}$$

For example, if the radius of a vessel is decreased by one-half, its resistance increases 16 times.

Alterations in Resistance

The lengths of the blood vessels are constant, and under normal circumstances the viscosity of the blood does not vary greatly. However, the radii of blood vessels can change. For example, the radii of the muscular arterioles vary according to the degree of constriction or relaxation of the arteriole muscles, and the arterioles are of major importance in determining the amount of resistance the blood encounters as it flows through the vessels.

The resistance to the flow of blood offered by the entire systemic circulation is called the *total peripheral resistance*. The resistance to the flow of blood offered by the entire pulmonary circulation is called the *total pulmonary resistance*.

Relationship among Flow, Pressure, and Resistance

The relationship among blood flow, pressure, and resistance can be expressed as:

$$\text{flow } (F) = \frac{\text{pressure}}{\text{resistance}} \tag{1}$$

The pressure term in this relationship is the pressure that drives the blood through a particular vessel or vessels. This pressure is represented by the pressure drop (ΔP) that occurs as the blood flows through the vessel, and it is equal to the pressure at the beginning of the vessel (P_1) minus the pressure at the end of the vessel (P_2). Thus, $\Delta P = P_1 - P_2$, and

$$F = \frac{\Delta P}{\text{resistance}} \tag{2}$$

The resistance term of this relationship includes the components of resistance discussed previously—blood viscosity, vessel length, and vessel radius. However, as previously pointed out, the component of resistance most likely to change and thus to affect the relationship is vessel radius, particularly arteriole radius.

The relationship among blood flow, pressure, and resistance expressed in equation 2 can be applied to the entire systemic or pulmonary circulation as well as to individual vessels or groups of vessels. For example, the volume of blood that flows through the systemic circulation in a given time is equal to the volume of blood ejected by the left ventricle during that time. If the time considered is one minute, then the flow term of the relationship is equal to

vis-koss'-i-tee

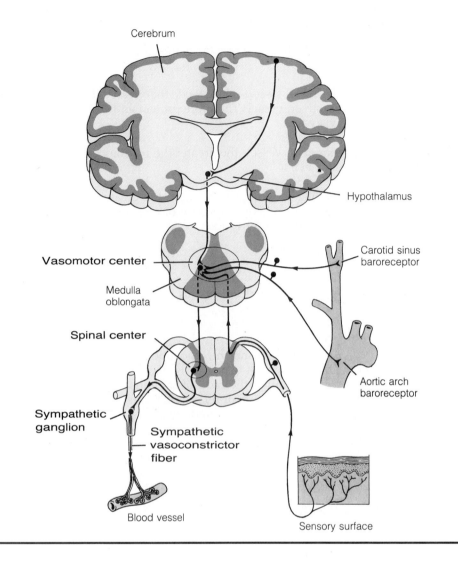

Figure 19.9
Diagrammatic representation of the neural pathways involved in the regulation of blood-vessel diameter.

the cardiac output (recall that the cardiac output is the volume of blood pumped per minute by a ventricle). Thus,

$$\text{cardiac output} = \frac{\Delta P}{\text{resistance}} \qquad (3)$$

The total resistance to the flow of blood through the entire systemic circulation is the total peripheral resistance. Thus, the resistance term of the relationship is equal to the total peripheral resistance, and

$$\text{cardiac output} = \frac{\Delta P}{\text{total peripheral resistance}} \qquad (4)$$

The pressure at the end of the systemic circuit (P_2) is the pressure within the veins at the right atrium. Since this pressure is almost zero, the pressure drop (ΔP) that occurs as blood flows through the systemic circuit is essentially equal to the pressure at the beginning of the circuit. That is, $\Delta P = P_1 - P_2 = P_1 - 0 = P_1$. This pressure ($P_1$) is the pressure in the aorta at the left ventricle due to the pumping action of the heart. However, as discussed in Chapter 18, the pressure within the aorta rises and falls with the beating of the heart. As such, the mean aortic pressure, rather than the aortic pressure during ventricular systole or diastole, provides the best indication of the pressure at the beginning of the systemic circuit. Thus,

$$\text{cardiac output} = \frac{\text{mean aortic pressure}}{\text{total peripheral resistance}} \qquad (5)$$

The resistance to flow in the large systemic arteries is only slight, and the pressure changes very little as the blood flows through them. Therefore, the mean aortic pressure is essentially equal to the mean arterial pressure, which can be approximated as

$$\frac{\text{systolic pressure} + 2 \text{ (diastolic pressure)}}{3}$$

and

$$\text{cardiac output} = \frac{\text{mean arterial pressure}}{\text{total peripheral resistance}} \qquad (6)$$

This relationship (equation 6) can be rearranged into the following form:

mean arterial pressure = cardiac output × total peripheral resistance   (7)

Thus, the mean arterial pressure is influenced by changes in either the cardiac output or the total peripheral resistance. Moreover, as discussed later, the body possesses receptors involved in sensing and regulating arterial pressure, and the activity of these receptors leads to changes in heart rate and contractile strength—which alter cardiac output—and to changes in vessel diameter—which alter total peripheral resistance.

ARTERIAL PRESSURE

The pressure within the large systemic arteries—which is essentially equal to the pressure generated by the pumping action of the left ventricle of the heart—is the immediate driving force for blood flow through the body's organs and tissues. Therefore, this pressure must be carefully maintained in order to ensure adequate organ and tissue blood flows. A number of factors affect the arterial pressure, through mechanisms that influence vessel resistance and cardiac function.

Neural Factors

Vasomotor Nerve Fibers

Vasomotor nerve fibers are efferent fibers that regulate blood vessel radii. With the exception of the capillaries, these fibers supply almost all types of vessels, particularly the arterioles. Changes in blood-vessel radii due to the activity of vasomotor nerve fibers alter the resistance of the vessels to blood flow, and this in turn can lead to a change in the mean arterial pressure (see equation 7).

Nerves of the sympathetic division of the autonomic nervous system contain both *vasoconstrictor* and *vasodilator* vasomotor fibers. The vasoconstrictor fibers are the most widely distributed and are the most important. At their junctions with blood vessels, they release norepinephrine, which stimulates the constriction of the vessels.

Vasodilator fibers—as well as vasoconstrictor fibers—are present in the sympathetic nerves that innervate skeletal muscles. At their junctions with blood vessels, the vasodilator fibers release acetylcholine, which leads to vessel dilation. The vasodilator fibers, however, are probably not very important in the overall control of vessel resistance. The major regulation of blood-vessel resistance is accomplished by the vasoconstrictor fibers, and most vessel dilation occurs as a result of diminished vasoconstrictor activity.

Vasomotor Center

In the lower third of the pons and the medulla oblongata of the brain is an area known as the **vasomotor center.** The vasomotor center plays an important role in the regulation of blood-vessel resistance and thus in the regulation of arterial pressure (Figure 19.9). Nerve impulses from this center are ultimately transmitted by vasoconstrictor fibers to blood vessels, and the vasomotor center is continuously, or tonically, active in promoting some degree of

F19.9

vasoconstriction, which is called *vasomotor tone*.

The activity of the vasomotor center is influenced by nerve impulses that arrive at the center from receptors associated with the cardiovascular system, as well as by nerve impulses from sensory surfaces of the body and from higher brain centers. Hormones and other substances in the blood also affect the activity of the vasomotor center. For example, carbon dioxide strongly stimulates the vasoconstrictor activity of the vasomotor center. When carbon dioxide levels increase, greater vasoconstriction is noted, and this fact has been used to explain the rise in arterial pressure seen early in asphyxiation.

Cardiac Center

The cardiac center, described in Chapter 18, influences arterial pressure by virtue of its effects on cardiac function. Many of the inputs that affect the vasomotor center also influence the cardiac center.

Baroreceptors

ba-ro-ree-sep'-ters
F19.9

In the arch of the aorta, in the slightly enlarged region where each common carotid artery divides into internal and external carotid arteries—that is, in the *carotid sinus*—and, to a lesser extent, in the walls of almost every large artery in the neck and thoracic regions are receptors that respond to the distension or stretch of the vessel walls. Since the degree of stretch of the vessel walls is directly related to the arterial pressure, these receptors function as pressure receptors, or **baroreceptors** (*baro* = pressure) (Figure 19.9). Nerve impulses from the baroreceptors ultimately inhibit the vasoconstrictor activity of the vasomotor center, and they also influence the cardiac center in such a manner that the activity of the parasympathetic nerves to the heart increases and the activity of the sympathetic nerves to the heart decreases.

When the arterial pressure rises, the rate of nerve-impulse transmission from the baroreceptors increases. As a consequence, blood vessels dilate, and the rate and contractile strength of the heart decrease, leading to a decline in cardiac output. These activities lower the arterial pressure.

When the arterial pressure falls, the rate of nerve-impulse transmission from the baroreceptors diminishes, resulting in an increased vasoconstriction and an increased rate and contractile strength of the heart. These activities raise the arterial pressure.

The arterial baroreceptors protect the cardiovascular system against relatively short-term changes in arterial pressure. However, they do little to protect the system from sustained long-term pressure changes. This situation is due to the fact that the baroreceptors display adaptation, and after a few days their rates of discharge return to normal levels regardless of continued high or low arterial pressures. The baroreceptors still function, but in essence they become "reset" to operate at a different pressure level.

Aortic and Carotid Bodies

The **aortic** and **carotid bodies,** which are located at the aortic arch and at the branchings of the common carotid arteries, contain receptors that are sensitive to arterial oxygen, carbon dioxide, and hydrogen ion concentrations. Like the baroreceptors, these receptors send impulses to the vasomotor center. Impulses sent in response to decreased arterial oxygen cause an increase in the arterial pressure. Changes in carbon dioxide and hydrogen ion concentrations can also alter the arterial pressure, but the effects of these substances by this pathway are relatively small.

Chemicals and Hormones

A number of chemicals and hormones influence the arterial pressure.

Angiotensin

an-jee-o-ten'-sin

Angiotensin is a powerful vasoconstrictor that raises the arterial pressure. When the arterial pressure falls, the formation of angiotensin, which is produced from precursors in the plasma by enzymatic action, increases.

Epinephrine

Epinephrine produced by the medullae of the adrenal glands causes a transitory increase in the systolic pressure within the arteries. Usually, when sympathetic nervous system activity causes widespread effects on blood vessels throughout the body, the adrenal medullae are stimulated, and epinephrine, as well as norepinephrine, is released.

Vasopressin

Vasopressin, or antidiuretic hormone, raises the arterial pressure by stimulating arteriole constriction. When the arterial pressure falls, the release of vasopressin from the pars nervosa of the pituitary gland increases.

Blood Volume

The volume of blood within the circulatory system affects the arterial pressure. In general, an increase in blood volume tends to raise the arterial pressure, in large part by increasing the venous return and ventricular filling, which in turn leads to an increased cardiac output. Conversely, a decrease in blood volume tends to lower the arterial pressure. Thus, mechanisms that alter blood volume can help control arterial pressure. For example, a rise in arterial pressure that leads to an elevation in capillary pressure favors the movement of fluid out of the capillaries and into the interstitial spaces. This movement of fluid decreases the blood volume and lowers the arterial pressure. Conversely, a fall in arterial pressure that leads to a decrease in capillary pressure favors the movement of fluid out of the interstitial spaces and into the capillaries. This movement of fluid tends to increase the blood volume and raise the arterial pressure.

The kidneys exert a significant effect on blood volume by virtue of their ability to regulate salt and water excretion, and they are particularly important in the long-term regulation of arterial pressure. A slight increase in the arterial pressure can cause a large increase in the formation of urine, which tends to decrease the blood volume and lower the arterial pressure. Conversely, a slight fall in the arterial pressure can lead to a substantial decrease in the formation of urine. As a consequence, fluids taken into the body tend to remain within the body, where they can increase the blood volume and thereby raise the arterial pressure.

Summary

From the previous discussion, it is evident that a number of different factors affect the arterial pressure. Consequently, the arterial pressure at any moment is usually the result of the combined influence of several factors. Moreover, different mechanisms work together to maintain the arterial pressure and, thus, to maintain a circulation that can meet the body's needs. For example, if the arterial pressure falls, neural mechanisms that involve baroreceptors, the cardiac and vasomotor centers, and vasomotor nerves, as well as chemical and hormonal mechanisms and mechanisms leading to changes in blood volume, may all act together to return the arterial pressure to normal.

Measurement of Arterial Pressure

The arterial pressure—which is commonly called the *blood pressure*— is usually measured indirectly (Figure 19.10). An inflatable cuff connected to a meter or mercury manometer is wrapped around the upper arm. The meter or manometer indicates the pressure within the cuff in millimeters of mercury (mm Hg). A pressure of 80 mm Hg, for example, is equal to the pressure exerted by a column of mercury 80 mm high. A stethoscope is placed below the cuff at the elbow in order to listen to the sounds in the brachial artery of the arm. The cuff is then inflated to a pressure sufficient to close the artery so that no blood flows past the cuff and no sound is heard with the stethoscope. The pressure is then slowly reduced until an intermittent thumping sound is heard. The pressure at which the sound is first heard is recorded as the *systolic pressure.* The sound is produced as follows: The contraction of the ventricles

F19.10

sis-tol'ik

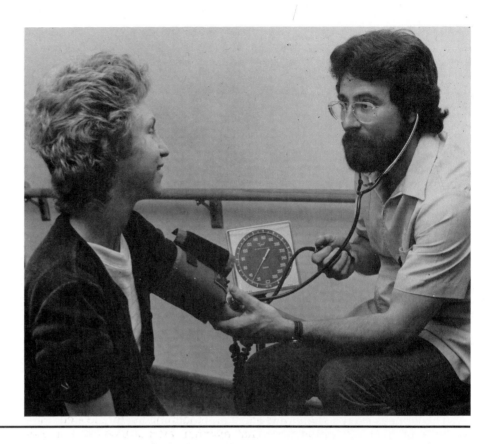

Figure 19.10
Indirect measurement of
blood pressure using a
manometer.

raises the pressure within the artery enough to overcome the resistance of the
cuff as it presses the artery closed. Consequently, blood flows past the cuff
into the lower arm in a turbulent fashion. It is the sound of this turbulence
that is heard with the stethoscope. When the ventricles relax and the pressure
in the artery drops, the cuff again closes the artery, no blood flows into the
lower arm, and no sound is heard with the stethoscope.

Upon further reduction of the pressure in the cuff, the intermittent
thumping sound becomes dull and muffled. The pressure at which this occurs is recorded as one diastolic end point (*diastolic pressure*). This point corresponds to the point at which the pressure in the artery, even during the
relaxation of the ventricles, is sufficient to overcome the resistance of the cuff,
and thus the vessel never closes. However, the cuff still exerts enough pressure on the artery to produce a turbulent blood flow that can be heard with
the stethoscope. If the pressure in the cuff is reduced further, the thumping
sound eventually disappears completely. The pressure at which this occurs is
recorded as a second diastolic end point. At this point, the flow of blood in
the vessel is smooth, and no intermittent sound of turbulence can be heard
with the stethoscope. Although the systolic pressure and both diastolic end
points may be recorded—for example, 120/85/80—generally only one diastolic end point is considered. Thus, arterial pressure is reported as a fraction
such as 120/80, with the numerator being the systolic pressure and the de-
nominator being the diastolic pressure. The difference between the systolic
and diastolic pressures is called the *pulse pressure*. The pulse pressure tends to
increase if the stroke volume of the ventricles increases or if the distensibility
of the arteries decreases.

BLOOD FLOW THROUGH TISSUES

Even when the mean arterial pressure is relatively constant, the blood flow
through many body tissues varies from time to time. Changes in the blood
flow through a particular tissue are primarily due to alterations in the resist-

ance of the tissue's blood vessels, and the resistance of the vessels is controlled by both local autoregulatory mechanisms and external factors.

Local Autoregulatory Control

Local autoregulatory mechanisms alter blood flow through a tissue according to the needs of the tissue. For example, in many tissues a decrease in the oxygen level leads to a local dilation of arterioles and to an increased relaxation of precapillary sphincters. These responses result in an increased blood flow through the tissue, which increases the delivery of oxygen. In addition, researchers have proposed that the metabolic processes of a tissue receiving an inadequate blood supply, as well as the low availability of nutrients and oxygen, lead to the formation and accumulation of substances that dilate arterioles and precapillary sphincters in the tissue. This dilation decreases the resistance to blood flow, leading to an increased flow through the tissue. Among the substances that have been proposed as local vasodilator substances are carbon dioxide, lactic acid, adenosine, adenosine phosphate substances, histamine, potassium ions, and hydrogen ions.

Some tissues have greater autoregulatory ability than others; this ability is particularly well developed in skeletal and cardiac muscle and in the gastrointestinal tract.

External Factors

The activity of external factors such as vasomotor nerve fibers and hormones can lead to the constriction or dilation of blood vessels in particular tissues and, therefore, to changes in blood flow through the tissues. However, external factors usually influence large segments of the vascular system, and they are generally more involved with mechanisms that function for the well-being of the entire body—such as those that maintain arterial blood pressure or body temperature—than they are with the control of tissue blood flow according to the needs of particular tissues. For example, in a warm environment, the activity of sympathetic vasoconstrictor fibers to skin vessels decreases, and the vessels dilate. This dilation results in an increased blood flow to the body surface, and thereby increases the loss of body heat from the blood to the environment.

CAPILLARY EXCHANGE

Capillaries are sites at which the exchange of materials between the blood and the interstitial fluid takes place. Most capillaries are permeable to water and small articles such as glucose, inorganic ions, urea, amino acids, and lactic acid; and these materials pass readily between the blood and the interstitial fluid. Permeability to protein, however, is usually quite limited, although it varies from tissue to tissue. It is highest in the liver, less in muscle, and very limited in the central nervous system.

Capillary Blood Flow

As previously described, blood flows slowly through the capillaries, allowing adequate time for the exchange of materials between the blood and the interstitial fluid. Moreover, the precapillary sphincter muscles at the entrances of the capillaries undergo cycles of contraction and relaxation at frequencies of 2–10 cycles per minute. Because of this cyclical activity, which is called *vasomotion*, capillary blood flow is usually intermittent rather than continuous and steady.

Local autoregulatory factors, particularly oxygen levels, have a strong influence on vasomotion. For example, in many tissues, when the oxygen level in a particular region is low, the intermittent periods of blood flow through the capillaries of the region occur more often, and each period of flow lasts longer. Thus, the blood flow through the region increases, leading to a greater delivery of oxygen and nutrients.

Although the blood flow through any one capillary is intermittent, there are so many capillaries that their function on an organ level becomes averaged. Therefore, it is possible to consider capillary function in terms of average capillary pressures, rates of flow, and rates of transfer of substances between the blood and the interstitial fluid. This approach is used in the following sections.

Movement of Materials between the Blood and the Interstitial Fluid

Several processes contribute to the movement of materials across capillary walls between the blood and the interstitial fluid.

Diffusion

Diffusion is the most important means by which substances such as nutrients and metabolic end products pass between the blood and the interstitial fluid. If a capillary is permeable to a particular substance, and if there is a higher concentration of the substance within the capillary than outside it, the net diffusion of the substance will be outward. If the substance is in higher concentration within the tissue fluid than within the capillary, its net diffusion will be inward. Lipid-soluble substances, including oxygen and carbon dioxide, can penetrate plasma membranes and move across capillary endothelial cells. Water-soluble substances, such as glucose and amino acids, can pass through water-filled clefts between adjacent endothelial cells. However, the size of the cleft limits the size of the particles that can pass through.

Endocytosis and Exocytosis

Substances are taken into one side of capillary endothelial cells by endocytosis and released from the other side of the cells by exocytosis. However, there is disagreement about the importance of this activity in the exchange of materials between the blood and the interstitial fluid.

Fluid Movement

A general movement of fluid—that is, water and dissolved particles to which a capillary is permeable—takes place across the capillary wall. As previously indicated, under certain circumstances, the movement of fluid between the blood and the interstitial fluid leads to alterations in blood volume (and arterial pressure). However, under normal circumstances, this movement causes little change in the volume of either the blood or the interstitial fluid. Several factors are involved in the movement of fluid across the capillary wall.

FLUID PRESSURES   The pressure of the blood within the capillaries (that is, the *capillary pressure*) tends to force fluid out of the capillaries and into the tissue spaces by filtration through the capillary walls. Opposing this movement is the pressure of the interstitial fluid (that is, the *interstitial fluid pressure*), which tends to move fluid out of the tissue spaces and into the capillaries. The interstitial fluid pressure, however, is generally much less than the capillary pressure, and there is considerable evidence that it is even below atmospheric pressure.

OSMOTIC PRESSURES   The presence of nondiffusable proteins in the interstitial fluid leads to the development of an osmotic force, called the *colloid osmotic pressure of the interstitial fluid*, which tends to draw water out of the capillaries by osmosis. This activity is opposed by the *colloid osmotic pressure of the plasma*, also called the *oncotic pressure*, which results from the presence of nondiffusible plasma proteins, particularly albumin. The oncotic pressure tends to draw water into the capillaries by osmosis.

on-kot'-ik

When osmosis occurs across a barrier like the capillary wall, which contains pores or clefts that are relatively wide compared to the width of a water molecule, substantial quantities of water move rapidly across the barrier, dragging along solutes to which the barrier is freely permeable. Thus, the

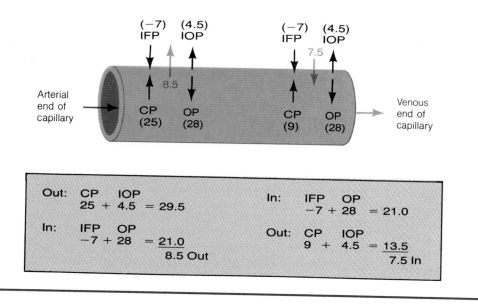

Figure 19.11
Forces involved in the movement of fluid into and out of capillaries. Values are general and may not apply to any particular capillary. They are expressed in mm Hg, with atmospheric pressure considered to be 0 mm Hg. Thus, negative values represent pressures below atmospheric pressure. CP = capillary pressure; OP = oncotic pressure; IFP = interstitial-fluid pressure; IOP = colloid osmotic pressure of interstitial fluid.

occurrence of osmosis across the capillary wall results in a movement of fluid similar to that due to the capillary and interstitial fluid pressures.

DIRECTION OF FLUID MOVEMENT   If the forces tending to move fluid out of a capillary (the capillary pressure and the colloid osmotic pressure of the interstitial fluid) are greater than the forces tending to move fluid into the capillary (the interstitial fluid pressure and the oncotic pressure), fluid leaves the capillary and enters the tissue spaces. If the inward forces are dominant, fluid enters the capillary from the tissue spaces.

The traditional view of fluid movement across the capillary wall is that under normal circumstances the capillary pressure at the arterial end of a systemic capillary is relatively high, and outward moving forces predominate (Figure 19.11). Thus, fluid moves out of the capillary into the tissues. As the blood flows through the capillary, the capillary pressure drops. Gradually the balance of forces favoring movement outward is reversed, and fluid moves back into the capillary at the venous end.

An alternative view of fluid movement across the capillary wall is that when a precapillary sphincter muscle relaxes during the rhythmic cycling of vasomotion, the capillary pressure rises and there is an outward movement of fluid all along the capillary. Then, when the precapillary sphincter constricts, the capillary pressure falls and there is an inward movement of fluid all along the capillary.

In either case, not all of the fluid that moves out of the capillaries returns to the capillaries. Some of the fluid, including any protein that escapes from the capillaries, enters the lymphatic system, to be ultimately returned to the bloodstream with the lymph.

VENOUS RETURN

Blood that leaves the capillaries flows into venules and then into veins on its way back to the heart. The greater the distance the blood travels, the greater the total resistance encountered and thus the greater the pressure drop. Blood in the venules and veins has traveled through the arteries, arterioles, and capillaries, and consequently the pressures within the venous circulation are quite low—for example, the pressures within the veins of the arms or legs average about 6–8 mm Hg.

The volume of blood that flows from the systemic veins into the right atrium of the heart in a given time is called the **venous return.** (Under normal circumstances, the same volume of blood flows from the pulmonary veins

F19.11

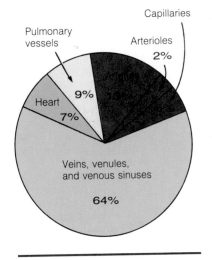

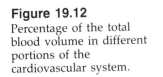

Figure 19.12
Percentage of the total blood volume in different portions of the cardiovascular system.

into the left atrium.) The magnitude of the venous return depends on the pressure driving blood through the veins and on the resistance of the veins to the flow of blood through them (recall that flow = ΔP/resistance). The pressure driving blood through the veins is represented by the difference between the pressure in the peripheral veins (the *peripheral venous pressure*) and the pressure at the right atrium of the heart. Consequently, factors that increase the peripheral venous pressure tend to increase the venous return, and factors that decrease the peripheral venous pressure tend to decrease it. For example, a decrease in the blood volume—as may result from hemorrhage—can lower the peripheral venous pressure and decrease the venous return, and an increase in the blood volume—as may result from a transfusion—can raise the peripheral venous pressure and increase the venous return. In a similar manner, factors that increase the right atrial pressure (which is normally close to 0 mm Hg) can also decrease the venous return. For example, a leaky right atrioventricular valve can cause an increase in the right atrial pressure and a decrease in the venous return.

The activities of skeletal muscles have an important influence on the venous return. The veins are quite flexible, and the movements of the skeletal muscles that surround them act in a pumping fashion to put pressure on the veins and compress them. This compression forces blood out of the veins and into the heart. The valves of the large veins contribute to the effectiveness of this pumping action by permitting blood leaving a compressed portion of a vein to flow only toward the heart and not in the opposite direction.

Breathing movements also influence the venous return. When a person inhales, the pressure within the thoracic cavity decreases, and the pressure within the abdominal cavity increases. These pressure changes facilitate the return of blood from the veins to the heart.

The veins have a large capacity, and they contain a substantial amount of the total blood volume. In fact, about 64% of the total circulating blood volume is normally in the systemic venous circulation, about 7% is in the heart, about 15% is in the systemic arterial vessels, about 5% is in the systemic capillaries, and about 9% is in the pulmonary vessels (Figure 19.12). Because of their large capacity, the veins serve as blood reservoirs. Although they are not heavily muscular, the veins can constrict and decrease their capacity, and even a slight venous constriction forces a good deal of blood into other portions of the cardiovascular system. Moreover, a constriction of the veins can decrease venous distensibility, raise the peripheral venous pressure, and increase the venous return. (In this regard it should be noted that the veins are relatively low-resistance vessels, and a moderate venous constriction does not greatly increase the overall resistance to blood flow.)

F19.12

EFFECTS OF GRAVITY ON THE CARDIOVASCULAR SYSTEM

Previous considerations of the pressures in various portions of the cardiovascular system generally assume the body to be in a horizontal position in which the blood vessels are at approximately the same level as the heart, and pressures due to the force of gravity acting on the blood are negligible. However, when the body is in an upright position, the weight of the blood affects these pressures. For example, when a person moves from a supine position to a position in which he or she is standing upright and perfectly still, the weights of the columns of blood within the blood vessels raise the pressure within the vessels of the lower regions of the body (Figure 19.13a). The increased pressure distends the veins, increasing their capacity so they contain a greater amount of the total blood volume. Moreover, the increased pressure results in an increased filtration of fluid out of the capillaries and into the tissue spaces. The accumulation of blood in the veins and the increased filtration of fluid out of the capillaries reduces the effective circulating blood volume. If no adequate compensatory adjustments occur, the arterial pressure falls, the blood flow to the brain declines, and the person becomes dizzy or

F19.13a

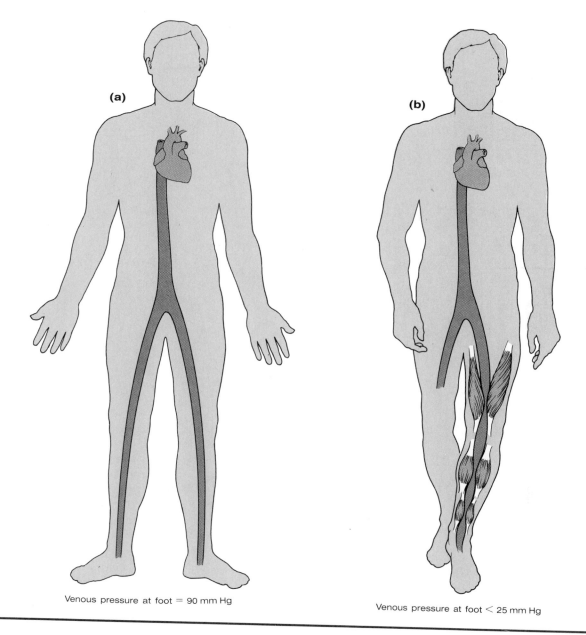

Venous pressure at foot = 90 mm Hg

Venous pressure at foot < 25 mm Hg

Figure 19.13

(a) When a person is standing upright and perfectly still, the weights of the columns of blood within the blood vessels raise the pressure within the vessels of the lower regions of the body. (b) During movement, muscular contractions compress the veins and force blood toward the heart. This activity reduces the pressure within the veins of the lower regions of the body, such as the veins of the feet.

faints. Under most circumstances, however, several mechanisms compensate for the effects of gravity when a person is upright.

One mechanism is the baroreceptor mechanism described earlier. For example, when a person sits up or stands after lying down, the force of gravity causes the blood to accumulate in the veins of the lower limbs and abdomen, and the arterial pressure in the head and upper body falls. However, the falling pressure is sensed by the baroreceptors, which initiate a generalized vasoconstriction and increase in heart rate and contractile strength, thus preventing a severe pressure drop that could result in diminished blood flow to the brain and a loss of consciousness.

A second mechanism that compensates for the effects of gravity is the pumping action exerted on the veins by the skeletal muscles of the legs. The contractions of the muscles compress the veins and force blood toward the

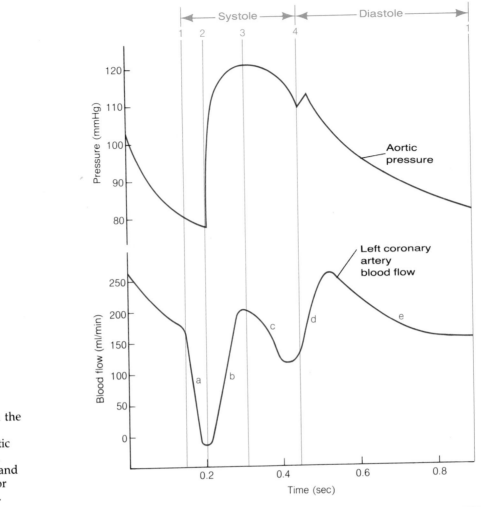

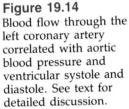

Figure 19.14
Blood flow through the left coronary artery correlated with aortic blood pressure and ventricular systole and diastole. See text for detailed discussion.

F19.13b   heart (Figure 19.13b). This activity enhances the venous return and reduces the pressure within the veins of the lower regions of the body. As a result, the upright position has less of an effect on venous distension and the accumulation of blood in the veins as well as on the capillary pressure and the filtration of fluid out of the cardiovascular system than would otherwise be the case. The importance of this mechanism is evident in the fact that when a person is standing perfectly still, the pressure within the veins of the feet can rise above 90 mm Hg, but the muscular activity of walking reduces the pressures within these veins to less than 25 mm Hg.

CIRCULATION IN
SPECIAL REGIONS

Certain regions of the body have special needs or present special problems that are met by various circulatory adaptations. In this section we consider some of the adaptations that occur in these regions.

Pulmonary Circulation

The pulmonary arteries and arterioles generally have larger diameters and thinner walls than corresponding vessels of the systemic circulation, and the resistance to blood flow through the pulmonary circulation is less than that through the systemic circulation. As a result, less pressure is required to move blood through the pulmonary circuit than through the systemic circuit, and the pressure within the pulmonary arteries (about 22 mm Hg systolic pressure, 8 mm Hg diastolic pressure) is considerably lower than the pressure within the systemic arteries (about 120 mm Hg systolic pressure, 80 mm Hg diastolic pressure).

The blood flow through different local areas of the pulmonary circulation varies with the oxygen levels in the areas. A decreased oxygen level causes vasoconstriction, and an increased oxygen level causes vasodilation. Although the local effect of oxygen on pulmonary vessels is the opposite of its effect on many systemic arterioles, it is consistent with the role of the lungs in providing oxygen to the blood. For example, if a portion of the lungs is not functioning effectively, the level of oxygen in the region falls. The decline in the oxygen level causes the vessels supplying the area to constrict, which results in a greater blood flow through the vessels in efficiently functioning areas of the lungs.

Coronary Circulation

The flow of blood through the coronary vessels is greatly influenced by the aortic pressure and by the pressures that build up in the walls of the heart chambers during systole—that is, the *intramural pressures*. The influence of these pressures is particularly evident in the left coronary artery (Figure 19.14). When the left ventricle contracts (line 1), the intramural pressure increases greatly, squeezing the coronary vessels and causing a significant reduction in the coronary flow (curve a). With the opening of the aortic semilunar valve (line 2), the effect of the intramural pressure of the ventricles is overcome by the increased aortic blood pressure. The increased aortic pressure causes an increase in blood flow through the left coronary artery (curve b). As ventricular systole progresses, the flow of blood into the aorta diminishes and the pressure within the aorta begins to fall (line 3). However, the left ventricle is still contracting at this time, and the intramural pressure remains high. As a result, there is a secondary slowing of the coronary flow (curve c). As ventricular systole ends (line 4), the intramural pressure drops, and the flow through the coronary vessels increases once again (curve d). During ventricular diastole, the flow through the coronary vessels is determined primarily by the aortic pressure, and the flow gradually decreases as the aortic pressure drops (curve e). Also, during this period, the atria contract and the pressure within their walls increases. This increase in the atrial intramural pressure also tends to decrease the flow of blood through the coronary vessels.

F19.14

Under resting conditions, the blood flow through the coronary vessels is approximately 250 ml per minute, which is about 5% of the cardiac output. However, during strenuous exercise, the coronary vessels dilate, and the coronary blood flow can increase to four or five times the resting level—that is, up to 1250 ml per minute. The dilation of the coronary vessels during exercise is believed to be primarily due to the influence of local autoregulatory factors (such as low oxygen levels resulting from increased cardiac activity).

An increased blood flow to the heart during exercise is particularly important because, even in the resting condition, the heart removes about 65% of the oxygen in the arterial blood that flows to it. The removal of this much oxygen is very efficient compared to most other tissues, where as little as 25% of the oxygen is removed. However, it means that during activity, the heart cannot greatly increase the oxygen available to it by increasing the percentage of oxygen removed from the blood. Therefore, increasing the coronary blood flow—and thus the delivery of oxygen—is the principal way in which additional oxygen is made available to the myocardium.

Cerebral Circulation

The neurons of the brain are unable to store glycogen, and the brain must have a dependable blood flow in order to ensure that sources of metabolic energy such as glucose are available to it. In addition, brain tissue must have a steady, adequate supply of oxygen. Irreversible damage occurs if the brain's

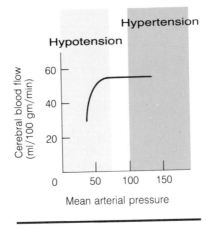

Figure 19.15
Relationship of mean arterial pressure and cerebral blood flow. The central area is the normal range of mean arterial pressure.

blood flow is stopped for over five minutes. The blood flow through the brain is about 750 ml per minute, which is 15% of the total cardiac output at rest. The overall metabolic rate of the brain varies little under widely different physiological conditions—for example, intense mental activity, muscular activity, or sleep—and under most circumstances, cerebral blood flow does not vary greatly.

The cerebral blood flow is relatively unaffected by changes in the systemic arterial pressure. Cerebral blood flow is remarkably constant over a very broad range of mean arterial pressures, and it does not decline significantly unless the mean arterial pressure is considerably below normal (Figure 19.15). This constancy of flow is believed to be due in large part to regulatory mechanisms that cause the large cerebral arteries to constrict when the systemic arterial pressure rises and dilate when the systemic arterial pressure falls.

The cerebral blood flow can be altered by changes in the concentrations of carbon dioxide or oxygen in the blood. An elevated concentration of carbon dioxide in the blood or a decreased concentration of oxygen causes the cerebral blood vessels to dilate, which increases cerebral blood flow.

The blood flow to different local areas of the brain increases in response to increases in the activities of the areas. For example, clasping the right hand is accompanied by an increased blood flow in the motor cortex on the left side of the brain. Local autoregulatory mechanisms are believed to contribute to changes that occur in local blood flow. An elevated concentration of carbon dioxide in an active local area of the brain or a decreased concentration of oxygen is believed to cause a dilation of arterioles in the area, which increases blood flow to the area. (In general, the mechanisms that regulate brain blood flow seem to be more sensitive to carbon dioxide concentrations than they are to oxygen concentrations.)

When there is a temporary interference with the cerebral blood flow, a person may faint, and factors that lower the arterial pressure below the level required to maintain an adequate cerebral blood flow can cause fainting. For example, the fainting that often accompanies severe emotional upset is due to a slowing of the heart and to a strong stimulation of the vasodilator fibers to skeletal muscle vessels. As a result, the arterial pressure declines, blood flow to the brain is impaired, and the person faints.

A more serious condition occurs when a cerebral vessel is obstructed or ruptured and the flow of blood to the region of the brain supplied by the vessel is reduced. Such a condition is called a *cerebrovascular accident*, or *stroke*. The reduced flow of blood results in an inadequate oxygen supply to the cells of the affected area and may cause permanent brain damage. The neurological effects of a stroke vary, depending on the site of brain damage.

Skeletal Muscle Circulation

During exercise, the energy requirements of skeletal muscles increase so greatly that they can be met only by a tremendous increase in the blood flow to the muscles. During strenuous exercise, muscle blood flow can be as much as 20 times greater than the flow at rest. The increased blood flow is made possible by the dilation of arterioles and the opening of previously closed capillaries within the muscles.

Two separate mechanisms are involved in increasing the blood flow through active skeletal muscles. The first mechanism involves the central nervous system. When signals are sent to a muscle to initiate muscular activity, sympathetic vasodilator fibers to the arterioles of the muscle are also activated by way of a pathway that originates in the cerebral cortex and passes downward to the hypothalamus. As a result, the arterioles of the muscle dilate, thus increasing the blood flow to the muscle. This vasodilation occurs even prior to exercise—that is, prior to any great decrease in the oxygen level within the muscle.

The second mechanism is the local autoregulatory response. The increased muscular activity during exercise decreases the oxygen level in the muscle and/or leads to an accumulation of local vasodilator substances. The

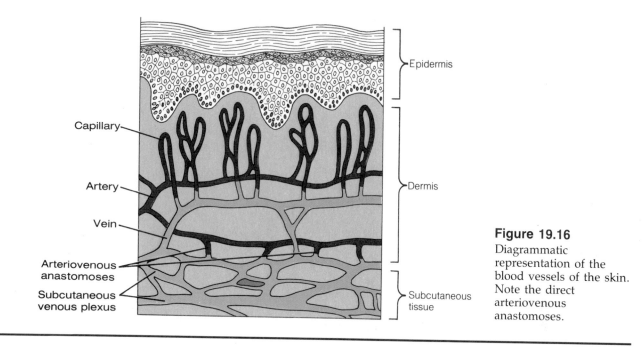

Figure 19.16
Diagrammatic representation of the blood vessels of the skin. Note the direct arteriovenous anastomoses.

decreased oxygen level and/or the local vasodilator substances produce a vascular dilation that increases the blood flow through the muscle.

Vasodilation and increased blood flow occur in skeletal muscles in which all sympathetic nerve fibers have been inactivated, and it appears that the local autoregulatory response is the more important of the two control mechanisms. The function of the sympathetic vasodilator fibers seems to be to increase the flow of blood through a muscle at the onset of muscular activity until the lack of oxygen within the muscle leads to vasodilation by way of the local autoregulatory response.

Cutaneous Circulation

The blood flow to the skin serves two functions: it provides nutrients to and carries metabolic end products away from the skin, and it brings heat from the internal structures of the body to the surface, where the heat can be lost from the body. The cutaneous blood flow is normally 20 to 30 times greater than is needed to meet the nutrient needs of the skin, and most of the flow is for purposes of temperature regulation.

The blood flow to the skin can vary tremendously. Under ordinary circumstances, cutaneous blood flow is approximately 400 ml per minute. However, under extreme conditions, the blood flow can increase to 2500 ml per minute. A unique arrangement of the blood vessels beneath the body surface makes such a large blood flow possible (Figure 19.16). In addition to the usual capillary beds, there are extensive subcutaneous venous plexuses (*plexus* = network) that can hold large volumes of blood. These plexuses are located close enough to the surface to allow heat to pass from the blood to the surface, and thus to be lost from the body. The blood flow through the subcutaneous venous plexuses varies with the degree of constriction or dilation of the vessels that carry blood to the plexuses, and a temperature control center in the hypothalamus acts through sympathetic nerves to regulate the diameters of these vessels. In some locations—for example, the hands, feet, and ears—blood can bypass the capillary system and flow directly from the arteries into the venous plexuses by way of vascular shunts called *arteriovenous anastomoses*. The walls of these arteriovenous shunts are muscular, and their constriction decreases the flow of blood into the venous plexuses and reduces the loss of body heat.

F19.16

CONDITIONS OF CLINICAL SIGNIFICANCE

Blood Vessels

High Blood Pressure

High blood pressure, or *hypertension*, can lead to heart attack, heart failure, brain stroke, or kidney damage. Approximately 20% of the population can expect to have high blood pressure at some time during their lives. In general, arterial blood pressures above 140/90 mm Hg are considered to be hypertensive. The degree of elevation of both the systolic pressure and the diastolic pressure is important in hypertension, since both pressures affect the mean arterial pressure.

Hypertension due to an unknown cause is called *idiopathic* or *essential hypertension,* and approximately 95% of the cases of high blood pressure are classed as idiopathic. Many researchers believe that high blood pressure is frequently caused by an excessive dietary intake of sodium or an excessive retention of sodium by the body, which leads to an increase in blood volume and cardiac output. Researchers have also proposed that an increased peripheral resistance due to vascular constriction contributes to high blood pressure, and a number of studies suggest a sequence of initially high cardiac output followed by increased peripheral resistance.

High blood pressure is sometimes associated with kidney diseases or conditions that impair renal blood flow. In some cases, the formation of renin in the kidneys leads to the formation of angiotensin and arteriolar constriction, although other factors are also likely to be involved. High blood pressure can also accompany the secretion of excessive amounts of norepinephrine and epinephrine by tumors of the adrenal medulla, and the secretion of excessive amounts of hormones by tumors of the adrenal cortex.

Atherosclerosis and Arteriosclerosis

As they grow older, a great number of people experience degenerative changes in the walls of their arteries that are known as *atherosclerosis.*

Atherosclerosis is characterized by deposits (plaques) of abnormal smooth muscle cells, lipid materials, and connective tissue in arterial walls. The atherosclerotic plaques can gradually narrow or occlude blood vessels. In addition, the plaques can promote the formation of blood clots within the vessels. In any case, the flow of blood through the vessels is often restricted or blocked. Frequently, the first indication of atherosclerosis is the dysfunction of an organ supplied by an affected vessel. For example, a restricted or inadequate blood flow to the heart can cause pain during exertion or emotional stress—a condition known as *angina pectoris*—and a blocked coronary vessel can lead to the death of a portion of the myocardium—that is, to a *myocardial infarct.* Similarly, the occlusion of a vessel supplying the brain may kill brain tissue.

In very advanced cases, atherosclerotic plaques can become calcified and there can be an extensive growth of fibrous tissue in arterial walls, giving rise to *arteriosclerosis,* or "hardening of the arteries." Arteriosclerosis can

greatly reduce the elasticity of the vessel walls. As the walls become less elastic, the vessels cannot properly expand and recoil in response to the pressure changes produced by the beating heart. Consequently, in severe cases, the pressure within the vessels rises quite high during systole and falls unusually low during diastole.

Edema

The accumulation of excess fluid in the tissues is called *edema.* Edema causes the tissues to swell, and it can impair tissue function because the excess fluid increases the distance that materials must diffuse between the capillaries and the cells. Many factors can contribute to edema. In general, any event that either enhances the movement of fluid out of the blood vessels and into the tissues or retards the return of fluid from the tissues back into the blood vessels can lead to edema. Thus, edema may result from (1) an increased capillary permeability, as may occur in certain allergic responses; (2) an elevated capillary pressure, due, for example, to blockage of a vein; (3) an increased interstitial fluid colloid osmotic pressure, caused, for example, by blocked lymphatic vessels; or (4) a decreased plasma colloid osmotic pressure, due, for example, to the loss of plasma proteins in the urine in certain kidney diseases.

Circulatory Shock

Circulatory shock is a condition in which the cardiac output is so reduced that the body tissues fail to receive an adequate blood supply. Any condition that decreases the cardiac output can lead to circulatory shock. Thus, circulatory shock can result from (1) diseases and weaknesses of the heart itself, caused, for example, by a heart attack; (2) reductions in blood volume, perhaps due to hemorrhage; or (3) vascular difficulties that reduce the venous return, due, for example, to extreme vasodilation and the accumulation of blood in the veins. In severe cases of shock, the inadequate blood supply to the tissues impairs tissue function so greatly that death results.

Aneurysms

An *aneurysm* is a localized dilation of an artery due to a weakness of the artery wall. The most frequent site of aneurysm formation is the abdominal aorta. The greatest danger from an aneurysm is that the affected vessel will rupture; however, aneurysms are also common sites of thrombus formation, and a portion of a thrombus may break off and block a smaller vessel. In addition, aneurysms can obstruct or erode neighboring organs or tissues.

Phlebitis

The inflammation of a vein is called *phlebitis.* Phlebitis can result from a number of conditions. One type of phlebitis is caused by bacteria that invade a vein, perhaps from an abscess or where the vein passes through an area of inflammation. In phlebitis, the inflammation of a vein can lead to the formation of a thrombus, with all of its accompanying dangers. This condition is called thrombophlebitis.

Varicose Veins

Varicose veins are veins that are dilated, lengthened, and tortuous. Varicose veins are often brought about by a congenital and inherited weakness of the walls and valves of the veins. Such veins are unable to efficiently carry blood toward the heart, thus allowing the blood to accumulate in the lower limbs when the body is in an upright position. Other varicosities may be caused by pressure on veins, such as might occur during pregnancy, in obesity, or from an abdominal tumor. In severe cases, varicose veins can interfere with the return of blood to such an extent that muscle cramps and edema occur. Hemorrhage, phlebitis, and thrombosis are also possible complications.

Effects of Aging

Aging apparently affects the heart and the blood vessels, since cardiovascular problems cause more than half the deaths of the population over age 65. However, these cardiovascular problems are not due to aging alone. Evidence suggests that in nonindustrialized societies there is little cardiovascular disease. By comparison, approximately 20% of the population of industrialized societies suffer from hypertension by the age of 65. Although age and genetic factors probably contribute to the changes in the walls of the blood vessels that lead to myocardial infarction or stroke, many investigators believe that in advanced societies environmental factors, and especially diet, play a more significant role than age or genetic inheritance.

Various studies have concluded that (1) a substantial lowering of plasma cholesterol would greatly reduce the number of heart attacks; (2) maintaining diastolic blood pressures of under 75 mm Hg would result in a threefold reduction in the number of heart attacks; and (3) regular moderate exercise would reduce the incidence of fatal heart attacks by approximately half. In addition, nonsmokers have half the incidence of heart attacks as do smokers. These studies suggest that aging may not be the major cause of cardiovascular disorders.

CARDIOVASCULAR ADJUSTMENTS DURING EXERCISE

During exercise, the dilation of skeletal muscle vessels due to local autoregulatory responses and to some extent to sympathetic vasodilator activity greatly increases the blood flow to skeletal muscles. At the same time, sympathetic vasoconstrictor activity causes a compensatory constriction of vessels elsewhere, particularly in the kidneys and gastrointestinal tract. The compensatory vasoconstriction does not fully offset the dilation of the skeletal muscle vessels, however, and the total peripheral resistance decreases.

Also during exercise, the activity of the sympathetic nerves to the heart increases and the activity of the parasympathetic nerves decreases. In addition, venous return is enhanced by the increased pumping effects of the contracting skeletal muscles and by the constriction of the veins due to sympathetic stimulation. As a consequence of these activities, both the heart rate and the stroke volume increase, resulting in an increase in the cardiac output. During most forms of exercise, the increase in cardiac output is somewhat greater than the decrease in total peripheral resistance, and the mean arterial pressure rises (recall that mean arterial pressure = cardiac output × total peripheral resistance).

It is believed that the neurally mediated cardiovascular changes characteristic of exercise—and also excitement, stress, or anger—can be initiated by nerve impulses that originate within the cerebral cortex and pass to the hypothalamus. The hypothalamus, in turn, acts on the brain-stem centers that influence cardiovascular function and directly on the sympathetic vasodilator neurons to skeletal muscle vessels.

ANATOMY OF THE VASCULAR SYSTEM

The blood of the circulatory system is carried throughout the body in a complex organization of vessels. Only the major blood vessels are named in this

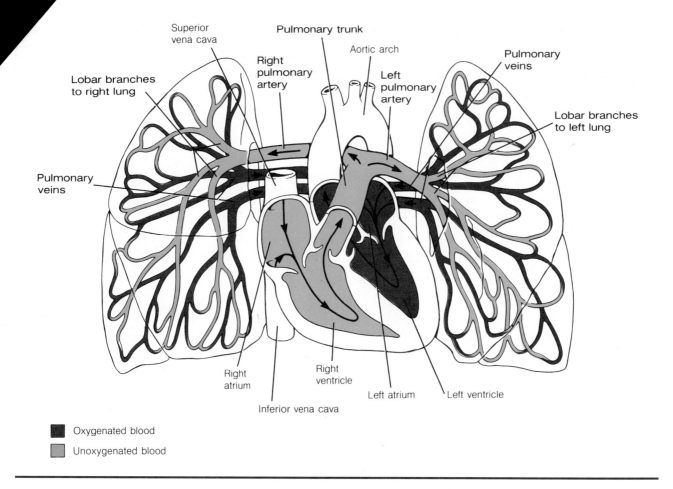

Oxygenated blood

Unoxygenated blood

Figure 19.17

Pulmonary circulation showing the pulmonary trunk dividing into right and left pulmonary arteries, which in turn divide into lobar branches. Blood from the lungs returns to the left atrium through the pulmonary veins. The arrows indicate the direction of blood flow.

F19.17

pull'-muh-na-ree

chapter, but you should bear in mind that there are many other smaller vessels that are not mentioned. The arteries branch into many small arterioles, which, in turn, branch into numerous microscopic capillaries where the exchange between tissues and blood occurs. From the capillaries, the blood enters tiny venules that join together to form the larger veins.

The vessels of the circulatory system can be divided into two separate circuits, each of which leaves and returns to the heart. The **pulmonary circuit** carries blood from the right side of the heart to the lungs and back to the left side of the heart. The **systemic circuit** carries blood that has just left the pulmonary circuit to the rest of the body and back to the right side of the heart.

Pulmonary Circuit

Blood enters the pulmonary circuit from the right ventricle of the heart through the **pulmonary trunk** (Figure 19.17). The short pulmonary trunk divides into **left** and **right pulmonary arteries,** which pass directly to the left and right lungs. As they enter the lungs, the pulmonary arteries divide into several **lobar branches,** one to each lobe of the lung. The right pulmonary artery divides into three lobar arteries; the left divides into two. (The right lung has three lobes and the left lung has two.) The lobar arteries divide into numerous orders of smaller arteries and arterioles and, eventually, into capillary plexuses located in the walls of the tiny air sacs (alveoli) of the lungs. It is through these capillary plexuses that gases are exchanged between the blood and the air.

From the capillaries, the blood is collected into venules and then into progressively larger veins that lead into left and right **pulmonary veins,** generally two from each lung. The pulmonary veins empty into the left atrium. The pulmonary arteries carry blood that has a high carbon dioxide content

FRONTIERS IN HEALTH
Wired Arteries

Degeneration of the muscle layer of an artery (the tunica media) can result in a ballooning of the inner and outer layers of the vessel wall, producing an aneurysm. An aneurysm in an arterial wall resembles a worn spot on a tire, in that when pressure builds up inside or when the wall thins too much, it can rupture. In a tire this is called a blowout. A similar violent event can occur in a blood vessel.

Blood spilling into the chest or abdominal cavity (in the case of a ruptured aortic aneurysm) or the brain (in the case of a ruptured cerebral aneurysm) often leads to death. In the United States alone, an estimated 30,000 people die each year from ruptured cerebral aneurysms, and 2500 die from ruptured aortic aneurysms.

The first line of defense against ruptured aneurysms is prevention. The best means of prevention is to reduce the two main causes, atherosclerosis and high blood pressure. The second line of defense lies in detecting aneurysms and treating them before they burst. Pain in the affected area generally alerts the physician to their possible presence, and they are fairly easy to detect by X-ray. Once a defective wall is found, surgeons must act fast. How they correct the aneurysm depends on the location and severity of the damage. For example, in cases of popliteal aneurysm, surgeons often remove the aneurysm and graft a vein in its place. To repair a cerebral aneurysm, which most often occurs in the cerebral arterial circle at the base of the brain, surgeons clamp or tie off the neck of the bulge on the arterial wall. This lowers the pressure in the bulge and prevents rupture. In larger arteries, however, surgeons use grafts of dacron or other synthetic materials to replace the damaged section of the artery. Regardless of which technique is used, the repair of aneurysms involves major surgery that is costly and risky.

Now, however, thanks to an innovative procedure, the surgical repair of aneurysms may become much less frequent. Dr. Andrew Cragg, a radiologist from the University of Minnesota, has been working with an alloy of nickel and titanium, called nitinol, that may someday be widely used to repair aneurysms quickly and safely, without the need for major surgery.

Nitinol is a metal with a "memory." When a fine nitinol wire is coiled around a cylinder and heated, it forms a tight spring. Cooling the spring causes it to revert to the form of a straight wire. When the wire is heated again, the coil reforms.

Dr. Cragg has developed a method for repairing aneurysms by capitalizing on nitinol's memory. He and his colleagues first make a nitinol coil the same size as the internal diameter of a damaged artery. They then push the wire through a specially cooled catheter, which is inserted into the damaged region of the artery. As the wire comes out the end of the catheter, body heat causes

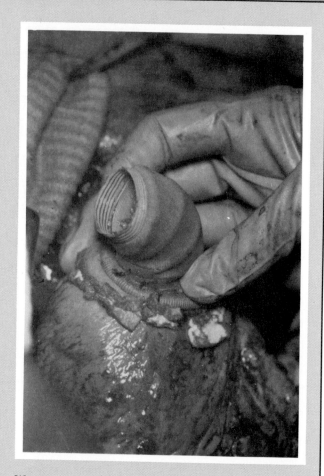

Woven dacron tubing positioned to replace an aneurysm.

it to coil back into the shape it was "taught." The nitinol coil becomes incorporated into the weakened portion of the arterial wall, strengthening the wall and preventing its rupture.

Nitinol has yet to be tested on humans, but experiments with dogs have been very encouraging. Dr. Cragg reports that four hours after emplacement, a fine layer of fibrin (a protein derived from the blood) covers the coil; after eight weeks, endothelial cells have grown over the coil, completely incorporating it into the arterial wall.

Because the insertion of wire coils can quickly strengthen the walls of arteries damaged by atherosclerosis and high blood pressure, without the need for major surgery, the procedure holds much promise for the thousands of people who suffer from aneurysms.

and is low in oxygen. The blood in the pulmonary veins, however, has a high oxygen content and is low in carbon dioxide. This pattern is the exact opposite of that for the systemic arteries and veins. In other words, the vessels are named according to the direction of the blood flow within them, not the condition of the blood they carry. Thus, any vessel that carries blood away

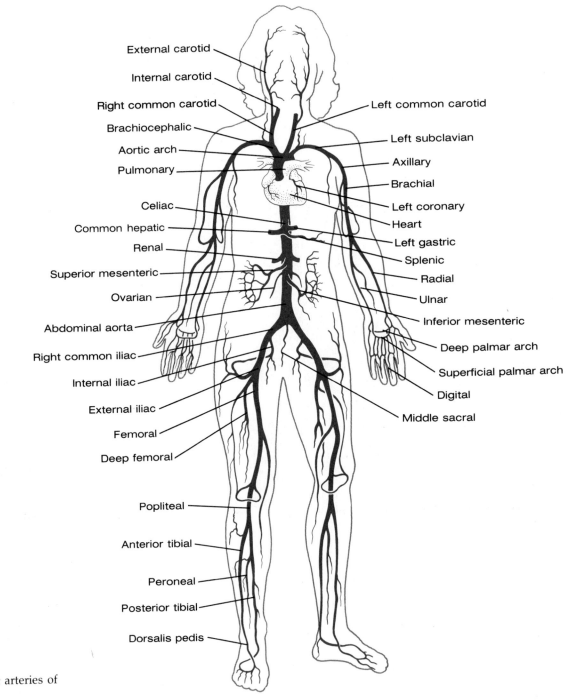

External carotid
Internal carotid
Right common carotid
Brachiocephalic
Aortic arch
Pulmonary
Celiac
Common hepatic
Renal
Superior mesenteric
Ovarian
Abdominal aorta
Right common iliac
Internal iliac
External iliac
Femoral
Deep femoral
Popliteal
Anterior tibial
Peroneal
Posterior tibial
Dorsalis pedis

Left common carotid
Left subclavian
Axillary
Brachial
Left coronary
Heart
Left gastric
Splenic
Radial
Ulnar
Inferior mesenteric
Deep palmar arch
Superficial palmar arch
Digital
Middle sacral

Figure 19.18
Major systemic arteries of
the body.

from the heart is an **artery** and any vessel that returns blood to the heart is a **vein.**

The vessels of the pulmonary circuit, unlike those of the systemic circuit, do not supply oxygen and nutrients to the lung tissues. The metabolic needs of the lungs are provided for by the small *bronchial vessels* of the systemic circuit.

Systemic Circuit

sis-tem'-ik Blood that is not in the pulmonary circuit is in the systemic circuit. The vessels of the systemic circuit transport blood to all tissues and organs of the body except the alveoli of the lungs.

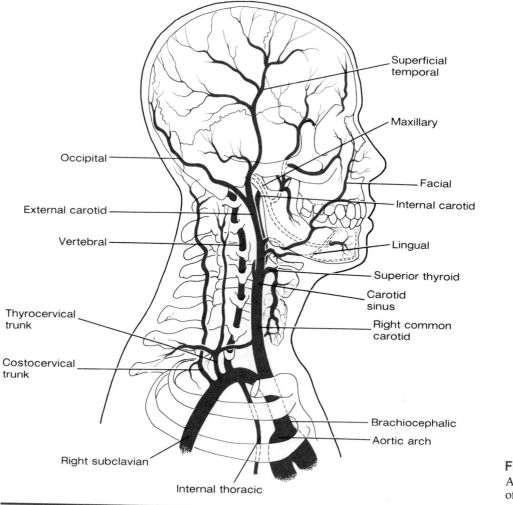

Superficial
temporal

Maxillary

Occipital

Facial

Internal carotid

External carotid

Vertebral

Lingual

Superior thyroid

Carotid
sinus

Right common
carotid

Thyrocervical
trunk

Costocervical
trunk

Brachiocephalic

Aortic arch

Figure 19.19
Arteries of the right side
of the head and neck.

Right subclavian

Internal thoracic

Systemic Arteries

Blood from the left ventricle enters the systemic circuit through the **aorta,**
from which all the arteries of the systemic circuit branch (Figure 19.18). For
descriptive purposes, it is convenient to divide the aorta into the **ascending
aorta,** the **aortic arch,** and the **descending aorta,** which has **thoracic** and **abdominal** portions. The branches of these various portions are summarized in
Table 19.1, p. 597.

F19.18

Table 19.1

BRANCHES OF THE ASCENDING AORTA   The ascending aorta is a short
vessel that passes in front of the right pulmonary artery. Its only branches are
the **right** and **left coronary arteries,** which arise from sinuses behind the
cusps of the aortic valve and supply the heart muscle (see Figures 18.7 and
18.8).

BRANCHES OF THE AORTIC ARCH   As the aorta leaves the pericardial
sac, it arches dorsally and to the left, forming the aortic arch. Three branches
arise from the aortic arch: the **brachiocephalic artery,** the **left common carotid
artery,** and the **left subclavian artery** (Figure 19.18). These arteries furnish all
the blood to the head, neck, and upper limbs.

F19.18

ARTERIES OF THE HEAD AND NECK   Shortly after arising from the arch
of the aorta, the brachiocephalic artery divides into a **right subclavian artery**
and a **right common carotid artery** (Figure 19.19). The right common carotid
artery and the **left common carotid artery** (which arises directly from the arch
of the aorta) supply most of the blood to the head and neck. The common

F19.19

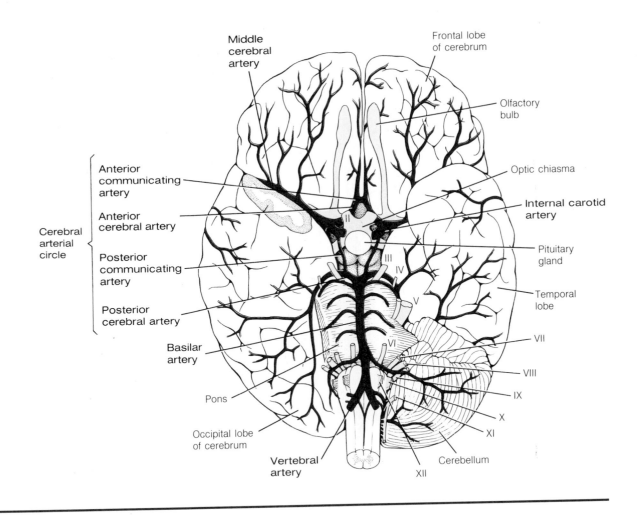

Figure 19.20

Arteries of the base of the brain, forming the cerebral arterial circle around the pituitary gland. To provide an unobstructed view, part of the right temporal lobe and the right cerebellar hemisphere have been removed. Roman numerals indicate cranial nerves.

carotid arteries course upward alongside the trachea a short distance before each bifurcates (divides) into an **internal carotid artery** and an **external carotid artery.** At this point of bifurcation, the vessels enlarge slightly, forming the **carotid sinus.** The sinus contains baroreceptors that assist in regulating blood pressure. Near this same region is a small oval **carotid body** that contains chemoreceptors sensitive to changing levels of oxygen, carbon dioxide, and pH in the blood that travels toward the brain.

The **external carotid artery,** which supplies most of the head and neck except for the brain, lies more superficially than the internal carotid artery. It passes through the parotid salivary gland and terminates as the **maxillary** and **superficial temporal arteries** (Figure 19.19). Just before terminating, the external carotid artery gives off the **occipital artery** to the posterior region of the scalp. Through these branches the external carotid artery supplies the muscles of mastication; the mucous membranes of the nose, the pharynx, and the palate; the teeth of the upper and lower jaws; some neck muscles; and the entire scalp region. While still in the neck, it sends branches to the thyroid gland **(superior thyroid artery),** tongue **(lingual artery),** and the skin and the muscles of the face **(facial artery).**

The **internal carotid artery** enters the cranial cavity through the carotid canal of the temporal bone. It travels forward in the canal and soon leaves the bone to pass forward and upward along the side of the sella turcica. The internal carotid artery gives off an **ophthalmic artery,** which follows the lower surface of the optic nerve and enters the orbit. The internal carotid artery supplies the brain through its terminal branches: the **anterior cerebral artery** and the **middle cerebral artery.** A small **anterior communicating artery** connects the right and left anterior cerebral arteries. These vessels also help form the **cerebral arterial circle (circle of Willis)** (Figure 19.20), which surrounds F19.20 the infundibulum of the pituitary gland.

The **vertebral artery,** which is the first branch of the subclavian artery, provides another major blood supply to the brain. The vertebral artery reaches the cranial cavity by passing through the transverse foramina of the cervical vertebrae and the foramen magnum. The right and left vertebral arteries join on the ventral surface of the pons of the brain, forming the **basilar artery** (Figure 19.20). The basilar artery continues forward and terminates as F19.20 **right** and **left posterior cerebral arteries,** which supply the posterior regions of the cerebral hemispheres of the brain. The basilar artery also gives off branches that supply the pons and the cerebellum. The **posterior communicating arteries** from the internal carotid arteries join with the posterior cerebral arteries to complete the cerebral arterial circle around the infundibulum.

ARTERIES OF THE UPPER LIMBS   The upper limbs are supplied by the **subclavian arteries.** As noted earlier, the right subclavian artery is a branch of the brachiocephalic artery, and the left subclavian arises directly from the aortic arch. The first part of each subclavian artery gives off branches that do not enter the upper limbs (Figure 19.21). Just lateral to the vertebral artery, F19.21 the **thyrocervical trunk** and **costocervical trunk** send branches to the lower neck, the dorsal scapular, and the intercostal regions. The **internal thoracic (internal mammary) artery** arises from the subclavian artery and passes downward beneath the costal cartilage. It supplies branches to the thoracic and intercostal muscles, the mediastinum, and the diaphragm.

After passing beneath the clavicle, the subclavian artery becomes the **axillary artery.** The axillary artery gives off branches to the lateral wall of the

---

CLINICAL CORRELATION

Traumatic Shock

CASE REPORT . . . .

THE PATIENT   A 26-year-old woman

REASON FOR ADMISSION   Multiple trauma

CLINICAL COURSE   The patient had been well before she was involved in an automobile accident. She was transported to the emergency room by helicopter, dazed but conscious. Her arterial blood pressure was 55/30 mm Hg (normal: 90–140/60–90 mm Hg), and her heart rate was 96/min (normal: 60–100/min). No head injury was indicated, and the heart and lungs were normal. However, an X-ray film showed widening of the mediastinum (which suggests injury to the aorta) and three broken ribs. The abdomen was swollen and tender (which suggests intra-abdominal hemorrhage). The right tibia appeared to be broken, and the right patella was severely lacerated. The hematocrit was 34% (normal: 37–48%), and the platelet count was 20,000/mm^3 (normal: 150,000–350,000/mm^3). An intravenous line was inserted and infusion of 10% glucose in saline was begun. Packed red cells, platelets, and fresh-frozen plasma were given. The urinary bladder was catheterized (for monitoring the production of urine).

Because of the evidence of internal bleeding, the patient was taken to the operating room, and an explor-

atory laparotomy (opening of the abdomen by way of an incision through the abdominal wall) was performed. One and one-half liters of blood and clots were removed from the peritoneal cavity, and several bleeding arteries were clamped and tied. The patient was given additional packed red cells, albumin, fresh-frozen plasma, and lactated Ringer's solution during the operation. Her arterial blood pressure stabilized in the range of 90/50 mm Hg. To assess the cause of the widened mediastinum, thoracic arteriography was done (fluroscopy of contrast medium injected at the root of the aorta). A traumatic rupture was found just distal to the origin of the left subclavian artery. The vessel was repaired through a left thoracotomy (opening of the chest by way of an incision through the thoracic wall), using a Dacron graft. Additional cells and fluids were given, and the patient's arterial blood pressure increased to 115/70 mm Hg.

OUTCOME   The patient was fortunate to receive prompt and expert care. After the bleeding was stopped and the arterial blood pressure had stabilized, her external wounds were treated and her broken bones were set. Recovery was uneventful.

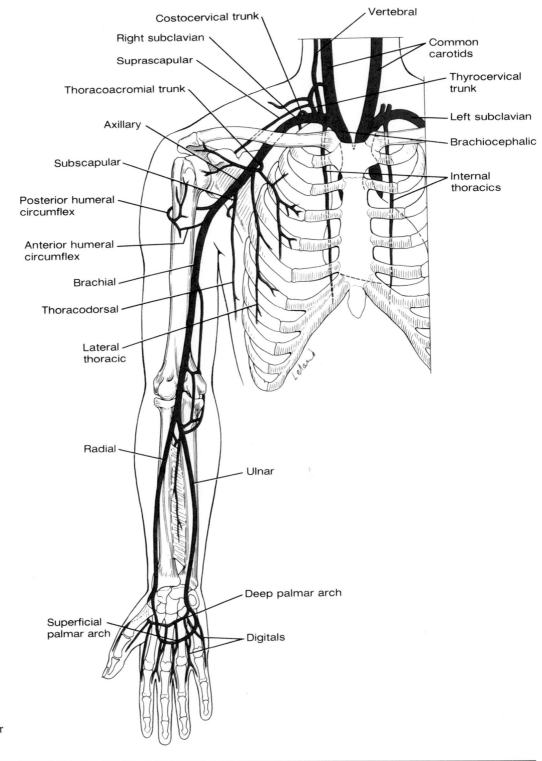

Figure 19.21
Arteries of the upper limb.

thorax (**thoracoacromial trunk, lateral thoracic artery, subscapular artery, and thoracodorsal artery**) and to the region around the proximal end of the humerus (**anterior** and **posterior humeral circumflex arteries**). It then continues into the arm as the **brachial artery,** which travels down the medial side of the humerus, supplying the muscles of the arm.

The brachial artery crosses the anterior surface of the elbow and divides into **radial** and **ulnar arteries.** Both vessels travel down the anterior surface of

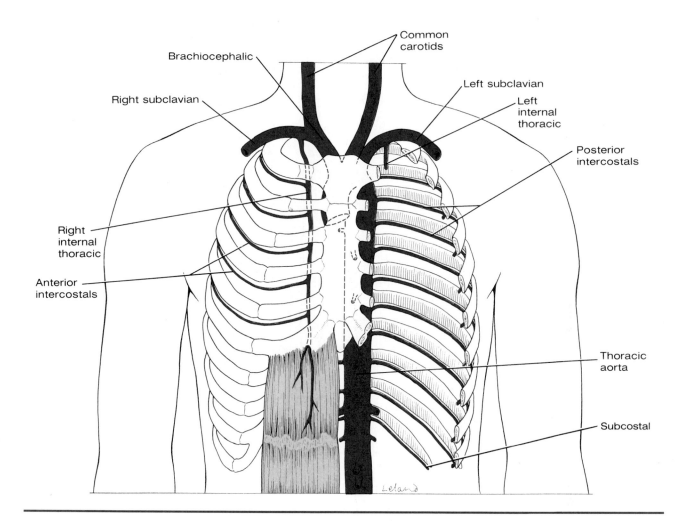

Common
carotids

Brachiocephalic

Right subclavian

Left subclavian

Left
internal
thoracic

Posterior
intercostals

Right
internal
thoracic

Anterior
intercostals

Thoracic
aorta

Subcostal

Leland

Figure 19.22
Branches of the thoracic
aorta. The anterior
portion of the rib cage
has been removed on the
left side.

the forearm—the radial artery alongside the radius, the ulnar artery along the
ulna. The radial artery is easily palpated on the lateral side of the wrist, where
it is used to take the pulse. The two arteries join in the hand through interconnecting branches of **superficial** and **deep palmar arches. Digital arteries** from
the palmar arches supply the fingers.

BRANCHES OF THE THORACIC AORTA   The arteries that supply the
thorax arise from the first part of the descending aorta, which lies to the left of
the vertebral column (Figure 19.22). All the branches from this region are
small. Most prominent are the **posterior intercostal arteries,** which travel between the ribs and anastomose with the **anterior intercostal arteries,** which
arise from each internal thoracic artery. In addition, **bronchial, esophageal,
subcostal** (beneath the 12th ribs), and **superior phrenic arteries** (which supply the diaphragm) also arise from the thoracic aorta.

BRANCHES OF THE ABDOMINAL AORTA   The descending aorta passes
through the *aortic hiatus* of the diaphragm and enters the abdominopelvic
cavity (Figure 19.23). It travels down the ventral surface of the vertebral column, supplying the posterior abdominal walls through four pairs of **lumbar
arteries.** The abdominal aorta terminates in front of the fourth lumbar vertebra by dividing into **right** and **left common iliac arteries** and a small **middle
sacral artery.** Immediately after entering the abdominopelvic cavity, the aorta
gives off a pair of **inferior phrenic arteries** to the underside of the diaphragm.
    A single **celiac artery** arises from the aorta just below the inferior phrenic
arteries, at about the level of the twelfth thoracic vertebra. The celiac artery is

F19.22

F19.23

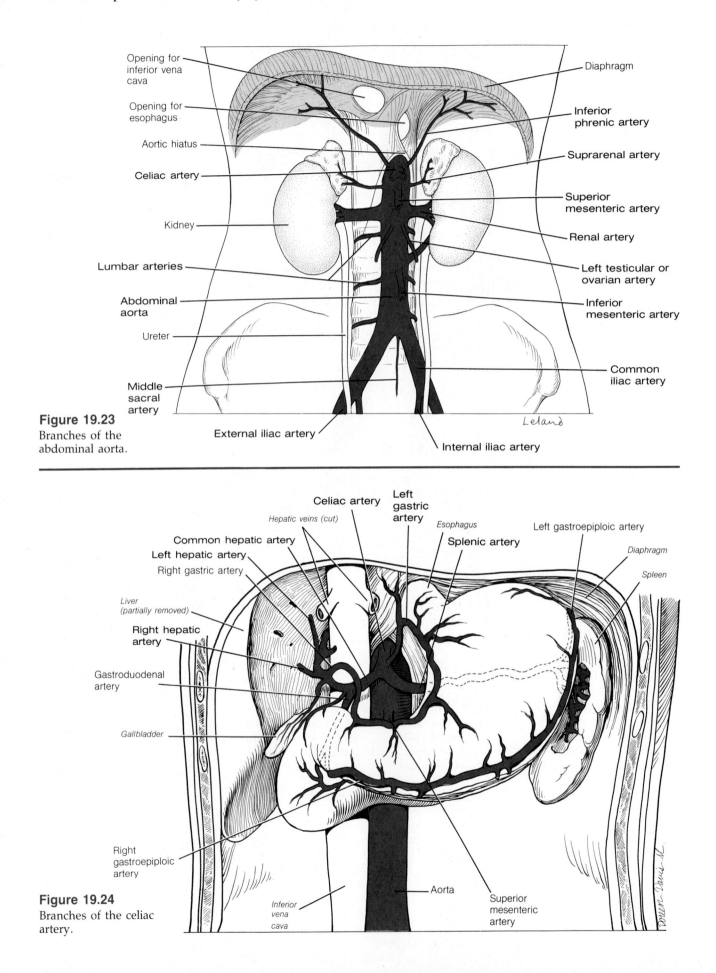

Figure 19.23
Branches of the
abdominal aorta.

Figure 19.24
Branches of the celiac
artery.

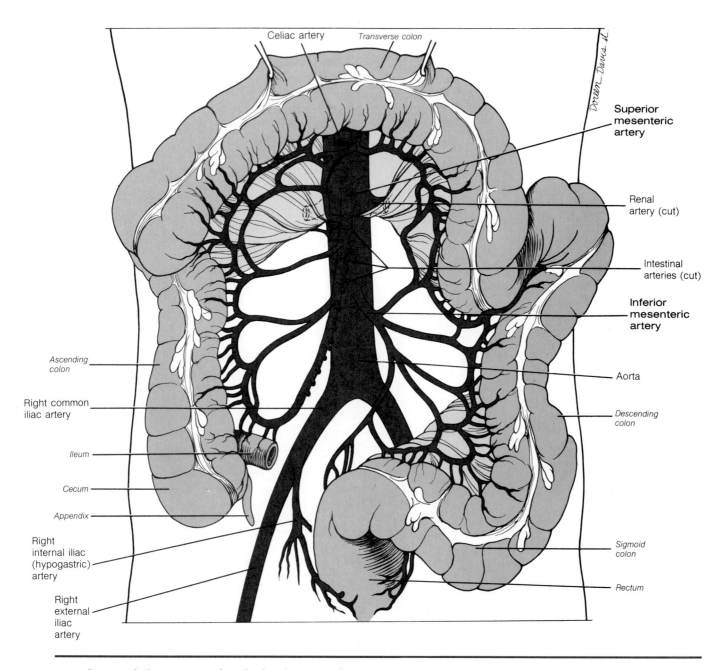

Celiac artery

Transverse colon

Superior mesenteric artery

Renal artery (cut)

Intestinal arteries (cut)

Inferior mesenteric artery

Aorta

Descending colon

Sigmoid colon

Rectum

Ascending colon

Right common iliac artery

Ileum

Cecum

Appendix

Right internal iliac (hypogastric) artery

Right external iliac artery

Figure 19.25
The superior and inferior mesenteric arteries. The transverse colon has been lifted up from its normal position.

very short and almost immediately divides into *left gastric, splenic,* and *common hepatic arteries* (Figure 19.24). The **left gastric artery** follows the lesser curvature of the stomach, supplying the lower part of the esophagus and part of the stomach. The **splenic artery,** which supplies the spleen, also gives off small branches to the pancreas and stomach, as well as the large **left gastroepiploic artery,** which supplies the greater curvature of the stomach. As it travels to the liver, the **common hepatic artery** sends branches to the stomach, duodenum, and pancreas through **gastroduodenal** and **right gastric** branches. The right gastric artery supplies the lesser curvature of the stomach, and the **right gastroepiploic artery**—a branch of the gastroduodenal artery—supplies the greater curvature. After giving off these branches, the common hepatic artery divides into **right** and **left hepatic arteries,** which supply the liver and gallbladder.

The single **superior mesenteric artery** arises from the aorta just below the celiac artery. It travels within the mesentery of the intestines, where it gives off several branches that form anastomosing loops (Figure 19.25). The superior mesenteric artery supplies all of the small intestine, as well as the cecum and the ascending colon and most of the transverse colon of the large intestine.

F19.23

Branching laterally from the aorta at the level of the superior mesenteric arteries are two **suprarenal arteries,** which supply the adrenal glands (Figure 19.23). These glands are also supplied by branches from the inferior phrenic and renal arteries. The paired **renal arteries** (Figure 19.23), which supply the kidneys, arise from the lateral margins of the aorta just inferior to the superior mesenteric artery.

F19.23

The arteries that supply the gonads (ovaries and testes) arise from the ventral surface of the aorta a short distance below the renal arteries (Figure 19.23). In the female, the **ovarian arteries** pass downward and laterally into the pelvic cavity to supply the ovaries; in addition, they give off branches to the ureters and uterine tubes. The **testicular (internal spermatic) arteries** of the male are longer than the ovarian arteries, since they pass through the inguinal canal and enter the scrotum, where they supply the testes.

F19.25

The **inferior mesenteric artery** is the final branch of the abdominal aorta (Figure 19.25). It is a single vessel that arises from the ventral surface of the aorta just above its bifurcation into the **common iliac arteries.** Branches of the inferior mesenteric artery supply part of the transverse colon, the descending colon, the sigmoid colon, and most of the rectum. This artery anastomoses with branches of the superior mesenteric artery.

F19.26

ARTERIES OF THE PELVIC REGION   Each of the common iliac arteries divides in front of the sacroiliac articulation into *internal* and *external iliac* arteries (Figure 19.26). The **internal iliac** *(hypogastric)* **arteries** enter the pelvic cavity and divide into branches that supply the pelvic viscera (urinary bladder, uterus, vagina, and rectum). In addition, branches from the internal iliac arteries supply the muscles of the gluteal and lumbar regions, the walls of the pelvis, the external genitalia, and the medial region of the thigh.

F19.26

ARTERIES OF THE LOWER LIMBS   The **external iliac artery** is actually a continuation of the common iliac artery. It travels downward and laterally through the iliac fossa and enters the thigh as it passes beneath the inguinal ligament. While in the pelvic cavity, the external iliac artery supplies branches to the muscles and skin of the lower abdominal wall.

Upon entering the thigh, the external iliac artery becomes the **femoral artery** (Figure 19.26). The femoral artery passes along the anterior medial region of the thigh. In the lower thigh, it passes to the posterior surface of the knee through an opening *(adductor hiatus)* in the tendon of the adductor magnus muscle, and is then known as the **popliteal artery.** While in the thigh, the femoral artery sends several branches to the skin and muscles of the region. The **lateral** and **medial femoral circumflex arteries** supply the region around the proximal end of the femur, whereas the **profunda (deep) femoral artery** passes posteriorly to supply muscles of the posterior compartment of the thigh.

F19.27

The **popliteal artery** is the continuation of the femoral artery. It passes behind the knee (through the popliteal fossa), supplying the muscles and skin of the area, and then divides into an *anterior tibial artery* and a *posterior tibial artery* (Figure 19.27).

The **posterior tibial artery** continues downward behind the tibia, supplying the muscles of the posterior compartment of the leg. Behind the ankle, it divides into **medial** and **lateral plantar arteries,** which supply the sole of the foot and form the **plantar arch. Digital arteries** arise from the plantar arch. Near its origin, the posterior tibial artery gives rise to the **peroneal artery,** which supplies the peroneal muscles in the lateral compartment of the leg.

The **anterior tibial artery** passes through the interosseous membrane that connects the fibula to the tibia. It then travels down the ventral surface of the membrane, supplying the muscles of the anterior compartment of the leg. The anterior tibial artery passes in front of the ankle and ends on the dorsum of the foot as the **dorsalis pedis artery.** Branches of the dorsalis pedis artery supply the dorsum of the foot and anastomose with the plantar arch on the sole of the foot.

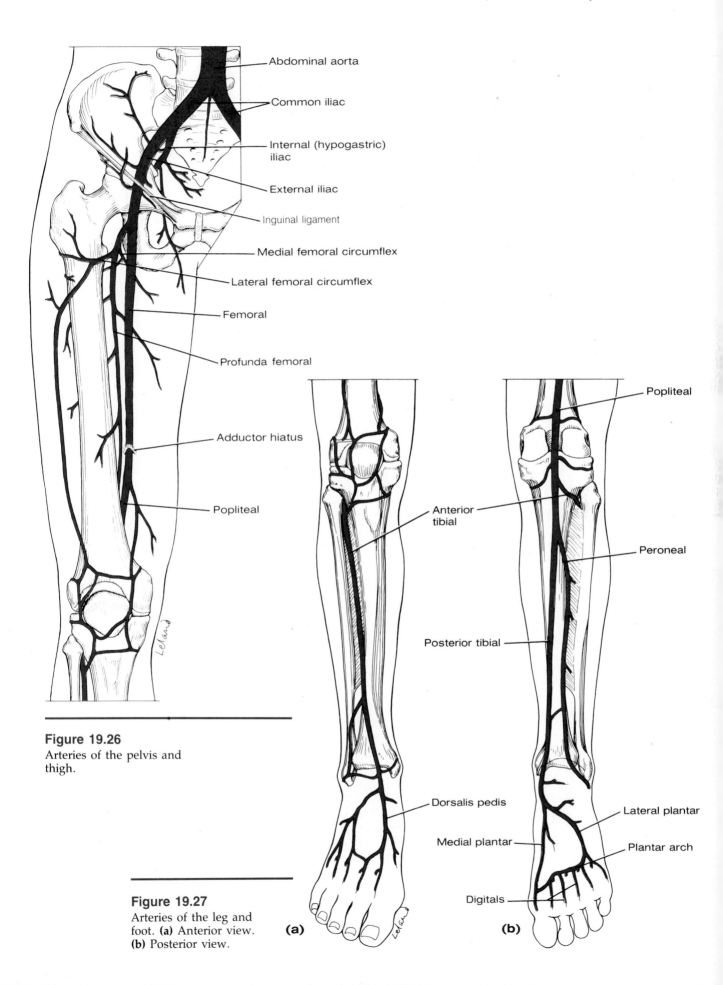

Abdominal aorta

Common iliac

Internal (hypogastric) iliac

External iliac

Inguinal ligament

Medial femoral circumflex

Lateral femoral circumflex

Femoral

Profunda femoral

Adductor hiatus

Popliteal

Popliteal

Anterior tibial

Peroneal

Posterior tibial

Dorsalis pedis

Lateral plantar

Medial plantar

Plantar arch

Digitals

Figure 19.26
Arteries of the pelvis and thigh.

Figure 19.27
Arteries of the leg and foot. **(a)** Anterior view. **(b)** Posterior view.

(a)

(b)

FRONTIERS IN HEALTH
Cleaning Out Clogged Arteries

Medical researchers are experimenting with a variety of techniques for cleaning out blood vessels that are clogged by blood clots and the buildup of atherosclerotic plaques. Several chemical agents have been found that dissolve the blood clots that often precipitate heart attacks. These substances help restore blood flow after an attack and significantly reduce heart muscle damage. Their use promises to spare thousands of lives every year.

In conjunction with the clot dissolvers, some medical researchers are using a small catheter with a tiny balloon attached to its tip. After the chemical clot dissolvers are administered, the catheter is maneuvered to the location of the clot and the balloon is inflated. The balloon causes the artery to expand and apparently loosens the atherosclerotic plaque from the wall. This technique, called balloon angioplasty, is not yet completely proven, but early results have been very promising.

Scientists are also looking for high-tech weapons to add to their arsenal against clogged arteries. Researchers at Stanford University Medical Center, for example, are experimenting to see if they can use lasers to break up atherosclerotic blockages in arteries. Clinical tests of this technique on human subjects have also been encouraging.

Medical researchers in France have used lasers to open up blocked arteries in patients undergoing open-heart surgery. A catheter containing fine glass fibers was inserted into the occluded arteries during surgery. Laser beams were then transmitted through the fiber-optic device, quickly burning away the blockage. However, even though the laser opened up the arteries, they all were blocked again within three months. In an attempt to determine why the arteries clogged again so quickly, workers are using a special catheter that not only delivers a laser beam but also uses a beam of light to illuminate the occlusion and a fiber to transmit the image back to the surgeon.

If reocclusion can be prevented, it is hoped that atherosclerotic blockages may be removed without opening a patient's chest. By inserting the laser catheter into the femoral artery and guiding it into the clogged vessel, surgeons may someday be able to remove atherosclerotic plaques with a fraction of the trauma incurred by open-heart surgery.

Medical opinions on the potential of lasers vary, however. Some cardiologists consider the technique ready for widespread use. Others believe cautious optimism is called for because of the potential risk of burning holes in arterial walls or weakening them, or even coagulating blood with the laser's heat.

One of the most exciting developments aimed at minimizing the potential damage is the "cool" laser, which emits ultraviolet light in short bursts, allowing the tissue to cool between pulses. This instrument could greatly reduce the possibility of tissue damage.

Scientists at the University of Florida at Gainesville are working on another technique to remove blood clots and atherosclerotic plaques from arteries. They have devised a special catheter, which is inserted into clogged vessels. Inside the catheter is a small revolving screw that, when turned, creates a mild suction. Tests in dogs indicate that the device is quite successful in vacuuming out blood clots and atherosclerotic plaques.

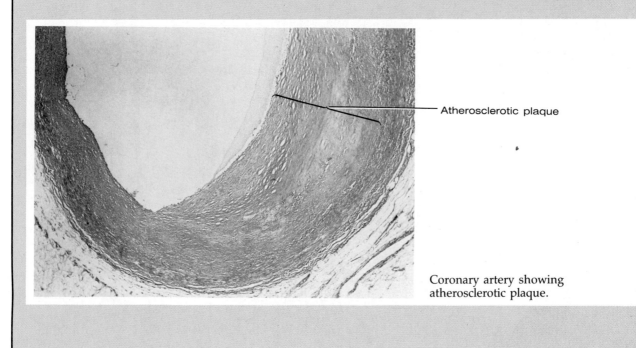

Atherosclerotic plaque

Coronary artery showing atherosclerotic plaque.

Table 19.1   Schematic Summary of the Systemic Arteries That Arise from Various Regions of the Aorta

ASCENDING AORTA
  Coronary arteries

AORTIC ARCH

Brachiocephalic
  → Right common carotid → Right internal carotid
                          → Right external carotid
  → Right subclavian → Axillary → Brachial → Radial
                                           → Ulnar
      → *Right vertebral*
      → *Right thyrocervical*
      → *Right internal thoracic*
      → *Right costocervical*

Left common carotid → Left internal carotid
                    → Left external carotid

Left subclavian → Axillary → Brachial → Radial
                                      → Ulnar
    → Left vertebral
    → Left thyrocervical
    → Left internal thoracic
    → Left costocervical

THORACIC AORTA
  Intercostals
  Bronchials
  Esophageals
  Superior phrenics

ABDOMINAL AORTA
  Inferior phrenics
  Celiac
    → *Common hepatic*
    → *Left gastric*
    → *Splenic*
  Superior mesenteric
  Lumbars
  Suprarenals
  Renals
  Testiculars or ovarians
  Inferior mesenteric
  Middle sacral
  Common iliacs (right and left) → Internal iliac
                                 → External iliac → Femoral → Popliteal → Posterior tibial → Peroneal
                                                                        → Anterior tibial

Systemic Veins

The blood that has been delivered throughout the body by the arteries is collected and returned to the heart through the veins. Since veins tend to be larger and more numerous than arteries, the capacity of the venous system is greater than that of the arterial system. The major veins, summarized in Table 19.2 (p. 604), tend to travel alongside the arteries. Although the general pattern of veins (Figure 19.28) is quite similar to the pattern of arteries, there are some important differences, which are explained in this chapter. Systemic veins are classified as *deep veins, superficial veins,* or *venous sinuses.*

Most of the **deep veins** travel alongside an artery and have the same name as the artery. Certain deep veins of the head and vertebral column are exceptions to this pattern, however.

The **superficial veins,** which lie just beneath the skin (in the hypodermis), return blood from the skin and the subcutaneous regions to the deep veins. Since there are no large superficial arteries, the names of the superficial veins do not correspond to the names of arteries.

Table 19.2

F19.28

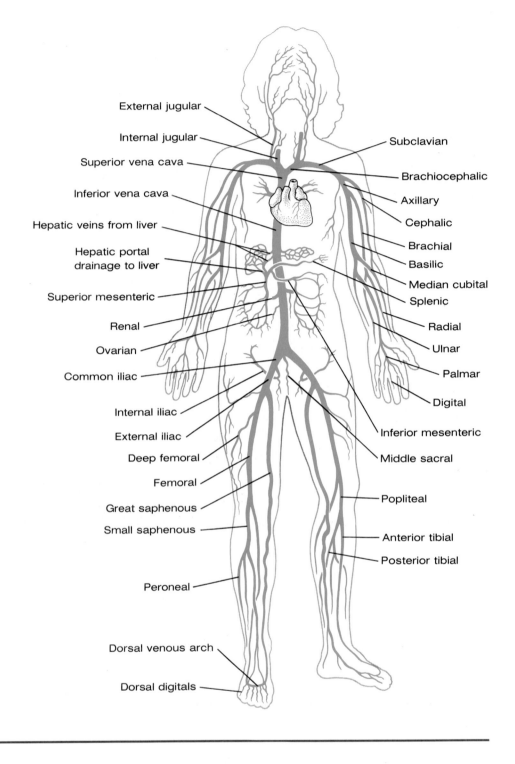

Figure 19.28
Major systemic veins of
the body.

The **venous sinuses** are not actually vessels, but are spaces that collect blood in certain regions and return it to the veins. The walls of venous sinuses are composed of connective tissue, with no muscle present, and they are lined with endothelium that is continuous with the endothelium of the capillaries and veins. The coronary sinus is a venous sinus that collects blood from the cardiac veins of the heart and returns it to the right atrium. Still larger venous sinuses are located in the dura mater, the outer meningeal covering of the brain (Figure 19.29). Most of the blood from the brain travels through these **dural sinuses** before entering the veins that return blood to the heart.

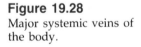

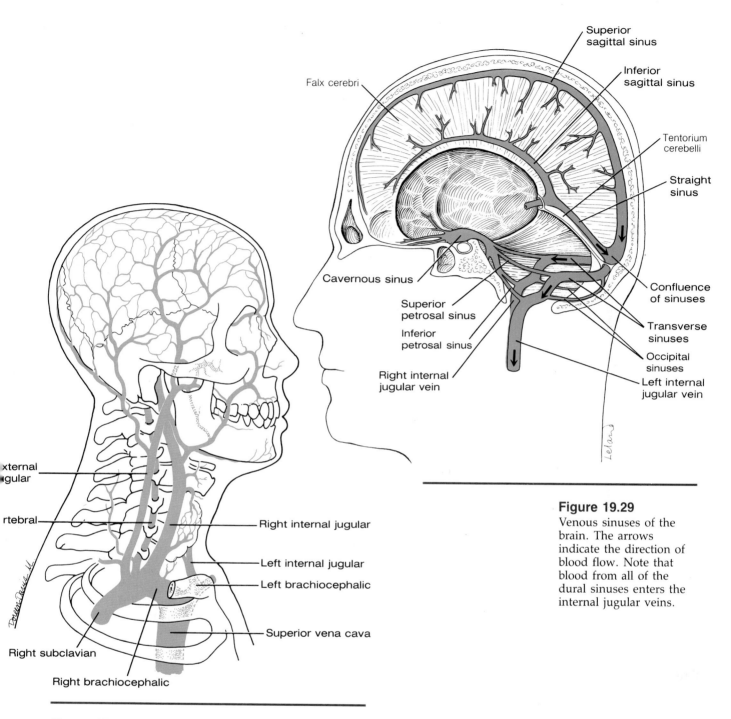

Figure 19.29
Venous sinuses of the brain. The arrows indicate the direction of blood flow. Note that blood from all of the dural sinuses enters the internal jugular veins.

Figure 19.30
Veins of the head and neck.

VEINS OF THE HEAD AND NECK   Most of the venous blood from the head and neck regions returns to the heart through the *internal jugular veins,* the *external jugular veins,* and the *vertebral veins* (Figure 19.30).

The **internal jugular veins** are larger and deeper than the external jugular veins. Each internal jugular vein drains a **transverse sinus** that receives blood from a **cavernous sinus,** the **superior sagittal sinus,** the **inferior sagittal sinus,** and the **straight sinus** (Figure 19.29). The internal jugular veins therefore serve as the major venous drainage of the brain. Each internal jugular

F19.30

F19.29

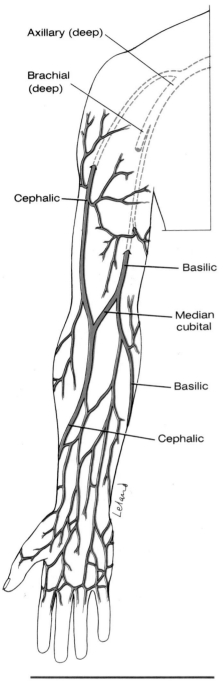

Axillary (deep)

Brachial
(deep)

Cephalic

Basilic

Median
cubital

Basilic

Cephalic

Leland

Figure 19.31
Superficial veins of the
upper limb.

vein leaves the skull through a jugular foramen, which is located in the junction between the petrous portion of the temporal bone and the occipital bone, and travels through the neck alongside a common carotid artery and a vagus nerve. The internal jugular vein joins with a **subclavian vein** to form a **brachiocephalic vein** (Figure 19.30).

The **vertebral veins** drain the posterior regions of the head (Figure 19.30). Each vertebral vein passes through the transverse foramina of the cervical vertebrae and joins a brachiocephalic vein.

The superficial regions of the head and neck, which are supplied by the external carotid arteries, are drained by the **external jugular veins** (Figure 19.30). Each external jugular vein passes superficially down the neck and empties into a **subclavian vein** lateral to the point at which the subclavian and internal jugular veins join to form a brachiocephalic vein. The right and left brachiocephalic veins join to form the superior vena cava, which empties into the right atrium of the heart.

VEINS OF THE UPPER LIMBS   The deep veins of the upper limbs follow the paths of the arteries and are given the same names: **axillary, brachial, radial,** and **ulnar veins** (Figure 19.28). The superficial veins (Figure 19.31) begin from venous networks that cover the dorsal and palmar surfaces of the hand. The **cephalic vein** arises from the lateral side of the dorsal veins of the hand, crosses to the ventral-lateral side of the forearm, and continues up the lateral side of the arm. At the shoulder, it goes deep and empties into the axillary vein. The **basilic vein** is the other major superficial vein of the upper limb. It arises from the medial side of the dorsal veins of the hand and ascends along the medial-posterior side of the forearm. Just below the elbow, the basilic vein travels to the front of the arm and goes deep above the elbow to join with the brachial vein, thus forming the axillary vein. There are several superficial branches between the basilic and cephalic veins. One of these, the **medial cubital vein,** which connects the two vessels in front of the elbow, is commonly used to give or receive blood. All the venous blood from the head, neck, and upper limbs is returned to the heart through the superior vena cava.

VEINS OF THE THORAX   The venous blood of the wall of the thorax also empties into the superior vena cava—in this case, by way of an *azygos system* of veins (Figure 19.32). The **azygos vein** begins in the upper right lumbar region and passes through the diaphragm via the *aortic hiatus* (along with the aorta and the thoracic duct of the lymphatic system) to enter the thorax. It continues up the posterior wall of the thorax, to the right of the vertebral column, and empties into the superior vena cava. The **hemiazygos vein** follows a similar course on the left side, passing in front of the vertebral column to empty into the azygos vein in the middle of the thorax. Most of the upper left region of the thorax is drained by an **accessory hemiazygos vein,** which also crosses the vertebral column to empty into the azygos vein. The upper three left intercostal veins drain directly into the left brachiocephalic vein through the left **superior intercostal vein.** The azygos system of veins also receives the **posterior intercostal, bronchial, esophageal,** and **pericardial veins.**

VEINS OF THE ABDOMEN AND PELVIS   Venous blood from the lower regions of the body is returned to the heart through the **inferior vena cava.** This large vessel is formed by the confluence of the **right** and **left common iliac veins** at the level of the fifth lumbar vertebra. It travels along the posterior body wall to enter the right atrium of the heart. As it passes through the abdomen, the inferior vena cava receives tributaries that correspond to most of the arteries that arise from the abdominal aorta—for example, **lumbar, renal, suprarenal, hepatic,** and **right testicular** or **ovarian veins.** (The **left testicular** or **ovarian veins** empty into the **left renal vein.**) The inferior vena cava does not receive blood directly from the digestive tract, pancreas, or spleen. The veins from these regions form the **hepatic portal system,** which is discussed in the next section.

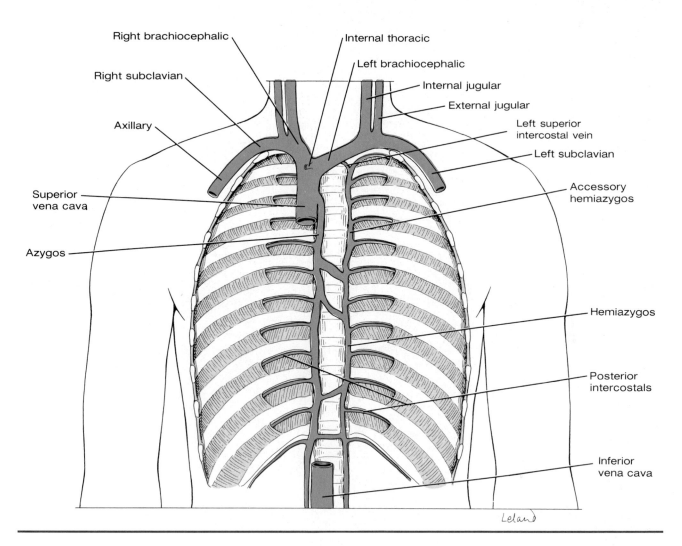

Right brachiocephalic

Right subclavian

Axillary

Superior vena cava

Azygos

Internal thoracic

Left brachiocephalic

Internal jugular

External jugular

Left superior intercostal vein

Left subclavian

Accessory hemiazygos

Hemiazygos

Posterior intercostals

Inferior vena cava

Leland

Figure 19.32
Veins of the posterior wall of the thorax. The anterior wall of the thoracic cage has been removed.

The veins of the pelvic region follow the pattern of the arteries and are given the same names. Most of them empty into the **internal iliac veins,** which join with the **external iliac veins** from the lower limbs to form the **right** and **left common iliac veins.**

THE HEPATIC PORTAL SYSTEM   Venous blood from the stomach, intestines, spleen, and pancreas is carried by small veins, most of which empty into three large vessels: the *splenic vein,* the *inferior mesenteric vein,* and the *superior mesenteric vein.* The **splenic vein** returns blood from the spleen. As it travels toward the midline of the body it receives tributaries from the stomach and pancreas. The **inferior mesenteric vein** returns blood from the rectum and the descending limb of the large intestine. It ascends beneath the parietal peritoneum and joins the splenic vein behind the pancreas. The common vessel formed by the junction of the inferior mesenteric vein and the splenic vein joins the superior mesenteric vein behind the neck of the pancreas. The **superior mesenteric vein** returns blood from the small intestine, the cecum, and the ascending and transverse limbs of the large intestine. The junction of the superior mesenteric vein and the splenic vein forms the **hepatic portal vein** (Figure 19.33). The hepatic portal vein ascends in the right border of the lesser omentum and enters the inferior surface of the liver, through which it follows a unique pathway that is discussed in Chapter 23. After leaving the liver, the blood travels through the hepatic veins to enter the inferior vena cava.

This modification of the circulatory system, whereby blood from the digestive tract, pancreas, and spleen passes through the liver before returning

F19.33

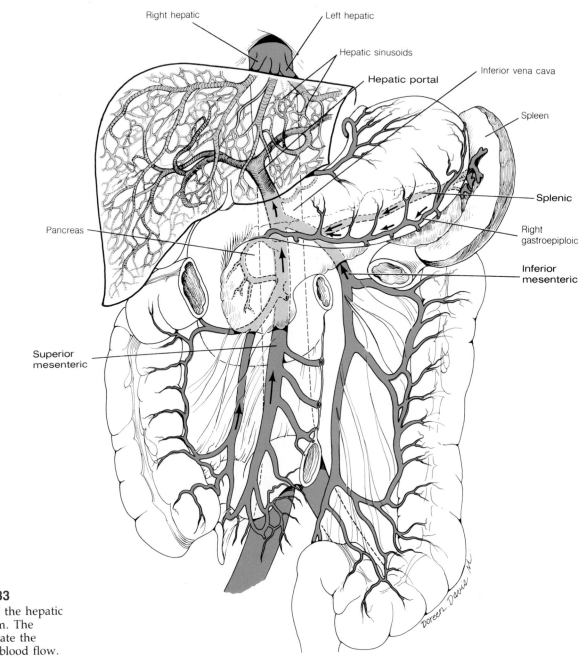

Right hepatic

Left hepatic

Hepatic sinusoids

Hepatic portal

Inferior vena cava

Spleen

Splenic

Pancreas

Right gastroepiploic

Inferior mesenteric

Superior mesenteric

Doreen Davis M.

Figure 19.33
The veins of the hepatic portal system. The arrows indicate the direction of blood flow.

to the heart, is very significant. The venous blood in the hepatic portal vein contains nutrients and other substances that have been absorbed from the digestive tract. The hepatic portal system carries these substances directly to the liver, where they can be removed from the blood and stored, metabolized, or, in the case of harmful substances, detoxified.

VEINS OF THE LOWER LIMBS   The deep veins of the lower limbs, like those of the upper limbs, travel along with the arteries and are given corresponding names: **external iliac, femoral, popliteal, anterior** and **posterior tibial,** and **peroneal veins** (Figure 19.28).

F19.28

Two large superficial veins of the lower limbs arise from a **dorsal venous arch** on the top of the foot (Figure 19.34): the *great* and *small saphenous veins.*

F19.34

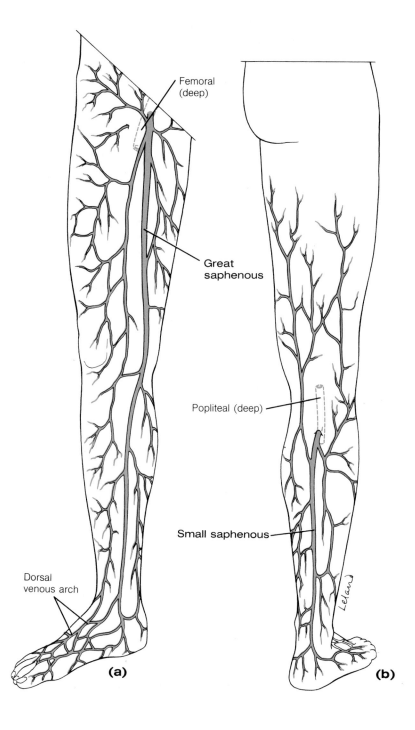

Femoral
(deep)

Great
saphenous

Popliteal (deep)

Small saphenous

Dorsal
venous arch

(a)

(b)

Figure 19.34
Superficial veins of the
lower limb **(a)** Anterior
view. **(b)** Posterior view.

The **great saphenous vein** is the longest vein in the body. It travels along the medial side of the foot, leg, and thigh, where it joins with the femoral vein just below the inguinal ligament. The **small saphenous vein** travels along the lateral side of the foot, crosses to the posterior surface of the leg, and joins the popliteal vein behind the knee. There are numerous connections between the great and small saphenous veins, as well as between them and the deep veins. Both the deep and the superficial veins of the lower limbs contain valves that assist in returning the blood to the heart against the force of gravity.

Table 19.2   Schematic Summary of the Major Veins That Empty into the Superior and Inferior Venae Cavae

SUPERIOR VENA CAVA
AZYGOS

    Hemiazygos

    Accessory hemiazygos

BRACHIOCEPHALIC
Internal jugular
Vertebral
Internal thoracic
Subclavian ←——————— Axillary ←—— Brachial ←Ulnar
    External jugular   Cephalic   Basilic Radial

INFERIOR VENA CAVA
Inferior phrenics

                    Inferior mesenteric
Hepatics←Liver←Hepatic portal ←Splenic
                    Superior mesenteric

Right suprarenal
Renals
    Left suprarenal
    Left testicular or ovarian
Right testicular or ovarian
Lumbars
Middle sacral
Common iliacs (right and left)
    Internal iliac
    External iliac← Femoral← Popliteal   *Posterior tibial*
                           Anterior tibial
                  Small saphenous
        Great saphenous

STUDY OUTLINE

TYPES OF VESSELS   p. 561

ARTERIES   Large vessels that carry blood away from heart; *arterioles* and *capillaries* are progressively smaller vessels.

VENULES, VEINS   Vessels that return blood to heart.

GENERAL STRUCTURE OF BLOOD VESSEL WALLS   p. 561

TUNICA INTIMA   Present in all vessels; composed of endothelium, connective tissue, and basement membrane.

TUNICA MEDIA   Composed of elastic fibers and smooth muscle; often separated from tunica intima by an internal elastic lamina and from tunica externa by an external elastic lamina.

TUNICA EXTERNA   Composed of collagen fibers.

STRUCTURE OF ARTERIES   pp. 561–563

ELASTIC ARTERIES   Large arteries; thick tunica media with many elastic fibers.

MUSCULAR ARTERIES (DISTRIBUTING ARTERIES)   Tunica media of smooth muscle cells; well-defined internal and external elastic laminae.

STRUCTURE OF ARTERIOLES   Diameter less than 0.5 mm; small lumen; thick tunica media of mostly muscle cells.   p. 563

STRUCTURE OF CAPILLARIES   0.5–1 mm long; 0.01 mm diameter; walls have endothelial layer only. Substances pass through walls through tight junctions, through cell membrane, within vesicles, or through pores between endothelial cells; sinusoids are large, thin-walled channels that receive blood from some arterioles.   pp. 563–565

STRUCTURE OF VENULES   Inner endothelium and some outer fibrous tissue; large venules encircled by a few smooth muscle fibers.   p. 565

STRUCTURE OF VEINS   Thin tunica media with few muscle fibers; thick tunica externa with little elastic tissue; larger lumen, thinner walls than arteries; some have valves and vasa vasorum.   p. 565

PRINCIPLES OF CIRCULATION   Efficient circulation requires adequate blood volume, blood vessels in good condition, and smoothly functioning heart.   pp. 565–569

VELOCITY OF BLOOD FLOW   In general, velocity of

flow in any segment of cardiovascular system is inversely related to total cross-sectional area of vessels of the segment; velocity of flow is lowest in the capillaries.

PULSE Elastic expansion and subsequent recoil of arteries that occurs in association with beating of heart; can be felt at various locations on body surface.

BLOOD FLOW Volume of blood that flows through a vessel or vessels in a given time.

PRESSURE Pumping action of heart imparts energy to blood; this energy is evident as pressure that drives blood through vessels.

RESISTANCE Essentially a measure of friction.

FACTORS THAT AFFECT RESISTANCE: Blood Viscosity, Vessel Length, Vessel Radius

ALTERATIONS IN RESISTANCE Vessel length is constant and blood viscosity normally does not vary greatly. Most alterations in resistance are due to changes in vessel—particularly arteriole—radius.

RELATION AMONG FLOW, PRESSURE, AND RESISTANCE Generally,

$$flow = \frac{pressure}{resistance}$$

and

mean arterial pressure = cardiac output
$\times$ total peripheral resistance

ARTERIAL PRESSURE Affected by a number of factors, through mechanisms that influence vessel resistance and cardiac function. pp. 569–572

NEURAL FACTORS

VASOMOTOR NERVE FIBERS Affect vessel resistance by altering vessel diameters. Sympathetic vasoconstrictor fibers release norepinephrine and cause vasoconstriction; sympathetic vasodilator fibers to skeletal muscle vessels release acetylcholine and cause vasodilation.

VASOMOTOR CENTER Tonically active brain-stem center that plays role in regulating blood-vessel resistance and, thus, in regulating arterial pressure; center influenced by inputs from higher brain centers, cardiovascular-system receptors, and receptors associated with sensory surfaces of body.

CARDIAC CENTER Can influence arterial pressure through effects on cardiac function.

BARORECEPTORS Influence vessel diameters, heart rate, and cardiac contractility, thereby affecting arterial pressure.

AORTIC AND CAROTID BODIES Contain chemoreceptors that can influence arterial pressure.

CHEMICALS AND HORMONES

ANGIOTENSIN Extremely powerful vasoconstrictor that raises arterial pressure.

EPINEPHRINE Causes transitory increase in systolic pressure within arteries.

VASOPRESSIN Raises arterial pressure by stimulating arteriolar constriction.

BLOOD VOLUME When increased, tends to raise arterial pressure; when decreased, tends to lower it. Kidney mechanisms that influence blood volume are particularly important in long-term regulation of arterial pressure.

SUMMARY A number of factors affect arterial pressure, and arterial pressure at any one moment is usually result of combined influence of several factors. Different mechanisms work together to maintain arterial pressure and thus to maintain a circulation that meets body's needs.

MEASUREMENT OF ARTERIAL PRESSURE Usually done indirectly using inflatable cuff and manometer. Blood pressure is reported as systolic pressure/diastolic pressure—for example, 120/80. Pulse pressure is difference between systolic and diastolic pressures.

BLOOD FLOW THROUGH TISSUES pp. 572–573

LOCAL AUTOREGULATORY CONTROL Local autoregulatory mechanisms alter blood flow in accordance with tissue needs. Oxygen deficiency can lead to local dilation of arterioles and increased relaxation of precapillary sphincters. Carbon dioxide, lactic acid, adenosine, adenosine phosphate compounds, histamine, potassium ions, and hydrogen ions may act as local vasodilator substances.

EXTERNAL FACTORS External factors such as vasomotor nerve fibers and hormones usually influence large segments of vascular system; generally more involved with mechanisms that function for well-being of entire body than with control of tissue blood flow according to needs of particular tissues.

CAPILLARY EXCHANGE Most capillaries are permeable to water and small particles such as glucose, inorganic ions, urea, amino acids, and lactic acid; and these materials pass readily between blood and interstitial fluid. Permeability to protein generally quite limited. pp. 573–575

CAPILLARY BLOOD FLOW Blood flows slowly through capillaries, allowing time for exchange of materials between blood and interstitial fluid. Precapillary sphincters contract and relax cyclically, and capillary blood flow is usually intermittent rather than continuous and steady.

MOVEMENT OF MATERIALS BETWEEN THE BLOOD AND THE INTERSTITIAL FLUID Several processes involved:

DIFFUSION Most important means by which substances such as nutrients and metabolic end products pass between blood and interstitial fluid.

ENDOCYTOSIS AND EXOCYTOSIS There is disagreement about the importance of this activity in exchange of materials between blood and interstitial fluid.

FLUID MOVEMENT Normally causes little change in volume of blood or interstitial fluid.

Fluid Pressures Capillary pressure tends to force fluid out of capillaries and into tissue space by filtration through capillary walls. Interstitial-fluid pressure opposes this movement.

Osmotic Pressures Colloid osmotic pressure of interstitial fluid tends to draw water out of capillaries by osmosis; colloid osmotic pressure of plasma tends to draw water into capillaries by osmosis.

Direction of Fluid Movement If forces tending to move fluid out of a capillary are greater than forces tending to move fluid in, fluid will leave, and vice versa.

VENOUS RETURN Influenced by blood volume, contraction of skeletal muscles, valves of veins, constriction of veins. **pp. 575–576**

EFFECTS OF GRAVITY ON THE CARDIO- VASCULAR SYSTEM When a person moves from supine position to standing upright and perfectly still, the weight of blood causes high pressures in blood vessels of lower regions of body. Baroreceptors and muscular contractions during movement help compensate for effects of gravity when a person is upright. **pp. 576– 578**

CIRCULATION IN SPECIAL REGIONS **pp. 578–581**

PULMONARY CIRCULATION Pulmonary circuit is a low-pressure circuit. Low oxygen levels cause constriction of pulmonary vessels by local autoregulatory mechanisms.

CORONARY CIRCULATION Greatly influenced by aortic pressure and by pressures that build up in walls of heart chambers during systole.

CEREBRAL CIRCULATION Total cerebral blood flow is remarkably constant over broad range of arterial pressures. Interference with cerebral blood flow can lead to fainting or stroke.

SKELETAL MUSCLE CIRCULATION During exercise, blood flow to skeletal muscles increases greatly, primarily due to local autoregulatory mechanisms, but sympathetic vasodilator neurons are also involved.

CUTANEOUS CIRCULATION Primarily for purposes of temperature regulation rather than skin nutrition. Blood flow can vary tremendously.

CARDIOVASCULAR ADJUSTMENTS DUR- ING EXERCISE Skeletal muscle blood flow increases greatly, due to dilation of skeletal muscle vessels. Other vessels constrict to compensate, but total peripheral resistance generally decreases. Heart rate and stroke volume increase, as does cardiac output. Mean arterial pressure usually rises. **p. 583**

CONDITIONS OF CLINICAL SIGNIFI- CANCE: BLOOD VESSELS **pp. 582–583**

HIGH BLOOD PRESSURE (HYPERTENSION) Most cases due to unknown causes. Many cases may be due to excessive dietary intake of sodium or excessive retention of sodium by body, leading to increased blood volume and cardiac output. An increased peripheral resistance due to vascular constriction may also be involved.

ATHEROSCLEROSIS AND ARTERIOSCLEROSIS *Atherosclerosis:* characterized by deposits of abnormal smooth muscle cells, lipid materials, and connective tissue. Deposits can narrow or occlude blood vessels or favor clot formation. *Arteriosclerosis* can lead to loss of elasticity of vessel walls.

EDEMA Accumulation of excess fluid in tissues. In general, can result from any event that enhances movement of fluid out of blood vessels and into tissues, or that retards return of fluid from tissues to bloodstream.

CIRCULATORY SHOCK Condition in which cardiac output is so reduced that body tissues fail to receive an adequate blood supply. Can be due to diseases or weaknesses of heart, a reduction in blood volume, or vascular difficulties that reduce venous return.

ANEURYSMS Localized dilation of an artery due to weakness of artery wall.

PHLEBITIS Inflammation of a vein that can lead to clotting.

VARICOSE VEINS Dilated, lengthened, and tortuous veins that can interfere with venous return and lead to clotting.

EFFECTS OF AGING Environment and lifestyle as well as age and genetic inheritance may contribute to cardiovascular problems associated with aging.

ANATOMY OF THE VASCULAR SYS- TEM **pp. 583–604**

PULMONARY CIRCUIT
1. From right ventricle to pulmonary trunk to pulmonary arteries to lobar arteries to capillary plexuses for gas exchange to venules to pulmonary vein to left atrium.
2. Transports oxygenated venous blood and oxygen-poor arterial blood.

SYSTEMIC CIRCUIT

SYSTEMIC ARTERIES Arise from aorta.

Branches of the Ascending Aorta Coronary arteries.

Branches of the Aortic Arch Brachiocephalic, left common carotid, and left subclavian arteries.

Arteries of the Head and Neck Common carotids, internal and external carotids, temporals, cerebral and vertebral arteries, and cerebral arterial circle.

Arteries of the Upper Limbs Subclavian, axillary, brachial, radial, ulnar, and digital arteries.

Branches of the Thoracic Aorta Intercostal, bronchial, esophageal, and superior phrenic arteries.

Branches of the Abdominal Aorta Lumbar, common iliac, middle sacral, inferior phrenic, celiac (gastric, splenic, common hepatic), suprarenal, renal, superior and inferior mesenteric, ovarian or testicular arteries.

Arteries of the Pelvic Region Internal and external iliac arteries.

Arteries of the Lower Limbs Femoral, popliteal, anterior and posterior tibials, peroneal, plantar, digital, and dorsalis pedis arteries.

SYSTEMIC VEINS Deep veins, superficial veins, venous sinuses.

Veins of the Head and Neck Internal and external jugulars, subclavian, brachiocephalic, superior vena cava, and vertebral veins.

Veins of the Upper Limbs Deep: axillary, brachial, radial, and ulnar veins. Superficial: cephalic, basilic, and median cubital veins.

Veins of the Thorax Azygos system: intercostal, bronchial, esophageal, and pericardial veins.

Veins of the Abdomen and Pelvis Inferior vena cava; common iliac, lumbar, renal, suprarenal, hepatic, right testicular or ovarian veins.

The Hepatic Portal System Hepatic portal vein formed by joining of inferior mesenteric, superior mesenteric, and splenic veins; carries blood from stomach, intestines, spleen, and pancreas; enters liver for removal of nutrients.

Veins of the Lower Limbs Contain valves. *Deep:* external iliac, femoral, popliteal, anterior and posterior tibials, and peroneal veins. *Superficial:* great and small saphenous veins.

SELF-QUIZ

1. The tunica media and tunica externa are not present in: (a) veins; (b) capillaries; (c) arterioles.

2. The endothelium is part of the: (a) tunica externa; (b) tunica media; (c) tunica intima; (d) adventitia.

3. The walls of the larger arteries and veins are so thick that they are supplied by small nutrient vessels called vasa vasorum. True or False?

4. Valves are typically found in arteries as well as in the larger veins. True or False?

5. Which of the following blood-vessel layers is present in vessels of all sizes? (a) endothelium; (b) tunica media; (c) tunica externa.

6. Compared to corresponding arteries, veins have: (a) a better-developed tunica media; (b) a generally larger lumen; (c) generally thinner walls; (d) both b and c; (e) all of these.

7. The velocity of blood flow is slowest in the: (a) arteries; (b) arterioles; (c) capillaries; (d) veins.

8. Blood pressure is generally lowest in the: (a) arteries; (b) capillaries; (c) veins.

9. In general, which of the following will cause an increased resistance to blood flow? (a) a decrease in vessel length; (b) a decrease in vessel diameter; (c) a decrease in blood viscosity.

10. The mean arterial pressure is equal to the cardiac output multiplied by the: (a) total peripheral resistance; (b) total systemic blood flow; (c) total venous return.

11. Which of the following is likely to lead to a fall in blood pressure? (a) decreased arterial oxygen concentration; (b) decreased activity of the carotid sinus baroreceptors; (c) stretching the wall of the aortic arch.

12. The kidneys exert a significant effect on blood volume by virtue of their ability to regulate salt and water excretion, and they are particularly important in the long-term regulation of arterial pressure. True or False?

13. When a person's blood pressure is reported to be 120/80, the 120 is the: (a) pulse pressure; (b) systolic pressure; (c) diastolic pressure.

14. Which of the following forces favors the movement of fluid into the capillaries from the tissue spaces? (a) colloid osmotic pressure of the plasma; (b) colloid osmotic pressure of the interstitial fluid; (c) capillary pressure.

15. Venous return is likely to be enhanced by a decreased peripheral venous pressure and an increased right atrial pressure. True or False?

16. The flow of blood through the coronary vessels is: (a) independent of the aortic pressure; (b) greatly influenced by the pressures that build up in the walls of the heart chambers during systole; (c) continuous and constant throughout the cardiac cycle.

17. Most of the cutaneous blood flow is for: (a) temperature regulation; (b) provision of nutrients to the skin; (c) removal of wastes from the skin.

18. During exercise (a) the stroke volume of the heart generally decreases; (b) the contractility of the cardiac muscle generally decreases; (c) the cardiac output generally increases.

19. The pulmonary arteries carry blood that has a low carbon dioxide content and a high oxygen content. True or False?

20. Match the terms associated with the systemic arteries with the appropriate description.

Aorta	(a) A major blood supplier to the brain; passes through the transverse foramina
Coronary arteries	
Internal carotid artery	(b) Supplies all of the small intestine
Vertebral artery	(c) Formed from bifurcation of abdominal aorta
Radial artery	
Superior mesenteric artery	(d) Vessel from which the pulse is usually taken
	(e) Branches of ascending aorta
Inferior mesenteric artery	(f) Supplies the muscles of the anterior compartment of the leg
Common iliac arteries	(g) Blood from left ventricle enters systemic circuit through this vessel
Anterior tibial artery	(h) Supplies the large intestine
	(i) Supplies the brain through its terminal branches—the anterior and middle cerebral arteries

21. The coronary arteries arise from the: (a) inferior vena cava; (b) superior vena cava; (c) right atrium; (d) ascending aorta.

LEARNING OBJECTIVES

After completing this chapter, you should be able to:

1. Describe the lymphatic system of vessels.

2. Discuss the functions of the lymphatic system.

3. Explain how excess interstitial fluid and plasma proteins are returned to the bloodstream by the lymphatic vessels.

4. Describe the structure of the lymph nodes, spleen, thymus gland, and tonsils.

CHAPTER CONTENTS

LYMPHATIC VESSELS

LYMPH NODES

LYMPHATIC DUCTS

MECHANISMS OF LYMPH FLOW

FUNCTIONS OF THE LYMPHATIC SYSTEM

LYMPHOID ORGANS

KEY TERMS AND DERIVATIVES

lymph (*lymph* = clear fluid) the watery fluid in the lymph vessels collected from the tissue fluid

lymphocyte (*cyte* = cell) a granular white blood cell that matures in the lymphoid tissue

interstitial (*inter* = between; *sistere* = to set) pertaining to or situated in the interspaces of a tissue

The Lymphatic System

The **lymphatic system** consists of (1) an extensive network of *capillaries* and larger *collecting vessels* that receive fluid from the loose connective tissues throughout the body and transport it to the cardiovascular system; (2) *lymph nodes*, which serve as filters of the fluid within the collecting vessels; and (3) the *lymphoid organs*, including lymphatic nodules, the spleen, the thymus gland, and the tonsils.

The lymphatic system is closely related both anatomically and functionally to the cardiovascular system. As indicated in Chapter 19, the volume of fluid that moves out of the capillaries and into the tissue spaces is slightly greater—by about 3 liters per day—than the volume of fluid that moves into the capillaries from the tissues. If this fluid were allowed to accumulate, the tissues would swell, producing edema. However, the lymphatic system returns this extra fluid (including any plasma protein that escapes from the capillaries) back to the bloodstream.

LYMPHATIC VESSELS

Interstitial fluid enters the lymphatic system by passing through the extremely thin walls of **lymphatic capillaries.** Once the fluid is within the vessels of the lymphatic system it is called **lymph.** Since the lymphatic system is a one-way system—only *returning* fluid to the bloodstream—the lymphatic capillaries are dead-end vessels (Figure 20.1). Their walls, like the walls of blood capillaries, are composed of a single layer of endothelium. However, lymphatic capillaries lack the surrounding basement membrane that ensheathes blood capillaries. Another difference between lymphatic and blood capillaries is that the edges of adjacent endothelial cells in lymphatic capillaries are only loosely attached and often overlap. This arrangement forms a functional one-way valve. The pressure of the interstitial fluid outside lymphatic capillaries pushes the edges of the endothelial cells inward, allowing the interstitial fluid to enter the capillaries. But once within the lymphatic capillary, this fluid cannot re-enter the intercellular spaces because pressure from within the capillaries forces the edges of the endothelial cells together, closing the "valve." Because of this structural arrangement, lymphatic capillaries are more permeable than most blood capillaries, and virtually all the components of the interstitial fluid, including proteins and other large particles (such as disease-causing organisms), can enter these vessels and are transported throughout the body. Most tissues contain plexuses of lymphatic capillaries located among the vascular capillaries. Special lymphatic capillaries called **lacteals** are located in the villi of the intestine. The lacteals aid in the absorption of fat from the digestive tract and carry it to the bloodstream as a milky fluid called *chyle*.

The lymphatic capillaries, which are widely distributed throughout the interstitial spaces of the body, join together to form progressively larger lymphatic vessels (Figure 20.2a). Generally, the larger lymphatic vessels, which are called *collecting vessels*, travel alongside the arteries and veins of the cardiovascular system and pass through one or more *lymph nodes* before emptying into either the *thoracic duct* or the *right lymphatic duct*, both of which return the

609

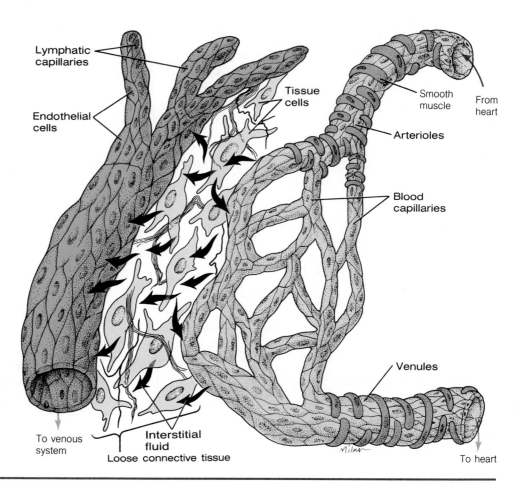

Figure 20.1
Schematic representation of the lymphatic capillaries showing their relationship to interstitial fluid, the blood vascular system, and tissue cells. The arrows indicate the directions of fluid movement. Note that the lymphatic capillaries begin as dead-end vessels.

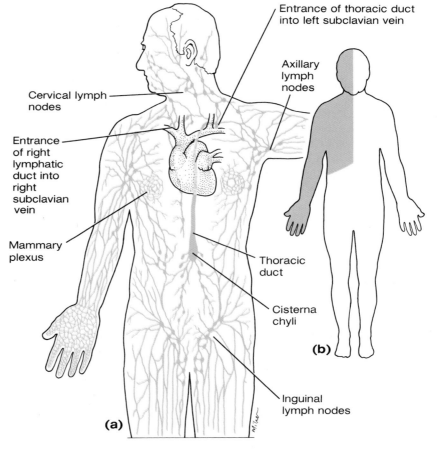

Figure 20.2
The lymphatic system. (a) Major lymphatic vessels and groups of lymph nodes. (b) Lymph from the colored area returns to the blood vascular system through the right lymphatic duct. Lymph from the remainder of the body travels through the thoracic duct.

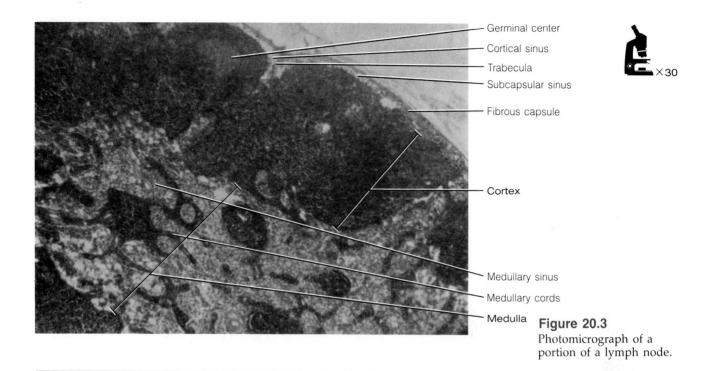

Germinal center
Cortical sinus
Trabecula
Subcapsular sinus
Fibrous capsule

Cortex

Medullary sinus
Medullary cords
Medulla

×30

Figure 20.3
Photomicrograph of a portion of a lymph node.

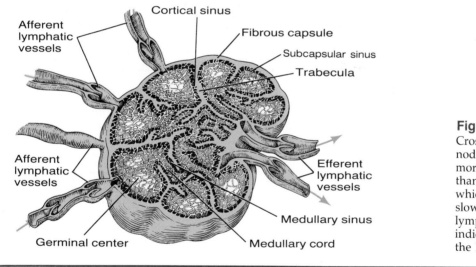

Afferent lymphatic vessels
Cortical sinus
Fibrous capsule
Subcapsular sinus
Trabecula

Afferent lymphatic vessels

Efferent lymphatic vessels

Germinal center
Medullary sinus
Medullary cord

Figure 20.4
Cross section of a lymph node. Note that there are more afferent vessels than efferent vessels, which has the effect of slowing the rate of lymph flow. The arrows indicate the direction of the lymph flow.

lymph to the bloodstream. The walls of the lymphatic collecting vessels are similar to the walls of veins, although they are thinner and, like veins, contain valves that occur in pairs, each on opposite sides of the vessel, with their free edges pointing in the direction of lymph flow. The valves aid in the movement of lymph by preventing its backflow.

LYMPH NODES

Lymph nodes are small round or bean-shaped organs distributed along the course of the lymphatic vessels (Figure 20.3). There are groups of lymph nodes in the groin, axilla, and neck, as well as in numerous other deeper locations.

Each node is enclosed in a **fibrous capsule** (Figure 20.4). Strands of connective tissue called **trabeculae** extend inward from the capsule and divide

F20.3

F20.4
tra-bek´-u-lah

the node into several compartments. These compartments are further subdivided by a network of reticular fibers that extend between the trabeculae. The node consists of an outer *cortical* region and an inner *medullary* region. Within the cortex of each node are separate masses of lymphoid tissue called **germinal centers,** which serve as a source of lymphocytes. *Lymphoid tissue* is a modified type of loose connective tissue, with a prominent network of reticular fibers between which there are many lymphocytes and macrophages. The cells of the medulla are arranged in the form of strands called **medullary cords.**

Lymph enters the convex surface of the lymph node through several **afferent lymphatic vessels** and filters slowly through irregular channels in the node, called **sinuses.** The sinuses are spanned by networks of reticular fibers. Lymph from an afferent lymphatic vessel enters a *subcapsular sinus* located just beneath the capsule. From there it percolates through *cortical sinuses,* which penetrate the cortex, and enters *medullary sinuses* located between the medullary cords. From the medullary sinuses lymph leaves by way of **efferent lymphatic vessels** at a small indentation called the *hilus.* Because there are fewer efferent vessels than afferent vessels, this has the effect of slowing the rate of lymph flow through the nodes, which allows phagocytic cells to remove foreign substances more effectively, thereby preventing their entry into the bloodstream. Generally, all lymph has passed through several lymph nodes before it is returned to the cardiovascular system.

Lymph that flows into the lymph nodes contains many foreign particles and microorganisms, some of which can cause diseases if they are not destroyed. As lymph percolates slowly through the sinuses within a node, large foreign particles such as bacteria become entrapped in the meshes of the reticular fibers that span the sinuses. The entrapped particles are soon attacked and destroyed by phagocytic cells, called *macrophages,* that line the sinuses. In addition, between the sinuses are masses of lymphoid tissue containing *lymphocytes* and *plasma cells* that produce antibodies for destroying certain foreign substances known as antigens. Unfortunately, not all disease-producing cells, such as cancer cells, are destroyed within the lymph nodes. Some cancer cells can survive and multiply within the nodes; thus the nodes serve as a site from which these cancer cells can spread throughout the body by way of the cardiovascular system. For this reason, swollen lymph nodes near cancer sites are often surgically removed.

The efferent collecting vessels from the lymph nodes of most regions of the body converge into larger vessels called *lymph trunks.* There are five major lymph trunks, four of them paired: (1) the unpaired *intestinal trunk,* which receives lymph from abdominal organs; (2) the *lumbar trunks,* which drain the lower limbs and some pelvic organs; (3) the *subclavian trunks,* which drain the arms and parts of the thorax and the back; (4) the *jugular trunks,* which drain the head and neck regions; and (5) the *bronchomediastinal trunks,* which drain the thorax. The lymph trunks, in turn, empty into either the thoracic duct or the right lymphatic duct.

LYMPHATIC DUCTS

F20.5   The **thoracic duct** arises from the **cisterna chyli,** which is a saclike enlargement that lies in front of the second lumbar vertebra (Figure 20.5). The cisterna chyli receives lymph from the lumbar and intestinal trunks, which drain the lower limbs and the viscera of the abdominopelvic cavity. As the thoracic duct passes through the diaphragm (along with the aorta) and travels upward through the thoracic cavity in front of the vertebral column, it receives lymphatic vessels that drain the left side of the thorax. Behind the brachiocephalic vein, the thoracic duct curves to the left and usually receives the left subclavian and left jugular trunks from the upper limbs and the left side of the head and neck before emptying into the *left subclavian vein* near its junction with the internal jugular vein.

As we have seen, the thoracic duct returns lymph to the bloodstream from the entire body *except* the upper right limb and the right side of the

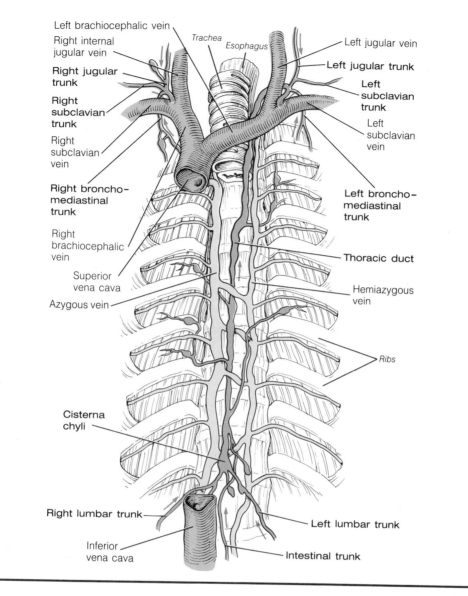

Left brachiocephalic vein
Right internal jugular vein
Right jugular trunk
Right subclavian trunk
Right subclavian vein
Right broncho-mediastinal trunk
Right brachiocephalic vein
Superior vena cava
Azygous vein

Trachea *Esophagus*

Left jugular vein
Left jugular trunk
Left subclavian trunk
Left subclavian vein
Left broncho-mediastinal trunk
Thoracic duct
Hemiazygous vein
Ribs

Cisterna chyli
Right lumbar trunk
Inferior vena cava
Left lumbar trunk
Intestinal trunk

Figure 20.5
Relationship of the thoracic duct and the right lymphatic duct to the blood vascular system.

thorax, neck, and head (Figure 20.2b). Lymph from these areas is returned to the *right subclavian vein* through a **right lymphatic duct.** The right lymphatic duct is a small vessel formed by the joining together of the right jugular, right subclavian, and right bronchomediastinal trunks, which drain lymph from these regions. The two lymphatic ducts thus convey all of the lymph that has been collected and filtered from throughout the body back into the cardiovascular system, and the lymph begins the circuit again as blood plasma.

F20.2b

MECHANISMS OF LYMPH FLOW

Lymph flows slowly; approximately 3 liters of lymph enter the bloodstream every 24 hours. The flow is slow because, unlike the cardiovascular system, the lymphatic system does not have a pump like the heart to keep it moving. Rather, lymph flow depends on more subtle forces, such as contractions of skeletal muscles, which apply pressure on the lymph vessels and compress them. This action forces lymph along the vessels. In a similar manner, the pulsing of nearby arteries can compress lymph vessels and move the lymph within them. In addition, when a section of a lymph vessel is distended by lymph it apparently can contract slightly and thereby propel the lymph along in the vessels. The valves within lymph vessels contribute to the effectiveness of these activities by permitting lymph in the vessels to flow only toward the bloodstream.

FUNCTIONS OF THE LYMPHATIC SYSTEM

The lymphatic system participates in several important activities, including the *destruction of bacteria*, the *removal of foreign particles from lymph*, the *specific immune responses*, and the *return of interstitial fluid to the bloodstream*.

Destruction of Bacteria and Removal of Foreign Particles from Lymph

Bacteria and other foreign substances are removed from lymph by phagocytes—primarily macrophages—that are present in the lymph nodes. During infections, the rate of formation of macrophages within the nodes is so great that the nodes enlarge and become tender.

Specific Immune Responses

In response to the presence of bacteria or other foreign substances, certain cells in the lymph nodes participate in specific immune responses, such as the manufacture of antibodies, that result in the destruction of the foreign substances. These responses are described in greater detail in Chapter 21.

Return of Interstitial Fluid to the Bloodstream

As mentioned earlier, edema would result if excess interstitial fluid were allowed to accumulate in the intercellular spaces. Moreover, the volume of blood would be reduced, since it is the major source of interstitial fluid. Perhaps of greater importance is the small amount of protein that is lost from the blood capillaries into the interstitial fluid but is normally returned to the blood by the lymphatic system. If this protein remained in the intercellular spaces, it would increase the osmotic pressure of the interstitial fluid and thereby affect the exchange of materials among the blood, the interstitial fluid, and the lymph.

LYMPHOID ORGANS

In addition to the lymph nodes, several organs are lymphoid in nature; these include the *spleen*, the *thymus gland*, and the *tonsils*. These organs, which have no direct association with the lymphatic system of vessels or with lymph, are an integral part of the body's immune system.

Spleen

The **spleen** is the largest lymphoid organ. It lies in the left hypochondriac region, between the fundus of the stomach and the diaphragm. The spleen is usually about 12 cm in length; however, its size and weight vary from person to person, as well as in the same individual under different conditions. The *diaphragmatic surface* of the spleen is smooth and convex, conforming to the undersurface of the diaphragm, with which it is in contact. The *visceral surface* is divided into gastric, renal, and colic surfaces that conform to the organs adjacent to it (Figure 20.6). Blood vessels enter and leave the spleen through a region on the visceral surface called the *hilus*.

F20.6

high´-lus

Like the lymph nodes, the spleen is covered by a strong fibrous capsule with trabeculae that extend into the organ and divide it into compartments. In the spleen, however, smooth muscle cells are also present in the capsule and the trabeculae. Two types of tissue, called *red pulp* and *white pulp*, are distinguishable within the compartments (Figure 20.7.) **Red pulp** is the more abundant. It consists of branching venous sinuses separated from each other by

F20.7

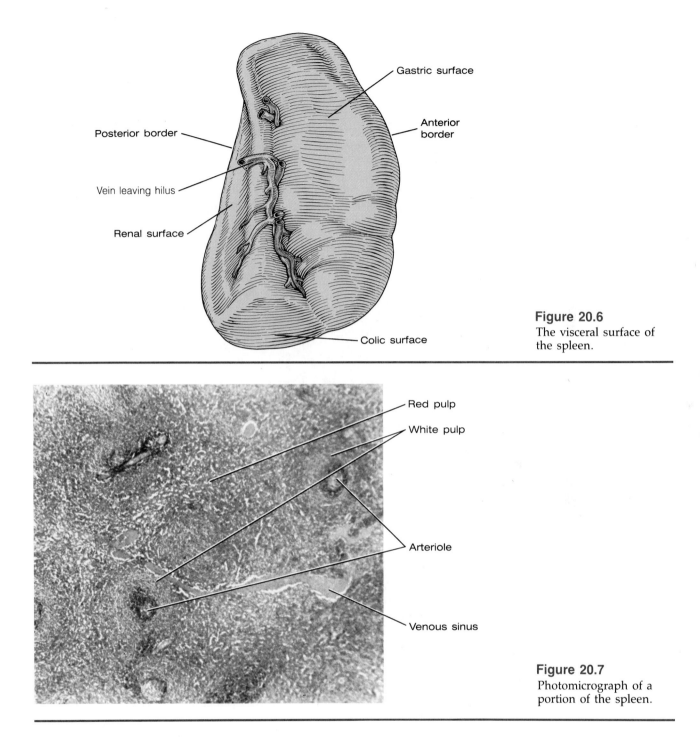

Figure 20.6
The visceral surface of the spleen.

Figure 20.7
Photomicrograph of a portion of the spleen.

columns of splenic tissue called *splenic cords*. Like other lymphoid tissues, the red pulp contains lymphocytes and macrophages; but it also contains red blood cells, which give the pulp its color. Scattered throughout the red pulp are round masses of **white pulp,** each surrounding an arteriole. White pulp, which does not contain red blood cells, is lymphoid tissue, and large numbers of lymphocytes are present in it. Blood enters the spleen at the hilus through the splenic artery. It may remain in the vessels that are surrounded by white pulp and enter the venous sinuses of the red pulp, or it may pass through the walls of the capillaries and filter between the cells of the spleen before entering the venous sinuses (Figure 20.8). From the venous sinuses, blood leaves the spleen through the splenic vein and travels within the hepatic portal vein to the liver.

F20.8

The spleen acts as a filter for the bloodstream much as the lymph nodes filter lymph. Like other lymphoid tissues, the spleen produces lymphocytes

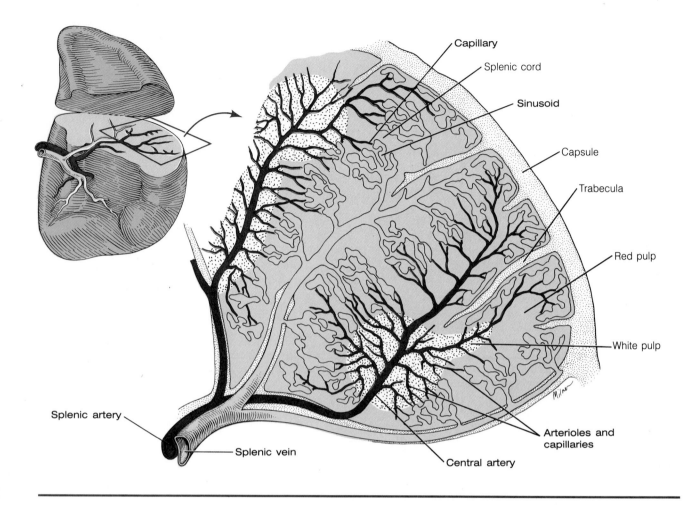

Capillary

Splenic cord

Sinusoid

Capsule

Trabecula

Red pulp

White pulp

Arterioles and capillaries

Central artery

Splenic vein

Splenic artery

Figure 20.8
The path of blood through the spleen. See text for explanation.

and plasma cells, which manufacture antibodies against foreign antigens. These activities are carried out primarily in the white pulp. In addition, the macrophages of the red pulp phagocytose old red blood cells as well as bacteria and other foreign particles. The spleen also serves to a limited extent as a blood reservoir. About 200 ml of the blood contained within the venous sinuses of the red pulp may be forced from the spleen into the general vascular system by contraction of the smooth muscles of the capsule. This can help to offset blood lost through hemorrhage. In the developing embryo, the spleen forms red blood cells; but except for some abnormal conditions that require the replacement of large numbers of red blood cells, this capability is lost following birth.

Thymus Gland

The **thymus gland** is a bilobed mass of lymphoid tissue located deep to the sternum in the anterior region of the mediastinum. It increases in size during early childhood, then begins to atrophy slowly, and diminishes following puberty. In the adult, it may be entirely replaced by adipose tissue. The thymus gland is covered by a connective tissue capsule that separates it into smaller lobules (Figure 20.9). Each lobule has an outer *cortex* and an inner *medulla*. The cortex is composed of densely packed lymphocytes. The medulla also contains lymphocytes, but in addition it has organized groups of cells referred to as **thymic corpuscles** (or *Hassall's corpuscles*).

The thymus confers on certain lymphocytes the ability to differentiate and mature into cells that can participate in specific immune responses (see Chapter 21). There is some evidence that the thymus also releases a hormone

F20.9

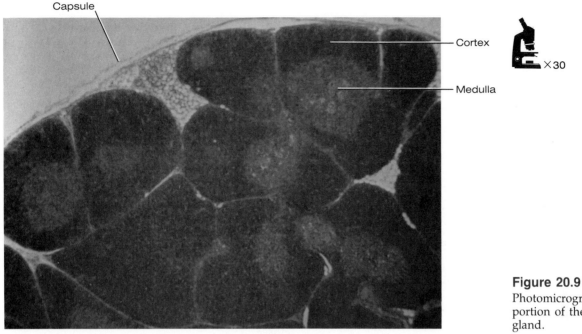

Capsule

Cortex

Medulla

×30

Figure 20.9
Photomicrograph of a portion of the thymus gland.

that can continue to influence the lymphocytes after they have left the gland.

Tonsils

The tonsils are small masses of lymphoid tissue embedded in the mucous membrane lining of the oral and pharyngeal cavities. The **palatine tonsils** are located on the posterior-lateral walls of the throat, one on each side. These are the tonsils that are noticeably enlarged when you suffer from a "sore throat." The **pharyngeal tonsils** are located on the posterior wall of the nasopharynx. They are at their peak of development during childhood, and when enlarged they are referred to as *adenoids*. On the dorsal surface of the tongue near its base is a group of **lingual tonsils.** Being composed of lymphoid tissue and surrounding the junction of the oral and nasal passageways, the tonsils serve as an additional defense against bacterial invasion of the body.

STUDY OUTLINE

LYMPHATIC VESSELS pp. 609–611

LYMPHATIC CAPILLARIES Return interstitial fluid to bloodstream.

LYMPH NODES Lymphatic tissue enclosed in fibrous capsule; afferent lymphatic vessels transport lymph into nodes; efferent lymphatic vessels transport lymph out of nodes; efferent collecting vessels form into five major lymph trunks, which empty into thoracic duct or right lymphatic duct. pp. 611–612

LYMPHATIC DUCTS pp. 612–613

THORACIC DUCT From cisterna chyli; drains most of body; transports lymph to left subclavian vein.

RIGHT LYMPHATIC DUCT Drains right upper portion of body; transports lymph to right subclavian vein.

MECHANISMS OF LYMPH FLOW Flows very slowly; depends on contractions of skeletal muscles squeezing on lymph vessels, the pulsing of nearby arteries, the presence of valves. pp. 613–614

FUNCTIONS OF THE LYMPHATIC SYSTEM p. 614

1. Destruction of bacteria and removal of foreign particles from lymph via phagocytes in nodes.
2. Specific immune responses, such as the manufacture of antibodies.

3. Return of interstitial fluid and plasma proteins to the bloodstream, thus preventing edema.

LYMPHOID ORGANS   Have no direct association with lymphatic system; are an integral part of the immune system.   **pp. 614–617**

SPLEEN   Fibrous capsule with trabeculae; contains red pulp and white pulp. Functions include: manufac-

ture of antibodies and phagocytosis of old red blood cells and foreign particles; site of red-blood-cell formation in the embryo.

THYMUS GLAND   Gives certain lymphocytes ability to participate in specific immune responses.

TONSILS   Palatine, pharyngeal, and lingual; provide defense against bacteria and other foreign particles.

SELF-QUIZ

1. Normally, slightly more liquid tends to enter the capillaries of the cardiovascular system than leaves it. True or False?

2. The walls of the lymphatic capillaries are composed of: (a) columnar epithelium; (b) adventitia; (c) a single layer of endothelium.

3. Lymphatic capillaries in the villi of the intestine are called: (a) lacteals; (b) cisterna chyli; (c) efferent lymphatic vessels; (d) trabeculae.

4. Rather large vessels that carry lymph to lymph nodes are called: (a) lymphatic trunks; (b) collecting tubules; (c) thoracic duct; (d) efferent lymphatic vessels.

5. Lymphoid tissue is a modified type of loose connective tissue, with a network of reticular fibers and containing many lymphocytes. True or False?

6. All of the major lymph trunks generally convey lymph into either the thoracic duct or the right lymphatic duct. True or False?

7. Which of the following assist in the flow of lymph within lymph vessels? (a) contraction of skeletal muscles; (b) pulsation of adjacent arteries; (c) valves within lymph vessels; (d) all of these.

8. Which is a function of the lymphatic system? (a) specific immune responses; (b) the return of tissue fluid to the bloodstream; (c) the destruction of bacteria and removal of foreign particles; (d) all of these.

9. The red pulp of the spleen consists of venous sinuses separated from each other by columns of splenic tissue called splenic cords. True or False?

10. The direction of blood flow through the spleen is from arteries into the venous sinuses, through arteries surrounded by white pulp, and out the hilus in veins. True or False?

LEARNING OBJECTIVES

After completing this chapter, you should be able to:

1. Describe the role of body surfaces in preventing the entry of potentially damaging agents or organisms into the body.

2. Explain the changes in blood flow and vessel permeability that occur during acute, nonspecific inflammation.

3. Describe the role and activity of phagocytes during acute, nonspecific inflammation.

4. Describe the roles of histamine, kinins, the complement system, and prostaglandins in acute, nonspecific inflammation.

5. List and explain the local and systemic symptoms of inflammation.

6. Explain the activities of interferon in resisting viral invasions.

7. Describe the functions of antibodies.

8. Describe the events that occur during a humoral immune response.

9. Describe the events that occur during a cell-mediated immune response.

10. Distinguish between active and passive immunity.

11. Explain the anaphylactic hypersensitivity response.

12. Describe the A-B-O and Rh blood group systems, and explain the difficulties that can arise when a person is transfused with an incompatible blood type.

CHAPTER CONTENTS

KEY TERMS AND DERIVATIVES

leukemia (*leuko* = white) a cancerous condition caused by an excessive production of leukocytes

leukocytes (*leuko* = white; *cyte* = cell) a white blood cell

macrophages (*macro* = large) a large phagocytic cell

Defense Mechanisms of the Body

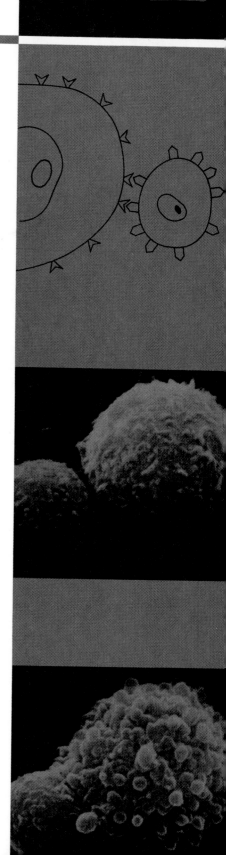

The body is continually exposed to a wide variety of potentially harmful factors, including bacteria, viruses, and hazardous chemicals. Pathogenic bacteria that invade the body often release enzymes that break down cell membranes and organelles or give off toxins that disrupt the functions of organs and tissues. Viruses can enter cells and utilize cellular facilities to reproduce more virus particles. They can kill cells by depleting them of essential components or by causing them to produce toxic substances.

Fortunately, the body is able to resist many organisms and chemicals that can damage tissues. This ability is called **immunity.** Some of the body's immunity is provided by nonspecific defense mechanisms that do not require previous exposure to a particular foreign substance in order to react to it. These mechanisms include the barriers formed by body surfaces and certain inflammatory responses that occur when tissues are injured. Immunity is also provided by specific immune responses that depend on prior exposure to a specific foreign material, recognition of it upon subsequent exposure, and reaction to it. These responses are mediated by specialized lymphocytes and cells derived from them.

BODY SURFACES

The skin and mucous membranes that form the body surfaces are the first barriers to the invasion of the body by potentially damaging factors. The intact skin prevents many microorganisms and chemicals from entering the body. The skin's sweat glands and sebaceous glands secrete substances that are toxic to many types of bacteria, and the outer portion of the epidermis contains a tough, water-insoluble protein called keratin that is resistant to weak acids and bases and many protein-digesting enzymes.

The mucous membranes lining body cavities that open to the exterior—for example, the digestive, respiratory, urinary, and reproductive tracts—secrete mucus, which entraps small particles that can then be swept away or engulfed by phagocytic cells. The mucous membrane of the upper respiratory tract contains ciliated cells that move mucus and its entrapped particles toward the throat (pharynx). The tears that bathe the mucous membranes of the eyes contain the enzyme *lysozyme*, which has antimicrobial activity, and the acidic gastric juice secreted by the stomach can destroy many bacteria and bacterial toxins.

Although the surface barriers are quite effective, from time to time potentially damaging agents or organisms do gain entrance to the body. It is then necessary for other defense mechanisms to attempt to neutralize or destroy them.

INFLAMMATION

In its most basic form **inflammation** is an acute, nonspecific, physiological response of the body to tissue injury caused by factors such as chemicals, heat, mechanical trauma, or bacterial invasion. Although some differences

621

Table 21.1   Events of Inflammation

Event	Significance	Contributing Mediators or Mechanisms
Dilation of small blood vessels	Increases blood flow to injured area; increases delivery of plasma proteins and phagocytic leukocytes	Histamine, kinins, complement components, prostaglandins
Increased permeability of small vessels (particularly capillaries and venules)	Plasma fluid and solutes, including proteins and other large molecules, move out of the circulatory system and into inflamed tissues	Histamine, kinins, complement components, prostaglandins
Slowing of blood flow	Associated with margination and pavementing of leukocytes	Increased blood viscosity and clumping of erythrocytes due to loss of fluid and solutes
Walling off of injured area	Limits spread of toxic products or bacteria	Formation of fibrin in injured tissues
Margination and pavementing of leukocytes	Associated with emigration of leukocytes	Clumping of erythrocytes and slowing of blood flow
Emigration of leukocytes	Leukocytes leave the circulation and enter injured tissues	Amebalike activity of leukocytes
Aggregation of leukocytes and macrophages at site of injury	Provides large numbers of phagocytic cells to engulf invading agents or organisms	Amebalike activity of leukocytes and macrophages attracted by chemotactic factors (for example, kinins, complement components)
Phagocytosis by leukocytes and macrophages	Removes invading agents or organisms and debris	Certain surface characteristics of materials to be phagocytized; opsonins

Table 21.1   may be evident—depending on the causative agent, the site of injury, and the state of the body—the fundamental events of the *acute, nonspecific inflammatory response* are similar in virtually all cases (Table 21.1). These events include changes in blood flow and vessel permeability, as well as the movement of leukocytes from the vessels into the tissues. The acute, nonspecific inflammatory response brings into the invaded or injured area plasma proteins and phagocytes that can inactivate or destroy the invaders, remove debris, and set the stage for tissue repair. The events of this response are mediated by a number of chemical substances that are released or generated in the injured area.

Changes in Blood Flow and Vessel Permeability

Although a transient vasoconstriction may occur immediately following an injury to the tissues, small vessels in the injured area soon dilate. This dilation leads to increased blood flow to the area, which increases the delivery of plasma proteins and phagocytic leukocytes. In addition, the permeability of capillaries and venules in the area increases, and plasma fluid and solutes, including proteins and other large molecules, move out of the circulatory system and into the inflamed tissues. This movement is aided by the vasodilation that occurs, because the dilation of arterioles in the area leads to an increased pressure within capillaries and venules that favors the filtration of fluid and solutes out of the vessels. Moreover, the movement of proteins from the blood into the interstitial fluid due to the increased vessel permeability diminishes the difference in protein concentration between the plasma and the interstitial fluid, and thus favors the accumulation of fluid within the tissues.

As fluid and solutes move out of the circulatory system and into the tissues during inflammation, the viscosity of the blood increases and erythrocytes clump together. This increases the resistance to blood flow, and as inflammation progresses, blood flow through small vessels in the injured area slows and sometimes even stops.

Walling-off Effect of Inflammation

One of the substances that moves from the blood into the tissue spaces during inflammation is fibrinogen. In the tissues, fibrinogen is converted to fibrin, forming a clot that walls off the injured area. This walling-off effect can delay or limit the spread of toxic products or bacteria.

Leukocyte Emigration

As erythrocytes clump together and the blood flow slows during inflammation, a process known as *margination* takes place, and leukocytes—particularly neutrophils and monocytes—are displaced to the periphery of the bloodstream, where they contact the endothelial linings of capillaries and venules in the inflamed area. At first the leukocytes slowly tumble or roll along the endothelial surfaces, but soon a phenomenon called *pavementing* occurs, and the leukocytes adhere to the vessel surfaces. The adhered leukocytes begin to exhibit an amebalike activity, and they squeeze between the endothelial cells of the capillaries and venules into the tissue spaces. The first leukocytes to arrive in the tissues are generally neutrophils, followed sometime later by monocytes. In the tissues, monocytes are transformed into macrophages, and some of the macrophages normally present in the tissues multiply and become motile.

Chemotaxis and Leukocyte Aggregation

Leukocytes and macrophages migrate through the tissues at varying rates of speed and aggregate at the site of injury or invasion. Their direction of movement is determined by chemical mediators released at the site of injury, and they are attracted to the highest concentrations of certain chemical substances. Such a movement of cells in response to chemical factors is called *chemotaxis*. Positive chemotaxis is a movement toward a chemical substance and negative chemotaxis is a movement away from a chemical substance.

kem-o-tak´-sis

In most acute, nonspecific inflammatory responses, leukocytes and macrophages aggregate at the site of injury or invasion in a fairly predictable sequence. Neutrophils usually predominate in the early stages of the response, and monocytes and macrophages predominate during the later stages. Neutrophils are present in greater numbers in the circulation and are more mobile than monocytes, and this probably contributes to their initial predominance. However, in the tissues, neutrophils die off faster than monocytes and macrophages. Moreover, neutrophils may produce or potentiate factors that facilitate monocyte emigration. These facts are believed to contribute to the predominance of monocytes and macrophages later in inflammation.

In contrast to the dominance of neutrophils, monocytes, and macrophages in most acute, nonspecific inflammatory responses, certain types of allergies and inflammatory responses to parasites are characterized by the presence of large numbers of eosinophils. In fact, a preponderance of eosinophils is often indicative of an allergic response.

Phagocytosis

A major benefit of inflammation is the phagocytosis of microorganisms, foreign materials, and debris by leukocytes—particularly neutrophils—and macrophages. The phagocytic cells identify the materials to be engulfed by "recognizing" certain surface characteristics that enhance the likelihood of phagocytosis, but exactly which characteristics are important in the acute,

fag-o-sigh-to´-sis

nonspecific inflammatory response, and how they trigger phagocytosis, is not certain. In any event, a phagocytic cell attaches to a particle and engulfs it by endocytosis, forming a membrane-bounded phagocytic vesicle that contains the engulfed particle within the phagocytic cell.

Enzymes associated with the membranes of phagocytic vesicles catalyze the production of substances that include hydrogen peroxide (H_2O_2), oxygen radicals (O_2^-), and halogenated oxides such as hypochlorite (OCl_2). The oxygen radicals and halogenated oxides are toxic substances that react with many organic molecules and inactivate many phagocytized microorganisms. In addition, lysosomes containing powerful digestive enzymes attach to phagocytic vesicles, forming digestive vacuoles, and the lysosomal enzymes are released into the vacuoles. Since the lysosomes appear microscopically as granules, this process is called *degranulation*. The lysosomal enzymes break engulfed materials down into products of low molecular weight that can be utilized by the phagocytic cell. Macrophages, but apparently not neutrophils, can also extrude residual breakdown products from the cells.

A phagocytic cell continues to ingest and digest foreign particles until toxic substances from the foreign particles and from the activity of the cell itself accumulate and kill the cell. Macrophages are larger cells that are less selective about what they engulf than are neutrophils, and they generally phagocytize more particles before they die. As a result, macrophages are particularly important in clearing an inflamed area of foreign material and tissue debris.

op´-se-ninz

The phagocytic activity of leukocytes and macrophages can be modified by many factors, including a number of proteins collectively called *opsonins*. Opsonins coat specific foreign particles and thereby render them more susceptible to phagocytosis.

In addition to their phagocytic activities, neutrophils can release lysosomal contents directly into the interstitial fluid. The lysosomal enzymes digest extracellular material, making it easier for macrophages to engulf it.

Abscess and Granuloma Formation

The ultimate goal of the inflammatory response is to overcome the injury or invasion of the body and clear the area for tissue repair. Occasionally, however, the inflammatory response is unable to overcome an invasion, and an abscess or a granuloma forms. An *abscess* is a sac of pus that consists of microbes, leukocytes, macrophages, and liquified debris walled off by fibroblasts or collagen. It does not spontaneously diminish and must be drained. A *granuloma* can form when the invading agents are microbes that can survive within the phagocytes or are materials that cannot be digested by the phagocytes. In such cases, layers of phagocytic-type cells form and are surrounded by a fibrous capsule. The central cells contain the offending agent. In this fashion, a person may harbor live bacteria such as tuberculosis-causing bacteria for many years without displaying any overt symptoms.

Chemical Mediators of Inflammation

Although tissue injury precipitates the acute, nonspecific inflammatory response, a variety of chemical substances mediate it. Some of these substances are released from tissue cells in the injured area, some are released from leukocytes, and some are generated by enzyme-catalyzed reactions.

Histamine

F21.1

Histamine is present in many cells, particularly mast cells, basophils, and platelets (Figure 21.1). It causes vasodilation (particularly of arterioles) and increased vascular permeability (particularly in venules). Several factors cause histamine release, including mechanical disruption of cells due to injury and chemicals secreted by neutrophils attracted to the site.

Kinins

kigh´-ninz

The *kinins* are a group of polypeptides that dilate arterioles, increase vascular permeability, act as powerful chemotactic agents, and induce pain (Figure

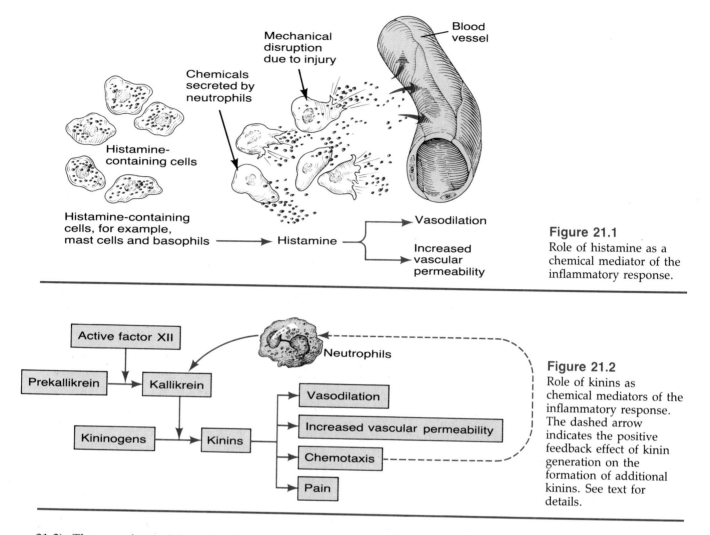

Figure 21.1
Role of histamine as a chemical mediator of the inflammatory response.

Figure 21.2
Role of kinins as chemical mediators of the inflammatory response. The dashed arrow indicates the positive feedback effect of kinin generation on the formation of additional kinins. See text for details.

21.2). They are formed from inactive precursors called kininogens that normally circulate in the plasma. The enzyme *kallikrein*, which is also present in the plasma in inactive form (prekallikrein), acts on kininogens to form kinins. Multiple factors can convert prekallikrein to kallikrein. Among them is active factor XII, which is also involved in blood clotting. Moreover, kallikrein is present in many body tissues, as well as in neutrophils, and once inflammation has begun, kallikrein from these sources contributes to kinin generation. In fact, the kinin-generating system displays a positive feedback in that kinins act chemotactically to attract neutrophils whose kallikrein can catalyze the production of still more kinins. In addition, tissue kallikrein cannot normally act on plasma kininogens because the kininogens are too large to pass out of the circulatory system. In inflammation, the kinins increase vascular permeability, allowing kininogens to enter the tissues and become activated by kallikrein.

Complement System

The *complement system* is particularly important in specific immune responses, but it can also contribute to acute, nonspecific inflammation. This system consists of a series of plasma proteins that normally circulate in the blood in inactive forms (Figure 21.3). The activation of the system's initial components initiates a cascading series of reactions that ultimately produce active complement molecules.

Active complement components can mediate virtually every event of the acute, nonspecific inflammatory response. Complement components enhance vasodilation and vascular permeability by direct effects on blood vessels, by stimulating histamine release from mast cells and platelets, and by

F21.2

F21.3

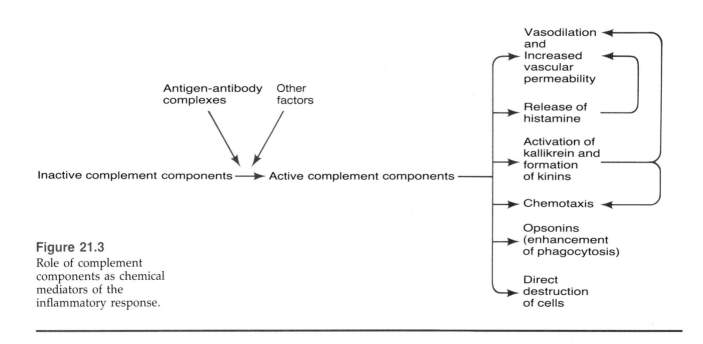

Figure 21.3
Role of complement
components as chemical
mediators of the
inflammatory response.

activating plasma kallikrein. They act as chemotactic agents in attracting neutrophils, and as opsonins that enhance phagocytosis by attaching to microbes, where they serve as structures to which phagocytic cells can bind. In addition, complement components directly attack invading microbes and kill them without prior phagocytosis. The complement components involved in this activity become embedded in the surface of a microbe and kill it by forming tunnels that make the microbe leaky.

As is discussed later in the chapter, structures called antigen-antibody complexes, which result from specific immune responses, are powerful activators of the complement system. However, other factors—for example, lipopolysaccharides in the cell walls of certain bacteria, protein A of staphylococcal microorganisms, and some lysosomal and bacterial enzymes—can also activate the complement system. Factors such as these may trigger the production of active complement components in acute, nonspecific inflammation.

Prostaglandins

pros-ta-glan´-dinz   *Prostaglandins* may also play a role in inflammatory responses. For example, prostaglandin PGE_1 induces vasodilation and increased vascular permeability, and it may stimulate leukocyte migration through capillary walls. Prostaglandins are believed to also contribute to the pain associated with inflammation, and they may mediate the fever-producing effects of a substance called endogenous pyrogen that is released from certain cells during inflammatory responses.

Leukotrienes

Leukotrienes are a group of substances produced by leukocytes. They are chemically related to the prostaglandins. Certain leukotrienes cause a relatively short-lived, transient constriction of arterioles that is followed by an increase in vascular permeability, particularly in venules. Some leukotrienes increase the adherence of leukocytes to endothelial cells of small venules, and some seem to be chemotactic agents that attract neutrophils. Leukotrienes are very potent constrictors of peripheral air passageways in the lungs, and they may play a significant role as mediators of asthmatic responses and certain allergic reactions. Researchers are actively studying the leukotrienes, and as more information about their actions is obtained, it seems likely that they will be found to play a significant role in inflammation and other activities.

Table 21.2 Local Symptoms of Inflammation

Symptom	Contributing Causes
Redness	Dilation of vessels, increased blood flow to injured area
Heat (increased temperature)	Dilation of vessels, increased blood flow to injured area
Swelling (edema)	Increased vessel permeability, and movement of fluid and solutes from the circulatory system into the tissue spaces
Pain	Increased pressure on sensory nerve endings in swollen tissues; effects of some chemical mediators of inflammation (for example, kinins) on afferent nerve terminals

Symptoms of Inflammation

Inflammation can produce both local and systemic symptoms.

Local Symptoms

The main local symptoms of inflammation are redness, heat, swelling, and pain (Table 21.2). The redness and heat (increased temperature) are due to the dilation of blood vessels and increased blood flow to the inflamed area. The swelling (edema) is caused by the increased vessel permeability and the movement of fluid and solutes from the circulatory system into the tissue spaces. The pain has been attributed to the increased pressure exerted on sensory nerve endings in the swollen tissues and to the effects of kinins and perhaps other chemical mediators of inflammatory responses on afferent nerve terminals.

Table 21.2

Systemic Symptoms

Fever is a common systemic symptom of inflammation, especially in inflammatory responses associated with the spread of organisms into the bloodstream. Monocytes and macrophages involved in the inflammatory response release endogenous pyrogen (believed to be a substance called *interleukin-1*), which influences the body's temperature-regulating mechanisms and induces fever. Although an extremely high fever can be harmful, phagocytosis is favored by higher body temperatures. Thus, fever (as well as the local increase in temperature at the site of inflammation) may enhance this activity. In addition to eliciting fever, the endogenous pyrogen also depresses the concentrations of iron and zinc in the blood. This action may retard the proliferation of bacteria that require a high concentration of iron to multiply.

Inflammation, particularly in response to many types of bacterial invasion, is often accompanied by an increased production and release of leukocytes. The leukocyte count in the peripheral blood may rise from about 7000 cells/mm^3 to 25,000 or more, and in most nonspecific inflammatory responses the increase is due to an absolute as well as a relative rise in the number of circulating neutrophils. In contrast, infections caused by viruses and protozoa are often associated with decreased numbers of leukocytes in the blood.

INTERFERON

The protein **interferon** provides some protection to the body against invasion by viruses, particularly until the more slowly reacting specific immune responses can take over. In general, when a virus invades the body, it enters a cell and reproduces more virus particles by utilizing cellular facilities to synthesize viral components, including nucleic acids and proteins. The new viruses are released to infect other, healthy cells and repeat the process. However, viruses (as well as other substances) also induce cells to produce interferon. One of the most effective inducers of interferon production is double-stranded RNA, and it has been suggested that viruses and other interferon inducers cause the production of double-stranded RNA within cells. The double-stranded RNA is believed to trigger the DNA of a cell to produce a messenger RNA that directs the production of interferon by the cell. The interferon leaves the cell and binds to receptors on the cell membranes of other cells. This binding triggers the synthesis of enzymatic proteins within the cells that can prevent viral reproduction by breaking down messenger RNAs and inhibiting protein synthesis. The newly synthesized enzymatic proteins are inactive until the cells are infected by a virus or exposed to double-stranded RNA. This activation requirement may help protect the normal nucleic-acid- and protein-synthesizing mechanisms of the cells from inhibition in the absence of viral infection.

It is thought that interferon inhibits the activity of some viruses by other mechanisms. Although interferon apparently cannot prevent the reproduction of certain viruses, the viruses are either not released from the cells or, if they are released, they are not capable of infecting other cells. Moreover, there is evidence that, at least in some cases, interferon-stimulated cells can transfer their interferon-induced antiviral activity to adjacent cells by mechanisms requiring cell-to-cell contact.

All cells can manufacture interferon when they are appropriately stimulated, and at least three different forms of interferon are produced. These different forms of interferon are named alpha interferon (also called leukocyte interferon), beta interferon (fibroepithelial interferon), and gamma interferon (immune interferon).

Interferon may play a role in protecting the body against some forms of cancer. Although viruses are suspected of causing some human cancers, interferon also seems to work against tumors thought to be caused by nonviral agents such as radiation and chemicals. In fact, interferon is increasingly viewed as a versatile agent whose effects on cells go beyond the inhibition of viral activity. For example, it slows cell division and inhibits the proliferation of both healthy and abnormal cells. It mobilizes and enhances the cell-killing properties of macrophages, and it stimulates the activity of a specialized group of lymphocytes called natural killer (NK) cells. Both macrophages and natural killer cells can attack tumor cells, and natural killer cells are believed to be important in *immune surveillance*—that is, in detecting and eliminating abnormal, potentially cancerous cells before they multiply and cause clinical cancer. In fact, it has been suggested that clinical cancer is due to ineffective immune surveillance mechanisms. Gamma interferon appears to be a more potent antitumor agent than alpha interferon or beta interferon.

SPECIFIC IMMUNE RESPONSES

The **specific immune responses** generally require previous exposure to a foreign agent or organism to be most effective. The specific immune responses are not always beneficial, however, and they are sometimes responsible for allergies and other harmful reactions.

Traditionally, two major aspects of the specific immune responses are recognized. One aspect, known as **humoral immunity,** is mediated by specialized proteins called antibodies. Antibodies are produced by plasma cells that are progeny of lymphoid cells called B cells. The second aspect, known as **cell-mediated immunity,** is mediated by certain populations of lymphoid cells called T cells.

Antigens

An **antigen** is a substance that has the capacity to elicit a specific immune response. (An antigen that actually elicits a specific immune response when it is present in a person's body is called an *immunogen*.) Substances that act as antigens generally have molecular weights of 8000 to 10,000 or more, and proteins, polysaccharides, complex lipids, and nucleic acids commonly act as antigenic materials. Lower-molecular-weight substances that are collectively called *haptens* can also act as antigens if they combine with larger molecules such as proteins.

A person generally does not generate specific immune responses to the antigens of his or her own body—that is, to *self antigens*. However, a person generally does generate specific immune responses to other antigens—that is, to *foreign antigens* (*nonself antigens*).

A group of antigens called **major histocompatibility complex (MHC) antigens** are particularly important in specific immune responses. (The MHC antigens of humans are also referred to as *human leukocyte antigens*, or *HLA*.) Different individuals—except identical twins—possess different MHC antigens. Thus, MHC antigens serve as important markers that distinguish a person's own cells (*self cells*) from other cells (*foreign*, or *nonself*, *cells*). There are two principal classes of MHC antigens: class I MHC antigens, which are present on the surfaces of virtually all nucleated cells of the body; and class II MHC antigens, which are present on certain cells involved in the generation of specific immune responses.

Antibodies

An **antibody** is a specialized protein that is produced in response to the presence of an immunogenic antigen. An antibody can combine with the specific antigen that stimulated its production to form an *antigen–antibody complex*. Antibodies belong to a family of proteins called globulins, and they are referred to as *immunoglobulins* (*Ig*). Five immunoglobulin classes of antibodies are distinguished. Immunoglobulin classes G and M (IgG and IgM) are involved in specific immunity against bacteria and viruses. Immunoglobulins of class E (IgE) are involved in specific immune responses to multicellular parasites and in certain allergic responses. Immunoglobulins of class A (IgA) are found predominantly in the gastrointestinal tract, in saliva, in sweat, and in tears. Class IgA antibodies are particularly important in protecting epithelial surfaces from bacteria and other microorganisms. Immunoglobulins of class D (IgD) are present in the body in relatively small amounts, and their precise function is unclear.

The basic structure of an antibody consists of two "heavy" and two "light" polypeptide chains (Figure 21.4). One end of each chain contains a variable region whose amino acid sequence differs for each kind of antibody. The particular conformation of the variable regions enables an antibody to bind to a specific antigen in a manner that is analogous to the interaction between an enzyme and its substrate. The remainder of each polypeptide chain is called the constant region. The constant regions contain binding sites for molecules and cells that function as effectors of antibody-mediated activities. The constant regions of the heavy polypeptide chains differ from one class of antibody to another, and these differences provide the basis for distinguishing the various classes of antibodies.

Antibodies inactivate or destroy foreign substances in a number of ways. They can combine with bacterial toxins or destructive foreign enzymes and inactivate them by inhibiting their interactions with target cells or substrates. They can also bind to surface components of viruses, preventing the attachment of the viruses to target cells and thereby keeping the viruses from entering the cells.

A major function of antibodies is to enhance the basic inflammatory response. Antigen–antibody complexes, particularly those involving antibodies of the IgG or IgM classes, are powerful activators of the complement system (Figure 21.3). As noted earlier, active complement components can mediate virtually every aspect of the acute, nonspecific inflammatory response. Moreover, if an antigen–antibody complex is on the surface of a foreign cell, active complement components attack the cell membrane to cause lysis and cell death. The utilization of complement components in this process is called complement fixation. Antibodies, particularly those of the IgM and IgG

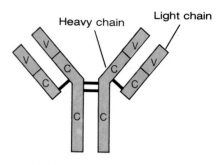

Figure 21.4

Diagrammatic representation of the basic structure of an antibody. The structure consists of four polypeptide chains, two heavy and two light. Many regions of the chains are of constant composition (C) among antibodies of a particular class. Only relatively small regions vary in composition (V) from antibody to antibody.

Ig
Classes

F21.4

F21.3

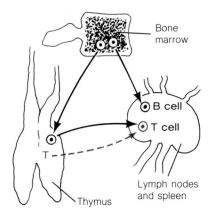

T = Thymosins

Figure 21.5
Formation and differentiation of B cells and T cells.

classes, also act as opsonins that enhance phagocytosis by attaching to foreign substances, where they serve as structures to which phagocytic cells can bind.

B Cells

B cells are lymphocytes that are committed to differentiate into the antibody-producing plasma cells involved in humoral immunity. Like all lymphocytes, B cells are derived from stem cell precursors in the red bone marrow (Figure 21.5). Stem cells that are destined to give rise to B cells proliferate and develop within the bone marrow (or within the liver and perhaps the spleen during fetal life). B cells continually circulate through the blood, tissues, and lymph, and at any one time large numbers of B cells are localized in the lymph nodes, spleen, and other lymphoid tissues.

T Cells

T cells are a heterogeneous group of lymphocytes that include those cells committed to participating in cell-mediated immune responses. Lymphocyte stem cells from the red bone marrow that are destined to give rise to T cells proliferate and differentiate in the thymus gland (Figure 21.5). In addition, hormonal substances from the thymus—for example, thymosins—are believed to influence T cells that have left the gland. Like B cells, T cells continually circulate through the blood, tissues, and lymph, and at any one time large numbers of T cells are localized in the lymph nodes, spleen, and other lymphoid tissues.

Accessory Cells

In addition to B cells and T cells, **accessory cells,** such as macrophages, play a significant role in specific immune responses. As is discussed in the following two sections, accessory cells present foreign antigens, together with self MHC antigens, to certain other cells involved in generating specific immune responses. The accessory cells present antigens by displaying on their surfaces both foreign antigens and self MHC antigens. For example, in many cases, a foreign antigen becomes bound to the surfaces of macrophages; then most of the antigen is endocytosed and digested by lysosomal enzymes of the macrophages. However, the macrophages either retain some of the unmodified antigen at their surfaces or, more likely, return some partially digested antigen (processed antigen) to their surfaces. As a result, the macrophages display the foreign antigen on their surfaces together with class I and class II self MHC antigens, which are normally present on macrophage surfaces.

Several kinds of cells besides macrophages may also serve as accessory cells that bind and present foreign antigens together with self MHC antigens, although the foreign antigens may be processed first by other cells. These antigen presenting cells include *interdigitating cells*, which are located in the T-cell regions of lymphoid tissues; *dendritic reticular cells*, which are located in the B cell regions of lymphoid tissues; and *Langerhans cells*, which are located in the skin and parts of the gastrointestinal tract. A foreign antigen that penetrates the epithelial surfaces of the skin or gastrointestinal tract can adhere to Langerhans cells and be carried by these cells to lymph nodes. Like macrophages, interdigitating cells, dendritic reticular cells, and Langerhans cells have both class I and class II self MHC antigens on their surfaces.

The antigen-presenting activities of the various accessory cells are important because some cells involved in specific immune responses generally recognize and respond to foreign antigens only when the antigens are presented together with self MHC antigens.

Humoral (Antibody) Immunity

Humoral immunity involves the production and release into the blood and lymph of antibodies to various antigens that the body recognizes as foreign. Humoral immune responses are important in providing specific immune resistance to bacterial toxins and to the extracellular phases of bacterial and viral infections. The humoral system responds to a foreign antigen as follows.

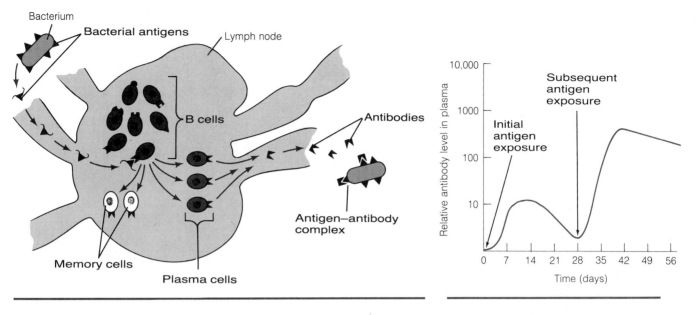

Figure 21.6
Diagrammatic
representation of the
events of a humoral
immune response. See
text for details.

Figure 21.7
Antibody production
following an initial
exposure to an antigen
and a subsequent
exposure to the same
antigen.

When a foreign antigen reaches lymphoid tissues such as lymph nodes or the spleen, a tiny percentage of the B lymphocytes is stimulated to undergo rapid cell division (Figure 21.6). Most of the stimulated cells develop into *plasma cells*, which produce antibodies that are released into the blood and lymph. The antibodies then combine with the specific antigen that stimulated their production, and responses such as those described in the earlier section on antibodies occur.

F21.6

Some of the stimulated B cells do not differentiate into plasma cells, but instead form *memory cells*, which provide the humoral system with a "memory" of the exposure to the antigen. The system responds rather slowly following an initial antigen exposure, and usually several days are required to build up substantial levels of antibodies (Figure 21.7). Because of the memory component, however, a subsequent exposure to the same antigen produces a very rapid outpouring of antibodies.

F21.7

Different foreign antigens stimulate different populations, or clones, of B cells (see Figure 21.6), and each clone can ultimately produce one specific antibody. The particular antibody a given clone of B cells can produce is displayed on the surfaces of the cells in the form of a receptor. When a foreign antigen enters the system, it binds with an antibody specific for it that is displayed on the surfaces of particular B cells. This binding contributes to the stimulation of the particular B cells so that they undergo division and differentiation into plasma cells that manufacture the antibody. There are sufficient clones of B cells within the body to allow for the production of antibodies to all of the millions of foreign antigens to which a person might ever be exposed.

F21.6

In actual fact, B cell stimulation and antibody production usually involve more than just the binding of a foreign antigen to antibodies displayed on the surfaces of particular B cells. In most cases, accessory cells, such as macrophages, and a class of T cells called *helper T cells* (T_H *cells*) are also involved. The accessory cells present the foreign antigen, together with self MHC antigens, to T_H cells. Certain T_H cells possess receptors that co-recognize both the foreign antigen and class II self MHC antigens displayed on the surfaces of the accessory cells. Interactions between the antigen-presenting accessory

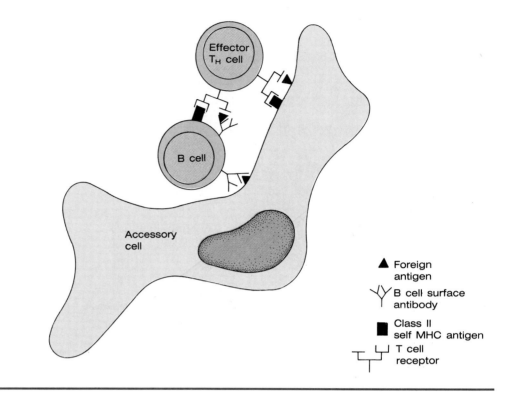

Figure 21.8
Stimulation of B cells by effector T_H cells. In many cases, receptors on the effector T_H cells are believed to recognize and interact with foreign antigen and class II self MHC antigens on the surfaces of accessory cells such as macrophages and also with foreign antigen and class II self MHC antigens on the surfaces of particular B cells.

▲ Foreign antigen

Y B cell surface antibody

■ Class II self MHC antigen

T cell receptor

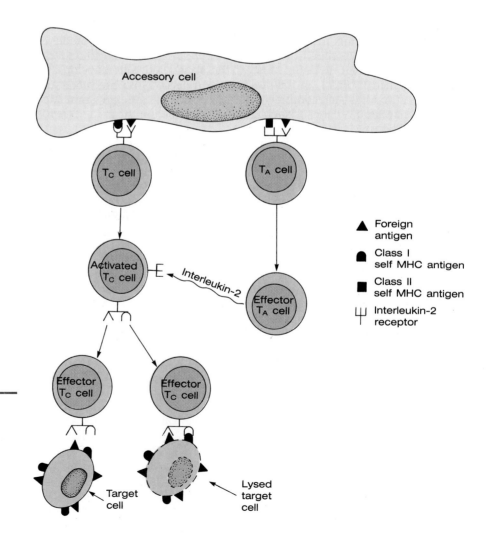

Figure 21.9
Diagrammatic representation of the activation and role of T_C cells during a cell-mediated immune response. See text for details.

▲ Foreign antigen

Class I self MHC antigen

■ Class II self MHC antigen

Interleukin-2 receptor

cells and these T_H cells activate the T_H cells. The activated T_H cells proliferate and differentiate, giving rise to effector T_H cells. The effector T_H cells, in turn, interact with B cells that have the foreign antigen bound to their surfaces. This interaction promotes the proliferation and differentiation of the B cells into antibody-producing plasma cells.

Some researchers believe that, in many cases, the promotion of B-cell activity by effector T_H cells also requires the participation of accessory cells such as macrophages. They propose that effector T_H cells recognize and interact with the foreign antigen and class II self MHC antigens displayed on accessory-cell surfaces and also with the foreign antigen and class II self MHC antigens on the surfaces of particular B cells (Figure 21.8). (Note that class II self MHC antigens are normally present on B-cell surfaces.) In addition, some studies suggest that, at least in certain cases, effector T_H cells secrete chemical factors that promote the proliferation and differentiation of B cells that have foreign antigen bound to their surfaces.

F21.8

Cell-Mediated Immunity

The cell-mediated immune responses depend on T cells, and particularly on groups of T cells called *effector cytotoxic T cells* (*effector T_C cells*). These responses are especially effective against cells that possess both foreign antigens and class I self MHC antigens on their surfaces. For example, most virus-infected cells acquire viral antigens on their surfaces, and they normally display class I self MHC antigens. (Recall that class I self MHC antigens are present on the surfaces of virtually all nucleated cells of the body.) Effector T_C cells produced by the cell-mediated immune system in response to the foreign viral antigens can attack and destroy the virus-infected cells. Thus, cell-mediated immune responses are important in providing specific immune resistance to the intracellular phase of viral infections. They are also important in providing specific immune resistance to fungi, parasites, and intracellular bacteria, and they play a major role in the rejection of solid-tissue transplants. The cell-mediated immune system responds to a foreign antigen as follows.

Like the B cells involved in humoral immunity, T cells are clonal, and they possess specific surface receptors. Specific clones of *cytotoxic T cells* (*T_C cells*) as well as a particular group of helper-class T cells called *amplifying T cells* (*T_A cells*) interact with accessory cells such as macrophages that display both foreign antigen and self MHC antigens on their surfaces (Figure 21.9). (T_C cells interact mainly with foreign antigen and class I self MHC antigens on accessory cell surfaces, and T_A cells interact predominantly with foreign antigen and class II self MHC antigens on accessory cell surfaces.) The accessory-cell–T_A-cell interaction causes the T_A cells to proliferate and differentiate into effector T_A cells, which produce and secrete a substance called *interleukin-2*. The accessory-cell–T_C-cell interaction activates the T_C cells and cause them to express a surface receptor for interleukin-2. The activated T_C cells, under the influence of interleukin-2, proliferate and differentiate into effector T_C cells, which participate in the cell-mediated immune response. (In addition, some T-cell progeny serve as a memory component of the system.) During a cell-mediated immune response, effector T_C cells leave the lymphoid tissues and travel throughout the body. When they encounter cells bearing both the foreign antigen that initiated their production and class I self MHC antigens, they combine with the cells (Figure 21.10). Effector T_C cells can lyse and thereby kill the cells.

F21.9

F21.10

Another group of T cells called *delayed hypersensitivity T cells* (*T_D cells*) are also involved in cell mediated immune responses, and they are important mediators of long-term, chronic inflammatory responses. In general, within lymphoid tissues, T_D cells interact with accessory cells that display both foreign antigen and class II self MHC antigens on their surfaces. This interaction activates the T_D cells, and they proliferate and differentiate into effector T_D cells (with some T-cell progeny apparently serving as a memory component). Effector T_D cells leave the lymphoid tissues and travel throughout the body. When effector T_D cells encounter cells such as Langerhans cells in the skin that display both the foreign antigen that initiated their production and class

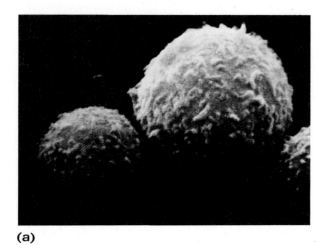

(a)

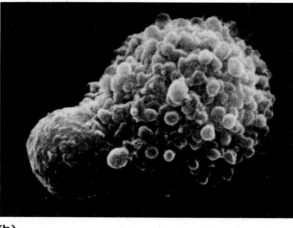

(b)

Figure 21.10
(a) Effector T cell (smaller sphere at lower left) attaches to a cell recognized as foreign.
(b) Destruction of the foreign cell is evidenced by the deep folds in its surface membrane. (×7250)

II self MHC antigens, they interact with them. The effector T_D cells are believed to release a number of different chemical factors that generate inflammatory responses and enhance phagocytosis. Among these factors are a chemotactic factor that attracts macrophages to the area and a cytotoxic factor that can damage or kill many cells except lymphocytes. (In addition to effector T_D cells, certain other groups of T cells may also release chemical factors that contribute to the generation of chronic inflammatory responses.)

Transplant Rejection

The cell-mediated immune responses play an important role in the rejection of solid-tissue transplants. The cells of different individuals who are not identical twins possess different MHC antigens (and also different antigens called *minor histocompatibility antigens*). When a tissue or organ from one individual is transplanted into another individual who is not the donor's identical twin, antigens on the cells of the donor's tissue or organ are recognized as foreign (or nonself) by the recipient's immune system, and effector T_C cells that can attack and destroy the cells of the transplanted organ or tissue are produced. This ability of effector T_C cells to attack the cells of a transplanted tissue or organ is somewhat puzzling because effector T_C cells generally do not recognize and attack cells bearing foreign antigens unless the cells also display class I self MHC antigens on their surfaces. However, the cells of a tissue or organ transplanted from a donor who is not an identical twin of the recipient display nonself MHC antigens. Perhaps the nonself MHC antigens are sufficiently similar to both self MHC antigens and foreign antigens that effector T_C cells produced by the recipient can recognize them in a manner similar to that by which they recognize self MHC antigens and foreign antigens (Figure 21.11). (Effector T_D cells may also be produced in response to the presence of foreign antigens from a transplanted tissue or organ, and the activities of these cells may contribute to the rejection of the transplant.)

F21.11

In transplant surgery, the likelihood that a specific immune response will lead to destruction and rejection of a transplanted tissue or organ is lessened by trying to select a tissue or organ whose antigenic composition closely matches that of the recipient and by utilizing drugs or antilymphocyte serum that weaken or suppress specific immune responses. In many cases, however, the drugs or antilymphocyte serum not only weaken or suppress specific immune responses directed against the transplanted tissue or organ, but they also diminish the ability of the immune system to provide protection against other foreign antigens, and the transplant recipient becomes very susceptible to infection.

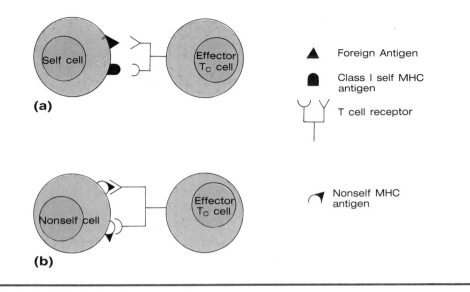

Legend for Figure 21.11:
▲ Foreign Antigen
● Class I self MHC antigen
Y T cell receptor
◖ Nonself MHC antigen

Figure 21.11
Diagrammatic representation of a possible method by which effector T_C cells could recognize nonself MHC antigens. **(a)** Recognition of a cell bearing self MHC antigens and a foreign antigen. **(b)** Recognition of a cell bearing nonself MHC antigens.

Active Immunity

A resistance to infection that results from an antigen-induced activation of an individual's specific immune responses is called *active immunity*. Most commonly, active immunity is acquired by infection. Alternatively, active immunity can be acquired through vaccination, in which a person is injected with a small amount of antigenic material. The material may be bacterial, a toxin, or some other substance. It is usually pretreated by drying, ultraviolet light, or some other means so that it is not strong enough to cause disease, but still acts as an antigen that stimulates specific immune responses.

Following the activation of an individual's specific immune responses by infection or vaccination, antibodies; effector T_C cells, and/or effector T_D cells may remain present to resist the disease-causing agent for some time. In addition, if an individual who has had an infection or has been vaccinated is exposed to the disease-causing agent at a later time, the memory components of the humoral and cell-mediated immune systems enable these systems to respond rapidly. It should be noted, however, that some microorganisms apparently do not activate the memory components of the systems, and the response to each exposure to one of these organisms is the same as the slowly developing response to the initial exposure.

Passive Immunity

Passive immunity is essentially "borrowed" immunity. For example, antibodies to a disease-causing agent can be transferred into an individual from sources (often nonhuman) that have been exposed to the agent. Passive immunity is of relatively short duration—generally only several weeks. Moreover, the recipient's own specific immune mechanisms are not stimulated to produce antibodies or effector T cells against the disease-causing agent. A common instance of passive immunity occurs during pregnancy. IgG-class antibodies formed by the mother move across the placenta to the fetus. These antibodies protect the infant during the period just after birth, when the infant's own ability to produce antibodies is relatively poor. Passive immunity can also be provided by injecting antibodies into an individual. For example, antibodies to the venom of a particular type of snake can be built up in a horse, recovered from the horse plasma, and injected into a person who has been bitten by that type of snake in order to provide immediate protection against the venom. A possible danger of such a procedure, however, is that injected antibodies may themselves be antigenic in the recipient's system—especially if they were obtained from nonhuman sources. Thus, the injected antibodies may stimulate the recipient to produce antibodies against them. This occurrence can lead to severe allergic responses.

Hypersensitivity

Hypersensitivities are immune responses that can be harmful to the body. Both the humoral and cell-mediated immune systems participate in these responses.

Role of Complement Components in Hypersensitivity

In some types of hypersensitivity—for example, immune-complex hypersensitivity and cytotoxic hypersensitivity—active complement components generated during a humoral immune response to an antigen damage body tissues and cells. For example, immune complexes of antigen, antibody, and complement can become deposited in blood vessel walls, where the activation of the complement system can initiate an inflammatory response that damages the vessels. The deposition of immune complexes in the kidneys can lead to an inflammatory condition known as glomerulonephritis (see Chapter 26).

Role of Effector T_D Cells and Effector T_C Cells in Hypersensitivity

In the type of hypersensitivity called cell-mediated hypersensitivity, or delayed hypersensitivity, the activities of effector T_D cells and effector T_C cells during a cell-mediated immune response cause damage to body tissues. For example, the toxin of poison ivy is a hapten that can bind to cell membranes in the skin. Consequently, exposure to poison ivy toxin can sensitize an individual and cause the production of effector T_D cells and effector T_C cells against the toxin. If such a sensitized individual is again exposed to poison ivy toxin, effector-T_D-cell and effector-T_C-cell responses (including the release of chemical factors) induce a tissue-damaging inflammatory reaction at the site of exposure that is characterized by itching, swelling, and vesication. Hypersensitivity reactions involving effector T_D cells and effector T_C cells usually take from one to three days to reach their peak intensities.

Role of Class IgE Antibodies in Hypersensitivity

Class IgE antibodies are involved in a type of hypersensitivity called anaphylactic hypersensitivity. This type of hypersensitivity is the cause of many common allergic reactions. In these reactions, the exposure of a susceptible individual to a particular antigen—for example, dust, pollen, or a particular food—sensitizes the individual and causes the production of class IgE antibodies from plasma cells. These antibodies bind to mast cells and basophils. When such a sensitized person is again exposed to the antigen, the antigen attaches to the antibodies bound to the mast cells and basophils. This attachment stimulates the release of substances that include histamine, leukotrienes, a chemotactic factor for eosinophils, and heparin. These substances induce inflammatory responses such as vasodilation and increased vascular permeability, as well as increased mucus secretion and the contraction of smooth muscles of the airways of the lungs.

Often, an anaphylactic-type allergic reaction is localized to a particular site in the body. For example, if the exposure to a particular antigen occurs in the nasal area, the result may be sneezing, runny nose, congestion, and other symptoms of hay fever due to the irritation, increased mucus secretion, increased blood flow, and increased protein leakage in the area. Occasionally (particularly in response to injected antigens), a systemic rather than a localized anaphylactic allergic reaction occurs, and severe hypotension and airway constriction quickly develop. This reaction is called *anaphylactic shock*, and it can be fatal if countermeasures such as the injection of epinephrine are not taken promptly. (Epinephrine prevents the release of substances from mast cells, and it antagonizes the actions of histamine and leukotrienes on smooth muscles.) Some sensitized people develop anaphylactic shock in response to the antigen in a single insect sting or in response to drugs such as penicillin.

The administration of antihistamines often provides some relief from anaphylactic-type allergic reactions, but the relief is frequently incomplete because chemicals other than histamine (for example, leukotrienes) also participate in these reactions. When the particular antigen to which a person is sensitive has been identified, desensitization therapy may be utilized. Desensitization consists of injecting an individual with small but increasing quantities of the offending antigen. Some researchers think this treatment results in the production of class IgG rather than class IgE antibodies to the antigen. According to this view, when the person again encounters the particular antigen, the class IgG antibodies attach to it and prevent the antigen from attaching to class IgE antibodies bound to mast cells and basophils. Alternatively, it has been proposed that desensitization causes a group of T cells called *suppressor T cells* (T_S *cells*) to suppress the synthesis of class IgE antibodies directed against the antigen. In either case, the antigen does not attach to class IgE antibodies that are bound to mast cells and basophils, and the anaphylactic-type allergic reaction does not occur.

Tolerance

Tolerance can be generally defined as the failure of the body to mount a specific immune response against a particular antigen. For example, the body is composed of proteins and other substances that are antigenic. Normally, however, the body exhibits self-tolerance and does not produce antibodies, effector T_C cells, or effector T_D cells against its own antigens.

Tolerance is due to the elimination, inactivation, or suppression of specific B cells and T cells as a result of appropriate antigenic exposure. Thus, under proper conditions, exposure to an antigen can prevent rather than initiate a specific immune response. Tolerance is more easily established in the fetus than in the adult. In this regard it should be noted that during fetal development a person is exposed almost exclusively to his or her own antigens. During this time, these antigens are believed to react with and paralyze or eliminate the particular clones of B cells and T cells that could produce specific immune responses against them. However, B cells and T cells continue to differentiate in the bone marrow and thymus throughout a person's lifetime, and the maintenance of tolerance to antigens such as self antigens requires an active and continuous inhibition of those B and T cells that could produce specific immune responses against them.

Two general mechanisms—clonal deletion and clonal suppression—are believed to be responsible for tolerance. In *clonal deletion*, exposure to an antigen eliminates or inactivates those clones of B cells and T cells specific for the antigen. Exactly how this occurs is not completely understood. However, under appropriate conditions, the exposure of B cells to an antigen apparently results in the removal or modulation of the receptors for the antigen on the surfaces of particular B cells. The removal or modulation of the receptors renders the cells unresponsive, or tolerant, to the antigen. Immature B cells appear to be much more sensitive to such an occurrence than mature B cells, and this may help explain why tolerance is more easily established in the fetus than in an adult.

In *clonal suppression*, an appropriate exposure to an antigen activates suppressor T cells, which in turn block the activity of helper-class T cells involved in initiating specific immune responses against the antigen. The suppressor T cells may also block the development of effector T_C cells and effector T_D cells responsible for cell-mediated immune responses against the antigen, and they may directly suppress the activities of B cells that could produce a humoral immune response against the antigen. Thus, in clonal suppression, B cells and T cells that can produce specific immune responses against particular antigens remain present, but their activities are blocked by suppressor T cells.

Autoimmune Responses

Occasionally, the body's tolerance to its own antigens breaks down, and *autoimmune responses* occur—that is, a person produces antibodies, effector T_C cells, and/or effector T_D cells that attack his or her own tissues. There are several causes of such occurrences. Drugs, environmental chemicals, viruses,

CONDITION OF CLINICAL SIGNIFICANCE

AIDS

In addition to conditions such as hypersensitivities and autoimmune responses in which specific immune responses cause damage to the body, problems can also arise when specific immune responses are deficient or absent. For example, infection by a virus called HTLV-III (human T-lymphotropic virus type III) or LAV (lymphadenopathy-associated virus) can damage the specific immune system and lead to a variety of disorders, including the condition known as AIDS (acquired immune deficiency syndrome). The HTLV-III virus can invade many body cells, but among its primary targets are a set of T cells known as T4 cells, which include the helper class of T cells. Following infection, the virus may remain dormant or inactive for long periods of time. However, once it becomes active and begins to reproduce new virus particles, it interferes with the normal function of and ultimately kills T4 cells, which carry out activities necessary for effective specific immune responses. The depletion of functional T4 cells produces a general immune suppression and deficiency that affects both humoral and cell-mediated specific immune responses and leaves the victim susceptible to a wide variety of infections and other problems that are ultimately fatal. At the present time, it is not certain what causes the dormant virus to become active and begin to reproduce.

The HTLV-III virus appears to be spread from an infected person to another person by direct exposure of the second person's bloodstream to virus-containing body fluids (such as blood or semen) from the infected individual. People infected by the HTLV-III virus generally produce antibodies against it, and the presence of these antibodies in a person's blood indicates that infection has occurred. However, the antibodies are not effective at eliminating the virus, and once infection occurs it is believed to be permanent.

Some reports indicate that the initial infection of a person by the HTLV-III virus can be accompanied by symptoms that resemble those of acute mononucleosis (for example, fever, sore throat, pain in muscles and joints). However, it is not known if these symptoms always—or even usually—occur on initial infection. As previously stated, following infection the virus may remain dormant for long periods of time, and many infected people apparently do not develop any overt signs of illness, at least for a number of years. Others eventually develop what is called AIDS-related complex (ARC), which is characterized by symptoms such as a generalized swelling of the lymph nodes, persistent fever, diarrhea, and a rundown feeling. After a period of time, ARC may resolve, or it may develop into AIDS. (AIDS may also appear without being immediately preceded by ARC). AIDS is characterized by an inability to successfully resist infections by organisms that would normally be resisted and cause no harm. (Such infections are called opportunistic infections.) For example, AIDS victims frequently develop a type of pneumonia that results from infection by a protozoan called *Pneumocystis carinii*, and they may suffer from a fungal infection of the mouth called thrush or candidiasis. They also have a greater than normal tendency to develop Kaposi's sarcoma, which is a cancer of the lining of blood vessels. AIDS victims ultimately die from such opportunistic infections and malignancies.

or genetic mutations may cause the formation of new or altered antigens on cell surfaces. The body treats these antigens as foreign, and generates humoral or cell-mediated specific immune responses against them. Alternatively, foreign antigens that are structurally very similar to some of the body's own antigens may stimulate the production of antibodies, effector T_C cells, and/or effector T_D cells that cross-react with body antigens. For example, certain streptococcal bacteria possess antigens that induce the formation of antibodies that cross-react with heart tissue, and severe, recurrent infections caused by streptococci sometimes lead to the development of rheumatic fever several weeks after the infection has subsided, suggesting an autoimmune response. Still another possibility is that certain body antigens are not normally exposed to the humoral or cell-mediated immune systems, and the body does not become tolerant to them. If, at some time, tissue disruption following, for example, injury or infection exposes these antigens, the body treats them as foreign and generates humoral or cell-mediated specific immune responses against them. For example, antigens of the cornea of the eye do not seem to circulate in the body fluids. Should damage to the cornea expose these antigens, specific immune responses against them may cause corneal opacity.

Suppressor T cells are thought to be particularly important in autoimmunity. If suppressor T cells are involved in tolerance (and particularly in preventing humoral and cell-mediated immune responses against a person's own antigens), as was previously suggested, then an inhibition or a deficiency of these cells could contribute to the development of autoimmune responses.

BLOOD GROUPS AND TRANSFUSION

The surfaces of erythrocytes contain particular antigens (agglutinogens) that can react with appropriate antibodies (agglutinins). These reactions form the basis of various blood-typing classifications.

a-gloo-tin´-a-jen
a-gloo´-tin-in

A-B-O System

The erythrocyte-surface antigens most often considered are those of the A-B-O system (Table 21.3.). The antigens of the A-B-O system, which are inherited, are designated A and B, and the lack of either A or B antigens on the erythrocytes is designated O. A given individual may have either antigen A or antigen B, both antigens A and B (AB), or neither antigen A nor B (O).

Table 21.3

Antibodies against antigens A or B begin to build up in the plasma shortly after birth. The antibody levels peak at about 8 to 10 years of age, and the antibodies remain present in declining amounts throughout the rest of life. The mechanism that stimulates the production of anti-A and anti-B antibodies is unclear, but it has been suggested that antibody development is initiated by small amounts of A- and B-type antigens that enter the body in the food, in bacteria, or by other means. A person normally produces antibodies against those antigens that are not on his or her erythrocytes, but does not produce antibodies against those antigens that are present on his or her erythrocytes. Thus, a person with antigen A has anti-B antibodies; a person with antigen B has anti-A antibodies; a person with neither antigen A nor B (O) has both anti-A and anti-B antibodies; and a person who possesses both antigens A and B has neither anti-A nor anti-B antibodies. The individual's blood type indicates the antigens he or she possesses and not the antibodies.

Transfusion

The mixing of incompatible blood types can cause erythrocyte destruction and other problems. For example, if a person with type A blood (antigen A on erythrocytes, anti-B antibodies in plasma) is transfused with type B blood (antigen B on erythrocytes, anti-A antibodies in plasma), the recipient's anti-B antibodies will attack the incoming type B erythrocytes. The type B erythrocytes will be agglutinated (clumped), and hemoglobin will be released into the plasma—that is, the cells will undergo hemolysis. Incoming anti-A antibodies of the type B blood may also attack the type A erythrocytes of the recipient, with similar results. However, unless large amounts of blood are transfused, this problem is usually not as serious because the incoming antibodies are diluted in the recipient's plasma.

Table 21.3 Summary of the A-B-O System

Blood Type	Antigens (Agglutinogens) on Erythrocytes	Antibodies (Agglutinins) in Plasma
A	A	Anti-B
B	B	Anti-A
AB	Both A and B	Neither anti-A nor anti-B
O	Neither A nor B	Both anti-A and anti-B

During a transfusion reaction resulting from the mixing of incompatible blood types, agglutinated erythrocytes can plug small blood vessels. When hemolysis is rapid, the released hemoglobin can precipitate in the urine-forming structures of the kidneys and contribute to kidney failure.

Since type AB individuals possess neither anti-A nor anti-B antibodies to attack incoming erythrocytes, they are often called universal recipients. On the other hand, since the erythrocytes of type O individuals will not be attacked by either anti-A or anti-B antibodies, these people are called universal donors. However, these terms are misleading because other erythrocyte antigens and plasma antibodies can cause transfusion problems. If possible, therefore, blood for transfusion should be closely matched to the blood of the potential recipient.

Rh System

Another example of an erythrocyte antigen–antibody system is the Rh system (so named because of its initial study in Rhesus monkeys). The Rh system consists of a group of surface antigens on erythrocytes, and certain Rh antigens are very likely to cause transfusion reactions. An individual who possesses these antigens is designated Rh positive and an individual who lacks them is designated Rh negative. The antibody components of the system—the anti-Rh antibodies—are not normally present in the plasma, but anti-Rh antibodies can be produced upon exposure and sensitization to Rh antigens. In general, an individual does not produce anti-Rh antibodies against Rh antigens that are present on his or her erythrocytes, but upon sensitization a person can produce anti-Rh antibodies against Rh antigens that are not on his or her erythrocytes.

Sensitization can occur if Rh positive blood is transfused into an Rh negative recipient. It can also occur when an Rh negative mother carries a fetus who is Rh positive (due to the presence of Rh antigens inherited from the father). In this case, some of the fetal Rh antigens may enter the maternal circulation and sensitize the mother so that she begins to produce anti-Rh antibodies against the fetal antigens. The most likely time for sensitization to occur is near the time of birth, but because it takes some time for the mother to build up anti-Rh antibodies, the first Rh positive child carried by a previously unsensitized Rh negative mother is usually unaffected. However, if an Rh negative mother who has been sensitized by transfusion or by a previous Rh positive pregnancy subsequently carries an Rh positive fetus, maternal anti-Rh antibodies may enter the fetal circulation and cause the agglutination and hemolysis of fetal erythrocytes. This can result in an anemic condition known as hemolytic disease of the newborn (erythroblastosis fetalis). In this condition, the breakdown of the hemoglobin released during hemolysis results in the formation of sufficient bilirubin to give the infant's skin a yellow color—that is, jaundice. In some cases, bilirubin is deposited in nerve cells, causing brain damage. The usual treatment for severe hemolytic disease of the newborn resulting from Rh incompatibility is to remove the infant's Rh positive blood and replace it with Rh negative blood from an unsensitized donor. The Rh negative erythrocytes will not be attacked by the maternal anti-Rh antibodies in the infant's system, and the replacement of the infant's blood reduces the levels of these antibodies.

Since the sensitization of an Rh negative mother to the Rh antigens of an Rh positive fetus usually occurs near the time of birth, it is common to inject Rh negative mothers soon after the delivery of an Rh positive child with agents that prevent or limit sensitization. These agents contain anti-Rh antibodies that bind to any fetal Rh antigens that have entered the mother's system. This inhibits the antigens from sensitizing the mother and causing her to produce antibodies against them.

METABOLISM OF FOREIGN CHEMICALS

Many toxic foreign chemicals—for example, drugs and chemical pollutants in air, water, or food—can enter the body by way of the gastrointestinal tract, lungs, or skin. A number of these chemicals, particularly organic chemicals, are metabolized in the liver and also to some extent in the kidneys, skin, and other organs.

Often, the metabolic transformation of a hazardous chemical decreases its toxicity and enhances its excretion. For example, a nonpolar, lipid-soluble chemical may be converted into a more polar, less lipid-soluble substance that can be eliminated from the body in the urine.

On the other hand, the metabolic processing of some foreign chemicals enhances their toxicities. For example, it is believed that many chemicals linked to the occurrence of cancer become carcinogenic only after metabolic transformation. Moreover, because the pathways involved in the metabolism of foreign chemicals generally process materials normally required by the body, the presence of foreign chemicals may upset the metabolism of necessary substances.

RESISTANCE TO STRESS

A stressful event can be thought of as any event that leads to an increased release of glucocorticoids such as cortisol from the cortices of the adrenal glands, resulting in glucocorticoid levels in the blood plasma that are higher than those present at the same hour of the day in undisturbed, normal individuals on a similar activity (sleep–wake) cycle. Among the events that can have this effect are noxious or potentially noxious occurrences such as physical trauma, prolonged heavy exercise, and various infections. Many stressful events are believed to enhance glucocorticoid release by either directly or indirectly promoting an increased release of corticotropin-releasing hormone (CRH) from neurons of the hypothalamus of the brain. The CRH, in turn, enhances the release of adrenocorticotropin (ACTH) from the pituitary gland, and the ACTH increases the release of glucocorticoids from the adrenal cortices. Moreover, there is evidence that certain infections cause the production of ACTH or an ACTH-like substance by extrapituitary sources (possibly monocytes or lymphocytes), and this substance may be at least partly responsible for an increased release of glucocorticoids in such cases. In addition, monocytes produce substances (for example, interleukin-1 and hepatocyte-stimulating factor) whose effects apparently include the ability to stimulate ACTH release from the pituitary.

Many investigators have suggested that the glucocorticoids are particularly important to a person's ability to combat physical stress. Since a major response to overwhelming stress is vasodilation and circulatory failure, it has been proposed that the glucocorticoids combat stress by permitting norepinephrine to induce vasoconstriction while at the same time preventing an excessive vasoconstriction that can lead to tissue ischemia (deficiency of blood) by acting directly on vascular smooth muscle and stimulating heart muscle. It has also been suggested that the metabolic effects of the glucocorticoids—for example, raising blood-glucose levels—are particularly important in mobilizing the body's resources to resist physical stress.

A common response to stress is the activation of the sympathetic nervous system and the adrenal medullae. This response elevates the levels of epinephrine and norepinephrine, which leads to responses such as increased blood-glucose levels, elevated blood pressure, increased heart rate, and increased blood flow to skeletal muscles. These responses can also be useful in meeting physical stress.

The levels of other hormones are also frequently affected by stress. For example, the release of antidiuretic hormone (ADH) and aldosterone can increase during stress, resulting in a decreased urine output and in increased retention of water and sodium that leads to an increase in blood volume.

A number of psychological situations that elicit emotional responses such as fear, anger, or anxiety—for example, final exams for college students or awaiting a surgical operation—are also associated with an increased release of glucocorticoids and stress responses. In this regard it has been suggested that, although stress responses can be of benefit to an individual in meeting physical stress, they may not be as useful in resisting psychosocial stress. In fact, it has been proposed that stress responses elicited by chronic psychosocial stress—for example, high-pressure employment or anxiety-provoking economic situations—actually contribute to pathological conditions such as high blood pressure or heart disease.

STUDY OUTLINE

BODY SURFACES   Skin and mucous membranes are first barriers to invasion of body by potentially damaging factors.   **p. 621**
1. Skin contains keratin; sweat and sebaceous glands secrete chemicals toxic to many bacteria.
2. Mucus secreted by mucous membranes entraps small particles that may then be swept away or engulfed by phagocytic cells.

INFLAMMATION   Most basic form is acute, nonspecific physiological response of body to tissue injury caused by factors such as chemicals, heat, mechanical trauma, or bacterial invasion.   **pp. 621–627**

CHANGES IN BLOOD FLOW AND VESSEL PERMEABILITY
1. Small vessels in injured area dilate, leading to increased blood flow that increases delivery of plasma proteins and phagocytic leukocytes.
2. Small-vessel permeability increases; plasma fluid and solutes (including proteins) move out of circulatory system and into inflamed tissues.
3. As inflammation progresses, blood flow through small vessels slows, sometimes stops.

WALLING-OFF EFFECT OF INFLAMMATION   In tissues, fibrinogen from plasma is converted to fibrin, forming a clot that walls off injured area; can delay or limit spread of toxic products or bacteria.

LEUKOCYTE EMIGRATION   As blood flow slows during inflammation, leukocytes marginate and pavement; leukocytes exhibit amebalike activity and squeeze between endothelial cells of capillaries and venules into tissue spaces. First leukocytes in tissues are generally neutrophils, followed later by monocytes, which become macrophages.

CHEMOTAXIS AND LEUKOCYTE AGGREGATION   Chemical mediators determine direction of movement of leukocytes. In acute, nonspecific inflammation neutrophils usually predominate initially; monocytes and macrophages predominate during later stages.

PHAGOCYTOSIS   Leukocytes—particularly neutrophils—and macrophages phagocytize microorganisms, foreign materials, and debris; opsonins can coat specific foreign particles and render them more susceptible to phagocytosis.

ABSCESS AND GRANULOMA FORMATION   Occasionally, inflammatory response is unable to overcome invasion and an abscess or granuloma forms.
1. *Abscess*: A sac of pus consisting of microbes, leukocytes, macrophages, and liquified debris walled off by fibroblasts or collagen.
2. *Granuloma*: Layers of phagocytic-type cells surrounded by fibrous capsule; central cells contain offending agent—for example, microbes that can sur-

vive within phagocytes or materials that cannot be digested by phagocytes.

CHEMICAL MEDIATORS OF INFLAMMATION

HISTAMINE   Present in mast cells, basophils, and platelets; released in response to mechanical disruption of cells by injury and chemicals from neutrophils; causes vasodilation and increased vascular permeability.

KININS   Group of polypeptides formed from kininogens by kallikrein; dilate arterioles, increase vascular permeability, act as chemotactic agents, and induce pain.

COMPLEMENT SYSTEM   Series of plasma proteins that normally circulate in inactive form. Active complement components can mediate virtually every event of acute, nonspecific inflammatory response; can attack and kill invading microbes without prior phagocytosis. Complement system can be activated by a number of factors.

PROSTAGLANDINS   May induce vasodilation and increased vascular permeability, stimulate leukocyte migration through capillary walls, and contribute to pain and fever of inflammation.

LEUKOTRIENES   Produced by leukocytes. Certain leukotrienes apparently cause increased vascular permeability; some seem to be chemotactic agents that attract neutrophils.

SYMPTOMS OF INFLAMMATION

LOCAL SYMPTOMS   Redness and heat due to vessel dilation and increased blood flow; swelling due to increased vessel permeability and movement of fluid and solutes out of circulatory system; pain due to increased pressure on nerve endings and effects of some chemical mediators.

SYSTEMIC SYMPTOMS   Fever (monocytes and macrophages release endogenous pyrogen); often, increased production and release of leukocytes.

INTERFERON   Provides some protection to body against viral invasion: Viruses induce cells to produce interferon, which then leaves cells to bind to receptors on cell membranes of other cells; this binding triggers synthesis of enzymatic proteins within cells that can act to prevent viral reproduction. Interferon may prevent release of viruses from cells or make viruses incapable of infecting other cells; may also help protect body against some cancers; increasingly seen as versatile agent with effects that go beyond the inhibition of viral activity.   **pp. 627–628**

SPECIFIC IMMUNE RESPONSES   Two major aspects: humoral immunity and cell-mediated immunity.   **pp. 628–639**

ANTIGENS Substances that can elicit specific immune responses; generally have molecular weights of 8000–10,000 or more; lower-molecular-weight substances (haptens) may elicit specific immune responses if they combine with larger molecules.

ANTIBODIES An antibody is a specialized protein produced in response to presence of immunogenic antigen; can combine with specific antigen that stimulated its production, forming antigen–antibody complex. Antibodies inactivate or destroy foreign substances by:
1. Inhibiting interactions of bacterial toxins or destructive foreign enzymes with target cells or substrates.
2. Preventing attachment of viruses to target cells.
3. Enhancing basic inflammatory response.

B CELLS Lymphocytes committed to differentiate into antibody-producing plasma cells involved in humoral immunity.

T CELLS Heterogeneous group of lymphocytes; include cells committed to participating in cell-mediated immune responses.

ACCESSORY CELLS Present foreign antigens, together with self MHC antigens, to certain other cells by displaying on their surfaces both foreign antigens and self MHC antigens.

HUMORAL (ANTIBODY) IMMUNITY Appropriate stimulation by a foreign antigen causes some B lymphocytes to undergo proliferation and differentiation; most develop into plasma cells that form antibodies, which are released and can combine with antigen that stimulated their production to inactivate or destroy the antigen. Some stimulated B cells form memory cells, and with subsequent exposure to the antigen, there is rapid outpouring of antibodies. In most cases, accessory cells (e.g., macrophages) and helper T cells are involved in stimulation of B cells by antigen.

CELL-MEDIATED IMMUNITY Under appropriate antigenic stimulation, certain T cells proliferate and differentiate into various effector T cells that participate in cell-mediated immune responses (and some T-cell progeny serve memory functions). Effector T_C cells leave lymphoid tissues to travel throughout body. When they encounter cells bearing both foreign antigen that initiated their production and class I self MHC antigens, they combine with the cells; they can lyse and thereby kill the cells. Effector T_D cells can also participate in cell-mediated immune responses; they release chemical factors that promote inflammatory responses and enhance phagocytosis.

TRANSPLANT REJECTION Cell-mediated immune system contributes importantly to rejection of solid-tissue transplants.

ACTIVE IMMUNITY Resistance to infection resulting from antigen-induced activation of individual's specific immune responses; may be acquired by infection or by vaccination.

PASSIVE IMMUNITY Essentially "borrowed" immunity. Examples: Antibodies to disease-causing agent can be transferred to individual from sources (often nonhuman) that have been exposed to agent; maternal IgG-class antibodies may pass to fetus during pregnancy.

HYPERSENSITIVITY Specific immune response that can do harm.

ROLE OF COMPLEMENT COMPONENTS IN HYPERSENSITIVITY Generation of complement components during humoral immune response to an antigen can sometimes damage body tissues and cells.

ROLE OF EFFECTOR T_D CELLS AND EFFECTOR T_C CELLS IN HYPERSENSITIVITY Activities of these cells during cell-mediated immune response can, in some cases, damage body tissues.

ROLE OF CLASS IgE ANTIBODIES IN HYPERSENSITIVITY Exposure of susceptible individual to particular antigen can sensitize individual and cause production of class IgE antibodies that bind to mast cells and basophils; on subsequent exposure to same antigen, antigen attaches to antibodies bound to mast cells and basophils, stimulating release of chemicals that induce inflammatory responses, increased mucus secretion, and contraction of smooth muscles of airways of lungs.

TOLERANCE Generally defined as failure of body to mount specific immune response against particular antigen. The body normally exhibits self-tolerance and does not produce antibodies, effector T_C cells, or effector T_D cells against own antigens.

AUTOIMMUNE RESPONSES Occasionally, body's tolerance to own antigens breaks down, and person produces antibodies, effector T_C cells, and/or effector T_D cells that act against own tissues.

CONDITION OF CLINICAL SIGNIFICANCE: AIDS p. 638

BLOOD GROUPS AND TRANSFUSION
Based on antigens on erythrocyte surfaces that react with appropriate antibodies. pp. 639–640

A-B-O SYSTEM Based on inherited antigens on erythrocyte surfaces. Antibodies against antigens of this system that are not present on surfaces of a person's own erythrocytes begin to build up in plasma shortly after birth.

TRANSFUSION Mixing of incompatible blood types during transfusion can cause agglutination and hemolysis of erythrocytes.

Rh SYSTEM Based on inherited antigens on erythrocyte surfaces. Antibodies can be produced against Rh antigens not on a person's erythrocytes upon exposure and sensitization to the antigens.

METABOLISM OF FOREIGN CHEMICALS
Many potentially toxic or poisonous foreign chemicals are metabolically transformed into less toxic, more easily excreted substances; metabolic transformations may sometimes enhance toxicity of foreign chemicals. pp. 640–641

RESISTANCE TO STRESS
Stressful event can be thought of as one that leads to increased release of glucocorticoids such as cortisol from cortices of adrenal glands, resulting in higher plasma-glucocorticoid levels than those present at same hour in undisturbed, normal individuals on similar activity (sleep–wake) cycle; glucocorticoids may help combat physical stress. Stress can also activate sympathetic nervous system and adrenal

medullae; their activities may also help combat physical stress. Some psychological situations elicit stress re-

sponses, but the responses may not help resist the stress. **pp. 641–642**

SELF-QUIZ

1. The skin and mucous membranes serve as barriers to the invasion of the body by potentially damaging factors. True or False?

2. During an acute, nonspecific inflammatory response: (a) the permeability of small vessels decreases; (b) plasma proteins are unable to leave the circulatory system; (c) vessels of the microcirculation dilate.

3. Which cells usually predominate in the early stages of most acute, nonspecific inflammatory responses? (a) neutrophils; (b) monocytes; (c) eosinophils.

4. Histamine: (a) kills invading cells directly; (b) causes dilation of arterioles and venules; (c) decreases vessel permeability, particularly in venules.

5. Kininogens are converted to kinins by: (a) kallikrein; (b) opsonin; (c) chemotaxin.

6. The complement system can be activated by: (a) kallikrein; (b) protein A; (c) histamine.

7. Active complement components can mediate virtually every event of the acute, nonspecific inflammatory response. True or False?

8. The fever that often accompanies inflammation may be due to: (a) endogenous pyrogen; (b) kinins; (c) properdin.

9. One of the most effective inducers of interferon production is double-stranded DNA. True or False?

10. Haptens: (a) commonly elicit specific immune responses in their free, uncombined forms; (b) are relatively low-molecular weight substances that can elicit specific immune responses if they combine with larger molecules such as proteins; (c) generally have molecular weights greater than 10,000 in their free, uncombined forms.

11. B cells: (a) are committed to participating in cell-mediated immune responses; (b) are formed from macrophages; (c) can differentiate into plasma cells.

12. The constant regions of the polypeptide chains of an antibody molecule contain binding sites that allow the antibody to attach to a specific antigen. True or False?

13. Antibodies are produced by: (a) effector T_C cells; (b) mast cells; (c) plasma cells.

14. Active immunity can be acquired through (a) the transfer of antibodies from mother to fetus; (b) the injection of a pretreated bacterial toxin; (c) the injection of antibodies obtained from another person.

15. Anaphylactic hypersensitivity involves: (a) activated complement components; (b) effector T_C cells; (c) class IgE antibodies.

16. Some relief from an anaphylactic-type allergic reaction may be provided by the administration of: (a) opsonins; (b) antihistamines; (c) mast cells.

17. Normally, antibodies are not produced against a person's own antigens. True or False?

18. An adult with type A blood would normally have: (a) anti-A antibodies in the plasma; (b) anti-B antibodies in the plasma; (c) neither anti-A nor anti-B antibodies in the plasma.

19. During stress, which substance(s) is (are) released by the cortices of the adrenal glands? (a) glucocorticoids; (b) ACTH; (c) CRH.

20. Glucocorticoids may help a person combat physical stress. True or False?

LEARNING OBJECTIVES

After completing this chapter, you should be able to:

1. Distinguish between the regions of the pharynx, and describe their respiratory functions.

2. Describe the framework of the larynx.

3. Describe the subdivisions of the bronchi.

4. Describe the gross anatomical structure of the lungs.

5. Describe the composition of the respiratory membrane.

6. Describe the factors responsible for the increased lung volume and the flow of air into the lungs during inspiration.

7. Describe the factors responsible for the reduced lung volume and the flow of air out of the lungs during expiration.

8. Discuss the factors that influence pulmonary airflow.

9. Explain the various lung volumes that are combined into the total lung capacity.

10. Describe the process by which gas exchange occurs between the lungs and the blood, and between the blood and the body tissues.

11. Explain how oxygen and carbon dioxide are transported in the blood.

12. Discuss the neural mechanisms involved in the control of respiration.

13. Describe the roles of oxygen, carbon dioxide, and hydrogen ions in the control of respiration.

CHAPTER CONTENTS

EMBRYONIC DEVELOPMENT OF THE RESPIRATORY SYSTEM

ANATOMY OF THE RESPIRATORY SYSTEM

MECHANICS OF BREATHING

TRANSPORT OF GASES BY THE BLOOD

CONDITIONS OF CLINICAL SIGNIFICANCE: PROBLEMS OF GAS TRANSPORT

CONTROL OF RESPIRATION

EFFECTS OF EXERCISE ON THE RESPIRATORY SYSTEM

CONDITIONS OF CLINICAL SIGNIFICANCE: THE RESPIRATORY SYSTEM

KEY TERMS AND DERIVATIVES

alveolus (*alveol* = a small cavity) an air sac in the lung

asphyxia (*a* = without; *sphyxis* = pulse) loss of consciousness resulting from a lack of oxygen

bronchitis (*itis* = inflammation of) an inflammation of the bronchi

dyspnea (*dys* = painful; *pnoia* = breath) labored, difficult breathing

epiglottis (*epi* = upon; *glottis* = the opening into the larynx) the membrane-covered cartilage at the back of the throat that guards the glottis during swallowing

hypoxia (*hypo* = under, below) a condition in which an inadequate amount of oxygen is delivered to tissues

pleura (*pleura* = side, rib) the membrane covering the lungs

The Respiratory System

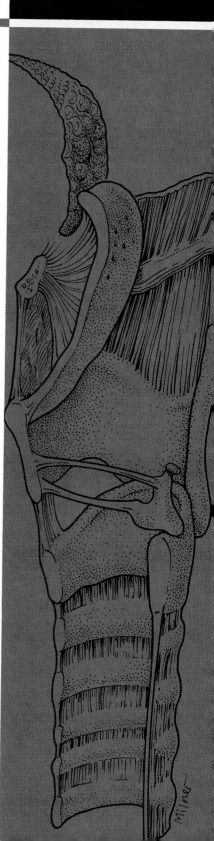

In order for the cells of the body to carry on their metabolic activities under aerobic conditions, they require a constant supply of oxygen and an efficient means of removing the carbon dioxide that their activities produce. Oxygen is supplied and carbon dioxide is removed by the **respiratory system,** with the assistance of the circulatory system. The respiratory system also makes vocalization possible. We are able to speak, sing, and laugh by varying the tension of the vocal folds as exhaled air passes over them.

Respiration is an integrated process, but for purposes of analysis it can be divided into five stages: (1) the movement of air into and out of the lungs; (2) the exchange of oxygen and carbon dioxide between the air in the lungs and the blood within the pulmonary capillaries; (3) the transport of oxygen and carbon dioxide throughout the body by the blood; (4) the exchange of oxygen and carbon dioxide between the blood and the interstitial fluid and cells; and (5) the utilization of oxygen and the production of carbon dioxide by metabolic processes within the cells. This chapter deals with the first four of these respiratory stages. The fifth is considered in Chapter 25.

The exchange of oxygen and carbon dioxide between the air and the blood occurs in the lungs. In order to reach the exchange sites in the lungs, the air must flow through a series of conducting passageways that branch from one another much like the branches of a tree. Air that enters the nose or mouth passes into the **pharynx** and is conveyed to the lungs by the **trachea,** which provides a branch called a **bronchus** to each lung. In the lung each bronchus divides many times into smaller and smaller tubules called **bronchioles,** which eventually terminate in small air sacs called **alveoli.** Gaseous exchange occurs in the alveoli.

EMBRYONIC DEVELOPMENT OF THE RESPIRATORY SYSTEM

The first indication of the development of the respiratory system in the embryo is the formation of an outpouch (*diverticulum*) from the ventral surface of the endoderm of the digestive tract, just behind the pharynx (Figure 22.1). This diverticulum, which appears in the four-week-old embryo, is called the **laryngotracheal bud.** As the bud elongates, the proximal portion develops into the *trachea*, and its distal end bifurcates, forming two buds that will develop into the *bronchi*. The bronchial buds continue to grow and rebranch, giving rise to many small tubes called *bronchioles*. The closed terminal portions of the bronchiole buds become dilated as they develop into the *alveoli* of the lungs.

It is clear from the pattern of embryonic development that the epithelium lining the entire respiratory tract is derived from endoderm. In the adult this lining is called the **respiratory epithelium.** With the exception of the lining of the smallest bronchioles and the alveoli, the respiratory epithelium is composed of pseudostratified ciliated columnar cells and scattered mucus-secreting goblet cells. The cartilage, muscles, and connective tissues of the trachea and the connective tissues of the lungs develop from embryonic mesoderm that becomes massed around the laryngotracheal bud.

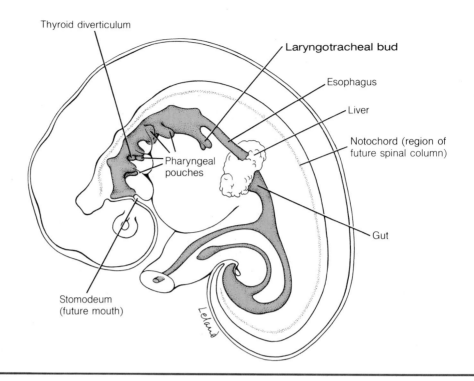

Figure 22.1
Lateral view of an embryo. The pharyngeal pouches, the thyroid diverticulum, and the laryngotracheal bud develop as outpouches of the pharynx.

ANATOMY OF THE RESPIRATORY SYSTEM

The respiratory system consists of the nose, nasal cavity, pharynx, larynx, trachea, bronchi, and lungs. The following sections describe the structures of these organs.

Nose and Nasal Cavity

Air enters the respiratory system through the **external nares (nostrils),** which lead to the **vestibule** of the nose. The lower part of the vestibule contains hairs that serve to trap the largest particles that might be drawn into the respiratory system during inspiration. The bridge of the nose is formed by the nasal bones. The rest of the framework of the nose consists of several plates of cartilage held together by fibrous connective tissue (Figure 22.2a). To form the **nasal septum,** one of the cartilages—the **septal cartilage**—joins with the nasal bones above, the vomer bone below, the perpendicular plate of the ethmoid posteriorly, and the maxillae inferiorly (Figure 22.2b). The septum divides the **nasal cavity** into right and left chambers. A deviation or deflection of the septum can interfere with the free flow of air through the nasal cavity, but this condition can be corrected surgically.

The bony roof of the nasal cavity is formed by the cribriform plate of the ethmoid bone (Figure 22.3). The lateral walls, which are irregular, are formed by the **superior** and **middle conchae** of the ethmoid bone and the separate **inferior concha** bones. Beneath the shelves formed by the conchae are recesses called the **superior, middle,** and **inferior meatuses.** The floor of the nasal cavity is formed by the bony **hard palate** (horizontal plates of the palatine bones and palatine processes of the maxillary bones) and the more posterior muscular **soft palate.** The palate separates the nasal cavity from the oral cavity. The nasal cavity opens posteriorly into the nasopharynx through the **internal nares** (*choanae*).

Passageways from the **paranasal sinuses** drain into the nasal cavities. Most of them open into the meatuses formed by the conchae. The paranasal sinuses are air spaces located in the frontal, maxillary, ethmoid, and sphenoid

F22.2a

F22.2b

F22.3

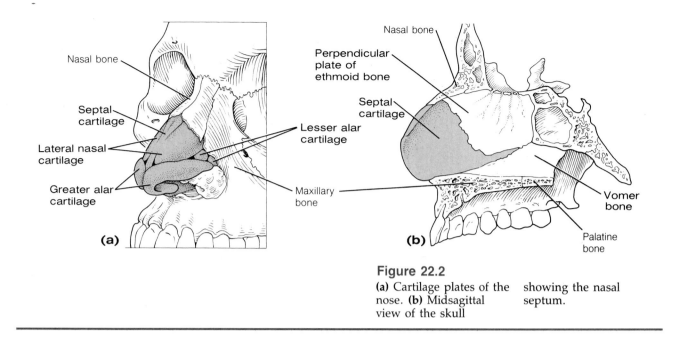

Figure 22.2
(a) Cartilage plates of the nose. **(b)** Midsagittal view of the skull showing the nasal septum.

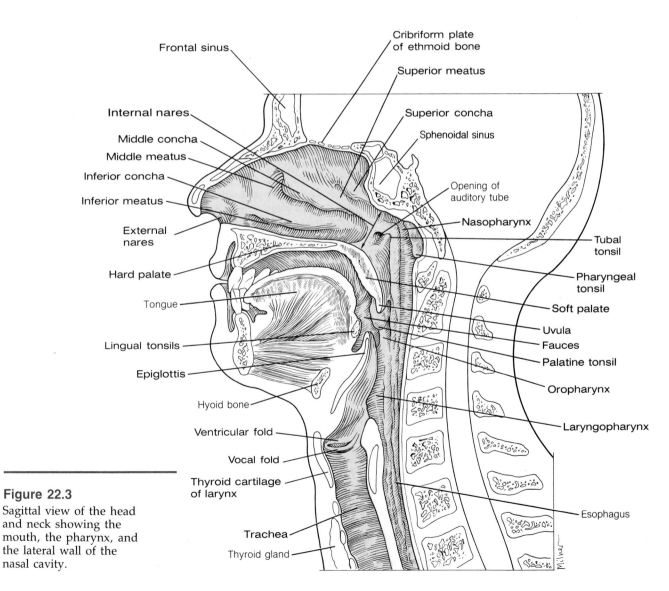

Figure 22.3
Sagittal view of the head and neck showing the mouth, the pharynx, and the lateral wall of the nasal cavity.

bones. The **nasolacrimal ducts,** which drain fluid (tears) from the surface of the eyes, also empty into the nasal cavity.

Lining of the Nasal Cavity and the Paranasal Sinuses

The vestibule of the nose is lined with stratified squamous epithelium that is continuous with the skin. The rest of the nose, the nasal cavity, and the paranasal sinuses are lined with a continuous mucous membrane of pseudo-stratified ciliated columnar epithelium that contains many mucus-secreting goblet cells. The part of the mucous membrane that is located at the top of the nasal cavity, just beneath the cribriform plate of the ethmoid bone, is a specialized epithelium called the **olfactory epithelium.** This region is supplied by the **olfactory nerve** (cranial nerve I), which passes through holes in the cribriform plate to reach the olfactory bulbs of the brain. Although the olfactory epithelium is sensitive to smells, it does not lie in the direct path of air flow; consequently, in order to detect odors it is helpful to sniff the air.

The mucous membrane has an extensive blood supply that warms the air as it is inhaled. Moreover, the mucous membrane saturates with water the air passing over it. The membranes of the nasal cavity and the delicate portions of the lungs are thus protected from becoming frozen or dried out. A sheet of mucus covering the mucous membrane further protects the respiratory system by trapping any small particles that get past the hairs guarding the external nares. The cilia of the membrane move in such a manner as to carry the particle-filled mucus toward the pharynx, where it can be removed by coughing or swallowing.

Infections of the Mucous Membranes

The mucous membranes of the nasal cavity may become inflamed because of infections (such as the common cold) or allergies. When inflamed, the blood vessels dilate, the membranes swell, and mucus secretion increases. The resulting congestion interferes with breathing and often causes a "runny nose." Such infection can spread into the mucous membranes of the paranasal sinuses, blocking their connections with the nasal cavity and causing them to fill with mucus. Since the sinuses act as resonance chambers, congestion changes the sound of the voice and can also cause a pressure increase in the sinuses such that a severe headache results. Infections of the mucous membrane of the nasal cavity can also extend through the nasolacrimal duct to the covering (conjunctiva) of the eye. Thus, it is not uncommon to have red, watery eyes along with a cold. It is also possible for infection to travel from the mucous membrane of the nasal cavity into the pharynx, thus producing a "sore throat." From the pharynx, the infection can spread into the bronchi of the lungs, causing coughing and possibly bronchitis, or through the auditory (eustachian) tubes into the middle ear. This occurs particularly in young children, in whom the lumina of the auditory tubes are relatively larger than in adults.

Pharynx

The **pharynx** is a tube that is used by both the digestive system and the respiratory system. It communicates with the nasal cavity (through the internal nares), the oral cavity (through the fauces), the middle-ear cavity (through the auditory tubes), the larynx (through the glottis), and the esophagus. The pharynx is muscular and is lined with a mucous membrane that is continuous with the mucous membrane of the structures with which it communicates. For descriptive purposes the pharynx is divided into three parts: *nasopharynx*, *oropharynx*, and *laryngopharynx*.

Nasopharynx

F22.3   The **nasopharynx** is located immediately behind the nasal cavity and is continuous with it through the internal nares (Figure 22.3). The mucous membrane of the nasopharynx, like that of the nasal cavity, is formed of pseudostratified columnar epithelium. On its lateral walls, the nasopharynx receives the **auditory (eustachian) tubes** that connect the nasopharynx with the cavity of the middle ear. Located near the openings of the auditory tubes are small masses of lymphoid tissue called **tubal tonsils.** On the posterior wall are the larger **pharyngeal tonsils.** When these tonsils become enlarged in response to an infection, they are called *adenoids.* Such enlargement can be chronic and can interfere with breathing through the nose, making it necessary to breathe through the mouth. The **soft palate** and **uvula** form the anterior wall of the nasopharynx.

Oropharynx

The **oropharynx** is a continuation of the nasopharynx, extending from the soft palate to the beginning of the laryngopharynx (Figure 22.3). It communicates with the oral cavity through the **fauces.** The oropharynx therefore receives food from the mouth and air from the nasopharynx. During exercise, air is also drawn into the oropharynx through the mouth, thus increasing the ventilation of the lungs. The mucous membrane lining of the oropharynx is stratified squamous epithelium, which is the epithelium typically found in the mouth and the upper portion of the digestive tract. This type of epithelium serves to protect the region from the abrasiveness of swallowed food. On the side walls are two **palatine tonsils.** Embedded in the base of the tongue is an aggregate of **lingual tonsils.** The tonsils are formed of lymphoid tissues and are therefore part of the body's immune system.

F22.3

ling'-gwal

Laryngopharynx

The **laryngopharynx** extends from the oropharynx above to the larynx and esophagus below (Figure 22.3). It communicates anteriorly with the larynx. Like the oropharynx, the laryngopharynx serves as a passageway for food and air and consequently is lined with a protective stratified squamous epithelium.

F22.3

Larynx

The **larynx** connects the laryngopharynx with the trachea, which continues below the larynx. Air passes through the larynx on its way into and out of the lungs. Any solid substance that enters the larynx, such as food, is generally expelled by violent coughing. The larynx forms a **laryngeal prominence** (*Adam's apple*) on the anterior surface of the neck. This prominence is particularly noticeable in males following puberty, when the larynx becomes larger than in females and the anterior region of its framework forms a more acute angle.

Framework of the Larynx

The larynx is formed by a total of nine cartilages—three unpaired and six that are paired together (Figure 22.4). These cartilages are held together, and attached to the hyoid bone above and the trachea below, by ligaments and muscles. The **thyroid cartilage** is the largest of the unpaired cartilages. It is formed by two broad plates that join anteriorly at an angle, producing the laryngeal prominence. The plates remain separated posteriorly, which leaves a wide opening in the laryngopharynx. Just below the thyroid cartilage is the ring-shaped **cricoid cartilage,** which is anchored to the thyroid cartilage above and the trachea below. The posterior region of the cricoid cartilage is longer than the anterior region. The third unpaired cartilage is the leaf-shaped **epiglottis.** The epiglottis is attached by its narrow end to the inner surface of the anterior region of the thyroid cartilage; its free upper portion projects like a flap behind the base of the tongue. During swallowing, the larynx is pulled upward, tipping the epiglottis so that it tends to deflect solids and fluids away from the opening of the larynx and into the esophagus.

F22.4

ep-e-glot'-tis

The **arytenoid cartilages** are the most important of the paired cartilages. Each arytenoid cartilage is shaped like a small pyramid and rests on the superior-posterior border of the cricoid cartilage. The posterior ends of the

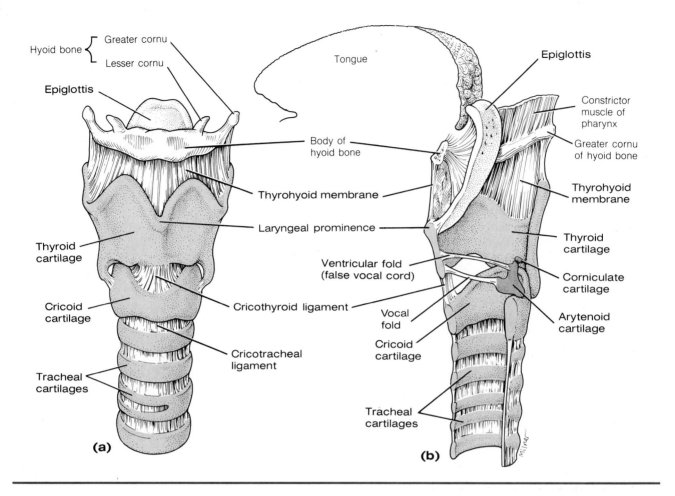

Figure 22.4
The larynx. **(a)** Anterior view. **(b)** Sagittal section.

F22.5

F22.5

lar-en-jigh'-tus

vocal folds are attached to the arytenoid cartilages, and movement of the cartilages is responsible for varying the tension on the folds. The other paired cartilages, the **cuneiform** and **corniculate cartilages,** are small and lie just above the arytenoid cartilages (Figure 22.5).

Mucous Membrane of the Larynx

The mucous membrane that covers the epiglottis and the upper parts of the larynx, where they are in direct communication with the laryngopharynx, is lined with stratified squamous epithelium. The remainder of the larynx is lined with pseudostratified ciliated columnar epithelium. The mucous membrane near the entrance to the larynx forms two pairs of horizontal folds that extend on each side from inside the angle of the thyroid cartilage to the arytenoid cartilages. The upper pair of folds are called the **ventricular folds** (*false vocal cords*). The lower pair are the **vocal folds** (*true vocal cords*). The opening between the vocal folds through which air enters the larynx is the **glottis** (Figure 22.5). Within the vocal folds are bands of elastic ligaments that connect to the thyroid, cricoid, and arytenoid cartilages. The elastic vocal ligaments can be tightened or slackened by the actions of certain intrinsic muscles of the larynx. These muscles also rotate the arytenoid cartilages, thus further varying the amount of stretch on the vocal folds. As a result of the actions of the intrinsic muscles, the glottis can be narrowed or widened. Air passing through the glottis causes the vocal folds to vibrate and produce a sound. The frequency of the vibrations, and therefore the pitch of the sound produced, depends on the tension on the vocal folds.

The mucous membrane that covers the vocal folds becomes swollen when irritated or inflamed. This swelling interferes with the ability of the folds to vibrate, and hoarseness may result—a condition referred to as **laryngitis.**

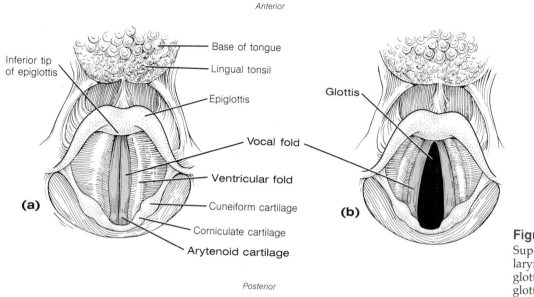

Figure 22.5
Superior view of the larynx showing **(a)** the glottis closed and **(b)** the glottis open.

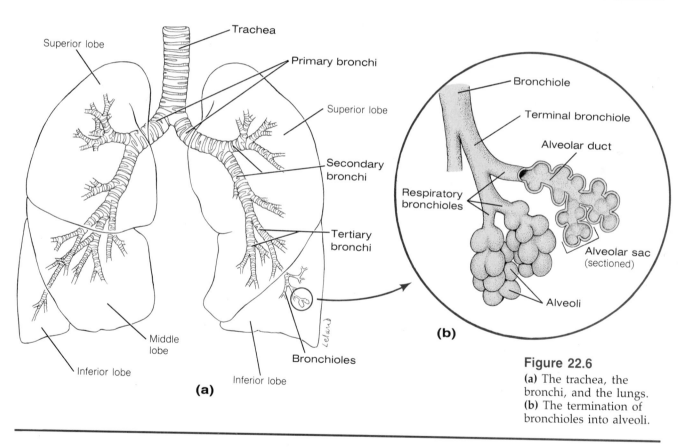

Figure 22.6
(a) The trachea, the bronchi, and the lungs. **(b)** The termination of bronchioles into alveoli.

Trachea

The **trachea** (windpipe) is a tube approximately 2.5 cm in diameter and 11 cm long. It extends from the larynx to the level of the sixth thoracic vertebra, where it divides into *left* and *right primary bronchi* (Figure 22.6a). The trachea lies against the anterior surface of the esophagus. The air passageway of the trachea is surrounded by a series of C-shaped hyaline cartilages that function to keep the trachea from collapsing, in the same manner as the wire rings in the hose of a vacuum cleaner. The cartilages are enclosed in an elastic fibrous

F22.6a

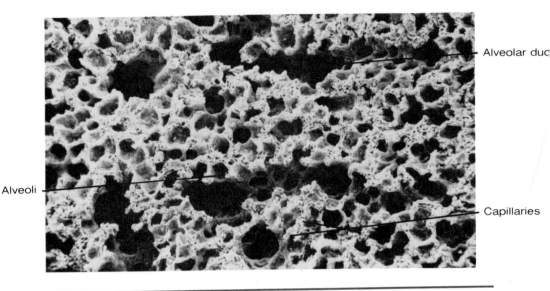

Alveolar duct

Alveoli

Capillaries

Figure 22.7
Photomicrograph of a lung section (×113). The numerous tiny openings are capillaries. (From *Tissues and Organs*: *A Text-Atlas of Scanning Electron Microscopy* by Richard G. Kessel and Randy H. Kardon. W.H. Freeman and Company. Copyright © 1979.)

membrane, and elastic fibers remain an important layer in the walls of all subsequent parts of the respiratory tract. Smooth muscles and dense fibrous connective tissue join the rings together posteriorly and close the opening of the C. When the passage of air through the upper airways is impeded, a direct airway into the trachea is sometimes created surgically through the anterior surface of the neck, between the second and third cartilagenous rings. This procedure is called a *tracheotomy*.

The trachea is lined with a mucous membrane of pseudostratified ciliated columnar epithelium that contains numerous goblet cells. Since the cilia beat upward, they tend to carry foreign particles and excessive mucus secretions away from the lungs to the pharynx, where they can be swallowed.

Bronchi, Bronchioles, and Alveoli

As the trachea passes behind the arch of the aorta, it divides into two smaller branches: the **left** and **right primary bronchi.** Each primary bronchus divides into still smaller **secondary bronchi,** one for each lobe of the lung. The secondary bronchi, in turn, branch into many **tertiary bronchi,** which further branch repeatedly, ultimately giving rise to tiny **bronchioles.** The bronchioles themselves subdivide many times, forming **terminal bronchioles,** each of which gives rise to several **respiratory bronchioles.** Each respiratory bronchiole subdivides into several **alveolar ducts,** which end in clusters of small, thin-walled air sacs called **alveoli** (Figure 22.6b). Often, several alveoli open into a common chamber called an **alveolar sac.** The bronchial tree is lined with pseudostratified ciliated columnar epithelium. However, in the respiratory bronchioles the epithelium loses its cilia and changes to cuboidal and then to squamous as the bronchioles extend distally. The many tiny alveolar air sacs and the bronchial system of tubes through which the air travels to enter or leave the alveoli give the lungs a spongy texture (Figure 22.7).

The walls of the primary bronchi, like those of the trachea, are supported by incomplete cartilage rings. In the lungs, the rings are replaced by small plates of cartilage that completely encircle the bronchus. Smooth muscles also encircle the bronchi. With further branching, the cartilage plates gradually become smaller, forming incomplete rings, and the smooth muscles that surround the air passageways become more common. The walls of the bronchioles contain no cartilage and are completely surrounded by smooth muscle. The fact that bronchioles have walls of smooth muscle explains one of the most disturbing symptoms of an asthma attack. In response to various allergens, the muscles of the bronchioles undergo spasms, and since there is no supporting cartilage, the air passageways are squeezed shut, making breathing very difficult.

F22.6b

F22.7

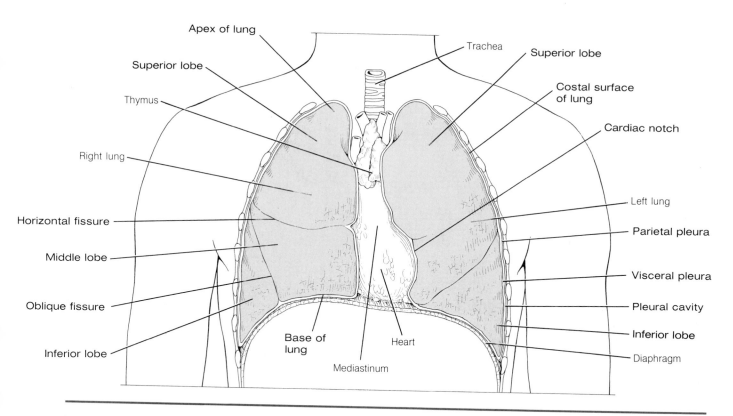

Figure 22.8
The lungs and the
thoracic cavity.

Lungs

Each lung is shaped somewhat like a cone, with the pointed **apex** fitting into the narrow space at the top of the thoracic cavity behind the clavicle (Figure 22.8). The **base** of each lung is broad and concave and rests on the convex surface of the diaphragm. A depression called the **hilus** is on the mediastinal surface of the lung. The hilus is the region where the structures that form the *root* of the lung—that is, the bronchus, blood vessels, lymphatics, and nerves—enter or leave the lung. The **costal surface,** which lies against the ribs, is rounded to match the curvature of the ribs. The left lung has a concavity for the heart, called the **cardiac notch,** on its medial surface. Each lung is divided into **superior** and **inferior lobes** by an **oblique fissure.** The right lung is further divided by a **horizontal fissure,** which bounds a **middle lobe.** The right lung therefore has three lobes, whereas the left has only two. In addition to these five lobes, which are visible externally, each lung is subdivided by connective tissue partitions into smaller units called **bronchopulmonary segments** (Figure 22.9). Each bronchopulmonary segment represents the portion of the lung that is supplied by a specific tertiary bronchus. The superior lobe of the right lung has three segments; the middle lobe has two segments; the inferior lobe has five segments. In the left lung, the superior lobe and the inferior lobe each have five segments, although some of them are not distinctly separated. The segments are important surgically because a diseased segment can be removed without having to remove an entire lobe or the entire lung. Also, disease does not spread so easily across the partitions that separate the segments, so pathology tends to be confined to one or several segments rather than spreading freely throughout the lung.

The two lungs are separated by a space called the **mediastinum.** Important structures are located in the mediastinum, including the heart, aorta, venae cavae, pulmonary vessels, esophagus, part of the trachea and bronchi, and the thymus gland.

The Pleura

Each lung is enclosed in a double-walled sac called the **pleura.** Both layers of the pleura are formed of serous membrane. The portion of the pleura that

ploor'-uh

F22.8

F22.9

Box 22.1
Smoker's and
Nonsmoker's Lungs

The ill-effects of tobacco smoke are clearly evident in this comparison between a smoker's lungs (top) and a nonsmoker's lungs (bottom). In the smoker's lungs, area A depicts a cancerous growth and the resultant narrowing of the lumen of the bronchus. The tumor has extended beyond the walls of the bronchus into the surrounding tissue (B). Lung cancer is typified by neoplastic growth—the presence of an abnormal mass (malignant tumor). The cancerous cells multiply and spread, destroying healthy tissue. Lung cancer is also characterized by metastatic growth, in which cancer cells spread throughout the body through the blood or lymphatic systems. Lung cancer can be treated with radiation, chemotherapy, or surgery to remove part or all of the lung.

Smoking is a contributing factor to heart disease, emphysema, and chronic bronchitis, as well as lung disease. Just one cigarette speeds up the heart rate, increases blood pressure, upsets the flow of blood and air in the lungs, and causes a drop in the skin temperature of fingers and toes. Less well known, but equally important, is the deleterious effect of secondhand smoke on the nonsmoker. It has been estimated that when a nonsmoker leaves a smoky environment, it takes four to five hours for 50% of the carbon monoxide to leave his or her body.

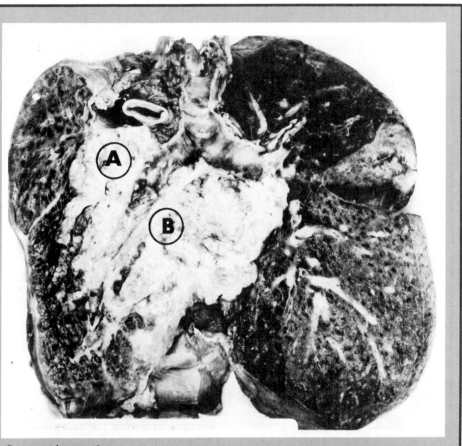

Lungs of a smoker

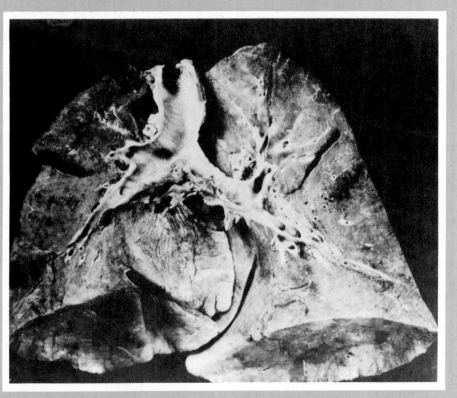

Lungs of a nonsmoker

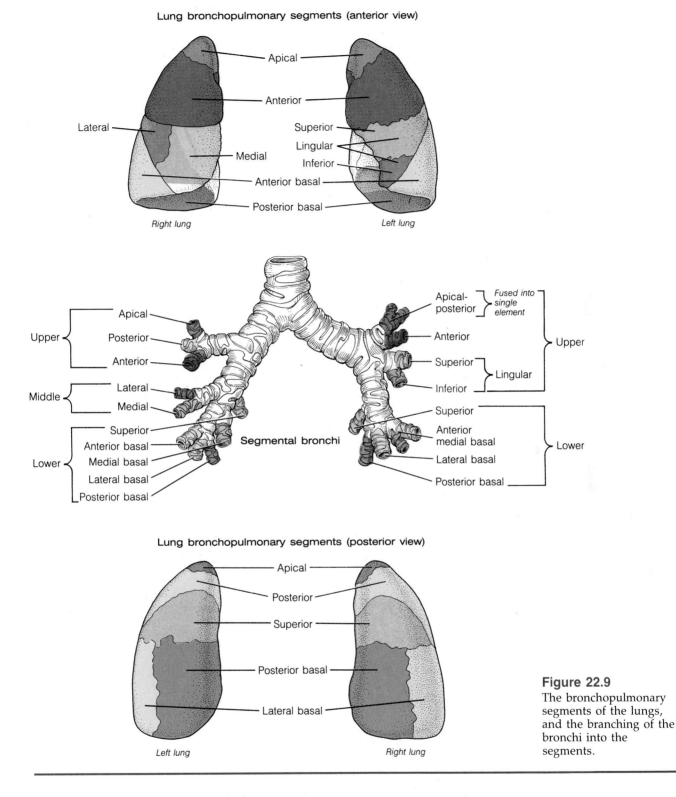

Lung bronchopulmonary segments (anterior view)

Apical
Anterior
Lateral
Superior
Lingular
Inferior
Medial
Anterior basal
Posterior basal

Right lung   *Left lung*

Apical
Posterior
Anterior
Upper
Lateral
Medial
Middle
Superior
Anterior basal
Medial basal
Lower
Lateral basal
Posterior basal

Segmental bronchi

Apical-posterior
Fused into single element
Anterior
Superior
Lingular
Inferior
Upper
Superior
Anterior medial basal
Lower
Lateral basal
Posterior basal

Lung bronchopulmonary segments (posterior view)

Apical
Posterior
Superior
Posterior basal
Lateral basal

Left lung   *Right lung*

Figure 22.9
The bronchopulmonary segments of the lungs, and the branching of the bronchi into the segments.

adheres firmly to the lungs is the **visceral pleura.** The portion that lines the walls of the thoracic cavity is the **parietal pleura.** The visceral and parietal layers are continuous at the hilus of the lung. Between the two layers of the pleura is an extremely narrow **pleural cavity,** which is filled with **pleural fluid.** The pleural fluid is secreted by the pleura, and it acts as a lubricant to reduce the friction between the two layers during respiratory movements. The pleural fluid also couples the visceral pleura and parietal pleura (and

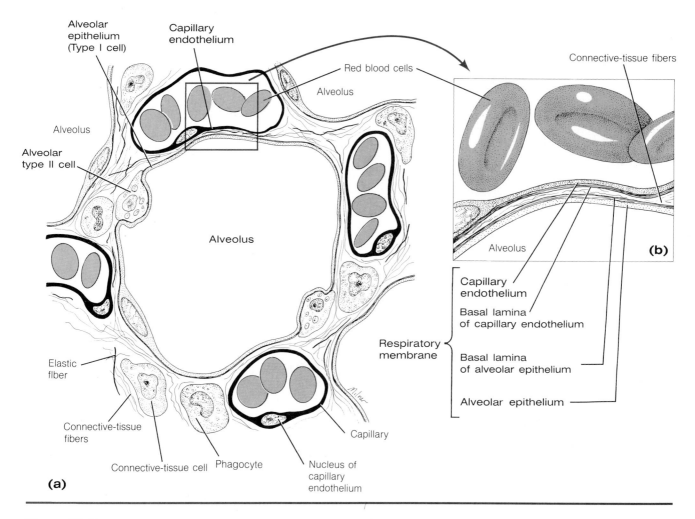

Alveolar epithelium (Type I cell)

Capillary endothelium

Red blood cells

Alveolus

Connective-tissue fibers

Alveolus

Alveolar type II cell

Alveolus

Alveolus

Elastic fiber

Connective-tissue fibers

Connective-tissue cell

Phagocyte

Nucleus of capillary endothelium

Capillary

(a)

Respiratory membrane

Capillary endothelium

Basal lamina of capillary endothelium

Basal lamina of alveolar epithelium

Alveolar epithelium

Alveolus

(b)

Figure 22.10

(a) An alveolus surrounded by capillaries, showing the respiratory membrane that separates the air in the alveolus from the blood in the capillaries. **(b)** The respiratory membrane of the lungs.

F22.10

therefore the lungs and the thoracic cavity structures) to one another. This coupling is analogous to the coupling of two sheets of glass by a thin film of water; it is extremely difficult to separate the sheets (the visceral pleura and the parietal pleura) by a force applied at right angles to their surfaces, even though the sheets can slide readily past one another. Because of this coupling, the movements of the lungs closely follow the movements of the structures that form the thoracic cavity.

Blood Supply of the Lungs

There is an important difference between the blood supplied to the alveoli and that supplied to the bronchi. The alveoli are supplied by branches of the pulmonary artery (which contains oxygen-poor blood), whereas small bronchial arteries from the descending aorta supply the bronchi (which contain oxygen-rich blood). Thus, the pulmonary arteries carry blood that will be aerated within the alveoli, whereas the blood within the bronchial arteries provides for the nourishment of the lung tissue.

The Respiratory Membrane

The air in the alveoli is separated from the blood by a very thin **respiratory membrane,** formed by the alveolar epithelium and its basal lamina and the endothelium of the pulmonary capillaries and its basal lamina. In some places there is a small amount of connective tissue in the form of reticular fibers and elastic fibers between the two basal laminae (Figure 22.10). It is obviously advantageous to have a thin respiratory membrane—since oxygen and carbon dioxide diffuse across it during respiration—while still maintaining the integrity of the blood vascular system. For the efficient diffusion of oxygen and carbon dioxide, the respiratory membrane must be moist, and the alveolar surfaces exposed to the air are coated with a thin layer of fluid.

The epithelium that lines the alveoli contains two types of cells. Most of the cells are of the simple squamous type through which gases diffuse readily. These cells are called *alveolar type I cells*. Scattered among the simple squamous cells are rounded or cuboidal cells—called *alveolar type II cells*—that may bulge into the lumen of the alveolus. The type II cells secrete a substance called *pulmonary surfactant*, which serves to lower the surface tension of the fluid that coats the alveoli, thus preventing the collapse of the alveoli and decreasing the muscular effort required to expand the lungs.

Phagocytic cells are also commonly present within the respiratory membrane of the alveolar walls, and they may be found free in the alveolar space. These phagocytes are sometimes called *dust* cells, since they often contain carbon particles obtained from inhaled air.

MECHANICS OF BREATHING

In order to maintain a concentration of oxygen and carbon dioxide within the alveolar air that is favorable to their diffusion across the respiratory membrane, it is necessary to constantly bring fresh air in and remove the air already within the lungs. The movement of air into and out of the lungs depends on pressure differences between the air in the atmosphere and the air in the lungs.

Atmospheric Pressure

At sea level, the gases of the atmosphere under the influence of gravity exert a pressure on the surface of the earth equivalent to that which would be exerted by a 760-millimeter-thick layer of mercury (Hg). Consequently, at sea level at any point on the surface of the earth, the pressure exerted by the column of atmospheric gases above that point is the same as the pressure that would be exerted by a column of mercury 760 millimeters high. Thus, *atmospheric pressure* at sea level can be expressed as equal to 760 mm Hg.

Pressure Relationships in the Thoracic Cavity

The lungs are stretched within the thoracic cavity, and the visceral pleura (which adheres to the lungs) is separated from the parietal pleura (which is attached to the structures forming the thoracic cavity) by only a thin layer of pleural fluid, which couples the visceral pleura and parietal pleura (and therefore the lungs and thoracic cavity structures) to one another. The walls and partitions of the lungs contain elastic connective tissue whose recoil tends to reduce the size of the lungs. In addition, as is discussed later, the surface tension of the fluid that coats the exposed surfaces of the alveoli also favors a reduction in the size of the lungs. Thus, the lungs tend to pull away from the structures that form the thoracic cavity and collapse. When the lungs are in the resting position following a normal exhalation, the collapsing force of the lungs is equal to a pressure of about 4 mm Hg (Figure 22.11).

F22.11

The pressure within the pleural cavity (the *intrapleural* or *intrathoracic pressure*) pushes against the outer walls of the lungs and also favors their collapse. However, the tendency of the lungs to pull away from the structures that form the thoracic cavity leads to a pressure within the pleural cavity that is somewhat less than the atmospheric pressure. When the lungs are in the resting position, the intrapleural pressure is approximately 4 mm Hg less than the atmospheric pressure—that is, approximately 756 mm Hg at sea level.

The alveoli and passageways within the lungs communicate with the atmosphere through the trachea, and the pressure within the lungs when they are in the resting position is equivalent to the atmospheric pressure.

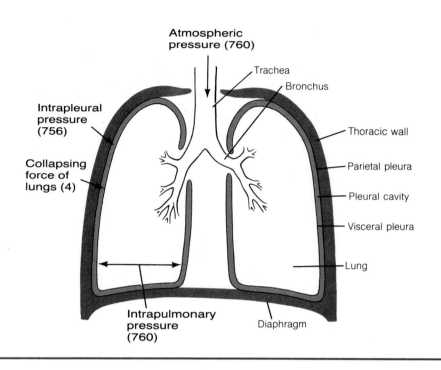

Atmospheric
pressure (760)

Trachea

Bronchus

Intrapleural
pressure
(756)

Thoracic wall

Parietal pleura

Collapsing
force of
lungs (4)

Pleural cavity

Visceral pleura

Lung

Intrapulmonary
pressure
(760)

Diaphragm

Figure 22.11
Intrapulmonary and
intrapleural pressures in
the resting position. All
pressures are given in
millimeters of mercury.

Thus, at sea level, the resting *intrapulmonary pressure* is 760 mm Hg. This pressure pushes against the inner walls of the lungs and opposes their collapse. Thus, under normal circumstances, the forces favoring the collapse of the lungs (the 4-mm-Hg collapsing force of the lungs and the 756-mm-Hg intrapleural pressure) are balanced by a force that opposes their collapse (the 760-mm-Hg intrapulmonary pressure). As a consequence, the lungs normally remain expanded within the thoracic cavity and do not collapse. However, if a lung or the thoracic wall is punctured or torn so that the pleural cavity is exposed to the atmosphere, air enters the pleural cavity and the intrapleural pressure becomes equal to the atmospheric pressure. As a consequence, the forces tending to reduce the size of the lung (the 4-mm-Hg collapsing force of the lung and the now 760-mm-Hg intrapleural pressure) become greater than the force opposing a reduction in lung size (the 760-mm-Hg intrapulmonary pressure), and the lung collapses. The presence of air in the pleural cavity is

noo-mo-thor'-aks   known as a *pneumothorax*.

Ventilation of the Lungs

Air flows from a region of higher pressure to a region of lower pressure. Thus, for air to flow into the lungs from the atmosphere, the pressure within the lungs—that is, the intrapulmonary pressure—must be less than the atmospheric pressure. Similarly, for air to flow out of the lungs, the intrapulmonary pressure must be greater than the atmospheric pressure. The pressure within the gas-filled spaces of the lungs can be altered by altering the volume of the spaces, and this can be accomplished by changing the volume of the thoracic cavity (recall that under normal circumstances, the pleural fluid couples the visceral pleura to the parietal pleura, and the movements of the lungs closely follow the movements of the structures that form the thoracic cavity). Thus, when the volume of the thoracic cavity increases, so does the volume of the lungs. The increase in lung volume lowers the intrapulmonary pressure below the atmospheric pressure, and air flows into the lungs until the pressure within the lungs again equals the atmospheric pressure. Conversely, when the volume of the thoracic cavity decreases, so does the volume of the lungs. The decrease in lung volume compresses the air within the lungs, and the intrapulmonary pressure rises above the atmospheric pressure. As a consequence, air flows out of the lungs until the pressure within the lungs becomes equal to the atmospheric pressure.

Inspiration

Inspiration, or *inhalation*, refers to the movement of air into the lungs. As previously indicated, air moves into the lungs when the volume of the thoracic cavity—and thus the lungs—increases, and the intrapulmonary pressure falls. During quiet breathing, the intrapulmonary pressure during inspiration decreases about 1 mm Hg below the resting intrapulmonary pressure, which is the same as the atmospheric pressure.

There are two ways of increasing the volume of the thoracic cavity during inspiration. One way is by contracting the diaphragm. When it contracts, the diaphragm flattens, lowering its dome. This movement increases the longitudinal dimension of the thoracic cavity. The second way is by elevating the ribs. In the resting position, the ribs slant downward and forward from the vertebral column. The contraction of muscles such as the intercostal muscles, which are located between the ribs, pulls the ribs upward. This movement increases the anterior-posterior dimension of the thoracic cavity. During normal, quiet inspiration, the contraction of the diaphragm is the dominant means of increasing the volume of the thoracic cavity and lowering the intrapulmonary pressure. The elevation of the ribs is most evident during forced inspiration.

Expiration

Expiration, or *exhalation*, refers to the movement of air out of the lungs. As previously indicated, air moves out of the lungs when the volume of the thoracic cavity—and thus the lungs—decreases, and the intrapulmonary pressure rises. During quiet breathing, the intrapulmonary pressure during expiration increases about 1 mm Hg above the resting intrapulmonary pressure.

During quiet breathing, the volume of the thoracic cavity is decreased by passive processes that do not involve muscular contractions. When the muscles involved in inspiration relax, the elastic recoil of the lungs, chest wall, and abdominal structures returns the ribs and diaphragm to their resting positions. This activity reduces the volume of the thoracic cavity and raises the intrapulmonary pressure.

During forced expiration, as occurs in exercise, muscles are involved in the further reduction of the volume of the thoracic cavity. The muscles of the anterior abdominal wall aid in forced expiration by exerting pressure on the abdominal viscera, thus forcing the diaphragm upward. The intercostals, the transversus thoracis, the quadratus lumborum, and the serratus posterior inferior muscles also assist in reducing the volume of the thoracic cavity by depressing the rib cage.

Traditionally, it has been thought that the roles of the intercostal muscles in respiratory movements could be distinguished, with the external intercostals assisting in forced inspiration and the internal intercostals assisting in forced expiration. However, recent studies have shown that both sets of intercostals are active in both forced inspiration and forced expiration, serving to control the distances between ribs.

Factors That Influence Pulmonary Airflow

The volume of air that flows in a given time between the atmosphere and the alveoli is influenced by the pressure that moves the air through the respiratory passageways and by the resistance the air encounters as it flows through the passageways. The relationship between these factors can be expressed as

$$\text{flow} = \frac{\text{pressure}}{\text{resistance}}$$

The pressure that moves the air through the respiratory passageways is represented by the difference between the atmospheric pressure and the pressure within the alveoli. A major factor influencing resistance is passageway diameter: the smaller the diameter of a passageway, the greater its resistance.

Smooth muscles associated with the respiratory passageways, particularly the bronchioles, can alter passageway diameters.

Under normal circumstances, the resistance of the respiratory passageways is slight, and a substantial airflow occurs in response to only a slight difference in atmospheric and alveolar pressures. During quiet inspiration, for example, the pressure difference between the atmosphere and the alveoli is only 1 mm Hg; yet approximately 500 ml of air enters the lungs with each breath.

Under certain circumstances, the constriction of the passageways (particularly the bronchioles) or the accumulation of fluid or mucus within the passageways increases the resistance sufficiently to impede the flow of air between the atmosphere and the alveoli. Since the passageways tend to widen as the lung volume increases during inspiration and narrow as the lung volume decreases during expiration, an increased resistance can particularly affect the flow of air out of the lungs. If the resistance increases, the pressure difference between the atmosphere and the alveoli must increase in order to maintain the airflow. Within limits, the pressure difference can be increased by an increased contraction of the respiratory muscles.

Parasympathetic nervous stimulation, histamine, and leukotrienes contract smooth muscles of the respiratory passageways and increase passageway resistance. Histamine also increases mucus secretion. Sympathetic nervous stimulation and epinephrine relax smooth muscles of the passageways and decrease passageway resistance.

Surface Tension and Pulmonary Surfactant

At an air–liquid interface, attractive forces between the molecules of a liquid such as water cause the molecules to be attracted more strongly into the liquid than into the air. This differential attraction gives rise to a tension at the surface of the liquid—the *surface tension*—that tends to reduce the liquid's surface area to the smallest possible value and to oppose any increase in the surface area. Thus, the surface tension of the fluid that coats the exposed surfaces of the alveoli tends to reduce the size of the alveoli and oppose the expansion of the lungs. If this fluid were pure water, which has a relatively high surface tension, the tendency of the alveoli to collapse would be so great that the expansion of the lungs during inspiration would require exhausting muscular effort. However, alveolar type II cells produce a phospholipoprotein complex called **pulmonary surfactant,** which lowers the surface tension of the fluid that coats the alveoli, significantly decreasing the muscular effort required to expand the lungs. In fact, even during heavy exercise, the ventilation of the lungs requires only about 3–4% of the total energy expenditure. Some children who are born prematurely do not produce sufficient quantities of pulmonary surfactant. The lack of adequate pulmonary surfactant production is believed to contribute to (or perhaps cause) an often fatal condition known as respiratory distress syndrome of the newborn, or hyaline membrane disease, in which breathing is difficult and labored.

Lung Volumes and Capacities

In young adult males, the lungs have a total capacity of approximately 5900 ml. However, a person cannot exhale all the air from the lungs, and about 1200 ml of air always remains no matter how forced the expiration. This remaining volume is called the **residual volume** (Figure 22.12).

F22.12

The volume of air that moves into and out of the lungs with each breath is called the **tidal volume.** During normal, quiet breathing, it is about 500 ml. The **inspiratory reserve** is the volume of air (approximately 3000 ml) that can be inspired in addition to the normal, quiet tidal volume. The sum of the normal, quiet tidal volume and the inspiratory reserve volume is called the **inspiratory capacity.**

Following a normal passive expiration of the quiet tidal volume, additional air can be forced out. This **expiratory reserve** is about 1200 ml. The sum

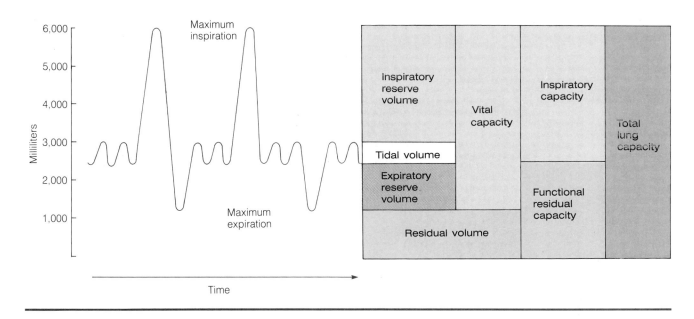

Figure 22.12
Lung volumes and capacities. The tidal volume indicated is the tidal volume during normal, quiet breathing.

Table 22.1 Average Lung Volumes in Young Males and Females (Age 20–30 Years)

| | Volume (in ml) | |
	Male	Female
Total lung capacity	5900	4400
Vital capacity	4700	3400
Inspiratory reserve	3000	2100
Tidal volume (at rest)	500	500
Expiratory reserve	1200	800
Residual volume	1200	1000

of the residual volume and the expiratory reserve volume is called the **functional residual capacity,** and the maximal amount of air that can be moved into and out of the lungs, from the deepest possible inspiration to the most forced expiration, is called the **vital capacity.** Therefore, the vital capacity represents the sum of the inspiratory reserve volume; the tidal volume during normal, quiet breathing; and the expiratory reserve volume. Note that it does not include the residual volume. In healthy young adult males the vital capacity is about 4700 ml. It is somewhat less in women because they tend to have smaller thoracic cages and smaller lung capacities (Table 22.1).

Table 22.1

Dead Space

The exchange of oxygen and carbon dioxide between air in the lungs and the blood occurs within the alveoli and, to some extent, within the small bronchioles and ducts that lead to the alveoli. However, all these gas exchange areas will be considered together as the alveoli. No gas exchange takes place within the remaining respiratory passageways (nose, trachea, bronchi, and so on). These air-filled passageways are referred to, therefore, as *anatomical dead space.* For healthy young adults, the volume of the anatomical dead space can be estimated to be approximately 1 ml for each pound of a person's "ideal" body weight. Thus, for a normal 150-pound male, the volume of the anatomical dead space is approximately 150 ml.

In some cases, oxygen and carbon dioxide exchanges between the air within certain alveoli and the blood are unable to take place efficiently because no blood reaches certain alveoli or because the ventilation of the alveoli exceeds the blood flow to them. Such occurrences have the effect of increasing the volume of dead space within the lungs beyond that of the anatomical dead space alone. The overall functional dead space within the lungs, including both the anatomical dead space and dead space attributable to the presence of alveoli that receive too little blood flow in relation to their airflow, is called the *physiological dead space*. In a healthy person, the volume of the physiological dead space and the volume of the anatomical dead space are the same or nearly the same. However, in certain disease conditions, the volume of the physiological dead space is significantly larger than that of the anatomical dead space.

Minute Respiratory Volume

The volume of air moved into the respiratory passageways in one minute is called the **minute respiratory volume.** It is equal to the respiratory rate (expressed as number of breaths per minute) multiplied by the volume of air that enters the respiratory passageways with each breath (the tidal volume). For example, during quiet breathing the respiratory rate is about 12 breaths per minute, and approximately 500 ml of air enters the respiratory passageways with each breath. Thus, the minute respiratory volume is 6000 ml/min (12 breaths/min × 500 ml/breath). The minute respiratory volume can be altered by altering either the respiratory rate or the amount of air entering the respiratory passageways with each breath.

Alveolar Ventilation

Because the exchange of gases between the lungs and the blood occurs within the alveoli and not within the respiratory passageways, the volume of air that moves into the alveoli from the atmosphere is more important than the volume of air that enters the respiratory passageways. With each inspiration, the first air to reach the alveoli is air that was left in the respiratory passageways—that is, in the anatomical dead space—from the previous expiration. Also, the last air to be inspired never reaches the alveoli. It remains within the respiratory passageways and is the first air to be exhaled. For a normal, 150-pound, young adult male, the volume of air within the respiratory passageways is approximately 150 ml. Therefore, of the approximately 500 ml of atmospheric air that enters the respiratory passageways with each inspiration during quiet breathing, only 350 ml reaches the alveoli.

Note that although only 350 ml of atmospheric air enters the alveoli with each inspiration during quiet breathing, a total of 500 ml of air actually moves into the alveoli (the first 150 ml is air that remained in the respiratory passageways from the previous expiration). Note, too, that the air within the alveoli is not completely replaced with each breath. For example, during quiet breathing, an expiratory reserve volume of 800–1200 ml and a residual volume of 1000–1200 ml remain within the lungs at the end of expiration, and only about 500 ml of air is taken into the lungs with each breath.

The volume of *atmospheric air* that enters the alveoli (either per breath or in one minute) and that can participate in the exchange of gases between the alveoli and the blood is called the **alveolar ventilation.** Thus, the alveolar ventilation represents the physiologically effective ventilation.

The alveolar ventilation *per minute* can be estimated as follows:

$$\text{alveolar ventilation} = \text{respiratory rate} \times (\text{tidal volume} - \text{physiological dead space volume})$$

To express the alveolar ventilation in ml/min, the respiratory rate is expressed as breaths/minute, and the tidal volume and physiological dead space volume are expressed as ml/breath. Thus, for a normal, 150-pound, young adult male during quiet breathing at a rate of 12 breaths/min, the alveolar ventilation is 4200 ml/min [12 breaths/min × (500 ml/breath − 150 ml/breath)].

Assuming the physiological dead space remains essentially constant, the alveolar ventilation can be increased by increasing either the respiratory rate or the volume of air inspired with each breath (that is, the tidal volume). However, the alveolar ventilation is generally increased more effectively by increasing the volume of air inspired with each breath than by increasing the respiratory rate. For example, if the respiratory rate doubles from its resting value (that is, doubles to 24 breaths/min) but the amount of air moved during each inspiration remains the same (that is, 500 ml/breath), the minute respiratory volume is 12,000 ml/min, and the alveolar ventilation is 8400 ml/min [24 breaths/min × (500 ml/breath − 150 ml/breath)]. If the respiratory rate remains the same (that is, 12 breaths/min) but the volume of air moved during each inspiration doubles (that is, to 1000 ml/breath), the minute respiratory volume is still 12,000 ml/min, but the alveolar ventilation is 10,200 ml/min [12 breaths/min × (1000 ml/breath − 150 ml/breath)]. During exercise and in most other situations where an increased supply of oxygen and an increased elimination of carbon dioxide are required, the increase in breathing depth—that is, the volume of air moved with each breath—is proportionally greater than the increase in respiratory rate.

Matching of Alveolar Airflow and Blood Flow

For efficient gas exchange, the airflow and blood flow to particular alveoli must be well matched. Poor gas exchange occurs when the alveolar airflow is either inadequate or excessive in relation to the blood flow. Local autoregulatory mechanisms contribute to matching alveolar airflow and blood flow. One of these mechanisms was considered in Chapter 19, where it was pointed out that pulmonary vessels constrict and blood flow decreases in areas where the oxygen level is low (and pulmonary vessels dilate and blood flow increases in areas where the oxygen level is high). In addition, the respiratory passageways dilate and the airflow to the alveoli increases in areas where the carbon dioxide level is high (and the respiratory passageways constrict and the airflow to the alveoli decreases in areas where the carbon dioxide level is low).

These autoregulatory mechanisms act in the following manner to match alveolar airflow and blood flow. In an area of alveoli whose airflow is inadequate in relation to their blood flow, the oxygen level will be low and the carbon dioxide level will be high. As a consequence, the pulmonary blood vessels in the area constrict and the blood flow decreases. In addition, the respiratory passageways dilate and the airflow increases. Thus, the alveolar airflow and blood flow become more closely matched. In an area of alveoli whose airflow is excessive in relation to their blood flow, the oxygen level will be high and the carbon dioxide level will be low. As a consequence, the pulmonary blood vessels in the area dilate and the blood flow increases. In addition, the respiratory passageways constrict and the airflow decreases. Again, the alveolar airflow and blood flow become more closely matched.

TRANSPORT OF GASES BY THE BLOOD

Partial Pressure

The atmosphere is a mixture of several gases, including nitrogen, oxygen, and carbon dioxide. The total atmospheric pressure at sea level is 760 mm Hg. According to Dalton's law of gases, the total pressure exerted by a mixture of gases is equal to the sum of the separate pressure exerted by each individual gas. Each individual gas in a mixture exerts a *partial pressure*, according to its percentage concentration in the mixture, that contributes to the total pressure exerted by the mixture. Thus, each atmospheric gas acts independently in generating the total atmospheric pressure, and each contributes partially to the total. For example, oxygen—which makes up approximately

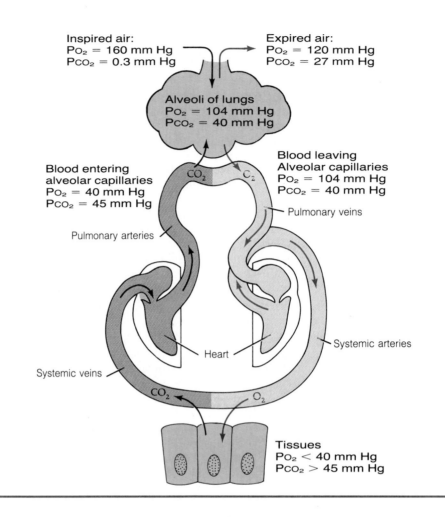

Figure 22.13
Partial pressures of oxygen and carbon dioxide in the alveoli of the lungs, in the blood entering and leaving the alveolar capillaries, and in the body tissues of a resting person at sea level.

21% of the atmospheric mixture of gases—contributes approximately 21% of the total pressure. Thus, the partial pressure of oxygen in the atmosphere at sea level is approximately 21%, or 160 mm Hg, of the total 760-mm-Hg sea-level atmospheric pressure.

Gases in Liquids

When a liquid is exposed to a gaseous atmosphere, gases enter the liquid and dissolve in it in proportion to their individual gas pressures. The number of molecules of a particular gas that enter a liquid, however, depends not only on the pressure of that gas in the atmosphere, but also on the solubility of the gas in the liquid. Nevertheless, by increasing the pressure of the gas, more molecules of it will dissolve in the liquid. A gas that is dissolved in a liquid can be treated mathematically under specific conditions as though it were in a free state and able to exert an independent pressure or partial pressure. Thus, each of the gases of the atmosphere can, while in solution, be considered to exert a pressure or partial pressure. Once a gas dissolves in a liquid, the gas molecules can diffuse, and the net diffusion is from regions of high pressure to regions of lower pressure. Moreover, if a liquid contains a particular gas at a higher pressure than the pressure of that gas in the surrounding atmosphere, molecules of the gas will leave the liquid and enter the atmosphere.

Oxygen and Carbon Dioxide Exchange Between Lungs, Blood, and Tissues

The partial pressure of oxygen (P_{O_2}) in the atmosphere at sea level, as previously noted, is approximately 160 mm Hg. The atmospheric partial pressure of carbon dioxide (P_{CO_2}) is only about 0.3 mm Hg. However, due to the humidification of the inhaled air and to the mixing of atmospheric air with air remaining in the lungs during each respiratory cycle, the gases in the lungs are not at the same partial pressures as in the atmosphere. The actual Po_2 and Pco_2 in the alveoli of the lungs, as well as the partial pressures of these gases in the blood entering and leaving the alveolar capillaries and in the body tissues of a resting person at sea level, are given in Figure 22.13. As can be seen from the figure, within the alveoli the Po_2 is about 104 mm Hg and the Pco_2 is approximately 40 mm Hg. These pressures remain fairly constant throughout the average respiratory cycle.

As the figure also shows, blood entering the alveolar capillaries from the tissues contains oxygen at a partial pressure of only 40 mm Hg and carbon dioxide at a partial pressure of 45 mm Hg. As a result of the partial-pressure differences between the oxygen and carbon dioxide in the blood and the oxygen and carbon dioxide in the alveoli of the lungs, there is a net diffusion of oxygen into the blood from the alveoli and a net diffusion of carbon dioxide into the alveoli from the blood. As a consequence of these activities, the partial pressures of oxygen and carbon dioxide in the blood leaving the alveolar capillaries are essentially equal to the partial pressures of these gases in the alveoli (Po_2: 104 mm Hg; Pco_2: 40 mm Hg).

Within the tissues, metabolic reactions utilize oxygen and produce carbon dioxide. Thus, the Po_2 in the tissues is approximately 40 mm Hg, whereas the Pco_2 is about 45 mm Hg. When the blood from the lungs, which contains oxygen at a relatively high partial pressure, reaches the tissues, there is a net diffusion of oxygen from the blood to the tissues. In a similar fashion, there is a net diffusion of carbon dioxide from the tissues to the blood. The end result is that the venous blood leaving the tissues contains oxygen at a partial pressure of about 40 mm Hg and carbon dioxide at a partial pressure of about 45 mm Hg. It is these partial pressures that were previously described for blood entering the alveolar capillaries of the lungs.

Thus, the exchanges of oxygen and carbon dioxide that occur between the blood and alveoli on the one hand, and the blood and tissues on the other, take place by simple diffusion and depend on the pressure differences of the respective gases that exist between the blood and the alveoli, and the blood and the tissues.

Oxygen Transport in the Blood

Since oxygen is relatively insoluble in water, relatively little actually dissolves in the fluid of the blood. In fact, under normal circumstances, each liter of systemic arterial blood contains only about 3 ml of dissolved oxygen. However, the total oxygen content of a liter of systemic arterial blood is about 197 ml. The additional 194 ml of oxygen is bound to the iron of the heme groups of hemoglobin molecules in erythrocytes.

The combination of hemoglobin with oxygen is reversible, and whether hemoglobin binds with or releases oxygen depends in large part on the oxygen partial pressure. The degree of saturation of hemoglobin with oxygen at any given Po_2 is indicated by the oxygen–hemoglobin dissociation curve (Figure 22.14). When the Po_2 is relatively high, hemoglobin binds with much oxygen and is essentially completely saturated. At lower oxygen partial pressures, hemoglobin binds with less oxygen and is only partially saturated.

The Po_2 is relatively high in the alveolar capillaries of the lungs. Here, oxygen from the alveoli diffuses into the plasma and then into the erythrocytes, where it binds with hemoglobin. Within the various body tissues, however, metabolic activities utilize oxygen, and there is a net diffusion of oxygen out of the blood. Thus, the Po_2 in the tissue capillaries is lower than that in the alveolar capillaries, and hemoglobin cannot bind with as much oxygen in the tissue capillaries as it can in the alveolar capillaries. As a result, the oxygen-rich hemoglobin that leaves the alveolar capillaries releases some of its oxygen when it reaches the tissue capillaries. This oxygen can then diffuse into the tissues for their use.

In addition to the Po_2, several other factors affect the binding of oxygen to hemoglobin. Under acidic conditions, the amount of oxygen that binds to hemoglobin at any given oxygen partial pressure is diminished. Thus, the higher the hydrogen ion concentration—that is, the lower the pH—the less oxygen is bound to hemoglobin at any given Po_2 (Figure 22.15). This effect is

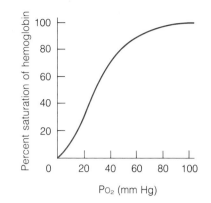

Figure 22.14
Oxygen–hemoglobin dissociation curve. The curve indicates the percent saturation of hemoglobin at different partial pressures of oxygen (Po_2) when the pH is 7.4 and the temperature is 38°C.

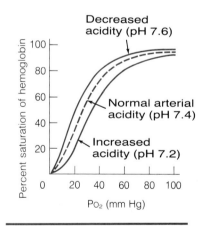

Figure 22.15
Graph indicating the percent saturation of hemoglobin at different partial pressures of oxygen, showing the effects of acidity.

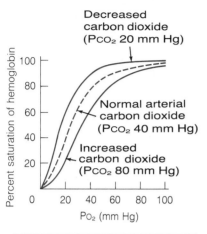

Figure 22.16
Graph indicating the percent saturation of hemoglobin at different partial pressures of oxygen, showing the effects of carbon dioxide.

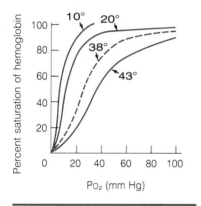

Figure 22.17
Graph indicating the percent saturation of hemoglobin at different partial pressures of oxygen, showing the effects of temperature (°C).

due to the fact that hydrogen ions bind with the hemoglobin molecules, altering their molecular structure and thereby decreasing the amounts of oxygen they can carry. The partial pressure of carbon dioxide has basically the same effect on the binding of oxygen and hemoglobin as the hydrogen ion concentration. That is, the higher the P_{CO_2}, the less oxygen is bound to hemoglobin at any given P_{O_2} (Figure 22.16). This effect is due in large measure to the fact that the P_{CO_2} can influence the pH in the following manner:

$$CO_2 + H_2O \rightleftharpoons H_2CO_3 \rightleftharpoons H^+ + HCO_3^-$$

carbon   water   carbonic   hydrogen   bicarbonate
dioxide   acid   ion   ion

As these reactions show, carbon dioxide can combine with water to form carbonic acid. (Although not absolutely required for this reaction to occur, an enzyme within erythrocytes, *carbonic anhydrase*, greatly increases the rate of the reaction.) The carbonic acid can dissociate into hydrogen ions and bicarbonate ions, and the hydrogen ions contribute to the hydrogen ion concentration. Thus, an increase in the P_{CO_2} tends to increase the hydrogen ion concentration and therefore lower the pH.

Temperature is another factor that influences the binding of oxygen and hemoglobin, and the higher the temperature, the less oxygen is bound to hemoglobin at any given P_{O_2} (Figure 22.17).

The influences of pH, P_{CO_2}, and temperature on the binding of oxygen by hemoglobin operate to ensure adequate deliveries of oxygen to active tissues that need it most. Active tissues tend to have higher hydrogen ion concentrations in their vicinities than do less active tissues (active skeletal muscles, for example, can produce lactic acid). They also tend to produce more carbon dioxide and to have higher temperatures as a result of their metabolic activity than do less active tissues. Consequently, hemoglobin binds less oxygen in active tissues (and thus provides more for delivery) than it does in less active tissues.

A substance produced by erythrocytes called DPG (2,3-diphosphoglyceric acid) binds reversibly with hemoglobin and alters the hemoglobin molecule so it releases oxygen. Thus, DPG promotes the release of oxygen from hemoglobin. DPG production increases in conditions associated with decreased oxygen supplies to tissues, and some investigators believe this response helps maintain oxygen delivery to the tissues. However, other investigators point out that an increased level of DPG not only favors the release of oxygen from hemoglobin in the tissues, but it also makes it more difficult for oxygen to combine with hemoglobin in the lung capillaries. Therefore, they question whether increased DPG production in conditions associated with decreased oxygen supplies to the tissues is always beneficial.

Carbon Dioxide Transport in the Blood

Carbon dioxide is transported by the blood in several different ways. A small amount of carbon dioxide—perhaps 8%—is transported in physical solution dissolved in the plasma. About 20% is transported in reversible association with various blood proteins. These carbon-dioxide–protein complexes are called *carbamino compounds*. Carbon dioxide can combine with many proteins in the blood. However, since the most abundant protein is the globin of hemoglobin, many carbamino unions form between carbon dioxide and the globin of hemoglobin. Thus, oxygen is carried by hemoglobin in association with the iron of the heme groups, and carbon dioxide is carried in reversible association with the globin protein portion of the hemoglobin molecules. Hemoglobin that is not carrying oxygen (*reduced hemoglobin*) is able to form carbamino compounds more readily than hemoglobin that is carrying oxygen (*oxyhemoglobin*). Thus, in the tissues, the presence of hemoglobin molecules that have given up their oxygen favors the combination of carbon dioxide with hemoglobin. In the lungs, the binding of oxygen to reduced hemoglobin molecules decreases their affinity for carbon dioxide and favors the displacement of carbon dioxide from the molecules. The displaced carbon dioxide can diffuse into the alveoli for elimination from the body.

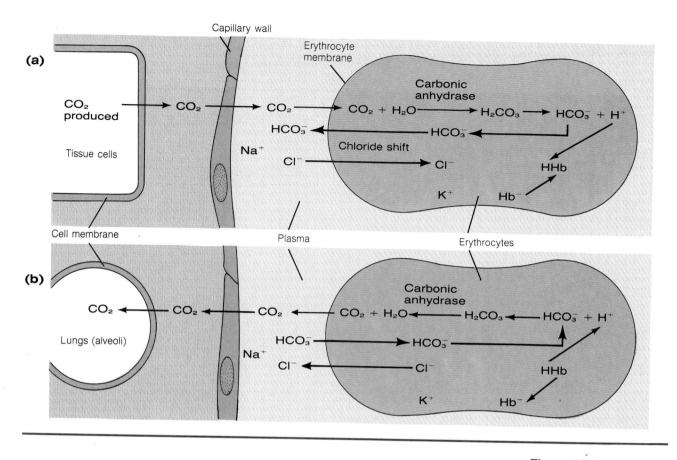

Figure 22.18
Transport of carbon
dioxide in the blood
as bicarbonate ions.
(a) Sequence of events
that occurs in the tissues.
(b) Sequence of events
that occurs in the lungs.
Hb is hemoglobin.

Most of the carbon dioxide in the blood—about 72%—is transported as bicarbonate ions (Figure 22.18). Carbon dioxide produced in the tissues diffuses into the plasma and from there into erythrocytes. Within the erythrocytes, the enzyme carbonic anhydrase facilitates the combination of carbon dioxide and water to form carbonic acid. The carbonic acid can dissociate into hydrogen ions and bicarbonate ions, and most of the hydrogen ions associate with hemoglobin molecules. This association eliminates free hydrogen ions and thus helps prevent a substantial increase in the hydrogen ion concentration of the blood. In fact, because of the ability of hemoglobin molecules (and other substances in the blood) to combine with hydrogen ions, the pH of venous blood leaving the tissues (pH = 7.35) is only slightly lower than that of arterial blood flowing to the tissues (pH = 7.4). The bicarbonate ions resulting from the dissociation of carbonic acid molecules diffuse out of the erythrocytes into the plasma, and in response to this movement of negatively charged bicarbonate ions, chloride ions from the plasma enter the erythrocytes. This exchange is known as the chloride shift. As a result of these events, carbon dioxide from the tissues is transported as bicarbonate ions in the blood.

At the lungs, the reverse of the preceding events occurs. Carbon dioxide diffuses out of the blood into the alveoli, and the concentration of carbon dioxide within the plasma and the erythrocytes declines. Within the erythrocytes, hydrogen ions combine with bicarbonate ions to form carbonic acid, which is split by carbonic anhydrase into carbon dioxide and water. As the concentration of bicarbonate ions within the erythrocytes diminishes, bicarbonate ions from the plasma diffuse in and chloride ions leave. The carbon dioxide that is produced leaves the erythrocytes and diffuses into the alveoli for elimination from the body.

The combination of hydrogen ions and bicarbonate ions to form carbonic acid and ultimately carbon dioxide and water is enhanced by the binding of oxygen to hemoglobin in the lung capillaries. When hemoglobin combines with oxygen, its ability to bind hydrogen ions decreases, and more hydrogen

CONDITIONS OF CLINICAL SIGNIFICANCE

Problems of Gas Transport

The transport of oxygen and carbon dioxide by the blood can be affected adversely by a variety of conditions. These include malfunctions of the cardiovascular system that interfere with the transport of gases by the blood, respiratory system problems that upset the gas exchange process, and deficiencies in the composition of the air breathed. The following are only a few of the possible conditions that can cause problems.

Cyanosis

Cyanosis is a bluish discoloration of the skin and mucous membranes. Reduced hemoglobin—that is, hemoglobin that is not combined with oxygen—appears darker than oxyhemoglobin, and cyanosis occurs when there is a significant concentration of reduced hemoglobin in the arterial blood. Cyanosis can result from inadequacies of either the respiratory or circulatory systems.

Hypoxia

Any state in which an inadequate amount of oxygen is delivered to the tissues is called *hypoxia*. Hypoxia can be caused by a number of factors.

In some cases, not enough oxygen reaches the blood, and the arterial P_{O_2} is below normal. Hypoxia due to such an occurrence is referred to as *hypoxemic hypoxia (arterial hypoxia)*. Hypoxemic hypoxia can result from insufficient oxygen in the air breathed or from respiratory problems—such as pneumonia, emphysema, or paralysis of the respiratory muscles—that prevent sufficient oxygen from reaching the blood within the lungs.

In other cases, enough oxygen reaches the blood, but there is not enough hemoglobin available to carry it to the tissues. Since the problem is often associated with anemia, hypoxia due to a deficiency of hemoglobin is referred to as *anemic hypoxia*.

In still other cases, the blood flow to the tissues is less than normal. The blood may be able to carry a normal amount of oxygen, but oxygen is not delivered to the tissues fast enough for the cells to maintain their normal metabolism. Hypoxia due to such an occurrence is called *stagnant hypoxia (hypokinetic hypoxia)*. Stagnant hypoxia can result from heart failure or the presence of emboli in the blood.

Carbon Monoxide Poisoning

Carbon monoxide (CO) is a colorless, odorless gas. It is present in the exhaust fumes of automobiles and is produced when carbon products such as wood and coal are burned. Carbon monoxide is hazardous because it and oxygen compete for the same hemoglobin binding sites. Hemoglobin has a much greater affinity for carbon monoxide than it does for oxygen, and when even small amounts of carbon monoxide are present in the air breathed, it preferentially occupies the oxygen-binding sites of the hemoglobin molecules. Moreover, the hemoglobin–carbon-monoxide bond is so strong that very little carbon monoxide is removed from the blood. Consequently, carbon monoxide causes drowsiness, coma, and ultimately death due to hypoxia.

---

ions leave the hemoglobin molecules than would otherwise be the case. These hydrogen ions can combine with bicarbonate ions to form carbonic acid and, ultimately, carbon dioxide and water.

As the preceding discussion indicates, carbonic anhydrase can catalyze the reaction:

$$CO_2 + H_2O \xrightleftharpoons[\text{anhydrase}]{\text{carbonic}} H_2CO_3 \rightleftharpoons H^+ + HCO_3^-$$

in either direction. The overall direction of the reaction depends on the partial pressure of carbon dioxide as well as on the pH and the bicarbonate ion concentration. In the tissues, where the P_{CO_2} is relatively high (and hydrogen ions are tied up by hemoglobin molecules), the formation of bicarbonate ions is favored. In the lungs, where the P_{CO_2} is lower (and hemoglobin molecules release hydrogen ions), the reaction moves from bicarbonate ions and carbonic acid toward carbon dioxide and water.

CONTROL OF RESPIRATION

The muscles responsible for inspiration–for example, the diaphragm and the intercostals–are skeletal muscles that require neural stimulation to initiate their contractions. When nerve impulses activate the inspiratory muscles, the thorax expands and the lungs fill with air.

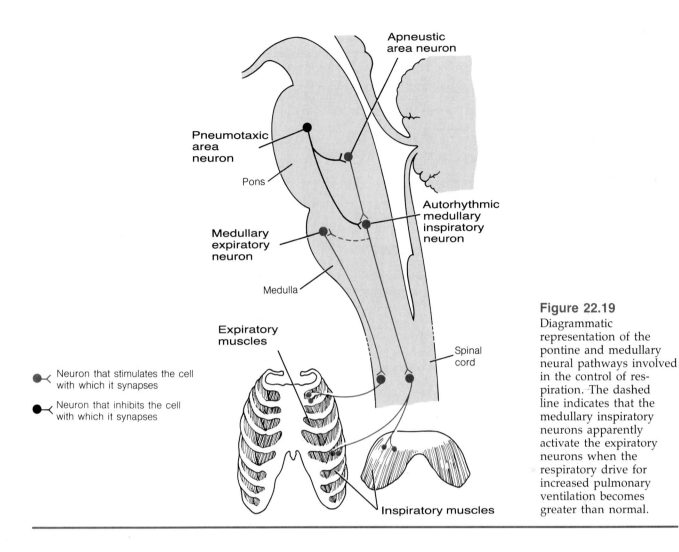

Apneustic
area neuron

Pneumotaxic
area
neuron

Pons

Autorhythmic
medullary
inspiratory
neuron

Medullary
expiratory
neuron

Medulla

Expiratory
muscles

Spinal
cord

● Neuron that stimulates the cell
with which it synapses

● Neuron that inhibits the cell
with which it synapses

Inspiratory muscles

Figure 22.19
Diagrammatic
representation of the
pontine and medullary
neural pathways involved
in the control of res-
piration. The dashed
line indicates that the
medullary inspiratory
neurons apparently
activate the expiratory
neurons when the
respiratory drive for
increased pulmonary
ventilation becomes
greater than normal.

Following inspiration, nerve impulses to the inspiratory muscles diminish greatly, and the diaphragm and other muscles involved in inspiration relax. When the inspiratory muscles relax, expiration occurs as a result of the elastic recoil of structures such as the lungs and chest wall, which returns the thorax to its resting position. During forced expiration, nerve impulses activate expiratory muscles—such as the muscles of the anterior abdominal wall—that help depress the ribs and diminish the size of the thoracic cavity.

Even when the thorax is at rest at the end of expiration, some nerve impulses are still transmitted to the respiratory muscles, and the muscles remain in a state of slight contraction (tonus), which helps maintain normal body posture.

Generation of Rhythmical Breathing Movements

The rhythmic pattern of inspiration and expiration characteristic of normal breathing depends on the cyclical stimulation of the respiratory muscles. This cyclical stimulation is due primarily to the activity of neurons called inspiratory neurons, whose cell bodies are found in the medulla oblongata of the brain (Figure 22.19). The medullary inspiratory neurons connect either directly or indirectly with neurons that supply the inspiratory muscles.

F22.19

Medullary Inspiratory Neurons

The discharges of the *medullary inspiratory neurons* are synchronized with inspiration, and these neurons cease discharging during expiration. It is currently believed that the medullary inspiratory neurons possess an inherent automaticity and rhythmicity. As a result, they undergo spontaneous, cyclical self-excitation and thus provide the underlying rhythm for respiration. Since expiration is basically a passive process that occurs when the muscles controlling inspiration relax, the cyclical activity of the inspiratory neurons can account for alternating cycles of inspiration and expiration.

Inputs from other neurons apparently modify the basic rhythm of the medullary inspiratory neurons. Among these inputs are connections with neurons of the pons, and the influence of afferent impulses from stretch receptors in the lungs.

Neurons of the Pons

Within the pons are two areas called the *pneumotaxic area* and *apneustic area*. The pneumotaxic area continually sends inhibitory signals to the medullary inspiratory neurons that tend to limit the period of inspiration. When the signals from the pneumotaxic area are strong, the period of inspiration is short; when the signals are weak, the inspiratory period lasts longer. The apneustic area sends stimulatory signals to the medullary inspiratory neurons that tend to prolong inspiration. However, the pneumotaxic area normally overrides the apneustic area.

Lung Stretch Receptors

The lungs contain stretch receptors that increase their activity as the lungs expand during inspiration. These receptors send afferent impulses to the medullary inspiratory neurons to inhibit the activity of the inspiratory neurons and aid in terminating inspiration. This pathway, however, does not appear to be of great importance in the control of respiration during normal breathing, and the response is weak unless the lungs are distended to a considerable extent. Thus, the lung stretch receptors are believed to be mainly a protective mechanism that prevents the lungs from overfilling.

Medullary Expiratory Neurons

The medulla also contains neurons called expiratory neurons. During most normal, quiet breathing, the *medullary expiratory neurons* are not active, and expiration is a passive process that occurs when the inspiratory muscles relax. However, when the respiratory drive for increased pulmonary ventilation becomes greater than normal, the expiratory neurons are activated—apparently by the medullary inspiratory neurons. The expiratory neurons send impulses to the expiratory muscles to facilitate expiration.

In summary, a major factor initiating inspiration is the spontaneous discharge of the inspiratory neurons in the medulla. This activity ultimately stimulates the diaphragm and other muscles involved in inspiration. Expiration occurs when the medullary inspiratory neurons cease firing as a result of their own self-limitation and also as a result of inhibitory impulses from such sources as the pneumotaxic area in the pons and lung stretch receptors. Following expiration, the inspiratory neurons spontaneously discharge again, and the cycle repeats itself. In some cases, expiration is facilitated by the activation of neurons that cause the contraction of expiratory muscles.

Control of Rate and Depth of Breathing

The rate and depth of respiratory movements are important because of their influence on alveolar ventilation. The rate of respiration is determined by the amount of time between bursts of nerve impulses to the respiratory muscles. The more often the respiratory muscles are activated, the more rapid will be the rate of respiration. The depth of respiration is determined both by the number of motor units of the respiratory muscles that are active and by the frequency of nerve impulses to the muscle cells of each motor unit. The more motor units that are active, the greater will be the expansion of the thoracic cavity and the more the lungs will expand and fill with air.

Breathing supplies the body with oxygen and eliminates carbon dioxide. Moreover, changes in ventilation can influence the plasma concentration of hydrogen ions because carbon dioxide can, as described earlier, undergo the following reaction:

$$CO_2 + H_2O \rightleftharpoons H_2CO_3 \rightleftharpoons H^+ + HCO_3^-$$

Thus, a change in ventilation that increases or decreases the partial pressure of carbon dioxide in the plasma is likely to affect the plasma hydrogen ion concentration. As a result, it seems reasonable to consider oxygen, carbon dioxide, and hydrogen ions as possible regulators of respiratory activity.

Arterial P_{O_2}

If the partial pressure of oxygen in the arterial blood is reduced below normal while the P_{CO_2} and pH are held constant, an increase in alveolar ventilation occurs (Figure 22.20). However, within the normal range of arterial partial pressures of oxygen (the normal arterial P_{O_2} is about 95 mm Hg), the effect of oxygen on alveolar ventilation is relatively slight. It is not until low partial pressures of oxygen are present that a major stimulatory effect of oxygen on alveolar ventilation is evident. This response is reasonable because hemoglobin remains almost completely saturated with oxygen until there is a considerable drop in the P_{O_2} (see Figure 22.14). Thus, the transport of oxygen to the tissues is not substantially diminished until the arterial P_{O_2} falls quite low.

The receptors sensitive to the partial pressure of oxygen are located at the bifurcations of the common carotid arteries, as well as at the arch of the aorta, in structures called the *carotid* and *aortic bodies*. Nerve impulses from these receptors are ultimately transmitted to the respiratory neurons to stimulate respiration. The receptors of the carotid and aortic bodies are sensitive to the P_{O_2} within their local tissue environment. However, the carotid and aortic bodies receive such an abundant blood supply that their oxygen needs are normally met without creating a significant difference between the P_{O_2} in their local environment and the P_{O_2} in the arterial blood. Thus, in essence, the receptors of the carotid and aortic bodies respond to the arterial P_{O_2}. A decrease in the P_{O_2} increases the discharge rates of the receptors, and the nerve impulses that are generated stimulate respiration.

Despite the presence of these receptors, conditions of moderate anemia generally do not alter ventilation even though the total oxygen content of the blood may be below normal. In anemia, the reduced oxygen content of the blood is due to a reduced number of functional hemoglobin molecules, and the arterial P_{O_2}, which is determined solely by the amount of dissolved oxygen, is not altered. Thus, as long as the carotid and aortic bodies receive a sufficient supply of oxygen, the P_{O_2} in their local tissue environment remains essentially normal, and the discharge rates of the receptors are not significantly altered. However, in certain conditions, such as severely low blood pressure, the oxygen supply to the carotid and aortic bodies is so deficient that the utilization of oxygen by the cells causes the P_{O_2} in the local tissue environment to fall below that in the arterial blood. In such cases, the receptors increase their rates of discharge even though the arterial P_{O_2} may still be normal.

Arterial P_{CO_2}

The partial pressure of carbon dioxide in the plasma is a major regulator of alveolar ventilation. An elevation of only 5 mm Hg in the arterial P_{CO_2}, with the P_{O_2} and pH held constant, increases ventilation by 100% (Figure 22.20). Conversely, a lower-than-normal arterial P_{CO_2} inhibits respiration.

The respiratory effects of the arterial P_{CO_2} are due to associated changes in hydrogen ion concentrations, particularly in the cerebrospinal fluid and the interstitial fluid of the brain. Carbon dioxide can readily diffuse from the plasma to the cerebrospinal fluid and the interstitial fluid of the brain, where it combines with water to form carbonic acid. The carbonic acid can dissociate into hydrogen ions and bicarbonate ions, thus increasing the hydrogen ion concentration of the cerebrospinal fluid and the interstitial fluid of the brain. The increase in hydrogen ion concentration stimulates respiration by acting

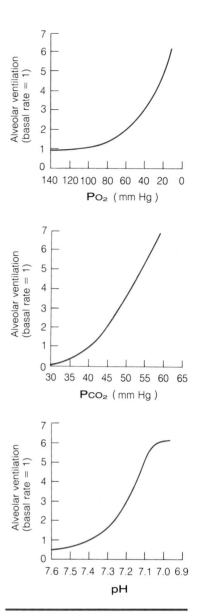

Figure 22.20
Effects of changing P_{O_2}, P_{CO_2}, and pH on alveolar ventilation in normal, healthy individuals when only one factor at a time changes and the others are held constant at normal levels.

F22.20

on chemosensitive cells in the medulla (central chemoreceptors) that have synaptic input to the respiratory neurons.

The carotid and aortic bodies are also sensitive to the arterial Pco_2, apparently because of the influence of carbon dioxide on the hydrogen ion concentration within their local tissue environment. However, the carotid and aortic bodies are much less important than the central chemoreceptors in mediating the respiratory effects of the arterial Pco_2.

Arterial pH

F22.20   The hydrogen ion concentration of the plasma can also affect respiration. Increasing the arterial concentration of hydrogen ions—that is, decreasing the arterial pH—with the Po_2 and Pco_2 held constant, substantially increases alveolar ventilation (Figure 22.20). Hydrogen ions can stimulate both the central chemoreceptors and the carotid and aortic bodies, but there is disagreement about which of these pathways is most important in mediating the respiratory effects of the arterial pH. In any case, compared to carbon dioxide, hydrogen ions diffuse rather poorly from the blood to the cerebrospinal fluid, and the arterial Pco_2 is generally more important than the arterial pH in normal respiratory regulation.

Interaction of Different Factors in the Control of Respiration

The various factors that influence respiration interact with one another to regulate breathing. For example, if the arterial Po_2 is decreased without keeping the Pco_2 or the pH constant, any increase in respiration due to the stimulatory effect of the low Po_2 can cause additional carbon dioxide to be expelled. The expulsion of additional carbon dioxide can lower the arterial Pco_2 and decrease the arterial concentration of hydrogen ions. Thus, the stimulatory effects of the decreased arterial Po_2 may be quickly counterbalanced by a lowered Pco_2 and a decreased concentration of hydrogen ions, both of which act to inhibit ventilation. At any instant, therefore, respiration is controlled by multiple factors, and the sensitivity to some factors, such as oxygen, may not be as great as the sensitivity to others, such as carbon dioxide.

EFFECTS OF EXERCISE ON THE RESPIRATORY SYSTEM

The muscular activity associated with exercise rapidly consumes oxygen and produces carbon dioxide. Thus, during exercise the body requires more oxygen and must eliminate more carbon dioxide than when at rest.

Increased Pulmonary Ventilation

During exercise both the rate and depth of breathing increase. During heavy exercise, the volume of air that reaches the alveoli may increase to as much as 20 times the resting volume.

The precise changes responsible for increased respiration during exercise are still not clearly understood, but several factors are thought to be involved. It is believed that when brain areas controlling movement transmit signals to exercising muscles, signals are also transmitted to the respiratory neurons to stimulate respiration. Consequently, increases in respiration occur even before muscular activity causes substantial changes in blood Po_2, Pco_2, or pH levels. Neural signals from joints and muscles involved in physical movements may also contribute to the stimulation of respiration during exercise.

If the neural mechanisms do not correctly adjust respiration to meet body requirements during exercise, the levels of chemical factors such as carbon dioxide may change, and these factors can then contribute to the overall control of respiration. The mechanisms that control respiration during exercise are quite effective, and in all but very heavy exercise the arterial Po_2, Pco_2, and pH remain almost exactly normal.

Increased Delivery of Oxygen to the Tissues

Both at rest and during exercise, the blood that leaves the lungs through the pulmonary veins is approximately 97% saturated with oxygen, and the factor that seems to be most important in determining how much oxygen can be delivered to the tissues during exercise is not the pulmonary ventilation, but the amount of blood the heart is capable of pumping. If a person can increase his or her cardiac output through training, the pulmonary ventilation appears to be capable of increasing enough so that the extra volume of blood is maximally saturated with oxygen.

CONDITIONS OF CLINICAL SIGNIFICANCE

The Respiratory System

Common Cold

The *common cold*, or *acute coryza*, is an inflammation of the mucous membrane of the upper respiratory tract that is familiar to most people. The initial inflammation is caused by various viruses and is often followed by bacterial infections of the sinuses, ears, or bronchi. When inflamed, the mucous membrane becomes engorged and swollen, causing discomfort and difficulty in breathing. Later the mucous membrane discharges a watery fluid that makes its presence known in the form of a "runny nose." Such discharges from the mucous membranes that line the paranasal sinuses can irritate the larynx and trachea, producing a cough.

Bronchial Asthma

Bronchial asthma is an allergic (antigen–antibody) response to foreign substances that generally enter the body by way of either air breathed or food eaten. It is characterized by episodes of wheezing and difficult breathing. In response to the foreign substances, the mucous membranes of the respiratory system secrete excessive amounts of mucus, and the smooth muscles that surround the smaller bronchi and bronchioles go into spasms. This narrows the passageways, making it difficult for air to move into or out of the alveoli.

Bronchitis

Bronchitis is an acute or chronic inflammation of the bronchial tree. It is caused by bacterial infection or by irritants in the inhaled air (such as smoke or chemicals). The mucous membranes of the respiratory system produce a sticky secretion that inhibits the normal protective function of the macrophages of the respiratory tract and hinders the self-cleaning actions of the cilia of the cells that line the bronchi. As the secretions accumulate within the bronchi, they are removed by coughing, which is annoying but serves the useful purpose of helping to keep the lungs clear.

Tuberculosis

Tuberculosis, an infection caused by the tubercle bacillus (*Mycobacterium tuberculosis*), can affect many parts of the body. However, because the bacterium most commonly enters the body by inhalation, tuberculosis of the lungs is the most prevalent form. Even if not inhaled, the bacilli can enter the lymphatic system or the bloodstream and thereby reach the lungs. When the bacilli reach the lungs, the lung tissue reacts by forming small clumps (tubercles) around the bacilli. Many of the bacilli are engulfed by phagocytes, and fibrous walls form around them. If the bacilli are not successfully walled off, the lung tissue is destroyed and the site of infection spreads. The process may continue until both lungs are extensively destroyed. Even if such a massive involvement of the lungs does not occur, the fibrosis in the affected portions interferes with the diffusion of gases and causes the lungs to lose their elasticity, thereby reducing the vital capacity.

Emphysema

Emphysema is a condition that develops slowly as a secondary response to other respiratory problems, such as chronic bronchitis and tuberculosis, or to environmental irritants, such as cigarette smoke and industrial pollutants. In persons suffering from emphysema, the alveoli become overdistended and the walls of the alveoli break down and are often replaced by fibrous tissue. This greatly reduces the surface area across which gaseous exchange can occur. As a consequence, the partial pressure of oxygen in the blood is lowered, and even mild exercise increases the breathing rate. In addition, the elastic tissues of the overexpanded lungs are reduced, making expiration difficult. A reduced expiratory volume is an early symptom of emphysema.

There are, therefore, two basic problems facing the victim of emphysema: (1) the lungs are "fixed" in inspiration; and (2) the respiratory surfaces of the lungs have deteriorated so much that they are no longer adequate to accomplish normal gas exchange. Unfortunately, the disease is progressive and irreversible.

Pneumonia

The inflammation of *pneumonia* causes a fibrinous exudate to be produced within the alveoli. The lung, or a part of it, becomes solid and airless, which makes it very difficult for gaseous exchange to occur within the alveoli. Most cases of pneumonia are probably caused by one of several viruses. Another common cause is the pneumococcus bacterium. However, pneumonia can also be caused by the inhalation of foods or other foreign bodies that obstruct a bronchus. This obstruction can lead to collapse of the lung, accumulation of fluids within the lung, and subsequent infection.

Pleurisy

Pleurisy is an inflammation of the pleural membranes that surround the lungs. The pleura are most commonly infected with the pneumococcus, the streptococcus, or the tubercle bacillus. In the early stages, the inflamed pleural membranes are "dry" and are covered with fibrous material. This condition causes pain during breathing. Adhesions between the layers of the pleura may develop as a result of pleurisy, and in severe cases, surgery may be necessary to remove them. In the later stages of pleurisy there is often excessive secretion of pleural fluid into the pleural cavity.

Effects of Aging

Tissue changes that occur in the respiratory system with increased age cause the chest wall to become more rigid and the lungs to become less elastic. Thus, although the total lung volume does not change with age, the rigidity of the chest wall and the loss of elasticity in the lungs results in a diminished ventilating capacity. Because of these changes, the vital capacity decreases in males from about 4700 ml at age 20 to about 4000 ml at age 70. Accompanying these changes is a decrease in the arterial Po_2, which is quite pronounced when a person is lying in a supine position, where breathing is more difficult. For this reason elderly people tend to become hypoxic during sleep, and they are often more comfortable if supported by several pillows. Carbon dioxide diffuses through tissues much more rapidly than oxygen; therefore, in contrast to the Po_2, the Pco_2 of the arterial blood is not affected much by age.

With increasing age, there is a decrease in the phagocytic activity of macrophages and in the activity of the cilia of the epithelial linings of the respiratory tract. This decreased activity makes the cleaning of the respiratory tract lining less efficient. The rigidity of the chest wall, the loss of elasticity in the lungs, and the diminished phagocytic activity and ciliary action all make elderly people more susceptible to pneumonia and other respiratory infections.

CLINICAL CORRELATION
Pulmonary Thromboembolism

CASE REPORT . . . .

THE PATIENT   A 45-year-old woman

PRINCIPAL COMPLAINT   Sudden onset of severe dyspnea (difficult breathing)

HISTORY   The patient had been healthy until four days previously when she was admitted to the hospital for treatment of injuries suffered in an automobile accident. Her left femur and ulna were fractured, and she had extensive intramuscular hemorrhage and other less serious cuts and bruises. In addition to medication for pain, she was given low prophylactic doses of an oral anticoagulant agent. Her leg and arm were in casts, but otherwise she recovered uneventfully until she developed a slight fever during the third night. The next day she complained that she could not breathe.

CLINICAL EXAMINATION   The patient's blood pressure was 135/90 mm Hg (normal: 90–140/60–90 mm Hg). Her heart rate was 110 beats/min (normal: 60–100 beats/min). Some rales (abnormal rattling or scraping sounds) and localized wheezes were heard over the lower lung field. The arterial Po_2 was 53 mm Hg (normal: >75 mm Hg), the arterial Pco_2 was 35 mm Hg (normal: 35–45 mm Hg), and the arterial pH was 7.46 (normal: 7.35–7.45). Spirometric tests showed that her minute respiratory volume was 9100 ml/min, and her alveolar ventilation was 4100 ml/min, or 46% of the minute respiratory volume (predicted: 60–70%). Her physiological dead space was 5000 ml/min, or 54% of her minute respiratory volume (predicted 30–40%). Alveolar Po_2 was calculated to be 100 mm Hg (with the patient breathing room air), and the alveolar-arterial Po_2 difference (that is, the alveolar-arterial oxygen gradient) was 47 mm Hg. Pulmonary thromboembolism was suspected because it is often associated with fractures or other injuries of the lower extremities.

Because of the severity and potentially lethal nature of the suspected disease, further tests were ordered. The patient was transferred to the cardiac catheterization laboratory, where a catheter was placed into the pulmonary trunk. Cardiac output was determined to be 3.8 liters/min (predicted 4.8 liters/min), and pulmonary arterial pressure was 60/45 mm Hg (normal: 12–28/3–13 mm Hg). The extent of vascular occlusion was determined by pulmonary angiography. A radioopaque material was injected into the pulmonary trunk, and the distribution of the material in the circulation of the lung was detected by X rays. This procedure showed decreased filling of two lobes of the right lung.

COMMENT   A major respiratory consequence of pulmonary emboli is an increase in the physiological dead space. The blood flow to alveoli in certain sections of the lungs may be completely or partially obstructed by clots, and ventilation of these alveoli is largely wasted. Because the patient frequently hyperventilates, arterial P_{CO_2} often is within normal limits, but alveolar P_{CO_2} is decreased (that is, closer to the P_{CO_2} of the inspired air). A decreased blood flow to a particular lung area also may decrease the availability of substrates for the synthesis of pulmonary surfactant, which causes alveolar instability and collapse.

The hemodynamic consequences of pulmonary emboli depend largely on the degree of vascular obstruction. Moderate obstruction may cause only an increased heart rate, and more serious problems may not arise if the patient remains quiet and the demand for oxygen does not increase. In this patient, the vascular obstruction was severe enough to produce arterial hypoxemia (a below-normal arterial oxygen tension), presumably due to a decreased functional area available for gas exchange at the lungs.

TREATMENT   Therapy was directed toward dissolving the blood clots and restoring the pulmonary circulation. Vigorous anticoagulant treatment was begun. Although anticoagulants do not cause the clots to dissolve, by preventing expansion of the clots they enable the fibrinolytic system of the body to remove them. The patient initially received 10,000 units of heparin intravenously and then continuous infusion of heparin at the rate of 1000 units every hour for seven days. She was switched to an oral anticoagulant several days before her discharge from the hospital, and this was continued for three months.

OUTCOME   The remainder of her recovery was uneventful.

STUDY OUTLINE

EMBRYONIC DEVELOPMENT OF THE RESPIRATORY SYSTEM   p. 647

LARYNGOTRACHEAL BUD   Diverticulum from ventral endoderm of digestive tract.

TRACHEA   From proximal part of laryngotracheal bud.

BRONCHI   From bifurcation of distal end of laryngotracheal bud.

BRONCHIOLES AND ALVEOLI   From bronchi.

RESPIRATORY EPITHELIUM   Derived from endoderm; lines entire respiratory tract.

ANATOMY OF THE RESPIRATORY SYSTEM
Respiratory system consists of nose, pharynx, larynx, trachea, bronchi, and lungs.   pp. 648–659

NOSE AND NASAL CAVITY   Air passes through external nares to vestibule of nose and to nasal cavity.

SEPTUM   Formed from septal cartilage, vomer bone, and perpendicular plate of ethmoid bone.

ROOF   Formed by cribriform plate of ethmoid.

LATERAL WALLS   Formed by superior, middle, and inferior conchae.

FLOOR   Formed by hard and soft palates.

VESTIBULE   Stratified squamous epithelial lining.

NASAL CAVITY, SINUSES   Mucous membrane of pseudostratified ciliated columnar epithelium. Blood supply warms air and saturates it with water; cilia trap small particles; internal nares connect nasal cavity with nasopharynx.

PHARYNX   Muscular structure lined with mucous membrane.

NASOPHARYNX   Receives internal nares and auditory tubes; contains tubal and pharyngeal tonsils.

OROPHARYNX   Food and air passageway; contains palatine and lingual tonsils; communicates with oral cavity through fauces.

LARYNGOPHARYNX   Food and air passageway; between oropharynx and esophagus and larynx.

LARYNX   Air conduction between laryngopharynx and lungs.
1. Framework of larynx consists of nine cartilages.
2. Mucous membranes of larynx are stratified squamous and pseudostratified ciliated columnar epithelium; form pairs of ventricular folds and vocal folds.
3. Glottis is opening between vocal folds.

TRACHEA (WINDPIPE)
1. Kept open by C-shaped hyaline cartilages.
2. Ciliated cells carry foreign particles away from lungs to pharynx.

BRONCHI, BRONCHIOLES, AND ALVEOLI
1. Primary, secondary, and tertiary bronchi→terminal and respiratory bronchioles→alveolar ducts→alveoli.
2. With branching, supportive cartilage gradually replaced by smooth muscle.

LUNGS
1. Three-lobed right lung; two-lobed left lung; each lobe further divided into bronchopulmonary segments.
2. Hilus—site where bronchi, blood vessels, lymphatics, and nerves pass into or out of lungs.
3. Mediastinum—space that separates lungs.

THE PLEURA
1. Visceral and parietal pleura—double-walled serous membrane sac that encloses each lung.
2. Pleural cavity between layers—filled with pleural fluid.
3. Pleurisy—inflammation of pleural membranes.

BLOOD SUPPLY OF THE LUNGS Pulmonary artery (oxygen-poor blood) supplies alveoli; bronchial arteries from descending aorta (oxygen-rich blood) supply bronchi.

THE RESPIRATORY MEMBRANE Thin membrane that separates alveolar air from blood—O_2 and CO_2 diffuse across membrane; consists of alveolar epithelium (and basal lamina) and capillary endothelium (and basal lamina); contains two types of cells: (1) *alveolar type I cells*—simple squamous cells that allow for gas exchange; (2) *alveolar type II cells*—rounded or cuboidal cells that secrete fluid.

MECHANICS OF BREATHING pp. 659–665

ATMOSPHERIC PRESSURE At sea level is 760 mm Hg.

PRESSURE RELATIONSHIPS IN THE THORACIC CAVITY
1. Elastic connective tissue of lungs and surface tension of fluid coating alveoli favor reduction in lung size, as does intrapleural pressure. These forces are balanced by intrapulmonary pressure, and lungs normally remain expanded and do not collapse.
2. If air enters pleural cavity, balance of forces maintaining lung expansion is upset and lung collapses.

VENTILATION OF THE LUNGS Air flows from high-pressure area to lower-pressure area; pressure within lungs is altered by altering volume of thoracic cavity.

INSPIRATION Volume of thoracic cavity increased by contraction of diaphragm and elevation of ribs. This increases lung volume, and intrapulmonary pressure drops below atmospheric pressure. Air then flows into lungs.

EXPIRATION When muscles involved in inspiration relax, volume of thoracic cavity—and lungs—decreases due to elastic recoil of respiratory structures such as lungs and chest wall. Decrease in lung volume raises intrapulmonary pressure above atmospheric pressure, and air flows out of lungs. During forced expiration, contraction of expiratory muscles further reduces volume of thoracic cavity.

FACTORS THAT INFLUENCE PULMONARY AIR-FLOW Volume of airflow per given time between atmosphere and alveoli is influenced by pressure that moves air through respiratory passageways and by resistance air encounters as it flows through passageways. Passageway diameter affects resistance; parasympathetic stimulation, histamine, and leukotrienes constrict smooth muscles of respiratory passageways and increase resistance; sympathetic stimulation and epinephrine relax smooth muscle of passageways and decrease resistance.

SURFACE TENSION AND PULMONARY SURFAC-TANT Pulmonary surfactant produced by alveolar cells lowers surface tension of fluid that coats exposed surfaces of alveoli, allowing lungs to be expanded with reasonable muscular effort.

LUNG VOLUMES AND CAPACITIES

RESIDUAL VOLUME About 1000–1200 ml of air that cannot be exhaled from lungs.

TIDAL VOLUME Volume of air moving in and out of lungs with each breath (about 500 ml during normal, quiet breathing).

INSPIRATORY RESERVE Extra volume of air that can be inspired in addition to normal, quiet tidal volume (about 2100–3000 ml).

INSPIRATORY CAPACITY Sum of normal, quiet tidal volume and inspiratory reserve volume.

EXPIRATORY RESERVE Extra volume of air that can be expired following normal, passive expiration of quiet tidal volume (about 800–1200 ml).

FUNCTIONAL RESIDUAL CAPACITY Sum of residual volume and expiratory reserve volume.

VITAL CAPACITY Sum of inspiratory reserve volume; normal, quiet tidal volume; and expiratory reserve volume (3400–4700 ml).

DEAD SPACE In some areas of respiratory system no gas exchange takes place; volume of anatomical dead space is about 150 ml.

MINUTE RESPIRATORY VOLUME Volume of air moved into respiratory passageways in one minute; equal to respiratory rate multiplied by volume of air entering respiratory passageways with each breath (tidal volume)

ALVEOLAR VENTILATION Volume of atmospheric air entering alveoli either per breath or in one minute that can participate in gas exchange between alveoli and blood.

MATCHING OF ALVEOLAR AIRFLOW AND BLOOD FLOW Local autoregulatory mechanisms are involved.
1. Low oxygen levels cause pulmonary vessels to constrict; high oxygen levels cause them to dilate.
2. Low carbon dioxide levels cause respiratory passageways to constrict; high carbon dioxide levels cause them to dilate.

TRANSPORT OF GASES BY THE BLOOD pp. 665–670

PARTIAL PRESSURE Oxygen, 21% of atmospheric gas mixture, contributes 21% of total atmospheric pressure.

GASES IN LIQUIDS Gases enter a liquid and dissolve in proportion to their individual gas pressures.

OXYGEN AND CARBON DIOXIDE EXCHANGE BETWEEN LUNGS, BLOOD, AND TISSUES

LUNG Partial pressure differences between O_2 and CO_2 in blood and in alveoli lead to net diffusion of O_2 into blood from alveoli and net diffusion of CO_2 into alveoli from blood.

TISSUES Net diffusion of O_2 from blood to tissues; net diffusion of CO_2 from tissues to blood.

OXYGEN TRANSPORT IN THE BLOOD Most is transported bound to iron of hemoglobin in erythrocytes. Binding of oxygen to hemoglobin is affected by oxygen partial pressure, hydrogen ion concentration, carbon dioxide partial pressure, temperature.

CARBON DIOXIDE TRANSPORT IN THE BLOOD
1. Dissolved in plasma.
2. Carbamino compounds (CO_2–protein complexes).
3. Bicarbonate ions.

CONDITIONS OF CLINICAL SIGNIFICANCE: PROBLEMS OF GAS TRANSPORT p. 670

CYANOSIS Bluish discoloration of skin and mucous membranes; occurs when there is significant concentration of reduced hemoglobin in arterial blood.

HYPOXIA Inadequate amounts of oxygen delivered to tissues.

CARBON MONOXIDE POISONING Carbon monoxide occupies oxygen-binding sites of hemoglobin.

CONTROL OF RESPIRATION
Inspiration occurs when nerve impulses activate diaphragm and intercostals; thorax expands; lungs fill with air. Expiration occurs when nerve impulses to inspiratory muscles diminish and respiratory structures such as lungs and chest wall recoil. During forced expiration, nerve impulses activate expiratory muscles. pp. 670–674

GENERATION OF RHYTHMICAL BREATHING MOVEMENTS

MEDULLARY INSPIRATORY NEURONS Believed to be inherently rhythmic.

NEURONS OF THE PONS

Apneustic Area Tends to prolong inspiration; normally overridden by pneumotaxic area.

Pneumotaxic Area Tends to limit period of inspiration.

LUNG STRETCH RECEPTORS May aid in terminating inspiration when lungs are considerably distended.

MEDULLARY EXPIRATORY NEURONS When respiratory drive for increased pulmonary ventilation becomes greater than normal, these neurons send impulses to expiratory muscles to facilitate expiration.

CONTROL OF RATE AND DEPTH OF BREATHING

ARTERIAL P_{O_2} Low P_{O_2} in arterial blood (with P_{CO_2} and pH constant) increases alveolar ventilation.

ARTERIAL P_{CO_2} Elevated arterial P_{CO_2} (with P_{O_2} and pH constant) increases ventilation.

ARTERIAL pH Increased arterial hydrogen ion concentration (with P_{O_2} and P_{CO_2} constant) increases ventilation.

INTERACTION OF DIFFERENT FACTORS IN THE CONTROL OF RESPIRATION At any instant, respiration is controlled by multiple factors and sensitivity to some factors, such as oxygen, may not be as great as sensitivity to other factors, such as carbon dioxide.

EFFECTS OF EXERCISE ON THE RESPIRATORY SYSTEM p. 674

INCREASED PULMONARY VENTILATION During exercise, both rate and depth of breathing increase.

INCREASED DELIVERY OF OXYGEN TO THE TISSUES Both at rest and during exercise, blood leaving lungs is approximately 97% saturated with oxygen; most important factor in determining how much oxygen is delivered to tissues during exercise seems to be not pulmonary ventilation but amount of blood heart is capable of pumping.

CONDITIONS OF CLINICAL SIGNIFICANCE: THE RESPIRATORY SYSTEM pp. 675–676

COMMON COLD Viral inflammation and sometimes subsequent bacterial infection of mucous membranes of upper respiratory tract.

BRONCHIAL ASTHMA Allergic response characterized by wheezing caused by excessive mucus and spasms of smooth muscles of bronchioles.

BRONCHITIS Acute or chronic inflammation of bronchial tree caused by bacterial infection or irritants.

TUBERCULOSIS Infection caused by *Mycobacterium tuberculosis*, which most commonly affects lungs.

EMPHYSEMA Progressive condition secondary to other respiratory problems; overdistended alveoli and reduced elastic tissue of lungs reduce expiratory volume.

PNEUMONIA Fibrinous exudate produced within alveoli, which causes part of lung to become solid and airless.

PLEURISY Inflammation of pleural membranes.

EFFECTS OF AGING
1. Chest wall rigidity.
2. Loss of lung elasticity.
3. Decreased ciliary activity of cells located on respiratory tract linings.
4. Decreased phagocytic activity, which increases susceptibility to infection.

SELF-QUIZ

1. In the embryo the respiratory epithelium develops from: (a) endoderm; (b) ectoderm; (c) mesoderm.

2. The nasal septum is formed from: (a) the vomer bone; (b) the ethmoid bone; (c) the septal cartilage; (d) all of these contribute to the nasal septum.

3. The floor of the nasal cavity is formed by: (a) the bony hard palate; (b) the membranous soft palate; (c) the inferior meatus; (d) both a and b.

4. The paranasal sinuses, like the nasal cavity, are lined with stratified squamous epithelium. True or False?

5. The pharynx communicates with the nasal cavity via the: (a) fauces; (b) glottis; (c) middle meatus; (d) internal nares.

6. Which of the following is located in the oropharynx? (a) pharyngeal tonsils; (b) palatine tonsils; (c) lingual tonsils; (d) both b and c; (e) all of these.

7. The largest of the unpaired cartilages of the larynx is the: (a) cricoid cartilage; (b) arytenoid cartilage; (c) thyroid cartilage; (d) cuneiform cartilage.

8. The vocal folds are attached to: (a) the thyroid cartilage; (b) the arytenoid cartilage; (c) the cricoid cartilage; (d) both a and b.

9. The trachea divides directly into: (a) tertiary bronchi; (b) primary bronchi; (c) respiratory bronchi; (d) terminal bronchi; (e) alveolar ducts.

10. Alveoli are associated with the: (a) trachea; (b) pharynx; (c) lungs; (d) nose.

11. The region where the bronchi and blood vessels enter and leave the lung is called the: (a) costal surface; (b) oblique fissure; (c) hilus; (d) cardiac notch.

12. Match the terms associated with the lung with the appropriate lettered descriptions.

Costal surface	(a) The space that separates the two lungs
Cardiac notch	
Oblique fissure	(b) A double-walled sac that encloses each lung
Mediastinum	
Pleura	(c) The portion of the pleura that adheres firmly to the lungs
Visceral pleura	
Parietal pleura	
Alveoli	(d) Lies against the ribs
	(e) Clusters of small, thin-walled air sacs
	(f) The portion of the pleura that lines the walls of the thoracic cavity
	(g) Divides each lung into a superior and inferior lobe
	(h) A concavity in the left lung

13. When air is moving into the lungs during inspiration: (a) the intrapulmonary pressure is higher than the atmospheric pressure; (b) the atmospheric pressure is higher than the intrapulmonary pressure; (c) the intrapleural pressure is higher than the atmospheric pressure.

14. During expiration: (a) the volume of the thoracic cavity increases; (b) the diaphragm contracts; (c) the intrapulmonary pressure is higher than the atmospheric pressure.

15. The volume of gas that remains in the lungs even after the most forceful expiration is the: (a) residual volume; (b) tidal volume; (c) anatomical dead space volume.

16. Which has the greatest volume? (a) expiratory reserve volume; (b) vital capacity; (c) inspiratory reserve volume.

17. Smooth muscles of respiratory system passageways are relaxed and the resistance to the flow of air through the passageways is decreased by: (a) parasympathetic nervous stimulation; (b) histamine; (c) epinephrine.

18. The minute respiratory volume is equal to the volume of air that enters the respiratory passages with each breath, multiplied by the: (a) respiratory rate; (b) alveolar ventilation; (c) anatomical dead space volume.

19. In general, the binding of oxygen to hemoglobin will be increased by: (a) an increase in the oxygen partial pressure; (b) a decrease in pH; (c) an increase in temperature.

20. Which of the following is a true statement about the transport of respiratory gases? (a) Most of the CO_2 is carried as bicarbonate ions within the blood. (b) CO_2 is never carried as bicarbonate ions. (c) On the average, 27% of CO_2 is carried as bicarbonate ions within the blood.

21. The activity of the apneustic area tends to shorten the period of inspiration. True or False?

22. In moderate anemia: (a) the carotid and aortic bodies generally provide a greatly increased stimulus to the respiratory center; (b) the amount of dissolved oxygen in the blood is abnormally low; (c) the reduced oxygen content of the blood is due to a reduced number of functional hemoglobin molecules.

23. An elevation of 5 mm Hg of arterial P_{CO_2}, with P_{O_2} and pH held constant, will: (a) inhibit ventilation; (b) increase ventilation; (c) have no effect on ventilation.

24. The respiratory rate increases in response to: (a) increased arterial P_{O_2}; (b) lowered arterial pH; (c) decreased arterial P_{CO_2}.

LEARNING OBJECTIVES

After completing this chapter, you should be able to:

1. List the principal components of the digestive system, and cite the chief functions of each.

2. Describe the microscopic and gross anatomy of the components of the digestive system.

3. List the types of teeth, according to shape, and cite the specific function of each type.

4. List the basic layers of the wall of the gastrointestinal tract.

5. Describe the modifications of the basic layers found in various regions of the gastrointestinal tract.

6. Describe the structure of the pancreas, liver, and gallbladder, and cite the functions of each.

7. Describe the factors involved in the control of chewing and swallowing.

8. Describe the factors that influence gastric motility.

9. Explain the movements that occur in the large intestine.

CHAPTER CONTENTS

EMBRYONIC DEVELOPMENT OF THE DIGESTIVE SYSTEM

ANATOMY OF THE DIGESTIVE SYSTEM

ACCESSORY DIGESTIVE ORGANS

MECHANICAL PROCESSES OF THE DIGESTIVE SYSTEM

CONDITIONS OF CLINICAL SIGNIFICANCE: THE DIGESTIVE SYSTEM

KEY TERMS AND DERIVATIVES

cecum (*cecum* = blind) the blind-end pouch at the beginning of the large intestine

chyme (*chyme* = juice) the semifluid contents of the stomach consisting of partially digested food and gastric secretions

esophagus (*eso* = carry; *phag* = eat) the muscular tube that carries food from the pharynx to the stomach

frenulum (*frenul* = restraint) the lingual frenulum is a membranous fold attaching the tongue to the floor of the mouth

gastric glands (*gastr* = stomach) minute glands in the stomach wall that secrete gastric juice

ingestion (*ingest* = put in; *ion* = the action of) the process of taking substances into the digestive tract

peristalsis (*peri* = around; *stal* = constrict) the progressive wave of contraction seen in body tubes (like the digestive tract)

peritoneum (*peri* = around; *tone* = stretch) the membrane lining of the interior of the abdominal cavity

pyloric region (*pylor* = gatekeeper) the final portion of the stomach preceding the duodenum

The Digestive System:
Anatomy and Mechanical Processes

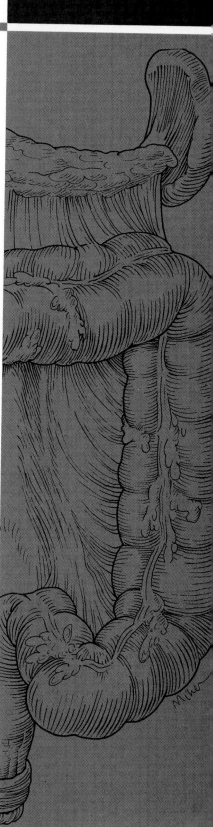

Every cell in the body requires a constant source of energy in order to perform its particular functions, whether these functions be contraction, secretion, synthesis, or any other. Ingested food provides the basic materials from which this energy is produced and new molecules are synthesized. Most food, however, cannot enter the bloodstream and be used by the cells of the body until it is broken down into simpler molecules. The *digestive system* alters the ingested food by mechanical and chemical processes so that it can ultimately cross the wall of the gastrointestinal tract and enter the blood vascular and lymphatic systems for distribution to cells throughout the body. After entering the cells, digested food molecules may be used in the production of energy to support body activity, or they may be incorporated into body structure.

The digestive system consists of a tube—called the **gastrointestinal tract, or alimentary canal**—extending from the mouth to the anus (Figure 23.1). As long as food remains in the gastrointestinal tract, it is technically still outside the body. To "enter" the body it must cross the epithelium that lines the wall of the digestive tract. Emptying into the digestive tube are the secretions of the salivary glands, gastric glands, intestinal glands, liver, and pancreas, all of which assist in the digestion of food. Although the gastrointestinal tract is a continuous tube, it is divided into specialized regions that each perform specific functions in the digestion of food. These regions include the mouth, pharynx, esophagus, stomach, small intestine, and large intestine.

The activities of the digestive system can be divided into six basic processes:

1. Ingestion of food into the mouth
2. Movement of food along the digestive tract
3. Mechanical preparation of food for digestion
4. Chemical digestion of food
5. Absorption of digested food into the circulatory and lymphatic systems
6. Elimination of indigestible substances and waste products from the body by defecation

In this chapter we are primarily interested in the anatomy of the digestive system and in the mechanical processes involved in preparing food for digestion. The chemical digestion of food and the absorption of digested food into the circulatory and lymphatic systems are discussed in Chapter 24.

EMBRYONIC DEVELOPMENT OF THE DIGESTIVE SYSTEM

Early in development, the embryo is a hollow cylinder covered on the outside with ectoderm. Its internal cavity, which is lined with endoderm, is the developing digestive tract (Figure 23.2). The portion of the tract that extends anteri-

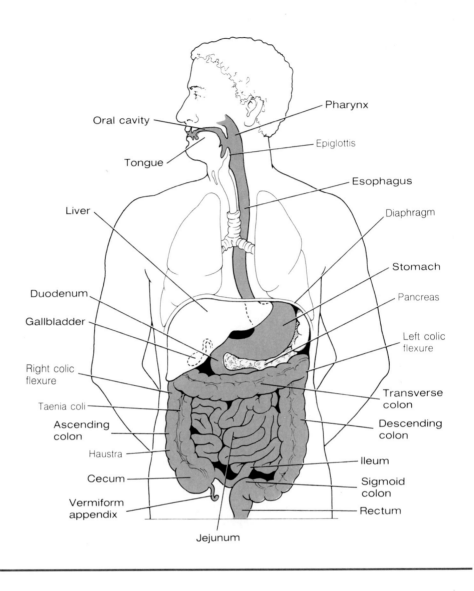

Figure 23.1
Structures of the
digestive system.

orly into the head region is the **foregut;** the portion that extends posteriorly is
the **hindgut;** and the central region of the tract is the **midgut.** Until the fifth
week of development, the midgut opens into a pouch called the yolk sac.
After the fifth week, the attached portion of the yolk sac constricts, sealing the
midgut.

Early in development, the anterior region of the foregut expands to form
the pouches of the pharynx. With continued growth, the foregut contacts the
surface ectoderm at the point where the ectoderm has formed a depression
sto-mo-dee'-um   called the **stomodeum** (the future mouth). The **oral membrane,** which separates the foregut from the stomodeum, breaks through during the fourth
week of development, and the foregut is then continuous with the outside of
the embryo through the mouth. In a similar manner, the hindgut contacts the
surface ectoderm at a depression called the **proctodeum** (the future anus).
With the rupture of the **cloacal membrane,** which separates the hindgut from
the proctodeum, the digestive tract forms a continuous tube from the mouth
to the anus.

With further development, hollow buds from the endoderm form at various places along the foregut. These buds will grow into the mesoderm that
surrounds the digestive tract and give rise to the thyroid gland, parathyroid
glands, salivary glands, liver, gallbladder, and pancreas. The thyroid and
parathyroid glands later lose their connections with the digestive tract and
function as endocrine glands. The salivary glands, liver, gallbladder, and

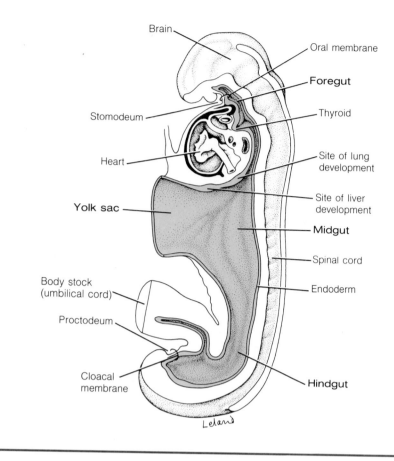

Brain

Oral membrane

Foregut

Thyroid

Stomodeum

Site of lung development

Heart

Site of liver development

Yolk sac

Midgut

Spinal cord

Body stock (umbilical cord)

Endoderm

Proctodeum

Cloacal membrane

Hindgut

Leland

Figure 23.2
Sagittal section of an embryo (approximately 22 days old) showing the future foregut and hindgut forming. Note that the midgut remains open to the yolk sac.

pancreas, however, retain their connections with the digestive tube by means of ducts. Consequently, they serve as accessory glands to the digestive system.

ANATOMY OF THE DIGESTIVE SYSTEM

It is important to keep in mind that the entire digestive tract is lined with **mucous membrane.** This membrane protects the underlying tissues and, at the same time, allows for the absorption of digested food in the intestine.

In order to function in absorption, the membrane must be thin and moist. The secretion of *mucus* by cells of the mucous membrane not only keeps the membrane moist, but because it is viscous, the mucus also serves as a protective mechanism. Thus, the thin membrane that lines the absorptive regions of the digestive tract provides adequate protection as long as it is coated with mucus.

Mouth

The **mouth** is the first part of the digestive tract. It extends from the lips to the oropharynx (Figure 23.3). The outer surfaces of the lips are covered with skin that has a relatively transparent surface layer of cells, allowing the underlying blood capillaries to show through. For this reason the lips appear red. Since the surface layer of the lips is not keratinized (horny), evaporation occurs from the lips. Consequently, the lips must be moistened frequently to prevent them from becoming dry and cracked. The inside surfaces of the lips and the rest of the mouth are lined with a mucous membrane that has a stratified squamous surface layer of nonkeratinized cells. No food is absorbed in the

F23.3

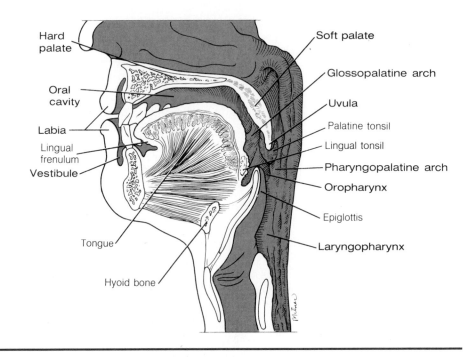

Figure 23.3
Sagittal section of the oral and pharyngeal cavities.

mouth, so the cells lining it need not be capable of absorption; and the stratified surface layer affords an extra degree of protection from abrasive food particles.

For descriptive purposes, the mouth can be divided into the *oral cavity*, which is the large space internal to the teeth that houses the tongue, and the *oral vestibule*, which is the small space separating the lips and the cheeks from the teeth and gums. The lips and cheeks aid in moving food between the upper and lower teeth during *mastication* (chewing). They also aid in speech.

The roof of the mouth is formed anteriorly by the **hard palate** and posteriorly by the **soft palate.** The hard palate is formed by the maxillary and palatine bones. The soft palate, which extends posteriorly from the hard palate, separates the oral cavity from the nasopharynx. It is composed primarily of muscles. The soft palate is pushed upward during swallowing, blocking the entrance from the pharynx to the nasal cavity and thus serving to prevent food and drink from entering the nasal cavity. The **uvula** is a small muscular flap that hangs down from the posterior margin of the soft palate. It serves as a frictional pad to prevent the soft palate from being pushed into the nasal cavity during swallowing. The soft palate is attached laterally to the tongue by the **glossopalatine arches;** it is connected to the wall of the oropharynx by the **pharyngopalatine arches.** The palatine tonsils, which are composed mainly of lymphoid tissue, are located in the fossae between the two arches, a region called the **fauces.**

Tongue

F23.3   The **tongue** forms the floor of the mouth (Figure 23.3). It is composed of interwoven bundles of skeletal muscles covered with mucous membrane. The **extrinsic muscles** of the tongue originate from the hyoid bone, the mandible, and the styloid processes of the temporal bones. These muscles protrude the tongue, retract it, and move it sideways. The **intrinsic muscles** originate and insert in the tongue. Their fibers run in various directions and modify the shape of the tongue in several different ways. Because of these two sets of muscles, the tongue is quite versatile and is used in manipulating food, swallowing, and speaking.

The mucous membrane covering the dorsum of the tongue is modified by
F23.4   the presence of numerous small projections called **papillae** (Figure 23.4). The

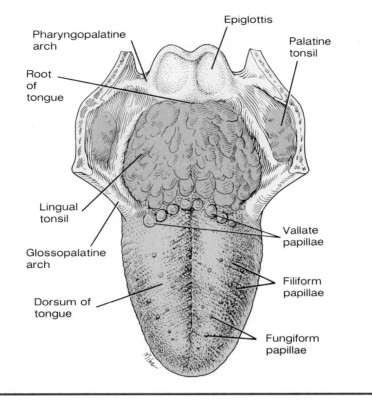

Pharyngopalatine arch

Epiglottis

Palatine tonsil

Root of tongue

Lingual tonsil

Glossopalatine arch

Dorsum of tongue

Vallate papillae

Filiform papillae

Fungiform papillae

Figure 23.4
Dorsal surface of the tongue.

papillae vary in shape; each type tends to be more common in certain regions of the tongue surface. The **filiform papillae,** which are small cones, are distributed in V-shaped rows over the entire dorsal surface of the tongue. Interspersed among the filiform papillae are flattened, mushroom-shaped **fungiform papillae** (Figure 23.5a). About a dozen large **vallate papillae** form an inverted V toward the back of the tongue (Figure 23.5b). **Taste buds** are found on the fungiform and vallate papillae. The filiform papillae make the surface of the tongue rough, allowing it to move food around during chewing. The posterior dorsal surface of the tongue contains an aggregate of small lymph nodules called the **lingual tonsil.**

F23.5a
F23.5b

The tongue is connected ventrally to the floor of the mouth by a fold of mucous membrane called the **lingual frenulum.** If the lingual frenulum is so short as to hinder the movement of the tongue, it interferes with speech. Such a condition, which can be corrected surgically, is called "tongue-tied."

Teeth

The teeth extend into the mouth from **alveoli** (sockets) located along the alveolar processes of the mandible and maxillary bones. **Gums** (*gingivae*), composed of stratified squamous epithelium and dense fibrous connective tissue, cover the alveolar processes. The sockets are lined with a fibrous membrane called the **periodontal membrane** (Figure 23.6). Destruction of the periodontal membrane by bacteria can lead to infection within a socket and loss of its tooth.

F23.6

The portion of each tooth that extends from the gum into the mouth is called the **crown.** One or more **roots** anchor the tooth to the alveolus. Between the crown and the root is a slightly constricted **neck.**

Each tooth is composed mostly of a hard, calcified substance called **dentin.** The dentin of the crown is covered by **enamel,** which is even harder than dentin. A bonelike substance called **cementum** covers the dentin of the roots and anchors the tooth to the periodontal membrane lining the alveoli. The central region of the tooth contains the **pulp cavity,** in which are found blood vessels, nerves, and a connective tissue called **pulp.** The pulp cavity extends

(a)

(b)

Figure 23.5
Papillae of the tongue.
(a) Fungiform papilla
(center) surrounded by
filiform papillae (×195).
(b) Vallate papilla (×606).
(From *Tissues and Organs:
A Text-Atlas of Scanning
Electron Microscopy* by
Richard G. Kessel and
Randy H. Kardon. W.H.
Freeman and Company.
Copyright © 1979.)

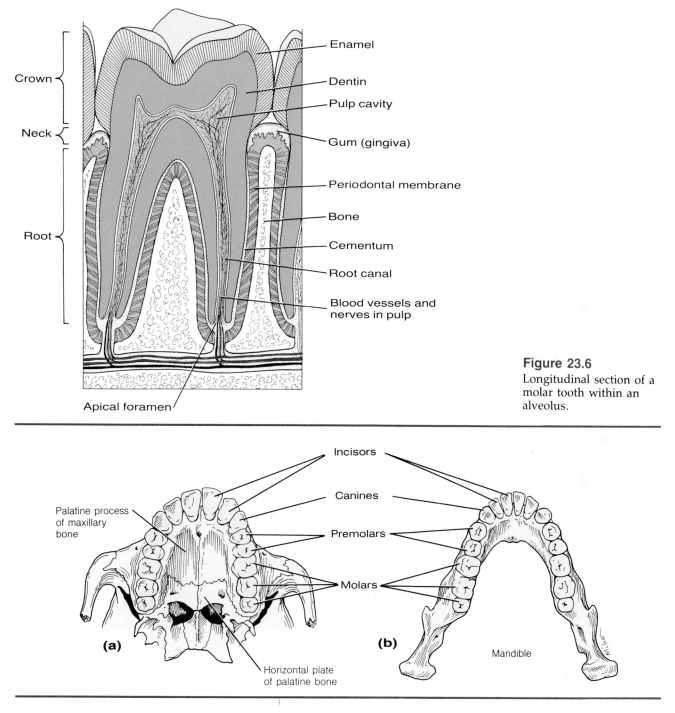

Crown

Neck

Root

Enamel

Dentin

Pulp cavity

Gum (gingiva)

Periodontal membrane

Bone

Cementum

Root canal

Blood vessels and nerves in pulp

Apical foramen

Figure 23.6
Longitudinal section of a molar tooth within an alveolus.

Incisors

Canines

Premolars

Molars

Palatine process of maxillary bone

Horizontal plate of palatine bone

(a)

(b)

Mandible

Figure 23.7
Permanent teeth in
(a) the upper jaw and
(b) the lower jaw.

down into each root as a **root canal.** At the end of each root canal is an **apical foramen,** through which blood vessels and nerves enter the pulp cavity.

Certain bacteria found in the mouth can produce enzymes and acids capable of breaking down tooth enamel. The points of enamel destruction are called *dental caries,* or *cavities.* Once the enamel has been penetrated, the dentin may also be destroyed by the enzymes and acids. If the decay reaches the pulp, it can irritate nerves in the pulp cavity, causing a toothache. The application of fluoride, either directly to the teeth or in drinking water, seems to make the enamel more resistant to bacterial enzymes and acids.

There are four types of teeth, named according to function or shape (Figure 23.7). Each type of tooth performs a specific function in preparing food for digestion. The front chisel-shaped teeth, called **incisors,** are especially

F23.7

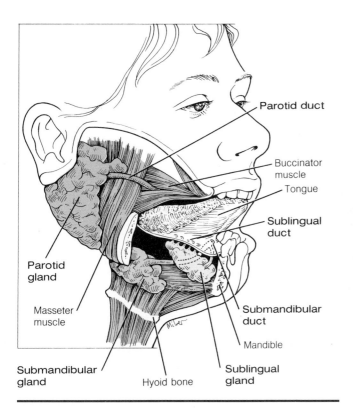

Parotid duct

Buccinator muscle

Tongue

Sublingual duct

Parotid gland

Submandibular duct

Masseter muscle

Mandible

Submandibular gland

Sublingual gland

Hyoid bone

Figure 23.8
The major salivary glands. The right side of the mandible has been removed to expose the submandibular and sublingual glands.

Table 23.1 Numbers of Specific Types of Teeth in Each Jaw

	Deciduous	Permanent
Incisors	4	4
Canines (cuspids)	2	2
Premolars (bicuspids)	0	4
Molars	4	6
TOTAL IN BOTH JAWS	20	32

adapted for cutting. On each side of the incisors are conical **canines** (*cuspids*), which serve to tear food. The **premolars** (*bicuspids*) and **molars** have broad crowns with rounded cusps, which aid in crushing and grinding food.

Normally, two sets of teeth develop during a person's lifetime. There are 20 **deciduous (milk) teeth,** which erupt through the gums at regular intervals, beginning with the incisors at about 6 months of age. All the deciduous teeth are usually present by age 2½ years. There are 32 teeth in the **permanent set.** The permanent teeth begin to replace the deciduous teeth at about 6 years of age and continue to do so until about age 17, when all of the temporary teeth generally have been replaced. The **third molars** (*wisdom teeth*) are the last to erupt, generally between the ages of 17 and 25. It is not unusual for the wisdom teeth to fail to erupt or to become wedged (impacted) in the jaw so that they are unable to erupt. Table 23.1 summarizes the numbers of each type of tooth present in each jaw, in both the deciduous and permanent sets.

In spite of extensive research, it is still uncertain precisely what causes teeth to emerge through the gums and to be replaced by other teeth in a fairly regular cycle. The development of a permanent tooth deep in the jaw puts pressure on the roots of the overlying deciduous tooth, causing the roots to disintegrate and the deciduous tooth to be shed. But this action has been shown not to be the complete answer. The presence of collagen and increased vascular pressure in the tissues surrounding the alveoli, and the regulation of hormonal levels by genes have all been suggested as being involved in the cyclic development and replacement of teeth.

Table 23.1

Salivary Glands

About 1000–1500 ml of saliva is secreted daily into the mouth. Some of this is produced by many small **buccal glands** located throughout the mucous membrane of the oral cavity. These glands secrete continuously and keep the mucous membranes moist. Most of the saliva, however, is secreted by three pairs of **salivary glands,** which are activated primarily by stimuli associated with food (Figure 23.8). These large salivary glands are of the compound tubuloalveolar type (see Chapter 4), being composed of blind-end tubules and alveoli. The largest of the paired salivary glands, the **parotid glands,** are located below and anterior to the ear, on the posterior surface of the masseter muscle.

buck'-kal

F23.8

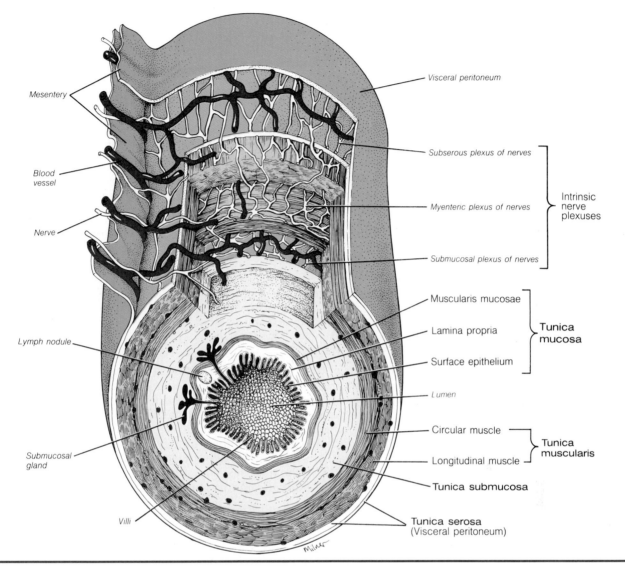

Mesentery

Blood vessel

Nerve

Lymph nodule

Submucosal gland

Villi

Visceral peritoneum

Subserous plexus of nerves

Myenteric plexus of nerves

Submucosal plexus of nerves

} Intrinsic nerve plexuses

Muscularis mucosae

Lamina propria

Surface epithelium

} Tunica mucosa

Lumen

Circular muscle

Longitudinal muscle

} Tunica muscularis

Tunica submucosa

Tunica serosa (Visceral peritoneum)

Figure 23.9
Basic structure of the wall of the digestive tract.

Each parotid gland has a duct that crosses the masseter muscle, pierces the buccinator muscle, and empties into the vestibule of the mouth opposite the second upper molar tooth. The **submandibular glands** (sometimes called the submaxillary glands) lie medial to the angle of the mandible. The ducts of the submandibular glands travel anteriorly under the floor of the mouth to open at the base of the frenulum of the tongue. The **sublingual glands** are located on the floor of the mouth within a fold of mucous membrane. Each sublingual gland has several small ducts that open onto the floor of the mouth and, often, a larger duct that joins with the duct of the submandibular gland.

Pharynx

Food that is swallowed passes from the mouth into the oropharynx and then on into the laryngopharynx. These two portions of the **pharynx** serve as common passageways for the respiratory and digestive systems; they are described in more detail in Chapter 22. Upon leaving the laryngopharynx, food enters the esophagus.

Gastrointestinal Tract Wall

Beginning in the esophagus and continuing all the way to the anus, the wall of the digestive tube has the same basic arrangement of four layers **(tunics),** with complex networks of nerves interconnecting the tunics (Figure 23.9). Although the structure of the wall is modified in various regions of the digestive tract, the four basic layers present are, from the lumen (cavity) of the gut

F23.9

outward, the *tunica mucosa*, the *tunica submucosa*, the *tunica muscularis*, and the *tunica serosa* or *adventitia*.

Tunica Mucosa

The **tunica mucosa** is the mucous membrane that lines the digestive tract. It consists of three layers:

1. An **epithelial layer** borders on the lumen. In the esophagus, it is stratified squamous epithelium. In the remainder of the tract, it is simple columnar epithelium.

2. The **lamina propria** is composed of loose connective tissue to which the epithelial cells are attached. Blood vessels, lymph nodules, and small glands are generally located in this layer.

3. Outside the lamina propria is a thin layer of smooth muscle fibers called the **muscularis mucosae.**

Tunica Submucosa

The **tunica submucosa** is a thick layer of either dense or loose connective tissue located deep to the mucosa. It contains blood vessels, lymphatic vessels, nerves, and, in some regions, glands.

Tunica Muscularis

In most regions of the digestive tract, the **tunica muscularis** is a double layer of muscle tissue. The muscle fibers of the inner layer are arranged circularly around the tube, whereas the outer fibers are oriented longitudinally along its long axis. At several points along the tract, the fibers of the circular layer are thickened, forming sphincters that control the movement of food from one region of the digestive tract to another. In the upper part of the esophagus, the tunica muscularis is composed of skeletal (voluntary) muscle fibers. Throughout the rest of the tract, it is smooth (involuntary) muscle. Rhythmic contractions of these muscles produce peristalsis, which moves food toward the anus.

Tunica Serosa or Adventitia

The outermost tunic of the digestive tube is composed primarily of a layer of connective tissue. In the esophagus, this connective-tissue layer merges into the connective tissue of surrounding structures and is called the **adventitia.** Along the rest of the digestive tract, the connective tissue is covered with a serous membrane consisting of a single layer of squamous epithelial cells, forming the visceral peritoneum. In this case, the outer tunic is called the **serosa.**

Intrinsic Nerve Plexuses

Between the four tunics that form the wall of the digestive tract, there are complex interconnections of neurons organized into intrinsic nerve plexuses. The plexuses are referred to as *submucosal, subserous,* or *myenteric,* according to their location. The myenteric plexus is located between the two layers of the muscular coat. The intrinsic nerve plexuses coordinate much of the activity of the digestive tract, and they make possible the occurrence of *intratract reflexes*. In these reflexes, the stimulation of a receptor within the wall of the digestive tract neurally influences—independent of the central nervous system—the activities of effector cells of the tract, for example, muscle or secretory cells. In addition, hormones and reflexes that operate by way of the central nervous system also influence digestive tract activity.

Esophagus

The **esophagus** is a muscular tube connecting the pharynx with the stomach. It is located behind the trachea. The esophagus travels through the mediastinum of the thorax and passes through the diaphragm by means of an opening called the *esophageal hiatus*. Food is moved through the esophagus by waves of contractions of the muscles in its wall (*peristalsis*). In the upper portion, near

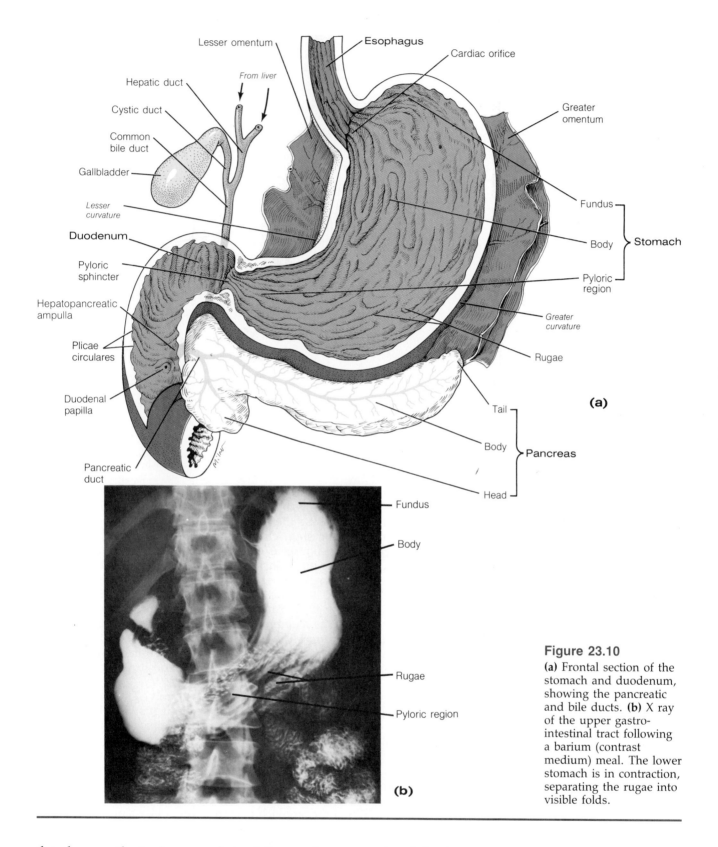

Figure 23.10

(a) Frontal section of the stomach and duodenum, showing the pancreatic and bile ducts. **(b)** X ray of the upper gastrointestinal tract following a barium (contrast medium) meal. The lower stomach is in contraction, separating the rugae into visible folds.

the pharynx, the tunica muscularis of the esophagus contains skeletal muscles. In the lower portions of the esophagus the muscles in the wall are smooth.

Stomach

Shortly after passing through the diaphragm, the esophagus empties into the **stomach** (Figure 23.10). The stomach lies to the left of the midplane just be-

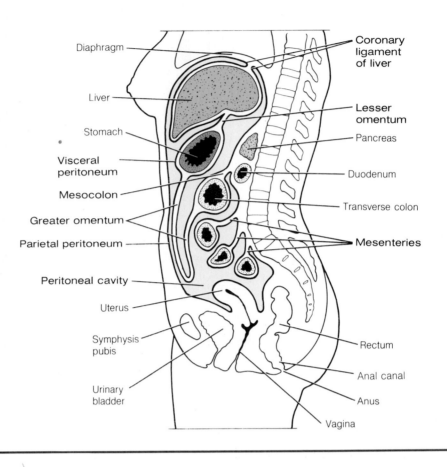

Figure 23.11
Sagittal section of the female abdominopelvic cavity, showing the peritoneum, the omenta, and the mesenteries.

neath the diaphragm. The opening from the esophagus into the stomach is called the **cardiac orifice.** The exit from the stomach, where it joins with the small intestine, is restricted by the **pyloric sphincter.**

The right border of the stomach, which is concave, is called the **lesser curvature.** The convex left border is called the **greater curvature.** The lesser curvature is attached to the undersurface of the liver by a mesentery consisting of a double layer of visceral peritoneum. This mesentery is called the **lesser omentum** (Figure 23.11). The two layers of the lesser omentum separate at the lesser curvature and form the serosa on the anterior and posterior surfaces of the stomach. The membrane layers rejoin at the greater curvature to form the **greater omentum.** The greater omentum is a folded mesentery forming an apron that covers the anterior surface of the transverse colon and the coils of the small intestine. It generally contains deposits of fat between its layers.

F23.11

The main portion of the stomach is called the **body.** The **fundus** is the portion that bulges above the entrance of the esophagus. The body of the stomach tapers inferiorly to form the **pyloric** region, which joins the duodenum of the small intestine.

The wall of the stomach is composed of the four basic layers (tunics) that are typical of the digestive tract. In the stomach, however, the epithelial cells of the mucosa are simple columnar rather than stratified squamous as in the esophagus. The epithelium of the mucosa remains simple columnar throughout the intestines. When the stomach is empty, the mucosa and submucosa form longitudinal folds called **rugae.** These folds flatten as the stomach fills.

A modification of the mucosa in the stomach is the presence of many **gastric glands,** which occupy the lamina propria. These glands, which secrete gastric juice, empty onto the surface of the mucosa through small invaginations called **gastric pits** (Figure 23.12). The gastric glands located in the fundus and body of the stomach (**fundic glands,** or **gastric glands proper**) contain several types of cells:

F23.12

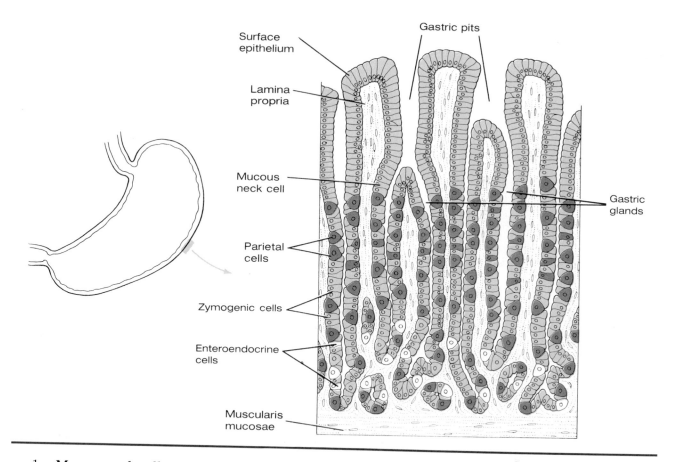

Surface epithelium

Gastric pits

Lamina propria

Mucous neck cell

Parietal cells

Zymogenic cells

Enteroendocrine cells

Muscularis mucosae

Gastric glands

Figure 23.12
Structure of the gastric glands.

1. **Mucous neck cells** are mucous-secreting cells located near the gastric pits.

2. **Parietal (oxyntic) cells** produce *hydrochloric acid.*

3. **Zymogenic (chief) cells** secrete *pepsinogen,* a precursor of the digestive enzyme *pepsin.*

The glands of the cardiac region **(cardiac glands)** and the pyloric region **(pyloric glands)** secrete mainly mucus. However, **enteroendocrine cells** that produce the hormone *gastrin* are present in the pyloric region.

Another modification of the tunics of the stomach is found in the tunica muscularis. Apart from the circular and longitudinal coats, the stomach wall contains an oblique layer of muscle between the circular layer and the submucosa. The additional layer of muscle in the wall makes extra strong contractions possible in the stomach and aids in its major functions—mashing the food and mixing it with digestive juices.

Small Intestine

The stomach empties into the **small intestine**—the longest (approximately 6 m) and most convoluted portion of the digestive tract. The small intestine joins the large intestine at the ileocecal valve. The small intestine is lined with simple columnar epithelium that contains cells specialized to absorb nutrients, which is a major function of the small intestine. On the basis of differences in microscopic structure, the small intestine can be divided into three regions that are not otherwise distinct from each other:

1. The **duodenum,** which is the first 25 cm or so of the small intestine, curves around the head of the pancreas (Figure 23.10a). The **common bile duct,** from the liver, and the **pancreatic duct,** from the pancreas, join together to form the **hepatopancreatic ampulla** (*ampulla of Vater*), which empties

doo-o-dee'-num or
doo-od'-de-num

F23.10a

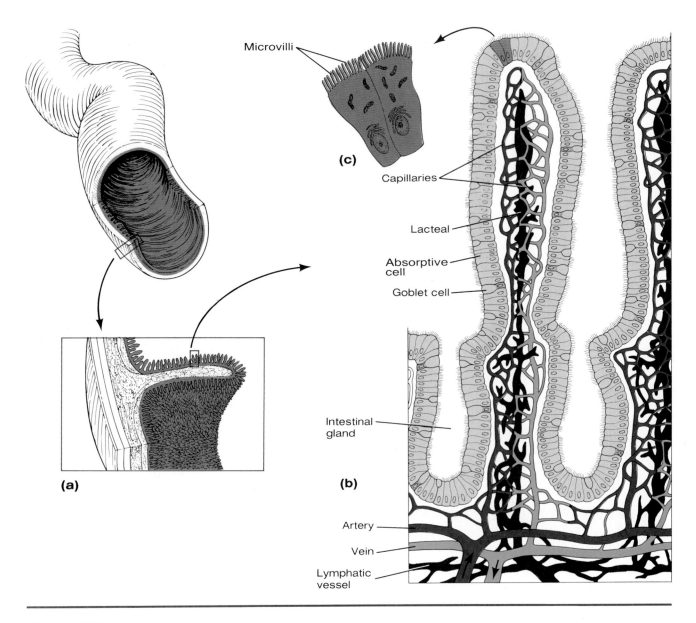

Microvilli

(c)

Capillaries

Lacteal

Absorptive cell

Goblet cell

Intestinal gland

(b)

(a)

Artery

Vein

Lymphatic vessel

Figure 23.13
Microscopic structure of the small intestine.
(a) Enlargement showing villi extending from a plica circulares.
(b) Higher magnification of a section through two villi, showing blood vessels and lacteals.
(c) Enlargement of two absorptive cells, showing microvilli on their free surfaces.

il'-ee-um

into the duodenum at the **duodenal papilla.** This opening is surrounded by a sphincter muscle called the **hepatopancreatic sphincter** (*sphincter of Oddi*). The common bile duct carries bile; the pancreatic duct carries digestive enzymes. The duodenum is retroperitoneal—that is, situated behind the peritoneum— and is tightly attached to the posterior body wall (Figure 23.11).

2. The next 2.5 m or so of the small intestine is the **jejunum.** This portion is suspended in the abdominal cavity by a mesentery.

3. The **ileum** is the remaining 3.5 m or so of the small intestine. The entrance of the ileum into the cecum of the large intestine is guarded by the **ileocecal valve,** which is composed of two folds of tissue. For 2 or 3 cm preceding the ileocecal valve, the muscle of the ileal wall is thickened, forming the **ileocecal sphincter.** The ileum, like the jejunum, is suspended from the posterior body wall by a mesentery. The mesentery allows the small intestine to move during the contractions of peristalsis, and it also provides support for blood vessels, lymphatic vessels, and nerves that supply the intestine.

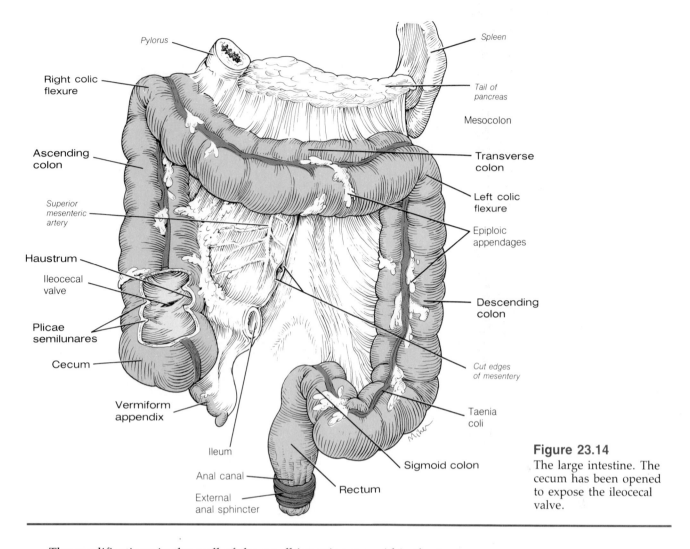

Figure 23.14
The large intestine. The cecum has been opened to expose the ileocecal valve.

The modifications in the wall of the small intestine are within the tunica mucosa and the submucosa. These layers form permanent circular shelflike folds **(plicae circulares)** that extend into the lumen of the small intestine and increase the surface area of the mucosa (Figure 23.10). The surface area is increased further by **villi,** which are tiny leaflike, tonguelike, or fingerlike projections of the mucosa into the lumen (Figure 23.13). In many areas, the villi are so numerous and so close together that the mucosa of the small intestine has a velvety appearance. Each villus is covered with a single layer of epithelial cells and contains a lymphatic capillary called a *lacteal*, which is surrounded by a network of blood capillaries. Two main types of epithelial cells cover the villi: *goblet cells*, which secret mucus, and more numerous *absorptive cells*, which participate in the absorption (and also the digestion) of food materials. The free surfaces of the absorptive cells are folded into microscopic projections called **microvilli.** Microvilli greatly increase the total surface area of the intestinal mucosa. The increased surface area provided by the plica circulares, villi, and microvilli aids in the absorption of digested foods, which must pass through the mucosa before entering blood capillaries or lacteals.

The mucosa contains many tubular **intestinal glands** (*crypts of Lieberkuhn*) located between the bases of the villi. In the submucosa of the duodenum are mucous glands called **duodenal** (*Brunner's*) **glands,** or **submucosal glands.** These glands generally empty into the intestinal glands.

ply'-ka
F23.10

F23.13

Large Intestine

The **large intestine,** which is about 1.5 m long, extends from the ileocecal valve to the anus (Figures 23.14 and 23.15). It is so named because its diameter in most regions is greater than that of the small intestine. Like the small

F23.14, F23.15

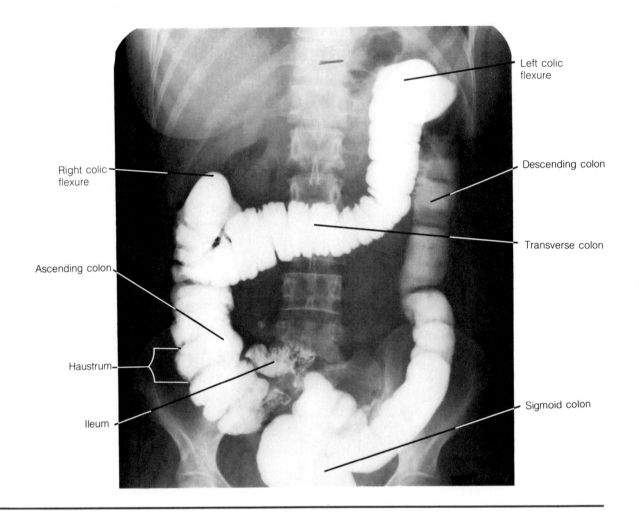

Right colic flexure

Left colic flexure

Descending colon

Ascending colon

Transverse colon

Haustrum

Ileum

Sigmoid colon

Figure 23.15
X ray of the lower gastrointestinal tract following a barium (contrast medium) meal.

intestine, the large intestine is lined with simple columnar epithelium with microvilli on their free surfaces. Both absorptive cells and goblet (mucous) cells are present, the latter being most abundant. Very few enzymes, if any, are produced by the epithelial cells. The large intestine begins as a blind pouch called the **cecum,** which receives the ileum of the small intestine. The **vermiform appendix** is a narrow blind tube that extends downward from the cecum. The wall of the appendix contains numerous lymphatic nodules. The large intestine extends upward from the cecum as the **ascending colon.** The ascending colon is not supported by a mesentery; instead, it lies tightly against the posterior wall of the abdomen. Just beneath the liver, the ascending colon bends sharply to the left **(right colic flexure)** and crosses the abdominal cavity as the **transverse colon.** This portion of the colon is suspended by a mesentery called the **mesocolon.** In the vicinity of the spleen, the transverse colon bends downward **(left colic flexure)** and forms the **descending colon.** The descending colon, like the ascending colon, is retroperitoneal. Where the descending colon reaches the left pelvic brim, it curves to the midplane via an S-shaped **sigmoid colon.**

Several variations occur in the four tunics of the colon. The tunica mucosa, for example, contains intestinal glands and a large number of mucous cells, but lacks villi. Moreover, while the longitudinal layer of the muscles of the tunica muscularis is continuous as a thin layer over the entire surface of the colon, most of the muscles of this layer are in the form of three bands—called **taeniae coli**—that run the length of the colon. The taeniae are shorter than the colon and therefore cause the colon to form small pouches called *haw′-strah* **haustra** (singular: *haustrum*). The mucosa between the haustra is thrown into **plicae semilunares** (semilunar folds), which extend into the lumen of the gut. Unlike the plicae circulares of the small intestine, the folds of the colon extend

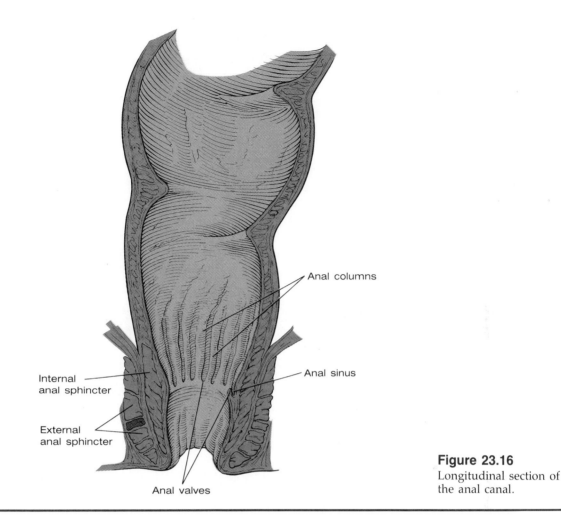

Anal columns

Anal sinus

Internal
anal sphincter

External
anal sphincter

Anal valves

Figure 23.16
Longitudinal section of
the anal canal.

only partway around the gut tube. Another unique characteristic of the colon is the presence on its external surface of small tabs called **epiploic append-ages.** These tabs are composed of folds of peritoneum filled with fat.

Rectum and Anal Canal

Beyond the sigmoid colon, the large intestine passes downward in front of the sacrum. This portion is called the **rectum** (Figure 23.14). The rectum has the same structure as the colon except that taeniae coli are not present, the longitudinal layer of muscles being again spread evenly over the entire sur-face.

F23.14

The terminal 3–4 cm of the large intestine is called the **anal canal** (Figure 23.16). This region is located below the pelvic diaphragm and thus is outside the pelvis.

F23.16

The mucosa of the anal canal forms a series of longitudinal folds known as **anal columns.** The anal columns are separated from each other by furrows called **anal sinuses,** which end distally in membraneous **anal valves.** The anal valves unite the lower ends of the anal columns. Within the anal canal, the epithelium changes from simple columnar, which is characteristic of the stomach and small and large intestines, to stratified squamous, which is typi-cal of the mouth and esophagus.

The anal canal opens to the exterior through the **anus.** The anal canal is surrounded by **internal** and **external anal sphincter muscles.** The internal sphincter is formed from a thickening of the circular layer of the smooth muscles of the tunica muscularis and is therefore not under voluntary control. The external sphincter is formed by skeletal muscle and is under voluntary control. Normally, therefore, a person can voluntarily control bowel move-ments.

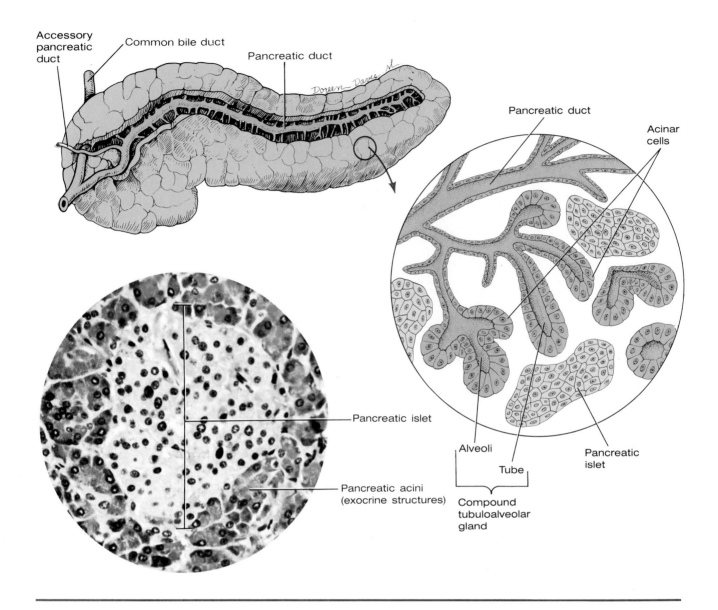

Figure 23.17
The pancreas. One inset shows the form of its compound tubuloalveolar glands, and the other is a photomicrograph of the acini and islets.

The circulation of blood through veins that travel the length of the anal columns is sometimes interfered with, causing them to become enlarged. This condition is called *hemorrhoids*. Hemorrhoids can be irritated by bowel movements, producing discomfort and bleeding, particularly if the feces are too dry, as in constipation.

ACCESSORY DIGESTIVE ORGANS

In addition to the many small glands situated in the wall of the digestive tract, there are large glands located outside the tract. The secretions of these glands, which are very important in the digestion of food, are carried to the digestive tract by way of ducts. The ducts and the secretory portions of the accessory digestive glands are derived from the endodermal lining of the gut tube of the embryo. These glands include the *salivary glands* (discussed earlier in this chapter), whose ducts empty into the mouth, and the *pancreas* and *liver*, both of which empty their secretions into the duodenum of the small intestine.

Pancreas

The pancreas is located behind the peritoneum and beneath the stomach (Figure 23.10a). The **head** of the pancreas is nestled in the curve of the duodenum, with the **neck, body,** and **tail** regions extending to the left. The tail reaches the vicinity of the spleen.

F23.10a

Microscopically, the pancreas resembles the salivary glands, containing many compound tubuloalveolar glands with their secretary cells arranged in short tubules or small sacs called **acini** (Figure 23.17). The acini consist of a single row of pyramid-shaped epithelial cells, with their apices converging toward the lumen. In acinar cells that are actively secreting, vesicles called *zymogen granules* accumulate toward the apex of the cells. The acinar cells secrete **pancreatic juice,** which contains digestive enzymes. The digestive enzymes of the pancreas are thought to be synthesized in the cytoplasm at the base of the acinar cells, where they enter the endoplasmic reticulum and are transported to the Golgi apparatus in the apical region of the cell. The Golgi apparatus concentrates the enzymes into zymogen granules, which are released in pancreatic juice.

F23.17

Pancreatic juice is transported to the duodenum by the **pancreatic duct** (*duct of Wirsung*). The pancreatic duct usually joins with the common bile duct, and they enter the duodenum together. An **accessory pancreatic duct** (*duct of Santorini*) often branches from the pancreatic duct and empties into the duodenum independently.

The pancreas does not just produce digestive enzymes, however; it also functions as an endocrine gland. Scattered among the acinar cells are clusters of endocrine cells called **pancreatic islets** (*islets of Langerhans*). The secretions of the cells of the pancreatic islets are not transported in ducts. Rather, they enter the bloodstream and travel throughout the body. The endocrine secretions of the pancreas are discussed along with the endocrine glands in Chapter 16.

Liver

The **liver** is a large organ that lies under the right side of the diaphragm. It is divided into two major regions: the **right** and **left lobes** (Figure 23.18a). On the inferior surfaces of the right lobe are smaller **caudate** and **quadrate lobes.** The right and left lobes are separated by a fold of parietal peritoneum called the **falciform ligament,** which attaches the liver to the anterior abdominal wall. In the free margin of the falciform ligament is a fibrous cord called the **ligamentum teres (round ligament).** The ligamentum teres is the remnant of the fetal umbilical vein that transported blood from the placenta to the liver. In postnatal life, it extends from the umbilicus to the undersurface of the liver. The falciform ligament is continuous on the superior surface of the liver with the **coronary ligament,** a fold of the parietal peritoneum that attaches the liver to the undersurface of the diaphragm (Figure 23.18b). The coronary ligament consists of anterior and posterior layers joined at their lateral margins by **right** and **left triangular ligaments.** Between the two layers of the coronary ligament is a region called the **bare area** of the liver. The bare area, which rests against the diaphragm, is the only portion of the liver not covered with visceral peritoneum. The under surface of the liver is anchored to the lesser curvature of the stomach by the **lesser omentum,** through which travel the hepatic artery, the hepatic portal vein, and the common bile duct. The inferior vena cava is partially embedded on the posterior surface of the liver (Figure 23.18c).

F23.18a

F23.18b

F23.18c

The liver receives blood from two sources: the **hepatic artery,** which supplies it with oxygenated blood from the aorta, and the **hepatic portal vein,** which carries venous blood from the digestive tract, pancreas, and spleen. The oxygen saturation of the blood in the hepatic portal vein is relatively low, but the blood contains a high concentration of dissolved nutrients absorbed from the intestines as a result of digestion. These nutrients are acted on in various ways as the blood passes through the liver. Approximately 1500 ml of blood flows through the liver every minute, of which 1100 ml arrives from the hepatic portal vein and 400 ml from the hepatic artery.

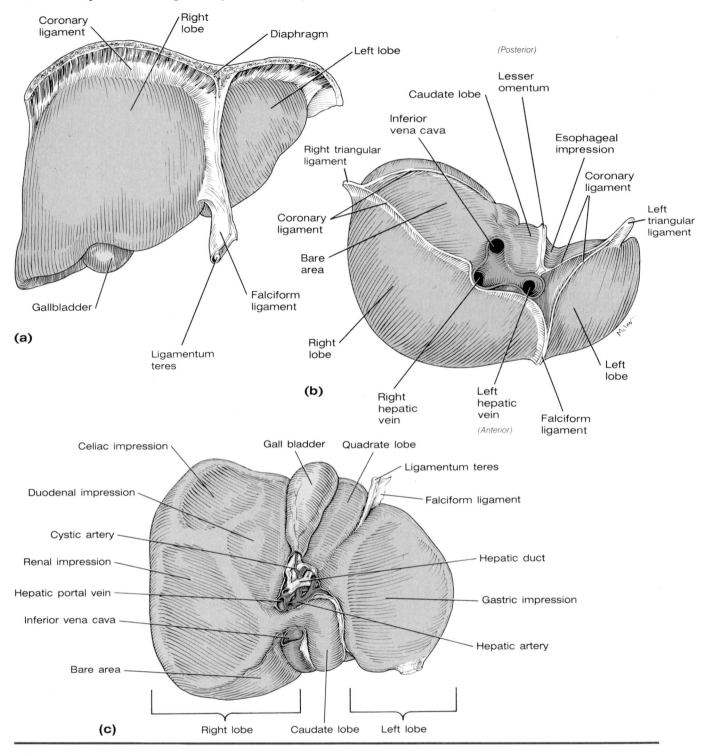

Figure 23.18
The liver and its supporting ligaments.
(a) Anterior surface.
(b) Superior (diaphragmatic) surface.
(c) Posterior-inferior surface.

F23.19

F23.20

The liver is composed of many tiny hexagonal compartments called **liver lobules,** with the corners of the compartments occupied by canals called **porta hepatis.** Branches of the hepatic artery and the hepatic portal vein, as well as a **bile duct,** travel in the porta hepatis. These three structures in close approximation within the porta hepatis are referred to as a **hepatic triad** (Figure 23.19). The most unusual aspect of hepatic circulation is that the branches of both the hepatic artery and the hepatic portal vein empty into the same **sinusoids,** which therefore contain a mixture of arterial and venous blood (Figure 23.20). The sinusoids of each lobule pass between the liver cells **(hepatocytes),** which are arranged in rows or sheets called **liver cords** or **plates,** and empty into a common **central vein** that passes through the center of each lobule, at right angles to the sinusoids. The central vein of each lobule drains into a **hepatic vein.** There are three hepatic veins, all emptying into the inferior vena

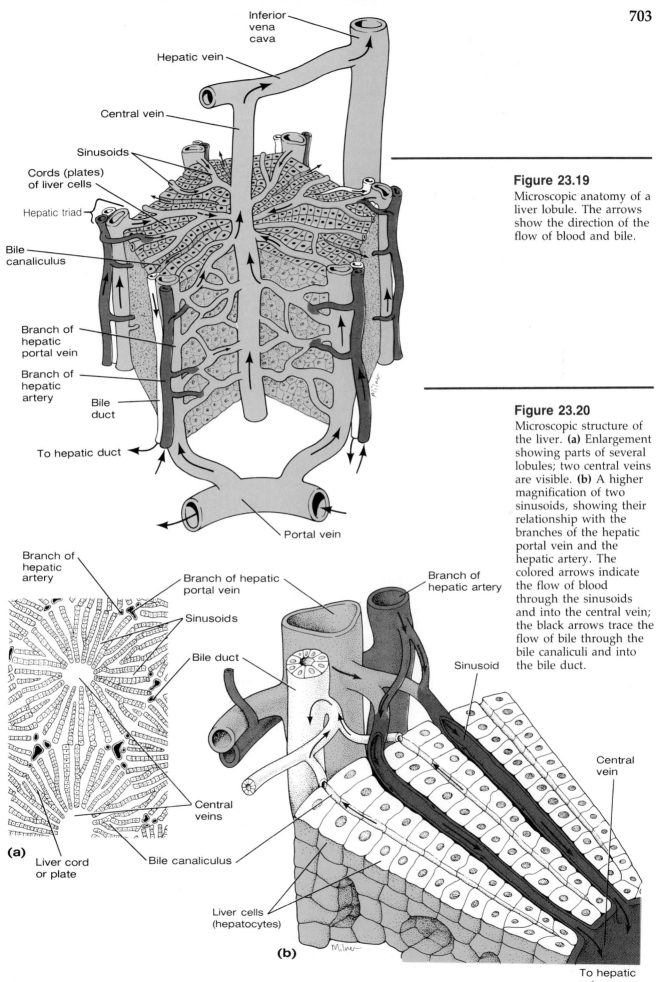

Inferior vena cava

Hepatic vein

Central vein

Sinusoids

Cords (plates) of liver cells

Hepatic triad

Bile canaliculus

Branch of hepatic portal vein

Branch of hepatic artery

Bile duct

To hepatic duct

Portal vein

Figure 23.19
Microscopic anatomy of a liver lobule. The arrows show the direction of the flow of blood and bile.

Figure 23.20
Microscopic structure of the liver. **(a)** Enlargement showing parts of several lobules; two central veins are visible. **(b)** A higher magnification of two sinusoids, showing their relationship with the branches of the hepatic portal vein and the hepatic artery. The colored arrows indicate the flow of blood through the sinusoids and into the central vein; the black arrows trace the flow of bile through the bile canaliculi and into the bile duct.

Branch of hepatic artery

Branch of hepatic portal vein

Sinusoids

Bile duct

Branch of hepatic artery

Sinusoid

Central vein

Central veins

(a)

Liver cord or plate

Bile canaliculus

Liver cells (hepatocytes)

(b)

To hepatic vein

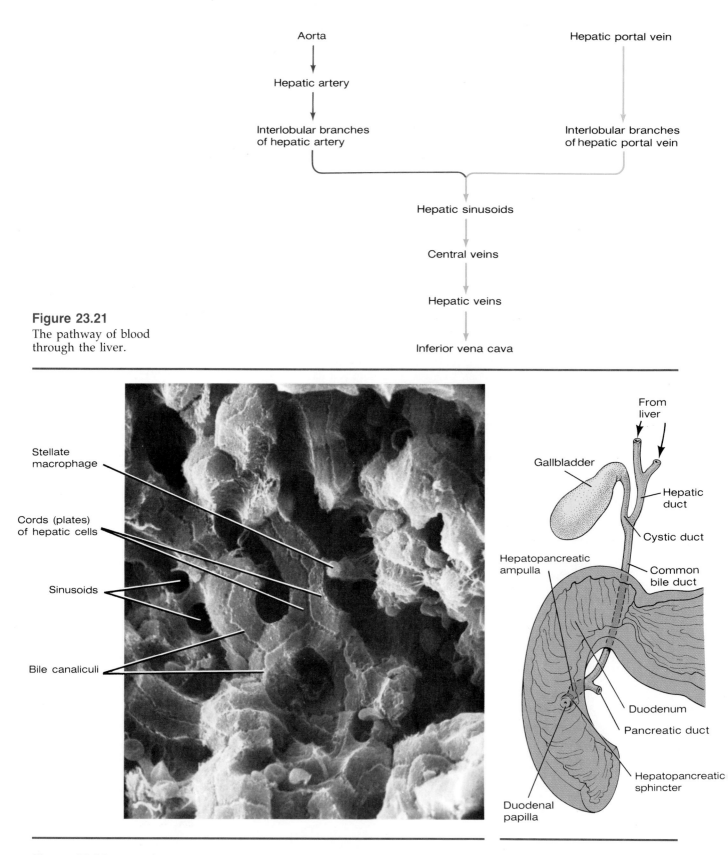

Figure 23.21
The pathway of blood through the liver.

Figure 23.22
Microscopic anatomy of the liver (×1716). (From *Tissues and Organs: A Text-Atlas of Scanning Electron Microscopy* by Richard G. Kessel and Randy H. Kardon. W.H. Freeman and Company. Copyright © 1979.)

Figure 23.23
The gallbladder and bile ducts.

the coordination of these events by the swallowing center fails, and food does enter the larynx or nasal cavity. The usual response to food that enters the larynx is a violent cough, which expels the food back into the pharynx.

The bolus is moved along the esophagus by a primary peristaltic wave—that is, by a peristaltic contraction of the muscle of the tunica muscularis of the esophagus that is basically a continuation of the peristaltic wave that began in the pharynx. About 2–5 cm above the cardiac orifice, the smooth muscle of the esophagus is normally in a state of tonic contraction, and it acts as a physiological sphincter (the *lower esophageal sphincter*) that keeps the lower portion of the esophagus closed. During swallowing, the lower esophageal sphincter relaxes, allowing the bolus to enter the stomach. Food generally requires between 5 and 10 seconds to move from the mouth to the stomach.

The distension of the esophagus by materials within it stimulates receptors in the esophageal wall and elicits secondary peristaltic waves that are not directly associated with the act of swallowing. These waves, which begin in the region of the upper esophageal sphincter, are of value in continuing the movement of any food material that may not have been moved completely through the esophagus by the primary peristaltic wave.

Gastric Motility

The mechanical activities of the stomach include: (1) storing ingested food until it can be utilized by the remainder of the gastrointestinal tract; (2) mixing the food with gastric secretions; and (3) moving the food into the duodenum of the small intestine at a rate consistent with efficient intestinal digestion and absorption.

Within limits, the stomach can expand to accommodate increasing amounts of food without a great increase in the pressure within it. This ability is due in part to neurally mediated reflex activity, and in part to the ability of gastric smooth muscle to lengthen, when stretched, without a marked increase in tension. As the food within the stomach is mixed with gastric juice, the mixture assumes a semifluid consistency and is referred to as *chyme*.

The opening between the stomach and duodenum is surrounded by the *pyloric sphincter*, which is generally partially open and thus offers only limited resistance to the movement of the stomach contents into the duodenum. However, this resistance is usually sufficient to prevent chyme from entering the duodenum in the absence of gastric muscular activity.

The stomach undergoes peristaltic contractions that begin in the body of the stomach and proceed toward the duodenum. The contractions are relatively weak in the upper region of the stomach, but they increase in strength and speed of transmission in the lower region, where the muscle of the stomach wall is quite thick. A strong peristaltic contraction in the pyloric region of the stomach forces some of the chyme out of the stomach and into the duodenum. In addition, the arrival of a peristaltic wave at the terminal pyloric region constricts this area and forces much of the chyme back toward the body of the stomach. This activity provides an effective mixing action.

The gastric smooth muscle contains pacemaker cells, which exhibit a basic electrical rhythm—that is, a spontaneous, cyclical depolarization and repolarization—that has a frequency of about 3 oscillations per minute. This rhythm is important in determining the frequency of the gastric muscular contractions, although not every cyclical depolarization necessarily results in a contraction.

Gastric motility (particularly the intensity of gastric muscular contractions) and the rate of gastric emptying are under the combined influence of a variety of excitatory and inhibitory mechanisms. For example, the volume of material contained within the stomach exerts a moderate influence on gastric motility and the rate of gastric emptying. As the stomach fills and the volume of its contents increases, the stomach wall is distended and mechanoreceptors in the wall are stimulated. The stimulation of these receptors initiates reflexes that enhance gastric motility, cause some relaxation of the pyloric sphincter, and promote gastric emptying. (Both the intrinsic nerve plexuses—that is, the intratract reflex pathway—and the vagus nerves and central nervous system

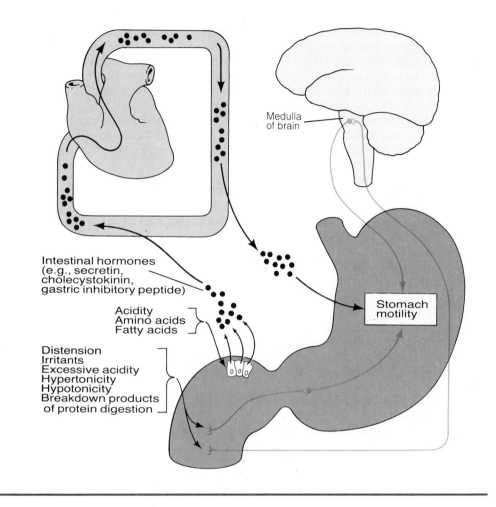

Figure 23.27
Diagrammatic representation of the neural (enterogastric reflex) and hormonal pathways by which the volume and composition of the chyme that enters the duodenum can inhibit stomach motility.

F23.27

are involved in these reflexes.) In addition, the hormone **gastrin** is released by cells in the pyloric region in response to distension (and also in response to the presence of partially digested protein and caffeine). Gastrin enters the bloodstream and ultimately returns to the stomach, where it stimulates gastric motility, relaxes the pyloric sphincter, and enhances gastric emptying. In general, the more liquid the contents of the stomach the more the rate of emptying increases with volume.

The volume and composition of the chyme that enters the duodenum exert a major influence on gastric motility and the rate of gastric emptying (Figure 23.27). As the duodenum fills with chyme, its wall is distended, and a reflex called the **enterogastric reflex** (*entero* = intestine) is initiated. In this reflex, nerve impulses that inhibit gastric motility, that produce a slight increase in the contraction of the pyloric sphincter, and that slow down gastric emptying are transmitted from the duodenum to the stomach both by way of the intrinsic nerve plexuses and by way of the vagus nerves and central nervous system. Irritants in the duodenum, an excessive acidity of the duodenal chyme (that is, a pH below 3.5), a relative hypertonicity or hypotonicity of the chyme, or the presence of breakdown products of protein digestion also activate the enterogastric reflex and inhibit gastric motility and emptying. Moreover, the acidity of the duodenal chyme and the presence of certain amino acids and fatty acids trigger the release of hormones such as **secretin, cholecystokinin,** and **gastric inhibitory peptide** from cells of the duodenum and jejunum. These hormones travel through the bloodstream to the stomach, where they exert an inhibitory effect on gastric motility and slow down gastric emptying.

The neural and hormonal duodenal mechanisms that inhibit stomach emptying tend to prevent additional chyme from entering the small intestine until the chyme already present has been effectively processed. The amount of gastric contents that enter the small intestine is therefore largely regulated by the small intestine itself.

Motility of the Small Intestine

Both segmentation and peristalsis take place in the small intestine. Segmentation mixes the chyme thoroughly with the digestive juices and exposes different portions of the chyme to the intestinal mucosa. In addition, segmentation causes some movement along the intestine. Peristaltic contractions also move the chyme along the small intestine, and they can contribute to the mixing of the intestinal contents. In general, the peristaltic waves of the small intestine are weak, and they die out after traveling only a short distance. Overall, the movement of the chyme through the small intestine tends to be relatively slow, allowing ample time for digestion and absorption.

Local mechanical and chemical stimulation by the intestinal contents is believed to be particularly important in the initiation and continuation of the movements of the small intestine. The motility of the small intestine is increased by distension; by hypertonic, hypotonic, or excessively acidic contents; and by the presence of certain products of digestion.

Like its gastric counterpart, the smooth muscle of the small intestine exhibits a basic electrical rhythm that is important in determining the frequency of intestinal muscular contractions. However, this rhythm varies in different portions of the intestine. It is most rapid in the duodenum—about 11 or 12 per minute—and decreases progressively along the length of the intestine to about 6 or 7 per minute in the terminal ileum.

The *ileocecal sphincter* is located at the end of the ileum near its junction with the cecum of the large intestine. The distension of the cecum initiates a reflex, by way of the intrinsic nerve plexuses, that stimulates the muscle of the ileocecal sphincter, and most of the time the sphincter is at least mildly constricted. The constriction of the sphincter facilitates absorption in the small intestine by slowing the movement of the chyme from the ileum into the large intestine. When food enters the stomach, a reflex called the *gastroileal reflex* intensifies peristalsis in the ileum. In addition, the hormone gastrin, which is released by cells in the pyloric region of the stomach, relaxes the ileocecal sphincter. These events enhance the movement of the chyme from the ileum into the large intestine.

The *ileocecal valve*, which is composed of two folds of tissue, is located at the opening of the ileum into the cecum of the large intestine. The folds of tissue are pushed apart and the valve opens when the contents of the ileum move into the cecum, but the folds are forced together and the valve closes when the contents of the cecum attempt to move backward into the ileum. Thus, the ileocecal valve helps prevent regurgitation of the contents of the cecum back into the ileum.

Motility of the Large Intestine

The movements of the large intestine are generally quite sluggish, and it frequently takes 18–24 hours for material to pass through the colon. Mixing movements of a segmental type occur within the colon at a much lower frequency than do those within the small intestine. At infrequent intervals, perhaps 3–4 times per day, substantial segments of the colon undergo several strong contractions, called *mass movements*, that propel material along the colon for considerable distances. Mass movements often occur following a meal, indicating that the presence of food within the stomach and the duodenum activates *gastrocolic* and *duodenocolic* reflexes, which increase colon motility.

The distension of the wall of the rectum by a mass movement of fecal material initiates the *defecation reflex*, which tends to move material out of the lower colon and rectum. Nerve impulses associated with the defecation reflex are transmitted by way of the intrinsic nerve plexuses, and these impulses are strongly reinforced by impulses that travel to and from the sacral region of the spinal cord. The defecation reflex stimulates peristaltic contractions of the descending colon, sigmoid colon, and rectum. It also causes the relaxation of the internal anal sphincter, which surrounds the anal opening. This combination of contraction and relaxation tends to propel the feces through the anus. However, a second sphincter, the external anal sphincter, also surrounds the anal opening. The external anal sphincter is composed of skeletal muscle, and it can be voluntarily controlled. If the external anal sphincter is voluntarily relaxed, defecation will occur. If the external anal sphincter is voluntarily contracted, defecation can be prevented. If this is done, the defecation reflex usually dies out after a few minutes, but it generally returns in several hours, when another mass movement propels additional fecal material into the rectum.

The defecation reflex is usually assisted through voluntary efforts. This assistance is accomplished by taking a deep inspiration and then closing the glottis and contracting the abdominal muscles. These actions raise the pressure within the abdomen, which aids defecation by increasing the pressure on the contents of the large intestine.

CONDITIONS OF CLINICAL SIGNIFICANCE

The Digestive System

Alterations in digestive tract motility can produce unpleasant changes in the digestive process, such as vomiting, diarrhea, and constipation.

Vomiting

Vomiting, which ejects the contents of the stomach through the mouth, is a complex reflex coordinated by a vomiting center within the medulla of the brain. During vomiting, the lower esophageal sphincter relaxes and the diaphragm and abdominal muscles contract. This increases the intra-abdominal and intragastric pressures, and forces the contents of the stomach through the esophagus and out of the mouth. The vomiting center receives input from many areas, both within and at the surface of the body. Thus, vomiting can result from dizziness, unpleasant odors or sights, and so forth, as well as from disturbances within the digestive tract. Excessive vomiting can lead to severe disturbances in the fluid, salt, and acid–base balances of the body.

Diarrhea

Diarrhea is characterized by watery stools, generally accompanied by frequent defecation. Diarrhea can be due to either an increased secretion of fluid into or a decreased absorption of fluid from the gastrointestinal tract. In either case, an increased volume of material within the large intestine can lead to more frequent defecation. Like vomiting, prolonged diarrhea can cause disturbances in the fluid, salt, and acid–base balances of the body.

Constipation

If the motility of the large intestine is decreased, digested materials remain within the colon and rectum for prolonged periods of time. The longer fecal materials remain within the colon, the more water is absorbed. As the feces lose water, they become drier and harder, making defecation more difficult and sometimes painful. This condition is referred to as *constipation*. Constipation can be caused by insufficient bulk in the diet, lack of exercise, or even by emotions.

CLINICAL CORRELATION

Spastic Colon

CASE REPORT . . . .

THE PATIENT   A 43 year-old-woman

PRINCIPAL COMPLAINT   Abnormal bowel movements

HISTORY   The patient related that her feces had been hard and pencil-thin and that she had experienced some

crampy discomfort in the upper left quadrant of the abdomen. These conditions were not clearly related to meals and were partially relieved by bowel movements. The patient also expressed fear that she had cancer of the colon. She remembered her bowel movements as being normal until she entered high school, at which age she began to notice occasional constipation, and her mother shared a laxative with her. When the patient went to college, she began having episodes of explosive diarrhea, with bowel movements three or four times a day, but not severe enough at night to disturb her sleep. She married shortly after graduating from college and again developed constipation, passing feces in small, rock-hard segments. After the birth of her second child, the patient began to experience aching in the lower abdomen. And at age 28 she had a hysterectomy (removal of the uterus). Her diarrhea and constipation had continued to alternate, with cycles of a few weeks to a few months. She developed dyspeptic symptoms (symptoms of disturbed digestion) in the upper right quadrant of the abdomen, particularly after fatty meals. Subsequently, at age 34, she had a cholecystectomy (removal of the gallbladder).

CLINICAL EXAMINATION   The patient appeared healthy, mildly obese, and slightly anxious. The results of the examination, which included proctoscopy (inspection of the anal canal and rectum) and barium enema (in which barium sulfate is introduced into the colon as a contrast medium for X-ray pictures), were

normal. A diagnosis of irritable bowel syndrome, or "spastic colon," aggravated by psychological stress, was reached.

COMMENT   Spastic colon is an intermittent condition that may occur even in psychologically healthy persons at times of stress, such as school examinations, marriage, divorce, and bereavement, but the precise mechanisms are not known. With this patient there was an abnormally close association between her emotions and the secretory and motor functions of her colon. Long-continued, excessive activity of the colonic muscle usually retards the passage of the contents, causing them to be excessively dry and hard. In diarrhea, the colon secretes more and contracts less; hence, the contents move faster.

OUTCOME   The patient was instructed to take a diet high in bran and other fiber, and her condition improved markedly. Unfortunately, not all patients who suffer from "spastic colon" are relieved by this kind of diet.

STUDY OUTLINE

EMBRYONIC DEVELOPMENT OF THE DIGESTIVE SYSTEM   pp. 683–685
1. Formed from endoderm.
2. Divided into foregut, midgut, hindgut.
3. Oral membrane breaks through to stomodeum, forming mouth.
4. Cloacal membrane ruptures through to proctodeum.
5. Endodermal buds along foregut, together with surrounding mesoderm, give rise to salivary glands, liver, gallbladder, pancreas.

ANATOMY OF THE DIGESTIVE SYSTEM
pp. 685–700

MOUTH
1. Lined with stratified squamous layer of nonkeratinized cells (except outer lips).
2. Lips and cheeks aid chewing and speech.
3. Roof formed anteriorly by hard palate; posteriorly by soft palate.

TONGUE   Forms floor of mouth.
1. Protruded, retracted, and moved sideways by extrinsic muscles.
2. Shape modified by intrinsic muscles.
3. Assists in food manipulation, swallowing, speech.
4. Papillae present on dorsal surface; taste buds present on some papillae.
5. Connected ventrally to mouth floor by frenulum, which can cause tongue-tied condition.

TEETH
1. Enamel-covered crown anchored by roots into alveolus (socket).
2. Pulp cavity, root canal, and apical foramen enclose vessels and nerves.

3. Incisors, canines, premolars, molars.
4. 20 deciduous teeth; 32 permanent teeth.

SALIVARY GLANDS

Tubuloalveolar Glands   Composed of small sacs (acini).

Parotid Glands   Below and anterior to ear.

Submandibular Glands   Medial to mandibular angle.

Sublingual Glands   On floor of mouth.

Saliva   Moistens mucous membranes.

PHARYNX   From oropharynx through laryngopharynx.

GASTROINTESTINAL TRACT WALL   Four tunics in wall of digestive tube; this basic structure modified in some regions of digestive tract.

TUNICA MUCOSA   Mucous membrane lining; consists of epithelial layer, lamina propria, and muscularis mucosae.

TUNICA SUBMUCOSA   Dense or loose connective tissue; contains blood vessels, lymphatic vessels, nerves, and glands.

TUNICA MUSCULARIS   Double layer of muscle tissue: inner layer circular, outer layer longitudinal.

TUNICA SEROSA (ADVENTITIA)   Outer layer of connective tissue.

INTRINSIC NERVE PLEXUSES   Interconnections of neurons; coordinate much of the digestive tract activity; make intratract reflexes possible.

ESOPHAGUS   Muscular tube connecting pharynx to stomach; upper portion contains skeletal muscles; lower portion has smooth muscles.

STOMACH From cardiac orifice to pyloric sphincter; body, fundus, and pyloric regions; greater and lesser curvatures.

TUNICA MUCOSA MODIFICATIONS

COLUMNAR EPITHELIUM

RUGAE

GASTRIC GLANDS
1. Mucous neck cells secrete mucus.
2. Parietal cells produce hydrochloric acid.
3. Zymogenic cells secrete pepsinogen.
4. Cardiac and pyloric glands secret mucus.

TUNICA MUSCULARIS MODIFICATION Includes an oblique muscle layer, as well as circular and longitudinal layers.

SMALL INTESTINE
1. Approximately 6 mm long; site of most digestion and absorption.
2. Three regions:

DUODENUM Retroperitoneal; receives common bile duct and pancreatic duct at duodenal papilla.

JEJUNUM Suspended by mesentery.

ILEUM Suspended by mesentery; entrance into large intestine surrounded by ileocecal valve; just preceding ileocecal valve is ileocecal sphincter.

3. Modifications in wall: Plicae circulares, villi, and microvilli increase surface area of small intestine. Contains goblet cells, intestinal glands, duodenal glands.

LARGE INTESTINE
1. Composed of cecum; ascending, transverse, descending, and sigmoid colons; rectum; anal canal.
2. Modifications in tunics: Tunica mucosa: has many mucous cells, lacks villi; tunica muscularis: longitudinal muscles form bands (taeniae coli) that cause colon to form pouches (haustra); epiploic appendages on external surface.

RECTUM AND ANAL CANAL
1. Rectum lies anterior to sacrum.
2. Anal canal mucosa forms longitudinal rectal columns.
3. Involuntary internal anal sphincter; voluntary external anal sphincter.

ACCESSORY DIGESTIVE ORGANS
pp. 700–705

PANCREAS
1. Cells in acini secrete pancreatic juice, which is transported to duodenum by pancreatic duct.
2. Endocrine secretions of pancreatic islets enter bloodstream.

LIVER
1. Right and left lobes separated by falciform ligament.
2. Ligamentum teres is remnant of fetal umbilical vein.
3. Coronary ligament attaches liver to undersurface of diaphragm.
4. Lobules consist of rows of cuboidal cells radiating outward from a central vein that empties into a hepatic vein.

5. Bile ducts collect bile from canaliculi that transport bile in opposite direction from blood flow.
6. Hepatic circulation: Hepatic artery supplies oxygenated blood to liver from aorta. Hepatic portal vein carries venous blood from digestive tract, pancreas, and spleen. Liver sinusoids contain mixture of arterial and venous blood. Sinusoids lined with highly permeable endothelium with attached phagocytic stellate macrophages. Hepatic veins empty blood from sinusoids into inferior vena cava.

GALLBLADDER AND BILE DUCTS
1. Gallbladder is small sac on inferior surface of liver; serves as bile storage site.
2. Cystic duct drains gallbladder, joins hepatic duct to form common bile duct, which joins with pancreatic duct to enter duodenum.

MECHANICAL PROCESSES OF THE DIGESTIVE SYSTEM   pp. 705–710
1. Segmentation—major mixing movement.
2. Peristalsis—major propulsive movement.

CHEWING Part voluntary, part reflex; tactile stimulation of food against teeth, gums, and anterior portion of roof of mouth leads to reflex relaxation of muscles that raise jaw; when jaw drops, stretch reflexes lead to muscle contractions that raise jaw.

SWALLOWING Movement of bolus into pharynx elicits swallowing reflex.
1. Peristaltic contraction of pharyngeal constrictor muscles.
2. Relaxation of upper esophageal sphincter.
3. Approximation of vocal cords, upward movement of larynx, elevation of soft palate.
4. Primary peristaltic wave in esophagus.
5. Relaxation of lower esophageal sphincter.

GASTRIC MOTILITY
1. Stomach can expand to accommodate food without great increase in intragastric pressure.
2. Stomach undergoes peristaltic contractions that move chyme into duodenum and provide effective mixing.
3. Basic electrical rhythm important in determining frequency of gastric contraction.
4. Gastric motility and emptying influenced by distension of stomach (via neural reflexes and gastrin) and by volume and composition of chyme in duodenum (via enterogastric reflex and intestinal hormones).

MOTILITY OF THE SMALL INTESTINE Both segmentation and peristalsis occur in response to local mechanical and chemical stimulation; gastroileal reflex intensifies peristalsis in ileum; gastrin relaxes ileocecal sphincter.

MOTILITY OF THE LARGE INTESTINE Movements, generally sluggish, include mixing movements of segmental type and mass movements; defecation reflex leads to emptying of lower portion of large intestine.

CONDITIONS OF CLINICAL SIGNIFICANCE: THE DIGESTIVE SYSTEM   p. 710

VOMITING Ejects contents of stomach through mouth; complex reflex leads to relaxation of lower esophageal sphincter and contraction of diaphragm and abdominal muscles.

DIARRHEA Watery stools generally accompanied by frequent defecation; can be due to either increased secretion of fluid into or decreased absorption of fluid from gastrointestinal tract.

CONSTIPATION Decreased motility of large intestine leads to increased water absorption and dry, hard feces.

SELF-QUIZ

1. The entire digestive tract is lined with: (a) cloacal membrane; (b) proctodeum; (c) mucous membrane.

2. The hard palate extends posteriorly from the soft palate, separating the oral cavity from the nasopharynx. True or False?

3. The hardest part of the tooth is the: (a) dentin; (b) enamel; (c) cementum.

4. The teeth specialized to tear food are the: (a) canines; (b) incisors; (c) premolars.

5. The largest of the salivary glands are the: (a) parotids; (b) sublinguals; (c) buccal.

6. Which one of the four layers of the gastrointestinal tract is responsible for peristalsis? (a) tunica mucosa; (b) tunica submucosa; (c) tunica muscularis; (d) tunica serosa.

7. Gastric glands secrete: (a) mucus; (b) cholecystokinin; (c) bile.

8. The point of exit from the stomach, where it joins with the small intestine, is guarded by a: (a) pyloric sphincter; (b) greater omentum; (c) lesser omentum.

9. The ducts from the pancreas and liver empty into the alimentary canal at the: (a) stomach; (b) duodenum; (c) ileum.

10. Duodenal glands produce: (a) amylase; (b) pepsinogen; (c) mucus.

11. As the descending colon reaches the left pelvic brim, it curves to the midline via an S-shaped sigmoid colon. True or False?

12. The internal anal sphincter is formed by a thickening of the circular layer of the smooth muscles of the tunica muscularis and is therefore under voluntary control. True or False?

13. The pancreatic islets are endocrine glands. True or False?

14. Bile is produced in the: (a) liver; (b) gallbladder; (c) pancreas.

15. When the hepatopancreatic sphincter relaxes, bile from the liver fills the common bile duct and then enters the gallbladder via the cystic duct. True or False?

16. Secondary peristaltic contractions in the esophagus: (a) move food from the stomach to the mouth during vomiting; (b) occur only during swallowing; (c) are generated in response to the distension of the esophagus.

17. Peristaltic contractions that occur in the stomach are generally strongest in the upper portion of the stomach and weakest in the lower portion near the duodenum. True or False?

18. The enterogastric reflex: (a) inhibits gastric motility; (b) stimulates the secretion of saliva; (c) causes defecation.

19. Both segmentation and peristalsis occur in the small intestine. True or False?

20. Mass movements characteristically move food through the stomach. True or False?

LEARNING OBJECTIVES

After completing this chapter, you should be able to:

1. Explain the functions and control of the salivary secretions.

2. Discuss the phases of gastric secretion.

3. Discuss the inhibition of gastric secretion.

4. List the principal digestive fluids that are present in the small intestine, and describe the factors that stimulate the secretion of each.

5. Describe the processes involved in the digestion and absorption of carbohydrates.

6. Describe the processes involved in the digestion and absorption of proteins.

7. Describe the processes involved in the digestion and absorption of lipids.

8. Discuss the absorption of vitamins.

9. Discuss the absorptive function of the large intestine.

CHAPTER CONTENTS

MOUTH AND SALIVA

STOMACH

SMALL INTESTINE AND ASSOCIATED STRUCTURES

LARGE INTESTINE

CONDITIONS OF CLINICAL SIGNIFICANCE: DIGESTION

KEY TERMS AND DERIVATIVES

digestion (*digest* = separate; *ion* = the action of) the bodily process of breaking down foods, chemically and mechanically, into substances capable of being absorbed by body cells

gastroenteritis (*gastr* = stomach; *entero* = intestine) an acute or chronic inflammation of the mucosa of the stomach and intestine

hepatitis (*hepat* = liver) inflammation of the liver

microvilli (*micro* = small; *villi* = shaggy) tiny projections of the free surfaces of certain epithelial cells

The Digestive System:
Chemical Digestion and Absorption

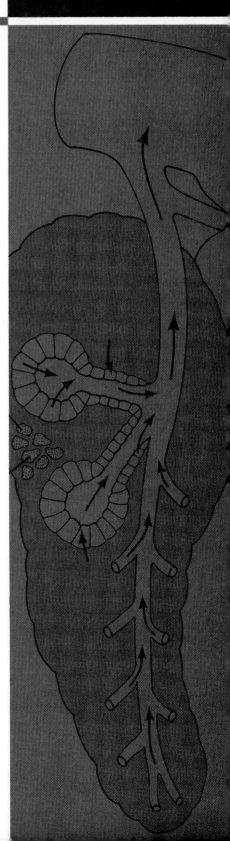

24

The chemical breakdown of the large molecules of foodstuffs into smaller molecules that can be absorbed from the gastrointestinal tract is accomplished by digestive enzymes. These enzymes function mainly by hydrolysis; that is, they split large molecules into smaller ones by introducing water into the molecular structures (Figure 24.1). In this fashion, large carbohydrate molecules such as starches are digested to monosaccharides such as glucose; proteins are split into their constituent amino acids; and fats are broken down primarily into monoacylglycerols and free fatty acids. The smaller molecules that result from digestion are absorbed from the lumen of the digestive tract, and nutrients ultimately enter the bloodstream to be carried to the cells of the body.

Many of the digestive enzymes and other secretory products of the gastrointestinal tract are produced and released by specialized epithelial cells that are commonly organized into exocrine secretory glands (Figure 24.2). These exocrine glands release their products by way of ducts onto the surfaces of the gastrointestinal tract.

MOUTH AND SALIVA

Three pairs of large salivary glands—the *parotid, sublingual,* and *submandibular (submaxillary) glands*—produce most of the saliva. The output of these glands is supplemented slightly by smaller buccal glands. Approximately two-thirds of the 1–1.5 liters of saliva produced per day by an average adult comes from the submandibular glands. Approximately one-quarter comes from the parotid glands, and the remainder is produced by the other glands.

Secretions of the Salivary Glands

The salivary glands produce two different secretions: (1) a mucous secretion that contains mucins; and (2) a serous secretion that contains the enzyme salivary amylase (ptyalin). The parotid glands produce serous secretions, the submandibular glands produce both mucous and serous secretions, and the sublingual and buccal glands primarily produce mucous secretions.

Mucins

The **mucins** are the major proteins of the saliva, and they have large polysaccharides attached to them. When mixed with water, the mucins form a highly viscous solution known as mucus, which lubricates the mouth and food.

Salivary Amylase

Salivary amylase splits starch molecules, which are composed of thousands of glucose units, into smaller fragments. Like other enzymes, salivary amylase has an optimal pH at which it functions best (6.9), but it is stable at pH values from 4 to 11. After food has been swallowed, the digestive action of salivary amylase continues in the stomach until the enzyme is inactivated by the acidic gastric juices.

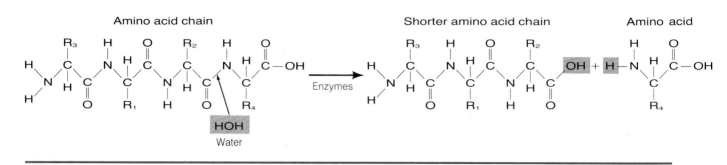

Amino acid chain   Shorter amino acid chain   Amino acid

Enzymes

HOH
Water

Figure 24.1
Hydrolytic cleavage of an amino acid chain to liberate a single amino acid. The peptide bond is broken by the introduction of water.

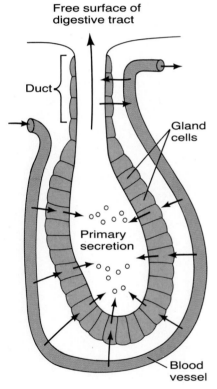

Free surface of digestive tract

Duct

Gland cells

Primary secretion

Blood vessel

Figure 24.2
Diagrammatic representation of an exocrine gland. Raw materials obtained from the blood are transformed by the gland cells into a primary secretion that is released into the glandular lumen. As the primary secretion moves along the glandular duct, active transport or diffusion across the duct walls may alter its composition.

Composition of Saliva

The exact composition of saliva depends on the glands from which it comes, on the secretion rate, and to some degree on the stimulus that evokes secretion. Generally, saliva is 97–99.5% water, and the pH of mixed saliva is usually from 6.0 to 7.0. Sodium, potassium, chloride, and bicarbonate ions are some of the main electrolytes of saliva. Saliva also contains kallikrein, as well as specific soluble blood-group substances of the sort responsible for blood type. Kallikrein acts enzymatically on plasma protein precursors to produce bradykinin, which is a vasodilator substance. The release of kallikrein from active salivary glands and the subsequent production of bradykinin is believed to increase blood flow to the glands by promoting local vasodilation.

Functions of Saliva

In addition to its lubricating and digestive functions, saliva assists in bolus formation and in swallowing by moistening food particles and holding them together. It also helps dissolve food so the food can be tasted. Saliva aids in speech by moistening the mouth and throat, and it has bacteriostatic properties.

Control of Salivary Secretion

Salivary secretion occurs in response to nerve impulses. Salivatory nuclei in the medulla-pons region of the brain receive impulses from the mouth and pharynx, as well as from higher brain centers. Nerve impulses travel to the salivary glands by way of the autonomic nervous system. Both parasympathetic and sympathetic stimulation cause salivary secretion, with parasympathetic stimulation having the greatest effect. The flow of saliva is enhanced by the presence of food in the mouth, as well as by the odor of food, the sight of food, the thought of food, or the presence of irritating food in the stomach and upper intestine. Chewing stimulates salivary secretion, whereas intense mental effort, dehydration, fear, and anxiety tend to reduce it.

STOMACH

Gastric Secretions

The gastric secretions include mucus, hydrochloric acid, and pepsinogen, which is a precursor of the protein-digesting enzyme pepsin.

Mucus

Cells that secrete a viscous, alkaline mucus line the surface of the stomach. The mucus, which adheres to the stomach walls in a layer 1–1.5 mm thick, lubricates the walls and protects the gastric mucosa. Even the slightest irritation of the mucosa directly stimulates these cells to secrete large amounts of mucus.

Glands that secrete mainly a thin mucus are present in the region of the cardiac orifice of the stomach (cardiac glands) and in the pyloric region (pyloric glands). Gastric glands in the body and fundus of the stomach produce some mucus, as well as additional secretions.

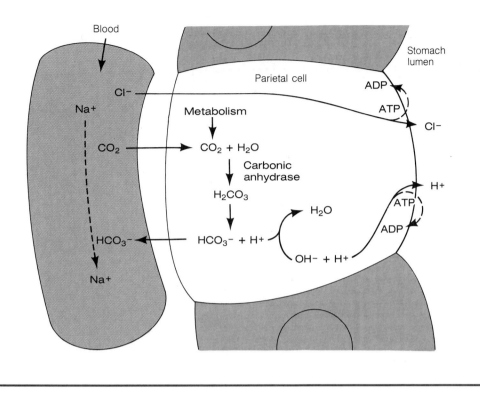

Figure 24.3
Potential mechanism for the secretion of hydrochloric acid (H^+Cl^-) by parietal cells.

Hydrochloric Acid

Cells in the gastric glands, called *parietal (oxyntic) cells*, produce hydrochloric acid, which is a strong acid that dissociates into hydrogen ions and chloride ions. The hydrochloric acid facilitates protein digestion in the stomach and also kills many of the bacteria that enter the digestive tract with the food. The parietal cells produce a hydrochloric acid solution that is about 0.16 molar (pH 0.8). Actual stomach acidity, however, depends on the rate of hydrochloric acid secretion, as well as on the rate of neutralization and dilution of the acid by other secretions.

The precise details by which the parietal cells produce hydrochloric acid are unknown. The prevalent theory holds that chloride ions are actively transported from the blood to the stomach lumen across the parietal cells, and hydrogen ions from dissociated water molecules are actively transported from the interior of the parietal cells to the stomach lumen, leaving behind hydroxide ions (OH^-) (Figure 24.3). Carbon dioxide from the plasma diffuses into the cells, where it (and also carbon dioxide from cellular metabolism) combines with water to form carbonic acid. This reaction is facilitated by the enzyme carbonic anhydrase. Carbonic acid can dissociate into hydrogen and bicarbonate ions. Thus, hydrogen, bicarbonate, and hydroxide ions are present in the cells. The hydrogen ions and hydroxide ions combine to form water, and the bicarbonate ions diffuse out of the cells into the blood. As a result of the basic bicarbonate ions entering the blood, the pH of the venous blood leaving the actively secreting stomach is higher than that of the arterial blood flowing to the stomach.

Pepsinogen

Cells in the gastric glands of the stomach called *zymogenic (chief) cells* produce pepsinogen, which is a precursor of the active enzyme pepsin. In the stomach, pepsinogen is converted to pepsin by contact with hydrochloric acid and previously formed pepsin. Pepsin, which is optimally active at acid pHs, digests proteins by breaking peptide bonds that involve certain amino acids, such as tryptophan, phenylalanine, and tyrosine. This activity results in smaller *peptide chains* of amino acids. The production of peptide chains is about as far as protein digestion in the stomach proceeds.

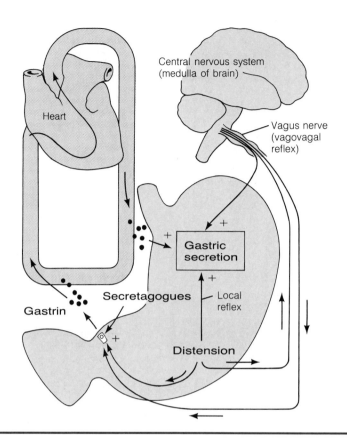

Figure 24.4
Gastric-phase pathways that result in increased acid secretion by the stomach.

Control of Gastric Secretion

The stomach glands produce as much as 3 liters of secretions (gastric juice) per day. Diet affects the amount of juice secreted, and a typical meal results in the secretion of up to 700 ml of gastric juice.

Gastric secretion is stimulated by both neural and hormonal mechanisms. Neural mechanisms involve the activity of both local, intratract reflexes and the vagus nerves from the medulla of the brain. Parasympathetic impulses transmitted by way of the vagus nerves to the stomach increase both pepsin and acid secretions and also cause some increase in mucous secretion. Hormonal mechanisms involve the hormone gastrin, which stimulates gastric secretion, particularly the release of hydrochloric acid. The neural and hormonal mechanisms of control are interrelated because vagal activity causes gastrin release.

The control of gastric secretion is divided into several regulatory phases.

Cephalic or Reflex Phase

Gastric secretion can be initiated by the sight, smell, or taste of food. This response is due to nerve impulses that travel from sensory receptors to the central nervous system and then by way of the vagus nerves to the stomach, where gastric secretion is stimulated. The secretory response elicited by the sight or smell of food is basically a conditioned reflex that is not elicited when a person is afraid, depressed, or has no desire for food (appetite). If the vagus nerves to the stomach are severed, secretion due to the cephalic or reflex phase ceases.

Gastric Phase

F24.4 The distension of the stomach by the presence of food stimulates the flow of gastric juice and also potentiates its acid and pepsin content (Figure 24.4). One way this response occurs is as a result of local reflexes that stimulate gastric secretion. The distension of the stomach also results in a vagovagal reflex by which impulses are sent to the brain stem and then back to the

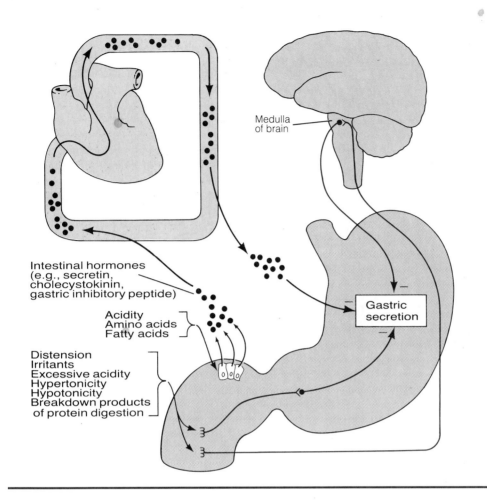

Medulla
of brain

Intestinal hormones
(e.g., secretin,
cholecystokinin,
gastric inhibitory peptide)

Acidity
Amino acids
Fatty acids

Distension
Irritants
Excessive acidity
Hypertonicity
Hypotonicity
Breakdown products
 of protein digestion

Gastric
secretion

Figure 24.5
Intestinal factors that
inhibit gastric secretion
(particularly hydrochloric
acid release) when the
stomach is actively
secreting and the
mechanisms that cause
gastrin release are
strongly stimulated.

stomach by way of the vagus nerves to stimulate the flow of gastric juice (and
the release of gastrin). In addition, local responses to the distension of the
pyloric region of the stomach or to the presence of substances called secreta-
gogues—for example, partially digested protein and caffeine—lead to the
release of gastrin.

Gastrin release is inhibited by high concentrations of hydrogen ions (acid)
in the stomach. This activity provides a negative feedback mechanism that
helps prevent excessive stomach acidity and helps maintain an optimal pH for
the function of the peptic enzymes. When the pH of the stomach contents
reaches 2.0, the gastrin mechanism for the stimulation of gastric secretion
becomes totally blocked.

Intestinal Phase

A slight stimulation of gastric secretion occurs in response to the presence of
chyme within the upper portion of the small intestine (particularly the duodenum) after the stomach has nearly emptied and gastric secretory activity is
relatively low following a meal. It is believed that this stimulation of gastric
secretion is mainly due to the production of small amounts of gastrin by the
duodenal mucosa in response to distension or chemical stimuli. However,
other hormones or reflexes are probably also involved.

Intestinal Factors That Inhibit Gastric Secretion

Although the presence of chyme in the intestine stimulates gastric secretion
during the intestinal phase, when the stomach is almost empty and gastric
secretory activity is relatively low, it frequently has an inhibitory effect on
gastric secretion during the gastric phase, when the stomach contains food
and gastric secretion is strongly stimulated (Figure 24.5). The distension of

kime

F24.5

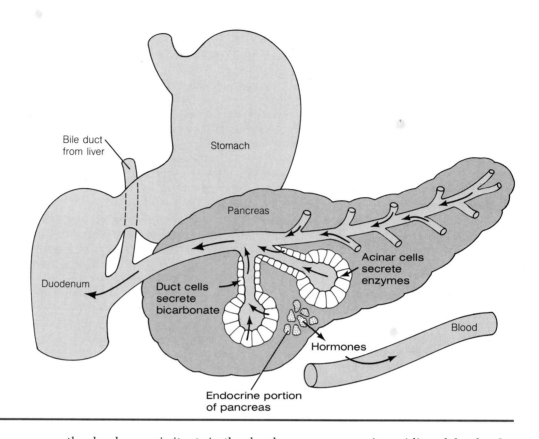

Figure 24.6
Diagrammatic representation of the pancreas. The glandular areas have been greatly enlarged.

the duodenum, irritants in the duodenum, an excessive acidity of the duodenal chyme, a relative hypertonicity or hypotonicity of the chyme, or the presence of breakdown products of protein digestion leads to a neural inhibition of gastric secretion by way of an enterogastric reflex. In this reflex, nerve impulses that inhibit gastric secretion are transmitted from the duodenum to the stomach both by way of the intrinisic nerve plexuses and by way of the vagus nerves and central nervous system.

kol-e-sis-toe-kine'-in In addition, the acidity of the duodenal chyme and the presence of certain amino acids and fatty acids lead to the release of intestinal hormones, such as secretin, cholecystokinin, and gastric inhibitory peptide, that enter the blood and travel to the stomach. When the stomach is actively secreting and the mechanisms that cause gastrin release are strongly stimulated, these hormones have an inhibitory effect on gastric secretion.

Protection of the Stomach

The mucosal surface of the stomach is composed of epithelial cells that are connected by tight junctions, and the stomach is believed to produce surface-active phospholipds that form an adsorbed hydrophobic layer between the gastric epithelium and the stomach contents. Moreover, the surface of the stomach is coated with a viscous, alkaline mucus. Together, these factors provide a protective barrier that prevents the hydrochloric acid and enzymes released into the stomach from digesting the stomach itself. In addition, damaged stomach cells are rapidly replaced, and the lining of the stomach is renewed about every three days.

SMALL INTESTINE AND ASSOCIATED STRUCTURES

The most important portion of the gastrointestinal tract as far as digestion and absorption are concerned is the small intestine and its associated structures. Food that arrives at the small intestine has not been completely digested and is not yet prepared for absorption. Some carbohydrate digestion is begun in the mouth by ptyalin, and the protein-digesting enzymes of the stomach pro-

tie'-a-lin

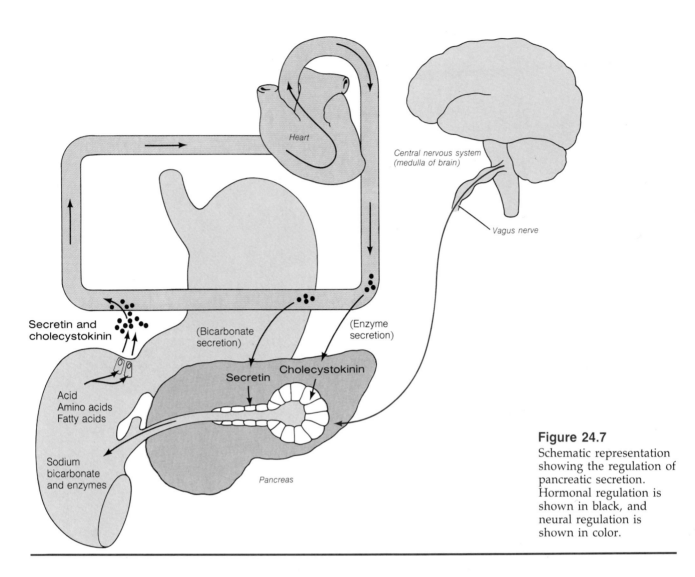

Heart

Central nervous system
(medulla of brain)

Vagus nerve

Secretin and
cholecystokinin

(Bicarbonate
secretion)

(Enzyme
secretion)

Secretin   Cholecystokinin

Acid
Amino acids
Fatty acids

Sodium
bicarbonate
and enzymes

Pancreas

Figure 24.7
Schematic representation
showing the regulation of
pancreatic secretion.
Hormonal regulation is
shown in black, and
neural regulation is
shown in color.

duce polypeptide chains of amino acids, but carbohydrate and protein digestion remain incomplete. Lipids that arrive at the intestine are largely undigested. Thus, most digestive activity occurs within the small intestine, as
does practically all of the absorption.

Three fluids are present in the small intestine: pancreatic juice, bile, and
intestinal juice. Substances within these fluids continue the processes of digestion. Many of the digestive processes are ultimately completed by enzymes associated with the membranes of the microvilli of the intestinal epithelium.

Pancreatic Juice

The pancreas is a dual structure that is both an exocrine gland that produces
digestive enzymes and an endocrine gland that produces hormones (Figure
24.6). The exocrine portion of the pancreas is composed of acinar cells and
collecting ducts. The ducts from the pancreas join and open into the duodenal
portion of the intestine, a short distance below the pyloric sphincter.

The exocrine portion of the pancreas produces an aqueous, isotonic fluid
with a high bicarbonate ion concentration and a basic pH (about 8.0). In the
intestine, this fluid, particularly the bicarbonate ions, helps neutralize the
acidic chyme from the stomach. The exocrine pancreas also produces carbo-
hydrate-, protein-, and lipid-digesting enzymes, as well as ribonuclease and
deoxyribonuclease, which are enzymes that digest RNA and DNA, respectively. The combined secretions of the exocrine portion of the pancreas make
up the pancreatic juice.

The exocrine secretory activities of the pancreas are controlled both hormonally and neurally (Figure 24.7). The intestinal hormone secretin, which is

F24.6

F24.7

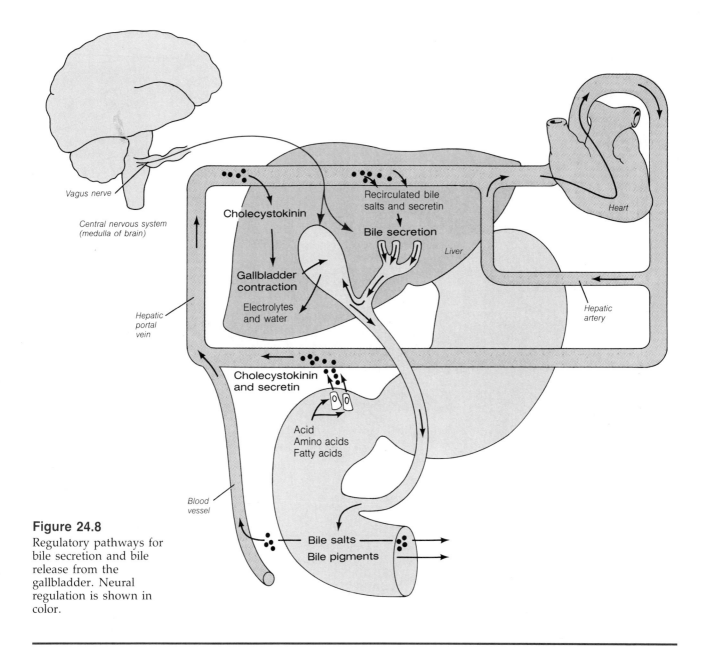

Figure 24.8
Regulatory pathways for bile secretion and bile release from the gallbladder. Neural regulation is shown in color.

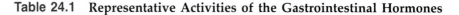

Table 24.1   Representative Activities of the Gastrointestinal Hormones

Gastrin Stimulates gastric secretion, particularly the secretion of hydrochloric acid

Secretin Inhibits gastric acid secretion when the stomach is actively secreting and the mechanisms that cause gastrin release are strongly stimulated
Stimulates the release of a watery fluid containing bicarbonate from the pancreas
Stimulates the secretion of bile

Cholecystokinin Inhibits gastric acid secretion when the stomach is actively secreting and the mechanisms that cause gastrin release are strongly stimulated
Stimulates the release of enzymes from the pancreas
Stimulates the contraction of the gallbladder

Gastric inhibitory peptide Inhibits gastric acid secretion when the stomach is actively secreting and the mechanisms that cause gastrin release are strongly stimulated

released primarily in response to the presence of hydrochloric acid, stimulates the release from the pancreas of a watery fluid that contains large amounts of bicarbonate ions. A second intestinal hormone, cholecystokinin, which is released principally in response to the presence of certain amino acids and fatty acids, mainly promotes the release of digestive enzymes from the pancreas.

Neurally, pancreatic secretion is stimulated by way of the vagus nerves, and the effect is mostly on enzymatic secretion. This response is most evident during the cephalic and gastric phases of stomach secretion.

Bile

The secretory and excretory activities of the liver continually produce bile in amounts averaging about 600–1000 ml per day. The bile travels by ducts from the liver, where it is produced, and the gallbladder, where it is stored, to the duodenum. Bile is an aqueous solution of water and electrolytes such as sodium and bicarbonate that contains bile salts (of cholic and chenodeoxycholic acids), bile pigments (such as bilirubin from the breakdown of hemoglobin), cholesterol, neutral fats, and lecithin. Approximately 94% of the bile salts released into the duodenum are reabsorbed in the ileum. They are then returned to the liver by the blood and resecreted. This cycle is known as *enterohepatic circulation*.

The rate of bile secretion is chemically, neurally, and hormonally controlled (Figure 24.8). Bile salts present in the plasma as the result of enterohepatic circulation are powerful choleretics—that is, chemical substances that stimulate bile flow. Stimulation of the vagus nerves can also increase the rate of hepatic bile secretion. The hormone secretin increases the secretion of bile, but it enhances primarily water and sodium bicarbonate secretion and not the rate of bile acid production. The hepatic blood flow also influences bile secretion, and within limits, the flow of bile increases as the hepatic blood flow increases.

F24.8

During periods when large quantities of bile are not required for digestion, bile is stored in the gallbladder, which has an approximate capacity of 40–70 ml. Within the gallbladder, water and electrolytes are rapidly absorbed, and the concentration of bile salts and pigments may increase five to ten times. Shortly after the ingestion of a meal, the gallbladder contracts and bile is released into the duodenum. The contraction of the gallbladder is stimulated primarily by the hormone cholecystokinin. Vagal stimulation can also cause weak contractions of the gallbladder. A summary of the activities of the gastrointestinal hormones (gastrin, secretin, cholecystokinin, and gastric inhibitory peptide) is presented in Table 24.1.

Table 24.1

Intestinal Secretions

Mucus

The first part of the duodenum contains *duodenal glands* (*Brunner's glands*), which secrete mucus. The mucus provides a protective coat for the intestinal mucosa. The duodenal glands produce mucus in response to tactile or irritating stimulation of the mucosa, vagal stimulation, and intestinal hormones, particularly secretin. Cells called *goblet cells* that are located on the surface of the intestinal mucosa also produce mucus in response to tactile or chemical stimulation of the mucosa. The intestinal glands (crypts of Lieberkuhn) also contain mucous-producing goblet cells whose secretory activity is probably controlled by local nervous reflexes.

Intestinal Juice

Intestinal glands, which are located over essentially the entire surface of the small intestine, produce an intestinal juice that has a pH of 6.5–7.5 and is isotonic with the plasma. The actual fluid secretions of the intestinal glands contain almost no digestive enzymes. Almost all of the enzymes in the intestinal juice are derived from the disintegration of enzyme-rich epithelial cells that are constantly being discharged from the tips of the intestinal villi.

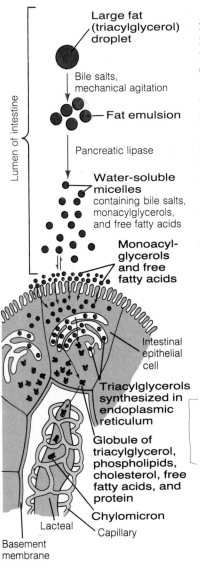

Figure 24.9
Processes of fat digestion
and absorption, and
chylomicron formation.

The main stimuli for the production of intestinal juice are local reflexes initiated by such things as tactile or irritating stimulation of the intestinal mucosa. Hormonal substances such as secretin and cholecystokinin also stimulate the secretion of intestinal juice, but these hormonal substances are not considered to be major controlling factors.

The membranes of the microvilli of the intestinal epithelium contain hydrolytic enzymes that are important in digestion. Among these are several different disaccharidases that participate in carbohydrate digestion and a number of peptidases that are involved in protein digestion. The microvilli also play an important role in the absorption of materials from the intestinal lumen.

Digestion and Absorption in the Small Intestine

Carbohydrates

Plant starches and sucrose are the major dietary carbohydrates. Small amounts of lactose (milk sugar) may also be present. Amylases in both the saliva (salivary amylase) and the pancreatic juice (pancreatic amylase) break starch into the disaccharides maltose and isomaltose (Table 24.2). Maltose and isomaltose are each composed of two monosaccharide glucose units. Sucrose is a disaccharide composed of the monosaccharides glucose and fructose. The disaccharide lactose is made up of glucose and galactose. The final digestion of these disaccharides to absorbable monosaccharide units is completed by the disaccharidase enzymes (maltase, isomaltase, sucrase, and lactase) of the microvilli of the small intestine.

Sugars are absorbed primarily in the duodenum and upper jejunum of the small intestine, and their absorption is practically complete by the time the chyme reaches the ileum. Glucose is absorbed by active transport, and galactose is transported by the same carrier as glucose. Fructose is apparently absorbed by passive processes.

Proteins

The pancreatic juice contains the inactive protein-digesting enzymes trypsinogen, chymotrypsinogen, and procarboxypeptidase. In the intestine, trypsinogen is converted to active **trypsin** by a substance called *enterokinase* from the intestinal mucosa and by previously formed trypsin. In turn, trypsin converts chymotrypsinogen to active **chymotrypsin,** and procarboxypeptidase to active **carboxypeptidase.** These active enzymes further the protein digestion begun in the stomach by pepsin (Table 24.2). Trypsin splits peptide bonds involving basic amino acids such as lysine and arginine, producing small peptide chains of amino acids. Chymotrypsin splits amino acid bonds involving aromatic amino acids such as tyrosine and phenylalanine, also producing small peptide chains. Carboxypeptidase frees the terminal amino acid from the carboxyl (acid) end of an amino acid chain. Together, the actions of these enzymes (as well as pepsin in the stomach) degrade proteins into an assortment of individual amino acids and small peptide chains.

A number of peptidase enzymes of the microvilli of the mucosal epithelium continues the process of protein digestion. Aminopeptidases liberate the terminal amino acid from the amino end of a small peptide chain. Tetrapeptidases split the final amino acid from tetrapeptides (chains of four amino acids), producing tripeptides and free amino acids, and tripeptidases split tripeptides into dipeptides and free amino acids.

Much protein digestion occurs in the upper portion of the small intestine, and protein is 60–80% digested and absorbed by the time the chyme reaches the ileum. During absorption, amino acids, dipeptides, and even tripeptides are taken into absorptive intestinal epithelial cells by active transport. Within the cells, dipeptidase and tripeptidase enzymes split the dipeptides and tripeptides into their component amino acids, and individual amino acids leave the cells and enter the bloodstream.

Lipids

Lipase enzymes in the stomach can digest some fat. However, dietary fat, which consists mostly of triacylglycerols, is digested primarily in the small intestine. The first step in the intestinal digestion of fat is emulsification—that is, the dispersal of large water-insoluble fat droplets into a suspension of fine droplets that provide an increased surface area on which the water-soluble, fat-digesting enzymes can act (Figure 24.9). (Bile salts produced by the liver and released into the intestine with the bile are important in this process.) The small, emulsified fat particles are digested by lipase enzymes, primarily from the pancreas (Table 24.2). This results mainly in the formation of monoacylglycerols and free fatty acids.

F24.9

Table 24.2

Bile salt molecules are amphipathic, and they can aggregate with one another to form small, water-soluble structures called micelles (see Chapter 2). The monoacylglycerols and free fatty acids formed during fat digestion associate with bile salt micelles, and it is in this form that they reach the intestinal epithelium. During absorption, the monoacylglycerols and free fatty acids are believed to leave the micelles and enter absorptive epithelial cells by simple diffusion. The bile salts, which can be reused, are ultimately absorbed in the ileum as part of the enterohepatic circulation process.

CLINICAL CORRELATION

Malabsorption

CASE REPORT . . . .

THE PATIENT   A 12-year-old boy

PRINCIPAL COMPLAINT   Failure to grow

HISTORY   The patient had an older brother who died of "pneumonia" at age 3 years; three other siblings are alive and well. He had been suffering from "asthma" since approximately 6 or 7 years of age, and had two episodes of pneumonia over the past three years. Between these episodes he had an increase in coughing and production of purulent sputum (which contains pus). He grew normally until the age of 7 or 8 years, but at the time of admission was in a lower percentile for height and weight. His parents considered him a sickly child, and he had not been able to maintain a normal play schedule with his friends, particularly in the summer. He had a large appetite, with no particular food intolerance. He described his bowel habits as normal. However, his feces were large in mass, light colored, poorly formed, somewhat foul smelling, and floated in water.

CLINICAL EXAMINATION   Laboratory examination of the feces revealed a fat content that comprised about 30% of the amount ingested (normal: less than 6%). Examination of his serum revealed a lowered protein concentration and a prolonged prothrombin time (a test of blood clotting ability). All of these findings are consistent with cystic fibrosis.

COMMENT   Cystic fibrosis is an inherited disease in which the central problem is the secretion of an unduly viscous mucus. Infection of the air passages (pneumonia) occurs because the cilia do not adequately clear the mucus toward the trachea. Other obstructed or partly obstructed tubes (for example, ureters, auditory tubes) usually grow excess bacteria. The sweat contains high concentrations of sodium and chloride ions; hence, in hot weather salt depletion is a problem. The patient had failed to grow normally because of a decreased absorption of food, although this problem was partly offset by his large appetite. The large amount of feces signifies malabsorption, and the foul smell is caused by bacterial degradation of unabsorbed foodstuffs. The low concentration of serum protein is a sign of severe undernutrition, specifically of failure to absorb adequate protein (amino acids). The prolonged prothrombin time is caused by deficient absorption of vitamin K. (Prothrombin is formed in the liver by a process that requires vitamin K.) The malabsorption evident in cystic fibrosis is related to blockage of the pancreatic ducts by the viscid mucus and replacement of the pancreatic acini by fibrous tissue. The digestive enzymes, amylase (which digests carbohydrate), lipase (which digests fat), and the proteinases, are not secreted. Thus, digestion of all three major classes of foodstuff is inadequate, and these nutritive materials are poorly absorbed.

OUTCOME   The patient was given pancreatic enzymes by mouth, protected by antacids, and he began to grow. However, because of the effect of the viscous mucus on pulmonary function, his life span may be shorter than normal.

During their entry into absorptive intestinal epithelial cells, many of the monoacylglycerols are digested to glycerol and fatty acids by an epithelial cell lipase. Within the endoplasmic reticulum of the epithelial cells, the free fatty acids combine with newly synthesized glycerol and small amounts of glycerol from the digested monoacylglycerols to form triacylglycerols. In addition, the cells synthesize phospholipids, cholesterol, and proteins. Small protein-coated globules of triacylglycerols, synthesized or absorbed phospholipids and cholesterol, and some free fatty acids leave the epithelial cells and enter the lacteals of the lymphatic system as minute droplets known as chylomicrons. Chylomicrons are, by weight, about 90% triacylglycerols, 5% phospholipids, 4% free fatty acids, 1% cholesterol, and a small amount of protein. Fat absorption usually takes place in the duodenum and jejunum and is completed in the ileum.

Ingested cholesterol occurs both as free cholesterol and as cholesterol esters (cholesterol combined with a fatty acid). The cholesterol esters are digested to free cholesterol and free fatty acids by a pancreatic cholesterol esterase (Table 24.2). Free cholesterol can associate with micelles, and it is ultimately absorbed at the intestinal epithelium.

Table 24.2

Vitamins

Fat-soluble vitamins such as A, D, E, and K associate with micelles, and they are absorbed in conjunction with fat digestion. Water-soluble vitamins, such as vitamin C and the B vitamins (with the exception of vitamin B_{12}), are absorbed by passive transport in the proximal portion of the intestine. Vitamin B_{12} combines with a glycoprotein substance, intrinsic factor, from the stomach and is actively absorbed across the intestinal wall at specific sites in the ileum.

Water and Electrolytes

The small intestine can absorb about 200–400 ml of water per hour. Five to ten liters of water derived from food, drink, and digestive secretions enter the small intestine each day, but only about 0.5 liter enters the large intestine. The rest is absorbed, primarily through the upper small intestine. Water can move across the intestinal wall in both directions, and the absorption of water from the intestine occurs osmotically in association with the absorption of particulate materials. In fact, the chyme within the duodenum becomes isosmotic with the blood plasma as a result of the osmotic removal or addition of water.

An acitve transport mechanism for sodium is present within the epithelial cells of the mucosa, particularly in the jejunum, and sodium is actively absorbed. Potassium, magnesium, and phosphate can also be actively absorbed through the mucosa. Chloride movement passively follows that of sodium in the upper part of the small intestine. However, chloride is actively transported in the ileum by a process in which the absorption of chloride from the intestine is coupled with the secretion of bicarbonate into the intestine. Calcium absorption occurs actively along the entire small intestine (particularly the duodenum) and requires vitamin D. The amount of calcium absorbed varies with body requirements and is controlled in part by hormones from the parathyroid glands.

LARGE INTESTINE

The free surface of the large intestine is covered with epithelial cells, but no villi are present. The mucosa possesses abundant mucous-secreting goblet cells and intestinal glands. The epithelial cells of the intestinal glands produce almost no enzymes, and the intestinal glands consist almost entirely of mucous-secreting goblet cells. The rate of mucous production within the large intestine increases in response to direct tactile stimulation of the surface goblet cells and as a result of local reflexes that involve the goblet cells of the intestinal glands. Extrinsic innervation also influences mucous secretion.

Table 24.2 Principal Enzymes Involved in the Digestion of Carbohydrates, Proteins, and Lipids[*]

Enzyme	Source	Action
CARBOHYDRATES		
Salivary amylase	Salivary glands	Splits starch into the disaccharides maltose and isomaltose
Pancreatic amylase	Pancreas	Splits starch into the disaccharides maltose and isomaltose
Disaccharidases	Intestinal epithelium	Split disaccharides into monosaccharides
Maltase		Splits maltose into glucose
Isomaltase		Splits isomaltose into glucose
Sucrase		Splits sucrose into glucose and fructose
Lactase		Splits lactose into glucose and galactose
PROTEINS		
Pepsin (Pepsinogen)	Zymogenic cells of stomach	Splits proteins into smaller peptide chains of amino acids
Trypsin (Trypsinogen)	Pancreas	Splits proteins and peptides into smaller peptides
Chymotrypsin (Chymotrypsinogen)	Pancreas	Splits proteins and peptides into smaller peptides
Carboxypeptidase (Procarboxypeptidase)	Pancreas	Frees terminal amino acid from the carboxyl (acid) end of an amino acid chain
Aminopeptidases	Intestinal epithelium	Free terminal amino acid from the amino end of an amino acid chain
Tetrapeptidases	Intestinal epithelium	Split final amino acid from tetrapeptides, producing tripeptides and free amino acids
Tripeptidases	Intestinal epithelium	Split tripeptides into dipeptides and free amino acids
Dipeptidases	Intestinal epithelium	Split dipeptides into free amino acids
LIPIDS		
Pancreatic lipase	Pancreas	Splits triacylglycerols mainly into monoacylglycerols and free fatty acids
Epithelial cell lipase	Intestinal epithelium	Splits monoacylglycerols into glycerol and fatty acids
Cholesterol esterase	Pancreas	Splits cholesterol esters into free cholesterol and free fatty acids

[*]Substances in parentheses are precursors of active enzymes.

Sodium is actively absorbed in the large intestine, and chloride absorption also occurs. Bicarbonate ions are actively secreted by the large intestine. In addition, the large intestine absorbs 300–400 ml of water per day as a consequence of sodium (and chloride) transport. Many microorganisms inhabit the large intestine, and some of the bacteria produce vitamins, such as vitamin K.

The fecal material that ultimately leaves the digestive tract consists of water and solids such as undigested food residue, microorganisms, and sloughed-off epithelial cells.

CONDITIONS OF CLINICAL SIGNIFICANCE

Digestion

In addition to conditions such as vomiting, diarrhea, and constipation, which are discussed in Chapter 23, other abnormal conditions of the gastrointestinal tract or its accessory structures commonly occur.

Peptic Ulcer

A *peptic ulcer* is an erosion of the wall of the gastrointestinal tract in an area of the tract exposed to gastric juice containing acid and pepsin. Peptic ulcers are most commonly found in the stomach (gastric ulcers) and the duodenum (duodenal ulcers). They may be caused by an excessive acid-pepsin secretion, or they may result from an insufficient secretion of mucus, which normally protects the gastrointestinal mucosa from being digested by the acid-pepsin gastric juice.

The most common symptom of peptic ulcer is pain. However, in some cases, the pain can be temporarily relieved by the ingestion of food, apparently because the food provides some protection by coating the ulcer or by acting as a buffer.

If the erosion due to a gastric ulcer is sufficiently severe, blood vessels in the stomach wall are damaged and bleeding occurs into the stomach itself (a bleeding ulcer). In extreme cases, a peptic ulcer can lead to a perforation—that is, to a hole entirely through the wall of the gastrointestinal tract. A perforation allows the contents of the gastrointestinal tract to pass into the abdominal cavity. Such an occurrence is very serious and requires immediate surgery.

Gastroenteritis

Gastroenteritis is an acute or chronic inflammation of the mucosa of the stomach and intestine. It is often caused by irritants such as excessive alcohol or cathartics (medicines that cause bowel movements), but it can have any of a wide variety of other causes, including viral infections, food allergies, overeating, and so forth.

Gallstones

Gallstones are particles—composed primarily of cholesterol or bile pigments—that sometimes form in the bile, particularly in the gallbladder. The stones can block the cystic duct from the gallbladder or the common bile duct to the duodenum. Gallstones are often painful, and those that block the cystic duct are especially so because the contractions of the gallbladder exert force on them.

In the case of cystic gallstones, bile is still able to reach the duodenum directly from the liver through the common bile duct. Blockage of the common bile duct, on the other hand, prevents bile from reaching the intestine and thus interferes with the proper absorption of fat. In addition, the bile pigments cannot reach the intestine to be excreted. These accumulate in the blood and are eventually deposited in the skin. This produces a yellow color in the skin, a condition that is called *jaundice*.

Pancreatitis

Pancreatitis is an inflammation of the pancreas that is often caused by the digestion of parts of the organ by pancreatic enzymes that are normally carried to the small intestine within the pancreatic ducts. In pancreatitis, the enzymes become activated within the ducts, and they may destroy the ducts and the pancreatic cells. Since pancreatic enzymes are very important in the digestion of carbohydrates, proteins, and fats, pancreatitis can produce severe nutritional problems.

Hepatitis

Hepatitis is an inflammation of the liver. Most commonly, it is due to a viral infection transmitted either by virus-infected blood or by contaminated food or water. The infected liver becomes enlarged, and its functioning is impaired, which can lead to jaundice. In some cases, liver function is depressed for a year or more.

Cirrhosis

Cirrhosis is a chronic inflammation of the liver that is progressive and diffuse. In the affected areas of the liver, some cells are replaced by fibrous connective tissue, thereby interfering with liver function. Cirrhosis can cause a reduction in the production of bile, in the excretion of bile pigments, and in the production of blood-clotting factors and plasma albumin. Cirrhosis also reduces the liver's ability to detoxify the blood, and toxins accumulate.

Appendicitis

The appendix is a blind-end pouch extending from the cecum. If it becomes obstructed—for example, with hardened fecal material—its venous circulation may be interfered with. Such interference reduces the oxygen supply to the area and permits bacteria to flourish. The appendix then becomes inflamed and filled with pus, a condition called *appendicitis*.

If an inflamed appendix is not removed soon enough it can rupture and release its contents into the abdominal cavity. This occurrence can cause peritonitis, which is an inflammation of the lining of the abdominopelvic cavity.

FRONTIERS IN HEALTH
Removing Gallstones the Easy Way

Jennifer has an odd, yellowish color to her skin, a condition called jaundice. She also experiences increasingly frequent episodes of severe, steady pain in the upper abdomen, accompanied by chills and nausea. Like many other people, Jennifer is suffering from gallstones.

Approximately 25 million Americans have gallstones. Each year a million Americans learn that they have them; about half of them will require surgery. Removal of the gallbaldder—an operation called a cholecystectomy—is one of the most common operations performed in the United States. It is estimated that gallstones cost more than a billion dollars a year in medical expenses and lost work time.

The most common gallstones are composed of cholesterol, with small amounts of calcium. Physicians have long been seeking a way to dissolve stones before they start causing pain by migrating into the cystic duct or the common bile duct, where they may obstruct the flow of bile. Such obstruction can cause severe infections in the gallbaldder and the pancreas.

Medical researchers may have a partial cure for gallstones. They have found a naturally occurring bile acid, chenodeoxycholic acid (CDCA) which, when taken orally, leads to the dissolution of cholesterol stones. CDCA is used in some 40 countries and has been approved for regular use in the United States by the Food and Drug Administration. It is sold under the name Chenodiol. However, CDCA must be taken for months, or sometimes for as long as 2 years, before even small gallstones disappear. For patients like Jennifer, whose conditions are acute, CDCA is too slow to provide relief from pain and possible infection. CDCA is also ineffective in individuals whose gallbladders no longer function properly. Nor does it seem to be effective against larger stones, or stones containing more than 4% calcium. Regardless of these disadvantages, thousands of people who cannot risk an operation, such as the elderly and those with heart disease, may find relief using this drug.

In addition to being very slow acting, Chenodiol has some side effects. Liver changes and diarrhea are the most common, but they are usually mild and disappear once treatment stops. In the search for a replacement for CDCA, medical researchers have begun testing a bile-type acid that has fewer side effects and seems to work faster than CDCA. The new drug is ursodeoxycholic acid (UDCA), and is called ursodiol.

Medical experts cautiously warn the public not to expect miracles from CDCA or UDCA. These drugs completely eliminate stones in only one of every five patients. Moreover, gallstones reform in up to half of all patients whose stones have been completely eliminated.

Suppose that a gallstone does move into the common bile duct, as it did in Jennifer's case. Can surgery still be avoided? Thanks to some pioneering research, there is a relatively painless way to dislodge stones that have become stuck in the common bile duct. This new method uses an instrument called a flexible endoscope. The endoscope is a flexible tube, which is swallowed by

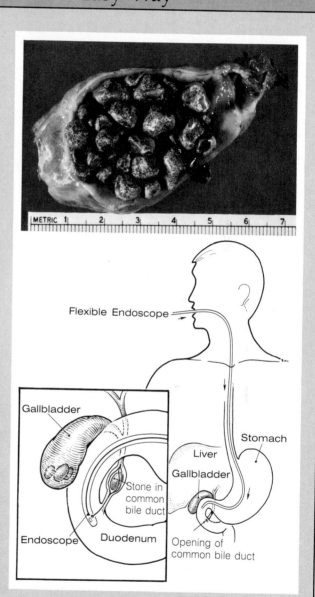

Top: Photo of gallbladder with gallstones. Bottom: Insertion of endoscope to the common bile duct.

the patient and guided by the physician through the stomach and into the small intestine. From there it is maneuvered into the common bile duct. Inserting the tube into the bile duct widens the duct, and this often suffices to flush the stone into the duodenum. In other cases, a small wire snare inserted through the tube is used to pull the stone out.

The endoscope contains thousands of tiny glass fibers that project light and convey a sharp image of the area being examined back to the physician. Unfortunately, the endoscope can only be used to remove stones from the ducts draining the gallbladder. If there are stones within the gallbladder, surgery is often required.

STUDY OUTLINE

MOUTH AND SALIVA   Saliva is produced by parotid, sublingual, submandibular, and buccal glands. pp. 715–716

SECRETIONS OF THE SALIVARY GLANDS

MUCINS   Major proteins of saliva; form mucus with water; mucus lubricates mouth and food.

SALIVARY AMYLASE   An enzyme that breaks down large starch molecules.

COMPOSITION OF SALIVA   97–99.5% water; pH 6.0 to 7.0; contains electrolytes (sodium, potassium, chloride, bicarbonate ions); kallikrein (leads to production of bradykinin, a vasodilator).

FUNCTIONS OF SALIVA   Lubrication; digestion; bolus formation; dissolves foods, aids taste; moistens mouth and throat; aids speech.

CONTROL OF SALIVARY SECRETION   Both parasympathetic and sympathetic stimulations cause secretion, with parasympathetic stimulation having greatest effect.

STOMACH pp. 716–720

GASTRIC SECRETIONS

MUCUS   Lubricates stomach walls; protects gastric mucosa; produced by cardiac, pyloric, and gastric glands.

HYDROCHLORIC ACID   Produced by parietal (oxyntic) cells; facilitates protein digestion; kills bacteria.

PEPSINOGEN   Produced by zymogenic (chief) cells; digests proteins when activated to pepsin.

CONTROL OF GASTRIC SECRETION

CEPHALIC OR REFLEX PHASE   Elicited by sight, smell, or taste of food.
1. Sensory receptors to CNS to vagus nerves to stomach for stimulation of gastric secretion.
2. Response elicted by sight and smell of food is basically a conditioned reflex.

GASTRIC PHASE   Elicited by distension of stomach; local reflexes; vagovagal reflex, gastrin.

INTESTINAL PHASE   Elicited by presence of chyme in upper portion of small intestine; gastrin production.

INTESTINAL FACTORS THAT INHIBIT GASTRIC SECRETION   An enterogastric reflex; under proper conditions, intestinal hormones.

PROTECTION OF THE STOMACH   Tall, columnar epithelial cells; alkaline mucous secretions.

SMALL INTESTINE AND ASSOCIATED STRUCTURES   Most digestion and almost all absorption. pp. 720–726

PANCREATIC JUICE
1. Basic—helps neutralize acid chyme from stomach.
2. Carbohydrate-, protein-, and lipid-digesting enzymes.
3. Ribonuclease and deoxyribonuclease.
4. Hormonal control by secretin, cholecystokinin.

5. Neural control by vagus nerves.

BILE   Produced in liver; stored in gallbladder; travels to duodenum.
1. Contains bile salts, bile pigments, cholesterol, neutral fats, and lecithin.
2. Chemical control by choleretics.
3. Neural control by vagus nerves.
4. Hormonal control by secretin.
5. Gallbladder contraction stimulated by cholecystokinin.

INTESTINAL SECRETIONS

MUCUS   Produced by duodenal glands (Brunner's glands), goblet cells, and intestinal glands; protects intestinal mucosa.

INTESTINAL JUICE   Produced by intestinal glands; isotonic with plasma; contains almost no digestive enzymes; local reflex control. Microvilli of intestinal epithelial cells contain digestive enzymes.

DIGESTION AND ABSORPTION IN THE SMALL INTESTINE

CARBOHYDRATES
1. Amylases break starch into disaccharides maltose and isomaltose.
2. Disaccharidase enzymes of microvilli break disaccharides into monosaccharides.
3. Glucose and galactose absorbed by active transport.

PROTEINS
1. Enterokinase and previously formed trypsin convert trypsinogen to trypsin.
2. Trypsin activates chymotrypsinogen and procarboxypeptidase.
3. Trypsin, chymotrypsin, and carboxypeptidase produce assortment of individual amino acids and small peptide chains.
4. Peptidase enzymes of intestinal epithelium continue process of protein digestion.
5. Amino acids absorbed by active transport.

LIPIDS   (Digested primarily in small intestine.)
1. Emulsification; bile salts are important.
2. Lipase enzymes digest small, emulsified fat particles mainly to monoacylglycerols and free fatty acids, which associate with bile salt micelles.
3. Absorption probably by simple diffusion.
4. Triacylglycerols resynthesized in intestinal epithelial cells.
5. Chylomicrons leave cells and enter lymph.

VITAMINS
1. Fat-soluble vitamins associate with micelles; absorbed in conjunction with fat digestion.
2. Water-soluble vitamins except B_{12} absorbed by passive transport in proximal intestine.
3. B_{12} combines with intrinsic factor; actively absorbed at specific ileum sites.

WATER AND ELECTROLYTES   Water absorbed mainly in small intestine; sodium can be actively absorbed; potassium, magnesium, and phosphate can also be actively absorbed; calcium absorption requires vitamin D.

LARGE INTESTINE   Epithelial cells, but no villi;

essentially no enzymes; mucous-secreting goblet cells. Absorption—sodium, chloride, and small amount of water. Fecal matter—composed of water, mucus, undigested food residue, microorganisms, and sloughed-off epithelium. Bacteria produce certain vitamins. p. 726

CONDITIONS OF CLINICAL SIGNIFICANCE: DIGESTION p. 728

PEPTIC ULCER   Erosion of wall of gastrointestinal tract in area of tract exposed to gastric juice containing acid and pepsin.

GASTROENTERITIS   Acute or chronic inflammation of mucosa of stomach and intestine; caused by irritants, viral infection, or food allergy.

GALLSTONES   Particles containing cholesterol and bile salts in bile; can block cystic or common bile ducts.

PANCREATITIS   Inflammation of pancreas; often caused by digestion of parts of organ by pancreatic enzymes.

HEPATITIS   Inflammation of liver, usually due to virus; causes enlargement of liver and impaired liver function.

CIRRHOSIS   Chronic liver inflammation; may result in reduced bile production, reduced bile pigment excretion, reduced blood-clotting-factor production, and blood toxin accumulation.

APPENDICITIS   Inflammation of appendix; can be due to obstruction that interferes with venous circulation.

SELF-QUIZ

1. Salivary amylase is produced primarily by which glands? (a) buccal; (b) duodenal; (c) parotid.

2. The optimal pH at which salivary amylase splits large starch molecules into smaller fragments is: (a) 6.9; (b) 4.2; (c) 8.4; (d) 11.7.

3. Saliva contains: (a) pepsin; (b) vitamin K; (c) kallikrein.

4. Chewing stimulates salivary secretion, whereas severe mental effort, dehydration, fear, and anxiety tend to reduce the flow of saliva. True or False?

5. The salivary substance that converts protein precursors to bradykinin is: (a) kallikrein; (b) salivary amylase; (c) mucin.

6. Parietal cells produce: (a) pepsinogen; (b) mucus; (c) hydrochloric acid.

7. Hydrochloric acid functions primarily to promote the absorption of vitamin B_{12} in the stomach and also kills many of the bacteria that enter the digestive tract with the food. True or False?

8. The chloride ions of the hydrochloric acid of the stomach are believed to come from: (a) chloride ions derived from carbonic acid; (b) chloride ions from plasma that are transported across parietal cells to the stomach lumen; (c) water molecules that combine with bicarbonate ions.

9. Protein digestion in the stomach proceeds to the stage of the formation of: (a) monoacylglycerols; (b) chylomicrons; (c) peptide chains.

10. Gastric secretion can be initiated by the sight, smell, or taste of food. True or False?

11. Gastrin release is inhibited by the presence of high concentrations of hydrogen ions (acid) in the stomach. True or False?

12. A slight stimulation of gastric secretion may occur in response to the presence of chyme within the upper portion of the small intestine (particularly the duodenum) after the stomach has nearly emptied and gastric secretory activity is relatively low following a meal. True or False?

13. Most digestive activity and practically all absorption occur within the: (a) mouth; (b) small intestine; (c) large intestine.

14. The secretion of pancreatic enzymes occurs in response to: (a) cholecystokinin; (b) salivary amylase; (c) mucin.

15. Pancreatic secretion, particularly enzymatic secretion, can be stimulated neurally by way of the vagus nerves. True or False?

16. Approximately 94% of the bile salts released into the duodenum are: (a) excreted with the feces; (b) reabsorbed in the ileum; (c) converted to bilirubin in the stomach by salivary amylase.

17. Match the following digestive agents with their appropriate lettered descriptions.

    Pancreatic juice
    Tetrapeptidases
    Amylases
    Bile salts
    Cholecystokinin
    Trypsin

    (a) Inhibits gastric acid secretion when the stomach is actively secreting
    (b) Aids in the emulsification of fats
    (c) Breaks certain peptide bonds that involve basic amino acids such as lysine and arginine
    (d) Enzymes of the microvilli of the mucosal epithelium
    (e) Contains the inactive protein-digesting enzymes trypsinogen, chymotrypsinogen, and procarboxypeptidase
    (f) Contained in both the saliva and the pancreatic juice

18. The absorption of which element occurs actively along the entire small intestine (particularly the duodenum) and requires vitamin D? (a) potassium; (b) calcium; (c) sodium.

19. Sodium is actively absorbed in the large intestine, and chloride absorption also occurs there. True or False?

LEARNING OBJECTIVES

After completing this chapter, you should be able to:

1. Outline the events of glycolysis and the Krebs cycle.

2. Describe the movement of electrons through the electron transport system.

3. Discuss the interconversion of carbohydrates, proteins, and triacylglycerols.

4. Explain the utilization of carbohydrate, protein, and triacylglycerols during the absorptive and postabsorptive metabolic states.

5. Distinguish between the two major classes of vitamins.

6. Describe the roles of minerals in the body.

7. Describe the concept of and the determination of the basal metabolic rate.

8. Explain three theories of nutritional regulation.

9. Explain three ways in which heat is transferred between the body and the environment.

10. Distinguish between heat stroke and heat exhaustion.

CHAPTER CONTENTS

UTILIZATION OF GLUCOSE AS AN ENERGY SOURCE

UTILIZATION OF AMINO ACIDS AND TRIACYLGLYCEROLS AS ENERGY SOURCES

OTHER USES OF FOOD MOLECULES

NUTRIENT POOLS

METABOLIC STATES

LIVER FUNCTION

VITAMINS

MINERALS

METABOLIC RATE

REGULATION OF TOTAL BODY ENERGY BALANCE

CONDITION OF CLINICAL SIGNIFICANCE: OBESITY

REGULATION OF BODY TEMPERATURE

CONDITIONS OF CLINICAL SIGNIFICANCE: THERMOREGULATION

KEY TERMS AND DERIVATIVES

coenzymes (*co* = with; *zyme* = cause to ferment) a nonprotein substance associated with and important for the activity of an enzyme

glycolysis (*glyc* = sugar; *lys* = break up) a metabolic pathway by which glucose is broken down into simpler substances

metabolism (*metabol* = change; *ism* = a process) the sum total of the chemical reactions that occur within the body

Metabolism, Nutrition, and Temperature Regulation

25

The chemical reactions that continually occur within the body are grouped together under the term *metabolism*. The metabolic reactions produce heat, and this heat is important in body-temperature regulation. Many metabolic reactions process absorbed nutrients such as glucose, amino acids, and triacylglycerols for use by the body. Portions of these substances are utilized for the synthesis of new structural materials, but the majority are used for the production of adenosine triphosphate (ATP), which provides energy to support the body's activities.

UTILIZATION OF GLUCOSE AS AN ENERGY SOURCE

One of the principal food materials utilized in the formation of ATP is the six-carbon sugar glucose. The breaking of the glucose molecule's chemical bonds releases energy that can be trapped in the form of ATP. In the body, glucose is broken down in a precisely controlled series of reactions that releases energy in a stepwise fashion. Each reaction or step is catalyzed by specific enzymes, often with the help of additional agents such as coenzymes or inorganic ions. The individual reactions are linked together into metabolic pathways in which the product of one reaction is a substrate for the next.

A metabolic pathway known as glycolysis breaks a glucose molecule into two molecules of pyruvic acid, and the energy released in this process is used to produce a small amount of ATP (Figure 25.1). In the absence of oxygen, glycolysis converts the pyruvic acid molecules to lactic acid. However, when oxygen is available, as is usually the case, pyruvic acid is broken down, and carbon dioxide and water are produced. This process, which is accomplished by a pathway called the Krebs cycle and a system known as the electron transport system, releases a considerable amount of energy, which is trapped in the form of ATP.

Glycolysis

The enzymes for the glycolytic reactions are located within the cytoplasm of the cells. (Figure 25.2 depicts the chemical transformations of glycolysis.)

The first step of glycolysis adds a phosphate group to the number-six carbon of a glucose molecule, producing a molecule of glucose-6-phosphate. The second step rearranges the glucose-6-phosphate into a molecule of fructose-6-phosphate. The third step adds a phosphate group to the number-one carbon of the fructose-6-phosphate to produce a molecule of fructose-1,6-diphosphate. The first and third steps each require a molecule of ATP, which provides a phosphate group and leaves behind a molecule of adenosine diphosphate (ADP). Thus, the initial stages of glycolysis do not produce energy, but actually use energy in the form of ATP.

The fourth step of glycolysis converts the six-carbon molecule of fructose-1,6-diphosphate to two three-carbon molecules. These are 3-phosphoglyceraldehyde and dihydroxyacetone phosphate. The fifth step transforms the dihydroxyacetone phosphate to 3-phosphoglyceraldehyde. Thus, each original glucose molecule forms two molecules of 3-phosphoglyceraldehyde, and the

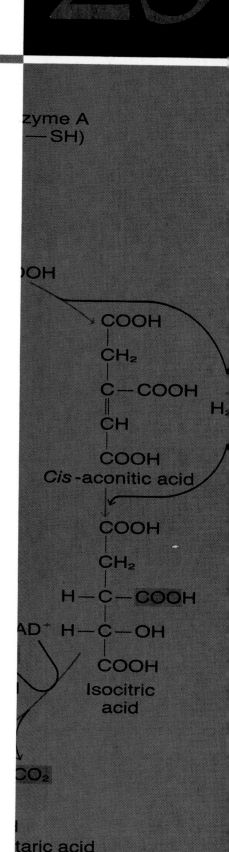

733

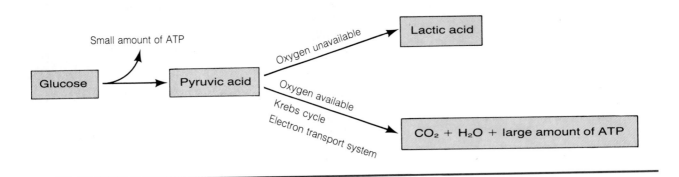

Figure 25.1

A metabolic pathway called glycolysis breaks glucose down to pyruvic acid, and the energy released in this process is used to produce a small amount of ATP. When oxygen is unavailable, glycolysis converts pyruvic acid to lactic acid. When oxygen is available, pyruvic acid is broken down, and carbon dioxide and water are produced. This process, which is accomplished by the Krebs cycle and the electron transport system, releases a considerable amount of energy, which is used to produce a large amount of ATP.

F25.2

remaining steps of glycolysis occur twice for every glucose molecule used. For simplicity's sake, however, each of the steps beyond 3-phosphoglyceraldehyde is shown only once in Figure 25.2.

The sixth glycolytic step converts 3-phosphoglyceraldehyde to a molecule of 1,3-diphosphoglyceric acid. In this step, an inorganic phosphate group is added to 3-phosphoglyceraldehyde. Also, two electrons and a proton (a hydrogen ion, H^+, is a proton) are transferred to the coenzyme nicotinamide adenine dinucleotide (NAD^+), forming NADH, and another proton is released as H^+. The loss of electrons by a molecule is called *oxidation*, and the gain of electrons is called *reduction*. Thus, in this step, the 3-phosphoglyceraldehyde is oxidized and the NAD^+ is reduced.

In terms of energy considerations, the structure of 1,3-diphosphoglyceric acid is especially important. Many high-energy compounds contain phosphate groups and have similar basic configurations. These can be written:

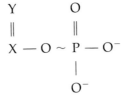

with the ~ indicating a high-energy bond. The Y is usually a carbon or an oxygen, and the X is a carbon or phosphorus. As can be seen from Figure 25.2, 1,3-diphosphoglyceric acid possesses this high-energy configuration. Thus, up to this point, the glycolytic reactions have produced from a single glucose molecule two three-carbon molecules of 1,3-diphosphoglyceric acid with high-energy configurations.

In the seventh glycolytic step, 1,3-diphosphoglyceric acid transfers its high-energy phosphate to a molecule of ADP, converting the ADP to ATP. A molecule of 3-phosphoglyceric acid results from the reaction.

Steps eight and nine of glycolysis rearrange the 3-phosphoglyceric acid into a molecule of phosphoenolpyruvate, which also possesses the high-energy configuration. In step 10, phosphoenolpyruvate transfers its high-energy phosphate to ADP, generating another molecule of ATP. A three-carbon molecule of pyruvic acid results from the reaction.

Up to this point, the glycolytic breakdown of a single glucose molecule to two molecules of pyruvic acid utilizes two ATP molecules and generates four, for a net profit of two. Along the way, two NADH + H^+ are produced. (Recall that the steps beyond 3-phosphoglyceraldehyde occur twice for each glucose molecule used.)

NAD^+ and other coenzymes are present in cells in only small amounts. Consequently, the NAD^+ that forms NADH + H^+ in the sixth glycolytic step must be regenerated if glycolysis is to continue processing glucose molecules. In order to regenerate NAD^+, NADH + H^+ must get rid of two electrons (and the accompanying protons) by transferring them to some acceptor molecule. When oxygen is unavailable, pyruvic acid acts as an electron acceptor, and the glycolytic pathway is completed by step 11, which converts pyruvic acid to lactic acid and regenerates NAD^+. Thus, glycolysis, which breaks a glucose

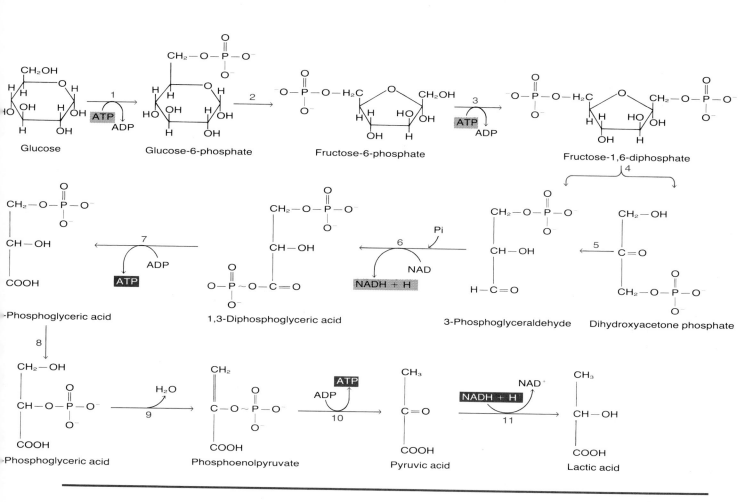

Figure 25.2
The steps of glycolysis by which a molecule of glucose is transformed into two molecules of lactic acid. See text for detailed discussion. "Pi" is inorganic phosphate.

molecule into two molecules of lactic acid, is an anaerobic pathway that requires no molecular oxygen. However, much of the energy contained within the glucose molecule remains within the lactic acid molecules, and the anaerobic breakdown of glucose to lactic acid by glycolysis yields a net profit of only two ATP.

Glycolysis alone cannot produce enough ATP to support the body's activities. However, it does allow cells to produce some ATP under anaerobic conditions. This ability is particularly important in skeletal muscle cells during heavy exercise.

To produce enough ATP to support the body's activities, oxygen-requiring, aerobic processes are necessary. When oxygen is available, as is usually the case, it acts as an electron acceptor by way of the electron transport system, and the pyruvic acid produced in step 10 of glycolysis is not converted to lactic acid. Instead it enters the Krebs cycle and is broken down. These aerobic events release a substantial amount of energy, which is used to produce a large amount of ATP.

Krebs Cycle

The enzymes of the Krebs cycle (or citric acid cycle, as it is also called) are contained within the mitochondria of the cells. In order to enter the Krebs cycle, a molecule of pyruvic acid enters a mitochondrion and combines with a molecule of coenzyme A to form acetyl-coenzyme A (Figure 25.3). During the reaction, one of the three carbons of pyruvic acid is lost as carbon dioxide, and an NADH + H$^+$ is formed. The two-carbon acetyl-coenzyme A molecule enters the Krebs cycle by combining with a four-carbon molecule of oxaloacetic acid to form a six-carbon molecule, citric acid. The Krebs cycle itself involves the stepwise rearrangement of citric acid to reform oxaloacetic acid. Along the

F25.3

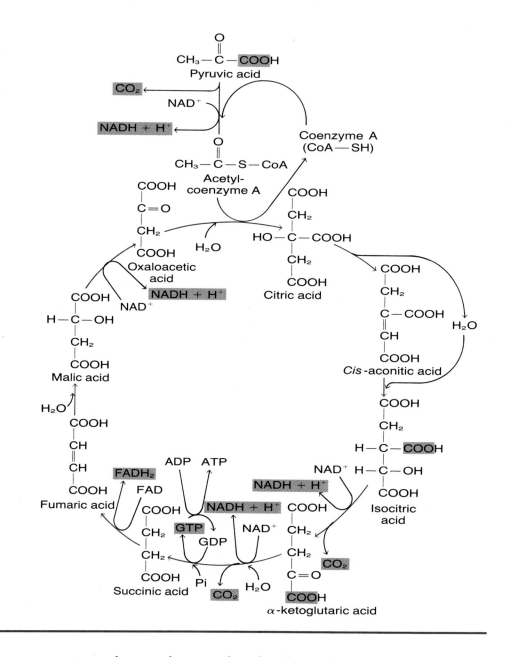

Figure 25.3

The reactions of the Krebs cycle.

way, two carbons are lost as carbon dioxide. At the conclusion of the Krebs cycle, three additional NADH + H$^+$ have been produced, and electrons and protons have been transferred to the coenzyme flavin adenine dinucleotide (FAD), forming FADH$_2$. In addition, a molecule of ATP has been formed from a molecule of guanosine triphosphate (GTP), which was itself formed from guanosine diphosphate (GDP).

Since two pyruvic acid molecules are produced from one glucose molecule, the results of two pyruvic acid molecules that each form acetyl-coenzyme A and enter the Krebs cycle are: six carbon dioxide molecules, two ATP, eight NADH + H$^+$, and two FADH$_2$. As is the case for glycolysis, the coenzymes NAD$^+$ and FAD must be regenerated from NADH + H$^+$ and FADH$_2$ if the Krebs cycle is to continue. This regeneration is accomplished by the oxygen-requiring processes of the electron transport system.

Electron Transport System

The final steps in the aerobic generation of ATP involve the transfer of electrons from the coenzymes NADH + H$^+$ and FADH$_2$ to oxygen, and the combination of protons (H$^+$) with the oxygen to form water and regenerate NAD$^+$ and FAD. These activities are accomplished by the electron transport system,

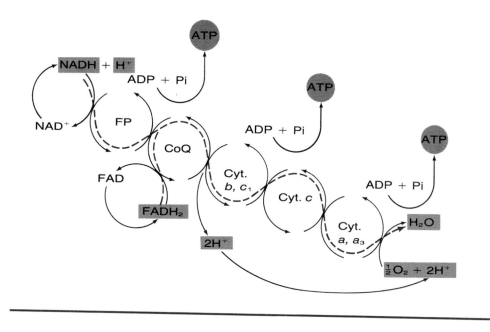

Figure 25.4
The electron transport system. Electrons are passed to various electron acceptors [flavoprotein (FP), coenzyme Q (CoQ), cytochromes b and c_1, cytochrome c, and cytochromes a and a_3]. Each electron acceptor is reduced when it receives electrons and reoxidized when it passes the electrons on to the next acceptor. The dashed line traces the path of electrons. At the end of the chain, oxygen is reduced and water is formed.

which is also located within the mitochondria. The electron transport system consists of a series of electron carriers, each of which can accept and donate electrons. Although not all the carriers are known with certainty, a likely sequence is shown in Figure 25.4. Some of the carriers are coenzymes that incorporate vitamins such as thiamine (vitamin B_1) or riboflavin (vitamin B_2). Others are proteins called cytochromes, which contain a heme group whose iron atom acts as an electron carrier. After being reduced by accepting electrons, each carrier becomes reoxidized by passing the electrons to the next carrier. Ultimately, the electrons are passed to oxygen, which combines with protons to form water.

Each NADH + H^+ produced within the mitochondria during the breakdown of pyruvic acid donates a pair of electrons to the electron transport system. As the electrons are passed from one carrier to another and finally to oxygen, energy is released. At three steps in this process, sufficient energy is released to synthesize ATP from ADP and inorganic phosphate. Thus, three molecules of ATP are produced from each of these NADH + H^+.

The NADH + H^+ generated in the cytoplasm during glycolysis cannot enter the mitochondria to donate electrons directly to the electron transport system. Instead, they donate their electrons to shuttle molecules that can penetrate the mitochondrial membrane and in turn donate the electrons to the electron transport system. One shuttle molecule, malate, donates the electrons before the first site of ATP synthesis, but another, glycerol phosphate, passes electrons to the carriers at a point beyond the first ATP-generation point. Thus, sometimes three and sometimes only two ATP are produced from each of the NADH + H^+ generated during glycolysis, and it is not yet certain which shuttle system predominates in normal living tissue. The electrons from $FADH_2$ also enter the electron transport system after the first ATP-generating step, and only two ATP molecules are produced from each $FADH_2$.

Overall, the aerobic breakdown of a single molecule of glucose produces a net profit of 36–38 ATP (Figure 25.5). Two ATP are generated from glycolysis and two from the Krebs cycle. The eight NADH + H^+ produced within the mitochondria yield 24 ATP in the electron transport system, and the two $FADH_2$ yield four ATP. The two NADH + H^+ from glycolysis yield four to six ATP, depending on the shuttle molecules involved. The process of forming ATP during glycolysis and the Krebs cycle is called *substrate-level phosphorylation,* and the formation of ATP during electron transport is known as *oxidative phosphorylation.*

The uncontrolled combustion of glucose in the presence of oxygen produces carbon dioxide and water and releases about 686,000 calories of energy

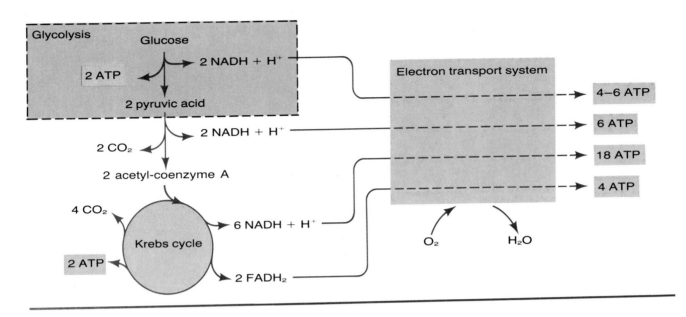

Figure 25.5
The aerobic breakdown of a single molecule of glucose produces a net profit of 36–38 ATP. A net profit of four ATP is produced by substrate-level phosphorylation during glycolysis and the Krebs cycle, and 32–34 ATP are produced by oxidative phosphorylation during electron transport. The glycolytic reactions occur in the cytoplasm of the cell. The other reactions take place within the mitochondria.

per mole (1 calorie is the amount of heat energy required to raise the temperature of 1 gram of water from 15°C to 16°C). Although dependent on pH, the production of a mole of ATP from ADP traps about 7300 calories of energy. Therefore, the net profit to the cell from the aerobic breakdown of a mole of glucose is 277,400 calories of energy ($38 \times 7300 = 277,400$). Thus, by breaking down glucose in a controlled series of reactions, the aerobic processes can trap in the form of ATP about 40% of the energy released. The other 60% appears as heat. Remaining at the end of the reactions are carbon dioxide and water. In actual fact, however, the ATP yield in living cells is often lower than the maximum (38) used in these calculations.

UTILIZATION OF AMINO ACIDS AND TRIACYLGLYCEROLS AS ENERGY SOURCES

In addition to glucose (which is a carbohydrate), molecules of other substances—most importantly amino acids and triacylglycerols—can undergo metabolic conversions to intermediates that can enter the glycolytic and Krebs cycle pathways and thereby provide energy for body activities.

Amino Acids

Amino acids can lose their nitrogen-containing amino groups and be converted to alpha-keto acids (α-keto acids) that can ultimately enter the Krebs cycle—for example, by way of pyruvic acid or the Krebs cycle intermediate oxaloacetic acid, both of which are α-keto acids. (An α-keto acid is similar to an amino acid except that it has oxygen rather than an amino group bonded to its alpha carbon.)

The process of converting an amino acid to an α-keto acid generally involves a reaction known as transamination in which the amino group of an amino acid is exchanged with the oxygen of an α-keto acid, most commonly α-ketoglutaric acid (Figure 25.6a). The acquisition of an amino group by α-ketoglutaric acid in exchange for its oxygen produces a molecule of glutamic acid (an amino acid), and the amino acid that provides the amino group is converted to an α-keto acid. Glutamic acid can then undergo a reaction called

F25.6a

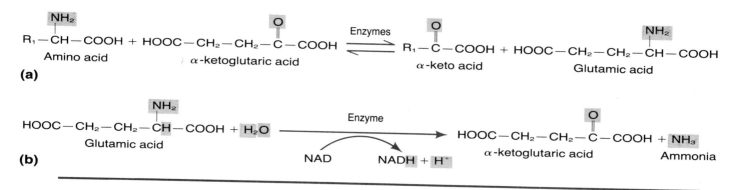

(a)

(b)

Figure 25.6

(a) Transamination reaction involving α-ketoglutaric acid and an amino acid. In this reaction, the amino group of the amino acid is exchanged with the oxygen of α-ketoglutaric acid. This exchange converts α-ketoglutaric acid to glutamic acid, and the amino acid to an α-keto acid. (Note that an α-keto acid is similar to an amino acid, except that it has an oxygen rather than an amino group bonded to its alpha carbon.) **(b)** Oxidative deamination of glutamic acid. This reaction forms α-ketoglutaric acid and ammonia. The overall result of the transamination reaction involving α-ketoglutaric acid and the subsequent oxidative deamination of the glutamic acid that is produced is to transform an amino acid into an α-keto acid, with the resultant production of ammonia.

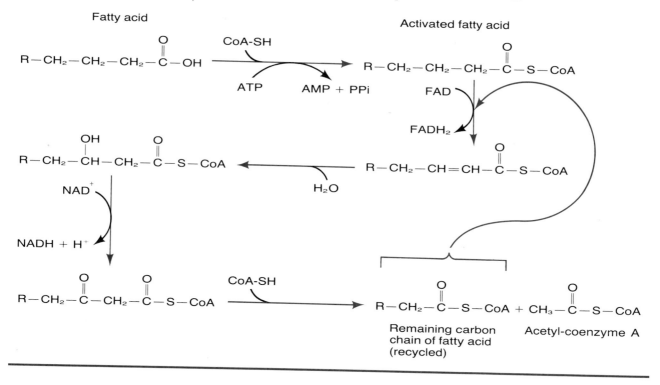

Figure 25.7

Breakdown of a molecule of fatty acid by beta oxidation.

oxidative deamination in which its amino group is removed, with the resultant production of ammonia, and α-ketoglutaric acid is reformed (Figure 25.6b). The overall result of the transamination reaction involving α-ketoglutaric acid and the subsequent deamination of glutamic acid is to transform an amino acid into an α-keto acid, with the resultant production of ammonia. As explained previously, the α-keto acids produced in this process can ultimately enter the Krebs cycle. The ammonia is converted to urea by the liver, and the urea enters the bloodstream and is excreted by the kidneys.

F25.6b

Triacylglycerols

Triacylglycerols (fat) can be broken down into their components, glycerol and fatty acids. The glycerol can be converted to 3-phosphoglyceraldehyde, which is one of the substances of the glycolytic pathway. The fatty acids can undergo a stepwise breakdown into molecules of acetyl-coenzyme A by the process of beta oxidation (Figure 25.7). In this process, a fatty acid molecule is

F25.7

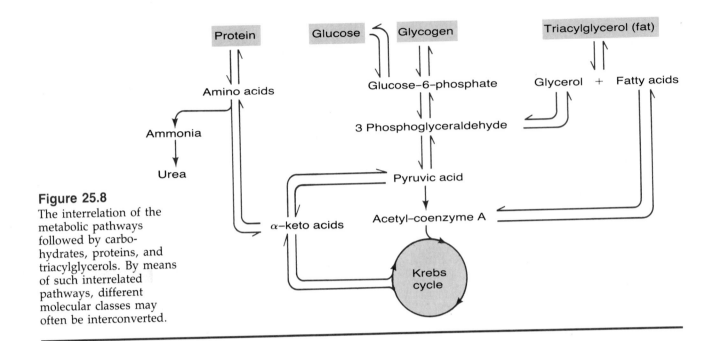

Figure 25.8
The interrelation of the metabolic pathways followed by carbohydrates, proteins, and triacylglycerols. By means of such interrelated pathways, different molecular classes may often be interconverted.

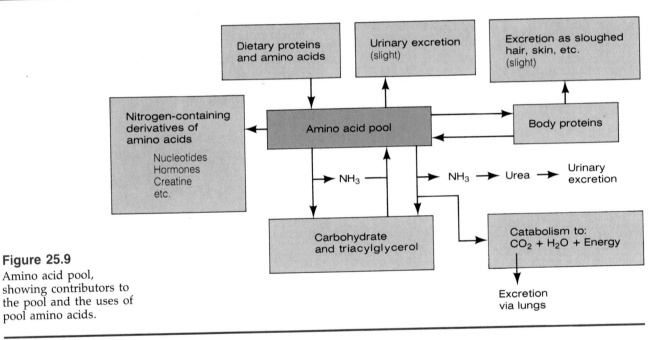

Figure 25.9
Amino acid pool, showing contributors to the pool and the uses of pool amino acids.

activated by combination with a molecule of coenzyme A. After a series of reactions, two carbons of the fatty acid carbon chain are broken away as a molecule of acetyl-coenzyme A, which can enter the Krebs cycle. The remainder of the fatty acid carbon chain can be recycled through the reactions and another two carbons can be split off as acetyl-coenzyme A. This process can be repeated until the entire carbon chain of the fatty acid is utilized.

OTHER USES OF FOOD MOLECULES

In addition to being broken down and utilized as sources of energy, food molecules are also built into larger structures. For example, amino acids can be assembled into proteins that function as enzymes or structural cellular

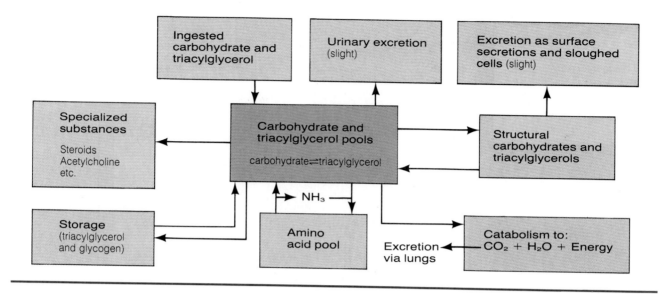

Figure 25.10
Carbohydrate and
triacylglycerol pools,
showing contributors to
the pools and the uses of
pool carbohydrates and
triacylglycerols.

components, and glucose molecules can be linked together to form large glycogen molecules, which are a storage form of carbohydrate. Moreover, when considering the uses of the various food molecules, it is important to realize that the metabolic pathways followed by carbohydrates (such as glucose), proteins (amino acids), and triacylglycerols (fat) are interrelated (Figure 25.8). Thus, pathways exist by which amino acids or glucose, for example, can be converted to triacylglycerols, or by which amino acids can form glucose. The liver is especially active in such interconversions; so is adipose tissue.

The ability to perform interconversions enables the body to synthesize many of the substances it needs. But some essential substances cannot be manufactured by the body, and these must be obtained from external sources. For example, the amino acids of proteins are essential sources of nitrogen. Although the body can form some amino acids, it cannot synthesize all of the amino acids it needs. Therefore, certain amino acids, called essential amino acids, must be obtained in the diet. In addition, some fatty acids cannot be synthesized, and these must also come from dietary sources. The same is largely true of vitamins, many of which serve mainly as coenzymes in chemical reactions.

NUTRIENT POOLS

In any consideration of metabolism, it must be remembered that the human body is not a static entity but rather exists in a dynamic steady state. The processes involved in the synthesis of protoplasm from simpler food molecules (anabolism) occur simultaneously with the breakdown of molecules to provide energy (catabolism). Thus, almost all the body's atoms and molecules are in a continual state of flux; even the atoms and molecules of bone are continually being exchanged.

In view of the continual turnover of the body's atoms and molecules, it is useful to consider the existence of nutrient pools within the body that can be drawn upon to meet its needs. For example, ingested protein (as well as the body's own protein) can be broken down to amino acids that contribute to an amino acid pool (Figure 25.9). Amino acids from this pool can be used for the synthesis of proteins and other nitrogen-containing structures required by the body. They can also have their nitrogen removed and be used as energy sources or be converted to carbohydrate or triacylglycerols.

Since carbohydrates are readily converted to triacylglycerols, carbohydrate and triacylglycerol pools are often considered together (Figure 25.10). Carbohydrates and triacylglycerols are major energy sources for the body.

F25.9

F25.10

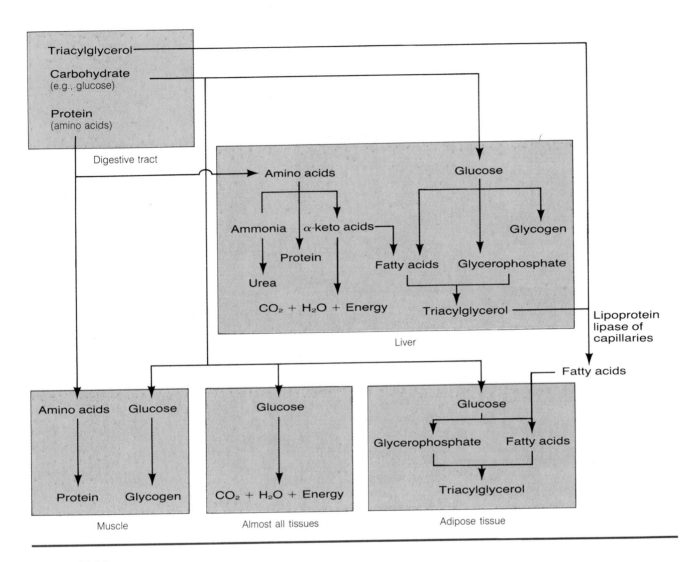

Figure 25.11
Absorptive state metabolic pathways. Although the diagram shows only muscle and liver tissue utilizing amino acids for protein synthesis, other tissues can also take up amino acids and synthesize protein from them.

The body is able to store excess triacylglycerol as well as some excess carbohydrate. Triacylglycerol stores, however, provide the major energy reserves for the body, and only comparatively small amounts of energy are stored as carbohydrate (glycogen).

METABOLIC STATES

The human body actually exists in two different metabolic states. After eating, the body is in an *absorptive state* in which food substances are being absorbed from the gastrointestinal tract into the bloodstream. When absorption is complete, the body is in a *postabsorptive*, or fasting, *state*. In the postabsorptive state, a person's needs must be met by the use of materials already present within the body.

Absorptive State

Carbohydrate

During the absorption of a typical meal (65% carbohydrate, 25% protein, and 10% triacylglycerol), the body usually utilizes carbohydrate (glucose) as its major energy source, although some energy is derived from absorbed amino acids and triacylglycerols (Figure 25.11). During the absorptive state, the body stores as glycogen some carbohydrate not used for energy, and it converts much to adipose tissue triacylglycerol.

F25.11

Carbohydrate is absorbed from the digestive tract largely in the form of the monosaccharides glucose, galactose, and fructose. However, glucose is the principal monosaccharide, and most of the galactose and fructose are either converted to glucose or enter essentially the same metabolic pathways as glucose. Therefore, the absorbed monosaccharides can be considered simply as glucose.

Glucose absorbed from the digestive tract enters the blood and is carried to the liver, where a large portion of it enters liver cells. The liver itself uses little glucose for energy during the absorptive state, and it converts much into triacylglycerol or glycogen. Some of the triacylglycerol is stored in the liver, but most is packaged into protein-containing particles called very-low-density lipoproteins (VLDL), which are released into the blood. As is discussed later, many of the fatty acids of the VLDL triacylglycerols are utilized in the formation of adipose tissue triacylglycerol. Moreover, adipose tissue cells can take up and convert to triacylglycerol absorbed glucose in the blood that is not taken up by the liver. Skeletal muscle and certain other tissues can also take up some glucose and store it as glycogen.

Protein

During the absorptive state, the body uses amino acids for protein synthesis, and amino acids are also converted into adipose tissue triacylglycerol (Figure 25.11). Some of the amino acids from digested protein are taken up by the liver, where their nitrogen is removed and they are converted to α-keto acids. The α-keto acids can ultimately be metabolized by the Krebs cycle to provide energy for the liver cells. They can also be converted to fatty acids, which can be synthesized into triacylglycerol, much of which is packaged into very-low-density lipoproteins and released into the blood. The liver also uses amino acids to synthesize proteins such as plasma proteins. Skeletal muscle cells and other cells take up many amino acids and synthesize proteins from them.

F25.11

Triacylglycerol

During the absorptive state, the body uses some triacylglycerol for synthesis, and some is used as an energy source. However, much triacylglycerol is stored in adipose tissue (Figure 25.11). Triacylglycerols from the digestive system enter the lacteals of the lymphatic system as a major component of chylomicrons. The lymphatic system delivers the chylomicrons to the bloodstream. The enzyme lipoprotein lipase, which is associated with the endothelium of capillaries, particularly in skeletal muscle and adipose tissue, breaks down the triacylglycerols of the chylomicrons (and also the triacylglycerols of the very-low-density lipoproteins from the liver), releasing free fatty acids. Some of the free fatty acids are taken up and utilized as an energy source by the muscle cells (not shown in Figure 25.11), and many are taken up and resynthesized into triacylglycerols by the adipose-tissue cells. The glycerol portion of the resynthesized triacylglycerols is derived from glucose.

F25.11

F25.11

When the absorptive-state uses of carbohydrates and amino acids as well as triacylglycerols are considered, it is evident that the components of adipose-tissue triacylglycerols are derived primarily from: (1) chylomicron triacylglycerols, (2) triacylglycerols produced as a result of the synthetic activities of the liver, and (3) glucose.

Postabsorptive State

During the postabsorptive state, the blood-glucose level must be maintained, especially for the support of the nervous system, even though no glucose is being absorbed. The level is maintained by using body-glucose sources and glucose-sparing, fat-utilizing metabolic pathways (Figure 25.12).

F25.12

Glucose Sources

The liver contains stores of glycogen that can be broken down to glucose. These glycogen stores, however, can support the body's activities for only about four hours. Other tissues, including skeletal muscle, also contain glycogen reserves, but these too are limited. Moreover, skeletal muscle lacks the

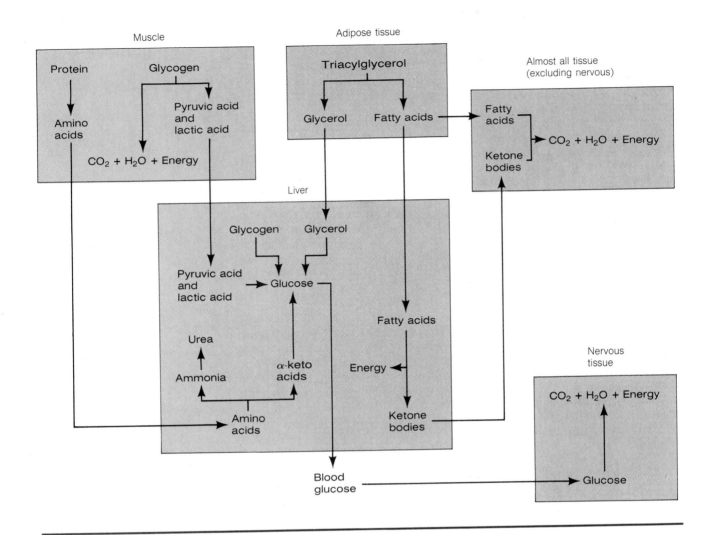

Figure 25.12

Postabsorptive state
metabolic pathways.

enzyme required to form free glucose from glycogen, and it cannot directly
release glucose to maintain the blood-glucose level. Instead, skeletal muscle
breaks glycogen down to glucose-6-phosphate, which it can catabolize for
energy by way of glycolysis and the Krebs cycle. In addition, the pyruvic acid
and, under anaerobic conditions, lactic acid that are produced by muscle glycolysis can enter the blood and travel to the liver, where they can be converted to glucose. As a result, muscle glycogen can serve as an indirect source
of blood glucose.

The catabolism of triacylglycerols to glycerol and fatty acids, particularly
in adipose tissue, can also provide glucose, because glycerol can be converted
to glucose by the liver.

Protein, however, is the major blood-glucose source during prolonged
fasting. Large quantities of muscle protein that are not essential to cellular
function (and to a lesser extent similar protein in other tissues) can be catabo-
lized to amino acids, which in turn can be converted to glucose by the liver.
During prolonged fasting, the kidneys as well as the liver produce substantial
amounts of glucose for the body.

Glucose Sparing and Triacylglycerol Utilization

During the postabsorptive state, almost all the organs and tissues of the body—
the nervous system is the major exception—diminish their use of glucose for
energy and depend primarily on triacylglycerol. This adjustment spares glucose from the liver for use by the nervous system, which normally requires

glucose as its principal energy source. During the postabsorptive state, adipose-tissue triacylglycerol is broken down and fatty acids are released into the blood. Almost all tissues except the nervous system can take up the fatty acids, which ultimately enter the Krebs cycle to provide energy for body activities.

During the postabsorptive state, the liver not only utilizes fatty acids as an energy source, but it also processes them into a group of substances called ketone bodies (acetoacetic acid, beta-hydroxybutyric acid, and acetone), which are released into the blood. Many tissues can take up these ketone bodies (specifically acetoacetic acid and beta-hydroxybutyric acid) and, by way of the Krebs cycle, utilize them as an important source of energy. Thus, the breakdown of adipose-tissue triacylglycerol during fasting ultimately provides energy for the body and spares glucose for the nervous system.

As a result of these activities, an average person can fast for many weeks, as long as water is provided. With prolonged fasting, however, a noteworthy change in brain metabolism occurs. The requirement that the brain must utilize glucose for its energy supply no longer seems to hold true after four or five days of fasting, and in addition to glucose, the brain begins to use large quantities of ketone bodies as energy sources. Since the major source of blood glucose during prolonged fasting is protein, this change conserves body protein that would otherwise have to be converted to glucose for the support of the nervous system.

Control of Absorptive and Postabsorptive State Events

Several hormonal and neural mechanisms control and regulate metabolism in both the absorptive and postabsorptive states. For example, when glucose is being absorbed following a meal—that is, during the absorptive state—the blood-glucose level may rise from about 85 mg per 100 ml to about 120 mg per 100 ml. The rise in the blood-glucose level causes an increased release of insulin, which influences both membrane transport and enzymatic activity. With the notable exception of most areas of the brain, insulin facilitates the entry of glucose into many cells, particularly skeletal-muscle and adipose-tissue cells. It favors the oxidation of glucose and the formation of glycogen within cells, and it inhibits glycogen breakdown. Insulin stimulates the active transport of amino acids into most cells, and it promotes protein synthesis and inhibits protein breakdown. Insulin also promotes the synthesis and inhibits the breakdown of triacylglycerols in adipose-tissue cells. Thus, in general, insulin promotes metabolic activities of the sort that occur during the absorptive state.

When absorption is completed and blood-glucose levels begin to fall—that is, during the postabsorptive state—insulin release diminishes. This favors the breakdown of glycogen, proteins, and triacylglycerols and the release of glucose, amino acids, glycerol, and fatty acids into the blood. A declining blood-glucose level also leads to an increased release of the hormone glucagon. Glucagon stimulates the breakdown of glycogen and the production of carbohydrate from noncarbohydrate precursors (such as amino acids) in the liver. Glucagon also stimulates the breakdown of triacylglycerols in adipose tissue. When the blood-glucose level is falling rapidly (but apparently not during prolonged fasting), glucose receptors in the brain are affected and sympathetic neural activity to the adrenal medullae and to adipose-tissue cells increases. The adrenal medullary hormones stimulate the breakdown of glycogen in liver and skeletal muscle as well as the breakdown of triacylglycerols in adipose tissue. The sympathetic neural stimulation to adipose-tissue cells also promotes the breakdown of triacylglycerols. The overall effect of these activities is to increase the concentrations of glucose, glycerol, and fatty acids in the blood. Thus, during the postabsorptive state, hormonal and neural mechanisms mobilize carbohydrates, proteins, and triacylglycerols for the maintenance of blood-glucose levels and the support of body activities.

Table 25.1   Vitamins

Vitamin	Examples of Functions or Importance	Some Results of Deficiency
Water Soluble Vitamins		
Vitamin B$_1$ (Thiamine)	Involved in the metabolism of carbohydrates and many amino acids	Deficiency leads to beriberi, which is characterized in some cases by cardiac muscle weakness and heart failure, and in other cases by peripheral neuritis, with muscle atrophy and occasionally paralysis
Vitamin B$_2$ (Riboflavin)	Component of FAD, which is involved in Krebs cycle reactions	Deficiency leads to cracking at the corners of the mouth
Niacin (Nicotinic acid)	Component of NAD$^+$, which is involved in glycolytic, Krebs cycle, and other reactions	Deficiency gives rise to pellagra, which is characterized by diarrhea, dermatitis, and mental disturbances
Vitamin B$_6$ (Pyridoxine)	Involved in amino acid and protein metabolism	Deficiency leads to dermatitis and perhaps convulsions and gastrointestinal disturbances in children
Vitamin B$_{12}$ (Cyanocobalamin)	Involved in erythrocyte maturation and nucleic acid metabolism	Deficiency results in macrocytic anemia, loss of peripheral sensation, and, in severe cases, paralysis
Pantothenic acid	Component of coenzyme A, which functions in the entry of pyruvic acid into the Krebs cycle and in the degradation of fatty acids	
Folic acid	Involved in the synthesis of purines and thymine, which are required for DNA formation; important in erythrocyte maturation	Deficiency leads to macrocytic anemia, which is characterized by the production of abnormally large erythrocytes
Biotin	Involved in amino acid and protein metabolism	Deficiency leads to muscle pain, fatigue, and poor appetite
Vitamin C (Ascorbic acid)	Important in the formation of collagen and in the maintenance of normal intercellular substances throughout the body	Deficiency leads to scurvy, which is characterized by poor wound healing, defective bone formation and maintenance, and fragile blood vessel walls
Fat-Soluble Vitamins		
Vitamin A	Important in the formation of photopigments; apparently needed for the maintenance of certain epithelial surfaces, such as the mucous membranes of the eyes	Deficiency results in night blindness and corneal opacity
Vitamin D	Needed for the proper absorption of calcium from the gastrointestinal tract	Deficiency leads to rickets in children and osteomalacia in adults
Vitamin E	Inhibits the breakdown of certain fatty acids	
Vitamin K	Required for the synthesis of prothrombin and certain other clotting factors in the liver	Deficiency leads to retarded blood clotting and excessive bleeding

LIVER FUNCTION

In addition to its important metabolic role, the liver performs several other important functions, including the following:

1. Produces albumin and other plasma proteins
2. Produces prothrombin and other blood-clotting factors
3. Excretes bilirubin, a breakdown product of hemoglobin
4. Produces bile, which aids in fat digestion
5. Carries out many reactions of carbohydrate, protein, and lipid metabolism
6. Stores glycogen
7. Synthesizes urea and ketone bodies
8. Stores iron (as ferritin), vitamins A and D, and other substances
9. Detoxifies many harmful drugs and toxins
10. Phagocytizes microorganisms and other foreign material

Many of these functions are discussed in other chapters.

VITAMINS

Normal metabolism and growth require not only adequate supplies of carbohydrates, proteins, and fats, but also minute amounts of other organic nutrients called *vitamins*. Vitamins do not provide energy or serve as structural components. Rather, most vitamins function either directly or after chemical alteration as coenzymes or components of coenzymes involved in various metabolic reactions.

In general, vitamins cannot be synthesized by the body; they must be obtained from external sources. The principal source of vitamins is ingested food, although some vitamins, such as vitamin K, can be produced by bacteria in the gastrointestinal tract. If raw materials, called provitamins, are provided, the body can assemble some vitamins. For example, the provitamin carotene found in foods such as spinach, carrots, liver, and milk can be used by the body in the production of vitamin A.

Vitamins are divided into two major groups: water-soluble vitamins and fat-soluble vitamins. Water-soluble vitamins, which include the B vitamins and vitamin C, are generally absorbed with water from the gastrointestinal tract. (Vitamin B_{12} is absorbed in association with intrinsic factor.) Fat-soluble vitamins, which include vitamins A, D, E, and K, are absorbed with digested dietary triacylglycerols. As a general rule, the storage of the water-soluble vitamins by the body is slight. The fat-soluble vitamins, with the exception of vitamin K, are usually stored to a greater extent. Vitamins A and D, in particular, are stored by the liver. Specific vitamins, examples of their functions in or importance to the body, and some of the results of their deficiencies are listed in Table 25.1. Recommended daily dietary allowances for selected vitamins are listed in Table 25.3.

Table 25.1
Table 25.3

MINERALS

The body needs not only organic compounds but also certain inorganic elements, or *minerals*. Minerals make up 4–5% of the body, by weight, and many of them are essential for life. Minerals are involved in numerous body func-

CLINICAL CORRELATION
Vitamin B_{12} Deficiency on a Vegan Diet

CASE REPORT ....

THE PATIENT A 46-year-old woman

PRINCIPAL COMPLAINTS Loss of appetite, weakness, sore tongue

HISTORY The patient had been well until about six months before admission, when she began to experience the symptoms that brought her to admission. She had experienced a psychological change five years earlier and become a "vegan" (one who eats no food of animal origin).

CLINICAL EXAMINATION The patient was pale, with a moderate yellowish tinge to the skin, but she did not appear acutely ill. The tongue was red and smooth, with a glazed appearance. The respiratory rate was 16/min (normal: 12–17/min). The chest was clear, no masses were felt in the abdomen, and kidney function was normal. The heart rate was 82 beats/min (normal: 60–100 beats/min), and the arterial blood pressure was 125/75 mm Hg (normal 90–140/60–90 mm Hg). The hematocrit was 34% (normal: 37–48%), and the hemoglobin content was 11 grams/100 ml (normal: 12–16 grams/100 ml). The mean corpuscular volume (the mean volume of a red blood cell determined by dividing the hematocrit by the red cell count) was 101 μm^3 (normal: 83–97 μm^3). The leukocyte count was 4500 cells/mm³ (normal: 4300–10,800 cells/mm³), with a normal differential (distribution of cell types).

The microscopic examination of the blood revealed macrocytosis (abnormally large erythrocytes) and hypersegmented neutrophils. Since these findings suggested pernicious anemia, the serum level of vitamin B_{12} was measured and found to be 48 picograms/ml (normal: 200–600 picograms/ml). Folic acid levels were normal. An examination of the bone marrow revealed large, immature red cells (megaloblasts), large leukocytes with bizarrely shaped nuclei, and decreased stores of iron. The concentration of iron in the serum was 89 micrograms/100 ml (normal: 50–150 micrograms/100 ml), and the total iron-binding capacity was 302 micrograms/100 ml (normal: 250–410 micrograms/100 ml). (Normally, only about 25% of the binding sites on transferrin are occupied by iron, and the remainder represent the latent iron-binding capacity of the plasma. If iron is defi-cient, fewer transferrin sites are occupied and the iron-binding capacity is increased.)

Because the principal cause of pernicious anemia is failure to absorb vitamin B_{12} due to loss of functional intrinsic factor associated with gastric atrophy, gastric secretory function was tested. Stimulation of the gastric parietal cells (which release both gastric acid and intrinsic factor) by gastrin revealed normal gastric secretion, and no antibodies to parietal cells or intrinsic factor were found. The absorption of B_{12} (as measured by the Schilling test) was found to be normal.

COMMENT The normal gastric secretory function and uptake of vitamin B_{12} indicated that the low concentration of vitamin B_{12} in this patient's plasma (which indicates inadequate stores of vitamin B_{12} in the body) was related to her diet. The body normally stores about 5000 micrograms of vitamin B_{12}, mostly in the liver, and several years may be required for its depletion, even when no vitamin B_{12} is absorbed. The patient had been on her exclusively vegetable (vegan) diet for about five years. The best sources of vitamin B_{12} in the diet are fresh meat and dairy products. Vegetable tissues contain no vitamin B_{12}.

Deficiency of vitamin B_{12} alters DNA synthesis in many tissues of the body, but its effects are especially evident in the hematopoietic (blood-forming) system. Impaired synthesis of DNA retards mitosis, and this retardation causes the production of large blood cells, both erythrocytes and leukocytes. Mitosis of epithelial cells also is affected, which probably accounts for the abnormalities of the mouth and the gastrointestinal system.

The patient's diet probably also accounts for the mild deficiency of iron that was observed, because iron from vegetable sources is not absorbed well.

TREATMENT The patient was given vitamin B_{12} supplement, 5 micrograms/day, and her condition improved dramatically.

OUTCOME Six weeks after the beginning of therapy, all of the symptoms relating to deficiency of vitamin B_{12} were gone, and the patient's hematologic profile was normal. Since she is committed to her vegan diet, she will have to continue to take supplementary vitamin B_{12}.

tions, and the activity of many enzymes depends on their presence. They are frequently found in chemical unions with organic compounds—for example, the iron of the hemoglobin molecule. The minerals used by the body are generally present in ionized form. Specific minerals and examples of their functions in or importance to the body are listed in Table 25.2. Recommended daily dietary allowances for selected minerals are listed in Table 25.3.

Table 25.2
Table 25.3

Table 25.2 Minerals

Mineral	Examples of Functions or Importance
Calcium	Needed for the formation of bones and teeth; required for blood clotting; necessary for normal skeletal and cardiac muscle activity; important in normal nerve function
Sodium	Exerts a major influence on the osmotic pressure of the extracellular fluid; important in muscle and nerve function; major cation (positively charged ion) of the extracellular fluid
Potassium	Important in normal muscle and nerve function; major cation of the intracellular fluid
Phosphorus	Needed for the formation of bones and teeth; a component of energy compounds (such as ATP) that are involved in energy storage and transfer; a component of nucleic acids (DNA, RNA)
Magnesium	Important in normal muscle and nerve function; activates a number of enzymes
Iron	A component of hemoglobin, which is important in oxygen transport; a component of myoglobin; involved in the formation of ATP as a component of cytochromes of the electron transport system
Iodine	Essential for the formation of the thyroid hormones thyroxine and triiodothyronine, which help govern metabolism
Copper	Required in the manufacture of hemoglobin and the pigment melanin
Zinc	A component of several enzymes, such as carbonic anhydrase and carboxypeptidase, and thus important for reactions catalyzed by these enzymes
Fluorine	Normally present in teeth and bones; appears to provide protection against dental caries (cavities)
Manganese	Needed for the formation of urea; activates some enzymes
Cobalt	A component of vitamin B_{12}, which is necessary for the normal maturation of erythrocytes

METABOLIC RATE

The *metabolic rate* is the total energy expended by the body per unit time. This energy is derived from the breakdown of carbohydrates, proteins, triacylglycerols, and other organic molecules that are originally obtained in the diet. The energy is used for biological work such as muscle contraction, active transport, or molecular synthesis, and it also appears as heat. Actually, only a relatively small amount of the energy is used for work; most appears as heat. However, the heat can be valuable in body-temperature regulation.

A reasonable estimate of the metabolic rate can be obtained by measuring the body's rate of oxygen consumption (Figure 25.13). For this purpose it is generally assumed that for every liter of oxygen consumed, the body produces 4.825 kilocalories of energy (1 kilocalorie = 1000 calories).

F25.13

Basal Metabolic Rate

Many factors—including eating, exercise, and exposure to extreme environmental temperatures—significantly influence a person's metabolic rate. Therefore, for purposes of comparison between individuals, the metabolic rate is often determined under specific, standard conditions, which eliminate many of these variables. The metabolic rate determined under these conditions is called the *basal metabolic rate*, and it is usually expressed in terms of kilocalories of energy expended per square meter of body surface area per hour.

The conditions under which the basal metabolic rate is determined include having the person awake but in a supine position at mental and physical rest in a room with a temperature between 20°C and 25°C (68°F to 77°F).

Table 25-3   Recommended Daily Dietary Allowances[a]

Age (years)	Weight (kg)	Weight (lb)	Height (cm)	Height (in.)	Protein (g)	Vitamin A (μg RE)[b]	Vitamin A (IU)[c]	Vitamin D (μg)	Vitamin D (IU)[c]	Vitamin E (mg α-TE)[d]	Vitamin C (mg)	Thiamin (mg)	Riboflavin (mg)	Niacin (mg NE)[e]	Vitamin B6 (mg)	Folic Acid (μg)	Vitamin B12 (μg)	Calcium (mg)	Phosphorus (mg)	Magnesium (mg)	Iron (mg)	Zinc (mg)	Iodine (μg)
Infants																							
0.0–0.5	6	13	60	24	kg × 2.2	420	1400	10	400	3	35	0.3	0.4	6	0.3	30	0.5	360	240	50	10	3	40
0.5–1.0	9	20	71	28	kg × 2.0	400	2000	10	400	4	35	0.5	0.6	8	0.6	45	1.5	540	360	70	15	5	50
Children																							
1–3	13	29	90	35	23	400	2000	10	400	5	45	0.7	0.8	9	0.9	100	2.0	800	800	150	15	10	70
4–6	20	44	112	44	30	500	2500	10	400	6	45	0.9	1.0	11	1.3	200	2.5	800	800	200	10	10	90
7–10	28	62	132	52	34	700	3300	10	400	7	45	1.2	1.4	16	1.6	300	3.0	800	800	250	10	10	120
Males																							
11–14	45	99	157	62	45	1000	5000	10	400	8	50	1.4	1.6	18	1.8	400	3.0	1200	1200	350	18	15	150
15–18	66	145	176	69	56	1000	5000	10	400	10	60	1.4	1.7	18	2.0	400	3.0	1200	1200	400	18	15	150
19–22	70	154	177	70	56	1000	5000	7.5	300	10	60	1.5	1.7	19	2.2	400	3.0	800	800	350	10	15	150
23–50	70	154	178	70	56	1000	5000	5	200	10	60	1.4	1.6	18	2.2	400	3.0	800	800	350	10	15	150
51+	70	154	178	70	56	1000	5000	5	200	10	60	1.2	1.4	16	2.2	400	3.0	800	800	350	10	15	150
Females																							
11–14	46	101	157	62	46	800	4000	10	400	8	50	1.1	1.3	15	1.8	400	3.0	1200	1200	300	18	15	150
15–18	55	120	163	64	46	800	4000	10	400	8	60	1.1	1.3	14	2.0	400	3.0	1200	1200	300	18	15	150
19–22	55	120	163	64	44	800	4000	7.5	300	8	60	1.1	1.3	14	2.0	400	3.0	800	800	300	18	15	150
23–50	55	120	163	64	44	800	4000	5	200	8	60	1.0	1.2	13	2.0	400	3.0	800	800	300	18	15	150
51+	55	120	163	64	44	800	4000	5	200	8	60	1.0	1.2	13	2.0	400	3.0	800	800	300	10	15	150
Pregnant					+30	+200	+1000	+5	+200	+2	+20	+0.4	+0.3	+2	+0.6	+400	+1.0	+400	+400	+150	†	+5	+25
Lactating					+20	+400	+2000	+5	+200	+3	+40	+0.5	+0.5	+5	+0.5	+100	+1.0	+400	+400	+150	†	+10	+50

[a]The allowances are intended to provide for individual variations among most normal persons as they live in the United States under usual environmental stresses. Diets should be based on a variety of common foods in order to provide other nutrients for which human requirements have been less well defined.
[b]Retinal equivalents.
[c]International units.
[d]Alpha tocopherol.
[e]Niacin equivalents.
†The increased requirement during pregnancy cannot be met by the iron content of habitual American diets nor by the existing iron stores of many women; therefore the use of 30–60 mg of supplemental iron is recommended. Iron needs during lactation are not substantially different from those of nonpregnant women, but continued supplementation of the mother for 2–3 months after parturition is advisable in order to replenish stores depleted by pregnancy.

Adapted from the National Research Council, National Academy of Sciences, 1980. *Recommended Dietary Allowances*, ninth edition. Washington, D.C.: National Academy of Sciences.

From J. L. Christian and J. L. Greger: *Nutrition for Living* 1986. Menlo Park: Benjamin/Cummings.

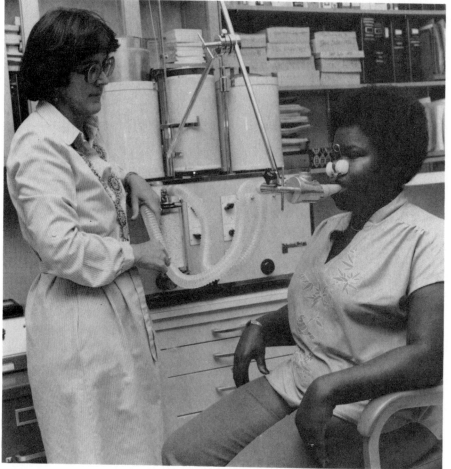

1 liter O₂ + Glucose, triacylglycerol or protein

↓

Liberates approximately 4.825 kcal energy

$$
\begin{array}{r}
O_2 \text{ consumed } = 15.4 \text{ liters/hr.} \\
\times \quad 4.825 \text{ kcal/liter} \\
\hline
74.3 \quad \text{kcal/hr.}
\end{array}
$$

Figure 25.13
Measurement of the metabolic rate by measuring oxygen consumption. The utilization of 1 liter of oxygen is approximately equal to the liberation of 4.825 kilocalories (kcal) of energy.

The person should also be in a postabsorptive state and have had nothing to eat for 12–18 hours. Under these conditions, no energy is expended in digesting, absorbing, and processing ingested nutrients, and none is expended by muscular contractions that do external work, such as moving objects in the environment. Moreover, the person is in moderate thermal surroundings.

The basal metabolic rate is not the body's lowest metabolic rate. The metabolic rate during sleep is 10–15% lower than basal, and this lower rate is presumably due, at least in part, to a more complete muscular relaxation during sleep.

Factors That Affect Metabolic Rate

Both sex and age influence metabolic rates. Basal metabolic rates tend to be lower in females than in males, and they tend to increase with age to approximately two or three years and decrease with age thereafter.

The ingestion of a typical meal containing carbohydrate, protein, and fat increases the metabolic rate 10–20% over basal. The effect of food consumption on metabolism is called the *specific dynamic action*. Protein ingestion has the greatest effect, and the ingestion of protein alone increases metabolism as much as 30%. The ingestion of carbohydrate alone may increase metabolism only 5%, and the ingestion of lipid alone increases metabolism approximately 8%. The energy expended in the digestion and absorption of food accounts for only a small portion of the increase. Most of the increase appears to be due to the processing of exogenous nutrients by the liver.

The greatest changes in metabolic rates are produced by muscular activity. Strenuous exercise can increase metabolism as much as 15 times.

REGULATION OF TOTAL BODY ENERGY BALANCE

The concept of energy balance implies a relationship between energy intake and energy expenditure, or outflow. If an energy balance is to be maintained, the energy acquired through food must equal the energy expended by the body. If the energy taken in exceeds the body's needs, some is stored—for example, as adipose-tissue triacylglycerol—and the person gains weight. If the energy taken in is less than the body needs, stored forms of energy are utilized, and the person loses weight. If the body's energy requirements and energy intake are balanced, body weight remains stable.

Since the body weights of most adults do remain relatively stable, there must be some sort of regulation to maintain a reasonable balance between energy intake and energy expenditure. In this regard, it appears that one of the major means of maintaining an energy balance is to control food (energy) intake in accordance with the energy expenditure requirements of the body.

Control of Food Intake

A number of brain areas, including certain areas of the hypothalamus, have been implicated in the control of food intake, and there are several theories about the nature of the signals that influence these areas and thus alter feeding activity. Some of the signals are believed to be concerned with maintaining normal levels of nutrient stores within the body. These signals are principally involved with *nutritional regulation*, and they could be particularly important in maintaining a balance between energy intake and energy expenditure. Other signals are believed to reflect the immediate effects of feeding on the digestive tract. These signals are principally involved with *alimentary regulation*, and they could be important in providing enough time for the gastrointestinal tract to process ingested food. In addition, *psychological factors* may have some influence.

Nutritional Regulation

GLUCOSTATIC THEORY According to the glucostatic theory of hunger and feeding regulation, the brain contains glucose receptor cells (neurons) that are sensitive to their own rate of glucose utilization. It is believed that when their rate of glucose utilization is high (as may occur when blood-glucose levels increase, perhaps as a result of eating), the glucose receptor cells increase their activity and feeding is depressed. Conversely, when their rate of glucose utilization is low (as may occur when blood-glucose levels decrease, perhaps as a result of fasting), the glucose receptor cells decrease their activity and feeding is enhanced. In support of this theory is the fact that a decrease in the concentration of blood glucose can often be associated with the development of hunger. In addition, an increase in the blood-glucose level increases the electrical activity of certain regions of the hypothalamus.

Increased levels of amino acids in the blood also tend to depress feeding, whereas decreased amino acid levels tend to enhance it. Although this effect is not nearly as great as that involving blood glucose, it may contribute to food intake regulation in a manner similar to that seen for blood glucose.

LIPOSTATIC THEORY According to the lipostatic theory of hunger and feeding regulation, some substance or substances (perhaps fatty acids and/or glycerol) are released from fat stores in adipose tissue in direct proportion to the total adipose-tissue mass. These substances act in a manner similar to that of glucose to depress feeding. Thus, the greater the total adipose-tissue mass, the less food is consumed. It is believed that such a mechanism, if it exists, would provide a reasonable form of long-term feeding regulation. In support of the lipostatic theory is the observation that increases in the quantity of adipose tissue in the body decrease the overall degree of feeding. In addition, the basal rate of glycerol release from adipose tissue appears to be directly related to the size of the adipose-tissue cells, and the long-term average concentration of free fatty acids in the blood is directly proportional to the amount of adipose tissue in the body.

THERMOSTATIC THEORY   According to the thermostatic theory of hunger and feeding regulation, brain areas involved in the control of body temperature interact with areas that control food intake. As a result of this interaction, exposure to a cold environment tends to enhance feeding, and exposure to a warm environment tends to depress it. The thermostatic theory is supported by the observation that animals exposed to cold tend to overeat, whereas animals exposed to heat tend to undereat.

HORMONAL INFLUENCES   The plasma concentrations of various hormones that affect the metabolism of glucose, amino acids, and triacylglycerols may also influence brain areas that control food intake. In support of this concept are the facts that insulin (which is secreted in increased amounts during food absorption) tends to suppress hunger, and glucagon (which is secreted in increased amounts during fasting) tends to stimulate hunger.

Alimentary Regulation

It has been proposed that the distension of the gastrointestinal tract, and especially the stomach, by food depresses feeding. Moreover, researchers have suggested that the gastrointestinal hormone cholecystokinin, which is secreted in association with the digestion of food, depresses feeding by either directly or indirectly activating vagal afferent nerve fibers. In addition, the body may be able to meter food intake in association with the acts of salivation, tasting, chewing, and swallowing. Such metering could limit food consumption. The regulation of short-term food intake—for example, the amount of food eaten at one meal—by such mechanisms as the distension of the gastrointestinal tract, the influence of the hormone cholecystokinin, and the metering of the amount of food eaten would help to coordinate feeding with the ability of the gastrointestinal tract to process the food.

Psychological Factors

It appears that caloric balance may be influenced by such reinforcing factors as the smell, taste, and texture of food. Moreover, some investigators have suggested that psychological factors not immediately related to energy balance may influence feeding, but other researchers dispute this contention.

CONDITION OF CLINICAL SIGNIFICANCE

Obesity

Obesity (the condition of being overweight) is a serious problem that often accompanies and contributes to a number of degenerative diseases, including high blood pressure, coronary artery disease, atherosclerosis, and diabetes. One view of obesity stresses that it is the result of consuming too much energy (food) in relation to energy expenditure (physical activity). Indeed, there is evidence that many sedentary people eat more in relation to their degree of physical activity than active people, and thus tend to become obese. Most cases of obesity, however, can be only partly explained by lack of exercise. It is only during the time that obesity is developing that energy intake exceeds energy expenditure, and in many obese people, body weight eventually reaches some higher-than-normal value at which it remains relatively stable. Thus, the mechanisms that control energy intake in relation to energy expenditure appear to be operative, but they maintain body weight at a higher-than-normal level. Perhaps the brain areas that control food intake malfunction or the incoming signals are somehow upset. Some researchers think habits and social customs as well as genetic factors contribute to the problem.

REGULATION OF BODY TEMPERATURE

Humans maintain a relatively constant body temperature, which is an important aspect of homeostasis because it allows the heat-sensitive chemical reactions of the body to proceed in a stable fashion. Under most conditions, the body temperature is higher than the environmental temperature. Thus, the heat produced by metabolism is of value in maintaining body temperature.

sir-kay'-dee-an

When determined in the morning under carefully controlled conditions, the oral temperature is approximately 36.7°C (98.1°F). However, not all parts of the body are the same temperature. The core or internal body temperature is usually considerably higher than the temperatures of the skin and external body parts. Moreover, the body temperature varies somewhat with activity, and there is even a 24-hour diurnal or circadian body-temperature cycle, with the lowest temperature generally occurring in the early morning after a night's rest, and the highest temperature occurring in the evening.

Despite these variations, however, body temperature does remain relatively constant. For this constancy to occur, heat gain must equal heat loss, and both the production of heat by the body and the loss of heat from the body are subject to some degree of control.

Muscle Activity and Heat Production

When a person is exposed to cold, muscle tone increases. This increase in muscle tone ultimately leads to oscillating, rhythmic muscle tremors (10 to 20 per second) known as shivering, which can increase body-heat production several times.

When a person is exposed to a warm environment, muscle tone decreases. This decrease in tone may not cause much of a decline in body-heat production, however, because muscle tone is normally quite low. Moreover, high environmental temperatures tend to raise body temperature, which increases the rates of the body's chemical reactions and consequently results in the production of additional heat.

Chronic exposure to cold causes some organisms to increase their heat production by means other than increased muscular activity. This phenomenon, called nonshivering (chemical) thermogenesis, is believed to be due mainly to an increased secretion of epinephrine, which substantially increases metabolism in certain tissues. However, it is not certain if nonshivering thermogenesis is a significant phenomenon in humans.

Heat Transfer Mechanisms

There are three principal means by which heat is transferred between the body and the environment: radiation, conduction and convection, and evaporation.

Radiation

All dense objects, including the body, continually emit heat in the form of infrared rays, which are a type of electromagnetic radiation. Thus, the body is continually exchanging heat by radiation with objects in the environment. The net direction of the heat exchange between objects by radiation depends on the surface temperatures of the objects, with the net direction being from objects with warmer surfaces to objects with cooler ones. Since the surface temperature of the body is usually higher than the surface temperatures of most objects in the environment, the net movement of heat by radiation is usually away from the body. However, the reverse may be true in a very warm environment or in the case of an exchange involving a warm individual object in the environment.

Conduction and Convection

Conduction is the transfer of thermal energy from atom to atom or molecule to molecule as the result of direct contact between two objects. In this transfer, heat moves from the object with the higher surface temperature to the object with the lower one. Most commonly, body surfaces are in direct contact with air, and conductive heat exchanges occur between air molecules and the body. Usually, the air-temperature is lower than the temperature of the body surface, and the body loses heat to the air by conduction. When heat is conducted from the body surface to air molecules, the air molecules are warmed. Since warm air is less dense than cool air, the warm air molecules rise away from the body surface and are replaced by cooler air molecules. This process, which is called *convection*, greatly enhances the conductive exchange of heat by constantly bringing cool air molecules into contact with the body surface. In fact, without convection, the conductive exchange of heat between the body surface and the air would be negligible. Convection, and thus the loss of heat from the body to the air by conduction, is enhanced by wind.

Evaporation

The evaporation of water requires the input of heat (heat of vaporization). Thus, when water evaporates from body surfaces, it carries heat with it, and the evaporation of 1 gram of water removes about 580 calories of heat from the body. Water is lost from body surfaces in two ways: by insensible water loss, which is not controlled for purposes of temperature regulation, and by sweating, which is an important thermoregulatory activity.

INSENSIBLE WATER LOSS   The skin is not perfectly impermeable to water, and some evaporative water loss continually occurs across external body surfaces. Moreover, respiration continually passes air in and out of the moist respiratory passages and the lungs. The inhaled air becomes humidified, and as a result, water is lost from the body with each expiration. Insensible water loss, which is so named because a person is not usually aware of its occurrence, can be as great as 600 ml per day.

SWEATING   Sweating requires active fluid secretion by sweat glands in the skin. Sweat is a dilute solution of sodium chloride that also contains urea, lactic acid, and potassium. In hot environments, sweat can be produced at rates up to 1.5 liters per hour. The sweat covers the body surface and evaporates, carrying heat with it. In fact, sweat must evaporate to produce its cooling effects. If it simply drips from the body, little benefit occurs. If the humidity (water-vapor concentration of the air) is high and the air is already saturated with water, no additional water can evaporate into the air. Thus, sweating is a more effective means of cooling in low-humidity environments than in high-humidity environments.

Blood Flow and Heat Exchange

The rate of heat exchange between the body and the environment by radiation and conduction depends in large measure on the temperature gradient (difference) between the body surface and the environment, and this gradient is influenced by the blood flow to the body surface (Figure 25.14). In warm environments, where the body must lose much heat in order to maintain a constant temperature, vessels to the skin dilate and a large amount of warm blood flows from the core of the body to the surface for heat exchange with the environment. This large blood flow increases the temperature of the body surface and creates a more favorable gradient for heat loss. Conversely, in cool environments, where heat conservation is important, blood vessels to the skin constrict, reducing blood flow to the body surface. This reduction in blood flow retains needed heat within the core of the body and reduces the temperature of the body surface, which decreases the gradient for heat loss to the environment.

F25.14

Adjusting blood flow to the body surface is a generally effective means of balancing heat production and heat loss within the moderate temperature range of 20–28°C (68–82°F). At temperatures lower than about 20°C, large heat losses require increased heat production—for example, by shivering—for compensation. At temperatures above about 28°C, additional heat loss by the body as a result of the evaporation of water—that is, sweating—becomes

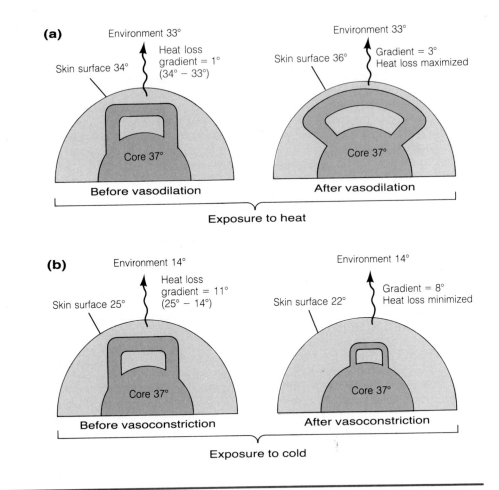

Figure 25.14
How the adjustment of blood flow to the body surface affects the thermal gradient for heat exchange by radiation and conduction.
(a) Vasodilation when exposed to heat.
(b) Vasoconstriction when exposed to cold. All temperatures are in °C.

necessary. In fact, when the environmental temperature is higher than the body temperature, the body gains heat by radiation and conduction, and evaporation is the only significant avenue of heat loss.

Humans do, of course, utilize other means of adjusting to high and low environmental temperatures. A person can, for example, reduce the body surface area over which heat loss occurs by curling into a ball. In addition, clothing helps to insulate the body from both heat and cold, as does shelter.

Brain Areas Involved in Thermoregulation

A number of brain areas, including certain areas of the hypothalamus, are involved in the integration of thermoregulatory activities (Figure 25.15). These areas receive input from temperature receptors in the skin and certain mucous membranes (peripheral thermoreceptors), as well as from receptors within the hypothalamus itself, the spinal cord, abdominal organs, and other internal structures (central thermoreceptors). The output of the integrating areas controls thermoregulatory activities such as adjustments in blood flow to the body surface, shivering, and sweating.

The operation of the body's temperature-control system appears to be analogous to that of a thermostat in which a particular temperature is set or called for (see Chapter 1). If inputs to the integrating areas from various thermoreceptors indicate that the actual body temperature differs from the called-for temperature, heat-gain or heat-loss adjustments are instituted as necessary to bring the actual temperature to the called-for temperature. Thus, if the body temperature is lower than called for, heat-conserving and heat-producing activities such as vasoconstriction and shivering are stimulated. If the

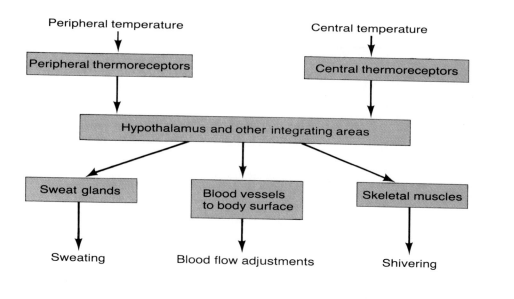

Peripheral temperature

Central temperature

Peripheral thermoreceptors

Central thermoreceptors

Hypothalamus and other integrating areas

Sweat glands

Blood vessels to body surface

Skeletal muscles

Sweating

Blood flow adjustments

Shivering

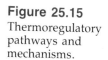

Figure 25.15
Thermoregulatory pathways and mechanisms.

temperature is higher than called for, heat-loss activities such as vasodilation and sweating are initiated.

Researchers have suggested that inputs from the peripheral thermoreceptors enable the body-temperature-control system to make thermoregulatory adjustments before there is a change in the central or core body temperature. For example, if a person is exposed to cold, inputs from the peripheral thermoreceptors are believed to cause vasoconstriction and shivering before there is a decrease in the core body temperature. This type of response would be particularly useful in maintaining a stable body temperature in environments where the temperature fluctuates.

CONDITIONS OF CLINICAL SIGNIFICANCE
Thermoregulation

Fever

The preceding view of temperature regulation envisions fever as a resetting of the body-temperature thermostat to a new, higher level. This resetting can be accomplished by substances called pyrogens, which are released during conditions such as infection or inflammation. The effects of the pyrogens are believed to be mediated by prostaglandins. (The synthesis of prostaglandins is inhibited by aspirin, which may explain why aspirin reduces fever.) When the body-temperature thermostat is set to a higher level in fever, the actual body temperature is initially below that level, and shivering, vasoconstriction, and chills can occur as the body attempts to raise its temperature to the new, higher setting. By the same token, when a fever breaks, the thermostat is reset to its original level and the body temperature is now above that level. Consequently, sweating and vasodilation can occur as the body brings its temperature back to the lower setting.

Heat Stroke and Heat Exhaustion

Occasionally, under heat stress, the body's temperature-control system breaks down. In such a case, the body temperature rises rapidly, accompanied by a dry skin and the absence of sweating. This condition is called *heat stroke*.

A more common condition experienced by individuals exposed to heat is *heat exhaustion*. In this condition, the body-temperature-control system remains functional, but as a result of extreme sweating (fluid loss) and vasodilation to lose heat, the individual collapses and has low blood pressure (hypotension) and a cool, clammy skin, with little rise in body temperature. Heat exhaustion is usually much less dangerous than heat stroke.

STUDY OUTLINE

UTILIZATION OF GLUCOSE AS AN ENERGY SOURCE
ATP is immediate source of energy for body's activity; glucose is one important food material utilized in ATP formation.   **pp. 733–738**

GLYCOLYSIS
1. Important pathway for glucose breakdown; enzymes in cytoplasm; no molecular O_2 required.
2. Glucose molecule broken into two molecules of pyruvic acid; 2 ATP spent, 4 ATP generated, 2 NADH + H^+ produced.
3. Pyruvic acid can be further degraded to lactic acid; uses 2 NADH + H^+, regenerating NAD^+.

KREBS CYCLE
1. Enzymes in mitochondria.
2. Two molecules of pyruvic acid each form acetyl-coenzyme A and enter Krebs cycle; produce: $6CO_2$, 2ATP, 8 NADH + H^+, and 2 $FADH_2$.

ELECTRON TRANSPORT SYSTEM
1. Transfer of electrons from NADH + H^+ and $FADH_2$ to oxygen, and combination of protons with oxygen to form water and regenerate NAD^+ and FAD; ATP generated.
2. Oxygen required.
3. Glycolysis, Krebs cycle, electron transport together yield net profit of 36–38 ATP per glucose molecule.

UTILIZATION OF AMINO ACIDS AND TRIACYLGLYCEROLS AS ENERGY SOURCES
Amino acids and triacylglycerols can undergo metabolic conversions to intermediates that can enter glycolytic and Krebs-cycle pathways, thereby providing energy for body activities.   **pp. 738–740**

AMINO ACIDS
Lose amino groups and form α-keto acids that can ultimately enter glycolytic or Krebs-cycle pathways.

TRIACYLGLYCEROLS
Split into glycerol and fatty acids; glycerol enters glycolytic pathway as 3-phosphoglyceraldehyde; fatty acids broken by beta oxidation to acetyl-coenzyme A, which can enter Krebs cycle.

OTHER USES OF FOOD MOLECULES
Built into larger structures; interconverted—for example, amino acids or glucose form triacylglycerols; essential amino acids, some fatty acids, and most vitamins must be obtained in diet.   **pp. 740–741**

NUTRIENT POOLS   **pp. 741–742**

AMINO ACID POOL
Protein synthesis; energy sources; conversion to carbohydrate or triacylglycerol.

CARBOHYDRATE AND TRIACYLGLYCEROL POOLS
Energy sources; storage.

METABOLIC STATES   **pp. 742–746**

ABSORPTIVE STATE

CARBOHYDRATE   Usually major energy source (glucose); converted to adipose-tissue triacylglycerol; some stored as glycogen.

PROTEIN   Some protein synthesis; converted to adipose-tissue triacylglycerol.

TRIACYLGLYCEROL   Some used for synthesis; energy; most stored as adipose-tissue triacylglycerol.

POSTABSORPTIVE STATE

GLUCOSE SOURCES   Liver glycogen supply; skeletal muscle glycogen supply; catabolism of triacylglycerols; protein provides major blood-glucose source.

GLUCOSE SPARING AND TRIACYLGLYCEROL UTILIZATION   Glucose reserved for nervous system; body energy largely from triacylglycerols.

CONTROL OF ABSORPTIVE AND POSTABSORPTIVE STATE EVENTS
Hormonal and neural mechanisms involved; insulin tends to promote absorptive-state metabolic activities; glucagon favors mobilization of stored carbohydrates and triacylglycerols.

LIVER FUNCTION
Storage and synthesis.   **p. 747**

VITAMINS
Most vitamins function either directly or after chemical alteration as coenzymes or components of coenzymes.   **p. 747**

WATER-SOLUBLE VITAMINS
Generally absorbed with water from gastrointestinal tract; examples: most B vitamins, vitamin C; vitamin B_{12} absorbed in association with intrinsic factor.

FAT-SOLUBLE VITAMINS
Absorbed with digested dietary triglycerides; examples: vitamins A, D, E, K.

MINERALS
Involved in numerous body functions; enzyme activity.   **pp. 747–748**

METABOLIC RATE
Total energy liberated by body per unit time; measured by determining rate of oxygen consumption.   **pp. 749–751**

BASAL METABOLIC RATE
Metabolic rate determined on awake, supine person; 20–25°C environment; postabsorptive state (12–18-hour fast period).

FACTORS THAT AFFECT METABOLIC RATE
Age, sex, specific dynamic action of food, muscle activity.

REGULATION OF TOTAL BODY ENERGY BALANCE
If energy requirements and energy intake are balanced, body weight remains stable.   **pp. 752–753**

CONTROL OF FOOD INTAKE
A number of brain areas implicated; several theories about signals that influence these areas.

NUTRITIONAL REGULATION   Theories include the glucostatic theory, the lipostatic theory, and the thermostatic theory.

Hormonal Influences   Plasma concentrations of hormones that affect metabolism of glucose, amino acids, and triacylglycerols may influence brain areas that control food intake.

ALIMENTARY REGULATION   Distension of gastrointestinal tract, influence of cholecystokinin, or ability to meter food intake may depress feeding.

PSYCHOLOGICAL FACTORS   Caloric balance may be affected by smell, taste, texture of food.

CONDITION OF CLINICAL SIGNIFICANCE: OBESITY   Serious problem; often accompanies and contributes to a number of degenerative diseases. **p. 753**

REGULATION OF BODY TEMPERATURE

Human body temperature generally relatively constant; thus, heat gain equals heat loss.   **pp. 754–757**

MUSCLE ACTIVITY AND HEAT PRODUCTION   In cold, shivering can increase body-heat production several times.

HEAT TRANSFER MECHANISMS

RADIATION   Emission of heat in form of infrared rays; net direction of heat exchange is from objects with warmer surfaces to objects with cooler ones.

CONDUCTION AND CONVECTION   *Conduction*: transfer of thermal energy as result of direct contact between two objects; aided by *convection*: air molecules exchange heat with body surface and move away to be replaced by other molecules.

EVAPORATION   Requires input of heat.

Insensible Water Loss   Achieved through exhaled air and through skin.

Sweating   Active fluid secretion; cools body by evaporation.

BLOOD FLOW AND HEAT EXCHANGE   Adjusting blood flow to body surface is generally effective in balancing heat production and heat loss within moderate temperature range (20–28°C).

BRAIN AREAS INVOLVED IN THERMOREGULATION

1. Receive input from peripheral and central thermoreceptors.
2. If inputs to integrating centers indicate an actual body temperature different from called-for temperature, heat-gain or heat-loss adjustments are instituted as necessary.

CONDITIONS OF CLINICAL SIGNIFICANCE: THERMOREGULATION   **p. 757**

FEVER   Body-temperature thermostat set higher; aspirin may inhibit synthesis of prostaglandins.

HEAT STROKE AND HEAT EXHAUSTION

HEAT STROKE   Failure of body-temperature-control system results in rapid rise of body temperature, dry. skin, and no sweating.

HEAT EXHAUSTION   Extreme sweating and vasodilation lead to collapse, with low blood pressure, cool skin, and little rise in body temperature.

SELF-QUIZ

1. The breakdown of glucose by glycolysis: (a) occurs within mitochondria; (b) does not liberate energy in the form of ATP; (c) requires no molecular oxygen.

2. When oxygen is not available, which substance serves as an electron acceptor to regenerate NAD^+ from $NADH + H^+$? (a) glucose; (b) pyruvic acid; (c) lactic acid.

3. Lactic acid molecules are degraded to carbon dioxide and water during glycolysis. True or False?

4. Unlike glucose, α-keto acids are unable to enter the Krebs cycle. True or False?

5. The fatty acids of triacylglycerols can be split by beta oxidation into molecules of: (a) glycerol; (b) acetyl-coenzyme A; (c) 3-phosphoglyceraldehyde.

6. During the absorptive state, the major energy source of the body is usually: (a) carbohydrate; (b) amino acids; (c) triacylglycerols.

7. Lipoprotein lipase breaks down chylomicron triacylglycerols, releasing free fatty acids. True or False?

8. During the postabsorptive state, glycogen in skeletal muscle cells is broken down to glucose, which is released from the cells into the blood. True or False?

9. Pyruvic acid and lactic acid produced by muscle glycolysis can be transported by the blood to the liver, where they can be converted to glucose. True or False?

10. During periods of fasting, which tissue has the greatest need for glucose to supply itself with energy? (a) nervous system; (b) muscular system; (c) liver.

11. Which hormone tends to promote metabolic activities of the sort that occur during the absorptive state? (a) epinephrine; (b) glucagon; (c) insulin.

12. Vitamins function primarily as: (a) energy sources; (b) structural components of the body; (c) coenzymes.

13. For a determination of the basal metabolic rate, a person must be: (a) in a postabsorptive state; (b) asleep; (c) in a room with a temperature of 5°C.

14. The greatest change in metabolic rate occurs as the result of: (a) protein ingestion; (b) muscular activity; (c) carbohydrate intake.

15. The basal metabolic rate is determined when a person is asleep and in a postabsorptive state. True or False?

16. In general, if the body's energy requirements and energy intake are balanced, body weight remains stable. True or False?

17. The transfer of thermal energy from atom to atom or molecule to molecule that occurs as the result of direct contact between two objects is called: (a) conduction; (b) radiation; (c) convection.

18. In warm environments, vessels to the skin: (a) constrict; (b) dilate; (c) remain essentially the same as when in a cold environment.

19. A resetting of the body's thermostat to a new, higher level may be accomplished by: (a) aspirin; (b) pyrogens; (c) thermostatin.

20. A person suffering from heat exhaustion is likely to exhibit: (a) an absence of sweating; (b) low blood pressure; (c) an extremely high body temperature.

LEARNING OBJECTIVES

After completing this chapter, you should be able to:

1. State the general functions of the kidneys.

2. Name the components of the urinary system.

3. Describe the embryonic development of the kidneys.

4. Describe the gross anatomical structure of the kidneys.

5. Describe the microscopic anatomy of a renal tubule.

6. Describe the filtration barrier that separates the blood in the glomerulus from the capsular space.

7. Explain the process of glomerular filtration.

8. Explain the processes of tubular reabsorption and tubular secretion.

9. Explain how the kidneys maintain a concentrated interstitial fluid in their medullary regions.

10. Explain how the kidneys produce different concentrations and volumes of urine.

11. List the physiological and neurophysiological chain of events leading to micturition.

CHAPTER CONTENTS

COMPONENTS OF THE URINARY SYSTEM

EMBRYONIC DEVELOPMENT OF THE KIDNEYS

ANATOMY OF THE KIDNEYS

PHYSIOLOGY OF THE KIDNEYS

URETERS

URINARY BLADDER

URETHRA

MICTURITION

CONDITIONS OF CLINICAL SIGNIFICANCE: THE URINARY SYSTEM

KEY TERMS AND DERIVATIVES

calyces (*calyc* = a small cup) cuplike divisions of the pelvis of the kidney

glomerulus (*glom* = little ball) a network of parallel capillaries in the kidney

juxtamedullary (*juxta* = near to) referring to the inner portion of the cortex of the kidney, adjacent to the medulla

nephron (*nephr* = pertaining to the kidney) part of a renal tubule

podocyte (*pod* = foot; *cyte* = cell) epithelial cells that form the inner layer of the glomerular

capsules; each has several foot processes extending from their cell body

renal papilla (*ren* = kidney; *papill* = nipple) the nipple-shaped tip of a renal pyramid that projects into a renal calyx

renin (*ren* = kidney) a substance released by juxtaglomular cells of the kidneys

uremia (*emia* = condition of the blood) a toxic accumulation in the blood of substances normally excreted by the kidneys

The Urinary System

If the cells of the body are to survive and carry out their functions effectively, they must be surrounded by a stable environment. As we have noted previously, the maintenance of a relatively constant internal environment in the body is called *homeostasis*. To maintain homeostasis the concentrations of such substances as water, sodium, potassium, calcium and hydrogen ions must remain relatively constant, as must the concentrations of a wide variety of cellular nutrients and products. Cellular metabolism constantly tends to upset the body's internal environmental balance by consuming some substances (such as oxygen and glucose) and producing others (such as carbon dioxide and urea). Also, substances may be added to the internal environment as a result of ingestion and removed from the internal environment by excretion.

Most body systems are involved in maintaining homeostasis. The digestive system, for example, supplies nutrients and also serves as a means of excreting some waste products. The lungs supply oxygen to the body and eliminate carbon dioxide. The skin also plays a minor role in excretion—sweat, for example, contains small amounts of urea and ammonia. The **kidneys**, however, are the main excretory organs, and they are critically important in maintaining the balance of substances required for internal constancy. The kidneys eliminate from the body a variety of metabolic products, such as urea, uric acid, and creatinine. Further, the kidneys conserve or excrete water and electrolytes as required so that the internal balance of these substances is maintained. In fact, kidney malfunction can cause severe and even fatal problems as a result of upsets in fluid and electrolyte balance. The kidneys also act as endocrine structures by producing a hormone called *erythropoietin*, which stimulates the production of red blood cells.

Because they are selective in what they excrete, the kidneys are able to maintain the internal environment within a range that is optimal for the survival of the body's cells. As a result of this selectivity, some substances that are vital to the cells may not be excreted at all, and other substances are excreted in varying amounts that depend largely on the needs of the body. Therefore, the kidneys have a major role in regulating the composition and pH of the interstitial fluid. Similarly, the amount of water removed by the kidneys from the blood, and thus from the interstitial fluid, varies with the needs of the body.

If the kidneys fail, there is no way for many of the substances they normally excrete to be removed from the blood. As a consequence, these substances accumulate in the blood and eventually in the interstitial fluid. Within a matter of days following kidney failure, the internal environment of the body can change so much that the body's cells no longer function. To prevent death, it is necessary to replace the nonfunctioning kidneys with healthy ones by means of a kidney transplant or to remove potentially harmful substances regularly from the blood with an artificial kidney machine (see Box 26.1).

COMPONENTS OF THE URINARY SYSTEM

The urinary system consists of the *kidneys*, which produce urine; the *ureters*, which carry urine to the *urinary bladder*, where it is temporarily stored; and the *urethra*, which transports urine to the outside of the body (Figure 26.1).

761

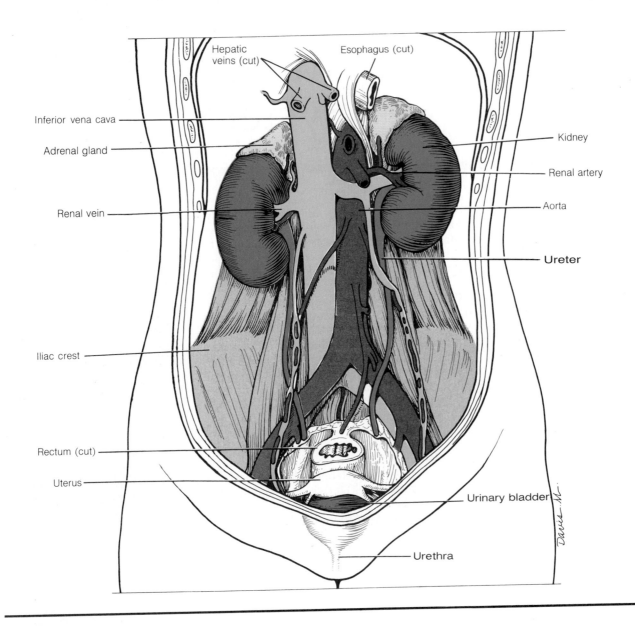

Figure 26.1
Organs of the urinary system. The anterior abdominal wall and most of the abdominal organs have been removed.

EMBRYONIC DEVELOPMENT
OF THE KIDNEYS

The kidneys develop in the embryo from columns of mesodermal cells called *intermediate mesoderm* that form just lateral to the somites. The somites give rise to the vertebrae that will form the vertebral column and to the trunk muscles. The columns of intermediate mesoderm begin to form in the superior trunk region of the three-week-old human embryo. As the more inferior regions of the intermediate mesoderm develop into kidneys, the older, more superior regions degenerate.

During embryonic development, three pairs of kidneys form in the intermediate mesoderm (Figure 26.2). The first and most superior kidney to form is called the **pronephros**. Even though the pronephros is never functional in the human, a **pronephric duct** develops and connects each pronephros with the exterior of the body through the embryonic cloaca. The pronephros begins to degenerate in the fourth week of embryonic development and is completely gone by about the sixth week. The pronephric ducts remain, however, and are utilized by the second pair of kidneys, the **mesonephros**.

F26.2
pro-nef'-ros

me-so-nef'-ros

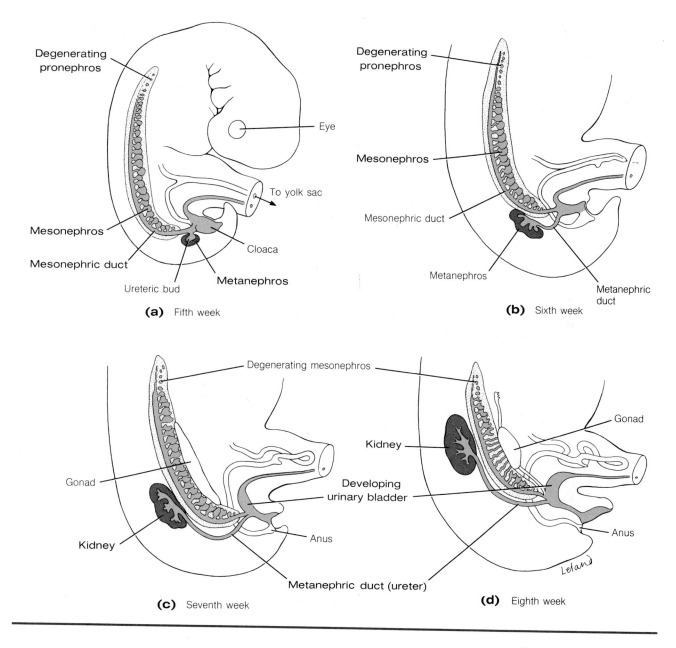

(a) Fifth week

(b) Sixth week

(c) Seventh week

(d) Eighth week

Figure 26.2

Embryonic development of the kidneys. The illustration shows the sequential development of the pronephros, the mesonephros, and the metanephros on one side of the embryo.

met-a-nef′-ros

As the pronephros degenerates, the mesonephros develops from the columns of intermediate mesoderm inferior to the region of the pronephros. Tubules from the mesonephros join with the pronephric ducts, which are then called the **mesonephric ducts.** By the sixth week, when the development of the mesonephros has reached its inferior limit, its superior portions begin to degenerate. By about the eighth week, only the most inferior regions of the mesonephros remain. Although there is no direct evidence that the mesonephros functions in the human embryo, it appears to be structurally capable of producing urine.

The **metanephros** is the third pair of kidneys to develop from the embryonic intermediate mesoderm. This pair develops into the adult kidneys. In about the fifth week of embryonic development, a hollow outgrowth called a **ureteric bud** arises from the distal end of each mesonephric duct, near its junction with the embryonic cloaca. The ureteric buds grow dorsally and cranially into the most inferior regions of the intermediate mesoderm. The upper ends of the ureteric buds, which come into contact with the intermediate mesoderm, enlarge. Each ureteric bud then elongates and carries its cap of

Box 26.1
Artificial Kidney

Effective kidney function is necessary for survival, and severe kidney disease can be fatal. The development of the artificial kidney machine, however, has greatly alleviated a number of kidney disease problems. The artificial kidney employs the principal of dialysis (see Chapter 2) to remove waste materials from the blood. In the artificial kidney, the patient's blood is passed through a dialysis tubing that allows urea, electrolytes, and other small molecules to move freely across its wall, but the tubing does not allow the movement of large protein molecules. The dialysis tubing is immersed in a bath that contains various substances. If urea or some other small molecule is present in the blood but not in the bath, it diffuses out of the dialysis tubing and into the bath and thus is

Dialysis unit of an artificial kidney. Blood enters and leaves the unit through the openings on the ends. The dialysis solution enters and leaves through the two openings on the side of the unit.

removed from the blood. If a specific low-molecular-weight substance is present in the bath at the same concentration as within the blood, then it diffuses into the tubing as fast as it diffuses out, and there is no net loss of the material from the blood. Thus, by regulating the composition

and concentration of materials within the bath, the types and amounts of materials that leave the blood by net diffusion out of the dialysis tubing can be regulated. In this way, waste products are removed from the blood while needed constituents are retained.

---

mesoderm cranially, eventually reaching what will be the permanent positions of the kidneys in the upper lumbar region. With further development, the *nephrons*, which are actively involved in the production of urine, form from the cap of intermediate mesoderm that covers each ureteric bud. The ureteric buds themselves develop into the *collecting tubules*, the *calyces*, the *renal pelvises*, and the *ureters*. The collecting tubules are also actively involved in urine production, and the calyces, the renal pelvises, and the ureters transport to the urinary bladder the urine that is formed in the nephrons and collecting tubules. These structures remain in the adult; they are described in the next section.

kay'-li-sees

ANATOMY OF THE KIDNEYS

F26.1

F26.3

The kidneys are paired reddish-brown organs situated on the posterior wall of the abdominal cavity, one on each side of the vertebral column (Figure 26.1). Each kidney is capped by an endocrine gland called the **adrenal gland.** The kidneys are approximately 11 cm long, extending from the level of the eleventh or twelfth thoracic vertebra to the third lumbar vertebra. Because of the presence of the liver, the right kidney is generally slightly lower than the left. The kidneys are located between muscles of the back and the peritoneal cavity (Figure 26.3). This retroperitoneal location makes it possible for the kidneys to be exposed surgically through the posterior body wall, without opening the peritoneal cavity.

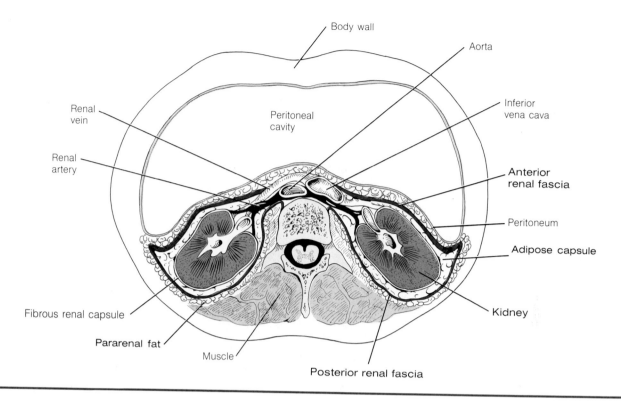

Body wall

Aorta

Renal vein

Peritoneal cavity

Inferior vena cava

Renal artery

Anterior renal fascia

Peritoneum

Adipose capsule

Fibrous renal capsule

Kidney

Pararenal fat

Muscle

Posterior renal fascia

Figure 26.3
Transverse section of the body trunk showing the retroperitoneal location of the kidneys and the renal fascia that surrounds them.

Tissue Layers Surrounding the Kidney

Each kidney is surrounded by three layers of tissues (Figure 26.3). The innermost layer, which covers the surface of the kidney, is the fibrous **renal capsule.** Surrounding the renal capsule is a mass of perirenal fat called the **adipose capsule.** The third tissue layer that covers the kidney is a double layer of fascia called the **renal fascia.** The renal fascia surrounds the kidney and the adipose capsule, completely enclosing them and anchoring the kidney to the posterior abdominal wall. There is an additional accumulation of fat (*pararenal fat*) outside of the renal fascia.

External Structure of the Kidney

The kidneys are bean-shaped, with convex lateral borders and concave medial borders. The medial border contains an indentation, called the **renal hilus,** through which the renal arteries enter the kidney and the renal veins and the ureter leave. The hilus opens into a space, called the **renal sinus,** in which the renal vessels and the renal pelvis are located.

Internal Structure of the Kidney

Three general regions can be distinguished in each kidney: the *cortex,* the *medulla,* and the *pelvis* (Figure 26.4).

F26.4

The **cortex** is the outer layer of the kidney, just deep to the renal capsule. Extensions of the cortex, called **renal columns,** pass into the medulla of the kidneys.

The **medulla** is located deep to the cortex and consists of several (up to 18) triangular **renal pyramids.** The pyramids are oriented so that their broad bases are covered by the cortex and their tips **(papillae)** project toward the renal pelvis. The pyramids are separated from one another by the cortical renal columns. Blood vessels that supply the cortex and medulla pass through the renal columns.

pa-pil'-lie

The papilla of each pyramid projects into a funnel-shaped chamber called a **minor calyx;** however, one minor calyx may receive two or more papilla. Several minor calyces join together to form a **major calyx.** There are generally

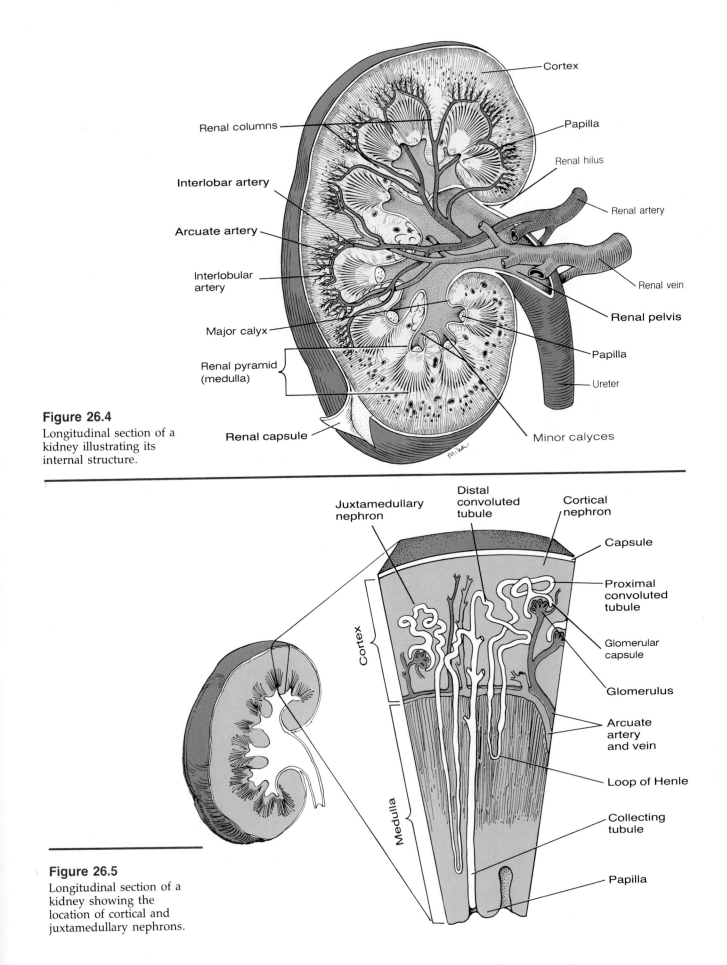

Figure 26.4
Longitudinal section of a kidney illustrating its internal structure.

Cortex

Renal columns

Papilla

Renal hilus

Interlobar artery

Renal artery

Arcuate artery

Renal vein

Interlobular artery

Renal pelvis

Major calyx

Papilla

Renal pyramid (medulla)

Ureter

Renal capsule

Minor calyces

Figure 26.5
Longitudinal section of a kidney showing the location of cortical and juxtamedullary nephrons.

Juxtamedullary nephron

Distal convoluted tubule

Cortical nephron

Capsule

Proximal convoluted tubule

Glomerular capsule

Glomerulus

Arcuate artery and vein

Loop of Henle

Collecting tubule

Papilla

Cortex

Medulla

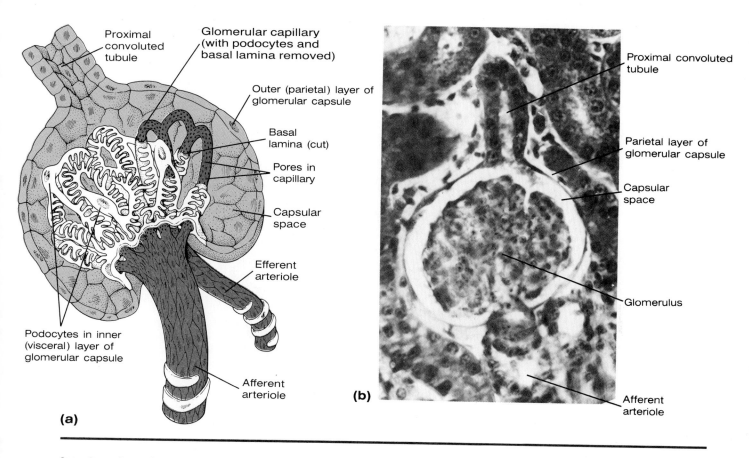

(a)

Proximal convoluted tubule

Glomerular capillary (with podocytes and basal lamina removed)

Outer (parietal) layer of glomerular capsule

Basal lamina (cut)

Pores in capillary

Capsular space

Efferent arteriole

Podocytes in inner (visceral) layer of glomerular capsule

Afferent arteriole

(b)

Proximal convoluted tubule

Parietal layer of glomerular capsule

Capsular space

Glomerulus

Afferent arteriole

Figure 26.6
Longitudinal section of a renal corpuscle.
(a) Drawing
(b) Photomicrograph.

2 or 3 major calyces and up to 13 minor calyces in each kidney. The major calyces join with one another to form the **renal pelvis,** which is the expanded upper end of the ureter. Urine passes as droplets from tiny pores in the papillae into the minor calyces. From there it travels into the major calyces, the renal pelvis, and finally into the ureter, which carries it to the urinary bladder.

The Renal Tubules

The **renal tubules,** which are the functional units of the kidneys where urine is formed, consist of **nephrons** and **collecting tubules.** There are estimated to be over 1 million nephrons in each kidney. Some nephrons—called **juxta-medullary nephrons**—extend deep into the medulla. Other nephrons—called **cortical nephrons**—do not penetrate as deeply into the medulla (Figure 26.5). Each nephron consists of two parts: (1) a network of parallel capillaries called a **glomerulus** and (2) a **tubule.** Various regions of the tubule differ from one another anatomically. Epithelial variations along the length of the tubule are related to variations in function. The proximal end of the tubule forms a double-walled cup known as the **glomerular capsule** (or *Bowman's capsule*), which surrounds the glomerulus. The capsule and the glomerulus together are called the **renal corpuscle** (Figure 26.6). The renal corpuscles are located in the cortical regions of the kidneys, and a process known as **glomerular filtration** occurs within them. In this process, which is an important aspect of urine production, some of the blood plasma (except for most proteins) passes out of the glomerular capillaries and into the space between the inner and outer layers of the glomerular capsule, forming a fluid called the **glomerular filtrate.**

The outer (parietal) layer of the glomerular capsule is composed of simple squamous epithelium that rests on a thin basal lamina. The inner (visceral) layer of the capsule is composed of specialized cells called **podocytes.** The podocytes have several processes that radiate from a central cell body and adhere to the basal lamina covering the squamous cells of the endothelium of

jux'-ta-med'-u-lair-ee

F26.5

glo-mer'-yoo-lus

F26.6

pod'-o-sites

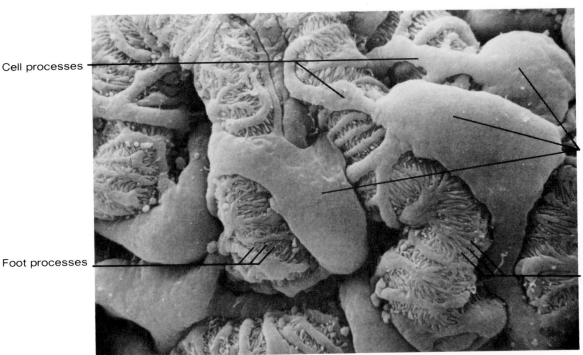

Cell processes

Foot processes

Podocyte
cell bodies

Filtration slits

Figure 26.7
Scanning electron
micrograph of podocytes
surrounding glomerular
capillaries (×3725). (From
*Tissues and Organs: A
Text-Atlas of Scanning
Electron Microscopy* by
Richard G. Kessel and
Randy H. Kardon.
W.H. Freeman and
Company. Copyright ©
1979.)

fen'-es-tray-ted

F26.8

F26.9

F26.10

the capillaries that form the glomerulus (Figure 26.7). These processes, in turn, branch into secondary and tertiary processes. The tertiary processes of the podocytes are called **foot processes,** or **pedicles**. The foot processes of one podocyte interdigitate with those of an adjacent podocyte, leaving an elaborate network of small clefts between them. These intercellular clefts are called **filtration slits,** or **slit pores**. A very thin **slit membrane** extends between the foot processes of adjacent cells, forming a diaphragm-like barrier that restricts the passage of molecules through the filtration slits. The capillaries that form the glomeruli are similar to other capillaries in that they consist of endothelium formed of a single layer of squamous cells; but they are a bit different due to the presence of many small, open pores that perforate the endothelial cells. Because of these pores, the endothelium of the glomeruli is referred to as *fenestrated endothelium* (*fenestra* = window).

The **filtration barrier** that separates the blood in the glomerular capillaries from the space in a glomerular capsule **(capsular space)** consists of only (1) the fenestrated endothelium, (2) the basal lamina, and (3) the slit membranes that cover the filtration slits (Figure 26.8). Consequently, many substances are able to pass through this barrier during glomerular filtration. Not all molecules are able to pass through the filtration barrier, however. The endothelial pores restrict the movement of blood cells and molecules larger than 16 nanometers in diameter, and the basal lamina and the slit membranes act as barriers to smaller molecules, allowing only those smaller than about 7 nanometers in diameter to pass through. As a result, the glomerular filtrate that enters a glomerular capsule includes most of the substances present in the blood plasma except most plasma proteins (Figure 26.9).

Negatively charged glycoproteins are associated with the filtration barrier. Consequently, positively charged molecules pass through the barrier faster than neutral molecules of the same size, and negatively charged molecules penetrate more slowly.

Beyond its glomerular capsule, each nephron forms a tightly looping tubule whose lumen is continuous with the capsular space. This region of the nephron is referred to as the **proximal convoluted tubule** because it is located in the cortex, close to the capsule, and is twisted (Figure 26.10). The wall of each proximal convoluted tubule consists of a single layer of thick cuboidal or pyramidal cells. The free surfaces of these cells, facing the lumen of the tubule, have many microvilli.

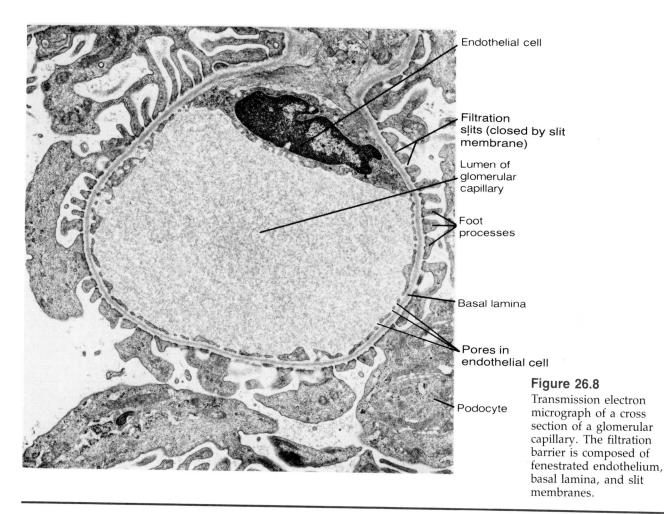

Endothelial cell

Filtration slits (closed by slit membrane)

Lumen of glomerular capillary

Foot processes

Basal lamina

Pores in endothelial cell

Podocyte

Figure 26.8
Transmission electron micrograph of a cross section of a glomerular capillary. The filtration barrier is composed of fenestrated endothelium, basal lamina, and slit membranes.

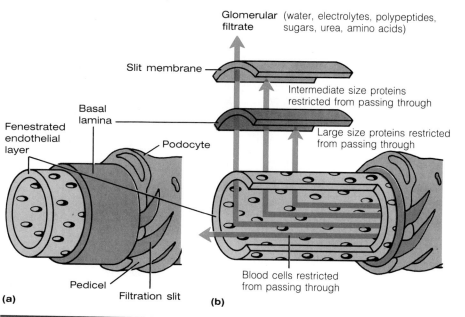

Glomerular filtrate (water, electrolytes, polypeptides, sugars, urea, amino acids)

Slit membrane

Intermediate size proteins restricted from passing through

Basal lamina

Fenestrated endothelial layer

Podocyte

Large size proteins restricted from passing through

Pedicel

Filtration slit

Blood cells restricted from passing through

(a)   **(b)**

Figure 26.9

(a) Structure of a glomerular capillary, with a surrounding podocyte. **(b)** Restrictions imposed by the various layers of the filtration barrier.

Beyond its proximal convoluted tubule, each nephron has a straight portion, the **proximal straight tubule,** and it then forms a **loop of Henle (ansa nephroni),** which passes into a pyramid of the medulla of the kidney (Figure 26.10). The loops of Henle of juxtamedullary nephrons are longer than those of cortical nephrons. As a consequence, the loops of Henle of juxtamedullary nephrons extend deeper into the medulla than do the loops of Henle of

F26.10

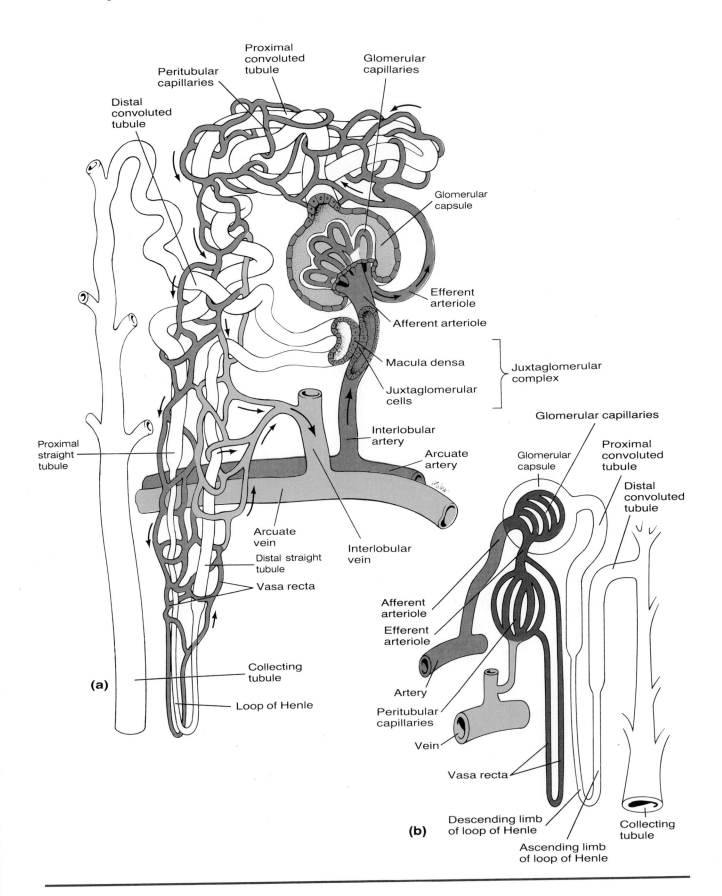

Figure 26.10

(a) A juxtamedullary nephron and its blood supply. (b) Schematic representation of the relationship between the juxtamedullary nephron shown in (a) and its blood supply.

cortical nephrons. The portion of each loop of Henle that descends into the medulla is called the **descending limb** of the loop of Henle. Because the epithelium of the tubule wall changes to thin squamous cells in the descending limb, this region is also referred to as the **thin segment** of the loop of Henle. Within the medulla the tubule makes a hairpin turn and ascends toward the cortex as the **ascending limb** of the loop of Henle, which passes out of the medulla and back into the cortex. The wall of the ascending limb is composed primarily of cuboidal cells, and this portion of the nephron is therefore referred to as the **thick segment** of the loop of Henle. The thick segment of the loop of Henle is also known as the **distal straight tubule.** In juxtamedullary nephrons the thin segment often extends around the hairpin turn and into the ascending limb. In these nephrons, therefore, the thick segment does not begin until the upper portion of the ascending limb.

Beyond its distal straight tubule, each nephron again becomes highly coiled. This portion is called the **distal convoluted tubule.** The wall of each distal convoluted tubule is composed of a single layer of cuboidal cells. The distal convoluted tubules of several nephrons empty into a common **collecting tubule** that transports urine back into the renal pyramids of the medulla. From 10 to 25 collecting tubules open on the papilla of each pyramid and empty into a minor calyx. The cells that make up the walls of the collecting tubules vary in shape from cuboidal to columnar.

Within the cortex of the kidney, the distal straight tubule contacts the blood vessel (the *afferent arteriole*) that carries blood to the glomerulus. At this point of contact, the cells of both the blood vessel and the tubule are modified. Among the smooth-muscle cells of the tunica media of the afferent arteriole are cells that contain prominent granules in their cytoplasm. These cells are called *juxtaglomerular cells*. Where the distal straight tubule contacts the juxtaglomerular cells, there is a concentration of nuclei and the cells appear higher than in other parts of the distal tubule. This region is the *macula densa.*

The juxtaglomerular cells plus the macula densa form a structure called the **juxtaglomerular complex** (or **juxtaglomerular apparatus**) (Figure 26.10). **F26.10** The juxtaglomerular cells are thought to secrete an enzyme called *renin* in response to lowered blood pressure. Through its effects on angiotensinogen, a plasma protein, renin indirectly raises the blood pressure in the kidney. This elevated blood pressure could play an important role in kidney functioning, since the blood pressure in the glomerulus must be maintained at a high enough level to cause filtration of substances into the glomerular capsule.

Blood Vessels of the Kidney

The importance of the relationship between the kidneys and the blood vascular system becomes apparent when one considers the large size of the **renal arteries,** which supply the organs. It has been estimated that, at rest, these vessels carry to the kidneys about 20% of the total cardiac output. In young adults, approximately 1100 ml of blood passes through the two kidneys every minute. Very little of this blood supplies the kidneys' nutritive needs. Rather, the large blood flow is related to the fact that the kidneys can maintain the homeostasis of the blood only if a considerable amount of blood passes through them.

Shortly after entering the hilus of the kidney, the renal artery divides into ventral and dorsal sets of vessels that pass on either side of the renal pelvis (Figure 26.4). These vessels travel between the pyramids and through the **F26.4** renal columns as the **interlobar arteries.** At the bases of the pyramids, which represent the junction of the medulla and the cortex, the interlobar arteries form arching branches that run parallel to the surface of the kidney. These are the **arcuate arteries.** At intervals, the arcuate arteries give off small **interlobular arteries,** which travel through the cortex toward the surface of the kidney. The interlobular arteries divide into several **afferent arterioles,** each of which supplies a renal corpuscle and forms a capillary network called the **glomerulus** (Figure 26.11). As mentioned earlier, it is at the glomerulus that blood **F26.11** comes in close contact with the cells of a glomerular capsule and the glomerular filtrate is formed.

Blood leaves each glomerulus through an **efferent arteriole** that passes out of the renal corpuscle. The efferent arterioles divide into networks of

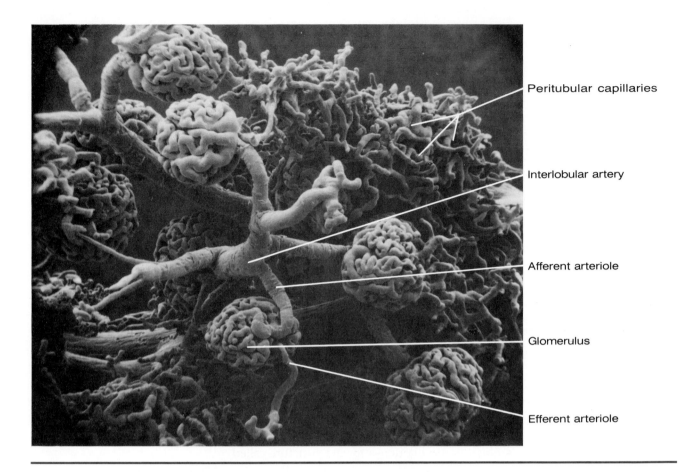

Peritubular capillaries

Interlobular artery

Afferent arteriole

Glomerulus

Efferent arteriole

Figure 26.11
Scanning electron micrograph of the blood vessels associated with glomeruli (×206). (From *Tissues and Organs: A Text-Atlas of Scanning Electron Microscopy* by Richard G. Kessel and Randy H. Kardon. W.H. Freeman and Company. Copyright © 1979.)

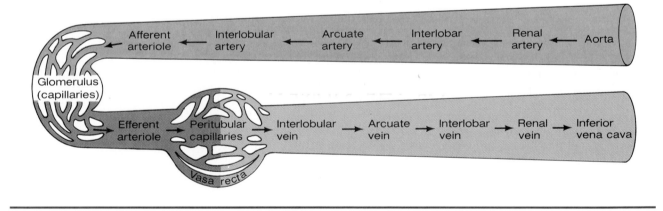

Figure 26.12
Summary of the pathway of blood through the kidney.

F26.10

capillaries that surround the proximal and distal convoluted tubules of the nephron as well as the cortical portions of the proximal and distal straight tubules and the collecting tubules. These capillaries are called the **peritubular capillaries** (Figure 26.10). Thin-walled vessels called **vasa recta** extend from the efferent arterioles of the juxtamedullary nephrons to supply the loops of Henle and the collecting tubules in the medulla of the kidney, particularly in the deep regions of the medulla. The vasa recta play a very important role in the formation of concentrated urine. The peritubular capillaries and vasa recta

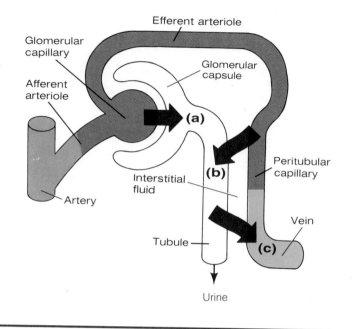

Efferent arteriole

Glomerular
capillary

Glomerular
capsule

Afferent
arteriole

(a)

Peritubular
capillary

Interstitial
fluid

(b)

Artery

Vein

Tubule

(c)

Urine

Figure 26.13
(a) As blood flows through the kidneys, some of the plasma is filtered out of the glomerular capillaries and into the glomerular capsules of the renal tubules. **(b)** Some materials within the blood that are not filtered into the renal tubules at the glomerular capsules can be secreted into other portions of the tubules. **(c)** As fluid flows along the renal tubules, water, electrolytes, glucose, and other substances required by the body are reabsorbed from the tubules and returned to the blood.

empty into **interlobular veins,** which in turn empty into **arcuate veins.** The arcuate veins lead into **interlobar veins,** which join with the **renal veins** that leave the kidney. The veins follow the same general course as the arteries of the same names. The pathway of blood through the kidney is summarized in Figure 26.12.

F26.12

There are several unusual and important features concerning blood flow through the kidney. As blood moves through the kidney, it flows through two sequential series of capillary beds—the glomerulus and the peritubular capillaries. The kidney is one of the few places in the body where this arrangement occurs. Generally, when blood leaves capillaries, it enters venules. In the kidney, however, efferent arterioles transport the blood from the first capillary bed (glomerulus) to the second capillary bed (peritubular capillaries). The presence of arterioles leading to and from the glomeruli has great functional significance. Like all arterioles, these vessels have smooth muscles in their walls that permit them either to constrict or to dilate in response to various types of stimulation. Thus, it is possible for a fairly constant blood pressure to be maintained in the glomeruli, even if the systemic pressure fluctuates. This constant blood pressure contributes to the efficient functioning of the kidney in spite of varying systemic conditions.

PHYSIOLOGY OF THE KIDNEYS

The excretory and regulatory activities of the kidneys depend in large measure on the efficient functioning of the glomeruli and the renal tubules (nephrons and collecting tubules). As blood flows through the kidneys, some of the plasma is filtered out of the glomerular capillaries and into the glomerular capsules of the renal tubules by the process of glomerular filtration (Figure 26.13). This process, which is primarily due to the pressure of the blood within the glomerular capillaries, is not very selective, and almost all plasma components—with the exception of large protein molecules—can enter glomerular capsules. From the capsules, the filtrate flows along the remaining portions of the renal tubules where water, electrolytes, glucose, and other substances required by the body are reabsorbed from the tubules and returned to the blood by both active and passive transport processes. Moreover, materials within the blood that were not originally filtered into the renal tubules at the glomerular capsules can be secreted into other portions of the tubules. The reabsorption and secretion processes are selective, and hormonal mechanisms enable the kidneys to exert a high degree of control over

F26.13

FRONTIERS IN HEALTH

New Surgery for Kidney Stones

Kidney stones are one of the most painful maladies that afflict humans. The sharp, often excruciating pain stabs the back and side or lower abdomen with such force that even powerful pain-killers are sometimes ineffective.

One in every ten men and one in every forty women will require medical attention because of kidney stones. Hundreds of thousands of Americans enter hospitals for treatment of kidney stones each year. Luckily, for the majority of them no surgery is required. The stones are allowed to pass through the urinary ducts and are excreted in the urine. But for some 20,000 to 50,000 people each year, surgery to remove the stones becomes necessary.

Until recently, kidney surgery was a major undertaking, and recovery was slow and painful. A 6- to 8-inch incision in the patient's side was required to allow surgeons to remove the stones, and a hospital stay of about 10 days was usual. After release from the hospital, another 8 weeks of recuperation at home were necessary before a patient could return to work.

Thanks to research that began in West Germany, many patients can now be up and walking the day after their surgery and can be back on the job in a week. The only visible sign of the surgery is a ¼-inch scar.

Instead of making a large incision to expose the kidney, surgeons now use a small hollow metal device, called a nephroscope. The nephroscope is inserted into the abdominal cavity through a tiny incision in the patient's side and, with the aid of fiber optics, guided into the kidney, from which it is possible to view the entire upper urinary tract. After locating the kidney stone the surgeon aspirates it if the stone is small. For larger stones, a tiny basket-like grabbing device is introduced through the nephroscope. The stone is clasped in the claws of this device and physically crushed. The crushed stone can then be sucked out through the hollow nephroscope. For still larger stones, however, a new method called percutaneous ultrasound lithotripsy is often used.

Dr. P. Alken and his colleagues at the University of Mainz Medical School in Mainz, West Germany were the first to use percutaneous ultrasound lithotripsy to pulverize kidney stones prior to their removal. This technique was introduced into the United States in 1981, and today it is used routinely in several major medical centers.

Lithotripsy is one of the many new medical uses of ultrasound—very high frequency sound waves inaudible to the human ear. To break up kidney stones, surgeons insert a hollow metal ultrasound probe through the nephroscope. The probe is then pushed into contact with the stone and turned on. Once the ultrasonic vibrations cause the hard outer "shell" of the kidney stone to crack, the remainder of the stone generally falls apart. Surgeons then suck out the fragments with a vacuum aspirator.

Clinical trials of lithotripsy have been very encouraging. A 1984 paper in the medical journal *Radiology* reports that surgeons were successful in removing all fragments of the kidney stone in 80% of the patients on the first attempt. The remainder of the cases required a second try. Overall, the success rate was 97%.

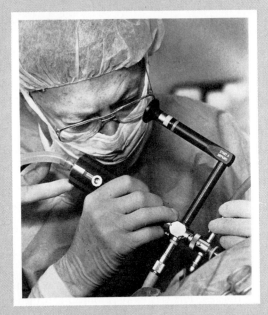

Surgeon using ultrasound through a nephroscope to break up kidney stones.

Although complications from ultrasound lithotripsy are about the same as for conventional kidney surgery, this new technique has several advantages for the patients. Discomfort and pain are drastically reduced, and recovery time and medical costs are cut to a fraction of what they were prior to the introduction of this technique.

As successful as ultrasound lithotripsy has been, however, the West German researchers have continued to look for new ways to remove kidney stones. A new method they are now working with may someday replace ultrasound lithotripsy.

Imagine a surgical technique that requires no surgery. A patient with a kidney stone is mildly sedated, given a local anesthetic, and placed in a warm water bath. Sitting perfectly still, he can listen to his favorite music while ultrasound waves, focused on the kidney stone, penetrate his skin and shatter the stone that is lodged in his kidney. When the procedure is completed, he returns home to pass the tiny fragments in the urine.

Sound like fantasy? At least 1000 West Germans have been treated with the technique and are now free of their painful stones. Six medical centers in the United States began testing the technique in the summer of 1984, and although the studies are not complete, early results are quite promising. However, despite an outstanding success rate and only minor side effects, this new technique may not become available to everyone. The machine alone costs $2 million, and only the largest medical centers will be able to afford it.

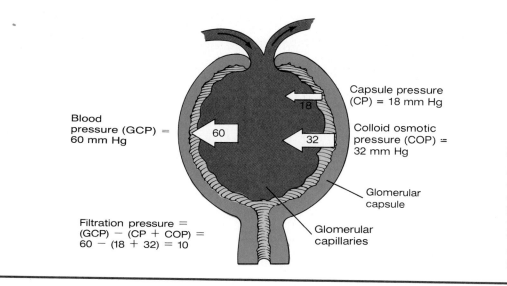

Blood pressure (GCP) = 60 mm Hg

Capsule pressure (CP) = 18 mm Hg

Colloid osmotic pressure (COP) = 32 mm Hg

Glomerular capsule

Glomerular capillaries

Filtration pressure = (GCP) − (CP + COP) = 60 − (18 + 32) = 10

Figure 26.14
The pressure of the blood within the glomerular capillaries (GCP) favors the filtration of materials into glomerular capsules. The pressure of the fluid within glomerular capsules (CP) and the osmotic pressure exerted by unfiltered plasma proteins that remain within the glomerular capillaries (COP) oppose filtration.

the amounts of many materials—particularly water and electrolytes—that are reabsorbed and returned to the blood or secreted into the tubular fluid.

When the reabsorption and secretion processes are completed, the fluid remaining within the renal tubules is transported to other components of the urinary system to be excreted as urine. Thus, urine consists of water and materials that were filtered or secreted into the renal tubules but not reabsorbed.

Glomerular Filtration

Approximately 16–20% of the blood plasma that enters the kidneys is filtered from the glomerular capillaries into the glomerular capsules of the nephrons as the glomerular filtrate. The remaining plasma continues into the efferent arterioles and on into the peritubular capillaries and vasa recta. Under normal circumstances, the kidneys produce approximately 180 liters (about 45 gallons) of glomerular filtrate per day. However, approximately 99% of the fluid volume of the filtrate is reabsorbed from the renal tubules and only about 1% (1–2 liters per day) is excreted as urine.

The filtration barrier between the plasma and the internal portion of a glomerular capsule is from 100 to 1000 times more permeable than a typical capillary, and water and solutes of small molecular dimension pass freely from the plasma into glomerular capsules. On the other hand, this barrier is relatively impermeable to large molecules, such as plasma proteins. As a result, the glomerular filtrate contains most plasma components at essentially the same concentrations as they are found within the plasma, but it is basically protein free. (Actually, a small amount of protein does appear in the glomerular filtrate, but this is normally reabsorbed by pinocytosis in the proximal convoluted tubules so that essentially no protein is present in the urine.)

Net Filtration Pressure

The pressure of the blood within the glomerular capillaries (that is, the glomerular capillary pressure) is a major factor in glomerular filtration. This pressure—about 60 mm of mercury (Hg) or more—is higher than the pressure in most capillaries, and it favors filtration out of the glomerular capillaries and into the glomerular capsules. The glomerular capillary pressure is opposed by the pressure of the fluid within glomerular capsules (about 18 mm Hg), and by the osmotic force exerted by unfiltered plasma proteins that remain within the glomerular capillaries (about 32 mm Hg). Thus, a *net filtration pressure* of about 10 mm Hg (60 − 18 − 32 = 10) favors the movement of materials out of the glomeruli and into the glomerular capsules (Figure 26.14). This net filtration pressure is responsible for glomerular filtration, which occurs as a bulk

F26.14

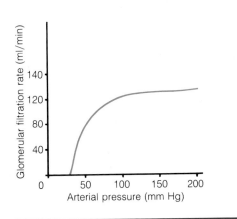

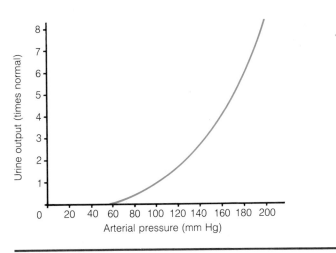

Figure 26.15

Graph indicating the effects of changes in arterial blood pressure on the glomerular filtration rate.

Figure 26.16

Graph indicating the effect of changes in arterial blood pressure on the output of urine.

flow of water and small dissolved particles from the glomerular capillaries into glomerular capsules.

Factors That Influence Glomerular Filtration

The rate of glomerular filtration at any moment depends on the relationship between the blood pressure within the glomerular capillaries that favors filtration and the forces that oppose filtration, which include the pressure of the fluid within glomerular capsules and the osmotic force exerted by the plasma proteins within the glomerular capillaries. Variations in any of these factors can alter the glomerular filtration rate.

INFLUENCE OF ARTERIAL BLOOD PRESSURE AND AUTOREGULATORY MECHANISMS   It might be expected that changes in the arterial blood pressure would lead to comparable changes in the pressure within the glomerular capillaries and, consequently, to alterations in the glomerular filtration rate. However, the influence of arterial blood pressure on the glomerular filtration rate is not as great as expected. The kidneys possess intrinsic autoregulatory mechanisms that maintain a fairly stable glomerular filtration rate over a rather wide range of systemic blood pressures (Figure 26.15). For example, when the arterial pressure rises, the afferent arterioles automatically constrict, preventing a major rise in the blood pressure within the glomerular capillaries. Thus, even when the arterial pressure increases from 100 to 160 mm Hg, the glomerular filtration rate increases by only a few percent.

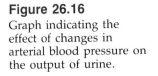

Even though autoregulatory mechanisms can minimize the effects of changes in arterial blood pressure on the glomerular filtration rate, arterial pressure changes nevertheless do have a substantial influence on the output of urine (Figure 26.16). This influence is due in part to the fact that autoregulation is not perfect, and an increase in the arterial pressure, for example, does cause some increase in the glomerular filtration rate. Since the kidneys normally form such a large volume of glomerular filtrate (about 180 liters per day), even a relatively small percentage increase in the glomerular filtration rate can cause a considerable amount of additional material to enter the renal tubules. Moreover, an elevation of the arterial pressure also causes a slight increase in the pressure within the peritubular capillaries, and this pressure increase tends to decrease the rate at which fluid is reabsorbed from the renal tubules. Consequently, an increase in the arterial blood pressure from 100 to 120 mm Hg approximately doubles the output of urine, and an increase in the arterial blood pressure from 100 to 200 mm Hg increases the urine output six to eight times.

INFLUENCE OF SYMPATHETIC NEURAL STIMULATION In addition to the influence of arterial blood pressure and intrinsic autoregulatory mechanisms, the activity of the sympathetic neurons that supply the kidneys can also influence the glomerular filtration rate. For example, an increase in sympathetic neural stimulation generally results in a preferential constriction of the afferent arterioles. This constriction results in a lowered glomerular capillary pressure (and in a decreased rate of blood flow into the glomeruli) and, as a consequence, in a lowered glomerular filtration rate.

Tubular Reabsorption

Many plasma components that are filtered from the glomerular capillaries into glomerular capsules are reabsorbed from the renal tubules to greater or lesser degrees (Figure 26.13). The reabsorption of a number of substances (especially water and many inorganic ions) is under hormonal control, and the amounts of such substances reabsorbed from the renal tubules vary with body needs.

Although filtration, as previously indicated, occurs by bulk flow, tubular reabsorption is the result of both diffusion and discrete transport mechanisms. Tubular reabsorptive transport can be either active or passive. Carrier molecules are often involved, and a single transport mechanism is sometimes responsible for the movement of several different substances with similar structures. Since carrier molecules can become saturated, tubular transport systems that utilize carriers—for example, the transport system for glucose—can transport only limited amounts of material per unit time. If the amount of glucose or some other substance transported by such a carrier system increases substantially in the plasma and subsequently in the glomerular filtrate, not all of the substance is reabsorbed and some appears in the final urine.

Active Tubular Reabsorption

A number of substances, including glucose, amino acids, and vitamins, as well as sodium, calcium, and chloride ions, are actively reabsorbed from the renal tubules. Active tubular reabsorption requires the expenditure of energy for the transport of a substance across the tubule wall. For example, in the proximal portion of a tubule, sodium diffuses passively from the tubular lumen into the tubular cells; then, energy-requiring active transport processes move sodium out of the cells into the interstitial fluid. From the interstitial fluid, the sodium enters the peritubular capillaries.

Note that in this example, the movement of sodium across the wall of the renal tubule requires the sodium to cross more than one membrane. Sodium must first enter the tubular cells through one membrane and then exit from the cells through another. Although only the exit of sodium from the cells is an active energy-requiring process, if any one step is active, the whole process is considered to be active. Thus, sodium is actively reabsorbed from the renal tubules. In subsequent sections of this chapter, the processes of tubular reabsorption (and tubular secretion) are considered as if only a single membrane is involved, and if any transport step is active, the whole process is considered to be active.

Passive Tubular Reabsorption

Some substances are passively reabsorbed from the renal tubules. For example, urea is freely filtered from the glomerular capillaries into glomerular capsules and appears in the glomerular filtrate at a concentration that is essentially equal to that of urea in the plasma. As the filtrate moves through the renal tubules, however, water is reabsorbed. The reabsorption of water increases the concentration of urea in the tubular fluid to a level above that of the urea concentration in the interstitial fluid surrounding the renal tubules. As a consequence, urea diffuses from the renal tubules to the interstitial fluid.

Thus, urea reabsorption is passive and depends on water reabsorption and also on the permeability of the renal tubules to urea. Since the renal tubules are generally not as permeable to urea as they are to water, only about 40–60% of the filtered urea is normally reabsorbed, even though about 99% of the filtered water is reabsorbed.

Many foreign chemicals are also passively reabsorbed from the renal tubules. The epithelial cells of the tubules are fairly permeable to lipid-soluble substances, and many drugs and environmental pollutants are lipid-soluble, nonpolar materials. Therefore, they are relatively easily reabsorbed from the tubules as a consequence of water reabsorption in a manner similar to that described for urea. As a result, such substances can be difficult to excrete. However, many of these substances are converted by the liver to progressively more polar, less lipid-soluble materials that are not as readily reabsorbed from the renal tubules. Such materials are excreted more easily.

Tubular Secretion

F26.13 Substances enter the renal tubules not only by glomerular filtration but also by tubular secretion (Figure 26.13). Tubular secretion can move substances that leave the peritubular capillaries into the renal tubules. As with tubular reabsorption, the processes of tubular secretion can be either active or passive, and they are often hormonally controlled. Hydrogen and potassium ions are secreted into the renal tubules, as are penicillin and many other chemicals not normally present in the body.

Excretion of Water and Its Effects on Urine Concentration

A person who has just consumed a large quantity of water must eliminate a considerable volume of fluid without losing excessive amounts of electrolytes and other vital substances if homeostasis is to be maintained. Under such conditions it is beneficial to produce a large volume of dilute urine. Conversely, a dehydrated individual who has had nothing to drink for some time must still produce urine for the elimination of wastes. In this case, it is advantageous to produce a small volume of concentrated urine, which requires the excretion of only small amounts of water. Thus, if the kidneys are to be effective as regulatory organs, they must be able to produce a range of urine volumes and concentrations according to the body's needs. In fact, the kidneys can produce urine as dilute as 65–70 milliosmoles per liter or as concentrated as 1200 milliosmoles per liter compared to the concentration of the blood plasma, which is about 300 milliosmoles per liter.

The ability to produce a low-volume, concentrated urine depends on the presence of a highly concentrated interstitial fluid within the medullary region of each kidney. Therefore, the mechanisms responsible for maintaining this high interstitial fluid concentration will be considered first, and then the processes involved in producing different volumes and concentrations of urine will be examined.

Maintenance of a Concentrated Medullary Interstitial Fluid

Within the medullary region of a kidney, the interstitial fluid concentration increases from about 300 milliosmoles per liter at the cortex to as much as 1200 milliosmoles per liter at the tips of the pyramids. The high medullary interstitial fluid concentration is due to the activities and anatomical arrangement of the loops of Henle of the nephrons (particularly those of the juxtamedullary nephrons), the collecting tubules, and the vasa recta.

ACTIVITIES OF THE LOOPS OF HENLE AND THE COLLECTING TUBULES The loops of Henle of the juxtamedullary nephrons extend deep into the medulla, and the collecting tubules pass through the medulla to the tips of the pyramids. As fluid flows along the loops of Henle and the collecting tubules, the reabsorption of solutes raises the concentration of the medullary

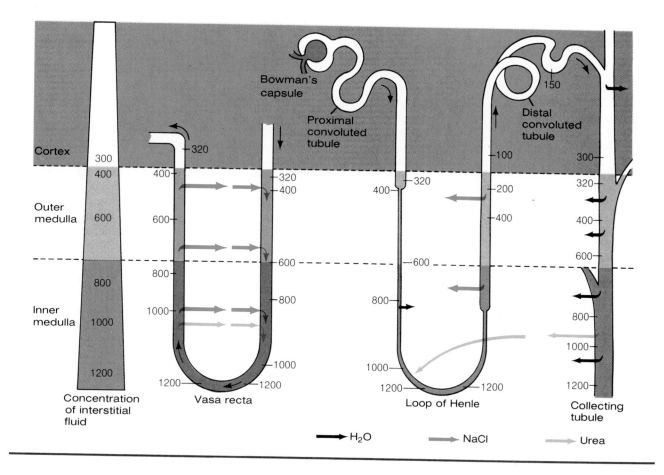

Figure 26.17
Mechanisms involved
in maintaining a
concentrated medullary
interstitial fluid and
producing a low-volume,
concentrated urine.
See text for detailed
discussion. The con-
centrations indicated are
in milliosmoles per liter.
They are the total
concentrations of all
solutes present, not the
concentrations of
individual substances.

interstitial fluid. For example, negatively charged chloride ions are actively transported out of the fluid in the upper portions (the *thick segments*) of the ascending limbs of the loops of Henle into the interstitial fluid of the outer medulla, and because of electrical considerations, positively charged sodium ions passively follow the chloride ions (Figure 26.17). This movement of sodium and chloride ions not only raises the concentration of the outer medullary interstitial fluid but, as discussed later, sodium and chloride ions are carried downward into the inner medulla by the blood flowing through the vasa recta. In addition, sodium ions are actively transported out of the fluid in the collecting tubules, with chloride ions following, and sodium and chloride ions can move out of the lower portions (the *thin segments*) of the ascending limbs of the loops of Henle into the interstitial fluid of the inner medulla. The reabsorption of sodium and chloride ions—particularly in the thick segments of the ascending limbs of the loops of Henle—contributes importantly to the high concentration of the medullary interstitial fluid.

Water can be reabsorbed from the fluid within the collecting tubules, and as discussed later, the amount of water reabsorbed is controlled hormonally by antidiuretic hormone. When a large amount of water is reabsorbed from the collecting tubules, a substantial amount of urea can diffuse out of the final portions of the tubules into the interstitial fluid of the inner medulla and thus contribute to the high concentration of the medullary interstitial fluid. This movement of urea occurs as follows.

The first portions of the collecting tubules are quite impermeable to urea. Consequently, when a large amount of water is reabsorbed, the concentration of urea in the tubular fluid increases considerably, and much of it diffuses out of the final portions of the collecting tubules into the interstitial fluid of the inner medulla (Figure 26.17). This activity raises the urea concentration of the interstitial fluid above that of the tubular fluid in the bottoms of the loops of Henle. As a result, some of the urea diffuses into the tubular fluid at the bottoms of the loops, raising its concentration. The tubular fluid is then carried through the ascending limbs of the loops of Henle and the distal convoluted tubules (both of which are essentially impermeable to urea) to the collecting tubules, where the reabsorption of water can further increase the fluid's urea concentration. Even more urea then diffuses out of the final portions of the collecting tubules, and the cycle is repeated. Thus, when much water is reabsorbed from the collecting tubules, this cycling process can multiply the urea concentration of the interstitial fluid of the inner medulla (and the urea concentration of the fluid within the final portions of the collecting tubules) to quite high levels.

Together, the mechanisms that raise the concentration of the medullary interstitial fluid are frequently referred to as the *countercurrent mechanism* of the loop of Henle because some aspects of the different mechanisms—for example, the cycling process involved in urea transport—depend on the fact that fluid flows in opposite directions in the essentially parallel ascending and descending limbs of the loops of Henle and the collecting tubules.

EFFECT OF VASA RECTA BLOOD FLOW   If a high medullary interstitial fluid concentration is to be maintained, the reabsorbed solutes (for example, sodium, chloride, and urea) that raise the concentration of the fluid must not be removed by the blood. The removal of the solutes is minimized by the pattern of blood flow in the vasa recta, which extend into the medulla.

The vasa recta form a countercurrent system. In this system, blood flows in opposite directions within the descending and ascending limbs of the vasa recta, and most solutes can pass readily between the blood and the interstitial fluid. The system operates essentially as follows.

F26.17   Blood with a concentration of about 300 milliosmoles per liter flows from the cortex into the medulla along the descending limbs of the vasa recta, which extend into regions of increasingly concentrated interstitial fluid (Figure 26.17). As the blood flows along the descending limbs, sodium, chloride, urea, and other solutes that are in higher concentration in the interstitial fluid than in the blood diffuse into the blood, raising its concentration. The concentrated blood then flows into the ascending limbs of the vasa recta, which extend into regions of decreasing concentrations of interstitial fluid. As the blood flows along the ascending limbs, sodium, chloride, urea, and other solutes that are now in higher concentration in the blood than in the interstitial fluid diffuse out of the vasa recta. As a result, by the time the blood leaves the vasa recta, its concentration is only slightly higher than it was when the blood initially entered the vessels. Thus, the blood flowing through the vasa recta removes only a small amount of solutes from the medullary interstitial fluid. In addition, the blood within the vasa recta contains unfiltered proteins that exert an osmotic force, the colloid osmotic pressure of the plasma. This force causes water that is reabsorbed from the loops of Henle and collecting tubules to be picked up and carried away by the blood, thus preventing it from severely diluting the medullary interstitial fluid.

Together, the activities and arrangement of the loops of Henle of the juxtamedullary nephrons, the collecting tubules, and the vasa recta maintain a high medullary interstitial fluid concentration that increases progressively from about 300 milliosmoles per liter at the cortex to as much as 1200 milliosmoles per liter at the tips of the pyramids. The kidneys utilize this highly concentrated interstitial fluid to produce a low-volume, concentrated urine.

Production of Different Urine Volumes and Concentrations

As previously indicated, the vast majority of the approximately 180 liters of glomerular filtrate formed per day is reabsorbed from the renal tubules and returned to the body. Moreover, the hormonal mechanism involving antidiuretic hormone that controls water reabsorption in the collecting tubules strongly influences the actual volume of filtrate reabsorbed. Consequently, this mechanism also influences the volume and concentration of the urine that is produced.

The concentration of the glomerular filtrate within glomerular capsules is essentially the same as that of the plasma—that is, about 300 milliosmoles per liter. As the filtrate flows along the proximal portion of a renal tubule, sodium ions are actively transported out of the tubule and into the interstitial fluid, with chloride ions following passively (Figure 26.18). In addition, a number of other substances, such as glucose and amino acids, are reabsorbed by various transport mechanisms. The proximal tubule is permeable to water, and water leaves osmotically, since the loss of sodium, chloride, and other substances transiently lowers the concentration of the tubular fluid and increases the concentration of the interstitial fluid. The net result of the removal of solutes and water from the proximal tubule is to reduce the volume of the original glomerular filtrate about 65–70%. Its concentration, however, remains at approximately that of the plasma. The materials that enter the interstitial fluid from the proximal portion of the renal tubule ultimately return to the peritubular capillaries and are carried away from the kidney with the blood.

The fluid remaining within the proximal tubule proceeds into the descending limb of the loop of Henle, which dips into regions of increasingly concentrated interstitial fluid (Figure 26.17). The descending limb is relatively permeable to water, and since the interstitial fluid outside the limb is more concentrated than the fluid within it, water moves out of the tubule. By the time the tubular fluid reaches the bottom of the loop of Henle, the loss of water has further reduced its volume and also raised its concentration above that of the plasma. In fact, at the bottom of the loop of Henle of a juxtamedullary nephron, the loss of water—together with the previously discussed entry of urea into the tubule—can raise the concentration of the tubular fluid to approximately 1200 milliosmoles per liter.

As already noted, within the ascending limb of the loop of Henle, sodium and chloride ions move out of the tubule and into the interstitial fluid (Figure 26.17). In the thin segment of the ascending limb, the outward movement of sodium and chloride ions apparently occurs passively by diffusion as the concentrated tubular fluid from the bottom of the loop of Henle flows up the ascending limb, which extends into regions of decreasing concentrations of interstitial fluid. However, some researchers believe the outward movement of sodium and chloride ions in this portion of the limb occurs by active transport. In the thick segment of the ascending limb, chloride ions are actively transported out of the tubule, and because of electrical considerations, sodium ions follow passively.

The ascending limb of the loop of Henle—particularly the thick segment—is essentially impermeable to water. Consequently, as sodium and chloride ions move out of the tubular fluid within the limb, the concentration of the fluid drops, but its volume is not greatly altered. By the time the fluid completes its passage through the ascending limb and enters the distal convoluted tubule, the reabsorption of sodium and chloride ions has lowered its concentration below that of the plasma.

The processes occurring within the distal convoluted tubule are complex. Certain ions are reabsorbed from the fluid within the tubule, whereas others are secreted into the fluid, and much of this activity is influenced by physiological regulatory mechanisms. In monkeys and presumably in other primates, such as humans, the distal convoluted tubule is relatively impermeable to water. In general, the concentration of the fluid within the distal convoluted tubule remains below that of the plasma (and it can even fall below that of the fluid leaving the ascending limb of the loop of Henle), but its volume does not change greatly.

Up to this point, much of the fluid volume of the glomerular filtrate has been reabsorbed—primarily in the proximal tubule. Moreover, many nutrients required by the body, such as glucose and amino acids, are almost completely reabsorbed in the proximal tubule. In the case of a juxtamedullary nephron, the sodium and chloride ions that move out of the ascending limb of the loop of Henle contribute to the high concentration of the medullary interstitial fluid, which is required for the production of a low-volume, concentrated urine.

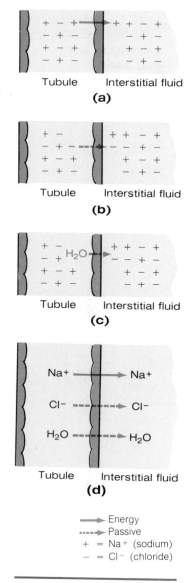

Figure 26.18
Sodium, chloride, and water movements in the proximal convoluted tubule. **(a)** Sodium ions are actively transported out of the tubule. **(b)** Because of electrical considerations, negatively charged chloride ions passively follow the positively charged sodium ions. **(c)** The reabsorption of sodium and chloride establishes an osmotic gradient that causes water to be reabsorbed from the proximal tubule. **(d)** Summary of the movements of sodium, chloride, and water in the proximal convoluted tubule.

Table 26.1 Representative Components and Characteristics of a 24-hour Sample of Urine

Component	Value
Calcium	0.01 to 0.30 g
Chloride	6.0 to 9.0 g
Creatinine	1.0 to 1.5 g
Potassium	2.5 to 3.5 g
Sodium	4.0 to 6.0 g
Urea	20.0 to 35.0 g

Characteristic	Value
pH	4.8 to 7.4
Specific gravity	1.015 to 1.022
Volume	1000 to 1600 ml

From the distal convoluted tubule, the fluid enters a collecting tubule, which extends from the cortex into the medullary region of the kidney where there are increasing concentrations of interstitial fluid. The permeability of the collecting tubule to water is controlled by antidiuretic hormone (ADH), which is manufactured in the hypothalamus of the brain and released from the pituitary gland. When the ADH concentration of the plasma is high, the collecting tubule is very permeable to water. As a result, water leaves the tubular fluid and enters the concentrated interstitial fluid. (Note that even though the permeability of the collecting tubule is high, the movement of water out of the tubule depends on the presence of the concentrated medullary interstitial fluid to establish the proper osmotic gradient.) The outward movement of water reduces the volume of the fluid within the collecting tubule and increases its concentration. This low-volume, concentrated fluid then leaves the tubule at the tip of a pyramid to be excreted as urine.

When the plasma-ADH concentration is low, the collecting tubule is not very permeable to water, and little water is reabsorbed from the tubular fluid. Consequently, the dilute fluid that flows along the distal convoluted tubule into a collecting tubule remains dilute, and its volume is not significantly reduced. (Moreover, ions can be actively reabsorbed from the collecting tubule, and this activity can lower the concentration of the tubular fluid even further.) In this case a high-volume, low-concentration fluid leaves the collecting tubules as urine.

As is evident from this discussion, the reabsorption of water from the renal tubules occurs by passive transport (osmosis), and depends on the reabsorption of solutes such as sodium and chloride ions to establish the proper osmotic gradient. However, once a gradient is established, the volume of water reabsorbed from the collecting tubules varies with their permeability to water, which is controlled by ADH. Consequently, ADH is important in regulating fluid volume excretion and in providing the kidneys with the ability to produce a range of urine volumes and concentrations according to the body's needs. Factors influencing ADH release are discussed in Chapter 27.

Minimum Urine Volumes

The body must excrete about 600 milliosmoles of waste products such as urea, sulfates, phosphates, and the like each day. The elimination of these materials, even at a urine concentration of 1200 milliosmoles per liter, requires a certain degree of water loss:

$$\frac{600 \text{ milliosmoles/day}}{1200 \text{ milliosmoles/liter}} = 0.5 \text{ liter/day}$$

Regardless of water intake or availability, this volume of fluid must be excreted simply to eliminate wastes. Beyond this requirement, however, physiological mechanisms come into play to vary the actual volume, concentration, and composition of the urine produced, in order to compensate for altered intakes of fluids, electrolytes, and other substances, as well as to compensate for varying losses of these materials by other pathways, including the lungs and sweat glands. The composition of a representative sample of urine is given in Table 26.1.

Table 26.1

Plasma Clearance

The *plasma clearance* of a substance is the rate at which the substance is eliminated or cleared from the plasma by the kidneys. By determining the plasma clearance of certain specific substances, the glomerular filtration rate and the rate of plasma flow through the kidneys can be determined. These types of information are useful in assessing kidney function.

The measurement of the plasma clearance of a substance requires the determination of (1) the rate of urine formation, (2) the concentration of the substance in the arterial plasma flowing to the kidneys, and (3) the concentration of the substance in the urine. With these values, the plasma clearance of

CLINICAL CORRELATION
Inappropriate Secretion of ADH

CASE REPORT . . . .

THE PATIENT   A 43-year-old man

PRINCIPAL COMPLAINTS   Headache, nausea, weakness, thirst, and abrupt loss of vision in one eye

HISTORY   At the age of 33 (ten years before this report), the patient was diagnosed as having multiple sclerosis (MS). He had developed numbness and weakness on the right side of the body that had progressed to quadriplegia (paralysis affecting the four extremities of the body). He was treated with adrenocorticotropic hormone (ACTH) and dismissed in improved condition. During the intervening years he experienced episodes of hemiplegia (paralysis of one side of the body) or paraplegia (paralysis of the lower limbs) that were ameliorated by ACTH.

CLINICAL EXAMINATION   Loss of vision was complete in the right eye, and the pupil was fixed. Deep tendon reflexes in the right leg were absent, and deep sensation was decreased in both legs. Sodium concentration in the serum was 110 mEq/liter (normal: 135–145 mEq/liter), potassium concentration was 4 mEq/liter (normal: 3.5–5 mEq/liter), and chloride concentration was 75 mEq/liter (normal: 100–106 mEq/liter). Blood urea nitrogen (BUN; nitrogen in the form of urea) was 7 mg/100 ml (normal: 8–25 mg/100 ml), and creatinine (a metabolic end product derived from creatine phosphate) was 0.3 mg/100 ml (normal: 0.6–1.5 mg/100 ml). The hematocrit was 33% (normal: 45–52%), and serum osmolality was 235 milliosmoles/kg water (normal: 285–295 milliosmoles/kg water). The patient weighed 60.4 kg (usual weight: 57.4 kg). The sodium concentration of the urine was 125 mEq/liter (normal: 50–130 mEq/liter), the potassium concentration was 45 mEq/liter (normal: 20–70 mEq/liter), the specific gravity was 1.020 (normal: 1.015–1.022), and the osmolality was 615 milliosmoles/kg water (normal: 500–800 milliosmoles/kg water).

COMMENT   The findings of normal osmolality and sodium concentration of the urine despite hypotonic plasma indicate inappropriate secretion of antidiuretic hormone (ADH). The concentrations of several constituents of the plasma—sodium, chloride, BUN, creatinine, and red cells (hematocrit)—were low, and the osmolality of the blood was low. Under these conditions, a dilute urine should have been excreted to increase the osmolality of the blood. For this to occur, the secretion of ADH should be low. Inappropriate secretion of ADH in this patient probably is related to demyelinating lesions of the hypothalamus (particularly in areas involved in the control and secretion of ADH) as one of the manifestations of MS. (Other effects of hypothalamic damage also have been seen in patients who have MS.) Consistent with this hypothesis, the development of hypotonic plasma in this patient occurred during worsening of the MS characterized by focal disorders of the central nervous system. The symptoms of headache, nausea, and weakness probably are related to the patient's low blood-sodium level.

TREATMENT   The patient was given hypertonic (5%) sodium chloride solution intravenously and his fluid intake was restricted to establish normal osmolality of the plasma. He was given both democlocycline—which antagonizes the action of ADH—and ACTH. He also was given cyclophosphamide, which suppresses the immune system of the body. Evidence indicates that MS is an autoimmune disease associated in some way with viral infection.

OUTCOME   The patient improved rapidly, and the vision in his affected eye began to return after two days. He produced large amounts of urine, and his weight returned to normal in three days. He was discharged after five days, with his blood chemistry normal and the MS in remission. However, MS is a progressive illness, and he will no doubt suffer relapse again.

the substance can be determined according to the following formula:

plasma clearance of A (ml/min) =

$$\frac{\text{rate of urine formation (ml/min)} \times \text{concentration of } A \text{ in urine}}{\text{concentration of } A \text{ in plasma}}$$

For example, suppose the concentration of urea in the plasma is 0.30 mg/ml. Suppose further than in one hour 66 ml of urine is formed (a formation rate that equates to 1.1 ml/min), and that this urine has a urea concentration of 17 mg/ml. The plasma clearance of urea would then be:

$$\text{plasma clearance of urea (ml/min)} = \frac{(1.1 \text{ ml/min}) \times (17 \text{ mg/ml})}{0.30 \text{ mg/ml}}$$
$$= 62 \text{ ml/min}$$

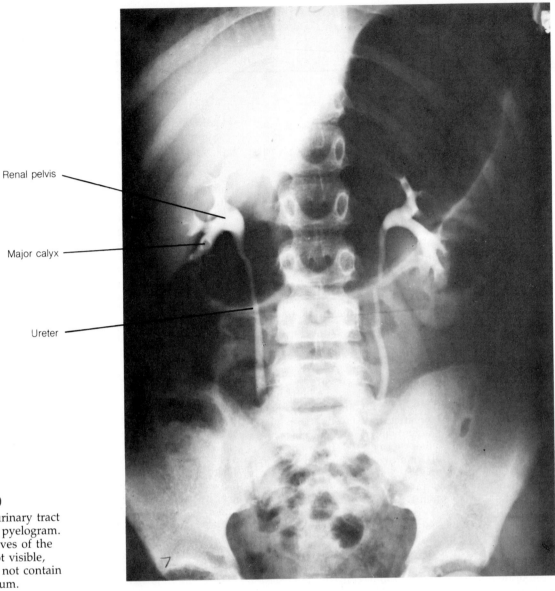

Figure 26.19
X ray of the urinary tract
by retrograde pyelogram.
The lower halves of the
ureters are not visible,
since they do not contain
contrast medium.

Renal pelvis

Major calyx

Ureter

This plasma clearance value means that in one minute an amount of urea is removed from the plasma that is equivalent to the amount of urea in 62 ml of plasma. (It does not mean that any single milliliter of plasma has all of the urea removed from it during one transit through the renal circulation.)

The clearance concept can be used to determine the glomerular filtration rate by measuring the plasma clearance of the carbohydrate inulin. Inulin is injected into the plasma, and its plasma concentration is determined. Inulin is freely filtered from the glomerular capillaries into glomerular capsules with the glomerular filtrate, but it is neither reabsorbed nor secreted by the renal tubules. As a result, the amount of inulin in a timed sample of urine is essentially equal to the amount of inulin filtered from the plasma into glomerular capsules in the same amount of time. For example, suppose that the one-hour, 66-ml urine sample of the previous example contained 1500 mg of inulin (22.7 mg/ml) and that the plasma inulin concentration was 0.2 mg/ml. Since inulin is neither absorbed nor secreted by the renal tubules, all of the inulin in this one-hour sample of urine must have been filtered into glomerular capsules from the plasma with the glomerular filtrate in the same hour. Moreover, because the plasma inulin concentration is only 0.2 mg/ml, in order to get 1500 mg of inulin into the urine in one hour, 7500 ml of plasma must have been filtered at the glomeruli in that hour. This activity equates to a glomerular filtration rate of 125 ml/min [(7500 ml per hr)/(60 min per hr)]. The glomerular filtration rate could also have been obtained using the plasma clearance formula, as follows:

$$\text{plasma clearance of inulin (ml/min)} = \frac{(1.1 \text{ ml/min}) \times (22.7 \text{ mg/ml})}{0.2 \text{ mg/ml}}$$
$$= 125 \text{ ml/min}$$

In the case of inulin, the plasma clearance (of inulin) and the glomerular filtration rate are essentially equal.

The concept of plasma clearance can also be employed to determine the rate of plasma flow through the kidneys. The substance utilized for this purpose is para-aminohippuric acid (PAH). PAH is both freely filtered at the glomerulus and secreted into the renal tubules, but it is not reabsorbed from the tubules. As a result, the plasma that passes through the kidneys is almost completely (actually about 91%) cleared of PAH. Suppose, to continue the previous examples, that sufficient PAH is injected into the plasma to achieve a plasma concentration of 0.01 mg/ml. Suppose also that the 1.1 ml of urine produced in one minute contains 6.25 mg of PAH (5.68 mg/ml). This means that a minimum of 625 ml of plasma [(6.25 mg per min)/(0.01 mg per ml)] must have passed into the kidneys in one minute to provide the amount of PAH that appears in the urine in one minute, even if all of the PAH is removed from the plasma. The same conclusion can also be reached by using the plasma clearance formula:

plasma clearance of para-aminohippuric acid (ml/min) =

$$\frac{(1.1 \text{ ml/min}) \times (5.68 \text{ mg/ml})}{0.01 \text{ mg/ml}} = 625 \text{ ml/min}$$

This value is called the effective renal plasma flow (ERPF). In fact, since only about 91% of the para-aminohippuric acid is in reality cleared from the plasma, the actual total renal plasma flow is about 687 ml/min [(625 ml per min)/0.91], or 989 liters per day. If the hematocrit is known (for example, 40%), then the total renal blood flow (1648 liters per day) can also be determined.

URETERS

The urine drips from the collecting tubules at the tips of the papillae and enters the minor calyces. The minor calyces join with the major calyces, which in turn join with the renal pelvis. From the renal pelvis, urine is transported to the urinary bladder by **ureters**, one from each kidney (Figure 26.19). The ureters descend between the parietal peritoneum and the body wall to the pelvic cavity, where they turn medially and enter the urinary bladder on its posterior lateral surfaces. Before opening into the bladder, the ureters travel obliquely through the bladder wall. As a result, contraction of the muscles of the bladder wall can compress the ureters and help prevent urine from flowing back into the ureters from the bladder, especially during bladder emptying (micturition). In effect, the muscles of the bladder wall act as sphincters on the ureters. Valvelike folds of the mucous membrane lining the bladder cover the orifices of the ureters and assist in preventing urine from flowing back into the ureters during micturition.

The walls of the ureters are composed of three layers: an inner *mucosal layer*, a middle *muscular layer*, and an outer *fibrous layer*. The mucosa has a surface layer of transitional epithelium, which is also typical of the urinary bladder and the urethra. The muscular layer of the ureter is capable of undergoing peristaltic contractions, thus propelling urine to the bladder.

F26.19

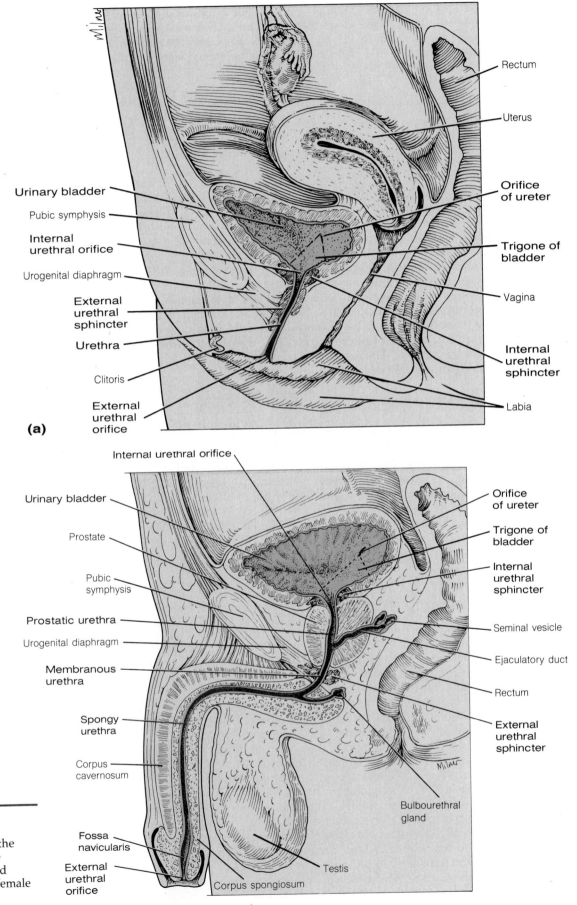

Figure 26.20
Sagittal section of the pelvis showing the urinary bladder and urethra in **(a)** the female and **(b)** the male.

URINARY BLADDER

The **urinary bladder** is a hollow muscular organ used to store urine (Figure 26.20). The bladder rests on the floor of the pelvic cavity, and like other urinary structures, it is retroperitoneal. The anterior surface of the bladder lies just behind the pubic symphysis. In males, the bladder is in front of the rectum. In females, it lies just anterior to the uterus and the superior portion of the vagina. When full, the bladder is spherical; but when it is empty, it resembles an inverted pyramid.

The urinary bladder, like the ureters, is lined with a mucous membrane of transitional epithelium. Covering the transitional epithelium is a tunic consisting of three layers of smooth muscle—inner and outer longitudinal layers on either side of a prominent circular layer. Throughout most of the bladder, the mucous membrane is loosely attached to the underlying muscular coat, and it appears wrinkled when the bladder is contracted. However, the internal opening of the urethra anteriorly and the openings of the two ureters laterally mark the corners of a triangular area in which the mucous membrane is firmly attached to the muscular coat and where it is therefore always smooth. This smooth, triangular region is called the **trigone** of the bladder.

As the bladder fills with urine, the pressure within it increases somewhat initially and then remains fairly constant up to a volume of about 300–400 ml (Figure 26.21). Beyond this point the pressure rises rapidly. The bladder can hold 600–800 ml of urine, but it is generally emptied before it reaches this capacity.

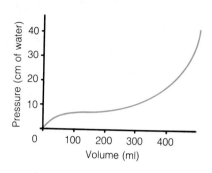

Figure 26.21
Graph indicating pressure changes within the urinary bladder as the bladder fills with urine.

F26.21

URETHRA

The **urethra** is a muscular tube, lined with mucous membrane, that exits from the inferior surface of the urinary bladder. It carries urine from the bladder to the exterior of the body. At the junction of the urethra and bladder, the smooth muscle of the bladder surrounds the urethra and acts as a sphincter (the **internal urethral sphincter**) that tends to keep the urethra closed. During bladder emptying, the contraction of the bladder and the resulting change in its shape pull the sphincter open. Thus, no special mechanism is required to relax this sphincter. As the urethra passes through the pelvic floor (**urogenital diaphragm**), it is surrounded by skeletal muscles that form the **external urethral sphincter.** When constricted, this sphincter—which is under voluntary control—is able to hold the urethra closed against strong bladder contractions.

In the female, the urethra is short (approximately 4 cm), and it runs along the anterior surface of the vagina (Figure 26.20a). It opens to the exterior at the **external urethral orifice,** which is located between the clitoris and the vaginal orifice.

F26.20a

The male's urethra is about 20 cm long and extends to the external urethral orifice at the tip of the penis (Figure 26.20b). In the male, the urethra is divisible into three parts—the *prostatic, membranous,* and *spongy urethrae,* which are named according to the regions through which the urethra passes. The **prostatic urethra** passes through the prostate gland and receives the ejaculatory ducts of the reproductive system on its posterior walls. Beyond the point of junction with the ejaculatory ducts, the male urethra is used for reproduction as well as urine transport. The short portion of the urethra that passes through the urogenital diaphragm is called the **membranous urethra.** The **spongy** (*cavernous*) **urethra** is the longest portion, extending from just below the urogenital diaphragm to the external urethral orifice on the glans penis. Within the glans penis, the urethra dilates into a small chamber called the **fossa navicularis.** A short distance below the urogenital diaphragm, the spongy urethra receives the ducts of the bulbourethral glands (Cowper's glands) of the reproductive system. In the penis, the urethra passes through the **corpus spongiosum,** one of three columns of erectile tissue.

F26.20b

MICTURITION

As the bladder fills with urine, its walls are stretched and mechanoreceptors within the walls transmit increasing numbers of sensory impulses to the sacral region of the spinal cord. These impulses ultimately stimulate parasympathetic neurons that supply the smooth muscles of the bladder wall and inhibit somatic motor neurons that supply the external urethral sphincter. As a consequence, when approximately 300 ml of urine has accumulated within the bladder, the muscles of the bladder wall contract, the external urethral sphincter relaxes, and **micturition** (urination, or voiding) occurs.

Although micturition as just described is essentially the result of a local spinal reflex, it is also influenced by higher brain centers, particularly centers in the brain stem and cerebral cortex. In addition to being transmitted to the spinal cord, sensory impulses from mechanoreceptors within the bladder walls are sent to higher brain centers. These impulses can lead to the sensation of a full bladder and to a feeling of a need to urinate. Moreover, impulses sent from the brain can either facilitate or inhibit the reflex emptying of the bladder, and with training it is possible to gain a high degree of voluntary control over micturition. As a result, urination can be either voluntarily induced or postponed until an opportune time. However, until control is developed and training is complete, the reflex response is the dominant factor governing bladder emptying. A baby, therefore, urinates whenever its bladder is sufficiently full to activate the spinal reflex.

CONDITIONS OF CLINICAL SIGNIFICANCE

The Urinary System

Pressure-Related Pathologies

Factors that upset the pressure relationships determining glomerular filtration rates can interfere with normal kidney function. (Recall that the glomerular filtration rate depends on the net filtration pressure, which equals glomerular capillary pressure minus capsular pressure minus colloid osmotic pressure of the plasma.)

Prostate Hypertrophy

In males, it is not unusual for the prostate gland, which surrounds the urethra just below the bladder, to hypertrophy (enlarge) and compress the urethra so it becomes difficult for urine to leave the bladder. As urine accumulates within the bladder, the pressure within the bladder rises. If the pressure increases enough, it causes a backing up of urine in the ureters, which produces a dilation of the renal pelvis and calyces and leads to an increased pressure within the glomerular capsules. The increased capsular pressure reduces the glomerular filtration rate and thus interferes with the kidneys' regulation of body-fluid composition.

Low Arterial Blood Pressure

Blood pressure changes associated with heart failure, hemorrhage, or shock can cause renal failure. In all of these conditions, there is a drop in arterial blood pressure, which reduces the ability of the kidneys to form glomerular filtrate. In addition, a substantial drop in blood pressure can activate reflexes involved in maintaining normal pressure within the major arteries of the body. These reflexes cause the afferent arterioles of the kidneys to constrict. The constriction of the afferent arte-

rioles helps elevate the general body blood pressure, but it diminishes the blood pressure within the glomeruli and further reduces the formation of glomerular filtrate. The reduction in the formation of glomerular filtrate diminishes the kidneys' ability to excrete wastes and regulate body-fluid composition.

Acute Glomerulonephritis

Acute glomerulonephritis is an inflammatory condition of the glomeruli that is the result of a hypersensitivity response in which immune complexes of antigen, antibody, and complement become deposited in the kidneys. It is frequently associated with streptococcal infections, such as throat infections. In acute glomerulonephritis, many of the glomeruli become blocked by the inflammatory response, and others become so permeable that they allow erythrocytes and large amounts of plasma proteins to pass into the glomerular capsules with the glomerular filtrate. In severe cases, total or almost total renal shutdown can occur. However, the vast majority of individuals afflicted with acute glomerulonephritis return to normal renal function within a few months.

Pyelonephritis

Pyelonephritis is a bacterial infection of the renal pelvis and the surrounding tissues of the kidney. The bacteria reach the kidneys from other sites of infection by way of the bloodstream or lymphatics, or the infection may spread up the ureters from the bladder. In pyelonephritis, the kidney may become swollen and congested and the pelvis may become inflamed and filled with pus.

Abscesses often develop on the surface of the kidney. Pyelonephritis generally responds well to treatment with antibiotics. In chronic cases, however, extensive scar tissue is formed in the kidney and renal failure becomes a possibility.

Proteinuria

In a number of renal diseases, the permeability of the glomerular capillaries may be increased to such an extent that large amounts of plasma proteins (mostly albumin) pass into the glomerular filtrate and are excreted in the urine. This condition is called *proteinuria*, or *albuminuria*. In severe cases, the loss of plasma proteins can be so great that the plasma osmotic pressure decreases substantially. As a result of the decrease in plasma osmotic pressure, there is an increased tendency for fluid to leave the systemic blood vessels and enter the tissue spaces—a condition that produces generalized edema (swelling) of the body.

Uremia

If the metabolic products (such as urea) that are derived from the breakdown of proteins are not excreted properly, they accumulate in the blood and produce a condition called *uremia*. Uremia affects several body systems, including the nervous system (convulsions and coma), digestive system (vomiting and diarrhea), and respiratory system (dyspnea, or labored breathing).

Kidney Stones

Kidney stones (renal calculi) sometimes form within the renal pelvis. The stones generally consist of uric acid, calcium oxalate, calcium phosphate, and certain other substances. What causes the stones to form is not known. However, there seems to be some correlation between their formation and kidney infections, a high concentration of salts in the urine, vitamin A deficiency, or hyperparathyroidism caused by a tumor.

A stone formed in the renal pelvis may remain there, or it may enter the ureter and pass to the bladder.

The stone often causes severe, painful contractions of the ureter as it travels through it. A more serious condition results if a stone becomes lodged within the ureter, obstructing the flow of urine to the bladder. In addition to the retention of urine, kidney stones can cause ulcerations in the lining of the urinary tract, which makes the tract more prone to infections. (See "Frontiers in Health: New Surgery for Kidney Stones" for information concerning the treatment of kidney stones.)

Cystitis

Cystitis is an inflammation of the urinary bladder accompanied by frequent and burning urination and by blood in the urine. In the acute form of cystitis, the mucous membrane lining the bladder becomes swollen and some bleeding occurs. In the chronic condition, the wall of the bladder can become thickened and its capacity reduced.

The bladder is generally quite resistant to bacterial infections, but under certain conditions bacteria become established in the bladder lining, thus producing cystitis. Cystitis can also be caused by chemicals or by mechanical irritation, such as catheterization. Women have a higher incidence of cystitis than men, probably due to their short urethra, which makes it easier for bacteria to reach the bladder from outside the body. It is not uncommon for *E. coli* bacteria from the woman's anal region to infect the urethra as a result of improper cleansing of the area.

Effects of Aging on the Kidneys

Kidney function declines progressively with age. At age 70 the glomerular filtration rate is only about 50% of the rate at age 40. Renal blood flow decreases from approximately 1100 ml per minute at ages 20–45 to only about 475 ml per minute at 80–89 years of age. There is a corresponding decrease in the function of the renal tubules and in their ability to concentrate the tubular fluid. However, the kidneys do retain their ability to regulate the acid–base balance of the body, although they respond less quickly to a sudden, large acid load.

STUDY OUTLINE

COMPONENTS OF THE URINARY SYSTEM
Kidneys, ureters, urinary bladder, and urethra.   p. 761

EMBRYONIC DEVELOPMENT OF THE KIDNEYS   pp. 762–764

FROM INTERMEDIATE MESODERM Three successive pairs of embryonic kidneys develop:
pronephros (with pronephric duct);
mesonephros (with mesonephric duct);
metanephros (becomes adult kidney).

URETERIC BUDS Develop into collecting tubules, calyces, renal pelvis, and ureters.

NEPHRONS Develop from cap of intermediate mesoderm that covers ureteric bud.

ANATOMY OF THE KIDNEYS
Paired reddish-brown, bean-shaped organs on posterior abdominal wall; retroperitoneal.   pp. 764–773

TISSUE LAYERS SURROUNDING THE KIDNEY Three layers: *renal capsule* (fibrous), *adipose capsule* (perirenal fat), *renal fascia* (double-layered).

EXTERNAL STRUCTURE OF THE KIDNEY Renal hilus is medial indentation; passageway for blood vessels and ureters. Renal pelvis located in renal sinus.

INTERNAL STRUCTURE OF THE KIDNEY
1. *CORTEX* Outer layer; also forms renal columns that pass into medulla.
2. *MEDULLA* Consists of renal pyramids.
3. *RENAL PELVIS* Expanded upper end of ureter; formed from joining of major calyces.

THE RENAL TUBULES Functional units of kidneys consist of nephrons and collecting tubules. Each nephron consists of a glomerulus (network of parallel capillaries) and a tubule (proximal end forms glomerular capsule).

RENAL CORPUSCLE Capsule and glomerulus; site of transfer between blood and nephron.
1. Visceral layer of glomerular capsule composed of podocytes; foot processes separated by filtration slits, which are covered by slit membrane.
2. Glomeruli have fenestrated endothelium.

FILTRATION BARRIER Formed of fenestrated endothelium, basal lamina, and slit membrane.

PROXIMAL CONVOLUTED TUBULE Single layer of cuboidal cells with microvilli.

LOOP OF HENLE Descending and ascending limbs.

DISTAL CONVOLUTED TUBULE Empties into collecting tubule.

JUXTAGLOMERULAR APPARATUS Distal convoluted tubule contacts afferent arteriole; composed of juxtaglomerular cells of afferent arteriole and macula densa of tubule; secretes renin.

BLOOD VESSELS OF THE KIDNEY

RENAL ARTERY Provides substantial blood flow, which allows kidneys to maintain blood homeostasis. Interlobar artery → arcuate artery → interlobular artery → afferent arterioles → glomerulus → efferent arteriole → peritubular capillaries → interlobular vein → arcuate vein → interlobar vein → renal vein.

TWO CAPILLARY BEDS Glomerulus and peritubular capillaries.

AFFERENT AND EFFERENT ARTERIOLES Maintain constant blood pressure in glomerulus.

PHYSIOLOGY OF THE KIDNEYS
pp. 773–785

GLOMERULAR FILTRATION
1. 16–20% of blood plasma entering kidneys is filtered from glomerular capillaries and into glomerular capsules.
2. Glomerular filtrate contains most plasma components at essentially same concentrations as plasma, but is basically protein free.

NET FILTRATION PRESSURE Filtration is favored by pressure of blood within glomerular capillaries and opposed by pressure of fluid within glomerular capsules and by osmotic force exerted by unfiltered plasma proteins in glomerular capillaries.

FACTORS THAT INFLUENCE GLOMERULAR FILTRATION

Influence of Arterial Blood Pressure and Autoregulatory Mechanisms Arterial blood pressure influences glomerular filtration rate, but effect is not as great as expected due to intrinsic autoregulatory mechanisms that maintain a fairly stable glomerular filtration rate over wide range of systemic blood pressures.

Influence of Sympathetic Neural Stimulation Increased sympathetic stimulation preferentially constricts afferent arterioles, lowering glomerular capillary pressure and glomerular filtration rate.

TUBULAR REABSORPTION Removes materials from renal tubules.

ACTIVE TUBULAR REABSORPTION Energy-requiring; glucose, amino acids, and vitamins, as well as sodium, calcium, and chloride ions are actively reabsorbed.

PASSIVE TUBULAR REABSORPTION Does not require energy; urea and many foreign chemicals are passively reabsorbed.

TUBULAR SECRETION Moves substances that leave peritubular capillaries into renal tubules; may be active or passive; hydrogen ions, potassium ions, penicillin are secreted into renal tubules.

EXCRETION OF WATER AND ITS EFFECTS ON URINE CONCENTRATION Kidneys can produce urine as dilute as 65–70 milliosmoles per liter or as concentrated as 1200 milliosmoles per liter.

MAINTENANCE OF A CONCENTRATED MEDULLARY INTERSTITIAL FLUID Activities and anatomical arrangement of loops of Henle of nephrons (particularly of juxtamedullary nephrons), collecting tubules, and vasa recta are important.

Activities of the Loops of Henle and the Collecting Tubules As fluid flows along loops of Henle and collecting tubules, reabsorption of solutes raises concentration of medullary interstitial fluid.

Effect of Vasa Recta Blood Flow Pattern of blood flow along vasa recta minimizes solute removal from medullary interstitial fluid.

PRODUCTION OF DIFFERENT URINE VOLUMES AND CONCENTRATIONS
1. Sodium ions actively transported out of proximal tubule; chloride ions follow passively. Substances such as glucose and amino acids are also reabsorbed. Water leaves osmotically. Volume of glomerular filtrate reduced 65–70%; concentration remains essentially that of plasma.
2. Water moves out of descending limb of loop of Henle. Tubular fluid is reduced in volume; its concentration is increased.
3. Sodium and chloride ions move out of ascending limb of loop of Henle, which is relatively impermeable to water. Concentration of tubular fluid is diminished; volume is not greatly altered.
4. Certain ions are reabsorbed from the fluid within distal convoluted tubule; others are secreted into the fluid. In monkeys and presumably other primates (such as humans), distal convoluted tubule is relatively impermeable to water. In general, concentration of fluid within distal convoluted tubule remains below that of plasma; its volume does not change greatly.
5. Permeability of collecting tubule to water controlled by ADH. If much ADH is present, permeability to water is high; water leaves and tubular fluid becomes concentrated and reduced in volume.

MINIMUM URINE VOLUMES Need to eliminate waste products such as urea, sulfates, phosphates requires daily excretion of about 600 milliosmoles of these materials.

PLASMA CLEARANCE Rate at which a particular substance is eliminated or cleared from plasma by kidneys. Plasma clearance of inulin indicates glomerular filtration rate.

URETERS p. 785

URINE TRANSPORT From renal pelvis to urinary bladder; retroperitoneal.

URINARY BLADDER p. 787

HOLLOW MUSCULAR ORGAN Urine storage.
1. Retroperitoneal.
2. Lined with transitional epithelium.
3. Three layers of smooth muscle.

URETHRA
Muscular tube lined with mucous membrane; carries urine from bladder to exterior of body; surrounded by internal urethral sphincter where it leaves bladder; surrounded by external urethral sphincter where it passes through urogenital diaphragm. p. 787

FEMALE Short (4 cm); runs along anterior surface of vagina.

MALE Urine passage and reproduction; long (20 cm).

MICTURITION p. 788
1. As bladder fills, local spinal reflex causes bladder to contract and external urethral sphincter to relax, resulting in micturition.
2. Impulses from brain can facilitate or inhibit reflex emptying of bladder; with training, high degree of voluntary control can be exercised.

CONDITIONS OF CLINICAL SIGNIFICANCE: THE URINARY SYSTEM pp. 788–789

PRESSURE-RELATED PATHOLOGIES prostate hypertrophy, low arterial blood pressure.

ACUTE GLOMERULONEPHRITIS

PYELONEPHRITIS

PROTEINURIA

UREMIA

KIDNEY STONES (RENAL CALCULI)

CYSTITIS

EFFECTS OF AGING ON THE KIDNEYS

SELF-QUIZ

1. The medulla of the kidney contains: (a) the adipose capsule; (b) glomeruli; (c) pyramids.

2. The expanded, funnel-shaped upper end of the ureter within the kidney is the: (a) renal pelvis; (b) renal hilus; (c) urethra; (d) nephron.

3. The capillary found in the glomerular capsule of a nephron is the: (a) glomerulus; (b) proximal convoluted tubule; (c) adipose capsule; (d) juxtaglomerular complex.

4. The space in each kidney that contains the renal vessels and the renal pelvis is called the renal: (a) capsule; (b) sinus; (c) cortex.

5. From the collecting tubules, urine enters the: (a) renal pelvis; (b) major calyces; (c) minor calyces; (d) distal tubules.

6. Podocytes are responsible for the formation of: (a) filtration slits; (b) fenestrated endothelium; (c) capsular space; (d) pronephros.

7. Renal corpuscles are located in the cortical region of the kidney. True or False?

8. The juxtaglomerular complex is formed from modification of the cells of: (a) the distal convoluted tubule; (b) the afferent arteriole; (c) both a and b.

9. Thin-walled vessels that supply the loops of Henle and the collecting ducts are called: (a) glomeruli; (b) vasa recta; (c) macula densa; (d) arcuate veins.

10. Essentially 100% of the plasma entering the kidneys is filtered from the glomerular capillaries and into the glomerular capsules of the nephrons as the glomerular filtrate. True or False?

11. Which force favors the movement of fluid out of the glomerular capillaries and into the glomerular capsules? (a) the osmotic force exerted by unfiltered plasma proteins within the glomerular capillaries; (b) the pressure of the fluid within the glomerular capsules; (c) the glomerular capillary pressure.

12. In general, an increase in the activity of the sympathetic neurons that supply the kidneys leads to a dilation of the afferent arterioles. True or False?

13. A rise in the arterial blood pressure: (a) reduces the glomerular filtration rate; (b) increases the output of urine; (c) has no effect on kidney function because of the influence of the kidney's intrinsic autoregulatory mechanisms.

14. Urea reabsorption from the renal tubules is passive and depends on water reabsorption. True or False?

15. At the proximal convoluted tubule of a nephron: (a) sodium ions are actively transported out of the tubule; (b) there is a net movement of water into the tubule; (c) there is a net movement of chloride ions into the tubule.

16. As fluid flows along the descending limb of the loop of Henle of a juxtamedullary nephron, there is a net movement of water out of the tubule. True or False?

17. Fluid at the bottom of the loop of Henle of a juxtamedullary nephron is: (a) less concentrated than the glomerular filtrate; (b) less concentrated than fluid within the distal convoluted tubule; (c) more concentrated than fluid within the proximal convoluted tubule.

18. With no antidiuretic hormone: (a) large volumes of urine are produced; (b) large amounts of protein appear in the urine; (c) glucose is not reabsorbed from the kidney tubules.

19. The urethra of the female is divisible into three parts—the prostatic, membranous, and spongy urethra, which are named according to the regions through which the urethra passes. True or False?

LEARNING OBJECTIVES

After completing this chapter, you should be able to:

1. List similarities and differences in the composition of the plasma, the interstitial fluid, and the intracellular fluid.

2. Describe the factors that influence the renal excretion of sodium.

3. Cite possible causes for a feeling of thirst.

4. Describe the factors that influence the renal excretion of water.

5. Explain how body potassium is regulated, and cite the regulatory factors involved.

6. Explain the actions of hormones involved in calcium regulation.

7. State what distinguishes a strong acid from a weak acid.

8. Explain the function of buffer systems in the body.

9. Explain how the respiratory system can have an important influence on the pH of the internal environment.

10. Explain how the kidneys act as essential regulators of acid–base balance.

CHAPTER CONTENTS

BODY FLUIDS

REGULATION OF THE BODY'S INTERNAL ENVIRONMENT

CONDITIONS OF CLINICAL SIGNIFICANCE: DISORDERS OF ACID–BASE BALANCE

KEY TERMS AND DERIVATIVES

dehydration (*de* = separation from; *hydra* = water) a condition resulting from an excessive loss of water

extracellular fluid (*extra* = outside) fluid within the body but outside the cells

intracellular fluid (*intra* = within) fluid within the cells of the body

acidosis (*aci* = acid; *osis* = a state of) an abnormal condition in which the hydrogen ion concentration of the blood tends to increase

alkalosis (*alka* = alkaline; *osis* = a state of) an abnormal condition in which the hydrogen ion concentration of the blood tends to decrease

Fluid and Electrolyte Balance and Acid–Base Regulation

27

As pointed out in the previous chapter, the body must regulate the concentrations of water, acids and bases, and electrolytes such as sodium, potassium, and calcium ions in its internal environment in order to maintain homeostasis. The kidneys are particularly important in the regulation of these substances, as are the lungs and the gastrointestinal tract.

BODY FLUIDS

There are three principal body fluids: (1) the *plasma*, or fluid within the blood vessels; (2) the *interstitial fluid*, or fluid that surrounds the cells; and (3) the *intracellular fluid*, or fluid within the cells (Figure 27.1). Together, the plasma and the interstitial fluid make up the *extracellular fluid*, or fluid outside the cells. It is this fluid that forms the body's internal environment.

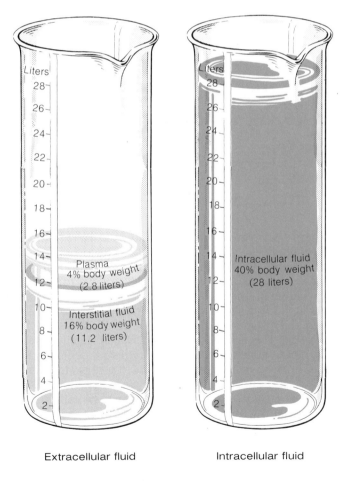

Extracellular fluid

Intracellular fluid

Figure 27.1
Distribution of body fluids.

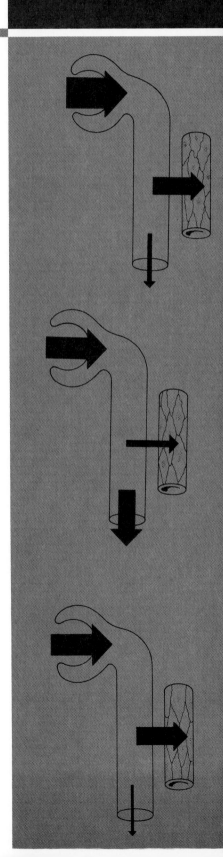

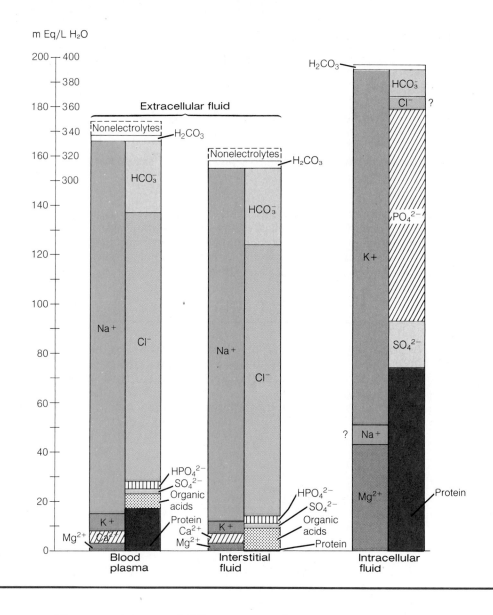

Figure 27.2
Composition of body fluids. Numbers on the left of the scale indicate amounts of cations or anions. Numbers on the right indicate the sum of cations and anions.

Composition of Plasma and Interstitial Fluid

The different body fluids are not isolated from one another, and a continuous exchange of materials occurs between them. As can be seen from Figure 27.2, the compositions of the plasma and the interstitial fluid are quite similar. These fluids are separated from one another by the walls of the capillaries, and the relatively free exchange of low-molecular-weight solutes that occurs across capillary walls accounts for the similarity between the two fluids. The major difference between the plasma and the interstitial fluid is that the plasma contains more proteins. This is due to the fact that the plasma proteins tend to remain within the blood vessels, since the capillaries are generally not very permeable to them. Under physiological conditions, these proteins normally exist as anions—that is, as negatively charged ions. For this reason, there is somewhat more sodium (a positively charged ion, or cation) and somewhat less chloride (an anion) within the plasma than within the interstitial fluid.

Since a relatively free exchange of materials does occur between the plasma and the interstitial fluid, the regulation of the fluid, electrolyte, and acid–base composition of the plasma by the kidneys (and to some degree by the lungs and gastrointestinal tract) also serves to regulate the fluid, electrolyte, and acid–base composition of the interstitial fluid.

Composition of Intracellular Fluid

The composition of the intracellular fluid is considerably different from that of the plasma and the interstitial fluid. The intracellular fluid is separated from the interstitial fluid by the membranes of the cells, which are relatively impermeable to proteins. Consequently, cellular proteins tend to remain within the cells. Moreover, the action of the cell membrane's sodium–potassium pump moves sodium ions out of the cells and accumulates potassium ions within them.

REGULATION OF THE BODY'S INTERNAL ENVIRONMENT

In order to maintain homeostasis, the addition of substances such as water and electrolytes to the body's internal environment must be balanced by the removal of equal amounts of these materials. Ingestion, of course, adds materials to the internal environment, whereas excretion by organs such as the kidneys removes materials. In fact, the kidneys' role in controlling the excretion of water and electrolytes is one of their most important regulatory functions.

Sodium Regulation

Sodium is the major extracellular cation, and because of its osmotic effects, variations in the body-sodium content can cause changes in the extracellular fluid volume, including the plasma volume. Changes in the plasma volume, in turn, can lead to changes in blood pressure, and mechanisms involved in plasma-volume and blood-pressure regulation are important in controlling body-sodium content.

Sodium Ingestion

Animal studies indicate that there are two components involved in determining the amount of sodium ingested (in the form of salt). One is the regulatory component, which governs the intake of sodium in such a way that a balance between sodium intake and outflow is maintained. The other is the hedonistic component, by which an animal demonstrates a strong preference for salt regardless of regulatory requirements. The degree to which regulatory and hedonistic components operate in humans is not entirely clear. Although persons severely depleted of sodium chloride often develop a desire for salt, people also seem to have a strong hedonistic salt appetite, and they consume large amounts of it whenever it is inexpensive and easily available. For example, in the United States, an average person consumes 10–15 g of salt per day even though less than 0.5 g per day is normally needed to meet regulatory requirements. Consequently, for humans it appears that the regulation of the body's sodium content is achieved mainly by controlling sodium excretion by the kidneys rather than by controlling sodium ingestion.

Renal Excretion of Sodium

The relationship between the glomerular filtration rate of sodium and the tubular reabsorption rate of sodium is very important in determining the amount of sodium excreted (Figure 27.3). For example, if much sodium is filtered into the glomerular capsules but only little is reabsorbed, then a good deal of sodium will be excreted.

F27.3

    Since sodium is freely filtered from the plasma into the glomerular capsules, any factors that alter the general rate of glomerular filtration will also alter the glomerular filtration rate of sodium. Among the factors that influence the glomerular filtration rate are (1) the arterial blood pressure, and (2) the activity of sympathetic nerves to the kidneys, which alters the diameters of the afferent arterioles.

    The reabsorption of sodium from the renal tubules is subject to physiological control, and the control of tubular sodium reabsorption is believed to be

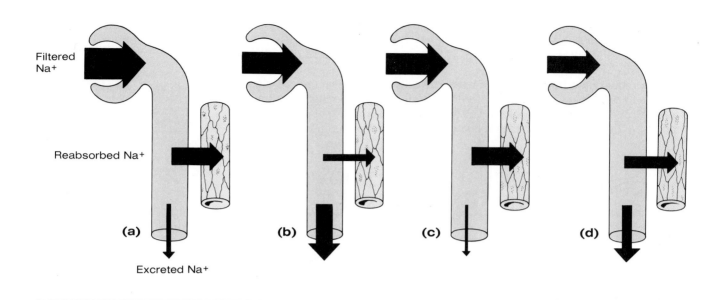

Filtered Na+

Reabsorbed Na+

(a)   (b)   (c)   (d)

Excreted Na+

Figure 27.3
The amount of sodium excreted by the kidneys may be altered by varying either the glomerular filtration rate (filtered sodium) or the reabsorption of sodium from the renal tubules.

particularly important in the long-term regulation of sodium excretion. A major factor in this control is the adrenal cortical hormone aldosterone. Aldosterone stimulates sodium reabsorption, particularly from the last portions of the distal convoluted tubules and the collecting tubules. When much aldosterone is present, almost all the sodium that enters the renal tubules is reabsorbed and as little as as 0.1 g of sodium is excreted per day. When little aldosterone is present, as much as 30–40 g of sodium can be excreted per day.

The release of aldosterone is enhanced indirectly by the substance renin, which is secreted into the blood by specialized cells (juxtaglomerular cells) of kidney arterioles in response to a number of factors. Although it is currently impossible to assign a precise quantitative role to each of these factors, a decrease in the renal arterial pressure and/or an increase in the activity of sympathetic nerves to the kidneys apparently stimulate renin secretion. Renin is necessary for the conversion of the precursor substance angiotensinogen, which is manufactured by the liver and is normally present in the blood, into angiotensin, and one of the effects of angiotensin is the stimulation of aldosterone release from the adrenal glands. Thus, an increased release of renin leads to an increased formation of angiotensin, which in turn enhances the release of aldosterone. The aldosterone then stimulates sodium reabsorption from the last portions of the distal convoluted tubules and the collecting tubules.

Factors That Influence Sodium Excretion

As previously indicated, a change in the body-sodium content can lead to a change in both the plasma volume and the blood pressure, and mechanisms involved in plasma-volume and blood-pressure regulation are important in controlling body-sodium content. For example, an increase in the systemic blood pressure, due perhaps to an increase in the plasma volume, acts directly on the kidneys to cause some increase in the glomerular filtration rate (Figure 27.4). Moreover, an increase in the systemic blood pressure (including the renal arterial pressure) acts on the kidneys to cause a decreased release of renin, which ultimately leads to a decrease in the level of aldosterone. Together, these responses increase the amount of sodium filtered into the glomerular capsules and decrease the amount of sodium reabsorbed from the renal tubules. As a result, the urinary excretion of sodium increases, and the total body mass of sodium declines. Because sodium, by virtue of its osmotic effects, has a direct influence on extracellular volume, the loss of sodium leads to a decrease in the plasma volume, which tends to return the blood pressure to normal.

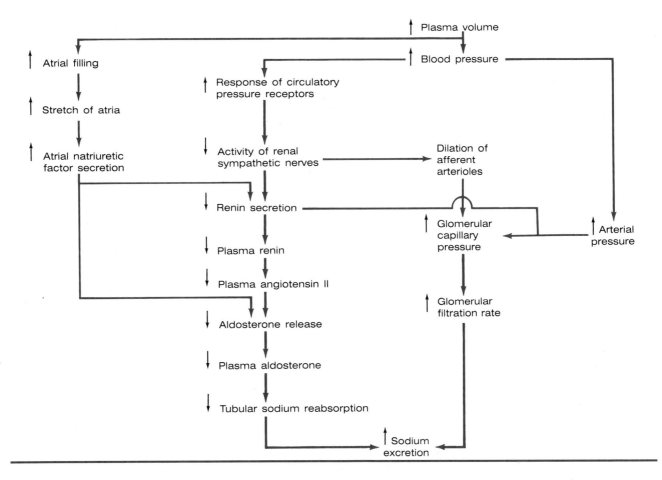

Figure 27.4

Effect of an increase in plasma volume on sodium excretion by the kidneys.

F27.4

Some researchers believe that both neural and hormonal mechanisms involved in plasma-volume and blood-pressure regulation help mediate the effects of changes in plasma volume and blood pressure on the renal excretion of sodium. According to this view (and to continue the preceding example), an increase in the systemic blood pressure causes an increase in the rate at which nerve impulses are transmitted to the central nervous system from circulatory pressure receptors such as the aortic arch and carotid sinus baroreceptors and receptors in the atria of the heart (Figure 27.4). The increased rate of nerve impulse transmission from these receptors leads to a decrease in the activity of the sympathetic nerves to the kidneys. The decrease in sympathetic activity, in turn, leads to a dilation of the afferent arterioles and, consequently, to an increase in the blood pressure within the glomerular capillaries, which contributes to the increase in the glomerular filtration rate. Moreover, the decrease in the activity of the sympathetic nerves to the kidneys also contributes to the decreased release of renin. In addition, an increase in the plasma volume can lead to an increased filling of the atria of the heart, which stretches the atrial walls. In response to stretch, atrial cardiac muscle cells release a polypeptide hormone called **atrial natriuretic factor.** Atrial natriuretic factor modifies the activity of the renin–angiotensin system by inhibiting the secretion of renin and also by directly inhibiting the adrenal secretion of aldosterone. It also acts directly at various sites in the kidneys to enhance sodium excretion.

Sodium Content and Sodium Concentration

The loss or gain of sodium by the body changes the total sodium *content* of the body, but it does not necessarily have a great effect on the actual *concentration* of sodium in the internal environment. As discussed in the next section, the body possesses receptors called *osmoreceptors,* which are sensitive to the osmotic pressure of the extracellular fluid. Regulatory mechanisms brought into play by these receptors compensate for changes in the osmotic concentration of the extracellular fluid (which is largely due to sodium) by stimulating or retarding water ingestion and water excretion in order to maintain the proper osmotic concentration. As a result, these receptors are of considerable importance in the regulation of the actual sodium concentration of the extracellular fluid.

Water Regulation

The water content of the body must be regulated because of its effect on both the osmotic pressure and the fluid volume of the internal environment.

Water Ingestion and Thirst

Although fluid intake is often influenced more by habit and social factors than by the need to regulate body water, water ingestion does seem to depend at least in part on the regulatory needs of the body. A feeling of thirst results from an increased osmotic concentration of the extracellular fluid and also from a substantial reduction in the plasma volume. It is thought that osmoreceptors in a hypothalamic thirst center in the brain, and perhaps also circulatory pressure receptors, stimulate thirst. Moreover, angiotensin stimulates thirst by a direct effect on the brain.

Often when a dehydrated person drinks water, the amount ingested approximates the amount required to bring the extracellular fluid back to the proper osmotic concentration. This fact is interesting because the person usually stops drinking before the ingested water has had time to be absorbed from the gastrointestinal tract and actually return the extracellular osmolarity to normal. Thus, the gastrointestinal tract may possess some sort of water-metering system that is able to regulate water intake in accordance with the needs of the body. Moreover, the control of thirst may have a learned or anticipatory component that enables a person to anticipate his or her fluid needs and therefore drink sufficient water to prevent dehydration. For example, the amount of fluid consumed in association with eating appears to be a learned activity that may prevent dehydration.

Renal Excretion of Water

Body water is also controlled by the excretory pathways of the kidneys, and the relationship between the glomerular filtration rate of water and the tubular reabsorption rate of water is of major importance in determining the amount of water excreted. For example, if much water is filtered into the glomerular capsules but only little is reabsorbed, then a good deal of water will be excreted.

Since water is freely filtered from the plasma into the glomerular capsules, any factors that alter the general rate of glomerular filtration will also alter the glomerular filtration rate of water. Some of these factors—particularly some associated with blood pressure regulation—were discussed in the previous section on sodium regulation. For example, an increase in the systemic blood pressure, due perhaps to an increase in the plasma volume, leads to an increase in the glomerular filtration rate. (Conversely, a decrease in the systemic blood pressure leads to a decrease in the glomerular filtration rate.)

As explained in Chapter 26, the amount of water reabsorbed from the renal tubules depends in large part on the collecting tubules' permeability to water, which is controlled by antidiuretic hormone (ADH). When the plasma-ADH level is low, the collecting tubules' permeability to water is low, and relatively little water is reabsorbed. Consequently, a large volume of urine is produced. Conversely, when the plasma-ADH level is high, collecting-tubule permeability is high, much water is reabsorbed, and only a relatively small volume of urine is produced.

Inputs to neural centers that influence ADH release come both from osmoreceptors located in the hypothalamus and from circulatory pressure receptors, particularly those in left atrium of the heart. The osmoreceptors, which are stimulated by increased extracellular osmolarity, stimulate the cells that secrete ADH. The pressure receptors, which are stimulated by increased atrial blood pressure (possibly resulting from increased plasma volume), inhibit the ADH-producing cells. Thus, considerable ADH release and substantial water reabsorption results from an increased extracellular osmolarity and a decreased blood pressure (plasma volume). Conversely, strong ADH inhibition and substantial water excretion results from a decreased extracellular osmolarity and an increased blood pressure. These responses, especially when coupled with the previously discussed changes in glomerular filtration rate that occur in response to changes in blood pressure, tend to return the extracellular osmolarity and the blood pressure (plasma volume) to normal when they either increase or decrease.

Factors That Influence Water Excretion

From the preceding discussion it is evident that the osmotic pressure of the extracellular fluid and the blood pressure (plasma volume) are major factors influencing water excretion. Moreover, the factors that influence water excretion are very closely related to the factors that influence sodium excretion, and both groups of factors are involved in the regulation of blood pressure as well as in the maintenance of the proper volume and concentration of the body's internal environment. Indeed, the activity of the mechanisms that influence water excretion and the mechanisms that influence sodium excretion can rarely be separated completely. For example, when an increase in the systemic blood pressure causes an increase in the glomerular filtration rate, both sodium and water are affected. Similarly, an increase in the plasma volume that increases the atrial and systemic blood pressure can decrease the reabsorption of sodium by way of the renin–angiotensin mechanism and reduce the reabsorption of water by way of the ADH mechanism. Moreover, a change in the body's sodium content due to a change in the excretion of sodium causes a change in the osmotic pressure of the extracellular fluid. The change in osmotic pressure is sensed by osmoreceptors that, in turn, cause the ADH mechanism to increase or decrease the reabsorption of water as necessary to reestablish the proper osmotic concentration of the extracellular fluid.

Potassium Regulation

Potassium is important in the excitability of nerve and muscle, and the potassium concentration of the extracellular fluid is closely regulated. Humans normally excrete potassium in amounts equal to ingested amounts, and thus potassium balance is maintained.

Potassium is freely filtered from the plasma into the glomerular capsules, and it can be both reabsorbed and secreted by the renal tubules. Normally, most of the filtered potassium is actively reabsorbed, and adjustments in the amount of potassium excreted depend mainly on how much is secreted into the renal tubules.

The amount of potassium secreted into the renal tubules is related to the potassium concentration in the tubular cells. When the potassium concentration of the extracellular fluid is high, more potassium will be present in the tubular cells, and as a consequence, more potassium will be secreted than when the potassium concentration of the extracellular fluid is low. A second factor that influences the elimination of potassium is aldosterone. Besides promoting the tubular reabsorption of sodium, this hormone enhances the tubular secretion of potassium. In fact, the potassium concentration of the extracellular fluid that bathes the adrenal glands is a major stimulus for the release of aldosterone from the adrenal cortex. When the potassium concentration of the extracellular fluid rises, so does the release of aldosterone. When the potassium concentration falls, the release of aldosterone decreases. Once aldosterone is released, however, it exerts both of its effects and enhances sodium reabsorption as well as potassium secretion.

Figure 27.5

Schematic representation of the carbonic-acid–sodium-bicarbonate buffer system. In solution, the sodium bicarbonate dissociates into sodium ions (Na^+) and bicarbonate ions (HCO_3^-). Because it is a weak acid, only a relatively small amount of the carbonic acid dissociates into hydrogen ions (H^+) and bicarbonate ions.

Calcium Regulation

The calcium level of the extracellular fluid is closely regulated, and low levels of calcium increase nerve and muscle excitability. Calcium regulation involves bone, which contains 99% of the body's calcium, as well as the kidneys and the gastrointestinal tract. The calcium-regulating activities of all three sites are influenced either directly or indirectly by parathyroid hormone (parathormone, or PTH) from the parathyroid glands.

Parathyroid hormone increases the plasma-calcium concentration (and decreases the plasma-phosphate concentration). It increases the movement of calcium and phosphate from bone into the extracellular fluid, and this activity is believed to be due at least in part to the ability of parathyroid hormone to stimulate the activity of cells called osteoclasts, which break down bone. In the kidneys, parathyroid hormone decreases the urinary excretion of calcium and increases the excretion of phosphate. Parathyroid hormone also enhances a step in the metabolic transformation of vitamin D_3 to a substance called 1,25-dihydroxycholecalciferol. This substance stimulates active calcium absorption from the intestine.

The release of parathyroid hormone is controlled by the calcium concentration of the extracellular fluid that bathes the parathyroid glands. When the calcium concentration is low, parathyroid hormone release increases, and this response tends to raise the extracellular fluid calcium concentration toward normal. Conversely, when the calcium concentration is high, parathyroid hormone release decreases, and this response tends to lower the extracellular fluid calcium concentration toward normal.

Calcitonin, a hormone from the thyroid gland, lowers the plasma-calcium level, primarily by inhibiting the removal of calcium from bone. Thus, the influence of calcitonin on the plasma-calcium level is opposite to that of parathyroid hormone. The secretion of calcitonin is controlled by the calcium concentration of the fluid that bathes the thyroid gland; high calcium concentrations enhance secretion. Calcitonin, however, appears to play a less important role in calcium regulation than parathyroid hormone does.

Hydrogen Ion and Acid–Base Regulation

Because hydrogen ions affect enzyme action, most metabolic reactions are sensitive to the hydrogen ion concentration of the fluid in which they occur. Consequently, the proper regulation of hydrogen ion concentration is essential for effective cellular function.

Acids and Bases

As explained in Chapter 2, acids are substances that liberate hydrogen ions, and bases are substances that accept them. The concentration of free hydrogen ions (H^+) determines the acidity of a solution, and a solution's acidity is measured on the pH scale, with lower pH values indicating higher hydrogen ion concentrations (acidity).

Acids can be grouped as either strong acids or weak acids. Strong acids are those that dissociate virtually completely in solution, providing large numbers of free hydrogen ions. Hydrochloric acid (HCl), which in solution dissociates into hydrogen ions and chloride ions, is an example of a strong acid. Weak acids are those that do not dissociate completely and do not provide as many free hydrogen ions when they are placed in solution. For example, when dissolved in water, some of the molecules of the weak acid carbonic acid dissociate into hydrogen ions and bicarbonate ions. Substantial numbers of carbonic acid molecules, however, do not dissociate into hydrogen ions and bicarbonate ions. Weak acids such as carbonic acid generally dissociate in a predictable fashion, with a certain proportion of the molecules dissociating to provide free hydrogen ions and a certain proportion not undergoing this dissociation.

pH of the Blood and Interstitial Fluid

The pH of the arterial blood is normally 7.4, whereas the pH of the venous blood and interstitial fluid is 7.35. However, the body's metabolic activities

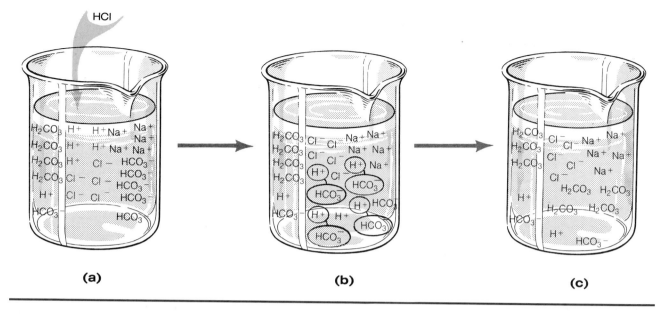

(a)   **(b)**   **(c)**

Figure 27.6

Effect of adding strong acid (HCl) to the carbonic-acid–sodium-bicarbonate buffer system. **(a)** The hydrochloric acid added to the solution dissociates into hydrogen ions (H^+) and chloride ions (Cl^-). This dissociation upsets the proportion that must be maintained between carbonic acid molecules that are dissociated into hydrogen ions and bicarbonate ions, and carbonic acid molecules that have not undergone that dissociation. **(b)** As a result, some of the hydrogen ions and bicarbonate ions combine into undissociated carbonic acid molecules **(c)** This combination ties up hydrogen ions from hydrochloric acid that would otherwise have been free to lower the pH of the solution.

generate acidic products that tend to raise the hydrogen ion concentration (and decrease the pH) of the internal environment. The phosphorus and sulfur of proteins, for example, are potential sources of phosphoric and sulfuric acids, and metabolic reactions produce organic acids such as fatty acids and lactic acid. Thus, under normal circumstances, the body must eliminate excess hydrogen ions in order to maintain the proper pH of the internal environment.

Buffer Systems

Although excess hydrogen ions must ultimately be eliminated from the body, principally by the kidneys, the body contains buffer systems that help stabilize the pH of the body fluids. A buffer system consists of a solution containing two or more chemical substances that can prevent extreme changes in the hydrogen ion concentration (pH) of the solution when either an acid or a base is added to it. A buffer system works by chemically combining with hydrogen ions as their concentration starts to rise and releasing them as their concentration starts to fall.

A solution of carbonic acid and sodium bicarbonate, for example, acts as a buffer system. When carbonic acid and sodium bicarbonate are placed together in solution, the sodium bicarbonate dissociates into sodium ions and bicarbonate ions. The carbonic acid, however, is a weak acid, and it does not dissociate fully into hydrogen ions and bicarbonate ions (Figure 27.5). Thus the carbonic acid–sodium bicarbonate buffer system contains undissociated carbonic acid, sodium ions, hydrogen ions, and bicarbonate ions. Moreover, a certain proportion of the carbonic acid is always dissociated into hydrogen ions and bicarbonate ions, and a certain proportion does not undergo this dissociation.

If hydrogen ions (for example, in the form of hydrochloric acid) are added to a buffer system such as the carbonic acid–sodium bicarbonate buffer system, many of the hydrogen ions do not remain in the free state to affect pH as they would if they were added to pure water (Figure 27.6). In the carbonic

F27.5

F27.6

acid–sodium bicarbonate buffer system, the addition of hydrogen ions (from hydrochloric acid) results in the presence of many hydrogen ions and many bicarbonate ions (from sodium bicarbonate) in the solution. These are the equivalent of carbonic acid molecules that have dissociated into hydrogen ions and bicarbonate ions. Consequently, the presence of these ions upsets the normal proportion that must be maintained between carbonic acid molecules that dissociate into hydrogen ions and bicarbonate ions, and carbonic acid molecules that do not undergo this dissociation. In order to reestablish the normal proportion, some of the hydrogen ions and bicarbonate ions combine into undissociated carbonic acid molecules. This combination removes from the solution some of the hydrogen ions resulting from the addition of the hydrochloric acid. These hydrogen ions would otherwise have been free to increase the hydrogen ion concentration of the solution and lower the pH. Instead, many of the hydrogen ions do not remain free, and the effect of the hydrogen ions from the hydrochloric acid on the hydrogen ion concentration (pH) of the solution is minimized.

Conversely, the removal of hydrogen ions from the carbonic acid–sodium bicarbonate buffer system—due, for example, to the addition of a base—also upsets the normal proportion that must be maintained between carbonic acid molecules that dissociate into hydrogen ions and bicarbonate ions, and carbonic acid molecules that do not undergo this dissociation. In this case, some of the carbonic acid molecules dissociate into hydrogen ions and bicarbonate ions in order to restore the normal proportion. This dissociation minimizes the effect of removing hydrogen ions on the pH of the solution.

In the body, the carbonic acid–sodium bicarbonate buffer system acts in the manner just described to stabilize the pH of the body fluids. Moreover, other body substances also act as buffers to resist changes in pH. Among these are large anions, such as plasma proteins, intracellular phosphate complexes, and hemoglobin molecules (reduced hemoglobin has a much greater affinity for hydrogen ions than oxyhemoglobin does).

Respiratory Regulation of Acid–Base Balance

The respiratory regulation of acid–base balance makes use of the fact that in solution carbonic acid exists in reversible equilibrium with carbon dioxide and water (dissolved carbon dioxide), as follows:

$$CO_2 + H_2O \rightleftharpoons H_2CO_3 \rightleftharpoons H^+ + HCO_3^-$$

Thus, the higher the concentration of carbon dioxide in the internal environment of the body, the more carbonic acid is formed. As more carbonic acid is formed, the normal proportion that must be maintained between carbonic acid that dissociates into hydrogen ions and bicarbonate ions, and carbonic acid that does not undergo this dissociation is upset, and some of the carbonic acid dissociates into hydrogen ions and bicarbonate ions. This dissociation tends to increase the acidity (lower the pH) of the internal environment. Conversely, the lower the concentration of carbon dioxide in the internal environment, the more carbonic acid forms carbon dioxide and water. This activity also upsets the normal proportion that must be maintained between carbonic acid molecules that dissociate into hydrogen ions and bicarbonate ions, and carbonic acid molecules that do not undergo this dissociation. As a result, some of the dissociated hydrogen ions and bicarbonate ions of carbonic acid combine into undissociated carbonic acid molecules. This combination removes hydrogen ions and tends to decrease the acidity (raise the pH) of the internal environment.

Since the carbon dioxide concentration of the body's internal environment influences the pH of the internal environment, and since carbon dioxide is eliminated from the body at the lungs, the respiratory system plays an important role in maintaining the pH of the internal environment. If much carbon dioxide is eliminated at the lungs, the CO_2 concentration in the internal environment decreases. This decrease results in a less acidic, more basic internal environment, with a higher pH. If only a small amount of CO_2 is eliminated at the lungs, its concentration in the internal environment increases. This increase results in a more acidic internal environment, with a

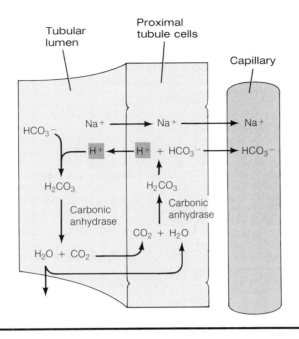

Figure 27.7
Secretion of hydrogen ions by cells of the proximal tubule. Also shown is the pathway by which hydrogen ions in the tubular lumen participate in the reabsorption of bicarbonate ions from the tubule.

lowered pH. Thus, the control of the CO_2 concentration in the internal environment of the body by the respiratory system serves as a kind of buffering mechanism. For example, if the pH of the internal environment falls below normal values, an increased elimination of carbon dioxide at the lungs tends to raise it. Conversely, if the pH of the internal environment rises above normal levels, a decreased elimination of carbon dioxide by the lungs tends to decrease it. In view of the ability of the respiratory system to influence acid–base balance, it is not surprising that respiration is itself sensitive to such factors as the carbon dioxide concentration and the pH of the system. The influence of these factors on respiration was discussed in Chapter 22.

Renal Regulation of Acid–Base Balance

Although the body's buffer systems can tie up some hydrogen ions and resist changes in pH to some degree, the continued production of acidic products by the body's metabolic reactions would eventually overcome this buffering capacity. Consequently, excess hydrogen ions must ultimately be eliminated from the body, and this is accomplished by the kidneys. The kidneys also regulate the extracellular concentration of bicarbonate ions, which is normally quite high.

SECRETION OF HYDROGEN IONS   Hydrogen ions are secreted into the renal tubules by tubular cells. Although the secretion mechanism is not well understood, it appears that the hydrogen ions are derived from carbonic acid (Figure 27.7). The enzyme carbonic anhydrase is present within the tubular cells, and it is thought to catalyze the formation of carbonic acid from carbon dioxide and water. Some of the carbonic acid dissociates into hydrogen ions and bicarbonate ions. The hydrogen ions are actively transported into the lumen of the tubule in exchange for sodium ions. The bicarbonate ions that remain in the tubular cells, as well as the sodium ions, pass out of the cells and into the blood.

REABSORPTION OF BICARBONATE IONS   The secretion of hydrogen ions by the mechanism described in the preceding section provides a way of recovering bicarbonate ions that are filtered from the plasma into the glomerular capsule (Figure 27.7). Bicarbonate ions within the tubular lumen combine with secreted hydrogen ions to form carbonic acid molecules. The carbonic acid molecules are then split into carbon dioxide and water by carbonic anhydrase associated with the membranes of the tubular cells in the proximal

CLINICAL CORRELATION
Metabolic Acidosis

CASE REPORT . . . .

THE PATIENT A 63-year-old man

PRINCIPAL COMPLAINTS Lethargy and confusion

HISTORY The patient had a long history of hypertension and congestive heart failure, for which he took digitalis (a cardiac stimulant) and a diuretic agent (an agent that increases the volume of urine). He also had moderate renal disease. The patient occasionally followed folk medicine, and on the advice of a friend, had begun to consume sulfur (Sublimed Sulfur, U.S.P., a fine powder of pure sulfur) for his malaise (general body weakness) and dyspnea (difficult breathing). He had been taking sulfur for a week before current admission, and it was estimated that he may have consumed as much as 75 grams/day during each of the last two days before admission. He reported no bowel movements for at least three days before admission. Because of his apparently deteriorated condition, his family brought him to the hospital.

CLINICAL EXAMINATION At admission, the patient's body temperature was normal, and the heart rate was 95/min (normal: 60–100/min). The arterial blood pressure was 165/95 mm Hg (normal: 90–140/60–90 mm Hg), and the respiratory rate was 22/min (normal: 12–17/min). The pH of the arterial blood was 7.18 (normal: 7.35–7.45). The serum-sodium concentration was 125 mEq/liter (normal: 135–145 mEq/liter), the serum-potassium concentration was 8.7 mEq/liter (normal: 3.5–5 mEq/liter), and the serum-chloride concentration was 103 mEq/liter (normal: 100–105 mEq/liter). The arterial

partial pressure of carbon dioxide was 28 mm Hg (normal: 35–45 mm Hg), the bicarbonate concentration was 10 mEq/liter (normal: 21–28 mEq/liter), and the arterial partial pressure of oxygen was 90 mm Hg (normal: 75–100 mm Hg). The pH of the urine was 4.9, the concentration of protein in the urine was 0.3 g/100 ml (normal: undetectable), and the numbers of red blood cells and white blood cells were excessive.

TREATMENT Sodium bicarbonate solution was infused and enemas were given to remove sulfur from the colon. After four days, the blood gases, plasma electrolytes, and pH of the arterial blood were normal, and the patient was discharged.

COMMENT Elemental sulfur is a relatively nontoxic substance, but it is converted to sulfide and then to sulfate by bacteria in the colon. Sulfur often produces diarrhea because of irritation produced by its metabolites, but such did not occur in this patient. Thus, acidic products of sulfur were absorbed, producing severe metabolic acidosis. The patient's renal disease also contributed to the condition, because the excretion of hydrogen ions was slow. The decreased arterial partial pressure of carbon dioxide reflects respiratory compensation for the metabolic acidosis, and the decreased bicarbonate concentration reflects the shift of the following reaction toward carbon dioxide and water.

$$CO_2 + H_2O \rightleftharpoons H_2CO_3 \rightleftharpoons H^+ + HCO_3^-$$

Complete correction of the condition requires the excretion of hydrogen ions by the kidneys.

convoluted tubule. The carbon dioxide enters the tubular cells, where carbonic anhydrase combines it with water to form carbonic acid. Some of the carbonic acid dissociates into hydrogen ions and bicarbonate ions. The hydrogen ions are actively transported into the lumen of the tubule in exchange for sodium ions. The bicarbonate ions that remain in the tubular cells, as well as the sodium ions, pass out of the tubular cells and into the blood.

CONTROL OF HYDROGEN ION SECRETION AND BICARBONATE ION REABSORPTION As can be seen from the preceding discussion, the tubular reabsorption of bicarbonate ions that are filtered from the plasma into the glomerular capsules depends on the presence of hydrogen ions within the renal tubules. Moreover, the rate of secretion of hydrogen ions into the renal tubules varies directly with the acidity of the internal environment.

Normally, the balance between the rate of filtration of bicarbonate ions into the glomerular capsules and the rate of secretion of hydrogen ions into the renal tubules is such that there are slightly more hydrogen ions secreted than there are bicarbonate ions filtered. As a result, essentially all the bicarbonate ions are reabsorbed, and the slight excess of hydrogen ions that remain after bicarbonate ion reabsorption is excreted in the urine.

If the concentration of hydrogen ions in the internal environment falls, the rate of secretion of hydrogen ions into the renal tubules diminishes relative to the filtration rate of bicarbonate ions into the glomerular capsules. As a consequence, not all of the bicarbonate ions are reabsorbed, and bicarbonate

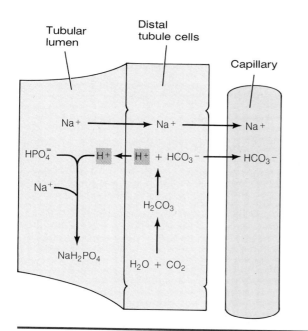

Figure 27.8
Secretion of hydrogen ions into the tubular lumen by cells of the distal tubule. The hydrogen ions are excreted in combination with phosphate compounds.

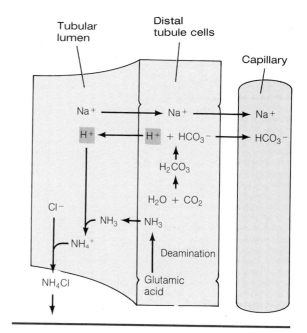

Figure 27.9
Secretion of hydrogen ions and ammonia into the tubular lumen by cells of the distal tubule. The hydrogen ions are excreted in combination with ammonia in the form of ammonium salts.

ions are excreted in the urine. The urinary excretion of bicarbonate ions diminishes the concentration of bicarbonate ions in the internal environment. The diminished concentration of bicarbonate ions leads to a decrease in the number of hydrogen ions tied up by the carbonic acid–sodium bicarbonate buffer system (as well as by other buffer systems), and the hydrogen ion concentration of the internal environment tends to rise toward normal.

If the concentration of hydrogen ions in the internal environment rises, the rate of secretion of hydrogen ions into the renal tubules increases considerably in comparison to the rate of filtration of bicarbonate ions into the glomerular capsules. Many of the secreted hydrogen ions are not needed for the reabsorption of bicarbonate ions, and they are excreted in the urine. Moreover, as previously described, the process of secreting hydrogen ions into the renal tubules provides sodium ions and bicarbonate ions that are returned to the body. The return of these ions increases the concentration of bicarbonate ions in the internal environment and thereby leads to an increase in the number of hydrogen ions tied up by the carbonic acid–sodium bicarbonate buffer system (as well as by other buffer systems). These responses tend to decrease the hydrogen ion concentration of the internal environment toward normal.

BUFFERING OF SECRETED HYDROGEN IONS   The mechanism that secretes hydrogen ions into the renal tubules can achieve a maximum concentration of free hydrogen ions in the tubular fluid that is equivalent to a pH of 4.5. At normal rates of urine flow this is insufficient to excrete all of the hydrogen ions that must be eliminated from the body. Consequently, buffers must be present within the tubular fluid to tie up secreted hydrogen ions so that the pH of the tubular fluid remains above 4.5, thereby allowing additional hydrogen ions to be secreted. Phosphate compounds (HPO_4^{-2}) and ammonia (NH_3) act as buffers to tie up hydrogen ions in the tubular fluid. The phosphate compounds are filtered at the glomeruli and are only poorly reabsorbed (Figure 27.8). They combine with hydrogen ions ($H^+ + HPO_4^{-2} \rightarrow H_2PO_4^-$) and are excreted in combination with a cation such as sodium (Na^+) in the form of a weakly acidic substance ($Na^+ + H_2PO_4^- \rightarrow NaH_2PO_4$). This same type of excretion can occur with other anions, such as lactate.

Ammonia is formed in the tubular cells by the deamination of certain amino acids, particularly glutamic acid (Figure 27.9). The ammonia diffuses

F27.8

F27.9

CONDITIONS OF CLINICAL SIGNIFICANCE

Disorders of Acid–Base Balance

Acidosis and Alkalosis

As previously indicated, the pH of the arterial blood is normally 7.4, whereas the pH of the venous blood and interstitial fluid is 7.35. However, certain conditions tend to alter the pH of the blood.

Acidosis is an abnormal condition in which the hydrogen ion concentration of the blood tends to increase and the pH of the blood tends to decrease. There are two categories of acidosis: respiratory acidosis and metabolic acidosis.

Respiratory acidosis is acidosis due to an increase in the carbon dioxide concentration, and consequently the carbonic acid concentration, of the body fluids. It frequently occurs as a result of asphyxia or as a result of diseases that interfere with the elimination of carbon dioxide by the lungs.

Metabolic acidosis is acidosis due to all other factors besides an excess of carbon dioxide in the body fluids. It can occur as a result of severe diarrhea or as a result of vomiting the lower intestinal contents. Either event causes the loss of bicarbonate-rich fluids from the gastrointestinal tract. This loss has the same effect as the excretion of bicarbonate ions by the kidneys—that is, a tendency to decrease the pH of the internal environment. Metabolic acidosis can also occur as a result of the production of an excessive amount of ketone bodies—for example, acetoacetic acid—during diabetes mellitus or as a result of the ingestion of acidic drugs.

Alkalosis is an abnormal condition in which the hydrogen ion concentration of the blood tends to decrease and the pH tends to increase. As was the case for acidosis, there are two categories of alkalosis: respiratory alkalosis and metabolic alkalosis.

Respiratory alkalosis is alkalosis due to a decrease in the carbon dioxide concentration, and consequently the carbonic acid concentration, of the body fluids. It can occur as a result of intense overventilation, which eliminates an excessive amount of carbon dioxide from the body. However, respiratory alkalosis is generally much less common than respiratory acidosis.

Metabolic alkalosis is alkalosis due to all other factors besides a deficiency of carbon dioxide in the body fluids. It can occur as a result of vomiting the acidic gastric contents alone, which rarely happens, or as a result of the ingestion of alkaline drugs.

Acidemia and Alkalemia

If the pH of the arterial blood falls below 7.35, a state of **acidemia** exists; and if the pH of the arterial blood rises above 7.45, a state of **alkalemia** exists. The occurrence of either acidemia or alkalemia is a serious problem, and a person generally cannot live more than a few hours if the pH of the arterial blood falls to about 7.0 or rises to about 8.0. Acidemia depresses the central nervous system, and a person suffering from severe acidemia usually enters a coma. Conversely, severe alkalemia produces an overexcitability of the nervous system that can lead to muscular tetany, to extreme nervousness, and in certain individuals, to convulsions.

from the tubular cells into the tubules, where it combines with hydrogen ions to form ammonium ions ($H^+ + NH_3 \rightarrow NH_4^+$). This combination removes ammonia (as well as hydrogen ions) from the tubular fluid and allows the diffusion of still more ammonia from the tubular cells into the tubular fluid. The ammonium ions are excreted in combination with anions such as chloride (Cl^-) in the form of very weakly acidic or neutral substances ($NH_4^+ + Cl^- \rightarrow NH_4Cl$) that do not significantly lower the pH of the urine. It is interesting to note that the production rate of ammonia increases when the hydrogen ion concentration has been elevated for two or three days. This increase in ammonia production provides additional buffering capacity for hydrogen ion excretion and allows the extra hydrogen ions to be excreted without exceeding the maximum allowable acidity of the tubular fluid.

STUDY OUTLINE

BODY FLUIDS Plasma, interstital fluid, intracellular fluid, and extracellular fluid. **pp. 793–795**

COMPOSITION OF PLASMA AND INTERSTITIAL FLUID Plasma and interstitial fluid are separated by capillary walls; are of similar composition.

COMPOSITION OF INTRACELLULAR FLUID Different from that of plasma and interstitial fluid.

REGULATION OF THE BODY'S INTERNAL ENVIRONMENT To maintain homeostasis, any addition of substances such as water and electrolytes to body's internal environment must be balanced by removal of equal amounts of these materials. **pp. 795–806**

SODIUM REGULATION Mechanisms involved in plasma-volume and blood-pressure regulation are important.

FACTORS THAT INFLUENCE SODIUM EXCRETION Mechanisms involved in regulation of plasma volume and blood pressure are important factors.

SODIUM CONTENT AND SODIUM CONCENTRA-TION Sodium loss or gain changes overall body-sodium content but not necessarily the actual concentration of sodium in internal environment.

WATER REGULATION Essential because of water's effects on osmotic pressure and fluid volume of internal environment.

FACTORS THAT INFLUENCE WATER EXCRETION Osmotic pressure of extracellular fluid and blood pressure (plasma volume) are major factors; factors that influence sodium excretion are very closely related to factors that influence water excretion.

POTASSIUM REGULATION Ingestion and excretion are normally in balance; most filtered potassium is actively reabsorbed; amount excreted depends mainly on secretion into renal tubules. Amount of potassium secreted is related to potassium concentration of renal tubular cells; influenced by aldosterone.

CALCIUM REGULATION Involves bone, kidneys, and gastrointestinal tract; influenced either directly or indirectly by parathyroid hormone. Calcitonin inhibits removal of calcium from bone.

HYDROGEN ION AND ACID–BASE REGULATION

ACIDS AND BASES Acids liberate hydrogen ions, bases accept them; concentration of free hydrogen ions determines solution's acidity.

PH OF THE BLOOD AND INTERSTITIAL FLUID pH of arterial blood normally 7.4; pH of venous blood and interstitial fluid normally 7.35.

BUFFER SYSTEMS
1. Solution containing two or more chemical substances that can prevent extreme changes in solution's pH when acid or base is added.
2. Body substances that act as buffers include carbonic-acid–sodium-bicarbonate buffer system, plasma proteins, intracellular phosphate complexes, and hemoglobin.

RESPIRATORY REGULATION OF ACID–BASE BALANCE Makes use of fact that in solution carbonic acid exists in reversible equilibrium with carbon dioxide and water. Large carbon dioxide elimination at lungs tends to reduce acidity (raise pH) of internal environment and vice versa. Respiration is sensitive to carbon dioxide content and pH of system.

RENAL REGULATION OF ACID–BASE BALANCE Hydrogen ions secreted into renal tubules; bicarbonate ions filtered into glomerular capsules can be reabsorbed by tubular cells; phosphate compounds and ammonia buffer hydrogen ions in tuberlar fluid.

CONDITIONS OF CLINICAL SIGNIFICANCE: DISORDERS OF ACID–BASE BALANCE p. 806

SELF-QUIZ

1. There are only slight differences in composition between the: (a) plasma and interstitial fluid; (b) intracellular fluid and plasma; (c) intracellular fluid and interstitial fluid.

2. It appears that humans normally regulate their body sodium content by carefully controlling the amount of sodium they ingest. True or False?

3. The arterial blood pressure can influence the glomerular filtration rate. True or False?

4. Angiotensin is a factor that stimulates the release of: (a) renin; (b) sodium; (c) aldosterone.

5. The loss or gain of sodium by the body does not necessarily have a great effect on the actual concentration of sodium in the internal environment. True or False?

6. A feeling of thirst results from: (a) an increased osmotic concentration of the extracellular fluid; (b) a substantial increase in plasma volume; (c) an increased volume of extracellular fluid.

7. The permeability of the kidney's collecting tubules to water is controlled hormonally by: (a) aldosterone; (b) ADH; (c) angiotensin.

8. ADH release is likely to increase in response to an increased: (a) blood pressure; (b) plasma volume; (c) extracellular osmolarity.

9. In most cases, an increase in the glomerular filtration rate of water is accompanied by a simultaneous decrease in the glomerular filtration rate of sodium. True or False?

10. Adjustments in the amount of potassium excreted depend mainly on how much is: (a) filtered into the glomerular capsules; (b) reabsorbed from the renal tubules; (c) secreted into the renal tubules.

11. Which hormone enhances the tubular secretion of potassium? (a) aldosterone; (b) ADH; (c) parathyroid hormone.

12. Parathyroid hormone decreases the: (a) movement of calcium from bone into the extracellular fluid; (b) movement of phosphate from bone into the extracellular fluid; (c) urinary excretion of calcium.

13. The normal pH of arterial blood is about: (a) 7.0; (b) 7.7; (c) 7.4.

14. The principal organ for the elimination of excess hydrogen ions from the body is the: (a) liver; (b) lungs; (c) kidneys.

15. When the CO_2 concentration of the internal environment rises, the pH of the internal environment also tends to rise. True or False?

16. If a large amount of carbon dioxide is eliminated at the lungs, the carbon dioxide concentration in the internal environment decreases, leading to a more acidic, less basic internal environment, with a higher pH. True or False?

17. When the pH of the internal environment of the body rises, the rate of secretion of hydrogen ions into the renal tubules: (a) diminishes in comparison to the rate of filtration of bicarbonate ions into the glomerular capsules; (b) increases in comparison to the rate of filtration of bicarbonate ions into the glomerular capsules.

LEARNING OBJECTIVES

After completing this chapter, you should be able to:

1. Describe the embryonic development of the internal and external reproductive structures.

2. Describe the structure and function of each part of the male internal reproductive structures and external genitalia.

3. Describe the process of spermatogenesis.

4. Discuss the hormonal control of testicular function.

5. Explain the functional value of the testes' location in the scrotum, outside the abdominopelvic cavity.

6. Describe the structure and function of each part of the female internal reproductive structures and external genitalia.

7. Describe the process of oogenesis.

8. Discuss the events of the ovarian cycle and the hormonal control of ovarian function.

9. Describe the development of the secondary sex characteristics during puberty in both males and females.

CHAPTER CONTENTS

KEY TERMS AND DERIVATIVES

gamete (*gam* = marriage) male or female reproductive cell

gonad (*gon* = seed) a gland or organ producing gametes; an ovary or testis

gubernaculum (*gubern* = to guide) a guiding, connecting cord structure; for example, the cord between the testis and the scrotal sac

spermatogenesis (*genesis* = origin) the process by which sperm cells are produced

uterus (*uter* = womb, bag) the organ that houses the embryo throughout its development

oogenesis (*genesis* = origin) the process by which ova are produced

The Reproductive System

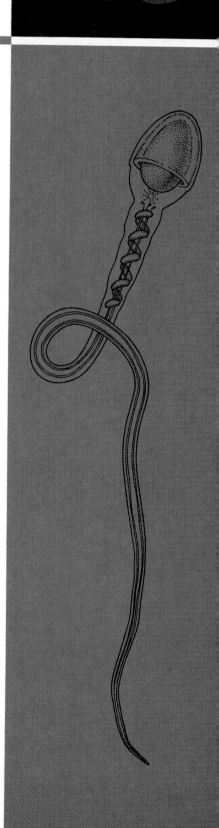

28

The organs of the male and female reproductive systems ensure the continuance of the human species. They do this by producing **gametes** (or *germ cells*) and by providing a method by which the gametes of the male (**spermatozoa,** or **sperm**) can be introduced into the body of the female, where one of the male gametes joins with a gamete **(ovum)** of the female. This union of an ovum and a spermatozoon is called *fertilization.* The organs of the female reproductive system provide a suitable environment in which the fertilized ovum **(zygote)** can develop to a stage at which it is capable of surviving outside the mother's body.

The organs that produce the gametes are referred to as the **primary sex organs,** or **essential sex organs.** These are the **gonads**—the **testes** in the male and the **ovaries** in the female. In addition to producing gametes, the primary sex organs also produce hormones that influence the development of male or female secondary sex characteristics and regulate the reproductive cycle. In the male, specialized interstitial cells in the testes produce a group of hormones called *androgens.* The principle androgen is *testosterone.* In the female, the ovaries produce *estrogens* and *progesterone.*

The structures that transport, protect, and nourish the gametes after they leave the gonads are called **accessory sex organs.** In the male the accessory sex organs include the *epididymis, ductus deferens, seminal vesicles, prostate gland, bulbourethral glands, scrotum,* and *penis.* Female accessory sex organs include the *uterine tubes, uterus, vagina,* and *vulva.*

EMBRYONIC DEVELOPMENT OF THE REPRODUCTIVE SYSTEM

The gonads are formed within retroperitoneal elevations called **genital ridges,** which protrude into the coelom (body cavity) just medial to the developing kidneys (mesonephros). Although the sex of the embryo is determined at the time of fertilization, the genital ridges do not begin to differentiate into testes until the seventh week or ovaries until the end of the eighth week of embryonic development. Gametes, which originate in the yolk sac, migrate to the genital ridges in the sixth week of development, prior to the differentiation of the ridges. During the undifferentiated period, the gonads develop in close approximation to the **mesonephric ducts,** which drain the embryonic kidneys. A second pair of ducts—the **paramesonephric ducts**—also develops. The paramesonephric ducts run parallel to the mesonephric ducts and empty into the urogenital sinus at the posterior end of the embryo (Figure 28.1a). Males retain the mesonephric ducts for transport of sperm, while females retain the paramesonephric ducts for the transport of ova and the nourishment of the fetus.

Development of the Internal Reproductive Structures

Following the undifferentiated period, during which all embryos have the same structure, male and female embryos take different developmental paths. As a result, unique internal reproductive structures are formed in each sex.

809

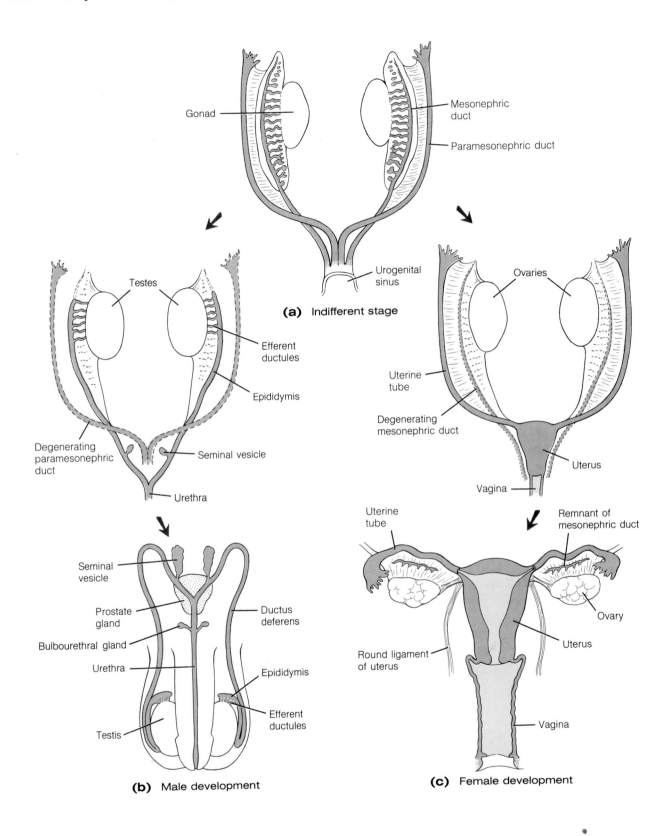

Figure 28.1

Embryonic development of male and female internal reproductive structures. Notice that the testes descend out of the abdominopelvic cavity into the scrotum, but the ovaries, which also descend, remain within the pelvic cavity.

Male Embryo

In the male embryo, the inner portion (**medulla**) of the undifferentiated gonads develops tubules that join with the mesonephric duct. These tubules become the **seminiferous tubules** of each testis, in which spermatozoa eventually are produced. In the male, the mesonephric duct is utilized for the transport of sperm from the testes to the exterior of the body. With further development, each mesonephric duct forms **efferent ductules,** an **epididymis,** and a **ductus deferens** of the adult (Figure 28.1b). Shortly after the gonads begin to differentiate into testes, the paramesonephric ducts degenerate.

sem-i-nif′-er-us

ep-i-did′-i-mis
F28.1b

Female Embryo

In the female embryo, the outer portion (**cortex**) of the undifferentiated gonads undergoes the greater development and forms follicles, in which ova develop. Coinciding with the differentiation of the gonads into ovaries, the distal ends of the paramesonephric ducts join together to form the **uterus** and **vagina.** The portion of the paramesonephric ducts between the ovaries and the uterus forms the **uterine tubes** (*fallopian tubes,* or *oviducts*), through which ova travel to the uterus (Figure 28.1c). In the female, the mesonephric ducts degenerate without contributing any functional structures to the reproductive system.

F28.1c

Effects of Hormones on Genital Duct Development

The development of the genital ducts into either male or female structures appears to be under the control of hormones produced by the testes. In the presence of testosterone that is produced after the development of the genital ridges into testes, the mesonephric ducts remain and develop further. At the same time the embryonic testes produce a peptide that inhibits the paramesonephric ducts and causes them to degenerate. If the gonads (testes) are removed from a genetically male embryo (XY), the embryo develops the reproductive structures of a female. That is, the paramesonephric ducts undergo further development and the mesonephric ducts degenerate. Similarly, the injection of testosterone into a genetic female (XX) causes the embryo to develop according to the male pattern. These results indicate that in the absence of the male hormone testosterone, all embryos, regardless of their genetic makeup, would develop into females. Under normal conditions, of course, embryos that are genetically male produce testosterone and therefore develop male reproductive structures.

Development of the External Reproductive Structures

The embryonic development of the external genitalia is also controlled by hormones produced by the gonads. The external genitalia, like the internal reproductive organs, remain in an undifferentiated state until about the eighth week. Before the production of hormones begins, all embryos develop a conical elevation—called the **genital tubercle**—at the point where the mesonephric and paramesonephric ducts open to the exterior of the body (Figure 28.2a). On the inferior surface of this tubercle is a shallow depression called the **urethral groove,** which opens into the urogenital sinus of the embryo. On either side of the urethral groove are slight elevations called the **urethral folds.** Bounding the folds laterally are rounded elevations called the **labioscrotal swellings.**

F28.2a

Male Embryo

If the embryonic gonads are producing testosterone, as occurs when the individual is a male, the genital tubercle elongates to form the **penis.** The urethral folds fuse, leaving an opening only at the distal end of the penis. The tube thus formed becomes the spongy portion of the urethra. The labioscrotal swellings develop into a pouch (**scrotum**) that eventually will receive the testes (Figure 28.2b).

F28.2b

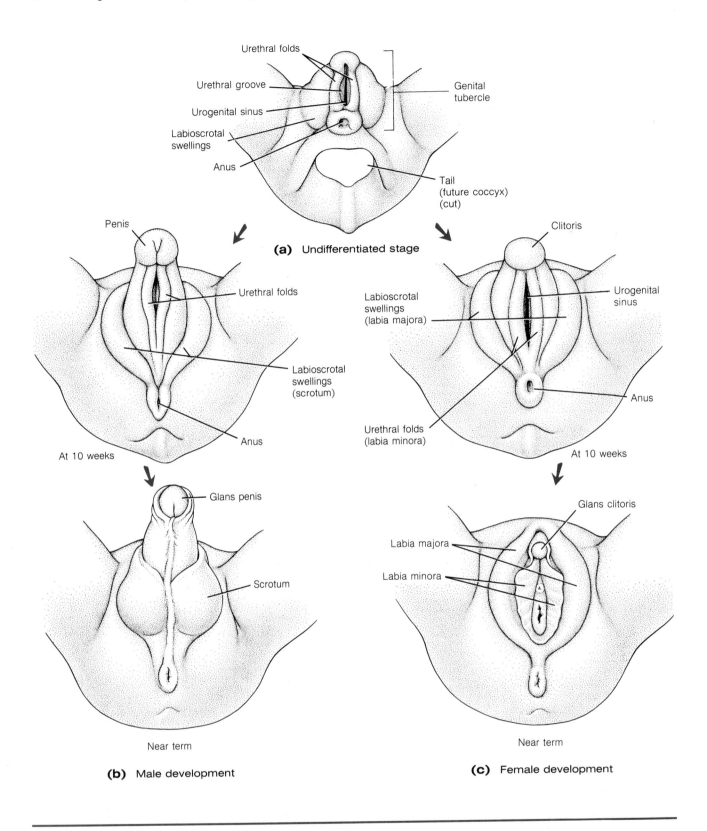

Figure 28.2
Embryonic development
of male and female
external reproductive
structures.

F28.2c

Female Embryo

In the female embryo, in the absence of testosterone, the genital tubercle does not elongate as much as in the male, and it becomes the **clitoris.** Nor do the urethral folds fuse in the female. Rather, they remain as the **labia minora,** surrounding the entrance into the vagina. The labioscrotal folds are not destined to receive gonads in the female. They remain as elevations called the **labia majora,** which flank the labia minora (Figure 28.2c).

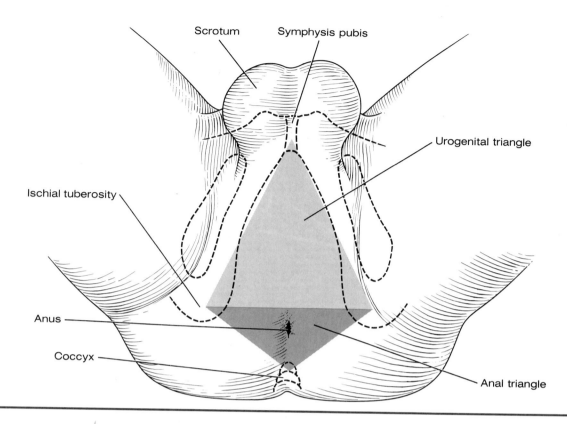

Figure 28.3
Divisions of the male perineum.

ANATOMY OF THE MALE REPRODUCTIVE SYSTEM

The male reproductive system includes the *testes,* which produce spermatozoa; a number of *ducts* that store, transport, and nourish the spermatozoa; several accessory *glands* that contribute to the formation of semen; and the *penis,* through which semen is conveyed to the exterior of the body.

The Male Perineum

The **perineum** includes all the structures that are located between the pubic symphysis anteriorly, the coccyx posteriorly, and the ischiopubic rami and the sacrotuberous ligaments laterally. The portion of the sacrotuberous ligament that bounds the perineum runs from the lateral margin of the sacrum and coccyx to the ischial tuberosity. The muscles of the perineum are described in Chapter 9. Here we are interested only in the surface anatomy of the region in males.

The male perineum can be divided into two triangles by a transverse line that passes between the ischial tuberosities (Figure 28.3). The anterior triangle, which contains the external genitalia, is called the **urogenital triangle.**

F28.3

CONDITIONS OF CLINICAL SIGNIFICANCE
Sexual Malformations During Embryonic Development

An understanding of the patterns of development of the reproductive structures makes it apparent how even slight malformations during embryonic development can cause any number of abnormalities in the adult. It is estimated that some degree of sexual abnormality occurs in one out of every thousand persons. In a male, for example, the urethral folds may fail to join completely, which leaves openings into the urethra on the undersurface of the penis. In a female with normal external genitalia, a ductus deferens may develop in addition to female internal sexual organs. Many other sexual abnormalities occur, but it is beyond the scope of this book to consider them.

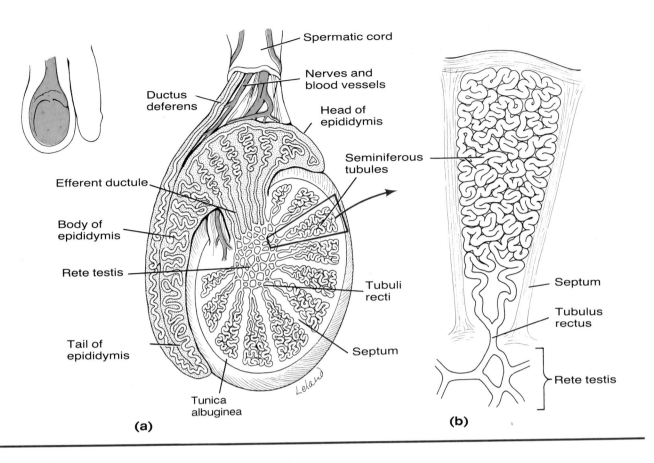

Figure 28.4
(a) Sagittal section of the ductus deferens, the epididymis, and the testis. (b) Seminiferous tubule within a single compartment of the testis.

The posterior triangle is called the **anal triangle** because it contains the anus.

Testes and Scrotum

The **testes** are the organs in which **spermatogenesis** (the production of spermatozoa) occurs. The testes are located in a skin-covered pouch called the **scrotum.** The scrotum consists of an outer layer of skin covering a thin layer of smooth muscle called the **dartos tunic.** Contraction of the muscles of the dartos tunic gives the scrotum a wrinkled appearance.

Each testis is an oval organ surrounded by a connective tissue capsule called the **tunica albuginea.** Inward extensions **(septa)** of the tunica divide the testes into compartments (Figure 28.4). Each compartment holds several highly convoluted **seminiferous tubules,** which contain gametes in various stages of development. Through the process of spermatogenesis, which occurs inside the seminiferous tubules, the gametes eventually develop into spermatozoa.

The seminiferous tubules in each compartment of the testis join together to form a short, straight tube called the **tubulus rectus** (*rectus* = straight). Toward the posterior portion of the testis, the tubuli recti from all the compartments form a network of tubes termed the **rete testis.** The tubes of the rete testis, in turn, empty into the **efferent ductules,** which leave the testis and enter the epididymis.

Clusters of cells called **interstitial endocrinocytes** (or *Leydig cells*) (Figure 28.5) are located in the loose connective tissue between the seminiferous tubules. These interstitial cells secrete the male sex hormone testosterone.

Descent of the Testes

As discussed earlier, the testes begin their development as retroperitoneal structures in the abdominopelvic cavity, just below the kidneys. As development proceeds, the testes move caudally toward swellings of the abdominal wall called labioscrotal swellings. These labioscrotal swellings, which are located just behind the penis in the anterior portion of the urogenital triangle of

F28.4

F28.5

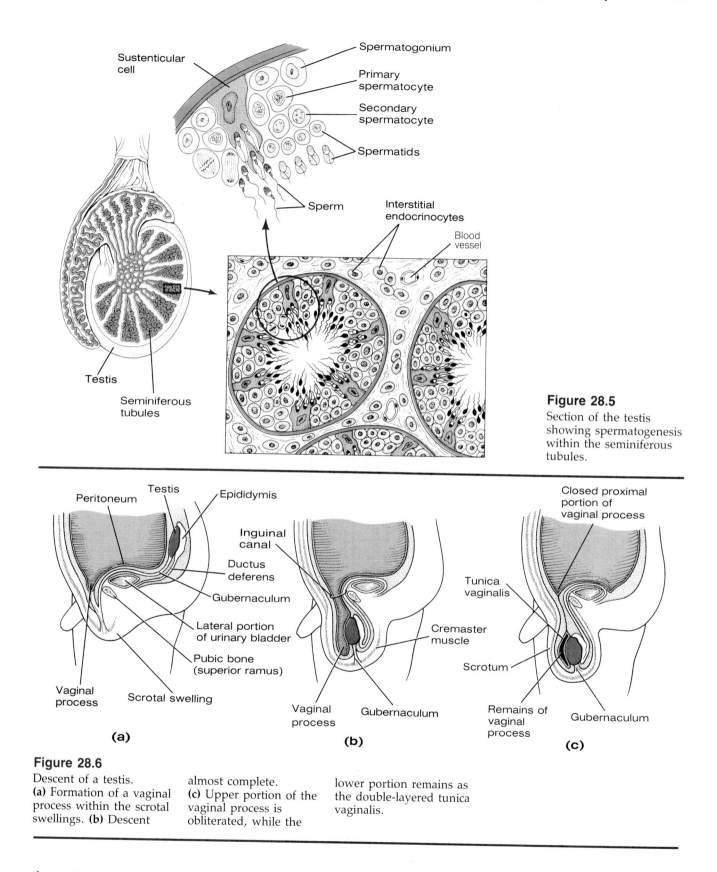

Figure 28.5
Section of the testis showing spermatogenesis within the seminiferous tubules.

Figure 28.6
Descent of a testis.
(a) Formation of a vaginal process within the scrotal swellings. **(b)** Descent almost complete.
(c) Upper portion of the vaginal process is obliterated, while the lower portion remains as the double-layered tunica vaginalis.

the perineum, develop into the scrotum (Figure 28.6). As the testes move toward the scrotal swellings, two outpouchings of the peritoneum called the **vaginal processes** protrude over the superior ramus of each pubic bone and extend through an **inguinal canal** into the two chambers of the scrotum. The testes, which remain behind the peritoneum, follow the vaginal processes out

F28.6

ing'-gwi-nal

CONDITION OF CLINICAL SIGNIFICANCE
Cryptorchidism

The location of the testes outside the abdominopelvic cavity in the scrotum is necessary for the normal development of sperm. Occasionally, one or both of the testes fail to leave the abdominopelvic cavity. This condition is called *cryptorchidism*. Eventually, the undescended testis will atrophy. Since sperm are not produced in cryptorchid testes, the person is sterile if both testes fail to descend. The problem in cryptorchidism is that normal spermatogenesis cannot occur at body-core temperature. The scrotum provides a means by which the testes can be maintained at a temperature lower than body-core temperature. Actually, by means of the **cremaster**

muscle the scrotum is able to regulate the temperature of the testes to some extent. The cremaster muscle, which is a continuation of the internal abdominal oblique muscle, extends down over the spermatic cord and the testes as a series of loops. When the environmental temperature is low, the cremaster muscle contracts, which brings the testes closer to the body and thus warms them. At higher environmental temperatures, the cremaster muscle relaxes and the scrotum becomes flaccid and elongates, thus removing the testes still further from the warmth of the body.

of the abdominopelvic cavity and through the inguinal canals into the scrotum. As the testes descend into the scrotum from their original position in the abdominopelvic cavity, their blood supply travels with them. Therefore, the testicular arteries leave the aorta—and the testicular veins join the inferior vena cava (the left testicular vein via the left renal vein)—in the region of the kidneys, near the original site of the testes, and follow their paths of descent into the scrotum.

After the testes have entered the scrotum, the inguinal canals narrow, constricting the upper portion of the vaginal processes. The lower portion of each vaginal process forms a doubled-layered sac that covers the testis. This portion is called the **tunica vaginalis.** If the upper portion of a vaginal process does not close off completely, it is possible for small loops of intestine to protrude into an inguinal canal. This condition is known as an *inguinal hernia.* Even if the vaginal process closes completely, this is an area of weakness in males, and thus a potential site of hernia. Because in females the gonads do not pass through the body wall and thus do not stretch and weaken the abdominal muscles surrounding the inguinal canals, inguinal hernias are less common in females than in males.

The actual cause of the descent of the testes into the scrotum is not known. However, it seems to be initiated by testosterone from the testes and certain hormones from the pituitary gland. A fibromuscular band called the **gubernaculum** is thought to assist in the descent, but the precise manner in which it does so is not certain. The gubernaculum extends from the caudal surface of each testis, passes through the body wall, and is attached to the floor of the scrotum. As the embryo grows, the gubernaculum shortens, but it is questionable whether it is strong enough actually to pull each testis toward the scrotum.

goo-ber-nak'-yoo-lum

Epididymis

The **epididymis,** which is located in the scrotum, is the first portion of the duct system that transports sperm from the testes to the exterior of the body (Figure 28.4). Each epididymis is an elongated structure that fits tightly against the posterior surface of the testis. It consists of a tightly coiled tube that receives the sperm from the testis through the efferent ductules. The tube is lined with pseudostratified columnar epithelium. The free surfaces of some of the columnar cells have long, nonmotile microvilli called *stereocilia.* The stereocilia serve to facilitate the passage of nutrients from the epithelium to the sperm. If uncoiled, the epididymal tube would be 4–6 m in length, and it is generally considered to serve as a storage site for sperm. The efferent ductules enter the upper region *(head)* of the epididymis. In its lower region (the *tail*), the tube becomes continuous with the ductus deferens. There are smooth muscles in the wall of the epididymal tube that contract during ejacu-

F28.4

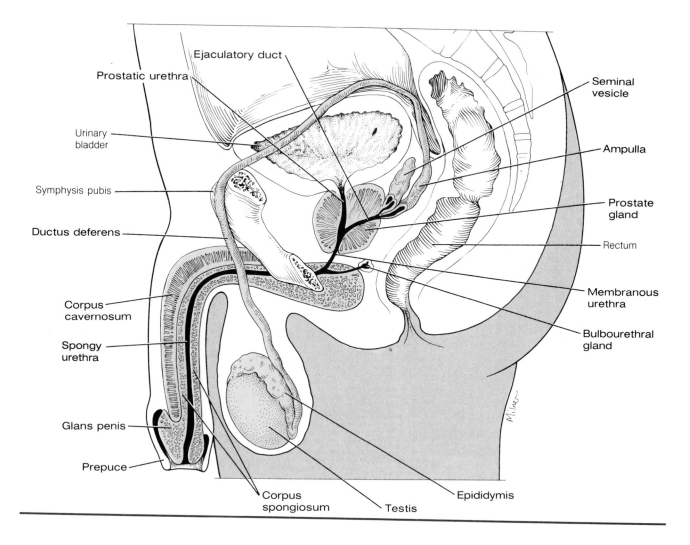

Ejaculatory duct

Prostatic urethra

Urinary bladder

Symphysis pubis

Ductus deferens

Corpus cavernosum

Spongy urethra

Glans penis

Prepuce

Corpus spongiosum

Testis

Epididymis

Seminal vesicle

Ampulla

Prostate gland

Rectum

Membranous urethra

Bulbourethral gland

Figure 28.7

Sagittal section of the male pelvis, with a portion of the left pubic bone attached to illustrate the path of the left ductus deferens.

lation. These contractions move sperm into the ductus deferens. During their slow passage through the epididymis, the sperm undergo a maturation process without which they would be nonmotile and infertile when they enter the female reproductive tract.

Ductus Deferens

The **ductus (vas) deferens** is a continuation of the duct of the epididymis (Figure 28.7). Each ductus deferens is a straight tube that passes along the posterior surface of the testis, medial to the epididymis, and ascends through the scrotum. The ductus deferens passes through the body wall by way of the inguinal canal and enters the abdominopelvic cavity. As it passes from the epididymis to the point where it enters the abdominopelvic cavity, the ductus deferens lies next to the vessels and nerves supplying the testis. All of these structures are enclosed in a sheath of fascia called the **spermatic cord.** Included in the spermatic cord along with the ductus deferens are the testicular artery, testicular vein, lymphatic vessels, and nerves. The veins returning from the testes form a network of connecting branches—the *pampiniform plexus*—around the testicular artery. This plexus is thought to absorb heat from the blood in the testicular artery, thereby lowering the temperature of the arterial blood and helping to maintain the temperature of the testes slightly below body-core temperature. This lower temperature is essential for normal spermatogenesis.

Inside the abdominopelvic cavity, the ductus deferens travels beneath the peritoneum along the lateral wall of the cavity, crosses over the top of the ureter, and then descends along the posterior surface of the urinary bladder,

F28.7

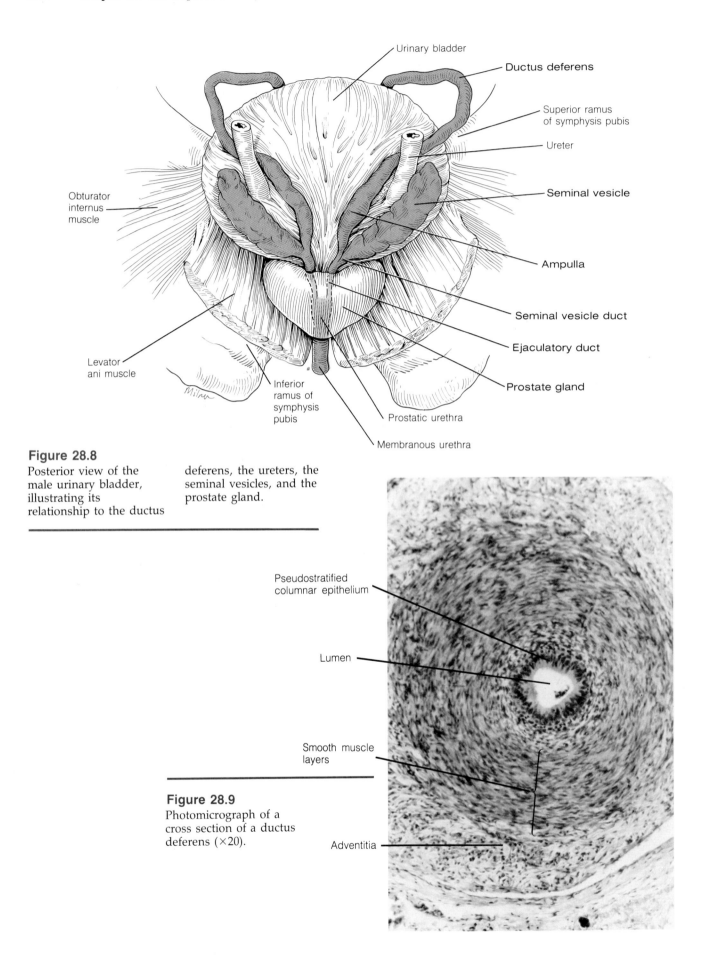

Figure 28.8
Posterior view of the male urinary bladder, illustrating its relationship to the ductus deferens, the ureters, the seminal vesicles, and the prostate gland.

Figure 28.9
Photomicrograph of a cross section of a ductus deferens (×20).

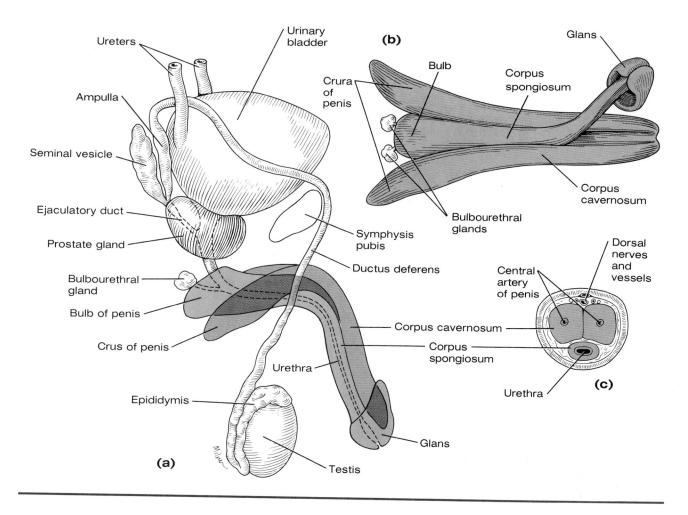

Figure 28.10
(a) Schematic representation of the male reproductive system. (b) Inferior view of a dissected penis, with the distal portion of the corpus spongiosum displaced to one side. (c) Cross section of a penis.

where it enlarges to form an **ampulla** (Figure 28.8). On reaching the inferior surface of the urinary bladder, each ductus deferens is joined by the duct of a seminal vesicle to form a short **ejaculatory duct** that passes through the prostrate gland to join the prostatic portion of the urethra. The ductus deferens, like the epididymis, is lined with pseudostratified columnar epithelium that has stereocilia on its free surface. The muscular coat of the wall of the ductus deferens is quite thick, consisting of three layers of smooth muscle (Figure 28.9). The muscle of the wall of the duct undergoes peristaltic contractions during ejaculation, propelling sperm into the ejaculatory duct.

Seminal Vesicles

The **seminal vesicles** are two membranous pouches located lateral to the ductus deferens on the posterior surface of the urinary bladder (Figures 28.8 and 28.10). The duct of each seminal vesicle joins with a ductus deferens to form an **ejaculatory duct.** The ejaculatory ducts penetrate the prostate gland and enter the urethra just below its point of exit from the urinary bladder. Contraction of the ejaculatory ducts propels spermatozoa from the ductus deferens, along with secretions from the seminal vesicles, into the urethra. The seminal vesicles secrete a viscous fluid that contributes to the formation of semen.

F28.8,
F28.10

Prostate Gland

The **prostate gland** (Figures 28.7 and 28.10) is a single organ that encompasses the urethra just below the bladder. Because it is located directly in front of the rectum, the prostate can be manually examined by means of a rectal examination in which a finger is placed in the anal canal and the gland palpated

F28.7,
F28.10

through the anterior wall of the rectum. The gland is composed of about 30 small tubuloalveolar glands from which a number of excretory ducts open independently into the urethra. It is enclosed in a capsule of fibrous connective tissue and smooth muscle fibers that extend into the gland and divide it into indistinct lobes. The prostate secretes a thin, milky, alkaline fluid that contributes to semen. The prostate tends to enlarge in older men and may cause difficulty in urination by compressing the prostatic portion of the urethra.

Bulbourethral Glands

F28.10

The **bulbourethral** *(Cowper's)* **glands** (Figure 28.10) are a pair of small glands located below the prostate on either side of the membranous portion of the urethra. Their secretion, which contributes to the semen, is carried to the urethra by a duct from each gland.

Penis

F28.10

pree'-puce

The **penis** is the copulatory organ by which spermatozoa are placed in the female reproductive tract. It consists of a shaft covered by loosely attached skin and an expanded tip, the **glans** (Figure 28.10). The skin continues over the glans as the **prepuce (foreskin).** In order to make it easier to keep the glans clean, the prepuce is often surgically removed shortly after birth in a procedure called *circumcision.*

The penis contains three cylindrical bodies, each of which is surrounded by a fibrous capsule. The three cylindrical bodies are held together by a connective tissue sheath that is covered by skin. Each cylindrical body is composed of highly vascular connective tissue called *erectile tissue*, which contains numerous spongelike spaces that fill with blood during sexual arousal, causing the penis to enlarge and become firm. This phenomenon is referred to as an *erection*. The two dorsal cylindrical bodies are called **corpora cavernosa penis.** The single ventral body is the **corpus spongiosum penis.** The urethra passes through the corpus spongiosum. The expanded distal end of the corpus spongiosum forms the glans of the penis. The proximal end of the corpus spongiosum is slightly enlarged, forming the **bulb** of the penis. The bulb is attached to the urogenital diaphragm, which forms the floor of the pelvic cavity. The two corpora cavernosa separate at their proximal ends and form the **crura** of the penis, which anchor it to the rami of the ischial and pubic bones.

Semen

Semen is a mixture of spermatozoa from the testes and fluids from the seminal vesicles, prostate gland, and bulbourethral glands. The secretion of the seminal vesicles, which contains fructose and prostaglandins, contributes about 60% of the bulk of the semen. Fructose is a major energy source for ejaculated sperm, which contain very little cytoplasm and therefore have only limited intracellular glycogen available from which to derive energy. The prostaglandins are thought to facilitate the process of fertilization by reacting with mucus at the cervix of the female uterus to make the mucus more compatible with sperm and by stimulating reverse peristaltic contractions that enhance the movement of sperm along the uterus and uterine tubes.

The alkaline fluid secreted by the prostate gland counteracts the mild acidity of other portions of the semen so that the semen itself is slightly alkaline, with a pH of about 7.5. The alkaline nature of the semen helps counteract the acid secretions of the female vagina (pH 3.5–4.0). This action is believed to be important because sperm do not become optimally active until the pH of their environment rises to about 6.0–6.5.

PUBERTY IN THE MALE

The time of life at which reproduction becomes possible is called **puberty.** Its age of onset varies considerably among individuals, but it most commonly occurs between 10 and 15 years of age. In males, puberty is marked by the first ejaculate of semen that contains mature spermatozoa, as well as by more gradual and subtle changes in personality and body form. The changes that occur in body form are called **secondary sex characteristics.** The penis, scrotum, and testes enlarge, and hair grows on the face, axilla, and pubic area. The voice deepens, and there is a tendency toward increased development of skeletal muscles, accompanied by a decrease in the amount of body fat.

The events that occur at puberty are the result of an increased production of testosterone by the testes, and they require the continued production of adequate levels of testosterone to sustain them. Prior to puberty, the hypothalamus of the brain secretes only small amounts of gonadotropin-releasing hormone (Gn-RH). Consequently, there is little release of the gonadotropins—follicle-stimulating hormone (FSH) and luteinizing hormone (LH)—from the pituitary, and the testes are not stimulated to produce large amounts of testosterone. At puberty, an alteration in brain function occurs that leads to an increased release of Gn-RH, which ultimately results in an increased production of testosterone. Some researchers believe that, prior to puberty, the hypothalamus is extremely sensitive to the negative feedback effects of testosterone. Thus, even though only low levels of testosterone are present, Gn-RH release is suppressed. But at puberty, the hypothalamus decreases in sensitivity to the negative feedback effects of testosterone, and Gn-RH is released in greater quantities. Alternatively, other researchers propose that prior to puberty the hypothalamic region that releases Gn-RH is inhibited by neural signals from the anterior hypothalamus, and Gn-RH release is suppressed. At puberty, this inhibition is relaxed, and Gn-RH is released in greater quantities.

SPERMATOGENESIS

The production of spermatozoa within the testes is called **spermatogenesis.** During spermatogenesis a unique type of cell division called meiosis takes place. (Meiosis is described in Chapter 3, and the specific events of this process should be reviewed at this time.) The purpose of meiosis, which is aptly called a reduction division, is to reduce the number of chromosomes in each gamete by half. The cells of the human body are *diploid;* that is, they contain 23 chromosomes from each parent, for a total of 46. Cells undergoing meiosis lose half of their total number of chromosomes, retaining 23 chromosomes. Such cells are said to be *haploid.* The formation of haploid spermatozoa (as well as haploid ova in the female) ensures that, following fertilization, when a male and a female gamete unite, the resulting cell will be a diploid cell that has 46 chromosomes.

sper-mat-o-jen'-e-sis

Spermatogenesis occurs within the seminiferous tubules of the testes. Located within the outer portion of each tubule are cells called **spermatogonia,** which have the diploid number of chromosomes (44 autosomes plus one X and one Y sex chromosome). The spermatogonia divide mitotically (see Chapter 3), thereby providing a continuous source of new cells that are used for the production of spermatozoa (Figure 28.11). Some spermatogonia move toward the lumen of the tubule and undergo a period of growth. These enlarged cells are called **primary spermatocytes.** Each primary spermatocyte undergoes a first meiotic division, which results in the formation of two **secondary spermatocytes.** One of the secondary spermatocytes contains 22 autosomes plus the X sex chromosome; the other has the remaining 22 autosomes and the Y sex chromosome.

F28.11

The second meiotic division divides each secondary spermatocyte into two **spermatids.** Thus, meiosis results in the formation of four haploid spermatids from one diploid spermatogonium. Each spermatid undergoes a complex series of changes by which it develops into a **spermatozoon.** During these changes, much of the cytoplasm of the cell is discarded and a tail containing a group of contractile proteins forms.

During spermatogenesis, the developing gametes are closely associated with **sustenticular** *(Sertoli)* **cells,** which extend from the outer portion of a

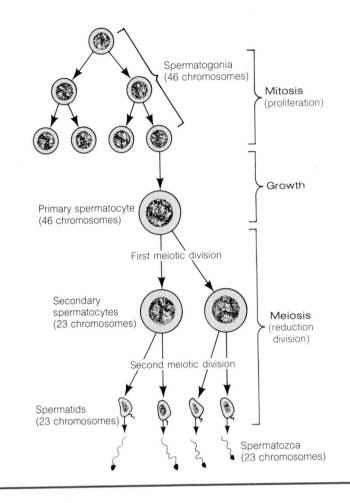

Spermatogonia
(46 chromosomes)

Mitosis
(proliferation)

Growth

Primary spermatocyte
(46 chromosomes)

First meiotic division

Secondary
spermatocytes
(23 chromosomes)

Meiosis
(reduction
division)

Second meiotic division

Spermatids
(23 chromosomes)

Spermatozoa
(23 chromosomes)

Figure 28.11
Diagrammatic
representation of
the process of
spermatogenesis.

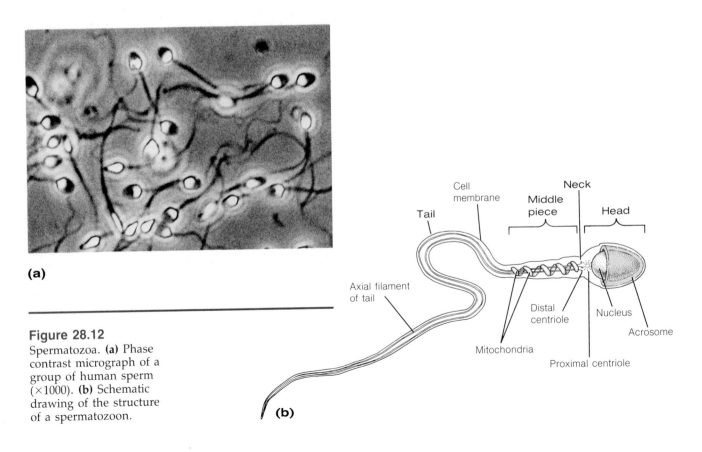

(a)

Figure 28.12
Spermatozoa. **(a)** Phase
contrast micrograph of a
group of human sperm
(×1000). **(b)** Schematic
drawing of the structure
of a spermatozoon.

Tail

Cell
membrane

Middle
piece

Neck

Head

Axial filament
of tail

Mitochondria

Distal
centriole

Proximal centriole

Nucleus

Acrosome

(b)

seminiferous tubule to the lumen (Figure 28.5). The sustenticular cells play an integral role in spermatogenesis, and they are believed to serve as a route by which nutrients and chemical (for example, hormonal) signals reach the developing gametes. In humans, the entire process of spermatogenesis, from spermatogonium to spermatozoa, requires about 72 days and occurs continuously beginning at puberty.

THE SPERMATOZOON

The **spermatozoon** is a flagellated cell that is divisible into **head, neck, middle piece,** and **tail** regions (Figure 28.12). The head consists almost entirely of the condensed nucleus, capped with a fluid-filled vesicle, called the **acrosome,** at its tip. The acrosome develops from the Golgi apparatus of the spermatid. A number of enzymes have been identified in the acrosomal fluid. In order for a sperm to fertilize an ovum, the acrosome must release its enzymes, which break down several layers that surround the ovum. Located in the neck region are two centrioles at right angles to one another. The distal centriole is oriented along the long axis of the sperm, and it extends into the tail, where it forms the flagellum (which gives the sperm motility). The middle piece is composed of mitochondria that meet end to end and spiral around the distal centriole. Most of the small amount of cytoplasm present in a mature sperm is located around the mitochondria of the middle piece. Because there is so little cytoplasm present, spermatozoa cannot survive very long on their own. Instead, they must derive nourishment from the semen in which they are suspended.

HORMONAL CONTROL OF TESTICULAR FUNCTION

The gonadotropins follicle-stimulating hormone (FSH) and luteinizing hormone (LH) are required for normal spermatogenesis and testosterone secretion. Moreover, testosterone is itself necessary for sperm production. FSH stimulates spermatogenesis by acting on the sustenticular cells, which are in close contact with the spermatogenic cells at all stages of development. LH stimulates testosterone secretion by acting on the interstitial endocrinocytes. In addition, since testosterone is required for sperm production, LH can be regarded as important in sperm production as well.

The release of FSH and LH from the pituitary is stimulated by Gn-RH from the hypothalamus. Testosterone from the testes inhibits LH release by way of negative feedback, but it has little effect on FSH release. The inhibitory influence of testosterone on LH release is believed to be exerted mainly at the level of the pituitary rather than at the level of the hypothalamus. The release of FSH appears to be inhibited by a glycoprotein hormone called **inhibin,** which is thought to be released by the sustenticular cells of the seminiferous tubules. Although not all interactions are fully established, the general relationship between Gn-RH, FSH, LH, testosterone, and inhibin is depicted in Figure 28.13.

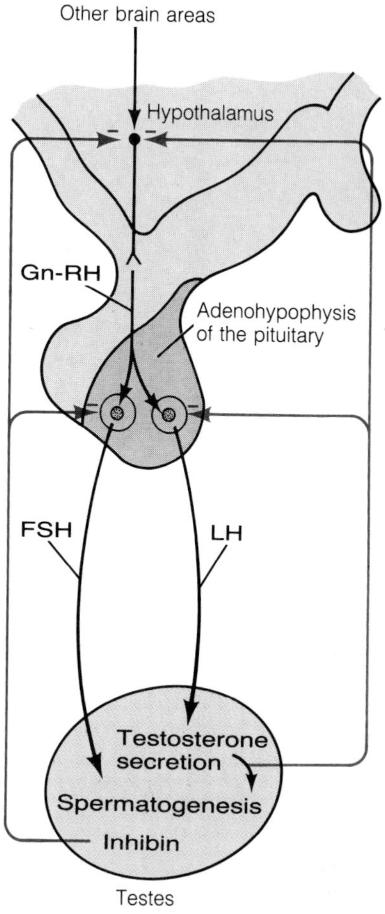

MALE SEXUAL RESPONSES

Erection

Sexual intercourse (coitus) is the usual method by which male sperm are deposited within the vagina of the female. In order to accomplish sexual intercourse, the male penis must become enlarged and firm, a process referred to as **erection.**

Erection is a vascular phenomenon in which the spaces within the corpora cavernosa and corpus spongiosum of the penis become engorged with

Figure 28.13
Relationship between Gn-RH, FSH, LH, testosterone, and inhibin in the male. The negative-feedback influences of testosterone (and presumably inhibin) are believed to be exerted mainly at the level of the pituitary rather than at the level of the hypothalamus. (With regard to the relationships depicted in this figure, it should be noted that some researchers believe there are separate releasing substances for FSH and LH. However, present evidence favors the existence of a single gonadotropin-releasing hormone [Gn-RH] that stimulates the release of both FSH and LH.)

blood. In the absence of sexual arousal, the arterioles that supply the corpora of the penis are partially constricted, which allows only limited blood flow into the penis. Under these conditions the penis is flaccid. During sexual excitation, parasympathetic nerve impulses travel from the sacral region of the spinal cord to the arterioles of the penis, causing them to dilate. The dilation of the arterioles allows more blood to enter the vascular spaces of the erectile tissues. As the erectile tissues of the penis expand due to the increased blood volume, the veins that drain the penis are compressed against the coverings of the penis. The compression of the veins reduces the flow of blood out of the penis, which futher increases the engorgement of the erectile tissues. As a result, the penis becomes enlarged and firm.

An erection can result from visual or psychic stimuli, including thoughts or emotions that originate in the brain, or from the physical stimulation of touch receptors on the penis and various other regions of the body. Moreover, psychic factors can greatly influence the effectiveness of physical stimuli. Indeed under adverse psychological conditions, the physical stimulation of the penis frequently will not cause an erection.

Emission and Ejaculation

In addition to causing the penis to become erect, continued tactile stimulation of the penis ultimately results in a forceful expulsion of semen. When sexual stimulation becomes sufficiently intense, sympathetic nerve impulses from the lumbar region of the spinal cord cause rhythmic contractions of the epididymis, the ductus deferens, the seminal vesicles, and the prostate gland. These contractions propel the contents of the ducts and glands into the urethra, where they form semen. Up to this point, the process is referred to as **emission.** Following emission, skeletal muscles at the base of the penis undergo rhythmic contractions that expel, or **ejaculate,** the semen from the urethra. Under normal circumstances a volume of about 2 ml of semen, which contains approximately 300 million sperm, is expelled during an ejaculation. The rhythmic contractions that produce an ejaculation are accompanied by intensely pleasurable sensations and widespread muscular contractions. This reaction is called an **orgasm.**

Following ejaculation, the arterioles that supply the penis return to their usual partially constricted condition, which decreases the blood flow into the penis. This decrease in blood flow relieves the vascular congestion within the corpora of the penis and allows the blood to return freely through the veins to the general circulation. As the accumulated blood leaves the corpora, the penis returns to a flaccid state.

ANATOMY OF THE FEMALE REPRODUCTIVE SYSTEM

The female reproductive system includes the *ovaries,* which produce ova; the *uterine tubes,* which transport, protect, and nourish the ova; the *uterus,* which provides a suitable environment for the development of an embryo, and the *vagina,* which serves as a receptacle for sperm.

Ovaries

The female gonads, or primary sex organs, are the **ovaries,** in which the female gametes **(ova)** are produced. Moreover, the *estrogens* and *progesterone*—hormones that influence the development of the female accessory sex organs and secondary sex characteristics—are secreted by the ovaries.

Each of the paired ovaries is oval and slightly smaller than a testis. Like all the organs of both the female and male reproductive systems, the ovaries are retroperitoneal. The ovaries do not migrate as extensively during embryonic development as do the testes. Rather, they descend only as far as the pelvis and lie against the lateral wall on either side of the uterus (Figures 28.14 and 28.15). The ovaries are held in this position by several ligaments. The largest

F28.14,
F28.15

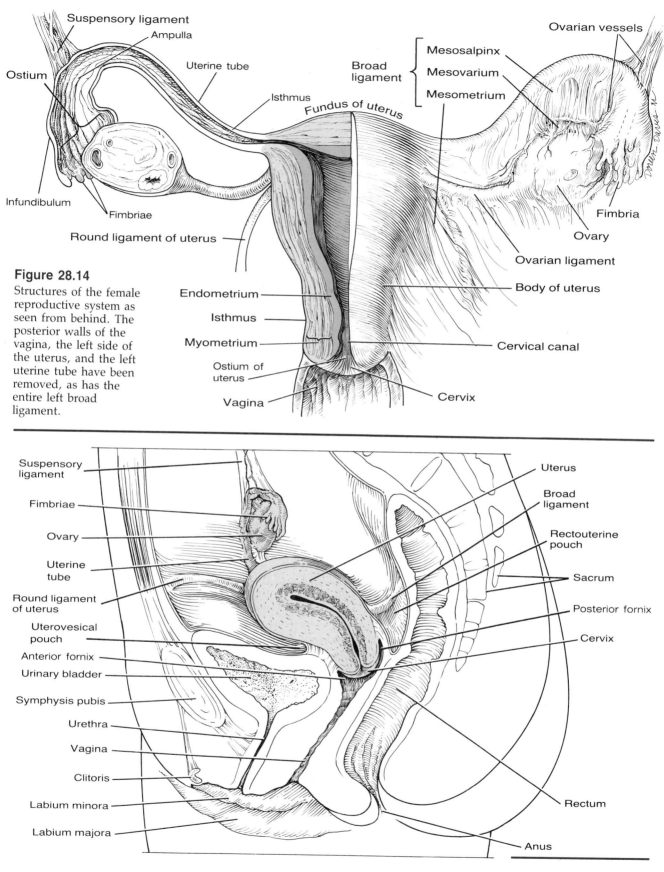

Figure 28.14

Structures of the female reproductive system as seen from behind. The posterior walls of the vagina, the left side of the uterus, and the left uterine tube have been removed, as has the entire left broad ligament.

Figure 28.15

Median section of the female pelvis.

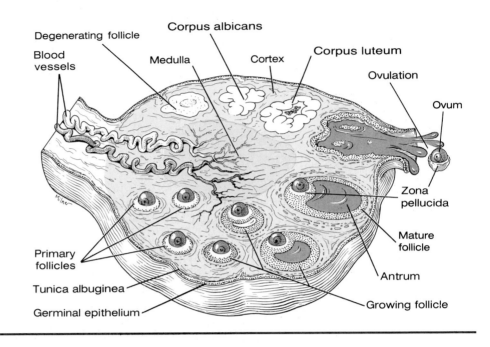

Figure 28.16

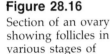

Section of an ovary showing follicles in various stages of development.

is the **broad ligament,** which also supports the uterine tubes, the uterus, and the vagina. The ovaries are suspended from the posterior surface of the broad ligament by a short fold of peritoneum called the **mesovarium.** A fibrous band called the **ovarian ligament** lies within the broad ligament and extends from the superior-lateral margins of the uterus to the ovary. The lateral margins of the broad ligament form a fold that attaches the ovary to the pelvic wall. This fold, through which the ovarian vessels reach the ovary, is known as the **suspensory ligament.**

F28.16   The peritoneum that covers the surface of the ovary is composed of relatively short, simple cuboidal cells rather than squamous cells as is typical of the rest of the peritoneum. This outer epithelial covering of the ovary is called the **germinal epithelium** (Figure 28.16). In spite of what the name implies, the germinal epithelium does not give rise to germ cells. Beneath the germinal epithelium is a fibrous connective tissue layer, the **tunica albuginea.** The ovary itself can be divided into an outer **cortex** surrounding a central **medulla.** The medulla is composed of connective tissue and contains blood vessels, lymphatic vessels, and nerves that enter or leave the ovary where it attaches to the mesovarium. At birth, the cortices of the two ovaries contain approximately 400,000 immature structures called **primary follicles.** Each primary follicle consists of an undeveloped ovum located within a small sphere that is composed of a single layer of cells, called *follicle cells.* Beginning at the time of puberty and continuing throughout the 30–40 years during which a female is capable of reproduction, one follicle usually matures fully and ruptures each month (approximately every 28 days), releasing an ovum that is then available for fertilization. This being the case, only about 400 of the total number of follicles reach full maturity during a woman's lifetime. The rest reach various stages of development and then regress and degenerate.

Uterine Tubes

F28.14   An ovum released at ovulation is carried to the uterus by a **uterine** *(fallopian)* **tube** (Figure 28.14). Each uterine tube, which extends from the vicinity of the ovary to the superior lateral angle of the uterus, lies between the layers of the broad ligament. The portion of the broad ligament that anchors each uterine tube is called the **mesosalpinx.** The medial constricted portion of the uterine tube is the **isthmus.** It opens into the uterus. The uterine tube is expanded somewhat in the region where it curves around the ovary. This region is called the **ampulla.** Fertilization generally occurs within the ampulla. The distal end of each uterine tube is called the **infundibulum.** The infundibulum opens into the abdominopelvic cavity very close to an ovary. This opening, called the **ostium,** is surrounded by small, ciliated, fingerlike projections called **fimbriae.** One of the fimbriae is generally attached to the ovary. The movements of the fimbriae and their cilia are thought to cause currents of peritoneal fluid to enter the uterine tube and thus carry the ovum that has been released from its follicle into the tube. Since the reproductive tract of the female opens into the peritoneal cavity through the two uterine tubes (one tube to each ovary) and to the exterior of the body through the uterus and vagina, infections of the lower reproductive tract can spread into the body cavity and cause peritonitis. (In males, there is no opening of the reproductive tract into the body cavity.)

The wall of the uterine tube is covered by the peritoneum. The *muscular layer* of the uterine tube, which is deep to the peritoneum, contains both circular and longitudinal layers of smooth muscles. The inner *mucosa* is thick and its epithelium is composed of simple columnar cells. The mucosa forms numerous branched folds that extend into the lumen of the uterine tube (Figure 28.17). Two types of cells are present in the epithelium of the mucosa: those with cilia and those without. Most of the cells have cilia that beat rhythmically toward the uterus. Thus, once within the uterine tube, the ovum is carried to the uterus by a weak fluid current caused by the beating of the cilia, possibly aided by peristaltic contractions of the smooth muscles in the walls of the uterine tube. Interspersed among the ciliated cells are cells that lack cilia. These cells are thought to be secretory cells that maintain a moist environment in the uterine tube and may serve as a source of nourishment for the ovum.

Uterus

The **uterus** is a single, hollow, pear-shaped organ that receives the uterine tubes at its upper lateral angles and is continuous below with the vagina (Figure 28.14). The upper portion of the uterus is called the **body.** Below its body, the uterus narrows into the **isthmus,** and where it joins with the vagina it forms the cylindrical **cervix.** The opening of the uterus into the vagina is the **ostium** of the uterus. The dome-shaped region of the uterine body above and between the points of entrance of the uterine tubes is the **fundus.**

The uterus lies in the pelvis, behind the urinary bladder and in front of the sigmoid colon and rectum. The peritoneum that covers the superior and posterior surfaces of the urinary bladder is folded back up the anterior surface of the uterus from the floor of the pelvic cavity. The space thus formed is called the **uterovesical pouch** (Figure 28.15). In a similar manner, the **recto-uterine pouch** is formed where the peritoneum that continues over the top of the uterus and down its posterior surface is folded back up the anterior surface of the rectum from the pelvic cavity floor.

The layers of peritoneum that cover the uterus anteriorly and posteriorly fuse along its lateral margins and extend to the lateral walls and floor of the pelvic cavity as the **broad ligament** (Figure 28.14). The portion of the broad ligament that is below the mesovarium and that anchors the uterus is called the **mesometrium.** Blood vessels and nerves reach the uterus and uterine tubes by traveling between the peritoneal layers of the broad ligament. The *uterine arteries*, which are branches of the internal iliac arteries, travel within the broad ligament to reach the cervical portion of the uterus. They then follow its lateral margin up to the isthmus of the uterine tube, where they give off tubal and ovarian branches.

Assisting the broad ligament in holding the uterus in place are the **round ligaments.** The round ligaments are fibrous bands that run within the broad ligaments from the lateral margins of the uterus, near the junctions with the uterine tubes, through the inguinal canals and into the labia majora. The ovarian and round ligaments form a continuum that is *homologous* to—that is, formed from the same embryonic tissue as—the gubernaculum of the male. Recall that the labia majora and the scrotum are also homologous structures.

The ligaments do not hold the uterus tightly in place; rather, they allow it to have limited movement. Normally, the uterus is tipped forward, its longitudinal axis forming an angle of about 90° with the longitudinal axis of the

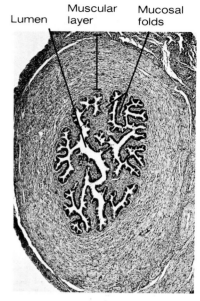

Lumen   Muscular layer   Mucosal folds

Figure 28.17
Photomicrograph of a uterine tube (×30).

F28.14

F28.15

F28.14

vagina. This angle varies, however, depending on how full the urinary bladder or the rectum is. The ligaments do not in fact provide much support for the uterus. Its chief support is provided from below by the muscles and membranes that form the floor of the pelvic cavity—the region known as the **pelvic diaphragm** and the **urogenital diaphragm.** These muscles all attach to a small, circular tendon located just posterior to the vaginal opening. This tendon is called the *central tendon* of the perineum. If the central tendon is weakened, which sometimes occurs as a result of childbirth, the uterus can move downward into the vagina. In this condition, the uterus is said to be *prolapsed.*

F28.14

The wall of the uterus is covered by a **serosa** formed by the peritoneum that is folded back over the uterus from the broad ligament. Beneath the serosa is a thick, muscular layer called the **myometrium** (Figure 28.14). The myometrium, which makes up most of the thickness of the wall of the uterus, is composed of bundles of smooth muscles running in various directions. The cavity in the uterus is lined with an epithelium of ciliated columnar cells. This surface epithelium plus an underlying connective tissue lamina propria called the *endometrial stroma* constitute the **endometrium.** The surface epithelium invaginates to form numerous tubular *uterine glands,* which extend down into the stroma. The endometrium consists of a thick superficial layer called the **functional layer** and a thin, deep, **basal layer.** The functional layer undergoes marked developmental changes during the menstrual cycle. Each month it thickens and becomes engorged with blood in preparation for receiving a fertilized ovum. If the ovum is not fertilized, the functional layer of the endometrium is sloughed off as the menstrual flow.

Vagina

F28.15

The **vagina** is the canal that leads from the cervix of the uterus to the exterior of the body (Figure 28.15). The smooth muscle tunic in the wall of the vagina is much thinner than the muscular tunic of the uterus. The mucosa lining the vagina has a protective surface layer of stratified squamous epithelium, as is typical of all the canals that open onto the body surface. The vaginal mucosa proliferates during the menstrual cycle in a manner similar to the endometrial changes of the uterus, but there is considerably less development in the mucosa of the vagina than in the mucosa of the uterus. The mucosa of the vagina contains numerous transverse ridges, or rugae. Near the entrance to the vagina the mucosa usually forms a vascular fold called the **hymen.** Generally, the hymen only partially blocks the vaginal opening, but in some cases it completely closes the orifice. The hymen, which is stretched and often torn during sexual intercourse, may also be torn by other means. Since the hymen may persist after sexual intercourse, its presence or absence is not a reliable method by which to prove or disprove virginity.

F28.15

The lumen of the vagina is generally quite small, and the walls that surround it are usually in contact with each other. The canal is capable of considerable stretching, however, as when the vagina receives the male penis during sexual intercourse or when it serves as the birth canal. The upper end of the vaginal canal surrounds the cervix of the uterus, forming a recess (the **vaginal fornix**) around the cervix (Figure 28.15). The largest portion of the recess lies dorsal to the cervix and is called the **posterior fornix.** The **anterior fornix** and the two **lateral fornices** are not as deep.

Female External Genital Organs

F28.18

When considered collectively, the external genital organs of the female are known as the **vulva,** or **pudendum** (Figure 28.18).

Under the influence of estrogens, there is a tendency in the female for fatty tissue to be deposited over the pubic symphysis. This deposition produces a mound called the **mons pubis.** (Because of the lack of estrogens, this fat deposit is lacking in males.) The skin over the mons becomes covered with hair after puberty.

Two rounded folds—the **labia majora**—extend backward from the mons. The outer surfaces of the labia majora are pigmented and covered with hair.

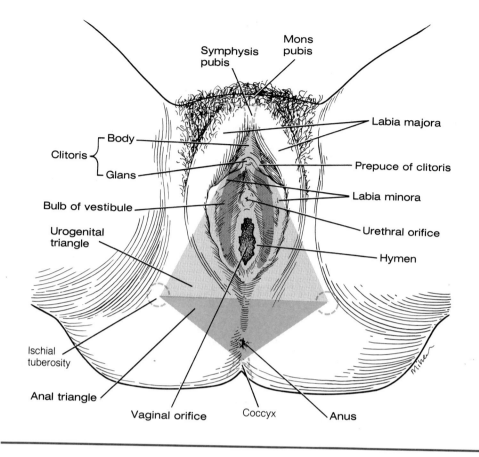

Figure 28.18
External genital organs of the female, showing divisions of the perineum.

Their inner surfaces are smooth, lack hair, and are moist because of the presence of large sebaceous glands.

The **labia minora** are two smaller folds located medial to the labia majora. Anteriorly they surround the clitoris. The labia minora are highly vascular and lack hair. They surround a space (the **vestibule**) into which the vagina and urethra open. The vaginal opening into the vestibule may be partially blocked by the hymen. A number of glands open into the vestibule and keep its walls moist. The ducts of the **greater vestibular glands,** which are homologous to the bulbourethral glands of the male, open into the vestibule on either side of the vagina. On each side of the external urethral orifice are the openings for the **lesser vestibular glands.** The vestibular glands lubricate the vestibule and thus facilitate sexual intercourse.

The **clitoris** is a small elongated structure located at the anterior junction of the labia minora. Most of the body of the clitoris is enclosed by a **prepuce** (foreskin) formed by the labia minora. The free exposed portion of the clitoris is called the **glans.** The clitoris is homologous to the dorsal portion of the male penis and, like the penis, is composed of erectile tissue. The clitoris contains two **corpora cavernosa clitoris,** but it does not contain a corpus spongiosum (which, in the male, surrounds the urethra). The female's urethra is not within the clitoris. Rather, it opens separately, posterior to the clitoris. The crura of the corpora cavernosa attach the clitoris to the rami of the pubic and ischial bones. The clitoris, which is very sensitive to touch, becomes engorged with blood when stimulated, contributing to the sexual arousal of the woman.

Just deep to the labia are two elongated masses of erectile tissue called the **bulb of the vestibule.** The bulb, which is homologous to the corpus spongiosum penis and bulb of the penis, passes on either side of the vaginal orifice. During sexual arousal, the bulb becomes engorged with blood, thus narrowing the opening into the vagina and squeezing against the male penis during sexual intercourse.

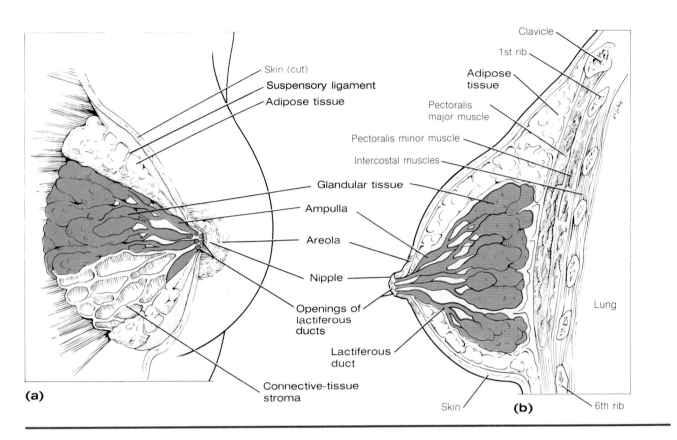

(a)

Skin (cut)
Suspensory ligament
Adipose tissue
Glandular tissue
Ampulla
Areola
Nipple
Openings of
lactiferous
ducts
Connective-tissue
stroma

Clavicle
1st rib
Adipose
tissue
Pectoralis
major muscle
Pectoralis minor muscle
Intercostal muscles
Lung
Lactiferous
duct
Skin
(b)
6th rib

Figure 28.19

The mammary gland.
(a) Anterior view of a
partially dissected left
breast. **(b)** Sagittal
section.

F28.18

F28.19

The Female Perineum

As in the male, the female perineum can be divided by a transverse line
between the ischial tuberosities into an anterior **urogenital triangle,** which
contains the external genitalia, and a posterior **anal triangle,** which contains
the anus (Figure 28.18). The region between the vagina and the anus is referred to as the *clinical perineum* because this area is sometimes torn by the
stretching that occurs during childbirth. Sometimes the tears even damage
the anal sphincter. To prevent tearing, an incision called an *episiotomy* is often
made in the perineum during delivery. This operation enlarges the vestibule
and makes delivery easier. Moreover, a surgical incision is easier to repair
than a tear.

Mammary Glands

Even though **mammary glands (breasts)** are also present in males, we discuss
their anatomy under the female reproductive system because it is only in the
female that the breasts are at all involved in reproduction. Even in the female
they are not involved in the actual reproductive process, however, but serve
to produce milk for the nourishment of the infant.

Each mammary gland is a skin-covered hemispherical elevation located
superficial to the pectoral muscles (Figure 28.19). Just below the center of each
mammary gland is a protruding **nipple** that is surrounded by a circular **areola.** The areola has many small elevations due to the presence of numerous
large sebaceous glands called **areolar glands.** The areolar glands produce a
waxy secretion that prevents chafing of the nipple during nursing. Both the
nipple and the areola are pigmented and have capillary beds located close to
their surfaces. The pigmentation becomes darker during pregnancy and diminishes somewhat afterward. Smooth muscles in the areola and nipple
cause the nipple to become erect as a consequence of stimulation.

Internally, the periphery of each mammary gland is composed of adipose
tissue held in place by a connective tissue stroma. Bands of connective tissue
extend from the anterior aspect of the stroma and attach into the dermis.

These connective tissue septa are called the **suspensory ligaments** of the breast. The septa subdivide the fat that lies superficial to the glandular tissue, and they give a smooth contour to the breast. Centrally, there are about 15–25 lobes of glandular tissue arranged radially around the nipple, each consisting of a separate compound tubuloalveolar gland. Each lobe is drained by a single **lactiferous duct,** which opens onto the nipple. The nipple, therefore, is perforated by numerous openings. Just before reaching the nipple, each lactiferous duct expands into a small milk reservoir called an **ampulla.** Glandular tissue also extends beyond the breast, into the axilla. Milk is released from the glands by a modified form of apocrine secretion.

PUBERTY IN THE FEMALE

On the average, females reach puberty a year earlier than males. In females, puberty is marked by the first episode of menstrual bleeding, which is called the **menarche.** At puberty, the vagina, uterus, and uterine tubes enlarge, and the deposition of fat in the breasts and hips causes them to become more prominent. In addition, the glandular portion of the breasts begins to develop. Pubic hair grows, and although the voice matures in the female, the change is not as dramatic as the voice change in males.

The events associated with puberty are largely the result of an increased production of estrogens by the ovaries, and they require the continued production of adequate levels of estrogens to sustain them. Prior to puberty, estrogens are secreted at very low levels, probably due mainly to the secretion of only small amounts of Gn-RH by the hypothalamus. At puberty, an alteration in brain function leads to increased Gn-RH release. The increased Gn-RH release stimulates the secretion of FSH and LH, which ultimately leads to an increased production of estrogens.

OOGENESIS

The production of ova within the ovaries is called **oogenesis,** and, as in the case of spermatogenesis, meiosis occurs during this process (Figure 28.20). Initially, precursor cells called **oogonia** divide mitotically and thereby provide the cells that will develop into ova. Each oogonium is a diploid cell containing 44 autosomes and two X sex chromosomes. By the time embryonic development is complete, several hundred thousand oogonia have undergone a growth phase and have become **primary oocytes.** The primary oocytes enter prophase of the first meiotic division, but do not complete it and are in first prophase at the time of birth. Each primary oocyte is located within a small sphere composed of a single layer of cells, thus forming a **primary follicle.**

It is not until puberty, when the process of follicle growth and maturation begins to occur, that a primary oocyte develops further. As a follicle matures, the primary oocyte within it completes the first meiotic division, producing two haploid cells that are of unequal size. The larger cell, called a **secondary oocyte,** receives half of the chromosomes and almost all of the cytoplasm. The smaller cell, called the **first polar body,** receives the other half of the chromosomes but very little of the cytoplasm. Thus, the secondary oocyte and the first polar body each contain 23 autosomes and one X sex chromosome. The secondary oocyte immediately begins the second meiotic division; however, it halts at metaphase and is released from the ovary in this stage when the follicle ruptures at ovulation.

The second meiotic division is not completed by the secondary oocyte unless it is fertilized. However, if fertilization does occur, meiosis is quickly completed. As was the case in the first meiotic division, the completion of the second meiotic division produces one large cell, the fully developed **ovum,** and one small cell, a **second polar body.** The first polar body also often undergoes the second meiotic division, producing two additional polar bodies. Therefore, during oogenesis, one ovum and as many as three polar bodies are

oh-oh-jen'-e-sis
F28.20

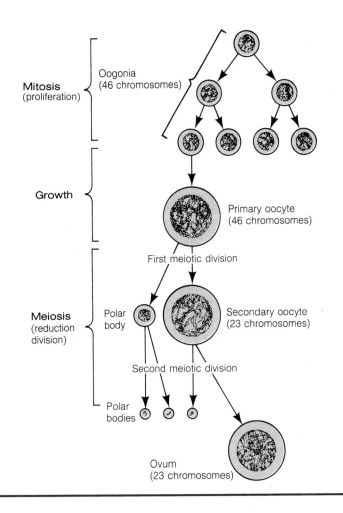

Figure 28.20
Diagrammatic representation of the process of oogenesis.

produced from an oogonium. Beginning at puberty, one primary follicle generally matures fully, and consequently one secondary oocyte (commonly referred to as an ovum) is released at ovulation each month (approximately every 28 days).

THE OVARIAN CYCLE

The ovarian cycle consists of a series of events that occur within an ovary, including the development of follicles, the release of an ovum from a mature follicle at ovulation, and the formation of a structure called the corpus luteum. The length of the ovarian cycle varies from about 20–40 days, averaging about 28 days. Therefore, the ovarian cycle is commonly considered to be a 28-day cycle, although in only about 30% of women is the cycle 27 or 28 days in duration. The ovarian cycle is closely correlated with another cycle, called the menstrual cycle, which involves a series of changes that occur within the lining of the uterus (and to a lesser degree within the vagina).

Development of Follicles

F28.16   Under appropriate hormonal stimulation, some primary follicles undergo futher development (Figure 28.16). The single layer of follicle cells that forms the wall of each primary follicle proliferates, creating a stratified layer of cells, now called *granular cells*. These follicles are then referred to as **growing follicles,** or **secondary follicles.** Some cells in the connective tissue outside the follicles condense into a layer around the follicle. The layer is called the *theca*. Theca cells join with the granular cells to become part of the follicle.

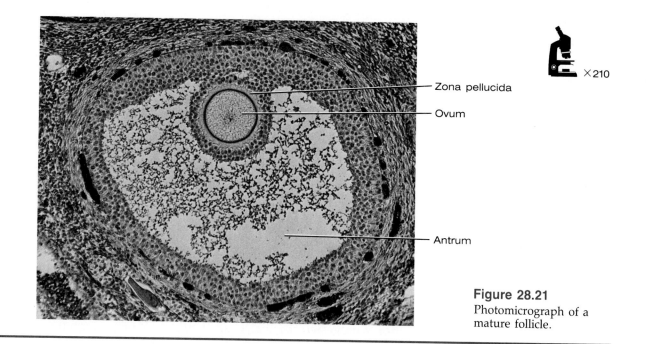

Zona pellucida

Ovum

Antrum

×210

Figure 28.21
Photomicrograph of a mature follicle.

With continued growth, a clear, noncellular region called the **zona pellucida** forms within the growing follicle. The zona pellucida separates the ovum developing within the follicle from the surrounding granular cells. As the solid, growing follicle becomes larger, a fluid-filled cavity called the **antrum** forms within it. The antrum displaces to one side of the follicle the developing ovum and a few layers of granular cells that surround it. As more fluid accumulates within the antrum, the follicle continues to enlarge and moves toward the surface of the ovary, where it produces a bulge. The follicle is then referred to as a **mature (Graafian) follicle** (Figure 28.21). Several follicles begin this series of changes each month, but generally only one reaches the mature phase. The rest degenerate. The process of degeneration of a follicle—and the ovum within it—is called *atresia*, and follicles undergoing atresia are referred to as *atretic follicles*.

F28.21

Ovulation

Under the proper hormonal conditions, the mature follicle ruptures and releases its ovum into the abdominopelvic cavity. This event is called **ovulation.** During ovulation, those follicle cells that directly surround the ovum remain with it. Therefore, the ovulated ovum is surrounded by a zona pellucida and a sphere of follicle cells that is now called the **corona radiata.** The ovum, with its surrounding zona pellucida and corona radiata, is swept into the uterine tube by the flow of liquid from the abdominopelvic cavity into the tube. This flow is produced by the beating of the cilia attached to the fimbriae that surround the opening of the uterine tube.

It generally takes 10–14 days for a primary follicle to develop into a mature follicle. During this time, the developing ovum completes the first meiotic division and reaches metaphase of the second meiotic division, as described previously. The ovum is in this stage when it is released from the ovary at ovulation.

Formation of the Corpus Luteum

Following ovulation, the ruptured mature follicle collapses. Shortly thereafter, the cells of the follicle increase in size and take on a yellowish color, due in part to the accumulation of lipid droplets. The resulting structure is called a **corpus luteum** ("yellow body"). The future of the corpus luteum depends on the fate of the ovum. If the ovum is not fertilized and pregnancy does not occur, the corpus luteum reaches its maximum development about eight to ten days after ovulation and then begins to degenerate. Eventually, all that remains is a white connective tissue scar called the *corpus albicans* ("white body"). However, if the ovum is fertilized and pregnancy does occur, the corpus luteum continues to develop, and it remains until near the end of the pregnancy.

HORMONAL CONTROL OF OVARIAN FUNCTION

Gonadotropin-releasing hormone (Gn-RH) from the hypothalamus, and follicle-stimulating hormone (FSH) and luteinizing hormone (LH) from the pituitary are involved in the control of ovarian function. The release of Gn-RH is influenced by the female sex hormones (and probably by inputs from other brain areas). Gn-RH, in turn, stimulates the release of both FSH and LH.

The female sex hormones—the estrogens and progesterone—are produced by the ovaries. Although estrogens are secreted to some extent by various types of ovarian cells, the primary sources of estrogens are follicle cells and cells of the corpus luteum. The corpus luteum is also the primary source of progesterone, and minute amounts of progesterone are secreted by follicle cells.

F28.22   The plasma levels of FSH, LH, estrogens, and progesterone during a typical ovarian cycle are illustrated in Figure 28.22. Follicle growth and maturation requires FSH, LH, and estrogens, and appropriate levels of these hormones initiate the ovarian cycle and follicle development. During the first portion of the cycle, the level of estrogens is relatively low. However, as follicle development proceeds, the secretion of estrogens progressively increases (recall that follicle cells are a primary source of estrogens). At relatively low levels, such as are present during the first portion of the ovarian cycle, estrogens are believed to limit the release of FSH and LH by acting in a negative-feedback fashion on the hypothalamus to inhibit the release of Gn-RH and perhaps also by exerting negative-feedback influences at the level of the pituitary. In addition, the hormone inhibin, which inhibits the release of FSH, has been isolated from the fluid of ovarian follicles, and researchers propose that it contributes to the decline in level of FSH that occurs during the second week of the cycle.

As follicle development continues and the level of estrogens increases rapidly, the effect of estrogens on the hypothalamus and pituitary changes. Relatively high levels—or perhaps rapidly rising levels—of estrogens, such as are present near the end of the second week of the cycle, stimulate the release of Gn-RH from the hypothalamus and also enhance the sensitivity of the LH-releasing mechanisms of the pituitary to Gn-RH. As a consequence, there is a sharp increase in LH secretion (and a lesser increase in FSH secretion). The high level of LH leads to the final maturation of the follicle, the rupture of the follicle, and ovulation.

Near the time of ovulation, the level of estrogens declines, and some follicle cells secrete small but progressively increasing amounts of progesterone. These events appear to be due to changes in the follicle that occur before ovulation, which are caused, at least in part, by the sharp increase in LH secretion. In fact, some researchers propose that the rise in LH stimulates progesterone secretion, and that the progesterone, in turn, initiates a series of events that leads to the rupture of the follicle and ovulation. For example, progesterone is believed to stimulate the production of protein-digesting enzymes that weaken the follicle wall.

Following ovulation, the corpus luteum forms and begins to secrete substantial quantities of estrogens and progesterone. When present together (or perhaps because of the dominant inhibitory effects of progesterone), these hormones act on the hypothalamus in a negative-feedback fashion to inhibit the release of Gn-RH and perhaps also on the pituitary to inhibit the release of

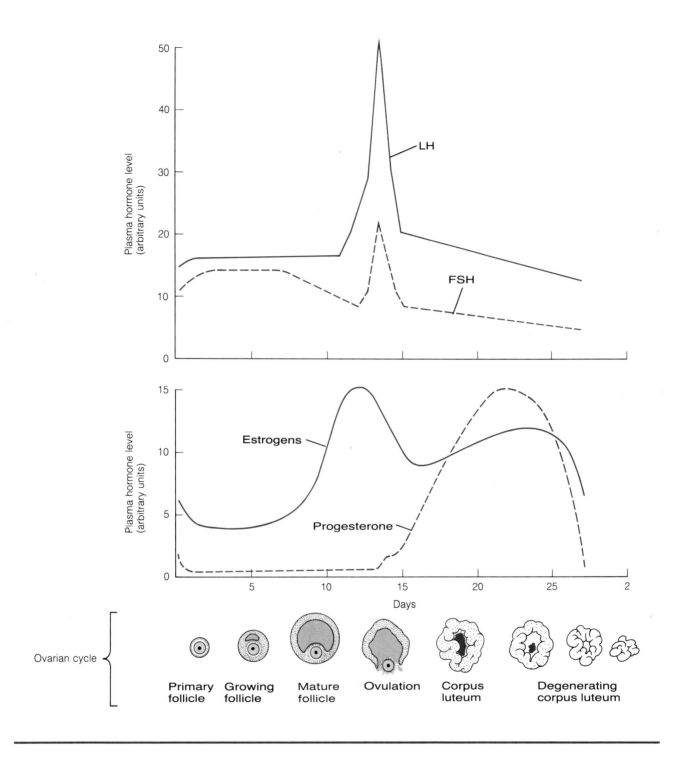

Figure 28.22
The plasma levels of FSH, LH, estrogens, and progesterone during a typical ovarian cycle.

FSH and LH. As a result, the levels of FSH and LH decline during the latter portion of the ovarian cycle. If the ovum is not fertilized and pregnancy does not occur, the corpus luteum reaches its maximum development about eight to ten days after ovulation and then begins to degenerate. As it degenerates, the levels of estrogens and progesterone fall, removing the inhibition of FSH and LH secretion. As a consequence, the levels of FSH and LH increase, other follicles are stimulated to develop, and another ovarian cycle begins.

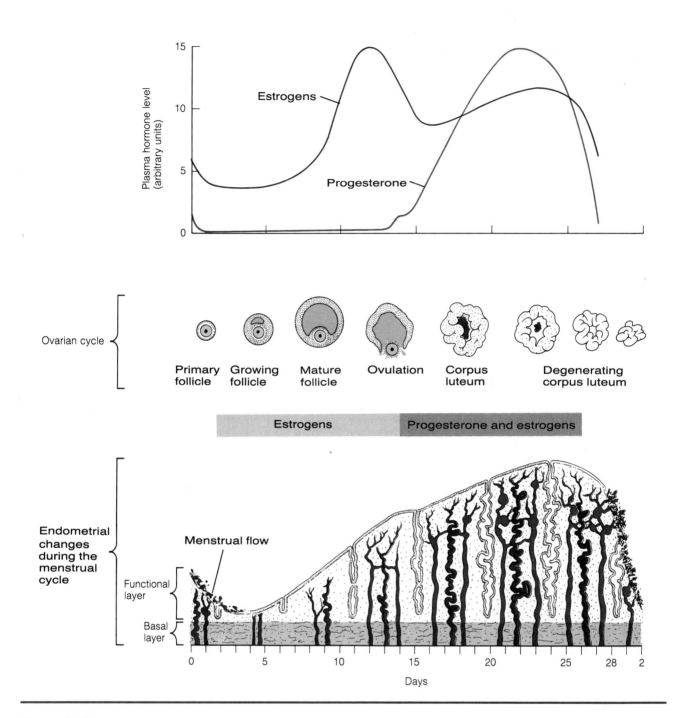

Figure 28.23
The menstrual cycle, illustrating cyclic changes in the endometrium of the uterus.

F28.23

THE MENSTRUAL CYCLE

The menstrual cycle consists of a series of changes in the endometrium of the uterus (and to a lesser degree within the vagina). It is closely associated with the ovarian cycle, and the estrogens and progesterone produced during the ovarian cycle control the events of the menstrual cycle (Figure 28.23).

During the first portion of the ovarian cycle, the estrogens produced stimulate the functional layer of the endometrium of the uterus, causing it to proliferate. Straight tubular glands form, and blood vessels invade the new endometrial epithelium. Thus, the thickness and vascularity of the functional layer of the endometrium increase.

FRONTIERS IN HEALTH

Premenstrual Syndrome: Everyone Has a Cure, But Is There One?

After years of telling women that their premenstrual irritability, depression, pain, and tension—symptoms of premenstrual syndrome—were "all psychological," medical scientists have changed their minds. Premenstrual syndrome (PMS), physicians now agree, is a legitimate medical problem.

No one knows quite how to classify this condition, which strikes four of every ten women of reproductive age. Complaints of those who suffer from PMS cover a wide range of physical and psychological symptoms. According to a recent report in the Harvard Medical School Health Letter, "PMS is easier to define by the timing of the symptoms than by the symptoms themselves." Almost without exception, the symptoms emerge just before menstruation begins. Luckily, most women suffer only a few of the discomforts, the most common being mood swings, nervous tension, and uncontrollable crying spells. For others there are fatigue, depression, headaches, bloating, swelling and tenderness of the breasts, and joint pain.

As difficult as it is to classify this disease, it is even more difficult to determine the cause of the complaints. Moreover, medical science is at a loss to describe how the psychological and physical symptoms are related to one another.

The cause of PMS is now under investigation at the National Institutes of Mental Health. Dr. David Rubinow, a psychiatrist who studies the effects of hormones on the brain, has begun a long-term study of the roles hormones and neurotransmitters may play in bringing about the many symptoms of PMS. Dr. Rubinow is studying the effects of estrogen and progesterone (ovarian hormones), luteinizing hormone and follicle-stimulating hormone (pituitary hormones that regulate the development of ovarian follicles and orchestrate the menstrual cycle), norepinephrine (an adrenal hormone), beta endorphin (the brain's own natural pain killer), and aldosterone (an adrenal hormone that regulates salt and water balance). Because PMS is such a complex condition, it will probably be a long time before medical scientists can truly pinpoint its cause or causes.

Ignorance of the cause of PMS has not deterred the health care industry from providing remedies. Dozens of "proven cures" now exist, ranging from massive doses of progesterone to vitamin B_6. Clinics specializing in PMS have opened, offering cures to suffering women. Because they are desperate, women will often pay large sums of money for a promised cure.

Let the buyer beware, though. Despite claims of success, there is very little scientific evidence to indicate that any of the "cures" really work. Consider the home remedies touted by popular magazines and health-food advocates. Exercise, stress management, and reductions in caffeine and sugar are among the most popular, but they have never been tested in the systematic way that is

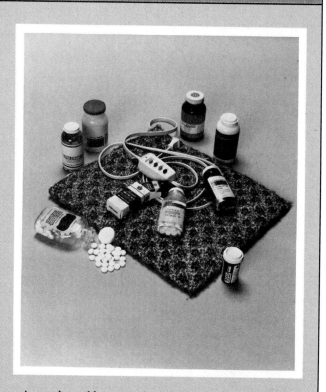

A number of home and medical remedies for PMS.

required for their acceptance by the medical profession. According to the Harvard Medical School Health Letter, "In all cases, these home remedies for PMS have received support from testimonials, but no solid research backs them up."

The medical remedies that physicians prescribe are, quite surprisingly, no more firmly established than the plethora of home remedies. Perhaps the most popular of all medically accepted treatments is progesterone, which physicians are now prescribing in large doses to patients suffering from PMS, despite the absence of any significant clinical evidence of its effectiveness. Prostaglandins have also recently begun to be used. Reports of success are encouraging but not yet sufficiently substantiated.

Work is presently under way to test the efficacy of these various treatments, but medical researchers will need several years to assemble a list of proven remedies for PMS. In the meantime, those who suffer from this condition should see a doctor to make certain that the symptoms are not the result of some other condition that has a proven cure.

CONDITIONS OF CLINICAL SIGNIFICANCE
The Reproductive System

Venereal Diseases

The most prevalent diseases of the reproductive systems of both males and females are the venereal diseases. These are infections that are spread through sexual contact.

Gonorrhea

Gonorrhea is an inflammation of the mucous membranes of the urogenital tract or rectum caused by the bacterium *Neisseria gonorrhoeae*. This inflammation usually results in a discharge of pus from the urethra and painful urination. In females, the cervix of the uterus and the uterine tubes can also become infected and inflamed, and *N. gonorrhoeae* is a frequent cause of pelvic inflammatory disease (see page 839). However, it is not unusual for a woman to have gonorrhea and transmit it through sexual contacts, yet have no apparent symptoms herself. In males, the spread of the gonococcus within the reproductive tract can lead to inflammation of the prostate, the seminal vesicles, and the epididymis.

Syphilis

Syphilis is a venereal disease caused by the bacterium *Treponema pallidum*. At the site where it enters the body, the bacterium usually produces a lesion called a *chancre*. Most commonly the chancre is on the penis or in the vagina. The chancre soon heals, and there may be no other symptoms for several weeks. However, during that time the infection spreads throughout the body by way of the bloodstream. After about six weeks, a skin rash accompanied by fever and aching joints develops. These secondary symptoms then disappear, and the disease enters a latent (inactive) period that can last for many years. During this latent period, the body may develop an immunity to the bacterium and destroy it, or the bacterium may spread to many different sites, including the nervous and vascular systems as well as various organs, causing damage to these structures. This latter occurrence produces severe and varied symptoms, depending on the structures affected. Syphilis can be detected by several different blood tests, one of which is called the Wasermann test.

Genital Herpes

Genital herpes is an increasingly common venereal disease. It is caused by the herpes simplex virus Type 2, which produces painful blisters on the reproductive organs. After some time the lesions heal, but they can recur periodically for years. The virus can also cause serious malformations of children born to infected mothers, and many researchers believe it is implicated in the occurrence of cervical cancer.

Male Disorders
Prostate Conditions

Although disorders of the male reproductive system are not restricted to the prostate gland, this is one structure that is frequently affected. *Prostatitis* is an inflammation of the prostate that is often due to bacterial infection. In this condition, the gland is swollen and tender, and in severe cases, abscesses can form. The prostate gland is also a common site of cancerous tumors, and an enlargement of the gland can compress the urethra, making urination difficult.

Impotence

Impotence is a fairly common male disorder in which a man is unable to attain an erection of the penis or to retain an erection long enough to complete sexual intercourse. Impotence can result from physical disorders of the vascular or nervous systems, and it is often a consequence of psychological or emotional problems.

Infertility

Male *infertility* is the inability of the male to fertilize the ovum. It is caused by either inadequate production of spermatozoa, the production of abnormal or nonmotile spermatozoa, or an obstruction that prevents the delivery of spermatozoa from the testes to the female vagina. The production of normal sperm can be interfered with by exposure to X rays, malnutrition, and certain diseases—including mumps.

Female Disorders
Abnormal Menstruation

Abnormalities of the menstrual cycle, which are among the most common disorders of the female reproductive system, can result from infections of the reproductive organs or malfunctions of the ovaries or the pituitary gland. Moreover, emotional or psychological factors are often involved.

Amenorrhea—the complete absence of menstrual periods—can be caused by disorders of the ovaries, the pituitary, or the hypothalamus. *Dysmenorrhea,* or painful menstruation, is usually associated with strong contractions of the uterus. Its cause is not known with certainty, but prostaglandins appear to be involved, and antiprostaglandin drugs are often effective in treating this condition. Moreover, dysmenorrhea often disappears following the first pregnancy.

Endometriosis

Some cases of severe dysmenorrhea and pelvic pain are caused by a condition called *endometriosis*. In this condition endometrial tissue develops in abnormal locations, often outside the uterus. The most common site for endometriosis is in the ovaries, but aberrant endometrial tissue has also been reported in the uterine ligaments, the pelvic peritoneum, and occasionally in various other locations. The tissue is thought either to enter the pelvic cavity from the uterus by way of the uterine tubes during menstruation or else to originate in the pelvic cavity as the result of abnormal embryonic differentiation of the epithelial cells that line the cavity.

Endometriosis becomes clinically important during menstruation, when the aberrant endometrial tissue, like the endometrium of the uterus, apparently reacts to hormonal levels in the body and undergoes cyclic bleeding. Unlike the endometrial tissue in the uterus, blood from the aberrant endometrial tissue generally has no means by which it can drain to the outside of the body. As a consequence, the blood collects in the aberrant tissue, causing pain and, in some cases, more serious complications. The condition is most common during a woman's reproductive life and declines in frequency after the age of 40.

Pelvic Inflammatory Disease

Pelvic inflammatory disease (PID) is an infection of the pelvic organs of the female reproductive tract, including the uterus and uterine tubes. It is caused by bacteria— usually gonococci, streptococci, or staphylococci—that generally reach the pelvic structures by passing through the vagina to the uterus and the uterine tubes. In some cases, bacteria from distant sites of infection reach the pelvic structures through the bloodstream.

Tumors

Tumors that develop within the female reproductive system can be either malignant (cancerous) or benign (non-

cancerous), and malignant tumors can metastasize— that is, spread to other locations within the body.

Ovarian tumors can be solid or they can occur as cysts (hollow sacs) that frequently contain fluid. Ovarian cysts are usually noncancerous, but a high percentage of solid ovarian tumors are malignant.

Tumors of the uterus and particularly the cervix are quite common. Fortunately, these tumors can be diagnosed early in their development by means of a simple test called a Pap smear. This test involves the removal (by a swab) of cells from the cervix and the surrounding area. The cells are then examined microscopically for signs of malignancy. With this procedure, cancerous cells can often be detected before any symptoms have appeared. If identified early, uterine tumors can be removed surgically or destroyed with radiation treatments. Since many women now have regular gynecological examinations that include Pap smears, the number of deaths due to uterine cancer has declined steadily.

Tumors are also common in the breasts, particularly after age 30. Breast tumors, whether malignant or benign, can be best detected by regular manual self-examination. Breast tumors are often removed surgically, and they are also treated by radiotherapy and chemotherapy.

Following ovulation and the formation of the corpus luteum, both estrogens and progesterone are produced. Under the influence of these hormones, the endometrial glands continue to grow and begin to secrete small amounts of a fluid rich in glycogen. In addition, the arteries within the endometrium become enlarged and spiraled. In this condition the endometrium is prepared to receive the embryo should fertilization occur.

If fertilization and pregnancy do not occur, the corpus luteum degenerates and the levels of estrogens and progesterone decline. As the levels of these hormones decrease and their stimulatory effects are withdrawn, blood vessels of the endometrium undergo prolonged spasms (contractions), which reduce the blood flow to the area of the endometrium supplied by the vessels. The resulting lack of blood causes the tissues of the affected region to degenerate. After some time, the vessels relax, which allows blood to flow through them again. However, capillaries in the area have become so weakened that blood leaks through them. This blood and the deteriorating endometrial tissue are discharged from the uterus as the menstrual flow. As a new ovarian cycle begins and the levels of estrogens rise, the functional layer of the endometrium undergoes repair and once again begins to proliferate.

FEMALE SEXUAL RESPONSES

During sexual stimulation, the responses of the female, like those of the male, are governed both by psychological stimuli and by tactile stimulation of the genital organs and other areas of the body. The neural pathways over which sexual stimuli are carried in the female are basically the same as in the male, and the motor impulses causing the various responses to sexual stimulation are carried over autonomic (that is, sympathetic and parasympathetic) nerves.

During sexual stimulation, the clitoris becomes engorged with blood in a manner similar to that of the penis, while eliciting widespread sexual sensations. Erectile tissue in the area of the vaginal openings also becomes engorged with blood. This narrows the vaginal orifice and facilitates the stimulation of the penis during intercourse. In addition, the epithelial lining of the vagina becomes lubricated through the copious secretion of mucus by glands located around the cervix of the uterus and the area of the vaginal opening. The lubrication provides for easy entrance of the penis into the vagina and facilitates rhythmic massaging stimulation of the female external genital organs as well as of the penis.

When sexual stimulation reaches sufficient intensity, the female undergoes an orgasm. The physiological changes that occur during an orgasm in the female are similar to those of the male, with the exception that there is no ejaculation in the female. The question as to whether a female orgasm in some way facilitates fertilization is not yet adequately answered. Certainly it is not necessary for a female to experience an orgasm for fertilization to occur.

MENOPAUSE AND THE FEMALE CLIMACTERIC

At about age 50, the ovarian and menstrual cycles gradually become irregular. Ovulation fails to occur during many of the irregular cycles, and in most women the cycles cease altogether over the next several months or, at most, a few years. The cessation of the menstrual cycle is referred to as **menopause**, and the entire period is called the **female climacteric.**

The female climacteric is thought to be caused by an inability of the ovaries to respond to hormonal signals, most probably due to a shortage of follicles that results from their ovulation or degeneration during the reproductive years. As a consequence, the production of estrogens and progesterone by the ovaries is quite low, and estrogen-dependent tissues such as the genital organs and breasts gradually atrophy. During this period, the production of FSH and LH by the pituitary is quite high, apparently due to a partial release of the Gn-RH, FSH, LH system from the negative-feedback influences of ovarian hormones. During the female climacteric some women experience such symptoms as sensations of warmth accompanied by sweating ("hot flashes"), irritability, fatigue, and anxiety.

STUDY OUTLINE

EMBRYONIC DEVELOPMENT OF THE REPRODUCTIVE SYSTEM pp. 809–812

EARLY DEVELOPMENT Genital ridges, paramesonephric ducts, and mesonephric ducts.

DEVELOPMENT OF INTERNAL REPRODUCTIVE STRUCTURES

MALE EMBRYO
1. Medulla of gonad gives rise to seminiferous tubules.
2. Mesonephric ducts form efferent ductules, epididymis, and ductus deferens; paramesonephric ducts degenerate.

FEMALE EMBRYO
1. Cortex of gonad becomes site of ova production.
2. Mesonephric ducts degenerate; paramesonephric ducts form uterus, vagina, and uterine tubes.

DEVELOPMENT OF EXTERNAL REPRODUCTIVE STRUCTURES
1. Genital tubercle forms penis or clitoris.
2. Urethral folds fuse in male but not in female.
3. Labioscrotal folds form scrotum in male, labia majora in female.

CONDITION OF CLINICAL SIGNIFICANCE: SEXUAL MALFORMATIONS DURING EMBRYONIC DEVELOPMENT Some degree of sexual abnormality occurs in one out of every thousand persons. p. 813

ANATOMY OF THE MALE REPRODUCTIVE SYSTEM pp. 813–820

THE MALE PERINEUM Divided into anterior urogenital triangle and posterior anal triangle.

TESTES AND SCROTUM

TESTES Spermatogenesis occurs in seminiferous tubules.

INTERSTITIAL ENDOCRINOCYTES Secrete androgens.

DESCENT OF THE TESTES Testes follow vaginal process through inguinal canal into scrotum.
1. Location of testes in scrotum necessary for normal sperm development.
2. Cremaster muscle assists in regulating temperature of testes.

EPIDIDYMIS First portion of duct system; storage of mature nonmotile sperm.

DUCTUS DEFERENS Continuation of duct of epididymis; heavy wall of smooth muscle; located in spermatic cord.

SEMINAL VESICLES Two membranous pouches lateral to ductus deferens; secrete viscid fluid that contributes to semen.

PROSTATE GLAND Encompasses urethra below bladder; secretes fluid that contributes to semen.

BULBOURETHRAL GLANDS Pair of glands below prostate; secretion contributes to semen.

PENIS Copulatory organ; contains three cylindrical cavernous bodies; deposits spermatozoa in female reproductive tract.

SEMEN Mixture of spermatozoa from testes and fluids from seminal vesicles, prostate, and bulbourethral glands; alkaline; fructose provides energy source for spermatozoa.

CONDITION OF CLINICAL SIGNIFICANCE: CRYPTORCHIDISM One or both testes fail to leave abdominopelvic cavity. **p. 816**

PUBERTY IN THE MALE **pp. 820–821**
1. The time of life at which reproduction becomes possible.
2. Events of puberty (for example, enlargement of penis, scrotum, and testes; characteristic hair growth, voice changes) result from increased testosterone production by testes.
3. At puberty, alteration in brain function leads to increased release of Gn-RH, which ultimately results in increased testosterone production.

SPERMATOGENESIS **pp. 821–823**
1. Occurs within seminiferous tubules of testes.
2. Spermatogonia (diploid) develop into primary spermatocytes (diploid).
3. Primary spermatocytes undergo first meiotic division, forming secondary spermatocytes (haploid).
4. Secondary spermatocytes undergo second meiotic division, producing spermatids.
5. Spermatids develop into mature spermatozoa.

THE SPERMATOZOON Flagellated; contains acrosome, mitochondria, centrioles, and little cytoplasm. **p. 823**

HORMONAL CONTROL OF TESTICULAR FUNCTION **p. 823**
1. FSH and LH release from pituitary is stimulated by Gn-RH from hypothalamus.
2. Testosterone inhibits LH release, mainly at level of pituitary.

3. FSH release believed to be inhibited by inhibin from sustenticular cells.

MALE SEXUAL RESPONSES **pp. 823–824**

ERECTION Penis becomes enlarged and firm.
1. Results from vascular engorgement of spaces within corpora cavernosa and corpus spongiosum of penis.
2. Can be caused by physical stimulation (a reflex phenomenon), visual or psychic stimuli.

EMISSION AND EJACULATION Results in forceful expulsion of semen from penis.

ANATOMY OF THE FEMALE REPRODUCTIVE SYSTEM **pp. 824–831**

OVARIES Production of ova; estrogen and progesterone production; cortex contains primary follicles.

UTERINE TUBES Carry ovum to uterus.
1. Fimbriae surround opening.
2. Thick mucosa of simple columnar cells; some ciliated.
3. Cilia and peristaltic contractions carry ovum to uterus.

UTERUS Composed of body, isthmus, and cervix.
1. Supported by central tendon of pelvic and urogenital diaphragms.
2. Endometrium thickens and becomes engorged to prepare for fertilized ovum each month.

VAGINA Canal from cervix to exterior of body; mucosa of stratified squamous epithelium; entrance partially covered by hymen.

FEMALE EXTERNAL GENITAL ORGANS (Vulva, or pudendum)
1. Labia majora and minora.
2. Greater and lesser vestibular glands lubricate vestibule.
3. Clitoris homologous to male penis.

THE FEMALE PERINEUM Divided into anterior urogenital triangle and posterior anal triangle. *Clinical perineum* is located between vagina and anus.

MAMMARY GLANDS
1. Nipple and areola pigmented.
2. Composed of adipose tissue and lobes of compound tubuloalveolar glands.
3. Ampullae serve as milk reservoirs.

PUBERTY IN THE FEMALE **p. 831**
1. Events of puberty (such as enlargement of vagina, uterus, and uterine tubes; deposition of fat in breasts and hips; characteristic hair growth) largely a result of increased production of estrogens by ovaries.
2. At puberty, alteration in brain function leads to increased release of Gn-RH, which ultimately results in increased production of estrogens.

OOGENESIS **pp. 831–832**
1. Oogonia (diploid) develop into primary oocytes (diploid).
2. Primary oocytes undergo first meiotic division, forming secondary oocytes (haploid) and polar bodies.
3. Secondary oocyte begins second meiotic division and completes it, if fertilized, to become a fully developed ovum.

4. Beginning at puberty, one secondary oocyte (commonly called an ovum) usually released at ovulation each month.

THE OVARIAN CYCLE   pp. 832–834

DEVELOPMENT OF FOLLICLES

1. Under appropriate hormonal stimulation, some primary follicles undergo further development.
2. A zona pellucida forms within growing follicle, as does a fluid-filled cavity called the antrum.

OVULATION

1. Under proper hormonal conditions, mature follicle ruptures and releases its ovum.
2. Ovulated ovum is surrounded by zona pellucida and sphere of follicle cells called corona radiata.

FORMATION OF THE CORPUS LUTEUM

1. Cells of ruptured mature follicle increase in size and take on yellowish color, forming corpus luteum.
2. If ovum is not fertilized and pregnancy does not occur, corpus luteum begins to degenerate about 8–10 days after ovulation.
3. If ovum is fertilized and pregnancy does occur, corpus luteum does not degenerate but remains until near end of pregnancy.

HORMONAL CONTROL OF OVARIAN FUNCTION   pp. 834–835

1. Appropriate levels of FSH, LH, and estrogens initiate ovarian cycle and follicle development.
2. As follicle development proceeds, secretion of estrogens progressively increases.
3. At relatively low levels, estrogens inhibit release of Gn-RH.
4. Relatively high levels—or perhaps rapidly rising levels—of estrogens stimulate Gn-RH release and enhance sensitivity of LH-releasing mechanism to Gn-RH, resulting in sharp increase in LH secretion, which leads to final maturation of follicle, rupture of follicle, and ovulation.
5. Estrogens and progesterone secreted by corpus luteum inhibit Gn-RH release and perhaps FSH and LH.
6. As corpus luteum degenerates, levels of estrogens and progesterone fall and levels of FSH and LH rise, stimulating other follicles to develop and the beginning of another cycle.

THE MENSTRUAL CYCLE   pp. 836–839

1. Estrogens produced during first portion of ovarian cycle stimulate functional layer of endometrium of uterus to proliferate, and thickness and vascularity of the layer increase.

2. Following ovulation, estrogens and progesterone from corpus luteum cause further development of endometrium.
3. As corpus luteum degenerates and levels of estrogens and progesterone fall, the functional layer of endometrium degenerates and menstruation occurs.

FEMALE SEXUAL RESPONSES   pp. 839–840

1. Clitoris becomes engorged with blood.
2. Epithelial lining of vagina becomes lubricated.
3. Erectile tissue in the area of vaginal opening becomes engorged with blood.

MENOPAUSE AND THE FEMALE CLIMACTERIC   p. 840

1. Sexual cycles gradually cease; believed to be due to an inability of ovaries to respond to hormonal signals.
2. Production of estrogens and progesterone is quite low and estrogen-dependent tissues gradually atrophy.

CONDITIONS OF CLINICAL SIGNIFICANCE: THE REPRODUCTIVE SYSTEM   pp. 838–839

VENEREAL DISEASES

GONORRHEA Inflammation of mucous membranes of genital tract or rectum caused by bacterium *Neisseria gonorrhoeae*.

SYPHILIS Infection caused by bacterium *Treponema pallidum*.

GENITAL HERPES Infection caused by herpes simplex virus Type 2.

MALE DISORDERS

PROSTATE CONDITIONS Such as bacterial inflammation (prostatitis) and tumors.

IMPOTENCE Inability to attain erection or maintain it long enough to accomplish sexual intercourse.

INFERTILITY Inability to fertilize ovum.

FEMALE DISORDERS

ABNORMAL MENSTRUATION May be due to infections or glandular malfunctions, as well as emotional and psychological factors.

ENDOMETRIOSIS Endometrial tissue in aberrant locations.

PELVIC INFLAMMATORY DISEASE Inflammation that can involve uterine tubes, ovaries, or peritoneum of abdominopelvic cavity.

TUMORS Can occur in ovaries, uterus, or breasts.

SELF-QUIZ

1. In the male embryo, which of these structures degenerate shortly after the gonads begin to differentiate into testes? (a) the paramesonephric ducts; (b) the seminiferous tubules; (c) the mesonephric ducts.

2. In the absence of the male hormones, all embryos, regardless of their genetic makeup, will develop into females. True or False?

3. Spermatogenesis occurs in the: (a) tunica albuginea; (b) seminiferous tubules; (c) sustenticular cells; (d) prostate gland.

4. The interstitial endocrinocyte cells produce: (a) sperm; (b) testosterone; (c) luteinizing hormone.

5. Occasionally, one or both of the testes fail to leave the abdominopelvic cavity, a condition called cryptorchidism, which results in sterility. True or False?

6. The scrotum regulates the temperature of the testes through the action of the: (a) gubernaculum; (b) cremaster muscle; (c) epididymis.

7. Match the structures of the male reproductive system with the appropriate lettered description.

   Epididymis
   Ductus deferens
   Seminal vesicles
   Ejaculatory ducts
   Prostate gland
   Bulbourethral glands

   (a) Structures that penetrate the prostate gland and enter the urethra just below its exit from the bladder
   (b) Encompasses the urethra just below the bladder
   (c) Structure that transports sperm to the urethra
   (d) The first portion of the duct system that transports mature sperm from the testes to the outside of the body
   (e) Located below the prostate on either side of the membranous portion of the urethra
   (f) Structures that join with the ductus deferens to form the ejaculatory duct

8. The acidity of the semen counteracts the alkaline pH of the vagina. True or False?

9. In the male, prior to puberty: (a) the testes continually produce large amounts of testosterone; (b) there is little release of follicle-stimulating hormone and luteinizing hormone from the pituitary; (c) the hypothalamus secretes large amounts of gonadotropin-releasing hormone.

10. Spermatids: (a) are haploid cells; (b) develop within the bulbourethral glands; (c) divide mitotically, giving rise to spermatogonia.

11. In the male, testosterone: (a) strongly stimulates FSH release; (b) inhibits LH release; (c) is not involved in sperm production.

12. The ligament that supports the ovary from the posterior surface of the broad ligament is: (a) mesosalpinx; (b) mesovarium; (c) round ligament; (d) ovarian ligament.

13. Ova mature in the: (a) uterus; (b) corpus luteum; (c) mature follicles; (d) cervix.

14. The vagina is lined with stratified squamous epithelium. True or False?

15. Collectively, the external genital organs of the female are known as: (a) the mons pubis; (b) the labia majora; (c) the vulva; (d) the pudendum; (e) both c and d.

16. The structure in the female that is homologous to the dorsal portion of the male penis is the: (a) hymen; (b) mons pubis; (c) clitoris; (d) labia majora.

17. The corpus luteum secretes: (a) luteinizing hormone; (b) follicle-stimulating hormone; (c) progesterone.

18. During the ovarian cycle, a high level of luteinizing hormone leads to the final maturation of the follicle, the rupture of the follicle, and ovulation. True or False?

19. During the first portion of the ovarian cycle, the functional layer of the endometrium of the uterus is stimulated to proliferate by: (a) progesterone; (b) estrogens; (c) endometriotropin.

20. During the female climacteric, the level of: (a) estrogens is relatively high; (b) FSH is relatively high; (c) progesterone is relatively high.

LEARNING OBJECTIVES

After completing this chapter, you should be able to:

1. Describe the events involved in the fertilization of the ovum and the implantation of the blastocyst.

2. Describe the formation and function of the placenta, and indicate the relationship between maternal and fetal blood in the placenta.

3. Describe how each embryonic membrane is formed and state its function.

4. Describe the process of parturition.

5. Discuss the development of the breasts and the hormonal activities involved in lactation.

6. Describe the unique features of fetal circulation.

7. Describe the circulatory changes that occur at birth.

8. Discuss how inherited traits are passed from one generation to the next.

CHAPTER CONTENTS

PREGNANCY

PARTURITION

CONDITIONS OF CLINICAL SIGNIFICANCE: PREGNANCY

LACTATION

FETAL CIRCULATION

HUMAN GENETICS

CONDITIONS OF CLINICAL SIGNIFICANCE: INHERITANCE

KEY TERMS AND DERIVATIVES

allantois (*allant* = sausage-shaped) the structure extending from the hindgut into the connecting stalk of the embryo

blastocyst (*blast* = germ; *cyst* = bladder) early stage of embryonic development during which cells form a hollow sphere

chorion (*chorio* = skin) the outermost fetal membrane; contributes to the formation of the placenta

fertilization (*fertil* = fruitful) the union of male and female gametes

trophoblast (*troph* = nourishment; *blast* = bud) outer layer of cells of the blastocyst; surrounds the inner cell mass and helps to nourish it

zygote (*zygot* = yoked together) the fertilized ovum before it divides; produced by the union of two gametes

Pregnancy, Embryonic Development, and Inheritance

We saw in Chapter 28 that two of the main functions of the reproductive system are to provide a means by which fertilization can occur and to furnish an environment that is conducive to the development of the fertilized ovum (*zygote*) to a stage at which it is capable of surviving on its own outside the mother's body. During the time that the zygote is developing within a woman, the woman is said to be **pregnant.** In this chapter we consider how fertilization occurs and discuss the major changes that occur in the woman during pregnancy. We also consider the embryonic development of the zygote and basic aspects of human genetics.

PREGNANCY

For pregnancy to occur following sexual intercourse, spermatozoa must reach the upper region of the uterine tube, where fertilization usually takes place. The fertilized ovum must then be transported to the uterus and become implanted within the endometrium, and a placenta must form.

Fertilization

An ovum remains capable of being fertilized for about 12–24 hours following ovulation, and sperm remain viable within the female reproductive tract for about 24–72 hours after ejaculation. Therefore, for fertilization to occur, sperm should be deposited within the female tract no earlier than 72 hours before ovulation and no later than 24 hours after ovulation.

Once within the vagina, spermatozoa travel through the uterus and reach the upper portion of the uterine tubes rather quickly, often requiring only minutes to do so. The rate at which the sperm are transported is too rapid to be due solely to the motility of the sperm, and it has been suggested that the prostaglandins found in semen assist sperm transport by stimulating reverse peristaltic contractions of the smooth muscles of the uterus and uterine tubes. The semen deposited in the vagina during sexual intercourse usually contains several hundred million sperm. However, sperm mortality is high, and only a few thousand reach the uterine tubes. This high mortality rate is one reason why large numbers of sperm are required for pregnancy to occur.

During their passage through the uterus and the uterine tubes, the sperm are believed to undergo a process called **capacitation,** which makes them capable of fertilizing an ovum. The process of capacitation is not clearly understood, but it apparently involves an alteration of the sperm membrane that allows the sperm to release enzymes upon contact with the corona radiata surrounding the ovum. The enzymes—for example, hyaluronidase and proteases—are believed to contribute to the breakdown of portions of the zona pellucida and the material that holds the cells of the corona together (hyaluronic acid), thus facilitating access to the ovum.

One of the sperm contacts the cell membrane of the ovum, fuses with it, and passes through into the cytoplasm (Figure 29.1a). Following this, the ovum completes the second meiotic division, and the nuclei (now called *pronuclei*) of the sperm and ovum unite to form a single nucleus. Thus, the ovum is fertilized, and the fertilized ovum, or **zygote,** contains the full complement of 46 chromosomes.

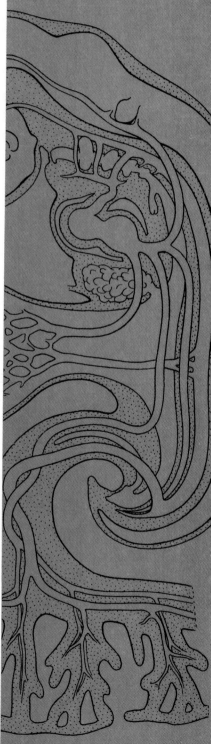

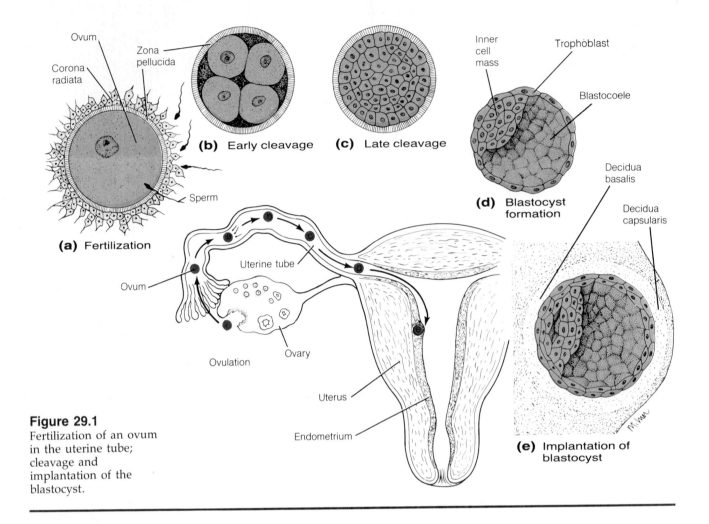

Figure 29.1
Fertilization of an ovum
in the uterine tube;
cleavage and
implantation of the
blastocyst.

The interaction between the sperm that enters the ovum and the cell membrane of the ovum brings about changes that prevent the entry of additional sperm. As a consequence of this interaction, chemical substances, including some polysaccharides, are released from the ovum. These substances cause the zona pellucida (and/or the cell membrane of the ovum) to become impenetrable by additional sperm.

Sex Determination

The genetic sex of an individual is determined at the time of fertilization, even though the individual's sex does not become apparent until about the eighth week of development. If an ovum is fertilized by a sperm containing an X sex chromosome, a female (XX) normally develops. If an ovum is fertilized by a sperm containing a Y sex chromosome, a male (XY) develops.

Development and Implantation of the Blastocyst

F29.1b,c

The zygote, which immediately begins to divide [undergo *cleavage* (Figures 29.1b and c)], passes along the uterine tube to the uterus—a journey that requires about three to four days. Indeed, if the zygote were to reach the lower end of the uterine tube much sooner, it could not enter the uterus, because this region of the uterine tube remains spastically contracted for about three days following ovulation. By the time the uterus is reached, or shortly thereafter, enough cell divisions have occurred for the zygote to have

F29.1d   developed into a fluid-filled sphere of cells called a **blastocyst** (Figure 29.1d).

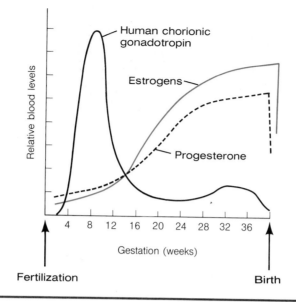

Figure 29.2
Relative amounts of human chorionic gonadotropin, estrogens, and progesterone in the maternal blood during pregnancy.

The outer sphere of cells of the blastocyst is called the **trophoblast** (*trophectoderm*). This layer will contribute to the formation of the placenta. The blastocyst contains an accumulation of cells called the **inner cell mass.** The embryo develops from the inner cell mass.

The blastocyst remains free within the lumen of the uterus for several days before it implants itself in the endometrium. During this period, it obtains nourishment from the uterine fluids and continues to undergo cell division. By this time, the zona pellucida has disintegrated, allowing the trophoblast to expand. About seven days after ovulation, the blastocyst undergoes implantation in the endometrium (Figure 29.1e). By this time the endometrium of the uterus has been prepared for implantation by estrogens and progesterone. Upon contact with the endometrium, the cells of the trophoblast release enzymes that digest cells of the endometrium, thus eroding the endometrial surface. This erosion makes additional fluids and nutrients available to the blastocyst and allows the blastocyst to burrow into the endometrium. As the blastocyst erodes deeper, the endometrium grows over it, and within a few days the blastocyst is completely implanted within the endometrium. For about seven days following implantation, the embryo obtains all of its nutrients from the destruction of endometrial cells by the trophoblast. Eventually, nutrients are obtained through the placenta, but erosion by the trophoblast continues to supply a significant fraction of the embryo's nutrients for the first two months of development, while the placenta is gradually enlarging and becoming increasingly functional.

Maintenance of the Endometrium

Because of the importance of the endometrium for the nourishment of the embryo, it is vital that the endometrium be maintained in a highly developed state so pregnancy can continue. Thus, the menstrual and ovarian cycles must be interrupted; menses must be prevented from occurring; and the development of additional follicles within the ovaries must be suppressed. In order to maintain the endometrium, there must be high levels of estrogens and progesterone within the blood (Figure 29.2). In addition to maintaining the endometrium, the estrogens and progesterone inhibit the secretion of FSH and LH, thus preventing the further development of any follicles.

During early pregnancy, the required estrogens and progesterone are produced by the corpus luteum. In the absence of pregnancy, the corpus luteum begins to degenerate about eight to ten days after ovulation. However, when pregnancy occurs, the corpus luteum remains and continues to secrete estrogens and progesterone.

go-nad-o-tro'-pin
kor'-ee-on

A major factor that causes the corpus luteum to remain and to continue to secrete its hormones is **human chorionic gonadotropin (HCG).** This hormone is produced by the **chorion,** a fetal membrane that develops from the cells of the trophoblast and later becomes involved in the formation of the placenta. HCG has properties that are very similar to those of LH. It is secreted as early as the second week of pregnancy, and its level reaches a peak during the third month and then sharply declines, remaining low throughout the remainder of pregnancy.

Another source of estrogens and progesterone during pregnancy is the placenta. By about the end of the second month, the placenta has developed to the point where it is secreting sufficient estrogens and progesterone to maintain the endometrium. Moreover, the substantial increases in the levels of estrogens and progesterone that occur as pregnancy progresses are due almost entirely to the secretion of these hormones by the placenta. Thus, beyond the second month, the continuance of the pregnancy no longer depends on the secretion of estrogens and progesterone by the corpus luteum.

Development of the Placenta

The **placenta** is an organ that, from the third month to the time of birth some six months later, serves to supply nutrients to, and remove wastes from, the embryo—which after the second month is called a **fetus.** Moreover, the placenta serves as an endocrine gland that produces a number of hormones, including estrogens and progesterone.

Because the superficial layer of the thickened endometrium is discarded by the body, either with the menses or at birth, it is called the **decidua** ("falling off"). The portion of the decidua that lies deep to the blastocyst is called **F29.1e** the **decidua basalis** (Figure 29.1e). The decidua basalis is involved in the formation of the maternal portion of the placenta. The portion of the decidua superficial to the blastocyst is the **decidua capsularis.** ·

As the blastocyst becomes implanted in the endometrium, the trophoblast separates into two layers: an outer **syncytiotrophoblast,** in which the cell boundaries disappear, and an inner **cytotrophoblast,** which is composed of **F29.3b** distinctly separate cells (Figure 29.3b). At this time a layer of mesodermal cells develops on the inside of the trophoblast. This layer of cells combines with **F29.3c** the trophoblast to form a membrane called the **chorion** (Figure 29.3c). As mentioned earlier, the chorion is the source of human chorionic gonadotropin. But it also forms an important part of the placenta. The syncytiotrophoblast proliferates and releases enzymes that erode the endometrium. At the same time the chorion develops many fingerlike projections **F29.3d** called **chorionic villi** (Figure 29.3d). Branches from the umbilical artery and vein of the fetus grow into the villi and form capillary beds. With continued development, the villi become surrounded by pools of the mother's blood, which has collected in sinuses, called *lacunae*, within the endometrium. These blood pools result from the syncytiotrophoblast's erosion of the decidua basalis region of the endometrium and the capillaries in the endometrium.

Thus, while maternal and fetal blood are separated by only a few layers of cells (chorion and endothelium of the fetal capillaries), there is normally no actual mixing of the two blood supplies. Oxygen, carbon dioxide, glucose, amino acids, fatty acids, various ions, and other substances can move in one direction or another between the maternal blood and the fetal blood either by diffusion or by active transport. The presence of a large number of villi provides an extensive surface area across which substances can pass. It is estimated that a fully developed placenta has a surface area of about 16 square meters.

It is clear, from the manner in which it is formed, that the placenta is a combination of maternal tissue (decidua basalis) and fetal tissue (chorion) **F29.4** (Figure 29.4). The fetal blood, which remains in the fetal vessels, enters the placenta through the **umbilical arteries** and passes through capillaries in the villi (where the exchange of substances occurs). The fetal blood then leaves the placenta—and returns to the fetus—through the **umbilical vein.** The

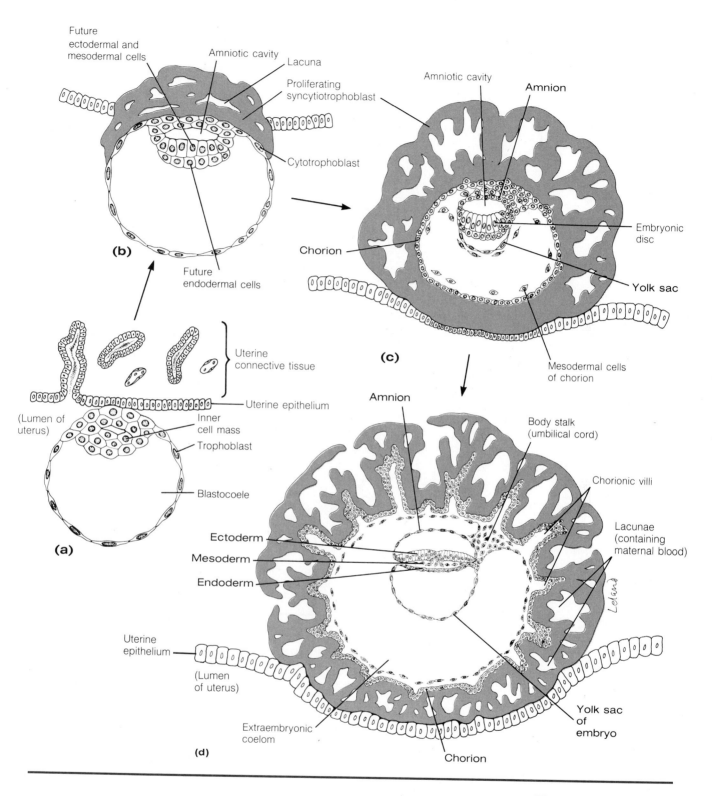

(a)

(Lumen of uterus)

Uterine connective tissue
Uterine epithelium
Inner cell mass
Trophoblast
Blastocoele

(b)

Future ectodermal and mesodermal cells
Amniotic cavity
Lacuna
Proliferating syncytiotrophoblast
Cytotrophoblast
Future endodermal cells

(c)

Amniotic cavity
Amnion
Chorion
Embryonic disc
Yolk sac
Mesodermal cells of chorion

(d)

Amnion
Ectoderm
Mesoderm
Endoderm
Body stalk (umbilical cord)
Chorionic villi
Lacunae (containing maternal blood)
Yolk sac of embryo
Uterine epithelium
(Lumen of uterus)
Extraembryonic coelom
Chorion

Leland

Figure 29.3
Successive stages of embryonic development, showing proliferation of the trophoblast, and early formation of the amnion, the chorion, the yolk sac, and the three primary germ layers.

umbilical vessels travel between the placenta and the fetus in the **umbilical cord.** The mother's blood enters the placenta through the **uterine arteries** (branches of the internal iliac arteries), flows through the blood pools of the endometrium (where the exchange of substances occurs), and leaves the placenta through the **uterine veins.**

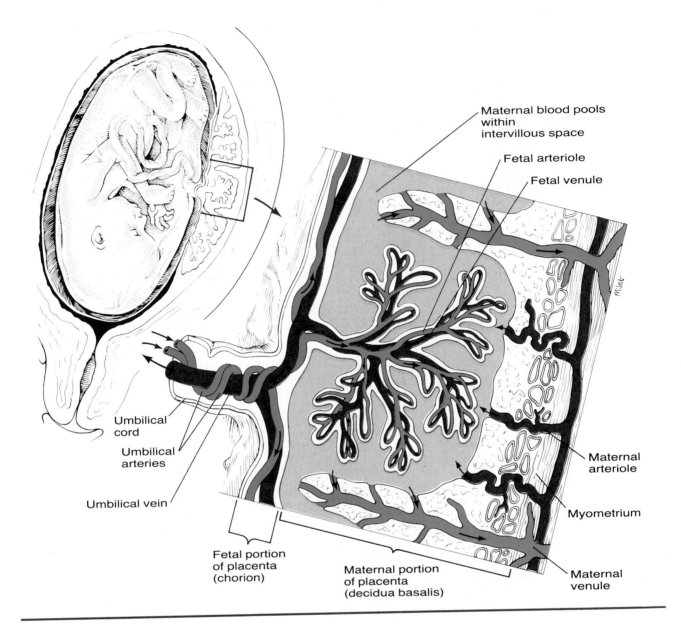

Figure 29.4
The vascular arrangement of the placenta. The arrows indicate the direction of blood flow.

Formation of Embryonic Membranes and Germ Layers

The placenta makes continued development of the embryo possible. Early in development, four embryonic membranes form from the embryo—the *amnion, yolk sac, allantois,* and *chorion*. These membranes assist in the embryo's development in various ways, although they contribute very little to the body of the embryo itself.

While the embryonic membranes are forming, the cells of the inner cell mass undergo movements that result in the formation of three germ layers: the *endoderm, ectoderm,* and *mesoderm*. The embryo—including all of its organs and structures—develops from these germ layers.

Embryonic Membranes

AMNION Following implantation, while the placenta is developing, the inner cell mass proliferates and a cavity forms that separates it from the trophoblast. The membrane that surrounds this cavity is the **amnion.** The amnion
F29.3 is a fluid-filled structure that eventually surrounds the fetus (Figures 29.3 and

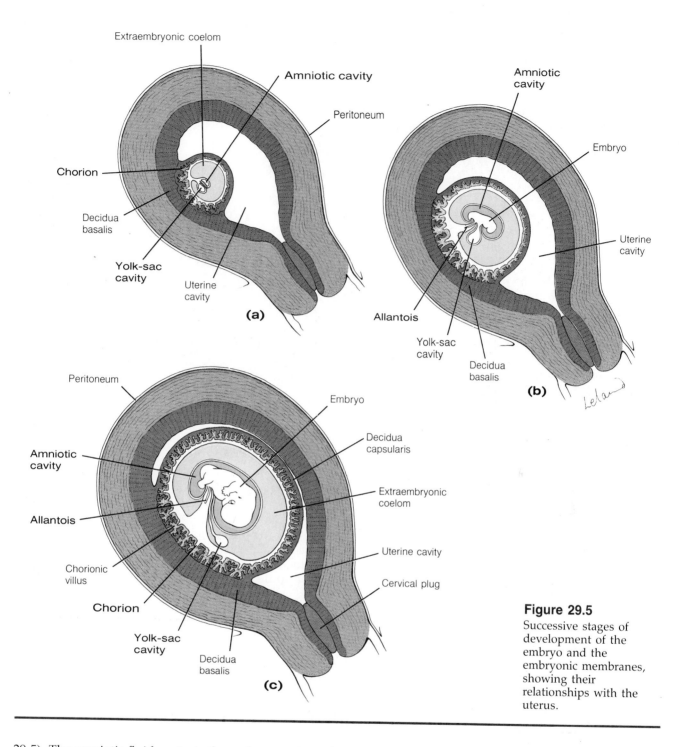

Figure 29.5
Successive stages of development of the embryo and the embryonic membranes, showing their relationships with the uterus.

29.5). The amniotic fluid protects the embryo against physical injury, helps to maintain the constancy of its temperature, and allows it to move freely. Those cells of the inner cell mass that form the floor of the amniotic cavity make up the **embryonic disc.** The cells of the embryonic disc, from which the embryo will be formed, separate into two layers: a layer of *ectodermal cells* that face the amniotic cavity and a layer of *endodermal cells* on the side toward the space (*blastocoele*) within the blastocyst. Included in the layer of ectodermal cells are cells that will contribute to the formation of mesoderm.

YOLK SAC   The endoderm cells of the embryonic disc undergo rapid mitosis and form the second embryonic membrane: the **yolk sac.** The yolk sac develops as a small cavity on the inner surface of the embryonic disc and eventually

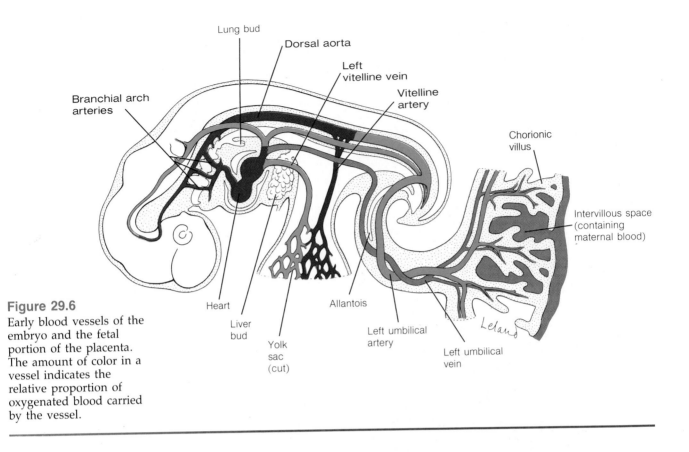

Figure 29.6
Early blood vessels of the embryo and the fetal portion of the placenta. The amount of color in a vessel indicates the relative proportion of oxygenated blood carried by the vessel.

extends into the umbilical cord. In the human, where very little yolk is present in the embryo, the yolk sac never becomes very large and serves no nutritive function. The yolk sac is, however, the source of the primordial germ cells that migrate to the gonads. With continued development, the endodermal cells on the undersurface of the embryonic mass that forms the upper portion of the yolk sac fold together, forming the gut tube of the embryo. The midgut region remains open to the yolk sac.

ALLANTOIS   Another diverticulum develops from the endoderm of the hindgut. This pouch is the **allantois,** the third embryonic membrane. The allantois serves as a storage site for embryonic waste products in most animals, though perhaps not in humans. Like the yolk sac, the allantois is in the umbilical cord. The blood vessels that supply the allantois (umbilical arteries and veins) carry the fetal blood supply to and from the placenta (Figure 29.6).

CHORION   The fourth embryonic membrane, the **chorion,** is formed from the trophoblast of the blastocyst plus the mesoderm that lines the inside of the trophoblast. The chorion surrounds the entire embryo and the other three embryonic membranes. As the amnion enlarges, it fuses with the inner layer (mesoderm) of the chorion. In combination with the umbilical blood vessels, the chorion forms the fetal portion of the placenta.

Germ Layers

In the second week of development, some of the surface cells of the embryonic disc settle deeper in the disc and spread between the **ectoderm** and **endoderm** layers that were formed by an earlier separation of the embryonic disc (Figure 29.7). These cells will form the **mesoderm.** As the sheet of mesoderm moves between the ectoderm and endoderm, it splits into two layers: *parietal* and *visceral*. The space between the layers becomes the **coelom,** or body space. Table 29.1 summarizes the main structures formed by the three primary germ layers.

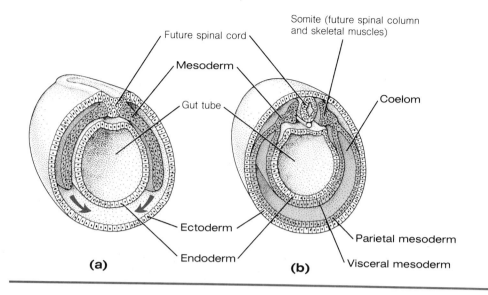

Figure 29.7
Schematic cross section of an embryo showing the development of the germ layers. **(a)** Early development: a sheet of mesoderm is moving between the endoderm of the gut and the ectoderm of the body surface. **(b)** Later development: the coelom has formed between the parietal and visceral layers of the mesoderm.

Table 29.1   Structures Formed by the Three Primary Germ Layers

ENDODERM

Epithelium of the digestive tract and its glands (e.g., the liver and the pancreas)
Epithelium of the urinary bladder and the urethra
Epithelium of the pharynx, the auditory tube, the larynx, the trachea, the bronchi, and the lungs
Epithelium of the tonsils, the thyroid, parathyroid, and thymus glands

MESODERM

Skeletal, smooth, and cardiac muscle
Cartilage, bone, and other connective tissues
Blood, bone marrow, and lymphoid tissue
Epithelium of blood vessels and lymphatics
Epithelium of the coelom and the joint cavities
Epithelium of the kidneys and the ureters
Epithelium of the gonads and reproductive ducts
Epithelium of the adrenal cortex
Dermis of the skin

ECTODERM

Epidermis of the skin
Hair, nails, and glands of the skin
Lens of the eye
Receptor cells of the sense organs
Epithelium of the mouth, the nostrils, the sinuses, and the anal canal
Enamel of the teeth
All nervous tissue
Adrenal medulla

Gestation

The *gestation* (pregnancy) period usually lasts about 280 days from the beginning of the last menstrual period. During this time the developing individual increases in size, and all the organs and structures of the adult are formed (see Box 29.1). During gestation, the total mass of the uterus increases to accommodate the growing fetus. In the later stages of pregnancy, the uterus occupies such a large portion of the abdominopelvic cavity that it can exert pressure on the rectum and the urinary bladder, causing constipation and frequent urination (Figure 29.8).

Box 29.1

F29.8

Box 29.1

Gestation of the Human Fetus

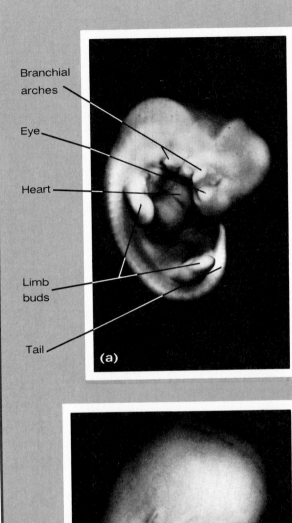

Branchial arches

Eye

Heart

Limb buds

Tail

(a)

The developing human at (a) 5 weeks, (b) 8 weeks, and (c) 28 weeks. At 5 weeks, the embryo has developed limb buds, which eventually develop into arms, legs, fingers, and toes. A rudimentary tail is present, as well as primitive branchial arches. At this stage, the brain is covered by only a thin layer of tissue, the eyes are just beginning to form, and the heart is larger in proportion to the body than at any other time in development.

By 8 weeks, the tail has degenerated, and the branchial arches have evolved into the jaw and various other structures. Gender is now determinable, and most of the major organs are present, including the liver, pancreas, and heart. By this stage, the fetus has acquired reflexes and is thus aware of touch, although it is unaware of sound (except for the mother's heartbeat).

By 28 weeks, eyelids are able to open and close, fingernails are present, and respiratory, circulatory, and nervous systems are established.

(b)

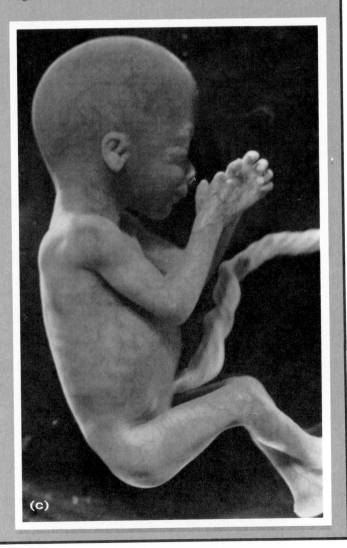

(c)

The presence of a growing fetus places extra demands on a woman's body systems. For example, a pregnant woman generally produces somewhat more urine than she did before pregnancy because her kidneys must process the excretory products of the fetus as well as her own excretory products. Diet is especially important during pregnancy, and an expectant mother's diet must include sufficient amounts of vitamins, minerals, proteins, and other substances to supply both her own needs and those of the fetus.

During pregnancy, the placenta produces a hormone called **human chorionic somatomammotropin (human placental lactogen),** and researchers believe this hormone affects the metabolism of the mother. Human chorionic somatomammotropin begins to be produced by about the fifth week of pregnancy, and its level increases throughout pregnancy. Human chorionic somatomammotropin decreases the mother's use of glucose; thus, more glucose becomes available for use by the fetus. This effect is believed to be useful in promoting the development of the fetus because the fetus uses glucose as a major source of energy. Human chorionic somatomammotropin also increases the mobilization of fatty acids from maternal adipose tissue. The fatty acids can be used in place of glucose as an energy source by the mother.

Twins

Twins occur in about one out of every 85 births. There are two types of twins, identical and fraternal, each the result of a different series of events.

Identical, or *monozygotic*, *twins* occur when an inner cell mass that develops from a single fertilized ovum separates into two masses, each of which gives rise to a complete embryo. Because identical twins are the result of the fertilization of a single ovum by a single spermatozoon, they always are the same sex and have identical genes. However, they may have separate placentae, depending on the stage of development at which the inner cell mass separates.

Fraternal, or *dizygotic*, *twins* occur when two ova are released during a single ovarian cycle and both are subsequently fertilized by separate spermatozoa. Fraternal twins, therefore, can be of the opposite sex and they can differ just as much as any siblings. Fraternal twins always have separate placentae.

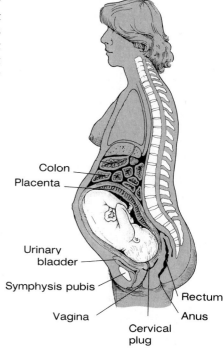

Figure 29.8
Sagittal section of a woman in a late stage of pregnancy. Notice how the rectum and the urinary bladder are compressed by the enlarged uterus.

PARTURITION

Parturition is the process by which the fetus is expelled from the uterus. In the last three months of gestation, the uterus undergoes weak, irregular contractions. During the final weeks of pregnancy, the contractions gradually become stronger and more frequent. Then, within a few hours, the contractions become quite strong and occur about every 30 minutes. This marks the beginning of labor.

par-toor-ish'-un

Labor

During *labor*, uterine contractions travel in peristaltic waves from the upper portion of the uterus downward toward the cervix, and each wave becomes progressively weaker as it approaches the cervix. As labor progresses, the contractions become stronger and more frequent, eventually occurring only one to three minutes apart. The uterine contractions propel the fetus toward the cervical canal and cause the cervix of the uterus to dilate. In addition, the uterine contractions exert pressure on the amniotic sac. When the pressure becomes great enough, the amnion bursts, releasing the amniotic fluid, which escapes through the vagina.

Delivery

In over 90% of births, the head of the fetus is downward so that the head is delivered first. This manner of delivery allows the head to act as a wedge, forcing open the cervical and vaginal canals as contractions of the uterus

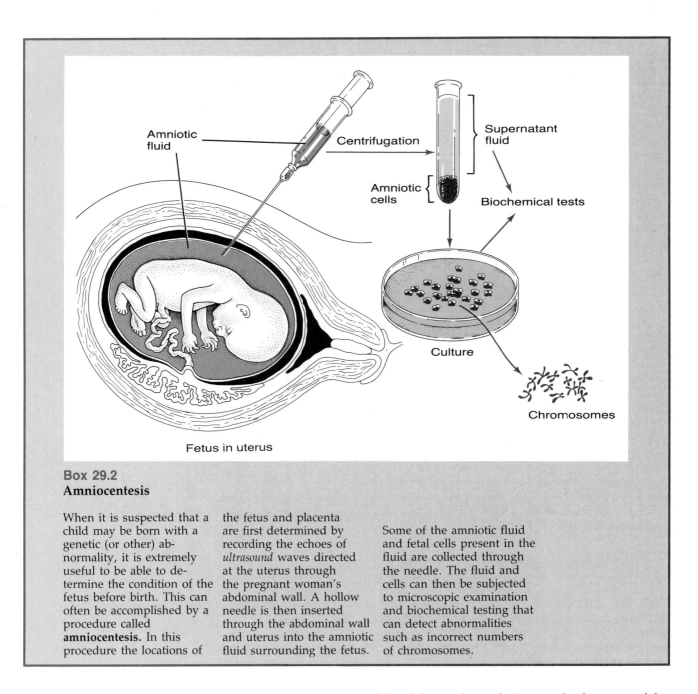

Fetus in uterus

Box 29.2
Amniocentesis

When it is suspected that a child may be born with a genetic (or other) abnormality, it is extremely useful to be able to determine the condition of the fetus before birth. This can often be accomplished by a procedure called **amniocentesis.** In this procedure the locations of the fetus and placenta are first determined by recording the echoes of *ultrasound* waves directed at the uterus through the pregnant woman's abdominal wall. A hollow needle is then inserted through the abdominal wall and uterus into the amniotic fluid surrounding the fetus. Some of the amniotic fluid and fetal cells present in the fluid are collected through the needle. The fluid and cells can then be subjected to microscopic examination and biochemical testing that can detect abnormalities such as incorrect numbers of chromosomes.

(assisted by contractions of the abdominal muscles) move the fetus out of the uterus. The head-downward position also has the advantage of allowing the fetus to breathe safely even before it is completely free of the birth canal.

Within a few minutes after delivery, the uterus contracts to a very small size. This contraction separates the placenta from the wall of the uterus and expels it as the *afterbirth*. Bleeding is generally minimal during this process because the placental blood vessels constrict and the contraction of the uterus squeezes closed the uterine vessels that supply blood to the placenta.

Within about five weeks following birth, the uterus returns to its normal size and the endometrial surface heals. If the mother is not nursing her child, menstrual cycles generally resume by about the end of the sixth week.

Factors Associated with Parturition

It is still not certain what initiates parturition. However, a number of different factors are believed to be involved, among which are the following:

1. The hormone oxytocin, which is released by the pars nervosa of the pituitary gland, is particularly effective at stimulating

uterine contractions late in pregnancy, although it is less effective early in pregnancy. Before the onset of labor, oxytocin levels are relatively low in both maternal and fetal (umbilical artery) blood. In very early labor, the oxytocin level in maternal blood remains low, but the level in fetal blood rises. In advanced labor, the oxytocin levels in both maternal and fetal blood are elevated.

2. Prostaglandins that can stimulate uterine smooth muscle are synthesized by the placenta, and their release increases during labor. Moreover, there is evidence that oxytocin stimulates prostaglandin production by the placenta.

3. The concentration of oxytocin receptors in the uterus and the placenta increases during pregnancy and is high at parturition. In experimental animals, it has been found that estrogens and uterine distension can increase the concentration of oxytocin receptors. During pregnancy, the level of estrogens gradually increases (see Figure 29.2), and near the time of birth the fetus grows rapidly, accelerating uterine distension.

F29.2

These observations have resulted in the following theory to explain the initiation of parturition. During pregnancy, the concentration of oxytocin receptors increases greatly, probably under the influence of rising estrogen levels and uterine distension. Although there is no substantial elevation in circulating oxytocin at the onset of labor, the rising concentration of receptors increases the oxytocin sensitivity of the uterus to the point where relatively strong uterine contractions are stimulated. In addition, oxytocin (much of which may come from the fetus) binds to placental receptors, stimulating prostaglandin production, and other substances released by the fetus at or near the time of birth may enhance prostaglandin production. Prostaglandins diffuse to the uterine muscle, where they act with oxytocin to produce even stronger uterine contractions. (In fact, some researchers believe that prostaglandins provide the major stimulus for uterine contractions at parturition.)

As uterine contractions move the fetus downward toward the birth canal, the fetus stimulates the cervix of the uterus. This stimulation triggers a neuroendocrine reflex that further strengthens the contractions. Nerve impulses transmitted from the cervix to the hypothalamus cause the pars nervosa of the maternal pituitary to release additional oxytocin, which enhances the contractions of the uterine muscle. Moreover, the stimulation of the cervix also seems to enhance uterine contractions by way of signals transmitted either neurally or along the uterine muscle to the body of the uterus. The enhanced uterine contractions push the fetus even more forcefully against the cervix, leading to the release of more oxytocin and still stronger uterine contractions. Thus, as described in Chapter 1, these activities act in a positive-feedback fashion to produce progressively stronger uterine contractions that expel the fetus from the uterus.

CONDITIONS OF CLINICAL SIGNIFICANCE

Pregnancy

Disorders of Pregnancy

Ectopic Pregnancy

The blastocyst normally becomes implanted in the upper portion of the uterine endometrium. However, in some cases, implantation occurs at a site other than the uterus. Such an occurrence is called an *ectopic pregnancy*. The most common ectopic pregnancy is a *tubal pregnancy*, which occurs within a uterine tube.

Ectopic pregnancies can endanger the life of the mother. They frequently cause internal hemorrhaging, since the trophoblast of the blastocyst destroys tissue at the site of implantation. The endometrium, which is the normal site of implantation, can withstand the destructive effects of the trophoblast, but most tissues simply break down. The lack of room for expansion is another danger in ectopic pregnancies. In the case of a tubal pregnancy, for example, the uterine tube may rupture as the embryo grows.

Placenta Previa

Occasionally, when implantation occurs in the lower regions of the uterus, the placenta covers the inner opening of the cervical canal. This condition is called *placenta previa*—that is, placenta leading the way. In this position, the placenta can irritate the cervix, which causes the uterus to contract and can result in a spontaneous abortion. If the pregnancy goes to term, the placenta is expelled before the baby is born, which causes the mother to hemorrhage severely. The expulsion of the placenta prior to the expulsion of the fetus can also stimulate the fetus to begin breathing while still within the birth canal.

Toxemia of Pregnancy

Toxemia (poison blood) associated with pregnancy continues to be a serious problem. Despite intensive research, the cause of toxemia is not known. It is probable that numerous factors produce toxemia—including malnutrition, metabolic disorders, endocrine imbalances, production of toxins, and a decreased blood supply to the uterus. As a prelude to toxemia, the woman often shows a strong tendency to reabsorb salt from the renal tubules, and with it, water. This excessive reabsorption of salt and water can usually be controlled with a low-salt diet or by the use of diuretic drugs to increase water loss.

If toxemia is unchecked, among other symptoms there may be excessive weight gain; edema of the lungs, kidneys, and brain; visual disturbances; and, eventually, convulsions. If the convulsive stage is reached, the condition is called *eclampsia*.

Birth Control

Over the years much effort has been expended in attempting to develop a convenient safe, and effective method of birth control, and a variety of birth control methods are now in widespread use.

Birth Control Methods Available to the Male

COITUS INTERRUPTUS   *Coitus interruptus* is the withdrawal of the penis just before the male orgasm so that ejaculation does not occur within the female tract. It is not a reliable method because it requires an unnatural response and perfect timing by the male.

CONDOM   A *condom* is a very thin sheath of rubber, plastic, or animal membranes that is pulled onto the erect penis just before intercourse. Condoms are an effective method of preventing fertilization as long as they do not tear or slip off after orgasm and allow semen to enter the vagina. They have the added advantage of providing protection against venereal disease.

VASECTOMY   *Vasectomy* is a form of surgical sterilization in which each ductus deferens is tied and cut. This procedure does not interfere with normal ejaculation, but spermatozoa are prevented from entering the semen.

Birth Control Methods Available to the Female

DETECTION OF OVULATION   Several methods of birth control rely on a woman's ability to determine the time of ovulation and to abstain from sexual intercourse for several days preceding and following ovulation. In one method, called the *rhythm method*, a woman keeps records of the lengths of her menstrual cycles and from these records predicts the length and likely time of ovulation for her current cycle. The rhythm method may be an effective method of birth control if the woman's menstrual cycles are regular and if the period of abstinence from sexual intercourse is long enough. However, menstrual cycles can vary in length, and the timing of ovulation can vary even when menstrual cycles are regular. Consequently, the likely time of ovulation during a current cycle can be difficult to predict accurately from records of past cycles.

Another method used to determine the time of ovulation is the measurement of body temperature. In this method, a woman determines her basal body temperature with a thermometer each morning upon awakening and prior to getting out of bed. Generally, a woman's basal body temperature will be somewhat lower prior to ovulation than it will be following ovulation. (The higher temperature following ovulation has been related to rising levels of progesterone.)

A third method for determining the time of ovulation is observation of the mucus secretions of the uterine cervix that are discharged at the vagina. Prior to ovulation, estrogens increase the quantity of alkaline cervical mucus that is secreted, and they decrease the viscosity and the cellularity of the mucus. These occurrences tend to favor the survival and transport of sperm. Following ovulation, progesterone causes the cervical mucus to be thick, tenacious, and cellular. This mucus contributes to the occurrence of conditions unfavorable for the penetration and survival of spermatozoa. At the vagina, following menstrual bleeding there are several days during which no mucus discharge is evident. These "dry days" are followed by the onset of "mucous symptoms" characterized by the appearance of increasing quantities of cloudy or sticky secretions that, in turn, are followed by the appearance of a clear, slippery, lubricative mucus. The clear, slippery, lubricative mucus is then followed by the occurrence of thick, tacky, opaque mucus. Generally, the period from the beginning of the mucous symptoms until the fourth day after the last day of appearance of clear, slippery, lubricative mucus is considered to be a woman's fertile period. (With this method the period of menstrual bleeding is also considered to be a fertile period).

DIAPHRAGM   A *diaphragm* consists of a thin hemispherical dome of rubber or plastic with a spring margin. It covers the cervix, thereby preventing the entrance of sperm into the uterus. A diaphragm, which is initially fitted by a physician, is generally used in combination with a spermicidal cream or jelly. Although a properly fitted diaphragm is quite effective, its use requires the woman to predict the occurrence of sexual intercourse so the diaphragm will be in place when needed, or else she must interrupt the prelude to the act while she inserts it.

INTRAUTERINE DEVICE An *intrauterine device* (IUD) consists of a small spiral, ring, or loop made of plastic, stainless steel, or copper that is inserted into the uterine cavity by a physician. Once in place, the IUD can remain within the uterus for a long time. An IUD does not affect the menstrual cycle, but it does prevent the implantation of the blastocyst into the endometrium. Some women's bodies cannot tolerate an IUD and expel it from the uterus. There have been so many health problems with certain forms of IUDs that they have been removed from public use.

ORAL CONTRACEPTIVES *Oral contraceptives*, better known as "the pill," are extremely effective methods of preventing pregnancy. There are several types of pills available, most of which contain a combination of synthetic estrogenlike and progesteronelike substances. When taken daily for the first 20 or 21 days of the menstrual cycle, these substances apparently prevent the rise in luteinizing hormone that leads to ovulation. Even if ovulation should occur (as it apparently does in some cases), oral contraceptives can still prevent pregnancy, presumably by thickening and chemically altering the cervical mucus so that it is more hostile to sperm and by making the uterine endometrium less receptive to implantation. In some women, oral contraceptives have

undesirable side effects, the most serious of which is a tendency to form blood clots.

CHEMICAL CONTRACEPTION Various types of douches, foams, suppositories, creams, and jellies that are introduced into the vagina either just before or after intercourse are available. Most of these prevent fertilization by acting as physical barriers to sperm and by serving as spermicides. They are most effective when used in combination with a diaphragm.

TUBAL LIGATION *Tubal ligation* is a form of surgical sterilization in which the uterine tubes are tied and severed. This procedure prevents spermatozoa from reaching the ovum to fertilize it, and it prevents the ovum from reaching the uterus.

INDUCED ABORTION In recent years, *induced abortion* of the embryo or fetus has become a more common method of birth control, although the moral issues involved continue to be debated. Induced abortion involves using various means to separate the implanted embryo or fetus from the wall of the uterus. The detachment of the embryo or fetus is generally accomplished either by scraping, by saline solution rinses, or by vacuum aspiration (suction).

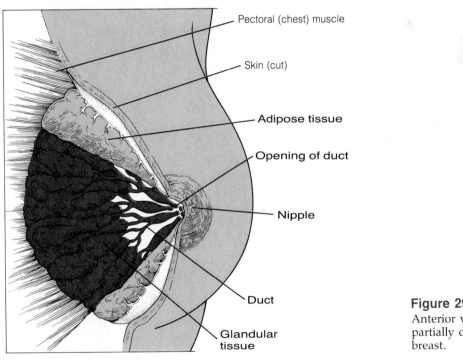

Figure 29.9
Anterior view of a partially dissected left breast.

LACTATION

Lactation, or the production of milk by the breasts (mammary glands), requires the interaction of several hormones.

A fully developed breast is composed of glandular tissue surrounded by adipose tissue (Figure 29.9). The glandular tissue consists of little sacs of   F29.9

Obstetrical Ultrasonics

Imagine having a window in the womb, through which physicians could view the largely secret world of the fetus. Such a window would be useful in detecting congenital birth defects, the position of the fetus just before delivery, and more, and could help save thousands of lives each year.

Is such an idea preposterous? In the strictest sense, perhaps it is. There is no serious thought of developing a porthole through which to observe the growing fetus. Medical scientists, however, now offer the next best thing, using a special technique called ultrasound imaging.

Ultrasound became quite popular in the 1970s as a "safe" substitute for X rays. X rays were once used to determine, among other things, the position of the fetus just before birth, but as more knowledge of the harmful effects of radiation became available, the use of X rays in obstetrics was greatly reduced.

The equipment needed for ultrasonic viewing of the unborn is relatively simple. A probe placed on the woman's abdomen gives off sound waves pitched so high that they escape detection by the human ear. Just as radar waves bounce off aircraft, ultrasound bounces off the tissues and travels back to a special receiving device in the probe. The information received is relayed to a small computer, which interprets the signals and creates a moving picture of the baby. Ultrasound images allow physicians to check for physical malformations such as missing limbs, determine the sex of the fetus, make measurements to determine whether growth is proceeding normally, and so forth.

Ultrasound imaging is also a valuable tool for physicians who specialize in surgery on the fetus. With the aid of this technique, surgeons have guided a needle into a fetus's bladder to drain urine that had accumulated there because of an obstruction in the urethra, and into the brain to drain the fluid-filled ventricles of a hydrocephalic fetus.

Ultrasound is proving to be a valuable tool in the medical laboratory as well. Scientists are using it to gain a better understanding of fetal motor development—determining what reflexes and what movements a fetus has and when they develop. The benefits of research along these lines are far reaching. Perhaps most important, a better understanding of fetal neural development will help physicians detect abnormal central nervous system development, which produces physiological symptoms that are much more subtle than gross physical defects. As an example, normal fetuses will turn away from the sound of a buzzer; a fetus with a hearing problem will not. Ultrasonic imaging allows the observation of such fetal movements. Early detection of deafness and other neurological defects offers hope of effective medical treatment.

Few people doubt the usefulness of ultrasound in medical diagnosis. Many medical scientists, however, warn that heat generated by ultrasound may have subtle, long-term effects on cells, perhaps causing cancer or genetic mutations.

Such concerns raise the question of just how safe ultrasound is. To date, at least 35 studies in animals and a smaller number in humans have failed to answer the risk question to the satisfaction of the medical community. While there is no evidence that ultrasound causes birth defects, retards physical growth, or induces cancer, more work is needed to assess fully the potential long-term effects. A number of laboratory studies which indicate that ultrasound can transform fibroblasts into tumor cells are cause for concern. Ultrasound has also been shown to produce changes in the surface properties and the motility of fibroblasts, and human lymphocytes undergo mutation when exposed to ultrasound in the laboratory. Whether this also happens in the body is not known.

The significance of these alterations must be determined before this promising technique can find a permanent niche in the field of obstetrics.

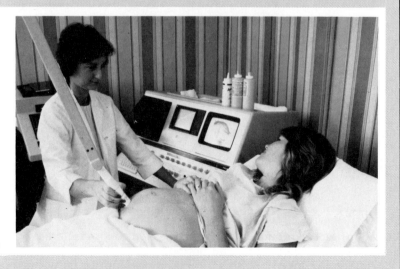

Ultrasound imaging.

secretory cells called *alveoli* that secrete milk into epithelium-lined ducts. The ducts carry the milk to the surface of the breast at the nipple. The alveoli and the initial portions of the ducts are surrounded by specialized contractile cells called *myoepithelial cells*.

Preparation of the Breasts for Lactation

Prior to puberty, the breasts are largely undeveloped. At puberty, the increased level of estrogens stimulates the growth and branching of the duct system within the breasts, and there is considerable breast enlargement due to fat deposition. The increased level of progesterone at puberty also contributes to breast growth. However, the alveoli do not develop completely, and the changes that occur at puberty do not result in milk production.

During pregnancy, the breasts enlarge still more, and the alveoli develop fully. These changes are due to the stimulatory effects of the high concentrations of estrogens and progesterone during pregnancy, as well as to the influence of the pituitary hormone **prolactin.** Prior to puberty, the prolactin level is low. However, estrogens stimulate prolactin secretion, and the prolactin level increases at puberty and rises still higher during pregnancy. (It is not known if estrogens influence prolactin secretion by way of the hypothalamus or by direct action on the pituitary.)

Production of Milk

Prolactin is the major hormone responsible for milk production. However, despite its relatively high levels (and the full development of the breasts) during pregnancy, significant milk production does not occur. This lack of milk production is due to the fact that the high levels of estrogens and progesterone present during pregnancy inhibit the milk-producing action of prolactin on the breasts (even though estrogens enhance prolactin secretion and act with it in promoting breast growth and development). At parturition, the loss of the placenta leads to a decline in the levels of estrogens and progesterone, and prolactin is able to stimulate milk production.

Following parturition, a reflex response is important in stimulating prolactin secretion. Mechanical stimulation of the nipples by the suckling infant initiates nerve impulses that are transmitted to the hypothalamus. The nerve impulses ultimately lead to prolactin release either by inhibiting the release of prolactin-inhibiting factor or by stimulating the release of a prolactin-releasing factor. Thus, even though the basal level of prolactin secretion returns to the prepregnancy level several weeks after parturition, when suckling is vigorous and frequent a substantial surge in prolactin secretion occurs each time the mother nurses her infant. The prolactin acts on the breasts to continue milk production. If suckling is less vigorous and frequent (as may occur after several months of nursing as the infant begins to eat significant amounts of other food), prolactin levels decline, and within about three months there is relatively little prolactin secretion in response to suckling. (In this regard, it is interesting to note that the rate of milk production by many nursing mothers declines substantially within seven to nine months after parturition.) If nursing is stopped (or if it never began), prolactin release in response to suckling does not occur and lactation ceases.

Milk Let-Down Reflex

The milk produced by the mammary glands tends to remain within the alveoli unless the specialized myoepithelial cells that surround the alveoli contract and force it into the ducts. The *milk let-down reflex*, which is initiated by suckling, leads to the contraction of the myoepithelial cells and the ejection of milk into the ducts of both breasts—not just the one being nursed—within a minute after suckling begins. In this reflex, the stimulation of the nipple by suckling sends nerve impulses to the hypothalamus, which leads to the release of oxytocin from the pars nervosa of the pituitary gland (Figure 29.10). Oxytocin causes the myoepithelial cells to contract, and this ejects the milk into the ducts, where it can be easily obtained by the nursing infant.

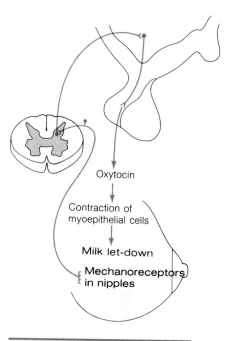

Oxytocin

Contraction of myoepithelial cells

Milk let-down

Mechanoreceptors in nipples

Figure 29.10
The milk let-down reflex. Stimulation of the nipples elicits the release of oxytocin from the neurohypophysis of the pituitary.

F29.10

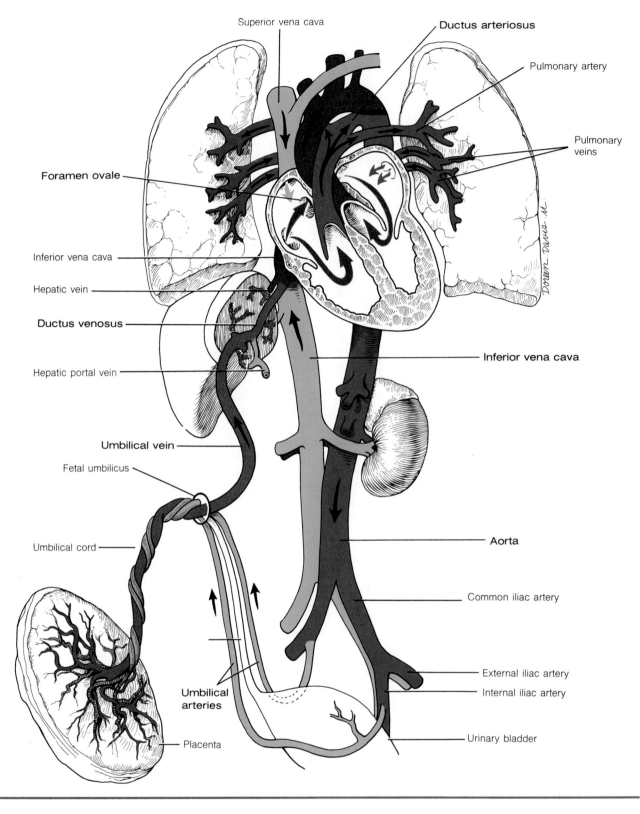

Figure 29.11

Fetal circulation. Blood that is highly oxygenated and rich in nutrients leaves the placenta and enters the fetus through the umbilical vein. After circulating through the fetus, the blood returns to the placenta through the umbilical arteries. The ductus venosus, the foramen ovale, and the ductus arteriosus allow the blood to bypass the fetal liver and lungs. The amount of color in a vessel indicates the relative proportion of oxygenated blood carried by the vessel.

Higher brain centers also influence milk let-down. For example, it is not unusual for a nursing mother to leak milk in response to the sound of a crying infant, even prior to physical contact between her and the child.

FETAL CIRCULATION

The circulatory system of the fetus contains some unique features that are well adapted to development in the uterus but must change dramatically following birth if the child is to survive.

Fetal Blood Pathways

Fetal blood travels to the placenta in the **umbilical arteries,** which are branches of the internal iliac arteries (Figure 29.11). After circulating through the capillaries of the chorionic villi, where exchanges of gases, nutrients, and wastes occur, the blood leaves the placenta in a single **umbilical vein.** The umbilical vein, therefore, carries blood that is rich in nutrients and has a high oxygen content. It is the only fetal vessel to do so. All other vessels of the fetus carry mixtures of arterial and venous blood. Between the placenta and the **umbilicus (navel)** of the fetus, the umbilical arteries and the umbilical vein travel within the umbilical cord. Part of the blood in the umbilical vein enters the liver and follows the usual path of blood through that organ and into the inferior vena cava. The remainder joins with blood from the hepatic portal vein and enters a bypass around the liver called the **ductus venosus.** The ductus venosus directly enters the inferior vena cava. Having much of the fetal blood bypassing the fetal liver causes no functional problems, since the mother's liver serves the needs of the fetus as the result of the exchange of materials across the placenta.

F29.11

In the fetus, blood from the inferior vena cava enters the right atrium, from which it can follow several pathways. It can follow the usual route, for example, entering the right ventricle and traveling to the lungs within the pulmonary arteries. Since the lungs are collapsed and nonfunctioning, however, they offer considerable resistance to blood flow. Therefore, most of the blood that leaves the right ventricle never reaches the lungs. Instead, it flows through a low-resistant shunt called the **ductus arteriosus,** which connects the pulmonary trunk to the arch of the aorta.

Another pathway is for the blood in the right atrium of the fetus to enter the left atrium directly by means of a flap-covered opening, called the **foramen ovale,** in the interatrial septum. The blood then passes into the left ventricle and the aorta. The foramen ovale is located in such a position that most of the blood from the inferior vena cava (which is highly oxygenated) is directed through it. Since the brain is supplied by arteries that branch off the aortic arch, the brain thus receives a good portion of this highly oxygenated blood immediately after it leaves the left ventricle. In contrast, much of the blood that enters the heart from the superior vena cava (venous blood) is directed into the right ventricle and pulmonary trunk. As we have noted, most of this blood passes through the ductus arteriosus, which joins the aorta beyond the arch. This blood is destined for the trunk and lower limbs of the fetus, where the oxygen demand is not so vital as that of the brain tissues.

Circulatory Changes at Birth

Separation of the infant from the placenta, which occurs when the umbilical cord is cut following birth, suddenly deprives the newborn of its oxygen source and its means of eliminating carbon dioxide. The resultant oxygen shortage and buildup of carbon dioxide cause the infant to take its first breath. The expansion of the lungs immediately lowers their resistance to blood flow, which allows as much as five times more blood to pass through the lungs after birth than before. This action also increases the return of blood from the lungs to the heart, which in turn increases the pressure in the left atrium. The increased pressure in the left atrium forces the flap of the foramen ovale against the interatrial septum, which immediately blocks the opening. All

F29.12

blood that enters the right atrium must then travel to the right ventricle and the pulmonary trunk. Eventually, the foramen ovale becomes permanently closed with fibrous connective tissue, leaving an indentation, called the **fossa ovalis,** on the interatrial septum (Figure 29.12). If the infant's foramen ovale fails to close, the higher pressure in the left atrium forces some of the blood that has returned from the lungs into the right atrium, thus recirculating the blood back to the lungs again. If the defect is large enough, the volume of blood flowing from the left atrium into the right atrium can cause pulmonary congestion and failure of the right ventricle. A severe defect in the interatrial septum must be surgically corrected.

With the lowered resistance in the lungs following birth, the pressure in the pulmonary arteries and the chambers of the right side of the heart is also reduced. As a result, blood is no longer forced through the ductus arteriosus into the aorta, where the pressure is now greater. In fact, for a short time following birth, blood may flow backward through the duct, from the aorta to the pulmonary artery. Within a few hours, however, the ductus arteriosus constricts, which prevents the reverse flow of blood from the aorta to the pulmonary trunk. Within a few weeks, the ductus arteriosus is generally completely obliterated by fibrous connective tissue and is then called the **ligamentum arteriosum.**

Once the umbilical cord is cut, blood no longer flows through the umbilical vessels or the ductus venosus. The proximal portions of the umbilical arteries continue to supply the urinary bladder—which has developed from the proximal portion of the allantois of the fetus—as **superior vesical arteries.** Beyond the bladder, the umbilical arteries fill in with connective tissue and travel on the inner wall of the abdomen to the umbilicus as the **umbilical ligaments.** In a similar manner, the umbilical vein becomes obliterated with connective tissue, and is then called the **ligamentum teres** (*round ligament of the liver*). Because blood is no longer carried in the umbilical vein, the ductus venosus gradually constricts, shunting more and more of the newborn's blood from the hepatic portal vein into the liver. Within a few weeks, the ductus venosus is permanently obliterated, remaining as the **ligamentum venosum.** With the closure of the ductus venosus the bypass around the liver is lost, so all the blood from the hepatic portal vein must then pass through the liver. This change is of vital importance because the hepatic portal vein drains the digestive system and the infant no longer has the mother's liver available to act on the various substances carried in the blood.

HUMAN GENETICS

The study of heredity in humans is called *human genetics*. A person's hereditary material is the DNA of the cells. The DNA provides a program, or set of instructions, that determines personal characters such as eye color or blood type. The DNA program is subdivided into informational units called **genes,** which control specific characters by presiding over the synthesis of chemical components of a person's cells. Each gene is a segment of a DNA chain, and most genes operate by specifying the sequence in which amino acids are joined to form polypeptides or proteins (see Chapter 3). Once formed, the polypeptides or proteins act as enzymes, as hormones, and in other ways to influence particular characters (the actual form of a character that is expressed in a person—for example, brown eyes or blue eyes—is referred to as a *trait*).

Human Chromosomes

Each human somatic cell (all cells except the reproductive cells) is a diploid cell that contains 46 chromosomes. Two of the chromosomes are **sex chromosomes** (two X chromosomes in females and one X and one Y chromosome in males). The remaining 44 chromosomes are called **autosomes.** The 44 autosomes can be grouped into 22 pairs of similar-appearing chromosomes (see Figure 3.21). One member of each pair contains genes derived from the person's father, and the other member of each pair contains genes derived from

F3.21

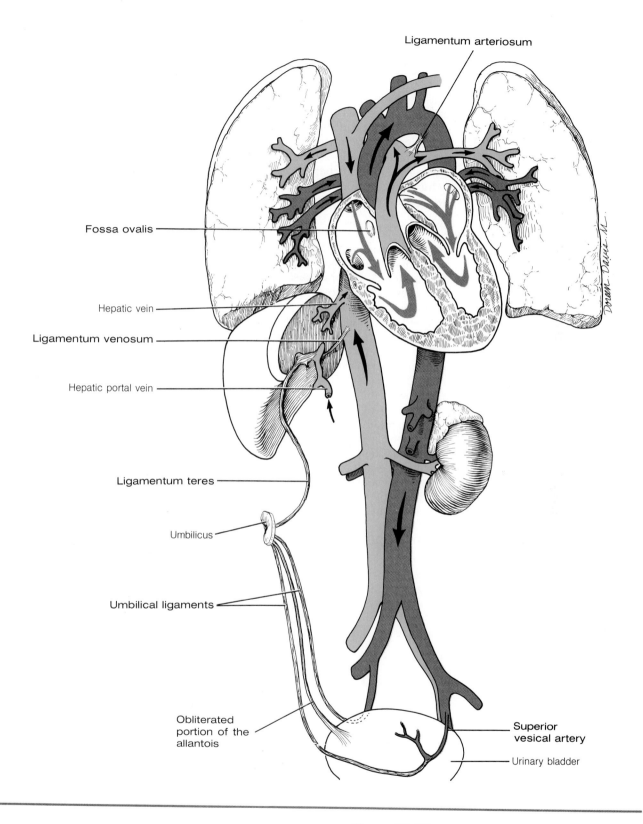

Ligamentum arteriosum

Fossa ovalis

Hepatic vein

Ligamentum venosum

Hepatic portal vein

Ligamentum teres

Umbilicus

Umbilical ligaments

Obliterated
portion of the
allantois

**Superior
vesical artery**

Urinary bladder

Figure 29.12

Changes in circulatory
pathways following birth.
These changes result from
the closure of the fora-
men ovale and the trans-
formation of the ductus
venosus, the ductus

arteriosus, the umbilical
vein, and the distal
portions of the umbilical
arteries into ligaments.
Color indicates vessels
carrying oxygenated
blood.

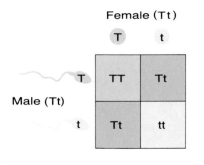

Figure 29.13
Use of a Punnett square to indicate possible genotypes of a fertilized ovum when a male and female who are both heterozygous for tongue rolling mate with one another.

the person's mother. Each pair comprises a set of **homologous chromosomes.** The two sex chromosomes of the female (XX) are also homologous, but the two sex chromosomes of the male (XY) are not.

Genes are arranged linearly along the chromosomes, and each gene occupies a specific position, or **locus,** on a particular chromosome. Homologous chromosomes each possess genes that control the same characters.

Genes that control the same character and that occupy the same locus on homologous chromosomes are called **alleles.** If a gene (an allele) for a particular character on one chromosome of a homologous pair is identical to the gene (the allele) for that character on the other chromosome of the pair, the person is **homozygous** for the character. If the two genes differ from one another, the person is **heterozygous** for the character.

Dominance

In some cases, one allele for a particular character can mask, or suppress, the effect of another allele for the character. In such a case, the first allele is said to be **dominant,** and the masked allele is referred to as **recessive.** In a homozygous person who possesses two dominant alleles for a particular character (a homozygous dominant), the effect of the dominant alleles is expressed, and the person is said to display a dominant trait with regard to the character. In a homozygous person who possesses two recessive alleles for the character (a homozygous recessive), the effect of these alleles is expressed, and the person displays a recessive trait. In a heterozygous person who possesses both dominant and recessive alleles for the character, the effect of the dominant allele is expressed, and the effect of the recessive allele is masked. Consequently, the person displays the dominant trait.

It is not always the case that one allele for a particular character masks the effects of a second allele for the character—that is, sometimes neither allele for a character on homologous chromosomes is completely dominant or completely recessive. In such a situation, the effects of both alleles are often expressed to some degree, and the person displays a trait that is intermediate between the traits that would be evident if one allele or the other were the only form of the gene present.

Genotype and Phenotype

A person's actual genetic makeup is called his or her **genotype.** The outward expression of the genotype—that is, the expressed traits—is the person's **phenotype.** In the case of dominant and recessive alleles, for example, homozygous dominant, homozygous recessive, and heterozygous individuals have different genotypes, but homozygous dominant and heterozygous individuals have the same phenotype.

Inheritance Patterns

Genetic traits can be passed from one generation to the next. For example, the ability to roll one's tongue is inherited in the manner expected for a dominant allele located on an autosomal chromosome. (The dominant allele for tongue rolling will be indicated by T and the recessive allele by t.) Thus, a person who is either homozygous dominant (TT) or heterozygous (Tt) will be able to roll his or her tongue, but a person who is homozygous recessive (tt) will not.

When a heterozygous individual who possesses both dominant and recessive alleles for tongue rolling forms haploid reproductive cells, the events of meiosis separate the two alleles in such a manner that half the gametes produced contain the dominant allele (T), and half contain the recessive allele (t). If a heterozygous male mates with a heterozygous female, there is a 50% probability that the sperm that fertilizes the ovum will contain the dominant allele (T) and a 50% probability it will contain the recessive allele (t). Similarly, there is a 50% probability that the ovum will contain the dominant allele and a 50% probability it will contain the recessive allele. This relationship and an indication of the possible genotypes of the fertilized ovum can be demonstrated in a *Punnett square,* as in Figure 29.13. From this Punnett square it is

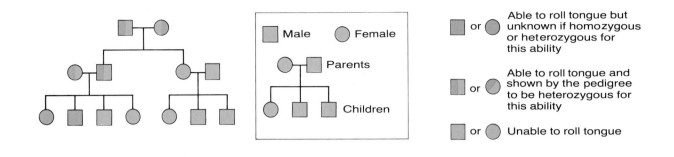

Figure 29.14
Pedigree of a hypothetical family illustrating the inheritance of tongue-rolling ability. The facts that parents with tongue-rolling ability can have children without this ability and parents without this ability have only non–tongue-rolling children indicates that tongue-rolling ability is inherited in the manner expected for a dominant allele. Moreover, the fact that parents who both have tongue-rolling ability have a child without this ability shows that the parents are heterozygous for this ability.

evident that there is a 25% probability that the person developing from the fertilized ovum will be homozygous dominant (TT) for tongue rolling, a 50% probability that the person will be heterozygous (Tt), and a 25% probability that the person will be homozygous recessive (tt). Phenotypically, there is a 75% probability that the person will display tongue-rolling ability (recall that a person who is either homozygous dominant or heterozygous for tongue rolling will be able to roll his or her tongue), and a 25% probability that the person will not.

The pattern of inheritance of a genetic trait in the members of different generations of a family can be recorded by constructing a *pedigree pattern*. Such patterns are used in human genetics to examine the manner in which particular traits are inherited. For example, the pedigree pattern for tongue-rolling ability illustrated in Figure 29.14 indicates that this ability is inherited in the manner expected for a dominant allele.

F29.14

Family pedigrees have proven to be useful for *genetic counseling*, during which prospective parents are advised concerning the chances of their children having a given abnormality. For some conditions, the counseling can rely heavily on pedigree patterns. For example, if one parent is heterozygous for the sickle cell anemia gene (see Chapters 3 and 17)—which causes abnormal hemoglobin in red blood cells—each child has a 50% chance of inheriting the gene. Most situations that require genetic counseling are too complicated to rely entirely on pedigree patterns, since they involve rare recessive alleles that remain masked in heterozygous individuals for many generations and are harmful only when present in the homozygous form. To detect such conditions it is often useful to perform biochemical analyses of cells and blood taken from the parents in an attempt to identify products of abnormal recessive alleles that may be present.

X-Linked Traits

Only a single X chromosome occurs in males. Consequently, the effect of a recessive allele located on the X chromosome of a male is generally expressed because there is no dominant allele to mask it. (The Y chromosome apparently has few or none of the same genes as the X chromosome, and it therefore does not mask recessive alleles on the X chromosome.) As a result, the effects of recessive alleles located on an X chromosome are more frequently expressed in males than in females. This situation is evident in the case of the blood-clotting abnormality hemophilia A (see Chapter 17), which is inherited in the manner expected for a recessive allele located on the X chromosome. To display this abnormality, females must have the recessive allele on both X chromosomes, but males must have the allele only on their single X chromosome. If a heterozygous female who has the recessive allele for hemophilia A

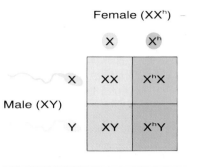

Female (XX^h)

	X	X^h
X	XX	X^hX
Y	XY	X^hY

Male (XY)

Figure 29.15
Use of a Punnett square to indicate possible genotypes of a fertilized ovum when a normal male who does not have hemophilia A mates with a female who does not have hemophilia A but does have the recessive allele for hemophilia A on one of her X chromosomes (indicated by X^h).

on one of her X chromosomes mates with a normal male, none of their female children will have hemophilia A. However, there is a 50% probability that their male children will have this disorder (Figure 29.15). In this case, neither parent has hemophilia A, but the mother is termed a *carrier* because she possesses a recessive allele for hemophilia A that can be passed on to her children.

Expression of Genetic Potential

In previous sections it was assumed that the effect of a dominant allele for a particular character is always expressed and, consequently, that a person who possesses such an allele will display the dominant trait with regard to the character. Although this assumption is generally true, cases are known in which a person possessing a dominant allele for a character does not display the dominant trait. These cases serve to emphasize the fact that individual genes operate as part of a person's overall genetic makeup and that, at times, the activities of genes for other characters may modify or otherwise influence the expression of the effect of a particular gene. Moreover, environmental factors can also influence the expression of the effects of particular genes. For example, a person may have the genetic capacity to attain great height, but the actual attainment of this height depends on the availability of the proper amounts and types of food. Thus, particular genes represent potentialities whose full effects and complete expression may be influenced by the activities of other genes and by environmental factors.

CONDITIONS OF CLINICAL SIGNIFICANCE

Inheritance

Many diseases and other abnormalities have genetic origins. Some of these are the result of abnormal numbers of chromosomes in cells, and others are due to malfunctioning or altered genes. A procedure called *amniocentesis* is often useful in detecting genetic disorders (Box 29.2).

Nondisjunction

A major cause of abnormal numbers of chromosomes in cells is a process called **nondisjunction,** which can occur during meiosis (Figure 29.16). Because of an abnormal segregation of chromosomes during the second division sequence of meiosis, a gamete may receive either both copies of a particular chromosome or neither copy. If one of these abnormal gametes participates in fertilization, the fertilized ovum can have either three chromosomes of one kind (a condition known as *trisomy*) or only one of that kind (*monosomy*).

Down's Syndrome

Down's syndrome (sometimes called "mongolism") is a genetic disorder characterized by a fairly consistent pattern of physical features and almost always by mental retardation. The most common and extreme form of Down's syndrome is due to the presence of three copies of autosome number 21, a condition known as *trisomy 21*. Trisomy 21 occurs most often in children born to mothers over 35 years old.

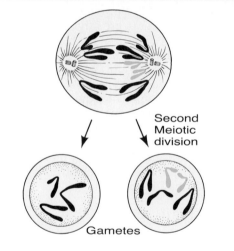

Figure 29.16
Nondisjunction occurring during meiosis. During the second division sequence of meiosis, both chromatids of a double-stranded chromosome go to one pole of the dividing cell. This results in the production of some gametes that have an extra chromosome and others without this chromosome. The nondisjunct chromosomes are shown in color.

Abnormal Numbers of Sex Chromosomes

When gametes with abnormal numbers of sex chromosomes are involved in fertilization, the result is a fertilized ovum with an incorrect number of X and/or Y chromosomes. A person developing from such a fertilized ovum is often abnormal. Table 29.2 lists some of the conditions associated with abnormal numbers of sex chromosomes.

Galactosemia

Galactosemia is a genetic disorder that is inherited in the manner expected for a recessive allele located on an autosomal chromosome. Babies afflicted with this condition cannot tolerate the milk sugar lactose. If they drink milk, they develop cataracts of the eyes and show symptoms of damage to the brain and liver. However, if newborn galactosemic babies are taken off milk soon after birth and fed special lactose-free formulas, they can develop normally.

Table 29.2   Normal and Abnormal Numbers of Sex Chromosomes in Humans

Chromosomes	Sex	Characteristics
XX	Female	Fertile
XY	Male	Fertile
XO(no Y)	Female	Turner's syndrome*
XXX	Female	Fertile or sterile
XXXX	Female	Fertile or sterile
XXY	Male	Klinefelter's syndrome*
XXXY	Male	Klinefelter's syndrome*
XYY	Male	Fertile

*Victims of Turner's and Klinefelter's syndromes fail to mature sexually (are sterile) and often show mental impairment.

STUDY OUTLINE

PREGNANCY   pp. 845–855

FERTILIZATION
1. Normally occurs in upper portion of uterine tube.
2. Sperm, as they pass through uterus and uterine tubes, are believed to undergo changes that make them capable of fertilizing an ovum.
3. When sperm contact corona radiata, they release enzymes that help break down portions of zona pellucida and the material that holds cells of corona together, thus permitting access to ovum.
4. One sperm passes through cell membrane of ovum and into cytoplasm.
5. Ovum completes second meiotic division, and nuclei of sperm and ovum unite.

SEX DETERMINATION
1. If ovum is fertilized by a sperm containing X sex chromosome, a female normally develops.
2. If ovum is fertilized by a sperm containing Y sex chromosome, a male develops.

DEVELOPMENT AND IMPLANTATION OF THE BLASTOCYST
1. Blastocyst is fluid-filled sphere of cells. Embryo develops from inner cell mass of blastocyst; outer sphere of cells is *trophoblast.*
2. After several days within lumen of uterus, trophoblast cells release enzymes that digest the cells of the endometrium, and blastocyst implants in uterine wall.

MAINTENANCE OF THE ENDOMETRIUM
1. High blood levels of estrogens and progesterone are required.
2. During early pregnancy, required estrogens and progesterone are produced by corpus luteum.

3. By about end of second month, placenta has developed to point where it secretes sufficient estrogens and progesterone to maintain endometrium.

DEVELOPMENT OF THE PLACENTA Provides fetal nutrient supply and waste removal; chorionic villi cover extensive surface area; connected to fetus by blood vessels in umbilical cord.

DECIDUA Portion of endometrium that is discarded at birth or with menses.

DECIDUA BASALIS AND FETAL CHORION Combine to form placenta.

FORMATION OF EMBRYONIC MEMBRANES AND GERM LAYERS

EMBRYONIC MEMBRANES

Amnion Fluid-filled, surrounds fetus; provides protection; permits movement; maintains temperature.

Yolk Sac Develops from endoderm of embryonic disc; provides nourishment.

Allantois Develops from endoderm of embryonic disc; allows for waste storage; allantoic blood vessels carry fetal blood to and from placenta.

Chorion Develops from trophoblast plus mesoderm; forms fetal portion of placenta.

GERM LAYERS Ectoderm and endoderm formed by separation of embryonic disc into two layers; mesoderm formed from cells of embryonic disc that spread between ectoderm and endoderm. *Coelom* formed when mesoderm sheet splits into parietal and visceral layers.

GESTATION Normally lasts about 280 days from beginning of last menstrual period.

TWINS

IDENTICAL TWINS An inner cell mass that develops from single fertilized ovum separates.

FRATERNAL TWINS Two ova fertilized by separate spermatozoa.

PARTURITION Process by which fetus is expelled from uterus. **pp. 855–857**

LABOR Uterine contractions propel fetus toward cervical canal; cervix of uterus dilates.

DELIVERY Contractions of uterus (assisted by contractions of abdominal muscles) move fetus out of uterus.

FACTORS ASSOCIATED WITH PARTURITION Though what initiates parturition is not certain, estrogens and prostaglandins are thought to be involved; increasing uterine sensitivity to oxytocin as pregnancy progresses may also be important.

CONDITIONS OF CLINICAL SIGNIFICANCE: PREGNANCY **pp. 857–859**

DISORDERS OF PREGNANCY

ECTOPIC PREGNANCY Implantation occurs at a site other than uterus.

PLACENTA PREVIA Placenta covers inner opening of cervical canal and is expelled before baby is born, causing mother to hemorrhage.

TOXEMIA OF PREGNANCY May be caused by numerous factors; often shows tendency for salt and fluid retention; results in lung, kidney, and brain edema.

BIRTH CONTROL

BIRTH CONTROL METHODS AVAILABLE TO THE MALE

Coitus Interruptus Withdrawal of penis before ejaculation.

Condom Cover pulled onto penis.

Vasectomy Ductus deferens is tied and cut.

BIRTH CONTROL METHODS AVAILABLE TO THE FEMALE

Detection of Ovulation Requires abstinence from sexual intercourse during fertile period.

Diaphragm Cover for cervix to prevent entrance of sperm into uterus.

Intrauterine Device Device inserted into uterus that prevents implantation of blastocyst.

Oral Contraceptives Pills containing synthetic estrogenlike and progesteronelike substances that inhibit ovulation, fertilization, or implantation.

Chemical Contraception Chemicals introduced into vagina that act as physical barriers to sperm and as spermicides.

Tubal Ligation Tying and severing uterine tubes.

Induced Abortion Removal of implanted embryo or fetus from uterus.

LACTATION **pp. 859–863**

PREPARATION OF THE BREASTS FOR LACTATION
1. At puberty, estrogens stimulate growth and branching of duct system within breasts, and fat deposition leads to considerable breast enlargement. Increased level of progesterone at puberty also contributes to breast growth.
2. During pregnancy, breasts become fully developed due to influences of high levels of estrogens and progesterone as well as prolactin.

PRODUCTION OF MILK
1. Prolactin is major hormone responsible.
2. Mechanical stimulation of nipples by suckling initiates reflex that stimulates prolactin secretion for some time after childbirth.

MILK LET-DOWN REFLEX Suckling reflexively causes oxytocin release and oxytocin causes myoepithelial cells to contract, ejecting milk into ducts of breasts.

FETAL CIRCULATION **pp. 863–864**

FETAL BLOOD PATHWAYS

BLOOD TO PLACENTA Carried by umbilical arteries.

UMBILICAL VEIN Carries high-nutrient, high-oxygenated blood from placenta.

FETAL VESSELS Most carry mixed arterial and venous blood.

FETAL CIRCUIT

Lungs Collapsed; therefore provide much resistance to blood flow.

Ductus Venosus Serves as liver bypass.

Ductus Arteriosus Carries blood from pulmonary trunk to aorta.

Foramen Ovale Opening in interatrial septum.

CIRCULATORY CHANGES AT BIRTH
1. Pulmonary circuit becomes functional.
2. Foramen ovale closes; forms fossa ovalis.
3. Ductus arteriosus is obliterated; forms ligamentum arteriosum.
4. Umbilical arteries become obliterated; form umbilical ligaments.
5. Ductus venosus is obliterated; forms ligamentum venosum.

HUMAN GENETICS The study of heredity in humans. **pp. 864–868**

HUMAN CHROMOSOMES Each human somatic cell is diploid cell containing 46 chromosomes: 44 autosomes and 2 sex chromosomes. Genes controlling same character and occupying same locus on homologous chromosomes are called *alleles*.

DOMINANCE In some cases, one allele for a particular character can mask, or suppress, effect of another allele for that character. First allele is *dominant*; masked allele is *recessive*.

GENOTYPE AND PHENOTYPE *Genotype*: a person's actual genetic makeup. *Phenotype*: outward expression of genotype—that is, person's expressed traits.

INHERITANCE PATTERNS Genetic traits can be passed from one generation to next, for example, inheritance of tongue-rolling ability.

X-LINKED TRAITS Only a single X chromosome occurs in males; effects of recessive alleles located on X chromosome more frequently expressed in males than females.

EXPRESSION OF GENETIC POTENTIAL Particular genes represent potentialities whose full effects and complete expression may be influenced by activities of other genes and by environmental factors.

CONDITIONS OF CLINICAL SIGNIFICANCE: INHERITANCE Many diseases and other abnormalities have genetic origins.   pp. 868–869

NONDISJUNCTION A major cause of abnormal numbers of chromosomes in cells. Occurs during meiosis.

DOWN'S SYNDROME Most common and extreme form is due to three copies of autosome 21. Characterized by consistent pattern of physical features, almost always by mental retardation.

ABNORMAL NUMBERS OF SEX CHROMOSOMES Cause a number of different abnormalities.

GALACTOSEMIA Genetically based inability to tolerate the milk sugar lactose.

SELF-QUIZ

1. Fertilization normally occurs within the: (a) ovary; (b) uterine tube; (c) uterus.

2. The interaction between the sperm that enters the ovum and the cell membrane of the ovum brings about changes that prevent the entry of additional sperm. True or False?

3. An ovum that is fertilized by a spermatozoon containing a Y chromosome normally: (a) develops into a male; (b) develops into a female; (c) has an equal probability of developing into either a male or a female.

4. During implantation, the cells of the inner cell mass release enzymes that digest the cells of the trophoblast. True or False?

5. During the early portion of pregnancy, the corpus luteum secretes: (a) progesterone; (b) human chorionic somatomammotropin; (c) human chorionic gonadotropin.

6. During pregnancy, the corpus luteum degenerates about eight to ten days after implantation. True or False?

7. The superficial layer of the endometrium is called the: (a) chorion; (b) decidua; (c) embryonic disc; (d) zona pellucida.

8. The placenta is formed from: (a) the decidua basalis; (b) the chorion; (c) a and b combined.

9. Maternal blood normally mixes with fetal blood within the placenta in order to exchange food materials and oxygen. True or False?

10. The placenta produces: (a) estrogens; (b) follicle-stimulating hormone; (c) luteinizing hormone.

11. The membrane surrounding the cavity that forms in the inner cell mass is the: (a) amnion; (b) yolk sac; (c) allantois; (d) chorion.

12. The layer of cells on the side of the embryonic disc toward the yolk sac is composed of endodermal cells. True or False?

13. The chorion forms when mesoderm lines the inside of the trophoblast. True or False?

14. The blood vessels that supply which of the following carry fetal blood to and from the placenta? (a) yolk sac; (b) amnion; (c) allantois; (d) chorion.

15. Which of the following are *not* formed from mesoderm? (a) muscles; (b) bone; (c) blood; (d) all of these are formed from mesoderm.

16. Oxytocin is particularly effective at stimulating uterine contractions early in pregnancy, but it is unable to do so late in pregnancy. True or False?

17. At puberty, the alveoli of the breasts develop fully and milk secretion begins. True or False?

18. Prolactin production: (a) occurs at a relatively high level prior to puberty; (b) ceases when the placenta is lost at parturition; (c) can be stimulated for some time following parturition by vigorous, frequent suckling.

19. The umbilical arteries of the fetus carry blood that is rich in nutrients and has a high oxygen content. True or False?

20. Most of the blood that leaves the right ventricle of the fetus never reaches the lungs. Instead, it enters the aorta through a shunt called the: (a) ductus venosus; (b) foramen ovale; (c) ductus arteriosus; (d) umbilicus.

21. Following birth the umbilical vein becomes obliterated with connective tissue and is called the: (a) fossa ovalis; (b) ligamentum teres; (c) ligamentum venosum; (d) umbilical ligament.

22. A person's actual genetic makeup is called his or her: (a) phenotype; (b) genotype; (c) heterotype.

23. In a male, the effects of recessive alleles located on the X chromosome are always masked by dominant alleles located on the Y chromosome. True or False?

Regional Anatomy

The best way to learn human anatomy is to dissect human cadavers. However, for a number of reasons, most undergraduate anatomy courses dissect some animal other than a human—generally a cat or fetal pig. Because figures in textbooks are of humans, what is observed in the laboratory animals must be extrapolated to the text illustrations.

One problem with relying so heavily on drawings is that they tend to be somewhat idealized. It is a simple matter to alter a drawing to include or exclude particular structures. Therefore, the drawings often do not exactly represent the spatial relationships that exist in the body. Students who rely entirely on drawings may assume that the body is just as neatly organized, with the various structures as clearly separated, as are the drawings.

The following photographs of actual human dissections clarify the relationships between the structures. The dissections are organized by region, according to the natural, main subdivisions of the body: head, limbs, thorax, abdomen, and pelvis. All of the labeled structures in the photographs are also illustrated and discussed in the text. The photographs should be used to reinforce the anatomical relationships seen in the drawings and the laboratory.

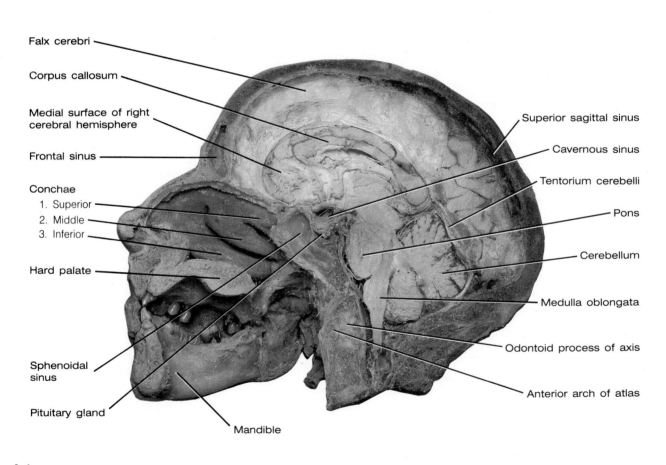

A.1

The head (midsagittal section).

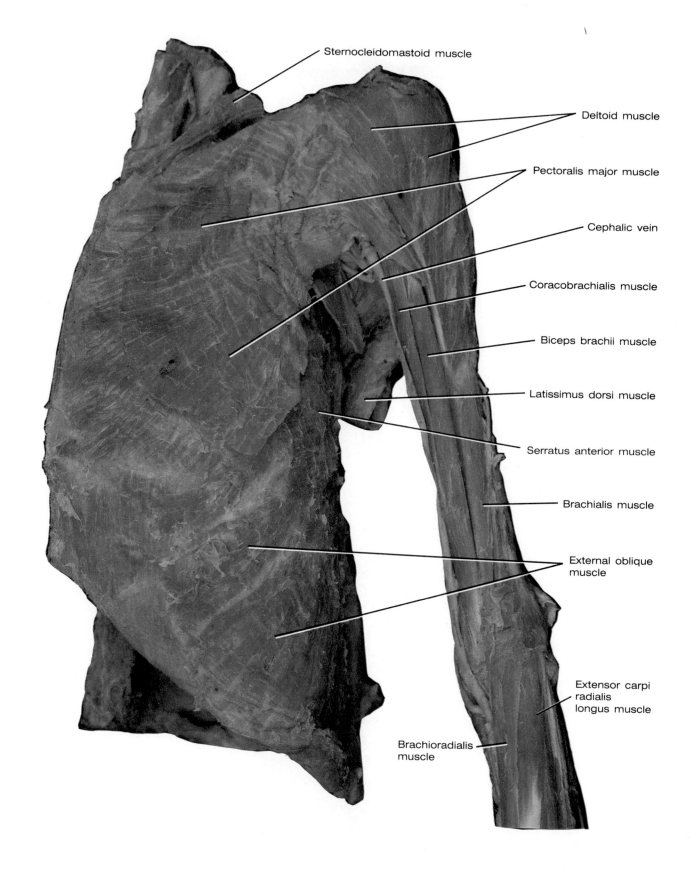

Sternocleidomastoid muscle

Deltoid muscle

Pectoralis major muscle

Cephalic vein

Coracobrachialis muscle

Biceps brachii muscle

Latissimus dorsi muscle

Serratus anterior muscle

Brachialis muscle

External oblique muscle

Extensor carpi radialis longus muscle

Brachioradialis muscle

A.2
Left chest, shoulder, and arm (anterior view).

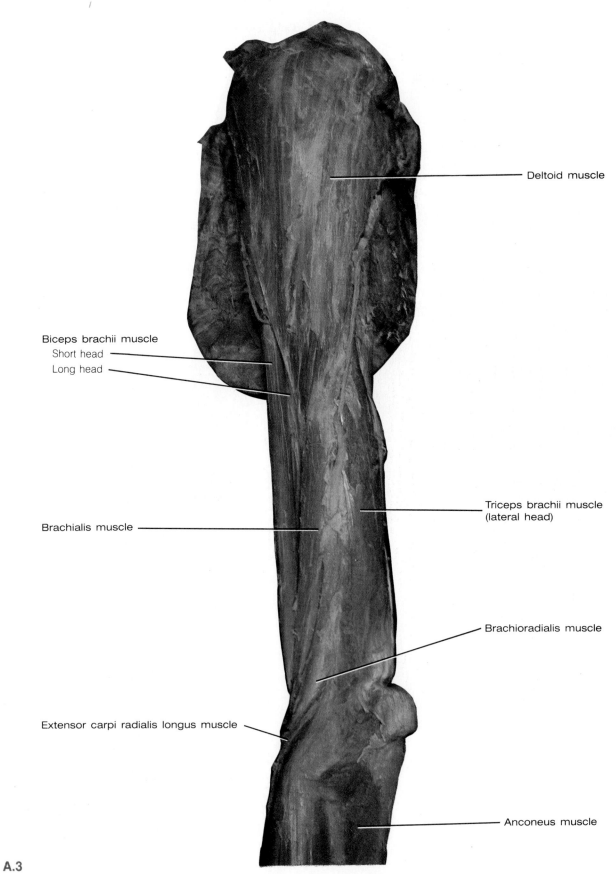

Deltoid muscle

Biceps brachii muscle
Short head
Long head

Triceps brachii muscle
(lateral head)

Brachialis muscle

Brachioradialis muscle

Extensor carpi radialis longus muscle

Anconeus muscle

A.3
Left shoulder and arm (lateral view).

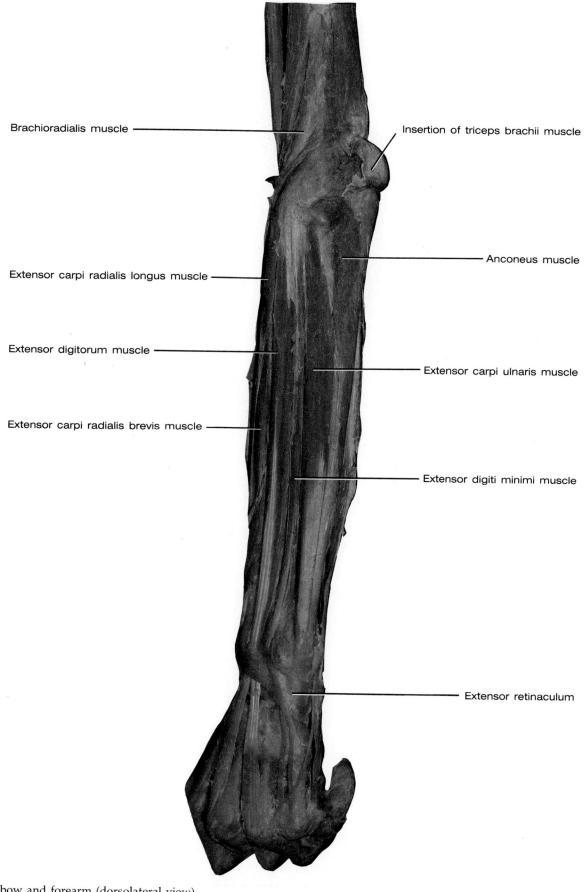

Brachioradialis muscle

Insertion of triceps brachii muscle

Extensor carpi radialis longus muscle

Anconeus muscle

Extensor digitorum muscle

Extensor carpi ulnaris muscle

Extensor carpi radialis brevis muscle

Extensor digiti minimi muscle

Extensor retinaculum

A.4
Left elbow and forearm (dorsolateral view).

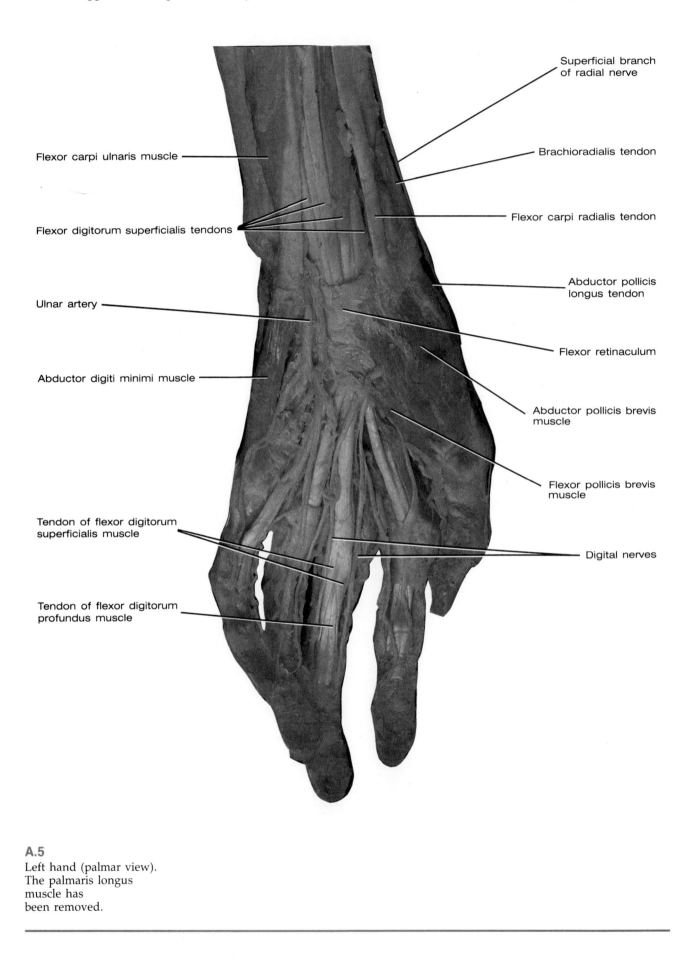

Superficial branch of radial nerve

Brachioradialis tendon

Flexor carpi ulnaris muscle

Flexor carpi radialis tendon

Flexor digitorum superficialis tendons

Abductor pollicis longus tendon

Ulnar artery

Flexor retinaculum

Abductor digiti minimi muscle

Abductor pollicis brevis muscle

Flexor pollicis brevis muscle

Tendon of flexor digitorum superficialis muscle

Digital nerves

Tendon of flexor digitorum profundus muscle

A.5
Left hand (palmar view).
The palmaris longus
muscle has
been removed.

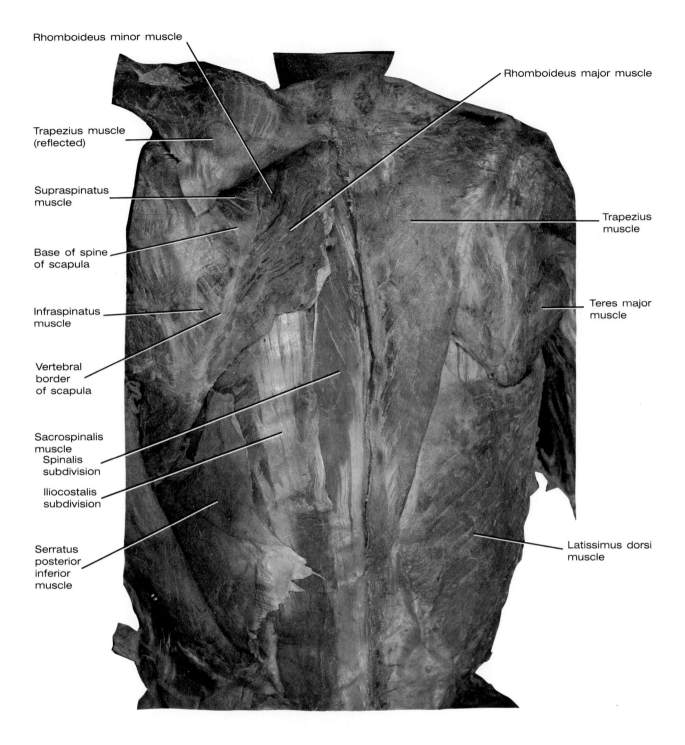

Rhomboideus minor muscle

Rhomboideus major muscle

Trapezius muscle (reflected)

Supraspinatus muscle

Base of spine of scapula

Trapezius muscle

Infraspinatus muscle

Teres major muscle

Vertebral border of scapula

Sacrospinalis muscle
  Spinalis subdivision

  Iliocostalis subdivision

Serratus posterior inferior muscle

Latissimus dorsi muscle

A.6
The back.

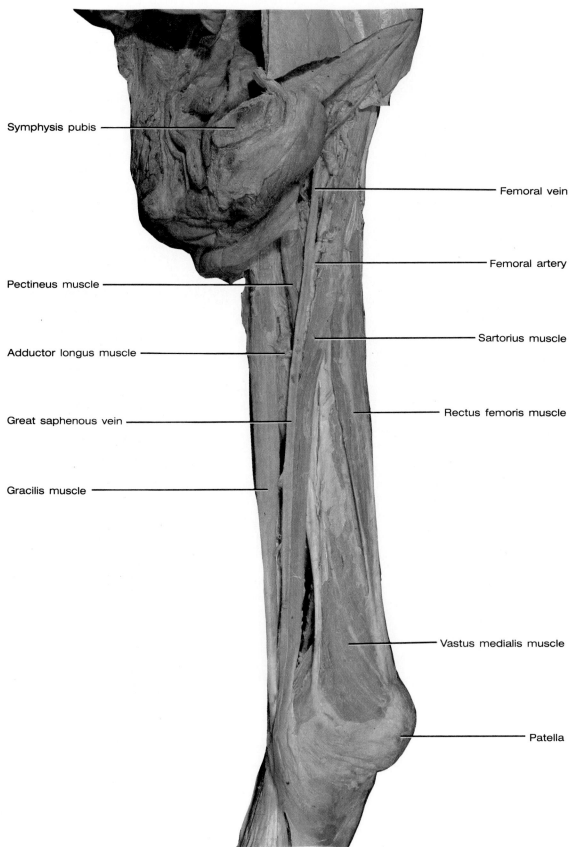

Symphysis pubis

Pectineus muscle

Adductor longus muscle

Great saphenous vein

Gracilis muscle

Femoral vein

Femoral artery

Sartorius muscle

Rectus femoris muscle

Vastus medialis muscle

Patella

A.7
Left thigh (medial view).

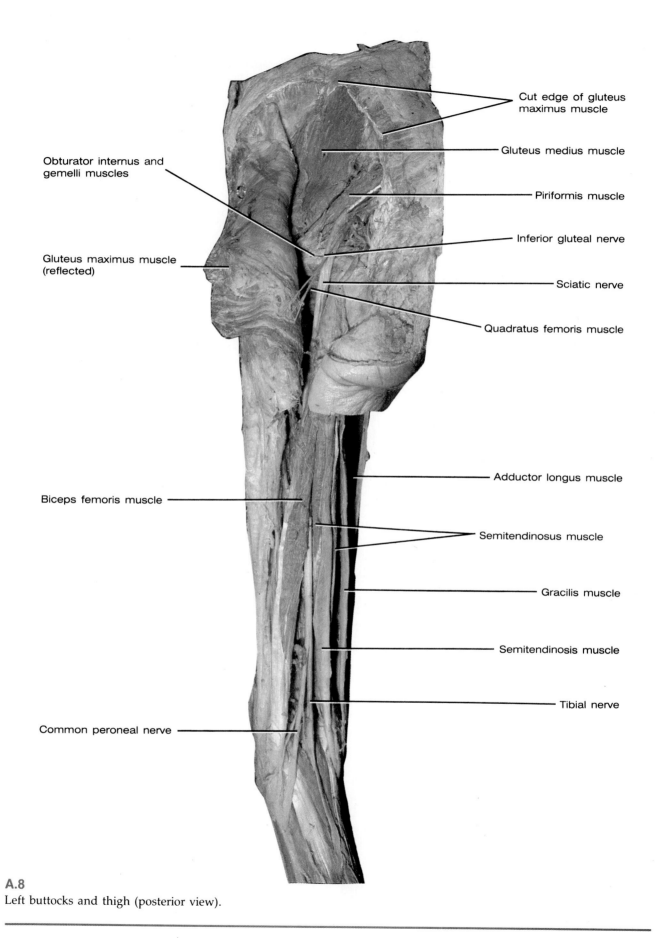

Cut edge of gluteus maximus muscle

Gluteus medius muscle

Piriformis muscle

Inferior gluteal nerve

Sciatic nerve

Quadratus femoris muscle

Obturator internus and gemelli muscles

Gluteus maximus muscle (reflected)

Adductor longus muscle

Semitendinosus muscle

Biceps femoris muscle

Gracilis muscle

Semitendinosis muscle

Tibial nerve

Common peroneal nerve

A.8
Left buttocks and thigh (posterior view).

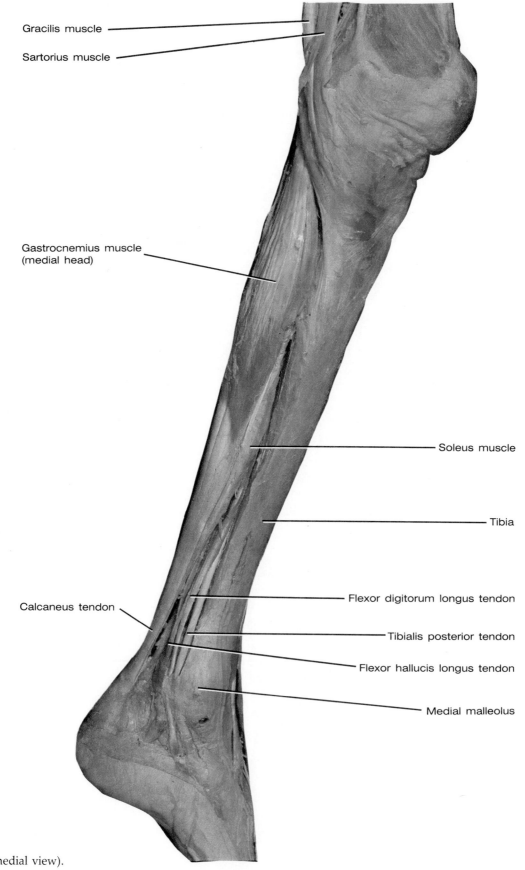

Gracilis muscle

Sartorius muscle

Gastrocnemius muscle
(medial head)

Soleus muscle

Tibia

Flexor digitorum longus tendon

Calcaneus tendon

Tibialis posterior tendon

Flexor hallucis longus tendon

Medial malleolus

A.9
Left leg (medial view).

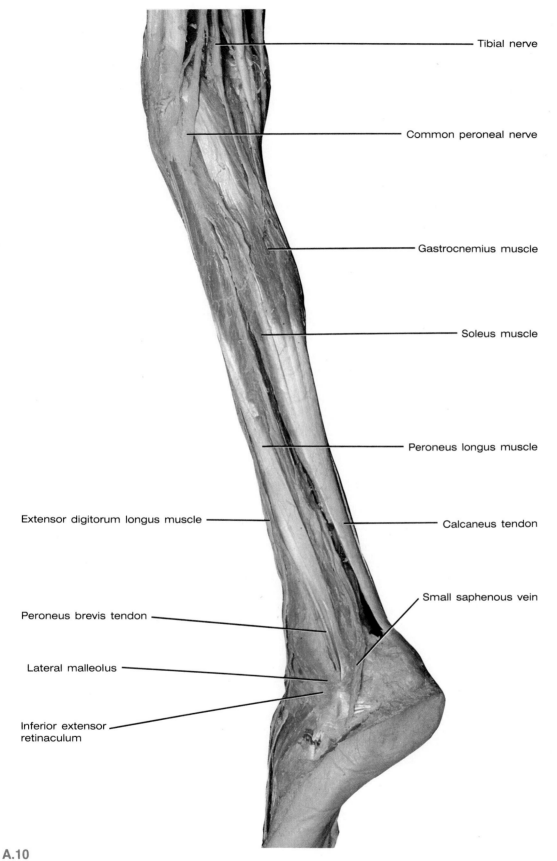

Tibial nerve

Common peroneal nerve

Gastrocnemius muscle

Soleus muscle

Peroneus longus muscle

Extensor digitorum longus muscle

Calcaneus tendon

Small saphenous vein

Peroneus brevis tendon

Lateral malleolus

Inferior extensor retinaculum

A.10
Left leg (dorsolateral view).

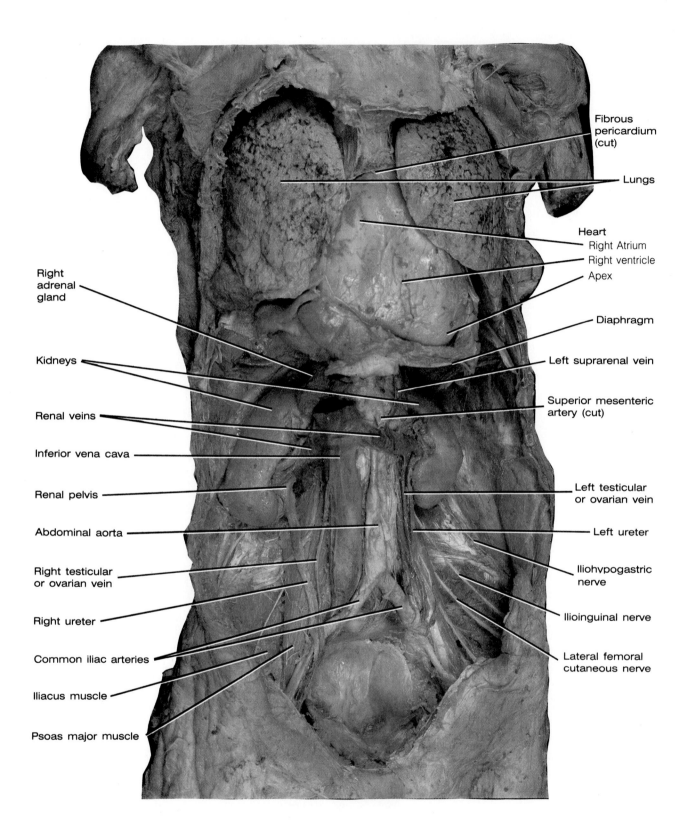

Fibrous pericardium (cut)

Lungs

Heart
Right Atrium
Right ventricle
Apex

Diaphragm

Left suprarenal vein

Superior mesenteric artery (cut)

Left testicular or ovarian vein

Left ureter

Iliohypogastric nerve

Ilioinguinal nerve

Lateral femoral cutaneous nerve

Right adrenal gland

Kidneys

Renal veins

Inferior vena cava

Renal pelvis

Abdominal aorta

Right testicular or ovarian vein

Right ureter

Common iliac arteries

Iliacus muscle

Psoas major muscle

A.11
Thoracic and abdominopelvic cavities (frontal section).
Most of the abdominal viscera and the fibrous pericardium
have been removed.

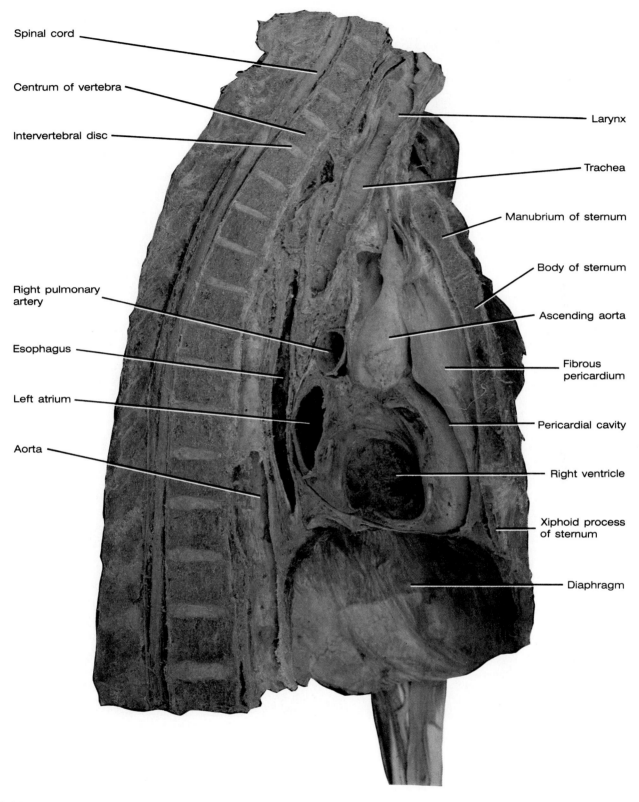

Spinal cord

Centrum of vertebra

Intervertebral disc

Larynx

Trachea

Manubrium of sternum

Body of sternum

Right pulmonary artery

Ascending aorta

Esophagus

Fibrous pericardium

Left atrium

Pericardial cavity

Aorta

Right ventricle

Xiphoid process of sternum

Diaphragm

A.12
Left thorax (midsagittal section).

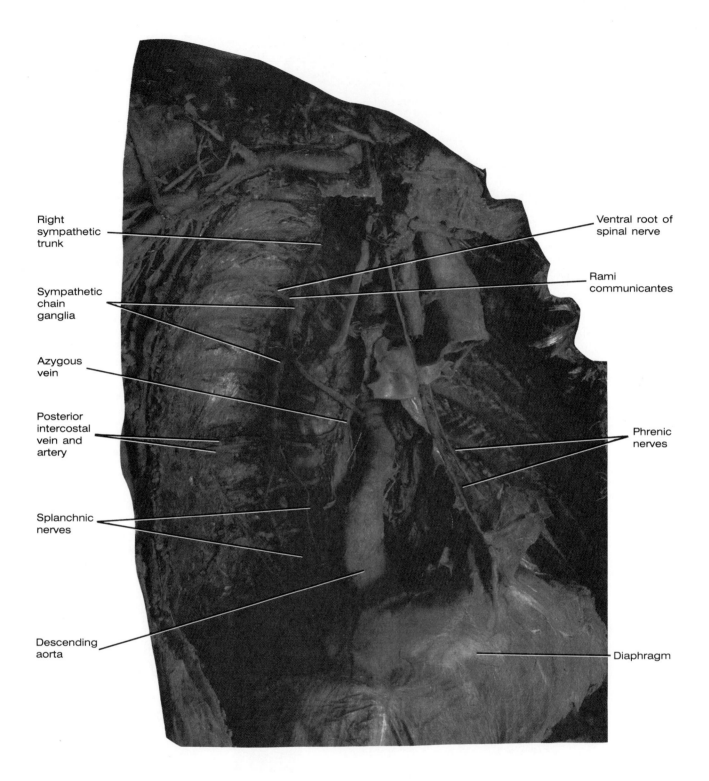

Right sympathetic trunk

Sympathetic chain ganglia

Azygous vein

Posterior intercostal vein and artery

Splanchnic nerves

Descending aorta

Ventral root of spinal nerve

Rami communicantes

Phrenic nerves

Diaphragm

A.13
Thoracic cavity, with lungs and heart removed.

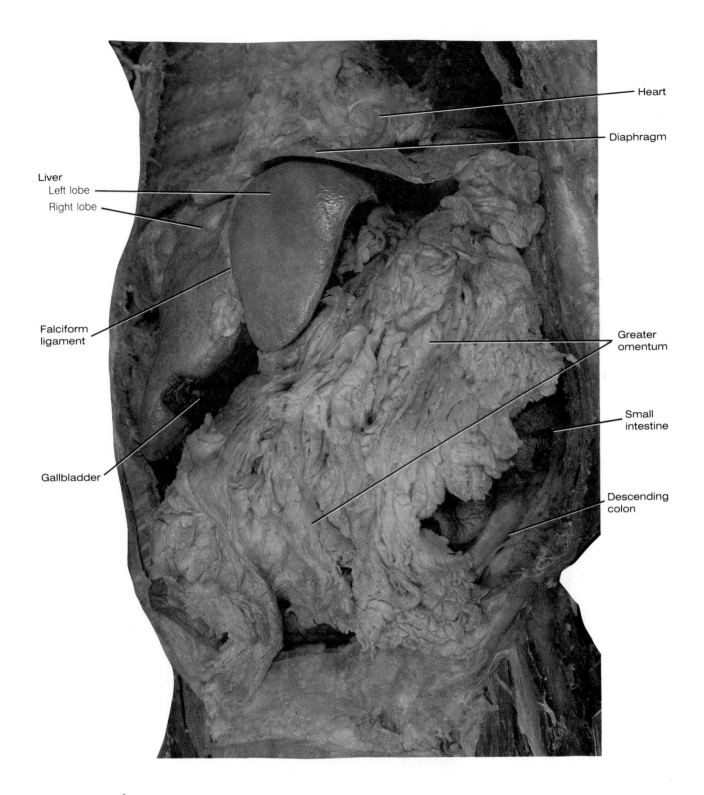

Heart

Diaphragm

Liver
    Left lobe

    Right lobe

Falciform
ligament

Gallbladder

Greater
omentum

Small
intestine

Descending
colon

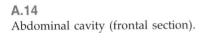

A.14
Abdominal cavity (frontal section).

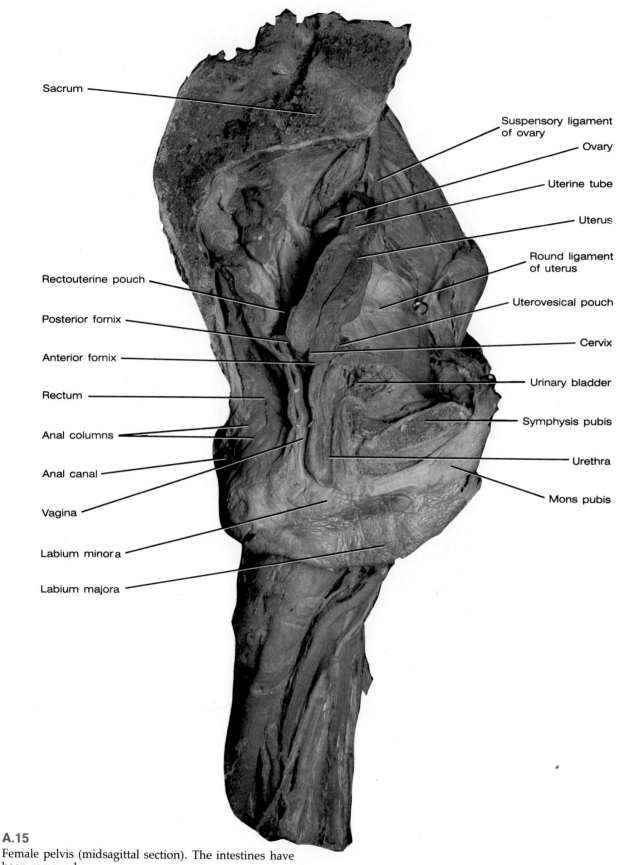

Sacrum

Suspensory ligament of ovary

Ovary

Uterine tube

Uterus

Rectouterine pouch

Round ligament of uterus

Posterior fornix

Uterovesical pouch

Anterior fornix

Cervix

Rectum

Urinary bladder

Anal columns

Symphysis pubis

Anal canal

Urethra

Vagina

Mons pubis

Labium minora

Labium majora

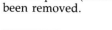

A.15
Female pelvis (midsagittal section). The intestines have been removed.

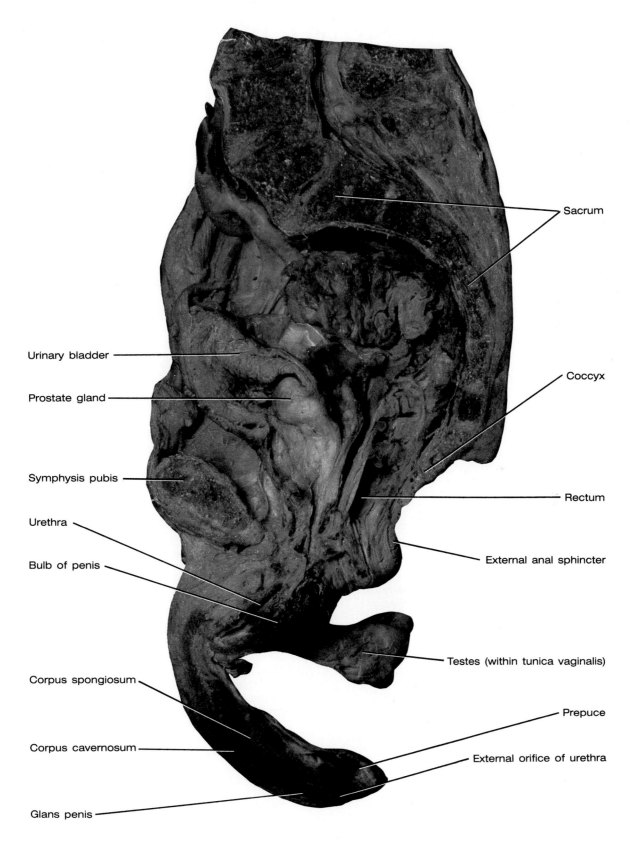

Sacrum

Urinary bladder

Prostate gland

Coccyx

Symphysis pubis

Rectum

Urethra

Bulb of penis

External anal sphincter

Corpus spongiosum

Testes (within tunica vaginalis)

Prepuce

Corpus cavernosum

External orifice of urethra

Glans penis

A.16
Male pelvis (midsagittal section). Most of the intestines
have been removed.

Appendix 2
Self-Quiz Answers

CHAPTER 1

1. c
2. F
3. a
4. b
5. c
6. Epithelial tissues: c, e
   Muscular tissues: a, g
   Nervous tissues: d, f
   Connective tissues: b, h
7. Prone position: k
   Anterior (ventral): g
   Posterior (dorsal): l
   Anatomical position: n
   Superior (cranial): a
   Medial: h
   Supine position: c
   Inferior (caudal): d
   Lateral: b
   Proximal: e
   Cervical: m
   Plantar: f
   Distal: i
   Lumbar: j
   Palmar: o
8. a
9. T
10. c
11. b
12. T
13. a
14. a
15. c
16. b
17. F
18. c
19. T
20. b

CHAPTER 2

1. T
2. c
3. a
4. b
5. F
6. a
7. T
8. b
9. Proteins: c
   Amino acids: g
   Peptide bond: b
   Dipeptide: e
   Polypeptide: i
   Primary structure: a
   Secondary structure: f
   Tertiary structure: h
   Quaternary structure: d
10. F
11. T
12. F
13. b
14. Solvent: c
    Solution: d
    Solute: a
    Suspension: b
    Colloid: e
15. c
16. F
17. b
18. a
19. c

CHAPTER 3

1. c
2. Specificity: b
   Saturation: d
   Competition: e
   Facilitated diffusion: a
   Active transport: c
3. b
4. b
5. T
6. F
7. b
8. T
9. a
10. c
11. b
12. a
13. b
14. Ribosomes: d
    Golgi apparatus: e
    Peroxisomes: a
    Basal bodies: b
    Microfilaments: c
15. F
16. c
17. T
18. Autosomes: h
    XY: g
    Homologous chromosomes: a
    46: f
    Diploid: c
    23: i
    Haploid: e
    Interkinesis: b
    Crossing over: d
19. F
20. c

CHAPTER 4

1. b
2. a
3. T
4. a
5. Squamous cells: c
   Stratified epithelium: f
   Simple columnar: e
   Columnar cells: a
   Cuboidal cells: d
   Simple squamous epithelium: b
6. b
7. c
8. T
9. Mucous membranes: d
   Lamina propria: c
   Serous membranes: f
   Mucin: g
   Mesothelium: b
   Endothelium: a
   Goblet cells: e
10. b
11. Fibroplasts: g
    Macrophages: f
    Ground substance: d
    Fibrocytes: a
    Fibers: c
    Collagenous: h
    Elastic: e
    Reticular: b
12. F
13. c
14. a
15. c
16. c
17. T
18. F
19. T
20. c

CHAPTER 5

1. b
2. T
3. Basal layer (stratum basale): e
   Stratum germinativum: f
   Keratohyalin: h
   Melanin: j
   Stratum corneum: a
   Eleidin: b
   Stratum granulosum: i
   Carotene: c
   Stratum lucidum: d
   Keratin: g
4. b
5. T
6. a
7. F
8. c
9. F
10. a
11. T
12. b
13. a
14. T
15. c
16. c
17. Acne: d
    Warts: g
    Psoriasis: f
    Impetigo: h
    Moles: c
    Herpes simplex: a
    Herpes zoster: i
    Cancers: e
    Eczema: b
18. b
19. T

CHAPTER 6

1. c
2. b
3. T
4. F
5. c
6. c
7. a
8. b
9. c
10. F
11. a
12. Process: i
    Trochanter: f
    Tubercle: j
    Condyle: d
    Sulcus: k
    Crest: h
    Facet: a
    Fossa: g
    Fovea: b
    Foramen: c
    Meatus: e
13. Fontanels: f
    Frontal bone: j
    Sinuses: d
    Parietal bones: a
    Occipital bone: h
    Ethmoid bone: b
    Temporal bones: l
    Maxillary bones: c
    Zygomatic bones: i
    Lacrimal bones: k
    Mandible: e
    Vomer bone: g
    An auditory ossicle: g
    Hyoid bone: n
    Cervical vertebrae: m
    Thoracic vertebrae: q
    Sacral vertebrae: o
14. a

15. c
16. a
17. d
18. b
19. c
20. a
21. c
22. c
23. Ilium: h
Ischium: k
Pubis: a
Acetabulum: m
False pelvis: b
Pelvic outlet: j
Femur: e
Greater trochanter: i
Condyles: c
Adductor tubercle: g
Tibia: d
Anterior crest: f
Fibula: l

CHAPTER 7

1. F
2. b
3. c
4. a
5. T
6. Synarthroses: d
Sutures: g
Syndesmoses: b
Synostosis: a
Amphiarthroses: f
Synchondroses: h
Symphyses: c
Synovial: e
7. b
8. F
9. F
10. a
11. Flexion: d
Extension: f, b
Abduction: c
Adduction: a
Plantar flexion: e
Dorsiflexion: g
12. F
13. c
14. b
15. b
16. Nonaxial: d, g
Gliding: g
Uniaxial: f, i
Hinge: i
Pivot: e
Biaxial: a, b, h
Condyloid: h
Saddle: b
Multiaxial: c
Sphenoid: c
17. a
18. a
19. b
20. Distal tibia/fibula: a, g
Pubic/pubic: b
Ulna/humerus: f
Sternum/clavicle: d, e
Radius/carpals: c
Tarsal/tarsal: d, e
Occipital bone/atlas: c
21. T
22. c

CHAPTER 8

1. T
2. a
3. Epimysium: c
Perimysium: h
Endomysium: f
Aponeuroses: d
Origin: a
Insertion: g
Unipennate: b
Bipennate: i
Multipennate: e
4. c
5. b
6. a
7. T
8. b
9. b
10. F
11. c
12. c
13. b
14. c
15. b
16. c
17. b
18. T
19. c
20. b
21. b
22. a
23. T

CHAPTER 9

1. Intercostalis: d
Rhomboideus: a
Pronator: f
Coracobrachialis: c
Gluteus maximus: b
Trapezius: a
Rectus abdominis: d, f
Pectoralis minor: b
Flexor carpi radialis: f
Biceps brachii: d, e
Quadriceps femoris: d, e
2. b
3. Depressor labii inferioris: d
Epicranius occipitalis: c
Corrugator: b
Orbicularis oculi: f
Orbicularis oris: g
Procerus: e
Risorius: a
4. a

5. T
6. b
7. T
8. Multifidus: c
Interspinales: d
Splenius capitis and cervicis: b
Spinalis thoracis and cervicis: a
9. c
10. T
11. F
12. a
13. b
14. Subclavius: b
Rhomboideus major: c
Pectoralis minor: a
15. Deltoid: a, b, d, e, f
Latissimus dorsi: b, c, e
Teres minor: f
Teres major: b, c, e
Coracobrachialis: a, c
Infraspinatus: c, f
16. a, b, e
17. Flexor carpi radialis: h
Extensor indicis: d
Extensor carpi ulnaris: e
Extensor pollicis longus: f
Extensor carpi radialis longus: b
Extensor digiti minimi: c
Flexor carpi ulnaris: g
Flexor digitorum superficialis: a
18. b, e
19. Gluteus maximus: d
Adductor magnus: e
Pectineus: b
Adductor longus: c
Gluteus medius: a
20. c
21. b
22. Tibialis anterior: c
Peroneous longus: b
Extensor digitorum longus: e
Flexor hallucis longus: a
Popliteus: d

CHAPTER 10

1. T
2. c
3. c
4. a
5. F
6. c
7. a
8. F
9. c
10. Nuclei: d
Soma: f
Center: h
Ganglia: a
Dendrite: b
Axon: i
Nerve fiber: c
Sheath of Schwann: e
Myelin: g
11. c
12. T
13. c
14. a
15. T
16. Motor neuron endings: e
Sensory neuron endings: a
Pacinian corpuscles: b
End-bulbs of Krause: f
Ruffini's corpuscles: d
Muscle spindles: h
Neurotendinous organs: i
Free nerve endings: c
Meissner's corpuscles: g
17. b
18. a
19. b

CHAPTER 11

1. a
2. b
3. c
4. b
5. b
6. b
7. c
8. F
9. a
10. a
11. F
12. b
13. a
14. T
15. T
16. T
17. a
18. c
19. F
20. F

CHAPTER 12

1. c
2. a
3. a
4. F
5. b
6. b
7. c
8. b
9. T
10. c
11. b

12. T
13. b
14. a
15. b
16. Cerebrum: i
    Projection tracts: e
    Basal ganglia: m
    Cerebral hemisphere: a
    Insula: k
    Primary motor area: p
    Visual area: b
    Olfactory area: l
    Thalamus: c
    Cerebral peduncles: n
    Metencephalon: o
    Pons: d
    Hypothalamus: g
    Ventricles: h
    Telencephalon: f
    Reticular formation: j
17. T
18. a
19. Cauda equina: g
    Meninges: d
    White matter: a
    Gray matter: j
    Ventral roots: c
    Fasciculus gracilis: i
    Spinothalamic tracts: b
    Spinocerebellar tracts: h
    Pyramidal tracts: f
    Extrapyramidal tracts: e
20. b
21. b
22. a
23. b
24. T
25. c
26. a

CHAPTER 13

1. c
2. a
3. c
4. a
5. a
6. b
7. a
8. c
9. c
10. a
11. b
12. a
13. d
14. c
15. T
16. T
17. F
18. c
19. a
20. c
21. T
22. T

CHAPTER 14

1. d
2. F
3. F
4. b
5. b
6. T
7. F
8. T
9. b
10. F
11. b
12. c
13. a
14. T
15. c
16. T
17. T

CHAPTER 15

1. b
2. c
3. a
4. b
5. F
6. c
7. a
8. b
9. a
10. c
11. a
12. c
13. a
14. T
15. b
16. a
17. T
18. c
19. b

CHAPTER 16

1. F
2. T
3. b
4. c
5. b
6. a
7. ADH: b
   Oxytocin: d
   LH: f
   TSH: e
   ACTH: a
   GH: c
8. c
9. F
10. c
11. a
12. F
13. Lack of ACTH: f
    Gonadotropin deficiency: c
    TSH deficiency: a
    ACTH overproduction: b
    Growth hormone deficiency: d
    Growth hormone overproduction: e
    Antidiuretic hormone deficiency: g
14. c
15. c
16. c
17. a
18. T
19. a
20. b
21. c
22. T
23. c

CHAPTER 17

1. b
2. F
3. F
4. T
5. a
6. T
7. Bilirubin: e
   Heme: b
   Erythropoietin: a
   Globin: c
   Ferritin: d
8. a
9. a
10. b
11. a
12. b
13. c
14. c
15. a
16. F
17. F
18. c
19. b
20. T

CHAPTER 18

1. b
2. Pericardial cavity: g
   Atria: d
   Ventricles: i
   Interatrial septum: a
   Inferior vena cava: f
   Aorta: b
   Epicardium: j
   Myocardium: c
   Endocardium: h
   Trabeculae carneae: e
3. c
4. c
5. a
6. c
7. F
8. a
9. a
10. T
11. T

12. b
13. a
14. T
15. c
16. b
17. F
18. T
19. c
20. a
21. a
22. b
23. b

CHAPTER 19

1. b
2. c
3. T
4. F
5. a
6. d
7. c
8. c
9. b
10. a
11. c
12. T
13. b
14. a
15. F
16. b
17. a
18. c
19. F
20. Aorta: g
    Coronary arteries: e
    Internal carotid artery: i
    Vertebral artery: a
    Radial artery: d
    Superior mesenteric artery: b
    Inferior mesenteric artery: h
    Common iliac arteries: c
    Anterior tibial artery: f
21. d

CHAPTER 20

1. F
2. c
3. a
4. b
5. T
6. T
7. d
8. d
9. T
10. F

CHAPTER 21

1. T
2. c

3. a
4. b
5. a
6. b
7. T
8. a
9. F
10. b
11. c
12. F
13. c
14. b
15. c
16. b
17. T
18. b
19. a
20. T

CHAPTER 22

1. a
2. d
3. d
4. F
5. d
6. d
7. c
8. d
9. b
10. c
11. c
12. Costal surface: d
Cardiac notch: h
Oblique fissure: g
Mediastinum: a
Pleura: b
Visceral pleura: c
Parietal pleura: f
Alveoli: e
13. b
14. c
15. a
16. b
17. c
18. a
19. a
20. a
21. F
22. c
23. b
24. b

CHAPTER 23

1. c
2. F

3. b
4. a
5. a
6. c
7. a
8. a
9. b
10. c
11. T
12. F
13. T
14. a
15. F
16. c
17. F
18. a
19. T
20. F

CHAPTER 24

1. c
2. a
3. c
4. T
5. a
6. c
7. F
8. b
9. c
10. T
11. T
12. T
13. b
14. a
15. T
16. b
17. Pancreatic juice: e
Tetrapeptidases: d
Amylases: f
Bile salts: b
Cholecystokinin: a
Trypsin: c
18. b
19. T

CHAPTER 25

1. c
2. b
3. F
4. F
5. b
6. a
7. T
8. F

9. T
10. a
11. c
12. c
13. a
14. b
15. F
16. T
17. a
18. b
19. b
20. b

CHAPTER 26

1. c
2. a
3. a
4. b
5. c
6. a
7. T
8. c
9. b
10. F
11. c
12. F
13. b
14. T
15. a
16. T
17. c
18. a
19. F

CHAPTER 27

1. a
2. F
3. T
4. c
5. T
6. a
7. b
8. c
9. F
10. c
11. a
12. c
13. c
14. c
15. F
16. F
17. a

CHAPTER 28

1. a
2. T
3. b
4. b
5. T
6. b
7. Epididymis: d
Ductus deferens: c
Seminal vesicles: f
Ejaculatory ducts: a
Prostate gland: b
Bulbourethral glands:
e
8. F
9. b
10. a
11. b
12. b
13. c
14. T
15. e
16. c
17. c
18. T
19. b
20. b

CHAPTER 29

1. b
2. T
3. a
4. F
5. a
6. F
7. b
8. c
9. F
10. a
11. a
12. T
13. T
14. c
15. d
16. F
17. F
18. c
19. F
20. c
21. b
22. b
23. F

Word Roots, Prefixes, Suffixes, and Combining Forms

PREFIXES AND COMBINING FORMS

a-, an- *absence or lack* acardia, lack of a heart; anaerobic, in the absence of oxygen

ab- *departing from; away from* abnormal, departing from normal

acou- *hearing* acoustics, the science of sound

acr-, acro- *extreme or extremity; peak* acrodermatitis, inflammation of the skin of the extremities

ad- *to or toward* adorbital, toward the orbit

aden-, adeno- *gland* adeniform, resembling a gland in shape

amphi- *on both sides; of both kinds* amphibian, an organism capable of living in water and on land

angi- *vessel* angiitis, inflammation of a lymph vessel or blood vessel

ant-, anti- *opposed to; preventing or inhibiting* anticoagulant, a substance that prevents blood coagulation

ante- *preceding; before* antecubital, in front of the elbow

arthr-, arthro- *joint* arthropathy, any joint disease

aut-, auto- *self* autogenous, self-generated

bi- *two* bicuspid, having two cusps

bio- *life* biology, the study of life and living organisms

blast- *bud or germ* blastocyte, undifferentiated embryonic cell

broncho- *bronchus* bronchospasm, spasmodic contraction of bronchial muscle

bucco- *cheek* buccolabial, pertaining to the cheek and lip

caput- *head* decapitate, remove the head

carcin- *cancer* carcinogen, a cancer-causing agent

cardi-, cardio- *heart* cardiotoxic, harmful to the heart

cephal- *head* cephalometer, an instrument for measuring the head

cerebro- *brain, especially the cerebrum* cerebrospinal, pertaining to the brain and spinal cord

chondr- *cartilage* chondrogenic, giving rise to cartilage

circum- *around* circumnuclear, surrounding the nucleus

co-, con- *together* concentric, common center, together in the center

contra- *against* contraceptive, agent preventing conception

cost- *rib* intercostal, between the ribs

crani- *skull* craniotomy, a skull operation

crypt- *hidden* cryptomenorrhea, a condition in which menstrual symptoms are experienced but no external loss of blood occurs

cyt- *cell* cytology, the study of cells

de- *undoing, reversal, loss, removal* deactivation, becoming inactive

di- *twice, double* dimorphism, having two forms

dia- *through, between* diaphragm, the wall through or between two areas

dys- *difficult, faulty, painful* dyspepsia, disturbed digestion

ec, ex, ecto- *out, outside, away from* excrete, to remove materials from the body

en-, em- *in, inside* encysted, enclosed in a cyst or capsule

entero- *intestine* enterologist, one who specializes in the study of intestinal disorders

epi- *over, above* epidermis, outer layer of skin

eu- *well* euesthesia, a normal state of the senses

exo- *outside, outer layer* exophthalmos, an abnormal protrusion of the eye from the orbit

extra- *outside, beyond* extracellular, outside the body cells of an organism

gastr- *stomach* gastrin, a hormone that influences gastric acid secretion

glosso- *tongue* glossopathy, any disease of the tongue

hema-, hemato-, hemo- *blood* hematocyst, a cyst containing blood

hemi- *half* hemiglossal, pertaining to one half of the tongue

hepat- *liver* hepatitis, inflammation of the liver

hetero- *different or other* heterosexuality, sexual desire for a person of the opposite sex

hist- *tissue* histology, the study of tissues

hom-, homo- *same* homeoplasia, formation of tissue similar to normal tissue; homocentric, having the same center

hydr-, hydro- *water* dehydration, loss of body water

hyper- *excess* hypertension, excessive tension

hypno- *sleep* hypnosis, a sleeplike state

hypo- *below, deficient* hypodermic, beneath the skin; hypokalemia, deficiency of potassium

hyster-, hystero- *uterus or womb* hysterectomy, removal of the uterus; hysterodynia, pain in the womb

im- *not* impermeable, not permitting passage, not permeable

inter- *between* intercellular, between the cells

intra- *within, inside* intracellular, inside the cell

iso- *equal, same* isothermal, equal, or same, temperature

leuko- *white* leukocyte, white blood cell

lip-, lipo- *fat, lipid* lipophage, a cell that has taken up fat in its cytoplasm

macro- *large* macromolecule, large molecule

mal- *bad, abnormal* malfunction, abnormal functioning of an organ

mamm- *breast* mammary gland, breast

mast- *breast* mastectomy, removal of a mammary gland

meningo- *membrane* meningitis, inflammation of the membranes of the brain

meso- *middle* mesoderm, middle germ layer

meta- *beyond, between, transition* metatarsus, the part of the foot between the tarsus and the phalanges

metro- *uterus* metroscope, instrument for examining the uterus

micro- *small* microscope, an instrument used to make small objects appear larger

mito- *thread, filament* mitochondria, small, filamentlike structures located in cells

mono- *single* monospasm, spasm of a single limb

morpho- *form* morphology, the study of form and structure of organisms

multi- *many* multinuclear, having several nuclei

myelo- *spinal cord, marrow* myeloblasts, cells of the bone marrow

myo- *muscle* myocardium, heart muscle

narco- *numbness* narcotic, a drug producing stupor or numbed sensations

nephro- *kidney* nephritis, inflammation of the kidney

neuro- *nerve* neurophysiology, the physiology of the nervous system

ob- *before, against* obstruction, impeding or blocking up

oculo- *eye* monocular, pertaining to one eye

odonto- *teeth* orthodontist, one who specializes in proper positioning of the teeth in relation to each other

ophthalmo- *eye* ophthalmology, the study of the eyes and related disease

ortho- *straight, direct* orthopedic, correction of deformities of the musculoskeletal system

osteo- *bone* osteodermia, bony formations in the skin

oto- *ear* otoscope, a device for examining the ear

oxy- *oxygen* oxygenation, the saturation of a substance with oxygen

pan- *all, universal* panacea, a cure-all

para- *beside, near* paraphrenitis, inflammation of tissues adjacent to the diaphragm

peri- *around* perianal, situated around the anus

phago- *eat* phagocyte, a cell that engulfs and digests particles or cells

phleb- *vein* phlebitis, inflammation of the veins

pod- *foot* podiatry, the treatment of foot disorders

poly- *multiple* polymorphism, multiple forms

post- *after, behind* posterior, places behind (a specific) part

pre-, pro- *before, ahead of* prenatal, before birth

procto- *rectum, anus* proctoscope, an instrument for examining the rectum

pseudo- *false* pseudotumor, a false tumor

psycho- *mind, psyche* psychogram, a chart of personality traits

pyo- *pus* pyocyst, a cyst that contains pus

retro- *backward, behind* retrogression, to move backward in development

sclero- *hard* sclerodermatitis, inflammatory thickening and hardening of the skin

semi- *half* semicircular, having the form of half a circle

steno- *narrow* stenocoriasis, narrowing of the pupil

sub- *beneath, under* sublingual, beneath the tongue

super- *above, upon* superior, quality or state of being above others or a part

supra- *above, upon* supracondylar, above a condyle

sym-, syn- *together, with* synapse, the region of communication between two neurons

tachy- *rapid* tachycardia, abnormally rapid heartbeat

therm- *heat* thermometer, an instrument used to measure heat

tox- *poison* antitoxic, effective against poison

trans- *across, through* transpleural, through the pleura

tri- *three* trifurcation, division into three branches

viscero- *organ, viscera* visceroinhibitory, inhibiting the movements of the viscera

SUFFIXES

-able *able to, capable of* viable, ability to live or exist

-ac *referring to* cardiac, referring to the heart

-algia *pain in a certain part* neuralgia, pain along the course of a nerve

-ary *associated with, relating to* coronary, associated with the heart

-atresia *imperforate* proctatresia, an imperforate condition of the rectum or anus

-cide *destroy or kill* germicide, an agent that kills germs

-ectomy *cutting out, surgical removal* appendectomy, cutting out of the appendix

-emia *condition of the blood* anemia, deficiency of red blood cells

-ferent *carry* efferent nerves, nerves carrying impulses away from the CNS

-fuge *driving out* vermifuge, a substance that expels worms of the intestine

-gen *an agent that initiates* pathogen, any agent that produces disease

-gram *data that are systematically recorded, a record* electrocardiogram, a recording showing action of the heart

-graph *an instrument used for recording data or writing* electrocardiograph, an instrument used to make an electrocardiogram

-ia *condition* insomnia, condition of not being able to sleep

-iatrics *medical specialty* geriatrics, the branch of medicine dealing with disease associated with old age

-itis *inflammation* gastritis, inflammation of the stomach

-logy *the study of* pathology, the study of changes in structure and function brought on by disease

-lysis *loosening or breaking down* hydrolysis, chemical decomposition of a compound into other compounds as a result of taking up water

-malacia *soft* osteomalacia, a process leading to bone softening

-mania *obsession, compulsion* erotomania, exaggeration of the sexual passions

-odyn *pain* coccygodynia, pain in the region of the coccyx

-oid *like, resembling* cuboid, shaped as a cube

-oma *tumor* lymphoma, a tumor of the lymphatic tissues

-opia *defect of the eye* myopia, nearsightedness

-ory *referring to, of* auditory, referring to hearing

-pathy *disease* psychopathy, any disease of the mind

-phobia *fear* acrophobia, fear of heights

-plasty *reconstruction of a part, plastic surgery* rhinoplasty, reconstruction of the nose through surgery

-plegia *paralysis* paraplegia, paralysis of the lower half of the body or limbs

-rrhagia *abnormal or excessive discharge* metrorrhagia, uterine hemorrhage

-rrhea *flow or discharge* diarrhea, abnormal emptying of the bowels

-scope *instrument used for examination* stethoscope, instrument used to listen to sounds of various parts of the body

-stasis *arrest, fixation*   hemostasis, arrest of bleeding

-stomy *establishment of an artificial opening*   enterostomy, the formation of an artificial opening into the intestine through the abdominal wall

-tomy *to cut*   appendectomy, surgical removal of the appendix

-ty *condition of, state*   immunity, condition of being resistant to infection or disease

-uria *urine*   polyuria, passage of an excessive amount of urine

Units of the Metric System

Unit	Metric equivalent	Symbol	English equivalent
LINEAR MEASURE			
1 kilometer	= 1000 meters	km	0.62137 mile
1 meter	= 10 decimeters	m	39.37 inches
1 decimeter	= 10 centimeters	dm	3.937 inches
1 centimeter	= 10 millimeters	cm	0.3937 inch
1 millimeter	= 1000 microns	mm	
1 micron	= 1/1000 millimeter or 1000 millimicrons	μ	
1 millimicron	= 1 nanometer = 10 angstrom units	nm, mμ	no English equivalents
1 angstrom unit	= 1/100,000,000 centimeter	Å	
MEASURES OF CAPACITY			
1 kiloliter	= 1000 liters	kl	35.15 cubic feet or 264.16 gallons
1 liter	= 10 deciliters	l	1.0567 U.S. liquid quarts
1 deciliter	= 100 milliliters	dl	0.03 fluid ounce
1 milliliter	volume of 1 g of water at standard temperature and pressure (STP)	ml	
MEASURES OF MASS			
1 kilogram	= 1000 grams	kg	2.2046 pounds
1 gram	= 100 centigrams	g	15.432 grains
1 centigram	= 10 milligrams	cg	0.1543 grain
1 milligram	= 1/1000 gram	mg	0.01 grain (about)
MEASURES OF VOLUME			
1 cubic meter	= 1000 cubic decimeters	m^3	
1 cubic decimeter	= 1000 cubic centimeters	dm^3	
1 cubic centimeter	= 1000 cubic millimeters	cm^3	
1000 cubic millimeters	= 1 milliliter (ml)	mm^3	

Glossary

Abdomen *ab'-do-men* the portion of the body between the diaphragm and the pelvis.

Abduct *ab-duct'* to move away from the midline of the body.

Abortion *a-bor'-shun* termination of a pregnancy before the embryo or fetus is viable outside the uterus.

Abscess *ab'-ses* a localized accumulation of pus and disintegrating tissue.

Absolute refractory period *ree-frac'-to-ree* period following stimulation during which no additional action potential can be evoked.

Absorption *ab-sorp'-shun* passage of a substance into or across a membrane or blood vessel.

Accommodation (1) adaptation or adjustment by an organ or organism in response to differences or needs; (2) adjustment of the eye for seeing objects at various distances.

Acetabulum *as-e-tab'-u-lum* the cuplike cavity where the femur fits on the lateral surface of the pelvic bone.

Acetylcholine *a-see-til-ko'-leen* a chemical transmitter substance released by nerve endings.

Acetylcholinesterase *es'-ter-ase* the enzyme that inactivates acetylcholine; present in muscle and nervous tissue.

Achilles tendon *a-kil'-leze ten'-don* the tendon at the back of the heel (calcaneus) that attaches the calf muscles.

Achondroplasia *a-kon-dro-play'-zee-ah* a condition, sometimes influenced by hereditary factors as well as hormonal levels, that produces a dwarf with short arms and legs but normal trunk and head.

Acid a substance that dissociates into hydrogen ions and anions when in an aqueous solution (compare with *base*).

Acidosis *ass-i-do'-sis* a condition in which the blood has a higher hydrogen ion concentration than normal with a decreased pH.

Acne inflammatory disease of the skin.

Acromegaly *ak-ro-meg'-a-lee* an abnormal pattern of bone and connective-tissue growth characterized by enlarged hands, face, and feet and associated with excessive pituitary growth hormone that is secreted after the epiphyseal cartilages have been replaced.

Acromion *ak-ro'-mee-un* the outer projection of the spine of the scapula; forms the highest point of the shoulder.

Acrosome *ak'ro-some* crescent-shaped structure molded to the nucleus and forming the anterior sperm head.

Actin *ak'-tin* a contractile protein of muscle fiber; myosin is another protein found in muscle.

Action potential an event occurring when a stimulus of sufficient intensity is applied to a neuron, allowing sodium ions to move into the cell and reverse the polarity.

Active state the development of tension by the contractile proteins of muscle.

Active transport net movement of a substance across a membrane against a concentration gradient; requires release and use of energy.

Adaptation (1) any change in structure, form, or habits to suit a new environment; (2) decline in the frequency of sensory nerve excitation when a receptor is stimulated continuously.

Addison's disease condition resulting from abnormal, deficient secretion of adrenal cortical hormones.

Adduct *ad-duct'* to move toward the midline of the body.

Adenohypophysis *ad-e-no-high-pof'-i-sis* the part of the pituitary gland that develops from Rathke's pouch.

Adenoids *ad'-e-noidz* enlargement of the pharyngeal tonsils in children.

Adenosine diphosphate (ADP) *a-den'-o-sene di-fos'-fate* the substance formed when ATP is split apart and releases energy.

Adenosine triphosphate (ATP) *a-den'-o-sene tri-fos'-fate* the compound that is an important intracellular energy source.

Adipose *ad'-i-pos* fatty.

Adrenal glands *ad-reen'-al* hormone-producing glands located superior to the kidneys; each consists of medulla and cortex areas.

Adrenalin *ad-ren'-a-lin* trademark name for epinephrine.

Adrenergic fibers *ad-ren-ur'-jick* nerve fibers whose terminals release norepinephrine or epinephrine upon stimulation.

Adrenocorticotropic hormone (ACTH) *a-dree'-no-kort-i-ko-tro'-pik* a hormone that influences the activity of the adrenal cortex and is released by the anterior portion of the pituitary.

Adventitia *ad-ven-tish'-yah* the outermost layer or covering of an organ.

Aerobic *ay-er-o'-bick* requiring oxygen to live or grow.

Afferent *af'-er-ent* carrying to or toward a center.

Afferent neuron *noo'-ron* nerve cell that carries impulses toward the central nervous system.

Afterbirth *af'-ter-berth* the expelled placenta after a baby has been born.

Agglutinin *a-gloo'-tin-in* an antibody in blood plasma that causes clumping of corpuscles or bacteria.

Agglutinogen *a-gloo-tin'-a-jen* (1) an antigen that stimulates the formation of a specific agglutinin; (2) an antigen found on red blood cells that is responsible for determining the ABO blood group classification.

Agonist *ag'-o-nist* a muscle whose contraction opposes the action of another muscle, its antagonist, which at the same time relaxes.

Albumin *al-bu'-min* a protein found in virtually all animal tissue and fluid.

Albuminuria *al-bu-min-oor'-ee-ah* the presence of albumin in the urine.

Aldosterone *al-dos'-ter-own* a hormone produced by the adrenal cortex that is important in sodium retention and reabsorption.

Alimentary *al-im-en'-ta-ree* pertaining to nourishment or the digestive organs.

Alkalosis *al-ka-lo'-sis* a condition in which the blood has a lower hydrogen ion concentration than normal with an increased pH.

Allantois *a-lan'-to-is* an embryonic membrane; its blood vessels develop into blood vessels of the umbilical cord.

Allele *al'-eel* one of two or more alternate forms of a gene.

Allergy *al'-ler-jee* a condition in which there is hypersensitive response to a particular substance or environmental element.

Alveolus *al-ve'-o-lus* (1) a general anatomic term referring to a small cavity or depression; (2) an air sac in the lungs.

Amenorrhea *a-men-or-ree'-ah* absence of menstruation.

Amino acid *a-mee'-no* an organic compound containing nitrogen, carbon, hydrogen, and oxygen; the building block of protein.

Amnion *am'-nee-on* the innermost fetal membrane; it forms a fluid-filled sac for the embryo.

Amniocentesis *am'-nee-o-sen-tee'-sis* a procedure for obtaining amniotic fluid. Amniotic fluid is removed (by inserting a needle through a pregnant woman's abdomen into the sac containing the fetus). Fetal cells in the fluid can be grown in a laboratory and subjected to chromosomal and/or biochemical analysis.

Amphiarthroses *am-fee-ar-throw'-seez* a type of articulation in which little motion occurs because of fibrocartilaginous connection of the articulating bony surfaces.

Amplitude *am'-pli-tude* largeness; fullness or wideness of extent or range.

Ampulla *am-pul'-lah* a saclike dilation of a duct or tube.

Anabolism *an-ab'-o-li-zem* the energy-requiring building-up phase of metabolism in which simpler substances are synthesized into more complex substances.

Anaerobic *an-ay-er-o'-bick* requiring no oxygen to live or grow.

Anastomosis *a-nas-to-mo'-sis* a union or joining of blood vessels or other tubular structures.

Anatomy the science of the structure of organs of organic beings.

Androgen *an'-dro-jen* a hormone or other substance that controls male secondary sex characteristics.

Anemia *a-nee'-mee-ah* a decreased number of erythrocytes or decreased percentage of hemoglobin in the blood.

Aneurysm *an'-yu-riz-em* a localized blood-filled sac in an artery wall caused by dilation or weakening of the wall.

Angina pectoris *an-jee'-nah pek'-tor-is* a severe, suffocating chest pain caused by brief lack of oxygen to heart muscle.

Angiotensin *an-jee-o-ten'-sin* a vasoconstrictor substance found in the blood.

Anion *an'-eye-on* an ion carrying one or more negative charges and therefore attracted to a positive pole.

Annulus fibrosus *an'-u-lus fi-bro'sus* firm, ring-shaped outer portion of the intervertebral disc.

Anorexia *an-or-ek'-see-ah* loss of appetite or desire for food.

Anorexia nervosa *ner-vo'-sah* a nervous condition in which an extreme loss of appetite leads to emaciation, malnutrition, and possible worse consequences.

Anoxia *an-ok'-see-uh* a deficiency of oxygen.

Antagonist *an-tag'-o-nist* a muscle that acts in opposition to an agonist or prime mover.

Anterior the front of an organ or part; the ventral surface.

Antibody a specialized substance produced by the body that can provide immunity against a specific antigen.

Antidiuretic hormone (ADH) *an-tee-dye-u-re'-tik* a pituitary gland hormone that controls the reabsorption of water by the kidney.

Antigen *an'-ti-jen* any substance—including toxins, foreign proteins, or bacteria—which, when introduced to the body, causes antibody formation.

Antithrombin *an-ti-throm'-bin* a substance in blood plasma that neutralizes thrombin, thereby inhibiting coagulation.

Antrum *an'-trum* an open space or chamber; a cavity, especially within a bone.

Anus *ay'-nus* the distal end of the digestive tract and the outlet of the rectum.

Aorta *ay-or'-tah* the major system artery; arises from the left ventricle of the heart.

Aortic arch *ay-or'-tik* the curved and most superior portion of the aorta.

Aortic body a receptor in the aortic arch sensitive to changing oxygen, carbon dioxide, and pH levels of the blood.

Aortic hiatus *high-ay'-tus* an opening behind the diaphragm giving passage to the aorta.

Aphasia *a-fay'-zhee-ah* a loss of the power of speech or a defect in speech.

Aplastic anemia *a-plas'-tic* anemia caused by inadequate production of erythrocytes resulting from inhibition or destruction of the red bone marrow.

Apocrine gland *ap'-o-krin* a gland in which the secretions accumulate toward the outer ends of the secreting cells; the cell ends pinch off and are released with the secretions.

Aponeurosis *ap-o-noo-ro'-seez* the fibrous or membranous sheet serving as a connection between muscle and the part it moves.

Appendix *a-pen'-dicks* a wormlike sac attached to the large intestine.

Aqueous humor *ay'-kwee-us hyu'-mer* the watery fluid of the anterior and posterior chambers of the eye.

Arachnoid *a-rak'-noid* weblike; specifically, the weblike, middle layer of the three meninges.

Areola *ah-ree'-o-lah* (1) any minute opening or space in a tissue; (2) the circular, pigmented area surrounding the nipple.

Arrector pili *ar-rek'-tor pih'-lee* tiny, smooth muscles of the skin which, upon contraction, pull the hair follicle, causing the hair to stand up.

Arteriole *ar-te'-ree-ole* a minute artery.

Arteriosclerosis *ar-te-ree-o-skle-ro'-sis* any of a number of proliferative and degenerative changes in the arteries.

Artery a vessel that carries blood away from the heart.

Arthritis *ar-thright'-us* inflammation of the joints.

Articulate *ar-tik'-u-late* to join together in such a way as to allow motion between the parts.

Ascites *as-sigh'-teez* accumulation of serous fluid in the abdominal cavity.

Asphyxia *as-fik'-see-ah* loss of consciousness resulting from a deficiency in the oxygen supply.

Asthma *az'-ma* a disease or allergic response characterized by bronchial spasms and difficult breathing.

Astigmatism *a-stig'-ma-tiz-em* a visual defect resulting from irregularity in the lens or cornea of the eye causing the image to be out of focus.

Ataxia *a-tak'-see-a* lack of muscular coordination.

Atelectasis *at-e-lek'-ta-sis* lung collapse or incomplete expansion of a lung.

Atherosclerosis *ath-er-o-skle-ro'-sis* changes in the walls of large arteries consisting of lipid deposits on the artery walls.

Atlas the first cervical vertebra; articulates with the occipital bone of the skull and the second cervical vertebra (axis).

Atom the smallest particle of an element; composed of electrons, protons, and neutrons; capable of existing individually or in combination with atoms of the same or another element.

Atresia *a-tree'-zhee-ah* (1) the abnormal closure of a body canal or opening; (2) degeneration of the ovarian follicle and the ovum within it.

Atrioventricular bundle (AV bundle) *a-tree-o-ven-trik'-yoo-lar* the bundle of specialized fibers serving to conduct impulses from the AV node to the right and left ventricles; failure of the AV bundle results in heart block; also called bundle of His.

Atrioventricular node (AV node) a specialized compact mass of conducting cells located at the atrioventricular junction in the heart.

Atrium *ay'-tree-um* a chamber of the heart receiving blood from the veins.

Atrophy *a'-tro-fee* a reduction in size or wasting away of an organ or cell resulting from disease or lack of use.

Auditory *aw'-di-to-ree* pertaining to the sense of hearing.

Auditory ossicles *aw'-di-to-ree oss'-i-kuls* the three tiny bones serving as transmitters of vibrations and located within the middle ear: the malleus, incus, and stapes.

Auditory (eustachian) tube *yoo-stay'shee-un* the connection between the middle ear and the pharynx.

Auricle *aw'-ri-kel* the external ear.

Auscultation *aws-kul-tay'-shun* the act of examination by listening to body sounds.

Autoimmune response *aw-to-im-myoon'* the production of antibodies or effector T cells that attack a person's own tissue.

Automaticity *aw-to-ma-ti'-si-tee* the ability of a structure, organ, or system to initiate its own activity.

Autonomic *aw-to-nom'-ik* self-directed; self-regulating; regulating; independent.

Autonomic nervous system the division of the nervous system that functions involuntarily and is responsible for innervating cardiac muscle, smooth muscle, and glands.

Autosome *aw'-to-some* a chromosome that is not a sex chromosome.

Axial skeleton *aks'-ee-al* the bones of the head and trunk: the skull, vertebral column, thorax, and sternum.

Axilla *aks-il'-ah* the armpit.

Axis (1) the second cervical vertebra; has a vertical projection called the odontoid process around which the atlas rotates; (2) the imaginary line about which a joint or structure revolves.

Axon *aks'-on* the process of a nerve cell by which impulses are carried away from the cell; efferent process; the conducting portion of a nerve cell.

Bacteria any of a large group of microorganisms, usually non-spore-producing and generally one-celled; found in humans and other animals, plants, soil, air, and water; have a broad range of functions.

Baroreceptor *ba-ro-ree-sep'-ter* a receptor that is stimulated by pressure changes.

Basal ganglia *bay'-zel gan'-glee-ah* gray matter structures deep inside each of the cerebral hemispheres.

Basal layer columnar cells in which mitosis takes place; also called stratum basale of the epidermis.

Basal metabolic rate *met-ah-bol'-ik* the rate at which energy is expended (heat produced) by the body per unit time under controlled (basal) conditions: 12 hours after a meal, at rest.

Base a substance that accepts hydrogen ions; capable of uniting with an acid to form a salt.

Basement membrane a thin layer of substance to which epithelial cells are attached in mucous surfaces.

Basophil *bay'-so-fil* white blood cells that readily take up basic dye; have a relatively pale nucleus and granular-appearing cytoplasm.

B cells lymphocytes that are committed to differentiate into the antibody producing plasma cells involved in humoral immunity.

Benign *bee-nine'* not malignant.

Biceps *bigh'-seps* two-headed, especially applied to certain muscles.

Bicuspid *bigh-kus'-pid* having two points or cusps.

Bifurcation *bi-fur-ka'-shun* division into two branches.

Bile a greenish-yellow or brownish fluid produced in and secreted by the liver, stored in the gallbladder, and released into the small intestine.

Bilirubin *bil-i-roo'-bin* the red pigment of bile.

Biliverdin *bil-i-ver'-din* the green pigment of bile.

Biofeedback an area of research in which subconscious feedback is raised to the conscious level.

Biopsy *bigh'-op-see* the removal and examination of live tissue.

Bipennate *bigh-pen'-nate* muscles in which the fibers are attached obliquely on both sides of the tendon.

Blastocyst *blas'-to-sist* a stage of early embryonic development.

Blood-brain barrier a mechanism that inhibits passage of materials from the blood into brain tissues and cerebrospinal fluid.

Bolus *bo'-lus* (1) a rounded mass of food prepared by the mouth for swallowing; (2) a concentrated mass of a pharmaceutical preparation, usually given intravenously.

Bowman's capsule *bo'-manz* the double-walled cup at the end of a nephron.

Brachial *bray'-kee-al* pertaining to the arm.

Bradycardia *brad-i-kar'-dee-ah* slowness of the heart rate; below 60 beats per minute.

Brain stem the portion of the brain consisting of the medulla, pons, and midbrain.

Bronchitis *brong-kigh'-tis* inflammation of the bronchi.

Bronchus *brong'-kus* one of the two large branches of the trachea leading to the lungs.

Buccal *buck'-kal* pertaining to the cheek.

Buffer a substance or substances that tend to stabilize the pH of a solution.

Bundle branch block a blocking of heart action resulting from a local lesion of the bundle of His; delayed contraction of one ventricle.

Bursa *bur'-sah* a small sac or cavity filled with fluid and located at friction points, especially joints.

Calcitonin *kal-si-to'-nin* a hormone released by the thyroid that lowers calcium and phosphate levels of the blood.

Calculus *kal'-ku-lus* a stone formed within various body parts.

Callus *kal'-us* (1) a bonelike material that protrudes between ends of a fractured bone; (2) a thickening of the skin caused by rubbing or friction.

Calyx *kay'-liks* a cuplike division of pelvis of the kidney.

Calorie *kal'-or-ee* a unit of heat; the amount of heat required to raise the temperature of 1 g of water 1° C is the small calorie; the large calorie is the amount of heat required to raise 1 kg of water 1° C—used in metabolic and nutrition studies.

Canal   a duct or passageway; a tubular structure.

Canaliculi   *kan-al-ik'-u-lee*   extremely small tubular passages or channels.

Cancer   a malignant, invasive cellular tumor that has the capability of spreading throughout the body or body parts.

Capacitation   *ka-pass-i-tay'-shun*   the process in which sperm undergo changes in the female reproductive tract making them capable of fertilization.

Capillary   *kap'-a-lar-ee*   a minute blood vessel connecting arterioles with venules.

Carbohydrates   *kar-bo-high'-drates*   organic compounds composed of carbon, hydrogen, and oxygen with the hydrogen and oxygen present in a 2:1 ratio; include starches, sugars, cellulose.

Carbonic anhydrase   *kar-bon'-ik an-high'-drase*   an enzyme that facilitates the combination of carbon dioxide with water to form carbonic acid.

Carcinogen   *kar-sin'-o-jen*   cancer-causing agent.

Carcinoma   *kar-sin-o'-ma*   cancer; a malignant growth.

Cardiac   *kar'-dee-ak*   pertaining to the heart.

Cardiac muscle   specialized muscle of the heart.

Cardiac output   the blood volume (in liters) ejected per minute by the left ventricle.

Carotene   *kar'-o-teen*   a yellow pigment; influences skin color.

Carotid   *ka-rot'-id*   the main artery in the neck.

Carotid body   a receptor at the bifurcations of the common carotid arteries sensitive to changing oxygen, carbon dioxide, and pH levels of the blood.

Carotid sinus   *sigh'-nus*   a dilation of a common carotid artery at its bifurcation; involved in regulation of systemic blood pressure.

Carpals   *kar'-puls*   the eight bones of the wrist.

Cartilage   *kar'-ti-lej*   white, semiopaque, fibrous connective tissue.

Caruncle   *kar'-un-kul*   a small fleshy protuberance.

Catabolism   *ka-tab'-a-liz-em*   the process in which living cells break down more complex substances into simpler substances; destructive metabolism.

Cataract   *kat'-a-rakt*   partial or complete loss of transparency of the crystalline lens of the eye or its capsule.

Catheter   *ka'-the-ter*   a narrow, hollow tube that can be inserted into a body cavity for withdrawal of fluids.

Cation   *kat'-eye-on*   an ion with a positive charge and therefore attracted toward a negative pole (cathode) in electrolytic cells.

Caudal   *kaw'dal*   in humans, the inferior portion of the anatomy.

Cecum   *see'kum*   the blind-end pouch at the beginning of the large intestine.

Cell   the basic biological unit of living organisms (except viruses), containing a nucleus and a variety of organelles; usually enclosed in the membrane.

Cell mediated immunity   immunity mediated by certain populations of lymphoid cells called T cells.

Cell membrane   the selectively permeable membrane forming the outer layer of most animal cells; also called plasma membrane or unit membrane.

Cellulose   *sel'-u-lose*   a fibrous carbohydrate that is the main structural component of plant tissues.

Cementum   *se-men'-tum*   the bony connective tissue that covers the root of a tooth.

Central nervous system (CNS)   the brain and the spinal cord.

Centriole   *sen'-tree-ole*   a minute body found in the nucleus of the cell; active in cell division.

Cerebellum   *ser-e-bel'-lum*   part of the hindbrain; controls movement coordination; consists of two hemispheres and a central portion (vermis).

Cerebral aqueduct   *ser-ee'-bral a'-kwe-duct*   the elongated, slender cavity of the midbrain that connects the third and fourth ventricles; also called the aqueduct of Sylvius.

Cerebrospinal fluid   *ser-ee-bro-spy'-nul*   the fluid produced in the cerebral ventricles; fills the ventricles and surrounds the central nervous system.

Cerebrum   *ser'-i-brum or se-ree'-brum*   the largest part of the brain; consists of right and left cerebral hemispheres.

Cerumen   *se-roo'-men*   earwax.

Cervical   *ser'-vi-kal*   refers to the neck or the necklike portion of an organ or structure.

Cervix   *ser'-vix*   (1) the cylindrical, inferior portion of the uterus leading to the vagina; (2) any necklike structure.

Chemoreceptor   *kem-o-ree-sep'-tor*   receptors sensitive to various chemical stimulations and changes.

Chemotaxis   *kem-o-tak'-sis*   movement of a cell, organism, or part of an organism toward or away from a chemical substance.

Chiasma   *kee-az'-ma*   a crossing or intersection of two structures, such as the optic nerves.

Cholecystectomy   *kol-e-sis-tek'-tom-ee*   removal of the gallbladder.

Cholecystokinin   *kol-e-sis-toe'-kine-in*   an intestinal hormone that stimulates gallbladder contraction and pancreatic enzyme release, but inhibits gastric acid secretion when the stomach is actively secreting.

Cholesterol   *ko-les'-ter-ol*   an organic alcohol found in animal fats and oil as well as in most body tissues, especially bile.

Cholinergic fibers   *ko-lin-ur'-jick*   nerve endings that, upon stimulation, release acetylcholine at their terminations.

Chondroblast   *kon'-dro-blast*   a cell that forms the fibers and matrix of cartilage.

Chondrocyte   *kon'-dro-site*   a mature cartilage cell.

Chorion   *kor'-ee-on*   the outermost fetal membrane; forms the placenta.

Choroid   *ko'-roid*   (1) skinlike; (2) the pigmented covering of the eye.

Chromosome   *kro'-mo-some*   the structures in the nucleus that carry the hereditary factors (genes).

Chyle   *kile*   a milky fluid consisting of fat globules in lymph formed in the small intestine during digestion.

Chyme   *kime*   the semifluid contents of the stomach consisting of partially digested food and gastric secretions.

Cilia   *sil'-lee-ah*   tiny, hairlike projections on cell surfaces that move in a wavelike manner.

Circadian   *sir-kay'-dee-an*   daily; on a daily cycle.

Circle of Willis   a union of arteries at the base of the brain.

Circumcision   *sur-kum-si'-zhun*   removal of the foreskin of the penis.

Circumduction   *sur-kum-duk'-shun*   circular movement of a body part.

Cirrhosis   *sir-o'-sis*   a chronic disease, particularly of the liver, characterized by an overgrowth of connective tissue.

Clitoris   *kli'-to-ris*   a small, erectile structure in the female, homologous to the penis in the male.

Clone   one or more offspring of like genetic constitution produced by asexual reproduction.

Cochlea *koak'-lee-ah* a cavity of the inner ear resembling a snail shell.

Coenzyme *ko-en'-zime* a nonprotein substance associated with and activating an enzyme.

Coitus *ko'-i-tus* sexual intercourse.

Colloid *kol'-loid* solute particles dispersed in a medium; particles do not settle out readily and do not pass through natural membranes.

Colloidal osmotic pressure (COP) *kol-loi'-dal os-mot'-ick* the pressure exerted on a membrane by the particles in a colloid; usually refers to the osmotic pressure of blood plasma and body fluids resulting from the presence of protein.

Colostrum *kol-os'-trum* the first milk secreted after parturition.

Coma *ko'-ma* unconsciousness from which the person cannot be aroused.

Complement system a system consisting of a series of plasma proteins that normally circulate in the blood in inactive forms. The activation of the system's initial components intiates a cascading series of reactions that ultimately produce active complement molecules, which can act as chemotactic agents, enhance phagocytosis, and exert other effects.

Compound a substance composed of two or more different elements.

Concave *kon'-kave* having a curved, depressed surface.

Concha *kon'-kah* a shell-shaped structure.

Condom *kon'-dom* a sheath worn over the penis during sexual intercourse to prevent conception and/or infection.

Conduction *kon-duk'-shun* the transfer of energy from molecule to molecule by direct contact.

Condyle *kon'-dile* a rounded projection at the end of a bone that articulates with another bone.

Cones one of the two types of photosensitive cells in the retina of the eye.

Congenital *kon-jen'-i-tal* existing at birth.

Conjunctiva *kon-junk-tigh'-vah* the thin protective membrane on the insides of the eyelids and the anterior surface of the eye itself.

Conjunctivitis *kon-junk-ti-vigh'-tis* an inflammation of the conjunctiva of the eye.

Connective tissue a primary type of tissue; form varies extensively, as does function, which includes support and storage.

Contraception *kon-tra-sep'-shun* the prevention of conception; birth control.

Contractility *kon-trak-til'-i-tee* a substance's ability to shorten or develop tension upon the application of a stimulus.

Contralateral *kon-tra-lat'-er-al* opposite; acting in unison with a similar part on the opposite side of the body.

Convection *kon-vek'-shun* the constant transfer of heat by movement of heated particles.

Convergence *kon-verj'-ence* turning toward or approaching a common point from different directions.

Convoluted *kon'-vo-lu-ted* rolled, coiled, or twisted.

Copulation *kop-u-lay'-shun* sexual intercourse.

Cornea *kor'-nee-ah* the transparent anterior portion of the eyeball.

Corona radiata *ko-ro'-nah ray-dee-aw'-tah* (1) the arrangement of elongated follicle cells around a mature ovum; (2) the crownlike arrangement of nerve fibers radiating from the inner capsule of the brain to every part of the cerebral cortex.

Corpus *kor'-pus* body; the major portion of an organ.

Cortex *kor'-teks* the outer surface layer of an organ.

Corticoid *kor'-ti-koid* a substance whose function and properties are similar to those of corticosteroids.

Corticosteroids *kor'-ti-ko-ste'-roidz* the steroid hormones released by the adrenal cortex.

Cortisol *kor'-ti-sol* a glucocorticoid produced by the adrenal cortex.

Costal *kos'-tal* pertaining to the ribs.

Cramp a painful, involuntary contraction of a muscle.

Cranial *kray'-nee-al* pertaining to the skull.

Creatine phosphate *kree'-uh-tin fos'-fate* a compound that serves as an alternative energy source for muscle tissue.

Crenation *kre-nay'-shun* the shriveling of an erythrocyte resulting from withdrawal of water.

Cretinism *kree'-tin-izm* a severe thyroid deficiency in the young that leads to stunted physical and mental growth.

Crista *kris'-ta* a crest or ridge.

Cryptorchidism *kript-or'-kid-izm* a developmental defect in which the testes fail to descend into the scrotum.

Crystalloid *kris'-tal-oid* a substance having a crystalline structure as opposed to a colloidal composition; particles can pass through a natural membrane.

Cubital *ku'-bi-tal* pertaining to the forearm.

Cupula *ku'-pu-la* a domelike structure.

Cushing's syndrome *ku'-shingz sin'-drome* a disease produced by excess secretion of adrenocortical hormone; characterized by adipose tissue accumulation, weight gain, and osteoporosis, for example.

Cutaneous *ku-tay'-nee-us* pertaining to the skin.

Cyanosis *sigh-a-no'-sis* a bluish coloration of the mucous membranes and skin caused by deficient oxygenation of the blood.

Cyst *sist* a sac with a distinct wall, containing fluid or other material; may be pathological or normal.

Cystitis *sis-tigh'-tis* inflammation of the urinary bladder.

Cytokinesis *sigh-to-ki-nee'-sis* the changes in the cytoplasm during cell division.

Cytology *sigh-tol'-o-jee* the science concerned with the study of cells.

Cytoplasm *sigh'-to-plaz-um* the protoplasm of a cell other than that of the nucleus.

Deamination *dee-am-i-nay'-shun* the removal of an amino group from an organic compound by reduction, hydrolysis, or oxidation.

Deciduous *dee-sid'-u-us* temporary.

Deciduous (milk) teeth the 10 temporary teeth replaced by permanent teeth; "baby" teeth.

Defecation *def-e-kay'-shun* the elimination of the contents of the bowels (feces).

Deglutition *dee-gloo-tish'-un* the act of swallowing.

Dehydration *dee-high-dray'-shun* a condition resulting from excessive loss of water.

Dendrite *den'-dright* branching; the branching neuron process that transmits the nerve impulse to the cell body; the receptive portion of a nerve cell.

Dental caries *den'-tal kar'-eez* tooth cavity.

Dentin *den'-tin* the calcified tissue beneath the enamel forming the major part of a tooth.

Deoxyribonucleic acid (DNA) *dee-ox-i-rye-bo-nu-klee'-ik* a nucleic acid found in all living cells: carries the organism's hereditary information.

Depolarization *dee-po-lar-i-zay'-shun* the neutralization to a state of nonpolarity; the loss of a negative charge inside the cell.

Depressor *dee-pres'sor* any substance that causes slowing, reduction of activity, or inhibition of another structure, organ, or substance.

Dermatitis *der-ma-tigh'-tis* an inflammation of the skin; nonspecific skin allergies.

Dermis *der'-mis* the deep layer of dense, irregular connective tissue of the skin; also called corium.

Desmosome *des'-mo-some* small, apposed, ellipsoidal plates in membranes of adjacent cells; also called macula adherens.

Diabetes insipidus *dye-uh-bee'-teez in-sip'-i-dus* a disease characterized by passage of a large quantity of urine of low specific gravity plus intense thirst and dehydration; a hypothalamic disorder is the cause.

Diabetes mellitus *mel'-li-tus* a disease caused by deficient insulin release, leading to failure of the body tissue to oxidize carbohydrates at a normal rate.

Dialysis *dye-al'-i-sis* the separation of substances from one another in a solution by taking advantage of their differing rate of diffusion through a semipermeable membrane.

Diapedesis *dye-a-pe-dee'-sis* the passage of blood cells through unruptured vessel walls into the tissues.

Diaphragm *dye'-a-fram* (1) any partition or wall separating one area from another; (2) a muscle that separates the upper thoracic cavity from the lower abdominopelvic cavity.

Diarthrosis *dye-ar-throw'-sis* a freely movable joint; a synovial joint.

Diastole *dye-ass'-ta-lee* a period (between contractions) of relaxation and dilation of the heart during which it fills with blood.

Diencephalon *dye-en-sef'-uh-lon* that part of the forebrain between the telencephalon and the mesencephalon including the thalami and most of the third ventricle.

Diffusion *di-fu'-zhun* the spreading of particles in a gas or solution with a movement toward uniform distribution of particles.

Digestion *di-jest'-yun* the bodily process of breaking down foods chemically and mechanically into compounds capable of being absorbed by body cells.

Dilate *dye'-late* to stretch; to open; to expand.

Distal *diss'tal* farthest from the center or midpoint of a limb structure.

Diverticulum *dye-ver-tik'-u-lum* a pouch or sac in the walls of a hollow organ or structure.

Dominant an allele, or the corresponding trait, that is expressed in all heterozygous cells or organisms.

Dorsal *dor'-sal* pertaining to the back; posterior.

Downs syndrome *dounz sin'-drome* an abnormality resulting from the presence of an extra copy of the genetic material contained in chromosome number 21. Characteristics include mental retardation and altered physical appearance.

Duct a canal or passageway.

Duodenum *doo-o-dee'-num* the first part of the small intestine.

Dura mater *doo'-rah may'-ter* the outermost and toughest of the three membranes (meninges) covering the brain and spinal cord.

Dysfunction *dis-funk'-shun* lack of normal function; disorder.

Dysmenorrhea *dis-men-or-ree'-ah* difficult, painful menstruation.

Dyspnea *disp'-nee-ah* labored, difficult breathing.

Dystrophy *dis'-tro-fee* a disorder caused by a defect or dysfunction of nutrition.

Ectoderm *ek'-to-derm* tissue forming the outer covering of the body and nervous tissues.

Ectopic *ek-to'-pik* not in the normal place; for example, in an ectopic pregnancy the egg is implanted at a place other than the uterus.

Edema *e-dee'-mah* an abnormal accumulation of fluid in body parts or tissues; causes swelling.

Effector *ef-fek'-tor* a motor or sensory nerve ending in an organ, gland, or muscle.

Efferent *ef'-er-ent* carrying away or away from, especially a nerve fiber that carries impulses away from the central nervous system.

Ejaculation *e-jak-u-lay'-shun* the sudden ejection of a fluid from a duct, especially semen from the penis.

Elastin *e-las'-tin* the main protein in elastic fibers of connective tissues.

Electrocardiogram (ECG) *ee-lek-tro-kar'-dee-o-gram* a graphic record of the electric current associated with heartbeats.

Electroencephalogram (EEG) *ee-lek-tro-en-sef'-a-lo-gram* a graphic record of the activity of nerve cells in the brain.

Electrolyte *ee-lek'-tro-lite* a substance that breaks down into ions when in solution and is capable of conducting an electric current.

Electron *ee-lek'-tron* a negatively charged particle in motion around the nucleus of an atom.

Embolism *em'-bo-liz-em* the obstruction of a blood vessel by a clot floating in the blood; may also be a bubble of air in the vessel (air embolism)

Embryo *em'-bree-oh* an organism in its early stages of development; in humans, the first two months after conception.

Emesis *em'-e-sis* vomiting.

Emphysema *em-fi-see'-muh* a condition caused by overdistension of the pulmonary alveoli or abnormal presence of air or gas in body tissues.

Enamel the hard, calcified substance that covers the crown of a tooth.

Encephalitis *en-sef-a-ligh'-tis* an inflammation of the brain.

Endocardial *en-do-kar'-di-al* pertaining to the inner lining of the heart.

Endocarditis *en-do-kar-di'-tis* an inflammation of the inner lining of the heart.

Endocardium *en-do-kar'-dee-um* the membrane lining the interior of the heart; endothelium and connective tissue.

Endochondral *en-do-kon'-dral* pertaining to the development of structures in cartilage.

Endocrine glands *en'-do-krin* ductless glands that empty their secretions directly into the blood.

Endoderm *en'-do-derm* tissue forming the digestive tube and its associated structures.

Endometrium *en-do-me'-tree-um* the mucous membrane lining of the uterus.

Endomysium *en-doo-mis'-ee-um* the thin connective tissue between the fibers of a muscle bundle.

Endoplasmic reticulum *en-do-plaz'-mik re-tik'-u-lum* a membranous network of tubular or saclike channels through the cytoplasm of a cell.

Endothelium *en-do-thee'-lee-um* the single layer of simple squamous cells that line the walls of the heart and the vessels that carry blood and lymph.

Energy the capacity to do work.

Enzyme *en'-zime* a substance formed by living cells that acts as a catalyst in bodily chemical reactions.

Eosinophil *ee-o-sin'-o-fil* a granular white blood cell whose granules readily take up a stain called eosin.

Epidermis *e-pi-der'-mis* the outer layer of cells of the skin.

Epididymis *ep-i-did'-i-mis* that portion of the seminal duct in which sperm mature and are transported from testes to body exterior.

Epiglottis *ep-e-glot'-tis* the elastic membrane-covered cartilage at the back of the throat; guards the glottis during swallowing.

Epimysium *e-pi-mis'-ee-um* the sheath of connective tissue surrounding a muscle.

Epinephrine *ep-i-nef'-rin* the chief hormone of the adrenal medulla.

Epiphysis *e-pif'-i-sis* the extremities of a long bone.

Epithelium *e-pi-thee'-lee-um* one of the primary tissues; covers the surface of the body and lines the body cavities, ducts, and vessels.

Eponychium *ep-o-neech'-ee-um* the fold of skin overlying the root of the nail.

Equilibrium *ee-kwi-lib'-ri-um* balance; a state when opposite reactions or forces counteract each other exactly.

Erythrocyte *e-rith'-ro-site* red blood cell.

Erythropoiesis *e-rith-ro-poi-ee'-sis* the formation process of erythrocytes.

Estrogen *es'-tro-jen* any substance that stimulates female secondary sex characteristics, female sex hormones.

Eupnea *yoop-nee'-ah* easy, normal breathing.

Excretion *eks-kree'-shun* the elimination of waste products from the body.

Exocrine glands *eks'-o-krin* glands that have ducts through which their secretions are carried to a particular site.

Exogenous *eks-og'-en-us* developing or originating outside the organ or part.

Expiration *ex-pi-ray'-shun* the act of expelling air from the lungs.

Exteroceptor *eks'-ter-o-sep-tor* an end organ that responds to stimuli from the external world.

Extracellular *eks-tra-sel'-u-lar* outside a cell.

Extracellular fluid fluid within the body but outside the cells.

Extrinsic *eks-trin'-zik* originating from outside an organ or part

Exudate *eks'-yu-date* material including fluid, pus, or cells that has escaped from blood vessels and has been deposited in tissues.

Facet *fas'-et* a smooth, nearly flat surface on a bone for articulation.

Fallopian tube *fal-low'-pee-an* the oviduct or uterine tube; the tube through which the ovum is transported to the uterus.

Fascia *fash'-ee-ah* the layers of fibrous tissue under the skin or covering and separating muscles.

Fasciculus *fa-sik'-yoo-lus* a bundle of nerve, muscle, or tendon fibers separated by connective tissues.

Feces *fee'-seez* material discharged from the bowel composed of food residue, secretions, and bacteria.

Fenestrated *fen'-es-tray-ted* pierced with one or more small openings.

Fetus *fee'-tus* the unborn young; in humans, from the third month in the uterus until birth.

Fibrillation *fib-ri-lay'-shun* irregular, uncoordinated contraction of muscle fibers.

Fibrin *figh'-brin* the fibrous insoluble protein formed during the clotting of blood.

Fibrinogen *figh-brin'-o-jen* a protein that is converted to fibrin during blood-clotting.

Filtration *fil-tray'-shun* the passage or straining of a solvent and dissolved substances through a membrane or filter.

Fissure *fis'-sure* (1) a groove or cleft; (2) the deepest linear depressions on the brain.

Fistula *fis'-tu-lah* an abnormal passage between organs or between a body cavity and the outside.

Fixator *fiks'-ay-tor* a muscle acting to immobilize a joint or a bone; fixes the origin of prime movers so that muscle action can be exerted at the insertion.

Flaccid *flak'-sid* soft; flabby; relaxed.

Flagella *fla-jel'-ah* long whiplike extensions of the cell membrane of some bacteria; serve as agents for locomotion.

Flexion *flek'-shun* bending; the movement that decreases the angle between bones.

Focal length *fo'-kal* the distance from the lens to the focal point.

Focal point the point at which light rays converge behind the lens.

Follicle *fol'-i-kal* a small sac or gland.

Follicle-stimulating hormone (FSH) a hormone produced by the anterior pituitary that stimulates ovarian follicle production in females and sperm production in males.

Fontanels *fon-tan-els'* the fibrous membranes at the body area where bones have not yet formed; babies' "soft spots."

Foramen *fo-ra'-men* a hole or opening in a bone or between body cavities.

Forebrain the anterior portion of the brain consisting of the telencephalon and the diencephalon.

Fossa *fos'-ah* a depression; often used as an articular surface.

Fovea *fo'-vee-ah* a pit, generally used for attachment rather than for articulation.

Frontal (coronal) plane a longitudinal section that runs at right angles to sagittal planes dividing the body into anterior and posterior parts.

Fulcrum *ful'-krum* the pivot point of a lever.

Fundus *fun'-dus* the base of an organ; that part farthest from the opening of the organ.

Funiculi *fu-nik'-ye-lee* (1) cordlike structures; (2) anterior and lateral divisions of white matter in the spinal cord.

Gallbladder the sac beneath the right lobe of the liver used for bile storage.

Gallstones particles of cholesterol or calcium carbonate that are occasionally formed in gallbladder and bile ducts.

Gamete *gam'-eet* male or female reproductive cell.

Gametogenesis *gam-e-to-jen'-e-sis* the origin and formation of gametes.

Ganglion *gan'-glee-on* a group of nerve-cell bodies, usually located in the peripheral nervous system.

Gap junction intercellular specialization with the cell membranes of adjacent cells only 20 Å apart; also called nexus.

Gastrin a hormone that stimulates gastric secretion, especially hydrochloric acid release.

Gastroenteritis *gas-tro-en-ter-i'-tis* an inflammation of mucosa of stomach and intestine.

Gastroesophageal sphincter *gas-tro-e-soff-a-jee'-al sfink'-ter* narrowing between the esophagus and the stomach.

Gene *jeen* one of the biological units of heredity located on chromosomes; transmits hereditary message.

Genetic counseling *jen-e'-tik* the counseling of parents about the chances of their children having particular genetic abnormalities.

Genetics *jen-e'-tiks* the science of heredity.

Genitalia *jen-i-tay'-lee-ah* the external sex organs.

Genotype *jen'-o-tipe* the genetic composition of an individual.

Gestation *jes-tay'-shun* the period of pregnancy; about 280 days for humans.

Gingiva *jin-jigh'-vah or jin'-ji-vah* the gums.

Gland an organ specialized to secrete or excrete substances for further use in the body or for elimination.

Glans a small glandlike mass of erectile tissue at the tip of the penis and the clitoris.

Glaucoma *glaw-ko'-mah* an abnormal elevation of the pressure within the eye.

Glia *glee'-a* see neuroglia.

Globin *glow'-bin* the protein component of hemoglobin.

Glomerulus *glo-mer'-yoo-lus* a knot of coiled capillaries in the kidney.

Glottis *glot'-tis* the opening between the vocal cords; entrance to the larynx.

Glucagon *gloo'-ka-gon* a hormone formed by islets of Langerhans in the pancreas; raises the glucose level of blood.

Glucocorticoids *gloo-ko-kor'-ti-koidz* the adrenal cortex hormones that affect metabolism of fats and carbohydrates.

Glucose *gloo'-kose* the principal sugar in the blood.

Glycerol *gliss-e-rol* an important alcoholic component of fat.

Glycogen *gligh-ko-jen* an animal starch; the main carbohydrate stored in animal cells.

Glycogenesis *gleye-ko-jen'-e-sis* the body's formation of glycogen from other carbohydrates.

Glycogenolysis *gleye-ko-jen-ol'-i-sis* the body's breakdown of glycogen to glucose.

Glycolysis *gligh-kol'-a-sis* the body's breakdown of glucose into simpler compounds, especially lactic acid.

Goblet cells the individual cells of the respiratory and digestive tracts that function as glands.

Goiter *goi'-ter* an enlargement of the thyroid gland.

Gonad *go'-nad* a gland or organ producing gametes; an ovary or testis.

Gonadotropins *go-nad-o-tro'-pinz* the gonad-stimulating hormones produced by the anterior pituitary.

Graafian follicle *graf'-ee-an fol'-i-kal* a mature ovarian follicle.

Graded response a response whose magnitude varies directly with the strength of the stimulus.

Gray matter the gray area of the central nervous system; contains neurons.

Groin the junction of the thigh and the trunk; the inguinal area.

Growth hormone a hormone that stimulates growth in general; produced in the anterior pituitary; also called somatotropin.

Gubernaculum *goo-ber-nak'-yoo-lum* a guiding, connecting cord structure; the cord between the testis and the scrotal sac, for example.

Gustation *gus-tay'-shun* taste.

Gyrus *jigh'-rus* a convolution on the surface of the cerebral cortex.

Hamstring muscles the posterior thigh muscles: the biceps femoris, semimembranosus, and semitendinosus.

Haustra *haw'-strah* pouches of the colon.

Haversian system or osteon *ha-ver'-zee-an, os'-tee-on* an organized system of interconnecting canals in the microscopic structure of adult compact bone.

Hay fever an acute allergic reaction of conjunctiva and upper air passages due to pollen sensitivity.

Heart block a defective transmission of impulses from atrium to ventricle.

Heart murmur an abnormal heart sound.

Heat exhaustion the collapse of an individual after heat exposure without failure of the body's heat-regulating mechanism.

Heat stroke the failure of the heat-regulating ability of an individual under heat stress.

Hematocrit *hem-at'-o-krit* the percentage of erythrocytes to total blood volume.

Heme *heem* the iron-containing pigment that is essential to oxygen transport by hemoglobin.

Hemiplegia *hem-i-plee'-jee-ah* paralysis of one side of the body.

Hemocytoblasts *hee-mo-sigh'-to-blasts* stem cells that give rise to all the formed elements of the blood.

Hemoglobin *hee-mo-glo'-bin* the oxygen-transporting component of erythrocytes composed of heme and globin.

Hemolysis *hee-mol'-i-sis* the destruction of erythrocytes.

Hemophilia *hee-mo-phil'-i-a* a clotting defect caused by an inherited genetic absence of a blood-clotting factor.

Hemopoiesis *hee-mo-poi-ee'-sis* the formation of blood.

Hemorrhage *hem'-o-ridj* the escape of blood from the vessels by flow through ruptured walls; bleeding.

Heparin *hep'-a-rin* a substance that prevents clotting found in many tissues, especially the liver.

Hepatic portal system *he-pat'-ik* the liver circulatory arrangement where the hepatic portal vein carries dissolved nutrients to the liver tissues.

Hepatitis *hep-at-eye'-tis* an inflammation and/or infection of the liver.

Hernia *her'-nee-ah* the abnormal protrusion of an organ or a body part through the containing wall of its cavity.

Herpes simplex *her'-peez sim'-pleks* a fever blister or cold sore; a virus-caused condition.

Herpes zoster *zos'-ter* an infection of the dorsal root ganglia of spinal nerves by a virus, causing pain and fluid-filled vesicles on the skin; also called shingles.

Heterosexuality *he-ter-o-seks-u-al'-i-tee* sexual interest in or desire for persons of the opposite sex.

Heterozygous *het-er-o-zigh'-gus* a situation in which two different alleles occur at a given locus on homologous chromosomes.

Hilum, hilus *high'-lum, high'-lus* the notched or depressed area where vessels enter and leave an organ.

Histamine *his'-ta-meen* a substance present in many cells which causes vasodilation and increased vascular permeability.

Histology *his-tol'-o-jee* the branch of anatomy dealing with the microscopic structure of tissues.

Holocrine glands *hol'-o-krin* glands that accumulate their secretions within their cells; secretions are discharged only upon rupture and death of the cell.

Homeostasis *hom-ee-o-stas'-sis* the state when the body organs function together to maintain a stable internal environment for the general well-being of the entire body; a state of body equilibrium.

Homeotherm *ho'-me-o-therm* an organism that produces its own heat and maintains a constant body temperature.

Homologous *ho-mol'-o-gus* parts or organs corresponding in structure but not necessarily in function.

Homosexuality *ho-mo-seks-u-al'-i-tee* sexual interest in or desire for persons of the same sex.

Homozygous *ho-mo-zigh'-gus* a situation in which the same allele occurs at a given locus on homologous chromosomes.

Hormones *hor'-mones* the secretions of endocrine glands; responsible for specific regulatory effects on certain parts or organs.

Humoral immunity *hyoo'-mer-al im-myoo'-ni-tee* immunity mediated by specialized proteins called antibodies.

Hyaline *high'-a-lin* glassy; transparent

Hydrocarbon *high-dro-kar'-bon* a molecule composed of only carbon and hydrogen.

Hydrochloric acid *high-dro-klo'-rik* HCl; facilitates protein digestion in the stomach; produced by parietal cells.

Hydrolysis *high-drol'-i-sis* the process in which a chemical compound unites with water and is then split into smaller molecules.

Hydrostatic pressure *high-dro-stat'-ik* the pressure of fluid in a system.

Hyperopia *high-per-o'-pi-a* farsightedness.

Hypertension *high-per-ten'-shun* high blood pressure.

Hypertonic *high-per-ton'-ik* excessive, above normal, tone or tension.

Hypertrophy *high-per'-tro-fee* an increase in the size of a tissue or organ independent of the body's general growth.

Hypodermis *high-po-der'-mis* the subcutaneous connective tissue; also called superficial fascia.

Hyponychium *high-po-neech'-ee-um* the thickened horny zone of the epidermis beneath the free border of the nail.

Hypothalamus *high-po-thal'-a-mus* the region of the diencephalon forming the floor of the third ventricle of the brain.

Hypothermia *high-po-ther'-mi-ah* subnormal body temperature.

Hypotonic *high-po-ton'-ik* below normal tone or tension.

Hypoxemic hypoxia *high-pock-see'-mik high-pock'-see-uh* a condition in which decreased oxygen is available to tissues because of decreased Po₂ in arterial blood.

Hypoxia *hi-pox'-ee-a* a condition in which a physiologically inadequate amount of oxygen is available to tissues.

Ileum *il'-ee-um* the lower part of the small intestine between the jejunum and the cecum of the large intestine.

Immune surveillance *im-myoon' sir-vail'-lantz* the body's immune response to cancer.

Immunity *im-myoon'-i-tee* the body's ability to resist many organisms and chemicals that can damage tissues.

Immunoglobulin *im-myoo-no-glob'-you-lin* an antibody.

Impetigo *im-pe-tee'-go* a highly contagious skin infection common in children.

Impotence *im'-po-tense* (1) lack of power, inability; (2) a male's inability to have sexual intercourse or maintain an erection.

In vitro *in vee'-tro* in a test tube, glass, or artificial environment.

In vivo *in vee'-vo* in the living body.

Infarct *in-farkt'* a region of dead, deteriorating tissue resulting from blood flow interference.

Inferior (caudal) pertaining to a position near the tail end of the long axis of the body.

Inflammation *in-flam-may'-shun* a physiological response of the body to tissue injury; includes dilation of blood vessels and an increase in vessel permeability.

Infundibulum *in-fun-dib'-yoo-lum* a funnel-shaped body part or passageway.

Inguinal *ing'-gwi-nal* pertaining to the groin region.

Inner cell mass an accumulation of cells in the blastocyst from which the embryo develops.

Innervation *in-ner-vay'-shun* the supply or distribution of nerves or nerve stimuli to a part.

Insertion *in-ser'-shun* the place or mode of attachment of a muscle; the movable part of a muscle as opposed to origin.

Inspiration *in-spi-ray'-shun* the drawing of air into the lungs.

Insulin *in'-su-lin* the hormone produced in the pancreas affecting carbohydrate and fat metabolism, blood glucose levels, and other systemic processes.

Integumentary system *in-teg-u-men'-tar-ee* the skin and its accessory structures.

Intercellular *in-ter-sel'-u-lar* between the cells of the body or part.

Intercellular matrix *may'-triks* the material between adjoining cells.

Interferon *in-ter-fer'-on* a chemical that is able to provide some protection against virus invasion of the body; inhibits viral growth.

Internal capsule the band of white matter between the basal ganglia and the thalamus.

Internal environment the environment within the body.

Internal respiration the exchange of gases between blood and tissue fluid and between tissue fluid and cells.

Interoceptor *in'-ter-o-sep-tor* a nerve ending situated in the viscera sensitive to changes and stimuli within the body's internal environment; also called visceroceptor.

Interstitial fluid *in-ter-stish'-al* the fluid between the cells or body parts.

Intervertebral discs *in-ter-ver'-te-bral* the discs of fibrocartilage between bodies of vertebrae.

Intervertebral foramina *fo-rah'-mi-nah* the openings between the pedicles of adjacent vertebrae through which the spinal nerves pass.

Intracellular *in-tra-sel'-u-lar* within a cell.

Intracellular fluid fluid within a cell.

Intramural pressure *in-tra-mu'-ral* the pressure built up in the walls of an organ.

Intrinsic factor *in-trin'-sik* a glycoprotein substance required for vitamin B₁₂ absorption.

Intrinsic muscle a muscle that has both its origin and its insertion in an organ.

Invert to turn inward.

Involuntary muscle a muscle not under control of the will; independent muscle.

Ion *eye'-on* an atom with a positive or negative electric charge.

Ionic bond *eye-on'-ik* a bond in which oppositely charged ions are held together by the attraction between opposite charges.

Ipsilateral *ip-si-lat'-er-al* situated on the same side.

Iris *eye'-ris* the pigmented, circular diaphragm in front of the eye's lens.

Ischemia *is-kee'-mee-uh* a local decrease in blood supply resulting from obstruction of arterial inflow.

Isometric *i-so-met'-rik* of the same length.

Isotonic *i-so-ton'-ik* having uniform tension under pressure or stimulation.

Isotope *i'-so-tope* a different form of a given element; isotopes have the same atomic number but different mass numbers.

Jaundice *jawn'-dis* an accumulation of bile pigments in the blood producing a yellow color of the skin.

Jejunum *je-joo'-num* the part of the small intestine between the duodenum and the ileum.

Joint the junction of two or more bones; an articulation.

Junctional complex the junction between cells in columnar and some cuboidal epithelium.

Juxtamedullary *jux'-ta-med'-u-lair-ee* referring to the inner portion of the cortex of the kidney, adjacent to the medulla.

Karyokinesis *kar-ee-o-ken-ee'-sis* see mitosis.

Keratin *ker'-a-tin* a fibrous insoluble protein found in tissues such as hair or nails.

Ketosis *kee-to'-sis* an abnormal condition during which an excess of ketone bodies are produced.

Kinetic energy *ki-net'-ik* the energy of motion.

Kinesthesia *ki-nez-thee'-zhah* the ability to perceive muscle movement.

Kinins *kigh'-ninz* group of polypeptides that dilate arterioles, increase vascular permeability, act as powerful chemotactic agents, and induce pain.

Krebs cycle the citric acid cycle; the series of reactions during which energy is liberated from metabolism of carbohydrates, fats, and amino acids.

Labia *lay'-bee-ah* lips.

Labor *la'-bor* the period characterized by strong, rhythmic uterine contractions that precedes the birth of a baby.

Lacrimal *lak'-ra-mal* pertaining to tears.

Lactation *lak-tay'-shun* the production and secretion of milk.

Lacteal *lak'-tee-al* the special lymphatic capillaries of the small intestine that take up chyle.

Lactic acid *lak'-tik* the product of anaerobic glycolysis, especially in muscle.

Lacuna *la-ku'-nah* a little depression or space; in bone or cartilage, lacunae are occupied by cells.

Lamina *lam'-i-nah* (1) a thin layer or flat plate; (2) the portion of a vertebra between the transverse process and the spinous process.

Laryngeal prominence *la-rin'-jul prom'-i-nense* the tubercle of the thyroid cartilage; Adam's apple.

Laryngitis *lar-en-jigh'-tis* an inflammation of the larynx.

Larynx *lar'-inks* the organ of the voice; located between the trachea and the base of the tongue.

Lateral *lat'-ur-al* away from the midline of the body.

Lateral sacs the reticular sites in muscle from which calcium is released.

Lens the elastic, doubly convex structure behind the pupil of the eye; focuses the light entering the eye on the retina.

Lesion *lee'-zhun* a tissue injury or wound.

Leukemia *loo-kee'-mee-ah* a cancerous condition in which there is an excessive production of leukocytes.

Leukocyte *loo'-ko-site* a white blood cell.

Ligament *lig'-a-ment* a band or sheetlike fibrous tissue connecting bones or parts.

Lingual *ling'-gwal* pertaining to the tongue.

Lipid *lip'-id* a substance that is almost insoluble in water but soluble in fat solvents; fatty acids and fats.

Lobe a curved, rounded structure or projection.

Locus *low'-kus* the position that a gene occupies on a chromosome.

Lordosis *lor-do'-siss* an excessive curve in the anterior lumbar spine; otherwise known as swayback.

Lumbar *lum'-bar* the portion of the back between the thorax and the pelvis.

Lumbar puncture a procedure involving insertion of a needle between the third and fourth lumbar vertebrae and into the subarachnoid space to sample cerebrospinal fluid.

Lumbosacral trunk *lum-bo-sak'-ral* a group of nerves that connects the lumbar plexus and the sacral plexus.

Lumen *loo'-men* the space inside a tube, blood vessel, or intestine.

Luteinizing hormone *lu'-tee-in-eye-zing* an anterior pituitary hormone that stimulates maturation of cells in the ovary and acts on interstitial cells of the male testis.

Lymph *limf* the watery fluid in the lymph vessels collected from the tissue fluids.

Lymph node a mass of lymphatic tissue.

Lymphatic system *lim-fat'-ik* a system of vessels carrying lymph closely related anatomically and functionally to the circulatory system.

Lymphocyte *lim'-fo-site* a granular white blood cell formed in the lymphoid tissue.

Lymphokines *lim'-fo-kines* substances involved in cell mediated immune responses that enhance the basic inflammatory response and subsequent phagocytosis.

Lysosomes *ligh'-so-somes* tiny organelles that originate from the Golgi apparatus and contain strong digestive enzymes.

Lysozyme *ligh'-so-zime* an enzyme capable of destroying certain kinds of bacteria.

Macula *mak'-yoo-la* the slightly yellow region lateral to the optic disc.

Macula adherens *mak'-u-la ad-hear'-uns* see desmosome.

Mammary glands *mam'-mar-ee* milk-producing glands of the breasts.

Malignant *ma-lig'-nant* life threatening; pertains to diseases that spread and lead to death, such as cancer.

Margination *mar-jin-ay'-shun* the adhesion of white blood cells to the walls of capillaries in the early stage of inflammation.

Mass number the combined number of protons and neutrons in the nucleus of an atom.

Mastication *mas-ti-kay'-shun* the act of chewing.

Matrix *may'-trix* the homogeneous intercellular substance of any tissue.

Meatus *mee-ay'-tus* the external opening of a canal.

Mechanoreceptor *mek-an-o-ree-sep'-tor* a receptor sensitive to mechanical pressures such as touch, sound, or contractions.

Medial *mee'-dee-al* toward the midline of the body.

Mediastinum *mee-dee-a-stigh'-num* the portion of the thoracic cavity between the lungs.

Mediated transport *mee'-dee-ay-ted* transport involving protein carrier molecules within the cell.

Medulla *ma-dul'-ah* the central portion of certain organs.

Meiosis *my-o'-sis* the last two cell divisions in gamete formation producing nuclei with half the full number of chromosomes (haploid).

Melanin *mel'-a-nin* the dark pigment responsible for skin color.

Melanocyte *me-lan'-o-site* a cell that produces melanin.

Menarche *me'-nar-kee* the onset of menstruation.

Meninges *men-in'-jeez* the membranes that cover the brain and spinal cord.

Meningitis *men-in-ji'-tis* an inflammation of the meninges covering the brain and spinal cord.

Menopause *men'-o-pawz* the physiological termination of menstrual cycles.

Menses *men'-seez* the recurrent monthly discharge of menstruation.

Menstruation *men-stroo-ay'-shun* the periodic, cyclic discharge of blood, secretions, tissue, and mucus from the mature female genital canal in the absence of pregnancy.

Merocrine glands *mer'-o-krin* glands that produce secretions intermittently; secretions do not accumulate in the gland.

Mesencephalon *mes-en-sef'-uh-lon* the midbrain.

Mesenteries *mes'-en-ter-eez* the doubled-layered membranes of the peritoneum that support most of the organs in the abdominal cavity.

Mesoderm *mes'-o-derm* tissue that forms the skeleton and muscles of the body.

Metabolic rate *met-a-bol'-ik* the energy expended by the body per unit time.

Metabolism *me-tab'-o-liz-em* the sum total of the chemical reactions that occur in the body.

Metabolize *me-tab'-o-lize* to transform substances into energy or materials the body can use or store by means of anabolism or catabolism.

Metacarpals *met-a-kar'-puls* the five bones of the palm of the hand.

Metastasize *me-tas'-ta-size* the spread of disease from one body part or organ into another not directly connected to it.

Metatarsals *met-a-tar'-suls* the five bones between the instep and the phalanges.

Metencephalon *met-en-sef'uh-lon* the anterior part of the hindbrain composed of the cerebellum and pons.

Microbodies *my-kro-bod'-eez* membrane-bound cytoplasmic structures containing oxidative enzymes; also called peroxisomes.

Microfilament *my-kro-fil'-a-ment* filaments associated with contractile activities of the cell and developmental modifications of cell and organ shape.

Microtubule *my-kro-too'-bule* cytoplasmic structures not bound in membranes having a support function.

Microvilli *my-kro-vil'-lee* the tiny protoplasmic projections formed on the free surfaces of some epithelial cells; appear brushlike when viewed through a microscope.

Mineralocorticoid *min-er-al-o-kor'-ti-koid* an adrenal cortical steroid hormone that regulates mineral metabolism and fluid balance.

Minerals the inorganic chemical compounds found in nature.

Mitochondria *my-to-kon'-dree-ah* the cytoplasmic organelles in the form of granules, rods, filaments responsible for generation of metabolic energy for cellular activities.

Mitosis *my-to'-sis* the division of the cell nucleus; often followed by division of the cytoplasm of a cell; also called karyokinesis.

Mixed nerves the nerves containing the processes of motor and sensory neurons; their impulses travel to and from the central nervous system.

Molar *mo'-lar* a solution concentration determined by mass of solute—one liter of solution contains an amount of solute equal to its molecular weight in grams.

Molecule *mol'-e-kewl* a very small mass of matter composed of atoms held together as a unit.

Moles elevations of the skin that are pigmented.

Monocyte *mon'-o-site* a large, single-nucleus white blood cell.

Monosomy *mo-no'-so-me'* the condition in which one of two homologous chromosomes is missing.

Mons pubis *monz pu-bus* the fatty eminence over the pubic symphysis in the female.

Motor nerve cells the nerves that carry impulses leaving the brain and spinal cord.

Motor unit a neuron and the muscle cells it supplies.

Mucus *mew'-kus* a sticky, thick fluid secreted by mucous glands and mucous membranes that keeps the free surface of membranes moist.

Mucous membranes the membranes that form the linings of the digestive, respiratory, urinary, and reproductive tracts.

Multipennate *mul-ti-pen'-nate* the muscles in which the fibers have a complex arrangement involving converging of tendons.

Multiple sclerosis *skler-o'-sis* a chronic condition characterized by destruction of the myelin sheaths of neurons in the spinal cord and the brain.

Multipolar neurons *mul-ti-pol'-ar* neurons that have one long axon and numerous dendrites.

Muscle fibers muscle cells.

Muscle spindles the complex capsules found in skeletal muscles that are sensitive to stretching.

Muscle twitch a single rapid contraction of a muscle followed by relaxation.

Muscular dystrophy *dis'-tro-fee* a progressive disorder marked by atrophy and stiffness of the muscles.

Mutation *mu-tay'-shun* an alteration in the genetic material.

Myelencephalon *mile-len-sef'-uh-lon* the lower part of the hindbrain, especially the medulla oblongata.

Myelin *my'-uh-lin* the white, fatty lipid substance forming a sheath around some nerves.

Myelinated fibers *my'-uh-li-nay-ted* axons (projections of a nerve cell) covered with myelin.

Myelitis *my-a-light'-us* an inflammation of the spinal cord.

Myocardial infarction *my-o-kar'-dee-al in-fark'-shun* a condition characterized by dead tissue areas in the myocardium caused by interruption of blood supply to the area.

Myocardium *my-o-kar'-dee-um* the cardiac muscle layer of the wall of the heart.

Myofibril *my-o-figh'-bril* a fibril found in the cytoplasm of muscle.

Myofilament *my-o-fil'-a-ment* the filamentous structures making up sarcomere consisting of thick and thin types.

Myogenic *my-o-jen'-ik* having the potential to contract automatically without nervous system stimulation.

Myoglobin *my-o-glo'-bin* muscle hemoglobin.

Myometrium *my-o-me'-tree-um* the thick uterine musculature.

Myopia *my-o'-pee-ah* nearsightedness.

Myosin *my'-o-sin* one of the principal proteins found in muscle.

Myotome *my'-a-tome* that part of a somite that differentiates into skeletal muscle.

Nares *nar'-eez* the nostrils.

Necrosis *ne-kro'-sis* the death or disintegration of a cell or tissues caused by disease or injury.

Negative feedback feedback that tends to cause the levels of a variable to change in a direction opposite to that of an initial change.

Nephron *nef'-ron* the functional part of the kidney.

Nerve fiber axon (nerve cell projection) together with certain sheaths or coverings.

Neuralgia *noo-ral'-jee-ah* a severe paroxysmal (spasm-producing) pain along the course of a nerve.

Neuritis *noo-righ'-tis* an inflammatory or degenerative condition of the nerves.

Neurofibril *noo-ro-fi'-bril* the fibril of a nerve cell, usually extending from the processes and traversing the cell body.

Neuroglia *noo-rog'-lee-uh* the nonneuronal tissue of the central nervous system that performs supportive and other functions; also called glia.

Neurohypophysis *noo-ro-high-pof'-i-sis* the portion of the pituitary gland derived from the brain.

Neuromuscular junction *noo-ro-mus'-kyoo-lar* the region where a motor neuron approaches skeletal muscle sarcolemma.

Neurons *noo'-ronz* the nerve cells that transmit messages throughout the body.

Neutron *noo'-tron* an uncharged particle located in the nucleus of an atom.

Neutrophil *noo'-tro-fil* the most abundant of the white blood cells.

Nexus *nex'-us* see gap junction.

Nitrogen balance the state in a normal adult in which the nitrogen excreted equals the nitrogen intake in the form of food.

Nondisjunction *non-dis-jungk'-shun* failure of separation of paired chromosomes during cell division.

Nuchal lines *nu'-kal* the three slight ridges on the external surface of the occipital bone.

Nucleoli *noo-klee'-o-lee* the small spherical bodies in the cell nucleus.

Nucleotide *noo'-klee-o-tide* a component of DNA and RNA consisting of a sugar, a base, and a phosphate group.

Nucleus *noo'-klee-us* (1) a dense central body in most cells containing the genetic apparatus of the cell; (2) the core of an atom.

Nucleus pulposus *noo'-klee-us pul-po'-sus* the central gelatinous part of the intervetebral disc.

Nystagmus *niss-tag'-mus* an oscillatory movement of the eyeballs.

Obesity *o-bee'-sit-ee* a condition of a person being overweight; often leads to other complications.

Occipital *ok-sip'-i-tal* pertaining to the back of the head area.

Occlusion *o-kloo'-zhun* closure or obstruction.

Olfaction *ol-fak'-shun* smell.

Oncotic pressure *on-kot'-ik* the osmotic force exerted by plasma proteins.

Oogenesis *oh-oh-jen'-e-sis* the process of origin, growth, and formation of the ovum.

Ophthalmic *of-thal'-mik* pertaining to the eye.

Optic *op'-tik* pertaining to the eye.

Optic chiasma *op'-tik kigh-az'-mah* the meeting of the optic nerves after entering the cranium.

Opsonins *op'-se-ninz* proteins that coat specific foreign particles and thereby render them more susceptible to phagocytosis.

Oral relating to the mouth.

Ora serrata *o'-ra ser-rat'a* the serrated margin of the retina.

Organ a part of the body combining two or more tissues to perform a specialized function.

Organelle *or-gan-el'* a specialized structure or part of a cell having a definite function to perform.

Organic compound any hydrocarbon or hydrocarbon derivative.

Orgasm *or'-gaz-um* the intense emotional and physical climax associated with sexual stimulation.

Origin the end of attachment of a muscle that remains relatively fixed during muscular contraction.

Osmoreceptor *os-mo-ree-sep'-tor* a structure sensitive to osmotic pressure.

Osmosis *os-mo'-sis* the passage (diffusion) of a solvent through a membrane from a dilute solution into a more concentrated one.

Ossicles *os'-si-kalz* the three bones of the middle ear: malleus, stapes, and incus.

Osteoblasts *os'-tee-o-blasts* the bone-forming cells.

Osteoclasts *os'-tee-o-klasts* the large cells that reabsorb or erode bone substance.

Osteocyte *os'-tee-o-site* a mature bone cell found in each lacuna.

Osteomalacia *os-tee-o-ma-lay'-she-ah* the softening of bone resulting from vitamin D deficiency in the adult.

Osteomyelitis *os-tee-o-my-a-light'-us* a disease in which the periosteum, the contents of the marrow cavity, and the bone tissue become inflamed.

Osteoporosis *os-tee-o-por'-o-sis* an increased softening of the bone resulting from a gradual reduction in the rate of bone formation while the rate of bone absorption remains normal; a common condition in older people.

Ostium *os'-tium* a small opening into a hollow structure.

Otic *o'-tik* pertaining to the ear.

Otitis media *o-tigh'-tis mee'-dee-ah* middle-ear infection.

Otolith *oh'-to-lith* one of the small calcareous masses in the utricle and saccule of the inner ear.

Ovarian cycle *o-va'-ree-an* the monthly cycle of follicle development, ovulation, and corpus luteum formation in an ovary.

Ovary *o'-va-ree* the female sex organ in which ova (eggs) are produced.

Ovulation *ov-u-lay'-shun* the maturation and release of an ovum.

Ovum *o'-vum* the female gamete (germ cell); an egg cell.

Oxidation *oks-i-day'-shun* the loss of electrons by a molecule; the process of substances combining with oxygen.

Oxygenated *oks'-i-je-nay-ted* the condition in which a substance is saturated with oxygen.

Oxygen debt *oks'-i-jen* the volume of oxygen required after exercise in excess of oxygen consumption in the resting state for an equivalent length of time.

Oxyhemoglobin *oks-i-hee'-mo-glo-bin* oxidized hemoglobin.

Oxytocin *oks-i-to'-sin* a hormone released by the posterior pituitary that stimulates contractility of smooth muscles of the uterus during labor and myoepithelial cells surrounding the alveoli of the mammary glands during lactation.

Palate *pal'-et* the roof of the mouth.

Palmar *pal'-mar* the anterior surface of the hands.

Palpation *pal-pay'-shun* examination by touch.

Palpebral fissure *pal'-pe-bral* the gap between the upper and lower eyelids.

Pancreas *pan'-kree-as* the gland located behind the stomach, between the spleen and the duodenum, producing both endocrine and exocrine secretions.

Pancreatic juice *pan-kree-at'-ik* a secretion of the pan-

creas containing enzymes for digestion of carbohydrates.

Pancreatitis *pan-kree-a-tigh'-tis* an inflammation of the
pancreas.

Papilla *pa-pil'-lah* a small elevation, nipple-shaped or
cone-shaped.

Papillary muscles *pa'-pil-lar-ee* cone-shaped muscles
such as those that project into the cardiac ventricular
lumen.

Paralysis *pa-ral'-i-sis* the loss of muscle function or of
sensation.

Paraplegia *par-a-plee'-gee-ah* paralysis of the lower
limbs.

Parasympathetic division *par-a-sim-pa-thet'-ik* a division of the autonomic nervous system; also referred
to as the craniosacral division.

Parathyroid glands *par-a-thigh'-roid* the several small
endocrine glands posterior to the capsule of the thyroid
gland.

Parathyroid hormone the hormone released by the
parathyroid glands that regulate blood calcium level.

Parenchyma *par-en'-ki-mah* the functional components
of an organ.

Parenteral *par-en'-ter-al* occurring through some route
other than the alimentary canal, such as intravenous.

Parietal *par-eye'-i-tal* pertaining to the walls of a cavity.

Parotid *pa-rot'-id* located near the ear.

Parturition *par-toor-ish'-un* the act of giving birth.

Patella *pa-tel'-ah* the kneecap.

Pathogenesis *path-o-jen'-e-sis* the development of a disease.

Pectoral *pek'-to-ral* pertaining to the chest.

Pedicle *ped'-i-kul* the portion of the neural arch between the centrum and the transverse process.

Peduncle *pe-dun'-kal* a stalk of fibers, especially that
connecting the cerebellum to the pons, mesencephalon,
and medulla oblongata.

Pelvis a basin-shaped structure, especially the lower
portion of the body's trunk.

Penis *pee'-nis* the male organ of copulation and urinary excretion.

Pepsin an enzyme capable of digesting proteins in an
acid pH.

Peptide bond *pep'-tide* a bond joining the amino group
of one amino acid to the acid carboxyl group of a second amino acid with the loss of a water molecule.

Pericardium *per-i-kar'-dee-um* the closed membranous
sac enveloping the heart.

Perichondrium *per-i-kon'-dree-um* a fibrous, connective-
tissue membrane covering the external surface of cartilaginous structures.

Perimysium *per-i-mis'-ee-um* the connective tissue enveloping bundles of muscle fibers.

Perineum *per-i-nee'-um* that region of the body extending from the anus to the scrotum in males and from
the anus to the vulva in females.

Periosteum *per-ee-os'-tee-um* a double-layered connective tissue that covers, invests, and nourishes the
bone.

Peripheral nervous system (PNS) *per-if'-er-al* a system
of nerves that connect the outlying parts of the body
and their receptors with the central nervous system.

Peripheral resistance the impedance to blood flow offered by the systemic blood vessels.

Peristalsis *per-is-tal'-sis* the progressive wave of contraction seen in tubes.

Peritoneum *per-i-ton-ee'-um* the membrane lining of
the interior of the abdominal cavity.

Peritonitis *per-i-ton-eye'-tis* an inflammation of the
peritoneum.

Permeability *per-mee-a-bil'-i-tee* that property of membranes which permits transit of molecules and ions.

Peroneal *per-o-nee'-al* pertaining to the outer side of
the leg.

Peroxisome *per-ox'-i-some* a membrane-bounded organelle in cells that contains oxidase and peroxidase;
also called microbody.

pH the symbol for hydrogen ion concentration; a measure of the relative acid or base level of a substance or
solution.

Phagocyte *fag'-o-site* a cell capable of engulfing and digesting particles or cells harmful to the body.

Phagocytosis *fag-o-sigh-to'-sis* the ingestion of foreign
solids by cells.

Phalanges *fay-lan'-jeez* the bones of the finger or toe.

Phantom pain a phenomenon whereby a person who
has undergone amputation continues to feel pain from
the amputated body part.

Pharynx *far'-inks* the muscular, membranous tube extending from the base of the skull to the esophagus.

Phenotype *fee'-no-tipe* the observable characteristics of
an individual as determined both genetically and envi-
ronmentally.

Phlebitis *fle-by'-tis* an inflammation of a vein.

Photoreceptor *fo-to-ree-sep'-tor* the specialized receptor
cells that can convert light energy into a nerve impulse.

Physiology *fiz-ee-ol'-o-jee* the science of the functions
of organic beings.

Pinna *pin'-nah* the irregularly shaped elastic cartilage
covered with skin forming the most prominent portion
of the outer ear.

Pinocytosis *pi-no-sigh-to'-sis* the engulfing of liquid by
cells.

Pituitary gland the gland located beneath the brain that
serves a variety of functions including regulation of the
gonads, thyroid, adrenal cortex, and other endocrine
glands.

Placenta *pla-sen'-tah* the organ to which the embryo attaches by the umbilical cord for nourishment and waste
removal; has an endocrine function as well.

Placode *plak'-ode* an ectodermal thickening in the early
embryo from which a sense organ or structure develops.

Plantar *plan'-tar* pertaining to the sole of the foot.

Plasma the fluid portion of the blood or lymph.

Plasma cells *plaz'-ma* cells that produce antibodies.

Platelet *plate'-let* one of the disc-shaped components
of blood; involved in clotting.

Pleura *ploor'-ah* the membrane covering the lungs.

Pleurisy *plur'-i-see* prolonged inflammation of the visceral and parietal pleuras, making breathing painful.

Plexus *plek'-sus* a network of interlacing nerves or
anastomosing blood vessels or lymphatics.

Plica *ply'-ka* a fold.

Pneumothorax *noo-mo-thor'-aks* the presence of air or
gas in a pleural cavity.

Podocyte *pod'-o-site* an epithelial cell located on the
basement membrane of the glomerulus, spreading thin
cytoplasmic projections over the membrane.

Polar body a minute cell given off by the ovum during
maturation divisions.

Polarized *po'-lar-ized* the state of an unstimulated neuron in which the inside of the cell is relatively negative
in comparison to the outside.

Poliomyelitis *po'-lee-o-my-a-light'-us* the viral destruction of nerve cell bodies within the anterior horns of
the spinal cord.

Polycythemia *pol-ee-sigh-theem'-ee-a* the presence of an abnormally large number of erythrocytes in the blood.

Polypeptide *pol-ee-pep'-tide* a small chain of amino acids.

Polyribosome *pol-ee-ribe'-o-some* a multiple structure composed of several ribosomes held together by a molecule of messenger RNA; also called polysome.

Polysome *pol'-ee-some* See polyribosome.

Pons (1) any bridgelike structure or part; (2) the structure connecting the cerebellum with the brain stem providing linkage between upper and lower levels of the central nervous system.

Positive feedback feedback that tends to cause the level of a variable to change in the same direction as an initial change.

Postganglionic (postsynaptic) neuron *post-gang-lee-on'-ik* a neuron of the autonomic nervous system having its cell body in a ganglion with the axon extending to an organ or tissue.

Preganglionic (presynaptic) neuron *pree-gang-lee-on'-ik* a neuron of the autonomic nervous system having its cell body in the brain or spinal cord and its axon terminating in a ganglion.

Prepuce *pree'-puce* the loose fold of skin that covers the glans penis or glans clitoris.

Pressoreceptor *pre-so-ree-sep'-tor* a nerve ending in the wall of the carotid sinus and aortic arch sensitive to vessel stretching.

Prime movers those muscles whose contractions are primarily responsible for a particular movement.

Process (1) a prominence or projection; (2) a series of actions or method of action for a specific purpose.

Proctologic *prok-to-loj'-ik* pertaining to the rectum or anus.

Progesterone *pro-jes'-ter-own* a hormone responsible for preparing the uterus for the fertilized ovum.

Pronation *pro-nay'-shun* the inward rotation of the forearm causing the radius to cross diagonally over the ulna—palms face posteriorly.

Prone refers to a body lying horizontally with the face downward.

Pronucleus *pro-noo'-clee-us* one of two nuclear bodies (one male and one female) of a newly fertilized ovum, the fusion of which results in formation of a cleavage nucleus.

Proprioceptor *pro-pree-o-sep'-tor* a receptor located in a muscle, tendon, joint, or vestibular apparatus of the internal ear; concerned with locomotion and posture.

Prostaglandin *pros-ta-glan'-din* a substance included in seminal vesicle secretion thought to facilitate fertilization; causes uterine contractions.

Protein *pro'-teen* a complex nitrogenous substance found in various forms in animals and plants as the principal component of protoplasm.

Proteinuria *pro-te-in-oo'-ree-ah* the passage of albumin or other protein in the urine.

Proton *pro'-ton* a particle carrying a positive charge; located in the nucleus of an atom.

Protrude *pro-trood'* to project or assume an abnormally prominent position.

Proximal *proks'-i-mal* toward the attached end of a limb or the origin of a structure.

Psoriasis *so-rye'-uh-sis* a chronic inflammatory skin disease characterized by development of red patches covered with silvery, overlapping scales.

Ptyalin *tie'-a-lin* a starch-splitting enzyme contained in saliva.

Puberty *pu'-ber-tee* the period at which reproductive organs become functional.

Pulmonary *pull'-muh-na-ree* pertaining to the lungs.

Pulmonary circuit the circulatory vessels of the lungs.

Pulmonary edema *e-dee'-mah* an effusion of fluid into the air sacs and interstitial tissue of the lungs.

Pulse the rhythmic expansion in arteries resulting from heart contraction; can be felt from the outside of the body.

Pupil an opening in the center of the iris through which light enters the eye.

Purkinje fibers *per-kin'-jee* the modified cardiac muscle fibers of the conduction system of the heart.

Pus the fluid product of inflammation composed of white blood cells, the debris of dead cells, and a thin fluid.

Pyelonephritis *pie-el-o-nef-rye'-tis* an inflammation of the kidney pelvis and surrounding kidney tissues.

Pyloric glands *pie-lor'-ik* the glands that secrete thin mucus; located in the region of the pylorus of the stomach.

Pyloric region the final portion of the stomach preceding the duodenum.

Pyramid any conical eminence of an organ, especially a body of longitudinal fibers on each side of the anterior median fissure of the medulla oblongata.

Pyrogen *pie'-ro-jen* an agent that induces fever.

Quadriplegia *quad-ri-plee'-jee-ah* the paralysis of all four limbs.

Radiate *ray'-dee-ate* diverging from a central point.

Ramus *ray'-mus* a branch of a nerve, artery, vein, or bone, especially a primary division.

Receptor *ree-sep'-tor* a peripheral nerve ending in the skin; specialized for response to particular types of stimuli.

Recessive *ree-seh'-siv* an allele, or the corresponding trait, that is expressed only in homozygous cells or organisms.

Reduction the gain of electrons by a molecule.

Reflex automatic, stereotyped reactions to stimuli.

Refracted bent.

Refracting media substances that bend light.

Refractory period *ree-frak'to-ree* the period of resistance to stimulation immediately after responding.

Relative refractory period the period following stimulation during which only a stronger than usual stimulation can evoke an action potential.

Renal *ree'-nal* pertaining to the kidney.

Renal calculus *ree'-nal kal'-ku-lus* a kidney stone formed in the renal pelvis or urinary bladder.

Renin *ren'-in* a substance released by the juxtaglomerular complex of the kidneys; involved with raising blood pressure.

Rete *ree'-tee* a network; often composed of nerve fibers or blood vessels.

Reticulocyte *re-tik'-u-lo-site* a nonnucleated, young erythrocyte.

Reticulum *re-tik'-u-lum* a fine network.

Retract to draw back, shorten, contract.

Rhinencephalon *rye-nen-sef'-a-lon* that portion of the cerebrum concerned with reception and integration of olfactory impulses.

Rhodopsin *ro-dop'-sin* the photopigment contained in the rods of the retina.

Ribonucleic acid (RNA) *rye-bo-nu-kle'-ik* the nucleic acid that contains ribose.

Ribosomes *rye'-bo-somes* the cytoplasmic structures at which proteins are synthesized.

Rickets *rik'-ets* a disease occurring in infants and young children characterized by softening of the bone caused by demineralization from malnutrition.

Rods   one of the two types of photosensitive cells in the retina.

Roentgenogram   *rent-gen'-o-gram*   an x-ray film.

Rotate   to turn about an axis.

Rugae   *ru'-je*   elevations or ridges, as in the mucosa of the stomach, uterus, and palate.

Sacral   *sa'-kral*   the lower portion of the back, just superior to the buttocks.

Sagittal plane   *saj'-i-tal*   a longitudinal section that divides the body or any of its parts into right and left portions.

Saliva   *sa-ligh'-vah*   the combined secretions of salivary and mucous glands of the mouth.

Salt   a compound that, when dissolved in water, dissociates into cations other than hydrogen ions and anions other than hydroxide ions.

Sarcoplasmic reticulum   *sar-ko-plas'-mik re-tik'-u-lum*   the membranous network running through skeletal muscle cells.

Sclera   *skleh'-rah*   the firm, fibrous outer layer of the eyeball; functions for protection and maintenance of eyeball shape.

Scoliosis   *sko-lee-o'-siss*   a lateral curve in the vertebral column.

Scrotum   *skro'-tum*   the two-layered sac enclosing the testes.

Sebaceous glands   *se-bay'-shus*   glands that develop from and empty their sebum secretion into hair follicles.

Sebum   *see'-bum*   the secretion of sebaceous glands; oily substance rich in lipids.

Secretion   *see-kree'-shun*   the passage of material formed by a cell from the inside to the outside of its plasma membrane.

Segmentation   *seg-men-tay'-shun*   the process of cleavage or splitting; the division of an organism into somites.

Semen   *see'-men*   the fluid produced by male reproductive structures; contains sperm.

Semilunar valves   *sem-i-loo'-ner*   valves that prevent blood return to the ventricle after contraction.

Seminiferous tubules   *sem-i-nif'-er-us*   highly convoluted tubes within the testes containing spermatocytes.

Sensory nerve   a nerve that contains only the processes of sensory neurons and carries nerve impulses to the central nervous system.

Sensory nerve cell   an initiator of nerve impulses following receptor activity.

Serous fluid   *ser'-us*   a clear, watery fluid secreted by the cells of the mesothelium.

Serum   *se'-rum*   the amber-colored fluid that exudes from clotted blood as the clot shrinks and then no longer contains fibrinogen.

Sesamoid   *ses'-a-moid*   denoting a small bone that is embedded in a tendon or in a joint capsule.

Sex chromosome   *kro'-mo-some*   chromosome that determines sex of the fertilized egg; X and Y chromosomes.

Sinuatrial node   *sigh-noo-ay'-tree-al*   the mass of specialized myocardial cells in the wall of the right atrium.

Sinus   *sigh'-nus*   (1) a mucous-membrane-lined, air-filled cavity in certain cranial bones; (2) a dilated channel for the passage of blood or lymph, which lacks the coats of an ordinary vessel.

Smooth (nonstriated) muscle   muscle consisting of spindle-shaped, unstriped (nonstriated) muscle cells; involuntary muscles.

Solute   *sol'-yoot*   the dissolved substance in a solution.

Solution   a homogenous mixture of two or more components.

Somatic nervous system   *so-mat'-ik*   a division of the peripheral nervous system; also called the voluntary nervous system.

Somite   *so'-mite*   a segment of the body of an embryo.

Sperm   that mature male germ cell, a spermatozoon.

Spermatogenesis   *sper-mat-o-jen'-e-sis*   the process of meiosis (cell division) in the male to produce mature male germ cells.

Sphenoid   *sfen'-oid*   wedgelike.

Sphincter   *sfink'-ter*   a muscle surrounding and enclosing an orifice.

Sprain   the wrenching of a joint producing stretching or laceration of the ligaments.

Squamous   *skway'-mus*   pertaining to flat and thin cells that form the free surface of epithelial tissue.

Stagnant hypoxia   *stag'-nant high-pock'-see-uh*   a condition marked by reduced available oxygen caused by slowed blood circulation.

Stasis   *stay'-sis*   (1) a decrease or stoppage of flow; (2) a state of equilibrium.

Static balance   *stat'-ic*   balance concerned with changes in the position of the head.

Stenosis   *sten-o'-sis*   constriction or narrowing.

Steroids   *ster'-oidz*   a specific group of chemical substances including certain endocrine secretions and cholesterol.

Stimulus   *stim'-u-lus*   an excitant or irritant; an alteration in the environment of a living thing producing a response.

Stomodeum   *sto-mo-dee'-um*   the primitive oral cavity of the embryo.

Strabismus   *stra-biz'-mus*   a squint; that abnormality of the eyes in which the visual axes do not meet at the desired objective point.

Stratum   *stray'-tum*   a layer.

Stress   any stimulus that directly or indirectly causes neurons of the hypothalamus of the brain to release corticotropin releasing hormone (CRH).

Striated muscle   *stry'-ay-ted*   muscle consisting of cross-striated (cross-striped) muscle fibers.

Stricture   *strik'-chur*   a contraction or inward pinching of a canal or duct.

Stroke   a condition in which a cerebral blood vessel is blocked.

Stroke volume   a volume of blood ejected by the left ventricle during a systole.

Stroma   *stro'-mah*   the supporting framework of an organ including connective tissue, vessels, and nerves.

Sty   an inflammation of the connective tissue of the eyelid, near a hair follicle.

Subcutaneous   *sub-ku-tay'-nee-us*   beneath the skin.

Sublingual   *sub-ling'-gwal*   located beneath the tongue.

Sulcus   *sul'-kus*   a furrow or linear groove; on the brain, a less deep depression than a fissure.

Summation   *sum-may'-shun*   the accumulation of effects, especially those of muscular, sensory, or mental stimuli.

Superficial (external)   located close to or on the body surface.

Superior   refers to the head or upper; higher.

Supination   *soo-pa-nay'-shun*   the outward rotation of the forearm causing palms to face anteriorly.

Supine   refers to a body lying with the face upward.

Surface tension   the contractile surface of a liquid or structure by which it tends to present the least possible surface; liquid meniscus formation.

Surfactant   *sur-fact'-ant*   a substance on pulmonary alveoli walls which reduces surface tension thus preventing collapse.

Suspension *sus-pen'-shun* a dispersing of particles throughout a body of liquid.

Suspensory ligament of an eye *sus-pen'-so-ree lig'-a-ment* fibrous ligament that holds the lens in place in the eye.

Sutures *soo'-churz* the immovable joints that connect the bones of the adult skull.

Sweat glands the glands that secrete a watery solution of sodium chloride (salt water).

Sympathetic division a division of the autonomic nervous system; opposes parasympathetic functions.

Synapse *sin'-apse* the region of communication between neurons.

Synaptic cleft *sin-ap'-tik* the space at a synapse between neurons.

Synaptic delay *sin-ap'-tik* the time required for an impulse to cross a synapse between two neurons.

Synarthrosis *sin-ar-throw'-sis* a fibrous joint; two types: sutures and syndesmoses.

Synergist *sin'-er-jist* a muscle cooperating with another to produce a movement neither alone can produce.

Synostosis *sin-os-tow'-sis* a union of originally separate bones by osseous material.

Synovial fluid *sa-no'-vee-al* a fluid secreted by the synovial membrane; lubricates joint surfaces and nourishes articular cartilages.

System a group of organs that function cooperatively to accomplish a common purpose; there are ten major systems in the human body.

Systemic *sis-tem'-ik* general; pertaining to the whole body.

Systemic circuit the circulatory vessels of the body.

Systemic edema *e-dee'-ma* an accumulation of fluid in body organs or tissues.

Systole *sis'-ta-lee* the contraction phase of a cardiac cycle.

Systolic pressure *sis-tol'-ik* the pressure generated by the left ventricle during systole.

T cells a heterogeneous group of lymphocytes that include those cells committed to participating in cell-mediated immune responses.

Tachycardia *tak-ee-kar'-dee-ah* the abnormal, excessive rapidity of heart action; over 100 beats per minute.

Tarsals *tar'-suls* the seven bones that form the ankle and heel.

Taste buds the receptors for taste on the tongue, roof of mouth, pharynx, and larynx.

Telencephalon *tel-en-sef'-uh-lon* the anterior subdivision of the primary forebrain that develops into olfactory lobes, cerebral cortex, and corpora striata.

Temporal summation *tem'-po-ral sum'-may-shun* the arrival of many nerve impulses at a synapse within a short period causing release of sufficient transmitters to initiate a nerve impulse in the postsynaptic cell.

Tendon *ten'-don* a band of dense fibrous tissue forming the termination of a muscle and attaching the muscle to a bone.

Testis *tes'-tis* the male primary sex organ that produces sperm.

Tetanus *tet'-a-nus* (1) the tense, contracted state of a muscle; (2) an infectious disease.

Thalamus *thal'-a-muss* the mass of gray matter at the base of the brain.

Theca *thee'-ka* a sheath.

Thermoreceptor *ther-mo-ree-sep'-tor* a receptor sensitive to temperature changes.

Thoracic *tho-ras'-ik* refers to the chest.

Thorax *tho'-raks* that portion of the trunk above the diaphragm and below the neck.

Threshold the lower limit of stimulus capable of producing an impression on consciousness or evoking a response in an irritable tissue.

Thrombin *throm'-bin* an enzyme that induces clotting by converting fibrinogen to fibrin.

Thrombocyte *throm'-bo-site* a blood platelet thought to be part of the blood-clotting mechanism.

Thrombocytopenia *throm-bo-sigh-to-pee'-nee-ah* a condition in which there is a decrease in the number of blood platelets below normal.

Thrombophlebitis *throm-bo-fle-by'-tis* an inflammation of a vein associated with blood-clot formation.

Thrombus *throm'-bus* a clot that is fixed or stuck to a vessel wall.

Thymus gland *thigh'-mus* a potential source of hormonal material; active in immune response.

Thyroid gland *thigh'-roid* one of the largest of the body's endocrine glands.

Tissue a group of similar cells and fibers forming a distinct structure.

Tolerance *tol'-er-antz* the failure of the body to mount a specific immune response against a particular antigen.

Tomography *toe-mog'-ra-fee* x-ray photography of a specific structure in a certain layer of tissue in the body, in which images of structures in other layers are eliminated.

Tonic *ton'-ik* refers to the state of continuous muscular or neuron activity.

Tonofibril *tone-o-figh'-bril* one of the fine fibrils in the cytoplasm of epithelial cells.

Toxemia *tok-see'-mee-ah* a condition in which blood contains poisonous products.

Toxic *tok'-sik* poisonous.

Trabecula *tra-bek'-u-lah* any one of the fibrous bands extending from the capsule into the interior of an organ.

Trachea *tray'-kee-uh* the windpipe; the cartilaginous and membranous tube extending from larynx to bronchi.

Tract a collection of nerve fibers having the same origin, termination, and function.

Trait *trate* a characteristic or quality of an individual.

Transducer *trans-doo'-ser* an agent that converts energy forms; for example, receptors of the nervous system transfer light into a nerve impulse.

Transmutation *trans-mu-tay'-shun* the change of one element into another.

Transverse processes the projections that extend laterally from each neural arch.

Trauma *traw'-mah* an injury, wound, or shock; usually produced by external forces.

Trisomy *trigh'-so-mee* the condition in which one chromosome is represented three times rather than the normal two.

Trochanter *tro-kan'-tur* a large, somewhat blunt process.

Trophic *tro'-fik* pertains to nutrition.

Trophoblast *trof'-o-blast* outer sphere of cells of the blastocyst.

Trypsin *trip'-sin* an active enzyme that splits proteins.

Tubal ligation *too'-bal li-gay'-shun* a form of surgical sterilization in which the uterine tubes are tied and severed.

Tubal pregnancy an ecotopic pregnancy that occurs within a uterine tube.

Tubercle *too'-bur-kul* a nodule or small rounded process.

Tuberosity *too-bur-os'-i-tee* a broad process, larger than a tubercle.

Tumor an abnormal growth of cells; a swelling; cancerous at times.

Tunica *too'-ni-kah* a covering or tissue coat; membrane layer.

Twitch a brief contraction of muscle in response to a stimulus.

Tympanic membrane *tim-pan'-ik* the eardrum.

Ulcer *ul'-ser* a lesion or erosion of the mucous membrane, such as gastric ulcer of stomach.

Umbilical cord *um-bil'-i-kal* a structure bearing arteries and veins connecting the placenta and the fetus.

Umbilicus *um-bil'-i-kus* the navel; marks site that gave passage to umbilical vessels in the fetal stage.

Unipennate *yoo-na-pen'-it* muscles in which all fasciculi insert onto one side of the tendon.

Unipolar neurons *yoon-i-pol'-ar* neurons in which embryological fusion of the two processes leaves only one process extending from the cell body.

Unitary smooth muscle muscle in which the cells contact one another at gap junctions and the impulses spread from cell to cell.

Unmyelinated fibers *un-my'-e-li-nay-ted* nerve axons that are not covered with myelin.

Urea *yoo-ree'-ah* the main nitrogen-containing waste excreted in the urine.

Uremia *yoo-ree'-mee-ah* a toxic accumulation in the blood of substances normally excreted by the kidneys.

Ureter *yoo-ree'-ter* the tube that carries urine from kidney to bladder.

Urethra *yoo-ree'-thrah* the canal through which urine passes from the bladder to the outside of the body.

Uvula *yoo'-vu-la* conical appendix hanging from soft palate.

Vacuole *vak'-yoo-ole* a clear space in a cell.

Valvular insufficiency *val'-vu-lar* a condition in which the cusps of the cardiac valves do not close tightly.

Varicose vein *var-uh-kos'* a dilated, knotted, tortuous blood vessel.

Vas a duct; vessel.

Vasa recta *va'-sa rek'-ta* capillary branches that supply loops of Henle and collecting ducts.

Vascular *vas'-ku-lar* pertaining to channels or vessels.

Vasoconstriction *vaz-o-kon-strik'-shun* the narrowing of blood vessels.

Vasodilatation *vaz-o-die-lay-tay'-shun* the relaxation of the smooth muscles of the vascular system producing dilated vessels.

Vasomotion *vaz-o-mo'-shun* an increase or decrease in caliber of a blood vessel.

Vasomotor center *vaz-o-mo'-tor* an area of brain concerned with regulation of blood vessel resistance.

Vasomotor nerve fibers *vaz-o-mo'-tor* the motor nerve fibers that regulate the construction or dilation of blood vessels.

Vasopressin *vaz-o-pres'-sin* another name for antidiuretic hormone.

Vein a vessel carrying blood away from the tissues toward the heart.

Ventral *ven'-tral* anterior or front.

Ventricle *ven'-tri-kal* (1) a small cavity or pouch; (2) the blood propulsion chamber of the heart.

Ventricles of the brain *ven'-tri-kalz* the cavities in the interior of the brain.

Venule *ven'-yool* a small vein.

Vertigo *ver'-tee-go* dizziness; the feeling of movement such as a sensation that the external environment is revolving.

Vesicle *ves'-i-kal* a small liquid-filled sac or bladder.

Viscera *vis'-ser-ah* the internal organs.

Visceral *vis'-ser-al* pertaining to the internal part of a structure or the internal organs.

Viscosity *vis-koss'-i-tee* the state of being sticky or thick.

Visual acuity *a-kyoo'-i-ty* the ability of the eye to distinguish detail.

Vital capacity the volume of air that can be expelled from the lungs by forcible expiration after the deepest inspiration.

Vitamins the organic compounds required by the body in minute amounts for physiological maintenance and growth.

Voluntary muscle muscle under control of the will; skeletal muscle.

Vulva *vul'-va* female external genitalia.

White matter the white substance of the central nervous system; the nerve fibers.

Yolk sac *yoke* endodermal sac that serves as the source for primordial germ cells.

Zonula *zon'-u-la* a small zone, usually beltlike.

Zonula adherens that part of a junctional complex of epithelial cells where the cell membranes are not modified and are separated by a gap of 200 Å.

Zonula occludens that part of a junctional complex of epithelial cells where the outer layer of cell membranes of adjacent tissues fuse.

Zygote *zy'-goat* the fertilized ovum before splitting (cleavage); produced by union of two gametes.

Index

*Page numbers in italics indicate pages with tables or illustrations.

913

Photo and Illustration Acknowledgments

Figure 1.1a, p. 4: Mark Tuschman.

Figure 1.1b, p. 4: Dr. J. de Groot, Department of Anatomy and Radiology, University of California, San Francisco.

Figure 1.2, p. 5: Courtesy of the Mayo Clinic.

Figure 1.4a, b, p. 11; 1.6, p. 14: Wayland Lee.

Figures 3.7, p. 63; 3.12, p. 69; 3.14, p. 70: J.W. Kimball, 1978. *Biology* 4/e, Addison-Wesley Publishing Company.

Table 3.1, p. 65: *From:* Luria, Gould, and Singer: *A View of Life*, 1981. Menlo Park, California: The Benjamin/Cummings Publishing Company. Table 3-3, p. 73.

Figures 3.10, p. 66; 3.11, p. 69: Compliments of Dr. Barry L. Batzing, Department of Biological Sciences, State University of New York, College at Cortland.

Figure 3.15, p. 71: T.E. Schroeder, University of Washington/BPS.

Figure 3.16, p. 71: B. King, University of California, Davis/BPS.

Figures 3.17b, c, p. 72; 3.21, p. 77: Courtesy of E. de Harven, Rockefeller University.

Frontiers box, p. 76: Media Vision/Peter Arnold, Inc.

Figures 4.2, 4.3, p. 90; 4.4, p. 91; 4.5, 4.6, p. 92; 4.8, p. 93; 4.12, p. 98; 4.14, p. 99; 4.16, p. 100; 4.20, p. 103; 4.21, 4.22, p. 104; 4.23, p. 105: Photos by Ed Reschke.

Figures 4.7, p. 93; 4.13, p. 39: Manfred Kage—Peter Arnold.

Figures 4.15, p. 100; 4.18, 4.19, p. 102: Biophoto.

Frontiers box, p. 106: Wide World Photos.

Figure 5.1, p. 112: Bloom and Fawcett, *A Textbook of Histology*, Phila., W.B. Saunders Co., 1975. Reprinted by permission.

Frontiers box, p. 115: Leonard Kamsler/Medichrome.

Figure 5.3, p. 118: Carroll H. Weiss, RBP, 1973.

Figure 5.5, p. 122: With permission of Dr. Robert Chase, M.D.

Figures 6.2b, p. 131; 6.35c, p. 168; 6.36b, p. 178: Lester Bergman & Associates.

Figure 6.7, p. 139: Dr. Henry Jones, Radiology, Stanford Medical Center.

Frontiers box, p. 158: Dr. Freiburger, Hospital for Special Surgery/Peter Arnold.

Figure 6.41c, p. 176; 6.43c, p. 178: Richard Humbert, Stanford University.

Box 6.2, p. 181: Dr. Steven Subotnick, DPM, MS.

Figure 7.8, p. 195: Wayland Lee.

Box 7.1, p. 197; Figure 7.13b, p. 203: Palo Alto Medical Foundation.

Figure 7.16, p. 205: Bobbe Wolfe/Time Magazine.

Figure 7.17, p. 206: Carroll H. Weiss, RPB, 1973.

Frontiers box, p. 207: Leonard Kamsler/Medichrome.

Figure 8.4, p. 217: Clara Franzini-Armstrong, University of Pennsylvania.

Figure 10.8b, p. 304: P. Schulz/Biology Media.

Figures 10.11, p. 306; 10.12, p. 307: © Manfred Kage—Peter Arnold.

Figure 10.14, pp. 310–11: Photos by Victor B. Eichler—Wichita, KS.

Figure 12.1b, p. 338: Lester Bergman & Associates.

Box 12.1, p. 339; Figure 12.17c, p. 357: Dr. J. de Groot, M.D., Ph.D., University of California, San Francisco.

Figures 12.2b, p. 340; 12.8b, p. 346; 12.9b, p. 348: M. Rotker/Taurus Photos, Inc.

Figures 12.6c, p. 342; 12.22, p. 362: © Manfred Kage—Peter Arnold, Inc.

Frontiers box, p. 370: Courtesy of Dr. Joseph H. Schulman, Neurodyne Corporation, and Pacesetter Systems.

Figure 12.32, p. 373: © William Thompson.

Figure 12.34, p. 374: © Dr. J.A. Hobson and Hoffman-LaRoche Inc., from DREAMSTAGE Catalog.

Figure 15.2b, p. 431: Don Fawcett/Photo Researchers/Science Source.

Box 15.1, p. 432: Owner JMC Eye Photo.

Figure 15.24, p. 449: Ivan Pieper, Ophthalmic Photographer, Division of Ophthalmology, Stanford University.

Figure 15.29c, p. 454: © Manfred Kage—Peter Arnold.

Frontiers box, p. 459: Diagram courtesy of 3M.

Figure 15.41b, p. 467: Lester Bergman & Associates.

Figures 16.2, p. 475; 16.3, p. 476: Adapted from "The Molecular Basis of Communication within the Cell," by Michael J. Berridge, *Scientific American*, October 1985, pp. 144, 146–7.

Figure 16.20, p. 495: Ed Reschke—Peter Arnold, Inc.

Frontiers box, p. 497: Cardiac Pacemakers, Inc.

Figure 17.3, p. 509: L. Weiss and R.O. Greep.

Frontiers box, p. 510: Dr. C.A. Hunt, University of California, San Francisco.

Figure 17.6, p. 516; 17.8, p. 519: © Manfred Kage—Peter Arnold, Inc.

Figures 18.4b, p. 531; 18.5b, p. 532: Lester Bergman & Associates.

Box 18.1, p. 535: Carroll H. Weiss, RBP, 1973.

Figure 18.12, p. 540: Ed Reschke.

Frontiers box, p. 555: Ed Reschke—Peter Arnold.

Figure 19.2, p. 562: P. Phelps and J. Luft, "Electron Microscopial Study of Relaxation and Constriction in Frog Arterioles," *Am. J. Anat.*, 125:399–428, 1969.

Figure 19.4, p. 564: Reproduced from the *Journal of Cell Biology*, 1974, 61:269–287 by copyright permission of the Rockefeller University Press.

Figure 19.5, p. 564: Courtesy of Ripon Microslides, Inc.

Figure 19.10, p. 572: © William Thompson.

Frontiers box, p. 585: Photo by Rod Cowe.

Frontiers box, p. 596: Ed Reschke—Peter Arnold.

Figures 20.3, p. 611; 20.9, p. 617: Ward's Natural Science Establishment, Inc., Rochester, NY.

Figure 20.7, p. 615: Biophoto/Science Source.

Figures 21.8, 21.9, p. 632; 21.11, p. 635: From Hood, Weissman, Wood, Wilson: *Immunology* 2/e, 1984. The Benjamin/Cummings Publishing Company, Inc.

Figure 21.10, p. 634: Andrejs Liepins, Sloan-Kettering Institute for Cancer Research.

Box 22.1, p. 656: Reproduced by permission of the American Cancer Society, Inc.

Figure 22.12, p. 663: Used by permission of Addison-Wesley Publishing Company, Inc., Redwood City, California.

Figure 23.17, p. 700: Ed Reschke—Peter Arnold.

Frontiers box, p. 729: Photo by Dianora Niccolini/ Medichrome; illustration by Harriet R. Greenfield.

Table 26.3, p. 750: Reprinted from "Recommended Dietary Allowances, Ninth Edition," 1980, with permission of the National Academy Press, Washington, D.C.

Figure 25.13, p. 751: © William Thompson.

Box 26.1, p. 764: Biological Photo Service.

Figure 26.6b, p. 767: Lester Bergman & Associates.

Figure 26.8, p. 769: Alan R. Liss, Publisher & Copyright Holder.

Frontiers box, p. 774: Dr. Irving H. Goldman, Crouse Irving Memorial Hospital, Syracuse.

Figure 28.9, p. 818: R. Knauft/Biology Media.

Figure 28.12a, p. 822: R. Yanagimachi, University of Hawaii/BPS.

Figure 28.17, p. 827: Bloom and Fawcett, *A Textbook of Histology*, Phila., W.B. Saunders Co., 1975. Reprinted by permission.

Figure 28.21, p. 833: © Manfred Kage—Peter Arnold, Inc.

Frontiers box, p. 837: Wayland Lee.

Box 29.1, p. 854: Dr. Robert Rugh and Dr. Landrum Shettles.

Frontiers box, p. 860: David York/Medichrome.